Understanding Surveillance Technologies

Spy Devices, Privacy,
History & Applications

Revised and Expanded
SECOND EDITION

Understanding Surveillance Technologies

Spy Devices, Privacy, History & Applications

Revised and Expanded
SECOND EDITION

J. K. PETERSEN

Auerbach Publications
Taylor & Francis Group
Boca Raton   New York

Auerbach Publications is an imprint of the
Taylor & Francis Group, an informa business

Auerbach Publications
Taylor & Francis Group
6000 Broken Sound Parkway NW, Suite 300
Boca Raton, FL 33487-2742

© 2007 by Taylor & Francis Group, LLC
Auerbach is an imprint of Taylor & Francis Group, an Informa business

No claim to original U.S. Government works
Printed in the United States of America on acid-free paper
10 9 8 7 6 5 4 3 2 1

International Standard Book Number-10: 0-8493-8319-6 (Hardcover)
International Standard Book Number-13: 978-0-8493-8319-9 (Hardcover)

Library of Congress Cataloging-in-Publication Data

Petersen, Julie K.
   Understanding surveillance technologies : spy devices, privacy, history & applications / J.K. Petersen. -- 2nd ed.
     p. cm.
   Includes bibliographical references and index.
   ISBN-13: 978-0-8493-8319-9 (alk. paper)
   ISBN-10: 0-8493-8319-6 (alk. paper)
   1. Electronic surveillance. I. Title.

TK7882.E2P48 2007
621.389'28--dc22

2006035040

Visit the Taylor & Francis Web site at
http://www.taylorandfrancis.com

and the Auerbach Web site at
http://www.auerbach-publications.com

CONTENTS

DEDICATION

This book is dedicated to Araminta Ross, who climbed every mountain and used whatever surveillance devices were at her disposal to further the quality of life and well-being of others. May we all make such wise choices.

ABOUT THIS BOOK

This was an enormous project. It required several years of research plus more than a year of 60-hour weeks to write, illustrate, and produce the first edition. The second edition, while essentially an update, required equally dedicated commitment because surveillance technologies are changing faster than books can be written.

One of the biggest problems encountered by the author while creating this reference was the discovery of fallacies and errors in supposedly reputable texts. While no text can be perfect, no matter how well researched (and much of the information about the inner workings of intelligence agencies is not public knowledge), there is a surprising amount of erroneous information in supposedly accurate references even about mundane and widely researched subjects related to surveillance.

For example, misinformation abounds about the invention and development of radar. The author encountered many references that unequivocally stated that radar was invented during World War II. As can be seen from the information included in the Radar Surveillance chapter, the concept for radar was discussed in the 1800s, radio-ranging instruments existed by the 1890s and were patented at least by 1904, long before WWII began. In fact, one of the early radar instruments was offered to the Germans for the war effort and rejected. While it is true that radar received priority development and funding during the war, especially in England, it is still important to recognize the developmental steps that led to the technology as we know it today. Similar incorrect information has been written regarding many other important inventions, including the telegraph and the telephone.

The author made an extra effort to locate original sources (patent diagrams, oral histories, engineering texts, etc.) to try to sort fiction from fact and has corrected a number of widespread inaccuracies (with references and patent numbers so readers can verify the information for themselves).

Another complication to producing a comprehensive text is the volatile political climate surrounding spies and widespread public use of spy technologies. The destruction of the World Trade Center and the subsequent creation of the department of Homeland Security wrought significant changes in focus and funding and the structure of the U.S. Intelligence Community. Allegations of unwarranted wiretapping by the government and illegal use of surveillance devices by regular citizens have both been much in the news lately. Some of the concerns are justified—criminals have unprecedented tools for committing crimes, but the potential for invasion of privacy and loss of constitutional freedoms stemming from the proliferation of surveillance technologies is enormous, as well, and must be considered as much of a priority as national security.

It is hoped that this text will help lawmakers, installers, government officials, educators, surveillance users, and journalists better understand surveillance technologies, so they can make proactive decisions to protect health and safety, while still safeguarding our basic freedoms.

PREFACE

This book sets the groundwork for understanding surveillance technologies by describing contemporary devices and current legislation in their historical context. These stepping stones will make it easier to understand more specialized texts on radar, sonar, video cameras, satellite imaging, and genetic profiling.

This book is suitable as a general reference for a variety of professions and also as a foundation text for post-secondary institutions offering surveillance studies courses.

Understanding Surveillance Technologies is the first comprehensive, introductory overview of the field of surveillance devices, first published in 2000 and remains the only comprehensive text on the subject. It comprises 18 chapters with more than 700 photos and illustrations and is suitable for college surveillance studies in sociology and political science, professional recruiting programs, and as a reference for beginning professionals in the fields of law enforcement, forensics, and military surveillance. It is also an indispensable reference for journalists, lawmakers, and community planners. It has been designed with a flexible, modular format so the chapters can be read in almost any order (chapters that share common topics are cross-referenced to alert the reader).

We are all being surveilled. It is no longer possible to avoid cameras, DNA tests, identity chips, border crossing cameras, highway monitors, ATMs, and other devices that record our movements, finances, and even our health. Many people don't realize that their activities are cataloged and stored in a multitude of databases, many of which are accessible on the Internet without the surveillee's knowledge or permission. Here are examples that illustrate this unsettling trend:

- Detailed information about people who have never logged onto the Net nor even used a computer is available to anyone with an Internet connection. Even ages and occupations are freely distributed by commercial sites and it is possible to acquire the names, addresses, and phone numbers, ages, and occupations of a person's neighbors quite easily, as well, which provides the data to create a composite picture of a neighborhood's social and economic characteristics. This book provides a better understanding of who is collecting this information, how they are doing it, and what they are doing with the data once they have it.

- Some hospitals now routinely take DNA samples of newborn babies. Similarly, the U.S. armed forces require submission of a DNA sample from new service members. This book explains the background and origins of DNA matching and possible social consequences of its use. In many instances, your DNA can reveal your gender, race, medical tendencies, and physical characteristics.

- Semi-nude and nude photos of unwary victims are sold on the Internet without the knowledge or permission of the surveillee. How is this possible? This book explains how these technologies work and why bootleg images may not be illegal. It further describes the ethical and social consequences of new forms of exploitation.

- Gaming centers, hotels, and trade shows use magnetic access cards to keep track of their guests. In casinos they can tell how often patrons play, how much they spend, and how frequently they visit the establishment. Even universities issue student cards that double as access cards. They allow access to vending machines, copy machines, and various retail services on campus. In some instances, this information is stored in sophisticated databases. This text describes a variety of user access and property surveillance technologies that provide travel suppliers, casinos, hotels, and retail outlets with detailed information on their patrons.

- Law enforcement agencies are consolidating their forensic and criminal databases and providing Internet access to authorized personnel from any part of the country. This provides new ways to solve serial murders and to catch felons who move from state to state, but it also makes a criminal less distinguishable from a law-abiding citizen because of the way databases are designed and used to store general information on law-abiding citizens. It is possible for law enforcement agencies to have good intentions and still unwittingly violate constitutional rights and basic freedoms. There are good ways and bad ways to structure databases so they help us distinguish between criminals and honest people, but many software programmers are unaware of how to incorporate these issues into their software design strategies. This text looks at databases that are used to fight crime and how we can take steps to support the efforts of law enforcement without turning the country into a repressive Big Brother society.

This is only a handful of the significant issues discussed in this book. There are also notes on the history and current state of intelligence-gathering in America, concerns about chemical and nuclear treaty surveillance and enforcement, and information about new technologies that make it possible to surveil space and other planets.

Surveillance devices are now used in every field of endeavor. From handheld magnifying glasses to sophisticated magnetic resonance imaging machines, 'spy' devices enable us to see beyond our basic senses in ways we wouldn't have imagined two hundred or even fifty years ago. This book is a fascinating journey through technology and provides significant food for thought and discussion as to how these devices can be used and whether we have the right to do so.

FORMAT

Understanding Surveillance Technologies is modular and follows the same format for each chapter. It starts with an introduction, describes some of the various kinds of devices that fall within the chapter's topic area, and describes the context in which the devices are usually used. This is followed by a historical overview of the major milestones associated with the technology.

After the history and evolution is a description of basic functions that have not been covered in the first three sections. Common applications for the technology are then described, followed by a discussion of some of the legal and ethical implications.

At the end of each chapter, there is an extensive annotated list of resources for further study short-listed for their relevance to the topic. The resources include bibliographies and selected media and online resources. Web addresses are provided for many of the important organizations, as are educational sites.

There are cross-references to alert the reader or instructor to information in related chapters, since some groups of chapters make more sense if they are read together. Chapter 1 should probably be scanned or read in its entirety first, and the following groups of chapters cross-reference. Otherwise, the chapters can be read in any order.

Acoustic Surveillance - These three chapters are strongly interrelated. It is a good idea to read Infra/Ultrasound Surveillance before reading Sonar Surveillance, as sonar is a specialized adaptation of acoustics that relies heavily on ultrasound.

The history section in the *Introduction & Overview* and the history section in the *Audio Surveillance* chapter describe a number of controversial wiretapping topics. These sections make more sense if they are cross-referenced and read in the above order.

Electromagnetic Surveillance - It is recommended that the reader cross-reference the Infrared, Visual, and Ultraviolet chapters. Together they comprise Light Surveillance. The Visual and Aerial Surveillance chapters make more sense if they are read together and much of the information in the Infrared Surveillance chapter is relevant to Aerial Surveillance.

Radar Surveillance uses radio waves, so it helps to read the Radio and Radar Surveillance chapters together.

Chemical & Biological Surveillance - The Biometric Surveillance chapter introduces a specialized subset of Chemical Surveillance, so it helps to read these chapters together. In turn, Genetic Surveillance is a subset of Biometrics.

Following the main body of the book is an extensive index.

ABOUT THE AUTHOR

The author is a technologist/consultant, writer, and university instructor (when the need arises) with a life-long fascination with gadgets, codes, and machines. When not writing technology references, Petersen enjoys outdoor activities, music, gourmet cooking, fiction writing, strategy games, and computer imagery.

Surveillance Technologies

Introduction

Surveillance Technologies

1

Introduction & Overview

1. Introduction

1.a. Scope and Focus

This book fills a significant gap in surveillance literature. There are thousands of books about spies, dozens of retail catalogs that list spying devices and, now, many books that discuss loss of privacy related to increasing levels of surveillance. However, this is the only foundation text that covers *surveillance devices* in a broader context so that readers can understand the origins and diversity of these devices, and how they are used in a wide range of fields.

Surveillance devices are increasingly used for domestic, governmental, corporate, retail, and residential security; intelligence and military applications; search and rescue operations; scientific research; skip tracing; and personal communications.

The page illustrated above was stamped with Base Censor and U.S. Censor Office stamps in 1943. It is a re-minder that communications are often monitored, opened, censored, returned, and sometimes even confiscated during times of political turbulence. [U.S. Army Signal Corp historic photo by McQuarrie, released.]

Specific Focus and Exclusions in this Volume

This book focuses on the ways in which surveillance devices are used and how their benefits must be balanced against their potential for harm and invasion of privacy. It does not explain how to build or configure spy devices (there are no circuit board or installation diagrams), and does not emphasize human-centered espionage, because there are a number of books already devoted to this specialized profession. Instead it focuses on how technology has changed our approach to information-gathering and local and national security.

Surveillance is a subset of the larger process of intelligence-gathering, and thus a key tool in intelligence operations, but it is also an equally important tool of wildlife conservation, weather forecasting, and institutional and residential security. At the end of each chapter, there is a carefully selected list of references for further study on each topic.

This text does not discuss technologies that may be used to extract information directly from humans through coercion or torture. Nor does it cover surveillance of nuclear radiation leaks except in a very general way. It does, however, provide a wealth of information on how technology is used to help humans hear, see, smell, and otherwise detect and monitor trends, people, buildings, vehicles, wildlife, military operations, contraband, and natural disasters.

Price and Availability of Surveillance Devices

Many surveillance devices are now portable and inexpensive—available through mail order catalogs, email, online vendors, and online auction companies. There are "spy shops" in most major cities. A week's wages is sufficient to set up a simple surveillance or security system whereas in the early 1990s, the same system would have cost thousands of dollars. Some examples illustrate this trend.

- Board-level pinhole cameras can be purchased in bulk for under $25 each. Outdoor color wireless bullet cameras with audio are now under $150.

- Video/audio transmitters are available for less than $33 each and an $80 VCR can be used to record the signal.

- Two-way radios with a two-mile range are now as low as $40.

- Basic computer systems can be purchased for less than $800 and fully functioning older models are available for less than $40.

- A high-speed Internet connection (which was $500 in 1997) now costs only slightly more than a telephone dialup connection. Depending upon whether it is DSL or cable, it may run from $25 to $39/month.

- Satellite images are free or available for reasonable fees on the Internet.

Political Openness

Democratic freedoms and specific U.S. governmental acts such as the *Freedom of Information Act* have created large repositories of "open source" software, some of which are used in information-gathering and surveillance applications. To balance the America-centricity of the examples in this book, the author has tried to select those that generically illustrate a class of technologies and trusts that the reader can extrapolate more general concepts from the information given.

While this text focuses chiefly on *devices*, there is also sufficient information to provide a sense of the *role* of these devices within the larger field of surveillance and the broad context of intelligence gathering.

1.b. Format

Understanding Surveillance Technologies is organized so the chapters can be read in any order. Cross-references indicate closely related sections. The only recommendation is that you read or scan the introductory chapter in its entirety, as it gives preliminary information relevant to all the chapters and will familiarize you with the history of surveillance and the general format of the rest of the book.

Much has changed since the first edition of this book was released in 2000, including legislative repercussions stemming from the World Trade Center attack, including the establishment of a *Homeland Security* directive within American government. Changes in legislation are reflected in this updated volume.

Chapters featuring related technologies have been grouped into sections. To make it easier to use this volume as a textbook or professional reference, each chapter follows the same basic format, consisting of

1. an *introduction,* providing the scope and focus,
2. *general types and variations of specific technologies,*
3. the *context* in which the technologies are most commonly used,
4. the *historical basis* for the evolution of the various technologies, including where they came from, how they were initially used, and how they evolved,
5. a *general description* of aspects not covered in previous sections,
6. *examples* of technologies used in practical applications and, in some cases, information about *commercial sources,*
7. *problems and limitations* inherent in a particular class of technologies,
8. *legislative restrictions, trends, and concerns* (note: the use of surveillance technologies for some purposes is highly illegal, with severe penalties)
9. *implications of use* of the technology—philosophical and ethical considerations.

Extensive resource information for further study is grouped toward the end of each chapter—enough to provide months' worth of reading. The author has tried to include the better references out of thousands extant. They are numbered and organized as follows:

10. **Resources**

    10.a. **Organizations** - some of the prominent agencies related to the topic

    10.b. **Print** - bibliographies of books, articles, and journals to aid the reader in locating intermediate and advanced print resources

    10.c. **Conferences** - some of the more significant conferences and workshops, with an emphasis on top industry conferences that occur annually

    10.d. **Online Sites** - selections of some of the more worthwhile sites on the Web

    10.e. **Media Resources** - a handful media resources, including films, museums, and television broadcasts

11. **Glossary** - a short list of words and abbreviations related to each subject area

It is hoped that the modular design will maximize your enjoyment of this book, allowing you to choose topics at will, while still getting an understanding of interrelationships.

1.c. The Impact of Surveillance Technologies

Important Trends

Surveillance technologies are developing rapidly and showing up in every aspect of daily life. They are changing our approach to military, border, retail, and residential security.

Surveillance technologies have altered the way in which we view ourselves and our neighbors. At one time, it was considered an individual's responsibility to curtail illegal or intrusive behavior against others. Now the onus is on the potential victim to protect him or herself and many are choosing technology for this task. Just as word processors superseded typewriters, surveillance devices are superseding many traditional methods of security and information-gathering. Video cameras are replacing security personnel, DNA profiles are used for traditional blood typing in parental custody lawsuits, and nanny monitors have replaced frequent trips to the nursery.

Military surveillance has changed as well. There was a time when trespassing a military zone could cost you your life if you stepped on a landmine. This caused many problems after conflicts were resolved, as the mines were difficult to remove and civilians would sometimes inadvertently step on them. Now audio/video cameras and motion sensors are often substituted for mines.

Motion Detectors Replace Land Mines. In 1961, the U.S. Naval Station at Guantanamo Bay, Cuba, began installing approximately 50,000 antitank and antipersonnel landmines in the buffer zone separating Communist Cuba and the Bay. Following a 1996 Presidential Order, these were excavated, transported to a demolition site (left) and prepared for destruction (right). Motion and sound detectors were substituted to monitor the base. [U.S. DoD 1997 news photos by R. L. Heppner, U.S. Navy, released.]

Shift In Access

The capability to unobtrusively observe other people's business conveys great power. With power comes responsibility, whether or not it is mandated by law. Until the mid-1990s, the power of surveillance was mainly in the hands of local and federal government agents and, to a lesser extent, private detectives. This is no longer true. It is now possible to purchase an aerial picture of your neighbor's back yard for less than $20/image, that is detailed enough to distinguish between a dog house and hot tub. You can purchase high-resolutions pictures of government buildings in foreign nations, refugee tankers, a controversial logging site, or the production yard of your chief business competitor on the other side of town. This globalization of geographic and socioeconomic information will have a widespread impact on commerce and international security.

Civilian access to spy devices has already resulted in some surprising developments. Citizens have become, in a sense, an extension of the government surveillance system. Nuclear installations in other countries, for example, are now closely scrutinized by private watchdogs who seek and disseminate information previously available only to government departments with limited budgets.

Web access is dramatically influencing surveillance, as well. Anyone with Internet access can log onto the World Wide Web and use reverse phone directories, public records databases, and genealogical databases to obtain large amounts of personal information. Marketing professionals are mining this data with considerable enthusiasm, judging by the significant increase in junk email and targeted postal mail in recent years. These publicly-accessible databases and search engines are discussed in more detail throughout this text.

Legislation

Technology changes faster than laws can be drafted to protect the vulnerable. It is impossible to anticipate all the ways in which a new technology may be abused. Currently, private citizens don't need special permission to purchase or own most types of video or photographic surveillance gadgets. The Federal Communications Commission (FCC) regulates these devices, but they cannot police every hobbyist who builds or uses radio-based gadgets for clandestine purposes. Satellites and network distribution channels have also put access to information in the hands of ordinary people. More than half the populace of developed nations can log onto the Internet and surf surveillance retailers or download instructions for building spy devices.

While military personnel and the intelligence community are the only ones with ready access to the highest resolution imagery, the gap between what civilians and those in traditional positions of power (e.g., national defense) can access has narrowed dramatically. This shift of access puts a great deal of power in the hands of an under-informed public which, in free societies, results in unethical and unscrupulous behavior on the part of a percentage of individuals willing or eager to take advantage of others. Lobbyists and lawmakers cannot afford to overlook the possible consequences of broad distribution of personal and private information and of potential 24-hour surveillance of every member of the populace. The increase in "upskirt" privacy violations, identity theft, and credit card fraud illustrates how access to personal information can have negative consequences.

Recent concerns about terrorism have been used as justification for increased governmental scrutiny of private citizens. The American Bar Association responded to concerns about allegedly illegal government surveillance by crafting ABA policy through their *Task Force on Domestic Surveillance in the Fight Against Terror*. This was drafted by a bipartisan panel, including former members of the U.S. intelligence and law enforcement communities, as well as national security law experts. The ABA approved these recommendations in February 2006.

At about the same time, the U.S. government became the target of increased criticism for allegedly surveilling without due process. To publicize their concerns, members of the ABA issued a news release on May 2006 calling on the Senate Judiciary Committee to carry out a thorough inquiry into the nature and extent of domestic surveillance conducted by the administration without warrants. The ABA expressed "deep concern" about possible violations of people's constitutional rights. In a letter to the Judiciary Committee, Michael S. Greco, President of the ABA, wrote, "... we are deeply concerned about the electronic surveillance of Americans without the express authorization of the Congress and the independent oversight

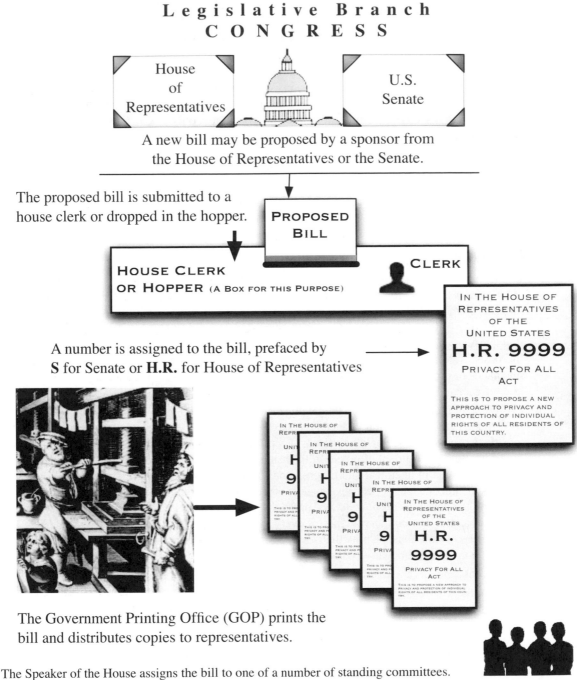

Legislative Branch
CONGRESS

House of Representatives

U.S. Senate

A new bill may be proposed by a sponsor from the House of Representatives or the Senate.

The proposed bill is submitted to a house clerk or dropped in the hopper.

PROPOSED BILL

HOUSE CLERK OR HOPPER (A BOX FOR THIS PURPOSE)

CLERK

IN THE HOUSE OF REPRESENTATIVES OF THE UNITED STATES

H.R. 9999

PRIVACY FOR ALL ACT

THIS IS TO PROPOSE A NEW APPROACH TO PRIVACY AND PROTECTION OF INDIVIDUAL RIGHTS OF ALL RESIDENTS OF THIS COUNTRY.

A number is assigned to the bill, prefaced by **S** for Senate or **H.R.** for House of Representatives

The Government Printing Office (GOP) prints the bill and distributes copies to representatives.

The Speaker of the House assigns the bill to one of a number of standing committees. Committee members listen to testimony regarding the bill and study it, in order to release/report out the bill with a recommendation to pass, revise, or lay it aside (table it). A released bill may be passed by unanimous consent or, in some cases, by a two-thirds vote. The bill is then read on the floor of the House and, if passed, moves to the Senate where it is announced by a presiding officer/Senator and, again, assigned to a standing committee.

Once released, the bill goes to the Senate floor to be voted upon and passed if it receives a simple majority. Then, a conference committee, with members from each House, resolves differeneces between House and Senate versions. Once approved, the U.S. GPO prints the document and the final version is certified by a clerk from the House that introduced the bill. Finally, the bill is signed by the Speaker of the House and the Vice President and forwarded to the President. At this point, the President has ten days to sign or veto the bill. If vetoed, a two-thirds vote of the Senate and of the House can still pass the bill into law.

of the courts."

The ABA also expressed concerns about government policies that appeared to support blanket authority for indefinite surveillance. This "wide net" was a radical departure from earlier tradition in which warrants were requested on a case-by-case basis and would expire within a stated time frame. The ABA urged that actions were carried out in a manner reinforcing "public respect for our system of checks and balances" in a way that would ensure national security without compromising constitutional guarantees.

United States Senate
Washington, D.C. 20510

Dear Chairman Specter & Senator Leahy:

As the Senate Judiciary Committee moves forward with its consideration of legislation regarding the use of warrantless electronic surveillance on American citizens, I write to express the views of the American Bar Association on this issue.

The ABA was deeply troubled by the revelations that our government is conducting domestic electronic surveillance outside of the process set forth in the Foreign Intelligence Surveillance Act ("FISA"). In response, I appointed the Task Force on Domestic Surveillance in the Fight Against Terrorism to explore the very difficult constitutional questions raised by unsupervised domestic surveillance. The Task Force was a bipartisan panel of distinguished lawyers that included a former Director of the Federal Bureau of Investigation, a former General Counsel of the National Security Agency and the Central Intelligence Agency, the National Institute of Military Justice General Counsel, and others with deep knowledge of national security law. Our position on these issues is based upon the unanimous report and unanimous expert recommendations of the Task Force that were adopted by a nearly unanimous voice vote of the ABA's 550-member House of Delegates. I refer to the ABA's policy below.

As your Committee considers legislation in this area, we urge you to act in a manner that reinforces public respect for our system of checks and balances and ensures that our national

Concerns About Possible Constitutional Violations of Unwarranted Surveillance. The ABA posted a publicly-accessible copy of a letter that was addressed to the Chair and to the Ranking Democrat of the Judiciary Committee of the U.S. Senate expressing "grave concerns" about domestic electronic surveillance "outside of the process set forth in the Foreign Intelligence Surveillance Act ("FISA")". [American Bar Association News Release, May, 2006.]

Responsibility and Social Evolution

Privacy erosion from increased use of surveillance devices is clearly evident. In the not-too-distant future, newborn infants may be routinely implanted with radio tracking devices, without their knowledge or consent, to 'ensure their safety' from wandering or kidnapping. These actions may be well intentioned initially but may subsequently be used to curtail freedom of choice and movement during a person's teenage years and beyond. Radio tracking systems can monitor movements, travel habits, buying habits, and even leisure-time activities. The potential for "branding" and profiling of individuals is significant. Once the information enters a computer database and is disseminated through a distributed system, it's almost impossible to recall it.

When freedoms are taken away quickly, people tend to fight back. When they are removed gradually, they tend to misjudge the potential harm until it becomes difficult or impossible to reverse the damage.

One or two generations of children who have never experienced our concept of freedom may not understand what they have lost until they reach adulthood, in which case it may be impossible to change their surrounding social structure. It is important to understand surveillance in its broader context so we can safeguard future freedoms as well as those we currently enjoy.

To resolve issues related to privacy and security, society must change and adapt, because eventually satellites, RF detectors, and unstaffed aerial vehicles may be tracking individual people on the move. Every action can be recorded with intelligent software and stored in immense databases. We can't allow a fascination with technology or entrepreneurial greed to cloud our judgment in such important matters.

Personal data is now widely available online. Switchboard and InfoSpace are just two of many Web-based services that provide addresses and phone numbers of almost every directory listing in America (some cover Canada and the U.K. as well). Most of these Web-based businesses include value-added pay services. For example, in a search for CRC Press, one service offered to provide detailed business and credit information, including employee size, sales volume, key executives, number of years in business, public information, and lines of business, etc., for a flat fee of $5. Many Web-based "phone" directories now allow you to view maps of a neighborhood, and even to look up the names and numbers of people in nearby houses. Such broad access to information was almost unknown before the late 1990s, and ease of data acquisition is changing the way private detectives carry out their investigations.

2. Kinds and Variations

2.a. Basic Terms and Concepts

Each chapter in this book includes basic terms and concepts related to the topic of the chapter. Some generalized surveillance concepts/terms are listed here, and additional terms are in the glossaries at the end of each chapter.

information In the context of surveillance, information consists of *knowledge, data, objects, events, or facts* which are sought or observed. It is the raw material from which intelligence is derived.

intelligence This is information which has been *processed and assessed within a given context*. Thus, the number of barrels of oil shipped by a nation in a year is *information* whereas the number of barrels of oil shipped by a nation in a year compared to other nations or compared to the previous year is *intelligence* if it can be used as an economic or political lever in comparative social contexts. It's difficult to know in advance what information may become part of a body of intelligence. Prior to their fatal accident, the videotaped sequence of Princess Diana and Dodi al Fayed leaving a hotel was information. After the accident, it formed part of an extensive investigation, especially of the driver, contributing to a body of intelligence.

surveillance Surveillance is the *keeping of watch* over someone or something. Technological surveillance is the use of technological techniques or devices to detect attributes, activities, people, trends, or events.

covert That which is masked, concealed, or hidden. Covert activities involve disguises, hidden equipment, camouflage, and shrouded activities intended to have a low probability of detection.

clandestine That which is secret, surreptitious, stealthy, sneaky, or furtive. Thus, a detective hiding behind a curtain in a window using binoculars to view someone from a distance is engaged in *covert* behavior, whereas a detective standing in plain sight in normal attire but secretly monitoring someone's activities is behaving in a *clandestine* manner. Similarly, a corporate agent wiretapping a competitor is engaged in covert surveillance while a corporate representative chatting at lunch with a competitor without revealing his or her agenda is engaged in clandestine surveillance (this is further explained in Section 2.c., following).

occult That which is hidden, concealed, secret, or not easily understood. Originally a generic term, this has gradually come to be associated with ghosts and psychics and has lost most of its practical use as a surveillance term. However, it still has some relevance with regard to reports of 'supernatural' events. It is included here because classified military exercises or tests are sometimes interpreted by uninformed onlookers as occult or *paranormal* events.

reconnaissance Reconnaissance is a preliminary or exploratory survey to gain information. Job-hunters often do reconnaissance on potential employers and vice versa. Law enforcement agents conduct crime scene reconnaissance in preparation for a full investigation. Military intelligence agents and strategists conduct reconnaissance of hostile territory before sending in troops.

2.b. Disclaimer and Regulatory Processes

Most surveillance technologies are, in themselves, neutral technologies—not inherently helpful or harmful. Their implementation, however, is rarely neutral. Surveillance technologies are used for reasons of distrust, fear, curiosity, sexual gratification, profit, exploitation, sales pressure, and sometimes as 'techie toys'. Once installed, the temptation to use them in unauthorized or inappropriate ways is substantial and often harmful.

Disclaimer: This text is intended to be educational and thus presents a broad view of surveillance devices and their implementation. The format inherently requires descriptions of technologies whose use *may be restricted or which may be illegal to own or use*. Many electronic eavesdropping devices may be built as hobby kits for learning about electronics and may even be legal to use in classrooms or homes, but may be *illegal for use* under other circumstances.

Legal restrictions vary greatly from country to country as do export criteria for the technologies described here. *This book does not encourage or endorse the illegal use of surveillance technologies.* Familiarize yourself with relevant regulations before purchasing surveillance equipment. Some general restrictions and regulations are listed toward the end of each chapter. Retail vendors will usually let you know if there are restrictions on the use of specific devices but less scrupulous Internet vendors may not.

Congressional Legislative Bills

The *U.S. Government Printing Office* (gpo.gov, gpoaccess.gov) gathers, produces, and preserves information about the work of the three branches of American government and makes it available to the public for purchase and, in some cases, at no cost, through the Federal Depository Library Program. Many of these documents govern the use of surveillance devices.

As an adjunct to interpreting surveillance and privacy bills, it is recommended that readers acquire a copy of the *Intelligence Reform and Terrorism Prevention Act of 2004* (S. 2845) and the accompanying conference report (H.Rpt. 108-796). You may also wish to reference the *Biographical Directory of the United States Congress, 1774–2005* (H.Doc. 108-222), which is available in print and digital format. This 2,236-page directory provides biographies of members of the Senate and House of Representatives who have served from the 1st through the 108th Congress. Members are listed by their full name, with nicknames or initials. H.Doc. 108-222 was compiled and edited under the direction of the *Joint Committee on Printing*, the *Senate Historical Office*, and the *House Office of History and Preservation*. These various documents are available for download, free of charge, from *www.gpo.gov* .

The GPO is constantly challenged to implement new technologies for distributing information to the public. On April 4, 2006, the GPO announced the release of a Request for Proposal (RFP) for a digital content management system (FDsys–Future Digital System) for providing permanent public access to federal government information. Digital content management systems in and of themselves make new forms of data surveillance possible.

2.c. Categories of Surveillance Activities

Surveillance technologies can be categorized in a number of ways according to

- the physical nature of the technology itself (infrared, X-ray, visual, etc.),
- the type of data derived (visual, aural, digital, etc.), or
- the nature of the surveillance with respect to the awareness of the person being surveilled.

This book is organized, chapter by chapter, according to the *physical nature* of the technology. Within each chapter, the more generic *data aspects* are described and cross-referenced to other chapters when appropriate, since there is overlap. The nature of the surveillance *with respect to the awareness of the surveillee(s)*, however, warrants further introduction as it is not covered in other chapters. This is important in the broader context of law and individual freedoms.

There are five general categories of surveillance activities:

implied surveillance *Surveillance that is mimicked or faked* with a variety of devices, including nonfunctioning cameras or empty camera housings and/or stickers claiming that an area is monitored, when in fact it isn't. Implied surveillance is a low-cost deterrent to theft and vandalism.

Retail Surveillance. An example of overt surveillance in a retail store. 1) The sign in the window alerts customers of the video security system, 2) cameras are clearly visible throughout the store, and 3) the video surveillance images are displayed in plain view near the cash register. (Note that there are still issues of storage and subsequent use of the videos that may impinge on privacy rights.) [Classic Concepts photos copyright 2000, used with permission.]

overt surveillance *Surveillance in which the surveillee has been informed of the nature and scope of the surveillance or in which the surveillance devices are clearly labeled* *and displayed.* Thus, an employee badge that constantly tracks workplace movement (assuming the employee has been fully informed of its role) or video camera surveillance in a department store, in which surveillees clearly see themselves on a monitor as they enter an area, are examples of overt surveillance.

Overt surveillance is often manifested in workplace or retail security systems in which employees or customers are informed that they are being watched. However, it is not sufficient to assume a person understands the function of a surveillance device because it is in plain sight. A wall-mounted camera that is visible to occupants of a room technically is *not* overt surveillance unless the surveillee explicitly knows a) that the camera is operating and b) that it is focused on the surveillee. If *both* these conditions are not met, then a device in plain sight is categorized as *implied surveillance*, if it is not functioning, or *clandestine surveillance,* if it is.

covert surveillance *Hidden surveillance. Surveillance in which the surveillance is not intended to be known to the surveillee.* Covert wiretaps, hidden cameras, cell phone intercepts, and unauthorized snooping in drawers or correspondence, are examples of covert surveillance. Most covert surveillance is unlawful and requires special permission, a warrant, or other 'just cause' for its execution. Covert surveillance is commonly used in law enforcement, intelligence gathering, espionage, and unlawful activities. The term 'black' is sometimes used to refer to covert operations—the deeper the black, the more secret it is. Some aspects of covert surveillance in retailing or the workplace are currently lawful, but are being challenged by privacy advocates who feel that prior notice of surveillance activities and clear identification of surveillance devices should be mandated by law.

Clandestine Surveillance. This dome-covered video camera is aimed at an outdoor ATM on a public sidewalk outside a financial institution, but does not qualify as overt surveillance because the public doesn't know where it's aimed or when it is active. In fact, several people outside the building mistook the surveillance camera for a light fixture, which it resembles. [Classic Concepts photos ©2000, used with permission.]

clandestine surveillance *Surveillance in which the surveilling system or its use is in the open but is not obvious to the surveillee.* Two-way surveillance mirrors above cash registers and entranceway cameras encased in aesthetically streamlined domes are recognizable as surveillance devices to professionals and the personnel who requisition and install them, but they are not always obvious to surveillees. The author queried customers outside a financial institution at which a dome camera had been installed overlooking the sidewalk and street. In every case the person queried was surprised to be told that the black and silver dome was a security camera not a light fixture. This type of clandestine surveillance, in which the device is not overtly hidden but is nevertheless inconspicuous, due to its placement, size, coloration, or design, is typical of surveillance devices in many public areas, including shopping malls, banks, and educational institutions.

Intrusions on personal privacy from clandestine surveillance devices are significant. Many video-plus-sound cameras are aimed at public squares, sidewalks, parking lots, and casual meeting places. Some are configured to broadcast live or almost-live over the Internet where anyone with a computer can capture and store the images without the knowledge or consent of the surveillees. Many surveillance systems that are claimed by their operators to be overt surveillance devices are actually *clandestine* surveillance devices. This has important ramifications for corporate and legislative policy-makers.

extraliminal/superliminal surveillance *Surveillance outside the consciousness of the person/entity being surveilled.* Extraliminal or superliminal means 'beyond consciousness'. Using video cameras or vital-sign devices to monitor an infant, a comatose hospital patient, or a mentally incompetent person who might be at risk of wandering or inflicting injury (on self or others) are examples of extraliminal surveillance. Extraliminal surveillance techniques, including tracking devices, may also be used for wildlife observation. Extraliminal surveillance is usually carried out to ensure the safety of the individual, or other people with whom the individual is interacting. It is primarily used in situations where informed consent is not possible.

It might be argued that anthropological observation of living primitive cultures or high-resolution satellite images of third-world cultures where technology is rare or nonexistent are forms of extraliminal surveillance, but since the surveillees are intellectually capable of understanding the concepts, surveillance of low-technology cultures is actually a form of covert or clandestine surveillance rather than extraliminal surveillance.

2.d. Categories of Surveillance Devices

Due to limitations of space, this text can't describe every existing surveillance technology, but it does cover representative examples of prevalent technologies. In overview, this text includes the following general sections:

1. *Surveillance Technologies*

   A general overview of surveillance history, devices, and intelligence-gathering. This book is modular, and the chapters can be read out of order, but it is probably helpful to read or scan the introductory chapter first.

2. *Acoustic Surveillance* - Audio, Infra/Ultrasound, Sonar

   Acoustic surveillance has been divided into three chapters. The first is *audio* technologies, those that operate within the range of human hearing; the second includes *infrasonic and ultrasonic* technologies, those which are primarily outside the range of human hearing; and the third is a specialized chapter for *sonar* because it is extensively used in marine surveillance and includes frequency ranges both inside and outside human hearing ranges.

3. *Electromagnetic Surveillance* - Radio, Infrared, Visible, Ultraviolet, X-Ray

Technologies that are based primarily on specific electromagnetic phenomena have been grouped together in this section. Infared, Visible, and Ultraviolet have been further subgrouped as *Light Surveillance* technologies (some people call them optical surveillance technologies, though they are not limited to optical devices).

Technologies that are not specifically electromagnetic but rely heavily on electromagnetic phenomena are included as well. *Radar and aerial surveillance technologies* rely heavily on radio and light phenomena. It is helpful to cross-reference the Visual Surveillance and Aerial Surveillance chapters.

4. *Biochemical Surveillance* - Chemical/Biological, Biometrics, Animals, Genetics

Chemical/Biological surveillance is a broad and highly technical field carried out largely in scientific laboratories, so it is covered mainly in its introductory and law enforcement aspects in this book. Genetic surveillance, an important subset of chemical/biological surveillance, and biometric surveillance, which is biochemical in origin, are discussed in separate chapters.

Animal surveillance is an important field, but it is not as prevalent as the other technologies and is given a correspondingly smaller amount of space in this text. It should be noted that it is a growing area of surveillance, and dogs and dolphins are used in many types of land and marine surveillance activities.

5. *Miscellaneous Surveillance* - Magnetic, Cryptologic, Computer

Magnetic surveillance is included in the miscellaneous section since it is not technically part of the electromagnetic spectrum. Some technologies are more difficult to categorize because they are not based on any one particular physical phenomenon, including *cryptology* and *computer surveillance* (only the basic user aspects of computer surveillance are introduced), and thus are grouped at the end.

2.e. Categories of Intelligence

This book is not about intelligence gathering, per se, but rather about devices that can aid in intelligence gathering specifically related to surveillance. However, it is helpful to have some idea of the general categories of intelligence, as it provides a framework for how particular classes of devices might be used. As described earlier, intelligence is information that has been processed and assessed within a given context. That context may require specific types of surveillance devices for effective information gathering. And, as was stated, we may not know in advance what information may later become part of a body of intelligence.

A few general categories of intelligence will be described here. (Note that these are generic categories and not the specific definitions used by the U.S. government—U.S. government definitions include stipulations about who might be the subject of the intelligence that may not apply in the general sense of the category.) The term *agent* as used in these descriptions means any agent (human, electronic, or otherwise) involved in gathering the information.

Note, some of the following INTs (forms of INTelligence), are dual-meaning in that they can refer to the technology used to gather information, or information gathered on the technology. For example, electronics intelligence can mean the use of electronics to gather various types of data for intelligence, or the use of various types of intelligence methods to gather information on electronics. For dual-meaning INTs, the meaning in practical use can usually be discerned from context.

biological/chemical intelligence (BICHEMINT) Intelligence derived from or by biological and/or chemical sources, such as biometrics, chemical stains, blood or saliva, hair, urine, gases, pharmaceuticals, etc.

communications intelligence (COMINT) Intelligence derived from communications that are intercepted by an agent *other than* the expected or intended recipient or which are not known by the sender to be of significance if overheard or intercepted by the COMINT agent. Oral and written communications, whether traditional or electronic, are the most common targets of surveillance for COMINT, but it may broadly include letters, radio transmissions, email, phone conversations, face-to-face communications, semaphore (flags or arms), sign language, etc.

In practice, the original data that form a body of COMINT may or may not reach the intended recipient. Data may be intercepted, or may reach the recipient at a later date than intended, or be intercepted, changed, and then forwarded. However, the definition of COMINT does not include the process of relaying delayed or changed information, but rather focuses on intelligence from the detection, location, processing, decryption, translation, or interpretation of information in a social, economic, defense, or other context.

computer intelligence (COMPINT) *Intelligence derived from or by computer networks, programs, algorithms, and data sources.* This is an important and growing source of information contributing to intelligence.

corporate intelligence (CORPINT) *General business intelligence.* A high proportion of CORPINT is corporate-competitor intelligence. General information on economic trends and imports and exports form part of CORPINT within ECONINT.

economic intelligence (ECONINT) *Intelligence related to business services, resource exploration, allocation, or exploitation with the potential for global or local economic impact.*

electronics intelligence (ELINT) *Intelligence derived from electronics-related noncommunications* (usually through electromagnetic, acoustic, or magnetic sources that are electronically generated or received) that are intercepted or derived by an agent *other than* the expected or intended recipient or that are not known by the sender to be of significance if overheard or intercepted by the ELINT agent. Radar signals, sonar pings, and magnetic disturbances are examples of ELINT information sources.

environmental/ecological intelligence (ECOINT) *Intelligence derived from observations of environmental patterns and characteristics, weather, pollution indicators, and ecological trends.* Weather intelligence (WEATHINT) is a subset of ECOINT, as is wildlife intelligence (WILDINT).

foreign instrumentation signals intelligence (FISINT) *A type of foreign TECHINT related to instrumentation within the broader category of SIGINT.* Mostly a military INT, this category isn't seen too often.

human intelligence (HUMINT) *Human-derived intelligence about activities, strategies, customs, etc.* This type of information is usually deliberately gathered by spies, agents, and operatives.

image intelligence (IMAGINT) *Video or photographic intelligence, which forms a large proportion of intelligence-gathering.* IMAGINT is varied and includes intelligence derived from remote-sensing technologies, aerial imagery, computer imagery, video footage, traditional photographs, infrared images, and more.

measurement and signature intelligence (MASINT) *The determination of characteristics related to identity which might include size, shape, volume, velocity, color, electrical characteristics, or composition.* Unique measurements can be used to determine 'signatures' for a fixed or moving object, such as an infrared or radar signature.

open-source intelligence (OPENINT) *Openly published or otherwise distributed, freely available sources of information such as books, journals, signs, lectures, ads, phone directories, genealogies, etc.* A large proportion of intelligence is acquired from OPENINT sources and these resources are increasing through the global Internet. Sometimes also called OSCINT, OSINT, or OPSINT, but the author recommends the less ambiguous OPENINT.

signals intelligence (SIGINT) Signals intelligence is commonly treated in the military as *a superset that includes COMINT, ELINT, FISINT, and TELINT.*

technical intelligence (TECHINT) *Intelligence derived from technical sources, usually machines, electronics, and instruments* as opposed to intelligence derived from human sources. Most of the surveillance technologies described in this volume are part of TECHINT.

telemetry intelligence (TELINT) *Intelligence derived from telemetric sources, that is, instruments that determine and calculate quantities or distances.* Telemetric data are often used to orient and control vehicles, projectiles, and satellites. Since telemetric data may be electronically generated or received, in many cases the data can be considered a subset of electronics intelligence (ELINT) and also of technical intelligence (TECHINT).

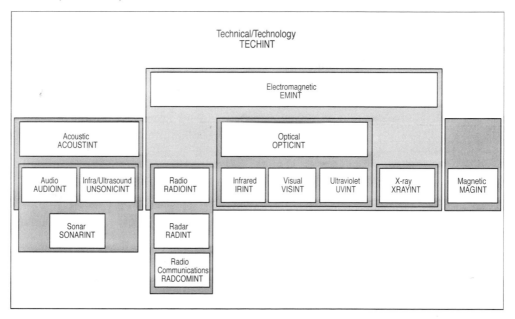

Sample Scheme for Organizing and Naming Common Technical Intelligence Categories

The above alphabetical list represents a selection of some of the common and older INT designations. Due to a two-syllable tradition for INT names, there is increasing ambiguity in naming schemes as technology grows and new INTs are added. We can continue the two-syllable tradition and tolerate the ambiguity or use three-syllable INTs where appropriate. The trend appears to be toward using three-syllable INTs to clarify the meaning.

Many organizations have developed the definitions and jurisdictions of their INTs over a period of decades and they have become intrinsically linked to many carefully developed policies and departments. For this reason, it's difficult to sort out and change existing INT designations to modernize them. However, for newer technologies, it may be useful to put them in some sort of logical order related to their physical properties. One possible scheme for organizing common TECHINT-related surveillance technologies is shown in the preceding chart.

Widely Used Technologies

The most widely used surveillance technologies are visual-, chemical-, and acoustic-based (particularly audio, radio, and sonar). Aerial surveillance is based primarily on electromagnetic technologies and is covered in a separate chapter due to its growing importance. Radar is also widely used, and infrared is steadily increasing in both aerial- and ground-based applications. This text concentrates more heavily on the more prevalent technologies, but does not overlook some less-used but valuable fields, including magnetic, animal, and cryptologic surveillance. Biometric surveillance is not yet a large field, but it is growing, and genetic surveillance may be the most significant technology of all. The technology that you ultimately choose depends upon what you need to know, what is lawful, and the type of environment in which the information-gathering devices will be used.

3. Context

Surveillance is a very context-sensitive field—the technology may be incidental or highly important. A homeowner checking on a housekeeper who is suspected of stealing might use a simple pinhole camera hidden in a smoke detector. A private detective observing a client's spouse in a crowded shopping mall, can usually accomplish the task with discretion and an unobtrusive camera. Some types of surveillance, however, require careful planning and make use of highly sophisticated technology.

In Sherlock Holmes' day, a detective would search mainly for visual clues (or convenient gossip) to solve a crime. Now, crime scene investigators and forensic labs utilize chemical dyes and powders, infrared and ultraviolet sensors, DNA sampling kits, bullet-analysis techniques and access to extensive databases of profiles, mugshots, fingerprints, and bullet casings to solve cases. Traditional low tech materials are sometimes combined with high tech. For example, if a news correspondent or foreign agent is sent to gather information on potential hostilities or human rights abuses in foreign territory, then specialized clothing, cosmetics, contact lenses, language classes, wireless recorders, aerial photographs of the region, maps, telegraph transmissions, satellite-modem-equipped notebook computers, accomplices, and a boat tucked away under a dock in a harbor, may all assist in achieving the desired ends.

Aerial images were once expensive and grainy ($4,000 per image in the early 1990s). At that time, they were primarily sold to researchers, large corporations, and military analysts. Now that high-resolution satellite images can be purchased for less than $25 per square mile, a dramatic shift is occurring in the applications for which these images are used.

The human nervous system is still the most important surveillance 'technology.' All the sophisticated inventions in the world are worthless without strategies, data analysis, and accurate interpretation of the results. The technological developments and devices introduced in this text are only effective if carefully selected and correctly installed. As they say in the computer programming industry, *garbage in, garbage out.* If the surveillance data is irrelevant or inaccurate (or absent altogether, due to faulty components or incorrect usage), it is of little value.

Planning is particularly important when using limited resources in armed conflicts. Left: A Wing Intelligence Officer with the U.S. Air Force, Capt. Muellner, updates a map of Entebbe in 1994. Middle: A member of the TAW target intelligene rach, TSgt. Olague, transfers drop-zone map coordinates to a satellite photograph during an exercise in 1991. Right: TSgt. Hawman from the 1st SOS Intelligence checks a map of southern Thailand before a 1996 preflight briefing in Japan. [U.S. DoD news photos by Andy Dunaway, H. H. Deffner, and Val Gempis, released.]

Maps and Aerial Photographs for Assessing Military Zones and Battle Damage. Left: An Air Force intelligence and targeting chief in the U.S. Air Force, Capt. Muellner, shows position on a map of Korea in 1993. Middle: An illustrator with a U.S. Mains Intelligence Company prepares a map for a 1996 orders brief in North Carolina. Right: A Navy intelligence specialist aboard an aircraft carrier in the Red Sea evaluates aerial photographs to assess 1991 battle damage. [U.S. DoD news photos by Michael Haggerty, Moore, and R. L. Kulger, Jr., released.]

3.a. Scientific Inquiry

Surveillance technologies are used in virtually every field of scientific inquiry. Archaeologists, anthropologists, geologists, meteorologists, marine biologists, zoologists, astronomers, geneticists, forensic pathologists, and sociologists all use surveillance, in one way or another, to monitor trends, conduct experiments, and gather scientific data.

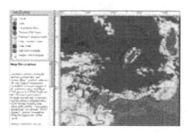 

Left: A sample fish-finding map from the OrbView-2 satellite. Satellite images are used by scientists, resource managers, and commercial harvesters. Right: An OrbView satellite image of Turkey taken in August 1999, after a major earthquake. [News photos ©2000 Orbimage, www.orbimage.com, now GeoEye, used as per copyright instructions.]

Aerial surveillance technologies are especially useful for astronomical study, archeological excavations, and ecosystems monitoring (weather, pollution, climate change).

3.b. Government Applications

Surveillance technologies are widely used by governments and corporations for national defense, local law enforcement, disaster assessment and relief, search and rescue, community planning, resource exploration, wildlife monitoring, property tax assessment, border patrol, camouflage detection, treaty negotiation and verification.

Military and search and rescue specialists monitor natural disasters and war zones to locate lost individuals. Left: Sgt. Lalit Mathias helps tourists into a Blackhawk helicopter after they were rescued near an avalanche in the Austrian Alps. Right: After severe flooding in Honduras following Hurricane Mitch, U.S. Army personnel from the 228th Aviation Regiment found and airlifted a child who was trapped on top of a house. [U.S. Army news photos by Troy Darr and Terrence Hayes, released.]

Rescue operations during natural disasters are often complicated by sustained bad weather and other sources of poor visibility (e.g., smoke). Finding victims who have been stranded on rooftops, in trees, in vehicles, and under debris can be a significant challenge.

Armed Forces Personnel Regularly Receive Intelligence Briefings. Left: Intelligence briefing during peacekeeping and humanitarian operations training. Austrian and Canadian participants are shown in this Partnership for Peace exercise in North Carolina, which operates according to NATO IFOR standards. Right: A U.S. Marine, Capt. Nevshemal, gives an intelligence brief to Hungarian and Central Asian platoons at Camp Lejeune, North Carolina. [U.S. DoD 1996 Released Photos by LCpl. R. L. Kugler, Jr. and LCpl. C. E. Rolfes, released.]

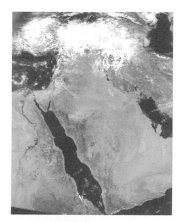

Government departments are making greater use of commercial image sources. Left: Satellite images of the islands of Japan, 9 May 1998. Right: Egypt, Saudi Arabia, and the Middle East, 3 April 1998. [News photos ©1998 OrbImage, now GeoEye as per licensing instructions.]

3.c. Commercial, Agricultural, and Government Applications

There are tens of thousands of legal surveillance applications. Examples include ATM security, public safety monitoring in tunnels and transportation systems, and environmental monitoring to preserve artifacts and artworks in museums and educational institutions.

Insurance adjustors, property value and damage assessors, marketing and promotional professionals, contractors, city planners, investigative journalists, weather forecastors, architectural planners, livestock and food inspectors, surveyors, fishers, crop yield assessors, forestry and fire-fighting personnel all use various forms of surveillance in their work.

Weather Surveillance. Left: This image of Hurricane Elena over the Gulf of Mexico in September 1985 clearly shows the spiral formation and elevated cyclonic cloud berm. Right: A SeaWinds radar data chart of Hurricane Floyd, one of the most destructive hurricanes of the 20th century, imaged in September 1999. Ocean wind speeds are indicated by adding colored arrows. SeaWinds is an orbiting imaging system managed by the Jet Propulsion Laboratories (JPL). [NASA/JSC and NASA/JPL news photos, released.]

Detailed satellite and aircraft-derived weather images of weather patterns, especially of impending storms, are of commercial interest to newscasters, boaters, commercial fishers, firefighters, production companies, insurance adjusters, and homeowners and businesses in the path of impending storms. Accurate weather forecasting can save hundreds, sometimes thousands of lives, and may help people minimize property damage by reinforcing levees, storm windows, and other protective structures.

Surveillance is also used by competitive businesses (and sometimes by governments) for unethical purposes. Seemingly benign technologies, when used covertly to gather information of a competitive nature, may give the surveilling parties an unfair or illegal advantage. It may also widen the economic gap between developed nations and nations that don't have access to technology.

Curran reports that the U.S. agricultural industry has gathered information on global agricultural production and mineral exploration in order to assess international markets.

> The first comprehensive inventory was called the Large Area Crop Inventory Experiment (LACIE) and was undertaken from 1974 to 1977 (MacDonald 1979). Wheat production of the USSR, Latin America, China, Australia, and India was estimated by multiplying the crop area derived from the Landsat sensor data by the estimated crop yield derived from meteorological satellite sensor data [Curran 1980a]. [Paul J. Curran, *Principles of Remote Sensing,* Longman Group, Ltd., 1985.]

The LACIE agricultural survey was not initiated as a spy project, per se. It was a joint effort by NASA, the National Weather Service, and the U.S. Department of Agriculture to "develop and test a crop forecasting system using Landsat data". At the time, it was found to be more difficult to identify and differentiate specific crops than was expected, but satellite imagery has improved significantly since 1974 and such systems have become more practical. The controversy centers around the fact that satellites circle the globe — forecasters monitor not only domestic crops, but the crops of foreign nations. This information is perceived by some as giving countries with access to satellite data unfair market advantages over others.

The development of dual-purpose satellites is not uncommon. The Television Infrared Observation Satellite (TIROS), was originally intended to provide both data for storm warnings and space-based reconnaissance for the Department of Defense (DoD). However, a number of changes occurred and Lockheed was contracted for a reconaissance-focused project that evolved into the Satellite Military Observation System (SAMOS). At about the same time, the Army Ballistic Missile Agency (ABMA) initiated Project Janus to test RCA's idea of putting a television camera in space for use as either a meteorology or surveillance system. When the DoD assigned satellite reconnaissance to the Air Force in 1958, the focus of Janus II became meteorology rather than reconnaissance, and Janus was transferred to the Research Projects Agency (ARPA), the organization that developed ARPANET, the forerunner to the Internet. Janus II was renamed TIROS.

TIROS was transferred to NASA with the understanding that they would cooperate with the Weather Bureau on its development. Interest in reconnaissance continued, however and, in the 1960s, the Air Force began developing a Defense Meteorological Satellite Program (DMSP) to meet military needs. For a while the technologies for military and meteorological satellites moved along separate paths, but were brought together again when the Weather Bureau and NASA parted company and the DoD offered launch services for Weather Bureau satellites. The Weather Bureau evolved into the National Oceanic and Atmospheric Administration (NOAA) and returned to a cooperative position with NASA (see the Aerial Surveillance chapter for more information on satellite surveillance).

3.d. Nonprofit and Public Welfare Applications

Surveillance technologies are used outside of governmental and corporate entities by public-concern watchdog agencies, as well. These organizations monitor government and corporate activities, pollution, investment patterns, natural resources, and price policies. In turn, governments use surveillance technologies to monitor those perceived as radical militant groups and individuals and groups suspected of terrorist intentions.

Nonprofit applications of surveillance technologies include monitoring environments, crops, wildlife, pollution, sociopolitical unrest, corporate management, manufacture of consumer goods, and military activities.

3.e. Personal Applications

Since the mid-1990s, surveillance technologies, particularly motion detectors, video cameras, and computer-related devices, have become consumer items and many surveillance images end up "in the wild" (available to anyone) on the Internet. Thus, there is a need to educate the public about the various devices, how they could or should be used, and implications for personal freedoms and privacy.

4. Origins and Evolution

Each chapter includes historical notes to put the technological development of a particular device or class of devices into its social context.

The historical sections in each chapter generally stand on their own, except that there is overlap between the chapters on radio devices and audio listening devices. It is worthwhile to cross-reference these. There is also some overlap between the infrared and aerial surveillance chapters, since infrared has become an important aspect of aerial surveillance. These, too, should be cross-referenced. They don't necessarily have to be read in sequential order, however.

Important milestones and events of a general nature regarding surveillance and the organizations and policies that have governed their use are described here.

4.a. Introduction

People have been spying on each other for thousands of years. However, more sophisticated surveillance operations became prevalent around the time of the Italian Renaissance.

The Formalization of Surveillance and Intelligence Gathering

Governments and individuals have surveilled each other since the dawn of competitiveness and political awareness. However, modern intelligence institutions in the west arose as a result of centuries-old conflicts between nations such as England and France. In 1573, when Queen Elizabeth I appointed Sir Francis, the Earl of Walsingham, as her Joint Secretary of State, she set the stage for a formalized secret service branch. Sir Francis became known for intercepting secret communications from Mary Queen of Scots that led to Mary's long imprisonment. Walsingham was a dedicated and meticulous administrator, who established a permanent peacetime intelligence service that established the British secret services.

By the 1600s, English documents by various writers and inventors, mention the development of submarine devices for surreptitiously approaching foreign surface vessels and "blowing them up."

In France, Armand-Jean du Plessis, Cardinal, Duc de Richelieu, used extensive intelligence services to protect his position and to influence the unity and future of France. He is quoted as saying "Secrecy is the first essential in affairs of the State."

Rivalries between the English and French continued to foment as Europeans migrated across the Atlantic to settle in the pioneer wilderness of North America. The *War of the Spanish Succession* erupted in 1702, followed by various other military outbreaks that motivated the continued surveillance of "foreigners". As borders began to solidify, tight security watches were put in place to prevent smuggling.

By the mid-1770s, George Washington was making regular use of coded messages, foreign agents, and other surveillance resources to further the aims of the American Revolution. In July 1789, Washington and the First Congress established the *U.S. Customs* service and the ports of entry it was entrusted to surveil. Border surveillance and search and seizure decisions relied heavily on human judgment until the late 1900s, when electronic devices were installed to aid border officials in scanning vehicles and license plates. It also made it possible to record names and identifying documents and characteristics in computer databases, and to patrol the borders at night.

4.b. Establishment of Surveillance-Related Agencies

As populations migrated westward across America in the early 1800s, crime and gangsterism increased. This resulted in a higher emphasis on law enforcement and surveillance, especially as farmers and settlers, including women and children, began to outnumber the original trappers and prospectors. The populace became increasingly concerned about law and order and family safety.

Smuggling wasn't the only problem that motivated lawmakers to approve surveillance practices in early America. A variety of regional currencies and a general wild west mentality had resulted in widespread counterfeiting and train robbing incidents. In 1806, the *Enforcement of Counterfeiting Prevention Act* was established to curb some of these problems. The act enabled U.S. Marshals and District Attorneys to officially police counterfeiting operations, a responsibility that was later transferred to the Secretary of the Treasury.

To curtail counterfeiting, the *National Currency Act* was enacted in 1863 to create and regulate a national currency.

Detective Agencies and Secret Service Agents

In 1846, a barrel-maker from Scotland, Allan Pinkerton (1819-1884), exposed a gang of counterfeiters, made history, and became the Deputy Sheriff. He thereafter became Chicago's first official city detective and founded Pinkerton's National Detective Agency in 1850—one of the oldest and most prominent detective agencies in history. In the decades that followed, Pinkerton's business and reputation grew as he and his agents investigated train robberies, embezzlement, assassination plans, spies, and much more. As a private establishment, Pinkerton had the flexibility to pursue criminals across state lines

In 1861, Abraham Lincoln appointed Pinkerton as his first secret service agent and, during the Civil War, Pinkerton organized a secret service within the U.S. Army.

General George McClellan employed Pinkerton to gather intelligence behind the lines of the Confederate forces during the Civil War. Pinkerton provided information on defenses, supplies, and transportation routes; he engaged in counterintelligence as well. After McClellan was demoted, Pinkerton went back to private detective work.

Pinkerton's Detective Agency. Left: Allan Pinkerton founded Pinkerton's Detective Agency in 1850. Right: Allan Pinkerton (also known at the time as E. J. Allen) in the field with Abraham Lincoln and Major General John A. McClernand. Detractors later claimed the 'plots' against Lincoln were fabricated to 'make work', leading Lincoln to establish his own secret service. [Left: Photo courtesy of Pinkerton Global Intelligence; Right: Photo courtesy of Library of Congress, by Alexander Gardner, October 1862, copyright expired on both images by date.]

The Secret Service in the American Civil War. Left: Members of the Secret Service meeting at Foller's House, Va. during the Peninsular Campaign. Middle: Allan Pinkerton, known as "E. J. Allen", on horseback during his time in the Secret Service at the main eastern theater of the Battle of Antietam. Right: George Banks, William Moore, Allan Pinkerton, John Babcock, and Augustus Littlefield. [Library of Congress May and September 1862 photos by George Barnard, James Gibson, and Alexander Gardner, copyrights expired by date.]

Pioneering Mugshots. Pinkerton's Detective Agency took advantage of advancements in photography to establish one of the first extensive collections of 'mug shots'—pictures of faces to identify and apprehend criminals. The tradition has continued to this day, with searchable computer databases gradually supplementing the print collections. From left to right: mug shots of famous criminals, including the Sundance Kid, Lena Kleinschmidt, and Alan Worth. [Photos mid-1800s, courtesy of Pinkerton Global Intelligence, copyrights expired by date.]

Pinkerton detectives institutionalized "profiling" by collecting criminals' pictures from posters and newspaper clippings, a practice that was eventually adopted by many local enforcement agencies. By the 1870s, Pinkerton's had developed the most extensive collection of 'mug shots' in America. The agency also became known for its logo, a picture of an eye, under which is the motto "We never sleep." This may have popularized the phrase 'private eye.'

Intelligence Documents. In 1863, officers of the *Bureau of Military Information Secret Service* were stationed in Bealeton, Virginia (left) and at Brandy Station in Feb. 1864 (right), which was established at the time George Sharpe was chief of the Bureau. Numerous books and pamphlets are housed in the tent; which are probably reference materials, including maps. [Library of Congress Civil War collection photos, copyrights expired by date.]

Scientific Input to Government

Scientists and inventors are often frustrated by the slow adaptation of new inventions by bureaucrats. Rear Admiral Charles Henry Davis, the great inventor Joseph Henry, and Alexander Dallas Bache suggested a commission or scientific organization to expedite the process of advising Union leaders, but no clear consensus was reached initially. Bache had wanted to institute a national science advisory body to aid the government in formulating policy and funding worthy projects. Finally, Joseph Henry helped establish the Navy's *Permanent Commission* in 1863, with himself, Davis, and Bache as members. John G. Barnard was later appointed to represent the interests of the Army.

April 28, 1895.

Dear Miss Sullivan:

My Private Secretary, Mr. McCurdy, has been absent in Chicago on Association business, and my time has been so occupied on account of the meeting of the National Academy of Science that I have been unable to write and thank Helen for the very beautiful poem she sent me. It has not yet been read before the Literary Society, because the meeting was postponed until May 6.

Helen's effort compares most favorably with the juvenile

Science Advisors to Government. Left: The "Academy of National Science" as it looked at the turn of the century in Philadelphia, Pennsylvania, not long after it was first established. A Permanent Commission of professionals had been founded in 1863 to provide expert advice to politicians regarding scientific inventions that could be used to further national interests. The founders hoped this effort would expedite the critical review and recommendations of new technologies. The Commission led to the establishment of the Academy of National Science which is today's National Academy of Sciences (NAS). Right: A copy of an 1895 letter from Alexander Graham Bell to Anna Sullivan (Helen Keller's teacher) mentioning his attendance at an early meeting of the Academy. [Library of Congress Detroit Publishing Company Collection and the Bell Family Papers collection, copyrights expired by date.]

By 1864, the Permanent Commission had created almost 200 reports, confirming its viability as an advisory body, and paving the way for establishment of the National Academy of Sciences (NAS). Bache became the Academy's first president, but due to illness leading to his death, Joseph Henry provided much of the leadership in its early days.

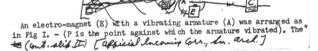

A. Graham Bell to Joseph Henry from 292 Essex St., Salem, Mass, 4/2/75:
"You were kind enough to express an interest in the experiments to which I directed your attention in Washington, and I trust that I do not take too great a liberty in addressing you again upon the same subject.
I have recently been led to the belief that an intermittent current of electricity creates a molecular vibration in the conductor through which it is passed; – and that this is the cause of the noise we heard proceeding from the empty helix of wire. I have repeated the experiment with coils of different resistances, and have satisfied myself that the intensity of the sound increases with the resistance employed. It is not enfeebled when the helix is stuffed with substances containing no oxygen.
I find further that a noise can be induced in a Ruhmkorff Coil without completing the circuit in which it is placed.

An electro-magnet (E) with a vibrating armature (A) was arranged as in Fig I. – (P is the point against which the armature vibrated). The

Scientific Advisement for Surveillance Technologies. Left: A copy of correspondence about electromagnetism from A. Graham Bell to Joseph Henry in 1875. Joseph Henry also studied electromagnetism and provided encouragement and counsel to many prominent inventors, including Charles Wheatstone and Samuel Morse. He significantly influenced the inventors of telegraph and telephone technologies. Along with Charles H. Davis (right), Henry and Bache helped establish the National Academy of Sciences based on the 'Permanent Commission.' Right: Rear Admiral Charles H. Davis in the early 1900s. [Library of Congress Bell Family Papers Collection; painting by F. P. Vinton. Copyrights expired by date.]

Surveillance of Coastlines

During the 1840s, Lt. Charles H. Davis (1807-1877) played a significant role in coastal surveillance. He worked with Alexander Bache on the *Coast Survey* and authored the *Coast Survey of the United States* in 1849. This was only one of many surveillance- and technology-related projects in which he participated over the next three decades.

Davis was promoted to Captain and later to Rear Admiral. In the 1860s, he was Chairman of the Western Navy Yard Commission and an advisor to President Lincoln. Under his direction, assistants conducted numerous surveys, including a reconnaissance survey of the area around Vicksburg. He also aided Joseph Henry and Alexander Bache in establishing the National Academy of Sciences. In the 1870s, he headed up the U.S. Naval Observatory and, from 1870 to 1873, was Shipyard Commander at Norfolk Naval Shipyard with four hulls honoring his name.

The U.S. Secret Service and Department of Justice

Within the Department of the U.S. Treasury, the Secret Service Division (SSD) operated from 1865 to 1879. One of its main concerns was the detection and apprehension of counterfeit rings. It also investigated forgery and securities violations.

The position of the U.S. Attorney General was first created in 1789. In 1870, the Attorney General was appointed the Director of the newly formed *Department of Justice* (DoJ). The Department of Justice is one of the oldest establishments to handle federal investigations and continues in this role today.

The armed forces were also establishing official intelligence departments at this time, with the *Office of Naval Intelligence* being created within the *Bureau of Navigation* in 1882.

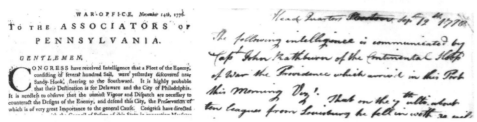

Historic Intelligence Documents. Left: In 1776, Congress received "Intelligence that a Fleet of the Enemy, consisting of several hundred Sail were yesterday discovered in Sandy Hook." Right: In 1778, George Washington indicated in correspondence that he had received intelligence from John Trathburn of the Continental Sloop of War, the Providence. Naval intelligence has an old tradition, leading up to the official establishment of an Office of Naval Intelligence in 1882. [Library of Congress, copyrights expired by date.]

In 1887, the Office of Naval Intelligence (ONI) directed naval forces to create a photographic record of coastal installations. Matthew Brady and Alexander Gardner were two important Civil War photographers who demonstrated the value of photo reconnaissance and the documentation of war (see photos following).

This photo provides a small surveillance challenge for readers who like solving real-life historic puzzles. The photo is labeled (among other things) "Bureau of Intelligence" in small letters at the bottom and "Commander" Chas. H. Davis is seated at the desk. The Library of Congress dates it as 1890 to 1901, but if Davis were a Navy Commander at the time, a rank just above Captain, it is more likely to have originated in the 1850s. But ... that was before the Civil War (which took place in the early 1860s) and an official intelligence office in the 1850s seems unlikely. Another possibility is that the term 'commander' is used loosely or that the image is mislabeled (which is not uncommon). Is this office possibly a forerunner to the *Bureau of Investigation*, which wasn't officially established until 1908? [Photo: Library of Congress collection, copyright expired by date.]

Strategic Photographs. Left: Photographer Matthew B. Brady, in 1889. Brady took many historic photos of political figures in the American Civil War. Middle: The shell-damaged deck of the Confederate gunboat "Teaser" captured by the U.S.S. Marantanza, as photographed by James Gibson in July 1862. Right: The war-damaged buildings at a Navy Yard in Virginia in 1864, documented by Alexander Gardner (1821-1882). Many strategic and documentary images of coastal installations were recorded during the War by these photographers. [Library of Congress (photo of Brady possibly by Levin Handy), copyrights expired by date.]

In 1890, the *Office of Naval Intelligence* was transferred to the *Office of the Assistant Secretary of the Navy* and later returned to the *Bureau of Navigation* in April 1898. Thus, many key agencies were taking form by the turn of the 20th century.

4.c. New Public Laws and the Rise of Technologies

Radio-Echo Detection

In Europe, a German inventor, Christian Hülsmeyer (1881-1957) sought to improve navigation by bouncing radio waves off of objects and detecting the returned signal. On 30 April 1904, he registered patent DRP #165546 for a *Telemobiloskop* (far-moving scope)—a radio device to help prevent marine collisions. This important innovation was the forerunner of modern radio-ranging techniques—an early radar system (see the Radio and Radar Surveillance chapters).

The "Progressive Era"

In the early 1900s, it was difficult to communicate over long distances in America. Most of the country was still wilderness. The northwest had barely been settled for fifty years, migrant shacks dotted the Mississippi River, and Native Americans still followed traditional hunting practices in forests and rivers along the Canadian border. Mechanical devices were uncommon. Most surveillance involved the use of simple telescopes, or the bribing of loose-tongued, eavesdropping telephone and telegraph operators. Community disturbances were handled by local law enforcement agencies. But, with technology evolving, and the population steadily growing, distance communications began to effect nationalism and the structure of the country.

By the early 1900s, the Department of Justice (DoJ) was using Secret Service agents ('operatives') to conduct investigations. In 1905, the Department created the *Bureau of Criminal Identification* with a central repository for fingerprint cards. Local law enforcement agencies began developing their own fingerprint repositories, perhaps in part because the DoJ was using convicts to maintain the federal print registry, a practice that was understandably controversial.

Attorney General Charles Bonaparte brought political pressure to bear on Congress to allow him to control investigations under his jurisdiction and, on May 1908, a law was enacted preventing the Department of Justice from engaging Secret Service agents.

Origins of the FBI

In 1908, during the latter part of Theodore Roosevelt's presidency, a *bureau of investigations* was established by Charles Bonaparte to investigate a variety of interstate, antitrust, copyright, and land fraud cases. Two years later when the *Mann Act* ("White Slave" act) was passed, bureau responsibilities broadened to include criminals who might not have committed federal violations, but who were evading state laws. After the outbreak of World War I, the *Espionage Act* and the *Selective Service Act* were passed and bureau responsibilities again increased. In the course of a decade, the bureau had grown from less than a dozen to over two hundred Special Agents. On completion of their terms in 1909, both President Roosevelt and Bonaparte recommended that the force of agents become a permanent part of the Department of Justice and General George Wickersham officially named this force the *Bureau of Investigation* (BoI).

War and Post-War Surveillance

In 1917, during Woodrow Wilson's presidency, the United States entered World War I. This spurred the deployment of surveillance technologies like never before. Submarine-spotting airships, airplanes, magnifying devices, code-breakers, etc. were all used in the arsenal to monitor hostile forces. As a result of the Great War, many new actions were taken to protect national interests.

The *Bureau of Investigation* acquired responsibility for the *Espionage Selective Service Act* and *Sabotage Act* and began to investigate enemy aliens. Specialization became an asset, with agents providing deciphering, foreign language, and problem-solving expertise. In 1919, the former head of the Secret Service, William J. Flynn, became the Director of the *Bureau of Investigation*.

In October 1919, the *National Motor Vehicle Theft Act* provided the BoI with tools by which it could prosecute criminals who tried to evade capture by crossing state lines.

Prohibition

The wages of war, combined with a liberal political climate in America (compared to the 'old countries' from which many had immigrated), created a number of smuggling and export problems, as well as civil disobedience that was in part attributed to the excessive use of alcohol.

Common civil and criminal problems at the time included street fighting, illegal drinking establishments, molestation of women, and an increase in organized crime and 'gangsterism'. These, along with "The Great War", led to *Prohibition*, the complete banning of a number of exports, social actions, and the sale and consumption of alcohol.* *Prohibition* granted Department of Treasury enforcement agents stronger powers for investigating and convicting criminals. Increased surveillance enabled law enforcers to find contraband and cleverly hidden stills and drinking establishments set up by those unwilling to accept the strictures or who were eager to cash in on the black market. Most of this surveillance was conducted by local authorities — the BoI had limited powers of jurisdiction.

In the early 1920s, the economy began recovering from World War I and the U.S. entered the "Roaring Twenties". New inventions, including gramophones, music boxes, telephones, and radios caused sweeping changes in communication, recreation, and surveillance methods.

Many men had been trained in the military during World War I and sought work in related fields after the war. A young law school graduate name J. Edgar Hoover, an assistant in the *General Intelligence Division,* was appointed Assistant Director of the Bureau of Investigation. In1924, the new President, Calvin Coolidge, appointed him Director, at a time when the BoI had grown to over four hundred Special Agents. When Hoover accepted the appointment, he made many changes. He had many agents replaced, overhauled promotions and work appraisals, and established inspections and training programs. Hoover further instituted a significant tool of surveillance, an *Identification Division*. It was an important agency—evidence has little value if it cannot be associated with the persons responsible for leaving the clues. Local police fingerprint files and federal files began to be amalgamated into a central resource.

UTTER DEMORALIZATION OF BOOZE GANG

After weeks of anxious, wakeful watching, the expected has come. The supreme court of the United States has held as constitutional, war-time prohibition. The rum gang had looked forward to, and sorter hoped for one more wet Christmas, but alas their expectations have gone a glimmering. Down in Kentucky they have on hand about 28,000,000 gallons of wet goods.

ONLY COLORED MAN APPOINTED PROHIBITION AGENT.

Roy D. Fowler, of this city, a former lieutenant in the 9th Ohio Battalion, National Guard, has received the appointment as Federal prohibition agent for the district comprising the states of Pennsylvania, Delaware, Ohio and Virginia. Fowler is the only Colored man in the district known to have secured an appointment as agent by the Internal Revenue Department. He was in Columbus this week, to which point all district agents had been called to receive instructions. The position pays $2,400 per annum.

Surveillance During Prohibition. Top Left: Announcement of 'Dry Christmas' after the Supreme Court upheld alcohol prohibition as constitutional, in 1919. Top right: Early efforts at prohibition didn't involve just alcohol; agents had been hired to uncover slavery and export violations, as well. Bottom Left: This clipping shows how the IRS was hiring Federal Prohibition Agents in January 1920 to detect and apprehend people breaking the newly implemented alcohol prohibition laws. Bottom Right: Federal Agents at the Customs House with confiscated liquor in Brownsville, Texas, in Dec. 1920. [Library of Congress clippings and photo from 1919 and 1920, copyrights expired by date.]

The Age of Communications

While Harding was the first President to greet Americans through the airwaves, radio truly came of age in the mid-1920s during the Presidency of Calvin Coolidge. Coolidge was the first U.S. President to make extensive use of press photography and radio technologies to further his campaign and administration goals. The devices would forever change the way political candidates and public officials communicated with the public..

---

*Total banishment of alcohol didn't work and was eventually repealed. Too many otherwise law-abiding citizens opposed it or flaunted the laws, but it had some lasting social effects nonetheless, as it resulted in restrictions that curbed the excessive use of alcohol which may have been contributing to crime at the time.

Political and Governmental Use of New Technologies. The campaign and administration of Calvin Coolidge in the 1920s was the first significant presidential period with extensive media coverage. Coolidge made liberal use of photography and radio broadcasting to promote his political goals. Left: Coolidge in the Oval Office. Right: Calvin Coolidge posing for press photos. More about this pivotal period of history is described in the Radio Surveillance chapter. [Library of Congress 1923 and 1924 Archive Photos, copyrights expired by date.]

The Great Depression

In 1929, the stock market crashed, the Great Depression descended, and up to a third of the population was unemployed at any given time over the next decade. It was an era of great need and crimes were committed by individuals who might have been law-abiding in better economic circumstances. It was also a time when people craved entertainment and a respite from hard economic realities. In spite of the financial upheavals, many great engineering feats occurred during the 1930s, including the construction of the Hoover/Boulder Dam, and the invention of cathode-ray tubes, better radar systems, and more efficient aircraft and radio communications.

Evolution of the BoI into the FBI

In July 1932, the *Bureau of Investigation* was renamed the *U.S. Bureau of Investigation*, solidifying its national focus. The same year, the Bureau established a Technical Laboratory to engage in a number of types of forensic intelligence-gathering, surveillance, and analysis activities (see the Biological/Chemical Surveillance chapter).

In 1934, responding to gangster activities and increased crime due to the poverty and hardships of the Depression years, the U.S. Congress granted greater powers to the *Bureau of Investigation*. By 1934, Bureau agents were authorized to make arrests and carry firearms. In 1935, the Bureau was renamed the *Federal Bureau of Investigation*. The FBI National Academy was established to train police officers in modern investigative methods.

Prior to World War II, science and technology were developing at a rapid rate. Cathode-ray tubes, essential elements of television, were incorporated into oscilloscopes and other kinds of display devices by the mid-1930s; radio ranging (radar) for air navigation was being promoted by 1937. Computers were invented independently by Konrad Zuse in Germany and Professor J. Atanasoff and his graduate student Clifford Berry in America. An explosion of technology appeared imminent when the war changed people's priorities; some projects were shelved and others developed more rapidly in different directions to serve the needs of national security.

In 1939, after several years of localized unrest, German-centered conflict broke out in Europe. Surveillance, espionage, and intelligence-gathering became major concerns for both nations at war and those seeking to avoid war. Due to the efforts of William Sebold and the FBI, the Duquesne Nazi spy ring in the U.S. was uncovered and defeated.

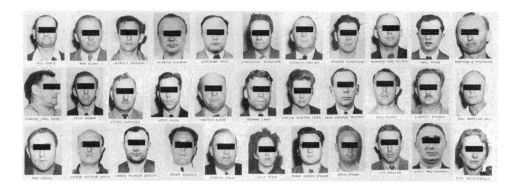

National Security Surveillance. Thirty-three members of the Duquesne Nazi spy ring were found guilty by jury in Dec. 1941 through the investigative efforts of the FBI, with assistance from William Sebold. [FBI news photo, released. Identities obscured in deference to their descendants.]

In 1940, many secret service and war-related measures were taken. The U.S. Congress passed the *Smith Act* and the draft was re-established. FBI surveillance was no longer limited to monitoring foreign powers, but also monitored American deserters and draft-dodgers. President Roosevelt established the *Special Intelligence Service* (SIS) in Latin America to monitor Axis activities and to disrupt Axis intelligence-gathering efforts.

4.d. Wartime Use of Surveillance Technologies

On 7 December 1941, the Japanese successfully attacked Pearl Harbor, in spite of U.S. intelligence reports warning of a break in diplomatic relations. The United States entered the war. A higher priority was put on the development and use of surveillance technologies, including radar defense and cryptological systems.

Military Intelligence on Foreign Nations. Left: In 1939, Albert Einstein wrote to the President warning of a possible German nuclear threat. Right: In 1942, the Manhatten Project was initiated to build a superweapon based on nuclear fission. Shown here is a portion of a June 1943 communication marked "Secret" from President Franklin D. Roosevelt to Robert Oppenheimer, head of the Los Alamos scientific team that was developing the atomic bomb. [Einstein drawing copyright 1998 by Classic Concepts used with permission; White House letter from the Library of Congress, J. Robert Oppenheimer Papers.]

Establishment of the CIA

When America entered the War, the number of FBI employees nearly doubled over the next three years and it was decided that a wider umbrella for national surveillance/reconnaissance services was needed. President Franklin D. Roosevelt asked New York lawyer William J. Donovan to draft a plan for an American intelligence service. In July 1941, Donovan was appointed as "Coordinator of Information". The *Office of Strategic Services* (OSS), the forerunner of the Central Intelligence Agency, was established, in June 1942, with Donovan as Director. The OSS collected and processed strategic information as required by the Joint Chiefs of Staff and conducted special operations not assigned to other agencies. At this time, the OSS did not have the wide-ranging jurisdiction of later organizations. The military forces and the *Federal Bureau of Investigation* still had significant responsibilities that included foreign matters.

These were turbulent times as the "European war" escalated into World War II. William Donovan formally suggested that the OSS be separated from the Joint Chiefs of Staff, and be put directly under the supervision of the President such that it might:

> ... procure intelligence both by overt and covert methods and ... at the same time provide intelligence guidance, determine national intelligence objectives, and correlate the intelligence material collected by all government agencies.

The War was reaching a climax and the need for inelligence was greater than ever, but political reorganization made the OSS short-lived. In August, 1945, the U.S. dropped an atomic bomb on Hiroshima. The day after the bomb exploded, Stalin appointed Lavrentii Beria to oversee a Soviet atomic bomb program.

In October 1945, the OSS was dismantled, and its functions transferred to the State and War Departments.

Three months later, in January 1946, President Harry S. Truman established the *Central Intelligence Group* (CIG), under the direction of the *National Intelligence Authority*, with a mandate to coordinate departmental intelligence to supplement existing services. Rear Admiral Sidney W. Souers, the Deputy Chief of Naval Intelligence, was appointed the first Director of Central Intelligence. The stage was set for a national, centralized approach to American intelligence operations in name, if not yet in fact.

Though public records are fragmentary, it is apparent that the Soviets, unnerved by the sudden atomic weapons superiority of the U.S., engaged in significant espionage at this time. Under pressure and time constraints to match American atomic explosives capabilities, spies and informants at Los Alamos may have contributed substantially to the development of the bomb that was tested by the Soviets in 1949.

U.S. reconnaissance was active in 1946. Historical notes from the 46th Reconnaissance Squadron indicate that Alaska-based personnel had learned to air navigate in polar regions in any season, making broad-based aerial surveillance possible. Fairbanks missions were initiated in July 1946 as reported by the Department of the Air Force:

> ... and in the following year flew more than 5,000 hours and 1,000,000 miles to test material and personnel. Its long-range project was to observe and photomap the Alaskan area for strategic location of defense installations. Other projects included ex-Magnetic Center Determined by Air Units. Exploratory flight for information necessary to establish regular air transport, service; and photomapping for oil formations, in cooperation with the Navy.... A system of reference-heading navigation which was worked out enables planes to fly anywhere and know their location to within one mile.

Thus, the combination of improvements in aerial reconnaissance and the deployment of the atomic bomb caused a shift in priorities in armed forces technologies. For hundreds of years, naval forces had been a dominant line of defense. Now air forces attained a new prominence in the overall strategic picture. For thousands of years, sheer manpower had been a significant aspect of warfare, now technology became an important means to enhance and perhaps someday replace human beings on the battlefield.

> From the earliest times through World War I, battles and wars were directed against people. The focus of effort was on killing enemy forces until the opposition withdrew or surrendered. Beginning with World War II and continuing through the Persian Gulf War, the main goal of battle made a transition from destroying people to destroying war machines. Tanks, airplanes, artillery, armored personnel carriers, air defense weapons and surface-to-surface missiles have been the prime objectives against which firepower is planned and directed. Now, however, there is a new era emerging–information. Information is the key to successful military operations; strategically, operationally, tactically and technically. From war to operations other than war, the adversary who wins the information war prevails.

> [General Glenn K. Otis, U.S. Army, Retired, Information Campaigns, 1991 quoted in *Military Review*, July-August 1998.]

4.e. Post-War Politics

Responsibility for Security and Related Policies - The CIA and FBI

Definitions and responsibilities related to surveillance in the U.S. were more explicitly defined following World War II and the trend toward nationalization and centralization continued. It was clear that the interception of radio and sonar signals and the decryption of sensitive communications significantly influenced the outcome of the War. Post-war vigilance included the sounding of air raid drill sirens, the building of bomb shelters and bunkers, and increased foreign intelligence-gathering.

In the wake of World War II, in September 1947, the *National Security Act of 1947* was issued. The act comprised a body of important policy guidelines for U.S. intelligence activities, some apparently stemming from recommendations made by Donovan three years earlier. It also represents another reorganization in which the *National Intelligence Authority* and the *Central Intelligence Group* were replaced by the *National Security Council* (NSC) and the *Central Intelligence Agency* (CIA).

> ... The function of the [National Security] Council shall be to advise the President with respect to the integration of domestic, foreign, and military policies relating to the national security so as to enable the military services and the other departments and agencies of the Government to cooperate more effectively in matters involving the national security.... The Director of Central Intelligence, who shall be appointed by the President, by and with the advice and consent of the Senate ... shall ... serve as head of the United States intelligence community; ... act as the principal adviser [sic] to the President for intelligence matters related to the national security; and ... serve as head of the Central Intelligence Agency.... [The National Security Act of 1947, July 1947 amended version.]

The CIA assumed responsibility for coordinating America's intelligence activities with regard to national security, and safeguarding intelligence methods and sources.

From 1945 to 1947, a nervous post-war society turned wary eyes on Communist sympathizers in the U.S., particularly those who might be in higher levels of U.S. government. Political activists monitoried the government and amassed classified U.S. documents. Secret service agents, especially those in the FBI were, in turn, monitoring those suspected of stealing U.S. secrets. In response, the 1946 *Atomic Energy Act* granted the FBI responsibility for determining "the loyalty of individuals ... having access to restricted Atomic Energy data." (This trend toward granting greater and greater powers for determining 'loyalty' to the United States continued for another decade.) The FBI's responsibility was broadened once again, to include background checks and monitoring of federal employees.

When Hitler died by his own hand, national fears turned from Germany to the U.S.S.R. and foreign intelligence became a higher priority than ever before. Americans feared a rumored atomic weapons program in Russia which, in turn, had been motivated by fear of American atomic weapons.

In July 1948, *National Security Council Intelligence Directive 9* (NSCID9) was issued to more clearly define national intelligence policies and responsibilities. This Directive remained in effect for some time.

In 1949, there were two significant milestones in national intelligence coordination within the U.S.:

- The CIA was granted broad powers of secrecy and funding exemptions through the terms of a new *Central Intelligence Agency Act.*

- On 20 May 1949, the *Armed Forces Security Agency* (AFSA) was formed within the U.S. Department of Defense, under the command of the Joint Chiefs of Staff.

AFSA was intended to direct the *communications intelligence* (COMINT) and *electronic intelligence* (ELINT) sections of signals intelligence units in the military. Over time, concerns were expressed from within the government and from within AFSA itself that the organization was not effectively carrying out its functions. Over the next several decades COMINT responsibilities were scrutinized and redefined a number of times.

Korea became a strategic outpost following Japan's defeat in the War. Soviet troops occupied much of northern Korea, while Americans occupied the region south of the 38th parallel. By September 1945, South Korea was under U.S. military rule. A U.N. *Temporary Commission on Korea* was tasked with drafting a constitution for a unified government for all of Korea. When denied access to the north, it continued with elections in the south, in May 1948. Shortly afterward, a constitution was adopted and the Republic of Korea was established. By late 1948, Soviet troops had withdrawn from North Korea, but maintained assistance; American troops similarly withdrew from the south in 1949, but maintained assistance.

Fears of a Soviet atomic bomb were confirmed on 29 August 1949 when the U.S.S.R. carried out a test detonation. Cold War vigilance and surveillance of foreign atomic capabilities became a U.S. priority. Global instabilities were further heightened when, in June 1950, North Korea invaded South Korea and U.S. troops became involved in defending South Korea.

Military Intelligence Acquires a Broader National Scope

In the early 1950s, technology flourished. New means of broadcasting radio, television, and telephone messages were developed. One of the most important aspects of microwave physics was that very-short waves were not reflected off the ionosphere, but passed right through, out into space. This spurred scientific research in astronomy, knowledge which was later applied to channeling Earth communications through orbiting satellites.

In March 1950, the *National Security Council Intelligence Directives* was established and defined the responsibilities of the *U.S. Communications Intelligence Board.* The Board provided coordination of government COMINT activities and advised the Director of the CIA in COMINT matters. Interestingly, COMINT activities were considered to be "outside the framework of other or general intelligence activities" and, conversely, other intelligence directives (e.g., electronic intelligence) were not considered directly applicable to COMINT. This point of view was later revised.

In 1952, a study was carried out under the chairmanship of Herbert Brownell to survey U.S. communications intelligence (COMINT), with a focus on the functions of AFSA. This became known as the Brownell Committee Report. The Report suggested that better national coordination and direction were needed, provoking action on the part of President Harry S. Truman.

In an October 1952 Memorandum, President Truman designated the Secretaries of State and Defense as a *Special Committee of the National Security Council for COMINT* to establish COMINT policies and to provide policy advisement, with help from the Director of the CIA, through the *National Security Council*. This document further established the *Department of Defense* (DoD) as the executive agent for the production of COMINT information. The 1948 *Intelligence Directive No. 9* by the *National Security Council* was revised in December 1952 in response to this Memorandum.

As a result of these studies and memoranda, security responsibilities were shifted from AFSA to the *National Security Agency* (NSA), with a broader mandate of national responsibility for communications intelligence. The responsibilities of the *U.S. Communications Intelligence Board* were revised as well. Electronic intelligence remained within the mandate of the armed services. Two decades later, the NSA again was granted broader powers.

The *National Security Act of 1947* was amended in 1953 to empower the President, with the support of the Senate, to appoint the Deputy Director of Central Intelligence (DDCI) who would stand in for the Director if he could not perform his duties or in the event of a vacancy in the position. Half a century later, the organizational structure would be changed again.

Post-War Prosperity and Baby-Boom Years

At this point in history, surveillance technologies were used almost exclusively by government and law enforcement agencies and were rarely seen in civilian applications. It was an era of trust and focus on personal responsibility. In the 1950s, in small and mid-sized towns in the U.S. and Canada, it was rare for people to lock their homes or their cars. Burglar alarms in businesses were uncommon and in homes, extraordinarily rare. Post office buildings stayed open 24 hours so people with rented boxes could retrieve their mail when it was convenient and small children played in parks unattended.

In sharp contrast, at the national level, policies on national security were being revised to increase security and surveillance, in the face of growing Cold War fears of Communist expansion.

In the past, the FBI and other national investigative organizations had the reputation of being rather shadowy entities, but with the proliferation of television and long-distance communications in general, government agencies came under greater public scrutiny. Then, in the mid-1950s, the FBI's name became more prominently associated with solving terrorist crimes when a high-profile case came to public attention. In 1955, a plane exploded in midair in Colorado, and it was the FBI that pieced together evidence to assist local law enforcement agents in winning a case against the person who planted the bomb.

U.S. Congress enacted laws to provide the FBI with greater powers to investigate racketeering, gambling, and civil rights violations. The powers of the NSA were also expanded. In September 1958,10 electronic intelligence (ELINT) was added by Directive (NSCID 6) to the responsibilities of the *National Security Agency*. This Directive was revised in January 1961.

At the same time, the perspective on intelligence functions was broadening and the motives and responsibilities of government were scrutinized [underlined for emphasis]:

> Information on the military strength and plans of a potential enemy is highly important, <u>but intelligence agencies cannot concentrate solely on the military because armed forces are only an instrument of policy. They must, at the same time, gather information on the aims and objectives of those who decide on the use of this instrument</u>. Military power is only one of many assets of the policy makers. Therefore, it is obvious that the collection of political intelligence is a matter of the first order, more so now than ever before, since the "hot war" has become a highly dangerous instrument of politics. The relative uselessness of the military instrument in the direct solution of political conflicts has extraordinarily enhanced the importance of the nonmilitary spheres of the international power struggle–diplomacy, economic and cultural policies, propaganda, psychological warfare, and subversive activity....
>
> [Dr. Georg Walter, "Intelligence Services", *Military Review,* August 1964.]

Public Announcements, Cuba, and Cold War Fears

Cars and TVs profoundly changed society in the 1950s and 1960s. As the population increased, it also became more mobile. During the Baby Boom years, automobiles made it possible for people to commute to work while living in less expensive houses in outlying areas now called suburbs. Due to increased mobility, neighbors and neighborhoods changed. Suspicions increased.

Public service announcers and insurance companies began using television to entreat people to lock their doors and their vehicles. This change of habits on a personal level would eventually fuel public demand for consumer surveillance devices.

The sixties were turbulent times. The availability of recreational drugs increased, civil rights issues were frequently on the news, and Cuba was invaded by Cuban exiles at the Bay of Pigs with support from the CIA, in April 1961. Then, in October 1962, the *Cuban Missile Crisis* erupted when intelligence services discovered Soviet-manufactured nuclear missiles in Cuba that were capable of reaching the U.S.

This touched a nerve in a generation sensitized to the devastation of World War II. It also illustrated an aspect of American politics that is often questioned—there is a general assumption by Americans that our democratic superpower knows what's best for the rest of the world and has a right to bear nuclear arms while keeping them out of the hands of other nations. Most people would acknowledge that this is a double standard and those opposed to the incongruity might even argue that the U.S., in global terms, is hanging onto its 'might and right' in the same way the Confederates held onto their 'might and right' to preserve their superiority and status quo in pre-Civil War days. Regardless of the interpretation, however, the fear of nuclear proliferation in the sixties resulted in an increase in foreign surveillance in general, and technologies designed to find or defeat nuclear weapons, in particular, flourished because the 'nuclear arms race' was no longer confined to the U.S. and the U.S.S.R.

New facilities and procedures were implemented to improve international intelligence-gathering. The modeling of foreign terrain, buildings, and weapons systems is one way to build a composite picture of a foreign nation's priorities and capabilities. In the mid-1960s, the CIA established a workshop for creating hand-crafted 3D models based on surveillance photos and intelligence reports. This facility operated for the next three decades. Computer modeling now serves much the same purpose.

At the same time, physics research led to the development of microprocessors which, in turn, made it possible to develop smaller, lower-cost personal computers. In military and large corporate circles, visionary computer engineers began to develop networking and time-share systems, the germ that grew into the ARPANET and, eventually, the Internet.

The Mid-1960s - Vietnam, Freedom of Information, and the Space Race

It's not clear how much of the social revolution and activism that occurred in the 1960s was due to increased news coverage and television viewing, but clearly it was a era of significant scrutiny of the government by the populace. Student council meetings at universities in the 1980s were generally poorly attended, with often less than a dozen people showing up to express their views. In contrast, in the sixties, as much as 80% of a student body would show up for meetings, with megaphones broadcasting debates to thousands of attendees.

The public demanded access to information and assumed a more active role in surveilling government officials. Government became more responsive and open to public opinion and thus, in 1966, the U.S. enacted the *Freedom of Information Act* (FOIA). This provided individuals with the right, enforceable in a court of law, to request access to records held by federal agencies upon submitting a written request. The scope of the Act, as it was written, was limited by nine exemptions and three exclusions. In spite of the exemptions, the FOIA was to become a driving force and an important tool of democracy. As government surveillance activities spread, so did those of the public.

In 1966, other changes occurred, as well. U.S. Coast Guard cutters were dispatched to aid Army and Naval surveillance forces involved in the Vietnam conflict. The U.S. Supreme Court granted the FBI greater leeway to investigate civil rights crimes and, in 1968, the *Omnibus Crime Control and Safe Streets Act* paved the way for FBI use of electronic surveillance to investigate specific types of violations. The FBI was concerned about terrorist activities and used investigative personnel and counterintelligence programs (CointelPro) to counteract these activities. However, Hoover is said to have discouraged intrusive methods such as wiretapping (see Chapter 2 for contradictions to this claim). These were eventually forbidden unless they were permitted under the terms of the *Omnibus Crime Control and Safe Streets Act.*

In the late 1960s, the world entered a new communications era with the introduction of geosynchronous satellites. By 1968, the U.S. was launching communications intelligence (COMINT) satellites (described in detail in the Aerial Surveillance chapter) specifically designed to carry out surveillance tasks. From this point on, both government and amateur satellite technologies developed rapidly. The following year, the Department of Defense (DoD) commissioned the *ARPANET*, the forerunner of the public Telenet network that was implemented in the mid-1970s.

In April 1971, J. Edgar Hoover formally terminated CointelPro operations.

4.f. Nationalization, Computerization, and Government Scandals

The Establishment of Research and Technology Labs

In the 1970s, a number of government departments established or upgraded laboratory facilities to reflect the increasing importance of advancements in science and electronics.

In 1970, U.S. Customs established a Research Laboratory to develop new analytical methods and to evaluate new instrumentation for use by Field Laboratories. The Laboratory would help maintain analytical uniformity among Field Labs and communication with other federal enforcement and technological agencies. It would further provide assistance to drug-screening and canine-enforcement programs.

 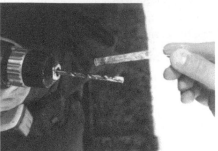

Increased High Tech Emphasis. Left: U.S. Customs Commissioner Kelly and Senator Campbell discussing high technology in law enforcement at a Senate Treasury Technology briefing and exhibit in Washington, D.C. Right: After a Customs canine made a positive hit on the hull of a ship inspected for drugs, the hull was drilled. The dust drawn out on the drill bit was placed in a chemical detector where it turned blue, testing positive for the presence of cocaine. Sixty-five kilos of cocaine were subsequently seized from the vessel. [U.S. Customs news photos by James Tourtellotte, released.]

National Intelligence and Personal Computers

In September 1971, the Kenbak-1, the first commercially-significant microcomputer, was advertised in Scientific American. The age of personal computers had begun, but few people took notice until the Altair was marketed in Popular Electronics and other magazines three and a half years later.

Up to this time, the NSA was specifically concerned with communications intelligence (COMINT). COMINT is part of a broader category called signals intelligence (SIGINT) which also includes electronic intelligence (ELINT) and telemetry intelligence (TELINT) which means that COMINT, as defined by the U.S. Department of Defense (DoD), was limited in scope. However, there were even further stipulations incorporated into the administrative definitions for COMINT that restricted NSA's jurisdiction beyond a general definition of COMINT.

In order to understand why the Intelligence Community was criticized for stepping outside its mandate over the next several decades, it should be understood that the general meaning of INTs, and the NSA-prescribed meaning of INTs, are not the same. The general definition of COMINT encompasses all types of communications (not just foreign). However, the NSA definitions for various types of intelligence inherently included limitations as to their use, as in this NSA definition for COMINT (communications intelligence):

COMINT in the early 1970s was defined as the collection and processing of "foreign communications" and was even further defined as restricted to communications that were "encrypted or intended to be encrypted." It did *not* include "unencrypted written communications, except ... [those] which have been encrypted or are intended for subsequent encryption." The NSA specifications for COMINT also explicitly excluded the interception of public broadcasts, except for "encrypted or 'hidden meaning' passages in such broadcasts."

These distinctions, in effect in the early 1970s, became important on several occasions in the 1990s, when NSA activities related to satellite and computer-network surveillance were questioned. They form the foundation for NSA statements that NSA personnel operate according to very specific guidelines.

In December 1971, NSA jurisdiction was increased. The Department of Defense issued a Directive prescribing the "authorities, functions, and responsibilities of the National Security Agency (NSA) and the Central Security Service (CSS)." This document identified the NSA as responsible for providing "the Signals Intelligence (SIGINT) mission of the United States and to insure secure communications systems for all departments and agencies of the U.S. Government." It further named the Central Security Service to conduct "collection, processing and other SIGINT operations as assigned."

In 1972, the National Security Council issued *Intelligence Directive No. 6* (NSCID6), "Signals Intelligence", prescribing the SIGINT activities and responsibilities of the Director of Intelligence. This document superseded NSCID 6 which had been in effect since September 1958.

In May 1972, J. Edgar Hoover died. He had been the Director of the FBI for almost 48 years. L. Patrick Gray was appointed by President Nixon as Acting Director the day after Hoover's death. Gray initiated progressive hiring practices by appointing the first women Special Agents to the FBI since the 1920s.

In September 1972, the Defense Security Service (DSS) was established to consolidate personnel security investigations within the Department of Defense (DoD).

The 1970s - The Watergate Scandal

In 1972, an incident at the Democratic National Headquarters in the Watergate Building involving illegal undercover operations, demonstrated the importance of understanding many aspects of surveillance, including wiretapping, investigative surveillance, and surveillance of government activities by the news media and the public.

In connection with Watergate, five men authorized by Republican Party officials were arrested for breaking in and allegedly photographing confidential Democratic documents. The following, blatant coverup took years to untangle. At a key period in history, just before the birth of microcomputer technologies, the behavior of President Nixon and his party members demonstrated to the country at large that trusted government officials were not above using surveillance devices, tape recordings, and wiretaps in illegal and unethical ways.

As turmoil erupted in the Nixon White House, computer networking began to take hold as distributed computing evolved on the ARPANET and the first network email programs were developed. These innovations were enthusiastically adapted by computer users. The ARPANET was demonstrated at the International Conference on Computer Communications (ICCC), and the *Internet Network Working Group* (INWG) was formed.

In July 1973, Clarence Kelley was appointed as the new FBI Director after Gray's opportunity was scuttled by the scandal in the Nixon White House. Kelley came from within the department, having been an FBI Agent for twenty-one years prior to his appointment. The following year, the FBI moved from a building shared with the Department of Justice to offices on Pennsylvania Avenue.

The post-Watergate era was a time of great public scrutiny and outcry, resulting in a number of important events and changes in U.S. policy directly related to surveillance and privacy.

The discovery of Watergate, and the subsequent obstruction of justice by members of the White House, turned a battery of eyes toward the Presidential administration. Vice President Spiro Agnew resigned after charges of tax evasion and, in August 1974, President Nixon resigned, as well.

Also, in 1974, the public and Congress scrutinized FBI methods to ensure that they didn't violate Constitutional rights. Further, the U.S. Congress passed the *Privacy Act* to ensure public access and the right to correct information about oneself held in federal files.

In spite of significant political upheavals, progress continued unabated in the computer networking world. In 1974, Vint Cerf and Bob Kahn published a document that was to became a significant milestone — "A Protocol for Packet Network Interconnection". This specified Transmission Control Program (TCP), a practical means of sending data over computer networks. Telenet, the first packet-based system for non-military use, introduced networking to the public. Telenet was essentially an open version of the ARPANET and it evolved rapidly.

Meanwhile, public and government housekeeping activities did not end with Watergate or investigations of the FBI. The NSA was now also under scrutiny.

In August 1975, the Director of the NSA admitted to the U.S. House of Representatives that the NSA "systematically intercepts international communications, both voice and cable". He acknowledged that domestic conversations were sometimes picked up in the course of gathering foreign intelligence, a fact that was later investigated and reported as 'incidental'. The exoneration didn't appear to alleviate watchdog concerns or reduce scrutiny of NSA activities. The public was demanding the right to monitor the actions of tax-supported intelligence agencies.

Birth of the Personal Computer

In 1975, a highly significant event occurred in electronics. Until this time, computers were very large, very expensive, and awkward systems primarily owned by governments, a few large corporations, and larger educational institutions. A few 'bleeding edge' personal computers had been unsuccessfully marketed since the Kenbak-1 was introduced in September 1971, but the Altair computer, introduced in kit form late in 1974, was a commercial hit that spurred the creation and growth of an explosive new consumer industry. COMINT would have to grow and adapt to handle the flood of information exchange that resulted from the proliferation of personal computers.

The FBI had been reasonably autonomous in its intelligence methods up to this time. As a result of scrutiny over the last several years, however, much more detailed specifications for its operation were handed down by Attorney General Edward Levi in March 1976 (foreign investigations) and again in April (domestic investigations).

As Director of the FBI, Kelley continued to carry out his responsibilities in spite of social turmoil. He implemented equal-opportunity hiring practices and established three clear national priorities consisting of foreign intelligence, organized crime, and white-collar crime.

The furor over internal intelligence activities continued. The Church and Pike Committees investigated these matters in 1975. The Church Committee focused on allegations of assassination plots and harassment of individuals and the overall structure and operations of the Intelligence Community. The Pike Committee focused more on the last decade or so of defense intelligence performance of the Intelligence Community. The resulting Pike Committee Report was withheld by a vote of the House (a draft was leaked to the press), but the Church Committee's Final Report was submitted to the public in April 1976. Responding to one of the Church report's main recommendations, a permanent *Senate Select Committee on Intelligence* (SSCI) was established by the U.S. Senate in May 1976.

In 1977, the Carter administration created the U.S. *Department of Energy* (DoE) as a response to the energy crisis and to act as an umbrella organization for energy-related enterprises and scientific research. Over the years, concerns were raised that security was not being handled as comprehensively and expertly as were scientific activities within the department. One of the cited examples was classified documents about advanced nuclear weapons that were available to the public in a Los Alamos library.

Responding to reports of surveillance abuse on the one hand and laxity on the other, President Carter signed Executive Order 12036 on January 1978, providing explicit guidance on the execution of intelligence activities, and reorganized the intelligence structure. The *Foreign Intelligence Surveillance Act of 1978* was a result of the findings and allegations of the Pike and Church Committee reports.

4.g. The Computer Age and Space Technologies

The Spread of Computers and Eyes in the Sky

In the mid- to late-1970s, in the computing world, personal computers established a new industry. The Radio Shack TRS-80 and Apple Computer were the two most prominent and successful products at the time. New stores, called 'Computing Centers' began to spring up throughout the country. Software programmers sold their software on cassette tapes and 8" floppy diskettes in plastic baggies with eight-page photocopied manuals for $30 to $200 per product, and made more money than they knew how to spend. Some of the most successful games programmers earned monthly royalties equal to the cost of a luxury car.

As computers evolved, computer users discovered new ways to engage in mischief and crime. A new breed of computer 'spy' emerged—a hacker was a computer techie who had the talent and ambition to 'break in' to other people's computers or software programs. They were capable of covertly examining system configurations and the content of private data files. When modems and computer bulletin board systems (BBSs) were established, hackers started breaking into a variety of government and financial institutions.

Computer electronics spread into other devices, as well. Toward the end of the 1970s, the U.S. implemented a new series of communications intelligence satellites as successors to the CANYON series. Sensitized members of watchdog agencies and the press were now concerned about satellite surveillance. It became more more difficult to carry out covert aerial surveillance without accountability and public relations efforts.

Then, in 1978, William H. Webster, a former federal Judge, became Director of the FBI, following the resignation of Clarence Kelley.

By the 1980s, the global population had grown to 5.5 billion people, cities swelled, people moved frequently, and safety and security became important concerns. In contrast to the 1960s, people began locking doors and installing security systems in increasing numbers. Video cameras dropped in price. It was a time of significant social and technological change.

Increased Governmental Surveillance of U.S. Persons

Emphasis on surveillance and countersurveillance increased as society moved into the 1980s, with a particular focus on new microcomputer technologies.

In March 1980, the Secretary of the Treasury delegated authority for technical surveillance countermeasures to the Assistant Secretary (Enforcement and Operations) under *Procedures for the Conduct of United States Secret Service Technical Surveillance Countermeasures* who reports to the *Under Secretary for Enforcement*. Most of the U.S. Treasury functions related to security and surveillance, including the U.S. Customs Service and the U.S. Secret Service, are currently under the jurisdiction of the Under Secretary, Department of the Treasury.

While it is hard to make generalizations based solely on unclassified documents, it appears that one of the trends over the past three decades, in the issuance of NSA-related intelligence directives, is that they were becoming wordier and more encumbered by the greater explicitness of the definitions.

In October 1980, the NSA/CSS issued the *United States Signals Intelligence Directive* (USSID), which prescribes policies and procedures, and assigns responsibilities related to the SIGINT System (USSS) with a stipulation to safeguard "the constitutional rights and privacy of U.S. persons." The Directive, titled "Limitations and Procedures in Signals Intelligence Operations of the USSS (FOUO)", superseded USSID 18 from May 1976.

The 1980 Directive specifically declares who or what might be considered 'foreign' persons and thus subject to SIGINT. Previous documents had indicated that U.S. citizens and permanent residents would not be considered foreign and implied that they were thus exempt from SIGINT; this document specified that an "agent of a foreign power" meant, among other things [underlined for emphasis]:

"b. Any person, <u>including</u> a U.S. person, who -

(1) Knowingly engages in clandestine intelligence gathering activities for, or on behalf of, a foreign power, which activities involved, or may involve a violation of the criminal statues of the United States; ....

(2) ....

(3) ....

(4) ....

c. A <u>U.S. person</u>, residing abroad, who holds an official position in a foreign government or the military forces of a foreign national and information about whose activities in that position would constitute foreign intelligence."

Later in the document *U.S. Person* is defined as:

"... a citizen of the United States, an alien lawfully admitted for permanent residence, an unincorporated association organized in the United States or substantially composed of United States citizens or aliens admitted for permanent residence, or a corporation incorporated in the United States.

a. The term "U.S. person" includes U.S. flag, nongovernmental aircraft or vessels. The term does not include a corporation incorporated in the United States that is openly acknowledged as a foreign government or governments to be directed and controlled by such foreign government or governments. [definition continues]"

As previously stated, it's difficult to make informed generalizations based only on declassified documents, but some trends can be extrapolated from the declassified U.S. Directives and the information at large. They indicate that

- surveillance was becoming more complex and technological in nature,
- NSA responsibilities were gradually broadening in scope,
- technology was increasing in importance in overall intelligence-gathering methods,
- secrecy and broad funding of U.S. secret services were being safeguarded,
- intelligence-related documents were becoming wordier and more complex,
- the scope of SIGINT had been explicitly broadened in the 1980 Directive to include possible surveillance of U.S. persons who the government considered to be security risks, and

- privacy advocacy, public desire for government accountability of surveillance activities, and the desire for access to information was increasing.

In October 1981, the Reagan administration revamped the President's Foreign Intelligence Advisory Board and named a number of distinguished citizens to serve on the Board. In December, clear goals were set and ambiguities clarified for the Intelligence Community.

In May 1982, William Casey, Director of Central Intelligence, issued a Directive that established a *Signals Intelligence* (SIGINT) *Committee* to advise and assist the Director of Central Intelligence (DCI) and the Director, National Security Agency (DIRNSA), within the Intelligence Community.

In June, President Reagan signed the *Intelligence Identities Protection Act of 1982* into Public Law 97-200. This imposed criminal penalties on anyone who wrongfully divulged the identities of covert intelligence personnel, a measure intended to protect vulnerable members of the Intelligence Community. In 1984, the President exempted the CIA from the search and review requirements of the *Freedom of Information Act* with respect to sensitive files through the *Central Intelligence Agency Information Act of 1984*.

Following a number of international terrorist bombings, the Director of the FBI added counterterrorism as a fourth national priority. The FBI was also granted greater jurisdiction along with the Drug Enforcement Administration (DEA) to handle narcotics investigations.

Evolution of the ARPANET and Network Addressing

By 1983, the ARPANET computer network had been split into two sections: ARPANET and MILNET (military network). MILNET continued to be used for nonclassified military communications while ARPANET came under the administration of the National Science Foundation.

In 1984, the development of computer network *name servers* and the evolution of routing technologies enabled computers to exchange messages without the prior declaration of an explicit end-to-end path. In less technical terms, this means that a message could be sent from A to F without declaring a specific route through B, C, D, or E. It's like sending a letter to your friend across the country with the understanding that the postal service will find the best way to route it (and that the routing might change, depending on what is currently most efficient). In the same way, a name-server system decides electronically on how to route email (or other data communications) on the user's behalf.

This approach to networking made systems more powerful and flexible; data "packets" were sent to their intended destinations regardless of how many computers were attached to a system and which ones were active at any given time. This made it necessary to establish a centralized authority to issue unique identifiers to each server. In other words, just a postal system needs a *street address* to deliver a letter, a networked computer needs an *electronic address* to deliver an electronic communication. Thus, the Domain Name System (DNS) was introduced, in 1984.

Computer communications came to rely on name-server concepts such that it didn't matter which route the communications packets took to reach their destination. They could be split up, sent through different routes, and reassembled at their destination. This made it more difficult to eavesdrop on electronic communications en route. Not knowing *where* to intercept the message or how much of the message could be retrieved at any particular point increased the security of distributed networks and made the task of surveilling them much more difficult. The weakest link in surveillance terms was no longer the network itself, but the less secure computers at each endpoint and the less security-conscious people using them.

Surveillance of Social Events and Government Consolidation

In 1984, Los Angeles hosted the international Olympic Games. Those associated with the Games wanted to prevent hostage situations as had occurred at the 1972 Munich Games. Surveillance to safeguard the athletes and the public were put into effect. The FBI established a *Hostage Rescue Team* to aid in public safety efforts. Since that time, many of the types of surveillance activities put into place for the L.A. Games are used in large-scale sports, music, and religious gatherings.

In 1984, the Assistant Chief of Staff for Intelligence established a new agency, the *Army Intelligence Agency* (AIA), to direct scientific, technical, and general intelligence (except for medical intelligence). In 1985, several departments were put under the direct control of the AIA, including the *Intelligence and Threat Analysis Center* (ITAC), the *Foreign Science and Technology Center* (FSTC), and the *Army Missile and Space Intelligence Center* (AMSIC).

Counterintelligence Efforts

1985 was dubbed 'The Year of the Spy' due to the discovery of penetration into America's sensitive agencies. The FBI uncovered a surprising number of espionage activities during this time and the CIA investigated internal security breaches that were traced, in part, to Edward Lee Howard and Aldrich H. Ames. Due to these revelations, counterintelligence became a higher national priority.

While U.S. intelligence agencies were investigating internal security concerns, the public was requesting increasing numbers of documents through the Freedom of Information Act (FOIA). It became apparent that a national clearinghouse for these documents would reduce redundancy and provide a valuable central resource for further research and analysis. Thus, in 1985, the *National Security Archive* was founded as a nonprofit, nongovernmental library, archive for storing information on international affairs and declassified U.S. documents obtained through FOIA and a public-interest law firm. Supported by public revenues, private foundations, and many documents donated by individuals, it has grown significantly.

In 1986, the Lawrence Livermore Laboratory established a formal counterintelligence program to identify trends in foreign intelligence threats against personnel, and to educate employees about counterintelligence issues.

In 1987, Judge Webster, head of the FBI, became the Director of the CIA. John E. Otto, Acting FBI Director in Webster's absence, established drug investigations as the fifth national priority after counter-terrorism. Judge William Steele Sessions then became Director of the FBI. Sessions began working with community institutions to implement drug-use reduction programs for the youth.

Two important political trends emerged in the mid-1980s: a greater emphasis on security within government circles and increased access to, and dissemination of, government information by private and independent agencies. These opposing trends of increased security and increased openness could ultimately be self-defeating or could indicate a healthy balancing dynamic for a democratic free society. Which interpretation is correct may not be knowable until a span of years has passed. What is known about the mid- and late-1980s is that investigations into domestic affairs and corruption in high levels of the government and financial institutions were turning up some unsettling facts.

FBI investigations at this time unearthed corruption and bribery within the U.S. Congress (ABSCAM) and other levels of government. Investigations of bank failures by the FBI and by investigative journalists gradually revealed a pattern of deliberate manipulation of real estate holdings and bank loans to further personal goals, especially those of certain directors associated with these institutions. In pursuit of money that became accessible through deregulation, the perpetrators falsified documents, committed bribery and, apparently, even committed murders. This internal looting resulted in the eventual collapse of several dozen savings and loan institutions.

Members of the Intelligence Community

It is not unusual for members of the Intelligence Community to transfer among different intelligence/security departments, or for senior members to be involved in special commissions and surveys.

Members of the Intelligence Community come from different walks of life, but many have military backgrounds. It is outside the scope of this chapter (and this book) to list prominent individuals in the Intelligence Community, since the focus is on technology and not human intelligence (HUMINT), but it is useful to read the profile of at least one intelligence professional with strong ties to the technology community. This puts a 'face' on the field and gives insight into the kinds of roles intelligence professionals play in government and industry.

Admiral Bobby R. Inman earned a Masters degree from the National War College in 1972 and subsequently held secret service positions for many years. He was Director of Naval Intelligence from 1974 to 1976, after which he became Vice Director of the Defense Intelligence Agency (DIA). He was the first naval intelligence specialist to reach a four-star rank. In 1977, he was appointed Director of the National Security Agency (NSA).

In 1980, Inman established a National Security Agency Director's Trophy to recognize outstanding performance by U.S. cryptologic mobile units that support military commanders. His career indicates a strong interest in cryptologic technologies and he has made public statements regarding the role of cryptology in security.

In January 1981, the Reagan Administration nominated Inman for the position of Deputy Director of the CIA.

In February 1982, *Aviation Week and Space Technology* printed an article by Inman titled "Classifying Science: A Government Proposal", which describes tension between the goals and climate of research and the government's national security aims, especially with regard to cryptologic research and applications. In the article, Inman suggests that scientists should include review of possible implications for national security before initiating a project:

> A potential balance between national security and science may lie in an agreement to include in the peer review process (prior to the start of research and prior to publication) the question of potential harm to the nation. The details of such a system would have to be resolved, of course, but cooperation will be better for all of us than confrontation.

> Included in such a system should be goals to simultaneously preclude harm to U.S. national security and to impose no unreasonable restrictions on scientific research, publication, or the use of the results. And when restrictions are judged necessary, speedy procedures for appeals, review and appropriate compensation should be included.

This is a tall order, given that the eventual results of any particular line of research are difficult and sometimes impossible to predict and that any government veto process would be strenuously opposed by many members of the scientific community. Inman continues:

> One example of this type of process is that recommended in the Public Cryptography Study Group. It is not easy to create workable and just solutions that will simultaneously satisfy the wide-ranging needs of national security and science, but I believe it is necessary before significant harm does occur which could well prompt the federal government to overreact.

Is this statement a veiled warning or a premonition of the heated government/private sector computer-encryption debates that erupted ten years later? (See the chapter on Cryptologic Surveillance for more information on encryption debates.)

In July 1982, Admiral Inman retired from the U.S. Navy, and his focus shifted from the government sector to business and technology. He accepted positions on the Boards of Directors of numerous companies and chaired a number of technology companies. Considering the challenge inherent in managing just one high-tech company, the diversity and overlap of Inman's corporate affiliations are surprising. Inman founded and became Chairman and CEO of the Microelectronics and Computer Technology Corporation (MCC)* from 1983 to 1986. From 1986 to 1989 he was Chairman, President, and CEO of Westmark System, Inc., an electronics holding company, and simultaneously held the position of Chairman of the Federal Reserve Bank of Dallas from 1987 to 1990.

In 1984, Inman convinced Douglas B. Lenat, a programmer with an interest in artificial intelligence and heuristic problem-solving algorithms, to join the newly formed MCC in Austin, Texas. Lenat would assemble a team to create "CYC", a computer program with common-sense reasoning skills. A new company, Cycorp, was spun off from MCC in 1994 and CYC showed promise in the area of intelligent searching, such as information retrieval from an image library in ways different from the traditional keyword-lookup structure.

In the mid-1980s, while chairing MCC, Inman was still involved in national security and reported on international diplomatic security through the *Inman Commission Findings*, submitted in 1985. These resulted from a panel set up by Secretary of State George P. Shultz.

The Findings made recommendations for improving international security, including modifications to physical structures. Recommendations included removing glass and blast-channeling corridors, moving people from the vicinity of windows, and creating high walls and longer building setbacks from streets (to protect from car bombs). Further recommendations involved improving surveillance and having guards inspect delivery vehicles at some distance from the embassy sites before accompanying them to the site. Inman's Findings further suggested that technological upgrades and increased operational countersurveillance could help deter terrorists.

In the mid-1990s, the Clinton Administration nominated Inman for the position of Secretary of Defense. To the surprise of many, Inman held a press conference in January 1994, withdrawing his nomination. Those who witnessed the conference expressed surprise. In a disjointed, rambling address, Inman cited personal attacks on him in the media among other reasons for withdrawing. Supporters speculated that perhaps Inman had doubts about returning to public service. William J. Perrey was replaced as nominee for the position.

Inman serves on the Boards of several prominent companies, including Science Applications International Corporation (SAIC). In March 1995, SAIC bought out Network Solutions Incorporated, the administrator of the domain name registry for the Internet, and a fee system was instituted for domain names through NSI. SAIC handles many government contracts that are directly or indirectly related to intelligence, defense, and law enforcement. By December 1997, SAIC was listed as the 41st largest private company in America with a high percentage of employee ownership.

---

*Among other products, MCC sells ExecuSleuth, a corporate surveillance software product with 'intelligent agents' designed to manage information by reading and understanding text, databases, and images. It is further designed to generate alerts when tactical and strategic changes occur. The product claims to track information on key competitors in realtime, to monitor changing market conditions, and to track changes in market conditions, technologies, and the infrastructure.

Although mainly involved in the private sector, Inman's opinion was still being sought with regard to the government Intelligence Community. In January 1996, Inman testified at a *Hearing of the Commission on the Roles and Capabilities of the United States Intelligence Community*. At the hearing, he offered his opinions on reorganization, made suggestions about priorities, emphasized the continued importance of human intelligence (HUMINT) activities, and expressed his opinion on the issue of disclosure of intelligence expenditures to the public.

Inman's affiliations are many. He holds the position of Adjunct Professor in the Department of Management at the University of Texas at Austin. He is listed as "Investor" on the Board of Directors of the Xerox Corporation and as an individual donor to RAND, a national security research organization. He serves or has served on the Boards of Directors of Fluor, Science Applications International, SBC Communications, Temple Inland, the Public Agenda Foundation, Southwestern Bell, and others. He further is a Trustee of the American Assembly, the Center for Excellence in Education, and the California Institute of Technology. He serves on the Executive Committee of the Public Agenda Foundation and is a member of the National Academy of Public Administration.

In spring 1999, Inman was interviewed about secret ECHELON global surveillance activities by a number of members of the press, despite the fact that he had not been officially affiliated with the NSA and the CIA since the early 1980s. Nevertheless, he is quoted as saying that the organization was not involved in intercepting and sharing economic secrets, but only "fair trade issues and trade violations...."

In 1998, U.S. embassies in East Africa were bombed. Following the bombings, Kenneth H. Bacon reported to a Department of Defense news briefing on the U.S. response to the bombings. In the brief, Bacon cited the 1985 Inman recommendations, stating that since 1986, many changes had occurred in the security of existing embassy buildings, and new buildings had incorporated 'Inman standards', that is, they were well fortified, with high walls and long setbacks from the streets.

Thus, Inman continued to influence aspects of security through the Inman Findings long after his departure from public service, and the press apparently perceives him as a liaison to the government, as it continues to seek out his perspective on Intelligence Community (IC) activities. This short profile indicates that specific individuals within the IC can have significant influence on both government and business developments over a long period of time.

4.h. Global Commerce, Government, and Communications

The Internet

In the field of technology, a computer revolution was underway in the 1980s. Cpt. Ralph Peters, in "Perspectives on the Future: The Army of the Future", *Military Review*, September 1987 wrote:

> At present, technology is outstripping the military imagination so swiftly that available hardware will continue to define tactics for a long time.

The growing global computer network was now known as the *Internet* and the security of the Internet, or the lack thereof, made a media splash when the Internet Worm spread through thousands of host computers in November 1988. To add to the commotion, the doctoral student who created the Worm turned out to be the son of the former Chief Scientist of the *National Security Agency* (NSA).

The use of computers for NSA intelligence-gathering was brought to public attention in

August 1988, when Duncan Campbell's article in *New Statesman* described a multinational electronic surveillance system called Project P415. Campbell reported that witnesses had been subpoenaed to provide the plans and manuals for the "ECHELON" system and described government projects for monitoring long-range radio and satellite communications.

Shortly thereafter, *The Cuckoo's Egg*, Clifford Stoll's autobiographical account of foreign agents infiltrating U.S. computer networks, lent further fuel to concerns about the potential for the misuse of electronic communications.

Late 1980s - Concerns About Personal and Computer Security

Airport security began to tighten up in the late 1970s after two decades of very low security and relative calm. However, plane hijackings and bombings had prompted spot-checks and use of metal detectors to be implemented. By the late 1980s, the sensitivity of gate surveillance devices had been increased and X-ray machines had been added to the security arsenal at airports. Later, with the destruction of the World Trade Center, airport security was increased once again.

Humans are social beings. They want to feel safe, but they also want to spend time with friends and relatives, so secured systems, like gated communities have to be designed to accommodate the social needs of their occupants. Gates that are monitored around the clock by human attendants are costly. To reduce costs, many gated neighborhoods began to substitute surveillance devices, including cameras, door openers, and motion detectors.

Security breaches of physical space were not the only concern in the 1980s and 1990s. Security breaches on the Internet affecting personal and business communications made people more aware of its vulnerability. *Firewalls*, a means of restricting network access or access to specific computers, were more widely established, and computer experts were called in by many of the larger business firms to provide security advice.

Vulnerability to attack was also of concern in official circles responsible for national security. DARPA (Defense Advanced Research Projects Agency) responded to network security concerns by forming the *Computer Emergency Response Team* (CERT) and the Department of Defense looked into the adoption of OSI instead of TCP/IP as a transmission protocol.

1990s - Provisions for Privacy, ECHELON, and Public Concerns

Internet growth exploded in the 1990s, from a quarter million hosts and less than 10,000 domains to more than 40 million hosts and two million domains. Much of this growth can be attributed to the development of the World Wide Web (WWW), a basic document-serving system released in 1991 that made graphical Web sites possible. The WWW, originally developed by Tim Berners-Lee, represented only a subset of the content of the Internet, but it was an important subset because it was comprehensible to businesses and ordinary users.

Most of the new growth on the Internet came from home and small business, people with no former background or technical expertise in computers. They were delighted to find a simple hypercard-like means of accessing computer files. The Internet was originally a small community of technically proficient programmers and researchers but now included a majority of people with minimal understanding of the vulnerability of computer systems.

There was tremendous growth in communications and the implementation of electronic surveillance technologies in the 1990s. The trends include

- greatly increased technological surveillance in both civilian and government activities, including search and rescue, domestic and business security, law enforcement, and national security;

- internationally, a trend away from larger, more generalized conflicts, to smaller localized conflicts fought with advanced tools of warfare;

- persistent rumors of a reported "Big Brother" ECHELON project;

- the expansion and evolution of the global Internet and its subset, the Web;

- government concerns about keeping up with the pace of technological change and effective deployment of new technologies;

- the establishment of a number of Net-based organizations for safeguarding privacy and freedom of information and expression on the Net, including the *Electronic Frontier Foundation* (EFF); and

- public concerns about use and possible abuse of surveillance technology by government agencies.

In the marketplace, detection systems and other surveillance technologies were becoming common. Car alarms were factory installed. Video cameras were added to home and office security systems to expand existing burglar alarms, and nannycams were installed in nurseries and daycare centers. Retail stores installed hidden or visible camera systems (sometimes both) in increasing numbers.

Another highly significant change occurred in the 1990s. Up to now, most communications technologies were analog, but digital technologies were on the rise. Digital systems made it easier to encode and encrypt broadcast communications and to replicate data. New means of encryption, spread-spectrum technology, and other ways to ensure privacy and anonymity were being developed by computer programmers and electronics engineers. As the technology changed from analog to digital, the interception and decryption of telephone conversations and satellite communications now required far more skill and dedication than ever before. Secure communications were available to the casual user in the form of encrypted messages.

ECHELON Rumors and Privacy Constraints

Since the early-1990s, rumors about a secret global surveillance system called ECHELON have circulated through the Internet community and some individuals have characterized it as a U.S. Government see-all/hear-all surveillance network that monitors and records every electronic communication on the planet. Given the massive, unwieldy volume of information that would result from such a mandate, and the poor signal-to-noise information-ratio that is inherent in most computer communications, it is more likely that ECHELON has specific operational foci and goals, and priorities that relate to national security. There are U.S. Government declassified documents that have been released that confirm the existence of ECHELON, but do not indicate the project's scope or focus. More telling perhaps are the privacy stipulations that are alluded to in other parts of documents that mention ECHELON.

In March 1991, the Central Intelligence Agency (CIA) moved into its new expanded headquarters and a 'time capsule' was built into the cornerstone with documents and photos representative of the period.

In September 1991, the U.S. Naval Security Group Command issued *NAVSECGRU Instruction C5450.48A* (cancelling C5450.48). The Instruction prescribed the missions, functions, and tasks of Naval Security Group Activity *Sugar Grove*, West Virginia, which was described as an active (fully operational) shore activity. The 544th Intelligence Group was included in the Tenant Commands. The Group was tasked with processing and reporting intelligence information and maintaining and operating an ECHELON site (the wording suggests there may be other

ECHELON sites), and [underlined for emphasis] to:

> "(3)(U) Ensure the privacy of <u>U.S. citizens</u> are properly safeguarded pursuant to the provisions of USSID 18.
>
> (4) ....
>
> (5)(U) Operate special security communications facilities, as directed."

Note that intelligence Directives from the late 1940s to at least the 1980s, including U-SID 18, had clauses that stipulated the safeguarding of the privacy of "U.S. Persons". It is not clear whether the wording of Instruction C5450.48A, which stipulates "U.S. citizens" rather than "U.S. Persons", indicates a change in direction or priorities, but from a legal point of view, such wording distinctions can be significant. Reading between the lines suggests that the ECHELON project mentioned in the same document has been defined and established in previous documents and may be just one installation in a multiple-site network.

The President's *Foreign Intelligence Advisory Board* (PFIAB) cited a classified document in reporting that listening devices were discovered in U.S. weapons-related facilities in the late 1980s. The PFIAB further reported in January 1998 that "Anecdotal evidence corroborates, and intelligence assessments agree, that foreign powers stepped up targeting of DOE" during the early part of 1992. The *Department of Energy* (DoE) is an important agency in the American resource and economic structure and its security is a high national priority. Perceived security weaknesses in the DoE and weapons facilities may have generated internal support for increased domestic and foreign security as well as projects like ECHELON.

In 1992, a *Memorandum of Understanding* aided in establishing a formal relationship between the FBI and the DoE with regard to counterintelligence activities.

In September 1992, Gerald E. McDowell of the U.S. Department of Justice received a response to his inquiries from Stewart A. Baker, General Counsel, representing the *National Security Agency/Central Security Service*. This document lends some support to the theory that there had been a gradual shift in priorities with regard to privacy, from 'U.S. persons' to 'U.S. citizens,' and from foreign communications to some that may have been domestic and thus were questionable. It further indicates an increased sensitivity to these privacy protections following reported abuses. Remember that it was stated and quoted from Directives in earlier decades that SIGINT and other intelligence aspects were explicitly restricted to *foreign* communications. Baker sought to allay concerns about government monitoring of the Banca Nazionale del Lavoro (BNL) and clarifies the agency's position on surveillance as a result of admitted abuses in the 1970s [underlined for emphasis]:

> ... The interception of communications by this Agency is extremely sensitive because of the danger it poses to the privacy of <u>American citizens</u>. In the early 1970s, this Agency improperly targeted the communications of a number of Americans opposed to the Vietnam War. In response to these abuses, uncovered by the Church and Pike Committees in 1975 and 1976, the Foreign Intelligence Surveillance Act of 1978, Executive Order 12333, numerous regulations now limit the targets of our collection efforts. As a result, NSA may only target communications for the purpose of producing <u>foreign</u> intelligence; we have no authority to target communications for law enforcement purposes. Our responses to you are thus based on a review of intelligence reports issued by NSA, and not a new review of intercepted raw traffic.

A declassified government document from December 1995, "Air Intelligence History", includes information on the "Activation of Echelon Units" which describes agreements to increase the Air Intelligence Agency's (AIA) participation by establishing new AIA units, with activation of these detachments in 1 January 1995. Portions of this document have been expunged, but in the ECHELON section, it mentions sites in West Virginia, Puerto Rico, and Guam.

In October 1996, the Center for International Policy (CIP), a watchdog agency and independent advocate for peaceful U.S. foreign policies, held a *Seminar on Intelligence Reform.*

4.i. Mid-1990s - Terrorism, Smuggling, and Technology Upgrades

In the mid-1990s, many organizations began installing surveillance technologies for both government and private use. Museums, service stations, department stores, grocery outlets, schools, and border stations added or upgraded surveillance systems, especially motion-detector and visual surveillance devices.

In January 1993, a foreign terrorist stood in front of CIA headquarters and shot at motorists, murdering two people and injuring three others. Then, in February 1993, the world was stunned when a great explosion damaged the World Trade Center in New York City, injuring over 1,000 people and killing six. Children were trapped for hours in the smoke-filled building.

Smuggling during the prohibition years resulted in many regulatory changes, yet smuggling is still a widespread problem. The influx of human cargo from developing or repressive nations received greater media coverage than ever before as news surveillance choppers flew over seized ships and cargo containers harboring human beings. When newer X-ray technologies were installed to inspect containers, they uncovered a surprising number of stowaways. Many of these refugees and illegal aliens had paid up to $60,000 per person to be transported to America, yet were found in poor health—hungry, thirsty, and diseased. Often the ships carrying them were unseaworthy. Sometimes the stowaways died before reaching their promised destinations. Major surveillance efforts by the DoJ working with the Secret Service, the IRS, and the Customs Service resulted in the apprehension of a large international ring of smugglers of human refugee and nonrefugee aliens in 1994. The smugglers were also subject to prosecution on other charges, including bribery, fraud, and money laundering.

Letter bombs disguised as holiday greeting cards were delivered to the U.S. and the U.K. Unfortunately, one bomb exploded in London, injuring two people. Three others were discovered at the U.S. Federal Penitentiary in Leavenworth. This prompted the offer of a reward leading to the apprehension of those responsible and increased surveillance of shipped packages by the U.S. Postal Service and private courier agencies. [U.S. Diplomatic Security Service news photos, released.]

Letter bombs have existed for a long time and there have been intermittent news reports about them, but in less than two months during the winter of 1996, more than a dozen letter bombs disguised as holiday greeting cards were delivered to U.S. and U.K. recipients, prompting the U.S. Postal Service to change regulations regarding the submission of packages for posting. Packages weighing more than a pound could no longer be conveniently dropped in postal slots, but had to be handed directly to postal employees during business hours. Private couriers also stepped up package surveillance.

4.j. Society in Transition

Greater Awareness and the Struggle for Balance

The Internet has given the public the opportunity to exchange information and to organize lobbying bodies in a way that has never before existed in human society. Never have modern governments been more closely surveilled, never have they been more accountable for their actions, nor have they encountered so much resistance to activities that they may deem necessary for the safety and security of citizens and which they might have carried out with impunity and secrecy in previous administrations.

In 1993, U.S. Attorney Louis J. Freeh was appointed as the new Director of the FBI.

In 1994, the U.S. established the *Civilian Career Management Program* (CCMP) which, in part, developed future civilian leaders for the Air Force intelligence community.

A number of government satellite imaging programs were declassified in 1995, increasing public awareness of aerial surveillance and creating an aura of general unease. A September 1995 memorandum from Daniel C. Kurtzer, Acting Assistant Secretary, *Bureau of Intelligence and Research*, to the Director of the NSA, J. M. McConnell, illustrates this struggle when the pursuit of security and privacy from different quarters of society shifts the balance:

> In response to NSA's request for a community reaction to the NRO proposal that the "fact of" SIGINT collection from space be declassified, we have looked at possible ramifications from a foreign policy as well as an intelligence perspective. We believe that official confirmation could have undesirable repercussions in those countries where _ are located -- _ -- and that certain steps are necessary before reaching any decision on the NRO proposal.
>
> In the cases of _ we strongly recommend that the US initiate formal consultations on the NRO proposal. As you are aware, the _ _ _ _ _ _ _ press and public already have concerns about activities at _ and there are aperiodic [sic] allegations in the press that the Americans are monitoring _ _ _ _ _ _ _ communications. _ that government will be particularly sensitive to unfavorable speculation with regard to the _ _ _ _ _ facilities. The consultations could take place in intelligence channels, provided the Department has an opportunity to review the record correspondence in advance.... [Underscores indicate sections that have been blotted out.]

Another significant struggle for power in the mid-1990s was the battle over computer encryption systems. When news of government proposals to restrict encryption and supply keys to the NSA were circulated over the Internet, the response from the public was swift, vocal, heated, and well-organized. The Internet was seen as a vehicle for fundamental change in society. Before the Internet, the government would have been able to establish the proposed encryption restrictions and to provide access by law enforcement and national security agencies to certain communications with far less debate and revision. But electronic communications forged a different relationship between government and the voting public. The paternalistic structure, in which public officials were entrusted to do 'what was best' for those they represented, was shifting to a more decentralized structure, in which the public was demanding equal access, equal privacy, and more direct input into the drafting of legislative bills through electronic communication. This shift became even more apparent when the issue of encryption came up again five years later (see more about the encryption debate and eventual outcome in the Cryptologic Surveillance chapter).

The *Electronic Freedom of Information Act Amendments of 1996* became Public Law 104-231, thus amending the act to provide public access to information in electronic format. This action, along with a number of subsequent actions on the part of the U.S. Government, suggested a trend to more openness in the late 1990s.

The Late 1990s - Reorganization, More Openness, Yet More Surveillance

By the mid-1990s, Global Positioning Systems (GPS), which permit the precise pinpointing of a location through satellite technologies, were showing up in discount catalogs at consumer prices. These and other types of tracking devices were now used on prison inmates on parole, or with limited movement privileges, and were available for children or older people who might wander unattended. By the late-1990s, they were available for less than $200. Pinhole cameras and digital camcorders came down in price, as well. Sophisticated surveillance devices had entered the mainstream.

The late 1990s were characterized by some remarkable changes in western society, stemming from information sharing made possible by the Internet. Changes in the media reflected these changes in society. Broadcast stations have always vied for 'ratings' (a measure of the number of viewers watching their shows). When ratings decline, broadcasters are sometimes willing to take greater risks. New spy devices made it possible to create a new genre of television programming similar to the Candid Camera concept. In this type of programming, real events and real people (as opposed to actors and scripts) are surveilled and recorded ('caught on tape'), often without the surveillee's knowledge. Small cameras were put in the hands and vehicles of various law enforcement personnel as they went about their jobs and the results broadcast to the viewing public either prerecorded or live. These 'reality' shows became popular evening viewing.

A similar change occurred in daytime TV, with talk shows gradually edging out many long-standing soap operas. Coincident with the rise of 'real' TV has been intercommunication on the Internet in a no-holds-barred, 'truth'-oriented, uncensored forums in which the most intimate details are described in unprecedented detail for anyone with an Internet connection to read (or see).

The result of this enormous open exchange of information is that by the turn of the millennium, the number of 'taboo' subjects in North American society significantly decreased and the sanctioning of personal and professional 'spying' on people's private lives increased. Many people who would have objected to having their private activities recorded and broadcast, didn't mind other people's private lives being recorded and broadcast for their entertainment. This double standard continues to pervade many decisions and policies on the use of surveillance devices.

A side-effect of the public appetite for open access to information, coupled with the general economic stability in the U.S. in the 1990s, appears to have been a greater willingness, on the part of government, to reveal information that hitherto would have been kept quiet, even if declassified. This trend was evident in the early 1990s and became more apparent by the end of the decade.

In 1997, the *Openness Advisory Panel* was established within the U.S. Department of Energy (DoE) to provide advice regarding classification and declassification policies and programs as well as other aspects of the *Openness Initiative*. It further was tasked with public relations through:

> "An independent evaluation of all DOE policies and procedures relating to enhancing public trust and confidence in the Department and its programs, with special emphasis on classification, declassification and openness policies;
>
> ....

ROLES AND RESPONSIBILITIES

> The Panel will focus on issues of primary interest to the public and on measures to ensure the Department remains responsive to public policy needs and continues to foster confidence with the public and the Congress...."

One example of this policy of 'openness' was the distribution of a 1998 organizational chart of the NSA Operations Directorate, a type of document that would previously have been kept secret. Such charts were not required to be released, even if unclassified, as per Public Law 86-36.

In some respects, 1998 was a banner year for the release of classified documents. The CIA alone responded to Executive Order 12958 by releasing more than a million pages of material, the largest amount ever declassified and released at one time.

In January 1998, The U.S. Energy Secretary, Federico Peña announced a reorganization of intelligence programs in the Department of Energy (DoE). The goal was to improve counterintelligence capabilities and coordination with the FBI and other law enforcement agencies. As a result, functions of the Office of Energy Intelligence were reformed into two new offices to ensure effective programs through the DoE complex and labs. Thus, foreign, local, and counterintelligence would now fall variously under the jurisdiction of the *Office of Intelligence* and the *Office of Counterintelligence*, both reporting directly to the Secretary and Deputy Secretary.

On 7 August 1998, U.S. embassy buildings in Kenya and Tanzania were devastated by terrorist bombs with the loss of many lives. On 20 August, the U.S. government launched missile attacks on installations in Afghanistan and the al-Shifa pharmaceutical factory, claiming it was being used for the creation of chemical weapons, a claim that stirred international controversy and was still disputed two years later.

In 1998, more than a hundred U.S. overseas diplomatic posts were outfitted with surveillance detection programs, and more were planned, but the embassy bombings, along with previous terrorist bombings such as the World Trade Center in 1993 and the Oklahoma Federal Building in 1995, prompted a reappraisal of global U.S. security. In April 1999, diplomatic security personnel expressed concern about diplomats stationed or traveling abroad and made recommendations to Congress for increased surveillance and warning systems, to guard against car bombs and other similar terrorist attacks. Such changes would be subject to approval by the Overseas Security Policy Board chaired by the Director of the Diplomatic Security Service. The recommendations included, among other things, the installation of security lighting and cameras and other unspecified items.

In June 1999, the President's Foreign Intelligence Advisory Board presented a report on security problems at the U.S. Department of Energy (DoE) titled "Science at its Best | Security at its Worst." Based on research and interviews, the Report described some of the persistent organizational and administrative security problems associated with the DoE, particularly with regard to nuclear weapons, some of which reflected policies and procedures and some of which reflected differences in philosophy and priorities. Openness and security are goals that do not easily coexist. The Board recommended some general organizational changes and specific changes related to "personnel assurance, cyber-security, program management, and interdepartmental cooperation under the Foreign Intelligence Surveillance Act of 1978."

The Internet, Global Surveillance, and Open-Source Information

The growth of the Internet is a milestone in terms of changing the dynamics of global surveillance. Within a period of five years, the Internet community changed from a relatively small group of government, academic, and computer professionals, to a society of over 30 million (and growing) personal, commercial, and governmental users.

The very nature of the Internet changed dramatically during this period of growth. What was originally a military communications medium became a research communications medium and then evolved into a dominantly personal and business communications medium, with research occupying a smaller percentage in the gold rush for entrepreneurial niches. Investment brokerage and investor habits also changed dramatically as the availability of personal and business information and service sites expanded dramatically. Some examples of personal information that became open to the public that were previously difficult or awkward to obtain, include

- voluminous personal genealogies of family relations, journals, and oral histories,
- course catalogs from academic societies, course outlines, lecture notes, research findings, and conference proceedings,
- balance sheets of public companies, made available online through the Securities Exchange Commission,
- ideas, opinions, and information sources of individuals on USENET newsgroups,
- reverse directories and public directories of almost everyone in North America with a phone number or address,
- public records that were previously difficult to find or access could now be easily retrieved without having to physically travel to the county of origin to request a copy,
- armed forces agencies' news photos and documents,
- patent documents that were previously difficult to view or search without professional assistance became easily accessible through searchable online databases.

This bonanza of information, originating mostly in North America and Western Europe, and to some extent, Japan, could now be downloaded in minutes by anyone, anywhere, with Internet access.

Changes in Communications Infrastructures

During the 1990s and 2000s, traditional methods of seeking and transmitting information were rapidly superseded by electronic technologies. Internet access began to supplement and replace traditional long-distance phone calls. Email, online forums, instant messaging services, and Internet phone services became prevalent. Commercial satellites capable of a variety of communications tasks, including data transfer and high-resolution aerial imaging, were placed in orbit. Computer graphics algorithms and displays were now used in many types of surveillance, including radar scopes, infrared imaging systems, satellite-interpretation programs, image and facial recognition programs, and 3D modeling of foreign territories (formerly crafted by hand in a CIA workshop for more than three decades).

Within the CIA, the *Directorate of Science and Technology* was assuming greater importance, with an increased emphasis on the gathering and processing of information through technology.

The age of easily accessed electronic open-source surveillance and computer imaging was in full swing by 1998.

Selective Security

New technologies are not always secure technologies.

International scandal erupted in northern Europe in 1997 when it was discovered that widely used computer correspondence and videoconferencing products purchased from U.S. vendors were not secure. A number of European government agencies and big businesses were using the products for sensitive communications, assuming that they provided 64-bit encryption. This wasn't true of all versions, however.

Because of U.S. software export laws, foreign-sold products were restricted to 40 bits, an encryption level that was easier to decode than 64-bit encryption. U.S. vendors didn't want to lose the European market. In order to get around export limitations, some U.S. software vendors, like Lotus, shipped their products with 64-bit keys, but designed them so that 24 bits of the code were broadcast along with the message. The U.S. National Security Agency (NSA) was then supplied with the key to decode the 24 broadcast bits, making it easier for them to decrypt the remaining 40 bits. Lotus initially expressed satisfaction with this solution, as it allowed them to sell products abroad while still satisfying export restrictions. Foreign purchasers were not so happy, however—the U.S. was the chief supplier of global commercial software programs and U.S. users were getting reasonably good encryption (64 bits), but foreign users were buying the same products and getting only limited encryption (in spite of the fact that it required some fairly substantial computing resources to decrypt a 40-bit scheme, some people felt that 40-bit encryption was insecure enough to be termed 'data scrambling' rather than 'data encryption'). European users felt betrayed that they had not been explicitly informed of the security weakness and NSA involvement.

This is just one example of challenges facing the U.S. Intelligence Community about how to curb illegal communications, and communications about illegal activities, that might affect U.S. national security. It is also provoked fear in some segments of society about U.S. Government intelligence agents monitoring private communications. The conflicts and debate were not immediately resolved and a few years later some surprising amendments were made to software export regulations as a result of the controversy.

Economic Espionage and ECHELON Revisited

Three years after the *Economic Espionage Act of 1996* was enacted, the FBI announced that they had convicted foreign company members accused of stealing trade secrets from an Ohio manufacturing firm. The case began when Avery Dennison's own internal corporate surveillance turned up evidence of possible espionage and turned the investigation over to the FBI. The perpetrators were found guilty in April 1999 by a federal jury—it was the first foreign company to be found guilty under the *Economic Espionage Act*.

By the late 1990s ECHELON rumors had attained folklore status on the Internet as a symbol of government use and abuse of surveillance technologies. By 1998, Internet users organized to 'fight back' against this as yet unconfirmed 'ear on the Net' by inserting politically volatile keywords in all their email messages, including *bomb, secret, Iran,* etc. The logic was that the government couldn't monitor everything and thus might be singling out sensitive documents for closer scrutiny by using keywords (it's reasonable to assume that there is selective filtering of data). By encouraging everyone to insert the conjectured 'hot' keywords into email, protestors intended to 'flood the spooks' with volumes of information too massive to process in order to subvert the surveillance system. Whether or not this grass roots protest was effective, it nevertheless illustrated that privacy advocates were concerned about the Big Brother implications of an extensive global surveillance net.

Even if ECHELON wasn't as pervasive as rumored, it emphasized the fact that it was now technologically possible to mount a global surveillance system with wide-ranging powers. By spring 2000, cable networks were running TV programs describing ECHELON as a top secret government project to monitor anything and everything that could be electronically seen or heard on the planet. This, in addition to the proliferation of video cameras in retail outlets, educational institutions, and voyeur sites on the Web, ignited a small, but vocal segment of the population to strongly scrutinize not just Big Brother but the burgeoning numbers of Little Brothers. By the year 2000, concerns about incremental loss of personal privacy were voiced in increasing numbers, concerns that were not entirely unjustified.

Interdepartmental Cooperation and Public Input

The 1990s was a time of substantial social and technological change, but it was also a time when long-distance intedepatmental communication increased and specialized facilities for deploying new technologies were established.

It was not unusual for federal agencies to cooperate in the investigation of crimes, but the means to do so were not always readily available. The global Internet has been a key factor in interconnecting various law enforcement agencies. Multiple-state arsons, serial killings, and kidnappings are easier to solve if investigators discover a pattern of crimes in other jurisdictions and work together to solve them. The Internet facilitated this type of cooperation and the trend to share project information was increasing in the mid-1990s.

In November 1997, integration of various *Department of Defense* (DoD) departments such as the *Polygraph Institute* and the *Personnel Security Research Center* into the *Defense Investigative Service* (DIS) resulted in the creation of the *Defense Security Service* (DSS) to reflect its broader mission. The three primary missions of the DSS are personnel security, industrial security, and security education/training.

One example of effective interdepartmental cooperation was the *National Church Arson Investigation* established in June 1996, which involved the combined resources of several federal agencies. Another was the May 1999 agreement between the *Bureau of Alcohol, Tobacco, and Firearms* (ATF) and the FBI to create the *National Integrated Ballistics Information Network* (NIBIN) *Board* to unify ballistics technology resources.

Integrated Forensic Databases

For many years, forensic scientists have helped law enforcement agencies in the identification of ammunition and firearms that are linked to various crimes—a function that has now been enhanced by computer databases. Steps were taken between 1996 and 1999 to make the ATF Integrated Ballistics Information System (IBIS) and the FBI DRUGFIRE system interoperable, thus providing comprehensive access to images of bullet and cartridge cases. The ATF also managed a gun-tracing system called Online LEAD.

The IBIS bullet-tracing system was capable of searching almost a million images for a match, using evidence associated with a crime. The integrated system, based on IBIS imaging, is now known as the National Integrated Ballistics Information Network (NIBIN) and is located in more than 200 federal state and local law enforcement forensic labs. Through this system, weapons used in seemingly unrelated crimes in different jurisdictions may be compared through digital photos and the information used as evidence.

To further support tracing of bullets and casings, in the early 2000s, the *National Institute of Standards and Technology* initiated a project to develop a virtual standard of digitized bullet profile signatures, based upon master bullets.

Left: The Reference Firearms Collection (RFC) aids the FBI in identifying firearms and their components. It provides a way to test and match firearms that might be associated with crimes. The Standard Ammunition File (SAF) is another collection maintained by the FBI, that includes whole and disassembled cartridges, shot wads, pellets, and other ammunition accessories, similar to those on the right, for study and comparison. [FBI Forensic Science Communications, released; Classic Concepts ©2000 photo, used with permission.]

Background Checks and Forensic Facilities

Another example of cooperative use of database identification techniques went into effect when the *National Instant Criminal Background Check System* began processing background checks on people seeking to purchase firearms. This system was established in conformance with the *Brady Handgun Violence Prevention Act* ("Brady Act") of November 1993, permanent provisions of which went into effect in 1998.

About half the states began participating in NICS in 1999, with the other half handling their own checks. It was found at that time that about 10% of the people applying for the firearms had outstanding warrants for arrest on a variety of charges. The computerized database system was developed through a cooperative effort of the FBI, the ATF, and state and local enforcement agencies, with the NICS Center housed at the FBI's Criminal Justice Information Services Division in West Virginia. Gun dealers could access the information in the system by personal computer over the Internet or by telephone inquiry.

In December 1999, the ATF began ground-breaking activities for a new National Laboratory Center, a complex of three labs, the ATF Alcohol Laboratory, the ATF Forensic Science Laboratory, and the Fire Research Laboratory. These labs provide a variety of investigative services including chemical surveillance, biometrics, explosives analysis, and various physical and instrumental analyses.

Another interesting foray into new strategies and alliances occurred in 1999 when the CIA established *In-Q-It*, a venture capital firm to fund promising Internet-related technologies to attract high-caliber expertise to the service. The director was a Silicon Valley entrepreneur who earned his money and reputation creating video games. The connection between the CIA and video games is not as distant as some might think. Simulations have always been a valuable tool of intelligence-gathering and strategic planning and video games require fast, multiple-media resources beyond those of almost any other type of microcomputer application: capabilities that are relevant to many types of visual and electronic surveillance.

In March 2000, the FBI announced that it had worked with Canada, Wales, and a number of banking and credit card firms to apprehend two 18-year-old males accused of international cybercrimes. Losses were estimated at over $3,000,000. In March 2006, they announced that suspects in the U.K. and the U.S. were arrested for widespread cybercrimes involving credit and debit cards. Suspects were openly selling stolen numbers and documents on the Web. The effort, called Operation Rolling Stone, was one of a number of Secret Service initiatives to combat identity theft and financial fraud.

Meanwhile, as cooperative alliances were being explored and tested in the U.S., satellite surveillance cooperation on an international scale was being critically scrutinized from some quarters, as well as turning up in news headlines.

Domestic and Foreign Concerns Regarding ECHELON

By March 2000, concerns about ECHELON had received worldwide attention and the international community began taking measures to sort out truth from rumors through official channels. The European Parliament (EP) discussed these matters on 5 July 2000, and set up a Temporary Committee on the ECHELON Interception System to inquire into ECHELON-related activities.

Both at home and abroad, people were concerned about rumors of ECHELON-based surveillance of nongovernmental humanitarian organizations and large overseas aerospace contractors. Thus, in April 2000, the NSA and the CIA reported to the House Intelligence Committee, explicitly denying accusations of surveillance of ordinary Americans or of industrial espionage to benefit U.S. firms.

In 2001, the European Parliament voted to acknowledge ECHELON as a global electronic surveillance network and cited examples in which it was believed the system was used to tap satellite communications and interfere with billion-dollar international contracts. The Parliament concluded in its report that Echelon was established by the U.S. in cooperation with Canada, the U.K., New Zealand, and Australia. This was felt, by the EP, to contravene human rights conventions and the U.S. was entreated to sign the Additional Protocol to the *International Covenant on Civil and Political Rights*.

The EP also called upon its member states to adopt the EU *Charter of Fundamental Rights* as a legally binding act "to raise standards of protection for fundamental rights, particularly with regard to the protection of privacy" and to establish a system for the "democratic monitoring and control of the automomous European intelligence capability".

When Technology Meets Opportunity

There is nothing surprising about attention to alleged U.S.-related surveillance activities. It is indeed true that the U.S. and other nations have been increasing their surveillance capabilities over the last several decades and that the U.S. has been involved, with a number of foreign allies, in various conflicts and peacekeeping activities. As a wealthy superpower, the U.S. is in a superior economic position to deploy the latest in electronic devices. As a nation with many world-class scientists, the U.S. is able to experiment with new technologies and techniques with relative freedom.

Thus, economic/scientific factors at the present time favor the U.S. as one of the dominant technological forces in surveillance technologies. American social factors also tend to further the proliferation of 'spy' technologies. The American public is humane and caring, but doesn't always have an in-depth understanding of the needs and concerns of people in foreign nations. Americans are also fiercely competitive and there are many who take for granted the democratic freedoms, and wealth of resources and technology, that are given to us from birth. These social factors may influence national policies and security activities in such a way that other countries may justifiably take issue with how we use surveillance technologies.

With new technologies there is always the 'temptation of opportunity' in which a technology put into use in good faith for one purpose may be used by unscrupulous individuals for another. Policing from within is as important as policing from without.

ECHELON has become a symbol and a focus for concerns over the increase in global surveillance technologies. The international furor may continue for some years until international checks and balances are established and sorted out. While this process continues, it is equally important that the use of surveillance technologies by private citizens and corporations be examined for their potential benefits and their potential to erode personal privacy and freedoms. It is the intent of this book to further the understanding of the prevalence and capabilities of these technologies so that informed choices and policies can be developed to serve everyone's needs.

Transportation Security

After the attack on the World Trade Center, in 2001, concerns about terrorist targeting of vehicles of mass transportation increased. In May 2005, the *Transportation Security Improvement Act* was reported in the Senate, with provisions for funding appropriations for the Secretary of Homeland Security for aviation, surface transportation, administration, intelligence, and research and development, as well as administrative authorizations for the Secretary of Transportation. In this bill, part of the funding for increased security would be paid directly by travelers, in the form of a "passenger security service fee" to be collected before boarding.

Measures for improving the security of bus, rail, marine, and air transportation have had mixed results. While the public may feel safer while traveling on trains and planes, the inconvenience of being searched and having to arrive hours earlier has reduced incentives to travel and affected the profitability of the industry as a whole.

The Department of Homeland Security now plays a significant role in U.S. lives. There have been a number of bills dedicated to its formation and role. In May, 2005, the *Department of Homeland Security Authorization Act for Fiscal Year 2006* (HR-817) was referred to the Senate Committee on Homeland Security and Governmental Affairs after its receipt from the House of Representatives. In it, $34 billion dollars was allocated for "necessary expenses" for 2006. A breakdown of some of these expenses yields the following picture of priorities and allocations

- U.S. Customs and Border Protection ($100 million of which was earmarked for technology to detect weapons of mass destruction and $134 million for the *Container Security Initiative*)—$6.9 billion (20%),

- a nuclear detection office and related activities—$195 million (0.57%)

- the Immigration and Customs Enforcement Legal Program and hiring and training of additional immigration adjudicators—$159.5 million (0.47%)

- chemical countermeasures development—$76.6 million (0.22%)

- the Office of Security Initiatives—$56 million (0.2%)

- terrorism preparedness—$2.04 billion (0.06%—$2 billion for grants to state and local governments as awarded by the *Office of State and Local Government Coordination and Preparedness* and $40.5 million for the *Office for Interoperability and Compatibility* with the *Directorate of Science and Technology* pursuant to the *Intelligence Reform and Terrorism Prevention Act of 2004*)

- cybersecurity R&D—$19 million (0.05%)

- directorates pursuant to subtitle G of title VIII of the *Homeland Security Act of 2002*—$10.6 million (0.03%)

- R&D for counter-threat technologies against human-portable air defense systems—$10 million (0.03%)

The bill also stipulates a consolidated background check process in which the Secretary of Homeland Security, in consultation with the Attorney General, would establish checks and, in some cases, IDs for transportation workers, hazardous materials carriers, the Free and Secure Trade program, the Registered Traveler program, and various border crossing programs. Significant about these background check processes is that they include provisions for collecting personal data and biometric information. The *Department of Homeland Security* would be tasked with implementing programs specified in the Authorization Act within 12 months of its enactment.

Regulations Governing the Use of Surveillance Devices.

The use of surveillance devices is governed by the U.S. Legislative Branch. Lawmakers regularly propose and amend bills that describe overall policy. Some bills are specifically focused on surveillance and some focus on aspects of surveillance related to national security. These bills also shape the structure and functions of various intelligence-gathering agencies.

The *Intelligence Reform and Terrorism Prevention Act of 2004* (S. 2845) has had a highly significant impact on the structure of the Intelligence Community and use of surveillance devices. A full conference report (H.Rpt. 108-796) was produced to accompany IRTPA. These are broad-ranging documents that describe extensive changes in surveillance authority. The following portions of the Act are particularly significant, and those marked with a sword (†) make more direct references to the use of surveillance devices.

Intelligence Community and Facilities Management

IRTPA mandates a number of important changes in intelligence administration, including

- reform of Intelligence Community management,

- revision of the definition of national intelligence,

- assigning responsibilities related to analytic integrity,

- structural organization of a number of intelligence and counterterrorism centers,

- revision of education and activities of the Intelligence Community, including military intelligence,

- conformation and amendment of a number of intelligence acts, officers, and organizations,

- staffing changes, and transfers, and

- improvement of FBI intelligence capabilities.

National Security and Terror Management

IRTPA further specifies screening, background checks, warning systems, and a variety of functions of Homeland Security, including the revision or implementation of

- transportation security, including the use of wireless communications monitoring, watchlists for maritime surveillance, checkpoint screening, biometric surveillance (or other technologies) for passenger and baggage security in the aviation industry, †

- border surveillance, including pilot programs for border security, †
- prevention of terrorist access to conventional and biological weapons, †
- strategies for identifying and dealing with terrorists, money laundering, and terror hoaxes, including pretrial terrorist detention,
- background checks for those seeking to become private security officers,
- economic policies that combat terrorism,
- counter-terrorist travel intelligence, †
- establishment of a human smuggling and trafficking center,
- Homeland Security funding and coordination,
- study of a nationwide emergency notification system, and
- a pilot study for digital warning systems. †

General Security, Surveillance, and Transborder Travel

IRTPA specifically calls for use of identification procedures and technologies, including

- the development of identification standards,
- changes in birth certificates, driver's licenses, and personal identification cards, and restrictions for the inclusion of social security numbers, †
- increased penalties for fraud, and a study on allegedly lost or stolen passports,
- biometric entry and exit systems, †
- establishment of visa and passport security programs,
- increased pre-inspection at foreign airports, †
- enhancement of public safety communications interoperability,
- conversion to digital television standards (for emergency communications),
- studies on telecommunications capabilities and requirements, † and
- Homeland Security geospatial information. †

Civil Liberties

IRTPA specifies civil rights and privacy issues, including

- privacy and civil liberties administration,
- protection of civil rights and liberties by the Office of the Inspector General,
- a privacy officer, and
- protection for human research subjects of the Department of Homeland Security.

The following IRTPA issues are less directly related to the deployment of surveillance technologies, but they form part of the underlying structure within which surveillance policies are implemented.

International Relations

IRTPA specifies a broad-based strategy to improve international relations, including

- assistance to foreign nations (e.g., Afghanistan),
- foreign relations and international policy,
- promotion of free media and "other American values",
- public diplomacy training,
- increase in foreign exchange, scholarships, and other forms of communication with Islamic nations,
- coalition strategies to combat terror, including international agreements to track and curtail terrorist travel with fraudulently-obtained documents,
- international standards for document and watchlist names transliteration (into the Roman alphabet), and
- international exchange of terrorist information.

National Intelligence Administration

The structure and staffing of various American intelligence agencies has been changed.

Significant amendments to a number of acts, including the *National Security Act of 1947*, the *Counterintelligence and Security Enhancements Act of 1994*, and the *Classified Information Procedures Act*, is that each instance of "Director of Central Intelligence" is specified by IRTPA to be replaced with "Director of National Intelligence" (IRTPA amendment to the *Foreign Intelligence Act of 1978*).

Additionally, in the *Counterintelligence Enhancements Act of 2002*, each instance of "President" is replaced by "Director of National Intelligence", thus shifting responsibility from the elected President to an appointed member of the intelligence community.

The Administration of Intelligence Offices

The Director of National Intelligence

The Director of National Intelligence (DNI) is a significant new position. The DNI is appointed by the President, by and with the advice and consent of the Senate and is governed by many of the requirements of IRTPA.

According to the terms of IRTPA, the DNI is an individual with national security expertise who, under the authority of the President, serves as head of the intelligence community and acts as principal advisor to the President, the *National Security Council*, and the *Homeland Security Council*. The DNI directs the National Intelligence Program from a location outside the Executive Office of the President, and may not simultaneously service as director of any other element of the intelligence community.

The General Counsel of the *Office of the Director of National Intelligence* (ODNI), appointed by the President in consultation with the Senate, is the chief legal officer of the ODNI.

As of October 2008, the Office of the Director of National Intelligence may not share the same location with any other element of the intelligence community.

DNI Responsibilities

The DNI disseminates intelligence to the President, legislative branches, and various departments and agencies, and is mandated by IRTPA to provide information that is "timely, objective, independent of political considerations, and based upon all sources available to the intelligence community and other appropriate entities".

The DNI participates actively in budget decisions and is responsible for protecting intelligence sources and methods from unauthorized disclosure. The DNI coordinates relationships with intelligence and security services of foreign governments or international organizations and provides incentives for personnel to fill domestic intelligence staff positions. The DNI coordinates the performance of elements of the intelligence community in consultation with heads of departments and agencies of the U.S. Government and the Director of the *Central Intelligence Agency*.

The DNI ensures that the National Intelligence Council "satisfies the needs of policymakers and other consumers of intelligence".

DNI Surveillance Authority

The DNI has no authority to direct or undertake "electronic surveillance or physical search operations" unless authorized by statute or Executive order. The DNI must see to it that any intelligence gathered through authorized means, pursuant to the *Foreign Intelligence Surveillance Act of 1978*, is disseminated so it "may be used efficiently and effectively for national intelligence purposes".

FBI Intelligence Capabilities

In the final report of the *National Commission on Terrorist Attacks Upon the United States*, it was urged that the FBI "fully institutionalize the shift of the Bureau to a preventive counter-terrorism posture".

IRTPA tasks the Director of the Federal Bureau of Investigation (D-FBI) with improving FBI intelligence capabilities and developing and maintaining a national intelligence workforce. The workforce shall comprise agents, analysts, linguists, and surveillance specialists trained in both criminal justice and national intelligence matters. The Bureau is directed to recruit individuals with backgrounds in "intelligence, international relations, language, technology", and other relevant skills.

The *Office of the Director of National Intelligence* assists the DNI and is composed of:

- the Director of National Intelligence and the Director's Principal Deputy Director (and any Deputy Directors appointed under section 103A or IRTPA),

- the National Intelligence Council, the General Counsel, and the National Counterintelligence Executive,

- other offices and officials, as may be established by law or designated by the DNI in the Office, including national intelligence centers,

- the Director of Science and Technology, and

- the Civil Liberties Protection Officer.

Civil Liberties Protection Officer

The Civil Liberties Protection Officer (CLPO) is appointed by the DNI and reports directly to the DNI. It is the responsibility of the CLPO to "ensure that the use of technologies sustain, and do not erode, privacy protections relating to the use, collection, and disclosure of personal information" and to ensure that personal information contained in a system of records subject to the "Privacy Act", is handled in compliance with fair information practices.

Director of Science and Technology

Surveillance technologies stem primarily from research and development in the science and engineering fields. Thus, the Director of Science and Technology (DST) would have a significant role in the application and development of surveillance technologies within the intelligence community.

The DST is appointed by the DNI and acts as the chief representative of the DNI for science and technology. The DST chairs the *Director of National Intelligence Science and Technology Committee* and assists the DNI in formulating a "long-term strategy for scientific advances in the field of intelligence". The principal science officers of the *National Intelligence Program* coordinate advances in R&D related to intelligence through the *Director of National Intelligence Science and Technology Committee.*

National Counterintelligence Executive

The *National Counterintelligence Executive* (NCE) is a component of the *Office of the Director of National Intelligence* and is tasked with performing the duties 1) prescribed by the DNI and 2) as delineated in the *Counterintelligence Enhancement Act of 2002.*

The National Intelligence Council

The *National Intelligence Council* (NIC) comprises senior analysts within the intelligence community and "substantive experts from the public and private sector" appointed by the DNI.

The NIC is responsible for:

- producing national intelligence estimates for the U.S. Government, including alternative views held by elements of the intelligence community,
- evaluating the collection and production of intelligence, and
- assisting the DNI.

Director of the Central Intelligence Agency

The Director of the *Central Intelligence Agency* (DCIA) heads the *Central Intelligence Agency* (CIA) and reports to the DNI. The DCIA is appointed by the President in consultation with the Senate. The Director is responsible for:

- enhancing analytical systems, human intelligence, and other CIA capabilities,
- developing and maintaining an effective language program,
- establishing and maintaining effective relationships between human intelligence and signals intelligence within the Agency, at the operational level,
- collecting intelligence through human sources and other "appropriate means",
- correlating, evaluating, and disseminating information to appropriate recipients, and
- coordinating relationships between elements of the intelligence community and the intelligence or security services of foreign governments or international organizations in matters involving intelligence related to national security or acquired through clandestine means.

The DCIA does not have police, subpoena, or law enforcement powers (or internal security functions).

The Central Intelligence Agency

The Central Intelligence Agency (CIA) assists the Director of the Central Intelligence Agency. For specifics, consult the *Intelligence Reform and Terrorism Prevention Act of 2004* (H.R. 108–796).

5. Descriptions and Functions

This introduction does not seek to duplicate the descriptions of individual technologies included in each chapter, but there are some general procedures that are common across a variety of surveillance technologies that are summarized here.

Surveillance technologies may be used globally or locally. "Scene investigation" is surveillance confined to a smaller area. Localized surveillance devices include ground-penetrating radar, magnetometers, handheld cameras, plaster casts, ultraviolet detection lights, and fingerprint powders. Scene investigation is useful in archaeology, wildlife conservation, insurance claims, and investigations of civil disputes, theft, violence, or arson.

Crime Scene Investigation

Crime scene investigation is an important example of localized surveillance. Incendiary chemicals, hair, skin, saliva, cigarette butts, clothing, notes, and other objects are often left behind. Sometimes the perpetrators remain near the site. Thieves and arsonists are sometimes caught because they stay to watch police investigate a scene or firefighters battle a blaze. Sometimes they are caught because they strike the same business or home more than once.

Crime scene investigation involves demarcating an area to protect it from contamination or disruption. It also important to visually record the scene before it is disrupted by investigators searching for clues and collecting potential evidence in whatever way is appropriate. The evaluation of clues includes lab analysis, deduction, and cooperation with everyone involved in the process. In cases of arson or homicide, it may be necessary to police the site to prevent the suspects from removing clues or striking again.

Surveillance techniques have become sophisticated enough to identify people from very small samples of human tissue, but this also makes it easy for evidence to be contaminated. Crime scene investigators are often dismayed when important clues are damaged. Footprints and fingerprints get trampled or smudged by bystanders, associates, or even by law enforcement agents themselves. Sometimes items associated with a crime are moved, removed, or lost. Most of the time, these actions are accidental, although, sometimes they may be deliberate.

Members of the Bureau of Alcohol, Tobacco, and Firearms (ATF) utilize a fleet of National Response Team (NRT) vehicles equipped to support arson and explosives investigations. Here, ATF agents sift through the debris at the scene of a fire looking for evidence of arson. Canine assistants are sometimes used to sniff out explosives or accelerants. The ATF cooperates with other law enforcement agencies in federal investigations involving alcohol, tobacco, and firearms. [ATF news photos, released.]

Archaeological Investigation

Archaeological surveillance at a dig site shares many similarities with the investigation of a crime scene (in fact, archaeologists are sometimes called to aid criminal investigators). Demarcation of the site to protect it from treasure-hunters or vandals is usually done first. Concerns about disturbing 'evidence' or contaminating a site are similar to those of law enforcement agencies investigating a crime scene.

Tape (or string) is now commonly used to mark a crime or anthropological/archaeological study site. The area may also be marked out into a grid for precise identification of materials that are removed from the site and sometimes reconstructed elsewhere. If there is concern about theft or vandalism, security guards or electric fences are sometimes raised.*

Many electronic devices are now available for site investigation. Ground-penetrating radar and magnetometers can provide information about underground objects (and prevent unnecessary digging or disturbance of a site). Geographic Positioning System (GPS) devices can show latitude and longitude. These types of devices are described in subsequent chapters.

Technology has made it possible to glean volumes of information from minute pieces of evidence. DNA samples have been taken from ancient skeletons and mummies. Gloves are routinely worn and face masks and body suits may also be worn to reduce falling hair and skin from investigators, particularly if sensitive DNA-profiling techniques are used on trace amounts of blood or hair. Environmental suits may also be worn to safeguard examiners working in hazardous or hostile situations. Environmental suits and biochemical clues are discussed in the Chemical & Biological Surveillance and Genetics Surveillance chapters.

6. Applications

Surveillance technologies are of little use unless applied toward a specific goal or task. Each chapter in this text includes a section on applications, to provide insight into how surveillance technologies are commonly used and who uses them. It isn't possible to include all the possible uses for every device, but there are enough to provide a better understanding of common applications.

6.a. Search and Rescue

Surveillance technologies are essential tools in search and rescue operations. When searching for individuals who have disappeared in lakes, oceans, avalanches, and forests, where hypothermia, hunger, or lack of breathable air can claim a life very quickly, technology can save precious time. Surveillance devices are also used to find people and animals when buildings collapse after bombings or earthquakes.

Surveillance devices serve many roles in search and rescue. After hurricanes and floods, surveillance devices not only help locate victims, but can provide data for assessing the scale of a disaster. They may aid in determining the character or status of chemical leakages, assess what it might take to rebuild, and aid insurance adjusters in surveying the damage and providing relief. Surveillance devices can predict natural disasters or warn citizens of impending danger from other sources. Infrared detectors, radar, satellite images of terrain, radio sets, and tracking beacons are all used in disaster prevention and lifesaving efforts.

---

*In one of history's ironic twists of fate, the famous Pinkerton Detective Agency recently supplied site security at an investigation scene when the alleged remains of Jesse James were exhumed to see if a positive identification could be made through DNA technology. Ironically, the notorious criminal was never apprehended by Pinkerton's Agency in the 1880s, when James was a wanted man, despite many attempts at his capture.

Strategies for Surveilance Deployment. Left: Surveillance technology is only useful when combined with good strategies. Here LtJ Seve Rutz, a controller with the Coast Guard Command Center in Juneau, is shown plotting a search pattern for a missing sailing vessel in January 1998. Right: C2PC, a computerized global command and control system, used by Pearson and Behner to plan a search. The C2PC is also useful for law enforcement activities. [U.S. Coast Guard news photos by Mark Hunt and Chuck Wollenjohn, released.]

The U.S. Coast Guard participates regularly in search and rescue operations using both air and marine vessels equipped with surveillance capabilities.

Search and Rescue Operations. Left: A burning fishing vessel from which the U.S. Coast Guard rescued five men and a dog off the coast of Unimak Island. Right: Coast Guard members assist a pilot and five tourists at the site where a helicopter crashed due to 'white-out' conditions on the Herbert Glacier near Juneau, Alaska. The search team brought food, shelter, and survival gear to help stranded victims make it through the night. [U.S. Coast Guard news photos by USCG and Mark Hunt, released.]

In the U.S., in 1989, the *National Urban Search and Research* (US&R) Response System was formed under the authority of the Federal Emergency Management Agency (FEMA). FEMA provides a framework for local emergency services to aid victims of structural collapse. FEMA is now familiar to many as an organization that was expected to respond to the severe flooding and other hurricane damage that destroyed the city of New Orleans. Members of the Task Force come from many disciplines—they include hazardous materials experts, structural engineers, search specialists, highly trained dogs, medical practitioners, and pilots. All of these specialists use surveillance technologies of one type or another in their work.

Examples of search and rescue technologies are covered more fully in the Sonar, Light, Aerial, and Animal Surveillance chapters. DNA identification of victims' remains is described in the Genetics Surveillance chapter.

Marine Search and Rescue. Left: Balloonists attempting to circle the Earth crash into the Pacific Ocean halfway through their journey. Search and rescue helicopters travel to the site from Air Station Barbers Point in Hawaii. Right: Coast Guard cutter crew members rescue three sport fishers adrift in a disabled vessel in eight foot seas near the island of Kauai. The location of marine victims is especially challenging if they have no radio or the radio has been disabled. [U.S. Coast Guard news photos by Marc Alarcon and Eric Hedaa, released.]

6.b. Border Patrol

Border patrol is an important aspect of national security. Customs and immigration officials seek to prevent abuses of human rights, trade laws, and the transport of hazardous materials, plants, and weapons. To achieve these aims, they regularly surveil people and vehicles crossing national borders.

The technology used in customs and immigration administration is becoming increasingly sophisticated. Infrared sensors, aerial surveillance, and X-ray machines have been used for some time, but high-resolution cameras, electronic databases, and intelligent recognition software (now available) will begin to play dramatic roles as they are incorporated into border surveillance.

By the end of the 1990s, cameras had been installed at many Mexican and Canadian border stations that formally were monitored only by humans. Cameras can be connected to computers to monitor when and how often specific vehicles or people cross the border and can alert border officials to unusual crossing. Patrolling borders away from official crossings is a greater challenge and technologies like aerial surveillance and infrared imagery are used from airplanes, helicopters, and unstaffed aerial vehicles on or over land or sea.

Marine patrols regularly ply the waters near border crossings to control smuggling. These benefiting from improved sonar systems. Remote-controlled marine and aerial vehicles, which greatly extend the surveillance 'reach' of patroling vessels, area also in use. Marine surveillance helps protect endangered species, prevents poaching or abuses of commercial fishing licenses, aids in search and rescue operations, and helps stem the flow of smuggled goods and refugees.

The human cargo trade, in which refugees pay up to $60,000 per person to be transported, in unsafe ships or containers, is of great concern to authorities. Refugees and stowaways often die or succumb to illness due to lack of food and sanitation, and they often bring diseases like tuberculosis and hepatitis into the country illegally entered. Customs and immigration officials have stepped up their surveillance of tankers or containers that may contain human cargo.

Sonar, X-ray machines, infrared sensors, cameras, and dogs are just some of the surveillance technologies now regularly used to detect border-runners and refugee claimants.

Border Surveillance. Left: An illegal alien found on board a fishing vessel by the Coast Guard is turned over to the U.S. Border Patrol. Various law enforcement agencies often cooperate to prevent smuggling, trafficking, or border-running. Right: Cuban refugees rescued from an unseaworthy homemade boat near the coast of Miami Beach, Fl. [U.S. Coast Guard 1999 news photos by Keith Alholm and Chris Hollingshead, released.]

6.c. Natural Resources Management and Protection

Surveillance Strategies. Left: Coast Guard members discuss tactics for searching fishing vessel spaces in 1990. Illegal fishing methods, catches, or contraband are sometimes found by the Coast Guard on routine checks. Right: A fish catch is inspected during a routine 1993 Coast Guard fishery patrol. Fishing vessels are also checked for compliance with regulations and minimum safety standards. [U.S. Coast Guard news photos by Robin Ressler, Ron Mench, released.]

The protection and management of natural resources, including food sources, are an important aspect of a nation's cultural and economic survival. Surveillance strategies and technologies help to monitor commercial harvesting, wildlife ecology, and poaching activities throughout the world. Without these protections, our resources would quickly be depleted, as has happened in the past in unregulated areas. DNA-monitoring, radio-collar tracking, sonar, and optical surveillance are technologies used to monitor natural resources and those who seek to abuse them.

Left: Coast Guard crew members successfully disentangle a humpback whale from a lobster trap. Humpback whale populations were on the endangered species list when their population dropped to only about 12,000 worldwide; in human terms, that's barely enough to populate one small town. A number of surveillance technologies, including radio tracking beacons and DNA matching, are being used to aid wildlife management and protection. Right: A highly endangered northern right whale found dead. Coast Guard personnel prepare lines to mark the site. [U.S. Coast Guard news photos, 1999 (Erb) & 2001 (Sperduto), released.]

6.d. Drug Laws Enforcement

Surveillance technologies are widely used by law enforcement agencies to uncover illegal activities such as drug manufacture and distribution. Infrared sensing, power-consumption monitoring, phone tapping, aerial photography, chemical sniffers, and canine scouts are all used to identify individuals, facilities, and vessels used in the drug trade.

Sometimes illegal drug-related suggling is discovered in the course of other activities, such as rescues from fire or violence indirectly related or unrelated to the drug activities. In these cases, several agencies may be involved in the search and seizure of drug caches and paraphernalia.

As an example of an accidental discovery of contraband drugs, the U.S. Coast Guard and the Mexican Navy responded to a fire onboard the vessel "Valera" off the coast of Mexico, in January 2000. The members of the Valera had abandoned ship and were rescued from the water by a small Coast Guard boat. While investigating the cause of the blaze, over three metric tons of cocaine were discovered and seized from two large compartments aboard the vessel. Cocaine is one of the more common drugs smuggled across maritime borders between Mexico and the United States.

Left: Bails of marijuana were found hidden below decks on a fishing vessel inspected by the U.S. Coast Guard. Right: Coast Guard personnel stand watch over 11.5 tons of cocaine seized in the eastern Pacific. [U.S. Coast Guard news photos 1997 & 2005 (Leshak), released.]

Once found, various methods are used by different agencies to determine the chemical makeup of suspected illegal drugs. Most enforcement agents are trained to make a preliminary guess by visual inspection, smell, and sometimes taste. Portable kits are available to assist in preliminary analysis and generally larger samples are then sent to a lab for confirmation or more extensive analysis in the case of mixed samples or blended drugs.

6.e. Military Applications

Surveillance technologies are prevalent in military reconnaissance and targeting operations. Many of the early satellite experiments were financed and developed for military purposes and radar technology, which was originally invented to improve marine safety, was developed for military surveillance in World War II. Intelligent and remotely-piloted vehicles are now also an important aspect of military operations.

Remote- or self-piloted aerial vehicles. Left: A Predator RQ-1 unstaffed aerial vehicle being handled by SSgt. Stroud at the Balad Air Base in Iraq. This is a long-endurance, medium-altitude vehicle used for econnaissance, surveillance, and target information acquisition. Right: Captain Songer maneuvers the Predator from a console stationed at the Balad Air Base. Bottom: A Global Hawk UAV by its hangar after flying a mission in Iraq. [DoD news photos by TSgt. Jensen, SSgt. Young, & TSgt. Gudmundson, 2004.]

The Boeing Corporation has been contracted to provide "persistent intelligence, surveillance, reconnaissance unmanned aerial vehicles services" in support of U.S. operations involving Iraq and globatl efforts to suppress terrorism. UAVs are unstaffed remotely-piloted or self-piloted aerial vehicles equipped with sensors, cameras, and communications equipment for reconnaissance and surveillance missions. Digital imagery from UAVs can now be monitored in realtime to track current events, troop movements, and changes in structures or terrain. UAV use has increase in Iraq and Afghanistan. The most prominent UAV systems, according to DoD news articles are the Predator and the Global Hawk (see the Aerial Surveillance chapter for more information on UAVs. The Audio, Sonar, and Radar chapters also have information on military applications.

6.f. National Intelligence Agencies

In 1992, Robert Gates, the Director of the Central Intelligence Agency (CIA), established an Office of Military Affairs in order to create a closer connection between the CIA and other members of the intelligence community. The main goal of this alliance was to provide intelligence that would enhance awareness during conflicts, particularly on the battlefield. The CIA gathers imagery and information from signals (radar, sonar, etc.), processes the information (e.g., through simulations) and provides it to battlefield commanders. The gathered data may include constant weather updates and continuous surveillance of the battle environment. John Deutch, Director of the CIA, in June 1995 stated "if the enemy does not have similar information, it means that victory will come more rapidly and therefore the casualties will be lower." Plans were in effect at the time to create a "national imagery agency" to serve in the collection, analysis, and distribution of intelligence imagery.

6.g. Commercial Products

Each chapter in this book includes a section on some interesting or representative commercial products and lists volume and price information on some.

Any commercial products listed are included for informational purposes only. Their inclusion does it imply an endorsement of the quality of their products or services. The included examples are intended as educational examples only, to provide an introduction to the types of products that may be purchased on the market as they relate to individual topics.

The following products are those of general interest to the topic of surveillance technologies; specific technologies are listed in individual chapters. In general, surveillance product vendors tend to fall into five main categories: personnel services (trained security officers equipped with surveillance skills and devices), training services (instructors and computer simulators), research and development services (firms that create and test new technologies), manufacturing services (firms that build products), and distribution services (firms that distribute and sell products).

Advanced Paradigms Inc. - Applications development, communications and training to federal agencies and commercial organizations. API has a number of prominent clients in the federal government and technology industries.

Alliant TechSystems - Markets to military, law enforcement, and the security industry. Products include quick-reaction products and high-speed cryptography.

Analytical & Research Technology Inc. - Systems development and integration. Provides the intelligence community with hardware, software, and integration services for data handling.

Applied Signal Technology - Designs and manufactures signal-processing equipment for a wide variety of telecommunications sources. The equipment is used for foreign signals reconnaissance by government and the private sector. Incorporated in 1984. Each office has a government-approved facility clearance.

Cloak and Dagger Books - A specialized bookstore with volumes on many aspects of intelligence, including military history, counterintelligence, codes and ciphers, espionage, and more. Bedford, New Hampshire.

Executive Intelligence Services - (EIS) An investigation company specializing in research and surveillance in workers' compensation, personal injury, fraud, and medical malpractice.

Executive Resource - A competitive intelligence consulting firm based in Montreal, Canada which provides corporate intelligence/strategic planning training and coaching services. Executive Resource also supplies a by-subscription online news journal, Competia Online.

Loyal Security, Inc. (LSI) - Leading-edge products for law enforcement, including access control, training, security consulting, counterintelligence, videoscopes, infrared illuminators, electronics belt-packs, and head-mounted displays. Based in North Carolina.

Mega Worldwide, Inc. (MWI) - A group of international companies which specialize in fields of high-threat security. MWI provides specialized equipment, training, and investigations services to state, federal, and foreign governments with experts in more than 50 countries.

National Security Archive - Microfiche and written publications based on declassified and unclassified government documents and scholarly research associated with these documents. Some of these are of general interest, some of academic interest, and some are priced for the library market.

Pacific-Sierra Research Corp - Founded in 1971 to carry out applied research. It markets to various defense and intelligence agencies and foreign and domestic clients. Primarily information technology, high-performance computing, software development, and submarine communications. Employee-owned.

The TEAL Team, Inc. (TTT) - Advanced security training and consulting in high-risk security tasks. Services are aimed at governments, law enforcement units, emergency response units, and large corporations. TTT claims that the staff includes senior experts drawn from organizations such as the Secret Service, FBI, SEALs, etc. Based in New Jersey.

In the surveillance industry, there are also 'shadow' organizations that are often loosely affiliated with large contractors, but which keep a low profile. These companies typically market covert services and technologies.

7. Problems and Limitations

Each chapter in this book has a section on problems and limitations that are specific to the technologies discussed in that chapter.

8. Restrictions and Regulations

The more relevant or interesting legal issues related to particular technologies are summarized in each chapter.

Some issues of general interest with regard to surveillance, information access, and privacy are described here. This is a sampling, it is in no way complete, and those seeking further information are encouraged to consult government and public legal archives on the Internet and in local libraries. (See also the Cryptologic Surveillance chapter for computer-related legislation.)

> *National Security Act of 1947*. USC, Title 50. Enacted and amended in 1947, this act was intended to promote national security by providing for a Secretary of Defense, a National Military Establishment, a Department of the Navy, and a Department of the Air Force, and for coordination of activities. It also established the *Central Intelligence Agency* (CIA). The act represented a significant reorganization of military establishments and foreign policy priorities and created new institutions.

Freedom of Information Act (FOIA). USC, Title 5, Subsection 552. Enacted by the U.S. Government in 1966 to provide individuals with the right to request access to information or records held or controlled by federal executive branch agencies upon submission of a written request. The Act stipulated nine exemptions and three exclusions. The exemptions generally cover issues relating to physical and financial security including national defense and foreign relations, inter/intra-agency communications, internal rules and practices, trade and financial secrets, law enforcement investigative information, and geological information on oil wells. The FOIA "does not apply to Congress, the courts, or the immediate office of the White House, nor does it apply to records of state or local governments." Most states enacted FOIA-type statutes which can generally be queried by writing to the state Attorney General. FOIA requests must be made in writing to the relevant federal agency. Search fees typically range from $10 to $30/hour plus copying fees with the first two hours of search and 100 pages of copying not charged for noncommercial requests. Response is up to 10 working days.

Privacy Act. Public Law 93-579. USC, Title 5, Subsection 552a. The U.S. Government acknowledges that it compiles federal records on individuals, including taxpayers, people in the military or employed by federal agencies, and those who receive social benefits such as student loans or social security. In response to this, in 1974, the U.S. Congress passed the Privacy Act to establish "certain controls over what personal information is collected by the federal government and how it is used." The act guarantees three rights to U.S. citizens and lawful permanent residents:

1) the right to see records about oneself, subject to the Privacy Act's exemptions;

2) the right to amend that record if it is inaccurate, irrelevant, untimely, or incomplete; and

3) the right to sue the government for violations of the statute, including permitting others to see your records, unless specifically permitted by the act."

The Privacy Act further establishes certain limitations on agency information practices and prohibits agencies from maintaining information describing "how an individual exercises his or her First amendment rights" unless there is consent or a statute permitting it or it is "within the scope of an authorized law enforcement investigation."

Like the Freedom of Information Act, the Privacy Act pertains to U.S. citizens and legal permanent residents and has certain exemptions which permit agencies to withhold information. The exemptions generally relate to information that could compromise national security or criminal investigations or which would identify a confidential source. Privacy Act requests must be made in writing to the relevant federal agency. Search fees are not charged, but copying fees may be charged. The response time is up to 10 working days.

Foreign Intelligence Surveillance Act of 1978 (FISA). Public Law 95-511, 50 USC §1805 (expanded in 1994), signed by President Carter. This Act establishes the procedures for an authorized government official to acquire a judicial order to authorize electronic surveillance or physical search in foreign cases. Probable cause must be shown that the target is associated with a foreign power and the surveilled premises are being used by the foreign power. Certain acquisition and disclosure requirements must be met. The Foreign Intelligence Surveillance Act Records System (FISARS) is the information system for FISA applications. Access to FISARS is restricted to personnel with TOP SECRET/SCI (Sensitive Compartmented Information) clearance with the Office of Intelligence Policy and Review (OIPR). The OIPR is not an investigative department, but rather a system for managing information received from the Intelligence Community.

Intelligence Oversight Act of 1980. Following allegations of wrongdoing by U.S. intelligence agencies, two committees were established, the Senate Select Committee on Intelligence (SSCI) in 1976 and the House Permanent Select Committee on Intelligence (HPSCI) in 1977. These, along with the Armed Services and the Foreign Relations and Foreign Affairs Committees, were to oversee and authorize the activities of the intelligence agencies. The Hughes-Ryan Amendment required that covert action notifications be given only to the two intelligence committees (other committees no longer had to be notified).

United States Intelligence Activities, Executive Order 12036, 24 January 1978. Signed by President Carter. This revoked Executive Order 11902 of 2 February 1976. It reorganized the intelligence structure and provided guidelines on the execution of intelligence activities. The Director of Central Intelligence (DCI) was given increased management authority over the Intelligence Community. The Secretary of Defense was designated to be the Executive Agent for Communications Security and the Director of the NSA was to execute the responsibilities for the Secretary of Defense. Senior officials of each agency were to report violations to the Attorney General.

Executive Order 12333, 4 December 1981. Signed by President Reagan. "Timely and accurate information about the activities, capabilities, plans, and intentions of foreign powers, organizations, and persons and their agents, is essential to the national security of the United States. All reasonable and lawful means must be used to ensure that the United States will receive the best intelligence available...." The Order describes goals, direction, duties, and responsibilities of the national intelligence effort to provide information to the President, the National Security Council (NSC), the Secretaries of State and Defense, and other Executive Branch officials. The Director of Central Intelligence is made responsible directly to the President and the NSC and duties of the CIA are put forth. The Department of Defense, Secretary of Defense, Department of Energy, and the Federal Bureau of Investigation (FBI) responsibilities are put forth. The Order further describes the execution of intelligence activities.

Intelligence Identities Protection Act, 23 June 1982. Signed into Public Law 97-200 by President Reagan. This imposed criminal penalties on anyone who wrongfully divulged the identities of covert intelligence personnel ("undercover intelligence officers, agents, informants, and sources"). In 1984, the President exempted the CIA from the search and review requirements of the Freedom of Information Act with respect to sensitive files through the Central Intelligence Agency Information Act of 1984.

Department of Defense Surveillance Countermeasures Survey Program, 23 May 1984. This is a Department of Defense (DoD) Instruction to update policies, responsibilities, and procedures for Technical Surveillance Countermeasures (TSCM) services which are included within the Defense Investigative Program.

Intelligence Organization Act of 1992. Organizational guidelines established by Congress for the Intelligence Community. This is one of the most significant Acts since the National Security Act of 1947. The oversight committees that had been established by the Intelligence Oversight Act of 1980 introduced intelligence reorganization bills leading to this Act. This basically established the actions the DCI had been taking to restructure the Intelligence community since 1980, establishing a legal framework, recognizing the DCI as the statutory adviser to the National Security Council and establishing the National Intelligence Council as the authority for intelligence analysis. It further defined the composition of the Intelligence Community.

Classified National Security Information, Executive Order 12958, 17 April 1995. A reform of the U.S. Government system of secrecy to create a uniform system for classifying, safeguarding, and declassifying information related to national security while supporting progress through the free flow of information. Sets out classification levels, standards, associated markings, guidelines for review, and associated authorities. Interpretation authority rests with the Attorney General. Signed by President Clinton. 12958 revokes 12356; it was later amended by EO 12972 (1995) and EO 13142 (1999).

Executive Order 12968, 2 August 1995. Describes access to classified information, financial disclosures, etc., which includes definitions, access eligibility, nondisclosure requirements, types of documents and reports that fall under this Order, and financial disclosure, including disclosure of codes and cryptographic equipment and systems, use of automated databases, access eligibility policy and procedure, standards, implementation, and general provisions.

Electronic Freedom of Information Act Amendments of 1996. Public Law 104-231. Established requirements for making information falling within Freedom of Information Act (FOIA) guidelines electronically available.

Economic Espionage Act of 1996. Public Law 104-294. A vehicle for prosecuting those who are found to be engaged in economic espionage, that is, the stealing of trade secrets and other business information. (Note that a war-related Espionage Act was enacted in May 1918.)

National Information Infrastructure Protection Act of 1996. Amends USC Title 18, the Computer Fraud and Abuse Act, to protect proprietary economic information. This amendment is Title II of Public Law 104-294 (see previous citation). It has important ramifications for 'computer hacking.'

Presidential Decision Directive No. 61 (DoE declassified version). A February 1998 restructuring Decision to form two independent intelligence offices within the Department of Energy (DoE), new counterintelligence (CI) measures, and a stronger cooperative relationship between the DoE and the FBI. Other measures include improved threat and vulnerability assessment and oversight and performance assurance. Coordination is provided by the National Counterintelligence Policy Board (NACIPB).

Act to Combat International Terrorism. Public Law 98-533. This in part establishes the Rewards for Justice Program in which cooperating individuals may be financially rewarded and/or relocated for providing information leading to the prevention of terrorism or the arrest and conviction of terrorists. This program is handled by the Diplomatic Security Service (DSS) of the U.S. Department of State.

United States Title 50 - War and National Defense, Chapter 36 - Foreign Intelligence Surveillance. This chapter covers electronic surveillance authorization, designation of judges, court order applications, order issuance, use of information, reports, criminal sanctions, civil liability, and authorization during time of war.

The law-making process is often slow, requiring three or more years for definitions and priorities to be sorted out before they pass through all the bureaucratic hurdles. In addition to general administrative and organization measures, there are also periodic acts and bills associated with funding, including:

Intelligence Authorization Act for Fiscal Year 1995, 14 October 1994. Public Law 103-359. This established a Commission on the Roles and Capabilities of the U.S. Intelligence Community who were to review and report on the efficacy and appropriateness of activities of the U.S. Intelligence Community by March 1996. Periodic authorization acts establish spending priorities and allocations for intelligence-related matters.

FY2000 Intelligence Authorization Bill. Budget appropriations for the fiscal year 2000 for U.S. Government intelligence-related activities. Similar to the above in that it is a periodic funds authorization act.

Department of National Homeland Security Act of 2001. This significant act was first introduced in the Senate exactly one month after the World Trade Center attack of 11 Sept. 2001. The primary focus of the bill was to establish a *Department of National Homeland Security.* Not surprisingly, considering the speed with which the original bill was drafted, it went through some changes over the next few months and was amended and submitted in November 2002 as the *Homeland Security Act of 2002* (H.R.–5005), emphasizing the *Department of Homeland Security,* information analysis and protection of the national infrastructure, terrorist countermeasures, emergency preparedness and response, and the security of transportation venues and borders. *www.dhs.gov/*

9. Implications of Use

Information on history, prices, and types of devices is useful, but not really complete unless the impact of the use of the technologies is also discussed, even if only briefly. Each chapter has a section that describes some of the social consequences and trends related to individual categories that provide a focus for thought and further discussion.

This book is about exciting new technologies that fulfill a range of political and social needs and, if used judiciously, can improve our quality of life. However, it is hoped that our increasing dependence on electronic devices doesn't completely supersede traditional methods (including common sense) because no technological device is immune from attack. Electronics are very sensitive to magnetic and high-voltage electrical fields. They can also be disrupted by powerful bursts of sound or coherent light (laser light).

A large predator, like a lion, doesn't have to fear smaller, weaker animals in its home range, but the large predator is the first to die if its range is diminished or food sources become scarce. The same vulnerability applies to electronics societies. Computers and surveillance devices might give Western nations an immediate political or economic advantage, but it also makes them extraordinarily vulnerable to electronic dependence. A high-energy radio-frequency (HERF) 'bomb' can instantaneously disable dozens or thousands of electronic devices with a concentrated burst of radio waves. This could destroy or significantly disrupt a fully automated infrastructure. It is possible to disable computerized power grids, transportation systems, and financial establishments with sophisticated weapons or with simple home-made bombs. Many destructive weapons cost less than a few thousand dollars to build and people who make these weapons aren't going to seek FCC approval before using them.

Little Brothers

Another important aspect of surveillance is increased use of viewing and recording devices by citizens and newshounds. There are often financial rewards for those who submit graphic and sensationalist examples to less ethical publications or Internet sites. One of the consequences of 'reality programming,' broadcasts based on the experiences of nonactors, detectives, law enforcers, search and rescue personnel, and others, is that the public gains a better understanding of how our society and our public safety systems work. One of the negative consequences is that the families of loved ones who are brutally killed in murders or kidnap attempts or who have died from terrible falls or accidents are subjected to a protracted mourning period in which wounds may be opened again and again when news shows, reality shows, and videos uploaded to the Internet are replayed. A viewer rarely knows when the images are going to be aired because the original videographers often license the footage to other broadcasters. Repeat airings can be painful and, in some cases, cruel. The definition of 'news' versus 'reruns' should be re-evaluated and the consequences of extended replays to surviving families should be considered when capturing and airing surveillance videos.

Media Exploitation of Private Information

There is a need for public discussion as to whether the news media may replay personal events with impunity or sell them to outside non-news broadcasters without permission from the victims or, if they are deceased, from their immediate families. The material may fit the definition of 'news' the first time it is played, but it can be argued that *replays* are no longer news (especially when they are licensed out to third parties). This could be considered commercial exploitation and should require the permission from family members or, if permitted, involve compensation for repeated broadcasts, in much the same way that actors get 'residual' payments for reruns.

Technological devices are rarely inherently good or bad. In the race for electronic superiority, it is hoped that the important ethical ramifications of using surveillance technologies will not be overlooked, since they form the basis for electing our representatives, safeguarding our freedoms, and enhancing and maximizing our quality of life and interactions with others.

This book is intended to provide a balanced view of complex issues, both bad and good, and a sampling of practical examples of interest, in order to broaden our understanding of surveillance technologies. Good references can aid in better decision-making, now and for the future.

10. Resources

10.a. Organizations

Each chapter has a section listing organizations of relevance to that particular category of technology. In addition, here are some references of general relevance to surveillance technologies and intelligence-gathering. No endorsement of companies is intended.

Advanced Technology Office (ATO) - Created as a result of a DARPA reorganization, ATO focuses on 'high payoff' maritime communications, early entry, and special operations. Communications projects include superconducting filters and secure large-scale wireless networks and mobile systems. www.darpa.mil/ato/

Air Force Foreign Technology Division (FTD) - One of several intelligence divisions in the U.S. Air Force, along with the Office of the Assistant Chief of Staff, Intelligence, the Air Force Intelligence Support Agency, the Air Force Electronic Security Command, and the Air Force Technical Applications Center. Originally established as the Foreign Data Section in 1917, the FTD publishes a regular bulletin.

Air Force Intelligence, Surveillance, and Reconnaissance (ISR) - Ensures U.S. military information superiority in partnership with other military services and national intelligence agencies. The U.S. Air Force operates a variety of ground sites and airborne reconnaissance and surveillance platforms around the world. Managed by the Director of Intelligence, Surveillance, and Reconnaissance which also handles the AIA. www.cia.gov/ic/afi.html

American Civil Liberties Union (ACLU) - A prominent, nonpartisan individual rights advocate providing education on a broad array of individual freedoms issues in the United States. Founded by Roger Baldwin in 1920, the ACLU seeks to assure preservation of the Bill of Rights which is associated with the U.S. Constitution. www.aclu.org/

American Institute of Physics (AIP) - Founded in 1931, the AIP has over 100,000 members worldwide in all branches of physics. AIP publishes a number of professional journals (*Acoustical Physics, Applied Physics Letters, Virtual Journal of Biological Physics Research, Computers in Physics, Computing in Science and Engineering, Journal of Applied Physics, Journal of Biomedical Optics, Journal of Electronic Imaging, Optics and Spectroscopy,* etc.) and provides searchable online access for subscribers. www.aip.org/

Army Intelligence Agency (AIA) - Established in the mid-1980s by the Assistant Chief of Staff for Intelligence to direct scientific (nonmedical), technical, and general intelligence. It was originally a counterintelligence and HUMINT agency which was rolled into the Intelligence and Security Command (INSCOM) in 1977 to unify intelligence services in the Army. It was then separated out again, but INSCOM reassumed command of the AIA in 1991. It was then discontinued and INSCOM created the National Ground Intelligence Center from the remaining Army units. See next listing.

Army Intelligence and Security Command (INSCOM) - A major operational intelligence agency of the U.S. Army, established in 1977 from the Army Security Agency (ASA), which dates back to 1930. INSCOM conducts intelligence and information operations in multiple disciplines for U.S. military commanders and national decision-makers. The organization also conducts a variety of production activities, including imagery exploitation, intelligence battlefield preparation, and science and technology intelligence production. It further engages in counterintelligence and force protection, electronic and information warfare, and support to force modernization and training. The European arm is located in Griesheim, Germany. www.vulcan.belvoir.army.mil/

Association of Former Intelligence Officers (AFIO) - AFIO is a nonprofit, nonpolitical educational association founded in 1975. Members comprise intelligence professionals engaged in promoting understanding of the role and functions of U.S. intelligence activities. AFIO publishes the journal *Intelligencer* and *Weekly Intelligence Notes* (WIN). www.afio.com/

Association for Crime Scene Reconstruction (ACSR) - Founded in 1991 to provide support to law enforcement investigators, forensic experts, and educators in the understanding of a crime scene, its reconstruction, and the gathering and preservation of evidence. www.acsr.com/

Atlantic Intelligence Command Joint Reserve Intelligence Program (AIC) - AIC is engaged in coastal studies, expeditionary support, evacuation-planning and other logistical services. It maintains the JIVA Operational Laboratory which processes SIGINT and IMINT and their integration. It also stands watch and aids in evacuation planning.

Bureau of Alcohol, Tobacco, and Firearms (BATF or more commonly ATF) - A U.S. Treasury Department law enforcement department dedicated to enforcing federal laws in order to prevent and suppress violent crimes, to collect revenues, and to protect the public in matters related to alcohol, tobacco, and firearms. BATF manages the Firearms Licensing Program to ensure compliance with federal laws. It also assists law enforcement agencies in the handling of violent crimes through the National Tracing Center. www.atf.treas.gov/

Bureau of Intelligence and Research (INR, sometimes called BIR in the popular media) - Originally the Interim Research and Intelligence Service, it has been designated as the Bureau of Intelligence and Research since 1957. It is now part of the U.S. State Department. INR publishes intelligence reports and collects normal diplomatic information and open-source intelligence. See State Department Bureau of Intelligence and Research.

Canadian Forces Intelligence Branch Association (CFIBA) - The professional organization of the Canadian Forces Intelligence Branch which fosters and promotes the traditions and well-being of its members. The Web site includes the "Intelligence Note Book" which provides an introduction to intelligence concepts and suggestions for further reading. www.intbranch.org/

Canadian Security Intelligence Service (CSIS) - Promotes Canada's national security and the safety of Canadian residents. CSIS provides information and assistance to safeguard scientific and commercial secrets. Founded in 1984 when an Act of Parliament disbanded the Royal Canadian Mounted Police Security Service which had carried out this mandate for 120 years. www.csis-scrs.gc.ca/

Center for Defense Information (CDI) - A national, independent military research organization founded in 1972 funded by public and foundation donations rather than government or military funding. The organization researches military spending, policies, and weapons systems. The military is studied from the public perspective and CDI provides information through the media and various publications. CDI provides assistance to the government by request. www.cdi.org/

Center for International Policy (CIP) - Founded in 1975 to promote U.S. foreign policies that reflect democratic values. CIP promotes a non-militaristic approach to international relations through education and advocacy. It promotes the restoration of democracy in oppressed nations and the removal of landmines. It sponsors an intelligence reform program. www.us.net/cip/

Central Intelligence Agency (CIA) - An independent agency established in 1947 through the signing of the National Security Act. The Director of Central Intelligence (DCI) was charged with coordinating national intelligence. The DCI is the head of the U.S. Intelligence Community as principal adviser to the President for intelligence matters. www.cia.gov/ (Office of the Director of Central Intelligence) www.odci.gov/

> **The Directorate of Intelligence** (DI) - The analytical arm of the CIA that provides intelligence analysis on national security and foreign policy issues.
>
> **The Directorate of Science and Technology** (DS&T) - Involved in science and technical innovation relevant to intelligence activities, such as imaging systems.
>
> **The Center for the Study of Intelligence** (CSI) - Provides research, publications, and a variety of educational programs. Houses historical materials.
>
> **The Electronic Document Release Center** (EDRC) - Provides document collection and information to provide an overview of access to CIA information.
>
> **Arms Control Intelligence Staff** (ACIS) - Arms control intelligence information.

Consumer Information Center (CIC) - A U.S. General Services Administration agency which publishes a free Consumer Information Catalog which lists more than 200 free or low-cost booklets on a wide variety of consumer-related topics. The Catalog can be downloaded. www.gsa.pueblo.gov/

Council of Intelligence Occupations (CIO) - A department of the Directorate of Intelligence (DI) which assesses skill levels against needed intelligence requirements and ensures a supply of expertise and talent in political, military, economic, scientific, leadership, imagery, and other intelligence fields. www.odci.gov/cia/di/mission/cioc.html

Counterintelligence Corps (CIC) - A branch of the U.S. Army descended from the Military Intelligence Service Counterintelligence Branch. The Corps was established in 1942 to handle both domestic and foreign missions. The 441st Military Intelligence branch of the CIC engages in surveillance of subversive activities, individuals and ideological movements, and security-related counterespionage. www.441st.com/

Counterterrorist Center (CTC) - A Directorate of Intelligence (DI) facility that tracks international terrorist activities to provide analysis and intelligence which assists in prevention and policy-making related to countering terrorist threats. It is now known as the DCI Counterterrorist Center. www.cia.gov/terrorism/ctc.html

Crime and Narcotics Center (CNC) - A department within the Directorate of Intelligence (DI) which provides intelligence and analysis on international organized crime, smuggling, and narcotics trafficking. Supports implementation of the Presidential International Crime Control Strategy for identifying and evaluating organized crime that affects U.S. economics and security. Computer databases are used to track the flow of goods and enforcement efforts. www.odci.gov/cia/di/mission/cnc.html

Defense Computer Forensics Lab (DCFL) - A Department of Defense (DoD) facility near Baltimore announced in September 1999. The DCFL employs agents and computers to engage in 'digital forensics' to trace and process computer-related activities, particularly viruses, files, corrupted, encrypted, or modified data to protect national interests. Information can assist military forces as well as aiding the FBI and local law enforcement organizations. www.dcfl.gov

Defense Intelligence Agency (DIA) - Established in 1961 to handle non-SIGINT, non-aerial, non-organic military intelligence activities. The DIA serves as a combat support agency for the Department of Defense. The DIA publishes a number of publications including the Defense Intelligence Estimates, Special Defense Intelligence Estimates, and Weekly Intelligence Summaries. www.dia.mil/

Defense Security Service (DSS) - A Department of Defense security agency which, among other things, conducts background investigations on Pentagon employees and contractors. The DSS handles the Personnel Security Investigations Program, the Industrial Security Program, and Security Education and Training. (Formerly the Defense Investigative Service (DIS)) www.dss.mil/

Defense Technical Information Center (DTIC) - A Department of Defense facility for providing access to and facilitating the exchange of scientific and technical information. It is part of the Defense Information Systems Agency (DISA). www.dtic.mil/

Department of Defense (DoD) - The U.S. DoD provides the U.S. military forces needed to deter war and ensure protection of the United States. It is headquartered at the Pentagon in Washington, D.C. The DoD provides a large archive of official press releases about defense agencies, priorities, and photos of activities of the combined forces of the U.S. at home and abroad. www.defenselink.mil/

Department of Energy (DoE) - This important U.S. federal body, founded in 1977 in response to the energy crisis, includes an Office of Intelligence and an Office of Counterintelligence which support DoE facilities and work in cooperation with the FBI and other law enforcement bodies. gils.doe.gov/

Department of Homeland Security (DHS) - Established to unify national organizations and institutions that are involved with national security, the DHS was established in response to the 11 Sept. 2001 attack on the World Trade Center and in conformance with the *Homeland Security Act of 2002*. www.dhs.gov/dhspublic/ www.whitehouse.gov/infocus/homeland/

Department of Justice (DoJ) - Under the direction of the Attorney General, the U.S. DoJ is charged with attaining and maintaining justice and fair treatment for Americans through the combined services of almost 100,000 attorneys, law enforcement professionals, and employees. Part of the DoJ responsibility involves detecting criminal offenders. It is headquartered in Washington, D.C. with almost 2,000 installations throughout the country. www.usdoj.gov/

Diplomatic Security Service (DSS) - A U.S. Department of State service that handles diplomatic security for U.S. diplomats stationed or traveling overseas. It also handles the "Rewards for Justice" counterterrorism program which provides rewards and relocation to people providing information leading to the arrest and conviction of those planning or perpetrating terrorist crimes. See also the Bureau of Diplomatic Security. www.heroes.net/ www.state.gov/m/ds/

Domestic Nuclear Detection Office (DNDO) - An office in the Department of Homeland Security tasked with developing and deploying a nuclear-detection system "to thwart the importation of illegal nuclear or radiological materials". www.dhs.gov/index.shtm

Electronic Frontier Foundation (EFF) - A prominent Internet-related organization established in 1990 by Mitch Kapor to serve as a lobbying body and information resource to safeguard public freedoms, particularly freedeom of expression and privacy, on and through the Net. www.eff.org/

Federal Bureau of Investigation (FBI) - This U.S. national investigative agency was first established as the Bureau of Investigation (BOI) in 1908. The FBI handles federal investigations and investigations of crimes that cross state lines. The FBI manages a lab with some of the most sophisticated forensic surveillance technologies and procedures in the world. More recently the FBI has been called in by NATO to aid in international investigations such as those which occurred in Kosovo. The National Infrastructure Protection Center is located at the FBI headquarters. fbi.gov/

Federal Information Center (FIC) - A U.S. General Services Administration agency which provides information to the public about the Freedom of Information Act and the Privacy Act. It can further aid an individual in locating the correct agency for information requests pertaining to the FOIA and the Privacy Act. Established in 1966. fic.info.gov/

Federation of American Scientists (FAS) - A prominent, privately funded, nonprofit organization which provides advocacy and analysis of public policy related to global security through science, technology, and education. Its distinguished membership includes many Nobel Prize Laureates. FAS evolved from the Federation of Atomic Scientists, founded in 1945 by members of the Manhattan Project. The site has extensive educational information on the U.S. Intelligence Community and global security topics. www.fas.org/

International Association for Identification (IAI) - A nonprofit organization for professionals engaged in forensic identification and scientific examination of physical evidence. The IAI provides a range of education and certification programs including latent fingerprint examination, crime scene certification, forensic artistry, etc. Descended from the International Association for Criminal Identification, founded in 1915. www.theiai.org/

International Centre for Security Analysis (ICSA) - A London-based consultancy research arm at King's College. Provides seminars, research, and publications on major topics in security and intelligence. www.kcl.ac.uk/orgs/icsa/

International Intelligence History Study Group - Founded in 1993 to promote scholarly research on intelligence organizations and their impact on historical development and international relations. The membership includes historians, scientists, cryptologists, former intelligence personnel, and politicians. The organization publishes a newsletter and provides excerpts online. As it is based in Germany, there is interesting information for historians from a European perspective. intelligence-history.wiso.uni-erlangen.de/

House Permanent Select Committee on Intelligence (HPSCI) - The CIA reports to this Committee as per the terms of the Intelligence Oversight Act of 1980 and pertinent executive orders. (See Restrictions and Regulations for more information on this.) The Website provides links to agencies within the U.S. Intelligence Community. intelligence.house.gov/

Marine Corps Intelligence Activity (MCIA) - Among other intelligence tailored to Marine Corps needs, the MCIA provides intelligence and educational materials related to urban warfare. www.quantico.usmc.mil/display.aspx?Section=MCIA

Mercyhurst College - An educational institution at which a professional intelligence/counterintelligence library is being organized. The college has been building an undergraduate intelligence studies program with assistance from R. Heibel (retired FBI), a member of the AFIO Board of Directors.

National Air Intelligence Center (NAIC) - Air Force intelligence production center within the Air Intelligence Agency. NAIC assesses foreign forces, threats, and weapons capabilities. It provides foreign air intelligence to Air Force operational units and the Department of Defense (DoD). Founded in 1993 from the amalgamation of the 480th Intelligence Group and the Foreign Aerospace Science and Technology Center. Members are also involved in community educational and humanitarian activities.

National Archives and Records Administration (NARA) - A significant archiving body for the United States which includes documents, histories, lists of agencies, artifacts, and images. The collection is extensive, including veterans' service records, federal records schedules, federal laws and Presidential documents. There is also an online exhibit featuring the Declaration of Independence and other historical artifacts. The Web site provides search capabilties. www.nara.gov/

National Association of Background Investigators (NABI) - A professional organization that provides education, newsletters, and other support services to members. www.background.org/

National Counterintelligence Center (NACIC) - Coordinates U.S. Government threats to national and economic security under the auspices of the National Security Council. NACIC personnel are drawn from various organizations related to the Intelligence Community. NACIC funds counterintelligence activities through courses and seminars and various public and private training programs. www.nacic.gov/

National Counterterrorism Center (NCTC) - The NCTC assists in integrating and analyzing domestic and foreign intelligence acquired from U.S. Government departments and agencies involved in counter-terrorist activities. The center "identifies, coordinates, and prioritizes the counterterrorism intelligence requirements" of U.S. intelligence agencies and "develops strategic operational plans for implementation". Vice Admiral Scott Redd became the first Director. www.nctc.gov/

National Drug Intelligence Center (NDIC) - Supports counter-drug operations by training, investigating and preparing strategic intelligence, and threat analysis/reporting. www.usdoj.gov/ndic/

National Foreign Intelligence Council (NFIC) - Created in 1982 to deal with budget issues and priorities related to foreign intelligence production. See also National Foreign Intelligence Program. baltimore.fbi.gov/nfip.htm

National Ground Intelligence Center (NGIC) - Through a number of separate buildings in Virginia, the NGIC provides scientific, technical, and general military assessments and recommendations including projects such as the identification and removal of landmines. See Army Intelligence and Security Command (INSCOM).

National Intelligence Council (NIC) - A council of National Intelligence offices which serves the DCI with strategic information and production to assist policy-makers in managing foreign policy. The Council draws on academic and private sector resources. www.odci.gov/ic/nic.html

National Military Intelligence Association (NMIA) - A professional forum for national and military intelligence personnel established in 1974 for the exchange of ideas and professional development. NMIA has approximately 200 members. It publishes the NMIA Newsletter and the American Intelligence Journal. www.nmia.org/

National Security Agency (NSA) - Originally descended from the U.S. Armed Forces Security Agency (AFSA) which was established in 1949, the organization was disbanded and then re-established as the NSA. The NSA has a broad set of responsibilities including aerial surveillance, SIGINT, cryptologic activities, computer communications, and counterintelligence strategies. www.nas.gov/

National Security Archive (NSA) - An independent non-governmental research institute and library located in the George Washington University in Washington, D.C. The Archive staff collects and publishes declassified government documents acquired through the Freedom of Information Act (FOIA). It receives non-governmental funding and donations of materials by private parties and foundations. www.gwu.edu/~nsarchiv/

National Security Council (NSC) - The NSC is an important unit that reports to the President and oversees intelligence activities, primarily national security and foreign policy. The statutory military advisor to the Council is the Chairman, Joint Chiefs of Staff. The NSC periodically issues National Security Intelligence Directives (NSCIDs) specifying definitions, duties, and responsibilities, some of which are described in the History and Evolution section earlier in this chapter. Created through the National Security Act of 1947. www.whitehouse.gov/WH/EOP/NSC/html/nschome.html

National Security Institute (NSI) - Product news, computer alerts, travel advisories, and a calendar of events for security professionals. NSI sponsors an annual forum on corporate and government security threats. www.nsi.org/

National Security Study Group (NSSG) - Also known as the Hart-Rudman Commission, this group has studied and documented national security in terms of threats and opportunities along with technology looking into the future. It has produced a series of reports for the NSSG Senior Advisory Board. NSSG sponsors the Future Tech Forum. www.nssg.gov/

National Technical Investigators Association (NATIA) - NATIA's members are surveillance professionals drawn from a number of fields, including law enforcement, the armed forces, and the government. Membership is not open to private detectives or vendors. The organization sponsors an annual seminar and exhibition. NATIA has one of the most interesting crests in the industry, an eagle holding two thunderbolts in its claws, with radio earphones and the motto "In God We Trust, All Others We Monitor". www.natia.org/

Naval Criminal Investigative Service (NCIS) - Uses and provides services and products for the investigation of crimes. Provides the NCIS System and Technology Threat Advisory that issues threat warnings related to foreign targeting of U.S. Navy critical program information. www.ogc.secnav.hq.navy.mil/

Navy Operational Intelligence Center (NOIC) - This evolved from the Navy Field Operational Intelligence Office (NFOIC) and functions below the Naval Intelligence Command. The NOIC monitors foreign marine-related vessel histories, locations, and activities.

Nonproliferation and International Security Division (NIS) - NIS centers report to the Associate Laboratory Director for Threat Reduction and respond to proliferation threats involving weapons of mass destruction (WMD) and develop and apply science and technology to deter proliferation to ensure U.S. and global security. www.lanl.gov/orgs/nis/

President's Foreign Intelligence Advisory Board (PFIAB) - An agency of the National Foreign Intelligence Program (NFIP), which also encompasses the DCI and the CIA.

Securities Exchange Commission (SEC) - The SEC Division of Enforcement surveils securities-related activities on the Internet. National Association of Securities Dealers (NASD) Regulation (NASDR) has an electronic surveillance department devoted to detecting and monitoring suspected insider trading violations, short selling, options trading, and 'drive-by manipulations.'

Senate Select Committee on Intelligence (SSCI) - The CIA reports to this Committe as per the terms of the Intelligence Oversight Act of 1980 and pertinent executive orders.

Society of Competitive Intelligence Professionals (SCIP) - International organization of competitive intelligence professionals. SCIP provides publications and educational seminars of particular interest to corporate intelligence personnel and business students. www.scip.org/

State Department Bureau of Intelligence and Research (INR) - INR was originally the Interim Research and Intelligence Service. The State INR established the Geographic Learning Site in 1998. It produces intelligence assessments related to foreign policy and international issues and educational funding management and training services. It is sometimes referred to as BIR in the popular media.

The **Strategic Intelligence** section of the Department of Political Science of Loyola University has a good list of links related to strategic intelligence, military intelligence, and economic intelligence. www.loyola.edu/dept/politics/intel.html

Terrorist Screening Center (TSC) - The TSC is tasked with consolidating terrorist watch lists and providing round-the-clock operational support to federal and other law-enforcement personnel.

Unified and Specified Command Intelligence Directorates (USCID) - Local installations of forces drawn from across U.S. military services. This intelligence is channeled to appropriate national agencies such as the NSA and the CIA.

10.b. Print Resources

Each chapter includes print resources that describe a few introductory texts and many intermediate and advanced texts pertaining to the topic of that chapter. Those of general interest to intelligence and surveillance are listed here.

The author has tried to read and review as many of these resources as possible before listing them, but sometimes had to rely on publishers' descriptions or the recommendations of colleagues. The annotations will assist you in selecting additional reading. It's a good idea to preview books before buying them or to seek out reviews by authors with interests similar to your own.

This list may include out-of-print publications. These can sometimes be found in local libraries and second-hand bookstores, or through inter-library loan systems.

Abrams, M.; Jajodia, S.; Podell, H. (editors), *Information Security - An Integrated Collection of Essays,* IEEE Computer Society Press, January 1995.

Adler, Allan, *Using the Freedom of Information Act: A Step by Step Guide,* Washington, D.C.: American Civil Liberties Union, 1987.

Bamford, James, *The Puzzle Palace: A Report on NSA, America's Most Secret Agency,* Boston, Ma.: Houghton-Mifflin, 1982. See also the 1983 Penguin reprint *The Puzzle Palace: Inside America's Most Secret Intelligence Organization.* Frequently cited by writers who specialize in the field of intelligence.

Berkowitz, Bruce D.; Goodman, Allan E., *Best Truth: Intelligence in the Information Age,* New Haven: Yale University Press, 2000, 224 pages. An analysis of the changing role of intelligence in the information age and a proposal to outsource or decentralize intelligence activities through new models and policies that include commercial resources.

Brin, David, *The Transparent Society: Will Technology Force Us to Choose Between Privacy and Freedom,* Reading, Ma.: Addison-Wesley, 1999. Brin, a physicist and science fiction author, provides nonfiction examples of trends and activities that erode our privacy along with a call for 'reciprocal transparency' to balance increasing technology tool use in government with that of other members of society.

Brookes, Paul, *Electronic Surveillance Devices,* Butterworth-Heinemann, 1996, 112 pages. Provides an overview of circuit diagrams and parts lists for a variety of electronic bugs. Does not go into detailed explanations of the circuits.

Brugioni, Dino A., *Eyeball to Eyeball: The Inside Story of the Cuban Missile Crisis,* listed as out of print. The author is a former CIA member who was a founding officer of the National Photographic Interpretation Center.

Brydon, John, *Best-kept Secret: Canadian Secret Intelligence in the Second World War,* Toronto: Lester Pub., 1993. A history of the Canadian contribution to Allied SIGINT operations and the evolution of the CSE.

Cain, Frank, *COCOM and its Intelligence Ramifications,* IIHSG 1996 annual conference presentation. How western industrial nations quelled exports of military technologies into Iron Curtain countries. The COCOM acted as a trade-control organization.

Campbell, Duncan, *Development of Surveillance Technology and Risk of Abuse of Economic Information (An appraisal of technologies for political control),* Working Document for the Scientific and Technological Options Assessment (STOA) Panel, Dick Holdsworth, editor, Luxembourg, April 1999.

Commission on National Security, *New World Coming: American Security in the 21st Century,* first installment of a 1999 report from the DoD-appointed panel to assess defense requirements and predict American vulnerability to terrorism. Available in Adobe PDF and HTML formats.

Committee on Government Reform and Oversight, U.S. House of Representatives, *A Citizens Guide on Using the Freedom of Information Act and the Privacy Act of 1974 to Request Government Records,* a booklet providing a detailed explanation of the Freedom of Information Act and Privacy Act. Available through the U.S. Superintendent of Documents.

Deforest, Peter; Lee, Henry C., *Forensic Science: An Introduction to Criminalistics,* McGraw-Hill, 1983, 512 pages. Suitable for technicians and law enforcement personnel.

Devereux, Tony, *Messenger Gods of Battle: Radio, Radar, Sonar: The Story of Electronics in War,* London: Brasseys, 1991. Introductory physical principles and history of technology in warfare.

Dorwart, Jeffrey M., *The Office of Naval Intelligence: The Birth of America's First Intelligence Agency 1865-1918,* Annapolis, Md.: Naval Institute Press, 1979. A history based on published and unpublished archival materials.

Dulles, Allen W., *The Craft of Intelligence,* New York: Harper & Row, 1963. Dulles (1893-1969) was a long-time Director of the CIA and is known for shaping the Intelligence Community in America. Princeton University has an extensive collection of his writings, speeches, and photographs documenting his life.

Eftimiades, Nicholas, *Chinese Intelligence Operations,* Arlington, Va.: Newcomb Pub. Inc., 1998.

Eliopulos, L., *Death Investigators' Handbook: A Field Guide to Crime Scene Processing, Forensic Evaluations, and Investigative Techniques,* Colorado: Paladin, 1993.

Ellit, S.R., *Scarlet to Green: A History of Intelligence in the Canadian Army 1903-1963,* Canadian Intelligence and Security Association, 1981, over 500 pages. Historical reference that includes the organization and activities of each military intelligence unit.

Feklissov, Alexandre, *Confession d'un Agent Soviétique,* Paris: Éditions du Rocher, 1999, 422 pages. The author, in his eighties, recounts his career in Soviet intelligence, including descriptions of the Rosenberg recruitment, development, and training in photographic techniques, and communications during the Cuban Missile Crisis.

Fisher, A. J.; Block, Sherman, *Techniques of Crime Scene Investigation,* Boca Raton, Fl.: CRC Press, 1998 (revised). Used as a reference for International Association for Identification's Crime Scene Certification, Level 1. Clinical analysis of crime scene investigation and real life examples.

Flaherty, David H., *Protecting Privacy in Surveillance Societies: The Federal Republic of Germany, Sweden, France, Canada and the United States,* Chapel Hill, N.C.; University of North Carolina Press, 1992.

Fuld, Leonard M., *The New Competitor Intelligence,* New York: John Wiley & Sons, 1995. Identifies information sources, includes analyses of case studies, discusses ethics. This is used as a text in business courses.

Garrison, D. H., Jr., *Protecting the Crime Scene,* FBI Law Enforcement Bulletin, September 1994.

Jacob, John, *The CIA's Black Ops: Covert Action, Foreign Policy, and Democracy,* Amherst, Prometheus Books, 1999, 350 pages. A discussion of CIA secret ("black") operations and their role in international politics and domestic policy.

Justice Department's Office of Information and Privacy, *Freedom of Information Act Case List.* Includes lists of cases decided under the Freedom of Information Act, the Privacy Act, the Government in the Sunshine Act, and the Federal Advisory Committee Act. Includes statutes and related law review articles. Updated every second (even) year.

Justice Department's Office of Information and Privacy, *Justice Department's Guide to the Freedom of Information Act Guide* and *The Privacy Act Overview.* Federal booklets that are updated annually. They are also available for download. www.usdoj.gov/oip/

Kahaner, Larry, *Competitive Intelligence,* New York: Simon & Schuster, 1996. This is used as a text in business courses.

Law Enforcement Associates, *The Science of Electronic Surveillance,* Raleigh, NC: Search, 1983.

MacKay, James A., *Allan Pinkerton: The First Private Eye,* Edinburgh: Mainstream Publishing, 1997, 256 pages.

McGarvey, Robert; Caitlin, Elise, *The Complete Spy: An Insider's Guide to the Latest in High Tech Espionage and Equipment,* New York: Perigee, 1983.

McLean, Donald B., *The Plumber's Kitchen: The Secret Story of American Spy Weapons,* Cornville, Az: Desert Publications, 1975.

Melton, H. Keith, *CIA Special Weapons and Equipment: Spy Devices of the Cold War,* New York: Sterling Publishing, 1993. Illustrates and describes tools of the trade. Melton has a personal collection of spy devices.

Melton, H. Keith, *OSS Special Weapons and Equipment: Spy Devices of WWII,* 1991, listed as out of print.

Melton, H. Keith, *The Ultimate Spy Book,* London & New York: Dorling Kindersley, Ltd., 1996. A visual encyclopedia of intelligence operations and equipment from about renaissance times to the present.

Melvern, Linda; Anning, Nick; Hebditch, David, *Techno-Bandits,* Boston: Houghton-Mifflin Co., 1984. Addresses issues of illegal technology transfer for military/industrial support of Communist nations.

Minnery, John, *CIA Catalog of Clandestine Weapons, Tools, and Gadgets,* Boulder, Co.: Paladin, 1990. Describes devices which Minnery claims are designed by CIA Technical Services.

National Security Archive, *Military Uses of Space: The Making of U.S. Policy, 1945-1991,* The Fund for Peace. See also the National Security Archive's *U.S. Espionage and Intelligence, 1947-1996.*

National Security Study Group, *The Hart-Rudman Commission Reports,* a series of reports regarding national security threats and opportunities and their relationships to future technologies, available in Adobe PDF format at www.nssg.gov/

Osterburg, James W.; Ward, Richard H., *Criminal Investigation: A Method for Reconstructing the Past,* Anderson Publishing Company, 1996.

Peake, Hayden B., *The Reader's Guide to Intelligence Periodicals,* Washington, D.C.: NIBC Press, 1992. Lists over 100 intelligence-related sources.

Peterson, J.; Mihajlovic, S.; Gilliland, M., *Forensic Evidence and the Police: The Effects of Scientific Evidence on Criminal Investigations,* National Institute of Justice Research Report, Washington, D.C., U.S. Government Printing Office, 1984.

Poole, Patrick S., *ECHELON: America's Secret Global Surveillance Network,* Washington, D.C.: Free Congress Foundation, October 1998.

Price, Alfred, *The History of U.S. Electronic Warfare: The Renaissance Years, 1946 to 1964,* Association of Old Crows, 1989. A history of the development and use of electronic warfare emphasizing intelligence-gathering and countermeasures.

Richardson, Doug, *An Illustrated Guide to the Techniques and Equipment of Electronic Warfare,* New York: Simon & Schuster, 1988.

Richelson, Jeffrey T., *A Century of Spies: Intelligence in the Twentieth Century,* Oxford University Press, 1997. A history of modern intelligence from the early days of the British Secret Service to present times including spies, agencies, and technological developments including aerial surveillance and ground station operations.

Richelson, Jeffrey T., *Foreign Intelligence Organizations,* listed as out of print, 1988.

Richelson, Jeffrey T.; Evans, Michael L. (assisting), *The National Security Agency Declassified: A National Security Archive Electronic Briefing Book.*

Richelson, Jeffrey T., *Sword and Shield: The Soviet Intelligence and Security Apparatus,* Cambridge, Ma.: Ballinger, 1986.

Richelson, Jeffrey T., *The Ties That Bind: Intelligence Cooperation Between the UKUSA Countries,* London: Allen & Unwin, 1985.

Richelson, Jeffrey T., *The U.S. Intelligence Community,* Boulder, Co.: Westview Press, 1999, 544 pages.

Richelson, Jeffrey T., *The U.S. Intelligence Community, 1947-1989,* Boulder, Co.: Westview Press, 1990. A comprehensive portrait of about two dozen U.S. intelligence-gathering organizations and the internal relationships and evolution of federal agencies including the FBI, the CIA, and military intelligence agencies. A related document, *The U.S. Intelligence Community: Organization, Operations and Management, 1847-1989* is available on microfiche, reproducing over 15,000 pages of documents from key intelligence organizations.

Saferstein, Richard, *Criminalistics: An Introduction to Forensic Science,* Englewood Cliffs: Prentice-Hall, 1997, 638 pages. Introductory text to the forensic sciences.

Science at its Best | Security at its Worst: A Report on Security Problems at the U.S. Department of Energy, A Special Investigative Panel, President's Foreign Intelligence Advisory Board (PFIAB), June 1999. A report on an inquiry requested by President Clinton on 18 March 1999 regarding "The security threat at the Department of Energy's weapons labs and the adequacy of the measures that have been taken to address it...." including counterintelligence security threat and its evolution over the last two decades.

Shane, Scott, *Mixing business with spying; secret information is passed routinely to U.S.,* Baltimore Sun, 1 Nov. 1996.

Shulsky, Abram N., *Silent Warfare: Understanding the World of Intelligence,* New York: Brassey's, 1993. A text that has been recommended for college courses on intelligence.

Smith, Lester; Rushour, Lester., *National Integrated Ballistics Information Network (NIBIN) for Law Enforcement, Denver DPD,* March 2005.

Smith, Walter Bedel, *Proposed Survey of Communications Intelligence Activities,* Report to the Secretary of State and the Secretary of Defense by a Special Committee, 1951.

Swanson, Charles R.; Chamelin, Neil C.; Terriro, *Leonard Criminal Investigation,* New York: Random House, 1981. Comprehensive text (used by some of the civil service programs) with practical introductory information on criminal investigation. Provides a framework for a variety of kinds of investigative activities. Includes illustrations and case studies.

Taylor, L.B., Jr., *Electronic Surveillance,* New York: Franklin Watts/Impact Books, 1987, 128 pages. A former NASA employee, Taylor highlights key aspects of electronic surveillance in a style suitable for teen readers and adults looking for an illustrated introduction to the field.

U.S. Government Manual, Washington, D.C. The official handbook of the U.S. federal government that describes offices and programs within the federal government, including top personnel and agency addresses and phone numbers. It is available for purchase for about $36 through the Superintendent of Documents or can be found in public libraries.

Wark, Wesley, *Canada and the Intelligence Revolution,* IIHSG 1996 annual conference presentation. Concepts relating to the Information Age and the Intelligence Revolution and Canada's experience and role in the intelligence community.

Weber, Ralph, *Spymasters: Ten CIA Case Officers in Their Own Words,* Wilmington, Del.: SR Books, 1999. A collection of interviews of almost a dozen CIA officials covering a range of topics from the early days of the NSC and the CIA to more recent events.

Westerfield, H. Bradford (editor), *Inside CIA's Private World,* London: Yale University Press, 1995.

Yost, Graham, *Spy-Tech: The Fascinating Tools of the Espionage Trade–What, How, and Who Uses Them,* New York: Facts on File, 1985.

Zegart, Amy B., *Flawed by Design: The Evolution of the CIA, JCS, and NSC,* Stanford: Stanford University Press, 1999, 342 pages. A critical analysis and discussion of the origins, development, and functions of the nation's primary intelligence agencies.

Articles

Aviation Week and Space Technology Editors, Electronic Countermeasures: Special Report, *Aviation Week and Space Technology*, 21 Feb. 1972, pp. 38-107.

Aviation Week and Space Technology Editors, Special Report on Electronic Warfare, *Aviation Week and Space Technology*, 27 Jan. 1975, pp. 41-144.

Caravella, Frank J., Achieving Sensor-to-Shooter Synergy, *Military Review,* July-August 1998. This is a personal account of the use of surveillance technologies (sensors, UAVs, intelligence reports, infrared satellite imagery, etc.) in armed conflict and the capabilities and limitations of the technology within the limits of current knowledge and procedures.

Clark, Robert M., Scientific and Technical Intelligence Analysis, *Studies in Intelligence*, V.19(1), Spring 1975, pp. 39-48.

Dumaine, Brian, Corporate Spies Snoop to Conquer, *Fortune*, 7 Nov. 1988, pp. 68-76.

European Parliament, *Report A5-0262/2001*, report on the existence of a global system for the interception of private and commercial communications, 11 July 2001.

Electronic Surveillance and Civil Liberties, Congress of the United States, Washington, D.C., Office of Technology Assessment, 1985.

Federal Government Information Technology: Management, Security, and Congressional Oversight, Congress of the United States, Washington D.C., Office of Technology Assessment Report Brief, 1986.

Hermann, Robert J., Advancing Technology: Collateral Effects on Intelligence, *American Intelligence Journal*, V.15(2), 1994, pp. 8-11. Hermann advocates a look at management of national systems and supports the exploitation of open-source intelligence and other information sources.

Hunter, Robert W., Spy Hunter: Inside the FBI Investigation of the Walker Espionage Case, Annapolis, Md.: U.S. Naval Institute, 1999. The recounting of years of Soviet espionage within the U.S. Navy. Hunter is a former FBI foreign counterintelligence agent involved in surveillance missions.

Logsdon, John M., Editor, et al, Exploring the Unknown: Selected Documents in the History of the U.S. Civil Space Program, undated, estimated mid-1990s.

Madsen, Wayne, Intelligence Agency Threats to Computer Security, *International Journal of Intelligence and Counterintelligence,* V.6(4), 1993, 413-488. Includes an international listing of computer-communications espionage capabilities of intelligence and law enforcement agencies.

Maiolog, Joseph A., I believe the Hun is cheating: British Admiralty Technical Intelligence and the German Navy, *Intelligence and National Security*, February 1996, V.11(1), pp. 32-58.

Woodward, Bob, Messages of Activists Intercepted, *Washington Post*, 13 Oct. 1975, pp. A1, A14.

Journals and Bulletins

This is just a small selection, as there are approximately 200 intelligence journals published regularly or semi-regularly in English alone.

AFIO Intelligence Notes, a weekly publication of the Association of Former Intelligence Officers. Includes news, reviews, and relevant Web information. Available by paper subscription, email subscription, and Web download.

Africa Intelligence, online by-subscription politcal/economic news journal published by Indigo Publications. www.indigo-net.com/

American Intelligence Journal, published by the National Military Intelligence Association, includes personal recollections and articles by leaders in the intelligence field. www.nmia.org/AIJ.htm

Bulletin of the Atomic Scientist, an analytic and often critical professional journal that includes articles about the intelligence community analysts. www.bullatomsci.org/

Competia Online, is an online by-subscription resource for news and articles on tools, analysis techniques, and strategic planning for corporate intelligence.

Competitive Intelligence Magazine, quarterly journal from SCIP with news and tutorial information on competitive intelligence (primarily business).

Competitive Intelligence Review, quarterly journal from SCIP and John Wiley & Sons, Inc. with practical research and analysis information on competitive intelligence (primarily business).

CovertAction Quarterly, www.mediafilter.org/caq/

Defense Intelligence Journal.

Director of Central Intelligence Annual Report for the United States Intelligence Community, an annual report of the accomplishments, priorities, and issues of concern to the CIA.

"Economic/Commercial Interests and Intelligence Services, Commentary No. 59", *Canadian Security Intelligence Service*, July 1995. Written by a strategic analyst, this focuses on protecting and pursuing a nation's economic/commercial interests.

Inside Fraud, U.K.-based professional bulletin on business fraud, its detection and management. Published by Maxima Partnering Ltd., London.

Intelligence and National Security, (I&NS) academically oriented journal.

Intelligence Newsletter, bimonthly journal of business intelligence, community watch, threat assessment, technology, people, etc., aimed at the needs of diplomats, officials, security companies, and academic researchers. Indigo Publications (founded in 1981), Paris, France.

Intelligencer, the journal of the Association of Former Intelligence Officers.

International Journal of Intelligence and Counterintelligence, (IJI&C) academically oriented journal.

Justice Department's Office of Information and Privacy, *FOIA Update,* a newsletter published quarterly with information and guidance for federal agencies regarding the Freedom of Information Act. Selected portions are available online. www.usdoj.gov/oip/foi-upd.html

Morning Intelligence Summary, an internal document produced by the INR each day for the Secretary of State, no matter where the Secretary may be stationed.

National Intelligence Daily, NID is a CIA secret bulletin that ceased publication in 1998 due to alleged press leaks. Its successor since 1998 has been the *Senior Executives Intelligence Brief (SEIB)* which has a more carefully monitored distribution.

National Military Intelligence Association (NMIA) Newsletter, lists news, activities synopses, upcoming events, book reviews and other information of interest to members. Available in hard copy, Adobe PDF format (from 1999), and through email subscription. www.nmia.org/

National Security Law Report, publication of the Standing Committee that is distributed eight times yearly to attorneys, government officials, and scholars. www.abanet.org/natsecurity/nslr/

Naval Intelligence Professionals Quarterly. Online access to members. http://www.navintpro.org/

On Watch: Profiles from the National Security Agency's Past 40 Years, Fort Meade, Md.: The National Cryptologic School, 1986.

Orbis, scholarly articles on theory and practice.

Periscope, a journal of the Association of Former Intelligence Officers.

Senior Executives Intelligence Brief (SEIB), CIA secret bulletin that replaced the *National Intelligence Daily* in 1998 in order to provide greater security from unauthorized distribution.

Studies in Intelligence, CIA publication that can be accessed through the CIA Electronic Document Release Center online.

Surveillant, Military Intelligence Book Center email newsletter. Ceased publication.

Weekly Defense Monitor, published by the Center for Defense Information, it includes information on world conflicts, the CIA, the NSA, arms agreements, weapons production, etc. from an independent public perspective. www.cdi.org/

World Intelligence Review, Heldred publication that went out of print in 1997. Back issues may be available. It was descended from *Foreign Intelligence Literary Scene.*

Your Right to Federal Records: Questions and Answers on the Freedom of Information Act and the Privacy Act, Washington, D.C.: U.S. General Services Administration and U.S. Department of Justice, November 1996.

10.c. Conferences and Workshops

Each chapter has a section listing conferences, workshops, and sometimes contests that are held to stimulate the technologies discussed in each chapter. In addition, there are some meetings of general interest listed below.

Many of these conferences are held annually at approximately the same time each year, so even if the conference listing is outdated, it may help you figure out the schedule for upcoming events. It is common for international conferences to be held in a different city each year, so contact the organizers for current locations. Many organizations announce upcoming conferences on the Web and some of them archive conference proceedings for purchase or free download.

The following conferences are listed in the order of the calendar month in which they are usually held.

The Digital Detective Workshop™, New York. The workshop covers acquisition tools and techniques for digital evidence. www.codexdatasystems.com/ddw.html

National Security Institute Forum. An annual forum that discusses emerging threats to corporate and government information.

Business Intelligence and Law Symposium II, sponsored by the Association of Former Intelligence Officers.

Competitive Technical Intelligence Symposium, held approximately every two years. In 2001, the symposium was held in San Diego and, in 2003, in New York City.

IIHSG Annual Conference, annual International Intelligence History Study Group conference.

InfowarCon, military operations, infrastructure protection, terrorism and espionage topics.

NASIRE Annual Conference, represents the concerns of Chief Information Officers (CIOs). *www.nasire.org*

AFIO National Convention, sponsored by the Association of Former Intelligence Officers, hosted by the Director NSA.

Police & Security Expo, is a commercial tradeshow for presenting professional products to law enforcement and security personnel. There is no admission fee, but attendees must present credentials.

National Intelligence Symposium, is an annual conference with secret status and restricted security check-in. In 2005, the symposium discussed the impact of the *Intelligence Reform Act of 2004* at the Northrop Grumman Corporation conference facility. Speakers include members of the intelligence community. The symposium is typically scheduled for mid-March.

SCIP European Conference and Exhibit, is an annual conference of the Society for Competitive Intelligence Professionals, London, U.K.

Forensic Investigative Conference, hosted by Jamerson Forensics of Fayetteville. Topics include forensic science, criminal profiling, and sex crimes.

Annual Academy of Behavioral Profiling, is an annual conference on behavioral evidence and criminal profiling for crime reconstruction and investigation.

Federal Information Assurance Conference, is an annual conference that focuses on government security culture and information assurance. *www.fbcinc.com/fiac/default.asp*

Eisenhower National Security Conference, this and a number of related events are planned by the Office of the National Counterintelligence Executive.

AFCEA Annual Conference, sponsored by the Armed Forces Communication and Electronics Association. The 2006 conference was held in Fort Bragg, in August. AFSEA also sponsors the *Biometrics Expo* (September) and the *Command, Control, Communications, Computers and Intelligence Systems Technology Exhibition,* (October).

Competitive Financial Intelligence Symposium, sponsored by SCIP, this symposium is usually held in the summer or fall in major cities such as New York (2000) and Chicago (2006).

Terrorism Awareness and Prevention Conference, includes government, corporate security, law enforcement, and academic speakers. In 2006, the two-day conference was held in Connecticut. Topics include intelligence gathering, terrorism trends and threats, and preparedness.

Terrorism Financing and Organized Crime, is an international conference focusing on practices and tools to combat organized crime, money laundering, and the financing of terrorist activities. The 2006 conference was organized by the FBI in cooperation with the U.S. Embassy Prague and Ministry of Interior of the Czech Republic.

10.d. Online Sites

This section lists Web sites of particular interest or relevance to this topic. In most cases, commercial sites are not included unless they have a particularly good educational focus or set of illustrations. The selected sites are generally well-maintained and up-to-date and are likely to remain at the same URL for some time. In the case of sites that change, keywords in the descriptions below can be used to try to relocate the site on a Web search engine (it is more likely that the site has been moved rather than deleted). Note that the list of organizations preceding this also includes many URLs for Websites.

Army Counterintelligence Discussion Group List. Initiated early in 1999 for active and former counterintelligence agents. Private, moderated list. Members must be approved. Covers counterintelligence trends, training, news, history, and current events.

Canadian Forces Intelligence Branch Association (CFIBA). This professional organization provides the Intelligence Note Book online, which provides an introduction to intelligence. It also hosts a Virtual Museum with information about Norse records dating back thousands of years and an illustrated history of Canadian contributions to the World Wars. There is also an annotated bibliography of suggested readings. www.intbranch.org

Center for Army Lessons Learned. A site with some interesting illustrated lesson-style articles that have arisen out of the real-world experiences of army personnel. While not directly related to the use of technology and surveillance devices, there are references to technological advances and their place within the overall frameworks of peacekeeping and armed conflicts. There is also access to Periscope, an online source for intelligence data and defense news, which is available to authorized users. call.army.mil/

CIA Electronic Document Release Center. A searchable archive of CIA documents that have been released or declassified under the Freedom of Information Act. Includes documents released since 1996. www.foia.ucia.gov/

Crime & Clues. This site includes links to forensics and crime investigation sites and provides a series of articles by various authors on crime scene protection, processing, and investigation. There is also signup for a crime scene investigation discussion list. crimeandclues.com

Cryptome. An online resource with articles and excerpts of communications regarding international surveillance and political developments. Includes information on wiretaps, ECHELON, cipher activities, Internet security, and more. cryptome.org/   www.jya.com/

Duncan Campbell IPTV Reports. This is an extensive, illustrated, news and investigative journalism site that covers global intelligence activities, including projects within the U.S. that have been haphazardly researched by others. While there is no way for anyone outside the secret services to authenticate much of the information, this site appears to be a better source of 'informed speculation' than many others on the Web. www.iptvreports.mcmail.com/

Granite Island Group Technical Surveillance Counter Measures. An extensive list of links to statutes and documents on intelligence and surveillance including the complete text of some of the more relevant Presidential Executive Orders. Maintained by James M. Atkinson, Communications Engineer. www.tscm.com/reference.html

Intelligence Online. "Intelligence Online: Global Strategic Intelligence" is a bimonthly online by-subscription journal of business intelligence, community watch, threat assessment, technology, people, etc., aimed at the needs of diplomats, officials, security companies, and academic researchers. It is a searchable, online version of the print journal *Intelligence Newsletter*. www.indigo-net.com/

Intelligence Resource Program. This resource site, with charts, articles, and links is compiled by the Federation of American Scientists. It includes an extensive complication of news reports (dating back to 1992), some with analyses, and programs, intelligence operations, documents, congressional material, and more. www.fas.org/irp/index.html

International Intelligence History Study Group Newsletter. The IIHSG publishes an abridged version of their print publication on their Web site. As this is focused on intelligence activities, it is of broad general interest to surveillance. intelligence-history.wiso.uni-erlangen.de/newsletter.htm

Law Reform Commission Publications. The Australian Law Reform site provides a series of papers on various aspects of law, including several on surveillance, e.g., What is Surveillance? Paper 12 (1997). www.legalaid.nsw.gov.au/nswlrc.nsf/pages/index

The Literature of Intelligence. *A Bibliography of Materials, with Essays, Reviews, and Comments.* J. Ransom Clark, J.D., a Faculty Dean at Muskingum College and former member of the CIA, has produced this extensive resource to fill what he considers to be a need for a "central, civilian-controlled entity for the collection, preparation, and dissemination of national-level foreign intelligence". The bibliography is well-organized, annotated, and cross-referenced according to topics. It includes search capabilities. intellit.muskingum.edu/

The National Instant Criminal Background Check System. A resource for Federal Firearms Licensees (FFLs) to request background checks on individuals attempting to acquire firearms. By filling out an appropriate form and phoning or accessing the database via the Internet, criminal background checks may be obtained 17 hours a day, seven days a week. www.fbi.gov/hq/cjisd/nics.htm

The National Security Archive. This is the world's largest resource of its kind, a nonprofit, non-governmental library, archive, and research resource founded in 1985. It includes information on international affairs, declassified U.S. documents obtained through the Freedom of Information Act and from a public-interest law firm. It also indexes and publishes documents in books, microfiche, and electronic formats. It is supported by public revenues and private foundations. The physical archive is located on the seventh floor of the George Washington University's Gelman Library in Washington, D.C. www.gwu.edu/~nsarchiv/

U.S. Senate Republican Policy Committee. This site has some political and legal interpretations of some of the intelligence-related Acts (e.g., the Foreign Intelligence Surveillance Act) and critiques of White House and Department of Justice interpretations and implementation of those Acts that are of interest. rpc.senate.gov/releases/1999/fr080699.htm

The **Strategic Intelligence** section of the Department of Political Science of Loyola University has a good list of links related to strategic intelligence, military intelligence, and economic intelligence. www.loyola.edu/dept/politics/intel.html

Vernon Loeb's IntelligenCIA Column. A biweekly column by the Washington Post reporter who specializes in intelligence topics. Available only in the online edition of the Nation section of the Post. www.washingtonpost.com/

10.e. Media Resources

A small selection of media resources is listed in each chapter, mainly popular feature films and videos. The media resources are not complete or comprehensive compared to other sections in this book, but are included to guide the reader to some visual representations and scenarios, both real and imagined, that provide a little extra insight into the field. They're included because some aspects of surveillance technology are hard to describe and illustrate in print.

The listed resources vary in quality from middle-of-the-road movies to science fiction thrillers and serious documentaries. They aren't necessarily included because they were good movies, but rather because they show devices and their use in a new or interesting way (e.g., Sliver isn't a five-star film, but the technology depicted in the film is representative of several important social and technological aspects of visual surveillance). Since the quality of the listed resources is uneven, short annotations are sometimes included in the description to alert the viewer. The author has viewed most of the media resources included, but not all of them. Some were included on the recommendation of colleagues and a few are based on publishers' descriptions.

Elizabeth, feature film biography of Elizabeth I, the wily Queen of England who firmly held her crown longer than most historic rulers, in part because of her use of the intelligence services of Sir Francis, Earl of Walsingham (not covered in detail, but the movie provides some background). The story is fictionalized and Hollywoodized, but gives a feel for the period and the fact that 'spying' has been around for a long time. Polygram, Cate Blanchett and Joseph Fiennes, 1998, 118 minutes.

The **German Historical Museum**, includes exhibits of interest and has been instrumental in establishing the Allied Museum for Berlin in which the Allied presence can be documented and preserved. The Allied Museum provides information on secret intelligence services.

The **Imperial War Museum** in London, England, has a permanent exhibit called the "Secret War Exhibition" that includes espionage history.

The **Nordic Museum** in Stockholm, Sweden, showed a special exhibition until 1 Sept. 2000 regarding espionage in and for Sweden. It included the prehistory of modern expionage and the conflicts between neutrality and espionage activities. The tools of espionage, such as cameras, listening bugs, and cryptographic equipment are featured. This exhibition has ended, but it may be possible to ask curators about access to parts of the exhibit and archival information regarding the exhibit.

Operation Solo, a July 1999 broadcast of the *History Channel,* based on the John Barron book *Operation SOLO: The FBI's Man in the Kremlin,* produced by Towers Productions, Inc. This covers a FBI counterintelligence operation from the early 1950s. The accuracy of some of the ideas presented in the program has been questioned, particularly the portrayal of J. Edgar Hoover, and the assertion that no communist connections to King were found. In general, however, this program is considered to be of interest.

The Oral History of the Office of Strategic Services, is a project to systematically interview the small number of living representatives of the original employees of the Office of Strategic Services (OSS) which was the predecessor to the U.S. Central Intelligence Agency (CIA). The project is sponsored by the CIA Center for the Study of Intelligence and initiated and directed by Christof Mauch. Unpublished memoirs of interviewees were also solicited as part of the project. It is archived in the Special Collections Division of the Lauinger Library at Georgetown University. Interview transcripts are also archived with the Library of Congress.

Red Files, a four-part Public Broadcast Service (PBS) documentary that provides a look at Soviet Union/U.S. Cold War rivalries. The episodes are titled "Secret Victories of the KGB", "Soviet Sports Wars", "Secret Society Moon Mission", and "Soviet Propaganda." There is a Web-based companion site to the series that includes story scripts, synopses, stores, and video clips. www.pbs.org/redfilesRed

The Rote Kapelle: 50 Years After, a documentary film written and directed by Yelena Letskaya, TROYKA Company, 55 minutes. Based on formerly classified documents in the KGB archives, this film includes interviews with former GRU resident agents and portions of German documentary images. The consultant is Professor Youri Zorya, who discovered some Nürnberg Trial (Nuremberg Trial) documents.

11. Glossary

Each chapter has a short glossary including terms, acronyms, and common abbreviations relevant to the chapter topic. For longer explanations, the reader is encouraged to consult more comprehensive dictionaries and encyclopedias.

Titles, product names, organizations, and specific military designations are capitalized; common generic and colloquial terms and phrases are not.

ACS/I	Assistant Chief of Staff/Intelligence
AFIC	Air Force Intelligence Command. Evolved from Electronic Security Command (ESC).
AFSA	Armed Forces Security Agency. Established in 1949 within the U.S. Department of Defense.
AIA	Air Intelligence Agency
AISC	Army Intelligence and Security Command
AITAC	Army Intelligence and Threat Analysis Center
ALoR	acceptable level of risk. An authoritative and carefully considered assessment that a system or activity meets certain minimum stated or accepted security requirements. Aspects that typically factor into ALoR include capabilities, vulnerabilities, potential threats, the possible consequences of those threats, and countermeasures options.
AMSIC	Army Missile and Space Intelligence Center
anomaly detection	The systems and processes used to assess deviant or unscheduled activities or presences which may indicate anomalous activities or unauthorized access. This interpretation assumes a baseline norm from which deviations are assumed to indicate some type of intrusion.
ASIM	Automated Security-Incident Measurement. A Department of Defense network access and traffic monitor.
BIR	Bureau of Intelligence and Research, better known as INR
breach	A successful disruption or intrusion that could indicate or result in penetration or vulnerable exposure of a system.
BS/I	Bachelor of Science degree with specialization in intelligence
C4I	Command, Control, Communications, Computers, and Intelligence
CBRN	Chemical, Biological, Radiological, and Nuclear
CCMP	Civilian Career Management Program
CDTD	Critical Defense Technology Division
CERT	Computer Emergency Response Team
CFIBA	Canadian Forces Intelligence Branch Association
CI	counterintelligence, competitive intelligence
CIFMP	Civilian Intelligence Force Management Program
CIM	Communications Identification Methodology
CIPMS	Civilian Intelligence Personnel Management System
CIR	Civilian Intelligence Reserve. A NIC-administered pilot program that was rolled into the Global Expertise Reserve.
CIS	Combat Intelligence System
CM	countermeasure(s)

CMRT	Consequence Management Response Team. An interagency response team which provides coordination and interagency communication for USG-Host Nation responses to CBRN events.
CSS	Central Security Service
CSSPAB	Computer Systems Security and Privacy Advisory Board
CIA	Central Intelligence Agency
countermeasures	Actions and systems intended to prevent, detect, or respond to a security threat.
CTC	Counterterrorist Center
CIWE	Center for Information Warfare Excellence
DCI	Director, Central Intelligence (Director of the CIA)
DCIM	Distinguished Career Intelligence Medal. Awarded by the Central Intelligence Agency (CIA).
DIA	Defense Intelligence Agency
DIC	Defense Intelligence College
DII	Defense Information Infrastructure. The phrase for the information assets of the U.S. Department of Defense.
DSS	Diplomatic Security Service
DSTC	Diplomatic Security Training Center
EAP	Emergency Action Plan
ECM	Electronic Countermeasures
espionage	This French term translates roughly to "spyage", the act of spying upon.
ferret	*n.* someone who provokes a response or illumination of a situation, not usually through confrontation but rather through subterfuge or calculated actions
FOIA	Freedom of Information Act
FTD	(Air Force) Foreign Technology Division
GDIP	General Defense Intelligence Program
GIITS	General Imagery Intelligence Training System
GITC	General Intelligence Training Council
GITS	General Intelligence Training System
HDBT	hard and deeply buried target
ICAC	(Canadian) Intelligence Collection and Analysis Centre
IIS	Intelligence Information System
In-Q-It	A venture-capital firm established by the CIA in 1999 to fund promising Internet/computer-related technological developments. *In* (intelligence) *Q* (Major Boothroyd, aka "Q" the fictional British secret service agent who created gadgets for Agent 007) *It* (information technology).
INSCOM	Intelligence and Security Command, within the U.S. Army
intelligence	The product resulting from the collection, analysis, integration, evaluation, and interpretation, within a given context, of a body of information, of which the information may variously include, but not be limited to, images, data, statistics, facts, objects, schedules, and figures.
ISRC	Intelligence, Surveillance, and Reconnaissance Cell
ITAB	Intelligence Training Advisory Board
JIVA	Joint Intelligence Virtual Architecture
JWICS	Joint Worldwide Intelligence Communications System. A computerized system which includes videoconferencing intended to facilitate communication among intelligence community agents.
misinformation	An important concept in investigations and covert activities, misinformation is the spreading of rumors, falsities, fraudulent claims or documents and the general creation of a deceptive persona or situation. When misinformation is used to discredit individuals or organizations, it can be potentially more devastating than a bomb.
MSD	Mobile Security Division. U.S. diplomatic protection division, the mobile arm of the DSS.

NAVSECGRU	Naval Security Group
NCE	National Counterintelligence Executive
NCS	National Cryptologic School
NFIB	National Foreign Intelligence Board
NIE	National Intelligence Estimates
NOIC	Navy Operational Intelligence Center
NSA	National Security Agency
NSC	National Security Council
NSG	Naval Security Group
NSTISS	National Security Telecommunications and Information Systems Security
NTIC	Naval Technical Intelligence Center
OIPR	Office of Intelligence Policy and Review, which is under the direction of the Counsel for Intelligence Policy, advises the Attorney General on national security activities and approves certain intelligence-gathering activities.
OOTW	Operations Other Than War
OPSEC	Operation Security. Securing an operation from discovery, infiltration, and intervention or compromise.
OSAC	Overseas Security Advisory Council
OSTP	Office of Science and Technology Policy
PFIAB	President's Foreign Intelligence Advisory Board
RSO	Regional Security Officer
SASO	Stability and Sustainment Operations
SC&DI	Surveillance, Control and Driver Information system
SCIP	Society of Competitive Intelligence Professionals
secret agent	colloquial term for individuals officially engaged in covert or clandestine surveillance or other intelligence-gathering activities
SIF	Securities Issues Forum
SIRVES	SIGINT Requirements Validation and Evaluation Subcommittee
SORS	SIGINT Overhead Reconnaissance Subcommittee
spoof	*n.* a trick, masquerade, decoy, or use of an imposter to control or divert attention, actions or events.
spook	*n.* a secret agent, spy, operative, or other (usually human) undercover, clandestine or covert agent.
spy	an agent (human, animal, robot) that overtly/clandestinely acquires secrets, information, or unauthorized entry to a facility, event, or nation
STIC	Science and Technology Information Center
sting operation	deliberate, planned, covert or clandestine manipulation of a situation to create an illusion intended to trap or entrap an individual or group
STOA	Scientific and Technological Options Assessment
TENCAP	Tactical Exploitation of National Space Systems Capabilities. A program for improving "combat-readiness and effectiveness of the U.S. Air Force through more effective military use of national space system capabilities."
TIARA	Tactical Intelligence and Related Activities
TRC	Terrorism Research Center
TSCM	Technical Surveillance Countermeasures. A TSCM survey is professional detection and reporting of the presence of technical surveillance devices, hazards, and security weaknesses.
UGIP	Undergraduate Intelligence Program. A Joint Military Intelligence College program for enlisted service members.
undercover	*adj.* concealed; covert or clandestine person, object, or operation
USCID	Unified and Specified Command Intelligence Directorates
USSS	United States Secret Service

101

Surveillance Technologies

Acoustic

Acoustic Surveillance

Audio

1. Introduction

Acoustic surveillance is a broad category that includes sounds within human hearing and sounds outside of human hearing. In this reference, acoustic surveillance has been divided into three sections: Audio Surveillance—technologies that are *within human hearing ranges*, Infrasonic/Ultrasonic Surveillance—technologies that are *outside of human hearing ranges,* and Sonar Surveillance—a specialized section that includes infrasonic, ultrasonic, and audio frequencies specifically used in sonar sensing.

Audio surveillance occurs over both wired and wireless systems. While wireless systems are discussed in this chapter in a general sense, the main focus is on wired systems. This is because many wireless audio communications are converted into radio-wave frequencies, which are discussed in greater detail in the Radio Surveillance chapter.

Encryption is often used to safeguard acoustic communications. Acoustic surveillance, computer encryption, and cryptology share a common history. Thus, you may wish to cross-reference the Cryptologic Surveillance and Computer Surveillance chapters for more background related to encryption and secured audio communications.

The Scoop Reporter II system is backup equipment enabling two-way radio signal transmissions. Here, SSgt. Stumm, a broadcast maintenance technician stationed in Japan, checks audio output levels through earphones. [DoD news photo by Gempis, 2005.]

The Phenomenon of Sound

Sound waves are periodic disturbances that occur in any material that can compress and decompress. Thus, sound waves can propagate through air, water, soil, oil slicks, wheat fields, and forests. They can occur in walls, mountains, and chunks of metal. If you've put your ear to the ground, or to a wall in a building, you've probably noticed that you can hear distant sounds through these media better than you can hear them through the air. Soil and plaster are denser than air and thus propagate the sound waves more readily. If the material is too dense, however, it becomes difficult for the material to compress and decompress and thus there's a point where sound travels through a dense material less readily.

 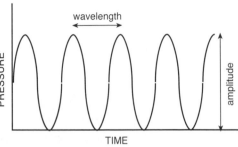

Left: The ripples in a pond that result from throwing in a pebble are often used to help people visualize how sound waves propagate in a medium. Imagine, however, that the sound waves travel outward from the source in all directions where there is a medium (not just along the surface of the water, but down and out, as well), and through the surrounding air, where they can't be seen. Sound travels at different speeds in different media, depending upon the density and composition of the media. Since water is denser than air, sound travels faster through water. Sound travels slower than light, an electromagnetic phenomenon, which is why we usually see lightening before we hear thunder, especially if the storm is far away. Right: Sound waves can be expressed symbolically as repeating sinusoidal waves. In this simple graph, the amplitude ('height') of the wave is represented on the *Y* axis and the length on the *X* axis (and repeats over time). [Copyright 1999, Classic Concepts, used with permission.]

When waves move through a medium, we say that they *propagate*. There are two common ways in which wave phenomena move through various gaseous or solid objects. There is actual movement *through* the medium (as radio waves penetrate through a wall) or there is a force *influencing* the medium that is intrinsically associated with the medium (as water is disturbed when a pebble penetrates its surface). There is an important distinction between these two types of wave phenomena:

- The first is similar to a small bird passing through the big holes in a chicken wire fence. Electromagnetic phenomena like radio waves can pass through a wall in somewhat the same way a small bird can pass through a wire fence.

- The second type of movement is like rapping on the wall, which has elastic properties, causing the sound to *propagate* longitudinally, using the wall as a medium. Pressure waves are built up in the wall as it responds to the energy from the pressure against it, alternately causing the molecular structure to compress and decompress. Without the wall, there would be no waves and, thus, no sound.

Our ears are specialized organs designed to respond to sound. If you put your ear on a wall that is being rapped, the disturbance is transmitted to the air and is guided down your ear canal, where it causes your ear drums to vibrate, behind which small bones in your skull carry the vibrations to your nervous system. The nerve impulses are then interpreted by your brain and matched to your brain's "database" of sounds to see if the sound is familiar. This is the phenomenon of hearing. In fact, having two ears set a little distance apart gives us even more information. Dogs, cats, and most other mammals take this process a step farther. They have directional ears. Not only can they turn their ears to "catch" the sounds (the same way we sometimes cup a hand by our ears to hear something better), but they can turn each ear in a different direction. An animal's brain does a little extra matrix algebra to process all this information. Humans are specialized to interpret human speech. Animals are better adapted to surveilling and interpreting a variety of environmental sounds. Since we can't move our ears independently of our heads, we have to turn our heads and 'measure' the relative loudness of the sound and the slight time delay it takes to reach each of our ears, to determine the direction from which the noise is coming.

Human hearing

Audio surveillance technologies are designed to enhance our sense of hearing, or to enhance other forms of stimuli and convert them to our hearing range.

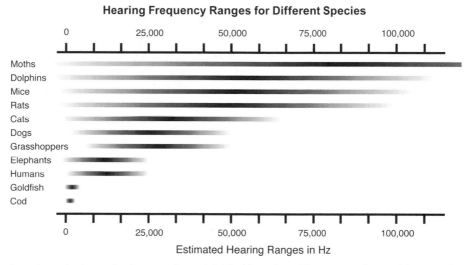

Insects, rodents, and cetaceans (e.g., dolphins) hear far wider ranges of sound frequencies than humans. Humans can hear up to about 18,000 Hz, but they are most sensitive to vibrations within about 1,000–4,000 Hz, which also corresponds to the frequencies of human speech. Thus, most communications devices are designed to be clear and sensitive within these ranges. Surveillance devices, on the other hand, often utilize frequencies above and below human hearing and some are even patterned after the hearing capabilities of other animals.

Human hearing is quite limited, as are some of our other senses (particularly smell). Raptors have better vision than humans, and most birds can see frequencies we can't see (and sense magnetic frequencies we can't sense), many mammals and insects have better senses of smell than humans, dolphins have exceptional sonar sensing, and a very large number of creatures have better hearing with respect to higher (and lower) frequencies. Some are able to hear repetitious sounds that we can't hear, as well. For example, birds can resolve repetitions that are much closer together than humans. The only reason we do well with such limited senses is because our brains have excellent capabilities for abstracting and processing the things we *do* hear.

The frequency chart indicates how human hearing compares to animal hearing in terms of frequencies. As can be seen, other animals not only have a broader range of frequencies, but their main frequency focus is different from ours. For example, rodents are tuned into higher frequencies.

We can feel sound waves with our skin, but we can't interpret those sounds very well, so we have specialized organs called ears. Sound waves stimulate the cilia and tiny bones on the other side of our eardrums. Then our brains interpret the intensity and characteristics of sound so that we may determine its source and direction. Sound waves also stimulate vibrations in various other materials, like minerals. These vibrations can be amplified or converted into other forms of energy or frequencies, with the use of electronic devices, so that we can hear things like radio waves.

Most children with normal hearing can hear frequencies from about 18 to 20,000 Hz. As humans age, or are subjected to various environmental stresses (illnesses, sudden loud noises, constant loud noises, or constant low-level noises), the sensitive mechanisms that allow us to hear are gradually damaged and both the sensitivity of our hearing and the frequency range gradually decrease. Adults typically hear frequencies of about 25 to 17,000 Hz and hearing may become impaired as we age.

Sound Surveillance

Almost all remote-sensing technologies involve *waves,* or phenomena that appear to behave like particles or waves, depending on how you study them, and since we can't see most of them, we resort to symbolic representations to measure and describe them. Since this volume focuses on the conceptual understanding and applications of surveillance technologies rather than on physics and engineering design, waves will be described with examples and a few basic diagrams rather than with mathematics. The Resources section lists more advanced texts that delve into the physics and math for those who want to better understand the physical theories related to sound.

The Character of Sound

Because sound is movement through a medium, it does not propagate in a vacuum. This characteristic seems to have inspired a lot of philosophical thought about whether sound even exists in our absence. If we momentarily borrow a concept from eastern thought, we can see how western thinking about sound sometimes has a Zen *koan*-like flavor. A *koan* is a statement or question intended to stimulate intuitive enlightenment. Koans typically seem paradoxical in the context of what we call 'reason'. They are to be 'understood' in a way that is different from the way in which we normally understand things through scientific investigation.* Common questions about sound with a koan-like flavor include "What is the sound of one hand clapping?" and "If a tree falls in the woods and no one is there to hear it, does it make a sound?" The Zen aspect of koans is outside the scope of this book (and perhaps any book), but these questions about sound are useful from a teaching perspective because they help stimulate inquiry into the characteristics of sound.

This text doesn't delve into the philosophical aspects of sounds, but from a physics point of view, the above questions help illustrate some of the basic terms and concepts associated with sound. The two characteristics of sound that are most relevant to surveillance are *frequency* and *volume*. Frequency is perceived by us as the *pitch* of the sound (the 'highness' or 'low-ness' of the note) and volume, the *loudness* of the sound. The sound of one hand clapping is considered to be *infrasonic*, a frequency *below* the range of human hearing (though elephants

---

*Tim Burton's movie *The Nightmare Before Christmas* gives a good illustration of the limitations of scientific reductionism when Jack Skellington systematically dismembers a Christmas bauble in order to try to discern its nature. In doing so, he perceives its parts, but loses the whole and, in fact, destroys the bauble.

might be able to sense it). One hand clapping is also considered to be *subaudible*, that is, a volume below which we are sensitive. For surveillance purposes, specialized instruments can be designed to pick up a low, soft sound and translate its frequency and volume into levels that we can perceive. Thus, if the sound of one hand clapping were both *amplified* and raised in *frequency*, we might hear a soft, rhythmic swishing from the disturbance of the air as the hand moves back and forth. These types of sound-processing techniques are useful in many aspects of audio surveillance.

Sound is not electromagnetic energy, but it has a lot of properties in common with electromagnetic technologies. If you place two tuning forks close together and hit one so that it vibrates, the other will begin to vibrate as well (though not as vigorously). This is called *resonance*. Resonance is a type of energy transfer. When objects are close together, they are more likely to resonate in response to those around them. Resonance is particularly noticeable if the objects are made of materials with good vibratory qualities. Tuning forks, guitar strings, and the surfaces of speakers are designed with materials known to resonate well at certain frequencies. Television antennas are designed with a particular shape and length so they resonate well in response to electromagnetic radio waves.

As with one hand clapping, sometimes a sound is not loud enough to be heard with the unaided ear. A guitar string that is not attached to the guitar can be stretched and plucked, but creates a soft, dead sound that isn't very appealing. When placed on an acoustic guitar, the vibrating string transfers energy to the air inside the guitar which bounces the sound around, giving it a warm, bright sound. The vibration (resonance) of the air around the string and in the wood itself aids in transferring sound from the string to the air inside the guitar, causing a melodious vibration we perceive as music. But what about an electric guitar? It doesn't have a sound box. There's no cavity inside to amplify the sound and no sound hole to direct the sound. In an electric guitar, the amplification is accomplished by transferring the sound via a 'pickup' attached to the body of the guitar. As the strings vibrate, the body vibrates and the pickup converts this to electrical energy that is transferred through a cord to an amplifier and speaker system. Electronic amplification can be substantial, which is why rock stars prefer to use electric instruments rather than acoustic instruments in large amphitheaters.

Sound Conversion and Display

Sometimes sounds (or other inputs like radar or computer data) are converted to other forms of energy so they can be 'heard' without creating noise or so that they can be analyzed in some other way. Some devices have a blinking light to indicate the pattern or frequency of a sound or input from other sources. Most computer dialup modems have a programmable *sound on* mode that allows you to listen to the flow of data over the line. This can help diagnose problems, but the sound is also annoying; it's a rough, screechy sound and most of the time the modem speaker will be set to 'off' with blinking lights indicating the flow of data instead. Similarly, most consumer radar detectors will indicate the presence of a radar signal by beeping, but this can be distracting if a conversation is going on in the car, so some models allow the sound to be turned off and will display blinking lights instead. This conversion of energy into different forms allows us to monitor data in whatever form is appropriate for a task. Audible sounds can be converted to other forms, and other forms of energy can be converted to audible sounds, as these brief examples illustrate:

- The stress levels in a human voice can be converted into a graph to indicate whether a person is nervous and possibly lying.

- Infrasound monitors, which detect sounds below the range of human hearing, can be electronically enhanced to convert the data into graphs or the sound itself can be raised in pitch and amplified to bring it into audio frequencies and levels that can be interpreted by humans. Infrasound is used in the surveillance of explosives detonations and the prediction of natural disasters or severe storms.

- In broadcasting, audible sounds are converted into radio signals to allow them to be transmitted without wires, and then converted back in order to be understood when they have been received. Radio and television are the most common examples, although satellite voice broadcasts are now becoming common.

Sound and the various electromagnetic technologies that are covered in subsequent chapters are important because together they comprise the great majority of surveillance devices. Seven chapters in this book are collectively devoted to sound and electromagnetic technologies and it will be seen that they have many aspects in common. But despite their similar features, sound vibrations and electromagnetic radiation are not the same physical processes.

Remember that sound requires a medium through which to direct energy. In fact, it could be said that sound *is* energy directed through a medium. In contrast, electromagnetic phenomena (like light) are energy with both wave- and particle-like properties which propagate *with or without a medium,* thus allowing electromagnetic energy to travel through a vacuum.

This chapter provides some background information on the nature of sound and then describes surveillance technologies that take advantage of the properties of sound within the human hearing range. It includes a range of audio technologies, primarily listening and recording devices, since they are inexpensive and widely used in many types of surveillance.

2. Kinds and Variations

While there are many technologies to surveil the Earth, its structures, weather, wildlife, and its surrounding atmospheric envelope, the majority of surveillance is directed at monitoring human activities including travel, agriculture, construction, politics, and communications. People are especially intent on surveilling other people's conversations, trying to discern their intentions and planned activities.

Many conversations occur in noisy or private settings, out of earshot of covert listeners (restaurants, bars, boats, private golf courses), yet a tremendous amount of human intelligence is still gathered from monitoring conversations that *can* be amplified or overheard. Consequently, much of this chapter focuses on telephone tapping and various types of remote listening systems.

In addition to face-to-face conversations, common means of audio communications include traditional telephones, digital voice technologies (including videoconferencing and Internet phone systems), cell phones and other wireless phones, voice pagers, radios, and telegraph-sounding keys. Sometimes security measures are taken to safeguard or disrupt audio communications including voice changing, voice scrambling, encryption, or jamming. (Wireless communications are covered more fully in the Radio Surveillance chapter and electronic communications such as email and Internet discussion groups are introduced in the Computer Surveillance chapter.)

The main categories of audio surveillance include

audio amplification/reduction - making a sound louder/softer so that it can more easily be detected, recognized, heard, or understood

audio filtering - separating out particular aspects of a sound, which most commonly involves filtering out noise or other forms of interference, but may also include filtering out particular voices or frequencies

audio location - detecting the location, origin, or direction from which the sound is coming

audio recognition - determining the source, type, or speaker of a sound

audio translation - converting a sound from one form to another, such as making the frequency higher or lower or translating human speech from one language to another

audio listening - eavesdropping on sounds, which may or may not include recording the sounds

audio logging - creating a record of administrative aspects of a sound including time of day, duration, frequency, etc.

audio recording - creating a replayable record of a sound in analog or digital format (or a hybrid format)

audio processing - analyzing a sound for a variety of characteristics including direction, origin, source, veracity, content, and sometimes even specific voices

Logging, *listening*, and *recording* have traditionally been the most common categories of audio surveillance. However, *audio processing* is increasing in prevalence because it can now be done with computer hardware/software that wasn't previously available.

Audio surveillance devices can also be subdivided into wired and wireless varieties, each with advantages and disadvantages:

wired *Wired devices are those which are physically connected through a solid medium to a speaker or recording device.* The connection is typically through common copper electrical wire, though it may also be through cable and fiber optic media. Wired listening devices are more difficult to use in covert operations because it is necessary to physically enter the premises for a period of time to install and hide the wire, which in some circumstances may be impossible. Wired bugs are not too difficult to detect, especially when they are in use. Activity on electrical wires can be found by current that 'leaks' through the insulation into the immediate surrounding area, while fiber optic is based on the transmission of light and doesn't create the same type of electrical emanation. Depending on the medium and the wire gauge (smaller wires are easier to hide), there may be limits to the effective length of the wire and the strength of the signal.

Fiber-optic cables provide greater security than wired connections, but they also require greater precision, as the end-couplings must be carefully spliced so as not to introduce kinks into the optical fiber, which would disrupt the travel of the light beam. Some installations use a wired connection that leads to a wireless connection elsewhere in a complex (the roof or a closet or wall) or vice versa.

wireless *Wireless devices are those which use air or water as the physical medium through which the sound travels* (remember that sound can't travel in a vacuum, so it can't be sent out into empty space unless it is first converted to another type of energy). With improvements in radio-frequency transmission technologies, wireless devices are increasing in prevalence.

Because it is not necessary to hide any wires, wireless transmitters are often favored for small hidden 'bugs.' The main disadvantage to wireless listening devices is that the transmission can be picked up by anyone in the vicinity with a receiver that can scan or tune to the transmitted frequency. Thus, they may be easy to detect when they are in operation and someone may be eavesdropping on the eavesdropper.

Another disadvantage to wireless listening devices is strict regulations about frequencies and ranges. If they have a range of more than 300 or 1,000 feet, they are typically subject to regulatory licenses and broadcast restrictions. Wireless communications are sometimes transmitted to a wired receiver which works as a relay station to further transmit the signal over a longer distance.

An understanding of the difference between analog and digital communications is important when choosing audio technologies. Communications used to be almost entirely analog, but now digital communications, which have some unique strengths, are being favored for more secure communications.

analog *Analog audio signals are those which are output as a continuous function.* The most common example used to describe an analog system is a clock with minute and hour hands that move smoothly through a 360° arc. As the hands sweep around the dial, *they pass through every point in the path of the arc.* Another example is an older dial radio in which AM stations are tuned by slowly moving the dial until the desired station comes in clearly. Old radios, clocks, tape recorders, and phones were based almost exclusively on analog technologies. By the 1980s, this was beginning to change.

digital *Digital transmissions are those which are output as discrete signals* such as high or low, on or off, in or out, etc. The common example of a digital display is a digital clock which, instead of sweeping through every second of the hour, 'jumps' from one second to the next, or one minute to the next without displaying intermediary moments of time. Another example is a pushbutton radio in which each button on the tuner 'snaps into' a specific frequency without scanning through individual stations as on older AM radios. (Besides the pushbutton concept, FM is used as an example because there are wide 'guard' frequencies around each broadcast frequency to guard against interference from nearby station frequencies and so is more illustrative of an 'on/off' system than AM broadcasts, which can overlap.) Many newer computerized electronics (including computers themselves) are based on digital logic and digital technologies. Digital devices provide two important advantages over analog devices:

replication Digital data can be perfectly replicated without degradation of the information. Data tend to degrade each time they are copied through an analog system. For example, a tape recording of a tape recording is never quite as perfect as the original. A photocopy of a photocopy of a photocopy gives a good illustration of how an image or signal loses information each time it is replicated. However, copying a computer software program from one computer to another, or a digital recording from one digital system to another or a digital CD from one CD to another makes a 'perfect' copy in the sense that the subsequent number of 'bits' equals the original number of 'bits' in quantity and content.

encryption Digital data are relatively easy to manipulate. You can move the bits around. You can change them into other bits to provide encryption, compression, or data processing. You can control them with computer electronics. Spread-spectrum technologies provide a means to 'hide' radio signals from eavesdroppers and encryption techniques make it difficult for eavesdroppers to interpret a signal even if they are able to capture it.

Eavesdropping and wiretapping of phone conversations appear to be exceedingly prevalent in business communications. Wiretapping and recording mechanisms are factory-built into some types of business phones. Telemarketing firms routinely monitor their employees' calls and many technical support lines are 'monitored for quality assurance' which means that either the supervisor listens in from time to time or the calls are recorded for later evaluation. The more common categories of devices used for monitoring purposes are listed here. Devices are further described in the *Description* and *Applications* sections.

pen register This was originally a device for analyzing *pulses* in a phone line to determine which number had been dialed. Since pulse phones are now rare, the phrase has come to mean a range of mechanical and electronic products that determine and display the number dialed, whether it is a pulse or tone signal. Pen registers may optionally be able to record the number for later review. Even more recently, the phrase has subtly changed to encompass numbers that have been dialed or 'otherwise initiated' since there are now autodialers and other electronic means of placing a call.

Pure pen registers are becoming less common. The trend in electronics is to bundle many features into one device. In other words, a device promoted as a 'pen register' might include other capabilities to log the time, frequency, and duration of a call. It is important, from a legal point of view, to define 'pen register' in its purest sense (of simply capturing/recording outgoing numbers), since there are laws governing who may use them and when they may be used. If the name 'pen register' comes to encompass other features, then laws are 'changed' without due process. Thus, a value-added pen register should be considered a 'pen register and logging device,' for example, to keep the legal distinctions clear. The logging aspect and guidelines for its use should be stipulated separately.

The proper interpretation of the name 'pen register' does not encompass call *content* or other *characteristics* of the call (e.g., attributes such as voice recognition or stress levels). Combination pen registers/loggers are often used for cost-accounting by businesses that bill calls to specific departments or budgets. The more limited pen registers are used in authorized law enforcement wiretaps. Unfortunately, pen registers don't capture just a number dialed. They can also capture PINs, credit card numbers, and other confidential information that is dialed in conjunction with the number.

scanning In terms of wired phone connections, this is the act of automatically dialing a series, range, or pattern of numbers in order to make a connection or determine characteristics of the line. A 'war dialer' is a type of scanner. It is an automatic-dialing device used by many telemarketing firms (sometimes illegally). War dialers have become more sophisticated. In the early days, a war dialer was simply programmed to dial numbers sequentially, one after the other. Now they can be hooked into electronic computer directories to dial specific names or people with particular personal profiles. Computer hackers sometimes scan numbers to see if a modem or fax tone responds to the call.* This is one way of seeking unauthorized access to computer systems or finding unpublished fax numbers that might provide document-delivery systems for employee or product information. Phreakers (phone hackers) sometimes scan for anomalies that might indicate special maintenance numbers or access points for long-distance calls or other services. (Scanning has a broader meaning in wireless communications which is described further in the Radio Surveillance chapter.)

tapping This is covert or clandestine access to or interception of a call. If you hook in an extension line, a specialized wiretapping device, a lineworker's set, a computer, or

---

* In the feature film *War Games,* a young hacker, played by Matthew Broderick, uses various techniques for locating dialup modems so he can break into restricted computer systems.

other device that is directly or closely associated with the physical line or the radio frequency, it is considered to be *tapping*. If you listen through some device not directly associated with the line or frequency, e.g., a microphone hidden in a wall near a phone, it's considered to be *bugging*. The listener may be a person or a recording device, or both. The euphemism 'monitoring' is sometimes used to mean overt or clandestine tapping of a call, as in certain business applications.

trapping This is the application of technology to 'seize' a call so the caller can't terminate the connection. In other words, someone places a call from a phone booth to a bookie several miles away. The call is 'trapped' so that the line remains active to allow it to be traced to its origin. The caller is unable to hang up the call, even by putting the headset in the cradle. This allows the time needed to locate the origin of the call. Trapping is often used in conjunction with tracing, so often in fact that in law enforcement it's called *trap-and-trace*, *lock-and-trace*, or *lockin-trace*.

tracing This is the process of determining the communications route of a call or points along that route. *Origin-tracing* is the determining source of the call. *Route-tracing* is determining the path through which the call is connected. *Terminal-tracing* is determining the points at which the call enters or leaves a network node or station.

A form of tracing is used by some 911 emergency systems to determine the origin of a call in case a frightened caller hangs up or experiences a medical emergency and is unable to provide an address. The term 'tracing' does not encompass the content, frequency, duration, or character of the call. In television programming, tracing is often depicted in kidnapping and ransom scenarios.

Traditional tracing occurs when a call is 'live', but with computerized logs and playback systems, tracing 'after the fact' is becoming easier all the time. The simplest means of low-tech tracing for long-distance calls is to look at a person's phone bill. The next simplest means is to access common carrier records that log both local and long-distance calls for billing purposes. The live tracing of cell phone calls can be done by triangulation, i.e., the comparison of signal strengths between several cell transceiver nodes. New cell phones equipped with Global Positioning System (GPS) capabilities now allow more precise tracing.

Caller ID Caller ID is a form of tracing. It is a pay service offered by telephone carriers which 'broadcasts' the number of the person calling to the person being called. By paying a monthly fee, the caller's number can be accessed from the line and displayed on a Caller ID-compatible device. Many answering machines and telephones now have Caller ID displays, but it is still necessary to pay for the service to get the number.

Call Blocking is a series of numbers that can be dialed to block Caller ID so the person being called doesn't know who is calling. Call Blocking is free. There was a lot of debate, when Caller ID was first introduced, about whether the caller or the callee should pay for Caller ID-related service.

From a public safety point of view, the person using Call Blocking should be charged and Caller ID should be free. This is analogous to a person freely looking through a peephole in the door to see a stranger before letting the stranger into his or her house. Unfortunately, the phone service was implemented the other way around, analogous to a stranger freely wearing a mask at the door and the home-owner having to pay a fee to see the stranger through his peephole. Communications carriers probably prefer the current system for economic rather than safety reasons.

common carrier These are communications service providers, including local exchange carriers (LECs), competitive-access providers (CAPs), cellular carriers, interexchange carriers, and providers of PCS and other mobile radio services. It further includes those cable and utility companies that provide telecommunications services. The concept of the common carrier is very important because law enforcement agencies have been lobbying Congress for legislation that requires common carriers to adapt their circuitry so that law enforcement agents can wiretap the lines.

3. Context

Because eavesdropping has been around for a long time, society has had time to debate many aspects of the practice and build a social framework around the legality of eavesdropping and its cousin wiretapping.

The contexts in which audio surveillance are most commonly used include

sound monitoring in the field, home, or workplace Listening devices are used to record the sounds of birds and whales, to monitor an infant crying or children playing in a daycare facility. It is used in hospitals to check on the needs of bedridden patients and on hiking trips to link partners or team members. It is also used in industrial yards to monitor safety around industrial equipment and in labs to check on employees working around hazardous chemicals or radiation. Many P.A. systems have a monitoring mode.

eavesdropping Audio surveillance of other people's conversations is common and some of it is illegal. It tends to occur in the context of relationship difficulties or business spying.

wiretapping Court-authorized audio surveillance of conversations is primarily used by law enforcement agencies to investigate and convict cases related to violent crimes and drug trafficking.

mobile listening/recording Gray area surveillance (an activity of questionable legality) is sometimes used to safeguard personal privacy. Hidden mobile microphones and transmitting or recording devices are sometimes used for the purpose of monitoring or saving the information and characteristics of a personal communication. This is called *wearing a wire*. An individual who has been granted a restraining order against a stalker or ex-spouse might wear a wire to alert others of violations of the restraining order or to gather evidence of those violations. Investigative journalists will often wear wires when seeking access to individuals or establishments that may be involved in alleged criminal activities, to record their responses and reactions. Employees and potential employees will sometimes wear a wire to record incidences of discrimination, sexual harassment, or workplace abuse.

Sound-Monitoring Devices

A listening device is any device which is designed to channel, focus, or amplify sounds to aid the listener in better recognizing the characteristics or content of the sounds. Sound tends to travel faster and more readily through solid objects than through air (try putting your ear to a wall or to the ground). Sound travels nearly five times as fast through water as it does through air. Thus, the context in which listening devices are used is important, as they depend not only on electronics design, but on their careful placement and alignment. For example, laser listening devices are precision instruments, requiring a steady base and the correct angles. If you don't have a solid, steady surface and a tripod for mounting the device, they're apt to record noise or unusable sounds.

Listening devices are generally designed to amplify a signal or to make it clearer (to remove static, noise, other voices, etc.). Amplification systems have been around for a long time; a stethoscope or glass can be placed against a wall, a parabolic dish can focus and thus amplify the sound from across a street, or an electronic amplifier can be embedded in a plant. These are all ways in which sounds are made louder and easier to understand.

Technologies for clarifying sound are more recent. Computer technology now makes it possible to analyze a signal and single out a particular voice, or to test it for stress levels that might be related to the truth of the information, or to fuse the tones from one 'frame' or moment in time to the next to algorithmically 'guess' at the content of a poor or faint recording. It is also possible to use computer technology to translate a conversation that is being carried out in a foreign language. Even if the translation isn't perfect, the context may be sufficient to make an educated assessment of the communication.

Tapping Devices

In most cases, phone tapping involves the installation of a device in close association with the phone being tapped. This is especially true now that electronic technologies make it harder to de-encrypt or intercept a call en route. Thus, tapping a phone usually involves gaining access to the premises, either inside or outside, depending on the type of tapping equipment and the configuration of the phone equipment. However, there are still circumstances where a call is tapped en route, usually at the premises of a local common carrier. Authorized law enforcement taps are usually done this way.

Single-line phones are the easiest to tap, and extension lines are not especially difficult. A tap can be set up on a specific line, or on all extension phones attached to that line, depending upon the placement of the tap on the circuit. Multiline phone systems are harder to tap, and phone-bugging devices for multiline systems such as private branch exchanges (PBXs) are harder to obtain and usually more expensive.

Society has become more mobile and so the demand for mobile communications continues to increase. Wireless phone conversations used to be relatively easy to access, when a single frequency was used, but spread-spectrum technologies are being built into consumer phones, making it much more difficult to access and decrypt a wireless conversation.

Because phone tapping is typically covert, most phone tap equipment is small, and thus portable, frequently no larger than a pocket pager, and sometimes as small as a matchbox or button, depending on its capabilities.

Wireline tapping devices are usually installed in buildings or on junction boxes or lines near the premises being tapped. They may also be installed on the local phone company premises, but these taps are in the minority. Portable or stationary recorders may be used in land vehicles, marine craft, or aircraft. Mobile recorders may be attached to moving conveyances or worn on the body.

With technology changing so rapidly, the term "wiretapping" has taken on a broader connotation and, in terms of tapping warrants, covers wire, oral, or electronic surveillance. In recent months, a number of key organizations have raised concerns about the U.S. intelligence community engaging in warrantless wiretapping. The American Bar Association expressed concerns, as did the New York Civil Liberties Union (NYCLU). In March, 2006, the NYCLU made a request to the Court of Appeals to open a secret lower court order about warrantless wiretapping by the NSA. The order was neither publicly released nor shown to the defendant in the case. It was the NYCLU's position that the American public has First Amendment rights to have access to the papers. In the words of the NYCLU Executive Director, "In this country we do not have secret courts".

Approved wiretaps are carefully monitored and the U.S. Courts release public information about the frequency of tapping requests and actual installations. In the court report for May 2006, it was stated that there were a total of 1,773 applications for approval to tap in 2005, an increase of four percent from the previous year. Tap-related arrests for the year totaled 4,674. The most common surveillance method reported was phone communications (95% of intercepts), including landline, cordless, mobile, and cellular phones.

4. Origins and Evolution

The history of audio surveillance involves four major aspects, the understanding of the science of sound, the evolution of technologies to project sound over distance, the evolution of devices to record sound, and laws to regulate the use of these technologies. Much of the current legislation regarding communications, recording, and wiretapping has its roots deep in the 19th century. (The aspects of audio surveillance that apply to wireless communications are described further in the chapter on Radio Surveillance.)

Humans have been recording events for at least 20,000 years, through images and later through text, but the recording of sound is a surprisingly recent event. Both sounds and images are now frequently recorded together on the same medium.

Coding systems enable sounds to be stored and replayed. Left: Carmen-Marsch music cylinders from a table-top music box were like cartridges—interchangeable. The pegs would hit the sound mechanism as the cylinder rotated, reproducing a tune. Right: A flat metal cylinder encoded with punched holes rather than raised pegs—from the American Radio Museum collection. [Classic Concepts photos ©1998, used with permission.]

The first visual recording devices were fingers and sticks, used by humans to draw images in dirt or sand. Later, the use of chisels, charcoal, and pigments provided a way to create a more permanent record, preserved for hundreds of thousands of years in petroglyphs and cave paintings. Thus, the earliest known efforts to record thoughts and communications were images that were carved or drawn by human hands. Later, written records on clay tablets and papyrus sheets provided a means to record histories, beliefs, people, and events. But no one had yet figured out how to record a human voice, the primary means of communication.

Replicating images was found to be easier than replicating speech. A pantograph was probably one of the earliest mechanical image-replicating devices. A pantograph is a zigzag 'arm' that holds a drawing implement in such a way that an image can be retraced with a stylus, with a drawing tool duplicating the motions, and hence the image, a few inches away. This concept of replicating an event or picture is the essence of recording technologies, whether they are audio or visual.

Music boxes are one of the oldest means of storing and replicating sound. They existed in Europe in the Middle Ages, although the idea may have originated from eastern trade with China. By the 1700s, there were many different music box mechanisms that could store and replay tunes. By placing pins in a rotating cylinder or holes in a platter or a strip of paper, a short, predetermined piece of music could be played many times.

The printing press is one of the most significant inventions in modern history. It was the first invention to provide a practical way to replicate and inexpensively disseminate multiple copies of the same document. This had important consequences for the permanence of written records. If some copies were lost, the information still had a good chance of surviving if multiple copies had been printed. The creation and replication of information also created a new tool of accountability. It was difficult to deny libelous statements when the proof had been 'captured' in the pages of a book. Similarly, when sound-recording devices were later invented, it became difficult to deny slanderous statements that had been captured on tape or another recording medium.

In 1904, Edwin Welte and Karl Bockish used the concept of music boxes to develop the Welte-Mignon reproducing piano, leading to a craze in 'player pianos' that could play 'by themselves' from notes punched in long rolls of paper. However, even sophisticated devices that played music were not yet able to record and reproduce the sound of a natural human speaking voice.

It may seem odd, but an ancient clay pot, handmade on an old-fashioned potter's wheel, is a type of crude sound-recording device. The grooves in a pot are like the grooves in a record platter, the vibrations of the fingers of the potter create indentations in the clay pot as it spins on a potter's wheel and the hands move upward, in somewhat the same way that the vibrations of a cutting stylus create indentations on a phonograph record as it spins on a revolving base. Unfortunately, the sensitivity of the potter's fingers and the signal-to-noise ratio results in a very poor recording, so we can't get much information about the past from 'playing' a pot, but it is interesting that some of the first phonograph 'records' were shaped like narrow, cylindrical pots, with fine grooves spiraling up the outside of the cylinder. Later, flat platters, called records, were used instead.

Eavesdropping - The Famous and the Infamous

Audio recording technology is one of the closest descendants of human spying and eavesdropping. The field has developed mainly to enhance or replace the human ear. People have rarely been able to resist the urge to eavesdrop. The impulse to listen in on other people's conversations is powerful and ubiquitous and has been around since long before the early days of the telegraph and telephone.

During the American Revolution, Lydia Barrington Darragh is reported to have eavesdropped regularly on the conversations of officers stationed at British headquarters near her home. She recorded these revelations in a simple code, hid them in large buttons on the clothing of messengers, and had them conveyed to her son, an officer in the Continental Army under the leadership of George Washington. Almost a century later, Allan Pinkerton, founder of the famous Pinkerton's Detective Agency, pointed out that many strategists had made the mistake of thinking that unenlisted women were not a serious threat to the Civil War effort.

One of the most remarkable stories of secret activities during the 19th century is that of "Harriet" Tubman (1820 or 1821-1913), a poor black slave born in a windowless one-room shack as Araminta Ross. From the age of six, she was often taken from her family and hired out to others. In 1844, she married John Tubman, a freed slave. When she became aware that her master's property was going to be sold (which would likely include her), she escaped, in 1849.

Her indomitable character led her to take great risks and to make many trips through dangerous territory in her efforts to free more than 200 other slaves. She frequently had to hide in haystacks, barns, and churches. Tubman is credited with great strategic and organizational skills. She used some surveillance technology to achieve her goals, as well, forging passes, writing messages in code, and carefully selecting costumes for the passengers she ferried through the "Underground Railroad".

UNVEIL "TUBMAN" MEMORIAL.

Auburn, N. Y.—Impressive ceremonies were held at the Fort Hill Cemetery by the Harriet Tubman Club of New York City, assisted by Empire State Federation of Women's Clubs at the grave of Harriet Tubman, the late famous conductor of the underground railway.

The "Civil War heroine" died last year at the age of ninety-six.

Harriet Tubman, who was born in slavery, fled to the North, and after regaining her freedom in this manner assisted 400 slaves to freedom. John Brown commissioned her "General

Left: Harriet Tubman successfully engaged in clandestine and covert activities for many years, freeing slaves through the "Underground Railroad" using codes, disguises, and forged passes. She also provided scouting and spying services to the Union Army during the Civil War. After the war, she raised money for black schools and created a home for the elderly. Middle: This Cleveland Advocate newspaper clipping has some errors (e.g., Tubman's age), but illustrates the respect that was held for her. Right: William H. Seward aided Tubman in acquiring this two-story brick home near Auburn, New York—an illegal transaction at the time. Tubman's home for the elderly became known as the "Harriet Tubman Home for Aged and Indigent Colored People". Her contemporaries greatly respected her, praised her valor and heroism and buried her with Military rites at the Fort Hill Cemetery. The rebuilt Harriet Tubman Home in Auburn, New York, is now a national historic site. [Harriet Tubman historical photo c1900; New York historical photo; Library of Congress clipping, 1915, copyrights expired by date.]

In 1850, Congress passed the Fugitive Slave act, making it a criminal offense to aid runaway slaves, but Tubman was so good at subterfuge that she was never caught, in spite of posted rewards for her apprehension. During the Civil War, in the 1860s, she worked as a nurse, a scout, and a spy, and received a number of official commendations from Union Army officers. Later in life, she produced letters from prominent dignitaries that substantiated their appreciation of her achievements.

Before long-distance communications, eavesdropping was an immediate and simple means of acquiring information, but only if the eavesdropper could get close to the conversation. Sometimes it was safer and more effective for an informant to enlist the talents of clothiers and chemists in creating a disguise. Sarah Emma Evelyn Edmonds (1841-1898) wrote in her memoirs that she worked as a spy for General George B. McClellan, carrying out eleven secret missions. It is difficult to substantiate these claims, due to a significant lack of corroborating evidence, but some historians have accepted that she enlisted with the army and traveled in disguise as "Frank Thompson". Indeed several women may have enlisted as 'men.'

THE FUGITIVE SLAVE LAW.

A bill to amend the act entitled "An act respecting fugitives from justice, and persons escaping from the service of their masters."

Be it enacted by the Senate and House of Representatives of the United States of America in Congress assembled, That the persons who have been, or may hereafter be, appointed commissioners, in virtue of any act of Congress, by the circuit courts of the United States, and who, in consequence of such appointment, are authorised to exercise the powers that any justice of the peace or other magistrate of any of the United States may exercise in respect to offenders for any crime or offence against the United States, by ar-

thus appointed, to execute their duties faithfully and efficiently. in conformity with the requirements of the constitution of the United States and of this act, they are hereby authorized and empowered, within their counties respectively, to appoint in writing under their hands, any one or more suitable persons, from time to time, to execute all such warrants and other process as may be issued by them in the lawful performance of their respective duties; with an authority to such commissioners, or the persons to be appointed by them, to execute process as aforesaid, to summon and call to their aid the bystanders, or *posse comitatus* of the proper county, when necessary to insure a

mentioned shall be conclusive of the right of the person or persons in whose favor granted to remove such fugitive to the State or Territory from which he escaped, and shall prevent all molestation of said person or persons by any process issued by any court, judge, magistrate, or other person whomsoever.

SEC. 7. *And be it further enacted,* That any person who shall knowingly and willingly obstruct, hinder, or prevent such claimant, his agent or attorney, or any person or persons lawfully assisting him, her, or them, from arresting such a fugitive from service or labor, either with or without service or labor, either with or without process as aforesaid; or shall rescue, such fugitive from the custody

ing such other duties as may be required by such claimant, his or her attorney or agent, or commissioner in the premises; such fees to be made up in conformity with the fees usually charged by the officers of the courts of justice within the proper district or county, as near as may be practicable, and paid by such claimants, their agents or attorneys, whether such supposed fugitive from service or labor be ordered to be delivered to such claimants by the final determination of such commissioners or not.

SEC. 9. *And be it further enacted,* That upon affidavit made by the claimant of such fugitive, his agent or attorney, after such certificate has been issued, that he has reason to apprehend that such fugitive

Political acts that threaten the status quo often result in civil unrest and an increase in surveillance activities at various levels of society. The Enactment of the 1850 Fugitive Slave law drove many people underground and caused others to become 'manhunters.' Critics declared that the Slave law deprived black people of their right to due process and changed citizens into 'bloodhounds' in search of slaves. It also made Harriet Tubman a more wanted 'criminal' than ever before, with rewards for her capture reaching as high as $40,000. Left: Rush Richard Sloane (1828-1908), a judge and mayor in Ohio, helped slaves escape after their own masters had arrested them. He was prosecuted for his actions under the Fugitive Slave law. Right: A signal lantern being raised on a flagpole at the John Rankin House in Ripley, Ohio. The lantern signaled to slaves that it was safe to cross the Ohio River. [Library of Congress American Time Capsule and Wilbur H. Siebert Collection, c1850; clipping from 1915, copyrights expired by date.]

Apparently pseudo-science accompanied technological science in the thinking of some of the military minds of the Civil War. Before being sent on a secret mission, Edmonds claims that the army gave her a *phrenological* examination. Phrenology is the association of certain mental and psychological traits with the physical characteristics of the skull. In those days it was believed that bumps on the head reflected the more developed portions of a person's brain. The phrenological examination of Edmonds revealed that she had well-developed "organs of secretiveness, combativeness, etc". Apparently these qualities, plus an oral interview, qualified her for a spying mission.

In her memoirs, Edmonds describes how she dressed up as a 'darky,' coloring her arms, head, face, and hands and acquiring 'a wig of real negro wool.' She then crossed through rebel lines and sought work in General Lee's camp. Edmonds recounts that she used nitrate of silver to preserve the color on her face when it began to fade. Later, she dressed as an Irish peddlar-woman. Edmonds apparently had a talent for accents, switching from 'pidjun-English' to an Irish brogue to fit her various disguises. Following the War, Edmonds received distinction and a Civil War veteran's pension.

Sarah Emma Edmonds published her "Female Spy" memoirs in 1894, claiming that she participated in the Civil War as enlisted man and male nurse "Frank Thompson". Right: Edmonds is illustrated in her book in one of her costume as a black 'contraband' working behind rebel lines. Bottom: The memoirs were reprinted in 1895 and again in 1999. Edmonds was inducted into the Military Intelligence Hall of Fame in 1988. A digitized copy of her book can be viewed online through the Making of America (MOA) site at the University of Michigan. [Images from *Nurse and Spy in the Union Army,* 1865, copyrights expired by date.]

Sometimes special 'surveillance talents' are revealed unexpectedly in emergency situations. Wiretapping without sophisticated equipment was demonstrated by an unusual traveler, not long before the Civil War. The story is told that in 1858, a Philadelphia telegraph operator named Anson Stager was taking a trip through the midwest when his train broke down. On finding out that the engine had failed, he asked the conductor if he was willing to order a replacement from the next station if Sager telegraphed the message. The conductor said yes and Stager "climbed a telegraph pole and lowered a wire to the ground. He thrust into the ground an iron poker from the coal stove in the coach, and tapped the end of the wire against it to order an engine". To receive the returning telegraph, Stager "stuck out his tongue, placed the wire upon it and received the electrical impulses". [G. Oslin, 1992.]

During the Civil War, Stager worked with the Telegraph Corps to create a wartime cypher-correspondence and set up a system of field telegraphs used during battle.

The assassination of President Abraham Lincoln brought together the Military Commission to determine the full extent and involvement of the 'conspirators' who had been captured in connection with the crime. Twenty-one-year-old Lewis Thornton Powell "Lewis Paine" or

"Payne" was indicted for assaulting Secretary of State William H. Seward and his household. At the time of the trial, the Military Commission was not aware that Powell was an agent of the Confederate Secret Service at the time he attempted to murder Seward. At the trial, all were found guilty and Powell and the others were sentenced to death by hanging.

Sarah Edmonds wasn't the only woman who claimed to have used the 'technology of disguise' to infiltrate the army. Loreta Janeta Velazquez has described numerous adventures in the Confederate Army as Lieut. "Harry T. Buford", and her experiences as a spy and secret service agent acting on behalf of General John H. Winder. In the course of her activities, she was arrested on numerous occasions and brought before General Butler (right) [Images from *The Woman in Battle*, 1876, copyright expired by date.]

Left: Anson Stager, a competent telegrapher, became an officer in the Federal Army, a member of the Telegraph Corps and eventually became a General. Right: A Chatanooga train and telegraph wires in c1864. [Library of Congress photos, copyrights expired by date.]

Anson Stager went on to help Gray and Barton establish the Western Union Teleraph Company. He is mentioned in connection with Western Union in some of the letters of Alexander Graham Bell's family in the late 1870s. Along with George Bliss, he paid a visit to Thomas Edison's laboratory in Menlo Park in February 1877 to discuss foreign rights to an 'electric pen' replicating system conceived by Edison in 1875. A few months after this visit, Edison was busy experimenting with an 'acoustic telegraph' system. The telephone age was imminent.

General John H. Morgan, a Confederate cavalry leader was known for his ability to tap Union lines. He could apparently mimic the telegraph style of Union telegraphers (a rare talent) and send misinformation and orders, signed with the names of Northern generals. The Union responded by encrypting their messages, providing the code only to generals and the War Department. (See Cryptologic Surveillance.)

In the American Civil War, General J. Stuart is reported to have carried a telegraph tapping device to intercept military communications, a slightly more sophisticated means than Stager's method of touching the wire with his tongue.

The telegraph had a substantial impact on the stock market. People who lived far from Wall Street could now participate. They could also eavesdrop on telegraph communications containing up-to-date market-related information. By 1864, there had already been at least one prosecution for the tapping of telegraph lines to obtain stock information [Dash, Schwartz, Knowlton, 1959]. A few years later 'stock ticker' machines (specialized printing telegraph receivers) were invented.

Left: Members of the 'Conspirators Court' Military Commission met for two months in spring 1865 to decide the fate of the prisoners accused of the plot that resulted in the assassination of President Lincoln. Lewis Thornton Powell (middle), a young, charismatic Secret Service agent for the Confederates, was included in the Lincoln conspiracy trial when he was apprehended for attempting to kill William H. Seward (right). Until then, he was not suspected of being a spy. [Minnesota Historical Society, 1865; National Archives photo by Alexander Gardiner, 1865; National Archives photo mid-1800s, copyrights expired by date.]

Another historic telegrapher who became renowned for surveillance activities was John E. Wilkie who became chief of the U.S. Secret Service in 1898. A 1906 edition of The Railroad Man's Magazine describes Wilkie as a newshound in his youth who 'scooped' big stories by eavesdropping on the 'fire-alarm-telegraph-wire'. The Secretary of the Treasury at the time apparently solicited Wilkie for the Secret Service without application. Wilkie began to tackle crime and made a long, stealthy stalk of a bold and substantial counterfeit gang, eventually convicting the counterfeiters. One of the members of the gang was found to be a former U.S. State District Attorney.

Left: The Western Union Telegraph Company grew and spread across the continent. This 1895 photograph shows the Western Union Telegraph office in Minneapolis in 1895. Middle: The Gulkana Signal Corps telegraph station in Alaska, 1910-1915. Right: A Western Union telegraph messenger, 1920. [Minnesota Historical Society photos; middle photo from the Alaska State Library, copyrights expired by date.]

Inventions Leading to the Telephone

Telephone technology is an evolutionary step arising from the revolutionary invention of the telegraph (both telegraph and telephone history are covered in more detail in the Radio Surveillance chapter).

The telegraph was the first technology that dramatically changed hand-to-hand physical communication, through objects and letters, to almost instantaneous communication over great distances. Even the carrier pigeon couldn't fly at the speed of sound or light. Telegraph concepts were developed in Europe in the late 1770s, and first put to practical use in England in 1836 by Wheatstone (1802-1875) and Cooke. In America, Samuel F. B. Morse (1791-1872) and his assistant Alfred Vail were developing telegraph equipment as well. In the days of telegraph systems, an eavesdropper had to learn the code to be able to listen in on messages. With the invention of the telephone in the late 1800s, eavesdropping became much easier.

In the late 1850s, Antonia Meucci, an Italian living in Cuba, attached wires to animal membranes to transfer sound through current, but due to his isolated location, news of his discoveries did not spread. In Europe and America, similar experiments were being conducted. Johann Philip Reis (1834-1874), a German inventor, demonstrated the transmission of a tone through wires in 1861. He reported in a letter that he could transmit words, but other evidence of his achievement has not survived. At about the time of his death, Elisha Gray (1835-1901), an American physicist, was experimenting with telegraphy and succeeded in transmitting tones, the basis of telephony. His inventions ran neck and neck with those of Alexander Graham Bell (1847-1922). Thus, the invention of the telephone was envisioned by a number of brilliant inventors at about the same time and it soon came to be a much-desired commodity in business and personal communications.

Now that inventors knew that wired telephony was possible, what about wireless communications? It didn't take long before they were competing to see who could develop the first wireless devices. In 1866, Mahlon Loomis (1826-1886) demonstrated that airborne kites could pass a signal from one kite to the other without a physical connection. Six years later, he applied for a patent for a wireless telegraphy system. Amos E. Dolbear, a university research professor and writer, was awarded an 1886 patent for a wireless telegraph system based on induction. In the mid-1890s, Aleksandr Stepanowitsch Popow (1859-1906) sent a wireless shipboard message to his laboratory in St. Petersburg. While not as quick to develop as the conventional telephone, these inventions eventually led to the development of radio and cellular telephone communications.

The Early Development of Recording Devices

A punch card can be used to store a weaving pattern so that it can be repeated at a later time. Music boxes use a similar system, storing songs on punched paper or metal. These were already in use by the early 1800s. However, recording technologies that could save human speech didn't exist until a few years before the outbreak of the American Civil War and didn't get firmly established until about a decade after the end of the War.

Sound-recording devices and stock ticker devices were developed around the same time, in the mid-1800s. While they might seem to be divergent technologies, they were held together by powerful social factors–the desire to know and record business transactions. The capability to record and transmit business information, whether by telegraph signals or by voice, made a permanent change in the economic structure of the world.

In the 1860s, a new type of telegraph was devised to transmit business information. In essence, the first stock ticker machines were 'printing telegraph' machines. Since they were a valuable business commodity that not everyone could afford, there was a temptation to spy on messages or access the telegraphic signals carrying stock information.

In 1857, Édouard-Léon Scott de Martinville developed one of the earliest known sound-recording devices, the *phonautograph*. He configured a funnel-shaped horn to channel the sound to a sensitive diaphragm which transmitted the sound vibrations to a stylus and recorded them on a rotating cylinder of blackened paper (charcoal or ink may have been used). This is the essential concept of cylinder and platter-based recording devices.

By 1877, Charles Cros in France and Thomas Edison in America had both invented devices to record sound. (Since Edison was both prolific and commercially astute, his devices are better known.) With the basic concept in hand, Edison set about trying to find a practical and readily available recording medium. He created early recordings on both waxed paper and tinfoil. The idea of using paper was not as unusual as it might seem, since certain music boxes and looms already relied on patterns punched in paper or metal.

Telegraphic stock tickers had an enormous impact on business dynamics since they made it possible to monitor and interact with the stock market from remote locations. Western Union first introduced stock tickers to brokerage firms in 1866. Left: The New York Stock Exchange building as it looked in the early 1900s. Middle and Right: Cartoon of ticker tape (a type of printed telegraphic message) being carefully studied by an investor. [Life Magazine cartoon, 1899; Library of Congress, Detroit Publishing Co., ca. 1904; copyrights expired by date.]

The success of these new sound-recording technologies meant that it was now possible to capture and replay nature sounds, music, and voices, with or without the knowledge of those being recorded.

As a surveillance technology, however, recording was still in its infancy. There were four big limitations to the early machines:

- The recorders were bulky and difficult to move around.

- They weren't truly a re-recording medium, as the wax had to be reheated or reapplied, or the metal had to be hammered out in order to be used again.

- The recording times were very short, generally only about three minutes.

- The quality of the sound was crude and scratchy, though probably adequate for identifying voices and speech. The bigger limitation was that microphones hadn't been invented yet and the speaker had to be near the horn that funneled the sound in order for a voice to register on the recording.

In spite of the limitations of early machines, Edison recognized the potential commercial value of a device that could record voice transactions over the telephone "so as to make that instrument an auxiliary in the transmission of permanent and invaluable records, instead of being the recipient of momentary and fleeting communication".

By 1881, there were dictating machines that freed the speaker from being present at the same location and time as the person transcribing the message. Recording technologies were of particular interest to business people trying to gain mobility or a competitive advantage. They provided greater flexibility in time scheduling. A recording could be shipped to another location or recorded off-hours and transcribed during working hours.

Phonographic Recordings

People who enjoyed phonograph records from the late 1800s until the late 1900s are familiar with them as a playback medium and most listeners had never seen a 'record-pressing' device. In the early days of recording, however, there weren't thousands of companies selling records, so there was very little to play on the phonograph machines. At that time, companies were selling phonographs as recording devices in much the same way they now sell tape recorders.

One of the ways in which phonograph recorders were marketed was for "Phonograph Parties". In trying to establish a new market, vendors encouraged people to throw parties in which they surreptitiously recorded their friends. The idea was to then surprise them with the fact that they had been recorded. The market for consumer phonograph-recording machines virtually disappeared, however, when record companies started mass-marketing prerecorded discs. People who wanted to make their own individual recordings no longer had easy access to phonograph recorders.

Meanwhile, the demand for prerecorded sound was booming. The early Edison recordings were sold on cylinders, but cylinders with variable-depth grooves were hard to mass-produce. Inventors looked for other solutions.

In 1887, Emile Berliner developed an important innovation in the honographic process in which the groove created by the recording stylus was etched side-to-side rather than up-and-down. This made it possible to mass-produce flat records by 'stamping' out platters in quantity and imprinting the grooves in a softer material. Berliner chose shellac because it was malleable when heated and hard when cooled. The Victrola company was built upon Berliner's U.S. patents. Edison's highly competitive spirit was roused and he responded by switching to discs, but they weren't compatible with the Victrola side-to-side-groove machines.

Advancements in Recording Technologies. Left: A Victrola could play but not record. Right: A phonograph and wire-recording machine could both play and record and led to the concept of tape recording devices. [Classic Concepts ©1997 photos, used with permission.]

Historic playing devices were limited, so inventors looked for more practical ways to record longer sequences and perhaps even re-record on the same medium.

In 1888, Oberlin Smith submitted a caveat for a patent (notice of intention to file) for a magnetic device for recording a phone conversation. But Smith never filed the actual patent, choosing instead to publish his ideas in the September issue of *The Electrical World*. Smith suggested recording on a metal-impregnated thread, an idea that led other inventors to experiment with wire.

By 1898, a Danish inventor, Valdemar Poulsen, had figured out a way to coil wire so that it could be rotated and passed under a magnetic pickup, thus creating a device that could record on fine wire. He demonstrated his invention at the international exposition in Paris, in 1900. Unfortunately, long spools of wire were awkward to handle and not ideal in terms of sound quality. Nevertheless, it was a breakthrough, and Poulsen patented his 'Telegraphone,' suggesting that it could be used for the unattended recording of phone messages. When he later added an automatic-answer feature; he had essentially developed the telephone-answering machine.

These basic recording technologies were improved over the next two decades by various inventors, to include amplification, better fidelity, a variety of recording media, and various mechanisms to turn the machines on and off. During this same time period, telegraph lines continued to reach out to the more remote areas of the continent and telephone switchboards were appearing almost everywhere. Business telegraphy, voice communications, and sound recording technologies were becoming well-established fixtures of society.

The basic aspects of sound recording were firmly in place by the 1910s, a fact that caused a stir in the motion picture industry. Up to now people had been getting their movie entertainment from 'silent pictures,' moving films without sound. By 1911, inventors were creating ways to synchronize sound with the pictures, creating a whole new genre called 'talkies' and putting a lot of the silent film stars out of work. This was an important development because, from this time on, many important technologies were able to simultaneously record images and sound. (See the Visual Surveillance chapter for further information on the recording of images.)

Left: Bales of wire blanket the rough terrain where the telegraph line crossed the Stikine River in B.C. at the turn of the century. Right: Bales of copper wire. Copper wire was used for almost every type of electrical installation from office buildings, to phone and telegraph wires, to wiring in automobiles. Most early telegraph and phone taps were set on copper wires. [B.C. Archives c1900 historical photo; Library of Congress 1925 Detroit Publishing Co. photo, copyrights expired by date.]

There followed a period of refinement in sound technologies. Optical sound tracks were added to films, broadcast sound fidelity was improved by modulation and, in 1917, E. C. Wente invented a condenser microphone to record clearer, more uniform sound. Edward Howard Armstrong developed the superheterodyne circuit, which improved the sensitivity and selectivity of radio receivers, so that amplifier tuning was no longer necessary. Technology was becoming more sophisticated and people were devising new ways to use it.

Regulation of Communications Technologies

In the latter half of the 19th century, developed nations began to organize their political economies around the new technologies. As telegraph systems unified nations, nations established regulations to control their use.

In 1835, in Britain, the Municipal Corporations Act called for individual regions to set up police forces overseen by local watch committees. In 1842, a Criminal Investigation Department was established. By 1849, the various district police stations were being interlinked and connected to Scotland Yard by telegraph lines.

Telephone services expanded across North America in the early part of the century, staffed by female operators. Left: Switchboard operators at work in Port Alberni, B.C. Right: Telephone operators in the Vancouver Hotel, B.C. [B.C. Archives ca. 1910s and 1916 historical photos, copyrights expired by date.]

America followed a similar path. The Boston and New York Police departments were established in 1838 and 1844, respectively. Within a year, the precincts in New York were interconnected by a police telegraph system. Due to political wrangling, the New York Metropolitan Police was created in 1857, with broader jurisdiction over New York City, Brooklyn, and Westchester County (coexisting with the New York Municipal police until 1870). By 1878, police telegraph boxes were being installed in the large cities of America.

At the federal level, changes were also occurring. The Internal Revenue Service was put in place in 1862. Then, in 1866, regulation of the telecommunications industry was initiated with the *Post Roads Act*, which granted the Postmaster General the authority to oversee rates for government telegrams and to assign rights of way through public lands.*

Public Officials and Charges of Wiretapping

While phone 'tapping' implies some electrical or mechanical connection to a phone line, in this historical section it will be used in its broadest context to mean any persons or recording devices 'listening in' on a wired or wireless phone conversation with whatever means are at their disposal. In other words, it encompasses most types of telephone eavesdropping.

In 1863, a 'draft riot' galvanized the New York Metropolitan Police to quell the disturbances.

---

*The telephone was not yet on the scene. The regulatory responsibility of the Post Roads Act eventually came under the jurisdiction of the Federal Radio Commission (through the Radio Act of 1927). Then, through the Communications Act of 1934, it became the Federal Communications Commission (FCC) which is still the main regulatory body.

The police were publicly commended:

> "The services of the Metropolitan Police, officers and men, during Riot Week, won for them the admiration and confidence of the community. Never did men meet an emergency so fearful with more promptness, unanimity, and courage, and never was hazardous and prolonged duty discharged with more willingness and fidelity".

> [David M. Barnes, "The Metropolitan Police: Their Services During Riot Week. Their Honorable Record", New York: Baker & Godwin, 1863.]

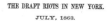
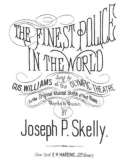

Prior to the investigation of the New York Police in the early 1890s, the actions and reputation of the Metropolitan Police were a source of pride for the city. Left: David M. Barnes authored "The Metropolitan Police: Their Services During Riot Week. Their Honorable Record" in 1863, commending police actions. Middle: Sheet music by Joseph P. Skelly exemplified the public perception of the Police force in 1875, prior to the Lexow Commission Report. Right: Theodore Roosevelt instituted reforms as a member of the Police Board from 1895 to 1897 (shown here as Governor in 1910). [Baker & Godwin, 1863; E.H. Harding sheet music, 1875; Library of Congress, Detroit Publishing Co., 1899, copyrights expired by date.]

New York enacted statutes to prohibit eavesdropping in 1892. Yet, to the astonished alarm of the residents of America's mightiest city, the first celebrated case of wiretapping involved the New York Police. They were charged with eavesdropping on sensitive communications and continued to tap as though they were exempt from the law. In 1894, a committee headed by State Senator Clarence Lexow released damning revelations of New York Police corruption, extortion, and wiretapping.

Reports of abuse in New York and elsewhere increased the population's awareness of wiretapping and made people more sensitive to the vulnerability of trusted officials and of conversations over public communications systems. In 1862, California enacted a law prohibiting the tapping of telegraph wires. In 1893, the police community at large founded the International Association of Chiefs of Police (IACP) to create a national forum for police organization, codes of conduct, and record-keeping. In 1899, the San Francisco Call accused the San Francisco Examiner of wiretapping reporters' communications and stealing exclusive stories, while thousands of miles away, the state of Connecticut enacted an electronic eavesdropping law that made it a state crime to listen to "the transmission of telegraphic dispatches or telephone messages to which he is not entitled". The California legislature responded, in 1905, to complaints of intrusions by extending the 1892 telegraph law to restrict wiretapping of telephone communications [Dash, Schwartz, Knowlton, 1959].

In 1895, Theodore Roosevelt was appointed to the New York Police Board and initiated changes that helped restore public confidence in the force that was lauded as "New York's

Finest" before the scandal. He reformed hiring practices, record-keeping, and identification procedures. The New York force had a small number of women 'matrons' but none had yet been hired to work at Police Headquarters. Roosevelt appointed Minnie Gertrude Kelly to a post in 1896. He also adopted the Bertillon system, a biometric bone-measuring identification system that predated fingerprinting. Roosevelt further reorganized the Detective Bureau and appointed Isabell Goodwin, who became a police matron in 1896 and went on to become an undercover agent. She was promoted to first grade detective in 1912.

Left: Thomas Edison's company recorded this film of the June 1899 Policeman's Parade in New York City. When Edison's company shot the film, the New York Police Department was still recovering from the corruption scandals of the early 1890s. The Lexow Committee, which investigated the Department, described serious criminal activity within the force which included wiretapping abuses. Right: The Policemen's Parade on Fifth Avenue, ca. 1903. The annual Police Parade was cancelled in 1895 due to the drop in public opinion but later restored with reforms that were spearheaded by Theodore Roosevelt. [Library of Congress, Thomas A. Edison, Inc.; Detroit Publishing Co., copyrights expired by date.]

The telegraph had a major impact on unifying the vast North American wilderness, yet the telephone patents of Alexander Graham Bell were a potential goldmine to any company who could use them to establish a commercial system to augment telegraph systems with something more natural. The Bell system was incorporated in 1878, based on Alexander Graham Bell's patents, less than 15 years after the telephone was first introduced in America. A telephone was quickly installed in the White House and President Hayes placed a call to the Bell company a few miles away. A telephone was also installed in the Washington, D.C. police station. By the early 1900s, the police departments in major cities had installed telephone systems that would eventually displace the telegraph lines.

Increased Mobility. Motorcycles and automobiles were adopted for police work all over the country at about the same time that telephones were superseding telegraphs and radio communications were being established. Top: The Police Department of the City of Bridgeport, Connecticut displaying motorcycles in October 1914. Bottom: Military police with motorcycles at division headquarters in Camp Zachary Taylor, Kentucky in front of the Western Union Telegraph Company building (center). [Library of Congress, Henry J. Seeley collection, 1914; Library of Congress, Caufield & Shook collection, ca.1918, copyrights expired by date.]

The quick installation of phones in the White House and the police station may, in part, have been a marketing/networking coup on the part of the Bell company. Under the astute business management of Theodore Vail, the Bell system did very well, so well, in fact, that much of the early legislation related to communications was in response to the activities of the Bell system and, later, AT&T.

Entrepreneurs wanted in on the growing communications profits. In the early 1890s, when the original Bell patents expired, thousands of independent telephone companies sprang up throughout the country. The Bell system responded by buying out, forcing out, and consolidating the independents. It regained its dominant position in just a few years. This situation had advantages and disadvantages. While innovation and local community-owned businesses were suppressed, consistency and stability were maintained under the direction of the corporate giant. This uniformity of hardware and services had a substantial impact on phone surveillance. A consistent national telephone infrastructure made it very easy to eavesdrop on telephone communications, especially before there were statutes in the United States to regulate wiretapping.

Other technological inventions were changing the country and the strategies of the people who were maintaining law and order. The invention of the automobile and the radio forever changed the way local law enforcement officers patrolled their beats and communicated with one another and with other precincts. It was a new age of transportation and communications. The concept of distance was to dramatically change over the next few years as automobiles and motorcycles superseded horses and bicycles.

Problems were still occurring in New York, however. In 1916, a New York mayor authorized wiretapping of Catholic priests in a charity-fraud investigation, without demonstrating that there was sufficient cause to initiate the action. In the course of investigating these police actions, the legislature discovered the police could tap any line in the New York Telephone Company and had listened to many confidential conversations. These continued accusations of wiretapping abuse brought more public and Senate attention to the issue.

Global Developments

Improvements in long-distance travel and communications allowed geographically diverse organizations to cooperate internationally. It now became possible to monitor criminals who fled to other countries. Up to this point, it was very difficult to capture a felon who melted anonymously into a foreign nation. In 1914, the first International Congress of Criminal Police came together in the Principality of Monaco to discuss global issues in policing including cooperative record-keeping and identification measures.

In 1923, at a meeting of the International Congress of Criminal Police, the International Criminal Police Commission was formed, where it gradually evolved into INTERPOL, an international data-collection agency that would assist with international investigations within certain guidelines.

Wiretapping Legislation in America

In America, it appeared as though wiretapping might eventually be outlawed altogether but for two strong social factors that caused the government to reconsider its use. The first was the outbreak of World War I and the fear of foreign infiltration. The second was Prohibition and the fear of gangsterism and general public opposition to some of the strictest prohibitions. Both of these events increased the prevalence of surveillance activities in general.

With the outbreak of World War I, the fear of foreign agents operating in America led to many acts intended to deal with foreign infiltration, including *The Act of October 29, 1918* in which Chapter 197, 40 Statute 1017 says in part that:

... whoever during the period of governmental operation of the telephone and telegraph systems of the United States shall, without authority and without the knowledge and consent of the other users thereof, except as may be necessary for operation of the service, tap any telegraph or telephone line, or wilfully interfere with the operation of such telephone and telegraph systems or with the transmission of any telephone or telegraph message, or with the delivery of any such message, or whoever being employed in any such telephone or telegraph service shall divulge the contents of any such telephone or telegraph message to any person not duly authorized or entitled to receive the same, shall be fined ... or imprisoned for not more than one year, or both.

This act established that tapping was to be used for counterespionage purposes only. It was not intended to endorse wiretapping of "U.S. Persons".

Delegates and speakers at the "National Conference on World Wide Prohibition" held in Columbus, Ohio, in November 1918. [Library of Congress Panoramic Photographs Collection, 1918, copyright expired by date.]

On 11 November 1918, the Germans signed the Armistice, ending World War I. From 19-22 November 1918, the "National Conference on World Wide Prohibition" was held in Columbus, Ohio with some significant consequences.

In 1919, the *18th Amendment* to the Constitution was ratified, prohibiting the "manufacture, sale, or transportation of intoxicating liquors within, the importation thereof into, or the exportation thereof from the United States and all territory subject to the jurisdiction thereof for beverage purposes ..". Enforcement of the measures was defined in the *Volstead Enforcement Act*.

SUPREME COURT DECISION ON THE PROHIBITION AMENDMENT AND THE VOLSTEAD ENFORCEMENT ACT ANXIOUSLY AWAITED

How the Supreme Court will pass upon the prohibition issues now before it, is a matter of daily concern. What the court will have to say on this vital issue is a matter of current interest to those for and against the liquor business. That a favorable decision in the interest of a better citizenship will be rendered, is

Left: Announcement of the Supreme Court Decision on the 18th Amendment and the Volstead Enforcement Act on alcohol prohibition, June 1920. Right: U.S. officials enforcing alcohol prohibition laws at the Brownsville Custom House in Texas, in Dec. 1920. [Ohio Historical Center Archives Library, 1920; Library of Congress South Texas Border Collection, 1920, copyrights expired by date.]

In 1927, Congress passed the *Radio Act of 1927*, Public Law 632. This important act established the Federal Radio Commission (FRC) to regulate use of the airwaves. Further, Section 27 stipulated that:

No person receiving or assisting in receiving any radio communication shall divulge or publish the contents, substance, purport, effect, or meaning thereof except through authorized channels of transmission or reception to any person other than the addressee ....

and no person not being authorized by the sender shall intercept any message and divulge or publish the contents, substance, purport, effect, or meaning of such intercepted message to any person; and no person not being entitled thereto shall receive or assist in receiving any radio communication and use the same or any information therein contained for his own benefit or for the benefit of another not entitled thereto ...

In 1927, wiretapping was outlawed in Illinois but, as had happened in New York, the police illegally continued the practice.

Constitution Issues and Wiretapping

The enactment of Prohibition brought many new cases into courts, some of which strongly challenged constitutional issues and the ability of the legal system to sort them out. Wiretapping featured prominently in one of these cases.

Wiretapping was used to catch liquor offenders during Prohibition. In a now-famous case, Roy Olmstead was caught running a $2 million a year smuggling operation out of the Pacific Northwest. Information about the illegal operation was largely obtained by four Prohibition Agents tapping through wires that were inserted by a lineman into the normal telephone lines outside the premises. The tapped dialog was transcribed into almost 800 typed pages that detailed illegal operations including the smuggling and sale of liquor, tax evasion, and bribes offered to local police officers. Olmstead and others were convicted of conspiracy to violate the National Prohibition Act, based partly on the wiretap evidence. Olmstead appealed. It went to the Supreme Court, but since no physical trespass had occurred, the Justices agreed that conversation was an intangible and thus did not constitute illegal search and seizure.*

In association with this string of events, Justice Louis Brandeis offered the following dissenting point of view:

The evil incident to invasion of the privacy of the telephone is far greater than that involved in tampering with the mails. Whenever a telephone line is tapped, the privacy of the persons at both ends of the line is invaded, and all conversations between them upon any subject, and although proper, confidential, and privileged, may be overheard. Moreover, the tapping of one man's telephone line involves the tapping of the telephone of every other person whom he may call, or who may call him. As a means of espionage, writs of assistance and general warrants are but puny instruments of tyranny and oppression when compared with wire tapping.

[Justice Louis Brandeis, *Olmstead v. United States*, 1928.]

In spite of this eloquent and impassioned plea, the U.S. Congress did not make wiretapping illegal, but neither was the matter permanently resolved. Brandeis's words would echo many times in cases through subsequent decades.

---

*Reference *Olmstead v. United States*, 277 U.S. 438, 462 (1928). Chief Justice Taft ascribed the rule to both the Fourth and Fifth Constitutional Amendments. Justice Brandeis and Justice Holmes dissented, offering the opinion that evidence admitted through violation of the Fourth Amendment in turned violated the Fifth Amendment.

Audio-Visual Developments

These days people typically own their own phones and are responsible for their phone lines from the junction box to the interior of the house or office. In the 1920s, however, except for a few communities or large corporations with independent phone systems, AT&T was in control of everything from the local office to the phone inside the house (which was leased to the customer). As the 1920s progressed, AT&T grew, technology improved, and recording technologies became more sophisticated and practical to use, but they weren't permitted on Bell system phones.

The telephone changed the way public service industries and businesses conducted their activities. Left: Forest-fire-reporting telephone lines provided a new way to report and track the progress of fires (Florida, 1937). Middle: A phone system provided a way for logging crews in different areas of the forests to intercommunicate (1941). Right: Telephones could now be used to direct construction crews from a good vantage point of a distance. Here a signal man instructs crews working at the Shasta Dam in California, in 1941. [Library of Congress FSA-OWI Collection photos by Arthur Rothstein and Russell Lee, public domain.]

AT&T had a virtual monopoly on telephone service in the 1920s and held jurisdiction over the phone lines and the phones they installed on customers' premises. They responded to developments in technology by banning the use of answering machines on the public networks. Thus, in America, recorders could only be used on private systems or those belonging to independent systems that permitted them.

By 1926, Dictaphone was selling the Telecord machine, an electronic telephone recorder that used wax cylinders. Synchronized sound movies, video technologies, and the earliest tape players were in development by the late 1920s.

The pressure on AT&T to permit telephone-answering machines increased. AT&T relented somewhat in 1930, allowing Dictaphone's Telecord machine to be used on private branch exchanges (PBXs). The machines caught on in America, in spite of the restrictions. People wanted the capability to monitor phone calls while they were away and to record conversations. Small businesses were eager for the machines, especially those that couldn't afford to hire a receptionist or who wanted to offer 24-hour telephone information. Europeans began using answering machines and the variety of types of recording media grew. In 1932, Loftin-White labs in New York announced a disc-based answering machine.

AT&T was concerned about how to maintain control and still meet public demand for answering machines and began to provide call-forwarding services so subscribers could arrange calls to be forwarded to live answering services.

As telephone service spread and improved, consumer magnetic-tape machines were soon to appear. In 1935, at the Radio Exhibition in Berlin, the *Magnetophon* was demonstrated, becoming available to the public in 1936. The Magnetophon provided a means to record on tape. A Swiss answering machine called the *Ipsophon* recorded on steel tape. You could even dial the fully automated Swiss machine to remotely retrieve phone messages. Semi J. Begun, a German immigrant working at the Brush Development Company in America, developed both steel tape and coated-paper tape recorders and sold the devices to the military. The machines could record on a variety of types of media including tapes, discs, and wire. Wire recorders of various types were being marketed in competition with the other formats and reached their peak in the late 1930s and early 1940s. Thus, there were now a number of practical formats to meet the growing demand for telephone call recording.

Tapping Phonelines. These photos illustrate the manual telephone systems still in use in the 1930s and 1940s. Their simple electrical/mechanical connections and uniformity made them easy to wiretap. Less scrupulous operators were also known to eavesdrop on calls, directly motivating the invention of automatic switching systems and phones with dials. Left: Switchboard operator at Littlefork, Minnesota in 1937. The phone jacks had to be manually plugged into the appropriate holes by a human operator to make a connection. Middle: Old-style home telephone with a hand crank, in Martin County, Indiana. Right: A telephone lineman using boot spurs to scale a pole to do maintenance on a line in 1940. Linemen's telephone sets were designed with alligator clips to hook temporarily to a line for testing. These were sometimes also used to tap lines. [Library of Congress photos by Roy Stryker, Arthur Rothstein, and Lee Russell. FSA-OWI collection, public domain.]

Further Legislation Related to Wiretapping

In 1934, the *Federal Communications Act* (FCA) prohibited the interception and divulgence of wire or radio communications; 47 U.S. Code, Section 605 reads:

> No person not being authorized by the sender shall intercept any communication and divulge or publish the existence, contents, substance, purport, effect or meaning of such intercepted communication to any person...

Even though the Federal Communications Act was aimed more at radio communications than wireline telephone communications, the Supreme Court interpreted the Act broadly. Thus, in a significant move, the FCA ruling was instrumental in reversing the famous *Olmstead* decision. On the basis of the new FCA law, the Court held in *Nardone v. United States* that wiretap information acquired by federal agents was not admissible as evidence. It further ruled in 1939, in *Weiss v. United States,* that this applied to federal tapping of intrastate as well as interstate communications. As a consequence of the various rulings, the Attorney General ordered a halt to FBI wiretapping. Ten years later, the pressure of World War II would cause the tide to turn once again.

The social and economic instability caused by the War rekindled the debate about wiretapping. America was emerging as a multicultural country such as had never existed before. Over the last half century, people from nations around the world, in numbers equal to Canada's entire population at the time, had immigrated to the United States. A large proportion of the population consisted of first generation Americans, with languages and cultural habits drawn from a wide spectrum. National security and local law enforcement agents felt that it might be relatively easy for a foreign spy to 'blend in' to this melting-pot landscape.

Since the *Nardone* case, wiretapping had been suppressed. Then, in 1940, J. Edgar Hoover, Director of the FBI, argued that the bureau needed broader wiretapping powers to detect and convict spies and subversives. Attorney General Robert H. Jackson decided that intercepting the communications would be permissible under the *Nardone* decision if the information accessed was not divulged, but kept within the federal government. President Roosevelt subsequently authorized the wiretapping of foreign agents for national security. He did, however, restrict approval to:

> ... persons suspected of subversive activities against the government of the United States, including suspected spies. You are requested furthermore to limit these investigations so conducted to a minimum and to limit them insofar as possible to aliens.

In response to the President's approval, in 1941, Attorney General Jackson instructed the Director of the FBI to maintain records of wiretapping, including cases, times, and places. While the requirement to keep a log might seem to establish accountability, the opposite may also be true. Jackson may, in essence, have asked the foxes to guard the henhouse, by not appointing an external, independent party to monitor the wiretaps.

Hoover continued his lobby for broader powers by opposing legislation requiring warrants for wiretapping.

In 1942, President Roosevelt established the Office of War Information to control news and propaganda which, among other things, generated a remarkable legacy of photographs detailing life (as they wanted it to be seen) in America.

Challenging the Olmstead Decision

The courts were busy, too, as another important case related to eavesdropping was debated that year.

In 1942, in *Goldman v. United States*, the plaintiff charged that law enforcement officers had violated the Fourth Amendment. They had bugged a wall adjacent to the plaintiff with a Detectaphone which could amplify and record the sounds in the next room. As there was no physical trespass of the bugged suite, the Court decided against the plaintiff, stating that the Fourth Amendment had not been violated. Justice Roberts voiced his opinion that there was no violation because there was no illegal search and seizure and the evidence obtained in this way was admissible in a federal court. This decision is important in another respect because it suggested a legal distinction might be made between bugs and wiretaps.

In 1943, a secret project, code-named VENONA, was quietly initiated within the U.S. Army Signal Intelligence Service (forerunner of the NSA) to monitor foreign communications.

Another prominent court case related to wiretapping occurred in 1945, in *Bridges v. Wixon, District Director, Immigration and Naturalization Service*. In its findings, the Court strongly denounced the Immigration and Naturalization Service (INS) and others for persistent hounding and attempts to deport Harry Bridges. Bridges was an Australian who had immigrated to America in 1920. Deportation proceedings were instituted against him on the grounds that he was affiliated with the Communist Party which, in turn, was allegedly seeking to overthrow the U.S. government. Statements in the case were passionate and direct, in part stating that:

.... The record in this case will stand forever as a monument to man's intolerance of man. Seldom if ever in the history of this nation has there been such a concentrated and relentless crusade to deport an individual because he dared to exercise the freedom that belongs to him as a human being and that is guaranteed to him by the Constitution...

Later in the statement, it comments on wiretapping of communications connected with Bridges:

Industrial and farming organizations, veterans' groups, city police departments and private undercover agents all joined in an unremitting effort to deport him on the ground that he was connected with organizations dedicated to the overthrow of the Government of the United States by force and violence. Wiretapping, searches and seizures without warrants and other forms of invasion of the right of privacy have been widely employed in this deportation drive.... The Immigration and Naturalization Service, after a thorough investigation of the original charges in 1934 and 1935, was unable to find even a 'shred of evidence' warranting his deportation and the matter officially was dropped...

At the federal level, eavesdropping was being employed on an international scale. A secret operation called SHAMROCK was initiated around the time of World War II. SHAMROCK was a telegraph-message-collection program in which three prominent international telegraph companies agreed to requests from the Government for access to certain international telegraph messages. The program was originally intended to extract telegrams relating to the communications of *foreign* targets (in compliance with national security guidelines), but later investigations indicate that the program gradually changed to include the communications of U.S. citizens and organizations, as well.

During World War II, all international telegraph traffic was screened by military censors, located at the companies, as part of the wartime censorship program. During this period, messages of foreign intelligence targets were turned over to military intelligence....

The Army Security Agency (ASA) was the first Government agency which had operational responsibility for SHAMROCK. When the Armed Forces Security Agency was created in 1949, however, it inherited the program; and, similarly, when NSA was created in 1952, it assumed operational control.

[U.S. Senate Select Committee reporting in "Intelligence Activities–The National Security Agency and Fourth Amendment Rights", 6 November 1975.]

Some people would argue that Operation SHAMROCK was a necessary wartime security measure. Others would oppose it on even those grounds. Most people, however, would agree that SHAMROCK should not have monitored the communications of U.S. citizens and should never have continued, as it did in various forms, for almost 30 more years. At the present time, it is difficult to reconstruct and study the operation because neither the Government nor the three companies kept a paper trail of these arrangements.

Until now, significant issues related to wiretapping had been decided in the courts, but Hoover saw an opportunity to repeat a wiretapping-related request to President Truman and the President signed it, perhaps not realizing its full implications. In 1950, George M. Elsey, the Assistant Counsel, expressed his concerns to President Truman that the wording of the Hoover memo was very broad, but Truman took no steps to reverse his approval. As far as the White House and the FBI were concerned, wiretapping could now be used in situations other than national security investigations of foreign agents.

Post-War Developments in Technology

World War II was winding down, families were reunited, the economy adjusted to Cold War and peacetime activities and the Korean conflict was stirring up in southeast Asia.

The end of the Second World War provided an opportunity for American and British technical investigators to 'discover' foreign technologies, including the Magnetophon recorder that was in use in German-occupied nations. The U.S. Alien Property Custodian seized the patent rights. Based on technological devices found overseas, the U.S. Department of Commerce published technical information about tape recording. John T. Mullin demonstrated a Magnetophon machine to the Institute of Radio Engineers in America. Magnetic tape recorders were poised to supersede the steel tape, 45 rpm, and wire recorders that had shared the market for the last several years.

By the late 1940s, tape recorders were beginning to win the 'format wars.' Even movie reels were changing from optical to magnetic sound. Meanwhile, 'Ma Bell' continued to face pressures from people who wanted to use telephone-answering machines. AT&T began leasing the machines to their customers.

In 1947, inventors at the Bell research labs invented the transistor, a development that was to have a stunning impact on the development of electronics devices from that point on.

Two things happened to dramatically affect AT&T's business practices in 1949. The Federal Communications Commission (FCC) had taken steps to relax regulations against the use of answering machines culminating in a landmark ruling in 1949, permitting consumers to use telephone-answering machines. The Department of Justice (DoJ), in turn, sued AT&T for antitrust violations (the second effort since 1913) to promote competition in the communications industry.

By 1950, AT&T was providing microwave relays in the east. Direct Distance Dialing was available by 1951; a human operator was no longer needed to connect long-distance calls.

While prosecuting AT&T, the DoJ had to cope with a scandal of its own. In March 1949, Judith Conlon, a DoJ employee, was caught ready to hand over confidential FBI documents to a United Nations employee from the Soviet Union. Conlon was arrested and charged with theft and later with conspiracy for the distribution of secret Department documents. The convictions were subsequently overturned, but the events brought internal security to the attention of outsiders, including Hoover's efforts to have the charges against Conlon dropped, presumably to prevent disclosure of FBI activities. Hoover is said to have instituted new filing procedures, placing FBI reports in separate files, depending on their content and sensitivity.

By 1950, the forest service and certain other emergency and search and rescue operations were beginning to use radio phones on a regular basis.

Continued Controversy Over Government Wiretapping

The early 1950s were years of nervous Cold War instability and conflicts in Korea. Public Law 513 was enacted, making it a crime to disclose classified information about American or foreign cryptography systems.

A secret operation called SOLO was initiated by the FBI, in the early 1950s, to monitor CPUSA, the Communist Party of the United States of America. SOLO amassed a large body of sensitive information through several presidential administrations over the next three decades.

Government wiretapping and issues of domestic versus national security continued to be controversial over the next few years.

- In 1950-1951, a Subcommittee on the Investigation of Wiretapping reported to Congress, generating four linear feet of records, now stored in the National Archives.

- In 1950, Presidential Directives authorized the FBI to investigate subversive activity but were not explicit as to how investigations were to be conducted. Government memos from the late 1940s to the late 1950s indicate an internal debate over whether the FBI should be limited to domestic national law enforcement or should be funded for "secret activities abroad".

- Attorney General Francis Biddle turned down applications he felt were not justified. This approval process was made explicit by Howard McGrath, the Attorney General in 1952 [Diffie and Landau, 1998].

- Wiretapping and internal security matters came up again at the December 1953 Legislative Leadership Conference.

Left: The Truman Cabinet members and various officials in 1950. President Truman is shown fourth from the right, with Attorney General J. Howard McGrath second from the right and Chairman of the National Security Resources Board, Stuart Symington, sixth from the left around the table. Right: Attorney General McGrath, President Truman, and Defense Secretary Louis Johnson in 1950. McGrath was in favor of strict approval procedures for authorIzed wiretaps and opposed the practice of wartime interment of immigrant Americans on the basis of race alone. [U.S. Dept. of Education archives (from NARA), public domain, photo on the left by Abbe Rowe.]

Another important legal case was tried in 1954 in *Irvine v. California*. The case discussed the concealment of listening devices in the walls of a residence. The Court upheld the bugging activities, but the Justices expressed "outrage[d]" at the "indecency of installing a microphone in the bedroom".

In 1954, Attorney General Brownell pressed for warrantless wiretapping to prosecute alleged Communists. The House Judiciary Committee accepted his argument, but the House of Representatives disagreed and no consensus was reached.

In 1956, Hoover briefed the cabinet about Communist efforts to influence civil rights movements. Some of the taps lasted for years, without specific prosecutions associated with the taps. Information that did not contain evidence of criminal activity was retained in files. It has been asserted that Hoover may have wiretapped prominent political figures and several Supreme Court Justices with little oversight of the tapping activities.

MEMORANDUM FOR: The Secretary of State
                   The Secretary of Defense

SUBJECT: Communications Intelligence Activities.

The communications intelligence (COMINT) activities of the United States are a national responsibility. They must be so organized and managed as to exploit to the maximum the available resources in all participating departments and agencies and to satisfy the legitimate intelligence requirements of all such departments and agencies.

I therefore designate the Secretaries of State and Defense as a Special Committee of the National Security Council for COMINT, which Committee shall, with the assistance of the Director of Central Intelligence, establish policies governing COMINT activities, and keep me advised of such policies through the Executive Secretary of the National Security Council.

On 24 October 1952, President Truman issued a Top Secret memorandum to the Secretary of State and the Secretary of Defense establishing a Special Committee of the National Security Council to handle COMINT (communications intelligence), to establish policies and provide advisement to the President. The resulting Directive was to replace NSCID No. 9. (Chapter 1 provides background information on NSCIDs, COMINT, and the NSA.) [NSA eight-page Top Secret, distribution status downgraded per NSC 28 Jan 1981.]

Left: In 1917, J. Edgar Hoover became a member of the legal staff with the Department of Justice. In 1924, he became Director of the Bureau of Investigation, retaining the position for almost 48 years. Hoover overhauled the Bureau in its early years and became extraordinarily influential in later years. Right: President Eisenhower decorated J. Edgar Hoover in 1955. Richard M. Nixon, later connected with illegal eavesdropping in Watergate, can be seen standing behind the President to the right. [Hoover portrait copyright expired by date; FBI Web site historical timeline news photos, released.]

In the world of technology in 1956, the U.S. Government and AT&T signed a consent decree that AT&T could only engage in common carrier communications services, excluding them from the computer industry so that competition could be maintained in the emerging industries. AT&T was further required to license the Bell patents on a royalty basis to independents.

Leasing an answering machine from AT&T in the mid-1950s was not inexpensive. It cost the equivalent of about twelve hours' clerical wages per month. In other words, in today's dollars, about $150 per month. That was far too high for the typical home user, but well worth it for a business that was looking for an option to paying secretarial wages. By 1957, AT&T had about 40,000 answering machine subscribers.

In 1957, the Wright Commission recommended federal legislation to support limited and authorized wiretapping that:

> ... would make admissible in a court of law evidence of subversion obtained by wiretapping by authorized Government investigative agencies. Wiretapping would be permissible only by specific authorization of the Attorney General, and only in investigations of particular crimes affecting the security of the Nation.
>
> [Commission on Government Security, "Report of the Commission on Government Security", Washington, D.C., 1957.]

Thomas F. Eagleton presented a report before the Subcommittee on Constitutional Rights opposing legal wiretapping, with records from between 1958 and 1960. His opposition was based largely on constitutional issues of privacy.*

Miniaturization and Emerging Wireless Systems

While American society was sorting out the legal implications of wiretapping, technology was changing dramatically, due to the development of transistor technologies. Small, light electronic parts could now be used to build portable radios, microphones, cheaper mainframe computers, and tiny surveillance devices. At the same time, wireless technologies continued to improve and business owners wanted to build private systems so they wouldn't have to rely on AT&T.

In 1959, private business owners applied to the FCC for permission to build their own private microwave systems. In the *Above 890* decision, the FCC ruled that there was sufficient bandwidth above 890 KHz to serve both private customers and AT&T. This challenge to its monopoly caused AT&T to step up development on its microwave communications systems. Above 890 also opened a crack in the door for entrepreneurs to develop new wireless technologies.

Seeing an opportunity, based on the Above 890 decision, Microwave Communications, Inc. (later MCI) applied to the FCC, in 1963, to build a microwave system between Chicago and St. Louis. They felt they could offer cheaper, better private service than what was available from AT&T. It took six years for the application to be approved, but this was a significant turning point in the communications industry, introducing competition that would eventually change the position of AT&T and the national telephony infrastructure.

Opposition to Listening Devices

The mid-1960s was a time of opposition to listening devices and an increased scrutiny of FBI wiretapping activities.

Listening devices were becoming smaller and more sophisticated. 'Bugs' were showing up everywhere. Laser listening devices were being prototyped and tested. In one significant case, law enforcement agents had driven a 'spike mike' into a wall under the apartment of a suspect. The wall wasn't punctured, but the vibrations through the building structure provided a good channel for sounds from the rooms that were being monitored. In spite of previous court decisions, in *Silverman v. United States*, the Court ruled in 1961, that the evidence obtained by bugging was inadmissible because it constituted a 'search' that had been carried out without a warrant. This was in spite of the fact that the wall wasn't punctured and no physical trespass had occurred. Thus, the Court took a stronger stand against unauthorized 'search and seizure' and provided a precedent for the protection of privacy.

---

*Papers with respect to this were collected as part of his correspondence between 1957 and 1964 and are housed in the Western Historical Manuscript Collection, Missouri.

In 1965, a Senate subcommittee studied electronic surveillance and the photographing of mail covers (envelopes), focusing a major part of their attention on the activities of the FBI. IRS activities were also scrutinized. The FBI was later accused of tapping the members of the subcommittee who were engaged in reviewing the FBI. Senator Edward V. Long was identified by the Bureau as unsympathetic. Not long after, Long was smeared in an article in a prominent magazine, linking him to a gangster. It is said he was pressured into signing a press release written by the FBI asserting that the FBI hadn't participated in uncontrolled tapping or eavesdropping. Long was subsequently defeated by Thomas Eagleton and retired.*

In 1965, Chief Judge Campbell reported to Congress that:

> My experiences have produced in me a complete repugnance, opposition, and disapproval of wiretapping, regardless of circumstances.... Wiretapping in my opinion is mainly a crutch or shortcut used by inefficient or lazy investigators.

In 1965, Attorney General Katzenbach, under a directive from President Johnson, tightened tapping requirements, imposing time limits on tap authorizations, stating:

> ... the record ought to show that when you talk national security cases, they are not really cases, because as I have said repeatedly, once you put a wiretap on or an illegal device of any kind, the possibilities of prosecution are gone. It is just like a grant of immunity.... I have dismissed cases or failed to bring cases within that area because some of the information did come from wiretaps. But here we feel that the intelligence and the preventive aspect outweigh the desirability of prosecution in rare and exceptional circumstances.

In 1967, a number of bills related to crime syndicates, admissibility in evidence of confessions, and wiretapping were submitted to the U.S. Senate. The President's Commission on Law Enforcement members claimed that a majority supported the authorization of law enforcement agents to use electronic surveillance.

By 1967, the distinction between wiretaps and bugs was disappearing, partly through improvements in technology and partly through subsequent legal decisions. In *Katz v. United States*, it was ruled that people have a reasonable "expectation of privacy" in using a public phone booth and a search warrant based on probable cause was required, "...a person in a telephone booth may rely upon the protection of the Fourth Amendment. One who occupies it, shuts the door behind him, and pays the toll that permits him to place a call is surely entitled to assume that the words he utters into the mouthpiece will not be broadcast to the world". The constitutional basis of the various privacy decisions was being strongly formulated by the growing number of cases related to listening devices.

Shifting the Focus to Law Enforcement

Numerous hearings that were critical of wiretapping were held around this time. Attorney General Ramsey Clark spoke against wiretapping, reflecting the public outcry, except for cases involving national security. He stated "I also think that we make cases effectively without wiretapping or electronic surveillance. I think it may well be that with the commitment of the same manpower to other techniques, even more convictions could be secured, because in terms of manpower, wiretapping, and electronic surveillance is very expensive".

If you have read Chapter 1, you will be aware that national security bodies were not authorized to focus surveillance on 'U.S. Persons,' and that hounding the citizenry was not to be a

---

*Long responded to the chain of events by telling the story from his perspective in "The Intruders, the Invasion of Privacy by Government and Industry", 1967.

goal of government intelligence-gathering. However, the FBI and the Department of Justice were concerned with domestic law enforcement and justice, as opposed to national security (the jurisdiction of the NSA), and thus considered intelligence-gathering of domestic activities to be within their mandate. Following this line of reasoning, in 1968, the *Interdivisional Information Unit* (IDIU) was consolidated under the administration of Attorney General Clark. This computerized system included files on organizations and individuals playing a role "purposefully or not, either in instigating or spreading civil disorders, or in preventing our checking them".

In 1968, organized crime was considered a serious problem and a number of studies concluded that the impenetrability of criminal groups justified wiretapping and bugs in law enforcement. The logic was that no specific victim was necessarily involved in these types of cases and that victims might be subject to threats, if they acted as 'stool pigeons' (informants), thus making apprehension and prosecution more difficult than in other types of crimes.

In 1968, the *Omnibus Crime Control and Safe Streets Act* was passed, establishing basic law for criminal investigation interceptions, generally in the cases of violent crimes, gambling, counterfeiting, and the sale of marijuana. It further set out requirements for telecommunications carriers to provide technical assistance and hardware adjustments to their equipment to aid law enforcement agencies in carrying out electronic surveillance, a move that provoked strong opposition that continues today (Title III was amended in 1970). The Act also created the National Institute for Justice, the research and technological development arm of the Department of Justice.

Watergate and The Computer Age

In the early 1970s, Government and press disclosures made it seem as though everyone was bugging everyone and this may have been true. IBM discovered that they had been bugged by Soviet agents. Attorney General Clark discovered, after denying the fact to a judge, that the FBI had been using electronic surveillance. News agencies found out they were being bugged. As soon as the miniature technology became widely available, it appears to have become widely used.

The 1970s are characterized by three important chains of events, one in communications delivery, one in national government, and one in technological development. These were

- the imminent breakup of AT&T,

- the loss of public trust of public officials with the Watergate Scandal, and

- advancing wireless telephony and emerging personal computers.

In 1974, the Department of Justice filed a comprehensive antitrust suit against AT&T, citing illegal actions in perpetuating its monopolistic business practices. The suit called for divestiture of some or all of the Bell Operating Companies and the further divestiture of Western Electric.

In 1974, in connection with the Watergate breakins, Judge John J. Sirica appointed six experts in audio technology to study one of the tapes made by President Richard M. Nixon in 1972. The June recording between H. R. Haldeman and Richard Nixon was obscured at a key point with a buzzing noise for about 18.5 minutes. The erasure appeared to be from a machine other than that on which the tape had been recorded. Investigators concluded that it was probably deliberate erasure produced on a Sony model 800B. By studying the magnetic signature, it was deduced that the erasure was done on a machine normally operated in Nixon's secretary's office. This is one of the more interesting stories in forensic investigation, as the experts (who were associated with IEEE, a respected electrical engineering organization), used spectrum analysis, waveform analysis, and digital signal-processing equipment to study the magnetic patterns on the tape. They prepared the medium by washing the tape in a fluid containing ferrite particles (which align to the patterns). The IEEE panel findings were presented and subsequently described in the April 1974 issue of *IEEE Spectrum*. The mystery of the original content of the tapes was not solved, however.

After a series of laws and amendments, in December 1975, Public Law 94-176 (89 Stat. 1031) established a *National Commission for the Review of Federal and State Laws Relating to Wiretapping and Electronic Surveillance* to study and review the operation and provisions of the chapter to determine their effectiveness. This Commission, comprising "competent social scientists, lawyers, and law enforcement officers" was to report to the President and Congress by April 1976.

As far as the communications infrastructure was concerned, AT&T was still fighting hard to maintain its control and market share of the telephone market. In the 1976 *Resale and Shared Use* decision, the FCC permitted unlimited resale and shared use of private line services and facilities. Interstate communication was more closely regulated, however. AT&T entreated Congress and the *Consumer Communications Reform Act* (the "Bell Bill") was enacted. This was possibly the most significant general communications regulation since the Communications Act of 1934 and would have favored AT&T. The issue wasn't fully settled, however, and aspects of the bill continued to be debated well into the 1990s. In a series of decisions called *Execunet*, the U.S. courts opened up the long-distance markets.

New Technologies and Common Carrier Obligations

An important trend in the late 1970s and early 1980s was the increasing sophistication and variety of communications and surveillance devices. Wiretapping was no longer the sole issue concerning the courts. Audio/visual "bugs", pen registers, surveillance cameras, and other devices were filtering out into the marketplace and being adopted by law enforcement agencies and the general public.

With technology changing so fast, investigators found themselves at a loss as to how to apply old laws and procedures to new devices and systems. They entreated Congress and the courts to help.

In *United States v. New York Telephone Co.,* in 1977, the Supreme Court found that telecommunications carriers were responsible for providing assistance to law enforcement agents "to accomplish an electronic interception". The technological details about how this should be achieved were not stipulated at the time. However, continuing improvements in computer technology made it clear that this issue would soon have to be addressed.

Until the age of modern electronics, wiretapping was easy. You chose a phone line, attached a device, and listened. Now it was getting complicated. A person could have more than one phone number; the phones could include a wireline phone, a cordless phone, a traditional cellular phone, or a PCS phone (or all four). The calls could be encrypted or sent over spread-spectrum frequencies to ensure privacy. The phone service might be accessed through several vendors at various times of the week or times of the day. Did the existing laws broadly apply to the new modes of communication?

Privacy advocates, in turn, were concerned that law enforcement agents were using the changes in technology to lobby for increased powers. If wiretapping powers were to be applied to new systems, there was going to be a stronger need than ever before for the communications providers to assist law enforcement officials in carrying out the tapping. Some of these providers had strong moral and financial reasons for opposing this requirement.

Bugs, Wiretapping, and Issues of Entrapment

It is worth pointing out that up to now, the regulations related to bugging and those related to wiretapping, while closely related, are not synonymous. Bugging appears to have been tolerated in some circumstances where wiretapping was not. In spite of this, they yield similar information and are often discussed together.

There have been some arguments as to whether the information obtained by wiretaps can be obtained by other less 'intrusive' methods and counter-arguments (e.g., by Louis Freeh, Director of the FBI) that important information is obtained by taps that can't be obtained any other way. The strongest argument in favor of taps is that the recordings provide strong evidence that is accepted more readily as 'truth' than the remembrances or hearsay of a human eavesdropper. It has also been argued that an electronic device might be less liable to entrap a suspect.

> But wiretaps and bugs enjoy two advantages over secret informants. First, the evidence they report as to what the defendant did or did not say is trustworthy. Second, and perhaps more important, a bug cannot encourage lawbreaking: It can neither advocate nor condone such conduct....

> In any event, for the purpose of my more general argument, it is enough to acknowledge that both legal tests of entrapment–objective and subjective–permit police to employ an enormous amount of routine deception, although the prevailing subjective test permits even more. [Jerome H. Skolnick, Deception by Police, *Criminal Justice Ethics,* 1982, Volume 1.]

Skolnick makes an interesting observation here, noting that a bug or wiretap cannot be used to entrap in the same way that a physically present undercover agent might entrap. In the context of Skolnick's argument, this is a good point. In the larger picture, outside of Skolnick's central theme, it would not be an appropriate argument for someone to use to justify the substitution of wiretapping information for that which might be obtained by undercover agents. After all, if entrapment is an issue, a law enforcement agent could use information obtained from a wiretap or bug *to enact a future entrapment* that might not have been possible without the information obtained from the tapping operation in the first place.

In other words, it should be remembered that surveillance techniques are not used in isolation and information obtained from one source is generally combined in a larger body of intelligence with information obtained from other sources.

Surveillance and Civil Rights

A lot of attention was focused on surveillance and information obtained from wiretaps in the late 1970s and early 1980s. This was in part a leftover from the Watergate investigations and in part a consequence of the increasing availability of surveillance devices. It was also a lingering legacy of the Cold War.

At this time, the Soviet Union was undergoing major changes, global commerce was opening up, and the Cold War was winding down. In 1980, Operation SOLO, a long-standing FBI surveillance operation, was terminated. In spite of global changes, Communist fears and general social intolerance were still evident in many of the public statements of Government officials which, in turn, were reflected in decisions about who or what the Government had a responsibility or right to surveil.

Dr. Martin Luther King, Jr. had been assassinated on 4 April 1968, but his name continued to come up again and again in the 1970s and 1980s, in part because of the extensive surveillance and wiretapping of his activities that had occurred while he was alive.

In January 1977, District Justice John Lewis Smith, Jr. ordered the FBI to purge its files of:

> ... all known copies of the recorded tapes, and transcripts thereof, resulting from the FBI's microphonic surveillance, between 1963 and 1968, of the plaintiffs' former president, Martin Luther King, Jr.; and all known copies of the tapes, transcripts and logs resulting from the FBI's telephone wiretapping, between 1963 and 1968, of the plaintiffs' office in Atlanta, Georgia and New York, New York, the home of Martin Luther King, Jr., and places of accommodation occupied by Martin Luther King, Jr....

> ... at the expiration of the said ninety (90) day period, the Federal Bureau of Investigation shall deliver to this Court under seal an inventory of said tapes and documents and shall deliver said tapes and documents to the custody of the National Archives and Records Service, to be maintained by the Archivist of the United States under seal for a period of fifty (50) years; and it is further ORDERED that the Archivist of the United States shall take such actions as are necessary to the preservation of said tapes and documents but shall not disclose the tapes or documents, or their contents, except pursuant to a specific Order from a court of competent jurisdiction requiring disclosure.

In 1983, Senator Jesse Helms made a series of inflammatory justifications to Congress regarding Dr. Martin Luther King, Jr., who had now been dead for 15 years. (The following quotes are brief so you are encouraged to read the full Congressional Record and form your own opinions.) In his statements, Helms remarked:

> Mr. President ... it is important that there be such an examination of the political activities and associations of Dr. Martin Luther King, Jr.... King associated with identified members of the Communist Party of the United States (CPUSA)....

> There is no evidence that King himself was a member of the CPUSA or that he was a rigorous adherent of Marxist idealogy or of the Communist Party line...

> [Jesse Helms, 3 October 1983 Congressional Record, Vol. 129, No. 130.]

In his statement, Helms asserted that King was vigorously entreated, by members of the Government, to sever all ties with the Communist sympathizers, which King did not do. He described how King continued to "address their organizations" and "invite them to his own organizational activities". Since Helms openly stated that King was not campaigning for Communism and was not a "rigorous adherent of Marxist idealogy or of the Communist Party line", it seems that the chief complaint against King was that "he had no strong objection to Communism" and, by implication, could be under their influence.

In public statements, King promoted a global humanity and a free country in which people were entitled to equality and a diversity of opinions, so it's not really surprising that King expressed a liberal view with "no strong objection to Communism". Helms asserted, for this reason, that he was a threat to be monitored and suppressed, and he and other detractors feared that "the Communist Party [would] infiltrate and manipulate King and the civil rights movement".

Those who oppose King's views have drawn parallels between his statements on eliminating racism and inequality with the Communist philosophy. They have generally failed to point out, however, that Communist philosophy and the militaristic Soviet implementation of that philosophy are fundamentally at odds with one another, with a repressive, coercive element that is anathema to most members of our free society, including those involved in civil rights movements. In retrospect, some might argue that criticisms of King were motivated not purely by fears of Communism but also by fears that his philosophy of a "new world order" would upset the traditional lines of power referred to as the 'old boys network' in our own political structure. Either way, it's difficult to assess the full facts in the case if the records are sealed until at least the year 2027, according to the Judicial decree.

Issues of Privacy and Constitutional Rights

By this time, wiretapping was outlawed in most states, but electronic eavesdropping, a more recent technology, was still largely permissible. Before this was sorted out, yet another class of devices was becoming available in the form of portable, wearable transmitters.

In 1963, the Chief Justice of the Supreme Court set the tone for many court decisions of the 'flower power' decade by noting that "... the fantastic advances in the field of electronic communication constitute a great danger to the privacy of the individual; [and] that indiscriminate use of such devices in law enforcement raises grave constitutional questions under the Fourth and Fifth Amendments ..." In *Lopez v. United States* (1963) the entrapment issues inherent in 'wearing a wire' were tested when an IRS agent used a pocket recorder to gather evidence. Since the wire was used in conjunction with the agent's activities, it was held that it was not listening in on conversations that could not otherwise have been heard and since there was no unlawful trespass, there was no instance of unconstitutional eavesdropping.

In 1967, in *Berger v. New York*, in which the plaintiff had been convicted of conspiracy based on eavesdropping evidence obtained from a recording device in an attorney's office, strong arguments were made both for and against the plaintiff. Justice Black offered a dissenting view that in part expressed sympathy for the position of law enforcement officials:

> Today this country is painfully realizing that evidence of crime is difficult for government to secure. Criminals are shrewd and constantly seek, too often successfully, to conceal their tracks and their outlawry from officers.... In this situation, 'Eavesdroppers,' 'Informers,' and 'Squealers' as they are variously called, are helpful, even though unpopular, agents of law enforcement....

> Since eavesdrop evidence obtained by individuals is admissible and helpful, I can perceive no permissible reason for courts to reject it, even when obtained surreptitiously by machines, electronic or otherwise. Certainly evidence picked up and recorded on a machine is not less trustworthy. In both perception and retention, a machine is more accurate than a human listener....

> The superior quality of evidence recorded and transcribed in an electronic device is, of course, no excuse for using it against a defendant if, as the Court holds, its use violates the Fourth Amendment. If that is true, no amount of common law tradition nor anything else can justify admitting such evidence. But I do not believe the Fourth Amendment, or any other, bans the use of evidence obtained by eavesdropping.

Justice Clark's opinion, on the other hand, was that:

> The claim is that the statute sets up a system of surveillance which involves trespassory intrusion into private, constitutionally protected premises, authorizes 'general searches' for 'more evidence,' and is an invasion of the privilege against self-incrimination.... We have concluded that the language of New York's statute is too broad in its sweep, resulting in a trespassory intrusion into a constitutionally protected area, and is, therefore, violative of the Fourth and Fourteenth Amendments...

In this ruling, Justice Clark made a distinction between bugs and wiretaps. Justice Douglas concurred with Justice Clark's opinion, stating:

... at long last, it overrules *sub silentio Olmstead v. United States,* 277 U.S. 438, and its offspring, and brings wiretapping and other electronic eavesdropping fully within the purview of the Fourth Amendment. I also join the opinion because it condemns electronic surveillance, for its similarity to the general warrants out of which our Revolution sprang and allows a discreet surveillance only on a showing of 'probable cause.' These safeguards are minimal if we are to live under a regime of wiretapping and other electronic surveillance.

Yet there persists my overriding objection to electronic surveillance *viz.,* that it is a search for 'mere evidence' which, as I have maintained on other occasions ... is a violation of the Fourth and Fifth Amendments, no matter with what nicety and precision a warrant may be drawn....

A discreet selective wiretap or electronic 'bugging' is, of course, not rummaging around collecting everything in the particular time and space zone. But even though it is limited in time, it is the greatest of all invasions of privacy. It places a government agent in the bedroom, in the business conference, in the social hour, in the lawyer's office–everywhere and anywhere a 'bug' can be placed.

If a statute were to authorize placing a policeman in every home or office where it was shown that there was probable cause to believe that evidence of crime would be obtained, there is little doubt that it would be struck down as a bald invasion of privacy, far worse than the general warrants prohibited by the Fourth Amendment. I can see no difference between such a statute and one authorizing electronic surveillance which, in effect, places an invisible policeman in the home. If anything, the latter is more offensive because the homeowner is completely unaware of the invasion of privacy.

The traditional wiretap or electronic eavesdropping device constitutes a dragnet, sweeping in all conversations within its scope–without regard of the participants or the nature of the conversations. It intrudes upon the privacy of those not even suspected of crime, and intercepts the most intimate of conversations...

Justice Clark's opinion included:

I would hold that the affidavits on which the judicial order issued in this case did not constitute a showing of probable cause adequate to justify the authorizing order. The need for particularity and evidence of reliability in the showing required when judicial authorization is sought for the kind of electronic eavesdropping involved in this case is especially great...

This case has lengthy arguments for and against the constitutionality of eavesdropping and evidence obtained by eavesdropping. The short quotes included here cannot fully convey its import and the judgment is worth reading in its entirety (it can be searched on the Web through www.findlaw.com and is on file online at Cornell University).

Telephony had continued to spread and evolve. Since the introduction of the Communications Act of 1934, legal interpretations of the regulations varied over the years with regard to wiretapping and related enforcement issues. Law enforcement activities were more specifically set out under the *Omnibus Crime Control and Safe Streets Act of 1968,* Title III. This Act defines the authority and specifies conduct and procedures of wiretaps by federal law enforcement agencies. Most of the states have enacted similar statutes supporting these restrictions. In 1970, the Act was amended to clarify the position and responsibilities of communications service providers for assisting law enforcement agents.

Public Disclosure of SHAMROCK

In November 1975, the *U.S. Senate Select Committee to Study Governmental Operations with Respect to Intelligence Activities* submitted "Intelligence Activities–The National Security Agency and Fourth Amendment Rights" which discussed an operation called "SHAMROCK" in which communications carriers were persuaded to work in conjunction with government officials. The report says in part:

> SHAMROCK was the cover name given to a message-collection program in which the Government persuaded three international telegraph companies, RCA Global, ITT World Communications, and Western Union International, to make available in various ways certain of their international telegraph traffic to the U.S. Government. For almost 30 years, copies of most international telegrams originating in or forwarded through the United States were turned over to the National Security Agency and its predecessor agencies.

> As discussed more fully below, the evidence appears to be that in the midst of the program, the Government's use of the material turned over by the companies changed. At the outset, the purpose apparently was only to extract international telegrams relating to certain foreign targets. Later the Government began to extract the telegrams of certain U.S. citizens.... There is no evidence to suggest that they ever asked what the Government was doing with that material or took steps to make sure the Government did not read the private communications of Americans.

In 1980, operation SOLO, the secret FBI operation that had been initiated in the early 1950s, was publicly disclosed and officially brought to an end.

Techies and Techie Toys

The late 1970s and beyond belonged to a new breed of technologically astute, intelligent, playful software programmers and electronics wizards. Fortunes were made by entrepreneurs, technologists began creating new communications channels, and subtle but significant changes in the fabric of the communications infrastructure were beginning to manifest through computers and computer networks.

In the late 1970s and early 1980s, it became a popular pastime for computer techies to build 'blue boxes,' handheld devices that could control a touchtone phone (usually a payphone) through tones. Blue boxes were designed to manipulate the phone line electronics to do mischief and make 'free' long-distance calls. Electronics buffs began to build and sell these illegal devices through the computer underground. Blue boxes became popular and the media began to release stories about the devices. It was assumed by the general public that most of the calls placed with blue boxes were made by university students and computer geeks, which was somewhat true, but investigators discovered that a large proportion of blue box thieves were well-paid professionals, including doctors and businessmen.

The significance of blue boxes was three-fold:

- the people who designed them were technologically capable of building various surveillance devices and marketing them as they had the blue boxes,

- the touchtone system through which they operated was shown to be easy to manipulate, and

- the concept of 'hacking' into the phone system could be applied to hacking into the new computer bulletin-board systems (BBSs) that were appearing in communities throughout the developed world.

Blue boxes made communications carriers and electronics designers more acutely aware of the security weaknesses of the existing phone system. Engineers were in the process of incorporating new computer technologies into many kinds of communications infrastructure systems at the time, including telephone signaling systems, with the result that Signaling System No. 7 (SS7) was introduced in the early 1980s. One difference between earlier systems and SS7 systems was that earlier systems carried the call control (signaling) information and the conversation (or computer data) on the same line. This is called an *in-band* system. SS7, on the other hand, carried the signaling information on a different line from the conversation or other data. This is called an *out-band* or *out-of-band* system. Not only was SS7 designed to be more flexible and powerful than earlier telecommunications systems but, as an out-band system, it was inherently more secure and less vulnerable to manipulation and unauthorized surveillance. In other words, blue boxes don't work on SS7 systems as they did on the older systems and neither do a number of other surveillance devices.

The early 1980s was also the time when many police departments began routinely taping all telephone conversations made on their lines, presumably with notice given to employees. Some departments provided an untapped line for the use of employees for personal calls, though there has been at least one incident [*Amati v. Woodstock* appeal, 1999] in which an untapped line was changed to a tapped line and the employees claimed they were not notified of the change.

By the 1980s, private branch exchange (PBX) phone systems were being equipped with 'back doors' that allowed phone service companies to access the equipment for service or maintenance, a concept that was adapted to software by computer programmers. In some phone systems, this access capability is provided through a DISA port. An external access port could also be used by off-site employees to dial into the system to make outgoing calls through the PBX for work-related communications. Understandably, such a system could be abused by employees committing fraud, or anyone wanting to place long-distance calls without paying the charges.

In the 1980s and 1990s, the digital switching systems began to supersede analog systems in most of the developed nations, and in the mid-1990s consumer 'Internet phones' with a telephone-style handset were able to interface with a computer keyboard to make international calls over the Internet for about $.05/minute, probably foreshadowing the next significant change in telephone dynamics, economics, and phone tapping.

The Breakup of AT&T and Implementation of the ECPA

In 1984, Judge Harold Greene divested AT&T of its Bell System regional operating companies. This has become known as the Modified Final Judgment (MFJ). Thus, with the breakup of AT&T and the growth of various independents, the uniformity of services and hardware changed to a proliferation of new technologies and means of providing services. At the same time that this important decision took place, analog technology was gradually changing to digital and wireless communications were beginning to increase in distribution. All of these factors greatly increased the technical expertise needed to surveil audio communications.

In 1986, Congress enacted the *Electronic Communications Privacy Act of 1986* (ECPA) and further amended the *Omnibus Crime Control and Safe Streets Act*, broadening the terms of the Omnibus act to include electronic communications.

In the ECPA, Congress acknowledged the capability of technology to intrude on personal privacy. The ECPA extended jurisdiction to wireless and non-voice communications and established rules for the use of pen registers and trap and trace devices in law enforcement. Freedom supporters followed the changes and the Senate commented on the ECPA as follows:

Most importantly, the law must advance with the technology to ensure the continued vitality of the fourth amendment. Privacy cannot be left to depend solely on physical protection, or it will gradually erode as technology advances. Congress must act to protect the privacy of our citizens. If we do not, we will promote the gradual erosion of this precious right. [Legislative committee report on the ECPA.]

Evolution in Phone Technologies and Call Security

With the growth of computer electronics, voice surveillance became more complex. As technologies evolved and diversified, it became necessary to first determine how the call was being placed (cell phone, PCS, cordless, Internet phone, etc.), and then to find out if the technology chosen was analog or digital (both were now common), and then to further determine whether the communication was being routed, spread, or encrypted, and finally to find an appropriate technology to intercept, decode, or record the conversation. Greater technological complexity sometimes confers greater security. Individuals using newer products do have a greater degree of privacy if the communications are encrypted. Source- or destination-level surveillance, however, is no more secure than before. In other words, fancy routing and decryption don't make the call secure if there's a bug in the pencil sharpener next to the phone or a spy with an ear to the wall.

Challenges for Law Enforcement

In terms of law enforcement, the technological revolution has greatly complicated the process of tapping conversations. Since there are more ways to place a call, there have to be more ways to access the calls. And since there are more ways to encrypt a call, it's more difficult to make 'en route' intercepts and to make any sense of the content of the calls. Phone numbers don't have to be tied to a specific physical address. Greater technical complexity usually results in higher costs for trained personnel and for the appropriate equipment. By the early 1990s, local and federal agencies were finding it hard to keep up with the pace of change and began to lobby for assistance from Congress.

Many users of the new cordless phone technologies assumed that existing wiretapping laws protected cordless conversations. The courts ruled otherwise, however, in 1992 in *United States v. David Lee Smith* by upholding the right of law enforcement to tap a cordless phone without a warrant. This situation was changed two years later when the *Communications Assistance for Law Enforcement Act of 1994* (CALEA), Public Law 103 414, required that the warrant requirements of earlier wiretap laws cover cordless phones as well, except in some instances of employers monitoring employee business communications.

The Communications Assistance for Law Enforcement Act (CALEA) was passed by the U.S. Congress in October 1994. This Act required telecommunications providers to assist law enforcement agencies, which meant providers would have to make changes to existing equipment to meet the call taps and trace needs of law enforcement agencies. This reinforced the 1970 Omnibus Crime Control and Safe Streets Act Amendment of 1970 common carrier obligation with the further responsibility to *modify equipment* to fulfill the terms of the Act. The Communications Assistance Act also authorized funds to reimburse direct costs to providers complying with the terms of the Act.

CALEA was an important piece of legislation. In its more specific terms, it required telecommunications providers to have the technical capability to isolate and access realtime calls and call identification information as well as the ability to provide this to law enforcement agencies offsite. However, the Act did not require that the carriers handle decryption except in cases where the target of the call had been provided with an encryption service.

The Recording of Personal Calls

In spite of attempts by developed nations to regulate and address the intricacies of wiretapping and decide who may or may not record conversations, the 'letter of the law' with regard to private citizens recording calls (especially their own) still remains somewhat subject to interpretation. In the U.K., the best answer is that both parties need to be informed of the action. In the U.S., state-by-state statutes differ as to whether one or both parties need to be informed. Clarification often does not come until a case is adjudicated through the court system and, even then, it may require a body of cases to establish the weight of priorities in one direction or another.

In August 1990, Senator Patrick J. Leahy, Chairman of the Senate Judiciary Subcommittee on Technology and the Law hosted a hearing that discussed Caller ID technology. He concluded the ECPA needed to be reviewed due to developments in communications technologies. He appointed a private sector task force which concluded that the new technologies were "challenging the existing statutory scheme for communications privacy".

The 1990s - Changeover to New Technologies

The early 1990s saw another change in hardware that was used to carry wired communications. For decades, copper wire was the medium of choice. By the early 1990s, however, substantial amounts of high-bandwidth optical fiber were beginning to coexist and, in some cases, replace copper wire. The volume and type of communications that were carried over fiber varied somewhat from what was being carried over copper wires and the means by which cable is 'tapped' differs from the tapping of copper wires.

In the mid-1990s, wireless communications and Internet voice capabilities were beginning to come of age. By this time, there were more than a thousand cellular switching networks in the U.S. alone and people were starting to use Internet phone systems to digitally place long-distance calls.

Another concept that began to take hold in the mid-1990s was the idea of 'number portability.' In the past, a phone number was associated with a physical location, just as a house number was associated with the physical location of the land on which the building was situated. With the increase in wireless communications and diversification of the industry, the idea of associating a number with a person rather than with an address, so that person could receive calls no matter where he or she might be located, began to appear technologically practical. While this would take years to fully implement, it would change audio surveillance technologies in a number of significant ways:

- A phone number would be tied to a person, like a social security number, rather than to an address.

- It might remain the same for the person's lifetime, rather than changing each time he or she moved.

- If people started to wear inexpensive wristwatch-sized wireless phones, the current system of phones might disappear, changing the way phones are tapped and tracked.

- If Global Positioning Systems (GPSs) were built into the tiny phones, the exact location of the person, to within about 60 feet, could be determined.

On the other hand, encryption techniques might become so sophisticated that tapping becomes impossible. That doesn't mean conversations will automatically be as safe as the communications media that carry them; there is still the potential for a conversation to be heard or recorded with tiny bugs. These could be designed to adhere innocuously to the bottom of a shoe, or the back of a lapel or could masquerade as a bird and fly along near a person to amplify and transmit a conversation to someone up to five miles away. Given the increasing sophistication of tiny remote technologies, such devices are now more real than science fiction.

Common Carrier Assistance Obligations

By October 1994, the *Wiretap Access Bill* had passed and was awaiting Presidential approval. This was originally proposed as the *Digital Telephony and Privacy Improvement Act of 1994*. It would legalize authorized surveillance of telecommunications systems and it stirred fears that law enforcement officials might require the common carriers to install systems that could be remotely monitored by law enforcement in a hands-off mode that would reduce the physical presence, and hence the accountability, of officials intercepting the conversations. In other words, in the past, a live phone carrier employee would be somewhat aware of who was tapping what, because they were involved in the activities, and could report anything that seemed out of the ordinary. With remote electronic boxes associated with the system, it would be difficult to establish the same checks and balances.

By 1994, bills were being proposed that would enable law enforcement agents to tap into the new digital communications technologies. On the one had, law enforcement officers argued this wasn't extending their jurisdiction, but rather continuing it in the face of new technologies. On the other hand, a gun-shy public, still sensitized to prior abuses on the part of trusted officials were opposing it. If communications carriers must have systems with tapping capabilities built in, engineers have to design them that way. But obsolescence and unsold goods are a commercial nightmare. Timothy Haight summed the situation up this way:

> ... it's expensive. The Feds have authorized 1/2 a billion dollars to pay for this, but the phone companies say it will cost a lot more... In the future, to avoid expensive retrofits, we can expect phone carriers to build in easy access at the outset. Allowing for wiretapping will become a design principle.

> These bills have a chilling effect on designers of technology. Design in security at your own risk. Next year it may be outlawed and you won't be able to sell it....

> [Timothy Haight, The Punishment of the Wise, *Network Computing*, November, 1994.]

In 1996, the first major overhaul of the *Telecommunications Act of 1934* occurred with the *Telecommunications Act of 1996*. This act essentially opened the doors to access and competition within the communications services industry.

Disrobing the Machine - Security Weaknesses

The FCC cited losses of over $400 million to fraud and security problems in 1996. Later that year, prohibitions against eavesdropping on wireless phone transmissions were tested in court in a politically sensitive headline case.

Around December 1996, a couple eavesdropped with a radio scanner on a cellular conference call between John Boehner, Newt Gingrich, and other Republicans discussing a House Ethics Committee investigation. They recorded the wireless call and subsequently turned it over to a Democratic Representative who released it to the media. It was then published in the *New York Times*. In April 1997, the eavesdroppers were fined $1,010 for violating FCC prohibitions.

In March 1997, Counterpane Systems and U.C. Berkeley jointly announced that their researchers had found a flaw in the privacy protection used in the most advanced digital cellular phones. The group described how an intrusion could be carried out in minutes using a personal computer. A digital scanner could pick up the numbers dialed on the key pad which might include PINs or credit card numbers. Ironically, the announcement came at the same time that legislators were scheduled to hold hearings on the *Security and Freedom Through Encryption* (SAFE) bill. The group criticized the 'closed door' design process as contributing to weaknesses in the resulting security systems associated with cell phones.

Wiretapping in Foreign Nations

Unfortunately, because of the complex legal and social issues associated with communications technology, there is not enough space in this volume to discuss international developments, but it is worth mentioning that not all countries permit wiretapping.

> Here in Japan, there is no crime problem that would seem to justify wiretapping. Nevertheless, the legalization of wiretapping is being proposed by the Ministry of Justice.... In this context, increased electronic surveillance reveals the state tendency to try to suppress autonomous people's movements by utilizing its police apparatus.
>
> [Toshimaru Ogura, Japan's Big Brother, The Wiretapping Bill and the Threat to Privacy, *Japan-Asia Quarterly Review,* V.28(1), 1997.]

The Ministry of Justice of Japan was considering permitting very broad wiretapping of telephones, cell phones, fax machines, and computers, upon issuance of a warrant. So it appears that the for-and-against debate is not restricted to the United States.

The Late 1990s - Variation and Sophistication

It hardly seemed possible that electronics evolution could go any faster, but by the late 1990s, the pace was still increasing, and new technologies were entering and exiting the market faster than consumers could figure out their practical applications.

By 1999, the cell phone industry was booming. Eager to provide value-added services in the competition for cell phone subscribers, hardware vendors requested authorization from the FCC to add global positioning system (GPS) capabilities to cell phone handsets. This was seen as a selling point, since the origin of a call could be used to deploy emergency services or to aid a lost caller in getting back on the right track. It could potentially also be interfaced with automobile computer-mapping systems to display local services and phone numbers.

In September 1999, the FCC agreed to allow the cell phone/GPS technology. GPS technology effectively turns a cell phone into a tracking device. Up to this point, the common way to locate the user of a cell phone was to ask him or her his location or to triangulate the position from the strength of signals reaching cell transceiver stations in the vicinity. Even then, it was only an approximation.

With integrated cell phone/GPS capabilities, not only could a person's location be known, but it could be followed continuously and logged to an accuracy of between 20 and 100 feet, depending on the terrain and speed of movement. While private individuals would probably object to such monitoring, employees using company cell phones might not have a choice and law enforcement officials could theoretically obtain warrants to access tracking information by providing 'just cause' for such an investigation. Even before the ethical aspects were resolved, commercial systems began to sell in mid-2000.

By June 2000, designers, vendors, and telecommunications carriers were required by the terms of the Communications Assistance for Law Enforcement Act (CALEA) to implement systems that would enable law enforcement officials to conduct approved wiretap operations. Implementation since 1994 had been slowed by a variety of technical factors including concerns by the communications carriers that the modifications were too costly and difficult (in spite of government funds allocations). The modifications were further hampered by the concerns of privacy rights activists who argued that FBI jurisdiction was being broadened rather than just maintained with regard to new technologies. In August 1999, the FCC issued a series of technical standards to facilitate the implementation of CALEA.

Increase in Wireless Telephony

By the turn of the century, the number of cellular subscribers exceeded the entire population of Canada. By 2000, AT&T was marketing wireless phone services that functioned in much the same way as regular wired telephone services, without roaming charges or complicated service agreements.

In the mid-1990s, Harris Communications began marketing "Triggerfish", a briefcase-sized device with a headphone jack designed to continuously monitor cellular phones and provide pen register numbers and "intercept documentation" for wiretapping.

With this type of technology proliferating and GPS capabilities being built into newer cellular phones, friends, enemies, competitors, stalkers, and law enforcement officials could listen to your calls, track you, or theoretically pinpoint your location in relation to the scene of a crime. Unlike traditional tapping equipment in which a physical connection may betray the presence of a tap or for which a court authorization is required before physical connections are made, a tapping system for wireless communications is 'invisible' and difficult to detect or monitor, especially after the fact.

Because it was known that scanners could be used to listen to wireless phone calls, the Electronic Communications Privacy Act of 1986 (ECPA) prohibited the monitoring of cell phone communications except for system administration and maintenance and authorized 'wiretapping.' The Triggerfish was thus marketed specifically to law enforcement agents, but that didn't mean there wouldn't be copycat vendors developing similar systems and selling them through the same underground that blue boxes were sold through a few years earlier.

Experts have argued that the best way to secure the privacy of vulnerable communications is through encryption. Spread-spectrum technologies, which move the conversation around through different frequencies, provide an added measure of security. Secure communications that protect the populace pose a continuing dilemma for law enforcement agents, so the debate hasn't ended yet.

Wiretapping, Implementation, and Opposition

In July 1998, the New York Times reported that Louis Freeh, Director of the FBI, had approached members of the Senate Appropriations Committee, asking them to approve an amendment that would

> ... provide police agencies with the precise location of cellular phone users, in some cases without a court order.

Attorney General Reno and the Federal Communications Commission then scheduled a meeting to discuss the FBI's case that "such legislation is needed if the agency is to stay current with an evolving technology that enables criminals to use mobile phones to avoid detection". The technology being proposed was similar to 911 emergency services that could use triangulation to track the location of a cell phone caller. Several privacy groups responded to this announcement by writing to the Senate Appropriations Chair, entreating him to reject the amendment.

Meanwhile, communications carriers were voicing serious concerns about modifying their equipment to comply with law enforcement wiretapping needs. The objections were raised for various reasons, including technological feasibility, cost, obsolescence, timeline and, in some cases, moral grounds related to privacy or law enforcement accountability. The result, in August

1998, was that a suit was filed in District Court by the United States Telephone Association (USTA) on behalf of about 1200 small, medium, and large providers of local telecommunications exchange and access services throughout the country (predominantly wireline). These companies represented over 95% of the nation's local access lines. The suit named the FBI and the Department of Justice (DoJ) as defendants and challenged the regulations requested by the FBI under CALEA. The suit cited problems with the cost-recovery regulations which were supposed to reimburse carriers for the required modifications. The implementation guidelines were described by the claimants as being "arbitrary, capricious ... contrary to law" and in excess of "the FBI's statutory authority".

The implementation deadline was originally set for 25 October 1998. In September, 1998, prominent communications carriers, including AT&T Wireless Services, Inc., Lucent Technologies, Inc., et al., filed a memorandum opinion and order before the FCC and were granted an extension until 30 June 2000 for complying with CALEA.

On 27 August 1999, the Federal Communications Commission (FCC) issued a news release regarding the adoption of technical requirements for wireline, cellular, and broadband Personal Communications Services (PCS) to comply with assistance capability requirements prescribed by the *Communications Act for Law Enforcement Act of 1994* (CALEA). It required that the capabilities requested by the Department of Justice (DoJ) and the Federal Bureau of Investigation (FBI) be implemented by wireline, cellular, and broadband PCS carriers. In other words, communications service providers would now be required to implement the Telecommunications Industry Association (TIA) interim standard (J-STD-025) and several "punch list" capabilities as well. The compliance deadline was set as 30 June 2000, with packet-mode communications capabilities to be in place by 30 September 2001.

At the time this is being written, many of these issues are not yet resolved and the body of court cases is not sufficient to provide guidelines in all instances. For further information, consult current Web sites that provide news on these matters and cross-reference the Radio Surveillance chapter for additional information on wireless technologies.

5. Descriptions and Functions

5.a. Listening Devices

Devices that direct or enhance sound vibrations are the most common kinds of listening devices. They provide a more effective path for vibrations to travel from the sound source (or near the source) to the ear or recording microphone or they channel or focus the sound in order to provide acoustic amplification. Some also provide electronic amplification, and the most sophisticated utilize computer processors to selectively choose or improve a sound. Some sound devices are designed to detect specific types of situations and to automatically trigger an alarm or sequence of events. There are five general categories of listening devices:

Sound-detecting devices are those that responds to sound (usually loud sounds). They can be designed to selectively detect the cry of a baby, the sound of an explosion or a bursting pipe, or a car accident at a freeway interchange or traffic intersection. The simplest devices respond to loud sounds or those of a particular frequency. The more sophisticated devices respond to specific types of sound (e.g., screeching brakes followed by a bang). Sound-detecting devices are often hooked up to alarms or other emergency indicators or may be programmed to shut down machinery, heating systems, to freeze traffic lights, or to turn on a camera.

Sound-channeling devices are those that direct sound. They range from water glasses costing a few cents to stethoscopes costing a couple of hundred dollars. They are frequently used to eavesdrop. Sound-channeling devices are also used to help diagnose or monitor traffic flow in digital networks or mechanical linkages in production lines. The flow of data in a digital circuit can sometimes be heard through a device attached to network cables or terminal points. Some technicians are so adept at monitoring the sounds in electrical equipment, that they can aurally detect switching points and traffic flow in banks of telecommunications devices.

Sound-focusing or acoustical-amplifying devices generally use cone shapes (like megaphones) or parabolic shapes to focus and enhance a sound. They are sometimes used in combination with electronic amplifiers. These range in price from $20 for a simple megaphone-style amplifier to about $500 for hand-held parabolic listening devices. The most sophisticated 'lapel-style' parabolic amplifiers use both acoustics to capture the sound and electronics to amplify the sound, which are more expensive. One common listening device that is often overlooked is a basic hearing aid. With electronic miniaturization, these are now so small, they can be hidden in the ear canal and are almost undetectable if the hair is worn over the ears. They can boost sounds or can be designed to selectively enhance sounds in particular frequency ranges (e.g., to listen to animal sounds).

Electronic sound-focusing and/or amplifiers are devices that capture sounds and transmit them to another location or a recording device and may also increase gain and enhance the volume; some will also improve the quality of the sound through computer processing. Computer-based amplification systems are capable of detecting a specific voice or set of keywords, of analyzing the sound, and of carrying out some complex processing. Electronic and computerized amplifiers/processors can range in price from $100 to thousands of dollars, depending on their features. The more sophisticated systems are usually desktop-based, rather than portable, and are generally used to process recordings made at another location (although van-based listening labs could be used to process sounds in realtime—at a cost of tens of thousands of dollars). Most electronic 'bugs' amplify ambient sounds or focus on particular sounds.

Sound converters are devices that respond to a sound and convert the stimulus into another form such as lights, motion, or text. Sound converters are useful in situations where sounds are being monitored by someone who is hearing impaired or who doesn't wish to have sounds in the area where the surveilled signals are being monitored, as in covert surveillance or the monitoring of wildlife that's easily spooked. The sounds of a dog barking, a car honking, an alarm, or a telephone ringing can be converted to a vibration or an illumination. Vibrating devices will sometimes be placed against the spine or the skull to increase bone-conduction and may have connections to earphones. More sophisticated devices can be designed to respond to specific sounds or to screen out ambient sounds (to reduce the chance of false alarms). The more recent programmable devices can be configured to learn to recognize a particular sound (since phones, for example, ring at different frequencies from phone to phone). It is probably only a matter of time before someone programs a listening device that can radio a pager and print a short text message such as "The smoke-alarm is ringing". or "A child is crying".

Microphones

Small microphones can be used to listen to conversations and other sounds and can often be interfaced with public address systems and recording devices. They are usually powered by lithium, AAA, or AA batteries.

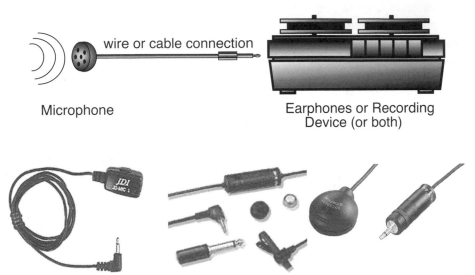

wire or cable connection

Microphone   Earphones or Recording
                              Device (or both)

Wired microphones. Left: This tiny microphone can be clipped to a lapel or pocket and at-tached to an amplifier or recording device, from Jing Deng Industrial Co. Ltd. Plug adaptors are available. Middle and Right: These electret condenser microphones are compact, high-sensitivity, omnidirectional, wide frequency-range 1.35-volt units that can be connected with public address systems or recording devices (e.g., for recording a conference), available from Yoga Electronics Co. Ltd. [Classic Concepts diagrams ©2000, used with permission. Supplier information courtesy of www.asia.globalsources.com.]

There are also extra-sensitive piezoelectric subminiature microphones which can pick up sounds to about 20 or 30 feet. Some of these tiny microphnes use the same technologies that are used in doctors' stethoscopes. Piezoelectric mics can be used as 'bugs' in walls or decorative furnishings and may be wired or wireless. A stethoscope itself is sometimes used to listen to sounds emanating from the next room by placing it against an adjacent wall. Small microphones range from $30 to $250.

Audio plug connections come in a variety of sizes and it's important to get one that fits correctly. Adaptors are readily available in electronics stores. Some plugs are monaural and others are stereo (stereo is now more common).

Olympus make a smaltelephone recording microphone that can record both voices of a telephone conversation, whether it's a wired phone or a cellular phone. The small device doubles as an earphone for listening to a radio or tape player. Jack sizes (with adaptor) are 2.5 mm and 3.5 mm. The microphone can be used to record to a microcassette recorder. Street price is about $30.

Suction-cup telephone pickups are designed to allow quick-and-dirty recording of a conversation in a location where a direct connection to the line isn't feasible (as on an airport payphone). The suction-base sound pickup is attached to the telephone mouth- or earpiece and connects with a wire that ends in a microphone jack. The jack can be plugged into a recording device or transmitter. The sound clarity isn't especially good, but if it's an important business transaction, a rough recording is probably better than nothing. Kits are about $15, assembled about $30.

BASIC COMPONENTS OF A WIRELESS SYSTEM

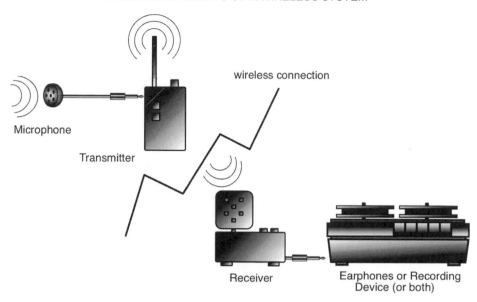

Wireless microphones usually transmit on FM, UHF, or VHF frequencies. Frequencies of 900 MHz and 2.4 GHz, which are widely used for short-range video and telephone transmissions, are also becoming more common for small wireless audio transmitters. Wireless microphones come in two basic models, a microphone with a separate transmitter connected by a cable or a microphone and transmitter built into the same housing. Wireless audio devices are often used for body-worn systems, e.g., 'wearing a wire.' [Classic Concepts diagrams ©2000, used with permission.]

Small, unobtrusive, wireless microphones come in a variety of shapes and sizes. Top Left: A tiny microphone which links to an FM transmitter and sends to a receiver up to 100 meters away, from Guangdon Takstar Electronic Co. Ltd. Top Middle: A tiny collar or lapel microphone sold together with a transmitter that can be placed in a pocket or on a belt to send audio to the receiver about 15 to 30 meters away from SCE Company Limited. Top Right: A condenser tie-clip or lapel-clip microphone which is sold separately from the VXM-168LTS wireless transmitter/receiver. Bottom Left: A wireless microphone that runs on AA batteries with 80-12000 Hz frequency response from Hisonic Audios Mfg. Group. Bottom Middle: This compact handheld microphone/transmitter works on a AA battery and weighs 210 grams, from Sekaku Electron Industry Co. Ltd. Bottom Right: This is an omni-directional electret condenser microphone with an FM transmitter/receiver that works in the 110-120 MHz frequency range, from Yoga Electronics Co. Ltd. [Classic Concepts diagrams ©2000, used with permission. Supplier information courtesy of www.asia.globalsources.com.]

Parabolic Microphone

Security, border patrol, military patrolling, wildlife biology, sportscasting, investigative reporting, and pivte detection are all professions in which parabolic listening devices to amplify sounds within a couple of hundred feet of the listener are used.

Parabolic microphones are sometimes also called 'umbrella' microphones due to their dish-like umbrella shape. They have highly reflective surfaces that use physics to 'capture' the sound and acoustics and electronics to equalize the sound and to actively filter it to produce the effect of amplification. If they are aimed carefully, they can be quite effective at bring the sound 'nearer' to the listener. The sound tends to amplify more at the higher frequencies.

Parabolic microphones are used by the news media to capture sports events, by field biologists and filmmakers to record animal sounds, and by law enforcement agents and private detectives to capture sounds from a distance. Most are portable and run on batteries, though some may have AC adaptors for stationary use. Prices vary from about $200 to $2,000+ depending on the size, model, and sound quality. Miniature versions are now available, but the majority are a foot or two in diameter. The range for common consumer models is about 100 to 250 feet.

Note, when using earphones with sound-amplifying equipment like parabolic microphones, it is important to use earphones or a microphone with a high-decibel shutoff system. This is a system which detects sudden loud noises and screens them out. Otherwise, if you try to use a normal headphone or earphone, you might either 'blow' the electronics by overloading them or, worse, damage your hearing with a sudden blast of amplified sound.

Left: A parabolic microphone uses a dish shape to collect and direct sound to the electronic components that process the sound and send it to the earphones. Right: A shotgun microphone is designed to pick up distant sounds directly in front of the microphone while minimizing surrounding noises. Both types are usually equipped with 'shutoff' circuits to prevent loud blasts of sound from damaging the hearing of the listener. [Classic Concepts drawings and photo ©1999, used with permission.]

Shotgun Microphone

A shotgun microphone resembles a long wand. It is designed to directionally pick up sound from a distance by attenuating the 'side' sounds. These microphones are used in conjunction with earphones, recorders, and video cameras and are especially popular for newscasting and detective work. They are a little less obvious than the umbrella-shaped parabolic microphones but need a bit of equalization (which also emphasizes unwanted noise) to the sound to get the same 'naturalness' as a parabolic microphone. Good quality shotgun microphones range from about $300 to $500. Small, handheld shotgun mics can be found for under $100.

Laser Listening Devices

A laser listening device is a piece of optical equipment intended to be aimed at a physical structure which is vibrating as a result of sounds near the structure (e.g., people talking inside or outside, near a window). The laser beam hits the structure and the sounds are then reflected back by influencing a change in the character of the beam. *This is a high-precision instrument and is only effective in ideal conditions.* In fact, in testing, the instrument is found to work better if it is focused on the debris and dust clinging to the surface of a window rather than the glass itself. Incidental vibrations from other sources, poor focusing, an incorrect angle, or vibration of the transmitter will all significantly degrade the returning signal.

For example, assume a conversation inside a distant building is being carried out over the noise of a television or radio broadcast, the complex vibrations hitting the window from both the conversation and the broadcast don't produce a 'clean' signal. Another limitation is that the window may be influenced by other vibrations, most often wind or traffic noise, which will confound the speech vibrations. Noise and interference are also likely to occur from poor weather with rain or hail. This is particularly severe if the precipitation is pelting the window. Finally, the conversants need to be near the window and speaking loudly enough to cause the surface to vibrate. If they are standing at the far end of the room and speaking in whispers, it's not likely that anything useful will be picked up by the system.

Laser listening devices come in two basic types:

reflecting laser The laser is aimed at a vibrating object, such as a window, which presumably is vibrating in response to the sounds nearby, such as a conversation. The laser beam hits the window, where it is influenced by the vibrations on the window from the inside, and acts as a 'carrier wave' to transmit the window vibrations back to the receiver.

interferometric laser Many of the limitations and characteristics of a reflecting laser also apply to the more sophisticated interferometric laser. A stable base and a good vibrating surface from which to surveil the sound are essential to its effective use. The precise angle of targeting is also essential as the beam needs to travel back to the exact receiving point in order to 'interfere' with the outgoing beam.

Since sound vibrations travel outward from the source in waves, the communication on one part of the window will be vibrating at a different part in the conversation than another part. As fast as sound travels, it's not instantaneous. Thus, the laser device must be firmly mounted on a very solid, unmoving base in order to precisely pinpoint one area of the window. One other limitation of laser devices is that they only work when aimed at surfaces that are firmly mounted and vibrate readily. Thick walls or moving objects do not make good laser targets. Laser listening devices are sometimes used in conjunction with other types of listening devices.

Laser listening devices are precision instruments requiring a tripod or other stable mounting base and are only effective under nearly ideal circumstances with clear line-of-sight, no precipitation, a precise focus and angle, and no interfering sounds. Commercial laser listening devices typically include a transmitter, a receiver, and an amplifier. Because they need to be solidly mounted and focused, they may be hidden inside camera or telescope housings to make them less obvious as listening devices, though there is still a likelihood that a camera aimed at suspect conversants may rouse suspicion. The mounting will usually have headphone/recorder connections and may also have a small speaker or a speaker connection. They are usually battery-powered, weighing about 15 pounds. The laser is usually tuned to the infrared spectrum (approx. 800 mm) in order to be invisible to unaided human eyes.

Sound Conduit Bugs

Sometimes bugs are cleverly installed against 'sound conduits.' In other words, the bug may not be in the room, but may be monitoring the sounds in the room through a heating duct or ventilation shaft some distance away. A metal vent can 'channel' the sound from a room quite well to a location several rooms away. Try talking through a long giftwrap tube to see how well sound will emanate from the far end of a tube or put a metal ruler against your ear and have someone whisper to the other end with his or her mouth close to the ruler. With some experimentation, you can get an understanding of how sound travels through ducts and materials. This, in turn, can aid in determining where someone might try to hide a bug.

Amplifying Microphones

Amplifying microphones are designed to pick up soft noises, whispers, animal sounds, machinery sounds, ticking, etc., in order to make them louder. These can be used to locate stowaways, burrowing animals, fugitives, leaking pipes, tunnels, bombs, and counterfeiting presses. Some amplifying microphones have special automatic 'gain' properties that allow them to boost the sound of a whisper without amplifying the other undesirable noises and loud sounds.

 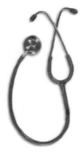 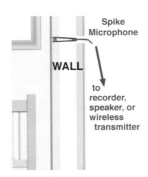

An acoustic listening device can be a simple glass that directs the sound to the ear or a more sophisticated device such as a stethoscope (middle) or contact/spike microphone (right). The Omnibus Crime Control and Safe Streets Act of 1968 has strict restrictions and penalties on listening devices used for unauthorized interception or disclosure. There are some exceptions for consenting parties and administrative or authorized law enforcement interceptions. [Classic Concepts photos ©2000, used with permission.]

5.b. Recording Devices

A recorder is any technology which provides a means to symbolize and imprint physical stimuli or events, so they can be reread or replayed. The more faithful the playback to the original event, the 'better' the recording. A recorder provides a way to document activities and store them for delayed or remote playback. Some recordings are transient or fragile, of only limited use, but most commercial recording devices are reasonably robust. Tapes are somewhat fragile, since the data can be damaged by magnetic interference; the data on newer 'hard' storage media such as CDs or DVDs are less likely to degrade over time.

Recorders are widely used to monitor activities, to provide a record of those activities, or to later analyze or evaluate the activities.

Audio recording is not synonymous with voice recording. Audio recording may include environmental sounds, footsteps, doors opening and closing, switches being thrown, or vehicles coming and going. While this information may not seem significant in itself, in the context of grounds security or an investigation, general activity levels, the timing of activities, or their presence or absence, may be important clues.

The most common recorders are audio and video recorders, though scent, tactile, and motion recorders have many applications, especially in scientific research. A seismograph is an instrument for recording earthquake characteristics and magnitude, but the same technology could be adapted to detect, track, and analyze movements within a building, within a vehicle, or along the ground.

Audio Recording Products

Commercial recording products come in many designs and price ranges, depending on their characteristics and features. As a general rule of thumb, the smaller the unit and the longer the recording times, generally the higher the price (except for professional desktop multitrack recorders). Typically, also, the greater the degree of automation, the higher the price, as they vary from manual to fully automatic. These are the basic types of recorders:

manual The recording occurs only when activated, usually by a human or a tripping mechanism.

manually activated, delayed The recording is manually activated, but delayed (as in setting a timer on a camera that causes the picture to be taken a few moments later). This allows the person to exit the scene without wasting recording time.

automatic timed The recording unit records on a timed basis, which may be scheduled, delayed, or random. Scheduled recorders may also be set to turn off at a certain time or after a certain interval.

automatic triggered The recording unit records in response to a signal or trigger, such as voice, speech (a particular spoken command), touch, motion, light, or a particular tone. In order to prevent constant on/off recording in a case of a conversation, there is often a built-in delay that the system will continue recording for a few seconds after the sound stops (since it may begin again after a pause).

Analog versus Digital Recorders

Digital recorders are becoming more popular, but are still limited by recording times, unless they are larger units equipped with hard drive storage. Removable PC Cards, similar to those used in digital cameras, can be used to increase recording times on smaller units, but have the same disadvantage as tape–the cards have to be exchanged. At the present time, handheld tape recorders usually record up to about 120 minutes, whereas digital recorders usually record up to about 20 or 60 minutes. With improvements in memory capacity, digital recorders will probably eventually supersede tape-based recorders and provide longer recording times.

Analog recording has the advantages of availability and low cost. Digital recording has the advantages of compression, quick upload to a computer system for storage, and opportunities for immediate or remote analysis of the data. Digital recording also creates an opportunity to edit the data. In the simplest case, unnecessary noise or information may be filtered out. *However, in terms of the integrity and admissibility of the information in court, it is very difficult, and sometimes impossible, to determine if digital data have been 'doctored,' that is, altered to serve the interests of the party doing the recording.*

Now that digital recording devices are becoming less expensive and more sophisticated, there is the probability that programmers will design 'smartcorders' that can selectively play back recordings according to a list of 'rules' based on priorities. Since humans are obsessively concerned with recording everything and since it's impossible to predict what might happen at any given moment in time, we store the recordings because a sound that's insignificant today might be highly significant tomorrow. However, there isn't necessarily a net gain. We may be creating a storage and playback nightmare in which we end up spending more time listening to tapes, managing archives, and searching for information, than being productive and encouraging people to take responsibility for their actions.

How can we alleviate these storage and retrieval problems? One way is to not to make the recordings in the first place. The gain in information (and prosecutions) in some cases may not gain enough to offset the expense, time, storage, and operations costs that are associated with constant monitoring. It's sometimes cheaper to hire a security guard to patrol the premises than to hire a technician to keep the system running and an archivist to swap tapes, put on labels, and manage a library full of recordngs. The second way is to create smart recording devices that can prioritize and play back the portions that are more likely to be significant (unusual patterns, schedules, sounds, or sound levels). This may not work in all circumstances, but if it worked in 80% of cases, it could save an enormous amount of clerical work and expense. With digital technologies, the idea is particularly feasible, since a digital recording device can selectively play back specific parts of a recording without winding through a long tape to find it.

Commercial Recorders

Cassette tape recorders are one of the oldest recording technologies, and still one of the most common in covert activities. A high proportion of gathered intelligence arises from verbal communications between individuals recorded on tape.

Portable recorders are usually $25 to $60. Miniature recorders range from about $30 to $400, depending on features. High quality desktop recorders can range from $350 to $5,000, with high-fidelity, multiple-track recorders at the high end.

Many miniature audio recorders are voice-activated, or activated by removing a pen, or other common implement, in order to hide the fact that the recorder has been activated.

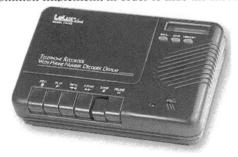

Left: This portable desk recorder begins to record automatically when the call is initiated. It has two speeds to provide longer recording times and works on AC or DC power. It decodes a number that has been dialed and records the information on the tape. Lelux Electronics Ltd. Middle: A palm-sized message recorder with LCD status display and timer alarm can randomly search and delete recorded messages. Headphones, external microphone, and IC memory cards are options. CASIL Research & Devel. Co. Ltd. Right: This tiny keychain digital recorder uses 4 button-cell batteries and records up to eight seconds for quick messages or important reminders. CASIL Research & Devel. Co. Ltd. [Classic Concepts ©2000, used with permission. Supplier information courtesy www.globalsources.com.]

Units that record on tape usually advertise the *total recording time*. Keep in mind that the *unattended* recording time is usually half of that claim. That is, if the manufacturer claims six hours of recording, it usually means turning over the tape after three hours. Some units are equipped to record on both sides without handling the tape, but these are in the minority and are usually a little bigger to accommodate the extra mechanisms. They also tend to be more expensive.

Telephone-answering machines are one of the most common devices used to record phone calls. They are equipped with a variety of features, including the ability to decode outgoing phone numbers and store them on the tape along with the recording or, if the Caller ID service is available, to store incoming numbers in association with the message. Time and date functions are usually supported. These features make it possible to create a record of business transactions, without having to type or handwrite notes about the time or party called.

Digital tape recorders are becoming more popular, but are still limited by recording times, unless they are larger units equipped with hard drive storage. Removable PC Cards, similar to those used in digital cameras, could be used to increase recording times on smaller units, the disadvantage being that they have to be manually exchanged.

Analog recording has the advantages of low cost and wide distribution. Digital recording has the advantages of compression, quick upload to a computer system for storage, and opportunities for analysis of the data. Digital recording also creates an opportunity to edit the data. In the simplest sense, unnecessary noise or information may be filtered out. In fact, many of the new tiny recorders have computer interfaces. You could theoretically upload the data, shuffle it around, delete, insert, and reload it to the recorder.

Many miniature audio recorders are voice-activated, or activated by removing a pen, or other common implement, in order to not call attention to the fact that the recorder has been activated. Others are hidden inside calculators, cell phones, pencil sharpeners, smoke detectors, clocks, and pagers.

Left: The Samsung SVR-P700 digital pen recorder weighs 41 grams and uses a AAA battery which lasts about 4 hours. It will record up to 70 minutes on flash memory and interfaces with other electronic devices. It can be reviewed with an earphone. The street price is around $200. Middle: This digital pen recorder looks like a gunmetal business pen and comes in models to record 40 or 20 seconds in two channels or 10 seconds of audio. Union Electric Corp. Right: The Samsung Voice Stick digital recorder records up to about 4 hours on 8 MB flash memory cards and is PC-compatible with a high-res LCD display for about $200. The voice-activated Micro-bar digital audio/phone/cell phone recorder with up to 8 hours recording is around $600. [Classic Concepts diagrams ©2000, used with permission. Supplier information courtesy www.globalsources.com.]

Remote Monitoring through a Phone Line

There are also systems that use the remote telephone itself as the listening device. In other words, you can attach a device to the phone line, call the number and listen to a conversation going on in the vicinity of the phone that was called or the device itself, if it is self-contained. Some of these systems will even defeat the ringer so that people at the remote location don't know that a phone call has come through and are not aware that the sounds in the room are being monitored. They can further be used to monitor the conversations of someone who has just hung up the phone. The listening range is usually up to about 20 to 30 feet from the phone depending on the design of the room and other ambient noises. These systems generally require that a device be attached to the remote phone line, necessitating access to the premises. There are legitimate and illegitimate uses of this technology. Some people use them to monitor a business after hours or to check on the activities of teenagers who have been left at home for a short while.

A similar device is a phone-hold monitor. This device monitors a remote phone conversation, but only when the hold button has been pressed at the other end. Since people often comment on the conversation at hand while on hold, the phone-hold tap can be revealing.

Most recorders work off the telephone power source, making them detectable by normal bug-sweeping procedures. They retail for about $200 to $500 with the more expensive units capable of monitoring more than one line. Many are self-contained, resembling a telephone junction box, and do not require a telephone (just a live telephone line) for operation.

When an audio or video recorder is placed covertly on a human, it is called 'wearing a wire' or 'being wired.'

Recorder Accessories

Switches that allow a regular tape recorder to be sound-activated can be purchased in kit form for about $10. They usually interface through the remote plug on the recorder.

5.c. Monitoring Phone Communications

Wiretapping involves the use of a device to access a conversation at some point in the physical connection related to the call, traditionally through a phone wire or cable. In recent years, the term is used more generically to include eavesdropping on both wired and wireless communications. The tools to tap a conversation have become more varied and sophisticated as electronic technologies provide ever-increasing ways to communicate using wires or radio waves in various forms.

Personal and business recording of phone conversations typically occurs at or near the phone being monitored, since the activity is usually consensual or illegal. Law enforcement tapping typically occurs adjacent the premises (in order to avoid issues of illegal trespass, search, or seizure) or at the local phone switching station (with the cooperation of the communications carrier).

The telephone in its most basic sense consists of two speaking/listening devices connected to each other by a string or wire, or other filament capable of conducting sound. For many years telephones were not much more than a power-amplified enhancement of this basic scheme. Thus, it doesn't take much effort or sophistication to attach a third listening device somewhere along the length of the conducting wire, to listen to the conversation at hand or to use a tape recorder to record it. This is the essence of phone tapping.

Pen Registers

The evolution of electronics has opened the door to other types of 'taps.' Sometimes the eavesdropper isn't listening to the content of the conversation, but rather is interested in who is being called or where or when. Since the advent of touchtone phones, this information is easy to determine, and there are handheld acoustical phone decoders (pen registers) that will log and display the numbers that are being called from a selected phone. Pen registers are used in business and law enforcement activities.

Communications Logging Devices

A logging device is one that keeps statistics on various aspects of calls, including the time, date, and duration of the calls. High-end pen registers, called pen register/loggers, sometimes have this information in addition to the numbers dialed. Some of these units can be combined with Caller ID services to provide additional information.

Basic Telephone Listening and Recording Devices

Analog phone technology, at its heart, is not complex, so listening to or *tapping* a phone conversation often doesn't involve high cost or complicated equipment. The conversations that occur on most systems consist of 'raw data,' words that are not scrambled or coded or spread over several frequencies. However, as phone technologies become increasingly digital (and increasingly wireless), tapping becomes more difficult. Digital technologies allow more opportunities for encryption and transmissions that can hide, obscure, or scramble the conversations.

The most common form of phone voice recorder is the pervasive telephone answering machine. Most answering machines will record two sides of an ongoing conversation simply by activating the record mode or memo mode with the touch of a button. There are laws protecting conversants from covert recording, so many answering machines are equipped with beeping tones to inform the person on the other line that the call is being recorded. In some regions one party must consent to the recording; in other regions, both parties must consent.

Some units have an option to disable the beep, as the beep is annoying if it is a legitimate call and both parties have agreed to the recording (e.g., a long-distance business transaction).

Wiretapping Devices

Some wiretaps are set up at the local switching office with the assistance of the service provider, but other (usually illegal) taps occur in the vicinity of the phone being tapped. Taps also occur legally in work environments at the discretion of a business owner.

Most nonanswering-machine sound recorders consist of a control circuit, which is either integrated into the recording device or housed in a separate box, and a transmitting or recording mechanism. These are essentially wiretapping devices. Most of the lower-priced consumer units are configured for single-line phones. Sometimes the control circuit and recording unit are integrated into the phone itself. Most voice recording units are advertised for their 'silent operation,' that is, no beeping tones are emitted while the call is taking place. Prices range from $25 to $300+. Units with good sound quality, dual-speed, long-play, and voice-activation features are in the $150 to $250 range. Those with extra features, such as call logging and enhanced CallerID displays are in the $300+ range.

Most wiretapping devices connect in series with the phone line, but there are some activating mechanisms that are connected in parallel. Some taps are not physically connected to the line. They are connected just adjacent to the line (or surrounding the line without touching it) and pick up electrical emanations from the line. This requires sophisticated equipment with noise filtering and gain to clean up and amplify the signal. The quality of the sound may not be as good as regular wiretapping devices, but the chance of detection is much lower and less expensive tap-defeating systems do not affect them.

Premium units that only record when actual conversation takes place, rather than during the entire duration of the call, provide longer recording times. Longer recording times can also be accomplished with combination digital/analog systems, where there is a digital buffer for the conversation which is then stored to analog tape without the long gaps or pauses that are common to conversations. Hybrid digital/analog systems can also interpret the touch tones to intercept and record the number dialed, and display them to a visual display incorporated into the unit, similar to a CallerID display. When hooked into a computer system, a high-end voice recording system can include databasing of the conversations, voice stress analysis, and other evaluations of the data. With banks of high-capacity storage devices, virtually unlimited recording is feasible.

A basic voice-recording model comprises a cigarette-package-sized box with a power switch, RCA audio jacks to hook to a recording device, and an RJ-11 phone jack to connect to the phone line. Longer recording times are possible with some simple modifications to the drive mechanisms of many recorders. Some systems are configured with two or three recording speeds. Slower speeds usually result in lower quality audio. Most tape-based surveillance units will record from one to twelve hours. Digital microcassette units usually record from 10 minutes to about an hour. Computer peripheral units usually connect through RS- serial or USB interfaces.

Most phone voice recorders are designed for analog systems, and are generally only able to detect a single line. Depending on the location of the unit they may or may not record calls taken on extension phones to that line. Thus, many of them don't operate on private branch exchange (PBX) lines in offices and institutions. To overcome this limitation, there are inexpensive units that connect between the phone and the handset to monitor the specific call taking place. Since these units can be seen by passersby, they are not suitable for covert recording.

Another device for digital PBX and other multiline systems is an adaptor that converts the signal from digital to analog, thus allowing recording devices, modems, and other analog equipment to be used. Adaptors are small enough to fit in carry-on luggage or in a laptop carrying case pocket, and sell in the $140 range.

The simplest phone taps include two leads that are hooked directly to the phone line on the one end, and attached to a listening device on the other end (like an earpiece or headset). A telephone lineworker's test set, which looks like a telephone handset with a dial and two wires hanging out, is a common piece of equipment that can be purchased almost anywhere. The test set can be connected to a phone box to monitor an ongoing conversation. Phone taps can be designed to record the conversation to an attached recorder, usually a simple cassette recorder, or can transmit the conversation without wires to a listener nearby, such as a receiver in a van half a block away, not unlike the 'spy vans' commonly seen on television shows. A surveillance van is usually equipped with a high quality antenna (sometimes several antennas) to intercept the signal.

As with all listening devices, there are ways of detecting that the phone is being tapped, and there is always the possibility, if the conversation is transmitted elsewhere, that some ham radio buff in his attic or basement may pick up the signal and blow the whistle on the eavesdropper.

Wiretap Accessories

It is possible to build components which attach to the remote and microphone jacks of standard or miniature tape recorders to cause the tape recorders to automatically start recording when the phone is picked up. Kit price is about $22.

Audio Transmitters

Wireless transmitters are those which convert the audio into radio signals and back to audio again at the receiving end. FM transmitters are attached to the phone line or hidden inside the phone. Most of the line transmitters are connected in series to one of the two copper wires that typically attach to a phone. The phone line is both the aerial and the power source (which makes it vulnerable to detection). These usually transmit from about 100 to 300 feet. An FM receiver tuned to the same frequency is needed to receive the signal. Kits sell for about $20 for both the transmitter and receiver. Longer range transmitter kits with tunable frequencies are about $30.

Sound transmitters are microphone/transmitting units that are essentially the same as intercom systems. They can be one-way or two-way, wired, wireless, or fiber optic. Wireless systems sometimes use the building wiring as an aerial. Intercoms are usually about $30 and kits, which may be smaller and less visible, about $20.

Basic fiber optic audio links consist of a microphone, a length of fiber-optic cable, and a speaker or connection to a recording device. While the microphone will still generate electrical disturbances, the cable itself does not and thus is less vulnerable to detection than electrical cable systems. Distance depends on the construction/style of the cable. Kits are about $35. Two-way fiber-optic system kits are about $62.

Stethoscope-style transmitters are based on the idea of using the listening portion of a stethoscope (the sensitive endpiece) in a suction-cup shape so it can be fastened on outer surfaces without entering a premises. It includes a transmitter (usually FM) to transmit up to about 500 feet.

Watch phones like those depicted in spy comics in the 1960s are now a reality. Tiny programmable phones that are worn on the wrist with digital displays that provide about 90 minutes of conversation and about 60 hours of standby are being distributed by Samsung featuring speech-recognition technology by Conversa, a Redmond-based company. [Conversa news photo 2000, released.]

Video and Digital Recorders

Video camcorders can be used as audio surveillance devices when other more specialized audio devices aren't available. The lens cap can be left on if video is not desired. Recordings on video tape have good quality sound and can be further enhanced by attaching a quality condenser microphone, parabolic antenna, or wireless audio receiver to the camcorder through a connecting cable.

5.d. Audio Changing or Jamming Devices

Sometimes confusing a bug is easier than finding and disabling it. Devices that protect against audio eavesdropping in this way usually consist of devices that generate a range of frequencies that run through the human audio spectrum to 'muddle' the vibrations and disturb any audio receivers that may be monitoring the location. The lower frequency usually starts around 5 to 20 kHz and the upper frequency is usually around 1,300 to 20,000 KHz (human hearing is about 20 to 18,000 KHz, depending on the age of the subject).

5.e. Institutional Phone Tapping

In movies and television shows, detectives and government agents are commonly shown tapping into personal and criminal conversations from receiving units in vans and stakeouts in abandoned buildings. In actual fact, there is a lot of paperwork involved in getting permission to tap. Timing is also of considerable importance in wiretapping for law enforcement purposes as permission to tap is specifically restricted to certain types of alleged crimes, usually violet crimes and drug-trafficking-related crimes. These crimes can often take two to five years to 'crack,' whereas permission to tap may last about 30 days. (See the history section in this chapter for the regulations and debates about the legality of wiretapping.)

In order to tap, law enforcement agents must go through a process more complicated than obtaining a general search warrant. In general, they must:

- obtain approval by the Attorney General, or his or her current acting agent,

- apply to a local U.S. Attorney to apply to the appropriate court,

- provide very specific information in the wiretap request. It must not only show probable cause for the tapping operation, but also must list some specific terms such as the identity of the person or persons being tapped, the location of the tap (with exceptions for roaming conversations), and even the types of conversations that might be expected to be accessed, and

- turn off monitoring if there are lengthy conversations on topics unrelated to the investigation.

All these precautions are in place because the tapping of a phone intrudes not just on the privacy of the suspect, who may be discussing issues unrelated to the suspected crime (and who is innocent until proven guilty), but also invades the privacy of the other people with whom the suspect is conversing, who may be innocent of any wrongdoing.

Once a law enforcement agency has permission to tap, the process is not over. It is usually necessary for the law enforcement agency and the phone service personnel to cooperate in establishing a tap, especially now that electronics have become more varied and complicated. With dozens of new ways to place calls on cell phones, Internet phones, and radio-based personal relay technologies, the technical challenges of tapping a call have greatly increased. Spread-spectrum phones also make it more difficult to access a conversation in progress, because the frequencies are changed and may further be encrypted. These systems are less vulnerable to both jamming and eavesdropping and they are now available to consumers for under $150.

Phone Conversations - Tracing a Call

Sometimes the source of a message is as important as, or more important than, the content of a message. 'Tracing' a line is another aspect of telephone surveillance which is commonly depicted in films and TV. Tracing involves identifying the source of the call. The methods for tracing wired communications are somewhat different from tracing wireless communications.

Prevalence of Law Enforcement Taps

The U.S. Department of Justice Criminal Division reviews a little over 1,000 wiretaps per year and provides assistance to federal enforcement agencies on the use of emerging technologies. The number of taps per year has increased very gradually over the last decade or so, but the increase is not statistically significant in relation to the increase in population in the U.S. The number of calls monitored with each tapping authorization varies greatly but may reach as high as 2,000. Historically, law enforcement has also contracted to outside detective agencies for various surveillance services and there are no records of how much information obtained from private detectives may be from tapping or bugging activities, if any.

There are two exceptions to the typical wiretap authorization requirements. The President can, following a declaration of war, authorize a wiretap for foreign intelligence for up to fifteen days without a court order through the Attorney General. Also, if the communications are exclusively between foreign powers or involve intelligence other than spoken communications from a location under the exclusive control of a foreign power there may be some leeway in tapping. (As the economy and politics of the world become more global, it is almost certain that U.S.-initiated foreign taps will be scrutinized and criticized by allies and other nations.)

At the present time, the Attorney General must inform the House Permanent Select Committee on Intelligence and the Senate Select Committee on Intelligence of all wiretap activity. The information is classified, but the Attorney General must provide the Administrative Office of the United States Courts with an annual report.

Who Gets Tapped?

Two-thirds of court-authorized law enforcement taps are related to drug-trafficking operations. The other third is mostly related to investigations of wide-scale fraud, such as Medicare/ Medicaid scams, military-contractor fraud, and situations in which law enforcement officers or politicians are found to be involved in illegal schemes. Wiretaps lead to just over 1,000 convictions per year, an average of about one conviction per non-FISA tap.

In the mid-1990s, the average cost of installation and monitoring of an institutional wiretap was almost $70,000.

An individual or corporate spy can probably find out what she or he wants to know through just listening in to phone conversations, but law enforcement agents need concrete evidence. They have to record the calls for the wiretap information to be useful in court. The calls must be recorded so there is no doubt as to its source or authenticity and no indication that the information has been altered in any way.

Given all this bureaucratic procedure, it is far more likely that you are being tapped by your Little Brothers: friends, neighbors, or business competitors, than by law enforcement agents. It is pretty easy for a neighbor to listen in to conversations on a party line or through a radio phone tuned to the same frequency as your cordless phone, or for a kid down the street to hook into your external line with parts from a new electronics kit he got for his birthday, or for an ex-spouse to attach an illegal tap to your line that was purchased for a few dollars on the Internet. In fact, in many cases, employers can legally tap into your calls at work with nothing more than a brief memo or mention that she or he might be 'monitoring' employee calls.

5.f. Covert Listening Device Countermeasures

People involved in high stakes activities, e.g., business deals, often suspect that they are being bugged, even when they aren't. People having relationship problems sometimes bug one another. Criminals seeking to avoid law enforcement agents sometimes use bugs, taps, and radio scans to monitor their movements to avoid capture. Criminals sometimes also bug a location in order to 'case the joint,' that is, to gather information, to determine when to enter the building without being detected or when security personnel are absent.

If there is reason to believe there are covert listening devices installed in a building or vehicle, there are a number of steps that can be taken to try to locate the bugs. Some people choose to hire a professional technical surveillance consultant who usually has the equipment to search for a variety of types of bugs; others try to do it themselves.

Success in detecting listening devices depends in part on making some good guesses on the type of device that is being used. Since most covert devices are electronic 'bugs' or wiretaps, most of the detection devices that are available are designed to locate changes in the electrical properties of a wireline, transmissions through the air, or electrical anomalies in the vicinity of an eavesdropping device. The process of locating bugs is called *bug sweeping* since many devices are swept through the air in much the same way that a metal detector is often swept back and forth along the ground.

Bug sweepers vary in type and sophistication. Some can only detect a bug when it is transmitting and are essentially scanners that seek a stronger signal within a range of frequencies, a pretty limited type of device, but since many bugs are purchased as hobby kits or through Internet dealers, they share common design features. These consumer bugs tend to send out signals that are in the standard FM broadcast bands or just below the frequencies of the standard FM broadcast bands. Knowing this makes it easier to detect this common type of bug. (There is more information about wireless transmitters/receivers and common broadcast frequencies in the Radio Surveillance chapter.)

Detecting Bugs

Because of legal restrictions and FCC requirements, the majority of audio transmitting devices that are used as small 'bugs' transmit in the FM broadcast ranges. Frequencies around 73 MHz are common. The biggest disadvantage of these as covert devices is that the transmission can be picked up by anyone scanning through channels on an FM radio or specialized receiver within the range of the transmitter. Transmitters which are designed for 'educational purposes' as electronics kits or from schematics or which are being used illegally to eavesdrop sometimes include FM transmitters that use frequencies outside the regular broadcast ranges. These require a special scanner or receiver tuned to the corresponding frequency. Nonbroadcast frequencies are less likely to be accessed by someone carrying a 'boom box' nearby, but they can be picked up by a nearby FM scanner. Since FM transmitters send a signal in all directions, they are highly vulnerable to detection.

Other frequencies that are typical include VHF, around 180 MHz (±50), and UHF, around 650 MHz (±300) with FM modulation.

There is a 'metalevel' of thinking that a lot of people overlook in surveillance and countersurveillance activities. Never forget that solving an eavesdropping puzzle is solving a logic puzzle and a psychology puzzle. Physically hunting for the bug isn't necessarily the best way to find it. Getting the person in the room responsible for planting a bug and watching their eye movements, or where they ask you to sit can be clues. Checking credit card statements for recent purchases from electronics retailers can be another. If you find out what type of devices the vendor sells, it's easier to choose the right type of bug-sweeping device (note, the legality of these actions varies with the circumstances). Many people shout 'hallelujah' with relief on finding a bug and overlook that fact that there may be others (or that they can be reinstalled). Never assume there is only one bug and never assume that multiple bugs are of the same type. Be suspicious of renovations, empty rooms next to a room that is suspected of being bugged, or furniture that has been shifted.

Because bugs may be difficult to find, some people choose to counter a bug, not by removing it, but by defeating its acoustical properties. Voice changers, voice scramblers, and noise generators are devices that alter or confuse the local sounds to make it difficult for them to be recorded or interpreted. High frequency radio waves can be used to temporarily or permanently disrupt nearby electronic devices (not feasible if sensitive components are nearby). Other people simply look for wide open spaces in which to carry out confidential conversations, like golf courses (and then make the mistake of hiring a questionable caddy or renting a bugged golf cart).

Sometimes metal-detecting wands and physical 'pat-down' searches are used to see if someone is 'wearing a wire' when entering a premises or the area where a transaction is about to take place. Public transportation systems (e.g., airlines) now routinely use walk-through access devices and wands to detect bombs and other weapons, but these devices can sometimes also be used to detect body-worn bugs and recording devices.

Detecting a bug that is hidden in a building or vehicle is generally a four-step process consisting of:

- making some common-sense preliminary guesses on where the bugs might be and what type they might be (if you're wrong, you back up and try other strategies)

- doing a preliminary visual search for changes or unusual aspects to the premises or vehicle (new paint, a small pile of dust, a crooked picture)

- discerning the frequency on which it is transmitting (wireless bug) or anomalies in the electrical signal strength or characteristics of the transmission (wireline bug), and

- finding the actual location of the bug or wiretap.

One type of bug that is more difficult to detect by these methods is a 'light bug' or fiber-optic device. Light doesn't emanate from a cable in the same way that electricity emanates from a cable. The weak point of fiber optic devices is that the cable itself has to be hidden somewhere and simple visual inspection, various ultraviolet lights, penetrating x-rays or other surveillance devices may detect the physical presence of a cable under a carpet, in a wall, or above the ceiling tiles (a very common place to run wires). Another light device is an infrared transmitter/receiver. Infrared transmitters can be designed to convert sound to light and the receiver can turn the signal back into sound again through a device that is worn or held near the head and which may have an earphone for quiet or covert listening. Such devices exist for the hearing-impaired and might also be appropriate for certain covert operations. These devices have three limitations: they are limited in range, they require an unobstructed line of sight, and the beam can be detected with an infrared sensor. They are useful, however, in noncovert surveillance in which radio waves are not an option (perhaps because of interference) and a clear line of sight is available.

A test sound is often generated in an area that is being swept for bugs to stimulate the microphone into electrical activity in order to detect its presence or measure its properties. Handheld computers are sometimes used in conjunction with bug sweepers to display statistics or mathematically analyze phase differences that can aid in finding the actual location of the bug.

Wiretap Countermeasures

There are a large variety of devices designed to detect wiretaps and bugs. In traditional landline phone conversations, the phone is powered by the line itself. Many tapping devices will use power from this line and the slight effect on the power in the line can be detected with the right equipment. Newer or more expensive tap devices are designed to create a minimum disturbance to the line to which they are attached. Bugs are usually powered by batteries.

A general understanding of phone system wiring and accessories is helpful in locating telephone taps. Many taps masquerade as common consumer jacks and accessories or are built into standard store-bought accessories.

Frequency counters, bug-sweeping devices for seeking out wireless transmitters are available as portable *scanners,* capable of scanning through a wide range of frequencies. Some come with LCD indicators, others with status lights. More sophisticated models can also keep a log of time and date and location through GPS and many personal computer interfaces. More expensive models usually scan over a wider range of frequencies.

Many telephone 'inside-premises' tapping devices are either built to look like normal phone accessories or they are hidden inside standard working phone accessories. These diagrams show some that are particularly vulnerable to tampering. Wall plates can be removed, the wiring altered, and replaced. Phone bugs are now so small, they can be hidden inside standard splitters and adaptors. Junction boxes and bell ringers are easy receptacles for hiding bugs, as they have room to spare and tampering is rarely noticed. [Classic Concepts photos copyright 1999, used with permission.]

Bug detection kits will often come with a variety of swappable sensors, called *sondes* to detect radio waves, infrared radiation, etc. More sophisticated systems with computerized readouts may require technical expertise to interpret the information. Professional systems will sweep through a wider range of frequencies, some as high as 4.5 GHz and can detect scrambled signals or spread-spectrum transmissions. They check for AM/FM radio transmissions, sub-carrier, carrier-only, SBB, and DSBSC signals. Most consumer models sweep up to about 2.4 GHz and may not be able to detect spread-spectrum transmissions.

Telephone handsets and telephone answering machines can also harbor bugs and taps. The mouth and earpieces of the older rotary phones are easy to unscrew in order to insert bugs. Deskset phones are usually accessible by screws in the bottom. Telephone answering machines can be readily opened and modified or equipped with a bug. [Classic Concepts photos ©1997, used with permission.]

High-end tap-detecting systems are significantly more powerful and versatile than most of the consumer bug-sweeping devices. Since the majority of taps are unsophisticated, they are not especially difficult to detect with experience and basic equipment. However, there are circumstances where the tap is miles from the premises and designed to be difficult to detect. It is difficult to physically trace the connection to this type of wiretap and it is sometimes necessary to use diagnostic equipment that can trace the phone line status from the premises all the way to the local telephone provider. This type of equipment can also detect anomalies at the junction boxes without physically inspecting them (though an inspection is recommended, whenever possible). But even with thee devices, there are limitations in trying to trace a line that is within a private branch exchange (PBX). For these, the status of the lines inside the exchange has to be tested and then the status of the lines leading into the exchange must also be tested. The price range for more sophisticated bug detectors is from $1,000 to $4,000.

Some cellular phones are now being equipped with detectors for transmitters that will sound a tone or cause the phone to vibrate to indicate there might be a bug nearby.

There are telephone 'guard' systems that are designed to detect and react to anomalies on the phone line, deactivating not only a large number of common wiretap devices but also deactivating the automatic recording feature on many phone recorders. Most of these systems are based on an electrical activity 'reference' and thus must be calibrated to a line when installed in order to detect future anomalies.

Warning Signs

How can you find out if your phone is bugged? There are general warning signs and specific measures you can take to prevent bugging.

- The first warning is if people seem to know things they shouldn't. This is not an electronic countermeasure, it's a common-sense countermeasure.

- Are there changes in the phone line while you are engaged in conversation? These can include unusual sounds, volume changes, popping noises, static or sound emanating from the phone when it's not in use.

- Do you get calls where the phone rings, but no one is on the line and you hear unusual sounds?

- If you have a radio near the phone that acts strangely, there may be a bug in the vicinity of the radio (or television).

Things to look for include junction boxes with extra wires, or messy wiring, decoy 'boots' (the dark protective sleeves on wires near telephone poles) that look like real ones but actually hide wiretap components and transmitters, and phone taps hidden inside common commercial phone components. [Classic Concepts photos ©2000, used with permission.]

Some tapping devices are wireless, transmitting the conversation using radio frequency (RF) signals to another location rather than directly recording or amplifying it at the source. These can be detected with 'scanners,' devices designed to scan through a range of radio frequencies to detect an outgoing signal.

Some wiretap detection devices can only detect a tap when it is active. There are also devices designed to detect audio bugs even when they are inactive, in walls or on phone lines. Non-linear junction detectors are one example. Some devices are based on detecting and analyzing harmonic levels to determine whether a signal is originating from an electronic or nonelectronic source.

Some private branch or multiline phone systems are sold with built-in security by including *telephone-line analyzers* as part of the hardware. These systems can scan up to a couple of dozen phones (or more) to check for taps or transmitters associated with the lines. They are not infallible, but they can serve as a deterrent.

Since phone lines are also used for other types of communications, e.g., fax machines, there are also systems for encrypting or otherwise scrambling the signal prior to transmission. Thus, voice scramblers and fax encrypters are available which may be built in or portable/handheld. Voice-changers are also available. They don't encrypt the content of the message, but they change the sound of the voice to make it difficult to identify the person engaged in the call.

To defeat wiretapping equipment, some systems will constantly broadcast noise or nonsensical speech through the line. Thus, if a recording device is attached, the tapes or other memory buffer will fill up with hours and hours of unusable signals.

Sometimes bug-detecting devices are permanently installed on a secure line, with LCD displays describing any unusual occurrences or likely causes of anomalies. Systems that provide maximum security with line-monitoring, readouts, and noise broadcasting sell in the $600 range.

This book does not provide a detailed description of telephone switching systems or the detailed mechanics of telephone tapping. Its purpose is to provide a broad understanding of the technology and there are many references off and on the Internet that give precise details of the functionings of various telephone systems. A few of the more common concepts and glossary terms are included at the end, to aid in your understanding, but you are urged to consult technical references if you want engineering information on the various technologies mentioned here. See also the Radio Surveillance chapter for more information on wireless technologies.

6. Applications

Listening devices can be used to monitor children, the sick, or the elderly, the sounds of wildlife, or business transactions. They can also be used to detect emergency situations (car accidents, explosions, etc.) and to activate alarms, warning lights, or other emergency responses. In military applications, they can be used to detect the presence of hostile forces, bombs, and hostile troop communications. In law enforcement they are used to monitor illegal gambling, racketeering, drug trafficking, violent crimes, and insider trading activities.

7. Problems and Limitations

Wireless Transmissions

The biggest problem with wireless devices is that most of them transmit radio waves which travel in every direction and can readily be intercepted with scanners or other receivers tuned to the same frequency as the outgoing signal. This makes third-party eavesdropping or detection more likely than with wired transmissions. Encrypted or spread signals make content of the conversation more difficult to detect, but the presence of the communication can still be detected.

Installation

The installation of surveillance devices is becoming easier, as tiny, self-contained consumer models are manufactured and sold to the general public. There are still circumstances, however, where technical expertise is needed to install components (particularly certain types of beacons and wireless systems, as well as laser-based listening systems). Access to the inside or near outside vicinity of the premises is necessary for the majority of listening devices and may be difficult in some circumstances.

Wiretap Detection

Many phone tapping devices can easily be detected and defeated, particularly those which make a physical connection to the phone line and use power from the line. Checking for anomalies in the power usage or differences between pretap characteristics and post-tap characteristics of the line are common ways in which taps are detected.

Recording Times and Vulnerability

The vast majority of recording devices have limited recording times, or significant degradation of the quality of the recording with longer recording times. Most require that a tape be turned over or a flash memory card swapped out. The smaller the devices, usually the shorter the recording time. Larger van-based units, or room-based units can be hooked up to large recording machines or large computer storage drives to extend recording times.

Because recording units are larger than basic listening units, they are more vulnerable to detection. A body-worn audio wire is very small and easy to hide in clothing, but the transmitter or recorder can often be detected by a visual inspection or pat-down search. Room-based recording units and their tapes or drives are vulnerable if there is access to the room. Thieves have often been known to break into recording rooms to take the tapes or recording devices that have captured a break-in or vandalism.

Emissions

One of the problems with electronic surveillance devices is that they emit radiation that may interfere with other nearby devices or may be compromised by radiation from other sources. Since the history of surveillance technologies is somewhat shrouded in myth and mystery and since many of the devices are illegal, not all of them have been manufactured to FCC standards or have been through FCC testing. This is also true of the kits and components sold for educational purposes which people sometimes try to use in offices or homes. Inadequate shielding or proximity to electronics devices is usually the reason for problems with buzzing, static, or erratic functioning.

Change of Use

One of the biggest problems with audio surveillance is that once the equipment is in place, intentions change, and monitoring occurs for longer than was originally planned or the purpose of the monitoring changes. The planned destruction or distribution of the information may also be changed after the fact. This tendency to change use or overstep the bounds of the original intention to tap is one of the valid reasons why privacy advocates oppose many types of surveillance.

8. Restrictions and Regulations

Because wiretapping and similar activities have been available to society for some time, restrictions to safeguard personal privacy have been put more firmly in place than for some of the newer surveillance technologies. In terms of the recording of personal phone calls in the U.S., in some states, one party must agree to the recording; in others, both parties must agree. Regulations for employee calls are different, with employers having a certain amount of leeway to protect business interests. It is wise, when making a legitimate record of a call, to ask the recipient on tape if she or he agrees to the recording of the call.

Some U.S. restrictions and laws of particular relevance to audio surveillance and privacy include

Communications Act of 1934. Public Law 416. U.S. Federal regulations established to organize and regulate interstate and foreign communications for national defense and to promote competitive communications technologies and services. The Federal Communications Commission (FCC) was established in accordance with the Act. The Act was amended by the *Omnibus Budget Reconciliation Act* (OBRA) to preempt state jurisdiction. It organized wireless communications into two categories: commercial mobile radio services (CMRS); and private mobile radio services (PMRS), including public safety and government services. (Note: 1992 Public Law 102-385 amends this Act to increase consumer protection and increase competition in cable television markets.)

Above 890 Decision. A 1959 Federal Communications Commission (FCC) decision which permitted private construction and use of point-to-point microwave links. Private companies could now utilize frequencies above 890 MHz for communications which might be useful on oil rigs, remote power plants, research stations, etc. As microwave communications technologies improved, the FCC was increasingly pressured for access to microwave broadcast frequencies. MCI was the first private commercial carrier service to take advantage of this Decision.

Omnibus Crime Control and Safe Streets Act of 1968. Public Law 90-351. This established procedure by which law enforcement agencies could obtain authorization to conduct electronic surveillance. It required telecommunications carriers to provide the "technical assistance necessary to accomplish the interception". It further created the National Institute of Justice (NIJ), the research and technological development agency of the Department of Justice (DoJ). Section 1212 of Public Law 91-452 repealed section 804 of this Act and was subsequently again repealed and amended.

Consumer Communications Reform Act of 1976. Also known as the "Bell Bill" because AT&T had lobbied for restoration of its monopolistic domination of the market and tried to reduce FCC regulatory authority over long-distance communications competitors. This provoked hearings into the Act and competition in the communications market. The result was the *Execunet Decision* in 1977, opening the long-distance market to competing companies.

Foreign Intelligence Surveillance Act of 1978 (FISA). This Act established legal standards for the use of electronic surveillance for counterintelligence and the collection of intelligence related to foreign activities within the U.S. It provided legislative authority for wiretapping and other electronic surveillance of foreign powers within and without the country. It further established the Foreign Intelligence Surveillance Court (FISC) to review and approve surveillance which could be used to monitor U.S. Persons. Amended in 1994 to provide limited authority for physical searches. Review of cases was conducted by a Committee beginning in the 104th Congress.

Electronic Communications Privacy Act of 1986. (ECPA). This amends Title III of the Omnibus Crime Control and Safe Streets Act of 1968. In essence, ECPA extends existing restrictions on unauthorized interception of communications over traditional media to cover electronic communications. It does not extend prohibitions in cases where one of the parties consents to the interception and does not extend the right into some work-related communications (there are employer exceptions that permit monitoring). There are also some exemptions for communications providers to permit system administrators to manage and troubleshoot the system. A number of state laws are patterned after the ECPA.

Communications Assistance for Law Enforcement Act of 1994 (CALEA). Public Law 103-414. This is sometimes referred to as the 'Digital Telephony' law. It requires that telecommunications providers ensure that law enforcement agents can execute court-authorized wiretaps. In many cases, this requires physical changes or upgrades to the providers' equipment.

Digital Communications and Privacy Improvement Act of 1994. 25 October 1994, signed into law by President Clinton. This was to ensure continued ability of law enforcement officials to conduct court-authorized electronic surveillance.

Comprehensive Counterterrorism Prevention Act of 1996. Signed into law by President Clinton on 9 Oct. 1996. This is a strategy to improve security in federal buildings and aircraft cargo holds. It also authorizes relocation of U.S. forces in foreign stations at high risk for terrorist attacks.

Communications Decency Act of 1996. A provision of the Telecommunications Reform Act that erupted in controversy as to definitions of 'lewd' or other materials that were being promoted as criminal because they might be objectionable to the general public, yet were considered acceptable within the more open climate of Internet communications. The Act was contested and, in June 1997, declared unconstitutional and in violation of individual rights of freedom of speech.

Telecommunications Act of 1996. The first substantive overhaul of the Communications Act of 1934. The intent of this Act is to enable open access to the telecommunications business and to permit any business to compete with any other telecommunications business. The primary impact was on phone and broadcast services with responsibility shifted away from state courts to the Federal Communications Commission (FCC), while much of the administrative workload remained with state authorities. It made it possible for Regional Bell Operating Companies to provide interstate long-distance services and for telephone companies to provide cable television services. Cable companies could now also provide local telephone services.

United States Title 18 - Crimes and Criminal Procedure, Part 1 (Crimes), Chapter 119 - *Wire and Electronic Communications Interception and Interception of Oral Communications.* This chapter covers the interception and disclosure of wire, oral, or electronic communications; manufacture, distribution, possession, and advertising of wire, oral, or electronic communications intercepting devices; confiscation; use-as-evidence prohibitions; authorizations; procedures; reports; recovery of civil damages; and injunction against illegal interception.

Unites States Title 18 - Crimes and Criminal Procedure, Part 1 (Crimes), Chapter 121 - *Stored Wire and Electronic Communications and Transactional Records Access.* This chapter covers unlawful access; disclosure of contents; requirements for government access; backups; delayed notice; reimbursement; civil action; exclusivity of remedies; counterintelligence access to records; and wrongful disclosure.

Communications Assistance for Law Enforcement Act. Public Law 103-414, enacted by the 103d U.S. Congress. Invokes assistance from the telecommunications industry to provide technological solutions for accessing call information and call content for law enforcement agencies legally authorized to do so.

There are many state laws of interest, so these are just examples:

- New Jersey Wiretapping and Electronic Surveillance Control Act
- Pennsylvania Wiretapping and Electronic Surveillance Act

See the Radio Surveillance chapter for information pertinent to wireless communications and the Introduction for information on laws of general relevance to surveillance.

9. Implications of Use

There appears to be an enormous market for bugs, recorders, bug-sweeping devices, amplifiers, and surveillance consultants judging by the number of vendors and products. The sheer sales volume of taping devices alone suggests that phone tapping is a widespread activity, despite abundant restrictions and regulations. This conclusion is further reinforced by the fact that vendors emphasize the 'quiet operation' of recording and listening units, such as the absence of warning beeps to notify a conversant that she or he is being recorded. The steadily decreasing cost of these devices makes it easy for people to consider their purchase.

The Dangers of Call Monitoring

The incidence of recorded calls, both covert and otherwise, is increasing. Many high-tech firms are now recording or listening in on customer product inquiry and technical support calls. Since the caller must be informed that the call may be recorded (in some areas, the caller must be explicitly asked if recording is OK), there is usually a message like, "This call may be monitored for quality control". Since this seemingly innocuous message is commonplace, many people no longer consciously realize that it means the call is being recorded, or may forget during the progress of the call, especially if they are made to wait for a long time while on hold.

This trend may lead to much broader use and acceptance of live monitoring or recording of calls with consequences that might not be in the best interests of the caller. Say, for example, that the call resulted in legal proceedings. Since the caller has tacitly agreed to the recording, but may not have been fully aware of the ramifications or consequences, the information could conceivably be used against him or her. The caller is almost never told what is going to be done with the monitored information, who will listen to it, or how long it will be kept on file. The caller also doesn't know if the call is being processed for voice stress or other psychological factors that might be used to manipulate a sales or business call. A copy of the call is in the hands of the callees, to use as they see fit, but not in the hands of the caller, who must try to remember the nature and contents of the call, which is difficult, especially if the matter comes up months or years later. Clearly, with new technologies that make it easy to integrate Call Monitoring into the phone system itself, the consumer's rights are not being protected in part because of the naive trust of consumers. If they don't understand the technology, they don't understand how it can be used to manipulate or compromise their security or safety.

Law Enforcement Monitoring of Calls

Privacy advocates have strongly objected to the monitoring of electronic communications by law enforcement agents. Law enforcement agents, on the other hand, are concerned about falling behind the technology curve and being unable to apprehend criminals who might use technologies in new ways to communicate with one another or might seek to commit new types of crimes made possible by the emerging technologies. It is difficult for law enforcement agencies to implement new policies or to use new technologies in crime prevention and detection without public support. This support will not be forthcoming unless these agencies stress accountability within the system and address the concerns of privacy and civil rights advocates which include

past abuses In the past, wiretapping abuses by law enforcement have been documented by Congress itself, causing the public to be 'gun-shy' of allowing the agencies any wider jurisdiction.

invisible access The adoption of newer remote network technologies can potentially allow law enforcement agents to tap without the same checks and balances that were in place when they had to physically enter a communications carrier's premises to cooperate in setting up a tap. If the equipment is in place and can be accessed remotely, and is decrypted after capture, then law enforcement activities essentially become 'invisible' and less subject to a public or corporate approval process and other traditional safeguards to prevent misuses or corruption.

repurposing There have been many examples of agencies taking information that was approved and gathered for one purpose and later using it for another, particularly when political administrations change. Thus, people are concerned that politically volatile or out-of-context information gathered on prominent public persons could be 'leaked' to the press to discredit someone with a different agenda from the mainstream or that sensitive economic information could clandestinely be used for business-related financial gain by friends or relatives of people within agencies that have access to the information.

discrimination There are concerns that surveillance targets and database lookup systems might be structured in a way that would unfairly target or marginalize minority groups.

Communications carriers and developers have also expressed concerns about law enforcement tapping capabilities being built into digital communications systems because it may build obsolescence into the systems, or may even be outlawed (and unable to be sold) if wiretapping legislation changes in the near future, thus threatening R&D and production expenditures within the telecommunications business community.

Given these concerns, it is important not just to lobby for use of new technologies and to learn to use and implement them well, but to give equal consideration to how accountability structures that protect the public can be built into the systems.

10. Resources

Inclusion of the following companies does not constitute nor imply an endorsement of their products and services and, conversely, does not imply their endorsement of the contents of this text.

10.a. Organizations

Alliance for Telecommunications Industry Solutions (ATIS) - Provides news and information on conferences, educational programs, software, and other support related to the telecommunications industry. ATIS sponsors the Electronic Communication Service Providers Committee in order to assist with compliance requirements for communications carriers. www.atis.org

Cellular Telecommunications Industry Association (CTIA) - CTIA provides member support and wireless products advocacy. The Web site provides news, commentary, information on law and public policy, statistics, consumer resources, and conference announcements. www.wow-com.com/

Center for Democracy and Technology (CDT) - This is an independent, nonprofit, public-interest policy organization which develops and implements public policy regarding liberty and democratic values. CDT is following and recording the debate over wiretapping legislation and privacy and providing research and study into this area of concern. Archives of reports are maintained on their Web site. www.cdt.org/

Central Computer and Telecommunications Agency (CCTA) - A United Kingdom government agency located in Norwich, which promotes good practices in information technology and telecommunications in the public sector.

Competitive Telecommunications Association (CompTel) - Provides representation for over 300 members before the FCC and Congress and supports the prosperity of the competitive telecommunications carriers and their suppliers in the U.S. and overseas. In 1999, CompTel joined with America's Carriers Telecommunication Association (ACTA). The group holds three conferences per year. www.comptel.org

Electronic Privacy Information Center - A public-interest research center located in Washington, D.C. EPIC was founded in 1994 to focus public attention of civil liberties and privacy issues associated with the electronic age. It works in association with Privacy International (U.K.) and others. www.epic.org/

Federal Communications Commission (FCC) - An important U.S. federal regulatory organization established in 1934 to regulate the broadcast industry by granting and administering licenses for radio communications. The FCC's responsibilities have been broadened since that time to include product emissions regulation and fair distribution of telecommunications resources. www.fcc.gov/

Fiber Optic Association, Inc. - An international nonprofit professional association representing the fiber-optic industry. It provides information, training, and certification. www.fotec.com/

Fibre Channel Association (FCA) - An organization supporting Fibre Channel technology which is capable of providing high-speed intercomputer communications for longer distances than the current popular SCSI standard, for example. www.fibrechannel.com/

Fibreoptic Industry Association (FIA) - A U.K.-based professional organization which includes educators, installers, and suppliers or fiber technologies. www.fibreoptic.org.uk/

Indiana State Archives - This resources includes seventeen volumes of telegraphic correspondence between Governor Morton and President Lincoln, Generals Sherman, Stanton, Grant, and others. The telegraph books and telegraphs not recorded in the books are stored on microfilm and the database index can be searched online. www.ai.org/icpr/webfile/archives/homepage.html

International Telecommunication Union (ITU) - The ITU is an important international organization that is based in Geneva, Switzerland and provides education and standards to the telecommunications industry. The ITU evolved from the Telegraph Union which was formed in 1865 (formerly CCITT). www.itu.ch/

Nathanson Centre for the Study of Organized Crime and Corruption - Provides historical and contemporary information on crime and corruption, including analysis and intelligence, alternatives to law enforcement, investigation, surveillance and undercover operations, and law and legislation. The focus is on Canada, but there are many generic references and an excellent annotated bibliography. www.yorku.ca/nathanson/

National Institute of Justice (NIJ) - NIJ was created as the research and technological development agency of the U.S. Department of Justice to sponsor special projects, research, and development to improve and strengthen the criminal justice system to reduce or prevent violent crime. www.ojp.usdoj.gov/nij/

Telephone Pioneers of America (TPA) - A nonprofit organization established in 1911 by the Bell system pioneers. TPA now includes more than 100,000 members. www.telephone-pioneers.org/

U.S. Department of Justice (DoJ) - Under the direction of the Attorney General, the DoJ is charged with attaining and maintaining justice and fair treatment for Americans through the combined services of almost 100,000 attorneys, law enforcement professionals, and employees. Part of the DoJ responsibility involves detecting criminal offenders. It is headquartered in Washington, D.C. with almost 2,000 installations throughout the country. www.usdoj.gov/

10.b. Print

Arrington, Winston, *Now Hear This! Electronic Eavesdropping Equipment Design*, Sheffield Electronics, 1997.

Berkel, Bob; Rapaport, Lowell, *Covert Audio Interception*, CCS Security Publishing, 1994, 720 pages.

Blum, Richard, *Surveillance & Espionage in a Free Society*, New York, London: Praeger Publishers, 1972.

Brookes, Paul, E*lectronic Surveillance Devices*, Butterworth-Heinemann, 1996, 112 pages. General descriptions, types of devices, circuit diagrams, and construction information. Of interest to hobbyists and corporate security technicians.

Brown, Robert M., *The Electronic Invasion*, New York: John F. Rider Publisher, Inc., 1967. A historical perspective on electronic bugging up to the time of publication. Discusses the evolution and miniaturization of the technology as well as distribution of tapping devices.

Bugman, Shifty, *The Basement Bugger's Bible: The Professional's Guide to Creating, Building, and Planting Custom Bugs and Wiretaps. Includes schematics, blueprints, photos, diagrams, anecdotes on phone taps, microphones, bugs and related audio surveillance devices*, Paladin Press, 1999. For academic study, 320 pages.

Campbell, Duncan, *Big Brother is Listening: Phonetappers and the Security State*, New Statesman, 1981, 70 pages. Campbell is one of the more vocal and credible of the high-profile Web journalists reporting on surveillance activities.

Carr, James G., *The Law of Electronic Surveillance*, New York: Clark Boardman Co., Ltd., 1977.

Chambers of Commerce of the State of New York, *Papers and Proceedings of Committee on the Police Problem*, City of New York, 1905, New York: Ayer Company Publishers, 1905. This title is still available. It describes investigations of police problems at the turn of the century, including the Lexow and Mazet hearings.

Chin, Gabriel J., *New York City Police Corruption Investigation Commission 1894-1994*, six volumes, New York: William S. Hein & Co., Inc., 1997.

Chin, Gabriel J., *Report and Proceedings of the Senate Committee Appointed to Investigate the Police Department of the City of New York* (Lexow Report), five volumes, New York: William S. Hein & Co., Inc., 1997.

Churchill, Ward; Vander Wall, Jim, *The COINTELPRO Papers: Documents from the FBI's Secret Wars Against Dissent in the United States*, South End Press, 1990, 468 pages. Includes statistics, information, and speculation on the documented and undocumented aspects of FBI wiretapping and mail openings during Hoover's administration.

Cook, Earleen H., *Electronic Eavesdropping,* 1983. Out of print.

Daley, Robert, *Prince of the City: The True Story of a Cop Who Knew Too Much,* Boston: Houghton Mifflin Co., 1978. A former New York Deputy Police Commissioner describes witnessing alleged corruption, theft, and perjury within the police department.

Dannett, Sylvia G. L., *She Rode with Generals: The True and Incredible Story of Sarah Emma Seelye,* Alias Franklin Thompson, New York: Thomas Nelson, 1960.

Dash, Samuel; Schwartz, Richard F.; Knowlton, *Robert E., The Eavesdroppers,* New Brunswick, N.J.: Rutgers University Press, 1959. Even though this is an older text, it is frequently cited by both writers and speakers. It was reprinted by Da Capo Press in 1971, 484 pages.

Diffie, Whitfield; Landau, Susan, *Privacy on the Line: The Politics of Wiretapping and Encryption,* Boston: MIT Press, 1998, 342 pages. Recipient of IEEE and other book awards, this takes the reader point-by-point through the history and politics of wiretapping, revealing the opposing opinions and complex issues involved.

Donner, Frank, *The Age of Surveillance: The Aims and Methods of America's Intelligence System,* New York: Vintage-Random House, 1981.

Doyle, Sir Arthur Conan, *The Man with the Watches,* a story in which Conan Doyle makes reference to the Lexow commission (which investigated the New York police in 1894). The story can be found in various Sherlock Holmes compilations, e.g., Doyle, Arthur C., Complete Sherlock Holmes, New York: Doubleday Books, 1960.

Edmonds, S. Emma E., *Memoirs of a Soldier, Nurse, and Spy: A Woman's Adventures in the Union Army,* Northern Illinois University Press, 1999. Previously published as Nurse and Spy in the Union Army: The Adventures and Experiences of a Woman in Hospitals, Camps, and Battle-fields, 1865 (384 pages) and originally published as The Female Spy of the Union Army, Boston: DeWolfe, Fiske & Co., 1864.

Erickson, William H., Chairman, *Electronic Surveillance - Report of the National Commission for the Review of Federal and State Laws Relating to Wiretapping and Electronic Surveillance,* NWC Report, U.S. Government Printing Office, 1976.

Fishel, Edwin C., *The Secret War for the Union. The Untold Story of Military Intelligence in the World War,* New York: Houghton Mifflin Co., 1996, 734 pages. Intelligence records usually 'disappear' after a War, but Fishel has uncovered documents that reshape our thinking about military strategists and how they influence the politics of war.

Fitzgerald, Patrick; Leopold, Mark, *Stranger on the Line: The Secret History of Phone Tapping,* London: Bodley Head, 1987. U.K. wiretapping history.

Garrow, David J., *The FBI and Martin Luther King, Jr. From Solo to Memphis,* New York, London: W.W. Norton & Co., 1981.

Goode, James, *Wiretap: Listening in on America's Mafia,* New York: Simon & Schuster, 1988.

Greene, Richard M., *Business Intelligence and Espionage,* Homewood, Il.: Dow Jones-Irwin, 1966.

Hall, Richard, *Patriots in Disguise,* New York: Paragon House, 1993. Women were not welcome in matters of war, but many got involved anyway, in male disguises. Illustrated history of some of the fascinating soldiers and 'male' nurses who eavesdropped during the Civil War.

Hartman, John Dale, *Legal Guidelines for Covert Surveillance,* Newton, Ma.: Butterworth-Heinemann, 1993, 235 pages.

Johnson, Pauline Copes, *City of Auburn Souvenir Celebration Booklet Commemorating 20 Years of History 1793-1993,* Cayuga County Historian's Office. Includes information on Harriet Tubman, Underground Railroad and Civil War spy.

Jones, R.; Taggart, R.; et al., *Electronic Eavesdropping Techniques and Equipment*, Washington, D.C.: National Bureau of Standards, Law Enforcement Standards Laboratory, 1977.

Lapidus, Edith J., *Eavesdropping on Trial*, New Jersey: Hayden Book Co., 1973, 287 pages.

Law Enforcement Associates, *The Science of Electronic Surveillance*, Raleigh, North Carolina: Search, Inc., 1983.

LeMond, Alan; Fry, Ron, *No Place to Hide: A Guide to Bugs, Wire Taps, Surveillance and Other Privacy Invasions*, St. Martin's Press, 1975, 278 pages.

Leonard, Elizabeth, *All the Daring of the Soldier: Women of the Civil War Armies*, New York: W. W. Norton, 1999, 320 pages, illustrated. Describes the eavesdropping and spy activities of women in the Revolutionary and Civil Wars in America.

Long, Edward V., *The Intruders, the Invasion of Privacy by Government and Industry*, New York: Praeger, 1967, 230 pages. Foreword by Hubert H. Humphrey. Long was unsympathetic to alleged FBI pressure to support their actions while holding a Senate seat, was subsequently defeated, and wrote this account of his experiences.

Marx, G.T., *Undercover: Police Surveillance in America*, L.A.: University of California Press, 1988.

Murphy, Walter F., *Wiretapping on Trial*, New York: Random House, 1965. Murphy is co-author of American Democracy, a widely used college text and a former member of the U.S. Marine Corps.

New Haven Board of Police Commissioners, Report on Wiretapping, 1978.

Office of Technology Assessment, *Electronic Surveillance in a Digital Age*, U.S. Government Printing Office, July 1995. Describes progress in electronic communications and surveillance and focuses on the work of law enforcement and telecommunications agencies to implement the Communications Assistance for Law Enforcement Act, Public Law 103-414 and other relevant laws.

Oslin, George P., *The Story of Telecommunications*, Macon, Ga.: Mercer University Press, 1992. The author, born in 1899, lived through many of the significant early developments in telecommunications and communicated directly with some of the pioneers in the industry. The book covers the technical and regulatory aspects, as well. He is credited with inventing the 'Singing Telegram.'

Paulsen, Monrad G., *The Problems of Electronic Eavesdropping*, American Law Institute American Bar Association Committee on Continuing Professional Education, 1977, 136 pages.

Pollock, David A., *Methods of Electronic Audio Surveillance*, Springfield, Il.: Charles C Thomas Publisher, Ltd., 1973.

Records of the Committee on the District of Columbia Subcommittee on the Investigation of Wiretapping. About four linear feet of records from the 81st to 92nd Congresses dating 1950 to 1951 housed as Record Group 46 in the National Archives and Records Administration.

Records of the San Francisco Field Office, 1950 to 1952. This is a series of clippings, affidavits, memorandums, pleadings, and telegrams about controversial program content related to investigations of the radio and television industry and wiretapping. Available through the regional National Archives and Records Administration office in San Bruno, California.

Richburg, Rod; Swift, Theodore N. (illustrator), *Wiretap Detection Techniques*, Austin Tx.: Thomas Investigative Publications, 1997.

Ruttledge, Hugh, *Everest 1933*, London: Hodder & Stoughton, 1934, 390 pages.

Schartz, Herman, *Taps, Bugs and Fooling the People*, The Field Foundation, 1977.

Schneier, Bruce; Banisar, David, *The Electronic Privacy Papers: Documents on the Battle for Privacy in the Age of Surveillance*, New York: J. Wiley, 1997, 747 pages.

Shannon, M. L., *The Bug Book: Wireless Microphones & Surveillance Transmitters,* with contributions by Kevin D. Murray, Boulder, Co.: Paladin Press, 2000, 168 pages. Awareness, detection, and countermeasures, including anecdotes.

Shannon, M. L., *The Phone Book: The Latest High-Tech Techniques and Equipment for Preventing Electronic Eavesdropping,* Recording Phone Calls, Ending Harassing Calls, Boulder, Co.: Paladin Press, 1998, 280 pages.

Swift, Theodore N., *Wiretap Detection Techniques: A Guide to Checking Telephone Lines, Boulder,* Co.: Paladin Press, over 100 pages, illustrated. Describes procedures for conducting eavesdropping countermeasures surveys. The author is a former counterintelligence officer who worked for the DEA for 11 years and has developed this book from his experience and his teaching of a law enforcement seminar. Topics include inductive wiretaps, testing for series devices, network schematics, line balance tests, spectrum analyzers, transmitter harmonics, and more.

Thomas, Ralph D., *The TSCM Bible: A Countermeasures Cookbook on Conducting Professional TSCM Services,* Austin, Tx.: Thomas Investigative Publications, Inc. Includes almost 300 pages and a computer resource disk. A comprehensive overview on conducting countermeasures sweeps and TSCM services including equipment and testing procedures.

Turner, WIlliam W., *How to Avoid Electronic Eavesdropping and Privacy Invasion,* Boulder, Co.: Paladin Press, 1972.

Velazquez, Loreta Janeta, *The Woman in Battle: A Narrative of the Exploits, Adventures and Travels of Madame Loreta Janeta Velazquez, Otherwise Known as Lieutenant Harry T. Buford, Confederate States Army,* Richmond, Va.: Dustin, Gilman & Co., 1876. The full text is available online through the University of North Carolina at Chapel Hill Libraries.

Whidden, Glenn H., *The Axnan Attack–A Detailed Composite Case History about How Corporate Electronic Eavesdropping is Accomplished,* Technical Services Agency.

Articles

The Center for Constitutional Rights, If an Agent Knocks: Federal Investigators and Your Rights, New York. Discusses FBI COINTELPRO activities and the rights of individuals who receive visits by FBI agents.

Delaney, Donald P.; Denning, Dorothy E.; et al., Wiretap Laws and Procedures: What Happens When the Government Taps a Line, Georgetown University, Sept. 1993.

Dempsey, James X.; Weitzner, Daniel J.; et al., Comments of the Center for Democracy and Technology in the Matter of Communications Assistance for Law Enforcement Act, CC Docket No. 97-213 before the FCC, 20 May 1998. See also Dempsey, James X., Statements before the Subcommittee on Telecommunications, Trade, and Consumer Protection of the House Committee on Commerce on the Wireless Privacy Enhancement Act of 1999 and the Wireless Communications and Public Safety Enhancement Act of 1999, 3 Feb. 1999 and Dempsey, James X., Communications Privacy in the Digital Age: Revitalizing the Federal Wiretap Laws to Enhance Privacy, *Albany Law Journal of Science & Technology,* 1997, V.8(1), available through CDT. www.cdt.org/

Denning, Dorothy E., Encryption and Law Enforcement, Georgetown University, Feb. 1994.

Dichter, Mark S.; Burkhardt, Michael S., Electronic Interaction in the Workplace: Monitoring, Retrieving and Storing Communications in the Internet Age, Morgan, Lewis & Bockus, LLP, 2000. Describes trends, laws, and historical precedents regarding employee electronic communications, including some references to corporate wiretapping and how the courts made distinctions between phone tapping and network 'tapping'. Article published as a PDF at www.morganlewis.com

Elder, Willie J., Jr., Electronic Surveillance: Unlawful Invasion of Privacy or Justifiable Law Enforcement, Yale-New Haven Teachers Institute Curriculum Unit, 1983.

Fillingham, David, Listening in the Dark - Wiretapping and Privacy in America, MIT paper for Ethics and Law on the Electronic Frontier, 1997. Discusses historical highlights and major legal decisions leading up to present-day encryption debates.

Freeh, Louis, FBI Director, Testimony in the Senate Judiciary Committee, Terrorism, Technology & Government Information Subcommittee, Senator Jon Kyl, Chair, 3 September 1997.

Guinier, Daniel, From Eavesdropping to Security on the Cellular Telephone System GSM, *ACM SIG-SAC Security Audit & Control Review,* 1997, V.15 (2), pp. 13-18.

Internal Revenue Service, Tax Professional's Corner: Handbook 9.4, Investigative Techniques, Chapter 6, Surveillance and Non-Consensual Monitoring, available online at www.irs.gov/bus_info/tax_pro/ in the Part 9, Criminal Investigation section.

Landau, Susan, Eavesdropping and Encryption: U.S. Policy in an International Perspective, published on the Harvard Information Infrastructure Project site. Discusses the historical role of government in communications policy and cryptographic concerns up to the present day from an international perspective.

Lyon, David, The New Surveillance: Electronic Technologies and the Maximum Security Society, *Crime, Law and Social Change,* 1992, V.18(1-2).

MacDonald, Fred J., Don't Touch That Dial: Radio Listening Under the Electronic Communications Privacy Act of 1986, Chicago: Nelson-Hall, 1989.

Matthews, Clark, Unanymous Nod for Wiretap Bill, *The Spotlight*, Nov., 1994. Discusses the Wiretap Access Bill, passed in Oct. 1994, and FBI lobbying that occurred prior to submission of the bill.

Merhav, N.; Arikan, E., The Shannon Cipher Systems with a Guessing Wiretapper, *IEEE Transactions on Information Theory*, Sept. 1999, V.45 (6), pp. 1860-1866.

Millman, Gregory, From Dragnet to Drift Net: Telephone Record Surveillance and the Press, Dudley Clendinen, Justice Dept. Gets Phone Records of the Time's Bureau in Atlanta, NYT, 6 Sept. 1980.

Miyazawa, Setsuo, Scandal and Hard Reform: Implications of a Wiretapping Case to the Control of Organization Police Crimes in Japan, *Kobe University Law Review*, V.23, pp. 13-27.

Morse, Wayne, Wiretapping proposals threaten historic gains, Washington, D.C., U.S. Government Printing Office, 1954.

Ogura, Toshimaru, Japan's Big Brother, The Wiretapping Bill and the Threat to Privacy, *AMPO, Japan-America Quarterly Review*, 1997, V.28(1). Discusses a controversial move to legalize wiretapping in Japan.

O'Neill III, Thomas F.; Gallagher, Kevin P.; Nevett, Jonathon L., MCI Communications Corporation, Detours on the Information Superhighway: The Erosion of Evidentiary Privileges in Cyberspace and Beyond, *Stanford Technology Law Review*, 1997, V.3. Provides a historical overview of wired/wireless communications up to the current Internet and legal responses to regulation and changes in the industry.

Saxbe, William B., Wiretapping and Electronic Surveillance, *Police Chief*, 1975, V.42(2), Feb., pp. 20-22.

Shuy, Roger W., Tape Recorded Conversations, *Criminal Intelligence Analysis,* Loomis, Ca.: Palmer, 1990.

Skolnick, Jerome H., Deception by Police, *Criminal Justice Ethics*, 1982, V.1(2). Skolnick is Co-Director of the Center for Research in Crime and Justice. The article discusses ethics, the concepts of legality, and the actions and theorized motives of dishonest individuals in positions of authority. Wiretapping in the Hoover years is discussed.

Steal, Agent, "Tapping Telephone Lines Voice or Data for Phun, Money, and Passwords Or How to Go to Jail for a Long Time", *Phrack magazine Number 16*, Aug. 1987.

Truman Library, "Oral History Interview with Joseph L. Rauh, Jr.", in the National Archives and Records Administration, transcribed by Niel M. Johnson, 21 June 1989. Includes references to Truman, the FBI, and wiretap documents of Tommy Corcoran.

Westin, Alan "Science, Privacy, and Freedom: Issues and Proposals for the 1970s", Columbia Law Review reprint, 1966, 47 pages. This article has been cited in a number of privacy-related legal judgments.

Journals

Safety and Security, by Carroll Publications, Ohio. Topics include fraud, fire protection, workplace surveillance, public safety, etc.

Wiretap Report, an annual publication by the Administrative Office of the United States Courts. Provides general and historic information on criminal wiretaps and procedures. There are tables breaking down major offenses by category and state. It can be downloaded in Adobe .pdf format. It does not provide specific information on the actual number of lines covered in each court order and does not specify statistical details for pen register and trap-and-trace activities, which comprise the majority. This type of information was collected by the FBI, however, for a period of two years, in order to estimate law enforcement needs incumbent upon communications carriers through CALEA.

The Department of Justice (DoJ) conducts surveys of Pen Register/Trap and Trace court orders.

See also the general privacy and security journals listed in Chapter 1 - Introduction and the computer communications journals listed in the Computer Surveillance chapter.

Reports and Testimony

Electronic Surveillance Task Force, Communications Privacy in the Digital Age, *The Digital Privacy and Security Working Group papers*, Interim Report, June 1997. Discusses electronic surveillance and the evolution of privacy protection and continued law enforcement lobbies for wider discretion in monitoring electronic communications. Lists and discusses key developments and related legislation. Available through the Center for Democracy and Technology. www.cdt.org/

Hearings on Wiretapping and other Terrorism Proposals, Testimony of David B. Kopel, Associate Policy Analyst, Cato Institute, Committee on the Judiciary United States Senate, 24 May 1995. Responds to fears provoked by the Oklahoma City bombing and describes the need for careful consideration of legislation and debates the issues over the expansion of wiretap authority.

Shelby, Richard C. (Chairman) et al., Special Report of the Select Committee on Intelligence, United States Senate, January 4, 1995 to October 3, 1996, 1997. A report contributing to public accountability that includes information on Committee review of the Foreign Intelligence Surveillance Act (FISA). Topics include reviews of foreign intelligence, legislation, arms control, counterintelligence, counter-terrorism, and counter-proliferation.

Subcommittee on Crime Committee on the Judiciary U.S. House of Representatives Oversight Hearing on The Implementation of the Communications Assistance for Law Enforcement Act of 1994, Dempsey Testimony, Testimony of James X. Dempsey, Senior Staff Counsel, Center for Democracy and Technology., 23 October 1997. This testimony includes substantial information on CALEA and its proposed implementation and pen register and other surveillance tools used by law enforcement. The reader is also encouraged to read other testimony related to this Oversight Hearing.

U.S. Congress House Committee on Education and Labor. Subcommittee on Labor-Management Relations, Hearing on House Rule 1900, Privacy for Consumers and Workers Act of 1993, Washington, D.C., U.S. GPO, 1994, 235 pages.

U.S. Congressional Record, Remarks of Senator Jesse Helms, 3 October 1983, V.129(130), pp. S 13452-13461. These remarks refer to surveillance of the activities and associations of Dr. Martin Luther King, Jr., and statements about communists and other 'extremists,' (the speaker included in this category those who were opposed to the Vietnam conflict). Wiretap evidence was ordered to be sealed for a period of several decades.

U.S. Department of Justice, Electronic Surveillance - Report on the FBI's Publication of the Second Notice of Capacity, January 1997, a report on the telephone system capacity that may be needed by law enforcement to carry out court-approved electronic surveillance with regard to implementation of the CALEA requirements of communications carriers. Later, further details were provided by the FBI on how the wiretap statistics were obtained and the capacity requirements calculated (Congressional Record, 9 February 1996).

U.S. National Commission for the Review of Federal and State Laws Relating to Wiretapping and Electronic Surveillance, Electronic Surveillance Report, Washington, D.C., 1976.

10.c. Conferences and Workshops

Many of these conferences are annual events that are held at approximately the same time each year, so even if the conference listings are outdated, they can still help you determine the frequency and sometimes the time of year of upcoming events. It is very common for international conferences to be held in a different city each year, so contact the organizers for current locations.

Many of these organizations describe the upcoming conferences on the Web and may also archive conference proceedings for purchase or free download.

The following conferences are organized according to the calendar month in which they are usually held.

Computer Telephony Expo, business and developer-oriented conference held annual in L.A.

ICASSP - IEEE International Conference on Acoustics, Speech, and Signal Processing, Annual, international technical conference.

ICC - IEEE International Conference on Communications, annual technical conference.

Association of Public Safety Communications Officials (APCO), annual conference usually held in late summer.

International Conference on Communication Technology - ICCT, international IEEE technical conference.

Mobile Battlefield Communications, international SMI conference.

DEFCOM - Defence Communications Exhibition, Nexus Media Ltd., annual international conference.

Military Satellite Communications, SMI international conference.

National Technical Investigators Association (NATIA), annual conference. The 2006 conference and trade show was held in August in San Antonio.

GLOBECOM - IEEE Global Telecommunications Conference, international IEEE technical conference.

10.d. Online Sites

Adventures in Cybersound. Milestones through 2500 years of communications history. This lists many of the interesting and important discoveries in microphones, sound recording, and wireless broadcasting technologies associated with audio/visual communications. Compiled by Dr. Russell Naughton. www.cinemedia.net/SFCV-RMIT-Annex/rnaughton/TV_TL_COMP_2.html

Center for Democracy and Technology. This organization has a specific section on wiretapping (see buttons at bottom of page) which lists developments in wiretapping activities and legislation and provides numerous links to prominent articles and excerpts of testimony related to the subject. There is also an overview of terms and devices. www.cdt.org/

DSL Sourcebook. The HTML version of "The DSL Sourcebook: Plain Answers on Digital Subscriber Line Opportunities", by Paradyne Corporation. This includes extensive information on Digital Subscriber Line technologies, including the existing copper-wire infrastructure, DSL concepts, emerging services, network models (including IP/LAN, ATM), and almost three dozen diagrams of performance descriptions, reference models, etc. www.paradyne.com/sourcebook_offer/sb_html.html#ch2

International Privacy and Wiretapping. On the Chicago-Kent Collect of Law Illinois Institute of Technology, there is an online Information Center. One page of particular interest is the U.S. Department of State Annual Human Rights Report (Privacy Report) which has been excerpted by Privacy International to list relevant information that briefly describes wiretapping authority and regulations in the various world nations. www.kentlaw.edu/ic3/islat/prvcysum.htm

The Sound Recording History Site. This site by David Morton includes a variety of topics relevant to sound recording including a historical chronology, information about recording devices, ideas that didn't work, dead sound-recording media, and thoughts on the surveillance society. www.rci.rutgers.edu/~dmorton/soundrechist.html

10.e. Media Resources

The Inspectors, a Showtime Network movie about the lives and works of two fictional federal law enforcement agents responsible for solving a mail bomb case. While this isn't directly related to audio surveillance, inspection of mail and eavesdropping on conversations historically have many parallels and this show includes other surveillance techniques of interest, as well.

11. Glossary

Titles, product names, organizations, and specific military designations are capitalized; common generic and colloquial terms and phrases are not.

AIN	Advanced Intelligent Network. A Signaling System No. 7-based telephone switching network which integrates ISDN and cellular services into PCS services. This may eventually be superseded by Information Network Architecture (INA).
AMPS	Advanced Mobile Phone Service. An analog cellular phone system introduced in the early 1980s.
ATM	Asynchronous Transfer Mode. A high-speed, cell-based, connection-oriented, packet transmissions protocol that can handle data with varying burst and bit rates. ATM evolved in the mid-1980s through standardization efforts by the CCITT (now ITU-T). It is an important digital data and communications format that is widely used in computer networking. Digital voice communications can be carried over ATM systems, with Voice over ATM being one of a number of models in development.

aural transfer — Defined by U.S. Code Title 18, Part I, Chapter 119 as "a transfer containing the human voice at any point between and including the point of origin and the point of reception".

bug — A covert or clandestine listening or viewing device that is noted for its small, inconspicuous (bug-like) size. Bugs used to primarily mean listening devices, small microphones that could be hidden in plants or phone handsets, but the term now is also used to describe tiny pinhole cameras that are as small as audio bugs used to be twenty years ago. A bug may be wired or wireless and may or may not be sending information to a recording device.

CCITT — Comité Consultatif International Télégraphique et Téléphonique. An important international standards body which is now known as the International Telecommunication Union (ITU).

CDMA — Code-Division Multiple-Access. A digital, wireless communications service based on spread-spectrum technology. Security can be provided through spread-spectrum modulation of the signal.

CELP — Code Excited Linear Predictive. This is a means of translating analog voice data to digital format so that conversations can be sent through digital networks like local area networks (LANs) and the Internet. When Internet phone services began to spring up, traditional long-distance carriers were concerned about loss of business due to people switching to less expensive Internet-based phone services. (While not necessarily CELP-based, there are many formats, videoconferencing systems with small video cameras and microphones can also be used with appropriate software to transmit conversation over the Internet, often without charges other than the Internet access itself.)

circuit switching — A type of end-to-end transmission system commonly used for phone connections in which the resources are allocated to a specific call for its duration and are not usually available for other purposes until the call is completed.

CPDP — Cellular Digital Packet Data. A packet-based open standard released in the early 1990s which is suitable for packet data services for mobile communications to extend landline services for mobile users. CDPD works over AMPS analog voice systems with transmission speeds up to 19.2 Kbps. CDPD is primarily an architectural structure and doesn't prescribe the kinds of service that can be carried over the system.

DECT — Digital Enhanced Cordless Telecommunications. A set of European wireless standards intended to unify digital radio standards for European cordless phones. It requires more cells than cellular, but supports higher densities.

DLC — Digital Loop Carrier. A telecommunications service provider type that emerged in the early 1970s which utilizes switches and multiplexers to concentrate low-speed services prior to distribution through a local central switching office. It is similar to a Local Loop Carrier in that it provides a physical connection between subscribers and a main distribution frame over copper wires, but of a digital nature.

DTMF — Dual-Tone Multifrequency. Sometimes also called TTMF (touch-tone multifrequency). A signaling method using two specific frequencies.

electronic communication

Defined by U.S. Code Title 18, Part I, Chapter 119 as "any transfer of signs, signals, writing, images, sounds, data, or intelligence of any nature transmitted in whole or in part by a wire, radio, electromagnetic, photoelectronic or photooptical system that affects interstate or foreign commerce, but does not include

(A) — the radio portion of a cordless telephone communication that is transmitted between the cordless telephone handset and the base unit;

(B) — any wire or oral communication;

(C) — any communication made through a tone-only paging device; or

(D) — any communication from a tracking device (as defined in section 3117 of this title)".

ESN	Electronic Serial Number. An identifier associated with wireless communications that is used for location and/or identification. In some devices, the ESN cannot be changed and may include information about the manufacturer. The ESN may further by keyed to services provided by the manufacturer, such as tech support or Web site access.
F-ES	Fixed End-System. A stationary data communications system (non-mobile) in which a mobile subscriber access landline telecommunications services. This is commonly used with computer-based modems.
FISA	Foreign Intelligence Surveillance Act
FWA	Fixed Wireless Access. A wireless radio-based telephone service that works in place of a local wireline loop, with common-carrier phone service. It is commonly used in regions where it's difficult to string wire, such as isolated areas, islands, or temporary installations.
HDSL	High Bit-Rate Digital Subscriber Line
HLR	Home Location Register. The basic service area list maintained on mobile communications subscribers.
IAP	Intercept Access Point. A point within the carrier's system where call information or communications of an intercepted call can be accessed.
ISDN	Integrated Services Digital Networks. A set of standards for digital data transmission that can work over existing copper wires and the newer cabling media. Thus, a traditional phone network can be used for high-speed data transmissions. Part of the difficulty of establishing ISDN services has been that the link to subscribers themselves is still mostly analog and appropriate terminal adaptors have to be installed to set up the service. In addition to this, cable modem services over existing TV cables have made data services available to many computer users at a lower cost than ISDN in many areas, thus hampering the implementation of ISDN services. ISDN is a flexible system allowing both voice and data communications over the same 'line.' There are two basic types of ISDN service, Basic Rate Interface (BRI) and Primary Rate Interface (PRI).
ITU	International Telecommunication Union. The ITU is based in Geneva, Switzerland and provides education and standards to the telecommunications industry. The ITU evolved from the Telegraph Union which was formed in 1865 (formerly CCITT).
J-STD-025	A Telecommunications Industry Association interim standard for defining services and features to support lawful, authorized electronic surveillance.
JATE	Japan Approvals (Institute) for Telecommunications Equipment. A Japanese regulatory agency, established in 1984, which is roughly equivalent to the Federal Communications Commission (FCC) in the U.S.
LEA	law enforcement agency
MFJ	Modified Final Judgment. This was a historic seven-year antitrust lawsuit that concluded in the mid-1980s between the U.S. Justice Department and AT&T in which AT&T was eventually divested. AT&T was permitted to retain Bell Laboratories and AT&T Technologies, but the Regional Bell Operating Companies (RBOCs) were banned from manufacturing. Local Access Transport Areas (LATAs) were created rather than retaining the existing local exchange boundaries.
MIN	Mobile Identification Number. An identifier associated with wireless communications.
MMCX	Multimedia Communication Exchange. A commercial phone/data server software developed by AT&T for providing multimedia services for private phone branch exchanges (PBXs).
MSC	Mobile Switching Center
MTSO	Mobile Telephone Switching Office

NGDLC   Next Generation Digital Loop Carrier. Evolved from Digital Loop Carriers in the 1980s, NGDLC provides telecommunications services based on very-large-scale integration technology intended for use over fiber optic or hybrid optic/copper wirelines.

NPRM   Notice of Proposed Rule-Making

oral communication

Defined by U.S. Code Title 18, Part I, Chapter 119 as "any oral communication uttered by a person exhibiting an expectation that such communication is not subject to interception under circumstances justifying such expectation, but such term does not include any electronic communication".

PBX   Private Branch Exchange. A local, internal multiple-line phone system that is commonly found in business and educational environments. You can generally tell you are using a PBX system, if you have to dial '9' before placing an outgoing call to a destination not included on the exchange. Exchange 'locals' are numbers within the exchange to designate individual phones.

PCS   Personal Communications Services

POTS   plain old telephone service. The basic analog communications over copper wires that existed for many decades.

roaming   Engaging in a wireless call while traveling outside of a subscriber's 'home area,' the area in which the basic service is established. Roaming charges, like long-distance charges on a wired line, tend to be higher than the basic service charges. By the year 2000, large carriers were starting to announce widespread coverage with reduced roaming restrictions.

SPCS   Stored Program Control Switch. A type of metropolitan communications switch introduced in the 1960s that is gradually being replaced by digital switches.

SXS   step-by-step. Used with regard to the old electromechanical communications switches used at the turn of the century to route traffic. A few rural SXS systems still exist but they are quickly disappearing.

TSCM   Technical Security/Surveillance Countermeasures

wearing a wire   The wearing of a hidden microphone and transmitting or recording device for the purpose of covertly monitoring or saving the information and characteristics of a personal communication.

wire communication

Defined by U.S. Code Title 18, Part I, Chapter 119 as "... any aural transfer made in whole or in part through the use of facilities for the transmission of communications by the aid of wire, cable, or other like connection between the point of origin and the point of reception (including the use of such connection in a switching station) furnished or operated by any person engaged in providing or operating such facilities for the transmission of interstate or foreign communications or communications affecting interstate or foreign commerce and such term includes any electronic storage of such communication, but such terms does [sic] not include the radio portion of a cordless telephone communication that is transmitted between the cordless telephone handset and the base unit".

wiretap   n. Strategies and equipment set up to access or intercept the contents of a communication, usually a voice call over a telecommunications network. Taps are most commonly made over copper wires, but means to tap directly into the phone unit or communications over fiber optic cable also exist. The term is now used in a broader sense to include 'listening in' on and recording wireless communications as well. Given the increasing complexity of communication and growth of wireless transmissions, the most effective taps are now usually at the source or destination or through switching systems enroute.

Acoustic
Surveillance

3

Infra/Ultrasound

1. Introduction

The previous chapter on audio surveillance focuses on sounds within the range of human hearing. However, there are sounds outside of human hearing that can be heard by many creatures including moths, dolphins, bats, cats, and dogs. These sounds can also be generated by machines and used to communicate signals, measurements, or conversations that are useful in surveillance.

This chapter introduces *infrasound*, sounds at frequencies below human hearing, and *ultrasound*, sounds above human hearing. Sonar detection used to be a manual process—an operator would listen to the pings through earphones—and thus used audible sound frequencies. Since sonar has been automated, however, it primarily uses ultrasound.

Sonar is a specialized area of acoustics, widely used in surveillance, and is covered in more depth in Chapter 4. Shock waves are somewhat like sound waves, and are discussed briefly in this chapter. Sound waves and shock waves are not the same, but devices to detect them are very similar. For an explanation of the distinctions between these phenomena, see Section 5 (Description and Functions).

Infra/ultrasounds have many advantages as surveillance tools. They can communicate or signal across distances without attracting attention. Compared to a number of other forms of

A severely damaged stretch of highway in Oakland, California, following the Loma Prieta earthquake. Infrasound detectors can aid in predicting many different types of natural disasters in order to warn residents to avoid roads or to take shelter underground. [U.S. Army Corps of Engineers news photo, public domain.]

communication, sound signaling is often economical—it is easier to generate certain sound frequencies and listening devices than it is to generate many types of electromagnetic frequencies and their related transmitters/receivers.

Sound technologies are less harmful, at normal intensities, than X-radiation or microwaves. They are also versatile. They can be used to detect objects, to measure distances, to map contours, and to create a 'picture' of objects within other objects or structures. Sound is used in applications as diverse as locating submarines, navigating across oceans or through caves, sending messages, testing the structural integrity of buildings, assessing quality on a production line, or imaging a growing fetus inside a womb. The ability to look through particulate environments or through certain opaque materials is a very useful capability of sound.

Much of our knowledge of infrasound and ultrasound comes from studying the anatomy and behavior of animals, insects, and fish. In some cases, humans have studied the abilities of various species of animals and used their findings to develop surveillance tools. Sometimes they even use animals directly. Dolphins and dogs can recognize and respond to signals above human hearing, and have been trained to perform certain surveillance tasks on behalf of humans. Humans have also made use of the sound-sensing abilities of other species as warning or searching aids.

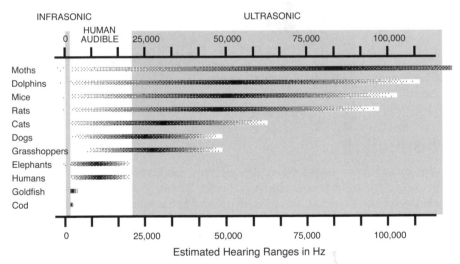

Estimated Hearing Ranges in Hz

Chapter 2 on audio surveillance and Chapter 4 on sonar surveillance provide most of the introductory and specialized information needed to understand acoustic technologies related to surveillance. This chapter is intended to introduce the general concepts related to inaudible acoustics—those below and above human hearing, respectively called infrasound and ultrasound.

- As can be seen from the two shaded regions in the above chart, infrasound is a very small portion of the sound spectrum and yet it is highly important as a monitoring technology for predicting and studying larger events such as storms, earthquakes, floods, and nuclear explosions. Infrasound is also used in compact pressure-vibration-based motion detectors.

- Ultrasound covers a much larger spectrum of frequencies than infrasound, and it is widely used in navigation, depth-sounding, and medical applications. Most aspects of medical imaging are outside the scope of this text, but it is worth mentioning that ultrasound can be used in forensic investigations of homicide or accidental death. Ultrasound is also used for robotics control. The marine applications of ultrasound, especially sonar, are many. The majority of sonar technologies now operate within ultrasound frequency ranges and are described in the following chapter.

1.a. Basic Concepts and Terminology

As illustrated by the shaded regions in the preceding chart, sounds with frequencies above human hearing are called *ultrasonic* and those below human hearing are called *infrasonic*. Human-audible sounds are called *sonic*.

Infrasonic vibrations are those with frequencies below 17.5 ± 2.5 Hz. Ultrasonic vibrations are those above $19,000 \pm 1,000$ Hz. Different texts and branches of science use different cutoff points, but most infra/ultrasonic designations are within these limits. In biological terms, there are no definite cutoff points, since hearing varies from person to person and changes as we age. Political bodies sometimes specify cutoff points in order to establish policies, such as those for detection devices associated with international disarmament.

Infrasonic vibrations can sometimes be converted to audible sounds by recording the vibrations with sensitive microphones and then speeding up the replay. The same can be done, to a limited extent with ultrasonic vibrations near the upper level of human hearing. However, we usually use other instruments to record and display infra/ultrasonic sounds, often converting the data into a graphical representation.

Frequency is separate from the intensity or 'loudness' of a sound (e.g., a 50,000 Hz ultrasonic vibration can have a high intensity and still would not be heard by humans). Intensity is expressed in units called *decibels,* which are described, along with an illustration of sound waves, in the following chapter.

Infrasound and Seismography

A *seismic* vibration is a disturbance or *shock wave* in a celestial body such as the Earth or the Moon. Events that cause shock waves generally create acoustic disturbances at the same time. Some people refer to a shock wave as an intense acoustic wave, although scientists prefer to make a distinction based upon the character of the shock wave, which travels very fast through an elastic medium, creating a conical wavefront. Since there are similarities between acoustic waves and shock waves and the detectors that sense them, seismic waves are discussed along with acoustic waves in this chapter. Seismic and acoustic vibrations can yield important information about large-scale events such as floods, tidal waves, slides, earthquakes, the impact of an asteroid, and volcanic eruptions. We describe them as shaking or rumbling sensations, often with low-level sounds that are more felt than heard. Acoustic waves can also result from chemical or nuclear explosions.

A *seismograph* is an instrument designed to record and display seismic vibrations. Seismographs are important tools for earthquake prediction and disarmament-treaty monitoring. Early seismographs used pens to trace lines on paper. More recent seismographs may use computer monitors to display digitally processed information. Some even have warning capabilities that will email data to a predefined address or set off alarms that will wake a scientist or graduate student interested in observing cataclysmic events such as the impending eruption of a steaming volcano.

Elephants are able to sense infrasonic vibrations down to almost the lowest end of the scale. They even communicate with one another using infrasonic vibrations that can't be heard by humans but may sometimes be felt as body vibrations. If you consider that elephants are large animals with broad feet that have a constant solid contact with the ground, it makes sense that they have evolved the ability to interpret some of the deeper vibrations that form part of their environment.

Ultrasound

The main applications for ultrasound are sonar, industrial materials testing, and medical imaging. Sonar is discussed in the next chapter. Medical imaging is a large and growing field beyond the scope of this text, but it's worth noting that most ultrasound systems are designed to convert acoustic information into a visual representation on a graph or computer monitor. This image can be a series of still frames or a realtime representation of movement, like an animation. The images can be inversed (like a negative), grayscale (to look more natural), or pseudo-colored (sometimes called false colored), to help to interpret the image. An ultrasound printout is called a *sonogram* or *sonograph*. Sonograms are sometimes printed on photographic film to resemble X-ray images.

The term *sonic* is sometimes incorrectly used to represent 'ultrasonic' (perhaps this came about for reasons of brevity). The term sonic correctly refers to sounds within the audible human hearing range.

1.b. Other Sound Concepts

The Sound 'Barrier'

When objects, bats, or birds fly, they create pressure waves in the air, but the disturbance is so minimal, it is hardly noticed. However, when objects travel very, very fast, they build up a region of pressure at the forefront of the nose or wings that becomes more and more difficult to overcome. Engineers thought the speed of sound might be the upper speed at which airplanes coul fly, so they labeled it the *sound barrier*.

With the right design and sufficient energy, however, it is possible to break the sound barrier. This happens, for example, when a bullet is fired from a gun. When a large object, like a Concorde jet, successfully breaks through the sound barrier, there is a dramatic pressure disturbance called a *sonic boom*, which is a high-intensity burst of sound that covers many frequencies, resulting from the *shock wave*. Sonic booms are so loud and disturbing, in fact, that the commercial use of fast jets is strictly regulated. They are only permitted to cross the sound barrier at certain times and locations (e.g., over the ocean). Sonic booms also limit the ways in which very-high-speed surveillance jets are utilized to carry out reconnaissance missions. They are flown on schedules and from locations that minimize the chance that the boom will betray their presence.

Sound Zones

The phrase *dead zone* is often used in connection with sound applications. Since sound is dependent upon the medium through which it travels, changes or variations in the medium will affect the pitch and speed of the sound and even the direction of travel. Some objects will 'block' sound, causing the sound waves to pass over or around the obstruction. In general terms, a dead zone is an unresolvable area, a region in which there is no sound, and so the sound probe cannot provide any useful information.

It is important to understand dead zones in terms of imaging and testing, so as not to confuse the dead region in the data or image as an actual object. For example, when you use sidescanning sonar on a boat, there will be a dead zone in the center between the right and left scans in which no data is being acquired. It should be interpreted as an unresolvable area rather than as a physical trench below the boat. That might seem obvious, but sometimes the effect is not so obvious. When the sidescanning sonar sweeps outward from the boat, the objects on the right and left will protrude to different degrees and cast sonar 'shadows' on the side of the objects that are away from the boat. It will appear as if nothing is there, when in fact there may be many protrusions within those unresolvable areas that aren't showing up on the scan because the sound never reaches them.

A dramatic moment was captured by photographer John Gay as an F/A-18 "Hornet" broke the sound barrier in July 1999. The light region represents the area where a massive pressure wave built up on the front of the jet, as it flew progressively faster. Then condensation formed as the air cooled to create a visible cloud as the jet broke through. When aircraft like military jets or the Concorde commercial jetliner break through the sound barrier, they create a sonic boom that can be heard for great distances. The industry is regulated so that the sound barrier is only broken away from populated areas. [U.S. DoD news photo, released.]

In industrial testing with ultrasound, there may be a dead zone between the resolving area of the probe and the material next to the area of the probe wherein the pulse cannot be processed, in other words, a region that is too close for the pulse to 'focus' to provide useful information. There may also be dead zones created by materials that cannot be penetrated by the sound.

Frequencies Used for Surveillance

To make a sound wave inaudible to humans, we generally manipulate its *intensity* and/or its *frequency*.

- Intensity is perceived by us as the loudness or softness of a sound. Intensity-control has some obvious applications in surveillance. A whisper is less likely to be heard by prying ears than normal speech and a tiny earphone held next to the ear is less likely to be noticed than a loudspeaker on a handheld radio or speakerphone.

- As has been mentioned, frequencies below 15 Hz or above 20,000 Hz are inaudible to humans. Our ears and brains aren't equipped to sense below and above those thresholds which means that these frequencies can be used to covertly convey information or to take measurements.

This chapter concentrates on technologies that use frequency to provide covert communications or which are designed to detect phenomena that we can't hear with our unaided ears.

More of the basic concepts and attributes of sound are described in the introduction sections of Chapters 2 and 4, and you may wish to scan those sections before continuing here.

2. Kinds and Variations

A number of technologies have been developed to detect non-audible sound waves. Infrasonic detectors can be land or water based.

- *Hydrophonic detectors* are commonly used to detect acoustic waves underwater and *seismographs* are commonly used to detect ground-level shock waves. Most hydrophones are general purpose; they detect sounds within a wide range of frequencies.

- Most seismographs are specialized to detect low-frequency pressure waves (so that they don't react to every honking horn or spoken word). Depending on their design, sensitivity, and location they may respond to slamming doors, moving feet, traffic rumbles, earthquakes, and strong tidal currents.

- *Vibrational motion-sensors* are somewhat like specialized seismic detectors. These small, infrasonic surveillance devices can be placed near entrances or restricted areas to detect approaching footsteps from people or animals and can sound an alarm or activate other surveillance devices.

Ultrasonic sonars are usually used in marine environments, though some are used for controlling robots on land. Ultrasonic detectors for medical and industrial purposes are usually used on land, but typically in conjunction with liquids or gels to aid in sound conduction between the detector and the material being checked.

3. Context

All surveillance technologies have to deal with clutter, *unwanted information of one sort or another, and infra/ultrasonic applications are no exception. Infra/ultrasound technologies are used in many different sciences and the means for interpreting the signals will depend in part on what type of information is desired and how accurate it needs to be.*

Since infrasound is used largely to monitor large natural events and explosions, there is a need to determine the source of the disturbance, the speed at which it is moving or dissipating, its magnitude and location, and its potential for harm. National weather and disaster-relief organizations are continually dealing with these issues and seeking to improve their surveillance technologies and forecasting abilities. The results are not always perfect, but there have been many strides in the last decade, particularly with satellite sensing, Doppler radar, and more sophisticated computer processing and display technologies providing additional data. With these tools it is becoming somewhat easier to predict major earthquakes, the path of a storm, avalanche activity, rock slides, and lava flow from a volcano. Devices are also used to monitor global political conflicts and compliance with disarmament treaties.

Seismographs can be used to detect infrasonic disturbances and shock waves, but distinguishing a natural event from a human event, such as a nuclear explosion, is both art and science. Identification patterns and criteria have been established to aid in the making of a 'best guess.' The *mb/Ms* method is one means of comparing the magnitude of *seismic body waves* to *seismic surface waves*. Thus, among other things, the influence from events above ground can be compared with events below ground to yield a profile of the event. In other words, a volcanic eruption may have different characteristics and a different mb/Ms profile than an earthquake or a tidal wave. When this information is compared with decades of collected data for various types of events, it provides a guideline for analysis within certain confidence levels. Local data on the surrounding geology can be combined with generalized data to narrow the evaluation even further. Data from other sensors can help round out the picture. Satellite photos, radiation detectors, broadband hydroacoustic detectors, and X-ray images can all be used in combination with seismographic detectors.

While no exact numbers are known, it is estimated that there are over 10,000 seismic monitoring stations worldwide, and the International Monitoring System (IRS) has almost 200 seismic stations. Many of these stations have a variety of detectors and can monitor shock waves and infrasound waves at the same time. There is considerable collaboration between different monitoring stations; data from several stations can be cross-referenced to reveal a better picture of the source and velocity of a disturbance. The data can also be used to cross-check individual stations to make sure equipment is working correctly and to eliminate data from faulty stations that might otherwise skew the overall results.

Simultaneous seismic events can sometimes be difficult to distinguish from one another or may make identification more difficult if one is deliberately used to mask others. For example, a nation testing nuclear explosions could detonate more than one device simultaneously, obscuring the exact nature of the activity and making it difficult to determine the magnitude of the event or whether there was one or more explosions. *Low yield tests*, those which are of lower magnitude, or tests conducted within insulating geographic structures can also be difficult to detect.

Industrial Testing and Safety

Ultrasound is an important aspect of flaw detection and quality assurance in industrial manufacturing. Computerized ultrasonic detectors are now routinely used to inspect welds, joints, cracks, laminates, and many other materials and fabrications. Ultrasonic detectors are usually *resonance systems* or *pulsed systems*. The basic ways in which ultrasound testing is conducted include

- *immersion testing* - materials are put in a liquid bath with fluids that are suitable for the type of material used and the testing probe is immersed as well, to send sound vibrations through the liquid medium and materials (sound doesn't travel well between air and liquid, as it will refract, hence the liquid medium), and

- *contact testing* - materials are coated with a liquid or gel to provide a good contact between the ultrasonic probe and the materials being tested. This method is not suitable for all types of materials, but is convenient and makes it possible to develop portable detectors.

Surveillance of infra/ultrasound is important in industrial work environments because workers cannot directly hear sounds outside of human hearing ranges which may, in fact, be intense enough or persistent enough to cause damage to hearing. For worker protection, various health and hearing agencies have established commercial and industrial guidelines for noise levels, ear protection, and worker exposure.

The U.S. Army Corps of Engineers, for example, has established guidelines for safety and health. The National Institute for Occupational Safety and Health (NIOSH) *Hazard Evaluation and Technical Assistance* (HETA) program responds to requests to examine hazards in the workplace, including acoustical hazards. HETA uses infrasound measurements along with audiometric testing and noise evaluation to assess an environment. (Other general health organizations that publish acoustic standards and guidelines are listed in the Chemical Surveillance chapter *Resources* section.)

4. Origins and Evolution

Note, ultrasound shares a common history with sonar technologies. It is recommended that you cross-reference the historical information in the following chapter with the information provided here.

Infrasonic and ultrasonic devices do not have a long history, as compared to optics, for example. They did not really become practical until there was a greater understanding of electronics and of methods of sending and detecting controlled acoustic resonances or pulses.

Infrasonics and Early Seismologic Studies

Records of cataclysmic events go back thousands of years and the ancient Greeks may have had some early earthquake-recoding technology that was sensitive to shock waves and infrasonic vibrations.

Chang Hêng, a Chinese philosopher, is reported to have invented the earliest known seismoscope, in 132 A.D. The instrument was a ceramic urn, adorned with eight open-mouthed dragons about its middle, each mouth sporting a small ball. If the urn was disturbed by a ground vibration, it would tilt and discharge one or more balls in the mouths of waiting ceramic frogs at the base of the urn, thus indicating the magnitude of the disturbance and its direction of origin [Needham, 1959]. News of this innovation does not appear to have reached the western world at the time, or if it did, the records were lost.

In 1783, an Italian clock-maker, D. D. Salsano reports his invention of a 'geo-sismometro,' a geoseismograph based on a brush suspended on a pendulum, that could make marks with ink on an ivory slab, a concept not far from later seismographs.

Robert Mallet, an Irish engineer responsible for the design of many London bridges, made some initial studies of seismic waves. Mallet had a novel approach. In 1851, he used gunpowder to create explosions and then studied the different geological effects of the pressure waves through the Earth, particularly the velocity of the waves.

The ruins of City Hall (left) and rubble-clearing (middle) after the 1906 San Francisco earthquake. Mass destruction from the quake prompted the improvement of earthquake-prediction technologies and more stringent building codes. The after-effects of a quake (right), such as floods and fires, can also have devastating consequences. [Library of Congress photos from the Detroit Publishing Co. and Pillsbury Photo Co., copyright expired by date.]

Other European scientists up and down the western coast of Europe were making similar studies, perhaps motivated by a series of strong earthquakes that had taken place in the previous century. Some came up wih ntresting ideas for recording earthquakes, including the use of mercury- or water-filled vessels.

By the mid-1800s, detection instruments were becoming more sophisticated, able to measure a number of different attributes of a phenomenon. As clock-making improved so did scientific instruments, which had many components in common with precision clocks. In 1856, a 'sismografo elettro-magnetico,' an electromagnetic seismograph, was installed by Luigi Palmeire in a volcanic observatory on Mount Vesuvius. It was intended to provide a variety of measurements, including direction, duration, and intensity. This device was also significant in that it measured both horizontal and vertical movement.

The Lassen Volcanic National Park was established in 1916 in response to eruptions of Lassen Peak around 1914 that continued to 1921. Sometime after 1916, the Loomis Seismograph Station was constructed in the park to sense shock waves and provide readings. [Library of Congress HABS/HAER collection 1976 photos by Fred English, public domain.]

In the late 1800s and early 1900s, there was much interest in the study of earthquakes in Japan, and both western and Japanese researchers based in Japan contributed to the emerging science of seismology. Some important quantitative scientific discoveries were made by Sikei Sekiya in the process of studying earthquakes and John Milne was instrumental in the founding of the Seismological Society of Japan in 1880. Milne later spearheaded the establishment of global seismic monitoring stations.

Ultrasonics - Early 'Reflectoscopes'

The *reflectoscope* was one of the earliest ultrasonic devices, but the term sometimes causes confusion because there were even earlier devices called 'reflectoscopes' that had nothing to do with sound; they were visual projection devices used to enhance lectures in the early 1900s. Visual reflectoscopes were often used in conjunction with other projection devices called *stereopticons*. Oldtimers may remember them as 'magic lanterns.' These evolved into present-day *opaque projectors* and the stereopticons are today's *slide projectors*. The Boston City Council authorized that schools be provided with electric 'stereopticons' and 'reflectoscopes' in 1902. Tuck Hall, the first business school in America, had its lecture room equipped with a stereopticon and reflectoscope at least as early as 1904. Thus, the term was in regular use before sonar and other modern acoustic-sensing devices were even invented.

As visual display devices evolved, they took on names more specific to their functions and the term reflectoscope was borrowed by inventors to describe acoustic sensing devices. The term is appropriate, since it is descriptive of the way sound is projected out from the projecting device, similar to the way light is projected out from a magic lantern. Just as the light is 'read' when it hits another object and reflects the image back to our eyes, most sound-detection instruments read reflected energy when it returns to its source or a receiving device at some other location.

Some of the earliest, most rudimentary acoustic 'reflectoscopes' began to emerge in the late 1930s but they didn't have the imaging capabilities of later systems. Up to this time, most sound technologies were within human hearing ranges because the technological means to generate high-frequency sounds weren't yet practical. Electronics needed to become more fully developed. (The emerging concepts of ultrasound were gradually applied to sonar, as well, as described in Chapter 4.)

Rail Expansion and Quality Control

As with so many other technologies, testing techniques for expanding rail systems arose, in part, because of a tragic accident. In 1911, in Manchester, New York, a train derailed, killing almost three dozen people and injuring dozens more. The U.S. Bureau of Safety investigated the tragedy and concluded that a rail with a defect that was not visually apparent was probably responsible for the event. When other tracks were subsequently inspected, it was found that the Manchester rail defect was widespread.

Heightened awareness provided an opportunity for inventors to create detection/inspection devices that could potentially save dozens if not hundreds of lives and prevent the damage of costly rail cars. A number of experimental techniques for rail inspection were tested, including sound probes and magnetic probes.

Top and Bottom Left: A cleanup crew surveys a train wreckage on the Intercontinental Railroad (ICRR) in Illinois, in October 1909. Bottom Right: A train wreck in 1914 illustrates the need for technologies to detect defective tracks and to maintain tracks once they are laid. The same technology can be adapted to check the structural integrity of the bridges and trains as well. [Library of Congress photos from International Stereograph Company, and photo from the South Texas Border collection by Robert Runyon, public domain by date.]

One of the inventors interested in tackling this challenge was Elmer Ambrose Sperry (1860-1930), who is best known for his invention of the gyroscope, an instrument now incorporated into gyrocompasses, which are an important tool of navigation. In 1923, Sperry initiated the development of a rail car especially designed to inspect lengths of track and to look for the type of transverse defect that had been discovered in the broken rail in Manchester.

Sperry's initial experiments involved the use of magnetic flux to seek out defects in the rails. These magnetic detectors were used for a time. Later experiments in the use of ultrasonic detectors showed significant promise, gradually overshadowing the magnetic detectors. (See the Magnetic Surveillance chapter for more information on Sperry's early detection devices.)

Display Technologies and Ultrasonics

The invention of the cathode-ray tube became a great boon to many detection technologies. Oscilloscopes, radar and sonar scopes, and scopes that could be used with future ultrasound devices began to be mass-produced around the mid-1930s, providing a means for visually representing sound and electromagnetic energy.

The early patents on ultrasonic sound detectors are reported to have appeared in the early 1940s and may, in part, have been inspired by the need for detectors for the construction and testing of large vessels during World War II. One of the first wartime applications was testing for flaws and de-laminations in the armored plating of tanks. It is likely that the technology was applied to the inspection of ships, as well.

Since the 1920s, Elmer Sperry had been involved with the detection of transverse defects in rail tracks, but it became apparent, as the science improved, that there could be many other types of defects that were not easily discovered with magnetic probes. Floyd Firestone and Elmer Sperry subsequently developed the 'supersonic reflectoscope,' which is essentially an ultrasonic flaw-detector, to detect internal discontinuities in industrial materials. Sperry continued over the decades to develop industrial products and Sperry reflectoscopes are now commonly used for inspecting materials and welds.

The invention of transistors, in 1947, contributed substantially to the development of smaller, more sophisticated, lower-power-consumption electronics devices. Other inventors became interested in the commercial potential of industrial detectors.

In Germany, after the war, inventors returned to peacetime inventions. Karl Deutsch, who rose to leadership in the field of industrial flaw detection, read a journal article about the reflectoscope and subsequently asked a young radio operator with electronics skills to build an ultrasonic flaw-detector similar to the reflectoscope. Before Branscheid completed the engineering task, other German inventors, Josef and Herbert Krautkrämer, announced that they had developed a similar device. In spite of the competition (or perhaps because of it), Deutsch continued his work, branching into other areas.

By the late 1940s, pulse-echo metal flaw-detectors had been developed in Germany by the Krautkrämers. The early models were gradually improved to operate at higher frequencies, with shorter pulse durations. These advancements helped improve the resolution of the devices and the reputation of the Krautkrämer company.*

Deutsch and Branscheid continued their work and developed an instrument called the Echograph, introducing it commercially in 1951. A patent dispute developed between the Echograph and the Sperry-reflectoscope, which was brought out in 1952. The Karl Deutsch company shifted its focus to other products, introducing the RMG in 1957, an instrument designed to measure crack depth. The Echograph was updated with new transistor technologies, resulting in a modular version that was designed for use with ultrasonic systems.

When sound detectors were first introduced, operators had to learn to interpret the acoustic information and to equate differences in pitch (frequency) and the pattern of echoes to make any sense of the data. As technologies began to move into ultrasonic frequencies, it was no longer possible to interpret the data auditorily unless the frequencies were brought down into

---

*The Krautkrämers subsequently wrote "Ultrasonic Testing of Materials". The 4th edition was published in 1990 by Springer-Verlag.

auditory ranges. Fortunately, the development of cathode-ray tubes (CRTs) made it possible to visualize the acoustic information. The only problem was that early CRTs featured low resolution, monochrome images, slow refresh rates, and a tendency to get 'out of tune;' in other words, they had to be adjusted all the time. Since it was difficult to capture the images on film by photographing the screen, more direct ways of installing a photographic camera were devised, and images of ultrasonic data were available by the mid-1950s.

In 1956, the pioneer C-scan images were created at the Automation Instruments Power Plant Inspection Division test lab with the Hughes Memotron. This large industrial instrument incorporated photographic imaging capabilities with sound pulse/receiver units.

Industrial Flaw-Detection

Since his initial work in the 1920s, Sperry had continued to tackle the challenge of rail line inspection.

Trolleys, trains, and cable cars were widely used throughout North America and Europe in the 1950s and it was important to inspect and maintain the tracks to ensure public safety. In 1959, the New York City Transit Authority (NYCTA) installed a Sperry Rail track-inspection car based entirely on ultrasonics.

'Portable' flaw detectors were being devised now as well, by both Sperry and Automation Instruments. The transition from tubes to transistors was making it possible to create instruments one-tenth the size of their vacuum-tube counterparts, but true miniaturization didn't emerge until the late 1960s and 1970s when microelectronics had a significant impact on both the size and processing capabilities of these industrial flaw-detecting machines.

Nuclear Science and International Regulations

Most of the acoustic surveillance technology in the early 20th century was focused on sonar and industrial ultrasonics, though seismic technology was being developed as well.

In 1945, at the end of World War II, the United States dropped atomic bombs on the cities of Hiroshima and Nagasaki. This provoked the development and test detonation of a Soviet atomic bomb in August 1949. The Arms Race had begun and concern was raised about the proliferation of atomic weapons and the means to detect their presence.

In the early 1950s, seismic stations to detect atomic explosions were installed in strategic locations, including Alaska, Germany, Greenland, Korea, and Turkey. Other technologies to detect radiation and atmospheric changes were also developed.

Up to this point, ultrasonic technologies were primarily analog, but this was about to change. In 1973, the Karl Deutsch company introduced the first digital ultrasonic flaw detector, called the Echotest.

Nuclear energy is not used only for building bombs. It is also used to power generators and to create explosions for geological surveys and resource exploration. Seismic measurements from various types of explosions are sometimes used to study geological formations. Unfortunately, the fallout and waste from nuclear explosions can have deadly consequences for plant and animal life, including death, radiation burns, some forms of cancer, immune-system damage, and a host of lesser health problems. The difficulty of disposing of nuclear wastes or avoiding nuclear fallout (radioactive particles) is part of the reason for international restrictions on nuclear testing and global surveillance of nuclear weapons development.

In May 1974, India exploded a "peaceful" nuclear device that was estimated on the basis of seismic and visual surveillance, to be about five kilotons or less (about half that announced by India). The Threshold Test Ban Treaty instituted that year set a limit of 150 kilotons for nuclear testing.

Technological Developments

At the end of 1974, the Altair desk-sized computer kit was introduced to hobbyists. Within two years, the microcomputer revolution was in full swing, forever changing electronics in dramatic ways. Microcomputer electronics were soon incorporated into a wide range of sensing devices including acoustic and seismic wave detectors.

In 1976, the Group of Scientific Experts (GSE) was given a mandate to study the scientific basis for monitoring a test-ban treaty from the CCD (now known as the Conference on Disarmament). The GSE was selected for this task on the basis that it had done a lot of scientific groundwork on seismic monitoring. The data would be transmitted, via satellite, to the International Data Center (IDC).

In 1976, the Karl Deutsch company, which had become a recognized leader in industrial detection technology, introduced a computer-controlled ultrasonic flaw-detecting system, called the Echograph 1160.

Ultrasound technologies for industrial testing and medical imaging continued to improve in the late 1970s and early 1980s. Microprocessors were now starting to be used in many electronics devices. The Karl Deutsch Echograph 1030, a microprocessor-controlled ultrasonic flaw detector, was introduced in 1983, in Moscow. The following year, the Echometer 1070 was demonstrated, equipped with a microprocessor-controlled ultrasonic thickness gauge. In 1987, Karl Deutsch, still a leader in industrial ultrasound technology, added more sophisticated data processing capabilities to its systems. By the early 1990s, improved very-large-scale-integration (VLSI) made it possible to produce smaller, more powerful detectors and portable versions of the industrial models.

From the time of the early ultrasonic reflectoscopes that were used for rail inspection, Sperry had also been continually improving and upgrading its products. The ultrasonic C-scanner was available in the 1980s, along with higher-end versions of the product with computer control and color display capabilities.

Digital seismographic instruments were becoming more prevalent by the mid-1980s, and the technology continued to improve, with low-noise seismometry, higher-bandwidth data rates, greater storage and processing capacities, and the increased ability to communicate through satellites and computers.

Political Policies and Technology

The United States sometimes sets domestic policies and carries out research that appear to contradict its foreign policies. At a time when the U.S. was seeking to monitor and suppress nuclear proliferation in other nations, it was also proposing internal testing procedures that could obscure America's own nuclear explosions. In the early 1980s, work in U.S. labs included design ideas for underground nuclear test facilities which could contain large-scale explosions to minimize the seismic impact, described by some as a 'quiet' bomb-testing installation where tests could be carried out without alerting the rest of the world. At that time and since, there have been many American initiatives for exposing, monitoring, and suppressing foreign weapons facilities, exposing the irony of the position of the U.S. Government with regard to quiet testing.

Supercomputers were emerging in the 1980s, providing new ways to compile and analyze complex data from multiple sensing stations. Powerful computers introduced new industrial testing and simulation capabilities as well. The improvement in computing technology made it possible for some aspects of nuclear testing to be simulated, providing developed nations with a means to carry out weapons some aspects of research with fewer actual detonations, options which some countries don't have.

In the early 1990s, the concept of 'nonlethal weapons,' began to get attention in political circles and in the press. Weapons designed to stun or disable, rather than kill, have been available as consumer items since about the 1980s and are regularly shown on science fiction series such as Star Trek™, but the use of these on a large scale was not given substantial attention until technologies improved enough to provide a great many more options. The development of lasers, electromagnetic-pulse devices and infrasound technologies made a broader range of nonlethal weapons possible.

On the one hand, there are definitely situations in which it is better to disable and apprehend than to kill. Fleeing criminals, wild animals loose in populated areas, and marauding forces are all examples in which nonlethal weapons can be a humane solution. On the other hand, some types of nonlethal weapons can do subtle, perhaps cumulative harm that is difficult to document and prove and thus difficult to trace to the time of the event or the persons responsible. There is no smoking gun in nonlethal attacks. Like chemical warfare, nonlethal warfare in the wrong hands is potentially one of the most frightening and horrific adaptations of science and should not be pursued without the most stringent of ethical judgments. Opponents to nonlethal weapons, which include certain infrasound devices, began to speak out against the technology in the early 1990s.

Global Monitoring

In May 1955, the Soviet Union proposed a ban on nuclear testing, initiating a process of negotiation that would span many decades. Some of the other key players were the United States, France, and Great Britain. In 1957, the Soviet Union made further suggestions for prohibitions that included international supervision of the process. In March 1958, the Soviet Union made an announcement that it would be discontinuing nuclear tests, with appeals to other superpowers to follow suit.

In the summer of 1958, the "Conference of Experts on Nuclear Test" convened in Geneva, Switzerland, to discuss issues related to the monitoring of a ban on nuclear tests. Confirmatory on-site inspections were also discussed to distinguish natural events from human-initiated explosions.

Beginning in 1959, the U.S. Air Force's Project VELA Program involved the installation of a major global network of seismographic stations, called the World-Wide Standard Seismological Network (WWSSN). This contributed substantially to the geosciences, especially plate tectonics, and created a new body of data from global monitoring. It also provided a means of detecting signs of nuclear detonations that could be instrumental in international monitoring.

In 1963, the Limited Test Ban Treaty (LTBT) was negotiated and subsequently signed and put into force. It prohibited nuclear explosions in the atmosphere, under water, or in outer space. Underground tests that might result in the spread of radioactive products were also banned.

In 1976, the Conference on Disarmament (CD) established the Ad Hoc Group of Scientific Experts (GSE), to develop and evaluate cooperative international actions toward enforcement of the Comprehensive Test Ban Treaty. The GSE has since been involved in the development of networked data-exchange systems and technical tests related to seismic monitoring.

Treaty Ratification and Enforcement

Despite disagreements, disarmament talks appeared to have gotten off to a reasonably good start in the 1960s and 1970s, especially on the issues of nuclear testing. But then, somehow, the negotiations dragged on. Large countries continued to develop new technologies that could conceivably be used as weapons, even if that was not the original intention of the technologies. Smaller countries continued to try to improve their weapons capabilities and were gradually arming themselves with nuclear capabilities. The Soviet Union collapsed and turmoil broke out in the nations bordering Russia and economic havoc descended on the former U.S.S.R. By the 1990s, disarmament had not come as far as people had hoped and the threat of nuclear war or terrorism in smaller nations increased. Progress in nuclear technology in China led to testing and the rest of the world relied on surveillance technologies to monitor the magnitude and progress of the testing activities.

In 1992, China conducted nuclear tests in May. The site location was recorded, a month after the blast, by the U.S. Landsat-4 imaging satellite. The aerial images suggested that the Chinese had prepared two test sites. Further nuclear tests were conducted by the Chinese Government in September. Seismic surveillance technologies, in conjunction with other data, were used to estimate the extent of the explosions.

Around this time, commercial satellite companies began marketing high-resolution images from Russian satellites. The resolution of commercial imagery at this time was sufficient to detect trucks and buildings (more information on satellite capabilities is described in the Aerial Surveillance chapter).

In October 1993, China is reported to have conducted an underground nuclear explosion at their Lop Nur test site. The Verification Technology Information Center (VERTIC) announced details of the test based on a variety of surveillance technologies and analyses, describing the range, location, and shaft orientation of the detonation. The conclusion that it was a nuclear explosion was based in part on the absence of surface waves, the 5.8-magnitude reading on the Richter scale, and surveillance monitoring of China's activities over the last several years. The news was communicated globally within the next couple of hours.

The results of the various reports of the nuclear test appeared to confirm that the global seismic network was suitable for monitoring larger-scale nuclear testing activities.

Progress in Disarmament and International Monitoring

By early 1996, the members of the Conference on Disarmament were making progress toward an international treaty on various weapons and surveillance technologies, primarily nuclear armaments, but there had also been discussion of an international network of seismic stations supplemented by a larger number of auxiliary stations. Networks of hydroacoustic and infrasound sensors were agreed to by most nations, while satellite data were still under discussion.

In May 1997, the international community was startled to discover that India had conducted nuclear tests. The limitations of seismic-sensing technologies were demonstrated when the nature and scope of these tests were not detected unequivocally by the *Comprehensive Test Ban Treaty* (CTBT) monitoring system, leading to concerns about enforcement of the treaty. It was known that explosions rated at less than one kiloton might be difficult to detect from a distance, but this was the first politically sensitive, unsettling occurrence that underlined limitations of both the technology and political process. In addition to the scientific ramifications, there were questions about whether the system should even be monitoring a country that was not yet party to the Treaty.

It's difficult to know what you have or haven't detected without concrete reference information. Controlled scientific tests can yield exact data. Observations and calculations based on second-hand information and carefully worded announcements from foreign nations are more difficult to assess. In a post-blast statement, the Prime Minister of India announced that a 43-kiloton thermonuclear device, a 12-kiloton fission device, and a 0.2-kiloton device had been detonated. India's leader stated that the 12- and 43-kiloton detonations were about a kilometer apart.

Initial data from the Incorporated Research Institutions for Seismology seemed to suggest only one event, with an estimated detonation that fell somewhere between the two larger devices, around 20-kilotons, from one source or simultaneous multiple sources very close together. This raised questions both as to the veracity of the political leader's statements and to the accuracy of readings and interpretation of seismic data.

Later data from more than 100 sources yielded evidence of a slightly larger magnitude, leading to estimates that the blast was about 30 kilotons, which still didn't match verbal reports from India. Seismic measures of the blast from different stations that reported to the International Data Center recorded magnitudes ranging from 4.7 to just over 5.0 which would indicate a blast of not much more than 15 ± 10 kilotons.

The Air Force Technical Applications Center (AFTAC) was commissioned to install twenty new seismic arrays around the world between 1997 and 2001, with the first to be installed in Peleduy, Russia, for operation by Russia's Special Monitoring Service. Data from these arrays would be scheduled to arrive at the IDC within one minute of their acquisition.

AFTAC was also planning the installation of new hydroacoustic arrays and a network of infrasonic arrays in the southern hemisphere to aid in CTBT monitoring.

In August 1997, a 'seismic event' in Russia stirred controversy over whether Russia had violated the CTBT and detonated a small nuclear explosion or whether the event was an earthquake. A seismic sensing array in Norway had picked up the ambiguous vibrations. This occurrence was significant in that it tested the global surveillance system's ability to pick up smaller events and whether it could distinguish between natural and human activities. The data gathered suggested it is not likely that it was a nuclear explosion, as it appears to have occurred out in the Kara Sea. That is not to say a nuclear test could not be carried out in this unlikely location; the Indian Sea is thought to have been the site for a possible earlier nuclear test explosion. But the body of data, taken as a whole, suggests it was a natural phenomenon.

By November, the U.S. Central Intelligence Agency (CIA), had released an assessment of the event, with input from outside experts, which stated that the event was not a nuclear explosion.

Nuclear weapons proliferation is a form of one-up-manship. The development of the U.S. atomic bomb was in response to rumors that the Germans were developing such a weapon during World War II. If indeed they were, the project was never successfully completed. In response to the U.S. detonation of two bombs in Japan, the Soviets set about creating their own 'atom bomb.' When India conducted nuclear tests in August 1997, neighboring Pakistan responded by conducting their own tests near the border of Afghanistan in May 1998. The Pakistan government claimed that five devices had been detonated. Seismic surveillance did not pick up five detonations and some think only one may have been detonated.

Assessments of nuclear detonations using a variety of surveillance technologies and human intelligence resulted in estimates that the U.S. and the Soviet Union had been responsible for about 85% of the nuclear detonations that had occurred up to this time. The total numbered over 2,000, about 26% of which were atmospheric explosions [Norris and Arkin, 1998].

In 1997, the Comprehensive Nuclear Test-Ban Treaty Implementation Act was put before the Canadian House of Commons to define Canada's obligations under the Treaty. The duties and functions of the National Authority included radionuclide monitoring, seismological monitoring, hydroacoustic monitoring, and infrasound monitoring. The Minister of Natural Resources was given responsibility for coordinating and, if necessary, operating verification measures as part of the International Monitoring System (IMS) "by means of seismological, hydroacoustic and infrasound monitoring".

By 1998, the CTBT had established guidelines for international monitoring that would include hydroacoustic monitoring and infrasound monitoring. The system was organized so that the data from these surveillance technologies would be collected and transmitted to an International Data Center (IDC) located in Vienna, Austria. Data could also be provided by nations through separate arrangements that were not formally within the IMS. Each party in the treaty would be provided with all data transmitted to the Data Center. The U.S. National Data Center would be responsible for consolidating the U.S. data contributed to the IDC as well as disseminating data from the IDC which would then be made available to government agencies and academic researchers.

By 1999, the U.S. was coming under international and domestic criticism for not having ratified the Comprehensive Test Ban Treaty (CTBT). Submitted for ratification by the Clinton administration in 1997, it was reported to have been held up by Foreign Relations Committee Chair Jesse Helms until he was given protocols for both the Anti-Ballistic Missile Treaty and the Kyoto Climate Change Treaty, which he is reported to oppose [Collins and Paine, 1999]. As a superpower, many smaller nations take their lead from the U.S. in treaty matters.

An international CTBT conference was scheduled to seek ways to move the Treaty into completion. At this point only Britain and France had ratified the Treaty, just two out of the 44 nuclear-capable nations that were needed before the Treaty could be in force.

5. Description and Functions

General descriptions and functions of acoustic devices are described in Chapter 2 and Chapter 4 and are not repeated here. The introductory sections of this chapter provide a good overview of the technologies, and the following Applications section rounds out the picture. Further information can be sought from the resources listed at the end of the chapter.

It was mentioned in the introduction that acoustic waves and shock waves are not the same. However, the devices that detect acoustic waves can be similar to those that detect shock waves, so this chapter hasn't made a strong distinction so far between shock waves and acoustic waves. However, in scientific circles, the two phenomena are considered distinct from one another.

Acoustic waves are *longitudinal*; that is, the vibrations occur in the same direction as the motion of the waves. (Electromagnetic waves, in contrast, are *transverse* waves.) The molecules that make up the acoustic medium, whether it's air, water, or some other elastic substance, will go through a series of compressions, with the particles alternately crowding together, and decompressions, where they 'relax' back into their previous positions. The particles in denser materials (wood, metal) generally resume their normal positions and less dense materials (air, water) tend to return to a similar state of relationships or equilibrium, if not the exact same positions, after the wave has passed through the medium.

Shock waves tend to be created by sudden, violent events, such as explosions, earthquakes, lightening strikes, and volcanic eruptions. (Strong acoustic waves are often generated by the same events at the same time, which is one of the reasons why they tend to be studied and measured together.)

A shock wave is a sudden violent pressure wave in an elastic medium (water, air, soil, etc.). Shock waves tend to travel faster than acoustic waves, but their intensity also tends to decrease more quickly as the heat generated by the wave is absorbed by the elastic medium. As shock waves die down, they lose intensity and the high pressure wavefront diminishes. It becomes difficult to distinguish them from events that originated as acoustic waves.

Some people claim that the difference between an acoustic wave and a shock wave is not so much a difference in the basic aspects of the phenomena, but rather a difference in intensity. In other words, some references will call a shock wave a high-intensity acoustic wave. Physics texts describe a shock wave as different in character from an acoustic wave. It consists of a conical wavefront that is produced when the *source velocity* exceeds the *wave velocity* of the wave.

From a visual point of view, a shock wave can sometimes be seen in a solid medium. (If you're close enough to see it, it's usually a good idea to leave, quickly.) If you have ever seen video footage of a violent explosion, you may have seen a shock wave. The effect is remarkable. If the ground is fairly flat and unpopulated, you can see it buckle up in waves that look as though a huge rock was thrown into the center of the 'pond,' except that the pond is sand or gravel or some other solid, elastic base that doesn't ordinarily move like water.

Shock waves move very quickly and disappear very quickly. Meanwhile, the acoustic rumbles from the explosion might reach you after the shock wave has knocked you right off your feet. In some cases the acoustic wave might never reach you, even if you feel the shock wave, because the sound may travel up and out. When Mt. St. Helen's in Washington State exploded in a volcanic eruption in 1980, many people 20 miles from the volcano didn't hear a thing, but were covered in ash from the eruption, while other people almost 200 miles away didn't see any ash, but heard the explosion so loudly it sounded like a large bomb exploding. Those who couldn't hear the explosion, even if they were nearby, are said to have been within the *cone of silence*, the region not affected by the acoustic waves.

Shock waves don't just travel through the ground, they may travel through air as well. When jets fly at supersonic speeds (faster than the speed of sound) they build up tremendous pressure on their forward surfaces, particularly on the wings and the tail. When they break through the *sound barrier*, there are significant shock waves and a loud 'sonic boom.'

For the purposes of an introductory text, the distinctions aren't as important as they are in scientific research, but it is a good idea to remember that seismic detectors and acoustic detectors are intended to record phenomena that are similar, but not exactly the same.

6. Applications

Low- and high-frequency acoustic waves and shock waves have many applications in surveillance. They have also been tested for their weapons capabilities, but these are only mentioned briefly as they are not directly concerned with surveillance and, in fact, the purpose of many surveillance devices is to detect and prevent the use of weapons or harm that might result from their use.

Historically, ultrasound transducers required a liquid contact, but more recently air-coupled transducers, developed by Dr. Schindel at Queen's University in Canada, and Dr. Hutchins at Warwick University in England, have improved to the point that many new sensing applications, particularly for industrial environments, are now possible.

Structural Assessment and Quality Assurance

In industrial applications ultrasound is used for non-destructive testing (NDT) and examination of materials, often in conjunction with other surveillance technologies, including infrared thermography and radiography. Ultrasound was found to be able to make a deeper assessment of some types of structures in some materials. Impact detection, cracking, and de-lamination are just some of the defects that may be detected with ultrasound. Ultrasound is favored for many applications because it does not require immersion or contact with the materials. The newer ultrasound technologies are beginning to replace some of the more expensive radiographic and computer tomography detectors for industrial applications.

Ultrasonic inspection of cracks, seams, and defects is now widely carried out in industrial manufacturing and quality assurance programs. By selecting specific types of transducers and arranging them in different ways, it is possible to gather data on a large variety of types of deect. Pulse-echo and pitch-catch are just two of the techniques that can be used. Industrial weld examinations are typically done at frequencies of 2.25 or 5.0 MHz.

Weather and Disaster Prediction and Monitoring

Seismographic detectors are often used in conjunction with Doppler radar weather imaging systems to predict the occurrence and magnitude of natural disasters, including earthquakes, floods, and volcanic eruptions. Left: A Pave Low IV helicopter about to refuel over severely flooded Central Mozambique, in March 2000. The U.S. military contributed to relief efforts after torrential rains and floods. Right: Spc. Rodney Porter unloads a shaken dog that was rescued from flood waters caused by Hurricane Floyd in September 1999. The National Guard provided humanitarian relief for people devastated by the hurricane. [U.S. DoD news photos by Cary Humphries and Bob Jordan, released.]

Infrasound is used for detecting and measuring earthquakes, tidal waves, volcano eruptions, explosions, large weather systems, and floods. The information can be applied to disaster prevention and relief and search and rescue efforts in a diverse number of situations.

The 'weather' outside our atmosphere is also monitored by a number of technologies including infrasound. When the Earth is hit by meteors we often notice their presence from the sparkles of light that occur when they fall to Earth and burn up in the atmosphere. During the day, however, it is harder to see these events, particularly if the meteors are small. Infrasonic vibrations from the disturbance can sometimes reveal their presence.

Avalanche Detection

With so many more people taking to the mountains for sports and sightseeing, avalanche prediction and detection are increasing concerns for park personnel and search and rescue professionals. Infrasonic instruments that continually monitor the ambient vibrations are used to detect avalanches and slides and communicate that information to the appropriate station, including the magnitude of the disturbance, the location, date, and time. The effective range of these systems varies with conditions and the local geography, but can reach up to about five kilometers.

Infrasound detectors are now used in many regions to predict and detect avalanches and rock slides. Here, a U.S. Army Blackhawk helicopter comes in for a landing after rescuing avalanche victims (right) who were stranded on a mountain in Austria in February 1999. [U.S. DoD news photos by Troy Darr, released.]

Collision Avoidance

One of the newer adaptations of infrasound is in guidance control and safety systems to aid in preventing collisions. Low-level sound vibrations at times or in environments where the vibrations shouldn't occur can serve as advanced-warning systems or as grounds security monitors. Infrasound has even been proposed as a means of reducing bird-aircraft collisions.

Communications Security

In efforts to make communications more secure, a number of technologies to change or scramble voices or to create confusing noise have been developed. Ultrasound is one of the means that has been used to disrupt microphones (such as pocket recorders or bodyworn wires), with mixed results. Effectiveness may depend on the type of microphone that is being used and the ability of the conversants to tolerate the ultrasound which can cause nausea and general malaise even if it can't be heard.

Grounds Security

There are many ways in which infrasound can be used in commercial, military, and personal security applications. Infrasonic ground vibrations can signal the presence of vehicles or foot traffic, the movement of air can indicate opening doors, windows, or drawers. If the sensor is triggered it can sound an alarm or turn on a microphone or video surveillance device. Infrasound burglar alarms are available for about $35. Ultrasonic applications for security are now increasing as well, with newer air-conducting ultrasonic technologies that are suitable for distance-sensing, room surveillance, and robotic imaging.

Ultrasound is sometimes combined in the same housing with infrared sensors. This combination can reduce the incidence of false alarms.

Ultrasound has been used along with infrared in guidance systems for autonomous robots designed for patrol and grounds security. Higher frequencies tend to provide more accurate assessments of distance in robotic navigation applications.

Microphonic Monitoring

Microphones designed to monitor sounds within and without human hearing ranges have become increasingly small, inexpensive, and powerful.

In September 1999, New Scientist reported that Emkay in West Sussex had developed a broadcast-quality microphone so tiny it would fit into a 2.5 millimeter cylinder. This broadband microphone can record ultrasound as well as audible sound, making it suitable for surveillance and wildlife research applications.

Medical Applications

Ultrasound allows internal biological structures to be detected and imaged in realtime and thus is highly valuable to the medical industry. The denser areas of cartilage and bone and some of the organs will show up more readily than fluid-filled cavities and tissues. Ultrasound is used to detect, assess, and monitor fetal development and a variety of disease conditions including heart disease, diabetes, and cancer.

Sandia Laboratories have developed some powerful three-dimensional ultrasound technologies that allow CAD and acoustic technologies to enhance one another to create detailed images that reveal internal structures in relation to one another. Three-dimensional ultrasound systems can be used in industrial non-destructive evaluation and testing, mine and bomb detection, and medical diagnostic imaging and prosthesis evaluation and fabrication.

Rail Inspection

Rail lines are subjected to great weights and temperature changes and require constant inspection and maintenance. Rails are now surveilled with a combination of ultrasound and electromagnetic sensors. The electromagnetic equipment tends to be bulkier than the sound-sensing equipment. These detection systems are usually housed in self-contained rail-inspection cars, with the probes mounted between the axles. Rail-inspection trucks are also used, but generally carry only the ultrasonic detectors.

Detection probes on moving conveyances can only move as fast as the technology can keep up with the data. For this reason, most sonar depth sounders have upper speed limits of about 20 knots and imaging systems rarely go above 1 to 4 knots depending on their resolution and sophistication. Rail inspection follows similar principles, with detection devices being moved along at speeds of up to about 12 ± 3 kilometers per hour.

Robotics Sensing

The Polaroid 6500 Ranging Module is an economical ultrasonic sensor that is used in many hobby and commercial ranging applications, especially for longer-range robot navigation. Distance calculation is not carried out by the Module, but can easily be calculated with interfaced microprocessors.

Weapons Detection

There are many detectors now used on public transportation systems to prevent possible harm from weapons and bombs. Airlines now regularly use X-rays and magnetic detectors. Ultrasound sensors are now being developed to remotely detect and image concealed weapons. For example, support from DARPA and the Concealed Weapons Detection Technologies grant program has resulted in the creation of an ultrasound detector that can detect both metallic and nonmetallic weapons under clothing up to a distance of about 5 to 8 meters. Information is available through www.jaycor.com .

7. Problems and Limitations

Passive infrasound, ultrasound, and seismic detectors do not pose any direct risks to physical structures or human health, since they are designed to detect rather than generate acoustic waves and shock waves. However, *active* infrasound and ultrasound are not completely harmless technologies. Most active acoustic technologies do not emit high intensity sound waves and have a low probability of doing harm, but since they involve generating an acoustic pulse that cannot be heard by humans and which can be emitted at high intensities, they must be used with a certain amount of caution.

Prior to the use of ultrasound for prenatal care, X-rays were sometimes taken of the fetus if the mother or fetus were determined to be at risk. The risk of using ultrasound for surveilling the womb to determine the health of the fetus is less than that of X-rays but is not known to be 100% safe. Many medical decisions are based on trade-offs. The risk of taking an ultrasound reading is considered to be much less than the risk of taking an X-ray. Taking an ultrasound reading is also considered to be less of a risk than not taking an ultrasound reading, as it is a valuable means to monitor the health and progress of a fetus.

Since infrasound has been evaluated as a potential weapon with the possibility of creating cardiac, respiratory, nervous system, or digestive system dysfunctions, it is clear that the application of the technology may not be safe in all circumstances, but generally, infrasound and ultrasound technologies are more beneficial than harmful.

8. Restrictions and Regulations

There are not a lot of regulations explicitly prohibiting or restricting the uses of infra/ultrasonic technologies for surveillance purposes, except perhaps in the medical industry, where it is used to assess fetal health and a variety of illnesses. The main concerns in industrial environments are monitoring the intensity of the acoustic waves and the provision of proper ear protection and education on the potential damage that can occur if protection is not worn.

The American Institute of Ultrasound in Medicine releases reports on a variety of subject areas including information on the medical use of ultrasound, safe usage guidelines, cleaning recommendations, and possible hazards. www.aium.org.

9. Implications of Use

The main concern in the use of infrasonic and ultrasonic technologies is keeping the intensity of the sound emanations to levels that are low enough to prevent damage to living tissues and to avoid prolonged exposure.

There has been very little controversy regarding the use of infrasound, ultrasound, and seismic detection technologies for surveillance. The two major concerns are the correct use of medical imaging for monitoring a growing fetus and the avoidance of high-intensity sound waves for military applications that may disrupt the hearing and health of marine animals.

10. Resources

Inclusion of the following companies does not constitute nor imply an endorsement of their products and services and, conversely, does not imply their endorsement of the contents of this text.

10.a. Organizations

Acoustical Society of America (ASA) - ASA is an international scientific society dedicated to research and dissemination of knowledge about the theory and practical applications of acoustics. ASA sponsors annual international meetings and annual Science Writers Awards for articles about acoustics. asa.aip.org/ www.acoustics.org/

The Federation of American Scientists (FAS) - A privately funded, nonprofit research, analysis, and advocacy organization focusing on science, technology, and public policy, especially in matters of global security, weapons science and policies, and space policies. Sponsors include a large number of Nobel Prize Laureates. FAS evolved from the Federation of Atomic Scientists in 1945. The site contains numerous surveillance, guidance systems, U.S. Naval documentation, and other radar-related information. www.fas.org/index.html

Incorporated Research Institutions for Seismology (IRIS) - This is a consortium of almost 100 research facilities, most at universities with strong research interests in seismology. It includes the Data Management System, which consolidates data from the collection centers conducting seismic recordings that make up the Global Seismographic Network (GSN). www.iris.washington.edu/

International Monitoring System (IMS) - This is a primary seismic network of stations that are designed to detect, locate, and identify underwater and underground events. There are also auxiliary stations contributing data to this system.

Naval Command, Control and Ocean Surveillance Center (NCCOSC) - Established in 1992, this center operates the Navy's research, development, testing, engineering, and fleet support for command, control, and communications systems and ocean surveillance. NCCOSC reports to the Space and Naval Warfare Systems Command. www.nswc.navy.mil/

Naval Research Laboratory (NRL) - The NRL conducts multidisciplinary programs of scientific research and advanced technological development of maritime applications and related technologies. The NRL reports to the Chief of Naval Research (CNR).

Naval Surface Warfare Center (NSWC) - Established in 1992, this center operates the Navy's research, development, testing, and fleet support for ship's hulls, mechanical, weapons, and electrical systems. NSWC reports to the Naval Sea Systems Command. www.nswc.navy.mil/

Ocean Mammal Institute (OMI) - A public awareness, research, and educational organization for the study and preservation of marine mammals. The Web site includes an information page on low-frequency active (LFA) sonar and its possible effects on marine life. www.oceanmammalinst.com/

Ultrasonics Industry Association (UIA) - A trade forum for manufacturers, users, and researchers in ultrasonic technology. The site includes descriptions of practical applications of ultrasonics, conference information, and other links. www.ultrasonics.org/

U.S. Office of Naval Research (ONR) - The ONR coordinates, executes, and promotes the science and technology programs of the U.S. Navy and Marine Corps through educational institutions, government labs, and a variety of organizations. ONR carries out its mandates through the Naval Research Laboratory (NRL), the International Field Office, the Naval Science Assistance Program (NSAP), and the Naval Reserve Science & Technology Program. ONR reports to the Secretary of the Navy and is based in Arlington, VA. www.onr.navy.mil/

U.S. Space and Naval Warfare Systems Command (SPAWAR) - This department designs, acquires and supports systems for the collection, coordination, analysis and presentation of complex information to U.S. leaders. agena.spawar.navy.mil/

10.b. Print

Hughes, Thomas Parke, *Elmer Sperry: Inventor and Engineer,* Baltimore: Johns Hopkins University Press reprint, 1993, 348 pages. Sperry was one of the early pioneers in the development of ultrasonic detectors for industrial purposes.

Krautkrämer, Josef and Herbert, *Ultrasonic Testing of Materials,* New York: Springer-Verlag, 1990 (reprint). The authors are pioneers of ultrasonic flaw-detection devices and Krautkrämer is now one of the largest companies in the industry.

Needham, Joseph. Needham wrote a series of books in the 1950s and 1960s on technology, industry, and science in China, published by Cambridge University Press in Cambridge. The topics include clockworks, metal construction, and chemistry. They also include discussions of technological information exchange between China and the western nations. Some of the volumes have been recently reprinted in shorter versions.

Sontag, Sherry; Drew, Christopher, *Blind Man's Bluff: the Untold Story of American Submarine Espionage*, New York: Public Affairs, 1998, 352 pages. Describes the Cold War submarine 'chess game' between the superpowers—the U.S. and the Soviet Union.

Articles

Cody, John D., Infrasound, Web article sponsored by the Borderland Sciences Research Foundation. It describes characteristics of infrasound and some of the events that create infrasonic shock waves, including earthquakes, floods, etc. www.borderlands.com/newstuff/research/infra.htm

Collins, Tom Z.; Paine, Christopher, Test Ban Treaty: Let's Finish the Job, *Bulletin of the Atomic Scientists*, July/August 1999, V.55(4). Discusses the political and some of the technological aspects of the Comprehensive Test Ban Treaty and its implementation.

Dewey, James; Byerly, Perry, The Early History of Seismometry (to 1900), *Bulletin of the Seismological Society of America*, February 1969, V.59(1), pp. 183-227. A very interesting article that describes some of the early inventions based upon primary sources, including some of the Italian inventions of the 1700s.

Georges, T. M., Infrasound from convective storms: Examining the evidence, *Reviews of Geophysics and Space Physics*, 1973, V.11(3), pp. 571-594.

Gilbreath, G. A.; Everett, H. R., Path Planning and Collision Avoidance for an Indoor Security Robot, *SPIE Proceedings: Mobile Robots III, Cambridge, MA,* Nov. 1988, pp. 19-27.

Jones, R. M.; Goerges, T. M., Infrasound from convective storms III Propagation to the ionosphere, *Journal of the Acoustical Soceity of America*, 1976, V.59, pp. 765-779.

Los Alamos Technical Report #LA-10986-MS, Observations of prolonged ionospheric anomalies following passage of an infrasound pulse through the lower thermosphere.

McHugh, R.; Shippey, G.A.; Paul, J. G., Digital Holographic Sonar Imaging, *Proceedings IoA*, V.13, Pt. 2, 1991, pp. 251-257.

Norris, Robert S.; Arkin, William M., NRDC Nuclear Notebook: Known Nuclear Tests Worldwide, 1945-98, *Bulletin of Atomic Scientists*, Nov./Dec. 1998, V.54(6).

Richards, Paul G., Seismological Methods of Verification and the International Monitoring System, Lamont-Doherty Earth Observatory of Columbia University. This describes general background on seismology and test ban monitoring and describes steps that can be taken to carry out the monitoring for the Comprehensive Test Ban Treaty.

Shirley, Paul A., An Introduction to Ultrasonic Sensing, *Sensors, The Journal of Machine Perception*, Nov. 1989, V.6(11). Describes ultrasonic ranging and detecting devices that work in air and the various environmental factors that can influence the signal.

Smurlo, R. P.; Everett, H. R., Intelligent Sensor Fusion for a Mobile Security Robot, *Sensor,* June 1993, pp. 18-28.

Thurston, R. N.; Pierce, Allen D., Editors, Ultrasonic Measurement Methods, *Physical Acoustics,* V.XIX, Academic Press, 1990.

U.S. Congress, Office Technology Assessment, Seismic Verification of Nuclear Testing Treaties, OTA-ISC-361, U.S. Government Printing Office, Washington, D.C., May 1988.

Journals

Bulletin of the Atomic Scientists, this professional journal includes references to infrasonic detectors. www.bullatomsci.org/

Bulletin of the Seismological Society of America, bimonthly professional journal.

The e-Journal of Nondestructive Testing & Ultrasonics, available from NDTnet. www.ndt.net/v05n06.htm

European Journal of Ultrasound, official journal of the European Federation of Societies for Ultrasound in Medicine and Biology. www.elsevier.nl/inca/publications/store/5/2/4/6/3/7/

International Network of Engineers and Scientists for Global Responsibility Newsletter, published in Germany, this provides information on various scientific and NATO activities. www.inesglobal.org/

Ultrasonic Imaging, published by Academic Press, available online up to 1996. www.idealibrary.com/cgi-bin/links/toc/ui

10.c. Conferences and Workshops

Many of these conferences are annual events that are held at approximately the same time each year, so even if the conference listings are outdated, they can still help you determine the frequency and sometimes the time of year of upcoming events. It is very common for international conferences to be held in a different city each year, so contact the organizers for current locations.

Many of these organizations describe the upcoming conferences on the Web and may also archive conference proceedings for purchase or free download.

The following conferences are organized according to the calendar month in which they are usually held.

IEEE International Conference on Acoustics, Speech and Signal Processing, IEEE international conference.

ICASSP, International Conference on Acoustics, Speech, and Signal Processing, annual conference.

New Millennium–New Vision for Ultrasonics, annual symposium.

Ultrasonic Transducer Engineering Conference, sponsored by the PSU College of Engineering.

ICANOV2000, International Conference on Acoustics, Noise, and Vibration., this was held at McGill University in August 2000 in conjunction with the *First International Conference on Acoustics, Noise, and Vibration.*

International Conference on Ultrasonics and Acoustic Emission, annual international conference.

Ultrasonics International Conference, current information not available, but conference proceedings for previous conferences have been published, some of which are archived on the Web.

10.d. Online Sites

The Early History of Seismometry (to 1900). This is an educational site of the U.S. Geological Survey center compiled by James Dewey and Perry Byerly which describes early seismoscopes, their invention, and development, along with some of the early scientific controversies. gldss7.cr.usgs.gov/neis/seismology/history_seis.html

Rail Inspection. There is a well-illustrated history of magnetic and ultrasonic testing of rail lines in America on the Center for Nondestructive Evaluation site at Iowa State University, contributed by Robin Clark of Sperry Rail Systems. www.cnde.iastate.edu/ncce/UT_CC/Sec.6.1/Sec.6.1.html

Ultrasonic History & Science Exhibition. An illustrated history of interesting practical applications for ultrasonics. www.tsc.co.jp/~honda-el/exhi_e.html

U.S. National Data Center. The Data Center provides information regarding U.S. monitoring related to the Comprehensive Test Ban Treaty (CTBT). It is a gateway between the U.S. and the International Data Center, providing data related to geophysical disturbances, particularly seismic data from nuclear explosions and other major events. The U.S. NDC is established at the Air Force Technical Applications Center (AFTAP). www.tt.aftac.gov/

U.S. Navy Marine Mammal Program and Bibliography. The U.S. Navy has an illustrated description of the use of dolphins and sea lions as members of a 'fleet' for a variety of marine tasks and studies. www.spawar.navy.mil/nrad/technology/mammals/

10.e. Media Resources

Cortland Historical Site. The Sugget House museum in Cortland contains personal and invention artifacts of Elmer Sperry, inventor of the gyroscope and of a number of important ultrasound technologies. www.cortland.org/

Earthquakes: Living on the Edge, a History Channel program in the 20th Century with Mike Wallace series. Details the history of some of the devastating earthquakes in California, Japan, and Mexico. Geologists, rescue workers, and city planners give their views on disaster preparation and prevention. VHS, 50 minutes. May not be shipped outside the U.S. and Canada.

Medical Imaging, from the History Channel *Modern Marvel*s series, this charts the history, from World War II submarine detectors to current fetal monitors, of ultrasound technology as it is used in medical imaging. VHS, 50 minutes. May not be shipped outside the U.S. and Canada.

11. Glossary

Titles, product names, organizations, and specific military designations are capitalized; common generic and colloquial terms and phrases are not.

acoustic spectroscope	
	an instrument for measuring acoustic attenuation or velocity as a function of acoustic frequency which aids in studying particle size distribution in situations where the attenuation spectra are already known. In practical applications, this instrument can be used to study mixed samples without having to dilute the samples (paint, blood, shampoo, etc.), as with light-based substance analysis, thus preserving their original form.
active tank	an ultrasonic tank stimulated to produce bubbles in a process called cavitation
blanketing	a limiting phenomenon in which the density of the cloud of bubbles in a cavitation field has reached a point where it doesn't become any denser with the application of additional energy. The intensity at this point is called a *blanketing threshold.*
cavitation	a phenomenon in which vapor bubbles form and collapse when a liquid is subjected to high frequency and/or intensity sound waves. Cavitation is a source of interference in sonar applications, but may actually be of positive benefit in certain infrasound and ultrasound industrial applications.
ECAH theory	a theory of ultrasonic propagation in dilute heterogenous systems
electroacoustics	the science of effects in interrelated electrical and acoustic fields, especially within a medium of charged particles
shear wave	a type of wave that propagates more readily in solids than in liquids

sonolysis	a process of applying acoustic energy to biological cells to disrupt their form and/or function
SVP	sound velocity profile
tone-burst	an acoustic pulse comprised of a number of cycles within a given frequency
transducer	a device to convert energy and project it as acoustic energy

Acoustic Surveillance Sonar

4

1. Introduction

The pinging sound of a sonar system is commonly used in movies depicting submarines and other underwater vehicles. This audible reference symbolizes the myriad sound surveillance technologies used to probe for vessels or structures underwater. Some sonar sounds are within human hearing ranges and some are not.

Chapter 2 Audio Surveillance describes information that is gathered through sound technologies within human hearing ranges, such as phone taps, listening devices, tape recorders, and amplifiers. Chapter 3 focuses on infrasound and ultrasound, frequencies below and above human hearing ranges. This chapter focuses primarily on acoustic-sensing and sonar devices that are designed for underwater detection, ranging, and imaging applications, many of which utilize frequencies outside of human hearing range.

Acoustic devices have several advantages as surveillance tools. It is easier to generate sound

The DSRV Deep-Submergence Rescue Vehicle is designed for rescue operations on disabled, submerged submarines. It can dive up to approximately 1500 meters, conduct a sonar search, and attach itself to the sub's hatch to transfer personnel to another location. [U.S. Navy photo, released.]

vibrations than it is to generate X-rays, for example, making it a less expensive technology in some respects. X-rays can cause damage if aimed at biological tissues, whereas sound waves, as they are typically used in surveillance devices, are not harmful (there are some high-intensity experimental sound devices that can harm biological systems, but these are less frequently used). In addition to being relatively safe, sound waves are versatile. They can be used to detect objects, to measure distances, and to create a 'picture' of objects within other objects. Sound is used in applications as diverse as locating submarines, sending messages, testing the structural integrity of buildings, and imaging the growth of a fetus inside a womb. The ability to look through murky environments or 'inside' certain opaque objects is a very useful capability of acoustical (sound) energy.

1.a. Basic Concepts and Terminology

Sonar is a sensing system based upon sound. It is recommended that you read the chapter on Acoustic Surveillance in conjunction with this one, so you understand the basic concepts associated with sound propagation.

Sonar is currently used as an acronym for *sound navigation and ranging*. The term has been ascribed various acronymic origins, but most historians agree it was popularized around the time of World War II, when detection of submarines and other underwater obstacles was an important concern. It replaced the World War I code-word ASDIC (which has been ascribed a number of meanings whose origins are in dispute). Sonar shares many concepts and basic terms with radar. In practical applications, radar tends to be used in nonaqueous environments whereas sonar tends to be used in *aqueous* (liquid) environments, due to the fast-moving properties of sound in water.

When sound is emitted from a transmitter, it creates pressure waves that move away from the source in many directions through elastic media (air, water, ground, etc.).

- In sonar technologies, the emitted sound pulse is called a *ping*. If there are obstructions encountered (less elastic objects) as the ping travels outward, the waves are reflected off the obstructions in various directions. Some may return to the transmitting area or another area to be picked up by a receiver (in most cases, the transmitter and receiver are near one another or housed in the same unit).

- As in radar, the intended deflecting structure is called the *target* and any irrelevant signals from other objects is termed *clutter*.

- Interference from the transmitting/receiving technology itself, especially at threshold levels, is called *noise*. The signal that returns from the target is called the *echo* although it is more accurately termed a *reverberation*.*

- The traditional term *echo* is used in this text to refer to the reflected signal. The degree to which a target shows up in a sonar scan and its 'visibility' characteristics comprise the sonar *signature*.

- The sending device in radio systems is typically called a transmitter, whereas in sonar it is more often called a *projector*.

- In sonar, the common basic receiving device is a *hydrophone*, since most sonar devices are used underwater.

---

*An echo is a repetition of a sound due to reflection, whereas a reverberation is the resulting effect of an impact which may result in a reflection.

- The device that converts sound waves to electrical impulses and vice-versa is more specifically called a *transducer*. The transducer, along with any mechanisms designed to amplify or focus the signal, comprises the projector.

- If more than one transducer is in the vicinity of a sonar transmission one system may pick up the signal emitted by another, resulting in a false signal called *crosstalk*. If more than one transducer signal is being intentionally generated, as in a multibeam scanning system, timing is introduced to avoid false detections by the receiver.

- A sonar image created through image processing, usually from scanning or multibeam arrays, is called a *sonograph* or *sonogram*. This image resembles the negative from a black and white photograph, but is usually computer-processed to inverse the image and to add 'false' color to aid in interpretation.

Signal/Noise

To interpret sound signals, you must be able to distinguish the target from the clutter. If the object that is being probed is very smooth and there are few nearby objects, signal scattering is minimized. If the object is very convoluted or rough, the signals will scatter in many directions and only a small number may return to the receiving unit. If the object absorbs most of the sound waves, the signal returning to the receiver may be weak and difficult to interpret. The density and composition of an object affects the amount and intensity of the returning signal. These characteristics may make it difficult to sense certain kinds of objects, but they can also be exploited to create devices and vessels with low sonar signatures—stealth technologies. The desired portion is called the 'signal' and the undesired, interfering, or spurious signals are called 'noise'. A high signal-to-noise ratio means the sensor is receiving a high proportion of good data to bad or undesired data. As instruments become more precise and sensitive, there is often a trade-off between signal and noise and a cutoff region at which the incoming data can no longer be trusted.

Active and Passive Systems

Sonar can be an active sensing technology (by emitting pings) or a passive sensing technology in that sound waves emanating from ambient sources can be sensed without initially sending out a signal. For example, the sound of a boat propeller echoing off the hull of a submarine may reach a passive sonar receiver.

In practical use, sonar is very similar to radar—signals are typically emitted and the returning signal analyzed. Sound surveillance differs from radar, however, in that sound is a transmission or *disturbance* through a medium (such as air or water), whereas radar, like light, does not require a transmission medium and can travel great distances through the empty regions of space. The essential characteristics of sound as they compare and contrast with electromagnetic sensing technologies are described in more detail in this and other chapters.

Typical Frequencies

Most sonar is designed to take advantage of certain sound 'windows' that have desirable properties and give good results. Sonar does not inherently have to use frequencies that are outside of human hearing, but most commercial sonar technologies use ultrasound, frequencies above the human hearing range. These sounds can readily be heard by dolphins and many types of whales. Most commercial depth-sounding devices work at a specific frequency, or at two frequencies, as in dual systems. There are a few 'tunable' systems that allow a frequency to be selected from a wider range of options.

The active frequency varies from product to product, but for reference, some common frequencies are shown in this chart.

Frequency	Common Applications
2 to 8 kHz	passive and active hull-mounted DSP sonar
20 kHz	*for reference, the upper range of human hearing*
27 kHz	some higher-speed towed sonars (e.g., 20 knots)
50 or 192 kHz	commercial fishfinders
dual 24 kHz & 200 kHz	depth sounding (bathymetry)
dual 33 kHz & 210 kHz	depth sounding
455 kHz	high-end, low-speed, high-res. bathymetry
variable, 3.5 kHz to 50 kHz	high-end, specialized applications

Higher and lower frequencies have different properties when used in sounding devices:

higher frequencies - generally used in shallower depths; narrower beam angle (sometimes called the *cone*). Tend to have a higher target definition, to be less susceptible to noise, and more susceptible to absorption and attenuation.

lower frequencies - generally used for greater depths and ranges, as they are not absorbed quite as readily; wider beam angle due to diffraction, and higher susceptibility to noise. The higher beam angle necessitates a wider aperture

1.b. Signal and Background Interaction

Sound waves generated above and below human hearing must compete with natural sounds. These natural sounds may make it more difficult to interpret a return signal from a sonar ping. Sound technologies fall between radar and infrared in terms of the difficulty of interpreting the signals. Since radar frequencies are mostly synthetic, there is very little background clutter, making radar targets relatively easy to detect and interpret. Infrared, on the other hand, is emitted by almost everything in our world, and it is technologically challenging to separate the desired information from irrelevant background information. Sound technologies fall between these two; aquatic environments have a moderate amount of natural background clutter, but not as much as most infrared-sensing technologies.

Like radar, sonar is often used to detect and locate objects. Sonar, radar, and infrared sensors can all be used as imaging technologies, to map terrain and objects in the surrounding environment. Unlike radar and infrared, which are primarily used above water and in space, sonar is primarily used below water and cannot be used in the 'empty' reaches of space that are insufficiently dense to propagate sound.

1.c. Biological Sonar Systems

Sound travels faster through water than through air, so it is not surprising that many sea creatures have senses that are very sensitive to sound vibrations and some even use biological sonar systems as a means of communication.

Humans are particularly well adapted to sensing the world through sight and sound. Sea creatures rely on other senses to navigate and survive. Light doesn't penetrate water as easily as air, especially at lower depths. Water also differs from air in that it can more readily suspend particles. Although rain and dust can obscure vision in air, they are rarely as thick as kelp or plankton or sandy underwater sediments. For these reasons, even though many sea creatures have good visual senses, there are many that have highly developed sound-sensing organ to aid in their survival.

Dolphins have sophisticated biological sonar detection, ranging, and communications systems that have served as models for many different aspects of communications and surveillance research. They have a sonar sensor called a 'melon' located inside the front of their heads. This organ enables them to seek prey and sense objects inside or behind other objects. A dolphin's sonar system like a cross between 'mind reading' (since dolphins can intercept each other's emitted signals) and 'X-ray vision' using sound waves rather than X-rays. Dolphins also use sonar to intercommunicate while traveling together. [Classic Concepts image, copyright 1996, used with permission.]

Dolphins are sociable, air-breathing mammals that live in the sea. They are highly intelligent with large complex brains, individual personalities, and a tight family structure. In the wild, a dolphin lives for about thirty years. Like humans they bear only about one child every two years. This low birth rate means they can't replenish their numbers as readily as fish and thus are vulnerable to environmental changes and depletion of their numbers.

Dolphins have a highly developed auditory region in the cortex and the auditory nerves have more fibers than human auditory nerves. The dolphin hearing range far exceeds that of humans. When we are young, we can hear sounds ranging up to about 18,000–20,000 Hz, whereas dolphins can hear higher than 100,000 Hz (perhaps as high as 150,000 Hz). Humans intercept most sounds through their external ears and bodily vibrations. Dolphins, because they live in the water, receive sounds more readily through their specialized jaws and a separate bone called the *auditory bulla*.

Dolphins have good eyesight and even better 'soundsight,' a type of biological sonar imaging that humans don't have. They don't just hear sounds, they appear to generate mental images of objects they sense through sound. Thus, they can 'see' inside things using sound. For example, a dolphin can see a developing fetus inside another dolphin or an anchor hidden inside a wooden box.

Dolphins intercommunicate with a wide variety of clicks and chirps and can use sound to determine their position and relationship to one another and other structures. The organ in their foreheads called a *melon* focuses the outgoing sound pulses, and the bony lower jaw vibrates when it receives the returning sound waves, transmitting them to the dolphin's inner ear.

Some of the sounds emitted by dolphins are audible to humans, but many must be detected with instruments and converted into human hearing ranges. A dolphin's biological *echolocation* system is similar to synthetic radar and sonar in that outgoing clicks are timed with incoming clicks so that the dolphin can interpret the returning signal to form a sensory picture of its environment. Since dolphins, like humans, are highly social animals, they have further developed a kind of 'mass communication' in which they visually and aurally share information while swimming in pods (family units) or schools. This shared sensory environment helps them to navigate and hunt for food and may in part explain why they become disoriented and confused when forced into close quarters within circular fishing nets.

Dolphins are trained in the U.S. Navy Marine Mammal Program to assist in mine detection and marking. Dolphins can carry out complex tasks and have been taught to attach neutralization charges to the mooring cables of marine mines. Dolphins and sea lions have also been taught to assist divers in search and salvage tasks and serve as underwater 'watch dogs' to provide warnings of intruders or unusual events. [U.S. Navy photo, released].

Dolphin sonar is sophisticated and has many practical applications. Dolphins have been trained to assist in mine detection, diver assistance, and other tasks. Sonar technologies have been developed along the lines of biological sonar. For example, multibeam sonar systems that send out a simultaneous swath of pulses are somewhat similar to a group of dolphins sending out pulses together.

Dolphins tend to group themselves in wide rather than long formations to enhance the efficiency of their sonar systems. In this way, each individual gets a part of a composite picture which is then communicated through sight and sound among members of the group. Similarly, sonar imaging systems can send out several beams in a swath to create a composite picture of the underwater environment. Multibeam sound-scanning or fast-moving-beam scanning can be digitally processed into a composite picture on high-resolution systems to create images that are very similar to normal photographs.

Left: A sidescanning sonar 'fish' in which the 'fins' are used to stabilize the device. The sonar transducers are located in the long black rectangle along the side of the body. The device is used to study the morphology of the sea floor. Right: A sidescanning sonar image of the sea floor near San Francisco showing the Taney Seamounts. These structures were discovered during the "EEZ Scan Project". [U.S. Geological Survey news photos, released.]

1.d. Components of Sound

There are three important components to a sound wave. A basic two-dimensional diagram can illustrate these essential characteristics and further description can help the reader understand the propagation of waves through 3D space.

Two-Dimensional Symbolic Representation of a Sound Wave

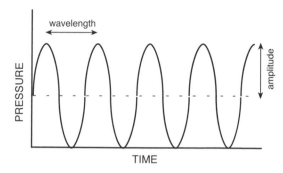

Acoustic waves are typically described in terms of amplitude, wavelength, and frequency. They are diagrammatically represented as periodic sine waves, with the peak indicating the highest pressure and the trough indicating the lowest pressure of the wave. Acoustic waves are longitudinal. (To picture this, imagine a quick lengthening tug on a spring that's attached at one end. You will see the 'pulse' travel straight through the spring from one end to the other without the spring bouncing up and down.)

amplitude Amplitude is the maximum distance that particles in a medium are displaced as a wavelike disturbance moves through the medium. People sometimes use the terms *magnitude* or *height* to describe wave amplitude. Amplitude symbolizes the phenomenon we perceive as loudness or softness. A higher amplitude sound tends to sound louder. Above certain amplitudes, damage can occur to biological structures—they can damage hearing and, at some levels, other organs or tissues. Amplitude is diagramed as the vertical height of the wave.

wavelength Waves represent a repeating phenomenon, sometimes called *cyclic,* or *periodic,* in which waves with similar characteristics succeed one another. Repetition of the wave is called *vibration* or *oscillation.* The distance between corresponding points on two successive wave disturbances, or *compressions,* is called a *wavelength.* Symbolically, a wavelength may be illustrated as the horizontal 'width' of a wave.

frequency The rate of repetition or *periodicity* of a succession of waves is the frequency. We usually describe sound wave frequency in terms of repetitions or *cycles* per second. Older texts use the term cycles to describe the frequency of sound and radio waves. Newer texts, in honor of Heinrich Hertz, use the term *Hertz* or its abbreviation *Hz.* Thus, a sound of 10,000 cycles per second would be expressed as 10,000 Hz or 10 kHz. On a graph, narrower wavelengths are used to symbolize higher frequencies. What we perceive as the *pitch* of the sound increases as the frequency increases and decreases as the frequency decreases. Not all humans are able to perceive frequency changes in sound. Tone-deafness is a phrase used to describe the inability to recognize or remember sound pitch differences. Many humans are tone-deaf to lesser or greater degrees. *Perfect pitch* is a term used to describe individuals who can recognize a specific pitch without a reference pitch; in other words, they know that a specific tone might correspond to 'A' which has been assigned to 440 Hz, for example. Good *relative pitch* is when individuals can 'hold a tune' once they have a reference note to get them started in the right key. In most symbolic diagrams, frequency is illustrated by the number of wavelengths repeated over a given unit of time.

AMPLITUDE VARIATION   FREQUENCY VARIATION

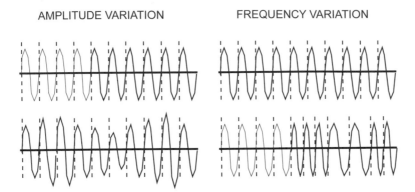

Reference waves (top) are shown with varying waves (bottom). Bottom Left: Sound waves that are varying in amplitude (loudness) are shown with varying peaks and troughs. Bottom Right: Sound waves that are varying in frequency (pitch) are shown with narrower and wider repetitions. Deliberate manipulation of the frequency or amplitude of a sound or electromagnetic wave is called modulation. Modulation is an important technique used in many types of communications technologies. [Classic Concepts copyright 1997, used with permission.]

1.e. Sound Characteristics

The speed of sound is related to the density of the material through which the sound waves are moving. Since water is denser than air, sound travels more quickly through water. The speed of sound in air at room temperature at sea level is about 344 meters/second. The speed of sound in seawater is more than four times faster, about 1500 meters/second (depending on temperature, depth, and salinity). Sound speed in a direction is termed *sound velocity*.

When sound moves from one medium to another, as from air to water, the sound is *refracted*; that is, it is bent. Since this is difficult to predict or control under conditions of moving waves, aircraft carrying sonar devices must get close to the surface of the water in order to put the sonar in the water. Surface vessels and submarines are suitable for deploying sonar, but aircraft are sometimes used for tactical purposes or to drop off self-contained sonobuoys with data-storage capabilities or radio transmitters.

Sound devices work well underwater because sound travels quickly through water. Basic sound-based vessel- and mine-detection systems work well in open waters where physical protuberances (turbulence, coastlines, reefs, wrecks, etc.) are less likely to attenuate or deflect the sound waves. Imaging systems, on the other hand, work well in clear, shallower waters where the reefs, shelves, coastlines, and other prominent features can be readily mapped.

Not all physical structures outside of the target will hinder a sound signal. Just as the ionosphere can be used to bounce radio signals around Earth in a desired direction, the ocean bottom or other physical structures or temperature regions can be used to bounce an underwater sound signal in a desired direction. Surrounding terrain can sometimes be used to channel or funnel sound communication.

Earth's atmosphere has a variety of layers that have different conductive and refractive properties. Marine environments also have layers created by the influence of sunlight, pressure associated with depth, sedimentation, plant life (kelp, plankton, etc.), and marine animals (fish, mammals, corals, etc.).

2. Kinds and Variations

Sonar sensing is based on sending out a 'beam' and analyzing a returning signal (reverberation or *echo*). A single beam is limited and is usually used for the more basic applications, including depth-sounding and pinging. Multiple beams in the form of sonar 'arrays' are powerful clusters or groupings of sonar beams that can be aimed in different directions to form a 'picture' of the surrounding environment. These are useful for identifying specific types of objects and for mapping terrain.

Sonar arrays are more complex than single-beam sonars and thus require more complex electronics to handle timing and processing. Computers are often used for this purpose, especially in high-resolution arrays used for mapping.

A *beamformer* is a spatial filter that aids in organizing and controlling the beams emitted by sonar arrays—it helps direct and focus the beam. Beamforming allows mathematical analysis of the results of multiple elements such that a narrow response in a particular direction may be analyzed. This aids detection of the direction from which the sound is returning. *Beamwidth* is expressed as the decibel-range of the beam, e.g., 3 decibels or 3 dB.

A *stave* is an electromagnetic element used to produce transducers. Staves can be used to create a curved array of hydrophone sensors, for example, so that the system doesn't have to rotate to sense in multiple directions.

Sonar systems may be 1) mounted on the operating vessel, as on a submarine, surveillance ship, or helicopter, 2) towed behind the operating vessel, or 3) dropped off and *scuttled* (or picked up later).*

The basic types of sonar schemes include

single-beam - the traditional scheme of sending out a single focused beam—the predominant method in the early decades,

sweeping - the beam is generated with spaced transducers, usually sweeping orthogonal to the direction of travel. This system is used in scanning sonars, and

multibeam - continuous beams spreading out from the source, usually in a fan-shape or swath. This is a more recent development as it involves more sophisticated timing and processing.

Passive and *active* sonar technologies have different strengths and weaknesses. Active sonar has a higher likelihood of detection and must balance the electronics and reverberating limits needed to send and receive signals from the same location. Passive sonar has a low probability of detection, and doesn't have to perform the same balancing act to receive strong signals, but it can't initiate a transmission.

Bistatic sonar uses a combination of active and passive technology. The transmission emanates from one location and is received at another. This takes advantage of the best characteristics of the different technologies and sometimes also provides tactical advantages.

Variable-depth sonar (VDS) is a means of overcoming some of the problems with sending signals at an angle through water that is 'layered'—that is, it has different temperature and particulate characteristics (sediment, salt, etc.) that might bend or *refract* the sound beam. Submarines have been known to 'hide' at greater depths to take advantage of the layered

---

*The term 'scuttle' comes from the name for a hole in a ship that is covered with a lid, which can be opened to admit a sailor or for tossing objects through. It is now also used more generically for things that are tossed overboard or punctured so they will sink.

characteristics of water to mask their presence. By using variable-depth sonar systems, it is possible to take a series of soundings at different depths and evaluate the information to correct for beam refraction. Fixed-depth sonars are those that are attached to a hull, drilling rig, or other single-depth platform. Variable-depth sonars are dropped or towed, and may be attached by electrical or fiber-optic cables.

Doppler systems are designed to exploit or compensate for the physics associated with movement in relation to the sound pressure wave. A Doppler sonar can detect and sometimes measure the change in frequencies of the sonar echo, based on the magnitude and direction of the change, relative to the motion of the sensor or transmitter. The phenomenon of Doppler shift can assist in target classification of a moving target (for a brief explanation of Doppler physics, see the Radar Surveillance chapter).

Sonobuoys are self-contained sonar systems that can be dropped off and picked up later to analyze the collected data, or they can be monitored regularly via radio signals that are transmitted to a receiver on a ship, sub, satellite, or coastal ground station. When sonobuoys are dropped in clusters, they can be used to compute directional information.

Expendable devices are those that decay, sink, or self-destruct when their useful life is over or, in covert operations, if they are in danger of being detected. These devices are usually self-contained and can store or transmit information until they are picked up or until they become nonfunctional. Sometimes the information is relayed through satellite communications systems. Some sonobuoys are designed to be expendable.

In commercial implementations, common ways to categorize sonar include

- **scanning sonar** (SS) - This is a stationary or mounted sonar that 1) sends simple signals in a general direction, usually to detect incoming objects (e.g., on a dock), or 2) that rotates slowly over some or all of a 360° arc to scan the surrounding region. Scanning sonars are used for navigation and monitoring applications. More complex data computations are required if the system is fixed to a moving vessel.

- **sidescanning sonar** (SSS) - This is the most common high-frequency sonar. SSS is used for object-detection (e.g., sunken wrecks), and marine surveying and mapping. The *along-track* beam-width is narrow and the *vertical* beamwidth is wide. Sidescanning sonar is towed. The beam usually scans to both sides of the vessel, producing a screen image or printout in which protuberances are recorded, though some may image only to one side. Objects as small as a coffee mug can be resolved with high-end systems. Commercial SSS systems typically use an operating frequency of about 100 kHz, though a few systems are multifrequency, and can be mounted on vessels or towed at speeds up to about 15 knots. Sidescanning sonars are optimized for imaging on one or both sides of the vessel (which is usually a surface vessel). This leaves a channel in the center, called a *water column,* that is not imaged (essentially a blind spot). Most SSS systems are linear and do not use focusing.

- **forward-looking sonar** (FLS) - FLS systems are useful for object and area surveillance. FLS systems can help detect the presence of vessels, marine life, mines, etc. A typical FLS is configured with pulsed arrays or continuously scanning arrays. More complex systems, with multiple beams and fast scanning rates, can scan a region within a single pulse period for 1D or 2D imaging. FLS systems operate at a variety of pulse rates, from about 100 kHz to about 600 kHz, depending upon the system. Some FLS systems are dual-axis—the scan can be set side-to-side or top-to-bottom.

- **downward-looking sonar** (DLS) - DLS, also known as bathymetric sonar, is less common than SSS and FLS. It is used for depth sounding, fish finding, and marine contour mapping. Single-beam systems scan over a wide, vertical swath, similar to a sidescanning sonar, while multibeam systems are more similar to multibeam FLS systems. Frequency ranges vary from about 12 kHz to about 1 MHz. Most DLS systems are vessel-mounted, though a few are towed.

- **sector-scanning sonar** (SSS) - This is similar to sidescanning sonar and may be towed on a platform. Sector-scanning sonar can be used in both active and passive modes and can be rotated through two axes. Operating frequencies for active systems are around 160 kHz. Passive modes have been used to detect frequencies up to over 300 kHz. Their range is typically up to about 250 meters.

Most sonar systems are used for one-dimensional or two-dimensional imaging, but scientists are studying ways in which the angle of the swath and the frequency of the pulses can be adjusted to yield three-dimensional images through computer processing. This is particularly effective when imaging stable structures (shipwrecks, reefs), but experimental systems have successfully imaged moving targets such as schools of fish and underwater vessels.

Beam direction and synchronization are important aspects of multibeam sonar arrays. The geometry of an array, consisting of the spacing and directional alignment of the beams, is called the *beam formation*.

Sonar targets are objects that are deliberately designed to show up well on sonar scans. They are installed temporarily or permanently to serve as reference points, to aid in adjusting sonar equipment and to provide reference information for the subsequent interpretation of sonar images. They are especially useful in areas that are being monitored for environmental changes, where precision may be important.

Towed arrays are versatile in the sense that they can be towed behind marine vessels or by air vehicles (usually helicopters hovering a short distance above the surface of the water).

Some sonar devices are called *towfish* because they are fish-shaped or because they utilize a fish-shaped depressor to weigh down the end of the cable. Deadweights and hydrodynamic depressors are commonly used.

Sometimes the towed sonar is cable-shaped and the cable may arc due to pressure from the water as it flows over the cable. A flexible cable assumes a curved *hyperbolic* shape called *catenary*.

Historically, *dipping* sonars, that could be dropped into the water by helicopters, became available in the 1950s. They incorporated a specialized compass that provided bearing information to the helicopter.

Multibeam systems are becoming more prevalent and are gradually superseding many single-beam systems. In multibeam sonar, several beams are activated at the same time, thus covering a wider area. The returning data makes it possible to model a three-dimensional contour image of the target area. A multibeam systems may be equipped with motion sensors and a gyrocompass.

Depth-Sounding

Depth-sounding is known in scientific jargon as *bathymetry*. In commercial marketing it is sometimes also called *echo sounding*. In its simplest sense, depth-sounding involves sending a signal downward and measuring the time it takes to reflect off the bottom and return. This information is used to calculate distance. In real applications, many soundings are taken, as debris, sediment and other obstacles could give a false impression of the depth.

Commercial bathymetric products generally have the following capabilities:

- The depths that can be sounded vary with the product, but are usually from about 100 to 1500 meters.

- Frequency ranges vary, but typically are around 24 or 200 kHz (or both, in dual systems).

- The number of beams may vary from 20 to 80 beams—some professional systems may have as many as 240 beams.

- The *swath coverage* (the angle of the beam-spread) is usually about 120° ± 30°. The adjacent beams fan out from a central point, like a peacock's tail.

- A signal series of sound waves is sent out, the return signal recorded, and another signal series sent, with an update rate of about 10 to 40 swaths per second.

- Newer systems have serial ports for transmitting data to a computer, and VGA or SVGA ports for displaying on a computer monitor.

Sonar Imaging

Imaging is a more recent application for sonar information. Pictures of the seabed, marine objects, reefs, wrecks, or vegetation can be created by sending out a series of sound signals along a series of swaths. This form of imaging incorporates many basic depth-sounding concepts, but typically requires a higher number of beams, more precise timing, and more complex computer processing than depth-sounding or target location applications. Imaging depths can range down to about 1500 meters.

3. Context

3.a. Applications Context

Because sound travels more readily through water, sonar has become an essential sensing, navigation, and imaging tool for surface and underwater vessels. Bodies of water cover almost 75% of the Earth's surface, so their importance to survival, transportation systems, defense operations, and communications should never be underestimated.

Most sonar technologies are used in lakes, rivers, and ocean environments. Since these bodies of water comprise almost 75% of the Earth's surface, underwater sensing technologies are extremely important to understanding our environment, our resources, and the security of our planet. [U.S. Geological Survey, public domain.]

The essence of most sonar applications is the use of sound reverberations to determine distance. Less often, sound is *modulated* to communicate messages. Sonar isn't specifically an underwater technology. However, since sound travels more slowly through air than through water, there are more efficient ways to determine distance or communicate through the air and, hence, the vast majority of sonar technology is used underwater.

Water is a complex environment. It is not as homogenous as may be assumed. The oceans, great rivers, and large lakes are richly layered—comprised of different temperature regions, sedimentary particles, plankton, kelp, reefs, and other objects that greatly influence the characteristics of sound travel through water. An awareness of this complexity and the general principles of sound, can aid sonar technicians, boaters, and imaging specialists in maximizing performance of sonar buoys, tow lines, depth sounders, and other sonar-related equipment.

Sonar is chiefly used for the following applications:

- surface or underwater navigation,

- detection of other bodies underwater,

- object and terrain mapping,

- underwater communications, and

- robotics ranging applications (both underwater and above water).

Sonar devices are often towed behind a vessel or below a helicopter, on a tether that may range from a few dozen yards to several kilometers in length. The towing vessel usually moves slowly, up to about 1 knot in speed for high-end imaging applications, and up to about 20 knots for basic depth-sounding. Future fast-processing or autonomous systems may be able to be towed faster. As pointed out by Robert Fricke in *Down to the Sea in Robots*, towing at very slow speeds in open water can be tedious for the surveillants and the crew, a job that could perhaps better be handled by robots.

Navigation

Many marine vessels use sonar to determine depth and the distance to marine structures and other vessels. These data make it possible to maneuver vessels and to select regions of water where it is deep enough and sufficiently free of dangerous obstacles to navigate safely.

Left: Sonar is essential to submarine navigation and reconnaissance objectives. This Trident ballistic missile submarine, the USS Ohio, has tallied over 50 strategic patrols. Right: Subs are now equipped with sophisticated electronic control centers. This nuclear-powered sub, the USS Seawolf (SSN 21), was commissioned in 1997 and has command centers devoted to displays of the data from various sensing apparatus and vessel systems. [U.S. Navy 1998 and 1997 news photos by Shawn Handley and John E. Gray, released.]

Mapping

Sound properties are suitable for measuring distance through water, which makes sonar a good technology for underwater mapping applications. Three-dimensional terrain mapping at its most basic entails making a series of depth soundings and consolidating the data into a mathematical or visual picture. By sending out many signals at precise locations, and creating a symbolic 'graph' of the results, a terrain map or map of submerged objects can be created. The image from this process resembles an inversed grayscale photograph (which can then be further enhanced with pseudo-coloring). This form of imaging works best with stationary objects.

Detection

Sonar is used to sense depth, debris, obstacles, reefs, fish, and whales. It is an essential tool in detecting hostile craft or foreign military vessels of unknown intentions. Passive sonar is widely used for national security operations.

Underwater vessels may also be spotted from the sky, but this is difficult if it's foggy, if the water is murky, or the vessel is deep under the water's surface. Since air vessels are faster and more flexible than sea vessels, they are frequently used in underwater-vessel spotting activities, but sonar is a more efficient way to detect certain types of vessels and the sonar devices need to be in contact with the water.

To accomplish this, an aircraft will fly low over the water and tow or drop a sonar device. The device then takes readings that are communicated directly or to the craft or recorded and analyzed later. To transfer the information immediately, the device will either send signals wirelessly or through a communications cable attached to the receiving system. The sonar device may also be left on its own to gather and store data to be picked up later, sometimes by a different vessel or craft.

Tactics

Deep-diving vessels, like submarines, will sometimes descend below the upper layer of the water to hide from sonar probes. Varying the angle of a sonar-detecting signal can sometimes aid in penetrating several layers, but may also introduce errors due to refraction. Each layer has different densities and particulate characteristics and thus may have different refractive indexes. Thus, many soundings and more than one type of sonar may have to be used to form a composite picture of the environment.

Sound travels relatively quickly through water, but it is not instantaneous. Thus, a passive sonar system may be used to determine if a marine vessel is being scanned by active sonar some distance away. The probing signal may only provide a few seconds of advance warning, but that small advantage could mean the difference between life and death if the probe originated from a hostile vessel.

When a fleet is traveling in hostile waters, the individual vessels may use spring and drift tactics to vary the composite radiant noise level to make it difficult to pinpoint an individual vessel or the overall size of a fleet.

Communications

Sonar is somewhat limited as a communications technology, compared to radio, but there are circumstances where it is practical to modulate sound waves to carry content, particularly where radio frequencies can't be used or are more likely to be detected.

The range at which sonar can be used for oceanic communications varies greatly with the conditions of the weather, the terrain, and the amount of sea traffic that is generating noise (propellers, engines, hulls, etc.). *Sound channels*, areas of terrain or sediment that can funnel and bounce sound over longer distances (somewhat in the way the ionosphere can bounce radio waves) may extend sonar ranges. Ships can intercommunicate to a distance of about 10,000 meters in good conditions. Submarines may communicate over longer ranges, sometimes up to 20 kilometers.

Robotic Ranging

Sonar is a useful way for robots to navigate, both above and below water. Since most of the sound technologies for robots are above ground (air-based) and use frequencies above the human hearing range, they are discussed more fully in the Infra/Ultrasonic Surveillance chapter.

Sonar Displays

Sonar technology is similar to radar in many ways, and sonar displays and imaging systems share many common characteristics with radar displays and imaging systems.

A sonar *scope* provides an image of the sonar echo when the signal is apprehended by the transducer. Scopes range from simple depth-sounding displays to complex imaging displays. Some of the most common displays include the following.

- Sonar *A-scan* displays are based upon basic Cartesian coordinate graphing systmes (somewhat like the heart-rhythm displays shown on TV medical monitors). As the beam traverses the screen, a vertical blip or *pip* (a raised section on the graph) indicates the presence of a 'hit' or target acquisition. Monitoring the beam to see if the target reappears on subsequent passes helps the operator distinguish real targets from false signals or passing debris.

- A *plan-position indicator* (PPI) display is familiar to radar technicians and anyone who has watched submarine movies. A PPI display has circular tick marks as reference points and a cross-hair dividing the screen into quadrants. As the beam sweeps through a 360° arc, it bright pips are displayed when targets are sensed.

- A sonar *mapping or imaging system* usually provides a grayscale or false color display of composite information on a computer monitor. The information may be a picture that has been built up from a number of 'passes' or sonar swaths.

4. Origins and Evolution

Sonar and submarines are almost always discussed together. Sonar enables submarines and surface vessels to navigate, to detect dangerous obstacles, and to locate other vessels. More advanced sonar systems also allow underwater terrains to be mapped and imaged. The history of sonar is closely allied to the development of submarines, surface marine craft, unstaffed remote vehicles, and autonomous sonobuoys and vehicles.

The concept of submarines has fascinated people for a long time. In early history, humans devised a way to stay underwater for longer periods by breathing through reeds or pipes. This simple concept led to the development of snorkels and submarines.

In 1578, William Bourne created plans for an enclosed wooden boat that could be rowed beneath the surface of the water. The whole vessel was bound in leather to make it waterproof. There is no evidence that Bourne ever built the boat, however.

John Napier, in 1596, mentioned the use of underwater sailing craft and his intention to direct their construction and use. A few years later, in 1605, a boat similar to the one designed by Bourne was actually built and tested, but it apparently ran into technical difficulties when it stuck to the river bottom.

The First Military Submarines. Left: The "Turtle", the first American submarine, was designed by David Bushnell and built in 1776, during the American Revolution. Powered by a hand propeller, it could stay mostly submerged, with only about six inches showing above the water. This drawing by Lt. Francis Barber, in 1875, has a few technical flaws, but gives an idea of how the system looked and worked. Right: The "Alligator", the first submarine purchased by the U.S. Navy for military use. [U.S. Navy Submarine Centennial Exhibit, drawings public domain by date.]

Cornelius van Drebbell, a Dutch physician and inventor of optical devices, was living in England when he created a greased leather submarine, in 1620. Apparently it could descend to depths of 12 to 15 feet. Bishop Wilkins mentions Drebbell's vessel in an English publication of 1648, noting that it was "already experimented here in England by Cornelius Dreble". Wilkins specifically mentions the surveillance and warfare applications of such a vessel and the fact that they could enable one's enemies to be "undermined in the water and blown up". David Bushnell, a graduate of Yale, built a historic submarine torpedo boat in 1776. It was intended that this new tool of warfare could provide a way to monitor and destroy British warships in the New York Harbor.

There were a lot of limitations and design problems associated with early submarines, but once the technology began to mature, it became clear that sonar could help the underwater vessels navigate and detect other vessels.

Understanding Sound

Developing practical sonar systems necessitated an understanding of the physics of sound. Although Leonardo da Vinci made some important early observations, most of our fundamental understanding of acoustics was developed between 1600 and the mid-1800s.

It's difficult to find a technological concept that wasn't anticipated by Greek philosophers 2,000 years ago, or illustrated by Leonardo da Vinci (1452-1519) in his notebooks, and our understanding of acoustics and sound devices is no exception. The Greek philosophers studied and discussed the properties of sound. Statements by Aristotle (384-322 B.C.) indicate that he may have grasped the concept of pressure or wave characteristics of sound moving through air.

Da Vinci's prolific imagination and gift for invention were far in advance of his time and he often was unable to put his hands on materials needed to realize his ideas. This didn't stop him from experimenting, however, with the materials at hand or from drawing mechanisms that couldn't be built. So perhaps we shouldn't be surprised that Leonardo suggested lowering air tubes into water as listening devices, an invention that is not unlike the early hydrophones developed for maritime sensing 400 years later. Da Vinci further contributed the important observation that sound travels in *waves*.

Around 1600, Galileo (1564-1642) began making systematic studies of sound and developed fundamental theories that would later aid in making devices to exploit the properties of sound. Galileo noted that *pitch* was related to the physical properties of the objects used to generate the sound.

The concept that sound was an inherent property of physical substances was further demonstrated in 1660 by Robert Boyle (1627-1691) when he removed air from a bell jar and showed that the sound diminished in relation to the removal of air from the jar. This experiment led to the conclusion and experimental confirmation that *sound required a medium for transmission* and thus could not exist in a vacuum (but not before several scientists proposed contradictory and incorrect theories that sound was emitted through particles).

Marin Mersenne (1588-1648) studied the harmonic qualities of sound and described tonal frequencies in *Harmonic universelle* in 1636. In 1640, he measured the speed of sound to be about 1,100 feet per second. These contributions have earned him the name of Father of Acoustics.

Isaac Newton (1647-1727) contributed essential mathematical understanding to the emerging theories about sound. He studied the action of sound through fluids, noted that the speed of sound in a medium was related to its physical properties of density and compressibility, and used those observations to mathematically calculate the speed of sound through the air. Newton published his pressure theories of sound in *Principia* in 1686.

By the late 1700s, scientists were conducting experiments to try to determine the properties of sound in water. It was known that sound traveled faster through water than through air, but human curiosity is rarely satisfied with generalities, and inventors eagerly sought to measure the speeds and characteristics of these sound waves. In 1822, Daniel Colloden lowered an underwater bell into Lake Geneva, Switzerland, in order to calculate the speed of sound. A bell on one vessel was rung underneath the water, while a vibrating sensor on another vessel some distance away indicated when the sound waves reached the other vessel. The time interval was then used to calculate the speed, with an accuracy that was quite good, considering the simplicity of the methods.

The original experiments in sound were mainly physical experiments with boats and reeds, gongs and horns, but now that mathematics of sound became important—the properties of sound could be explored more precisely through theoretical calculations and then tested through physical experiments. Thus, the 1800s heralded an era when sound was explored as much with a pencil and paper as it was with horns and funnels.

As the symbology for mathematics evolved, so did the mathematical representation of sound. Sound could be expressed on paper with repeating sine waves in a two-dimensional coordinate system. Pressure and time could be assigned to the axes of a graph. The repetition of the sine waves could represent the periodicity of the sound waves. The stage was set for more advanced manipulations by mathematicians such as Johann Karl Friedrich Gauss (1777-1855) and Jean Baptiste Joseph Fourier (1768-1830).

In the 1820s, Fourier applied the superposition of sines and cosines to time-varying functions and showed that they could then be used to represent other functions. Fourier studied heat conduction through materials (a concept somewhat related to sound conduction) and his analytical techniques later led to many new techniques of mathematical modeling. Fourier's theories, when applied to sound, showed that the analysis of the harmonic qualities of sound could provide a fuller understanding of the phenomenon.

In 1826, French mathematician, Jacques Sturm, provided more accurate measurements of the speed of sound in water than were calculated by Daniel Colloden. The experiments of Sturm and other scientists confirmed that density and elasticity were important attributes that contributed to the speed of sound through various materials.

The Beginnings of Sonar

In 1880, scientists Pierre and Jacques Curie discovered the piezoelectric effect. Piezoelectricity is a form of electromagnetic polarity arising from pressure, particularly in crystalline substances. If this effect could be exploited, it might lead to the invention of new devices. Thirty-six years later it would contribute to the development of acoustical detection devices.

In 1900, the U.S. Submarine Force was established with the purchase of a 'modern' sub, the Holland VI, commissioned in October as the USS Holland (SS-1). The role of the early subs was coastal and harbor defense. One of the more important visual surveillance devices, the *periscope*, began to be used on subs at this time. A periscope was basically a mounted adaptation of the telescope that became specialized for covert, submersible situations.

The sinking of the Titanic in 1912 had a significant influence on the development and evolution of surveillance technologies. Not only were the rules for maritime telegraphy changed to prevent future accidents, but water microphones, called *hydrophones* were developed for detecting icebergs and other obstacles, in order to improve marine safety. Thus, early sonar systems emerged at about the same time and for the same reasons as the original radar devices, to prevent collisions at sea. But whereas radar was initially designed to avoid collisions with other ships and large structures along the large rivers and coastlines of western Europe, sound-ranging was originally designed to help prevent collisions with dangerous icebergs and to navigate in regions where there were lightships.

A lightship is somewhat like a lighthouse. It is a ship-mounted beacon to indicate position and the possible presence of dangerous obstructions. By equipping a lightship with a foghorn or other sounding device, and a bell below the waterline which could be sounded at the same time, information could be conveyed to nearby ships equipped with hydrophones. The sounds were within human hearing ranges so sailors could manually interpret the acoustic signals.

Underwater Mines and Submarine Warfare

Entire wars have been won and lost at sea. For centuries, naval fleets were an essential aspect of a country's military arsenal and explosives delivered by harpoons or barrels could be used to disable a fleet. This led to a new idea, underwater mines that would only explode when sufficiently disturbed. Mines are one of the most devilish inventions of man and, by the time World War I erupted, they were being planted underwater by ships and subs to protect coastal territory and to wreak havoc on enemy shipping lanes. Since no one appreciated hitting a mine hundreds of miles from land, surveillance technologies for detecting mines were avidly sought and tested. The submarines and ships that were planting the mines were also targets of sonar sensing devices.

In times of war, peacetime technology is adapted to serve the needs of national security, federal funding is diverted to further its development, and improvements for war-related purposes are the natural consequences of the increased emphasis on research and development. Rudimentary hydrophones were already used for maritime safety by the time World War I broke out, but now hydrophones were seen in a different light, as not just a means for avoiding icebergs and ships, but also as a tool for detecting submarines, avoiding minefields, and perhaps even charting a course through hazardous booby-trapped waters.

Thus, it may have been the outbreak of the war in 1914 that motivated Constantin Chilowsky, a Russian scientist living in Switzerland, to propose the idea of using high-frequency sound for detecting submarines [Hunt, 1954.].

Military Submarine Ships. Left: Pre-World War I submarines resembled small ships more than modern subs and didn't have any of the deep-diving capabilities of later subs. Nevertheless, their low-slung, less-conspicuous design made them suitable for patrols, mine detection, and mine laying. Right: The commissioning of the USS Holland, in 1900, represented the beginning of the U.S. Submarine Force. The Spanish-American War had prompted Theodore Roosevelt to make the purchase of the Holland VI in 1898. [Library of Congress Detroit Publishing Company Collection ca. 1910, copyrights expired by date.]

Thomas Edison was one of the pre-eminent inventors in America at the time World War I broke out. When the U.S. Navy set up the Naval Consulting Board (NCB) to advise the Secretary of the Navy in 1915, Edison was established as its director. Even before the U.S. took an aggressive position in the War, the NCB was engaged in research and development for devices that could detect submarines and other marine vessels. It soon became clear that the German *U-boat* submarines that were covertly attacking ships and laying minefields were a dangerous threat.

It was around 1915 that Chilowsky began collaborating with Paul Langevin (1872-1946), a gifted French physicist, to develop hydrophones for underwater detection purposes. In 1916, Langevin, a contemporary of Einstein, made a discovery that was important not just for the future development of sonar, but also for many important aspects of electronics. He found that the piezoelectric effect in quartz crystals could be applied to his research on acoustics.* By 1917, Fernand Holweck (1890-1941), who later became the director of the Curie Laboratory, was also collaborating with Langevin on his hydrophone experiments. The result of the work of these scientists was the development of a mosaic of thin slices of quartz crystals sandwiched between two steel plates. By applying an electric current, the quartz slices could be influenced to change shape and resonate at a specific frequency. This was an important milestone for electronics.

The technology of anti-submarine warfare (ASW) was being debated in American literature at least as early as 1917 and that year a special Committee was set up to tackle the challenges of detecting and destroying German *U-boat* submarines, further leading to the establishment of the Naval Research Laboratory. The *National Research Council* contributed by arranging an international conference in June 1917, bringing together British, French, Swiss, and American experts to discuss the science and development of U-boat detection systems [Weir, 1997].

---

*Compressing a quartz crystal can cause its polarity to reverse. Langevin's discovery of this property in quartz, which has a constant vibrational interval, provided a way to generate sounds far above the range of human hearing. His subsequent experiments, and those of other scientists, led to active sonar systems, ultrasound transducers, and the use of quartz in various timing applications.

The details have been lost with time, but sometime during World War I, the British Royal Navy, which was closer to the front, established an Anti-Submarine Division to study applied acoustics and solve some of the challenges of marine warfare. Surviving references from about 1918 indicate there may have been an Allied Submarine Detection Investigation Committee in Britain by that time. For the next two decades, active hydrophones, the first modern sonar systems, were developed (called ASDICs at the time).

Charles Max Mason (1877-1961), a mathematics professor, worked on submarine detection devices as a member of the NRC submarine committee during World War I. He was instrumental in the development of a naval multiple-tube, passive submarine sensor, a device that focused sound so that an operator, equipped with earphones, could use the sound levels to discern source and direction. Mason has been credited with inventing acoustical compensators.

High-frequency sound projectors had not yet been developed. The projectors used at this time tended to be tuned to the upper ranges of human hearing, around 18 kilo*cycles* ± 4 (the tradition of using *kilohertz* had not yet been established). If more than one projector was used at the same location, it was important to make sure they weren't transmitting at the same frequency or the operator would get a blast on his listening frequency that was many times louder than a normal echo.

Early hydrophones were better suited to depth-sounding than to submarine detection, but were apparently used to some extent to locate U-boats in the Atlantic. Some claim that the UC-3 was the first German submarine to be located by hydrophones and sunk in April 1916, while other sources claim that the UC-3 was laying mines a month later and therefore could not have been sunk. However, the UC-49, a mine-laying submarine, is reported to have been pursued by the HMS Opossum using hydrophones, successfully attacked with depth-charges, and sunk in August 1918 [Perkins, 1999].

Echo-ranging in those days was called *supersonics*, a term that now refers more generically to sound or other phenomena traveling faster than the speed of sound in air (thus a supersonic jet is one that breaks the sound barrier by traveling faster than the speed of sound).

The early, passive hydrophone systems had very little range, and thus were limited in use. However, vacuum-tube electronics were providing new ways to amplify electrical signals and scientists correctly speculated that this might be combined with a hydrophone to improve the signal, leading to experiments with active sound sensors. The actual application of these experimental systems on crowded, bobbing ships at the time was probably very limited if, in fact, they were even used before the end of the War.

Post-War Developments

There are many problems to deal with in post-war societies. Mines don't just disappear when wars end—they pose dangers to vessels for many decades if they are not removed, necessitating the development of peacetime mine-detection strategies and removal devices. The detection and clearing of mines were as important after the War as they were during the War. With the resumption of commercial shipping, there were many ships and crews who were unfamiliar with mines and ill-equipped to deal with them. To complicate matters, mine-laying activities by hostile forces continue even during peacetime.

Like mines, submarines don't just automatically disappear at the end of a war. After World War II, submarines continued to patrol the oceans, wary of of another global conflict. Crews trained to use sonar during the war sought employment in related fields after the War, or looked for ways to use apply wartime technologies for other purposes. They began to put more energy into adapting sonar foroceanographic sciences, resulting in the first rudimentary, underwater terrain-mapping technologies. In the 1920s, hydrophones continued to be used for submarine-spotting, but were adapted for other uses, as well, and practical active sonar systems slowly emerged. The earliest bathymetric charts—images of the seafloor terrain—were produced around 1923.

Improvements in Submarine Technologies. Left: After World War I ended, research and development on submarines and sonar systems continued and some were diverted to scientific uses. War usually creates, as an aftermath, a nervous society, concerned about defense readiness and the prevention of future conflicts. The development of longer-range, more powerful, deeper-diving subs were priorities following the War. The V-4 was authorized in 1925 for mine-laying operations and was renamed the "Argonaut". Right: The S-class submarines were used in the 1920s and 1930s as test platforms for higher-frequency sonars with smaller, trainable transducers. Narrower-beam, lower-interference systems were being developed. [U.S. Navy historical photo, public domain; U.S. Navy historical photo from the Lt. Oscar Levy collection, released.]

Other seafaring nations were making similar post-war adjustments. In 1927, in Britain, an ASDIC research and development unit was established at the Portland Naval Base at HMS Osprey to study hydrophones, ASDIC, and other aspects of submarine detection. In 1929, this unit was renamed the *Anti-Submarine Experimental Establishment* (A/SEE).

The *National Academy of Sciences* in America emphasized the importance of oceanographic research in 1927 and recommended a permanent facility for this purpose. The result was the *Woods Hole Oceanographic Institute* (WHOI), founded in 1930 in a lab on the east coast.

In 1933, the *Washington Navy Yard* manufactured 20 sets of echo-ranging systems, considered a major development in sonar. While the author could find only the briefest reference to these systems, it appears that their design may have been based upon the steel-and-quartz-sandwich transducers first developed by Paul Langevin and his contemporaries at the end of World War I. Sound-echo-ranging equipment was installed on American destroyers in 1934, including the USS Rathburne (DD-113). Thus, sound-ranging systems, as we now understand them, had been established by this time.

By the mid-1930s, the use of radio or sound signals for civilian and commercial ranging applications was documented in engineering texts, and aircraft began to use radar (radio ranging), which is similar to sonar ranging, for navigation. Not long after, scientists began to understand that salinity and water temperature had important effects on underwater acoustical transmissions.

Sonar training centers began to spring up around the world. In 1939, the *West Coast Sound School* was opened at the San Diego Destroyer Base. The same year, the *Atlantic Fleet Sound School* opened at the submarine base in New London, Connecticut (in 1940, this school was transferred to Key West, Florida). In Canada, a small submarine-detection school was established in Halifax.

When surveillance technologies improve in effectiveness, those who wish to remain unseen adapt countermeasures to maintain their secrecy. This was as true for sonar as for any other technology. Since visual spotting at the surface was effective in detecting submarines in World War I, postwar submarine designers devised ways for subs to dive deeper and to stay under for longer periods of time. As subs dove deeper and longer, sonar was improved to detect them under the water rather than at the surface of the water. As World War II progressed, the German U-boat crews realized that the current sonar systems were optimized for finding them in deeper water and were less effective when the subs attacked near the surface. This spurred the development of more specialized sonar systems and more effective radar systems, thus countering the countermeasures.

World War II

By the late 1930s, submarine-spotting and sonar navigation systems were no longer peace-time technologies. The world was again entangled in global conflict and both sonar and radar were important technologies contributing to the outcome of the War.

After Britain's anti-submarine establishment on the south coast at Portland was bombed in 1940, it was moved to Fairlie, near Glasgow, and research and development continued through the War, including the design of mechanisms for sending out depth charges.

From 1940 to 1967, the Raven (AM 55) swept the oceans for mines. It was the lead ship of 93 minesweepers serving the U.S. and Britain in World War II.

During World War II, sonar had evolved to the point that submarines could no longer lurk in the dark undetected under the surface of the water. Anti-sonar measures became as important as sonar itself. In order to try to reduce their sonar signatures, the Germans experimented with synthetic rubber as a skin to counter Allied sonar probes. Special materials are now regularly used on subs and aircraft to selectively deflect or absorb sonar and radar probes.

Greater scientific research was being applied to acoustic sensing at this time. The U.S. forces were benefiting from private institutions and setting up new labs of their own. The *Woods Hole Oceanographic Institute* contributed important defense-related research to the U.S. forces during this time.

The bathothermograph, an early sonar imaging technology that became available during the war, was incorporated into submarines to aid them in covert patrol and attack operations. The USS Herring (SS-233), left, and the USS Scorpion (SS-278) were equipped with the submarine bathothermograph (SBT) system. The Herring spotted and sank the German U-163 in March 1943. The Scorpion was lost, perhaps to a mine explosion, soon after a rendezvous with the Herring in February 1944. The Herring was presumably sunk from Japanese attacks to the conning tower in 1944. [U.S. Navy historical photos, public domain.]

In 1941, Waldo K. Lyon established the *U.S. Naval Arctic Submarine Laboratory*. Since the arctic was a particularly challenging environment in which to navigate, sonar research and development were essential to carrying out successful surveillance in icy regions.

In October 1941, the 'bathothermograph' was introduced to the naval fleet. The training of specialists and assignment of crew to take bathymetric observations from aboard ships were begun so the technology could be used in patrol and attack operations. Bathothermography helped a submarine hide in thermal layers to avoid detection by enemy sonar systems.

The ASW Patrol Ships and Destroyers

In December 1941, after the attack on Pearl Harbor, the importance of the U.S. Submarine Force increased. Subs were used to hold the line in the Pacific and made hundreds of patrols. They even used FM-based sonar to pursue the Japanese into their own waters.

Sonar equipment is generally installed or towed underwater, and thus is vulnerable to damage from debris, collisions, ice, and mine blasts. Retractable sonar domes existed before this time but, by the early 1940s, most of the larger ships were fitted with retractable domes.

The new sonar equipment on military ships necessitated new training and procedures in technical and strategic handling of the technology. Two ships of interest that were involved in anti-submarine activities were the USS Jacob Jones and the USS Roper.

In February 1937, The USS Jacob Jones (DD 130) participated in minesweeping training and in 1940 joined the Neutrality Patrol, which had been formed in September 1939 to patrol the western hemisphere. After two months of duty, the ship returned to training operations. In September 1940, the Jacob Jones sailed to New London, where the crew took acoustics training for anti-submarine warfare (ASW), after which she continued to Key West for further training before rejoining the Neutrality Patrol. Sometime in late 1941 or early in 1942, the Jacob Jones detected an underwater submarine and began to attack with depth charges. Contact with the submarine was lost, however, and the Jacob Jones continued on her way. A month later, while heading south from Iceland, she again detected a submarine, but depth charges were apparently ineffective.

In February the Jacob Jones became a member of a roving ASW patrol and soon detected and gave chase to a submarine. Depth charges yielded oil slicks, but no confirmation of sinking a sub. On 27 February 1942, the vessel searched for survivors around the wreckage of a torpedoed tanker off the coast of Delaware, then set course south. By the light of the following morning, an undetected German U-boat (U-578) fired at Jacob Jones, surprising her and ramming her with at least two torpedoes. The ship was destroyed and sank rapidly, and only 11 made it to shore alive. Wreck divers have reported that the Jacob Jones lies in pieces with its torpedoes intact near the Indian River. A month after the sinking of the Jacob Jones, the U.S. Navy established the *Submarine Chaser Training Center*.

The USS Jacob Jones, the historic anti-submarine training vessel, was sunk by U-boat torpedoes from an undetected submarine in February 1942, off the coast of Delaware. The USS Roper (DD 147) was also used as an ASW training ship, but was converted to a transport in 1943, and decommissioned in 1945 to be sold for scrap. [U.S. Navy historical photos, public domain.]

Submarine Surveillance. Left: A 110-foot Patrol Coastal (PC) submarine harrier in the construction yard in Stamford, Connecticut, March 1942. The responsibilities of Patrol Coastal and Patrol Sub Chaser (PCS) vessels were to patrol various waterways and conduct interdiction surveillance. Right: Current patrol ships like this Cyclone-class Patrol Coastal also provide support to Navy SEAL operations. [Library of Congress FS/OWI photo by Howard Liberman, U.S. Navy news photo, both public domain.]

The other vessel reported to have been used for ASW training was the USS Roper (DD 147), which also formed part of the Neutrality Patrol. This vessel was luckier than the Jacob Jones. In April, 1942, she sighted a German U-boat at the surface off the coast of North Carolina. The Roper pursued the submarine and succeeded in sinking the Nazi vessel designated U-85. The U-85 was the first German U-boat to be sunk in American waters. She sunk to a depth of about 95 ± 10 feet in waters with strong currents, with all hands lost. The location is now a designated German grave site.*

Patrol Coastal (PC) and Patrol Sub Chaser (PSC) vessels handled many of the surveillance and ASW tasks from the time of World War II. Rushville (PSC-1380) was put into service in 1943. She was a Patrol Sub Chaser vessel assigned to ASW training duty at the *Fleet Sonar School Squadron* in Key West. In December 1944, the Rushville was outfitted with special experimental acoustic gear.

Other Acoustic Applications

Acoustic devices are used for many purposes. A strong enough wave, such as a boom from a large explosion, can deform the ground in rolling waves (technically a shock wave rather than a sound wave), blast buildings into matchsticks, and knock a person off his feet. Thus, the pressure characteristics of sound can be used in other ways besides the detection of seafaring vessels. In 1942, U.K. scientists created a device to generate sounds that could clear a minefield by detonating the mines. This *hammer box* was usually lowered into the water well ahead of the path of a minesweeper specially constructed to withstand the subsequent explosion.

The use of the term ASDIC for sound-ranging applications was specifically tied to submarines, but sonar was now being used in a broad range of military and commercial applications. The more familiar term *sonar* has been attributed to American underwater acoustics specialist and director of the wartime *Harvard Underwater Sound Laboratory*, Frederick V. (Ted) Hunt, to provide a euphonious analog to 'radar'. Others maintain the term was coined to represent

---

*Mike Leonard has created a built-from-scratch model of the German submarine U-85. A photo of the model can be viewed at the simplenet.com site. warship.simplenet.com/images/Leonard/U85.jpg

'sonic, azimuth, and range'. Either way, it eventually replaced the British term ASDIC. Currently, the term sonar is used as an acronym for 'sound *na*vigation *r*anging' or 'sound *na*vigation *a*nd *r*anging'.

Frequency adjustments on sonar systems at this time were still somewhat crude. If the system could operate at more than one frequency (or if the frequency needed tuning), the job was usually done with a screwdriver rather than a switch or knob as on modern systems. Secrecy further limited the number of people who were qualified to repair or calibrate a sonar system. In fact, quartz, which was increasingly used for its piezoelectric properties, was referred to by the codeword *asdivite* until about the time the term *sonar* caught on and the term ASDIC was relegated to history.

The State of Sonar Technology

By the mid-1940s the variety of sound-sensing devices had increased significantly. *Automatic* and *recording* sonars had been used during the war, and many continued to be used in military vessels after the war. These were used chiefly for navigation, submarine spotting, and automatic firing systems. During periods of hostile contact or attacks, *range recorders* could be used to make a visual record of sounds from a sonar receiver and could be used to activate depth charges or thrown charges. *Bearing recorders* for monitoring gun bearings, were sometimes used in conjunction with range recorders.

Although sonar technology improved between World War I and World War II, by the end of World War II, some aspects of sonar were still essentially the same as prior to the war. The frequencies used were still primarily within audible hearing ranges and slightly above, and their range had not significantly improved. Sophisticated sonar systems emerged sometime after the conclusion of the War.

Post-World War II

After the war, a number of wartime vessels were put into service as training vessels or decommissioned for use as salvage or targets. The Canadian government acquired a British L-class submarine (L-26) and scuttled it a few miles off Pennant Point on the east coast. It was used as a *sonar target* for location and identification training purposes for at least the next twenty years.

The Fairlic anti-submarine research center near Glasgow was returned to its original site in Portland on the south coast in 1946. During post-war organizational changes in 1947, the name was changed to *HM Underwater Detection Establishment*.

Before the end of the War, a U.S. team had the opportunity to study German U-boat technology. As a result of this, both during and after the War, the advanced design features of the German subs were incorporated into U.S. subs, taking the best features from both worlds. Thus, Greater Underwater Propulsive Power (GUPPY)sytems and streamlined hulls were developed to increase cruising speeds while submerged, snorkel systems were added, and array sonar systems were incorporated to keep pace with improvements in the subs themselves.

Post-War and Cold War Sensing Applications

The *Woods Hole Oceanographic Institute*, founded in 1930, had by this time established an *Underwater Sound Lab* (USL). In 1948, the U.S. Navy began working with the USL on countermeasures to the Soviet submarine force. This was apparently the origin of Project KAYO, which explored the use of submarines with low-frequency, passive, bow-mounted sonar arrays for ASW tasks [Cote, Jr., 1998].

The effective range of sub-spotting sonars at this time was about 3500 ± 300 meters, using analog broadband detectors. The expertise of the operator was crucial to interpretation of the signals and good pitch discrimination was an asset.

In the late 1940s, the U.S. began to install a more-or-less permanent system of 'underwater ears' in strategic ocean locations on the Atlantic continental shelf. These cable-connected hydrophones enabled detection of any unidentified vessels approaching the seacoast, particularly submarines. The system also provided a way to monitor and keep in touch with U.S. submarines or detect a vessel that might be acting erratically, indicating that it might be in distress and unable to radio for help. The system, called SOSUS (*sound surveillance system*) wasn't good at sensing very slow-moving vessels, since they didn't create enough of a disturbance to register on the system, but anything traveling more than about eight knots was vulnerable to detection. The original SOSUS system was based upon single-beam sensing recorded on paper. Later, computers would be used to process and analyze the acoustical data.

There were many important improvements in sonar technology following World War II. The invention of transistors in 1947 was in part responsible for significant improvements in electronics capabilities and electronics miniaturization. Left: In 1951, the Guavina (SS 362) was equipped with an experimental searchlight sonar that made it possible to distinguish the sound signatures of specific vessels (a concept that was also applied to radar identifiers for aircraft around this time). This new sonar system could differentiate signals far better than previous systems. Right: The Albacore (AGSS-569) was an experimental submarine with a streamlined hull designed to create less noise in the water and thus avoid registering on acoustic surveillance systems. It was also equipped with the first fiberglass sonar dome, in 1953. [U.S. Navy historical photos, public domain.]

By the late 1940s, sonar was increasingly used for resource studies and environmental-sensing purposes, especially for studying gases, minerals, and fish behaviors and habitats—applications that have continued to grow.

Post-War Adjustments to Acoustics Research

Cold War politics in the early 1950s resulted in a high priority being placed on sonar research and development. At the same time, training for sonar operators, which had been very practical and hands-on up to this time, began to include some of the theory and mathematics associated with sonar. The math was extremely valuable as it could be used to create lookup tables for predicting detection ranges, a step toward modern acoustics-sensing technologies.

By the 1950s, sonar was used for more comprehensive and detailed mapping of ocean terrain. Sonar images revealed impressive features, including extensive mountains and valleys that had never before been seen. Later developments of multibeam sonars would greatly enhance this process.

By this time, the U.S. was building nuclear subs. The first nuclear sub, the Nautilus (SSN-571), was commissioned in 1954 and was underway by January 1955. It was the first true U.S. submersible submarine and the first ship to reach the North Pole. After decommissioning in 1980, Nautilus was put on display at the *Submarine Force Museum*.

Improved Diving and Acoustic Sensing

The Soviet launch of the Sputnik satellite in 1957 was a wake-up call for politicians and an inspirational shot-in-the arm for scientists. As a result of international competition for space-based resources, there were many technological firsts in the early 1960s. Deep diving, improved sonar, and new space technologies all contributed to navigation, exploration, and communications in the 1960s and beyond.

The Mariana Trench. The Mariana[s] Trench, also called Challenger Deep, is considered to be the deepest part of the Earth's oceanic terrain and thus provides a significant challenge to diving technology. NOAA-NGDC provide a computer-animated model of the trench based upon bathymetry-topography data. [Animation by Dr. Peter W. Sloss, public domain materials courtesy of NOAA/NGDC.]

In 1960, Don Walsh and Jacques Piccard descended to the record-breaking depth of 10,912 meters (sometimes reported as 10,915 meters) in the Trieste II bathyscaphe, to study the depths of the Mariana[s] Trench. This inspired the development of research submarines with greater maneuvering capabilities. Even forty years later, this is a remarkable achievement, considering only a handful of vessels (e.g., the NR-1) can dive deeper than 7,000 meters and none has yet beaten the Trieste II's record.

In the U.S.S.R. and the U.S., submarines were becoming basic strike force vessels and some military analysts in the U.S.S.R. suggested that surface vessels might eventually be superseded by underwater forces for strike activities. At this time, the Soviets had a force of over 400 subs, about half of them long-range vessels. The U.S. estimated that the Soviets were also constructing nuclear subs.

The operating range of early sonar systems was limited, but as transducers gradually replaced older quartz oscillating systems, the range was increased by a factor of about three, depending on conditions—a considerable improvement.

Technological One-Upmanship

The problem of detecting and avoiding underwater mines was now critical. Mines had been improved to the point where contact was no longer necessary to set them off. Sensitive non-contact detonators that could react to the physical emanations of marine craft made them deadlier than ever—longer-range, more sophisticated detection systems were needed to avoid them.

At about the same time, variable-depth sonar systems began to be used to overcome some of the limitations in trying to track underwater vessels that were using water layers to hide from sensors—layer characteristics could effectively refract a surveillance beam and provide misleading data. Variable-depth sonar (VDS) technology was in part developed in Canada.

Postwar research labs in the early 1950 and 1960s were actively improving sonar technology, resulting in more powerful and flexible systems. More sophisticated submarine countermeasures, weapons-deployment, and improved sidescanning and dipping sonars for use with helicopters, were all initiated or developed during this time.

During the Korean War, in the late 1950s and early 1960s, U.S. Naval submarines were used for both sonar and photo reconnaissance during surveys of Korean and Soviet mine fields, shipping lanes, and coastlines. Submarine hunter-killer groups were also dispatched. In 1964, the transport subs *Perch* and *Sealion* were recommissioned to support SEAL operations, to collect intelligence, and to aid in search and rescue operations.

In the early 1960s, the CIA, interested in Soviet military warfare communications, was intercepting secret Soviet publications such as *Voyennaya Mysl* (Military Thought) published by the Soviet Ministry of Defense, which discussed various tools and strategies of warfare, including naval surface vessels. These articles were translated into English and distributed to the Defense Intelligence Agency (DIA) and various defense and intelligence directors.

Innovation and Disaster

The lead ship in a new class of stealth submarines was the USS Thresher (SSN 593). It was a Permit-class vessel—quieter, more streamlined, and equipped with advanced sonar and weapons systems. In April 1963, only three years after it was commissioned, tragedy struck the U.S. Navy when the USS Thresher (SSN 593) ruptured and sank off the coast of New England, taking down the entire crew. Turning tragedy into technology, the Navy began research and development on vessels that could aid submarines in distress, in hopes of offloading their personnel and bringing needed emergency equipment and supplies. This led to the establishment of the *SubSafe* certification program and the *Deep Submergence Rescue Program* and vessels that could carry out rescue dives to thousands of feet.

Submarine Strengths and Weaknesses. Left: The USS Timosa (SSN 606) is one of the Permit-class nuclear subs designed for quieter operation and greater stealth. These subs could also dive deeper than older models. Right: The USS Thresher (SSN 593) was lost with all hands in 1963, possibly due to a defect in the piping system. Ironically, acoustics might have aided in detecting a possible defect, by more thorough application of ultrasonic testing to the joints and systems. [U.S. Navy news photos, public domain.]

In 1964, the U.S. Navy built Alvin, a submersible vessel capable of taking two scientists down to a depth of about 4,000 meters. Alvin was equipped with measuring instruments, cameras, and mechanical arms for collecting samples. It was delivered to the *Woods Hole Oceanographic Institute* for civilian research.

By the 1970s, computer and satellite technologies were being utilized to create some interesting new sonar inventions. The ALACE system of temperature-measuring floats was developed by Scripps and Webb Research to gather information underwater and surface about once a month to transmit data to a satellite receiver. It would then resubmerge to continue its mission. This innovative concept could be applied to sonobuoys designed to take sonar readings and autonomously transmit the data through a satellite.

Sonar Search Technologies. In May 1968, the USS Scorpion (SSN 589), a U.S. nuclear submarine first commissioned in 1960 disappeared southwest of the Azores with all hands, with no clear answers as to why she sank. She was found in October 1968 by a towed deep-submergence vehicle. The image on the right was probably taken at the time the deep-sub vehicle located the wreck on the ocean bottom late in 1968. [U.S. Navy historical photos, released.]

Deep-Submergence Craft

In 1969, the U.S. Navy launched its first deep-submergence vessel, a nuclear-powered craft, the NR-1. The NR-1 is able to stay underwater for extended periods at depths of about 700 meters and is designed for object recovery, geological surveying, oceanographic research, and the installation and maintenance of various types of underwater equipment. The vessel is equipped with sonar, special manipulators, cameras, and a TV periscope that enable it to be used for high-resolution mapping and searching operations. In 1986, the NR-1 had the unenviable task of searching for wreckage from the space shuttle Challenger which exploded shortly after launch. In 1997, it was engaged in the second of two archaeological expeditions in the Mediterranean in search of ancient Roman merchant ships along with the JASON remotely operated vehicle from *Woods Hole Oceanographic Institute*.

Left: The NR-1 is a unique nuclear-powered deep-sea vessel first used by the Navy in 1969. It can dive to more than 700 meters for extended periods to carry out sophisticated military, commercial, and scientific search and mapping operations. Right: A Deep-Submergence Rescue Vehicle (DSRV) designed to perform rescue operations on disabled, submerged submarines. The first DSRV was launched in 1970. This 15-meter Lockheed Missiles and Space, Co. vessel can dive up to about 1500 meters to conduct a search via sonar. It utilizes both search and navigation sonar systems. [U.S. Navy news photos, released.]

The NR-1 is equipped with visual viewing ports with external lighting, and color and grayscale video and still cameras, manipulator arms for grasping and cutting, and sonar searching and navigation systems. The vessel also sports the Obstacle Avoidance Sonar (OAS) developed by the University of Texas.

The development of personal computers in the mid-1970s eventually led to ways for sonar equipment to be interfaced with computer processing systems and improved displays, and provided new means to coordinate the timing and processing of multibeam sonar and sonar array sensors two decades later.

The 1970s - New and Improved Sonar Applications

Acoustic systems in the early century were used mainly for navigation, submarine-spotting and mine detection. However, as the technology improved and the use of torpedoes increased, acoustics began to be used for other purposes, including homing devices and torpedo detection. The frequencies at which sonar were used had also become ultrasonic. Whereas early sonar systems operated at about 18 kHz, newer systems could generate frequencies of 100 kHz and beyond. These systems were not yet realtime, however. Like early radar, which relied on sending out a pulse and waiting to sense a returning pulse, with an interval between pulses, sonar didn't provide an up-to-date picture, but rather a picture of things as they were just a moment before the signal registered on the printout or display. Since microcomputer electronics were not yet fully developed in the early 1970s, an additional delay was introduced by limited signal-processing capabilities.

By the 1970s, sport fishing was changing from a quiet, contemplative pastime to a high-tech search for the biggest schools and the best fish. Like hunters seeking trophy stags, fishing enthusiasts began using sonar to track, identify, and catch fish big enough to match their fish stories. *Fish-finders*, as they came to be called, weren't very sophisticated at first—the displays were rudimentary and the systems bulky and expensive—but they provided a new way to determine the presence and depth of fish below a boat. Distance calibrators could convert the data into feet or meters. The early systems were costly and were used more for commercial fishing than sport fishing but, with the development of microelectronics in the mid-1970s, the situation began to change.

In 1970, the U.S. Navy launched the USNS Hayes as one of 28 special missions ships. The Hayes was converted to function as an acoustic research ship in 1986 and completed and reclassified as T-AG in 1992. The Hayes now transports, deploys, and retrieves acoustic arrays and conducts acoustic surveys and testing operations. The catamaran design provides a wide operating deck and a sheltered region between the hulls. Since high-end imaging applications generally require slow towing speeds, the Hayes has two engines specially equipped to maintain a speed of two-to-four knots.

By the mid-1970s, various systems with widely spaced sonar devices were being developed to provide more accurate estimates of range and location. These essentially used the concept of 'triangulation' (the combination of data from more than one sensor targeting the same general area) to provide a more accurate picture of the incoming signals. Improved digital signal-processing systems (DSPs) were also being developed, which transformed many aspects of electronics processing.

As computer technologies improved, many sonar systems were upgraded to be more portable and powerful. Existing systems were replaced, overhauled, or upgraded. The SOSUS system first implemented in the early 1950s was increasingly computerized. New systems, such as the Surveillance Towed Array Sensor (SURTASS) were developed.

By the 1980s, the sensitivity and power of sonar had greatly improved over the systems used in the War and the early 1960s, when transistors and computer components were coming on the scene. Improvements included the extension of sensing ranges from less than 1,000 meters to almost 200,000 meters (200 kilometers) in good conditions. In the future, that range would be extended even further, with new communications technologies.

Important improvements occurred at this time in Doppler sonar systems. The early Doppler systems were not easy for untrained personnel to use and were not suitable for distinguishing objects that were traveling at varying speeds within moving currents. To improve on existing systems, the U.S. Navy devised ways for narrow-band filters to detect the Doppler frequency in order to combine the data with compensating circuits and reference measurements. This enabled water velocity to be compared with incoming data to create a more accurate sonic picture of detected objects [Skoures, Farace, 1970].

The 1980s - Increasing Sophistication and Computerization

In the early 1980s, the U.S. Navy began to develop a mine countermeasures (MCM) force made up of two new classes of ships and minesweeping helicopters. The Avenger- and Osprey-class ships were designed to detect, classify, and destroy moored mines and bottom mines using conventional minesweeping techniques combined with newer sonar and video systems, cable cutters, and remote-controlled detonating devices.

Improved diving equipment and techniques, combined with video and sonar surveillance technologies made the discovery of previously 'unrecoverable' wrecks a possibility. Many wrecks given up for lost were found and explored or salvaged with advanced technologies in the 1980s and 1990s.

In June 1985, divers discovered the wreck of a German sub that had been taken over by the U.S. after the War and subsequently sunk during explosives testing. The U-1105, dubbed the 'Black Panther,' had been commissioned for the war in June 1944. It was an experimental German sub incorporating a dark, synthetic rubber skin that was code-named 'Alberich.' At one point, the U-1105 managed to hide for 31 hours after attacking the HMS Redmill with acoustic torpedoes. The modified Type VII-C Kriegsmarine sub was able to evade detection for the duration of the war, but was surrendered to the U.S. at the end of the War. The rubber skin was studied by the Naval Research Laboratory and MIT's Acoustic Laboratory and the vessel later became Maryland's first underwater dive preserve.

In 1989, the *Key West Fleet Sonar School* was closed, so that just the *Fleet ASW Training Center* in San Diego remained to serve the anti-submarine needs of the U.S. Navy.

Acoustic-Sensing Challenges

The 'littoral environment' is the coastal region within which tidal changes occur. Littoral environments in many regions are complex and craggy. They are filled with rocks, debris, fish, plants, corals, and thick layers of shifting sediment. The difficulty of distinguishing targets from the clutter in these coastal regions is significant. Scientific vessels, diving boats, and special-ized fishing boats were demanding more sophisticated systems that could aid in navigating and surveying these tidal environments. The same was true for military defense surveillance and commercial shipping safety. Mines that were planted in tidal waters were difficult to dis-tinguish from debris and outcroppings. Sensing systems could prevent damage to boats and, in some cases, loss of life.

Littoral environments also provided a place for new, smaller, remote-controlled and au-tonomous craft developed in the late 1980s and 1990s to hide.

In 1990, the Office of Naval Research established the *High Area Rate Reconnaissance* (HARR) program to develop both shallow and deep water technologies for countering mine threats. HARR used both *side-looking sonar* (SLS) imaging and *toroidal volume search sonar*

(TVSS) or etecting unburied mines. The TVSS system is donut-shaped, providing omnidirectional beaming perpendicular to the direction of the tow. It was tested in the mid-1990s in the Gulf of Mexico, in water depths from 28 to 160 meters. The SLS was tested in the late 1990s, generating sonar images that could be processed and viewed in realtime on a display screen.

Left: A U.S. Navy Ocean Surveillance ship equipped with an advanced linear towed-array system for underwater detection. Right: Three Stalwart-class Ocean Surveillance ships were refitted for above-surface surveillance using radar and advanced communications systems. These are now being used in support of Navy counter-drug-trafficking surveillance. [U.S. Navy news photos, released.]

In the early 1990s, with a reduced Soviet threat, the U.S. Navy lowered the priority on certain ocean surveillance vessels that had been commissioned since the mid-1980s. Three of these Navy ships were converted to handle narcotics-trafficking detection rather than underwater acoustics surveillance. Their underwater acoustic arrays were removed and replaced with an above-water surveillance system, an air-search radar, and up-to-date communications systems.

Other ships in the fleet continued their ocean surveillance activities to support the anti-submarine efforts of other vessels. The *Surveillance Towed-Array Sonar System* (SURTASS) used on the ships is comprised of listening devices, computerized processors, and electronic satellite communications. Due to the needs of high-resolution sonar, the ships are designed to be stable at slow hull speeds, even under adverse weather conditions, to handle the towed linear-array system. The *Impeccable*-class ships have modules that house pairs of high-powered active-sonar transducers which can be used with either monostatic or bistatic receivers.

There are three computer-related technologies that were emerging at this time that were particularly important to surveillance technologies, including sonar.

- The development of fiber optics permitted the manufacture of longer, lighter sonar-towing and communications cables, and enabled higher-bandwidth applications that were especially useful for sonar arrays with many elements, such as imaging systems.

- By the mid-1990s, the development of high-resolution Global Positioning System (GPS) technologies was beginning to have a substantial impact on the evolution of sonar technologies. The ability to pinpoint (or even more closely approximate) the location of sonar devices such as sonobuoys with the added GPS data, permitted applications that had not been previously possible. Satellite-derived gravity imagery of ocean structures also became an important adjunct to Earth-based oceanography. Scientists could combine satellite and sonar imagery, sometimes in real-time, to develop better quality underwater terrain maps.

- Digital signal processing (DSP) made inroads in the 1990s and contributed to the development of more sophisticated computerized sonar systems with greater levels of automation.

Increasingly sophisticated sonar systems provoked counterdevelopment of increasingly 'quiet' submarines—vessels equipped with special skins, low sonar signatures, quiet propulsion, and other features to improve their stealth capabilities. More than ever, the U.S. submarine force had earned the name of "Silent Service".

Upgrading to New Technologies

In the late 1990s, computer electronics and software made some remarkable advancements. Virtual storage devices jumped in capacity and dropped in price, processors continued to become faster, and software became increasingly sophisticated. The technology of instrumentation was advancing far more rapidly than the design of the vessels themselves. Submarines and ships were improving, but the basic designs were little different from those that existed around the time of World War I. The concepts of seaworthiness had been developed and refined over centuries, whereas modern computer electronics disciplines were less than thirty years old and still in their growth and development stages.

This technological incongruity created a dilemma for vessels equipped with large, built-in sensing systems. Newly-installed electronic consoles were becoming obsolete in months or even weeks, compared to the vessels themselves, which could be used in one capacity or another for decades. Thus, during the late 1990s, many surveillance craft, from helicopters to submarine-chasers, were upgraded or retrofitted with advanced sensing systems, instead of retiring the vessels. One example of this is the *Acoustic Rapid COTS Insertion* (ARCI) upgrade for submarines. It was initially installed for evaluation in the USS Augusta in 1997 and subsequently planned for fitting into other vessels in the fleet over the next decade.

In September 1998, new U.S. Navy attack submarines were equipped with advanced stealth features, and sonar systems for anti-submarine and mine warfare. They could also launch un-staffed underwater or aerial vehicles for mine reconnaissance and intelligence-gathering.

Scientific Applications

The emphasis on peacetime scientific surveillance missions increased in the 1990s. Submarines and surface vessels equipped with advanced sonar systems were used for search and rescue, salvage, ecological studies, and scientific expeditions to the north and south poles. The information gained was valuable for both national defense objectives and civilian research. Like the early dual-purpose weather/defense satellites, dual-purpose use of vessels helped justify their continued funding.

Multisensor Imaging

With the computerization of tracking systems, data from different sensing technologies may be combined into one visual 'map'. Sonar signals may be included with data from radar and visual surveillance technologies. The days of discrete modes of sensing are probably over for many applications.

In 1998, the U.S. Navy, due to priority changes and funding cutbacks, gave up the maintenance and operation of its deep-submergence vehicles. Both the Sea Cliff and the Turtle were decommissioned and retired from service. The Turtle was put into storage and the Sea Cliff transferred to the *National Deep Submergence Facility* at the *Woods Hole Oceanographic Institute* for alterations that would make it suitable for scientific research.

Aircraft Wreckage Search, Rescue, and Recovery

Surveillance technologies have evolved to the point where many types of vessels are equipped for search and rescue operations. The use of these technologies in the recovery of downed aircraft is described both here and in the Applications section.

The careful coordination of video, sonar, radio, and computer technologies for search and rescue was demonstrated by Canadian forces when their sonar was used to locate a jetliner that disappeared near Nova Scotia, in September 1998.

When Swissair Flight 111 crashed off of Peggy's Cove on the east coast, one of the systems used in the search for wreckage was the Canadian Towed-Array Sonar System (CANTASS), along with a number of remotely operated vehicles (ROV). The Deep Sea Inspection System (DSIS) is equipped with both video and sonar surveillance systems. The Bottom Object Inspection Vehicle (BOIV) is a medium-sized remotely operated vehicle and the Phantom, is a small ROV. The Phantom, equipped with color video and high-definition, scanning sonar, created images of the ocean floor that revealed the contours of the stricken jetliner. In addition, the HMCS Okanagon, an Oberon-class submarine, used passive sonar to listen for the locating signal from the airliner's flight data recorder.

The data collected by the remote vehicles were transmitted by radio from the supply vessel HMCS Anticosti to military oceanographers stationed at the Bedford Institute in Halifax who were aboard a Coast Guard cutter. It was then fed to computer software that cross-referenced it with a record of the airplane's seating plan. This coordinated effort enabled the identification of all the victims, on behalf of the airline and the bereaved families.

In the process of searching for the Swissair flight, the searchers got an unexpected surprise—a large, old object, perhaps a ship, perhaps a submarine, lying on the ocean floor.

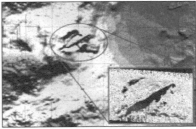

Left: The HMCS Kingston, used at the scene of the Swissair Flight 111 crash as a sidescanning survey ship (alng wih the HMCS Anticosti). Right: A sonar image of what appears to be an old vessel was taken off the coast of Nova Scotia while searching for wreckage of the Swissair craft by the Canadian Towed-Array Sonar System (CANTASS) in conjunction with the Phantom video- and sonar-equipped remotely operated vehicle. [Canadian Department of National Defence 1998 news photos, public domain.]

The Matthew is a survey and research vessel that was dispatched for sonar search operations when Swissair Flight 111 crashed into the ocean. Right: Diving Tenders Grandby and Sechelt were used for diving operations in the recovery. [Canadian Department of Defence 1998 news photo, public domain.]

Acoustic Homing Sensors

Acoustic homing sensors are used on underwater torpedoes for precise targeting, but still have limitations that require accessory guidance systems. Acoustic homing torpedoes are typically launched from submarines or advanced mines.

Acoustical homing systems have been in use at least since World War II, when the Royal Air Force was equipped with acoustic homing torpedoes and used them along with depth charges to sink the U-954. In recent years, acoustic homing systems have become more practical and accessible and it was clear by the mid-1990s that the U.S. and her allies were not the only countries using this technology. American intelligence indicated that Iran had been testing wake-homing and wire-guided acoustic homing torpedoes on their submarine fleet [Baus, 1996].

Technological Advances

Rapid-scanning sonars, laser sonars, and within-pulse electronic-sector-scanning sonars are just some of the technologies that have been more recently introduced.

One of the most challenging applications for sonar ranging and detection is navigating beneath ice floes at the Earth's poles. Even with advanced sensing equipment and the best sonar operators, the challenge of navigating in tight quarters beneath massive, deep, jagged ice protrusions is not for the faint of heart. To add to the challenge, very little oceanographic data exist on regions permanently shrouded in ice.

In March 1999, the Sturgeon-class attack submarine USS Hawkbill dove beneath icefloes in the Bering Sea to begin an eight-week arctic research mission. The vessel had been specially outfitted with sensing systems for navigating under arctic ice, including a forward-looking ice-finding sonar for detecting the protruding icebergs and high-frequency sonar for constantly monitoring ice draft (extent below sea level). The Hawkbill was dispatched to collect scientific data and samples to further research on ocean warming, and geophysical and chemical structures. For surfacing maneuvers, an upward-looking video display was used in conjunction with sonar guidance.

One of the significant experiments from this mission was the Arctic Climate Observations Using Underwater Sound (ACOUS) project in which a vertical hydrophone array was dropped through a hole in the ice to receive transmissions from a 20-Hz underwater sound signal emitted from some distance away. This provided a measure of oceanic temperature along the range of a 2800-kilometer path.

Satellite Communications

During the late 1990s, with improvements in commercial and military satellite communications, a number of sonar-satellite technologies were invented.

Since sonobuoys were often self-contained units, left unattended for long periods of time, they were sometimes difficult or expensive to relocate. One solution was to automatically scuttle them at the end of their useful life. Thus, for surveillance missions, the data had to be transmitted to vessels or ground stations. Encrypted radio-wave communications with satellites provided a practical solution to this problem. Not only could the data be sent to a remote station for analysis, but the buoy itself could be controlled to some extent with telemetric data relayed through the satellite.

This general idea could also be applied to communications between a submarine and a buoy. Using sonar signals, information could be sent to the buoy floating at the surface, which, in turn, could use radio signals to communicate with the satellite. This way, the submarine wouldn't have to surface to communicate with air or ground stations. The major disadvantage was that the submarine had to be near the buoy to connect to it through a communications cable (see the Magnetic Surveillance chapter for some proposed improvements to this scheme).

Automation and the Future

By the late 1990s, in spite of all the automation and improvements in sonar technology, the main 'brain' of the system was still the human being who examined and analyzed the sonar screen or sonar images. The difference between a mine, a wreck, or a submerged sub was still largely determined by a subjective evaluation by a trained professional. Strides were being taken, however, to automate some of the recognition capabilities of the software, in an effort to create 'smart' sonar systems. This brings together the accumulated knowledge of many fields, including artificial intelligence, vision systems, robotics, and even medical imaging.

While electronics technology continues to improve in leaps and bounds, many of the breakthrough developments in sonar are not in hardware, but rather in software that provides new ways to mathematically model and analyze the signal data.

In 1997, Nelson and Tuovila, with the U.S. Navy, developed a fractal and nearest-neighbor clustering technique for identifying the clutter in sonar images—that is, a system to help identify and screen out unwanted or unimportant information. While this technology was developed to help with mine detection, it clearly applies to many other types of sonar images and further can be adapted to other imaging systems, including radar, X-ray, and MRI.

In the 2000s, the emphasis has been on sonar simulation and training systems, especially for use with submarines. Geographical mapping of ocean data, combined with sound profiles enable shore-based simulation of sonar systems and the development of new approaches. Advanced concepts and mathematical data manipulation techniques will probably result in future sonar systems far beyond anything we can currently imagine.

5. Descriptions and Functions

5.a. Sound Units and Characteristics

While an in-depth knowledge of the mathematics of sound is not necessary to appreciate the information in this chapter, it is helpful to know some of the basic units and terminology associated with the physics and representation of sound. Here are the most essential sound measurement concepts and units (see also the other chapters within the section on Acoustic Surveillance).

Loudness

Sound intensity or what we perceive as 'loudness' is described in *decibel* (dB) units. *Deci* refers to ten and *bel* is derived from Alexander Graham Bell. Our perception of changes in intensity as it becomes higher or lower is that the differences are not equal but progressively larger or smaller. The decibel scale is not a linear scale that increases in equal steps, but rather a *logarithmic* scale, in which changes are *exponential* as you move up or down the scale. Thus, 40 decibels is not 4 times 'louder' than 10 decibels, but rather 10^4 (40) times 'louder' and the magnitude of difference continues to increase as you go up the scale.

What we call 'loudness' corresponds roughly to the concept of *intensity,* which has been more objectively defined in math and physics for use in calculations.

As is discussed further in the sections on electromagnetic media, *wavelength* and *frequency* are mathematically related. In electromagnetics, the wavelength is equal to the speed of light divided by the frequency. Similarly, in acoustics, the *wavelength* is equal to the *speed of sound in the medium* divided by *frequency* or, expressed symbolically, $\lambda = c/f$.

It is usually easier to understand the concept of intensity with a few examples common to everyday life. Here are some examples of approximate sound amplitudes/intensities of various occurrences as measured in air in *decibel* units:

decibels	action or phenomenon
0	*human threshold - for reference*
25	soft whisper nearby
60	human conversation
80	public places, restaurants, subways
85	*level at which hearing protection should be worn*
100	carpentry shop, power tools; uncomfortable, cumulative damage
125	next to jackhammer; painful and harmful to human hearing
140	jet engine, arm's-length gunshot; damaging to human hearing
160	explosions; highly deleterious to human hearing
200	serious and permanent hearing damage

You cannot directly translate sound levels in air to sound levels in water in terms of decibels. Not only have sea creatures evolved different types of sound-sensing biology, but the physics of sound travel through water (speed, propagation, absorption, distance, etc.) is different from that through air (see Impedance immediately following). There is, however, a basic calculation that can provide a rough idea of correspondence, at least from the point of view of human perception. Large cargo ships generate about 190-200 dB underwater and whales typically communicate at about 160 to 175 dB. If you want to convert the sound levels in air to sound levels in water, add 62 dB. Thus, a very loud sound like a 140 dB jet engine in air might equate roughly to a 202 dB roar in the water.

Impedance

The concept of *impedance* is familiar to electricians and electronics engineers because they have to deal with the impedance levels of various materials in circuits and wiring distributions. Impedance is usually less familiar to lay readers who may not have encountered the concept and may not even be familiar with the term. Yet it helps to have a basic understanding of impedance because *decibel* measurements describing the 'loudness' or intensity of sound in one medium do not necessarily equate to the same levels in another medium. This is because of the *pressure* and *impedance* characteristics related to the medium.

To impede basically means to hinder. In electricity, impedance refers to a hindrance, an opposition, to current flowing through a circuit. In a sound-carrying medium, impedance refers to the *ratio of the pressure exerted* to the *volume displacement* within that medium. Water has a higher impedance level than air. What this means is that a 100 dB sound in air *isn't directly equivalent* to 100 dB in water. (To get a rough idea of the conversion, subtract 62 when converting air to water. Thus, 100 dB in air is about 48 dB in water and 180 dB in air is about 118 dB in water.)

To give some practical examples of sound travel in water that can be compared to sound travel in air as illustrated in the chart above, the intensities at which whales communicate underwater are typically from 175 to 190 dB and large ships (e.g., tankers) tend to generate louder sounds in the 190 to 200 dB intensity range. In fact, some scientists have suggested that whales may have been able to communicate for distances of more than 2,000 miles in the days before the shipping noise. Their communications range is now estimated to be a few hundred miles or less.

Speed and Velocity

The speed of sound depends on the characteristics of the medium through which it moves, since sound is essentially a disturbance in a medium—a series of pressure waves. A wavelength of sound of a particular frequency in water is approximately four times as long as a corresponding wavelength in air. The speed of sound in water is approximately 1500 m/s whereas the speed of sound in room temperature air is approximately 340 m/s. Stated as an example, a message shouted from a beach to a boat 3,000 meters away takes about eight seconds to arrive at the boat, whereas the same message shouted from below the surface of the water to a dolphin or submerged hydrophone 3,000 meters away takes about two seconds.

Many things affect the speed of sound. If the medium through which sound travels is reasonably homogenous, like air or water, it is not too difficult to determine or predict how fast it will travel. If there are other substances present, however, especially those that might change periodically (dust or water vapor in air, kelp or sediment in water), the speed will be affected in less predictable ways and there will be increased scattering and absorption.

Temperature also affects sound. In water, as the temperature increases, the speed of sound decreases. As the depth, and hence the pressure, increases, the speed of sound increases. Sometimes the two effects counteract one another, as in deep oceans where pressure increases and temperature decreases.

When we talk about the *velocity* of sound, we are describing it not only in terms of its speed but also in terms of its direction. It's a handy way to describe two concepts together in one word. The concept of velocity is commonly used in physics to describe sound travel.

Pressure

Sound pressure is expressed in terms of the force it exerts at one moment of time on a specified area—in most cases, the force it instantaneously exerts on one square meter. Sound pressure units are most often expressed as *micropascals* (μPa). One *pascal* (Pa) is the pressure of a force of one *Newton* exerted over one square meter.

Since this is just a basic introduction, it is recommended that you consult acoustics and physics texts listed in the Resources section for more advanced (and more precise) information. There is also a recommended site "An Introduction to Underwater Acoustics" that gives a good, short introduction at a level beyond the introduction presented here. newport.pmel.noaa.gov/whales/acoustics.html

5.b. Basic Sonar Systems

Most of the aspects of basic sonar systems have been described in *Section 2. Types and Variations* and *Section 3. Context*. This section provides just a little extra information on some example systems and more are described in *Section 6. Applications*.

Most marine sonar systems are hull-mounted, towed, or self-contained. Sometimes a variety of sensing systems are used together.

Hull-Mounted Systems

Hull-mounted sonar systems may be either active or passive and commonly have one or two transducers. The transducers are usually tuned to different frequencies, since range and noise-susceptibility vary with frequency. Hull-mounted transducers are usually not the highest in resolution but they can be used on a vessel that is charting an irregular course, whereas towed sonar works better if the course is even, slow, and straight. Hull-mounted systems are most often used for depth-sounding and navigation but may also be used for some detection purposes.

Towed-Array Sonar

Towed systems tend to be higher in resolution and more often used in covert or scientific imaging applications. They are better at resolving small targets and are less subject to noise emanating from the towing vessel than a hull-mounted system. However, they can provide erroneous signals on turns or irregular moves and may not be suitable when vessels are used in evasive maneuvers.

Bearing-ambiguity-resolving sonar (BARS) is a system manufactured by British Aerospace to improve torpedo detection. The towed-array system provides rapid single-sensor location of incoming threats independent of the ship's maneuver or convergence-delay requirements. In 1997, the U.S. Department of Defense selected a BARS project for the Foreign Comparative Testing (FCT) program.

Sonobuoys

Sonobuoys can be dropped off and picked up later, contacted through sonar or through a cable for communications purposes, or can be designed to relay with communications satellites. They can collect data over a period of time and relay the data at regular intervals, or store it until the buoy is retrieved. Sonobuoys are most often used for environmental research or tactical communications.

5.c. Remote-Controlled and Autonomous Vehicles

With improvements in fiber-optic communications and radio transmitters, there has been increasing interest in self-contained, remote-controlled, and autonomous vehicles for underwater applications. Since the vessels are often less expensive to build and operate than piloted vehicles, and since there is less danger of loss of life, they are suitable for some of the more dangerous search, rescue, and recovery missions.

Sonar-equipped *remotely operated vehicles* (ROVs) and *autonomous underwater vehicles* (AUVs) are used for oceanographic research, insurance assessments, wreckage detection and recovery, structural engineering surveys and inspections, underwater repairs, underwater tunneling and drilling activities, cable-laying, and mineral and gas exploration.

Remote-Controlled Vehicles

The U.S. Patent Office lists quite a number of designs for remotely-controlled vehicles, many of which are specifically designed for surveillance tasks. For example, Silverman, Simmons, and Croston have developed a surveillance vehicle resembling a small tank for navigating via radio control in hazardous environments. The top deck is designed so that it can include an articulated arm for taking surface samples. The remote station can monitor the vehicle through TV cameras, sound, and a variety of environmental sensors (U.S. Patent # 4709265, 1987).

A more recent system patented by Weinstein incorporates analog video cameras, microwave technology, and the TCP/IP networking protocol for transmitting data through an Ethernet transceiver. Combined with wireless video, it can be used to visually surveil a remote location and send the information through common computer standards for data transmission. This is a very generalized system that can apply to many different hardware configurations (U.S. Patent # 7,051,356, 2006). It is so generalized, in fact, that its patentability may be disputed.

These systems do not specifically use sonar as their sensing mechanism, but sonar sensing systems are often combined with other technologies in remote-controlled systems, or are developed for sound sensing based upon similar concepts as other systems.

Examples of remotely operated vehicles. Left: The Phoenix LW is a deep-diving inspection and light work ROV. It can be used in inspecting underwater oil and tunnel work and mine countermeasures applications. It is controlled through fiber-optic telemetry. Right: The Phantom HD2+2 is a long-umbilical, multiple-sonar ROV suitable for contraband and evidence recoveries, inspections, and tunnel penetrations. It includes built-in video, auto-depth and heading, and accommodation for adding more sensors. [News photos courtesy of Deep Ocean Engineering, www.deepocean.com/]

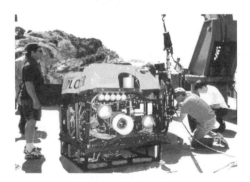 

Left: The HySub 25 built for Eastern Oceanics uses the ISE Advanced Telemetry Control System, a color camera, and 360-degree sector-scanning sonar. Right: The HySub 50 is a deep-seabed intervention system (DSIS) that has been supplied to the Canadian Department of Defence and was used at the Swissair 111 crash site. It features a cage, winch, and robotic manipulators, along with an ISE Advanced Telemetry Control System. [news photos courtesy of ISE Ltd., www.ise.bc.ca/]

Robots and Autonomous Vehicles

Robots are not really a distinct category from remotely operated and autonomous vehicles, but we tend to ascribe the term robots to mechanical/electronic devices that are smaller or friendlier or cuter, or more creature-like in one respect or another. The term *automaton* is also sometimes used. The more specific term *android* is reserved for robots that are distinctly human-like in their physiological and psychological makeups.

The Autonomous Lagrangian Circulation Explorer (ALACE) Float is an underwater robot developed by Davis and Webb that is neutrally buoyant and drifts for a month or so at depths up to about 680 meters, measuring the temperature of the marine environment. When its tour is complete, it rises to the surface and transmits its present location, and findings on temperature and salinity, to a satellite. It then resubmerges and continues gathering information in this cyclic pattern for up to five years.

Left: The ARCS project was initiated to serve a need for an under-ice vehicle for hydrographic surveys for the *Canadian Hydrographic Service*. In 1986, it underwent obstacle-avoidance sonar trials. It was subsequently acquired by the *Canadian Department of National Defence* and became the forerunner of the larger Theseus Autonomous Underwater Vehicle. The vehicle uses a Data Sonics Acoustic Telemetry link, EDO 3050 Doppler Sonar, and has a depth range of about 400 meters. Right: The Dorado was developed to meet the growing need for systems to deal with deadly advanced mines. It is used by the *Canadian Department of National Defence* and is being sold with forward-deployed sidescanning sonar by DCN International of France. The system can pull a sonar-equipped towfish at speeds up to 12 knots and depths up to 200 meters. Lower depths provide stability for sidescanning sonar and the data can be transmitted through a 2.4 GHz radio transmission. [News photos courtesy of ISE Ltd., www.ise.bc.ca/]

6. Applications

6.a. Depth Sounding

Acoustic devices are commonly used to measure the depth from a floating rig or marine vessel to a lake or ocean bottom. Depth sounders may also detect submerged objects that move between the sounder and the bottom. Since tides can dramatically depth, depending on the season or time of day, it is important vessel safety to monitor depth in order to avoid dangerous obstacles or the chance of 'bottoming out'. A depth sounder can also be used to scout fishing or diving locations.

Depth sounders are used for reconnaissance, hydrographic surveys, dredging operations, bridge building, and mineral prospecting. They are widely used by marine forces, the U.S. Army Corps of Engineers, the National Oceanic and Atmospheric Administration (NOAA), corporate marine industrial companies, and others.

Depth sounders may be analog, digital, or both. Some can switch between analog and digital displays and others use separate screens for the two displays. Some units can interface with printers and some have memory buffers for storing a sequence of information, or ports for uploading data to a computer. More sophisticated systems can use the data to create an image called a *bottom profile*.

When combined with computer processing and Global Positioning Systems (GPS) information, it is possible to take depth soundings from one location and compare them with another or to take soundings from one particular location and compare them with soundings taken at the same place at a different time. These could be overlaid on a computer screen or compared side-by-side.

Some of the variations on depth-sounding equipment that are of interest include

- Many new digital depth sounders can interface with existing analog depth sounders (or stand alone). These can further be connected to computer systems.

- A sounder may be equipped with an alarm to indicate that the depth has dropped below a certain predetermined level or user-selected level.

- Some depth sounders are self-contained, incorporating a transducer, others can be interfaced with the transducer as a separate unit, allowing a choice of transducers.

- Most depth sounders work on a single frequency or are switchable between two frequencies, but tunable sounders are also available. Dual-frequency sounders will sometimes display (or print) the results of the selected frequency as either a different shade or color.

- Some sounders must be used at slower speeds, others can be used at higher speeds.

- Depth sounders now come with a variety of ports for printers, computers, and video units, including VGA/SVGA, parallel, serial, keyboard, and floppy drive interfaces.

- More recent interferometry equipment is being used for some depth-measuring applications.

6.b. Commercial and Sport Fishing

Sonar can be used to locate fish and other marine prey. Phased-array sonar is used in commercial 'fish finders' as well as in modern nuclear submarines.

Identifying small objects on the move requires sophisticated technology. Many workplace computer technologies have arisen from the development of resource- and speed-intensive video games and, similarly, many sonar technologies have arisen from the development of commercial products like fish-finders. After all, fish are faster and smaller than submarines and just as anxious to avoid being caught. If you can resolve the technical problems of distinguishing small, moving fish in cluttered tidal regions, you can apply some of the knowledge to systems designed to locate larger, slower-moving underwater structures and vessels.

6.c. Marine Mapping

Terrain-mapping is used by marine geologists, biologists, and ecologists to study the characteristics and evolution of marine environments. It is also useful to prospectors seeking shipwrecks, minerals, and gas deposits. Military strategists map underwater routes, strategic locations, and good sound channels for sending sonar messages.

The *U.S. Geological Survey* (USGS) is one of the prominent departments engaged in surveying, mapping, and researching Earth's environment, including the marine environment. Many USGS surveys use digital sidescanning sonar, which collects data in swaths that are subsequently assembled into regional views. The images are further processed to correct for distortions and signal noise. When combined in sequence, the images can be viewed further as 'flyby' animations such as those seen in video games.

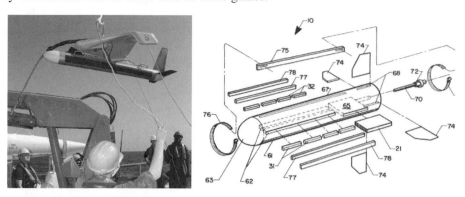

Sonar Towfish. A "towfish" is a roughly fish-shaped sensing system that is towed behind a marine vessel or beneath a low-flying helicopter. Left: A Klein 5000 sidescanning sonar used aboard a mine-countermeasures ship to survey marine terrain for man-made objects. The towfish has five beams on each side that emit high-frequency sound pulses. Right: This sidescanning towfish is designed with slots to hold the electronics and attachments, it is sealed with urethane and bonded with acoustic systems that couple with acoustic signals. The unit is designed to work with reduced power requirements. [U.S. Navy news photo by Lippmann, 2002, released; U.S. Patent # 5,142,503 by Wilcox & Wilcox, 1992.]

The USGS projects are funded with taxpayer dollars and an extensive library of images and MPEG animations is available to the public. The images can be downloaded through the Web or by *anonmous ftp*. Here is an example of a composite digital mosaic.

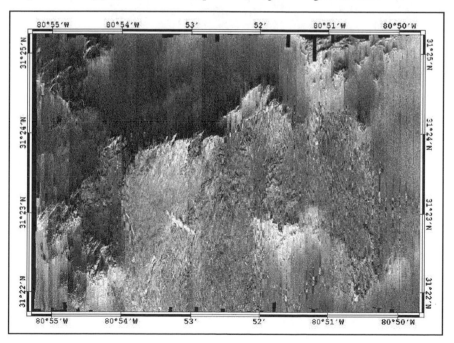

Sidescanning sonar grayscale image of *Gray's Reef National Marine Sanctuary*, Georgia, August 1994. A bathymetric map (map of depths) of this region is also available with 'false color' assignments to indicate depth. [U.S. Geological Survey, public domain.]

Sonar mosaic and bathymetric images can be superimposed over one another to provide an image that incorporates both textural and depth information.

An important function of marine mapping is to identify and monitor underwater hazards. When ships, submarines, and houseboats sink, they sometimes disappear into the silt and mud. Natural disasters or unusual tidal activity will sometimes uncover and move these objects to an area where they pose hazards to shipping traffic. The U.S. Coast Guard marks these obstructions with buoys and will sometimes use sidescanning sonar mapping (sometimes through contractors) to locate potentially dangerous hazards.

The U.S. Coast guard contracted Aquatic Systems Corporation (ASC) to conduct a sidescan survey with their survey vessel, which was placed in a U.S. Army Corps of Engineers barge. The Coast Guard cutter Osage then placed buoys on identified obstructions to alert shipping traffic to the potential hazards. This particular survey of the Ohio River, after a major flooding episode, took seven days.

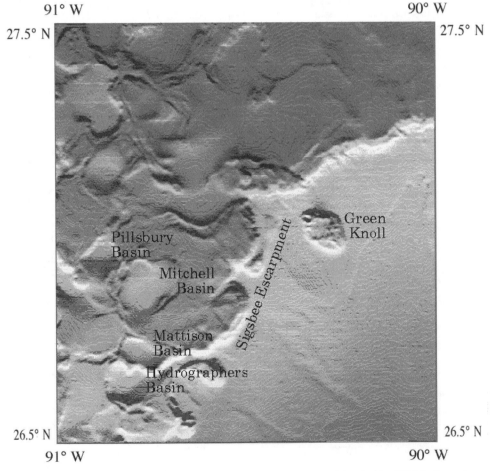

Sea-Beam Shaded-Relief Bathymetry. A shaded relief image combined with a NOS bathymetry false color image to provide a colored relief map (seen here in grayscale) that makes topological features clearer and easier to interpret. [U.S. Geological Survey, public domain.]

6.d. Active Submarine and Underwater Vessel Location

There are a number of marine vessel, coastal, and airborne surveillance systems especially designed for detecting submarines. The Canadian Forces Sea King is a ship-borne undersea surveillance and warfare (USW) helicopter that is carried aboard maritime frigates, replenishment ships, and destroyers. It is equipped to detect, locate, and optionally destroy hostile submarines. It is further equipped for surface surveillance, transport, and search and rescue missions. The Sikorsky Sea King uses forward-looking infrared radar and passive/active sonar.

Top Left: A U.S. Navy Seasprite helicopter is a ship-based helicopter with anti-submarine and anti-ship surveillance and targeting capabilities. Top Right: An SH-60 Blackhawk helicopter with a Helicopter Anti-Submarine Squadron hovering off the bow of the carrier USS Enterprise. Left: This Seahawk helicopter is from a Helicopter Anti-Submarine Squadron (HS-5) equipped with a dipping sonar that is submerged as the helicopter hovers only about 60 feet above the water. Right: Another Seahawk from the "Nightdippers" stopping for fuel on a range-reconnaissance mission. Seahawks are also used in search and rescue missions. [U.S. Navy 1996 news photos by Pat Cashing, Timothy Smith, Jim Vidrine, and Chris Vickers, released.]

Sometimes subs are sought because they are in distress. In 1963, the USS Thresher was lost with all hands, at a time when rescue vehicles couldn't reach subs that were very deep. Since that time, rescue vehicles have become more sophisticated and can locate and rescue crews that would have been unreachable in earlier times.

The U.S. Navy fleet of attack submarines is specifically designed to seek out and destroy enemy submarines and surface ships. The Navy emphasizes the role of technology in this mandate. "The concept of technical superiority over numerical superiority was and still is the driving force in American submarine development". The attack submarines are also used for collection missions in intelligence operations. In 1989, the Navy began constructing their Seawolf-class submarines. These vessels are designed to be "exceptionally quiet" with advanced sensing capabilities.

Other technologies, such as magnetic surveillance, in the form of magnetic anomaly detection (MAD), are sometimes used together with sonar.

The Deep-Submergence Rescue Vehicle (DSRV) helps rescue disabled or submerged submarines. It is designed to be easily transported and deployed by trucks, aircraft, or marine vessels, or by an appropriately fitted submarine. This 15-meter Lockheed Missiles and Space, Co. vessel can dive down to about 1500 meters, conduct a search via sonar, and attach itself to the sub's hatch in order to transfer up to 24 personnel to another location. It utilizes both search and navigation sonar systems. [U.S. Navy photo, released.]

6.e. Special Underwater Reconnaissance Missions

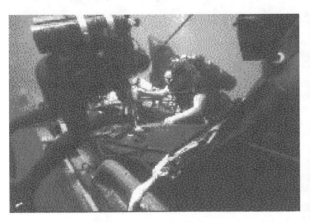 

Left: U.S. Navy SEALs undergo rigorous training in underwater and parachuting operations. The underwater operations include a variety of reconnaissance, demolition, and clandestine operations in support of U.S. military objectives. Right: Divers involved in the detection and salvage of wreckage often carry handheld sonar devices and video cameras. [U.S. Navy photos, released.]

The U.S. Navy Benjamin Franklin-class subs are equipped for special operations and support the work of the Navy SEAL (Sea, Air, Land) groups (which evolved in 1962 from Combat Swimmer Reconnaissance Units). SEALs perform a wide variety of special functions and clandestine operations, as well as some that are more routine, including minesweeping. In 1983, the Underwater Demolition Teams (UDTs) were redesignated as SEAL teams. Hydrographic reconnaissance and underwater demolition were added to various SEAL responsibilities.

6.f. Undersea Wreckage Location and Recovery

Sonar surveillance is an integral tool for locating underwater wreckages, including sunken ships, submarines, and downed aircraft. Divers have used sonar, acoustic listening devices, ultrasound, and remote-controlled vehicles to locate sunken vessels and voice and data recorders (black boxes) from a number of commercial airline disasters. Often sonar is used in conjunction with cameras and radar. Here are just a few examples in which sound technologies have been used with other surveillance technologies to locate and assess wreckages.

Submarine Recovery

The Confederate submarine H. L. Hunley is credited as the first submarine to sink a warship in battle. In the 1864 exchange, the Hunley itself sank. The Hunley has been evaluated in the murky waters in the past (e.g., 1996), but new technology made it possible to learn more about the submarine's structural status to aid in its recovery. Of special interest was whether any major corrosion or damage had occurred. A Coastal Inspection Inc. representative, Denis Donovan, worked with engineers to develop a custom ultrasonic transducer capable of measuring the thickness of civil war wrought iron to an accuracy of hundredths of an inch. This technology has applications in other aspects of surveillance as well.

Assessment and recovery of the Hunley also provide an example of how 'low tech' techniques are used. Molding putty was pressed against the hull in the murky waters to record impressions of surfaces, rivets, and seams. This molding technology can be applied above ground, as well, in crime scene assessment, where putty or plaster may be used to record surfaces, footprints, or objects of interest.

Using submarines for underwater surveillance was relatively new technology when the H. L. Hunley submarine (left) located and sank the USS Housatonic (right), an Osipe class war sloop, in 1864, submarine technology was so recent, in fact, that the sub was powered by eight people turning a hand-crank. The wreck of the Housatonic was dynamited at the turn of the 20th century to clear shipping lanes, but the Hunley was located by Clive Cussler with the National Underwater and Marine Agency (NUMA) in relatively good condition using surveillance devices that could probe the murky waters. [U.S. Navy historical archive photos of paintings by R. G. Skerrett.]

Aircraft and Flight Recorder Location and Recovery

Left: A U.S. Navy-contracted representative, Andy Sherrell, of Oceaneering Advanced Tech is shown with headphones interfaced to an underwater acoustic locator. Sherrell was listening for pings from the downed EgyptAir Flight 990 flight recorder (black box). Right: The U.S. Navy Deep Drone 7200 remotely operated vehicle was used in the search and recovery of the EgyptAir flight. [U.S. Navy 1999 photo by Isaac D. Merriman.]

A support ship from the U.S. Navy fleet, the USS Grapple (ARS 53), provides remotely operated vehicle (ROV) and diving assistance at crash sites, including Swissair Flight 111 and EgyptAir flight 990. The rescue and salvage vessel is equipped with a variety of surveillance devices, including the Mobile Underwater Debris Survey System (MUDSS), synthetic-aperture sonar, and the Laser Electro-Optics Identification System, with which it can image details of the ocean floor. It further transports a Deep Drone remotely operated vehicle (top right) and provides equipment and support for divers. FBI Agent Duback is shown tagging the recovered cockpit voice recorder from EgyptAir Flight 990 (bottom left) which was located by the Deep Drone ROV and then flown to Washington, D.C., for analysis. [U.S. DoD 1998 and 1999 news photos. USS Grapple by Todd P. Cichonowicz. Divers by Andy McKaskle. FBI-tagging photo by Isaac D. Merriman. ROV photo by Tina M. Ackerman.]

The SCORPIO (left) is the remotely operated marine surveillance vehicle (ROV) that successfully located the flight data recorders from downed Alaska flight 261 in February 2000. The pilots of the SCORPIO use video cameras and control panels to remotely monitor and steer the SCORPIO from on board the MV Kellie Chouest, a Submarine Support Ship. [U.S. DoD 2000 news photos by August C. Sigur, and Spike Call.]

TWA 800, which went down in the ocean on 17 July 1996, was located by naval divers in approximately 150 feet of water off the coast of Long Island, New York, with handheld sonar sets, a towed pinger locator, GPS systems, and sidescanning sonar. A remote-controlled MiniROV vehicle which can dive to a depth of about 1,000 feet was also used.

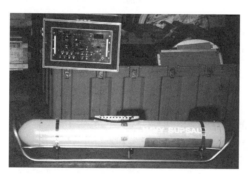

Two types of technology used to probe oceans are sensitive microphones and remote-controlled vehicles equipped with sonar. Left: a U.S. Navy 'pinger locator'—a high-sensitivity underwater microphone, called a hydrophone, that is towed in the water to pick up marine signals. Right: a sonar- and camera-equipped remote vehicle. Both were used to assist in the search and recovery of TWA flight 800, in 1996. [Photos: U.S. DoD 1996 news photos by the U.S. Navy.]

A U.S. three-person, torpedo-bombing Douglas TBD-1 Devastator aircraft that crashed into the ocean in about 800 feet of water in 1943 was recently located by a diver and subsequently claimed by a private collector/salvager. This salvage operation is of interest because the collector staked a claim based on a videotape of the wreck by the diver and because there is a dispute over ownership claims between the collector and the federal government, with the initial judgment in favor of the collector. Similar cases are likely to occur as surveillance technologies improve. Many historic wrecks have been abandoned because the technologies did not exist to find or recover them. Now that location systems and diving/recovery equipment have improved, private collectors are claiming ownership over the artifacts as 'abandoned' and the original owners are disputing the claims because they realize it is now possible to regain what they once thought was unrecoverable.

The National Underwater Marine Agency (NUMA) has used sonar to locate the wreck of the Carpathia, the ship that came to the rescue of the Titanic survivors in April 1912. The Carpathia was subsequently sunk by German U-boat torpedoes.

The 16th century wreck of the Mukran warship was located in the Baltic Sea in 1994 with a towed, color-imaging, sector-scanning sonar used in combination with radar buoy locators, GPS, and a video camera housed in a waterproof plastic ball that was towed at about two knots.

The USS Grapple (ARS 53) is a U.S. Navy rescue and salvage vessel that is dispatched to disaster sites in which planes, ships, or other wreckage are being sought underwater. It is equipped with synthetic-aperture radar (SAR) and a laser electro-optics identification system to create detailed images of the underwater terrain. It also has a number of diving systems and robotic vehicles to aid in search and recovery operations.

Historic Artifacts Recovery

Many smaller items lost from sailing vessels, but not directly associated with shipwrecks have been located and recovered and many more remain to be found. Thousands of items old and new have been thrown from boats deliberately, or accidentally during storms. Sometimes smugglers and poachers have unloaded contraband to avoid being caught, intending to come back later for the items (in shallower waters), but not always succeeding in relocating them. Many of these items are valued for their precious metals or gems or because they are historic collectors' items. Others have cultural and historic value for the stories they tell and the myths they confirm or deny. When items are lost in marine environments, sonar is one of the most important technologies used to locate them:

- Sidescanning sonar aided in the 1999 Underwater Atmospheric Systems, Inc. recovery of a ship's anchor identified by J. Delgado, Director of the Vancouver Maritime Museum, as being from the era of Captain Vancouver's explorations of the Pacific Northwest.

- Years of sweeping the bottom of Lake Michigan with sidescanning sonar enabled Harry Zych to discover the scattered debris of the Lady Elgin, sunk in 1860, with many historic artifacts of interest.

6.g. Scientific Research

Sonar has many applications in scientific research. The newer imaging systems can create realistic underwater terrain maps that resemble conventional grayscale photos. The stronger reflectors are actually imaged as dark areas on most systems, which give the appearance of a grayscale negative, but computerized systems will generally reverse the images so the 'highlights' are light and appear more natural to humans and thus are more easily interpreted as three-dimensional objects.

The Remotely Operated Platform for Ocean Science (ROPOS) is a fiber-optic-controlled remote-sensing robot for deep sea research. It is Canada's primary means of reaching deep ocean terrain to a depth of almost 5,000 meters. The vehicle includes illuminators, mechanical arms, low-light video cameras, color imaging sonar, and instrumentation to carry out a variety of scientific and commercial tasks. ROPOS was lost at sea in 1996 during a severe west coast typhoon that partially disabled the American research vessel from which it was being operated. Surveillance aircraft were used to locate the lost robot vessel, but by the time recovery craft arrived, it had disappeared.

Left: The submarine USS Pogy (SSN 647) is primarily a frontline warship, but a portion of the torpedo room was converted into a research laboratory. It is shown here surfacing through an arctic ice flow during a 45-day research mission to the North Pole in which data were collected on the chemical, biological, and physical properties of the Arctic Ocean. Geophysics, ice mechanics, and pollution were also studied. Right: A survey boat used for environmental missions by the Army Corps of Engineers. Sonar is used for a variety of surveys, including dredging and condition reports. [U.S. Navy 1996 photo by Steven H. Vanderwerff, released and U.S. Army Corps of Engineers 1995 photo, released.]

The Coast Guard Cutter Polar Sea (WAGB 11) has been providing support to scientific exploration in the polar regions since the 1970s. Its sister ship, the Polar Star (WAGB 12), has five laboratories and can accommodate up to 20 scientists for at-sea studies in fields such as geology, vulcanology, oceanography, and sea-ice physics. Here, Bill Chaney of Arctic Inc. is preparing to send a sonar measurement probe through a hole in the ice floe. [U.S. Coast Guard 1985 news photo by C. S. Powell.]

In 1998, scientists used sonar and computer programs to survey the bottom of Lake Tahoe, which is gradually filling in and becoming more cloudy, due to erosion and the proliferation of algae. The results showed the lake to be shallower than expected, with a flat, highly sedimented bottom. Sonar imaging provided a set of maps of the lake bottom topography that can help ecologists and planners budget and administrate a program to maintain the health of the lake.

6.h. Mine Detection

There are many types of mine-detection technologies because there are many different types of mines. Metal detectors are not suitable for locating plastic mines and acoustic-triggered mines have to be located in different ways from vibration-triggered mines. Marine mines have different characteristics and visibility from land mines. There are moored mines (that trigger on contact), rocket-propelled mines, and vertical-rising mines.

Sonar is one of the more important surveillance technologies for detecting underwater mines or navigating through them. Sidescanning and forward-looking sonar are two of the common types of sonar configurations used in mine operations. A safe route through a marine minefield is called a *Q-route*. The process of detecting and disabling mines as they are encountered en route is known as *in-stride breaching*.

Mines have been used by many nations over the years, not only for warfare but, in some cases, to aggressively assert territorial rights over precious coastline resources. Unrecovered mines from previous wars still lurk in the ground and in the water and both military and commercial vessels need to be aware of their presence and danger. Almost three-quarters of the damage to U.S. Navy ships in the late 1980s and early 1990s is reported to have been caused by mines [Kaminski, 1996]. With the breakup of the Soviet Union, a large quantity of sea mines, estimated at almost half a million, became potentially available for purchase by various other nations. It has been estimated that more ship casualties can be attributed to mine detonations than to missile, submarine, and air attacks since 1950 [Baus, 1996].

Responding to the challenge of marine mines, the U.S. armed forces have formed a number of centers and strategies over the years, including the Army's *Mine Warfare Center*, and the Navy's *Mine/Countermine Command Center.* Centers like these, along with Marine Corps systems, provide coordination and communication for land and underwater mine detection and clearing strategies.

- New detection technologies deployed from ships and submarines include unpiloted aerial vehicles like the Predator. The Predator can communicate with a host vessel many miles away through a super-high-frequency-gain flat-plate antenna mounted on the controlling vessel. A computer controls the direction of the antenna and a control console rebroadcasts the video through a joint deployable intelligence support system (JDISS) terminal inside the host.

- The Falcon (MHC 59) and the Cardinal (MHC 60) are examples of U.S. Navy Osprey-class coastal mine hunters. They are equipped with high-definition, variable-depth sonar systems and remotely operated submarines for neutralizing mines. The vessels were commissioned along with the Shrike, in 1997, to form a fleet of 12 Osprey-class ships. There have been two previous Navy minesweepers with the name Falcon, one which conducted rescue and salvage during World War II, in operation from 1918 to 1946, and the AMS 190, which operated from 1954 to 1976. Three previous Cardinal minesweepers have also patroled coastal waters.

- In 1945, a ship called Robin helped to clear the mine barrage in the North Sea. In 1996, a minesweeper of the same name was commissioned as the fourth of 12 Osprey-class ships authorized by Congress. These large mine-hunting ships are built entirely of fiberglass. They are used for reconnaissance, classification, and neutralization of moored and bottle mines in harbors and coastal waterways. Robin uses a high-definition, variable-depth sonar and a remotely piloted vehicle for reconnaissance activities.

- In 1998, the USS Raven (MHC 61) was added as a U.S. Naval Osprey-class minehunter. The vessel is designed to clear harbor, coastal, and ocean waters of mines using a high-resolution sonar system and a remotely piloted mine-neutralization vehicle.

- Minesweepers are typically equipped with hulls with very low magnetic and acoustic signatures, to reduce the chance of detection, and are constructed with fiberglass to absorb the shock of a mine explosion.

Marine minesweepers now typically use a combination of ships and remote-controlled or autonomous vessels to locate and disable mines. Recent improvements in high-resolution sonar systems have resulted in significantly improved signal and image quality, making it easier to identify and locate mines, which can be quite small and yet still deadly.

 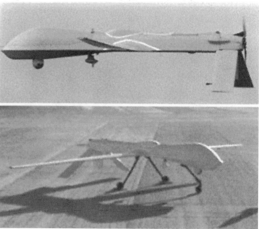

Sonar for locating or navigating through mines. Left: In a training exercise, U.S. Navy divers attach an inert satchel charge to a marine mine. Right: The Predator unpiloted drone can be launched from a submarine or ship to provide aerial reconnaissance of surface and underwater objects and hostile weapons-launching systems along coastlines. [U.S. DoD 1997 news photos by Andrew McKaskle, Jeffrey Viano, and Linda D. Kozaryn, released.]

The dolphin squad, described in more detail in Chapter 18, is also used to detect mines through natural sonar. Dolphins are highly intelligent and very social and are interested in interacting with people. One task they've learned is to attach neutralization devices to mines. Sea lions have good underwater senses as well, and have been trained to a limited extent, but they are a little less serious about their work than dolphins, and appear to be more interested in pursuing extracurricular activities.

6.i. Military Applications and Intelligence-Gathering

Information contributing to a body of intelligence is rarely gathered with just one device or class of technology. Sonar is only one category of device that is used to surveil bodies of water; most naval reconnaissance vessels use a combination of sonar, radar, and optical technologies (especially visual and infrared). Marine intelligence vessels are routinely equipped with multiple antennas that interface with a variety of display and control consoles.

There are a variety of types of seagoing intelligence-gathering vessels operated by nations such as Russia. Top Left: The Soviet Moma-class intelligence collector Seliger underway during a NATO exercise in 1986. Top Right: A Soviet Okean-class intelligence collection ship in a NATO exercise in the North Atlantic in 1986. Bottom: Balzam-class Russian intelligence collector, Belmorye (SSV-571) patrolling near Cuba and the U.S. east coast in 1993. [Photos: U.S. DoD news photos by Jeff Hilton (top) and uncredited photographer (bottom).]

Military Applications

Military forces make extensive use of acoustic sensing devices for oceanographic terrain mapping, navigational charting, submarine detection, search and rescue, targeting, tracking, and wreckage detection and recovery. To counter hostile threats over the decades, the U.S. Navy has been particularly active in developing and improving acoustic-sensing technologies. Examples of some of the more recent and relevant projects include

Advanced Deployable System - A short-term, undersea, rapidly deployable surveillance system for monitoring submarine and other underwater vessels in shallow littoral regions. It is designed to use a large-area field array of passive acoustic sensors interconnected and tethered through fiber-optic transmissions cables.

Distant Thunder - A project for the development of advanced signal-processing techniques for shallow water coastal environments that uses computer algorithms to process the target echo data received from low-frequency active (LFA) sources.

AUTEC range systems - The Undersea Test and Evaluation Center employs a variety of acoustic beacons and underwater acoustic telemetry to create tracking systems for use both above and beneath the surface as firing platforms and targets. These systems can further be adapted for two-way communications systems. The Underwater Digital Acoustic Telemetry (UDAT) system provides modem-based, bidirectional communications. Outgoing data from submarines are transmitted through a broadband, low-frequency transducer that doubles as an acoustic tracking device (a range pinger).

Autonomous and Remotely Operated Vehicles - The USNS Mohawk, for example, is equipped with a mini-rover MR-2, a remotely operated vehicle, a passive pinger-locator that is towed behind the vessel, and a shallow-water intermediate search system (SWIS) sidescanning sonar. The pinger generally sounds the first alert. Then the sidescanning sonar can comb the area more thoroughly, with the mini-rover following up with a camera and, finally, the divers enter the water with handheld sonar devices that can detect signals from reflective objects and pinger-locators.

Weapons Guidance Simulations - Since it is costly and sometimes dangerous to conduct live weapons training with new recruits, simulations are increasingly used for orientation and initial preparation. One such system at the Warfare Analysis Facility (WAF), uses a supercomputer to simulate the underwater acoustic environment and torpedo-target geometry. The software generates a representation of what the torpedo 'hears' through its travels. Acoustic input stimulates the torpedo's guidance system which, in turn, responds with appropriate steering commands to the computer system.

The Navy oceanographic program studies the influence of oceans on weapons and sensors in terms of both design and performance. The oceanographic ships aid in this research.

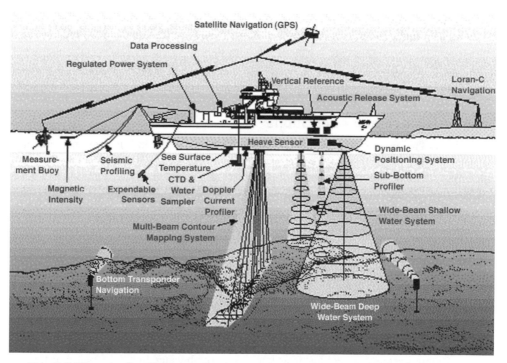

This U.S. Navy drawing illustrates the wide range of sensors that are used on the oceanographic survey fleet. Almost every type of data is collected, from acoustical information to magnetic intensity and seismic profiling. Both narrow- and wide-beam sonar are installed, along with measurement sonobuoys and satellite communications systems. [U.S. Navy news photo, released.]

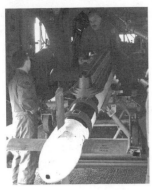 

Left: A high-resolution sidescanning sonar is loaded onto a helicopter by members of a Helicopter Mine Countermeasures Squadron. Right: A Knighthawk helicopter conducting a low-frequency sonar test and evaluation operation in the Caribbean Sea. [U.S. Navy news photos by Grisham, left, 2005, and uncredited photographer; released.]

7. Problems and Limitations

Since sound travels slowly through air (compared to electromagnetic energy), but quickly through water, its use in above-ground surveillance is somewhat limited.

The most common problems with acoustic technologies are 1) *attenuation*, 2) the gradual loss of signal, and 3) *interference*—noise from undesired sources or from the inherent characteristics of the medium. There are many other limitations associated more specifically with sonar, and not all are discussed here, but enough are included to illustrate the main technical difficulties associated with the technology.

7.a. Interference and Aberrations

Marine environments vary greatly in texture, pressure, heat, and the presence and composition of living organisms. Movement can cause waves, surges, eddies, and bubbles, all of which may attenuate or alter or create a *discontinuity* in a solar signal. Here are some common sources of problems:

- The rhythmic jouncing motion of a vessel with a mounted sonar or a towfish is known as *heave* and may create artifacts in the sonar image. Significant interference with the signal resulting from water that is rough at the surface can increase *clutter*.

- Unwanted aberrations, called *kiting,* can result from the side-by-side motion of a long cable or trailing towfish.

- Bubbles can persist for some time, particularly in rough water or in the wake of a large vessel, and can significantly diminish a sonar signal in a process called *cavitation* that adds *noise* to the signal.

- Cavitation can arise not just from bubbles in rough water, but also from acoustic interference when transducers are placed close together. With the increasing prevalence of sonar arrays with finer and finer elements, cavitation alone can be a serious problem. The U.S. Navy has been working on array-organization and the use of rubber separating materials to help minimize this problem.

When sonar transducers are fixed to a boat, it is generally advisable to place them in line with the azimuth of the vessel. The azimuth is the horizontal direction, expressed as the angular distance between the heading of the vessel and the direction of the transducer. This is recommended in vessels that do not permit beam-steering (the beam is the broadest point across the width of the boat).

Other types of interference, from signals bouncing off terrain or noise generated by other vessels, can also be a problem. For broadband sonar, systems of multiple receivers arranged in an array have been proposed for determining the strength and direction of the signals based on time-of-arrival at different receivers in the array, to process out the spurious information (e.g., Bourdelais interference-canceling, 1999).

7.b. Depth Sounding and Navigation

In many cases, an echo sounder must be set to point straight down, limiting the range of 'visibility.' In depth-sounding maneuvers where a vessel is moving back and forth along a region to map the terrain below, there can be aberrations in the data each time the vessel turns and there may be channels in the center (blank spots), and gaps in the track between successive tracks. Sidescanning sonar can overcome some of the limitations of channels and track gaps if the successive tracks (swaths) are overlapping. This takes longer, as some of the data will be redundant, but it also provides a more accurate reading of the terrain.

New interferometric technologies are overcoming some of the limitations of basic depth-sounders. These can survey a region up to eight times faster than a traditional depth-sounder, provide good three-dimensional data, and have a much wider angular coverage. They also require more careful installation and calibration, and a higher degree of computer processing.

Ambient noise problems are gradually being analyzed, predicted, and corrected-for with computer algorithms and digital signals processing, so there have been many improvements in the last half decade. Some inventors have even sought out ways to use the noise and clutter as reference points rather than screening them out, yielding some interesting results, particularly in littoral environments.

Ice poses special challenges for sonar, so polar ice prediction models (PIPS) have been developed to create energy disruption fields to indicate pressure ridge formations. Radar is sometimes used with sonar to provide more information about obstacles and their characteristics.

7.c. Ecosystem Disruption and Preservation

Undersea Life

It is important to keep in mind that the oceans are home to many sensitive forms of life. Over the last century, the noise from shipping alone has significantly reduced the effective communication distance of whales, and may damage the hearing/vibration-sensing organs of a variety of marine creatures. It may also disrupt reef organisms that are highly sensitive to change.

Sonar is an important communications and sensing capability of many marine mammals. Synthetically generated low-frequency active (LFA) sonar from marine vessels may significantly endanger the hearing and physical well-being of human divers and the hearing, communication, breeding, and social structures of marine mammals such as whales. LFA sonar involves sending high-decibel, low-frequency sound waves through bodies of water. It is one of the technologies used to detect submarines. Current LFA systems operate at about 180 dB with potential up to about 260 dB, with a field range of about 2 kilometers (with source level at the transmission point at about 250 dB).

The *U.S. Marine Mammal Commission* has released reports describing evidence of trauma and disruptions to marine life from LFA sonar testing and use and the *Natural Resources Defense Council* has issued statements regarding the potential dangers of LFA to marine life. www.nrdc.org/

Waste Monitoring

Another environmental aspect in which sonar is important is in the detection and monitoring of garbage and other wastes that are deposited by lake- and ocean-going vessels.

Sonar is used to help monitor industrial and radioactive wastes that have accidentally or intentionally been deposited on seabeds and lake bottoms. It is not uncommon for industrial wastes to be stored in drums and dropped overboard to rest on the bottom of the sea. On other occasions, tankers with drums of PCBs, industrial effluent, medical waste, and nuclear waste sometimes lose parts of their loads or sink, leaving toxic materials at the mercy of the corrosive seawater. Sonar allows the containers and wrecks to be located and monitored. It is extremely difficult to recover or neutralize these oceanic waste dumps, but at least sonar can provide information on when the materials might leak or become dangerous to nearby shipping and marine life.

Resource Exploration, Conservation, and Management

Sonar can be used to find new forms of life deep under the sea, to study thermal emissions, coral, and caves. It is further used to prospect for resources such as minerals, oils, and gas.

The monitoring of marine populations, particularly fish and endangered species, is carried out with help from sound-sensing systems. Conservation organizations can track the progress of protected and returning species and keep watch for violations and problems. Fisheries organizations use sonar to monitor and estimate fish migrations and numbers, and use the information to manage fisheries permits, quotas, and species restrictions to help ensure resources for the future.

7.d. Image Interpretation and Analysis

It is important to take into consideration the limitations of sound travel through water when interpreting sonar data and images from mapping operations. Differences in the properties of thermal or sediment layers in the water can distort an image. Shadows can be misleading if their origins and characteristics are not understood. For example, sidescanning sonars not only will commonly have a 'black spot' in the center of the channel between right and left scans, but the shadows on either side of the scan will fall away from one another. These 'shadows' are not the result of light from the sun or other illumination, but rather the result of lower reflectance or complete absence of reflectance off the sides of obstacles not facing the outgoing pulses of the scan.

A towed sonar system often has to be swept back and forth over an area. Care must be taken to interpret the information gathered during a turn, because misleading angles or anomalies in the signal caused by the turn might look like actual structures in a sonar image. It is best to sweep so that the turns are reimaged in another straight swath to provide good data, if possible.

7.e. Satellite Communications

Tactical satellite communications from submarines and other underwater vessels can be accomplished at the water's surface, using radio waves or, for covert operations, under the water's surface, using electrical signals sent to a sonobuoy through a tether. This is less visible and more flexible, but also has disadvantages. Typically the communications buoy must be connected to the underwater vessel, limiting the physical range of the host vessel and increasing the chance of detection when the buoy is communicating with a satellite or other receiver. The vessel is also more likely to be spotted if it is making noise near the surface or is visually spotted by patroling aircraft.

If the vessel is engaged in untethered communications using sound (similar to using a wireless modem, except with sound pulses), then there is the danger that the sound signal itself may be intercepted by passive listening devices. Other limitations with sound communications include the slower data rates and greater susceptibility to noise (see the Magnetic Surveillance chapter for some alternate solutions to submarine/satellite communications).

8. Restrictions and Regulations

Export Restrictions

Sound-sensing, more than ever, is dependent upon computer processing. This is especially true of acoustic guidance systems, sound communications, and high-resolution imaging systems. The export of software from the U.S. is restricted in many ways. Thus, source codes for target detection and analysis and other acoustic technologies may not be exportable. In other words, American vendors can't sell some of their sound-related technologies on the global market.

Underwater Diving and Salvage Operations

Acoustic surveillance technologies are frequently used to locate downed vessels and old sunken wrecks. There are a variety of local and federal laws and regulations concerning the preservation and salvage of sunken wrecks, especially those belonging to current commercial organizations and the Armed Forces. A sampling of the federal acts and regulations of relevance include

Archaeological and Historic Preservation Act of 1974 (16 U.S.C. 469)

Abandoned Shipwreck Act of 1987 (43 U.S.C. 2101)

Abandoned Shipwreck Act Guidelines (55 FR 50116)

Protection of Historic Properties (36 CFR 800)

Recovery of Scientific, Prehistoric, Historic, and Archaeological Data (36 CRF 66)

These laws have been debated in the courts quite vigorously since the advent of new technologies. Divers using new sonar devices can now find and recover artifacts and vessels that were apparently abandoned, while those who have abandoned them are suddenly claiming prior rights. It is best to look up some of the recent cases on wreck diving and ownership, because the laws in this area are by no means clear or easy to anticipate with certainty.

9. Implications of Use

Because sound ranging equipment is mainly used for underwater navigation, scientific research, and national defense, there has not been a lot of political conflict over its use, compared to telephone wiretapping, for example. It is considered a valuable tool, much like a flashlight or hammer, and is not perceived as an immediate threat to the security of daily life.

One area in which there is continuing controversy, however, is over the use of high-decibel sound as a potential underwater weapon or sensing device. Testing a high-decibel sound wave underwater is somewhat like testing high explosives above ground. The noise and shock-waves can be disturbing or very dangerous to nearby life, including humans. By the same reasoning, powerful soundwaves under the water have the potential to significantly disrupt aquatic flora and fauna, particularly the acutely sensitive hearing of aquatic mammals, including dolphins, whales, and sea lions. The blast of a bomb near a person's ear can cause irreparable damage. Clearly it is possible that the blast of a sound wave near a dolphin, whose hearing is far more sensitive and essential to its survival than a human's, may also create irreparable damage.

10. Resources

Inclusion of the following companies does not constitute nor imply endorsement of their products and services and, conversely, does not imply their endorsement of the contents of this text.

10.a. Organizations

Acoustical Society of America (ASA) - ASA is an international scientific society dedicated to research and dissemination of knowledge about the theory and practical applications of acoustics. ASA sponsors annual international meetings and annual Science Writers Awards for articles about acoustics. The site includes digitized whale calls and other underwater sounds. asa.aip.org/ www.acoustics.org/

Centre for the History of Defence Electronics - In addition to general information on defense, the organization holds colloquia on specific topics, including a Sonar History colloquium in 1996 (information or contacts from this may still be available). Bournemouth University, School of Conservation Sciences. old.britcoun.org/eis/profiles/bourneuni/bmthfacl.htm

Coastal Ocean Acoustic Center - Involved in the development of the Pt. Sur Ocean Acoustic Observatory for undersea research which came about when the Navy was downsizing and preparing to abandon much of the Sound Surveillance System that stretched 25 miles out to sea, it has now been converted to a research platform. In addition to Acoustic Thermometry studies, there is a hydrophone operating at a depth of more than a mile which can be used to study the communications and migration patterns of whales. Information is available through the Naval Postgraduate School.

Defence Evaluation and Research Agency (DERA) - A U.K. Ministry of Defence agency involved with non-nuclear research, technology, testing, and evaluation. DERA, with a staff of about 12,000, is one of Britain's largest research facilities, involved with research and testing of aviation technologies, electronics, command and information systems, sensors, weapons systems, and space technologies. DERA provides research data on airborne radar, target recognition techniques, and active/passive microwave ground radar systems. www.dra.hmg.gb/

The Federation of American Scientists - A privately funded, nonprofit research, analysis, and advocacy organization focusing on science, technology, and public policy, especially in matters of global security, weapons science and policies, and space policies. Sponsors include a large number of Nobel Prize Laureates. FAS evolved from the Federation of Atomic Scientists in 1945. The site contains numerous surveillance, guidance systems, U.S. Naval documentation, and other radar-related information. www.fas.org/index.html

Large Cavitation Channel - The world's largest underwater acoustic research and development facility being developed by the David Taylor Research Center in Tennessee. Information is available through the Oak Ridge National Laboratory. www.ic.ornl.gov/HTML/ic94156.html

Marine Corps Intelligence Association (MCIA) - MCIA supports U.S. Marines and their counterparts who participate in collecting, processing, and analyzing intelligence data and products. MCIA is based in Quantico, VA. and publishes INTSUM, a quarterly journal, for its members. mcia-inc.org/

Military Sealift Command (MSC) - One of three commands reporting to the U.S. Transportation Command, MSC is associated with national security and specifically involved in marine salvage, oceanographic surveys, and replenishment. Sonar is one of the important technologies used for accomplishing these goals and the MSC handles a number of submarine support ships through the Washington, D.C. Navy Yard. The Special Missions Program provides services for the U.S. military and federal government including underwater surveillance, flight data collection and tracking, acoustic research, submarine support, and oceanographic and hydrographic surveys. Multibeam sonar is one of the technologies used by MSC to chart the ocean floor and most of the world's coastlines. These data provide information that helps ensure safe marine surface navigation, supports submarine navigation and cables, and aids in weapons testing and the detection of hostile vessels. There is additional information and many photographs of the fleet on MSC's extensive Web site. www.msc.navy.mil/

National Oceanographic Atmospheric Administration (NOAA) - This important organization provides a great deal of practical and educational oceanographic information. NOAA serves as a sales agent for the National Imagery and Mapping Agency (NIMA) and provides nautical charts available to the public. NOAA also includes the National Marine Fisheries Service (NMFS). MapTech is a free online resource that houses the largest source of NOAA digital charts and USGS topographic maps in the world. www.noaa.gov/ www.maptech.com/

National Sonar Association (NSA) - A nonprofit organization founded by Frank Crawford in 1982 to bring together the various sonar and sound technicians who worked in a variety of fields, without regard to time in service or career status. Publishes The Ping Jockey. The organization sponsors reunions to keep alive the knowledge, experiences, and social contacts of the organization's members. www.sonarshack.org/

National Underwater and Marine Agency (NUMA) - A nonprofit, volunteer and membership-supported foundation, founded in 1979. NUMA investigates, reports, and preserves our marine heritage by locating sunken vessels using a variety of search and surveillance technologies. For example, NUMA used *scan sonar* to pinpoint the location of the sunken Carpathia, the ship that rescued the survivors of the sunken Titanic but which succumbed to German U-boat torpedoes in 1918. www.numa.net/

Naval Command, Control and Ocean Surveillance Center (NCCOSC) - Established in 1992, this center operates the Navy's research, development, testing, engineering, and fleet support for command, control, and communications systems and ocean surveillance. NCCOSC reports to the Space and Naval Warfare Systems Command (NAVSEA). www.nswc.navy.mil/

Naval Historical Center (NHC) - The NHC has an Underwater Archaeology Branch which uses a number of surveillance technologies to locate, assess, and document sunken vessels of archaeological interest. Projects have included the location and excavation of battleships and submarines going back to the Civil War.

Naval Intelligence Foundation (NIF) - A nonprofit organization founded in 1989 in response to a 1986 bequest given to the Navy and Marine Corps Intelligence Training Center (NMITC). NIF seeks to preserve, advocate, and support the culture and heritage of naval intelligence and to recognize and reward achievement on the part of active duty intelligence personnel. NIF is closely affiliated with NIP and has no membership per se (see next listing)

Naval Intelligence Professionals (NIP) - Founded in 1985, this nonprofit organization aims to further knowledge of the art of maritime intelligence and to further camaraderie among naval intelligence professionals. Members include retired and active duty service members and civilians who serve within the naval intelligence community. NIP publishes a quarterly journal and maintains an email discussion list. www.xmission.com/~nip/

Naval Research Laboratory (NRL) - The NRL conducts multidisciplinary programs of scientific research and advanced technological development of maritime applications and related technologies as well as Internet security projects. The NRL reports to the Chief of Naval Research (CNR). The Navy also has an Acoustic Research Detachment in Bayview, Idaho.

Naval Surface Warfare Center (NSWC) - Established in 1992, this center operates the Navy's research, development, testing, and fleet support for ship's hulls, mechanical, weapons, and electrical systems. NSWC reports to the Naval Sea Systems Command. www.nswc.navy.mil/

Naval Undersea Warfare Center (NUWC) - Established in 1992, this center engages in research, development, testing, and fleet support for submarines and other underwater vessels and weapons systems. NUWC reports to the Naval Sea Systems Command (NAVSEA). www.nswc.navy.mil/

Ocean Mammal Institute (OMI) - A public awareness, research, and educational organization for the study and preservation of marine mammals. The Web site includes an information page on low-frequency active (LFA) sonar and its possible effects on marine life. www.oceanmammalinst.com/

U.S. Army Corps of Engineers (USACE) - This corps is involved in research and development, navigation, flood control, environmental programs, and military construction. These tasks often involve work in marine environments involving the use of sonar. www.usace.army.mil/

U.S. Geological Survey (USGS) - The Coastal and Marine Geology Program department of the USGS conducts systematic geological mapping of coastal and marine environments and provides a seafloor mapping server online. Sidescan-sonar surveys are described and illustrated. Track maps can be downloaded in PostScript (vector) or GIF (raster) formats. kai.er.usgs.gov/

U.S. Naval Intelligence - Part of the Central Intelligence Agency, Naval Intelligence supports the Department of the Navy and the maritime intelligence requirements of various national agencies. It is located primarily in the National Maritime Intelligence Center in Maryland. www.odci.gov/ See also the Space and Naval Warfare Systems Center. www.nosc.mil

U.S. Naval Salvage and Diving - Uses Global Positioning System (GPS) and stationary and handheld sonar devices to located sunken marine vessels, aircraft, and debris.

U.S. Office of Naval Research (ONR) - The ONR coordinates, executes, and promotes the science and technology programs of the U.S. Navy and Marine Corps through educational institutions, government labs, and a variety of organizations. ONR carries out its mandates through the Naval Research Laboratory (NRL), the International Field Office, the Naval Science Assistance Program (NSAP), and the Naval Reserve Science & Technology Program. ONR reports to the Secretary of the Navy and is based in Arlington, VA. www.onr.navy.mil/

U.S. ONR Center for Autonomous Underwater Vehicle (AUV) Research - Founded in 1987, through an interest in using unpiloted underwater vehicles for covert mine countermeasures, the Center has since expanded to include other types of surveillance and commercial monitoring. These craft are especially useful in shallow waters where it's difficult or dangerous to operate other types of vessels. www.cs.nps.navy.mil/research/auv/

U.S. Space and Naval Warfare Systems Command (SPAWAR) - This department designs, acquires, and supports systems for the collection, coordination, analysis, and presentation of complex information to U.S. leaders. agena.spawar.navy.mil/

Woods Hole Oceanographic Institution - This prominent private, nonprofit research and education organization founded in 1930 includes an Ocean Acoustics Lab with information and publications on acoustic tomography, acoustic scattering, acoustic-mode coupling, and many other acoustic-related topics. www.oal.whoi.edu/

10.b. Print

Au, W.W.L., *The Sonar of Dolphins,* New York: Springer-Verlag, 1993, 277 pages. Comprehensively summarizes the research on the physiological, acoustical, mathematical, and engineering characteristics of the dolphin's sophisticated sonar system. It is currently superior to human-engineered systems in noisy environments and there is still much to learn from studying dolphin sonar and applying the information to the design of synthetic systems.

Bramley, J. Murton; Blondel, Phillippe, *Handbook of Seafloor Sonar Imagery,* John Wiley and Sons, 1996.

Burdic, William S., *Underwater Acoustic System Analysis,* Englewood Cliffs, N. J.: Prentice-Hall, Inc., 1991, 466 pages.

Bureau of Naval Personnel, *Introduction to Sonar,* NAVPERS 10130-B, Washington, D.C., 1968. (The Bureau also publishes manuals for Sonar Technicians.)

Coates, Rodney F., *Underwater Acoustic Systems,* New York: Halstead Press, 1989, 188 pages.

Cohen, Philip M., *Bathymetric Navigation and Charting,* United States Naval Institute, Anapolis, MD, 1970.

Cox, Albert W., *Sonar and Underwater Sound,* Lexington, Ma.: Lexington Books, 1974.

Curry, Frank, *War at Sea,* includes insights into the challenges of becoming a good submarine detector during World War II. Curry kept lengthy diaries of his experiences.

Edgerton, H. E., *Sonar Images,* Englewood Cliffs, N. J.: Prentice Hall, 1986.

Fish, J. P.; Carr, H. A., *Sound Underwater Images: A Guide to the Generation and Interpretation of Side-Scan Sonar Data, Orleans,* Ma.: Lower Cape Publishing, 1990.

Heppenheimer, T. A., *Anti-Submarine Warfare: The Threat, The Strategy, The Solution*, Arlington, Va.: Pasha Publications, Inc., 1989.

Hervey, John, *Submarines,* Brassey's, 1994, 289 pages. Illustrated description of state-of-the-art submarine sonar, electronic warfare, weapons systems, communications, and navigation.

Hunt, Frederick V., *Electroacoustics: The Analysis of Transduction and Its Historical Background,* originally published in 1954, reprinted by the Acoustical Society of America in 1981, 260 pages. This text describes the development of acoustics and transducer systems of the time and the early history, including some primary sources, of the development of sonar.

Koch, Winston E., *Radar, Sonar, and Holography - An Introduction,* New York: Academic Press, 1973.

Leary, William M., *Under Ice,* College Station, Tx.: Texas A&M University Press, 1999. The biography of Dr. Waldo K. Lyon who established the Arctic Submarine Laboratory in 1941.

Mazel, C., *Side Scan Sonar Record Interpretation,* Oxford: Academic & University Publishers Group, 1990.

National Defense Research Committee, *Principles and Applications of Underwater Sound,* Washington, D.C., 1976.

Sontag, Sherry; Drew, Christopher, *Blind Man's Bluff: the Untold Story of American Submarine Espionage,* New York: Harper Perennial Library, 1999, 514 pages. Describes the Cold War submarine 'chess game' between the superpowers, the U.S. and the Soviet Union.

Stefanick, Tom, *Strategic Antisubmarine Warfare and Naval Strategy,* Institute for Defense and Disarmament Studies, 1987.

Wilson, O.B., *Introduction to Theory and Design of Sonar Transducers,* Los Altos, CA: Peninsula Publishing, 1988.

Articles

Au, W.W.L., Echolocation in Dolphins with a Dolphin-Bat Comparison, *Bioacoustics*, V.8, pp. 137-162. Describes echolocation and reception systems in dolphins and bats, with comparisons and contrasts necessitated by the different characteristics of sound when traveling through water or air.

Au, W.W.L.; Martin, D.W., Insights into Dolphin Sonar Discrimination Capabilities from Broadband Sonar Discrimination Experiments with Human Subjects, *Journal Acoust. Society of America*, 1980, V.68 (4), pp. 1077-1084. Indications are that dolphin-like signals can be played back at a slower rate to humans who were then able to make target discriminations about as well as dolphins.

Au, W.W.L.; Snyder, K.J., Long-Range Target Detection in Open Waters by an Echolocating Atlantic Bottlenosed Dolphin, *Journal Acoust. Society of America.,* 1980, V.68 (4), pp. 1077-1084. Dolphins were able to detect small stainless-steel objects at 113 meters in open water.

Baus, Thomas C., Forward ... From the Sea: Intelligence Support to Naval Expeditionary Forces, Marine Corps Command official document written in fulfillment of a requirement for the Marine Corps Command and Staff College, 1996.

Brussieux, Marc; Martin-Lauzer, Francois-Regis, Harbor Surveillance and Mine Disposal Using ROVs, *Sea Technology*, April 1998. The authors discuss terrorist mine threats within harbors and populated areas and the use of surveillance on board a remotely operated vehicle with a sidescanning sonar and a remote camera, systems used by the French Navy.

Burroughs, Chris, Portable chemical sensor system promises new way of detecting underwater explosives, *Sandia Lab News,* October 1999, V.51(20). Describes how an underwater vehicle with sonar and chemical sensors can be used to detect underwater ordnances.

Cote, Jr., Owen R., Innovation in the Submarine Force: Ensuring Undersea Supremacy, *Undersea Warfare,* Autumn 1998, Issue 3.

Cutrona, L.J., Additional characteristics of synthetic-aperture sonar systems and a further comparison with non-synthetic-aperture systems, *Journal of the Acoust. Soc. of America,* May 1977, V.61 (5), pp. 1213-1217.

Cutrona, L.J., Comparison of sonar system performance achievable using synthetic-aperture techniques with the performance achievable by more conventional means, *Journal of the Acoust. Soc. of America,* 1975, V.58, No. 2, Aug., pp. 336-348.

Fricke, Robert J., Down to the Sea in Robots, *MIT Technology Review*, October 1994. The author describes a scientific ocean sonar mission for collecting data and how future, tiny, autonomous vehicles to be used will teach more about the ocean environment.

Gilbreath, G. A.; Everett, H.R., Path Planning and Collision Avoidance for an Indoor Security Robot, *SPIE Proceedings,* Mobile Robots III, Cambridge, Ma., Nov. 1988, pp. 19-27.

Gough, P. T., A Synthetic Aperture Sonar System Capable of Operating at High Speed and in Turbulent Media, *IEEE Journal of Oceanic Engineering,* April 1986, V.11(2), pp. 333-339.

Hackmann, Willem, Seek & Strike - Sonar, Anti-surmarine Warfare and the Royal Navy 1914-1954, *HMSO,* 1984, 487 pages. Illustrated history of sonar and anti-submarine warfare, with technical data.

Hobbs, C. H.; Blanton, D. B.; et al., A marine archaeological reconnaissance survey using side-scan sonar, *Journal of Coastal Research*, 1994, V.10(2), pp. 351-359.

IEEE Journal of Oceanic Engineering, V.17 (1), Jan. 1992. Special issue on synthetic-aperture sonar (SAS).

Kaminski, Paul D., Three Musts for Affordable Naval Mine Warfare, *DoD Defense Viewpoint*, V.11(69), 1996.

Kraeutner, P.H.; Bird, J.S., A PC-Based Coherent Sonar for Experimental Underwater Acoustics, *IEEE Trans. Instrum. and Measurement,* V.45(3), June 1996.

Mann, Robert, Field Calibration Procedures for Multibeam Sonar Systems, Surveying and Mapping Program technical report, c1997.

McHugh, R.; Shippey, G.A.; Paul, J.G., Digital Holographic Sonar Imaging, *Proceedings of the Institute of Acoustics*, V.13, Pt. 2, 1991, pp. 251-257.

Moore, P.W.B.; Bivens, L.W., The Bottlenose Dolphin: Nature's ATD in SWMCM Autonomous Sonar Platform Technology, *Proceedings of Autonomous Vehicles in Mine Countermeasures Symposium,* Naval Postgraduate School, Monterey, CA, April, 1995. A discussion of U.S. Navy marine mammal systems.

Parent, M. D.; O'Brien, T.F., Linear swept FM (Chirp) sonar seafloor imaging system, *Sea Technology*, pp. 49-55.

Perkins, David J., German Submarine Losses from All Causes during World War One, University of Kansas *World War One Naval Site,* September 1999.

Preston, J. M.; Gorton, P.A., Stability Criteria for platforms for multi-beam sidescan sonars, *Proceedings of the Institute of Acoustics,* V.15, Part 2, Bath University, April 1993, pp. 373-380.

Ramsdale, Dan J.; Stanic, Stephen J.; Kennedy, Edgar T.; Meredith, Roger W., High Frequency Environmental Acoustics Research for Mine Countermeasures, *Proceedings of the Symposium on Automomous Vehicles and Mine Countermeasures*, April 1995.

Schock, S. G.; LeBlanc, L. R., Chirp Sonar: A New Technology for Sub-Bottom Profiling, *Sea Technology*, pp. 35-39.

Smurlo, R. P.; Everett, H.R., Intelligent Sensor Fusion for a Mobile Security Robot, *Sensor,* June 1993, pp. 18-28.

U.S. Navy, Quicklook: Low Frequency Sound Scientific Research Program Phase III, discusses the use of low-frequency sound in a several-phase testing program that included underwater acoustics.

Weir, Gary E., Surviving the Peace: The Advent of American Naval Oceanography, 1914-1924, *Naval War College Review,* Autumn 1997. Describes how the naval forces set in motion in World War I and peacetime pressures to justify their continued existence led, in part, to the development of the oceanographic sciences. Includes footnotes and historical references and information on the origins of military sonar.

Journals

All Hands, monthly U.S. Navy publication from the Naval Media Center.

Applied Acoustics, monthly U.K. journal.

Journal of the Acoustical Society of America, monthly professional journal.

Naval War College Review, a collection of articles and essays about Naval history, politics, technology, and strategies. 1992 to present are archived online. www.nwc.navy.mil/press/Review/revind.htm

The Ping Jockey, the official publication of the National Sonar Association nonprofit organization.

Sea Technology, Compass Publications, Inc. A recognized authority for ocean design and engineering applications. www.sea-technology.com/

Undersea Warfare, the official professional magazine of the U.S. Submarine Force. The general public may subscribe through the Superintendent of Documents.

10.c. Conferences and Workshops

Many of these conferences are annual events that are held at approximately the same time each year, so even if the conference listings are outdated, they can still help you determine the frequency and sometimes the time of year of upcoming events. It is very common for international conferences to be held in a different city each year, so contact the organizers for current locations.

Many of these organizations describe the upcoming conferences on the Web and may also archive conference proceedings for purchase or free download.

The following conferences are organized according to the calendar month in which they are usually held.

International Conference on Electronics for Ocean Technology, an annual conference at least until 1987, with proceedings and articles by various contributors available for 1987 (5th conference).

Oceanology International, international hydrographic and marine survey equipment conference.

Connecting Technology: The Vision of the Future, jointly sponsored by the Department of the Navy Chief Information Office and the DON Information Technology Umbrella Program. A government-sponsored international conference for Navy IT and Marine Corps technology and policy.

Oceanic Imaging Conference, with topics including side-looking sonar, active synthetic-aperture sonar, and radar surface-scattering techniques.

Symposium On Autonomous Underwater Vehicle (AUV) Technology, Sponsored by the Oceanic Engineering Society, IEEE.

IEEE Symposium on Autonomous Underwater Vehicle Technology, held every two years in the 1990s—conference proceedings are available online.

International Autonomous Underwater Vehicle Competition, sponsored by the Office of Naval Research (ONR) and the Association for Unmanned Vehicles System International (AUVSI),. Competitors design and build completely autonomous underwater systems capable of traversing a body of water, navigating a series of gates, returning to a designated recovery zone, and determing the maximum depth of the recovery zone.

International Conference on Shallow-Water Acoustics, Beijing, China, April 1997 (Proceedings have been published). Funded by ONR.

European Conference on Underwater Acoustics, technical conference on acoustic science and the development, testing, and application of sonar sponsored by the European Commission and European Acoustics Association. The conference has been held every other year since 1992.

MCIA 2000 Convention and Reunion, sponsored by the Marine Corps Intelligence Association.

SPAWAR/NDIA. PowerPoint™, presentations from this conference are available to the public. agena.spawar.navy.mil/pubinfo.nsf/$defaultview/8A73AB731D4F714A88256818006506C4/$File/spawar_ndia_99_industry_days.html

10.d. Online Sites

alt.sci.physics.acoustics. A USENET newsgroup initiated by Angelo Campanella which is the principal newsgroup for acoustics discussions. Another group, comp.dsp is also a good resource for digital signal processing discussions.

American Pioneer Sonars. This informational/historical site is hosted by H. B. Pacific Pty Ltd. It focuses somewhat on fish-finding, but there are also many general-interest diagrams and informational/historical references covering different aspects of sonar. www.hbpacific.com.au/american.htm

American Underwater Search and Survey, Ltd. This is a commercial site with considerable educational content including many pictures and a history and background to sonar and sonar imaging. www.marine-group.com/

Basic Method of Conserving Underwater Archaeological Culture. An extensive description of the handling and preservation of a variety of artifacts and remains that are apt to be found in underwater environments. Of interest to divers, salvagers, treasure hunters, and those studying underwater archaeology. www.denix.osd.mil/denix/Public/ES-Programs/Conservation/Underwater/archaeology.html

Jerry Proc's ASDIC, Radar and IFF Systems Aboard HMCS Haida. This site is recommended. It provides a ten-part, well-illustrated historical and current highlights of sonar, radar, and minesweeping operations through the various wars and up to the present. It took the author a year to compile and present this material and it's worth reading. webhome.idirect.com/~jproc/sari/sarintro.html

Lowrance Sonar Tutorial. The Lowrance and LEI Extras, Inc. sites are commercial sites, mainly devoted to sport fishing, but this tutorial has some good diagrams and general information on sonar technologies (cone angles, frequencies, etc.), thermoclines, and some specific information on various transducers which are all worthwhile. www.lowrance.com/marine/tutorial/transducers.htm www.lei-extras.com/tips/sonartut/Default.htm

The Mukran Wreck: Preliminary Report. An illustrated report on the Mukran Wreck with a short description of the technologies used to locate the 16th century Nordic warship in the Baltic Sea. Springmann, Maik-Jens, April 1999. atle.abc.se/~m10354/mar/publ/mukran.htm

Reson Multibeam Sonar. Reson is a commercial supplier of echosounders and imaging sonars which has a good online information site that includes a glossary, FAQ, and numerous charts comparing relative characteristics, range, and depth information for a variety of sonar devices operating at different frequencies. www.reson.com

U.S. Navy Marine Mammal Program and Bibliography. The U.S. Navy has an illustrated description of the use of dolphins and sea lions as members of a 'fleet' for a variety of marine tasks and studies. www.spawar.navy.mil/nrad/technology/mammals/

There is also an extensive annotated bibliography with over 200 references of sonar-related scientific dolphin/porpoise/whale studies listed at this site. www.spawar.navy.mil/sti/publications/pubs/td/627/revd/ch1sound.html

U.S. Navy Submarine Centennial. The Navy is featuring a centennial celebration with a site devoted to submarine technology through the years, the history of the force, a chronology of important events and various campaigns. There is also a bibliography with further references. www.chinfo.navy.mil/navpalib/ships/submarines/centennial/subs.html

Welcome to Oregon Reference. This is one of the best personal sites the author has encountered, organized by Sam Churchill. It includes well-illustrated educational sections on the use of remotely operated vehicles, including underwater vehicles used to salvage plane and ship wrecks off the Oregon coast. Of particular interest is "Under the Oregon Coast: An Underseas Adventure". The site has many links to sea vessels, aircraft, relevant news clips (some with sound), charts, and ROV companies which use sonar in their vessels. www.teleport.com/~samc/ and www.teleport.com/~samc/seas/deep3.html

10.e. Media Resources

Commander Little Interview. Commander Charles Herbert Little CD RCN, a former head of Naval Intelligence is interviewed on his experiences in setting up a Canadian naval intelligence organization by John Frank in 1995. From the *Seasoned Sailors* video series by Policy Publishers Inc., Ottawa, 56 minutes.

Fast Attacks and Boomers: Submarines in the Cold War, is a Smithsonian National Museum of American History exhibit which began 12 April 2000. It was scheduled to coincide with the centennial anniversary of the U.S. Navy's Silent Service. The exhibit features recently declassified information and equipment from decommissioned U.S. nuclear-powered subs. Visitors can walk through the various submarine compartments, including the attack center, maneuvering rooms, and others. It will be open until April 2003. The U.S. Navy is also sponsoring smaller exhibits in other locations.

The Hunt for Red October, feature film starring Sean Connery and Michelle Pfeiffer, Director John McTiernan. It takes place in 1984 with a Soviet sub heading for the U.S., 1990, 134 minutes.

Frank Curry Diary. Leading Seaman Frank Curry RCNVR, A Canadian Naval ASDIC (sonar) operator was assigned to HMCS Kamsack, a new corvette. He kept a diary about his experiences in this vessel and later, on the HMCS Caraquet minesweeper. Video interview by Peter Ward in 1997. From the Seasoned Sailors video series by Policy Publishers Inc., Ottawa, 56 minutes.

Naval Undersea Museum. The museum includes naval history, undersea technology, and marine science in a new building with 20,000 square feet of exhibits including deep-submergence vessels, Confederate mines, and a simulated control room. It holds a large naval undersea history and science artifacts collection. Keyport, Washington.

Sea Hunt, television series from the late 1950s and 1960s starring Lloyd Bridges. There were 155 episodes, two of them in color. It depicts the adventures of a submarine and its divers. Episode 1127 is called the Sonar Story in which sonar is used to foil smugglers. In Episode 1132, a scientist is testing an underwater detection device.

Submarine Force Museum and USS Nautilus Historic Landmark. Opened in 1986, this official Navy submarine museum traces the history of the 'silent service' from the time of the Turtle submarine to modern submarines. Exhibits include the attack center of a WWII submarine and operating periscopes. Short films are also shown. The museum includes a research library. Groton, Connecticut.

The U.S. Naval Memorial Museum. Located in the Navy Yard in Washington, D.C., this museum includes the 1866 Intelligent Whale, one of the earliest naval submarines as well as ship models, working periscopes, uniforms, medals, art, and photographs.

11. Glossary

Titles, product names, organizations, and specific military designations are capitalized; common generic and colloquial terms and phrases are not.

ACS	Advanced-Concept Submarine. An exploration study by MIT graduate students to analyze and evaluate new technologies that can be used to design relatively low-cost state-of-the-art submarines. The ACS design includes a CAVES sonar system.
ALFS	Airborne Low-Frequency Dipping Sonar usually deployed from helicopters
ASW	Anti-Submarine Warfare
AUV	autonomous underwater vehicle
BARS	bearing-ambiguity-resolving sonar
batfish	a towed body containing instruments whose depth can be controlled
bathymetry	measurement of depths of water bodies such as lakes, oceans, etc.
bathysphere	a vehicle built for deep-sea exploration which is sturdy enough to withstand high pressure at great depths.
bioacoustics	the use of acoustic sciences and technologies to study biological organisms or systems (e.g., using fish-finders to study migration paths) and the study of acoustical systems in biological organisms (such as dolphin sonar, bird calls, grasshopper singing, etc.)
BSP	Barra Side Processor/Barrel Stave Projector
BT	bathymetry

CANTASS	Canadian Towed-Array Sonar System
CAVES	Conformal Acoustic Velocity Sonar
COMSEC	Communications Security
CTFM	Continuous Transmission Frequency Modulation. A type of system found in scanning devices such as forward-looking sonar.
DSV	deep submersible vehicle
EAD	expendible acoustic device. A sound-sensing device that may sink, decay, or scuttle itself when its useful life is finished or if there is a possibility of detection.
echo	the repetition of a sound resulting from reflection of the sound waves
FOTA	fiber-optic towed array
GLORIA	a wide-aperture, sidescanning, sonar-array, towfish imaging system that is being used in oceanographic research
IDR	initial detection range
LBVDS	lightweight broadband variable-depth sonar
LFA	low-frequency active
MAD	magnetic anomaly detection. A means of locating objects by their magnetic disturbance. While not a sonar method, it is one of the means for locating underwater vessels from under the water or from the air.
MCM	mine countermeasures
MIUW	Mobile In-shore Undersea Warfare program of the U.S. Navy which uses a variety of visual, thermal, radar, and sonar surveillance devices installed in vans to surveille shorelines to monitor and defend coastal regions.
NRL	Naval Research Laboratory
NSS	Naval Simulation System
ONR	Office of Naval Research
PSA	planar sonar array
QMIPS	a sidescanning sonar system
reverberation	an effect or result of an impact that resembles an echo
ROPOS	Remotely Operated Platform for Ocean Science. This is a robot system remotely controlled through a fiber optic cable built by International Submarine Engineers of Port Coquitlam.
ROV	remotely operated vehicle
SAD	sonar acoustic data
SAS	synthetic-aperture sonar
SIGINT	Signals Intelligence
SMB	submarine message buffer
SOFAR	sound fixing and ranging. A means of channeling sound through natural structures as those which exist in deep ocean regions.
SOSUS	Sound Surveillance System/Sound Surveillance Under Sea. A U.S.-deployed system of sound detection devices placed in strategic underwater locations throughout the world's oceans.
SURTAS	Surveillance Towed-Array Sensor/Sonar
SURTASS	Surveillance Towed-Array Sonar System. A low-frequency active sonar being tested and deployed by the U.S. Navy for international marine surveillance.
SVP	sound-velocity profile
SWARM '95	1995 Shallow Water Acoustic Random Medium experiment. An oceanographic-acoustic field study.
SWIS	Shallow-Water Intermediate Search, a type of sidescanning sonar
TACTAS	tactical towed-array sensor/sonar
TASS	Towed-Array Sonar System
TSP	tactical sonar performance
ULQ-13(V)	a countermeasures signal simulator system
VDS	variable-depth sonar

Electromagnetic Surveillance 5

Radio

1. Introduction

Radio is one of the most significant surveillance technologies ever developed. Radio waves are invisible to humans. They can pass through many objects, including biological systems, walls, windows, and fences. Radio waves can travel great distances—they can be sent around the globe by bouncing them off the ionosphere, or they can be projected far out into space. Radio waves operate over a range of frequencies and can be used at low or high intensities. Like other electromagnetic technologies, they do not require a medium through which to travel. Unlike sound waves, which require an elastic medium for transmission, radio waves can pass through a vacuum.

Radio waves comprise a broad segment of the electromagnetic spectrum, far broader than the visible spectrum. Because radio signals can be sent on many different frequencies at the same time without necessarily interfering with each other, radio waves can be used in many different types of applications, including television and radio broadcasting, phone and computer communications, and radar sensing and imaging.

Radio waves coexist at different wavelengths. The tuner in your AM or FM radio enables

A small helmet-mounted microphone enables Capt. Draper to maintain hands-free communication with an Apache attack helicopter near Tal Afar, Iraq. [U.S. Army news photo by Bailey, 2006, released.]

you to select the station of your choice from hundreds that are broadcasting at the same time. The same goes for broadcast television. The *tuner* is a radio-wave-receiving instrument inside the radio or TV that is designed to resonate and respond to the frequency selected with a pushbutton or dial.

Older AM radios have a tendency to "drift" and sometimes to overlap signals from frequencies that are close together, but most of these problems have been overcome with more recent technologies.

Radio waves are fast, flexible, invisible, and relatively easy to generate, but they are not unlimited in availability. If you have a low-power radio transmitter sending out a signal at a particular frequency and your next door neighbor has a high-power radio transmitter transmitting at the same frequency, the strong signal will clobber the weak signal just as surely as a semi-trailer will clobber a motorcycle if it tries to inhabit the same physical space at the same time. These constraints must be taken into consideration when designing and using radio systems and have necessitated a system of regulatory agencies to allocate *bandwidth* and licenses to various commercial, government, and amateur organizations.

Radio waves are relatively easy to intercept. A radio transmission typically radiates in all directions and the sender has no control over who tunes into the signal with a radio receiver tuned to the broadcast frequency. For this reason, signals may be encrypted in a variety of ways to increase privacy. Sending frequencies may be changed (either between transmissions or during transmissions), or the signals may be sent out at short random (or apparently random) intervals. These are common strategies for sending signals, and can safeguard a majority of communications, but they are not infallible. Every time a new way to hide the content or character of a transmission is invented, someone finds a way to 'crack' the code or strategy. Sometimes this is done for reasons of national defense or to enable private or institutional eavesdropping, and sometimes it is done for the thrill of solving a problem. There are many avid chess-players and puzzle-solvers who are also amateur or professional radio hobbyists who enjoy the challenge of figuring out encryption schemes.

This chapter focuses mainly on radio technology as it is used in broadcasting, amateur radio, and government communications. It puts a greater emphasis on radio receiving than it does on radio transmission, because this is the aspect that is important to surveillance, but a certain amount of attention to transmitting is necessary because the character of the transmission will determine the strategy that is needed to apprehend the signal. It also puts greater emphasis on the auditory aspects of radio technology, as visual technologies using radio waves (e.g., television transmissions) are covered more fully in Visual Surveillance.

Radar (radio direction/detection and ranging) is a specific aspect of radio technology that is so prevalent in surveillance that it warrants a separate chapter. It is recommended that you read the Radio Surveillance chapter first, if you are planning to study radar surveillance.

2. Kinds and Variations

Basic Terms and Concepts

Radio waves are a form of radiant energy in the electromagnetic spectrum, together with light, X-rays, and gamma-rays. Radio waves are longer than light waves and quite a bit longer than X-rays. We cannot hear, see, or smell them, so we must convert the energy into something we can sense in order for the information to be useful for most applications. Radio frequencies are often converted to audio frequencies within human hearing ranges and, in the case of television, into light frequencies that create images we can interpret through our visual senses. Sometimes we translate radio signals into beeps, buzzes, or alarm signals.

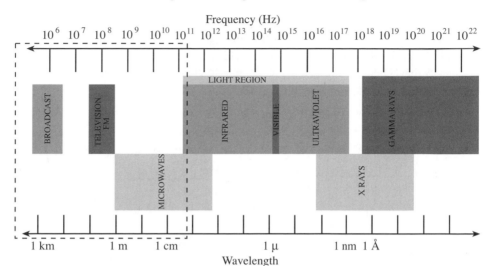

Radio waves occupy a large portion of the electromagnetic spectrum, compared to visible wavelengths. They range from microwaves at the longer-wave end of the infrared spectrum to extremely long wavelengths used by broadcast technologies. Because radio waves of the same frequency originating from different sources can interfere with one another, radio transmissions are carefully regulated and administered by the *Federal Communications Commission*. The FCC assigns licenses for broadcasting that stipulate the frequencies and power levels that may be used.

Radio technologies are highly varied, so it's difficult to organize them into just a few categories, but for the purposes of surveillance, radio technologies are commonly used for:

broadcasting (transmitting and/or receiving) - Radio waves are used to communicate instructions, locations, intentions, and activity reports. Broadcasting is done through broadcast stations, amateur systems, cell phones, remote controls (for components, appliances, and robots), and small-scale radio transmitters.

signaling - Radio waves are used to indicate that alarms, perimeter detectors, traps, or other objects have been activated, changed, moved, or deactivated in some way, allowing them to be used in many types of security systems, obstacle or hazard detection/avoidance systems, and wildlife management programs.

tracking - Radio waves can be used to indicate the position, movement, and speed of an object or living organism. Radio beacons have been attached to ships, cars, aircraft, spacecraft, people, animals, birds, bees, bicycles, kites, whales, and just about anything else that can carry a small electronic device and power cell.

A radio wave doesn't inherently carry communications information. In order to carry information, a radio wave, called a *carrier wave,* is *modulated* to add the informational content. Sometimes only part of the wave is sent and the rest (usually a symmetric portion) is mathematically reconstructed at the receiving end to save transmission resources.

The two prevalent schemes for modulating radio waves are *amplitude modulation* (AM) and *frequency modulation* (FM). Amplitude modulation is the more common and the more technically accessible means of modulating a radio signal. Amplitude modulation was discovered long before frequency-modulation. In fact, many engineers claimed that frequency-modulation was impossible. Well, it wasn't impossible, but it took a brilliant engineer a decade of devoted work to prove them wrong. Frequency modulation is now an essential technology used not only in FM radio broadcasting, but in many phones, tracking beacons, and other small-scale electronic devices. The chief advantage of FM over AM is the clarity of the signals. This comes at a price, however. FM signals are sent with *guard bands*, which are 'empty' frequencies on either side of the frequency that is being broadcast, to prevent interference. This uses up precious bandwidth, but is still preferred for applications where signal fidelity is important.

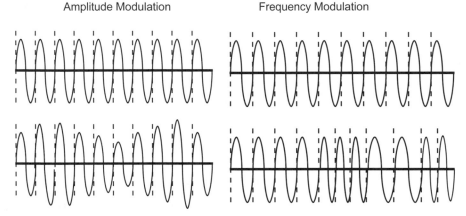

Radio waves don't inherently carry communications; it is necessary to add information to the waves as they are broadcast. The top signal is unmodulated. Amplitude modulation (bottom left) and frequency modulation (bottom right) are the two primary means by which 'carrier waves' are modulated to add information to the signal, usually in the form of speech, music, or data. [Classic Concepts ©1998 diagram, used with permission.]

3. Context

Radio wave energies are used everywhere, in broadcasting, consumer appliances, telephones, satellites, vehicles, alarm systems, electronic article surveillance systems, appliances, tracking tags, radio collars, leg collars, and wireless networks. Radio technologies are one of the most prevalent tools of our society. The capability to communicate without wires has many advantages.

Radio technologies that are used for surveillance are not tied to any particular industry. They are used just as readily in home and commercial applications as they are in professional or military applications.

For surveillance uses, radio technology has pros and cons. The most obvious advantage is the ability to send information invisibly without wires, over distances—sometimes considerable distances. The biggest disadvantage to radio waves is that they are indiscriminate. They don't care where they go or who intercepts them, making them vulnerable to detection and interception. While it's difficult to hide the presence of radio waves, it is becoming easier, with advanced hardware and software technologies, to hide the content of radio communications.

4. Origins and Evolution

By the late 1700s, emigration to the "New World" had greatly increased and people were searching for more efficient ways to communicate over long distances. All non-oral communi-

cations in those days were delivered by hand, animals, or birds (e.g., carrier pigeons). Carrier pigeon communications had some advantages over messages carried by people, but they also had some drawbacks. Pigeon-mail didn't work well over oceans and there was no guarantee the pigeon would reach its destination. There was also a limit to how much information a pigeon could carry. Inventors wanted to find ways to send longer messages over greater distances, preferably in two directions. Thus, telegraphs, telephones, and wireless communications evolved over the next two centuries.

4.a. Distance Communications Through Wires

The Telegraph

Invention depends upon dreaming new dreams. The dream of radio grew out of the eye-opening invention of the telegraph. The very fact that information could be transmitted almost instantaneously over great distances is probably the single most revolutionary discovery in the history of communications. The telegraph demonstrated that communications could be sent at astounding speeds over wires, so inventors thought "Well, how can it be done *without* wires?" Since the invention of the telegraph, most new forms of communications, including radio, have been evolutionary rather than revolutionary in nature.

France and Switzerland led the way in conceptualizing early telegraph systems, but it took about 80 years before practical electrical telegraphs, as we know them, were implemented in England and, soon after, in America.

The concept of electric telegraphy is recorded in a letter from 1753, but the initials "C.M." do not reveal the identity of the inventor. The first *frictional telegraph*, which used wires to electrically stimulate movement in pith balls at the receiving end, was introduced in Switzerland twenty years later, in 1774, by George Louis Lesage.

The first 'wireless' *optical telegraph*, using a system of towers and semaphores, was invented in the early 1790s by Claude Chappé (1763-1805), with assistance from his brothers. Stations were established across France but the system was somewhat cumbersome and impractical in bad weather.

Electromagnetic energy is the basis for the great majority of surveillance technologies and yet, surprisingly, electromagnetism was not demonstrated until 1819 by Hans Christian Ørsted (1777-1851). Ørsted was a Danish scientist and professor who discovered the *electromagnetic effect* while demonstrating how current could influence the behavior of a magnetic needle to a class of physics students in Kiel. His reports on the electromagnetic effect enlightened many scientists and established the foundation for electric telegraphs. Both André-Marie Ampère (1775-1836) and Johann Karl Friedrich Gauss (1777-1855) subsequently demonstrated electromagnetic telegraph systems, but they were not yet fully practical.

Europeans invented the first telegraph systems, but Americans were equally interested in developing the technology. In the 1830s, Joseph Henry (1797-1878) was experimenting with electromagnets in America and also actively assisting inventors on both continents. There was a lot of emigration in the early 1800s and the Atlantic was a formidable obstacle that separated families and colleagues—the desire to find a way to communicate across the ocean must have been very powerful. Finally, in England, in 1836, Charles Wheatstone (1802-1875) and William Fothergill Cooke succeeded in making the first practical working telegraph. The next year, in America, Samuel Finly Breese Morse (1791-1872) developed a working telegraph. Joseph Henry had encouraged and nurtured both these inventors, so it may not be coincidental that their discoveries occurred at approximately the same time.

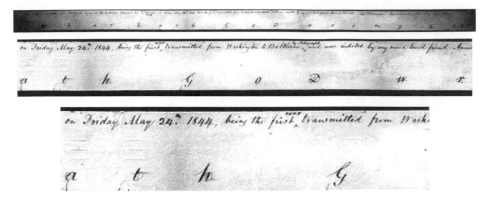

This is the historic paper tape from the first public demonstration of the Morse telegraph on the Baltimore line, on 24 May 1844, with the coded, and subsequently handwritten, message "What hath GoD wrought?" The message was sent by Morse from the U.S. Supreme Court to Vail in Baltimore and subsequently given by Mrs. George Inness to the Library of Congress. [Library of Congress Prints & Photographs Division, copyright expired by date.]

Telegraphic Surveillance

Like so many aspects of surveillance, we don't have records on how often early telegraph messages were intercepted. Social expectations and codes of conduct were different in those days. Telegraph operators had a professional responsibility not to repeat confidential communications and most of them respected this code of honor. Yet, even then, there were pieces of news that were whispered to family or friends or news reporters willing to pay for information on deaths, disasters, or business transactions. Since telegraph technology was relatively simple, surveillance didn't require highly technical skills. To eavesdrop, all you had to do was learn Morse code and tap into a wire. Even learning Morse code could be avoided if the eavesdropper had an informant working in the telegraph office.

Morse code wasn't invented to hide communications, however, but rather to enable communications on the simple on/off electrical/mechanical systems of the time. Samuel Morse had originally developed a somewhat cumbersome 'lookup book' for coding and decoding messages. This would have made telegraph communications harder to translate but also harder to intercept, as it would require possession of the relatively large lookup volume. However, history took a turn when Morse's assistant, Alfred Vail (1807-1859), came up with a simpler idea. The younger Vail was from an ambitious, mechanically apt manufacturing family. He was responsible for constructing and adjusting many of Morse's mechanical devices. While Morse developed the lookup code, Vail was altering the telegraph key from horizontal to vertical to allow easier finger motion and to permit spaces on the paper tape. This suggested an idea for a simple dot-dash code that would be easier to learn. A colleague who worked with Vail reports in Century Magazine that he and Vail visited typesetters to see which letters of the alphabet were more frequently used and that Vail used this information as the basis of the system that became Morse code. Since Vail had a contract with Morse to turn over his inventions in exchange for a percentage of commercial profits, his employer received the credit and the system became known as American Morse Code.

In terms of communication, the Vail approach greatly streamlined the sending of messages. From a surveillance point of view, it also made it easier to eavesdrop on telegraph signals anywhere along the line, since a lookup reference wasn't needed. People who were determined to intercept the communications of others could tie into a wire somewhere out of sight, or even out in the wilderness, without worrying about being detected. These were the first examples of *wiretapping* in the surveillance industry. If the eavesdroppers weren't mechanically inclined, their options were more limited but they could still try to interpret or acquire the paper tape on which the message was recorded in code, or strike a deal with a telegraph operator to reveal the contents. No matter how simple or sophisticated the technology, the human element then, as now, was usually the weakest point in the chain.

When *sounders* were developed, telegraph receivers emitted a *click* instead of creating a paper tape. While paper-tape telegraph receivers didn't disappear right away (actually, they evolved into teletype machines), many telegraph offices replaced tape with sounders.

 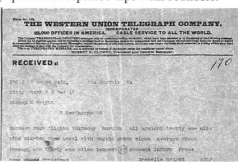

Telegraphy was a milestone technology—the first to enable fast distance communication. Left: A historic telegraph key. Right: The historic Western Union telegram from Orville and Wilbur Wright to their father, announcing the success of their first airplane flights at Kitty Hawk, in December 1903. [Classic Concepts ©1998, American Radio Museum collection, used with permission; Library of Congress Words and Deeds Collection, copyright expired by date.]

With sounders (audio telegraph systems), an eavesdropper who understood Morse code could listen to an incoming message from a nearby room or window, without having to physically hook into the electrical line. No doubt many communications were 'overheard' this way. The telegraph sounder also changed other aspects of telegraph surveillance. With the original paper tape machines, there were at least two records of a transaction, the original Morse code on the tape, and the operator's written transcript of the message. When sounders replaced paper tape receivers, someone had to be staffing the station at all times to receive an incoming message. In addition, there was no way to verify the accuracy of the message as transcribed by the telegraph operator. The operator could make mistakes or could deliberately alter the contents of the message. This was especially easy in large telegraph offices where many sounders were operating at the same time and the resulting noise would obscure the actions of any one person. Sounding systems were thus less accountable than paper tape systems, but also harder to intercept 'after the fact' by stealing the paper transcript of the message, since there was now only a translated message rather than a translated message and an original paper tape message.

Laying Down the Lines

Inventors were already looking for ways to send wireless telegraphs, but they hadn't yet discovered radio technology. Consequently, the early telegraph systems needed a physical medium through which to send electrical impulses and laying telegraph wires became an international passion. By 1851, France and England had been connected by an underwater cable and efforts were underway to connect England and Canada. Gutta-percha, a rubbery substance that could protect deep-sea cables, was a key element in establishing oceanic connections.

Cyrus W. Field is remembered for heading the project that resulted in the laying of the transatlantic telegraph cable, a project that had its disappointments and ultimate success. In fact, the two attempts to lay the cable in 1858 (and a previous attempt in 1857) were unsuccessful. It took experiences from the failed missions and the procurement of gutta-percha (an insulating, protective substance) to make the project viable and successful.

Left: This 1858 sheet music cover commemorates the laying of the transatlantic telegraph cable. At the top of the cover is the cable, showing the various layers over a copper core. In the center is a portrait of Cyrus Field, the coordinator (inadvertently reversed by the printer) and below are the ships used to lay the cable. Right: Cyrus W. Field was a young, wealthy New York merchant who came out of early retirement to tackle the project. It was an ambitious venture that ended up taking many years of effort on the part of the 'retired' Field and the others involved in the project. [Library of Congress Historical American Sheet Music Collection and America's First Look into the Camer Collection, copyrights expired by date.]

Transcontinental Cables

Overland connections were being established, as well. By October 1861, through the remarkable efforts of Hiram Sibley, Ezra Cornell, and a small crew, the first transcontinental telegraph spanned America from east to west. This feat was remarkable not only because it was accomplished in six months under budget, but because America was still a wilderness with many regions unsettled. The transcontinental telegraph was completed when buffalo still freely roamed the plains, only eight years after white pioneers settled in the Lynden area of the Pacific Northwest.

For businesses, and families separated by great distances, the telegraph was a wonderful invention. People in Europe and America no longer had to wait half a year for a message to travel by boat from one continent to the other and back. But it still required specialized knowledge to directly receive a message. You either had to learn Morse code or depend on the veracity of the message as interpreted by the operator.

Like a tree sending root tendrils into fertile ground, new inventions seem to always spark more inventions in ever-increasing numbers. Now that the viability of distance communications had been established, the public was eager for more scientific wonders. They wanted a way to speak directly to friends and business associates, without the impediment of Morse code and they didn't like those annoying (and unsightly) wires.

Service providers didn't like the wires either—they were expensive to install and maintain and buffalo were constantly knocking over the telegraph poles in their quest for better scratching posts. Inventors and entrepreneurs were eager to find solutions, since the commercial potential for wireless communications was huge.

By 1864, the telegraph had shown its value as a wartime communications tool. Most Union headquarters could communicate with the leaders, coordinating troop movements and receiving progress reports. However, the lines were vulnerable to wiretapping and false messages could be sent to mislead the unwary. Union forces apparently issued decoy messages that enabled General Sherman to draw Confederate troops out into the open on the pretense of sending a

Union troop to Savannah.

By 1866, Mahlon Loomis (1826-1886) had discovered that the Earth could act as a conductor, which meant a telegraph could potentially be devised with one wire rather than two. He had also demonstrated with kites that a signal could travel through the air without direct contact between kite strings (or wires). By the 1870s, several breakthroughs were on the horizon. Loomis received a patent for a wireless telegraphic system in 1872 and the telephone was poised to make its debut. The convergent development of wireless telegraphy and voice telephony gradually led to the birth of radio communications.

Setting up the first telegraph lines was a significant challenge. The landscape was rough and untamed, supplies were scarce, and everything had to be done by hand. Here, laborers perched atop telegraph poles like crows as the poles were being erected and the lines slung, in April 1864. Once the wires were in place, using them wasn't easy either. The typical 'telegraph station' was little more than a 25-square-foot covered wagon posted near a telegraph pole. [Library of Congress photo by Timothy O'Sullivan, copyright expired by date.]

Once their value had been demonstrated, the nation began laying communications lines anywhere they could travel. A transcontinental line was laid in 1861 and local lines were erected in the more distant communities, gradually being connected to the national infrastructure over the next several decades. The men shown here were installing the Juneau-Skagway telegraph cable. [Alaska State Library Winter and Pond Collection photo, copyright expired by date.]

Left: This giant wooden structure was the Butler's Signal Tower at Cobb's Hill, near New Market, Virginia in 1864. Right: A similarly constructed signal tower in Jacksonville, Florida, photographed at about the same time as the tower on the left. Towers are still used for radio communications. [Library of Congress Civil War Photographs, copyrights expired by date.]

Pioneer Radio Communication. Left: Irwin M. Ellestad at his wireless telegraph station in his home in Lanesboro in 1909. Right: Western Electric cable wound on huge reels. There was a tremendous expansion of transportation and communications technologies in the 1920s with a need for great quantities of cable for cable cars, bridges, electricity, and telegraph lines. The telephone didn't supersede the telegraph system, however, as they came to be used for different types of applications. [Minnesota Historical Society photos, 1909; Library of Congress gift from the State Historical Society of Colorado, ca. 1920; copyrights expired by date.]

The Telephone

The telephone was originally conceptualized as a 'harmonic telegraph' (a telegraph able to convey tones) by early inventors trying to find a way to send tones, music, or voice over telegraph installations.

The radio telephone is an important tool of modern life, but most telephone communications for the first century were sent over wires. The telephone was invented at almost the same time in both Europe and America, primarily by Johann Philip Reis (1834-1874), Elisha Gray (1835-1901), and Alexander Graham Bell (1847-1922). A little-credited Italian inventor in Cuba, Antonia Meucci, preceded Reis and Gray by a number of years with an acoustic membrane device but, unfortunately, his distant location put him at a disadvantage. When he traveled to America with information on his important invention, some of his papers were apparently stored at the Western Union offices. It has been said that both Bell and Gray had access to Meucci's papers, so perhaps it's not a coincidence that both Bell and Gray, both ambitious inventors, rushed to the patent office on the same day.

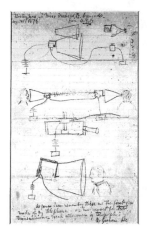 

Left: Bell's drawing shows a 'harmonic telegraph' that evolved into the Bell telephone. This drawing shows one person speaking into a cone and another listening at the other end. It is dated 21 March 1876. The date may have been estimated later as there is a subsequent note by Bell at the bottom of the page that states, "As far as I can remember these are the first drawings made of a telephone ..". Right: A model built to reproduce Bell's telephone. It was photographed around 1920. [Library of Congress Alexander Graham Bell Family Papers; Library of Congress Detroit Publishing Company Collection, copyrights expired by date.]

Compared to a telegraph key, a telephone was easy to use, so it's understandable that telephone technology caught on quickly with the public and spread rapidly. By 1878, Connecticut had a commercial telephone exchange based on human operators, batteries, and hand-placed fiber-wound cords for establishing electrical connections. The era of the 'Hello Girls' and a new class of telephone surveillants was born ... almost.

In fact, the first switchboard operators were predominantly men. Women were barred from most professional occupations at the time (including most clerical work) and even when they managed to find jobs, they were usually required to quit as soon as they were married. In those days, in many locations, there were town ordinances that a woman couldn't walk down the street unchaperoned by a man, so a single woman trying to seek work or travel to or from work was extraordinarily hampered by social restrictions. As it turned out, however, men didn't do well as switchboard operators. There surely were men who were good operators, but there are many reports that male operators were impatient, undisciplined, and rude, often cursing out the callers. This motivated business owners to try women instead. These female operators became known as *Voice with the Smile*, *Hello Girls*, and *Call Girls*, the latter taking on a different connotation over the years that may or may not have originated in the early days of telephone technology.

Anyone who has watched television shows from the forties and fifties has probably seen a clip of a telephone operator listening in on people's conversations. In some cases, the eavesdropping was known and expected, since operators at a central point in a marginal connection sometimes helped interpret the voices of the calling parties when they were hard to understand (crackling lines full of static and noise were common in early telephony). As the connections improved, however, the operator was supposed to turn a deaf ear during the conversation, listening only if he or she was summoned or needed to end the connection or etablish a new one. Nevertheless, anecdotes of phone operators eavesdropping are numerous.

Historic switchboards consisted of a series of jacks, and holes in which to place the jacks. When a call came in on a line associated with a jack, it was electrically connected to the line belonging to the callee by manually plugging the jack into the appropriate hole. A headset worn by the operator enabled the operator to monitor the call in order to ensure the connection and terminate it when appropriate. Later systems added lights to alert the operator of incoming calls and disconnections so that each call need not be monitored (which ensured better privacy). These switchboards eventually evolved into multiline or private branch exchange (PBX) telephone systems that were more difficult to tap than single-line phone systems. [Switchboard ca. 1909, public domain by date.]

Coherers

Wired telephone automation and technology improved and spread while inventors were seeking ways to devise practical wireless systems.

A number of individual discoveries in the late 1800s led to the development of simple but effective radio transmitters and receivers. One of the most important of these was the invention of the *coherer*.

David Hughes (1831-1900) emigrated from Britain to America where he became a schoolteacher and an inventor. He is responsible for a number of audio and telegraph devices that furthered the development of radio and telegraph technology. In 1877, Hughes invented a carbon microphone. By 1878, he had developed a coherer, a device in which particles clump together in response to an electrical discharge. He was also one of the first to devise a printing telegraph.

During this time, research into wireless systems of communication continued and the electrostatic wireless telephone was patented by Amos Dolbear, a professor and science writer, in 1886.

In 1890, a French inventor, Édouard Branly, developed the Branly detector, a radio-wave detector that incorporated a coherer. This detector was later adapted by Guglielmo Marconi. (Marconi was also known to have adapted the Castelli coherer for transatlantic experiments and to have designed some coherers of his own.) The coherer was an important component of the detector because it formed a sort of simple 'switch'. The particles in the tube could exhibit two *states*, loose and clumped. The electrical conductivity was lower or higher, depending upon the state.

 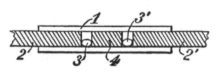

The Marconi coherer from the American Radio Museum collection (left) is constructed of a thin glass tube with an ivory base and two fiber-wound wires to make the electrical connection. On the right is a diagram of a Castelli coherer used by Guglielmo Marconi in transatlantic communications experiments. The conductor plugs on the ends (2,2') are separated from an iron plug in the middle (4) by two mercury pockets (3,3'). [Classic Concepts ©1997, used with permission. Diagram: Scientific American 4 Oct. 1902, copyright expired by date.]

In 1894, Oliver Joseph Lodge (1851-1940) did a number of experiments with coherers. His research contributed to knowledge on tuning in radio waves and transmitting over longer distances.

Automatic Telephone Switching and Marine Radio Systems

Technological surveillance as a field was about to become a powerful force in society. Now that radio technology was emerging and telephone systems were being improved, significant opportunities for monitoring newsworthy information or sensitive financial communications presented themselves. Some people recognized the potential for abuse and explored ways to reduce eavesdropping.

The invention of the *direct-dial* telephone system specifically came about because a businessman, Almon B. Strowger suspected his competitor of enlisting a phone operator (apparently the competitor's wife) to aid him in getting customers. Strowger was an American mortician with a talent for invention. When he suspected that his business was being undermined by a competitor working in collusion with the phone operator, he developed an automatic telephone switching system—a *step-by-step switch* that could connect a local call without assistance from a human operator. He patented the step-by-step system in 1889 followed by a *dial-switch* system in 1892.

At about this time, Romaine Callender, a Canadian music teacher and instrument maker, began patenting telephone systems based on a different model. By 1895 he had developed a different type of automatic switching system, aided by the expertise of the Lorimer brothers. Automatic switching systems began to remove the human element from local calls.

As land-based telephone switching systems evolved, so did marine radio communications. Even though the technology was very new, British warships were being equipped with radios, by 1899, that were developed by the young Italian inventor, Guglielmo Marconi (1874-1937). The systems were limited in range, less than 100 miles, but they were important tools of strategic cooperation. Only two years later, Marconi demonstrated that it was possible to send signals across the Atlantic Ocean from Canada to England, a distance of more than 2,000 miles.

In 1901, Strowger founded Automatic Electric and marketed his equipment to independent phone companies that were in competition with the Bell Telephone company. Surprisingly, automatic systems didn't become prevalent in small communities until almost 50 years later, and in the 1960s there were still manual switchboards in small businesses and small communities throughout North America and Europe.

In spite of their slow introduction, automatic telephone switching systems changed the ways in which eavesdroppers tapped into phone conversations and a variety of wiretapping systems and devices were developed (or commandeered from telephone technicians). Many early incidences of wiretapping were carried out with telephone lineworker phonesets that consisted basically of a phone handset with clips to hook into a phone line. While it was a rather crude form of surveillance, this device was widely used.

Assertion of Rights and Priorities

At the turn of the century, the U.S. government made a strong political move to acknowledge the importance of wired communications by granting broad rights of way for the installation of communications systems through public lands. In a U.S. Statute at Large published in 1901, known as the *Right of Way Act*, the Secretary of the Interior was authorized to "permit the use of rights of way through the public lands, forest and other reservations of the United States, and the Yosemite, Sequoia, and General Grant national parks, California" for electrical power, telephone, and telegraph communication, irrigation, and water supply.

4.b. Wireless Communications and the Birth of Radio

Crystal Detectors

The *crystal detector* was the next significant step in the development of radio communications. A crystal detector was an elegantly simple way of harnessing the natural resonating properties of a crystal and amplifying the vibrations so they could be heard by humans. Some types of crystals have natural rectifying properties. This means that they permit current to flow through them more readily in one direction than another. This is extremely useful in electronics, because it provides a certain amount of control over energy flow.

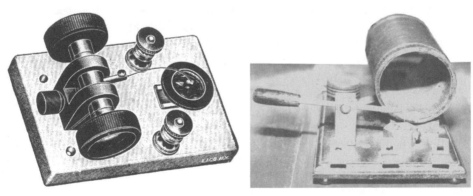

Left: A diagram of a crystal detector showing the round mounting base that holds a crystal, the catwhisker wire that makes physical contact with the crystal and transfers the vibrations to the main unit, and two leads for the headphones. Right: A crystal detector with a tuning coil, wound around a hollow cylinder, from the American Antique Radio Museum. [Diagram public domain by date; photo by Classic Concepts, ©1997, used with permission.]

Galena and carborundum were two substances that act as natural rectifiers. By mounting galena or carborundum crystals in a base so that they were constrained, but not too tightly (they needed to be able to vibrate), it was possible to detect their vibrations with a fine wire called a *catwhisker* and direct the vibrations to a pair of sensitive (high impedance) earphones or an amplifying component (amplifiers as we know them had not yet been invented). In other words, crystals could function as basic radio receiving units. The device didn't require any electricity and could be fairly easily transported. By carefully selecting the crystals, different radio broadcast frequencies could be tuned in.

In 1908, Greenleaf Whittier Pickard filed a patent for a crystal detector (they had been marketed as early as 1906) and in 1911 he patented a crystal detector with a catwhisker or *feeler*. Tuning coils, which consisted of specific lengths of wire wound carefully around hollow cores, were added to some crystal detectors. In time, crystal detectors superseded cohering detectors, though some hobbyists still enjoy collecting and using them.

Electron Tube Technology

Crystal detectors (sometimes retroactively called crystal radios) had some great advantages. They didn't require power and they had new and exciting capabilities. They were sufficiently useful that the U.S. Armed Forces ordered quantities of 'industrial strength' crystal detecting boxes for work in the field. Radio-operator training began to emerge. The detectors also had some significant disadvantages. The user had to select good crystals, keep them carefully oriented, adjust the contact tension of the catwhisker, and listen with earphones. The broadcast couldn't easily be heard by anyone else in the room. This was good for security, but not so good for conferencing or public information broadcasts.

Scientists sought diligently for ways to amplify the signal so that radio transmitters could send stronger signals and receivers could amplify the incoming signals. The solution came in two stages: the first was the Fleming valve and the second was the Audion triode.

John Ambrose Fleming (1849-1945) was an English engineer who had been investigating the *Edison effect*, discovered by Thomas Alva Edison (1847-1931) in 1883 (patented in 1884). Many significant discoveries are based on small observations. While working with light bulbs, Edison sealed a wire inside a glass bulb near a filament and noticed that a spark could jump the gap between the hot filament and the metal wire. Fleming had been trying to improve wireless receivers and also had been investigating the Edison effect in 1904. In the course of his research, he developed a two-electrode bulb that could be converted to direct current (DC) when attached to a radio-receiving system. The system didn't provide much improvement over previous methods but the concept opened up a new class of devices. Just as the crystal in a crystal detector acted as a rectifier, the elements in Fleming's *diode* also acted as a rectifier. The electron bulb just needed one more thing to make it a truly great invention. It wasn't Fleming, however, who made the breakthrough.

Lee de Forest (1873-1961) was an ambitious, prolific inventor. He was not quite as brilliant in physics as a few of his remarkable contemporaries and he was never blessed with good interpersonal skills, but de Forest was highly gifted and exceptionally tireless and persistent in his search for new discoveries and their practical applications.

De Forest acquired a Fleming valve and experimented with it, adding a third element, thus creating the first *triode*. This third element was the key to the future of electronics. In a Fleming valve, the electrons flowed at will from the cathode to the anode, but there was no way to harness their power. In the triode, the third element acted as a 'controlling grid', providing a way to influence the flow. This was a significant achievement, and de Forest recognized its commercial value, naming it the *Audion*. He patented his most important invention in 1907. He subsequently devised many more inventions and developed a number of practical applications based on the Audion, including wireless telegraph receivers.

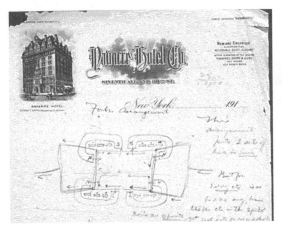

Left: A portion of Lee de Forest's notes on hotel stationery from around 1915. Right: A selection of electron tubes from the great variety that evolved from the original triode patented by de Forest in 1907. These examples are from the American Antique Radio Museum collection. [Library of Congress, copyright expired by date. Classic Concepts ©1997, used with permission.]

The age of electronics had begun. Electron tubes provided designers with a whole new way to build radio sets and thousands of other types of devices. The control and amplification of radio signals were greatly enhanced by this technology. The simple, tubeless crystal detectors were soon superseded by radios that included tubes. With a means to amplify radio wave signals, it was now possible to broadcast over longer distances to a bigger audience, ringing in the age of radio broadcasting and the age of radio surveillance.

4.c. Radio Broadcasting

Nathan Stubblefield broadcast his voice through the air without wires in a public demonstration in 1902, in Philadelphia. He had apparently first made it work in 1892, but did not obtain patents until 1908. Stubblefield's name has almost been forgotten, because he never managed to successfully market his invention [Kane, 1981].

Radio broadcasting essentially began in 1906 with the aircasts of inventor Reginald Aubrey Fessenden (1866-1932). From that time on, there were many amateurs and professionals who were excited about the technology and eager to try it out, but they discovered that airwaves had to be shared. Cooperation was essential in the orderly evolution of the field and not everyone was 'playing fair'. The government responded by passing an act of Congress to regulate amateur broadcasts in order to prevent interference with government stations and marine communications.

The sinking of the Titanic focused the nation's attention on the necessity of 24-hour monitoring of communications at sea. The Carpathia went to the aid of the stricken passengers, but the Carpathia was not the ship that was closest to the Titanic when it was sinking. The nearer ship had failed to respond to her distress calls because the wireless operator was off-duty and hence unaware of her plight.

As a result of the tragedy of the Titanic, by 1912, all ships were required to have wireless equipment, and more stringent regulations as to how and when the wireless systems were to be monitored were stipulated.

With the *International Radio Convention* and the *Radio Act of 1912*, the U.S. Department of Commerce was given regulatory authority by the U.S. Congress and unregulated use of the airwaves came to an end. It also marked a time when commercial, government, and amateur use began to become distinguished by wavelength regions in which they were permitted to operate. In terms of surveillance, this made designated 'government' frequencies of particular interest to foreign nations. It also made each group protective of its 'territory', a situation which still exists today.

The first radio stations were often improvised out of whatever materials and space were available but, gradually, over the next few years, facilities and their organization improved. Top Left: Minnesota's first broadcasting station improvised at the side of a road, in 1914, with James A. Coles (left) speaking into a microphone. Top Right: The WLAG broadcast station's transmitter room in the Oak Grove Hotel, Minneapolis in 1920. Bottom Left: The WCCO broadcast station in 1920. Bottom Middle: The WBAH broadcast station in Dayton, 1922. Bottom Right: The Dayton Company Broadcast room in 1923. [Minnesota Historical Society photos, copyrights expired by date.]

The American Civil War and World War I

One of the outcomes of the Civil War was the belated recognition of the rights of black Americans (in principle if not fully in practice). As a result, in 1887, African-American citizen Edward William Crosby, a writer and telegraph operator, was publicly acknowledged as the telegraph editor for the Buffalo Sunday Times.

Portable radios had some obvious advantages. In 1912, Ralph van Deman implemented the first U.S. Army mobile intercept van, called a Radio Tractor unit. In 1916, Brig. General John Pershing used a number of these radio systems to communicate among units crossing the Rio Grande, on the way to Columbus, New Mexico.

In 1917, General John Pershing used major newspapers to recruit unmarried female telephone operators for Signal Corps duties in France. They were required to be healthy, college-educated, and able to fluently speak French and English. Four hundred and fifty were selected for training and, in the spring of 1918, the first 33 operators were transported to France. Because they were near the front lines, the women were issued gas masks and steel helmets, and worked shifts that sometimes lasted as long as 48 hours.*

Top: Military telegraph operators worked under less than ideal circumstances. They are shown here stationed in tents at Bealeton (left) and in a Military Telegraph battery wagon in Petersburg (right) during the American Civil War. Bottom: The Military Telegraph Corps in fallen Richmond, Virginia in 1865. [Library of Congress Prints & Photographs Division, mid-1860s, copyrights expired by date.]

  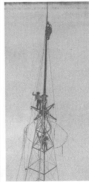

A wireless broadcasting tower being erected at Fort Brown, in September 1914. Middle and Right: The erected wireless tower with men on the mast. [Library of Congress South Texas Border Collection of the Robert Runyon Photograph Collection, copyrights expired by date.]

---

*Upon return to the United States, the female Signal Corp operators were not accorded veteran status or benefits, because Army Regulations were written in the male gender.

Commercial Broadcasting and Political Alliances

Commercial broadcasting began in Europe in 1913 and, by 1919, in America, Charles 'Doc' Herrold and Frank Conrad had provided broadcasts for local listeners out of primitive 'stations'. In 1920, KDKA became the first commercial station in America. The Radio Corporation of America (RCA) was also founded in 1920. David Sarnoff, who was long associated with RCA, was instrumental in forming the National Broadcasting Corporation (NBC), in 1926. In 1927, the Columbia Broadcasting System (CBS) was formed. Commercial radio had been born, and with it, competition for airwaves became big business.

Radio technologies, from the original telegraph that preceded radio, to modern wireless telephones, have played an increasing role in all government activities since then—from campaigning to running the Presidency.

One important milestone in government use of radio during this period of growth, was the broadcast of the Presidential election returns of 1920, but the full impact of radio on communications and the political system was not felt for another three or four years. The Coolidge papers illustrate how radio was further incorporated into political activities and how big business promoted the activities of selected politicians and vice versa. E. McDonald, President of the National Association of Broadcasters, wrote to President Calvin Coolidge, on 5 Nov. 1923, with a compelling argument that radio was a better way to reach voters than 'stumping' (foot-campaigning). An excerpt from the letter states:

> ... I can see where you, for instance, may relieve yourself of the most malignant of all the breaking down processes to which you will be subject, the speech-making tour. This is the worst of presidential tensions.... I need hardly mention how a speech-making tour preys upon one's health. Talks from train platforms under all manner of weather conditions, talks in congested auditoriums, countless interviews between times, meals under more or less strain, ... all are incidents of the speech-making tour and all have a telling physical effect, and are unnecessary.

> Science comes to the rescue and offers radio as a pleasant substitute for the old nerve-racking method. A speech over radio can be made with no more tax than the reading of a paper into the delicate microphone. You could calmly, unobserved if you wished, certainly undisturbed, deliver your talks under the ideal conditions of your study in the White House and be heard by every citizen of the United States who chose to give ear by means of a radio set...

Calvin Coolidge addresses a crowd through large radio-broadcast megaphones in July 1924 (left) and stands next to a radio-equipped automobile with his hand on the speaker, 14 Aug. 1924, during his campaign for the Presidency. Right: Coolidge posing in a group shot with Hoover and members of the Radiomen's Association in Oct. 1924. [Library of Congress National Photo Company Collection, copyrights expired by date.]

McDonald wasn't entirely correct about personal campaigning being unnecessary, but he was a catalyst in an important chain of events that resulted in far greater use of the airwaves by those seeking to communicate with the public. Coolidge used radio to campaign and to broadcast messages to the nation's radio-listening population during the mid- and late-1920s. It may also be true that covert political communications were carried out to a greater extent from this point on, given increasing familiarity with the technology in White House circles. The radio-listening audience greatly increased over this period, partly in response to the political broadcasts.

A 1925 report on radio broadcasting prepared by C. Coolidge Parlin, a manager with the Curtis Publishing Company, indicates revenue growth from about $2 million in 1920 to over $350 million in 1924. The number of radio-tube-based receiving sets increased over the same period from about 60,000 to almost 4 million with over 560 licensed stations in operation. Thus, there was now a compelling way to send the same information to everyone across the continent at the same time.

Capitalism abhors a vacuum and the tremendous interest in radio communications created a market for companies supplying radio services and equipment. In the 1930s, radio broadcasting was in the hands of big broadcast corporations and small amateur stations; consumers primarily purchased listening equipment (radio receivers) and most amateur radio transmitting systems were home-brewed from parts or kits. Regulation of the broadcast industry and the requirement to demonstrate competency further narrowed the field to the more technically inclined. For this reason, many of the radio products were sold in electronics stores, rather than department stores. Radio Shack got its start in 1934 and is still one of the better known sources of hobbyist supplies and consumer-level surveillance devices, including metal detectors, radio scanners, intercoms, video cameras, and phone recording devices.

By 1932, the New York Police Department was using mobile radio sets in patrol cars, receiving broadcasts from WPEG, the police department's radio station. More than 5,000 radio calls were made in the first six months of their operation. By 1937, the force was experimenting with two-way radios, but full implementation was delayed for a little over a decade by World War II.

By the late 1930s, television was being advertised to radio hobbyists (more about the history of television, a radio-wave technology, is covered in the Visual Surveillance chapter).

Regulation and Growth of the Radio Industry (Air Wars)

Airwaves are not unlimited and, in the name of competition, larger stations drowned out smaller ones in the early days of radio. Regulation of airwaves is handled by different agencies in different nations and a certain amount of international cooperation is essential because broadcast waves don't stop at international borders. In the United States, the jurisdiction for regulating the broadcast industry was passed from the U.S. Secretary of Commerce to the Federal Radio Commission (FRC) in 1927.

At first, mobile radio communications were strictly restricted to VHF broadcasting for aircraft operations only by the FRC, but the FRC eventually opened up "portable mobile" communications to broader use.

In 1933, the first radio 'scanners' in the form of *all-band receivers* were issued by the FRC to selected agencies to monitor radio communications. These were the first government-sanctioned general-purpose radio surveillance devices. Monitoring stations were responsible for monitoring communications for a couple of hours a day to check compliance with FRC broadcast regulations.

In 1934, the FRC responsibility was handed over to the Federal Communications Commission (FCC) and essentially remains there still. The FCC continued monitoring radio transmissions and acquired oscilloscopes help carry out their regulatory functions. Radio surveillance tools were becoming more specialized and more sophisticated and the U.S. Congress accorded the FCC broad-ranging powers to listen to a wide spectrum of broadcasts. By this time there were hundreds of broadcast stations and about 60,000 amateur radio operators around the world, 75% of whom were in the United States and Canada.

Radio Weathercasting and Search and Rescue Surveillance

A substantial segment of radio surveillance is dedicated to monitoring the airwaves for emergency calls, often resulting in the radio and television rebroadcasting of severe weather warnings or the coordination of search and rescue operations. The Amateur Relay Radio League (ARRL) Emergency Corps was established in the mid-1930s to provide emergency communications, as a result of storms, floods and, more specifically, the loss of a plane in the Adirondacks, in December 1934.

By 1938, FCC regulations for "portable mobile" were expanded to permit maritime mobile communications (with frequency and location restrictions).

The Evolution of Radio Devices

Alternating current (AC) systems gradually began to replace direct current (DC) battery-operated systems in the early 1930s, a trend that continued until the early 1960s, when portable radios (and hybrids) became popular again. The use of alternating current permitted larger, more complex radio systems to be manufactured.

The development of electron tubes was a significant part of radio history. These components continued to evolve until, in the early 1930s, higher-gain screen-grid tubes were beginning to replace the early triode tubes for radio applications.

By the late 1930s, regenerative receivers, which had been a mainstay, were being replaced by superheterodyne receivers.

The Doppler effect has been mentioned in a number of chapters in this text. Briefly, it refers to the changes in frequency (and thus pitch) that occur when an object generating a pressure wave is moving relative to the receiver. This important concept was being investigated for its potential use in radio systems. By the 1940s, Doppler concepts were used to create radio direction-finding systems and, in the late 1940s, Servo patented a quasi-Doppler direction-finding system. Doppler concepts have also been applied to sonar and synthetic-aperture radar (SAR).

Up to this point, radio broadcasts used a method of modulating carrier waves called *amplitude modulation* (AM). *Frequency modulation* had been discussed in scientific circles, but the general consensus was that it was impossible. Almost everyone held this point of view except a gifted and persistent inventor named Edwin Howard Armstrong (1890-1954). In the early 1940s, after a decade of dedicated effort, Armstrong proved that FM broadcasting was not only possible, but had some important advantages over AM broadcasting.

Radio Communications in Wartime (Air Wars II - The Foreign Menace)

On 24 October 1940, J. Edgar Hoover described wartime espionage operations in a memo to President Truman. Hoover's memo included information on infiltration of a German Intelligence Service radio station:

> The Federal Bureau of Investigation has been operating for a period of many months on the eastern seaboard a shortwave radio station which is utilized by the German Intelligence Service for transmission of reports of German Agents in the United States to Germany. The Directors of the German Secret Service in Germany also communicate

with this station furnishing instructions and requests for information to the operators of this station for transmittal to German Agents in the United States. Needless to say, no one knows that this German communication system is actually controlled and operated in the United States by Special Agents of the Federal Bureau of Investigation, who are considered both by German Intelligence Services in Germany and in the United States to be actual members of the German espionage ring. Through this station the Federal Bureau of Investigation has been able to develop voluminous information concerning the identity of German agents in the United States, their movements, interests and program...

.... Arrest is considered inadvisable except in extraordinary cases because counter espionage methods of observation and surveillance result in a constantly growing reservoir of information concerning not only known but also new agents of these governments...

[John Edgar Hoover, FBI, *Memo to Major General Edwin M. Watson,* 24 October 1940, declassified 3 April 1975.]

Not long after receiving this memo, a Presidential Directive established the Foreign Broadcast Monitoring Service within the FCC, in February 1941.

In December 1941, as World War II security measure, the U.S. government virtually shut down amateur radio.

Radio Security Innovations

During the World War II, many American celebrities offered their services as volunteers and entertainers to support the efforts of the U.S. Government and American service members. Hedy Lamarr (Hedwig Eva Maria Kiesler 1913-2000) was best known as a movie star, but she had a keen analytical intelligence, as well. She worked out a system of radio security called *frequency hopping*—a *spread-spectrum* technique in which a radio transmission is 'hopped' through a series of frequencies in order to escape detection and, in some cases, to create a clearer signal by using good channels. Lamarr developed frequency hopping in her search for a better system for guiding torpedoes—one that was less likely to be jammed by opposing forces. Antheil's contribution was to suggest a means to synchronize the frequencies at either end with coded paper tape (similar to a player piano roll), but it was not an easy method to implement at the time and didn't become practical until later, when it could be handled automatically by electronics. Lamarr and Antheil received a patent for their important innovation on 11 August 1942.

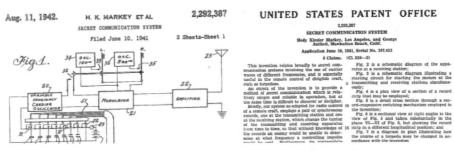

Frequency Hopping for Radio Control of Remote Craft. Hedy Kiesler Markey (Hedy Lamarr) developed a secured guidance system for remote-controlled devices based on the concept of frequency hopping and George Antheil suggested a means of coding the hops with perforations on paper rolls. Thus, an enemy would be unable to anticipate the pattern of radio frequency changes without the coded pattern. [U.S. Patent #2,292,387, August, 1942.]

Frequency hopping became an essential aspect of military communications, and is now widely used for wireless communications. Since 1972, more than four hundred U.S. patents specifically dealing with frequency hopping have been awarded. Unfortunately, by the time the world caught up with Lamarr's idea, her patent had expired and she never received any monetary benefit. She didn't receive credit either, until decades later, when women began to be acknowledged for their scientific contributions.

Meanwhile, surveillance of foreign communications continued to be a high national priority. In July 1942, the *Foreign Broadcast Intelligence Service* was established to succeed the *Foreign Broadcast Monitoring Service* by an order of the *Federal Communications Bureau*. Its responsibility was to record, translate, and analyze foreign broadcast programs.

Military Use of Radio. Radio Operator Robbins at his radio post in a hut in New Guinea. [U.S. Army May 1943 Signal Corps Historical Archive photo by Harold Newman, released.]

In December 1945, the FCC *Foreign Broadcast Intelligence Service* was transferred to the *Military Intelligence Division, War Department*, by an order of the Secretary of War and then, in August 1946 to the Central Intelligence Group (CIG), *National Intelligence Authority*. Following the war, it was succeeded by the *Foreign Broadcast Information Service*.

Wireless communications over distance improved but, even with improving technologies, it was found that radio waves still required good lines of sight, relay stations, or good reflective surfaces from which to bounce signals. In armed conflicts, these aids to radio communications were often absent. In the Korean War, in the early 1950s, for example, it was discovered that rugged terrain significantly hindered radio signals. At the time, handheld portable FM radios (AN/PRC-*x*), with broadcast ranges varying from 1 to 8 miles (less in rough terrain) were favored for communications between service members. Since satellite relay stations were not yet available, ground-based radio relay stations had to be improvised and, to escape detection, often had to be moved, under less than ideal circumstances.

Innovations in Radio Communications

In 1946, Signal Corps technicians in New Jersey successfully managed to bounce radio signals off the Moon. Thus, it was demonstrated that radio waves could be used to communicate through space, a concept that would later be applied to communications satellites.

In 1947, scientists at Bell Laboratories developed a significant new electronic component—the transistor. Transistor technology would eventually supersede large, fragile, power-consuming vacuum tubes for most applications.

Demand for radio waves began to exceed the supply by the late 1940s, so the *Federal Communications Commission* (FCC) imposed a temporary freeze on applications for TV broadcasting

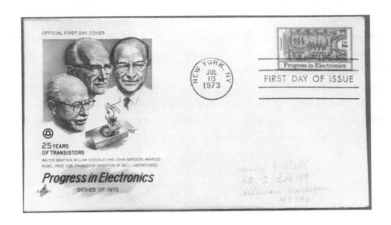

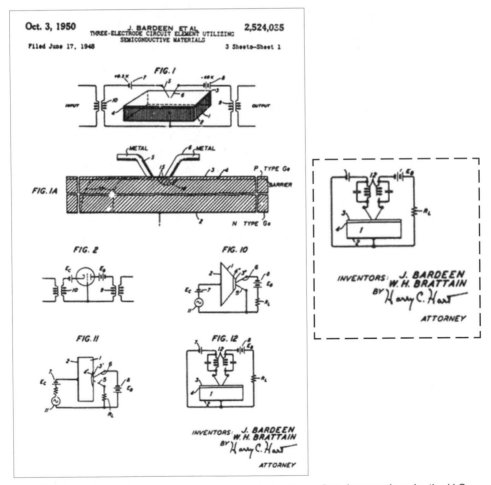

Top: The inventors of the transistor were acknowledged on a first-day envelope by the U.S. Postal Service. Bottom: Diagrams from the original patent for the transistor, developed at Bell Laboratories in 1947. The patent document is signed by Bardeen and Brattain, although William Schockley and Ralph Brown are also credited with contributing to the technology. [First Day Cover from the collection of the author, used with permission; U.S. Patent and Trademark Office, public domain.]

licenses. This doesn't seem to have discouraged innovation, however, and inventors continued to develop the technology. Meanwhile, the moratorium, which was only supposed to last a few months, was finally lifted by the FCC four years later.

Because it is a visual medium, people often forget that television broadcasting is a radio technology (described in the Visual Surveillance chapter). By 1950 more than four million televisions were estimated to be in use in the United States and Canada, a number that more than doubled by the end of the year.

Radio use was increasing as well, and transistors made it possible to fabricate small, handheld *transistor radios* and transportable radio systems that could be mounted in cars.

In the 1950s, some innovative radio technologies were developed by the Stoddart Aircraft Radio Company, which had supplied a number of radio technologies to Howard Hughes and the military during the war. Headed by Richard R. Stoddart (1900-1972), the company produced what has been called the first *audio-spectrum* receiver. We typically classify radio and audio spectrums as different phenomena. Radio is an electromagnetic wave-particle phenomenon that we can't sense, whereas audio is a longitudinal pressure-wave phenomenon that we can hear within audible frequencies. The Stoddart audio-spectrum receiver (NM-40A) was a detection instrument that could detect frequencies down to about 30 Hz. Up to this time, sets typically tuned to frequencies ranging from about 100 MHz to about 1 GHz. The NM-40A could be used as either a narrowband or a broadband receiving system, separately receiving the electrical and magnetic components of a wave, characteristics that made it suitable for scientific research [Layer, 1993].

Another milestone of the early 1950s took transistor technology to the next level. Inventors began creating electronic components as solid blocks, called *solid-state technology*, an innovation proposed by Geoffrey W. A. Drummer. This was the beginning of integrated circuits and large-scale integration (LSI) technology, which contributed dramatically to the evolution of electronics and the development of miniature components and microcomputer systems.

4.d. Modern Electronics

Miniaturization and Memory

The development of the transistor is one of the most significant inventions in all of electronics history. The evolution of large-scale integration (LSI) and very-large-scale integration (VLSI), following the invention of the transistor, eventually put small computers and handheld radios within economic reach of the general public. It also linked radio electronics and computers in important ways. From this point on, their evolution would be inextricably intertwined, with computer technology converging with almost every aspect of electronics and radio communications.

Prior to the development of VLSI electronics, computers and radio scanners were limited devices. Computers were programmed by manually changing wires, somewhat like the old telephone switchboards. Radio scanners were tuned by replacing crystals that would resonate at specific frequencies. Without computer memory, everything had to be hard-wired and hand-tuned. Choices were limited to a few computer programs or a few radio frequencies. The ability to store and retrieve electronic information is, in itself, as important as the invention of the transistor to every aspect of surveillance devices as we know them today.

Global Politics and Increased Surveillance

Following World War II, bomb shelters, air raid siren tests, and war drills created a background to people's lives in what came to be known as the *Cold War*, the political unrest between the Soviet Union and western nations. The McCarthy witch-hunt for communist sympathizers was chilling evidence of repressed fears and the willingness of people to sell each other out when faced with threats of government reprisal. Surveillance of the U.S.S.R. was considered an American priority and surveillance devices and techniques flourished.

In 1963, the Emergency Broadcasting System (EBS) was established. Television viewers were periodically reminded of its existence by short interruptions in scheduled TV programming with a sustained tone followed by an announcement about emergency instructions. Baby boomers in western nations all came to recognize the announcement, "This is a test of the emergency broadcast system.... This is only a test...".

The Birth of Satellite Communications

Arthur Charles Clarke (1917-), an English scientist and science fiction writer, demonstrated remarkable foresight when he described in considerable detail, in the late 1940s, the future of geostationary satellites as communications devices. A decade later, the Russians took the first step toward practical implementation of Clarke's ideas when they launched Sputnik I and, a month later, Sputnik II. This began the competitive decade of the 'Space Race' and exploration of the Moon.

The Score satellite was the first voice-capable satellite launched into space by the American space program. Both voices and code could now be sent over tremendous distances without underwater cables or dozens of relay stations. In 1958, President Eisenhower sent a Christmas greeting through the Score satellite. The world of global wireless communications through satellites opened up through Russian and American developments.

Left: An early 100-foot satellite named "Echo", designed by the Space Vehicle Group at the Langley Research Center in a test trial, in June 1961. It was sent into space uninflated, to be inflated when it reached the desired position. Right: The evolution of radio and television broadcasting brought the American people and their government into a new relationship. It was now possible for taxpayers to see how their money was being spent, and the development of the space program, among other government-funded programs, could be shared through radio broadcasts with the viewing public. This is a WHK radio interview in December 1963 at the Space Power Chamber with Centaur in the background. [NASA/Langley Research Center news photo; NASA/GRC news photo by Reidel, released.]

Even their proud inventors will admit that satellites are some of the funniest things you've ever seen. Some look like garbage cans, others resemble umbrellas, still others look like oversized children's party balloons. The first amateur satellite looked like a small homemade bomb and even some of the most sophisticated satellites used today look like they were built in an abandoned garage. Nevertheless the basic physics in these devices is extraordinary and their capabilities go far beyond their homely exteriors. Both passive satellites (that reflected radio waves) and active satellites (that could transmit their own radio waves) came to be developed, although the early passive satellites have been superseded, for the most part, by active satellites.

A few years later, astronauts on the Moon began transmitting images and sound that ushered in a whole new era of exploration and communications. By the mid-1970s, media programming could be broadcast across the nation by satellite. The Turner network and PBS took advantage of using the new technology.*

The Microcomputer Revolution

In the early 1970s, Intel introduced tiny computer chips that would spur the evolution of the first desktop computers. By the mid-1970s, the quiet revolution in electronics miniaturization and memory storage resulted in programmable microcomputers and radio scanning devices. Prior to the mid-1970s, only the government and large corporations could afford the huge operating rooms, equipment, and staff that were necessary for complex computing or radio surveillance activities. Suddenly this was no longer true. By early 1975, some significant hobbyist electronics from the previous couple of years came into the mainstream (or at least the mainstream of consumer electronics buffs) and the commercialization of computing and surveillance industries was picking up speed.

Robotics benefited greatly from the creation of tiny chip technologies. It was now possible to design small, 'intelligent' teaching, learning, detection, and surveillance devices that could autonomously or semi-autonomously explore their environments, gather objects and information, and report back to a central database or controlling console. Radio technology made it possible to remotely-control and communicate with robots, making new forms of industrial production and scientific and military exploration possible.

Radio-Based Long-Distance Remote Control. Left: Rocky the Robot, a radio-controlled prototype for planetary exploration, is designed to navigate over rough terrain and pick up samples along the way. The 56-pound robot shown here in Sept. 1991 was a test robot for future lighter robotic rovers that could be sent to the Moon or Mars. Rocky III was designed and built at the Jet Propulsion Lab (JPL) in California. Right: A radio-frequency ACR location beacon weighing only 190 grams was designed to be part of a survival kit taken into space by Mercury astronauts in 1962. [NASA/JPL news photos, released.]

---

*The evolution of satellite technologies is discussed further in the Aerial Surveillance chapter.

Radio Communications and New Industries

By the early 1980s, business owners, educational institutions, and technophiles were arming themselves with radio scanners, sophisticated calculators, and programmable personal computers, and using them to listen in on rest of the world. Marine radio and radar systems got smaller and cheaper and could now be purchased by small craft owners. Truckers and law enforcement personnel were no longer the only people using radios while driving—ordinary motorists could now afford them. The Citizens Band (CB) radio craze swung into high gear, computer modems enabled people to intercommunicate in new ways, portable scanners could suddenly handle hundreds of channels, and portable radio players and headsets pervaded beaches, schoolyards, and shopping malls. The reprogrammable miniature electronics revolution of the 1970s may someday overshadow the industrial revolution in terms of its importance in our history books.

The value of radio technologies for emergency ambulance, firefighting, and law enforcement services had been recognized for many decades, but small, efficient, portable systems didn't really become prevalent until the 1970s and 1980s. Handheld radios would now become a great boon to field workers, rescue professionals, and scientific and military users. As the systems became less expensive, events organizers and crowd control professionals began using them, as well. Portable radio technologies were now used by private detectives and private security agents on behalf of their clients, making it possible for them to communicate with colleagues and accomplices at a distance.

As cell phones caught on, it became clear that motorists could call for help, report crimes, or relay information about unexpected news events. Reporters began transmitting news features to their editors from their laptops, using radio-frequency modems.

The Pressure for Airwaves

In the 1970s and 1980s, electronics underwent major evolutionary changes and our understanding of radio wavelengths improved. Many radio frequencies were once considered 'junk frequencies', in other words, we knew the frequencies existed but didn't know how to harness their capabilities. Prior to the 1980s, ultra-high frequency (UHF) and very-high frequency (VHF) were used for a high proportion of radio broadcasting applications. Microwave frequencies were considered useful only for radar applications and many of the 'junk' frequencies were allocated to amateur radio buffs. But those amateur radio buffs were (and are) a bright bunch of people. They used their lemon frequencies to make lemonade. They were responsible for a number of important advancements in the use of radio technologies (and radio-related satellite technologies) that now utilize formerly unwanted wavelengths.

Unfortunately, amateur radio buffs were rewarded for their ingenuity by having to fight to keep their new discoveries. The situation has, at times, been similar to the struggle for indigenous treaty rights. History is full of examples of people being forced out of their homelands onto land that was considered undesirable by incoming settlers. Then, if valuable resources like oil, gold, or timber were subsequently discovered on those lands, the indigenous people were again relocated to less valuable lands.

On a smaller scale, amateur radio 'territory' is sometimes difficult to safeguard. As soon as new technologies evolve to make use of 'junk' frequencies, commercial or government lobbying agencies apply to the FCC and vie with one another for access to the new resources. Because wavelengths are limited and wireless communications demands are increasing as the population increases, various battles over radio bandwidth continue, and the situation may not be resolved in the near future, if ever.

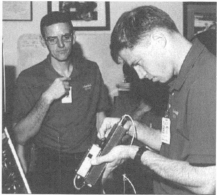

Left: Amateur radio hobbyists have been involved in many of the experiments and discoveries related to radio technologies and satellite communications. Left: Lou McFadden, an amateur radio operator at JSC, controls a console knob supporting the Space Shuttle Amateur Radio Experiment (SAREX) operations in July 1985. Right: Members of the STS-58 Spacelab Life Sciences crew training with amateur radio equipment. William McArthur and Richard Searfoss were scheduled to carry the Shuttle Amateur Radio Experiment (SAREX) payload along on the Shuttle mission in 1993. [NASA/JSC news photo; NASA/GRC news photo, released.]

Left: Astronaut Scott Carpenter receiving a wireless radio call from President John F. Kennedy through a radio telephone while on board the USS Intrepid spacecraft, on his four-hour and 56-minute space mission (July, 1962). Right: President Gerald Ford uses a radio telephone to communicate with the crew aboard the orbiting Apollo spacecraft on 18 July 1975. [NASA/JSC news photos, released.]

Civilian Use of Radio Technologies

Wireless phone systems were not new in the 1980s—many rural and small island populations were using wireless radio phones in the 1970s—but large-scale implementation of wireless networks and, in particular, cellular technologies, began in the early 1980s and picked up momentum in the mid-1990s. But inexpensive, wide-coverage wireless phones *were* new, as were commercial satellite systems that could carry the signals across continents, if desired.

Once again, the users of wireless technology began knocking on the FCC's door, requesting more bandwidth. In 1985, the FCC opened up three spread-spectrum frequency bands to unlicensed commercial use and regulatory adjustments continue to be made on a somewhat regular basis.

Miniaturization continued in the mid-1980s. Computers, radio sets, and components were still being made smaller and more efficient. There didn't seem to be an end to the process yet. Tiny microphones for a variety of purposes were designed and wireless microphones became more prevalent.

Local computer bulletin board systems (BBSs) made it possible for people to interact through computer networks. It became possible to access a number of computers with just one dialup phone call, instead of having to make individual calls to each system. ARPANET and other networks were beginning to evolve into a global network that would eventually become the Internet. By the mid-1980s, radio engineers were developing ways to use computers and radio transceiving stations together and 'packet' radio was evolving. The marriage of radio technology and data storage and processing was well underway by the early 1990s.

As communications technologies evolved, so did the devices through which the services were supplied. Television and radio programs were broadcast for many years using radio frequencies, but now cable was starting to emerge as a data transmissions medium. By using light signals over fiber-optic media, it was possible to deliver crisp, clean, fast programming that far exceeded the quality that most people were getting with their home television antennas. Cable also made it possible to deliver airwave broadcasting from around the world to local subscribers. The television broadcast industry began to consolidate itself.

Modern Use of Radio Technologies

Throughout the 1980s and 1990s, as technology advanced, it became clear that frequencies that were once considered unimportant or unusable were very important indeed. Microwave frequencies could now be used effectively for broadcasting, a fact that was once in dispute, and could even be used to cook food. In fact, the early microwave ovens were called radio ranges and radar ranges until the newer term "microwave oven" caught on. Because of the shorter wavelengths used, microwave broadcasting required different shapes and sizes of antennas from the traditional long-wave broadcasting frequencies.

These photos show how radio/television broadcasting has provided ground-based viewers with a look into space and how the technology has improved in three decades. Left: A historic, telecast of Astronauts Armstrong and Aldrin walking on the surface of the Moon during an Apollo II mission, in July 1969. Right: An Earth-based radio station interviewing space-based astronaut, Susan Helms (holding a tiny microphone) through a wireless communications link. Helms, aboard the Endeavour Orbiter Vehicle, was in orbit with her colleagues, in Jan. 1993, when this video image was captured. [NASA/JSC news photos, released.]

The explosive proliferation of radio technologies in the 1990s was due mainly to the decreasing size and price of transmitters and receivers. In the 1970s and 1980s many people expressed interest in having satellite television services, but most considered the parabolic dishes to be too large and expensive. When the size and cost of the systems dropped, satellite services began to seriously compete with cable for a portion of the viewing audience.

The same was true of wireless phone services. When the cell-phone handsets became smaller and less expensive, and the service regions became broader, people began using cell phones for business and research. As the price of the technology dropped, the number of personal users increased.

Modern Surveillance Tools

There were three important trends that contributed to the dramatic increase in surveillance technologies in the mid- and late-1990s. These were decreased size, decreased cost, and increased availability. Radio technologies specifically related to surveillance had been somewhat restricted to law enforcement, detective, and espionage applications up until this time—now they were available to anyone.

- In the mid-1990s, increased miniaturization was possible, and the price of components dropped dramatically. A color, wireless, pinhole camera dropped from over $1,000 in the early 1990s, to less than $200 by the year 2000 and dropped again to $120 by 2006.

- The birth of the Internet created two significant dynamics in the marketing of radio surveillance devices. 1) Educational articles explained how these devices worked and how to set them up, making people more willing to take a chance on buying or building them, and 2) small electronics manufacturers who had good technical skills but no marketing skills or distribution rights suddenly had a new way to sell their products by opening 'retail stores' on the Internet. Thus, cutting out the 'middle man' greatly increased the number of vendors who sold direct to users. As packaging and distribution costs decreased, so did prices.

Cellular phones are a good example of a radio technology that started as a basic consumer item and evolved into what is essentially a spying device. As digital imaging and video technologies continue to miniaturize, manufacturers have equipped cell phones with cameras and audio/video recorders. Cell phone users can now instantly record events going on around them, without the knowledge or consent of those in the vicinity and, further, can effortlessly send them to remote locations.

Cell phones also pose a danger to the privacy of the person using the phone. If the phone is issued by an employer, the calls may be monitored. This is a problem if managers are monitoring subordinates but subordinates are not permitted to monitor managers, in turn. It creates a lopsided system of double standards that could unfairly affect promotions, performance evaluations, and job security. In addition, many mobile phones are now equipped with Global Positioning System (GPS) components that can determine the location of the phone at any given time. This might be useful in an emergency where people are stranded or injured and searchers are trying to find them, but it also gives the service provider a means to track a person's movements. This information may have to be turned over to law enforcement agencies upon issuance of a subpoena and the cell phone user may not be immediately informed of the transaction.

Now that cell phones are essentially small computers, equipped with memory and sometimes extended keyboards, data integrity can be a problem. If confidential personal or business information is stored on a cell phone and the phone is lost or stolen, the data may be compromised as well. A vandal wouldn't even have to steal the phone to get the data. They could either offload it to another device or upload it to the Internet. It is a good idea to password-protect data-capable radio phones.

FCC Developments

The *Federal Communications Commission* has the unenviable task of trying to regulate an industry that changes by the minute. With demand for radio frequencies at an all-time high, the competition for bandwidth among commercial, governmental, and amateur interests is increasing.

New uses for ultra-wideband (UWB) technology include radar imaging and high-speed data transmissions. In 2002, the FCC adopted and announced a report and FCC Order permitting the marketing and operation of certain new products that exploit UWB frequencies. This was described by the FCC as a "cautious first step" due to the risk of interference with government UWB-based operations and was issued subject to further review. Many of the initial UWB systems would be restricted to use by law enforcement personnel, fire and rescue organizations, scientific research institutions, commercial mining companies, and construction companies. Wall-imaging systems would be further restricted for use only by law enforcement, and fire and rescue organizations. The same restrictions would apply to UWB surveillance systems, such as "security fences," for detecting intruders. UWB medical imaging systems were to be supervised by licensed health care practitioners.

The FCC also recognized the role of UWB frequencies for use in home and business networking applications and radio-wave-based measurement devices.

Radio surveillance equipment has quickly become part of the social landscape, with a number of consequences that are discussed later. Radio-controlled visual and auditory surveillance devices are showing up everywhere. The implementation of radio-frequency security and tracking systems is probably still in its infancy, with many more inventions and installations expected over the next decade. The prevalence of radio transmitters for wearable surveillance systems is still developmental, but is expected to increase. In 2000, these systems cost between $1,000 and $3,000, but they have dropped in price. We probably haven't even begun to anticipate all the changes these devices will eventually bring.

5. Description and Functions

5.a. Consumer Technologies

The electronics revolution has resulted in novel ways to use frequencies that were previously considered impractical. It has also created new classes of computer peripherals that rely on radio wave technology. There are now home and office security systems, cordless phones, stereo transmitters, wireless intercoms, and wireless-USB printers that use radio waves to transmit within 300-foot or 1000-foot limits. Regulations for low-power, limited-range devices are less stringent than for higher-powered devices, and low-power products are proliferating in consumer applications.

5.b. Improving Security and Efficiency

Security

Radio broadcast communications are inherently open to anyone within broadcast range who has a receiver that can tune to the specific frequency being broadcast, but this is not a secure means of exchanging information. In the quest for confidentiality, a number of strategies have been adopted, including encryption, short-range communications for short periods of time, and spread-spectrum technologies. Spread-spectrum techniques are increasing in prevalence due to their effectiveness.

Code-division multiple-access (CDMA) is a digital, wireless communications system that was originally incorporated into military satellites. CDMA was favored for its resistance to jamming and its greater security. CDMA provides access to multiple users with less interference than is encountered with some other techniques. Authentication of the source transmitter is possible, further limiting the possibility of eavesdropping.

Traditionally, radio communications were contained within narrow-bandwidth frequency ranges that rarely varied during any individual communication. However, it was clear during times of political instability that more secure forms of communication were needed, and the same is true for transmitting business transactions of economic importance. Thus, a number of ways of modulating the signal through a variety of frequencies, none of which could be predicted by an eavesdropper, substantially reduce the risk of being overheard.

Spread-spectrum techniques are commonly categorized as follows (and hybrid systems exist).

frequency hopping (FH-CDMA) - A group of changing frequencies modulated by the information bits in a two-step process. The carrier frequency is modulated first, then these modulated frequencies are modulated further, while keeping them independent. The receiver must be synchronized to the transmitter, which hops among available frequencies. Both are tuned to a reference center-frequency.

direct sequence (DS-CDMA) - Codes are used to modulate information bits such that each code is assigned to prevent the overlap of signals from user to user. The carrier is modulated to contain the information and the modulated signal is further modulated to spread it across spectrums. The code is regenerated by the receiver, which uses the information to demodulate the transmission.

time hopping - A carrier is on/off keyed with the speed of the keying determining the amount of signal spread.

chirp - A specialized form of spectrum spreading in which a carrier sweeps through a range of frequencies. It is primarily used in radio detecting and ranging (radar) applications.

Spread-spectrum technology has some interesting properties that make it suitable for secure transmissions. Spreading a transmission over many frequencies makes it harder to intercept or to jam. It also tends to be more difficult to pick up, the farther it travels from the transmitter. In other words, a listener some distance from the transmitter who is not expecting the signal is less likely to be aware of its presence.

For those involved in radio signal surveillance, spread-spectrum is more difficult to detect and decipher than traditional radio broadcast technologies.

Increasing Efficiency

The FCC limits access to radio bandwidth, so developers must find ways to use frequencies more efficiently, especially as more individuals switch to cell phone technologies and wireless computer data networks. Many schemes have been developed to achieve this end. While the following basic examples are not specifically designed to increase security (most were developed to improve capacity), some of them do add a measure through their specific implementation.

frequency-division multiple-access (FDMA) - a means for increasing capacity on communications channels. Available frequencies can be subdivided to allow more simultaneous links. The technique is common in cell phone and satellite communications.

time-division multiple-access (TDMA) - a digital technology that improves efficiency by assigning time slots to several calls within one bandwidth region, thus increasing capacity. *Extended-TDMA* (E-TDMA) is a further advancement on this general idea. TDMA is similar to CDMA and is used by some of the large wireless service providers. There are variations on TDMA in Japan and Europe.

5.c. Radio Transmitters

Covert Transmitters

Radio transmitters are widely used in surveillance activities. 'Wearing a wire' (being equipped with a microphone and recorder or radio transmitter) is an integral part of many law enforcement investigations and of investigative journalism. Employees who are harassed by employers sometimes wear wires to acquire evidence of the harassment. Undercover agents wear wires to collect evidence of criminal activities such as intent to kill a spouse, to rob a bank, to kidnap a child, to traffic drugs, to acquire illegal weapons, etc. Investigative journalists often wear wires when documenting embezzlement, corporate abuse of resources or employees, employee theft, or insider trading.

One area for potential innovation, now that radio transmitting units and microphones are so small, is to incorporate them into clothing, rather than attaching them as separate components. For surveillance activities, this would be particularly advantageous as there would be no outward sign of the electronics, and no bulges or lumps to give away the system's location in a 'pat search'. That is not to say that clothing-incorporated microphones and transmitters couldn't be found—metal detectors and other surveillance devices might reveal the presence of anomalous threads or fabrics—but a visual or cursory physical inspection wouldn't immediately reveal their presence.

Radio Tags

Radio tags are tiny transmitting or receiving/transmitting units that can be attached to almost anything to provide information about the location, movements, or audio signals associated with their use. Since the tiny size of radio tags usually means their transmission distance is limited, it may be necessary to have many receivers spread over an area. This is not necessarily difficult. Small receivers can be built and deployed over a wide geographic area, like a receiving array. The data from these receivers can, in essence, create a visual picture, through computer processing, of the movement of tags within the receiving region. Another means of tracking the tags is to bring a hand or vehicle antenna and receiver into the region where the tags are expected to be. A further use of the tags is on vehicles and other objects that regularly move in the vicinity of receiving stations such as bridges, weigh stations, or other checkpoints.

Tag readers or receivers are sometimes called *interrogators*. Radio tags are used in wildlife tracking, industrial yards and construction sites, assembly lines, and military operations. They could potentially also be used to track livestock, individuals who are prone to wander and who might injure themselves or get lost, competing athletes (expedition racers, long-distance runners, or cyclists), inventory, or employees at large events such as fairs or rock concerts.

Vendors promoting the use of radio tags for identification refer to them as *RF/ID* (radio-frequency identification) and *RF/DCI* (radio-frequency data collection) systems. Some of the manufacturers of these systems are familiar names in the electronics industry, including Texas Instruments, Philips Semiconductors, Brady, and Gemplus.

RF/ID technology has improved significantly since the mid-1990s. Not only has the cost of the tags dropped to below $1 per tag, but realtime tracking systems (as opposed to systems that report the location of the tag at the last checkpoint) are now available. This opens up the possibility of individual object tracking that could be used for shipping or luggage services. It also opens up the more controversial possibility of monitoring store purchases of individual items or equipping children with tracking tags so that parents have an idea of where the children may be on their way to schools, playgrounds, or the houses of friends.

Because of FCC emissions regulations and the small size of radio tags, most of them are short-range devices, usually from about half a meter to about six meters. Those used for wildlife tracking have longer ranges, due to the difficulty of getting close to wildlife in wilderness areas, and may have ranges up to about half a mile or more, depending on the size and power of the tag and the type of battery used. Tags for specialized purposes may be heat- or water-resistant.

Most radio tag systems used in ID and inventory systems work in the 900 MHz and 2.4 GHz regions of the spectrum—the same frequencies commonly used for short-range cordless phones, small home/office wireless transceivers, and local wireless computer network links.

Sometimes radio tag systems are used in conjunction with video cameras. When the tag passes a receiver or checkpoint, the video camera is activated and creates a record of the animal, bird, object, mall shopper, worker, intruder, or vehicle that triggered the camera.

In July, 2006, the FCC issued a notice of proposed rule-making to allocate bandwidth in the 401–406 MHz range for radio communication with implanted and body-worn medical devices. Up to this time, the Medical Implant Communications Services (MICS) band from 402–405 MHz had been used for radio-based implants. The new proposal would extend this by 1 MHz at the outer ranges and provide a way to transmit to the devices, as well, making it possible to remotely administer aid or medications via radio signals or to control artificial limbs. Thus, radio-controlled body-worn or implanted devices could have a positive effect on a person's quality of life. Implants could also have negative consequences, as well.

Recognizing the potential for abuse from implanted radio tracking devices, the state of Wisconsin took a proactive stance by instituting a law making it illegal to require a person to have a microchip implant that could be used for identification or tracking. The concern came as a response to the commercial promotion of chips that were designed to be injected under the skin to uniquely number and identify people.

5.d. Radio Direction Finders

Direction finders are used in search and rescue, covert surveillance, wildlife tracking, and for locating unauthorized radio-frequency transmitters.

Radio direction finding is an important technology for navigation as well. Boats and vehicles use radio direction finding to chart a course or determine a position. Radio beacons are usually used together with direction finders. Several beacons make it possible to more quickly and precisely determine location and relative bearings.

Basic Concepts

There are a variety of types of direction finders, including Doppler, Adcock, and various multi-element arrays. Most of them depend upon line-of-sight for effective transmissions. Basically, the direction finder uses a transmitted signal as a reference signal for determining the bearing to a target transmitter by evaluating the angle of the radio wave in relation to the antenna site. Theoretically, this is a fairly simple physical/mathematical determination. In real life, however, radio waves don't usually travel unimpeded. There are often trees, buildings, particles, and other reflective or absorbent surfaces that scatter and impede the signals so that what reaches the receiving antenna isn't a perfect wave.

To understand the technical issues, imagine throwing a pebble into the center of a still, smooth pond. Waves radiate out in clean, predictable lines to an 'antenna' near the shore. It is relatively easy to calculate the direction from which the pebble was dropped based upon the direction of the waves hitting the antenna. Now imagine a few rubber duckies, children's boats, and swans floating around the pond. When you throw in the pebble, some of the waves will be impeded, deflected, or absorbed by the various objects in the water. By the time the 'signal' or wave reaches the 'antenna' (the edge of the pond), there are complex wave patterns that must be processed for the information to be of any value. The directions and relative intensities of radio waves must be assessed over time to make a good estimate of which waves are the significant ones.

It's not unusual to see wildlife or search and rescue teams walking around in the wilderness holding up spindly devices that resemble old TV antennas, with several limbs sticking out from the main shaft. As the searcher moves this receiving antenna around, a beeping sound is usually emitted by the device to indicate the relative strength of an incoming radio signal. By rotating and waving the antenna, closer approximations to the right direction can be auditorily estimated by the loudness of the beep (this is similar to manually interpreting sonar pings). With skill and a little bit of luck, the searchers can eventually locate the avalanche victim, whale, or grizzly bear equipped with a radio beacon that emits a regular pulse that can be picked up by a receiving antenna.

New Systems

Direction-finding systems have been used in marine navigation for decades and are now becoming very important in land-based systems, as well. Combined with Global Positioning System (GPS) consoles and databases of city or terrain maps, a direction finder becomes a very sophisticated positioning or tracking system. With onboard vehicle displays, these hybrid systems can show you where you are on the map, where you've been, and where you might like to go (based on previous trips or stated preferences, e.g., back roads). Intelligent vehicle systems that essentially drive themselves use adaptations of these technologies for navigation.

Direction finders can be used to track rental cars, police vehicles, or truck convoys. In fact, some new vehicles are being sold with radio tags or beacons that can be located with direction-finding technology as security measures against theft and, if stolen, to aid in the apprehension of the thief.

Packet radio engineers have found ways to interface direction finders to computer networks. This makes it possible to tag and track marine or ground-based vehicles from any location with an Internet connection. Thus, commercial industries can monitor their vehicle resources and shipments and law enforcement agents can cooperate with state or federal authorities.

5.e. Radio Receivers

Roaming Radio

There has been a great deal of interest in developing powerful portable radio transceiving units with good sound quality. Radio receivers enable news, instructions, data, and other information to be relayed to remote users, while transmitters enable the user to report back, ask questions, or upload information. Portable systems also provide a way to control various types of electronic devices through radio signals or transmitted audio commands. Radio receivers are useful in robotics applications, as well, for control, navigation, and data acquisition. Portable radio-wave systems are may be handheld or attached to the body or clothing. They are sometimes referred to as "untethered" systems, especially in military operations.

One project of interest is the Nomadic Radio developed at MIT. Nomadic radio is a body-worn *roaming radio* receiver system. Through radio signals, the system regularly downloads information from the transmitter. As a message system, it incorporates speech synthesis and speech recognition to provide a multi-information source. It can input news, voicemail, and email. Thus, a person on the move doesn't have to carry a computer or other handheld system to keep up to date on what is going on. Broadcasts are automatically downloaded to the system and transmitted auditorally to the user. Sportscasters, investigative journalists, field workers, undercover agents, stock traders, handicapped individuals, and marketing and sales representatives are all potential users of hands-free systems.

The essential concept has many possibilities. A GPS location-base could allow location-relevant information to be downloaded, including information on nearby transportation venues, shops, restaurants, hotels, police stations, or phone booths anywhere in the world. A heads-up system could add visual output on a small screen attached to a helmet or a pair of glasses.

A patented sound neckset developed at Nortel for hands-free telephony could just as easily be used with a number of roaming radio technologies.

Interest in body-worn systems continues. In 2006, the Institute of Electronics, Communications and Information Technology (ECIT), sponsored a three-year research grant for investigating new techniques for conformable body-worn antennas, radio frequency circuit modeling, antenna construction, and measurement validation based upon high-frequency radio technology.

Field Radios. Left: In May 2006, the *National Institute of Standards and Technology* (NIST) announced a *Defense Advanced Research Projects Agency* (DARPA) program to test wearable sensors for use by soldiers. A variety of types of sensors would capture GPS positional data, sounds, and images. Data would then be extracted from sensors and recordings and combined with soldier observations to create a multimedia representation of the information collected. Right: Spc. Bonina uses a radio while conducting perimeter security at Guantanamo Bay, Cuba. [NIST news photo, 2006; Army news photo by Staker, 2005, released.]

5.f. Satellite Communications

Satellite signals are now used for many types of communication, including television and radio broadcasts, satellite phones, and computer data transfer. Depending on the position of the satellites, their number, and the degree of ground-based support, communication over any particular satellite system may range from a few hours a day to round-the-clock. The four basic components of satellite systems are transmitters, antennas, relay stations/repeaters, and receivers. In smaller systems, the transmitters or receivers may be housed in the same unit with the antenna. In larger systems, the antennas may be separate and may even be spaced out over several acres.

Satellite Transmitting and/or Receiving Units

Left: Satellites can now be used to relay voice messages around the globe. The U.S. Armed Forces use portable satellite phones that can be carried in a backpack and set up in the field. Here Cpl. Carlos Rivera, a radio operator, uses a satellite phone to send a message from the village of Skugrici, Bosnia/Herzegovina, Right: An Omni Tracs satellite transmitter is installed in a Humvee by R. Troxell. The Omni Tracs system reports the vehicle position to the Global Command and Control System in near-realtime. [U.S. DoD 1999 Army news photo by James Downen, Jr., and 1998 news photo by Christina Horne, released.]

A satellite phone is used outdoors in the Republic of Gabon to receive information about incoming aircraft and their cargos. This was part of a contingency plan for evacuating U.S. citizens from Zaire, if the need arose. Phone operator Air Force TSgt. Darrin Brown is an aerial porter for an Air Mobility Operation Squadron. [U.S. DoD 1997 news photo by Andy Dunaway, released.]

GPS

The Global Positioning System (GPS) is a space- and ground-based navigational system that was originally designed and used by the U.S. military as part of the Navy Navigation Satellite System. Satellites and ground stations intercommunicate to determine the location of a particular spot on or around the Earth. Three or more satellites can be used to 'triangulate' a position based on a point relative to the satellites. The satellites orbit in 12-hour cycles and provide 24-hour service.

When GPS satellite technology was opened to the commercial market, designers immediately found hundreds of uses for the technology, including vehicle and aircraft navigation, search and rescue operations, wildlife tracking, and much more. GPS receivers are now routinely built into cell phones and other electronic devices. Until recently, the resolution of the military GPS was much higher than civilian GPS frequencies, but the gap continues to narrow. As of May 2000, civilian frequencies were made accurate to about 10 to 20 meters.

GPS. Left: Handheld Global Positioning System (GPS) units used by the U.S. Air Force. The GPS units provide navigational reference point, time, and date information. Here, SSgt. Bozeman demonstrates an operational check. To conserve memory space, the GPS clocks reset to zero after 1,023 weeks. However, not all manufacturers anticipated the consequences of clocks resetting. Right: Sgt. Husen describes the Land Warrior's GPS which combines global positioning data with video, night vision, infrared targeting, and a heads-up display. [U.S. DoD 1999 news photo by Lance Cheung; Army news photo by Triggs, 2002, released.]

Depending upon how it is implemented and which electronics are included with the GPS technology, GPS can be used to determine longitude, latitude, altitude, and velocity. This information can further be cross-referenced with databases containing street maps and topographical maps, vehicle monitoring systems, and even radio tracking beacons.

Satellite Antennas

Radio stations, portable command sets, and mobile stations are often equipped with receiver, transmitter, and relay capabilities in one unit. In the past, the need for a large, curved surface antennas to receive satellite signals limited their utility and portability. However, during the 1990s, a number of inventors, including Choon Sae Lee, a former Hughes aircraft engineer and professor at Southern Methodist University, were developing units with smaller, flatter surface areas that could work effectively and be less expensive to produce. Just as flat screen monitors are superseding bulky cathode-ray-tube monitors, compact satellite antennas are superseding many large, dish-shaped antennas.

A 20-foot Quick Reaction Satellite Antenna is installed by U.S. Air Force personnel in Southwest Asia as part of an Air Mobility Command Tanker Task Force. The antenna provides worldwide communications. [U.S. DoD 1998 news photos by Efrain Gonzalez, released.]

Wireless Radio-Frequency Satellite Communications. Left: Two U.S. Army 58th Signal Company soldiers work at a field console in the confined space of the AN/TSC-85B satellite communications van near Kaposvar, Hungary, to support the NATO Implementation Force. Right: An Army paratrooper sets up a satellite radio during a patrol in Al Fllujah, Iraq. [U.S. DoD 1996 news photo by Larry Aaron; Army news photo by Johnson, 2004, released.]

5.g. Remote-Controlled Vehicles and Robots

The remote transmission of radio control signals has many applications in surveillance. They enable machines to go where people can't, including deep in the ocean, out into space, or into hostile or hazardous environments. Remotely operated vehicles (ROVs) typically use radio-frequency controllers, although some are tethered with fiber-optic cables.

Mine Detection Technologies versus Brute Force. Left: There are many strategies for locating dangerous mines that involve detection and removal, but sometimes it's easier to drive a heavy vehicle over the transportation route to set off the mines ahead of other vehicles and people. In this case, a remotely controlled Panther, based on a modified M-60 tank hull, is used to detonate contact mines and magnetic mines. It is remotely controlled by the U.S. Army staff riding in the vehicle behind it. Right: Xavier is a Carnegie Mellon robot, equipped with a large variety of sensors, that can communicate with the Internet through a radio-frequency Internet link. [U.S. DoD 1996 news photo by Jon Long; Carnegie Mellon news photo, released.]

Carnegie Mellon University's Learning Robot Lab, has a mechanism named Xavier that looks like a cross between a trash can and a high-tech podium. Xavier is actually a sophisticated robot that wanders the halls of the university, learning and interacting with its environment. The base is a standard four-wheel synchro-drive, and the top is custom-equipped, built by the researchers. Xavier is sensor-rich with a sonar ring array with two dozen sensors, a laser light striper, and a Sony color camera mounted on a pan-and-tilt head. There are onboard microprocessors and a laptop attached to the top of the 'podium'. One of the most unique aspects of the robot is a radio link through Ethernet to the Internet that enables Internet users to send command controls to Xavier through a console on the Web, when the robot is online and active. Xavier has speech processing capabilities—it can both talk and understand spoken commands.

Carnegie Mellon's CyberATV program is developing a number of processing systems that can be integrated into surveillance robots. Shown here is a finished vehicle (left) and a 'look under the hood' (right) at the various sensors and components that make up the system. [Carnegie Mellon Institute for Complex Engineered Systems news photo, released.]

At the Institute for Complex Engineered Systems at Carnegie Mellon, there are also some distributed robotics projects specifically aimed at surveillance applications. One of these is the *Perception of Visual Surveillance* project within the CyberATV program, which includes research activities for object detection, object classification, object correspondence, and object tracking, all essential components of good surveillance systems. For example, the vision system can focus on a person, pick out relevant features, and then look up these patterns in a database. Other robotics projects at CyberATV include vehicular control devices.

6. Applications

ID and Inventory Tracking

There was a time when all inventory tracking was done by clerks with clipboards. Now, computerized databases interfaced with cash registers are becoming prevalent and systems that incorporate Radio Frequency ID tags (RF/IDs) are on the rise. A tiny radio-frequency chip that can be imbedded in most consumer items can signal a computer with a wireless receiver or could even "talk" with items embedded with similar chips. In the year 2000, RF/ID chips were about the size of a fingernail. Now some are smaller than a pinhead. Retailers don't even have to worry about inserting them—they will probably be embedded in the product at the assembly-line stage. An electronic product code (EPC) provides the chip with a unique ID—an identifier that may eventually replace UPC barcodes.

RF/ID chips differ from barcodes in many ways. They can be read electronically by computers operating 24 hours a day, and can be tracked all the way from the manufacturing plant to the home or office in which the product is eventually used. In contrast to UPC codes, you can't just pull a sticker off to remove the identification, the chips are difficult to remove.

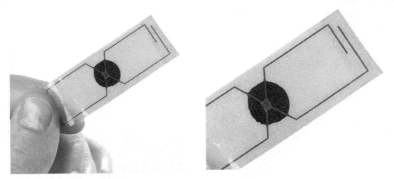

Not much longer than a thumb and much thinner, this radio-frequency (RF) tag can be used for identification and tracking applications. Originally developed for inventory monitoring systems, RF tags are now smaller and less expensive and can be used for shipping, vehicle monitoring, toll monitoring, consumer tracking, and public events applications (e.g., season's ticket holders). Depending upon the system, they can be individually programmed, or programmed collectively within an RF field. [Courtesy of the Pacific Northwest National Laboratory.]

Radio tags are used in a wide variety of surveillance and monitoring programs. Here are just a few examples from a variety of industries:

- The *Heavy Vehicle Electronic License Plate* (HELP) program is a Canadian automatic truck weigh-in system in which radio tags are attached to license plates and scanned when the trucks pass through weigh stations and border entry points. The data from the tags is sent to a central processing facility and is used to improve flow of the trucks by automatically identifying and weighing them.

- The U.S. Postal Service, in conjunction with Savi Technology Inc., has been installing a system of radio tags to automatically track mail containers. In combination with Internet data communications, this system can provide realtime computer reports on the status of the mail containers, providing their location at any particular time. The programmable radio tags and automatic readers are designed to work with the robotic tray-loading and sorting systems that are already in place.

- The TNO Institute of Applied Physics has developed an electronic ammunition identification system that incorporates radio tags. The Ammunition Registration System (ARES) identifies dummy ammunition as it is loaded, using a detector on the simulator's firing chamber. Each piece of ammunition is fitted with an electronic label transponder with a unique identification code. The transponders receive their energy from the detector, so that individual batteries for the transponders are not required. Data transmitted by the transponder are sent to an instruction console where an instructor can monitor and evaluate the actions of students.

- A number of ski resorts, amusement parks and other businesses that issue seasons passes are beginning to incorporate radio tags into the systems for access control at lifts, gates, and rides. This allows them to monitor such aspects as location and frequency of use, in order to manage traffic flow and safety mechanisms.

Employee Tracking

Radio tag technologies are used to track employee movements while on the job (and some employers have even tried to enforce it off the job). It is possible to use radio tracking devices and computer processing to monitor an employee as he or she passes through checkpoints or to monitor them on a realtime using computerized systems. GPS technology is sometimes used to monitor employees who work outside of the office space, such as truck drivers and on-the-road sales and marketing representatives. Most of the current tracking systems are tied to vehicle movements, but more and more, the tracking of individual employees is being implemented in the workplace. For example, certain city governments have proposed using GPS technology to track employees.

Employee tracking has always been controversial. It's a one-way system, in which senior employees track subordinates, but subordinates are rarely permitted to track their seniors. As was mentioned earlier, this system has a high potential for abuse when it comes to fair competition for promotions, raises, and company tenure. Superiors could use tracking data to cast their subordinates in an unfair light.

There is also concern about privacy invasion. When first introduced, some employers used tracking systems to track employees everywhere, even in the bathroom, and some even began to insist that employees wear the systems off the job. To further complicate the picture, vendors have developed RF/ID implants and are promoting them as a way to monitor movements, IDs, and entry and exit points. Some states have stepped forward and instituted laws to prevent employers from requiring RF implants. However, voluntary implants are still permitted and what is a worker to do if he or she needs a job and the person who is willing to submit to an implant has an unfair advantage of being hired? Financial coersion is often used as a way to force people to give up basic rights and a problem that hasn't yet been addressed in the courts.

Inmate and Parolee Tracking

Systems have been developed to attach radio transmitters to pre-trial, medium-risk defendants, and to parolees and prison inmates. The radio units vary in size and style. RF chips can be inserted into pagers, armbands, and legbands. Some proponents have suggested using radio implants, that are injected under the skin, or nanotech beacons injected into the bloodstream. All of these programs are controversial.

Critics of judicial tracking systems have expressed concerns about 'branding' or dehumanizing individuals beyond what a free and civil society would consider just or proper. Nevertheless, these programs continue to increase in number for a variety of economic and safety reasons. Here are examples of some current and proposed monitoring systems:

- There are now programs for sexual offenders that grant parole in lieu of incarceration, if the offenders will wear a radio tracking system that ensures that they don't go within a certain distance of specified environments, e.g, within 500 yards of a school ground. These programs permit the offender to work and interact with society instead of remaining in overcrowded prisons.

- Pre-trial defendants, out on bail, are monitored with radio tracking devices so that if they remove the device or leave town, their last position can be known and reported, within minutes, to the appropriate authorities.

- Sometimes biometrics are combined with radio technology. With offender voice-page systems, the offender's voice is first recorded on a reference system. He or she is then equipped with a pager that is beeped on a random or scheduled basis. Offenders must respond within a designated time limit, usually about 10 minutes, or steps will be taken to apprehend them. With current systems, the offender has to find an available phone. However, by incorporating a cell phone system with the pager, it wouldn't be necessary to hunt for a working telephone to make the callback. The caller's voice is then verified against the voiceprint. If the cell phone is equipped with GPS technology, the offender's whereabouts can be recorded at the same time the call is made. If the voice doesn't match, or the location is out of bounds, steps can be taken.

- Another system combines radio-frequency monitoring, GPS location information, and Web-based mapping to track the tagged person's movements within thirty feet. Programs to use these systems with criminal offenders are considered cost-effective compared to the cost of housing and feeding them in prison. A similar system was tested in eastern Canada, in 2006, to make it possible to release offenders awaiting trial. Offenders were outfitted with ankle units, GPS transceivers, and a beeper. If they wandered into a zone that was off-limits, the beeper would sound and a text message warned them that they were off-limits.

Search and Rescue

Radio technologies can be very valuable in search and rescue operations. A radio beacon can aid search crews in locating a buried avalanche victim. Handheld radios enable people to intercommunicate in dense woods or rough terrain where they can't see one another but need to keep in touch. Radios can warn of impending floods, hurricanes, and earthquakes. Radio communications systems further allow the coordination of rescue activities and communication between search craft (helicopters, planes, boats, snowmobiles, etc.) and foot crews.

Left: The Thailand Rescue Coordination Center and the U.S. Air Force have been engaged in cooperative search and rescue training in Khao Na Ting. Royal Thai Air Force Sgt. Jankeeree (left) and U.S. Air Force MSgt. Sitterly from the Pacific Rescue Coordination Center are shown here communicating on PRC-90 HF handheld radios. Right: A closeup of the PRC-90 HF radio used here by Air Force Major Jean Trakinat. [U.S. DoD 1996 news photos by Gloria J. Barry, released.]

The *National Oceanographic and Atmospheric Administration* (NOAA) carries an emergency surveillance system in the form of a *Search and Rescue Tracking System* (SARSAT) on its polar-orbiting satellites. The satellites can receive emergency transmissions from persons in distress and relay them to ground stations in the U.S. and abroad. The signals are forwarded to the nearest rescue-coordination center which, in turn, dispatches the information to emergency rescue personnel.

Humanitarian assistance training in Virginia using a radio setup that can be carried in a backpack and set up quickly in the field. P1C Eric Perez uses the radio to communicate with other Marines as part of the Marine Expeditionary Unit Service Support Group 26. [U.S. DoD 1998 news photo by J. T. Watkins, released.]

Rush Robinette of the Sandia National Laboratories has pointed out that strategies and algorithms for locating the point source of a chemical or biological attack can be applied to the location of a skier buried in an avalanche, thus providing another tool to aid search and rescue professionals. Following this line of reasoning, he and his group have developed computer programs to provide a form of 'group intelligence' to a swarm of mini-robots that could enable them to rapidly locate a source of contagion. Similarly, search and rescue workers with GPS receivers and radio equipment could be teamed with robots to locate avalanche victims faster. Since suffocation and hypothermia can claim a buried skier or hiker in a short period of time, any technology that speeds up the location of the victim can have life-saving consequences. Many skiers and hikers now routinely carry radio beacons, allowing them to send out a distress signal that can aid searchers in locating their position, even if they are buried or otherwise obscured.

Remote-Controlled Robots. Left: Spc. Schakey works an arm-mounted radio controller to remotely move a "tough-bot" surveillance robot that surveys an area before soldiers move in. Right: A robot 'team' has been set up here to illustrate how radio-equipped group robots might be used to aid search and rescue operations to locate avalanche victims faster, thereby increasing their chances of survival. [Army news photo by Bailey, 2005, released; Sandia National Laboratories 2000 news photo.]

The 'swarm algorithm' used in the Sandia simulations is called *distributed optimization* and was found in trials to locate theoretic victims four times faster than any previously published search scheme. In 'rough terrain' simulation tests, the results were even better. In their primary applications for the U.S. Department of Defense, the tiny robots intercommunicated through radio transmitters, thus enabling them to share information in order to home in on the desired target. Each robot broadcasts the strength of the radio beacon signal, from its current position, to the other robots, allowing the robots, as a group, to refine the area of search. Searchers equipped with palm-sized computers could be kept informed of the progress of the search and receive instructions as to where to go.

Remotely-controlled combat robots. SSgt. Tordillos uses a joy-stick console for remotely operating an armed robotic system. The screen enables the operator to monitor the position of the robot. [U.S. Army news photo by Jewell, 2004, released.]

Tiny Autonomous Robots. Adkins and Heller at Sandia labs have developed robots so small they can perch on a coin, complete with motors to control the wheels, and a temperature sensor. Future versions will include communications devices and additional sensors. The robots will be able to navigate in small spaces and fit where larger robots and people wouldn't be able to go. The units could be useful for sensing chemical and biological weapons and for locating land mines. Tiny robots could also be useful for 'swarming' or neural net applications and reconnaissance missions. [Sandia National Laboratories news photos, 2001, released.]

Robots technology continue to improve—robots are becoming more powerful, more autonomous, more sophisticated and, in some cases, smaller. Sandia labs has created autonomous robots that weigh less than one ounce that are powered by watch batteries. They include a pair of motors to control the wheels, a tiny ROM processor, and a temperature sensor. Development is underway to add a camera, microphone, chemical sensor, and communications device.

Natural Resources Management and Protection

Wildlife tracking collars are some of the more familiar radio technologies that are used in the study and management of resources, but radio technologies can also be used to study geophysical and astronomical phenomena.

The FORTE (Fast On-orbit Recording of Transient Events) satellite, developed jointly by the Los Alamos National Laboratory and Sandia National Laboratories, has a sophisticated radio-frequency receiving unit and 30-foot antenna that can record individual lightning strokes, providing new information in the atmospheric sciences. The recordings can further be linked with measurements taken by ground-based instruments. This in turn can help in weather surveillance. Meteorologists and weather forecasters can identify storm systems that might pose a danger through high winds or hail. The radio receiver can sample frequencies at a very high rate and store the information until it is transmitted to ground units. The system isn't just for weather surveillance, however, as it can further serve as a research platform for early warnings of nuclear detonations, which emit electromagnetic pulses similar to those generated by lightning. The measurements gathered by FORTE can aid in building up a knowledge base that could help scientists discriminate more easily between the different phenomena.

Another innovative use of radio frequencies is for the cleanup of contaminated sites. Just as a microwave oven uses radio frequencies to cook food, radio waves can be used to help sterilize materials in a number of industrial and commercial settings.

National Defense

Wireless portable or transportable radio and satellite links are an intrinsic part of national defense and military operations in the field.

Left: U.S. Marines from Communication Company and the 1st Light Armored Reconnaissance Battalion assemble a high-frequency radio communications antenna at a training exercise in Alaska. Right: A Troppo Scatter antenna protruding from camouflaged 'terrain' at Fort Bliss, Texas. The Air Control Squadron uses antennas for long-range communications by bouncing radio signals off the troposphere. [U.S. DoD Released 1998 news photo; 1995 photo by Mike Doncell, released.]

Scanning

Devices which are designed specifically to surveil radio signals in order to locate broadcasts of interest are called scanners. Some are sold with police and fire frequencies preprogrammed to make it easy to locate and listen to these frequencies. News reporters and those who are simply curious will often listen to the emergency frequency bands. Scanners can range in price from less than $100 to thousands of dollars. Most consumer scanners are designed to be portable and easy to use. Some of them include memory for storing frequencies of interest for quick access later on, somewhat like the push-buttons on a car radio that let you quickly select your favorite stations.

Scanners usually have an LCD display to display the frequency currently set and some tabletop models have outputs for displaying information on computer systems or for hooking into better quality room speakers.

Many scanners are designed to jump over frequencies that are specifically allocated to certain applications, such as cell phone communications. Hobbyists will occasionally modify scanners to overcome this situation, so the device will scan continuously. Not all circuits can be modified to scan over the full spectrum.

Some of the emergency frequencies are allocated on a national basis, and some are suballocated on a regional basis. It's a good idea to look up local regulations.

Scientific Instruments

Radio technology has been used to enhance or extend a number of other surveillance technologies, such as optical spectrometers. An acousto-optical tunable filter (AOTF) is a tiny solid-state crystal with a radio-frequency driver attached. The driver allows a specific frequency to be applied to the acoustic driver, which sets up a wave (grating) within the crystal. Light can then be shone through the crystal, which is diffracted. Since radio frequencies can be quickly changed, the instrument can be stimulated at various parts of the spectrum with the narrowband AOTF prefilter extending the capabilities of the spectrometer [Baldwin and Zamzow, 1997].

7. Problems and Limitations

The main limitations in the use of radio wave technologies are power consumption, availability of bandwidth, and detection/interception of private messages.

Power Consumption

Some progress is being made on the problem of power consumption. Smaller systems and longer-life batteries have alleviated this problem to some extent. In some cases, rechargeable batteries and solar power can help, but these are not practical in all situations related to surveillance. Surveillance is sometimes carried out in remote areas where access to power is limited or absent.

Availability of Bandwidth

Because radio waves are a physical phenomenon which must share space in the physical world, they must be allocated and used wisely in much the same way as any renewable resource. The airwaves are not unlimited. More powerful signals will overwhelm less powerful signals of the same frequencies. The Federal Communications Commission has stringent bandwidth allocations and licensing requirements for the frequencies used and the strength of the signals. Radio systems used for surveillance must respect these regulations or those of foreign nations if the surveillance takes place abroad. With the increased prevalence of home security and control systems based on radio frequencies, along with cordless phones, radios, and roaming computer data systems, the pressure for more bandwidth is increasing, with commercial interests constantly clamoring for more.

Message Security

A great deal of progress has been made on safeguarding the privacy of messages. New spread-spectrum and encryption techniques help hide the presence of a transmission to some extent, and help protect the content of a transmission. This helps safeguard the confidentiality of communications related to surveillance. However, it also makes it more difficult to surveil suspicious communications by potentially hostile nations or individuals. Law enforcement agencies have been concerned about their eroding ability to tap into conversations of a criminal nature through authorized wiretaps, now that there are so many new ways in which criminals can communicate with one another through PCS, cells phones, handheld radios, laptop computers, etc. Congress has tried to balance the pressure for public privacy in radio communications with law enforcement requests for access to encryption keys and radio communications with no clear resolution at this time.

8. Restrictions and Regulations

When the U.S. Congress passed the Communications Act of 1934, regulation of the radio industry passed from the Federal Radio Commission (FRC) to the Federal Communications Commission (FCC). The FCC was made responsible for wired and wireless electrical and radio communications. The FCC is the chief regulatory body for radio communications. There are strict fines and penalties for violating FCC regulations and some even include prison terms. Radio broadcasting is big business and big businesses don't like intrusions into their operations, so the industry is policed as much by the people who use it as by the various justice agencies.

Some of the regulations related to audio communications and privacy are covered in the Audio Surveillance chapter. The FCC Web site is the best source of information for questions regarding broadcast, usage, and manufacturing of RF-related devices.

> *Wireless Privacy Enhancement Act of 1999.* H.R. 514, a proposal to amend the Communications Act of 1934 to strengthen and clarify prohibitions on electronic eavesdropping, and other purposes. Referred to the Senate committee in March 1999.

> *Wireless Telephone Protection Act.* Public Law 105-172 was passed in April 1998. It amends the Federal criminal code to prohibit the knowing use, production, trafficking, control, or custody of hardware or software that has been configured to insert/modify telecommunications identification information such that it can be used without proper authorization.

Originally secure communications were only available for military purposes, but gradually the technology has opened up to commercial markets. Consequently, there are some frequency-hopping and spread-spectrum regulations. Here is one example:

> *U.S. Title 47, Chapter 1, Subchapter A - Part 15, Subpart C* This Subpart describes compliance provisions, including channel separation parameters, the number of hopping frequencies, permitted time of occupancy within a frequency, maximum peak output power of transmitters, and the processing gain of direct-sequence and hybrid systems. Priority of critical Government requirements are also noted. Government systems using these bands are typically airborne radio-locations systems.

9. Implications of Use

Short-Range Systems and Security

When homeowners or small business owners use common 900 MHz or 2.4 GHz cordless phones, wireless video transmitters, or wireless intercoms to talk to friends or business partners, they may not realize they're broadcasting to everyone within half a city block or so. In many cases, all that is needed to listen in on other people's conversations is a radio scanner, a device that can scan and tune a range of frequencies.

Similarly, if a homeowner buys a wireless video transmitter so the kids can watch cable TV on a computer monitor in the playroom while the adults watch the same show on the television set in the living room, the person next door, who might not have a cable hookup, can get 'cable for free' by watching the same broadcast through a wireless receiver tuned to the same frequency. While this probably isn't a common scenario, given the different viewing tastes and schedules of different families (and hopefully some commitment to honesty), it does illustrate how easy it is to spy on people's conversations or video broadcasts without their knowledge.

Radio frequencies are not like car keys. While the auto industry admits there is some duplication in car keys, the chances of finding a car that will open with your key in the same neighborhood or a nearby parking lot are very low. In contrast, the chance of many people in the same area using some of the same radio frequencies is very high. Anyone who has used a wireless handheld radio has probably talked or listened to dozens of strangers in the vicinity who are using the same frequencies.

To give another example of the ease with which people can 'tune in' on each other, suppose you have a wireless security system installed in your home or small office that has a range of up to about 300 or 1000 feet using a common frequency. Let's say it includes video cameras in the nursery, bedroom, hallway, office, or garage. When the system is activated you are broadcasting images of you in your private setting to anyone with a receiver that can be tuned to the same frequencies in the near vicinity. This might be a neighbor, business, or scanner in a mobile van. The transmission doesn't stop at the walls of the building.

Very few consumer-level wireless security systems are encrypted or otherwise protected from offsite receivers. The only 'protection' you have is selecting a 'house code' from one of a few choices, usually between 4 and 15 frequencies. All the house code is, is the choice of the specific frequency on which the transmission is occurring. In other words, it isn't very secure. What you see or hear on your own monitors, anyone within the range of the transmitters can also see or hear, if they are sufficiently interested. If they know what brand of security system you are using (many owners have stickers or signs giving away the brand), it makes it even easier to intercept the signals because they can narrow down the range of frequencies to check.

An unscrupulous person with some technical background could not only look in on what you are doing by scanning a wireless video or audio system, but might even be able to determine the layout of the house, the location of the cameras, your schedule, and the approximate value of your possessions. If the scanner is a thief rather than a voyeur, she or he may have enough information to enter your premises and still avoid the cameras. It isn't likely to happen, since it's far easier to break into a building without a security system, but technically it can be done right now with off-the-shelf products.

Radio Tracking Beacons

Radio tracking beacons are now so small that they can be used to track cargo, retail goods, and parolees. They can even be attached to tiny birds and bees to chart their movements. As the technology improves, it has become possible to transmit more than a simple radio beacon. Inventory systems are being designed which will be capable of transmitting a variety of information about the product with a 'smart tag.' Thus, it is only a matter of time before the checkpoints in retail stores can receive information about the products you are carrying in your shopping bag as you walk out of a store. This would include not just the purchases you made in that store, but purchases you made elsewhere, since most retail systems are standardized. Removing the tags is not always an option, since some of the products are being installed with 'source tags' which are imbedded in the wrapping or in the product itself and cannot be easily removed.

What can be done to safeguard privacy? Besides the option of not shopping in stores with smart tags, the tags could be designed so that there is a checkpoint code associated with the tags, so that each retailer is scanning at a different code and thus could not read the data from other retailers. This could be set up on a simple key encryption system in much the same way that computer data is currently protected. Retailers would be motivated to encrypt their tags so that they wouldn't be giving away valuable marketing information to other retailers.

Another consequence of the availability of tiny radio tracking beacons is the possibility that some people will implant them in children. One might expect parents to be strongly opposed to such an idea, but unfortunately, catastrophic events can cause people to take steps that they might otherwise oppose. If a series of child murders or kidnappings were to occur in a region, some parents would support the use of radio beacons in their efforts to try to protect their children. However, as the children grew older and were more able to fend for themselves, the motivation for the beacons may change on the part of the parents. Just as parents have been known to read their children's diaries to see what they are doing as teenagers (a questionable substitute for good communication and spending time with a child), there will be some parents willing to insist that their teenagers wear radio beacons. Children who resist this parental requirement might select some unpleasant options, including trying to remove the (implanted) beacon themselves.

If this scenario seems unlikely, consider that fact that there are already companies that market wearable beacons intended to help manage older people who might wander and injure themselves and for prison parolees as an option to incarceration. There are also companies who are beginning to market Global Positioning System (GPS) trackers for keeping track of children. Given the fact that these electronic gadgets are currently attached to arm or leg bands or incorporated into jewelry, it's not a long jump to assume that some company will eventually market implants for their convenience (and because they can't be easily removed). Some advocates of tracking have even suggested injecting nanites (microscopic robotic technologies) into a person's bloodstream.

New technologies open up new possibilities, some of which may have long-term negative effects on our social structure and relationships. There will always be companies willing to market these technologies, whether or not they have any positive benefit, if there is a profit to be made. It is hoped that the long-term implications of the use of tiny radio technologies will be considered carefully so that we don't create a future in which we gradually but irrevocably lose the freedoms that we currently enjoy.

10. Resources

Inclusion of the following companies does not constitute nor imply an endorsement of their products and services and, conversely, does not imply their endorsement of the contents of this text.

10.a. Organizations

American Communication Association (ACA) - The ACA is a not-for-profit organization which promotes academic and professional research and applications in the principles and theories of human communication. www.americancomm.org/

American Relay Radio League (ARRL) - One of the oldest and most significant radio organizations in the world with tens of thousands of members worldwide. The AARL was founded in 1914 by Hiram Percy Maxim, assisted by Clarence D. Tuska and their colleagues. The ARRL cooperates with various radio groups and governing authorities, including the FCC and the ITT. Their monthly publication QST has been published for over 80 years. www.arrl.org/

Association for Unmanned Vehicle Systems International (AUVS) - AUVS promotes the advancement of unmanned vehicles systems and publishes Unmanned Systems magazine. It hosts regional seminars and trade shows along with an aerial robotics competition. www.erols.com/auvsicc/

Federal Communications Commission (FCC) - The FCC has an enormous amount of information related to radio broadcasting and consumer devices and is worth visiting. There are historical sections, various bureaus, depending on your needs and interests, current news releases, information on various cable and wireless services, and a variety of FAQs. www.fcc.gov/

IEEE (formerly the American Institute of Electrical Engineers) - The AIEE was originally founded in 1884 to represent the EE profession and to develop standards for the industry. Bell and Edison were among the first vice-presidents. AIEE merged with the Institute of Radio Engineers (IRE) in 1963 to form the IEEE one of the foremost engineering organizations in the world. The IEEE hosts many professional gatherings and publishes a wide variety of documents and journals. www.ieee.org/

International Radio and Television Society Foundation, Inc. (IRTS) - Based in New York, the society provides educational programs, awards, and information on broadcasting-related events. www.irts.org/

National Association of Broadcasters (NAB) - The NAB has represented the technological and governmental interests of radio and television industries for over 75 years in order to promote the broadcasting industries. www.nab.org/

Robotic Industries Association (RIA) - An industry association which recommends robotics resources, books, videos, seminars, and other information of interest in the practical and industrial applications of robotics. www.robotics.org/

Society of Broadcast Engineers (SBE) - A professional organization providing support for education, certification, and seminars in the various broadcasting fields. Based in Indianapolis, Indiana. www.sbe.org/

Note that a number of universities and technical institutes have robotics labs, including MIT, Carnegie Mellon, Berkeley, Georgia Tech, Indiana University, and the University of Washington.

10.b Print Resources

The author has endeavored to select these carefully, reading and reviewing many of them, and consulting reviews by reputable professionals for many of them, as well. In a few cases, it was necessary to rely on publishers' descriptions on books that were very recent, or difficult to acquire. It is hoped that the annotations will assist the reader in selecting additional reading.

These annotated listings include both current and out-of-print books. Those which are not currently in print are sometimes available in local libraries and second-hand book stores, or through interlibrary loan systems.

Curtis, Anthony R., *Monitoring NASA Communications*, Lake Geneva, Wi.: Tiare Publications, 1992.

Directory of North American Military Aviation Communications, Hunterdon Aero, 1990.

Eisenson, Henry L., *Scanners & Secret Frequencies*, San Diego: Index Publishing Group, 1997, 318 pages. A well laid-out, accessible introduction to scanning and radio hobby activities using scanners. The author is an ex-Marine and electronics technical specialist.

Helms, Harry L., *How to Tune the Secret Shortwave Spectrum*, Blue Ridge Summit, Pa.: TAB Books, Inc., 1981. Out of print.

Kane, Joseph Nathan, *Famous First Facts*, New York: H. W. Wilson, 1981, 1350 pages.

Mann, Steve; Niedzviecki, Hal, *Cyborg: Digital Destiny and Human Possibility in the Age of the Wearable Computer*, Doubleday Canada, 2001. This describes how wearable computers can be used to monitor and filter our environment and to integrate the information with digital databases, thus becoming, in essence, an extension of our sensory system and turning us into surveillants in a broad sense.

Parlin, Charles Coolidge, *The Merchandising of Radio*, Philadelphia: Curtis, 1925. Available through the Library of Congress.

Yoder, Andrew R., *Pirate Radio Stations: Tuning in to Underground Broadcasts*, Summit, Pa.: TAB Books, 1990. Out of print.

Articles

Advanced Electronic Monitoring for Tracking Persons on Probation or Parole, NCJ 162420, *National Institute of Justice Technical Report,* 1996. Discusses electronic monitoring of pretrial-released offenders and of those on alternative imprisonment regimes.

Baldwin, D. P.; Zamzow, D. S., Emissions Monitoring Using an AOTF-FFP Spectrometer, *Talanta.* 1997, V.45, pp. 229-235.

Baldwin, David, Development of a Multielement Metal Continuous Emissions Monitor, *Characterization, Monitoring, & Sensor Technologies Reports,* September 1998.

Ballard, Nigel, Radio Surveillance (Bugging) in the U.K., *The Hacker Chronicles,* September 1998.

Diehl, Christopher P.; Saptharishi, Mahesh; Hampshire II, John B.; Khosla, Pradeep K., Collaborative Surveillance Using Both Fixed and Mobile Unattended Ground Sensor Platforms, *Proceedings of AeroSense '99*, SPIE, 1999.

Dugan, James, Using Nano-Technology to Rack Inmates and Parolees, Tunxis Community College, Program Class IV, 1998. The author describes the insertion of nanoprobes into inmates' and parolees' bloodstreams to provide a means to constantly track their whereabouts.

Trebi-Ollennu, A.; Dolan, John M., An Autonomous Ground Vehicle for Distributed Surveillance: CyberScout Internal Report, Institute for Complex Engineered Systems, Carnegie Mellon University. April 1999.

Wood, Christina, Privacy and the Wearable Computer, PCMag.com, 18 Dec. 2001.

Journals

Amateur Radio Newsline, a written transcript prepared from the Newsline Radio scripts, an audio news service distributed by telephone, provided by Newsline Internet Services. www.arnewsline.org/

The IEEE Radio Communications Committee (RCC) sponsors and promotes technical papers and tutorials on the engineering aspects of communications systems and equipment. www.comsoc.org/

QEX, An ARRL-sponsored forum for communications experimenters.

QST, official publication of the Amateur Relay Radio League (ARRL). www.arrl.org/

10.c. Conferences and Workshops

Many of these conferences are annual events that are held at approximately the same time each year, so even if the conference listings are outdated, they can still help you determine the frequency and sometimes the time of year of upcoming events. It is very common for international conferences to be held in a different city each year, so contact the organizers for current locations.

Many of these organizations describe the upcoming conferences on the Web and may also archive conference proceedings for purchase or free download.

The following conferences are organized according to the calendar month in which they are usually held.

INTIX EUROPE, sponsored by the International Ticketing Association. This is not specifically aimed at radio technologies, but a number of ticketing tracking systems using radio technologies are now displayed by exhibitors at these types of conferences.

International Robots & Vision Show, a biennial conference with practical solutions to robotics applications.

Unmanned Systems, robotics and radio-controlled vehicles show and competitions.

Epigenetic Robots, an annual workshop for discussing original research on robotics that brings together the fields of neuroscience, artificial intelligence, and developmental sciences.

Forum on Wildlife Telemetry. A report on this conference is available on the U.S. Geological Services site. www.npwrc.usgs.gov/announce/press/wildtele.htm

Intelligent Vehicles Symposium and Autonomous Vehicle Cooperation and Coordination, international symposium sponsored by IEEE

Ionospheric Radio Systems & Techniques, annual international conference, sponsored by the Institution of Engineering and Technology.

NAB Radio Show, sponsored by the National Association of Broadcasters, this large trade show includes a technical program for broadcast engineers.

PREC, the annual Public Radio Engineering Conference includes presentations on radio technologies.

10.d. Online Sites

Frequency-Hopping and Spread-Spectrum Information. This surveillance supply site provides an article on radio technologies by A. Gil, A. Freeman, and M. Levin that explains frequency-hopping and there is also an explanation of digital spread-spectrum. www.midniteyes.com/freqhopping.html

NARTE. This is the Website of the National Association of Radio and Telecommunications Engineers includes information on certification and FCC testing of products, industry news, employment information, and radio engineering events calendar. www.narte.org/

Porthcurno Museum of Submarine Telegraphy. Virtual Tour. This site shows the floorplan, exhibits, and history of exhibits related to submarine telegraphy and cable communications at Porthcurno Valley in the U.K. Many interesting stories and glimpses of the physical museum are available. www.porthcurno.org.uk

QST. The official journal of the American Radio Relay League (ARRL), founded in 1914. The ARRL site includes editorial articles, downloadable propagation charts, and many useful links to information about amateur radio events, licensing, and other resources. www.arrl.org

Note that there are also many radio technology discussion groups on the Internet on USENET, including *alt.ham-radio*, *alt.radio*, and *rec.radio*.

10.e. Media Resources

Marconi: Whispers in the Air, a Biography Channel show on the precocious inventor, Guglielmo Marconi, who developed many of the pioneering telegraph and radio technologies that emerged a century ago. VHS, 50 minutes. May not be shipped outside the U.S. and Canada.

Radio: Out of Thin Air, a History Channel show from the Modern Marvels series that documents how the invention of radio changed society. Casey Kasem and Larry King explore the qualities of the medium and its development. VHS, 50 minutes. May not be shipped outside the U.S. and Canada.

Sneakers, is a Universal City Studios feature film that stars Robert Redford, Ben Kingsley, and Dan Akroyd. It is an action thriller that features a wide variety of surveillance strategies and devices. 1992, just over 2 hours long.

11. Glossary

Titles, product names, organizations, and specific military designations are capitalized; common generic and colloquial terms and phrases are not.

ABS	automatic broadcast scheduler
ACC	access control center
ACI	adjacent channel interference
ACS	alternate control station
ACTS	Advanced Communications Technology Satellite
ADPCM	adaptive differential pulse code modulation
AF	audio frequency
AFSAT	Air Force Satellite System
AGC	automatic gain control
AJ	anti-jam
ALE	automatic link establishment
AM	amplitude modulation/modulator
AMP	amplifier, automated message processing
ASC	automatic switching center
ASK	amplitude shift keying
AT&T	American Telephone & Telegraph
BCST	broadcast
CDMA	code-division multiple-access
C3	Command, Control Communications
CCC	clear channel capacity
CCS	communications control system
CDM	code-division multiplexing
CDMA	code-division multiple-access
cellular	low-power radio coverage with transceiver 'cell' units
COMSAT	communications satellite
DDS	digital data service
DII	Defense Information Infrastructure
DMS	Defense Messaging System
DPCM	differential pulse code modulation
DPSK	differential phase shift keying
EAM	Emergency Action Message
EAS	Emergency Alert System (replace Emergency Broadcast System)
EM	electromagnetic
EMI	electromagnetic interference
FCC	Federal Communications System, chief radio regulating body
FIPS	federal information processing standard
FM	frequency modulation/modulator
FSK	frequency shift keying
GEOS	geostationary satellite
GLORI	Global Radio Interface
GPS	Global Positioning System
heterodyne	to produce a beat between two frequencies. A steady introduced current can be used to selectively create and control an electrical beat which can then be amplified. A

	heterodyne repeater can be used to create an intermediate frequency which can be amplified, modulated, and retransmitted.
I&A	identification and authentication
IEEE	Institute for Electrical and Electronic Engineers
IETF	Internet Engineering Task Force
JRSC	jam-resistant secure communications
JTRS	Joint Tactical Radio System, a programmable military radio system
LEIS	Law Enforcement Information System
MBA	multi-beam antenna
MFSK	multiple frequency shift keying
MILSATCOM	military satellite communications
MILSTAR	Military Strategic and Relay (satellite)
NAVCAMS	U.S. Naval Communications Area Master Station
NAVTEX	Navigational Text, international marine information broadcast at 518 Hz
NBAM	narrowband amplitude modulation
NBFM	narrowband frequency modulation
NCA	National Command Authority
PCM	pulse code modulation
PM	phase modulation
POTS	plain old telephone/telecommunications service
PR	packet radio
PRT	packet radio terminal
RF	radio frequency
Rx	receiver
S/N	signal-to-noise ratio
SAMPS	semi-automated message-processing system
SARSAT	Search and Rescue Satellite
SATCOM	satellite communication
scanner	a system for detecting transmissions across a range of frequencies in order to locate transmissions of interest
SCF	satellite control facility
SDLS	satellite data-link system
SDMA	space division multiple access
SDN	secure data network
SDS	satellite data system
SRF	superconducting radio frequency
SS	spread spectrum, solid state
SST	spread-spectrum technology
SVN	secure voice network
TDM	time-division multiplexing
TDMA	time-division multiple-access
TH	time hopping
TRANSEC	transmission security
TS	top secret
Tx	transmitter
UNI	user-to-network interface
VOBRA	Voice Broadcast Automation
VSAT	very small aperture terminal

Electromagnetic Surveillance

6

Radar

1. Introduction

Radar is an acronym variously described as *ra*dio *d*irection *a*nd *r*anging and *ra*dio *d*etection *a*nd *r*anging. It is a remote-sensing technology for detecting and interpreting radio signals, usually in the microwave frequencies. Radar systems can be passive or active. In *active* radar, radio signals are emitted toward an object or structure with the intent of intercepting and interpreting the reflected signals (the incident electromagnetic energy). The examples that follow will help make this more clear.

Imagine it's foggy and you can't move forward, but you have a bag of tennis balls and you want know if there's anything ahead of you. You can try listening to see if you can hear any echoes (passive sensing) or you can throw a tennis a ball and note whether it bounces off a nearby object and returns. Now imagine a whole bag of tennis balls thrown at a constant rate of speed. With a few calculations based on when the balls return, you could determine the distance to an object. By throwing a large number of balls in an ever-widening arc, you could even get some idea of the size and shape of the object. This is the general concept behind radio detection and radio ranging.

Now imagine you're an animal with sonar generators and sensors, like a bat or a dolphin, and you can send out ultrasonic vocalizations (clicks or chirps) that bounce off nearby rocks

U.S. Air Force airborne surveillance, command, control, and communications staff monitoring the radar scopes on an E-3 Sentry aircraft. [U.S. DoD news photo 1992, released.]

353

or trees. As the chirps echo and bounce, you build up a "picture" in your mind of the surrounding terrain—even if it's dark. This is the general idea behind radar imaging in biological systems—a concept that has been adapted to electronic systems. The returning radio signals can be apprehended and interpreted into X-ray-like grayscale images.

Radar is similar to animal sonar in many ways, but differs in that it uses radio waves–a type of electromagnetic radiation–rather than sound. Radio energy travels at the speed of light, much faster than sound, and radio 'waves' and other electromagnetic energies have particle-like qualities that enable them to travel in a vacuum.

The radar signals used in most surveillance technologies are electronically synthesized, and are invisible to humans, but can be sent and sensed with specialized devices, and translated into visual or auditory information that can be more easily understood. Radar is a very useful technology for surveillance applications.

Radar Basics

Radio waves are considered to be at the 'lower' end of the electromagnetic energy spectrum because the wavelengths are longer. For the purposes of illustration, let's accept that electromagnetic waves travel 300,000 kilometers (km) per second in a vacuum. They are known to travel more slowly when impeded by water or thick particulate matter, but the differences are slight, or can be calculated in order to compensate for them. By relating time and the speed of the wave, the distance can be calculated.

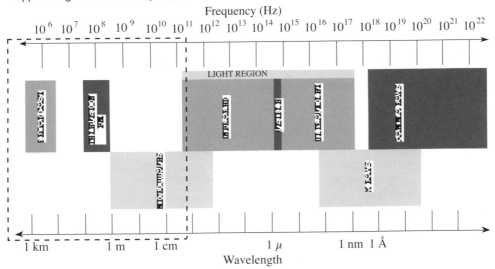

When symbolically represented as part of the electromagnetic spectrum, radio waves are illustrated as the longer waves next to the optical region. The wavelengths used for most broadcast applications range in length from around one meter to one kilometer. Those used for radar vary, but are typically in the microwave frequencies, closer to the infrared portion of the optical spectrum, and measure up to about one meter in length. [Classic Concepts ©1999 diagram, used with permission.]

When radar signals are emitted from a radio wave transmitter, they will continue away from the source indefinitely (gradually diminishing through *attenuation*), unless they encounter obstructions. An obstruction reflects the signals in various directions—usually returning some to the radar receiver. The returning "echoes" can yield information about the deflecting object or structure, usually termed the *target*. If the obstruction is very smooth, in relation to the length of the waves, the scattering of radio waves will be minimal (assuming it is not a highly absorbent material). If the obstruction is very rough or convoluted (as in rough terrain or radar-defeating chaff), the signals will be scattered in many directions and only a small portion of the signal may reflect back to the receiving antenna. If the object is very small and longer wavelengths are used, the signals may pass right by, making an object 'invisible' to the radar.

The reflected signals are collectively called the radar *echo*. When intercepted and correctly interpreted, different aspects of the echo, such as polarization, spectral reflectivity, and time of arrival, can be visually interpreted and displayed, and mathematically analyzed to yield information about the size, shape, location, and velocity of the radar target. If the radar is mounted on a moving object, like an aircraft, the angle of the radar signal, relative to the trajectory of the aircraft and the velocity of the aircraft, is considered in the calculations.

Sending out a radar signal does not guarantee its return, even if it hits the desired target—radar signals are affected by terrain, some types of weather (depending upon the frequencies used), radar-absorbing or -jamming systems, and other sources of *attenuation* (gradual diminution of the signal). Factors that interfere with radar signals are discussed further in Section 7 (Problems and Limitations).

Visualizing Radar Signals

We cannot see radio waves, but radar echoes can be intercepted, interpreted, and displayed in a number of ways. When a radar receiver intercepts incoming radio waves, they can be converted to electrical impulses which can then power a number of other electronic devices including radar displays. Specialized cathode-ray tubes (CRTs) intended for displaying radar 'blips' or *pips,* and computer monitors, are commonly used to represent ground-based radar target data. Digital or film images taken from aircraft or satellites using radar imaging sensors are similar to traditional grayscale photographs and are discussed further in Chapter 9. This chapter focuses mainly on ground-based radar data that are displayed symbolically, rather than photographically, on traditional cathode-ray tubes or digital viewing systems.

Radar systems range from small, portable devices to large Earth or marine stations. They can be deployed on the ground, or from high-altitude planes or orbiting satellites.

Radar was invented in the early part of the 20th century for marine navigation and adapted for air navigation by 1936, at about the same time that commercial availability of cathode-ray tubes was increasing. Because the concepts of radar ranging and sound ranging (sonar) are similar, radar and sonar technologies share some common history and terminology. Like sonar, radar came into regular military use during World War II and was further enhanced for various applications when the development of the transistor in 1947 introduced solid-state technology. Radar is now an essential tool of marine and air navigation and is widely used in aerial and planetary imaging.

Surveillance Applications

Radar is extensively used in surveillance activities, including military targeting, tracking, and defense, and civilian monitoring of moving vessels, hazards, threats, environmental changes, and weather systems. Radar is popular because of its versatility. Certain short-range, low-power systems for specialized purposes are priced in the consumer range. They can be used in light or dark and in many kinds of weather (some frequencies work better in wet weather than others). More expensive medium- and long-range radar tracking stations are stationed throughout the world and are used for navigation, international surveillance, and defense.

Economics of Radar Systems

Compared to other surveillance technologies such as video surveillance, genetic surveillance, or audio surveillance, high-end radar surveillance can be expensive. Video surveillance systems can now be installed for a few thousand dollars or a few tens of thousands of dollars. DNA profiles can cost as little as $30 for a basic dog-breeder's canine DNA profile and a commercial human DNA workup is now less than two thousand dollars. The cost of tapping a phone for employer monitoring of telemarketing calls ranges between a few hundred and a few thousand dollars.

High-end radar systems, in contrast, often cost millions of dollars, depending on their complexity, range, and the extent of computer processing incorporated into the systems. They are mainly used in law enforcement, weather forecasting, military surveillance, and astronomical research. Because radar systems tend to be commissioned by large corporations, universities, and government agencies, this chapter has a stronger emphasis on companies and contractors than in other chapters. If you desire more detailed information on radar you can contact the organizations listed in the *Resources* section at the end of this chapter, most of which have Web sites on the Internet.

2. Kinds and Variations

There are many kinds of radar systems but they all work on similar general principles. The cost associated with radar varies according to the range and variety of frequencies that can be transmitted, the strength of the pulse, and the complexity of the receiving system. The most expensive systems are those that have high-power, long-range transmitters and computerized artificial intelligence-equipped (expert system) receivers.

Radar signals travel in two general directions (assuming that a target is encountered)—from the sender toward the target and then away from the target (ideally, back to the sender, depending upon the angle of reflection). Radar systems can be *send-and-receive*, *receive-only*, or *send-only*. Send-and-receive systems and receive-only systems are the ones most often used for surveillance activities.

- *Send-and-receive* **systems** represent the great majority of surveillance-related radar. These are *active* systems, that aim a signal at a target and interpret the returning signal in order to derive information about the target. Send-and-receive systems are sometimes used in receive-only or send-only modes.

- *Receive-only* **systems** are usually defensive or covert systems in which the stealth vessel or target has a receiver equipped to detect an incoming radar signal, but no transmitter to betray its presence with radar signals that can, in turn, be detected by others. Car-mounted radar detectors are examples of small-scale receive-only systems. Most of them beep and/or blink when a radar signal is detected. Receive-only systems, in general, are the least expensive kinds of radar since they do not require the specialized electronics or power needed to send out radar pulses. However, more sophisticated receive-only systems may be used on stealth vessels that have radar-defeating features—such as chaff or radar-resistant paint—to absorb, avoid, or scramble the incoming signal by deflecting it in many directions. Even though the signal is detected, the reflected signal gives an obscured 'picture' of the vessel that was encountered by the radar probe. Stealth vessels can be designed to have 'small' radar signatures, and decoys are often designed with reflectors to look bigger. Receive-only systems are *passive* systems.

- ***Send-only* systems** are usually *beacons*, to provide warnings or location information, or *decoys* for use in situations of armed conflict. As a decoy, a radar signal may be transmitted to simulate a significant vessel or force stationed at the source of the signal, when in fact it may just be a portable transmitter or an unstaffed, automated system. The decoy is intended to confuse, to draw fire, or to draw attention away from other forces or installations. Send-only systems are *active* systems, because they generate radio energy, but are without receiving components.

Most of the systems discussed in this chapter are send-and-receive, and may be designed with single antennas or with multiple antennas. When the same antenna is used for both transmitting and receiving, it is called *monostatic radar*. Monostatic systems are very common. If there are two antennas, it is called *bistatic radar*. If there are more than two antennas, it is usually called *polystatic radar* or a *radar array*. The antennas must be spaced to minimize interference with one another.

Radar Sending

Radar signals can be *pulsed* or *continuous*, with the vast majority being pulsed. A pulse is a discreet burst of radio waves. The pulses are repeated at carefully timed intervals. The *pulse-repetition rate* is a count of the number of successive pulses per unit of time. Since radar waves travel at a more-or-less constant speed (most synthetic radar signals are sent through the Earth's atmosphere), it takes more time for radar signals to travel longer distances. Thus, longer radar ranges typically require slower pulse-repetition rates.

Radar Receiving

Radar sensors are categorized in a number of ways, and there are many hybrids and variations. The basic varieties include altimeters, scatterometers, and *synthetic-aperture radars* (SARs). Radar altimeters are commonly used in aircraft. Additional information about SAR and airborne radar systems is provided in the Aerial Surveillance chapter. Radar data are often used in conjunction with data from other technologies, including cameras, sonar, and seismic detectors. Since radio energy doesn't travel as effectively as sound through water, it's not commonly used for below-water sensing.

Modulation

Frequency modulation is a means of sending radio signals by altering their frequency. It is based on the same general principles as public FM broadcasting—the main difference is that broadcast frequencies use longer wavelengths than most radar applications. In radar, if frequency modulation is used in conjunction with continuous signals, then more information, including the distance to the target, can be calculated.

Variations

There are many different varieties of radar systems. Here are some common schemes and concepts of interest:

automatic detection and tracking radar (ADT) - This is a form of track-while-scanning system in which each rotation of the antenna yields data on the targets within its range. The visual display shows 'streaks' to represent the paths of the objects (assuming that they are moving) rather than pips. The movement and direction of vehicles or marine vessels can be monitored on this type of system.

phased-array radar (PA) - Not all antennas are designed to rotate while transmitting radar beams. For example, electronic phased-array radars can scan the radar beam back and forth within a certain *swath* to provide rapid updates of events within the sweep of the beam.

synthetic-aperture radar (SAR) - SAR is a more recently developed system, now commonly used in aerial and satellite sensing, in which one dimension represents the *range* (the *along-track* distance from the radar transmitter to the target) and the other represents the *azimuth* (the *cross-track* perpendicular to the range). The distance traveled by the craft using the radar becomes the 'synthetic' aperture in place of a very long physical antenna.

continuous-wave radar (CW) - the majority of radar systems are pulsed, but some specialized radars, such as weapons or scientific research systems, use continuous-wave technologies. CW radar transmits and receives at the same time, taking advantage of the Doppler effect, in which frequencies shift and thus are different for outgoing and incoming waves. This contrasts with pulsed radar which receives signal between transmission pulses to prevent interference.

Doppler radar - The Doppler principle is widely used in designing and evaluating radar signals. A *Doppler shift* occurs when an object is in motion relative to a reference point such as a radar receiver. If you are standing near a moving train that's blowing a whistle at an unchanging frequency, the pitch that you hear will change as the train approaches and passes you. This is due to compression and expansion of the sound waves relative to your position. Similarly, a tracked object which is moving toward or away from a radar receiver will cause a shift in the transmitted frequency that subsequently reaches the receiver. By measuring the frequency, information about the velocity of the object can be calculated, thus Doppler radar typically scans a target in terms of speed rather than range. Doppler concepts are widely used in weather-tracking and traffic radar systems. Sometimes two Doppler radar systems are used in tandem for calculating additional spatial information.

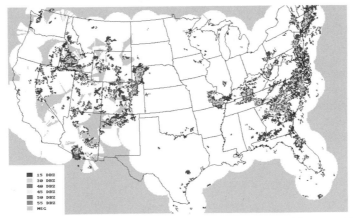

Radar Imaging. By sensing radar signals over a wide area, it is possible to combine the data to create a weather map. Areas of precipitation are indicated with color coding to make it easier to interpret the image. Higher levels of precipitation are expressed with 'hotter' colors and state lines have been added for reference. [U.S. National Weather Service summer 2000 weather photo, released.]

Radar is used throughout the transportation industry for homing, tracking, altitude sensing, object detection and identification, and land-, sea-, and air-traffic monitoring and enforcement.

Radar antennas can be readily seen on the decks and masts of watercraft, especially ferries and coast guard cutters. Radar systems on aircraft are usually enclosed and may be located 1) in the nose (especially receiving systems), 2) in a radome (a large, streamlined, rotating antenna cover), or 3) under the bely of the plane (side-looking radar).

Left: A weather officer with the All Weather Service (AWS) demonstrates weather surveillance systems, in 1980. She is pointing out views of severe storms that were photographed from ground level and imaged with a Doppler radar system. Right: The P-3C Orion aircraft is equipped with several surveillance systems including an AN/APN-227 Doppler radar and sonar tracking system (shown here in a counter-narcotics flight, in 1995). [U.S. DoD 1980 news photos by William D. Boardman and Paul J. Spiotta, released.]

Left: Contractors to the U.S. Air Force lift a Next Generation radar transmitter/receiver trailer for installation at the NEXRAD site in the Azores. Right: Domes protect radar antennas from weather damage in a U.S. Navy installation in Antarctica. [U.S. DoD 1996 news photo by Lemuel Casillas, 1995 photo by Edward G. Bushey, Jr., released.]

Identification Systems

The terminology and procedures for training radar engineers and operators were developed in the late 1930s, and evolved considerably in the 1940s when commercialization of radar and military use of radar increased. One important development was the creation of *identification friend or foe* (IFF) and later *identification friend, foe, or neutral* (IFFN) systems. These made it possible to distinguish various radar targets.

An IFF system is like a password. A signal 'query' is transmitted to the radar target which, in turn, transmits back a coded pulse. The code can be prearranged or dynamically assigned during operation. Commercial airlines and military aircraft are equipped to transmit various basic identifiers plus, as systems became more sophisticated, additional information such as origin, airline, flight number, mission, etc. On symbolic radar displays, this information can be assigned to an individual target so it can be identified by name, shape, or color. Thus, an aircraft carrier might be assigned a blue rectangle, while a foreign submarine might be assigned a red circle, thus aiding the tasks of tracking and interpretation. Symbolic systems can be particularly important in the identification and tracking of fast-flying surveillance craft that have small radar signatures intentionally designed to avoid identification on radar.

Visual Displays

There are many kinds of radar displays and the visual interpretation of the images requires various degrees of training and skill. Radar is used by air traffic controllers to guide take-offs and landings of commercial and military aircraft. It is used by aircraft and marine vessels to detect hazards or incoming projectiles. Clearly, since many lives may be at stake, good training and skill are necessary for personnel to evaluate three-dimensional air traffic or various navigational hazards based on their interpretation of pips or symbols on a two-dimensional screen. Radar display systems are described in more detail in Section 3 (Context).

Traditionally, radar technology has depended upon frequencies associated with radio waves, but the invention of lasers has provided other ways to apply basic ranging concepts associated with radar. An optical radar, called *lidar* (*li*ght ra*dar*) or *ladar* (laser radar), transmits very narrow, coherent light beams rather than radio waves. The returning data can be used to create high resolution images (see the Infrared Surveillance chapter for related information).

3. Context

A radar beam is essentially cone-shaped, like the light from a flashlight. It becomes broader as it moves away from the source and gradually 'fades' or *attenuates* as various particles and objects absorb or reflect the beam. Newer, higher-frequency narrow-beam radar systems have been developed that do not spread in the same way, which can provide more precise targeting. However, there is usually a trade-off that limits the range. The high-frequency beams don't usually travel as far when they have to pass through Earth's atmosphere, which contains particles and vapor. An unobstructed radar beam travels in a straight line (on a cosmic scale, gravitational forces can bend the path of electromagnetic waves, but for terrestrial purposes, assume the line of travel is straight).

A radar beam is *absorbed* by 'spongy' objects (dirt, leaves, fog), *refracted* (bent) by materials with different densities (e.g., as it passes from air into water), and is *reflected* by reasonably solid objects. Some of the reflected beam is *scattered* in various directions and some of it is reflected back to the source. It is the reflected signals that are called the returning *echoes*.

Very short electromagnetic wavelengths, which include ultraviolet, X-rays, and gamma rays, can be harmful to humans — shorter wavelengths are associated with higher energy levels than longer ones. Ultraviolet can cause sunburn and sometimes cancer. X-rays can cause substantial damage to human tissue and should be used judiciously. However, the relatively long wavelengths of radio waves, when used in normal ways in broadcasting and remote sensing, do not appear to harm human tissues.*

Radar signals are versatile because they can be used day or night in a variety of weather conditions and, depending on the transmitter and the frequencies used, can travel great distances.

3.a. Range

Radar range to a target is somewhat limited by *line of sight*. Certain reflective objects can interfere with both the outgoing and returning radar signals. When the signal deflects off of many objects other than the *target*, not only does the signal attenuate more quickly, but relevant targets become more difficult to distinguish from radar *clutter*.

Higher frequencies (shorter wavelengths) tend to have shorter ranges, as they can be scattered by atmospheric moisture and particles. Lower frequencies (longer wavelengths) tend to have longer ranges. However, the power of the transmitter is also a factor, so these are only general guidelines.

There are many varieties of radar systems, but a general rule within categories of radar systems is the longer the radar range, the more power the system must generate. Most radar systems are surface-based (land, air, and water), in the sense that they are used within the Earth's atmosphere; however, a number of radar systems are now installed in probes, orbiting satellites, and high-flying aircraft. These typically send and receive signals through the Earth's atmospheric envelope, though some also send signals out into space. Atmospheric particles reduce radar range through attenuation (scattering and absorption). Adjustments for rotation, angle, and time of day are made on space-based systems to maximize performance from limited power sources (satellites are dependent upon limited-power batteries and solar energy converters).

Interference from other sources of radio waves at or near the same frequencies as a specific radar may occur and may further limit range.

Very short-range radars, e.g., those used in traffic enforcement, may scan up to a few miles or a few hundred feet, depending upon terrain. With a clear line-of-sight, short-range radars may scan up to about 50 miles. Long-range radars may scan up to a few hundred miles, and very long-range radars, such as those used on satellite systems, may scan in excess of 500 miles. There are also very long-range radars used in radio astronomy that send signals out into space where there are vast areas without atmospheric particles to attenuate the signals—these may travel for millions of miles. One of the first times radar was used for exploring and mapping the heavens was when experimenters bounced a signal off the Moon, in the mid-1940s.

3.b. Radar Receivers

Our environment is full of radio signals of different frequencies that travel constantly through air and walls (and our bodies), outside of our conscious awareness. Consumer radio broadcasts that are sent through the air as inaudible waves, are converted to audio frequencies so we can enjoy them in our homes. Similarly, traditional TV aerials or satellite dishes capture invisible television broadcasts that are converted by a TV tuner and displayed on a monitor. Since humans cannot see or hear radar pulses, which are a type of radio wave, these signals must be converted to a form that enables us to interpret the information. The most common means of viewing radar echoes is on a radar scope, where a reflected signal is indicated as a bright area. Auditory signals are sometimes used for less sophisticated applications like vehicle radar detectors.

A radar *scope* provides an *image* of the radar echo at the moment the signal is apprehended by the receiver. Since most radars are pulsed (a signal is sent and the system waits for its return before sending the next one), there are 'blank' moments between pulses in which echoes will

---

*There are some concerns that microwave frequencies in radio-based communications devices (e.g., cell phones) could be harmful, but there isn't firm evidence to substantiate these concerns at this time. However, high-power microwave communications systems are dangerous, as are unshielded microwave ovens.

not be displayed on the scope. Since the mid-1930s, most radar systems have been designed to display on cathode-ray tubes (CRTs) in which the inside of the viewing surface (the inside front glass portion of the monitor) is coated with phosphors that are excited when the beam emitted from the back of the CRT hits the front. When they get excited, they glow, creating an image on the radar screen. On many systems, the brightness of the glow can be controlled. In pulsed radar, this glow gradually fades and is replaced by the image from the next received signal, providing the illusion of a continuous signal.

If the radar system uses a low pulse rate, in which there is a longer time delay between radar signals, a *long-persistence-phosphor* CRT can be used. On these displays, the phosphors glow longer to maintain the image longer. This helps provide a sense of the target image as long as the viewer remembers that it is a stop-action picture with a built-in delay. In other words, the lingering glow shows the scene as it 'was', not as it 'is', but that is usually more useful to the operator than a blank screen.

Vector and Raster Displays

There is more than one way to control the travel of a CRT beam, so display devices are further divided into *vectors*, in which the beam is aimed only at the part of the screen that is illuminated, and *raster* displays, in which the beam sweeps continuously across the full screen surface as on television sets or computer monitors, creating a *frame*.

Traditional 'pip' radars have *vector displays* similar to those used in traditional oscilloscopes. The beam becomes a visual reference—it traces a predetermined path that is altered in one or two directions if a radar signal is received. Thus, a repeating beam, traveling from left to right, may be altered in the vertical direction to indicate the presence of a signal representing a radar echo. Similarly, a circular beam following a spiral path from the center to the outside might be expressed as 'bright spots' to represent radar echoes, thus providing a two-dimensional representation of a three-dimensional space scanned by the radar. Vector displays have been in use for several decades.

1   **2**   **3**

Vector displays commonly trace a single beam at whatever portion of the screen needs to be lit at a particular time (1) or, on some radar scopes, trace a predetermined path and illuminate relevant targets (2) if they are present. Raster displays trace a frame that covers every part of the screen, usually from left to right and top to bottom. The frame refresh rate is usually about 60 times per second and is continuous, whether or not a target is detected (3). Raster displays are used on many computer-based radar systems with resolution typically expressed in pixels (e.g., 1024 x 768). [Classic Concepts illustration, ©2000, used with permission.]

On monochrome displays, the beam is turned on to illuminate the screen when 'cued' by a signal. On color displays, the signal may be displayed as another color or another intensity against a selected background. Computer-controlled raster displays provide great flexibility in the types of images that can be displayed, and symbols can be dynamically substituted by the system or the operator to represent objects or situations. Raster computer displays have been in common use for about 20 years (and have been used for TV displays for half a century).

Pip and *brightness* radar displays provide 'approximate' images due to delays and the limits of spatial resolution. The range of sizes of objects in the real world is much greater than the range of sizes of objects that can be represented on a small screen. Since nothing can be displayed smaller than the pip or pixel itself, or bigger than the total area of the screen, compensations in interpretation must be made. In other words, if the imaging resolution of the display is such that a blip symbolizes something the size of a horse, then smaller objects such as bicycles or go-carts would appear to be as big as a horse, or won't show up at all.

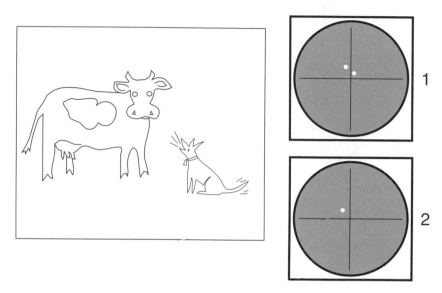

The spatial resolution of a TV-sized radar display cannot match the spatial resolution of acres of land, so operators must learn to adjust their thinking to interpret the images. Depending upon the distance and resolution of the system, the dog in the scene will either appear to be the same size as the cow (1) or will not be displayed at all (2) because it is too small to register. 'Tuning' the system to display smaller objects doesn't solve the problem because too many objects on the screen clutter the image and large objects would be so large they would fill up the screen or overpower the smaller ones. Thus, a radar system must be set up to balance the spatial relationships of the target and background, an example of the context-sensitive nature of radar system design, tuning, and interpretation. [Classic Concepts illustration, ©2000, used with permission.]

It should be noted that despite limitations of operator interpretation and resolution, radar detection systems can still be remarkably accurate. Radar can be used to track fast-moving aircraft and incoming projectiles using computer processing to correlate the data, and may even be set up to control the launch of a countermeasures device.

In portable systems, cathode-ray-tube (CRT) displays are impractical unless the viewing screen is very small. The electron beams in a CRT need to travel some distance in order to illuminate the full surface of a screen, so traditional CRT monitors are long and bulky. Newer display systems, especially those used with portable computers, utilize a number of technologies to reduce the bulk of the screen, including display systems such as *gas plasma*, *liquid crystal diode* (LCD), and *thin film transistor* (TFT). While initially not as crisp as CRT displays, these more recent display technologies are improving rapidly and are very practical in the field, as the monitors may be only about 1/2" thick, weighing less than two pounds for a 14" viewing area.

The most common categories of display systems are

- *A-scope* - displays a signal as a time-varied *blip*, suitable for monitoring the progress of a moving target.

- *B-scope* - displays range and bearing information in rectangular coordinates.

- *plan-position indicator* (PPI) - displays part or all of an arc, as the radar antenna rotates through some portion of an arc, up to 360°. The echoes appear as bright spots on a long-persistence monitor (a monitor in which the lighted areas retain their intensity before fading). The fade can be minimized with computer processing, but it is important for the operator to remember that the image is only current at the moment it displays, like a stop action photo that is retained until the next photo is snapped [Beam, 1989]. The revolution of the antenna may be as slow as six revolutions per minute. The refresh rate of the radar image is related to the speed of revolution of the scanning antenna and the refresh rate of the image display. There is a relationship between the pulse rate and the scanning speed of the radar. If the scanning rate is high and the pulse rate is low, fast-moving objects might evade detection. The radar range, in turn, will impact the pulse rate.

Photographs of different radar displays and operations consoles in use by air traffic controllers and military personnel are included later in this chapter to give the reader a sense of the variety of radar terminals that are in existence.

Radar Antennas

Since the directional character of radar beams is important, both for targeting and interpretation, most radar antennas are mounted on movable mechanisms. Radar antennas typically rotate around their vertical axes, and some can be angle-adjusted as well.

The SPS-49 radar antenna (left) and a radar dome (right) are just two of several systems used on board the aircraft carrier USS Abraham Lincoln (CVN-72). Military marine vessels are also routinely equipped with a variety of radar countermeasures devices that may include noise and jamming systems and chaff (a type of thread-like reflective material) to scatter incoming signals. [U.S. DoD 1990 news photo by Don S. Montgomery, released.]

Most ground-based aerial tracking radar systems use a *conical scan*. Because this type of system can be jammed, multi-input *monopulse tracking* radars were developed to receive data as a single pulse, which resists detection [Beam, 1989].

Military ground equipment with scanning radar antennas are incorporated into IFFN systems (Identification Friend Foe or Neutral). Identification systems are also used in civilian applications, but are less likely to incorporate key-coding to thwart imposters.

Left: A variety of types of radar antennas are visible on the Soviet-built USNS Hiddensee, including a High Pole-B EW antenna, a Square Head Identification Friend or Foe (IFF) receiver on the back, a Plank Shave missile-targeting antenna on the front, a Kivach 3 surface-search antenna on the bridge, and a radome with a Bass Tilt gunfire control antenna to control two types of gun systems. Right: The U.S. Navy cruiser USS Normandy (CG-60) is equipped with a cone-shaped SPQ-9A surface gun control radar on the fore mast, an SPG-62 radar illuminator to the left, an SPS-29(V)6 air-security radar on the main mast with a small dome-shaped SPS-64(V)9 navigational radar above and forward on the mast. On the top aft deck are two SPG-62 radar illuminators and on the right is a QE-82 antenna for WSC-3 UHF Satcomm. On the aft deck house are two panels of the SPY-1A fixed antenna radar array. [U.S. DoD 1993 and 1994 news photos by Don S. Montgomery, released.]

The U.S. Navy guided-missile destroyer USS Scott (DDG-995) is equipped (left) with an SPG-51 missile radar illuminator on top of the bridge, above which is a dome-shaped SPQ-9A surface gunfire-control radar, above which is an SPS-49(V)5 EW air search radar. Near the top of the mast is the SPG-55 surface-search antenna. To the aft end of the bridge is the SLQ-32(V)5 EW. The large screen (right) is the SPS-48E 3D search radar, below which is the SPG-60 gun/missile control. On the far right is a QE-82 antenna for WSC-3 UHF Satcomm. [U.S. DoD 1994 news photos by Don S. Montgomery, released.]

Radar Frequencies

Radar operates in a wide variety of frequency ranges, with the choice of frequencies related to the character of the application, the cost, and whether the sensing is of a covert nature. Civilian radar applications often operate at fixed-center frequencies, whereas covert applications may use frequency-hopping schemes to escape attention or interference.

As wavelengths become shorter, the problem of attenuation increases, so narrowband radar systems tend to be used for short-range applications, such as traffic radar detectors. Broadband detectors, that detect over a range of frequencies, usually have a sensitivity trade-off. There is also a trade-off between stationery radars (which are usually ground-based) and moving radars. Moving radars usually have about a 25% loss in range, depending on the system and the frequency used. However, if they gain a clearer line-of-sight between the transmitter and the target, as on an aircraft, the good line-of-sight more than compensates for the loss in range.

It took a while for electronics engineers to figure out how to exploit the higher frequencies for radar applications, just as it took time to figure out how to generate ultrasound radiation for sonar applications. Consequently, many early radar systems emitted high frequency (HF) and very-high frequency (VHF) energy. As electronics science improved, radar was applied to L-band frequencies and beyond. Each frequency range has advantages and disadvantages.

Note, the following frequency ranges are not 'fixed'. They are administrative designations that sometimes change (or are subdivided) as radio science advances and new applications emerge to take advantage of their various properties. The designations are relatively stable at the longer wavelengths, but are still being developed at the higher energy microwave frequencies and have changed somewhat over the last several years.

3 to 30 MHz, high frequency (HF or *shortwave*)

> Suitable for long-range applications; can be 'bounced' off the Earth's ionosphere to extend the range, allowing for *over-the-horizon* radar signals. Attenuation from particulate matter in the atmosphere is minimal. However, HF also requires long antennas and has to compete with other radio wave traffic. HF waves are more economical to generate than microwaves.

30 to 300 MHz, very-high frequency (VHF)

> The pros and cons of VHF for radar transmissions are similar to those for HF, although the antennas don't need to be quite as long. VHF waves are more economical to generate than microwaves. Some satellite imaging radars are in this region, including some environment-sensing SAR applications.

300 to 1,000 MHz, ultra-high frequency (UHF)

> Many surveillance activities, including air traffic control and military applications, are conducted in the UHF frequencies. UHF is suitable for long-range radar applications, has moderate attenuation, and is moderately economical.

400 MHz to 1 GHz (P-band)

> A long-distance experimental radar band that has been used for a variety of applications, including radar interferometers and synthetic-aperture radar. There are a variety of weather, satellite, military, and forest service radar applications that operate within the 1200 to 1700 MHz frequency range. P-band is also used for astronomical radar sensing. Interference from certain of the applications that have been approved to use in these frequencies (e.g., certain satellites with strong signals) makes it difficult for them to be used for anything else.

1 to 2 GHz (L-band)

> A long-distance frequency range used for experimental, satellite, weather, and military operations. The NASA/JPL synthetic-aperture imaging satellites do much of their imaging in L-band and C-band frequency ranges.

2 to 4 GHz (S-band)

A medium-high-distance frequency range used for air traffic control, weather monitoring, consumer short-range transmitters, and consumer cordless telephones. There are some three-dimensional radar imaging applications carried out at this frequency range and at higher frequency ranges. Some identification systems work at this range, as do some atmospheric boundary-layer radars.

4 to 8 GHz (C-band)

A medium-high-distance frequency range used in a variety of surveillance activities and military operations, including weapons control systems. This range has medium-long-range capabilities and is also reasonably good for precision applications. Some synthetic-aperture radars used for remote sensing operate in this region. A number of air-traffic control and air defense applications operate within about 3 to 6 GHz. Some altimeters operate in this region, as well.

8 to 12 GHz (X-band)

A medium-distance, higher-precision range used in professional applications, traffic radars, military applications, civilian marine applications, and weather forecasting. X-band radar is used to assess sea-surface characteristics, including temperature and air disturbances. Missile-targeting radars operate in this region. Most synthetic-aperture radars operate in this frequency range. Nose-mounted aerial radar systems often operate in the X-band and K-band regions. The surveillance of airport runways is carried out in this range (and in the K-band range). Some vehicle radar detectors work at about 10.5 GHz, as do some focused phased-array imaging radars.

above 12 GHz

The frequency designations above this region are currently somewhat fluid, and different maximum and minimum ranges have been published by various organizations, depending on their specific applications needs. In general, these higher frequencies are used for shorter-distance, high-precision detection systems, landing systems, and airborne radar. Some special-purpose weather radars operate in this region, e.g., air particle sensors. Some vehicle radar detectors operate at about 24 GHz, with some of the wide-band radar detectors at about 34 GHz. The designations found in this higher-frequency region include K-band, Ku-band, Ka-band, Q-band, V-band, and W-band (around 100 GHz) in that general order, almost all of which are used for satellite radar, although some frequencies at the lower end (e.g., around 20 GHz) have been set aside for mobile voice communications and guidance systems and there is an allocation within the Ka-band for non-geostationary fixed-orbit satellite services as well as subregions for various satellite and multipoint communications distribution services.

infrared (optical region)

Portions of the infrared region use specialized very-short-distance, high-precision, optical radar, mainly for distance-determination and weapons-control systems.

As wavelengths become progressively smaller, they become more susceptible to attenuation from particles and vapors. Some scientists have capitalized on this 'limitation' by using high-frequency radar to map the structure and shape of cloud formations.

4. Origins and Evolution

There is a widespread misconception, even among professionals, that radar was invented during World War II. This is probably because U.K. military strategists in World War II recognized the importance of radar and developed and used it extensively, influencing the outcome of the war. However, radar didn't originate in the U.K. or during wartime. It was invented in Germany more than thirty years earlier, at the turn of the century, for the purpose of preventing marine collisions. Radar evolved gradually and became an important marine and aerial navigation technology by the mid-1930s. With the outbreak of World War II, radar became one of the significant technologies used to defeat the German forces, which is ironic, considering the German military administration rejected the first radar inventions offered for their use by German scientists.

Basic Concepts

One of the most important scientific concepts that has been applied to modern radar was developed in 1842, by Johann Christian Doppler (1803-1853), who studied the way in which motion could compress sound waves and thus alter the frequency of sound relative to the perspective of the observer. This principle was later found to apply to other wavelike phenomena, like light. Since radio energy has wavelike properties, Doppler's observations were eventually applied to designing radar systems that utilized the Doppler effect to judge speed.

In essence, radar involves the sending and receiving of reflected electromagnetic waves. Thus, initial discoveries which led to the development of radar were made in the 19th century by inventors such as Heinrich R. Hertz (1857-1894), a German physicist employed at the University of Karlsruhe. Hertz demonstrated the reflection of electromagnetic waves by other electric inductors, in 1886.

Radar Ranging

In the late 1800s and early 1900s, searchlights were being used on marine craft to help prevent collisions with other vessels and to illuminate hazardous obstacles and floating debris. However, the system was severely limited by range and weather. Christian Hülsmeyer (1881-1957) sought a way to improve navigation by using radio waves, and on 30 April 1904, registered patent DRP #165546 for a *Telemobiloskop* (far-moving scope), a radio device to aid marine craft in preventing collisions. Hülsmeyer's use of radio waves as a remote-sensing technology was a significant advancement. His invention provided greater range and utility in bad weather than searchlight systems. The German engineer's device, the first to fully embody the basic principles of radar, was demonstrated on 18 May 1904.

Amateur radio broadcasting flourished between 1906 and 1920 and many hobbyists were eager to study the characteristics of radio transmissions and to experiment with practical applications. Their experiments included bouncing radio signals off various surfaces to further study phenomena such as reflection, refraction, and attenuation.

During the early century in Europe, Hans Dominik, a science fiction writer, and Richard Scherl, developed the *Strahlenzieler* (raypointer), which used radio wave echoes as a detection system. Scherl produced a working model and offered it to the Germans for use in World War I, in February 1916, but it was rejected by German administrators as 'unimportant' to the war effort. Soon after, in the United States, Nicola Tesla, an eccentric genius who did hundreds of experiments with electricity and radio technology, described radar concepts in the *Electrical Experimenter* (1917).

In the 1920s, military personnel began taking an interest in using radar technology for navigation and remote sensing. A. H. Taylor of the U.S. Naval Research Laboratory observed, in 1922, that a radio echo from a steamer could be used to locate a vessel. He may have independently come up with the idea, or he may have had conversations with colleagues resulting from the lectures given that year by Guglielmo Marconi, a radio pioneer, who described the use of radar for navigation.

Up to this time, radar was used to sense the presence of an object, but it was not yet used for deriving further information about the object. In 1926, American researchers Gregory Breit (1899-1981) and Merle Anthony Tuve (1901-1982) bounced a pulse-modulated radio signal off the conducting layer of Earth's high atmosphere (termed the ionosphere by R. Watson-Watt) in order to determine its distance. While the high frequencies commonly used for radar would not be practical for measuring the ionosphere (high-frequency waves pass right through it), this experiment nevertheless demonstrated concepts fundamental to radar, i.e., bouncing a signal off a wave-reflecting object and analyzing the resulting echo to calculate the distance.

In the 1920s and 1930s, radar technology was put to wider practical use. Radio signals were bounced off marine vessels to detect their presence. Ships show up well on radar, in contrast to their water environment, and they move slowly, in contrast to aircraft. However, as the technology was improved, aircraft were tracked, as well, and surveillance radars began to gradually evolve into other areas. In Italy, during this time, pioneer experiments in weapons-detection were carried out. While these early systems didn't yield sophisticated information, due to the lack of display devices and analytic methods and equipment, they did embody basic radar detection and ranging capabilities by sensing the presence and approximate direction of movement of a vessel or weapons charge.

Radar Display Devices

Karl Ferdinand Braun (1850-1918), a German physicist, developed a number of technologies that were later to become important in radio communications and radar systems. He created a crystal rectifier that was used in early crystal radio sets and, in 1897, developed the oscilloscope — a cathode-ray display system that was later adapted for imaging radar signals.

In the late 1920s, the research of Vladimir Kosma Zworykin (1889-1982) led to significant improvements in radar technology. Zworykin, a Russian-born American researcher, experimented with beams emitted in an electron tube and, in 1923, patented the *Iconoscope* television tube, a variation on the cathode-ray tube. Display tubes have been an indispensable component in radar scopes ever since the early 1930s, and modern versions are regularly used in vector displays and computerized raster display systems.

The three-element electron tube was a mainstay of electronics from the early 1900s until the transistor age emerged in the late 1940s. Based on the two-element Fleming tube, Lee de Forest added a third element, a controlling grid, to create the *Audion,* in 1906. Thus, electron tubes could now be used to build transmitters, amplifiers, receivers, and many other electronic components.

Practical Radar Systems

Robert Watson-Watt (1892-1973), a Scottish physicist, is credited as a radar technology pioneer for a patent for a device he called a *radiolocator* (1935), designed to detect aircraft. Early systems were known as radio direction finders (RDFs). Watson-Watt was an advocate for narrow-beam radar designs. Based on his research, by the following year, the British Royal Air Force (RAF) was operating a radar-based air warning network for aircraft detection.

By 1937, radio waves were being used for direction-finding, homing, and ranging for commercial aircraft and marine vessels, and were well-documented in engineering texts (e.g., Terman's *Radio Engineering*). Direction-finding was used to help aircraft and marine vessels stay to a course. Homing devices came to be used to guide various craft to specific locations, like guiding an aircraft to a landing base on the ground or on an aircraft carrier. Radio-ranging systems consisting of beacons designed to lay in a course in a predetermined direction, were implemented to mark primary air routes throughout the U.S.

A letter to the Chief Signal Officer in Washington, D.C. on 5 May 1937, requesting radio equipment "...to detect the presence of aircraft by reflected signals ..." was penned by Lt. Col. M. F. Davis, Air Corps, who was stationed in the Panama Canal Zone at the time. By summer, negotiations with RCA and Westinghouse for high-power VHF triode tubes were underway, with Westinghouse eventually winning a bid to supply radar systems to the U.S. military [Helgeson, www.bwcinet.com/acwrons/equip/SCR-270.html].

Experimental designs, improved electronics, and more comprehensive radar systems incorporating CRT technology were in development in a number of locations by the late 1930s. Leo Young developed a radar duplexer, with an integrated antenna for transmitting and receiving. A 200-MHz version was completed late in 1937 and tested on the USS Leary. An improved version was demonstrated on the battleship USS New York in 1939, heralding the age of marine radar for strategic warfare and defense. In 1942, the USS New York was used for a time as a radar training vessel.

As might be expected for machines developed prior to the invention of microelectronics, early radar systems were bulky and limited in frequency and range. Marcus Laurence Elwin Oliphant (1901–2000), a physicist in the labs of the University of Birmingham, secured a grant to build Europe's biggest cyclotron. It was in these labs that a significant new radar was invented, the first electron-tube *cavity magnetron*, built by British physicist Henry Boot (1917-1983) and biophysicist John T. Randall (1905-1984). The magnetron led to significant improvements in radar accuracy in the early 1940s, when high-power magnetrons, capable of generating microwaves suitable for radar transmissions, were developed.

Zworykin's cathode-ray tube display technology coupled with Boot and Randall's transmission systems provided key elements in the design and manufacture of improved, modern radar systems. By 1940, the British east and south coasts had been installed with high-tower transmitting aerials. Given the German aircraft superiority in World War II, these radar improvements and installations were important to the outcome of the Battle of Britain.

Military Development and Use of Radar

With the War raging in Europe and fears of it spreading, the early 1940s became a time of intense military development and deployment of radar in the United States.

In June 1940, the U.S. President established the National Defense Research Committee (NDRC) to support scientific research on military technology, which included contributions to the development of airborne radar systems. In July, the NDRC Radar Division held its first meeting, defining its mission and developing microwave technologies for radar-based aircraft early warning systems and interception. The exchange of information with the British, generally remembered as the Tizard mission, resulted in many new ideas and advancements. In August, information was exchanged on important developments including British detection of German aircraft, British ship and airborne radar systems, identification systems, and news of the cavity magnetron—an advanced means of generating radio waves in the microwave range. The Chief of Naval Operations subsequently requested samples of various British radar systems.

With a clear goal, and information on British systems, the Navy began developing radar identification systems, including airborne surface-detection radar, and ship-based detection radar.

In November 1940, the *Massachusetts Institute of Technology* (MIT) *Radiation Laboratory* was the site for the first general meeting of the Radar Division and became involved in many of the development projects that resulted from the meeting. The Chief of Naval operations authorized the use of the acronym "RADAR", and consolidated various radio ranging names into the phrase "Radio Detection and Ranging Equipment".

By March 1941, the Navy was reporting that they were able to track aircraft up to a distance of about 100 miles, and recommended the installment of aircraft identification devices. By May, Project Roger had been established at the *Naval Aircraft Factory* to support the *MIT Radiation Lab* and the *Naval Research Laboratory* radar projects. Project Roger aided in the installation and testing of the systems, which included aircraft radio control and search and blind bombing capabilities. In July, the first Identification Friend Foe (IFF) systems were installed in aircraft, along with some British ASV radar systems. Radar Plot systems for ship-borne air warning were approved for installation into aircraft carriers.

In August 1941, radar systems similar to those we currently use, were installed in U.S. aircraft. The AI-10, a microwave radar developed by the *MIT Radiation Lab*, was tested in an XJO-3 aircraft.* The tests led to the development of the ASG and the AN/APS-2. The success of radar projects prompted the *Bureau of Aeronautics* to request radar guidance equipment for their assault drones. They were interested in having automatic homing devices and a means to transmit target information to a control operator. The homing devices were an important concept that could potentially be used to guide missiles to their targets. The Bureau established a plan for the installation of long-range radar into patrol planes and short-range search radar in a torpedo plane. The installation of radio altimeters was also planned. By November, intercept radars had been successfully designed and were eventually installed in some of the F4Us.

Left: This F4U-5P "Corsair" aerial reconnaissance plane was equipped with radar and aerial surveillance equipment (there is a camera hatch below the cockpit). It was a variant on the F4U-5. The F4U-1P was equipped for photo-reconnaissance. This image was taken around the summer of 1950. The earlier F4U was one of the first aircraft to be equipped with interception radar, in the summer of 1941. Right: The USS Valley Forge transporting aircraft in April 1949. Some of the aft planes are F4Us. [U.S. Navy All Hands and Historical Center Collection historical archives photos, public domain.]

Toward the end of the year, the Navy was contracting out the construction of airborne search radars that had been developed by the *Naval Research Lab*, while research and development of other innovations continued. One of the most important of these was a transceiving antenna. ~~Up to this time conventional, separate, Yagi-Uda antennas (a common branching type) had~~
*The XJO-3 was among the first twin-engine aircraft to successfully execute aircraft carrier take-offs and landings with tricycle wheels, in August 1939. and could detect surface ships at up to about 40 miles.

been used for the transmitting and receiving of radar pulses. This system was cumbersome, so research on a 'duplex' system, in which a switch could toggle between the input and the output, allowed a single antenna to be used where two previously were necessary — making the radar systems more streamlined and reliable.

Improvements in radar had an important impact on strategy and the administration of military resources, and changes were now rapidly made in how personnel and equipment were deployed to take advantage of the new capabilities.

Overseas Developments

Letters from *RCAF Overseas Headquarters* indicate that Canada was also establishing radar surveillance posts. Some of these were overseas, where it was not easy to hide and haul around bulky radar equipment. By the time the U.S. entered the War, in December 1941, a secret Canadian radar installation, north of Singapore, had detected an incoming enemy raid and provided warning of the impending attack. When threatened, the radar post was moved to Singapore Island and later relocated to another part of the island. The members are reported to have improvised when necessary, sometimes using palm trees instead of masts for installing their radar antennas.

By December 1941, some radar defense systems had been put into service in the Continental U.S. and some of the U.S. possessions. Not surprisingly, these early installations were described as ineffective by U.S. Army personnel and historians:

These [radar defense systems] consisted mainly of obsolete SCR 268 searchlight control sets, sited to attempt to utilize it [them] for early warning and also the original version of the SCR 270 early warning set. The coverage provided was inadequate for two reasons. One was a misconception at first of the technical qualifications of a good radar site. This conception was that generally the more altitude the radar was set on, the greater the coverage. This proved false and practically all the radars had to be relocated on new sites. The other reason for the weakness of the system and one difficulty faced by the entire program of radar development, was the lack of trained operating personnel and technicians immediately available. An extensive and intensive training program was under away [sic] both in the United States and the United Kingdom but initially the quantity and quality of personnel could not be produced fast enough. Another difficulty encountered was the maintenance and constant modifications to the design of the sets. Spare parts could not keep up with aqnd [sic] ground observer units...

Rudimentary technology and the lack of trained operators were not the only impediments to early warning systems. The effective use of radar was hindered by the lack of effective administrative channels for conveying the warnings that resulted from the analysis of radar and radio communications data. For example, radio messages indicating a break in Japanese diplomatic relations just before the attack on Pearl Harbor were apprehended by U.S. military personnel, but the intelligence wasn't processed and forwarded in time to establish defenses against the attack.

Radar systems improved substantially during the course of the War and thereafter. The historian quoted above describes how radar could be used more effectively on the front lines during and after wars:

... as the war progressed more and more thought was given to the offensive use of radar.... This made it possible to effect interception of Stuka raids well before the enemy reached the front line. Interceptions were controlled and made from radar plots by a ground controller and visual contact by the pilot with the enemy flight....

The Navy is a great user of radar and control both offensively and defensively. Ships carry radar and control equipment for detecting enemy flights and making the interception. Their control ships controlled all army air activity initially in the island operations. They also use it for control of all types of automatic and heavy gun firing, detection of enemy and identification of friendly shipping.

Strategic air operations use radar extensively. They do not, however, incorporate the use of ground control such as does the Tactical Air. They are linked to the ground as their airborne radar APQ-13, APS-15 and others all pierce the clouds and reveal the landfalls and terrain features over which they fly.... One of the mostly highly prized activities utilizing radar and control is that of GCA, Ground Control Approach. This equipment has demonstrated its effectiveness in both ETO and in the Pacific areas. It picks up aircraft which are either in distress and in overcast, or lost in zero conditions and guides...

[Colonel Rex J. Elmore, Development of Radar and Control in Air Operations, *Military Review*, August 1946, unclassified.]

THE WHITE HOUSE
WASHINGTON

January 17, 1942.

MEMORANDUM FOR THE PRESIDENT

Last night you "wondered" about two items. Here is what information I have been able to get today.

RADAR for Small Craft. You specifically mentioned the possibility of using aircraft RADAR on small surface craft. The Bureau of Ships tells me that to be effective for as much as two miles on a periscope three feet in height above the water, the radar antenna (weight slightly above 100 pounds) must be at least thirty feet above the water. Most small craft, of course, will not have masts which can carry that weight, nor are they susceptible of alteration to permit carrying such top side weight. BuShips states that they are developing radar equipment for all types of vessels in excess of 75' in length, and hope to be in production with same about July 1942.

The nub of the problem seems to be that when the antenna is placed but slightly above the water's surface, the more difficult the problem becomes. Aircraft, medium and large surface craft, use comparatively low frequencies, whereas in small surface ships very high frequencies are necessary.

Cedar Point, Maryland. As I recall it, you said "Tiney Point" - and that the activity there was being undertaken by the Bureau of Aeronautics.

This January 1942 Memorandum for the President from John L. McCrea to President Franklin D. Roosevelt responds to the President's inquiries about placing radar on marine craft. It explains, in simple terms, some of the difficulties with installing radar systems on small craft, including weight and height requirements that are impractical for smaller vessels. [National Archives historical documents, declassified public domain.]

Communications terminology and protocols evolved substantially in the 1940s with World War II (and the following Cold War) as motivating factors. Identification friend or foe (IFF) systems improved at this time as well. Homing devices, based on homing in on a radar beam reflected from the intended target, came into use around the summer of 1942. Discussions about the establishment of early warning radar stations began about this time as well, based upon the desire to extend the range of 'sight' to beyond the horizon.

INFORMATION ABOUT RADAR

EXPLANATION

CXAM-1   Radar designed primarily for detecting aircraft at long ranges (antenna approximately 18' X 18').

IE   Radio homing device.

TBS   Ultra high frequency radio transmitter.

ZB   Part of homing device.

IFF   Radar identification device "Friend and Foe."

MCW   Modulated continuous wave radio.

SG   Radar designed primarily for detecting aircraft at medium ranges (antenna approximately 7' X 8').

An excerpt from a February 1942 letter from J. B. Dow to President Roosevelt indicates that FDR was interested in keeping up-to-date on how radar technology was used during World War II. This excerpt lists a few of the common codes in use at the time. [National Archives historical documents, declassified public domain.]

The aviation industry scored one especially big bonus from radar research and development in January 1943, when a new aircraft approach radar, that had barely been tested, prevented an emergency situation. A snowstorm shut down an airfield at NAS Quonset Point before the arrival of a flight of PBYs. The ground-control approach crew were able to detect the incoming planes on the search radar and relayed landing instructions through the control tower to the aircraft for contact landing, thus bringing in the planes with the first Ground Control Approach (GCA) system.

New Strategies for Warfare

Radar evolved and supplemented hydrophones in the detection of submarines and other underwater targets. Helicopters were equipped with radar and dipping sonar systems to aid in submarine detection. In the Battle of the Atlantic, radar provided a way to locate surfaced U-boats from the water or from the air.

In November 1943, a radar-equipped Avenger was used to guide two Hellcat fighter planes to enemy aircraft during nighttime. On the second reconnaissance, the team successfully detected and engaged the enemy.

Warfare changed, as radar provided a way to 'see' in the dark and at distances not previously possible. But as radar evolved, so did anti-radar technologies. Materials with lower radar 'signatures' and radar-jamming techniques were developing alongside the radar technologies, creating a competitive environment in which scientists were always trying to keep a step ahead of the counter-technology.

Aerial Early Warning Systems

Up until the time World War II broke out in the 1940s, World War I was known as "the Great War". World War II was initially called "the European War" until Japan and the U.S. got involved. After the eruption of World War II, the Great War was retroactively named World

War I. Concerns about preventing a third global conflict were probably already surfacing before the conclusion of World War II, and resources were put into early warning systems using new radar technologies.

Stationary radar warning platforms were effectively used during World War II, but military planners felt there was a need for radar-equipped aircraft for early warning systems. In 1944, the U.S. Navy established such a program through the *Naval Research Lab* and the *Radiation Laboratory* (later known as the *Lincoln Lab*). Experimenting with a combination of radio ranging and radio communications technologies, scientists developed an airborne radar system which could transmit radar video data to sea or ground platforms. Thus, aircraft carriers, for example, could dispatch a reconnaissance aircraft which could subsequently transmit the data to a ship-based combat information center (CIC). The concept was sound, but the technology was lacking. The range of the video data link just wasn't sufficient at the time to provide effective early warning capabilities. The solution proposed in 1945 was to establish the CIC on board the aircraft itself [Bouchard, 1999].

The honor guard from the 11th Airborne Division Reconnaissance Battalion presents arms as Allied representatives arrive for the ceremonies commemorating the surrender of Japan in Tokyo Bay, 2 September 1945. Reconnaissance and surveillance of Japanese radio communications and marine vessels were significant factors in countering Japanese offensives in the Pacific during the War. [U.S. Army Signal Corp Collection, public domain.]

When World War II ended, scientists looked for new ways to exploit radar technology. In 1946, radio waves were bounced off the Moon and back, demonstrating that the energy not only could penetrate the ionosphere, but that FM modulation, developed by Armstrong, had great potential as a measurement and communication technology. That same year, a group of engineers from *Douglas Aircraft* proposed the use of satellites as observation posts, though actual implementation did not occur until many years later.

Following the War, the Federal Communications Commission (FCC) allocated certain frequency bands for radar, including X-band and K-band frequencies for commercial use, and other bands for commercial aviation and military use.

The development of airborne radar systems led to the modification of a number of land-based aircraft to accommodate air search radar systems, which were put into service in 1946. Later that year the Patrol Bomber Squadron flying the planes was transferred to Rhode Island and redesignated the Airborne Early Warning Development Squadron Four (VX 4). Another move to Maryland in 1948 established it as the center for airborne naval early warning. The result of equipping and flying the modified aircraft led to the development of specialized radar-equipped reconnaissance planes.

Radar had proved its worth as a detection and navigational tool and was expected to continue to be used for other purposes at the conclusion of the War.*

Gradually, destroyers were outfitted with radar systems, with some of the ships specifically assigned as *radar pickets*. The Sullivans (DD-537), for example, served on radar picket duty in the Pacific in 1945.

During the summer and fall of 1948, a U.S. submarine, the Spinax, was modified for radar and communications capabilities by adding equipment similar to that used in the naval destroyers. It thus became the first radar picket submarine—redesignated SSR 489. The Spinax Electronic Counter-Measures (ECM) equipment was used to detect radar or similar electronic radiation. On the port side of the periscope wells was the submarine-spotting radar.

The Sullivans, was one of the earliest U.S. naval destroyers assigned to radar picket duty in the Pacific region in the mid-1940s, shown here in October 1962. Right: The USS Valley Forge is shown in April 1949. Also note atop the tripod mast there is a large SX radar antenna. [U.S. Navy historical photos, released.]

In the late 1940s, Arthur C. Clarke predicted and described in detail the process of putting geostationary satellites in orbit around the Earth. His predictions were not only remarkably accurate, but probably in part inspired the development of satellite technology.

The Transistor Revolution

The invention of the transistor in 1947 at Bell Laboratories heralded the next significant stage of growth in surveillance technologies and, once the technology became established, had a dramatic effect on the evolution of radar devices. Transistors and the evolution of semiconductor technology represented an important shift from mechanical to electronic components and systems, and from large-scale to small-scale radar sets that were equal to or better than their older large-scale counterparts. That is not to say that transistors completely superseded tubes. For certain high-frequency radar applications, vacuum tubes continued to be important system components but, overall, the design and efficiency of radar systems improved with

*There are many books that provide histories and details on Allied radar installations in World War II, some of which are listed in the *Resources* section at the end of the chapter.

transistorized parts.

In 1947, scientists at Bell Laboratories introduced a milestone development in electronics, the transistor, which made microminiaturization and a new generation of devices possible. [U.S. Postal Service first day cover from the author's collection, used with permission.]

Adding Radar to the Fleets

As radar technology advanced, and became more practical and affordable with the addition of transistor technologies, planes and ships were increasingly outfitted with a variety of types of radar systems.

By the early 1950s, several airplane designs had evolved from the original B-25 Mitchell bombers that were first flown in 1940. These descendants included the TV-25K and TB-25M, which were subsequently modified to serve as training aircraft for teaching Hughes E-1 and E-5 *fire control radar* operation skills. The addition of radar to many aircraft and ships continued through the 1950s, along with training for radar operators—prompted in part by the Korean War.

Helicopter Surveillance. Left: Sikorski helicopters from the First Marine Division Reconnaissance Company in Korean operations, September 1951. Helicopters were first equipped with some of the modern radar technologies in the mid-1940s. Right: Part of the instrument panel of a Navy Seahawk helicopter. The Seahawks first entered U.S. Navy service in 1983. The crew includes the pilot, an airborne tactical officer and a sensor operator ("senso") who operates the radar and magnetic anomaly detector, as well as interpreting acoustical data. [U.S. Navy All Hands Collection and news photo, public domain.]

By the 1950s, new radar technologies were spreading around the globe. Australia, for example, shifted its defense policies to become more self-sufficient, adding to its air force and establishing its own system of air defense and air traffic control radars.

Establishing Early Warning Systems

The beginnings of extensive U.S. early warning radar in Canadian territory began with the "Pine Tree Line". This was an air surveillance radar system intended to safeguard the northern approach to the U.S. It stretched across southern Canada, becoming operational in 1951. However, due to its proximity to the U.S. border, it was considered to be insufficient warning, and patrols off the coasts supplemented the Line.

In 1952, to further address the desire for early warning systems for national defense, Project Lincoln was initiated at MIT, which had been an important center for radar development during the War. The result of the Lincoln research was a recommendation that an early warning system be established across northern Canada. However, there were objections about the difficulties of establishing and maintaining radar stations in the cold northern reaches of Canada. In spite of the inherent difficulties, Project 572 became the Distant Early Warning (DEW) Line.

Radar picket lines along the coasts were established in the 1950s, including the Inshore and Contiguous Barriers. Extensions of the DEW Line were also planned. The radar picket stations were originally patrolled by radar picket destroyer escorts, but gradually were patrolled by radar picket ships and were later supplemented by airborne early warning craft.

Left: Early warning radar picket aircraft flying near Korea in August 1951. These Douglas AD-3W "Skyraider" planes were based aboard the USS Boxer (CV-21) as Squadron VC-11. Right: Ships supporting the Alaska DEW Line in September 1955. The USS Harris County and the USS San Bernardino County are shown unloading at Point Barrow, Alaska. Bottom: The USS San Bernardino County supporting the DEW Line operations in an ice belt of the shore of Alaska in September 1955. [U.S. Navy historical photos, public domain.]

The Alaskan portion of the northern DEW Line was completed in 1953, and was extended across northern Canada by 1956, coming into operation the following year. The DEW Line had been extended into the Atlantic as well, and test patrols of the Atlantic Barrier began in the summer of 1956.

Radar Automation

Despite all the advancements in radar during the War, and following the invention of the transistor, in 1947, there were still many aspects of radar that had not been automated. One of these was the plotting of trajectories. Since bright pips on a radar screen are only there for as long as the target can be seen and resolved by the radar, the radar operator must have a good memory and sense of what is happening to track activities within a region under surveillance. Hence, manual plotting systems kept track by recording reference marks and tracking changes on accompanying charts, providing a picture of what was happening to analysts and strategists evaluating the information. With aircraft and other vessels traveling progressively faster as technology improved, this system quickly became obsolete.

IBM was one of the companies working on solutions to automate plotting and introduced its Semi-Automatic Ground Environment (SAGE) system in the 1950s, continuing development on the system into the 1960s. Since the changeover from vacuum tubes to transistors was not yet complete, SAGE's 'brain' was based on a gigantic, old-style computer technology. By the time SAGE was fully functional, newer, smaller electronics were superseding historic components and SAGE became an expensive behemoth with limited processing capabilities.

Establishing Surveillance Systems

America was building radar stations across northern Canada and it became clear that better coordination between the defense forces of the two nations was needed. As a result, the *North American Air Defense Command* (NORAD) was established in September 1957.

Sputnik was a milestone technological achievement that took the public and many members of the U.S. government by surprise, but it wasn't a surprise to amateur radio enthusiasts. In the summer of 1957, the Russian Federation announced, in print, to the world's radio technologists, that they would be launching a satellite into orbit to study the ionosphere. They provided information on the radio wave frequencies that would be used for the satellite broadcasts so that communications received by listeners around the world could be logged to help track the progress of the satellite. The frequencies selected were just above those used for global standard time signals. In October 1957, the Federation successfully launched Sputnik I, the world's first artificial satellite. A month later, they launched Sputnik II, with a passenger—a dog named Laika—to test radiation levels and to investigate the feasibility of putting a human in space. The craft was equipped with a slow-scan TV camera for broadcasting images to ground-based receivers. Competitive space programs and modern surveillance and communications satellites were spurred by these momentous events in the 1950s and early 1960s. In response to the launch of Sputnik, America announced that they would be the first to put a man on the Moon.

In the late 1950s, a Minitrack system was developed for the *National Research Laboratory* (NRL) Vanguard Satellite Program, which detected signals from Sputnik and subsequent satellites to monitor their orbital positions. Since not all satellites could be expected to broadcast signals that could be interpreted by American systems, the idea of reflecting signals off the satellites to track their positions led to a more elaborate system called the Space Surveillance System (SPASUR) that became operational in 1961, using a frequency of 108 MHz.

In January 1959, Fidel Castro seized power in Cuba and established ties with the Soviet Union, raising U.S. concerns over national security. A number of U.S. destroyers were converted to radar pickets and surveillance efforts increased overall. U.S. intelligence agents reported the installation of Soviet missiles in Cuba. In 1960, the *Central Intelligence Agency* was involved in a plot against Fidel Castro. Meanwhile, Americans had been flying spy missions over Soviet territory and, in May 1960, Khruschev announced having shot down an American plane (described further in the Aerial Surveillance chapter).

In January 1961, President Eisenhower's administration broke off relations with Castro when Castro ordered a reduction in U.S. embassy staff.

In 1961, an invasion of Cuba at the Bay of Pigs was planned, and only days later aborted. A U.S. radar picket station was established at the southern tip of Florida, to monitor the region between the U.S. and Cuba. Political negotiations and military pressure on shipping and submarine activity in the vicinity of Cuba eventually resulted in the withdrawal of Soviet offensive weapons, closely monitored by U.S. forces in the area. In November and December 1962, the political situation with Cuba was very unstable and many reconnaissance, patrol, minesweeping, and anti-submarine ships were dispatched to the region. The event, which fortunately didn't erupt into a major nuclear war, has been remembered as the Cuban Missile Crisis. Radar was extensively used to patrol the region.

By the early 1960s, radar technology had advanced to the point where it could be used to remotely land an aircraft. At about the same time, satellite technology had taken to the skies and radar was used not just to warn of impending enemies in wartime, but to conduct global surveillance in peacetime as well. One of the early radar surveillance systems was a U.S. installation in Japan that scanned adjoining Soviet territory.

Space Surveillance

In 1965, the U.S. SPASUR system was upgraded to the Naval Space Surveillance System (NAVSPASUR) and the transmissions frequency was changed from 108 MHz to 216.88–217.08 MHz. The system was essentially a high-powered, continuous-wave, bistatic radar that sent out a latitudinally fanned radar beam, which is often referred to as a radar 'fence'. When the beam encounters a reflective object in orbit, the returning beam can be intercepted and analyzed by ground receiving stations, providing information on the existence of a satellite and collectively calculating its position. The receiving stations used interferometers, Doppler data, and triangulation to calculate information about the orbiting object detected by the sensors. Once a general position was known, more targeted surveillance could be conducted with other sensors if desired, to determine the location and characteristics of the satellite.

By the late 1960s and early 1970s, radar imaging was contributing to our ability to map terrain and investigate the unique 'signatures' of various types of materials. Radar frequencies suitable for probing beneath the Earth in soil and sand were studied and certain lower frequencies were found to be capable of deeper penetration (with some trade-offs in resolution).

Multisensor surveillance vessels were developed in the early 1970s, first by converting existing systems, and then by designing specialized stealth and surveillance craft that could carry several sensing systems. One of the early multisensor platforms was based on a modified B-25 "Mitchell" (a common World War II bomber plane) which was outfitted with side-looking radar (SLAR), an infrared scanner, radiometers, and aerial cameras. Many of the B-25s were modified to become training craft in the 1950s. Radomes were fitted in the noses and radar equipment was installed in the bomb bays.

A series of reconnaissance and fighter aircraft were developed over the next three decades, starting with early achievements such as the highly classified U-2 to more recent designs like the F-18 Super Hornet and the F-22 Raptor. These fast, high-flying aircraft gathered information from out of the reach of conventional weapons. The U-2 was equipped with long-rang cameras, radar-intercept receivers, and recording equipment, in order to gather information on missile sites, troop deployment, and nuclear installations, especially in the U.S.S.R. during the Cold War years [Rowan and Deindorfer, 1967].

In October 1983, the Naval Space Command (NSC), a component of USSPACECOM (the U.S. Space Command), began operations. The NSC, among other things, handles satellite-based space surveillance and early warning systems for defense.

Technological Advancements

Computer technology and microminiaturization were well established by the mid-1980s and the components used in many types of radar systems, making the systems smaller, more economical, and more powerful.

By the late 1980s, digital signal processing (DSP) was incorporated into radar systems to aid in automation and signal interpretation [Mardia 1987]. Miniaturization was also improving, enabling smaller, lower power-consumption systems to be developed. It was also becoming more practical to use higher frequencies, thus broadening the range of products and improving consumer radar products.

In the early 1990s, many of the new microelectronics technologies that had been developing over the last decade were tested in military situations. Homing devices, missile-seeking systems, radar imaging systems, and countermeasures had all evolved significantly and were gradually incorporated into the military arsenal.

Left: Naval F-14 "Tomcat" aircraft were equipped with long-range AWG-9 radar systems which, when engaged, would apparently cause Iraqi MiGs to turn away. The Tomcat is shown here launching from an aircraft carrier in August 1999. Right: A Tactical Air Reconnaissance Pod System (TARPS) being removed from an F-14. [U.S. Navy news photos, released.]

Left: The EA-6B Prowler was used for reconnaissance missions in Operation Desert Storm. Shown here are Prowlers patrolling the "No Fly" zone over southern Iraq in February 1998. The Prowlers provide 'cover' for accompanying planes and ships by jamming enemy radar systems, data links, and communications. Middle and Right: The radome, a protective radar cover, is clearly visible on this E2-C Hawkeye Airborne Early Warning aircraft. Hawkeyes provide long-range radar and communications support for aircraft carriers. [U.S. Navy photos by Chuck Radosta and Tedrick Fryman, III, Johnny Grasso, released.]

Radar systems along borders were continually monitored and gradually updated (and sometimes renamed) as the technology evolved. The NAVSPASUR warning system came to be known as the NAVSPACECOM fence, in 1993, though many continued to call it NAVSPASUR even into the year 2000. In Operation Desert Storm, Navy EA-6B Prowler aircraft used reconnaissance and sensing devices to determine the location of a threat. They would then jam and destroy enemy radar systems. In early 1991, the Prowlers damaged enemy early warning systems and attacked critical radar sites with high-speed anti-radiation missiles (HARM).

During hostilities, fire-control radars were used by the Iraqis, along with Silkworm missiles in Kuwait. The Gulf marine region was defended by several underwater mine fields that were detected by American surveillance. Extensive mine-clearing operations were carried out and the Silkworm radar systems targeted to disable the mines.

Electronic reconnaissance. Specially equipped EP-3Es. An S-3B worked reconnaissance in conjunction with the USS Valley Forge (CG 50) in the Persian Gulf, using forward-looking infrared and inverse synthetic-aperture radar ISAR). Left: An S-3B Viking and ES-3A Shadow. The Shadow is with the Fleet Air Reconnaissance Squadron shown here in support of Operation Southern Watch over the Gulf. Right: An S-3B Viking conducting air operations with the Japanese Maritime Self-Defense Force in the Sea of Japan in November 1999. [U.S. Navy 1999 photo by Michael Pendergrass, released.]

The RIM-7 NATO Sea Sparrow air-to-air missiles are designed with radar guidance systems to counter offensive missiles. In other words, the air-to-air missile is used to track and destroy a previously fired incoming missile. Note the various radomes on the ship's tower (left). [U.S. Navy 1999 and 2000 photos by Johnnie Robbins and Brett Dawson, released.]

Mapping the Earth from Space

One of the most significant aerial mapping missions using radar, was the spring 2000 flight of the U.S. Space Shuttle *Endeavour*. This mission obtained 3D topographical images of about 80% of the Earth's terrain using two large radar antennas. One was located in the Endevour, the other on a 197-foot mast. The area covered was about 43.5 million square miles, extending from Canada to Cape Horn. Over 300 digital tapes (about 12 terabytes) of radar data resulted from the mission. The tapes were copied for study and evaluation by NASA and the *National Imagery and Mapping Agency*, with the originals stored in environmental chambers for safekeeping and backup. The data were gathered for military navigation and targeting, with less precise charts available to scientists, pilots, and search and rescue groups. (Charts are degraded to make them less useful if they are acquired by hostile forces.)

Radar is now routinely used in navigation, geographical surveying, law enforcement, international surveillance, air traffic control, intruder security systems, missile tracking, scientific research, especially astronomy, and the forensic sciences.

5. Description and Functions

5.a. Basic Radar Systems

The basic concepts associated with radar have been described in the introductory sections of this chapter. This section fills in a few gaps and provides some sample images of radar console operators at work in a number of different environments.

A typical radar send-and-receive system consists of radio wave-generating equipment, an antenna to direct the waves, an antenna to receive the returning waves (which may be the same as the sending antenna, or separate), a receiver to interpret the waves, and a display system—which may or may not include computer algorithms for plotting movement over time, iconizing the data, or storing it for subsequent replay.

Most radar systems are pulse systems in which a series of pulses follows one after another, each carefully timed before the next pulse is transmitted.

In its most basic form, a send-and-receive radar consists of

- a transmitting system with a *generator* to create a pulse, a *modulator* to process the pulse, an *oscillator* (usually a vacuum tube) to create high frequencies, an *amplifier* to 'enlarge' the signal and an *antenna* to focus and direct the pulse outward toward the intended target(s), and

- a receiving system with an *antenna* to intercept and direct the returning pulse, an *amplifier* to increase the strength of the signal by increasing the voltage, a *processor* to automate aspects of data processing, and a *display device* to convert the signal into a visual representation that indicates the presence and relationships of reflective objects encountered by the transmitted beam.

Radar displays can be roughly grouped into *range indicators* and *plan-position indicators*. There are hybrid systems as well.

Radar systems can range from handheld, to large-scale arrays covering a few acres. Ground-penetrating radars are one of the more interesting technologies used in law enforcement and archaeological investigations.

Radar-based landing systems for planetary probes can help provide data on elevations, roughness, distribution of slopes, and bulk density of the surface—information needed to carry out a landing. A reflective surface that echoes back the radar signal expedites a good landing.

In the U.S., radar frequencies are allocated and monitored by the *Federal Communications Commission* (FCC). In Europe, cooperative arrangements between nations have existed over the decades and are gradually being brought under the umbrella of the European Union. Frequencies of 2.7 to 3.4 GHz are used in civil and military radar applications in Europe. This allocation came under discussion and review in the late 1990s, with some concerns about changes expressed by NATO.

5.b. Radar Displays

As mentioned earlier, the most common way to represent and display radar signals is on a specialized cathode-ray tube (CRT) called a *radar scope* though, increasingly, radar signals are imaged on more compact computer monitors including transistor- or gas plasma-based display systems. Radar scopes generally have alignment and measurement references superimposed over the display field to help the operator interpret the information. Since radar signals travel through three-dimensional space, and a CRT represents the signal on a two-dimensional coordinate system, various means of symbolic representation and interpretation are used. Depending on what type of information is sought, different aspects of the echo are selected or emphasized. Newer systems can model three-dimensional terrain.

Different types of radar console stations in the combat information center (left) and the combat direction center (right) of the aircraft carrier USS Kitty Hawk (CV-63). [U.S. DoD 1987 and 1991 news photos by James R. Gallagher, released.]

Left: A radar plot station on the Soviet-built USNS Hiddensee acquired by the Federal German Navy in 1991. Right: An air traffic control surveillance radar console on which Lou Scarozza monitors approaching aircraft at a U.S. Naval Air Station. [U.S. DoD 1993 photos by Eric A. Clement and by Don S. Montgomery, released.]

Radar is an essential tool for navigation on ferries, commercial tankers, cruise liners, larger boats, and military vessels. On aircraft carriers, radar is also used for monitoring aircraft that are associated with the carrier, and foreign vessels (both surface and airborne). At times, radar is used for directing combat operations. Aircraft carriers are equipped with many different radar antennas communicating with a variety of different radar console centers.

5.c. Digital Technology

Traditionally, radar screens display sweeping or bouncing lines or bright blips (pips), but radar display technology is changing with the introduction of computer electronics.

Computer technology now makes it possible to take traditional radar bouncing lines and bright blips and assign symbols or icons to individual aspects of the display. This can make the information more user-friendly by, for example, assigning a boat symbol to a large, slow object, and a plane symbol to a fast-moving target, once the source of the data has been determined. Radar operators must understand that symbolic mapping is never a perfect match with the original three-dimensional space, as proportions and sometimes even interpretations of the signal may be misleading. High-tech displays must be monitored for errors as carefully as any traditional display.

Analysis and evaluation of information by a human operator are still essential aspects of radar technology and a radar operator requires training, practice, and the ability to concentrate and interpret visual information well. On iconic systems, in order to provide checks and balances for ambiguous signals, pips and other symbolic images may be displayed side-by-side (on a split screen or on separate monitors), or the display may be toggled from one to the other. On some computerized systems, the 'raw' radar pips may not be accessible.

Computer technology has provided other improvements to radar displays beyond iconic representations of information. Here are some of the digital processing and display innovations provided by computerization:

tracking and plotting - This can be used to analyze the progress of targets over time, somewhat like the radar weather maps that are shown on TV, in which successive images of a storm are played one after another in rapid succession to create and animated sequence. This task was traditionally done by hand, using a pencil and graph paper.

storage and replay - Digital radar systems with memory caches and hard drives can be used to 'replay' previous image segments, or to freeze or analyze a particular segment of information, if a review of recent or past events is desired.

comparison and analysis - Expert systems can simultaneously compare current information against past events, to call attention to patterns or changes that might be significant, and to situations that might be important.

display customization - Computer displays can also provide dynamic assignment of colors, or selective display of information, to aid in interpretation and representation.

As with all technologies, there are trade-offs. A traditional display, used well, may still be better than a computerized display used by an inexperienced operator. A computerized display that demands a lot of processing power (or electricity) will have trade-offs in terms of speed and portability, and a symbolic display is only as good as the symbol assignments provided by the operators. Nevertheless, in spite of limitations (and complications) provided by computers, their storage, replay, and signal-processing capabilities have improved radar, and not just in terms of display systems. The capability to send out carefully timed beams to create sophisticated radar arrays and three-dimensional images is another important benefit of computer technology.

5.d. Synthetic-Aperture Radar

There are many kinds of radar, but one of the more recent and important is synthetic-aperture radar (SAR), a technology that an resolve a fairly broad area of terrain in high resolution, making it suitable for radar imaging applications like mapping the Earth's terrain, tracking ships, or monitoring weather systems. Satellite SAR scans of the ocean, for example, can create images about 100 kilometers square, providing good coverage of coastal features or shipping traffic.

The basic idea of a *synthetic* aperture is that distance can be 'substituted' for the length of an antenna. Radio waves can be a kilometer in length or more and antennas have been traditionally constructed so they are mathematically proportionate to the length of the wave so they will resonate in tune with the signal. Since there are practical limits to the size of physical antennas, especially when they are mounted on aircraft or small ships, some clever substitutions have been developed.

SAR is calculation-intensive. In its simplest form, SAR provides a 2D image in which one dimension represents the *range* (the *along-track* distance from the radar transmitter to the target) and the other dimension represents the *azimuth* (the *cross-track* perpendicular to the range). These concepts of range and azimuth are also used in sonar.

Now imagine that a SAR is mounted on an aircraft moving in a straight line. Since long antennas are needed to resolve long wavelengths and the length and shape of antenna transmitters and receivers are related to the dimensions of the radio wave (and its associated frequency), a 'synthetic' aperture can be achieved by associating *the distance flown* with what would normally be the physical antenna length, thus creating a *synthetic aperture*. In other words, the distance flown (as opposed to a physical aperture) makes it possible to create the effect of a long antenna.

The Doppler effect, described earlier, is used in SAR to resolve the shifted frequencies associated with the movement of the craft. Like other aspects of radar, there are trade-offs between resolution, speed, and the length of the radar pulses, but some of the most interesting radar images are generated with this technology.

5.e. Magnetrons

While microprocessors have replaced many types of vacuum-tube technologies, tubes are still important components in high-frequency applications, and radar transmitters commonly use vacuum-tube pulsed oscillators called *magnetrons,* developed in the late 1930s by Boot and Randall. A high-voltage pulse is applied to the magnetron which, in turn, emits a microwave-frequency radar pulse. While the noncoherent pulse of a crystal-controlled system cannot be perfectly locked, it has the advantages of small size and conversion efficiency. Coherent systems, which may be more costly and less efficient than a magnetron, are still useful, however, for special applications.

5.f. Space Systems

Radar signals on Earth are monitored with specialized, fast, high-altitude planes and orbiting satellites. These systems are commonly used in surveillance because they are too distant to be shot down with conventional missile systems. Information can be sent through radio frequencies, but since these signals can be intercepted, physical storage media such as cassette tapes or film in protective wrappings are sometimes used instead, and ejected and intercepted as they fall to Earth. Since they miss their landing targets, they are sometimes equipped with dyes or transmitters to increase the chance of recovery. The recovery of physical storage media can be costly and somewhat hit-or-miss, however, so encrypted and spread-spectrum radio communications are increasingly favored (these are discussed in more detail in the Radio and Cryptologic Surveillance chapters).

One nation's satellite is another nation's reflector. In the late 1950s and early 1960s, when Earth-orbiting satellite technology began to evolve, unsettled political relations between the U.S. and the U.S.S.R. motivated military and scientific personnel to make use of opposing forces' space technology whenever possible. The interception of signals from a transmitting satellite was an obvious objective, but engineers didn't overlook the fact that an orbiting body, even if it was not actively transmitting, could provide a reflective surface useful for calculating positional information. Ground-tracking stations were developed to transmit signals to both passive and active space-based objects and receivers were used to intercept the information for interpretation. Thus, a 'picture' of objects in the sky could be created from this information and new or unusual movements or objects could be detected and scrutinized more closely.

5.g. Ground-Penetrating Radar

Most radar signals are directed through Earth's atmosphere from ground stations and aircraft, although satellite radar from orbiting platforms is becoming more prevalent. However, ground-penetrating radar (GPR) has been found to be a useful technology for detecting the composition of the environment for a short distance under the ground.

Ground-penetrating radar was originally developed as a way to measure depths through ground structures such as glacier ice (this was easier than trying to drill hundreds of holes to try to assess depths of a particular region of ice). However, it was found that radar could yield more than just depth information. It could also image areas with varying dielectric properties, yielding a highly useful set of data for locating geophysical structures (water, silt, rock strata, etc.) and synthetic objects buried underground.

5.e. Stealth Radar

Stealth radar is a system that is virtually undetectable due to the characteristics of the radar signal. As an example, in 2006, Ohio State University announced that it had developed a system of radar that is "invisible" to conventional radar detection schemes because the signal resembles random noise. Thus, it would be screened out by the signal processing subsystem of many radar-detection devices and yet would not interfere with broadcast communications signals. This has potential applications for radar sensing behind walls for military applications, search and rescue, traffic safety, and possibly even for medical imaging. The price of the system is very competitive, and its range is from a few inches to many thousands of miles.

6. Applications

As has already been shown, radar has many commercial, scientific, and political applications. The examples in the previous sections demonstrate it as an integral tool to military surveillance. This section presents a broader view of radar applications, describing some of the scientific and civilian uses and interesting experimental radars that may become more common in the future.

6.a. Traffic Enforcement

Radar has been used in traffic enforcement for several decades and systems have increasingly become smaller and more accurate. Radar detectors have also evolved in step with traffic radar systems in a cat-and-mouse game that has caused the regulations regarding radar and radar detectors to be scrutinized and adjusted on a regular basis.

One of the more common technologies for speed enforcement is a single-shot radar 'gun' that only sends a signal when the vehicle is in visual range of the operator and the trigger is pulled. Conventional radar detectors don't provide advance warning of this type of instrument. However, some low-power, semiconductor-based, dash-mounted radar-jamming transmitters can interrupt the signal, sending back unusable data. These jamming systems use a cavity resonator to generate microwave signals powered by the car's 12-volt battery.

Police radar detectors generally operate in the X-band at about 10.5 GHz or the K-band at about 22 GHz. They are directional, but since many of the gun-style radar systems used in law enforcement are used in front or behind, it handles most applications. There is some potential for spurious signals, as reflections off large objects may trigger other radar devices in the vicinity, but the problem is not serious.

Certification is necessary to operate and calibrate radar speed control devices. Courses are offered, which usually require about four hours of classroom lectures and hands-on instruction.

On the left is an older gun-style police radar system, on the right a somewhat newer one, though speed radar systems are upgraded every few years to take advantage of new technologies. A typical system consists of one or two radar transmitters, a controlling console, tuning forks (for calibrating the system), and a remote control. [Classic Concepts photos copyright 1999, used with permission.

This Speed Watch vehicle-monitoring system, set up by local police in a 25 MPH zone two blocks from a school zone, shows how an unstaffed radar traffic monitoring system can provide a visual reminder to speeding motorists to slow down. Over the half hour period during which these photos were taken, motorists approached the system at an average speed of 33 MPH (with highs of 37) but they consistently slowed to 22 MPH by the time they passed the unit. [Classic Concepts photos, copyright 2000, used with permission.]

Photoradar, in which the radar system senses an approaching or retreating vehicle and snaps a picture of the license plate, has been used in Asia and Europe since the 1980s, and has also been implemented in British Columbia and a number of states in the U.S. It is mainly used for speed monitoring and traffic light enforcement. Traffic enforcement tickets are sent to the owner of the vehicle, as determined by the number on the plate, regardless of who was driving at the time of the infraction. Commercial photoradar systems use radio waves in the Ka-band—a choice that was made in part to help defeat radar detection systems.

Those trying to avoid prosecution for speeding have come up with various schemes to interrupt this system, including sprays and plastic covers that scatter electromagnetic radiation so that it cannot be effectively photographed. These systems are generally illegal.

Another traffic enforcement strategy is to inform motorists of their speed so they can make adjustments voluntarily. These systems are usually placed in locations where there is a habitual speeding problem, in areas where there are particular hazards such as a blind intersection or curve, or children playing near parks or schools. As motorists approach the system, the radar sends out a series of pulses to detect the vehicle speed and updates a large sign to display the speed. As vehicle speed changes, so does the display.

6.b. Civilian Applications

The capability of radar to 'see in the dark' and 'peer into structures' allows it to be used for many security and diagnostic applications. Computerization has made it possible to design smaller, less expensive, more powerful devices that can now be marketed for smaller-scale commercial uses. It is likely that these uses will increase over the next several years.

 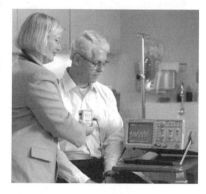

Left: A concealable radar security detector can sense intruders up to a distance of six meters. Designed at the *Lawrence Livermore National Laboratory* (LLNL), the repeat radar signal from this micropower impulse radar (MIR) detects motion in the room by responding to changes in the returning echo. Right: The same impulse radar technology can be used to peer into novel places, as shown with this micropulse radar stethoscope used by Elaine Ashby, M.D., to measure inventor Tom McEwan's heart rate. [LLNL news photos by Jacqueline McBride and James E. Stoots, Jr., released]

6.c. Scientific Research

Radar has been a great boon to scientists studying almost every branch of science, but especially environmental studies, geology, archeology, anthropology, and astronomy.

Archaeology and Anthropology

Radar can often detect physical features, patterns, and trends that can't be seen any other way. It has helped us to locate ancient burial sites and villages, unusual rock formations, and changing geological formations. Recent radar images have aided archaeologists and historians in finding out more about the technology and habits of ancient civilizations.

The remains of old villages and cities may not be noticed with the unaided eye, but when you image them with radar from the air or with ground-penetrating radar on the surface, suddenly building foundations, fields, squares, and other structures can be distinguished from more recent structures, because they differ in composition or density. In some cases, radar images have revealed structures for the first time, as in archaeological finds in the Sahara Desert.

The Great Wall of China has been constructed and reconstructed over a period of many centuries, giving a glimpse into the turbulent political history of the region, its battles, its wall-builders, its economy, and its industry. Visual and radar aerial imaging of the wall has revealed that its boundaries have changed many times. This information has helped scientists discover regions of the wall that could not be seen from the ground.

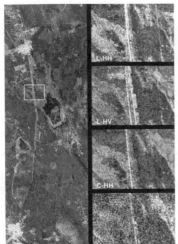

Left: This is an aerial view of a segment of the Great Wall of China in north-central China. The details on the right show close-ups of particular sections of the wall. The satellite photos were taken by the Jet Propulsions Lab Spaceborne Imaging Radar-C (SIR-C) in the L-band. The image shows the younger portion of the wall, built during the Ming Dynasty (about 600 years ago) and also the ruins of the older wall, built 900 years earlier, during the Sui Dynasty. The L-band can also reveal vegetation as bright areas. The images can help researchers trace the path of the old sections of the wall, providing a glimpse into the turbulent political history of China. Right: This satellite image of Angkor, Cambodia, taken in 1994 from the Space Shuttle Endeavor, shows 60 temples, some dating back as far the 800s. Apparently undiscovered structures were revealed by this image, which combines L-band and C-band radar data. Angkor had a population of up to a million residents before it was abandoned in the 1400s. [NASA/JPL news photos, 1996, released.]

Earth Monitoring and Mapping

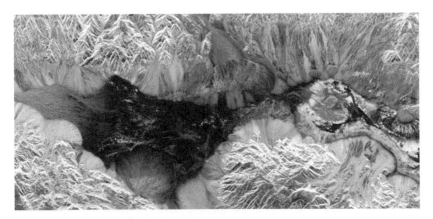

This image of Death Valley, California was imaged with the SIR-C/X-SAR NASA Spaceborne Imaging Radar. The synthetic-aperture system is equipped with L-band, C-band, and X-band microwave radar capabilities. The picture above was imaged by the *Jet Propulsion Lab* using the L-band instrument. It shows how radar imaging is not only useful but often very beautiful. [NASA/JPL 1995 news photo, released.]

Many aerial images of the Earth are created with radar technologies. From weather-mapping to space shuttle-based terrestrial mapping, radar has contributed important information to our understanding of our planet.

Left: A digital radar image of the Sierra Nevada Mountains acquired during a 1994 NASA space shuttle mission using the Spaceborne Imaging Radar C/X-Band Synthetic Aperture Radar system. It has been superimposed with a photograph taken by the astronauts on the shuttle. Color values are added via software to the grayscale radar images to aid in interpretation. Right: This computer-generated simulation of the spring 2000 Shuttle Radar Topography Mission (SRTM) shows how the space shuttle is used to collect 3D measurements of about 80% of the Earth's land surface (excluding the poles) using radar-sensing equipment with a resolution of about 16 meters. A 60-meter mast was installed on the shuttle, along with a C-band radar imaging antenna. The mission used the same radar system that was twice flown in the Space Shuttle Endeavour missions in 1994, called the Spaceborne Imaging Radar-C (SIR-C). This was a cooperative project between the *Defense Mapping Agency* of the U.S. *Department of Defense* and NASA's *Jet Propulsion Laboratory* (JPL). [NASA/JPL 1996 and 1994 news photos, released.]

Human-caused and natural environmental changes are now monitored with radar by many organizations, such as the *Space Radar Laboratory* (SRL), which has carried out space shuttle missions since the early 1990s. The systems are versatile, with the ability to take radar measurements of any type of region in any kind of weather. They are used to study geographic features such as rock formations, volcanic activity, ocean currents, vegetations, snow and ice packs, and wetlands. By comparing and contrasting data collected on different missions, small and large-scale changes and patterns may be discerned. The payloads have included the Spaceborne Imaging Radar *Synthetic-Aperture Radar* (SAR), which used C-band and X-band frequencies (SIR-C/X-SAR), and the *Measurement of Air Pollution from Satellite* system (MAPS). The data from the missions are distributed by NASA to the international scientific community.

Oil Spill Detection and Monitoring

Radar is used to detect whether tankers stray or deliberately navigate into restricted zones. It can also be used to detect unauthorized dumping or spills if the substance coats the water (e.g., oil) and is thick enough to alter the radar-scattering pattern of the surface of the water. The radio waves may be impeded, however, if water currents are especially rough. Side-looking radar (SLAR) systems mounted on aircraft are useful for this type of surveillance.

Underground Investigations

Ground-penetrating radar can be used to measure the depths of glaciers and ice floes, as well as locating old septic tanks, oil drums, and wells. It can help map geophysical structures and mineral veins. It can even be used to detect underground bunkers or covert storage facilities.

Ground-penetrating radar provides a way to initially survey a site without digging things up and disturbing the area. This not only saves time and money that might be spent on extensive excavations, but can help protect gravesites, private property, or sensitive environmental areas.

Surveillance with ground-penetrating radar is suitable for scientific geophysical research, resource exploration, forensic searches for old gravesites or possible victims of homicide, and hunting for fossils and artifacts. The radar probe, which resembles a large handheld metal detector, can create an image that shows changes in density at different depths up to several feet. On a computer display, a grid may be overlaid on the radar pattern to help interpret the data.

Space Exploration and Astronomy

Radio waves can travel for vast distances through space and, thus, radar is an important tool of astrophysical research. It has been used, for example, as a measurement tool to pinpoint distances between the Earth and solid celestial bodies like planets and asteroids. There are many uses for radar in astronomy and interplanetary exploration and the reader is encouraged to consult other resources, as well as the National Aeronautics and Space Administration (NASA) at www.nasa.gov for information on radar technologies and pictures of radar imaging.

Weather Monitoring and Forecasting

Weather forecasting, an important application of radar, has already been mentioned and illustrated in earlier sections. In addition to systems that were discussed previously, some of the more recent applications for weather surveillance include the forecasting of flash floods. Next Generation Radar (NEXRAD) systems, for example, can monitor wind velocity, storm movement and intensity, data that can be used to warn residents of impending danger.

In 2004, *Environment Canada* announced completion of the installation of 31 Doppler radar systems as part of the national weather-forecasting system. The system can help detect blizzards, tornadoes, and heavy precipitation to a distance of about 250 kilometers.

Even air-traffic controllers are benefiting from radar-derived weather information. In 2002, the *Federal Aviation Administration* FAA/NWS radar-based weather network data was included in radar displays showing aircraft positional data. The system enables air-traffic controllers to assess aircraft movement in relation to weather systems and to reroute air traffic, if necessary.

6.d. Instrumentation and Navigation

Navigation

Radar is widely used for marine and air navigation, and experimental vehicle navigation systems are under development that may support autonomous vehicles and collision-avoidance systems.

Radar is used in marine navigation to locate landmasses, harbors, and docking areas, potentially dangerous obstacles, and moving surface-marine or sub-marine craft.

Radar is also used in the aviation industry to safely get aircraft on and off the ground without colliding with other aircraft, buildings, or terrain that might be obscured by severe rain, snow, or fog. The radar systems used in air traffic control are almost entirely pulsed systems divided further into *primary radar* and *secondary surveillance radar* (SSR).

- High-resolution *primary radar systems* are used to track the movement of airport ground traffic, aircraft approaches, and en route air travel. They developed from military Identification Friend or Foe (IFF) systems, adding the Selective Identification Feature (SIF), a code that was crucial to tracking individual aircraft.

- *Secondary surveillance radar* (SSR) can intercept and interpret more information than a primary system, including altitude, hijacking, communications difficulties, and unspecified emergencies. The ground interrogator and the aircraft transponder communicate on different frequencies to reduce interference between incoming and outgoing signals. SSR systems are beginning to use tracking systems derived from missile guidance technologies for angle measurement.

Radar altimeters are used for both ground and air applications, including navigation and weather monitoring. They are also used for extraterrestrial applications, such as interplanetary space explorers such as the Mars lander.

Left: The F-16 Fighting Falcon is equipped with a Dispenser Weapons System 39 (DWS) which includes an onboard computer and a radar altimeter to automatically guide it to its target with data transmitted from the aircraft. Right: A Chinook (CH-47) helicopter brings relief supplies to victims of Hurricane Mitch in Honduras. The Chinook "Super D" models are equipped with a number of new instruments, including radar altimeters. [U.S. DoD 1994 news photo by Cindy Farmer, 1998 news photo by Thomas Cook, released.]

Personal Navigation

Solid-state miniaturization has resulted in smaller and smaller radar systems, to the point where a type of radar system has been incorporated into canes for the blind. For covert operations, such a system could be used for navigating in areas that are dark, or where light might increase the possibility of detection.

6.e. Security, Surveillance, and Defense

Radar is used in many ways to warn of movement, vessels, projectiles, and unidentified objects. From commercial security systems to satellite early warning, and ballistic missiles defense, radar is a versatile technology that can provide good information without a lot of interference from other types of signals.

Long-range radar systems warn of incoming missiles and aircraft. SPADATS (Space Detection and Tracking System) is a joint program of the United States and Canada used to monitor artificial satellites.

Some radar plotting is still done by hand but many of the plotting activities associated with radar monitoring are now automated by computer systems. Computerization enables the creation of 'animations' that can illustrate movement over time and the monitoring of individual targets (much the same way weather radar can show a storm system approaching the coast of Florida by playing a series of stop-action frames in quick succession).

Left: A synthetic-aperture radar (SAR) receiving information from the airborne Joint Surveil-lance Target Attack Radar System (J-STARS). A J-STARS Boeing E-8 aircraft was used to monitor vehicle and troop movements and relay them to stations in Hungary as part of the NATO Implementation Force (IFOR) in Bosnia and Herzegovina. Right: A navigation officer on a Boeing 707 plots a course on a laptop computer as part of J-STARS. In flight, the radar can detect and track more than 120 miles of terrain. A 40-foot radome under the forward fuselage houses a phased-array radar antenna which feeds information to Army and Air Force operators with access to realtime large-screen graphics consoles. [U.S. DoD 1996 news photo by L. Aaron, released.]

Aircraft patrols and satellite links have become an intrinsic part of early warning and defense systems utilizing radio waves for communications, and radar systems for specific sighting and imaging tasks.

Left: On a goodwill visit in 1989, various radar systems could be seen on the Soviet Slava-class cruiser Marshall Ustinov (CG-088). A Top Steer surveillance radar is mounted on top of the bridge. A Top Pair (Top Sail and Big Net) surveillance radar is situated further aft. Right: A U.S. Navy Aircraft Approach Controller monitoring an aircraft marine carrier radar display on the USS George Washington. The radar console aids the controller in managing air traffic flow and takeoffs and landings in the vicinity of the carrier. This mission to the Adriatic Sea was in support of the NATO Implementation Force (IFOR). [U.S. DoD 1989 news photo by Don S. Montgomery, 1996 news photo by Joe Hendricks, released.]

 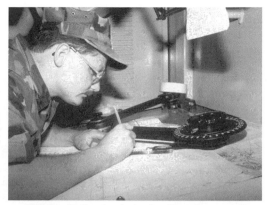

Left: A U.S. Navy Mobile Inshore Undersea Warfare Unit set up in Jordan for radar surveillance in 1985. Right: Inside a Mobile Unit control van, an Operations Specialist plots the location and progress of enemy ship contacts during an exercise in North Carolina in 1992. [U.S. DoD news photos by Mike Moore and William G. Davis, III, released.]

Left: An Air Defense Command radar tracking station in Trinidad. Right: An Air Defense Command radar on a snow-covered vantage point in Nova Scotia, Canada. [U.S. DoD 1968 news photos by Ken Hackman, released.]

Left: A Joint Service team assembles an Army-Navy Transportable Ground 75-Air Control and Warning Radar System on the McGregor Test Range in New Mexico. Right: An Airborne Early Warning E-2C Hawkeye lands on an aircraft carrier. Note the streamlined radar-protecting radome on top of the craft. [U.S. DoD 1996 news photo by Regina Height and 1990 news photo by Don S. Montgomery, released.]

Pave Paws, an Early Warning phased-array radar system designed to detect incoming sea-launched ballistic missiles (left, in 1995). The display console (middle) and computer room (right) of the An/FPS-115 Pave Paws system in 1986. Data storage and analysis are important aspects of radar monitoring and imaging. High-storage-capacity hard drive and tape systems have now superseded most of the older computer formats. In 1986, Random-Access Memory (RAM) sold for $700 MByte compared to $1.50 MByte in 1999 and hard drives sold in 1986 for hundreds of dollars for only 20 megabytes whereas $200 would buy over 4 GBytes in 1999. Thus, substantial changes in capabilities have occurred not only in storage capacity and data access, but in the speed and size of display consoles (especially raster-based digital consoles). [U.S. DoD news photos by Ken Wright and Don Sutherland, released.]

Projectile Detection and Guidance

Radar is used to to detect and track incoming missiles, and to guide outgoing missiles. Radar was used to counter Iraqi Patriot missiles after they had been fixed. The Patriot missiles themselves were guided by radar and used a local radar to increase targeting precision.

Fire-control computers are used to help determine the path of a guided missile and Missile Guidance Radar (MGR) is used for homing guidance. Missile Tracking Radar (MTR) can be replaced or supplemented by a high-power continuous-wave illuminator (CWI) radar that is interfaced with the Target Tracking Radar (TTR) to home in on the Doppler frequencies generated by the target.

An AN/MPQ-53 radar set for the MIM-104 Patriot tactical air defense system deployed during Operation Desert Shield. It uses a multifunction, phased-array device for surveillance, tracking, and guidance. Note that the sandy 'terrain' around the radar unit is actually camouflage. [U.S. Air Force news photo, released.]

Radar can be used to track a variety of incoming projectiles, including hostile fire.

The AN/TPQ-36 Fire Finder radar system is used by the U.S. Army to monitor incoming projectiles (rockets and mortars as small as .50-caliber), calculating their origin and possible landing location. The Target Acquisition Battery Bravo 25th Field Artillery Regiment shown here was deployed as part of a NATO Implementation Force (IFOR) in Bosnia and Herzegovina. Similar AN/TPQ-27 Fire Finding radars were also used. [U.S. DoD 1996 news photos by Lisa Zunzanyika-Carpenter, released.]

Mine detection

In spite of technological evolution in developed nations, there are still many places in the world where the chief mine-detection tool is a long stick. Success at finding a mine often results in severe consequences for the person carrying the stick. The technique is inefficient and impractical in regions of active combat. The next best commonly available, low-cost tool, a metal detector, is not suitable for locating plastic explosives. As a consequence, a number of low- to medium-cost solutions have been proposed, some of them incorporating radar technology. Ground-penetrating radar (GPR) systems have been tested in a number of soil types using a variety of frequencies, including X-band, C-band, and L-band ranges [Hanson et al., 1992].

One mine-detection strategy of interest is a micropower impulse radar (MIR) developed by *Lawrence Livermore National Laboratory* (LLNL). A self-contained, compact, low power, ultra-wideband radar impulse can be assembled into arrays and the data used to create two-dimensional and three-dimensional images of objects in sand, soil, or even concrete. Both metallic and plastic landmines have been detected in 5 to 10 cm of moist soil in test runs, with depths of up to 30 cm for dry soils. Other researchers, investigating the practicality of detecting landmines with radar have suggested that winter conditions may significantly hinder detection, depending upon the type of road surface. Radar sensor data can be combined with infrared sensor data to provide a more complete picture of the topography. Mine-detection technology can also be applied to other ground-penetration surveillance activities, such as bunker or structure detection. The systems can be aircraft- or vehicle-mounted, depending upon the circumstances. GPR technology is somewhat generic and can be applied to other fields of study, including archaeological exploration and crime scene investigation.

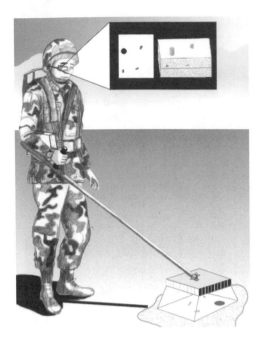 

Left: A handheld radar-based mine detector developed by the *Lawrence Livermore National Laboratory*. It employs micropower impulse radar technology to provide three-dimensional images of underground objects. Objects are intended to be identified by the size, shape, and mixture of reflected microwave frequencies. More than a dozen radars can be incorporated into the system, which also includes a waist-mounted computer and a 'heads-up' display for viewing the radar images. Top Right: One of several types of land mines. Bottom Right: A computer image showing lighter areas that indicate buried objects that might be mines. [LLNL news photo by Marsha Bell, released.]

6.f. Commercial Products

There are thousands of radar products, ranging from tiny car-mounted or handheld radar detectors to acreage-wide, large-scale-array, transceiver radar stations. There isn't room to catalog them all here, so instead, here is an annotated selection of some major companies supplying some representative products, especially medium- and large-scale national and international radar surveillance systems. These companies can help answer questions about more specific types of radar technologies, their applications, and their cost. Note: inclusion in this list does not imply endorsement of their products—the names are provided for information purposes only.

ARCTEC (a joint venture of ATCO Frontec Services Inc. and Arctic Slope World Services, Inc.) - Selected by the U.S. Air Force to manage the strategic northern network stations known collectively as the *Alaska Radar System* (ARS). The ARS control center is located in Anchorage. ARCTEC develops and operates the Solid State Phased Array Radar System (SSPARS), international stations that track and monitor orbiting satellites and ballistic missiles and Canada's North Warning System, which stretches across the Canadian arctic. The system is used for surveillance and tracking of commercial aircraft.

BAe Australia - Development, maintenance, and operations of Jindalee's Over The Horizon Radar (OTHR). This system, supplied to the Australian Defence Force (ADF), is a unique wide-area surveillance system.

The Boeing Company - Boeing is well-known for aircraft design and manufacture, including the manufacture of autonomous and remotely-controlled unstaffed aerial vehicles, but it also oversees multifirm contracts that include aircraft radar design and installation.

British Aerospace Defence Systems, in collaboration with the U.K. Defence Evaluation and Research Agency - Developers of the SAMPSON multi-function, active-array radar, which can be used in ground-based missile defence systems and in advanced air defence systems on naval destroyers. The SAMPSON can handle both tracking and surveillance with data from multiple incoming sources, and is highly resistant to interference from sea clutter and jamming.

California Microwave - Contracted by the U.S. Army Communications & Electronics Command for A-Kit surveillance equipment for Airborne Reconnaissance Low Multi-function (ARL-M) aircraft, to be completed by 2001.

Celsius Group - The Celsius Tech Electronic division supplies chaff, flare dispensers, and second generation radar warning receiver systems such as are being installed in the JAS 39 Gripen multirole combat aircraft. The systems are designed to detect, analyze, and evaluate hostile radar threats.

Hughes Aircraft Company - Provides APG-73 radars to projects led by The Boeing Company. See next listing.

Hughes International Corporation - This company is well-known for the design and manufacture of radar systems for command, control, and communications, including the AN/MPQ-64 Sentinel Air Defense Systems 3-D air defense radars that assist the U.S. Military in locating and identifying hostile targets such as aircraft, unstaffed air vehicles and cruise missiles. Designed with radiation and electronic countermeasures, these systems began production in the mid-1990s. Hughes has been involved with satellite sensing systems for many years.

Japan Radio Co., Ltd. (JRC.Co) - Established 1915, Japan Radio is a supplier of marine electronics and communications equipment, including ground surveillance radars, radar simulators, wireless systems, mobile earth stations, GPS-based vehicle location systems, graphic workstations, direction finders, and Doppler sounders. Tokyo, Japan.

Lockheed Martin Corporation - A well-known aerospace contractor, the Ocean, Radar & Sensor Systems division provides radar systems for the APS 145 which is installed in E2-C aircraft. Lockheed Martin, in conjunction with Boeing and Pratt & Whitney, under contract to the U.S. Air Force, rolled out the first F-22A Raptor fighter jet in spring 1997. The F-22 is gradually replacing the F-15.

Nav Canada - Provides products to Canada's civil air navigation services, including air traffic control systems, electronic navigation aids, and various computer and radar systems.

Northrup Grumman Corp. - Well-known in aerospace, the company was contracted by the Wright-Patterson Airforce Base, in 1997, to supply a radar microwave automatic test equipment shop for testing and repairing APG-68 fire control radar used in F-16 aircraft that are sold to foreign nations. Northrup Grumman also manufactures radar system improvement program (RSIP) kits through Boeing contracts.

Raytheon Company - Contracted by The Boeing Company to provide F-15 C/D aircraft radar upgrades as well as F/A-18E/F Super Hornet active, electronically scanned array (AESA) radar systems. These systems are intended to increase detection and tracking ranges and to improve image resolution in phased steps up to about 2004. Raytheon has cooperated on a number of radar-related projects with Northrup Grumman Corporation and Boeing. Raytheon has been contracted to provide ALR-67(V)3 radar warning receiver systems for F/A-18E/F Super Hornet aircraft, including countermeasures receivers. Features include advanced threat-detection capabilities, pulse Doppler and continuous-wave tracking, and identification of radio frequency and millimeter-wave threat emitters.

Textron Systems - Developers of ladar, laser radar systems for long-range detection, tracking, and imaging of a variety of types of targets.

Thomson-CSF Airsys - Developers of RAC 3D multimode, warning and coordination radar systems for use with very-short-range to medium-range surface-to air weapons systems. The system can also be used as a defense surveillance system for troops and conflict zones. It is useful in marine, land, and airforce operations. The RAC 3D system is operational in a number of NATO member nations.

7. Problems and Limitations

Attenuation (the gradual loss of signal) is an integral limitation of radar systems used over land through Earth's atmosphere. Severe weather conditions, mountains, intervening buildings or vessels other than the target vessel, etc. are all obstacles that may degrade or deflect the signal before it reaches its target or return destination. Attenuation tends to increase as the length of the waves decreases, while the corresponding frequency increases.

Interpretation of Displays and Radar Data

The effective use of radar displays takes training and experience. The interpretation of a radar signal is largely dependent upon the mode of representation (rectangular grid, conical scan, etc.), the type of display (raster or vector), the speed at which the display is updated, and the resolution of the display. The quantity of objects on the display and their speeds can also influence how easy or difficult it might be to assess radar targets. Here are some of the more common concerns in interpreting radar data:

- The actual size of objects relative to one another cannot be fully represented on a small screen. The difference in size between the smallest pixel or pip on a radar screen and the entire screen image is much less than the difference in size between a dog and a house. Even common objects, if they are small, require a resolution of about 300 dots per inch to clearly represent details and relative sizes, whereas most raster monitors used on computerized radar systems have a resolution of about 72 or 96 dots-per-inch.

- Fast movement can be missed on a radar display. For example, movie film frame rates that simulate 'natural' movement range from about 24 to 30 frames per second (about 1600 per minute). In other words, 24 to 30 still frames presented in succession can simulate a horse running or a plane flying through the clouds so they appear natural to us. Since most radar systems are pulsed, a radar 'frame' is similar to a film frame, in that it is a 'snapshot' of a moment in time. It would seem logical to pulse the radar at a rate of 24 to 30 frames per second for the convenience of human viewers, but this is technologically difficult and expensive. The rate of the pulse is related to the signal range and the speed of movement of the transmitting unit, which may be a rotating antenna in some systems. An antenna rotation may be as low as six revolutions per minute.

- A radar display is typically 2D (although 3D systems are becoming more prevalent). There are always trade-offs when 3D information is represented on a 2D display.

- Radars can be 'fooled' by objects made with highly absorbent materials or materials that scatter the radio waves in many directions. Chaff is a long, stringy, reflective material that flutters in the wind to scatter electromagnetic radiation. This is a common radar countermeasure intended to confuse the sensor and, in particular, to influence the homing direction of incoming projectiles.

- Radar data can provide erroneous information if the target is deliberately using highly reflective materials spaced out over an area to make something look bigger than it actually is. For example, reflecting posts and surfaces can be put outside the corners of a small encampment to make it appear larger.

- While radio waves travel quite readily through air and space, they are refracted and impeded in water. Sonar, using acoustical waves rather than radio waves, is usually used in liquid environments.

Antennas

Since the length or diameter of an antenna must mathematically related to the length of the radio wave being apprehended, longer radio waves typically require larger antennas. In domestic defense installations, large antenna arrays can be installed on domestic or foreign ally property. But, since many surveillance activities require that the sensing system be hidden, not easily recognized, or mounted on a moving vehicle, a smaller antenna is generally preferred. Thus, the frequencies that can be used will, in part, be limited by the size and shape of the transmitting and receiving antennas.

Since antenna size and frequency are closely related, and most antennas are of a fixed size, many radar systems will operate on only one frequency, or on a limited number of frequencies. This imposes a further limitation on the covert nature of radar use. When scanning repeatedly on the same frequency, it is harder to hide the signal and to prevent jamming.

Radar Countermeasures

There are many ways to counter a radar probe, including radar-reflecting barricades, radar-absorbent or radar-scattering paints and filaments, and frequency jamming. All are used in military operations.

Radar countermeasures are tied to the type of radar, the frequency, and the type(s) of antennas used. Monopulse tracking radar antennas are designed with multiple inputs to the receiver so that calculations can be derived from signal pairs. Thus, a single unobtrusive pulse, aimed at a target, can provide information about that target. This is especially valuable in covert operations where transmitted signals are best kept to a minimum. However, because it depends upon a minimum level of transmissions, without repeat verification pulses, this type of system, if detected, is easily countered with a series of pulses designed to confuse the receiving system [Beam 1989].

The Blackbird SR-71, developed in the late 1950s and flown in 1964, is still one of the most unique, fastest, highest-flying stealth aircraft in the world. The Blackbird, or "habu" as it is sometimes called, was the first generation of U.S. military "stealth" aircraft utilizing radar-absorbent materials to yield a radar cross-section or *signature* of less than ten square meters.

Radar-resistant design characteristics, including absorbent and reflective materials, were also used on the classified F-117 Nighthawk fighter bomber plane that was developed in the 1970s (but not shown to the public until the late 1980s). Similarly, the Tacit Blue Technology Demonstration Program stealth aircraft with a low radar signature was flown in the early 1980s and unveiled to the public in 1996.

The B-2 'batwing' stealth bomber is another U.S. military craft that is radar-resistant. The B-2 was used in combat in March 1999.

TheF-22 Raptor fighter jet is a more recent stealth fighter aircraft with a low radar signature. It uses a radar-resistant 'skin' to lower its radar signature. Manufacturing these new stealth aircraft is a high-precision job. Tolerances for attaching the skin are very fine—0.009 inches. The Raptor has a passive radar receiver to hide the transmission signature—the active transmitter is operated offboard. This F-22 is the replacement for older F-15 models. The F-22s use radar systems similar to those retrofitted into the older F-15s.

Left: The Tacit Blue program demonstrates low-detection properties of stealth aircraft. The Tacit Blue was designed with low-detection-probability structure and materials and a low-probability-of-intercept radar and other sensors. From the side, it resembles an aerodynamically-sleek motorhome with very thin, tapered wings and tail. It was a surveillance craft for covertly monitoring battle lines providing intelligence and target information in realtime to ground command centers. This aircraft flew for three years, beginning in February 1982. Right: The F-117 "Nighthawk" was designed as a stealth plane with a particular type of design and radar-absorbing materials built into the structure of the plane to significantly lower the radar signature (the infrared signature is lowered as well). There are a few aperture openings for sensors, but the apertures are basically hidden in this model. [U.S. DoD 1996 news photo and 1999 news photo by Mitch Fuqua, released.]

More and more, as the technology evolves, automatic countermeasure capabilities are built into radar systems. For example, under Northrop Grumman, the Signal Technology Corporation is contracting to build internally installed Tactical Electronic Warfare (TEWs_ systems for the AN/ALQ-135(V), intended to automatically detect and jam hostile radar signals.

Bandwidth and Frequency

In the early days of radio wave experimentation and broadcasting, frequency overlap and crowding weren't significant problems, but as the technology improved and was adapted to commercial enterprises, the demand for and limitations of the airwaves became apparent. High-power stations overpowered low-power stations and amateur radio operators. Signals transmitted at the same frequency interfered with one another, and other problems arose. To manage the demand for airwaves, the *Federal Communications Commission* (FCC) was established as a result of the *Communications Act of 1934*. It was implemented to regulate the American broadcast industry by allotting frequencies, time slots, and call signs. Later, it was further charged with controlling consumer products emissions from computers, telephones, etc.

Anyone manufacturing or operating radar devices must comply with the restrictions set out by the FCC. Certain low-power, short-range devices may be used in certain circumstances, but these are generally in the hobbyist/educational category and not widely used for surveillance activities, and most low-powered devices that are suitable for surveillance are regulated by various privacy laws, even if the frequencies are loosely enforced. As a result of federal regulation, the bandwidths assigned to radar are defined, with adjustments from time to time (usually when technologies improve and formerly 'useless' frequencies become valuable commodities).

In addition to regulatory limitations, radar frequencies have inherent physical limitations.

Radar operates over a wide variety of frequencies, but not all are suitable for use in adverse weather conditions. In long-range surveillance activities, scattering can be a significant problem in wet conditions when frequencies above 10 GHz are used. Then again, what limits one application might benefit another. If you are specifically trying to detect turbulent weather conditions, then scattering becomes a useful detection device rather than a liability, and frequencies below 10 GHz would be inappropriate because they would pass right through the storm without echoing back.

Incursions/Interference with Amateur Radio

Amateur radio enthusiasts have pioneered many of the technological advances in broadcast and remote-sensing applications. They launched some of the first satellites, developed telemetry technologies, and showed how former 'junk' wavelengths could be used in practical applications. As a consequence, there have been many debates between amateur radio operators and the FCC regarding radio frequency allocations.

The military and commercial satellite industries have grown significantly since the mid-1990s, but they require radio bandwidth to control their systems' locations and orientations through *telemetry* and also need it to communicate their data back to Earth. The developers of low-orbit satellites (LEOs), for example, wanted frequencies set aside for their use that were already in use by amateur radio operators.

Wind-profiler radar systems operating near 50, 449, and 1000 MHz have the potential to interfere with amateur radio. These systems are used by weather forecasters to chart wind patterns in the higher atmosphere. Another problem for amateur radio operators (called 'hams') is increased interference on some amateur UHF allocations from Earth Exploration Satellites (EES) that arc used for mapping by synthetic-aperture radars (SARs).

8. Restrictions and Regulations

Radio frequency resources available for radar are already severely stretched, especially since some of the frequencies are also used by television and radio networks, ham radio operators, educational institutions, and scientific research labs. They are increasingly needed for consumer items, as well, including cordless phones, cell phones, personal radios, and many kinds of intercoms and security systems.

Because radar is based on the use of radio waves, nearly all aspects of radar are regulated in the U.S. by the FCC. There may be a few short-range hobby and educational applications that are loosely regulated but, for the most part, FCC approval must be obtained before radar units can be sold commercially and FCC broadcast regulations must be followed, including appropriate FCC licensing, for their use.

A stronger radio signal will overpower a weaker signal if the competing systems are near the same frequency. AM stations can overlap or drift and produce fuzzy audio. FM stations avoid some of these problems and provide a cleaner signal by maintaining 'guard bands', unused frequencies, on each side of the transmission frequencies. For these reasons, specific frequencies are designated for particular types of tasks, licenses are allocated by the FCC for broadcast stations, and the whole industry must be carefully organized and monitored to make sure there are enough radio frequencies to go around. With the steadily increasing world population, now over 6.4 billion, this is becoming more difficult than ever.

FCC licensing is required for the calibration of radar units and radar operator training and certification programs are in place to provide minimum competency skills for the use of radar (particularly for traffic or air navigation radar systems).

The *Institute of Electrical and Electronic Engineers* (IEEE) has established standards for the letter-band terminology used with reference to radar (Standard 521-1984) and the U.S. *Department of Defense* (DoD) lists the letter-band terminology in its *Index of Specifications and Standards*.

International regulations also apply to radio frequencies used by commercial and military aircraft and various satellite transmitters and receivers.

Vehicle Radar Detectors

The use of vehicle-mounted radar detectors is prohibited or strictly regulated in many parts of the world. In general, speeding is seen as a danger to innocent victims, including passengers and other motorists, and law enforcement officers are hired to uphold traffic laws. Radar is one of the tools commonly used to detect and record excessive vehicular speed. Speeding is also seen as a problem by insurance companies, who have to pay out large sums of money as a result of damage caused by speeding, a cost that is partially borne by other drivers. Since speeding is a voluntary action, accidents due to speeding are considered by insurance and law enforcement agents (and innocent victims) to be unnecessary and preventable.

There are laws and guidelines determining how police radar must be calibrated, used, and maintained, along with regulations on the use (or prohibition) of radar detectors. By the mid-1960s, it was generally accepted that stationary traffic-monitoring radar, when properly calibrated and operated, could yield accurate vehicle speeds. By the mid-1970s operating guidelines for moving radar were in development, and minimum training standards for operators were gradually established.

Vehicle radar systems are not used only for traffic enforcement. Radar navigation systems may be incorporated into surface vehicles. As a result of requests for bandwidth for experimental systems, the FCC, in 1998, temporarily restricted amateur access to 76 to 77 GHz to provide bandwidth for commercial development and use of frequencies for applications such as vehicle radar collision-avoidance systems. This occurred to the dismay of amateur radio operators, in spite of reports that incompatibility would not be a problem.

The FCC often applies compensatory trade-offs when bandwidth is taken from one set of applications and reallocated for another. In one case, the FCC upgraded amateur and amateur-satellite allocation of 77.5 to 78 GHz from secondary to coprimary.

9. Implications of Use

Since radar cannot be seen or heard, it is widely used as a surveillance technology, but it is rarely used in the day-to-day surveillance of personal or business activities in the same way that audio or video devices. For this reason, there is much less controversy over the use of radar than of many other technologies. Radar has been applied in many safety and security applications, particularly air navigation and weather forecasting, and we all benefit from the technology in these ways. FCC regulation of radio frequencies has helped to prevent unethical uses of the technology by requiring that radio devices be approved before sale and that specific frequencies be used for specific types of tasks.

On the whole, up to this point, radar appears to have provided more benefits than negative consequences.

10. Resources

Inclusion of the following companies does not constitute nor imply an endorsement of their products and services and, conversely, does not imply their endorsement of the contents of this text.

10.a. Organizations

Army Research Office (ARO) - This North Carolina-based U.S. Department of Army office considers funding bids for research and development in leading-edge fields such as communications, surveillance and countersurveillance, including radar and lidar technologies. www.aro.ncren.net/research/baa99-1/baa99.htm

ATC Radar Tracker and Server system (ARTAS) - A EUROCONTROL project involved in the development of ARTAS V3U installed at Schiphol in use as a primary surveillance data source since 1998 with further versions in development since December 1997 the most current being ARTAS 1 V5. ARTAS comprises tracker, server, recorder, and system manager stations along with associated FDDI-ring computer network architectures. A router bridge is used to interface with external sources of data, including radar data sources.

Automatic Dependent Surveillance (ADS) - A EUROCONTROL project for implementation and operation of ADS in Europe, including evaluation of ADS-B and ADS-C technologies in terms of cost efficiency and safety.

Defence Evaluation and Research Agency (DERA) - A U.K. Ministry of Defence agency involved with non-nuclear research, technology, testing, and evaluation. DERA, with a staff of about 12,000, is one of Britain's largest research facilities, involved with research and testing of aviation technologies, electronics, command and information systems, sensors, weapons systems, and space technologies. DERA provides research data on airborne radar, target recognition techniques and active/passive microwave ground radar systems. www.dra.hmg.gb/

EUROCONTROL - *European Organisation for the Safety of Air Navigation.* An international organization founded in 1960 for overseeing air traffic control in the upper airspace of almost 30 European member states. www.eurocontrol.be/

EUROCONTROL Surveillance Team - A working team of EUROCONTROL since 1998 for the development of long-term surveillance strategies which include evaluation of Mode S Surveillance strategy/technology. Meets primarily in Brussels, but there have been meetings in Italy and Germany as well.

The Federation of American Scientists - A privately funded, nonprofit research, analysis, and advocacy organization focusing on science, technology, and public policy, especially in matters of global security, weapons science and policies, and space policies. Sponsors include a large number of Nobel Prize Laureates. FAS evolved from the Federation of Atomic Scientists in 1945. The site contains numerous surveillance, guidance systems, U.S. Naval documentation, and other radar-related information. www.fas.org/index.html

High Frequency Active Auroral Research Program (HAARP) - Studies the properties and behavior of the ionosphere, with particular emphasis on communications and surveillance systems for civilian and defense purposes.

Joint Surveillance and Target Attack Radar System (J-STARS) - A joint effort of the U.S. Air Force and Army program. The goal is common battle management and targeting capability for detecting, locating, classifying, and tracking stationary and moving targets. Thus, radar-based craft transmitting to ground stations via secure surveillance and control data links can serve as warning systems, and as a means to attack long-range targets. Test systems were deployed starting in the mid-1990s.

Surveillance Analysis Support System (SASS-C) - An ATC center-based Surveillance Analysis Workbench for ATC Radar and Tracker performance measurement. A EUROCONTROL project, SASS-C embodies radar data acquisition, classification, and object detection. It also includes other radar-related technology including radar plot feeders and radar plot filters.

U.K. Defence Research Agency - Includes Naval Radar Research division. Involved in the development of multifunction radar systems for weapons systems.

U.S. Naval Space Command - This arm of the military includes a Fleet Surveillance Support Command and a Satellite Operations Center which are involved in surveillance and warning activities through the monitoring of sea and airspace and orbiting satellites. Field stations along the U.S. southern regions, along with tracking and communications systems, comprise an electronic 'fence.' Data are gathered, cataloged, and evaluated on a more-or-less continuous basis.

10.b. Print Resources

Barton, David K.; Ward, Harold R., *Handbook of Radar Measurement,* Dedham, Ma.: Artech House, 1984, 426 pages.

Beam, Walter R., *Command, Control, and Communications Systems Engineering,* New York: McGraw-Hill Publishing, 1989, 339 pages.

Blackman, Samuel S., *Multiple-Target Tracking with Radar Applications,* Artech House, 1986, 449. While a little out of date, this is still considered a good reference for those needing to understand and solve target tracking problems.

Brown, R. Hanbury, Boffin: *A Personal Story of the Early Days of Radar, Radio Astronomy and Quantum Optics,* Bristol, New York: Adam Hilger, 1991, 192 pages. A personal account of the development of ground and airborne detection radar, and early radar applications in astronomy.

Buderi, Robert, *The Invention that Changed the World: How a Small Group of Radar Pioneers Won the Second World War and Launched a Technical Revolution,* New York: Touchstone Books, 1998, 576 pages. A lively, detailed account of radar as it was used in World War II.

Chrzanowski, Edward J., *Active Radar Electronic Countermeasures,* Artech House, 1990, 246 pages. Written to accompany an intensive course on the subject, this book describes radar systems designed to interfere with hostile radar systems.

Curlander, John C.; McDonough, Robert N., *Synthetic Aperture Radar: Systems and Signal Processing,* New York: John Wiley & Sons, 1991, 647 pages. Handbook for SAR imaging systems theory and design.

de Arcangelis, Mario, *Electronic Warfare: From the Battle of Tsushima to the Falklands,* Poole, Dorset, U.K.: Blandford, 1985. Discussions of electronic intelligence (ELINT) and electronic countermeasures (ECM) with an introductory history.

Devereux, Tony, *Messenger Gods of Battle. Radio, Radar, Sonar: The Story of Electronics in War,* London: Brassy's, 1991. Historical information and basic physical principles of electronics used in warfare.

Dillard, Robin A.; Dillard, George M., *Detectability of Spread-Spectrum Signals,* Artech House, 1989, 149 pages. Reference for engineering designers for intercept-reduced communications.

Hovanessian, S. A., *Radar Detection and Tracking Systems,* Dedham, Ma.: Artech House, 1978.

Jelalian, Albert Y., Editor, *Laser Radar Systems,* Artech House, 1992, 292 pages. Engineering design reference.

Levanon, Nadav, *Radar Principles,* New York: John Wiley & Sons, 1988, 320 pages. Includes technical theory, techniques, practical examples, and academic exercises.

Long, Maurice W., *Airborne Early Warning System Concepts,* Artech House, 1992, 519 pages. General reference on systems and trends.

Lothes, Robert N.; Wiley, Richard G.; Szymanski, Michael B., *Radar Vulnerability to Jamming,* Artech House, 1990, 247 pages. Engineering course on jamming, deception, and radar analysis.

Mardia, Hemant Kumar, *Digital Signal Processing for Radar Recognition in Dense Radar Environments,* Ph.D. thesis, The University of Leeds, 1987. Discusses the use of DSP in radar surveillance systems and describes the design of a radar warning receiver and an electronic surveillance measures receiver.

Price, Alfred, *The History of U.S. Electronic Warfare: The Renaissance Years, 1946 to 1964,* Alexandria, VA: Association of Old Crows, 1989. History of the development and application of U.S. electronic warfare, including the gathering of electronic intelligence and countermeasures.

Rowan, Richard Wilmer; Deindorfer, Robert, Secret Service: 33 Centuries of Espionage, New York: Hawthorn, 1967, 786 pages.

Skolnik, Merrill I., *Introduction to Radar Systems,* New York: McGraw-Hill, 1986 (2nd edition). In spite of being somewhat dated, this text is considered to have good coverage of radar technology at a high undergraduate or beginning graduate level.

Skolnik, Merrill I., *Radar Handbook,* New York: McGraw-Hill, 1989 (2nd edition). A comprehensive updated version of a 1970 reference for radar engineers which includes digital and Doppler radar, radar subsystems, and more.

Stevens, Michael, *Secondary Surveillance Radar,* Artech House, 1988, 300 pages.

Stimson, George W., *Introduction to Airborne Radar,* El Segundo, Ca.: Hughes Aircraft Company. There is also a second edition by Scitech Pub., 584 pages. Lucid, updated, well-illustrated text of radar techniques.

Wehner, Donald R., *High-Resolution Radar,* Norwood, Ma.: Artech House, 1994 (2nd edition), 593 pages. Spatial radar systems theory and design.

Wiley, Richard G., *Electronic Intelligence: The Analysis of Radar Signals,* Dedham, Ma.: Artech House, 1993 (2nd edition), 337 pages. Practical reference for systems design.

Wiley, Richard G., *Electronic Intelligence: The Interception of Radar Signals,* Norwood, Ma.: Artech House, 1985, 284 pages.

Articles

Bouchard, Joseph F., Guarding the Cold War Ramparts: The U.S. Navy's Role in Continental Air Defense, *Military Review,* Summer 1999. An interesting brief history of the establishment of early warning radar systems in air and ground stations during and following World War II.

Easton, R.L.; Fleming, J.J., The Navy Space Surveillance System, *Proceedings IRE,* April 1960, Vol. 48, p. 663-669.

Grossnick, Roy, Naval Aviation? A Century of Evolution, on the U.S. Navy site NANews. This is not specifically about radar, but it discusses the impact of technology on naval strategy and technological developments over the years. The author is head of the Aviation History Branch of the Naval Historical Center and Hal Andrews serves as technical advisor.

Hanson, J. V.; Evans, T. D.; Hevenor, R. A.; Ehlen, J., Mine Detection in Dry Soils Using Radar, *U.S. Army Topographic Engineering Center Report No. R-163*, March 1992.

McDonell, Michael, Lost Patrol, *Naval Aviation News*, June 1973, pp. 8-16. This is a sort-fact-from-fiction story about five TBM Avengers that disappeared in 1945 somewhere in the Bermuda Triangle.

McLean, J.W.; Murray, J.T., Streak tube lidar allows 3-D surveillance, *Laser Focus World*, Jan. 1998, p. 171-176.

Journals

IEEE has a number of collections of articles in proceedings from various engineering conferences, including *Radar and Signal Processing* and *Radar, Sonar, and Navigation*.

Journal of the Optical Society of America, includes articles on optical simulations of radar and other topics.

Quarterly Journal of the Royal Astronomical Society, includes articles on radar research on the upper atmosphere and ionosphere.

See also the journal listings in the Radio Surveillance and Sonar Surveillance chapters, as there is considerable overlap in the topics.

10.c. Conferences and Workshops

Many of these conferences are annual events that are held at approximately the same time each year, so even if the conference listings are outdated, they can still help you determine the frequency and sometimes the time of year of upcoming events. It is very common for international conferences to be held in a different city each year, so contact the organizers for current locations.

Most organizations describe upcoming conferences on the Web and may also archive conference proceedings from previous events for purchase or free download.

The following are organized according to the calendar month in which they are usually held.

IMDEX Asia Includes multifunction radar and surveillance keynotes and seminars. Sponsored by EDS.

European Radiocommunications Committee (ERC) Conference, held in Poland in 1998 at the invitation of the National Radiocommunications Agency of Poland.

International Radar Symposium (IRS). Includes radar surveillance topics.

EUSAR - European Conference on Synthetic Aperture Radar, Germany. All aspects of SAR, including sensors, signal processing, data management, and imaging.

IEEE International RADAR Conference, annual international conference.

International Conference on Radar Systems, annual international conference.

International Conference on Microwaves, Radar, and Wireless Communications, annual international conference held since the mid-1980s.

GPR, International Conference on Ground-Penetrating Radar, biennial international conference.

International Laser Radar Conference, annual international conference since the early 1980s.

European Conference on Radar Meteorology, annual international conference since 2000.

International Military Operations and Law Conference, annual conferences. In 2000, merged with the

USPACOM Legal Conference. Topics included the discussion of international rules for reconnaissance and surveillance of the coasts of other nations near military establishments.

ATCA, The Air Traffic Control Association annual meeting,. Numerous topics of radar use in air traffic services are included, especially upgrades and expansions in numerous countries. The use of satellite technology for safety and navigation, in expanses where radar surveillance is impractical or impossible, are also discussed.

10.d. Online Sites

Defence Systems Daily: The Internet Defence & Aerospace News Daily (DSD). Topical articles and searchable archives with numerous news items on specific commercial products utilizing radar technologies. defence-data.com/

The Federation of American Scientists' Military Analysis Network. This site has Navy documents online which can be read on the Web. Two sections in particular, Fundamentals of Naval Weapons Systems, and Laser Fundamentals are of interest to surveillance. www.fas.org/man/dod-101/navy/docs/index.html

The Radar Operator Course. This is a 100-image traffic radar slide show summarizing radar history, legal issues, Doppler principles, radar theory, beam reflection, shadowing, and phenomena that can interfere with the radar signals. The course was organized by Ian Wallace of the Washington State Criminal Justice Training Commission, based on the national Highway Traffic Safety Administration radar courses. www.wa.gov/cjt/radar/index.htm

The USS Spinax (SSR 489). This site is devoted to stories and pictures (lots of pictures) about the crew and the operations of the Spinax, the first U.S. radar picket submarine. www.spinax.com/

10.e. Media Resources

Air Defense, from the History Channel *Weapons at War* series. The story of radar, rockets, and anti-aircraft guns that defend against airborne attacks. Shows the British labs where radar was developed during the War. Includes combat footage shot aboard ships in the Pacific. Shows anti-aircraft technology in the Korean, Vietnam, and Gulf wars. VHS, 50 minutes, cannot be shipped outside the U.S. and Canada.

Radar, from the History Channel *Modern Marvels* series. Radar as a pivotal technology in World War II, particularly the Battle of Britain. Shows the cavity magnetron and provides inside accounts on the development of radar during the War. VHS, 50 minutes, cannot be shipped outside the U.S. and Canada.

Stealth Technology, from the History Channel *Modern Marvels* series. Introduces the stealth technology of the F-117 Nighthawk and larger B-2 bomber aircraft. Describes how these stealth planes avoid detection by radar and other technologies. VHS, 50 minutes, cannot be shipped outside the U.S. and Canada.

11. Glossary

In the field of radar, which is used throughout military operations, acronyms and abbreviations are so numerous, they cannot all be included here, but there are some of interest.

Titles, product names, organizations, and specific military designations are capitalized; common generic and colloquial terms and phrases are not.

AASR	advanced airborne surveillance radar
ABF	adaptive beam-forming. An error-compensation system for antenna arrays.
ADAR	air defense radar
ADS-B	Automatic Dependent Surveillance - Broadcast
AESA	active electronically scanned array. A type of radar array being adapted for use in stealth fighter jets.
AEW	Airborne Early Warning
ARPA	automatic radar plotting aid
ARS	airborne radar system, aerial reconnaissance and surveillance
ARTS	Automated Radar Terminal Systems. See STARS.
ASARS	advanced synthetic aperture radar system
ASR	airport surveillance radar
ATAR	advanced tactical radar, automatic target recognition
AWACS	Airborne Warning and Control System. Electronic surveillance aircraft. In 1999, The Boeing Company retrofitted a number of the aircraft through Daimler-Chrysler Aerospace with Radar System Improvement Program (RSIP) kits. These kits upgrade the radar computers and control maintenance panels to detect smaller, stealthier craft. In E-3 aircraft, the upgrade improves the sensitivity of the pulse Doppler radar and provides greater anti-jamming capabilities. RSIP kits have also been used to upgrade U.K. Sentry AEW1 aircraft. Upgrades to older aircraft also often include GPS improvements as this technology has evolved substantially in the last 5 years. See Joint STARS.
Bragg scatter	Scatter of radar echoes resulting from very small fluctuations relative to the length of the radar waves, usually occurring in the atmosphere.
BRWL	bistatic radar for weapons location
BMEWS	Ballistic Missile Early Warning System. A U.S. government radar-based missile warning system installed along the arctic from Greenland to England.
BWER	bounded weak echo region. A region of weak radar reflectivity which is bounded by a differently reflecting environment, as might occur in a storm with variations in updraft and precipitation.
CANOPUS	Canadian satellite observing systems incorporating advance imaging radar, optical, and sensing instruments.
CANTASS	Canadian Surveillance Towed Array Sonar System. A long array of small hydrophones which is towed. Used in submarine sensing and warfare.
chaff	stringy materials, often made of metal, designed to influence radar probes, often as a radar countermeasure. Naval ships often have chaff 'tubs' on their decks. Chaff is sometimes also dropped by balloons. It is sometimes used to test radar systems. A chaff 'corridor' is sometimes called 'confetti.'
DAR	defense acquisition radar
EA-6B Prowler	U.S. Air and Naval electronic warfare aircraft equipped with radar communications jamming systems. However, new technologies are always being developed, and enemy use of frequencies beyond those of the EA-6B is likely.
ECDIS	electronic chart display and information system
EFIS	electronic flight instrument system

ESM	electronic surveillance measures
ETRAC	enhanced tactical radar correlator
FAAR	forward area alerting radar
FFR	fire finder/finding radar
FIRESTORM	Federation of Intelligence, Reconnaissance, Surveillance, and Targeting, Operations and Research Models
FLIR	forward-looking infrared radar
FLTTD	Field Ladar Technical Transition Demonstration. A laser radar program of the U.S. Army.
FOPEN	foliage penetration
FOSIC	Fleet Ocean Surveillance Information Center
FSS	frequency surveillance system
GBR, GSR	ground-based radar, ground surveillance radar
GCA	ground control approach. Aircraft landing systems which typically include radar systems monitored by air traffic controllers.
GPR	ground-penetrating radar
GSTS	ground-based surveillance and tracking system
HIPAR	high power acquisition radar
IEC 60872–1	an international standard prepared by the IEC technical committee related to automatic radar plotting aids (ARPA) used in maritime navigation and radio communications equipment
IEC 61174	an international standard prepared by the IEC technical committee related to maritime navigation and radio communications equipment
IES	A machine vision, radar signal detection, and military reconnaissance system which interprets imagery to provide information on movement, as of troops and other objects (Matzkevich, 1995).
IFSAR	interferometric synthetic-aperture radar
ISAR	inverse synthetic-aperture radar
ISTA	intelligence, surveillance, and target acquisition
Joint STARS, J-STARS	Joint Surveillance and Target Attack Radar System. An air-to-ground E-8C aircraft surveillance system for locating, classifying, and tracking ground targets. See AWACS.
JTIDS	Joint Tactical Information Display System. Tactical information provided through a digital link. This is a passive display system that can provide information about the surrounding environment without generating a transmissions signature. Suitable for use in stealth and fighter aircraft to indicate the positions of other aircraft or proximate objects.
ladar/lidar	laser radar/light radar. Can be used for weather information and for the detection of airborne chemical and biological agents. Suitable for long-range applications such as detection, tracking, and imaging hard and soft objects. In addition to the range and velocity information provided by traditional radar, ladar can be used to calculate length through a 'soft' object, for example, and can provide information about complex motion. The length-measuring capabilities are highly valuable in studying and imaging specific objects in order to determine their type and dimensions (tanks, helicopters, decoys, etc.).
LOPAR	low power acquisition radar
LRS	long-range surveillance
MIR	micropower impulse radar
MSTAR	man-portable surveillance and target acquisition radar
NASOG	North American Special Operations Group

NAVSPASUR	Naval Space Surveillance System. NAVSPASUR makes use of continuous wave (CW) multistatic radar systems to reflect energy beams to receiving stations where position and trajectory information is calculated. In 1965 the frequency was changed from the SPASUR system's 108 MHz to 216.9-217.1 MHz. Called NAVSPACECOM fence since the early 1990s.
NORAD1	North American Aerospace Defense Command
OTHR	over-the-horizon radar
PAR	precision-approach radar
PHALCON	phased-array L-band conformal radar. A commercial aircraft long-range radar system manufactured in Israel. Diplomatic feathers were ruffled when Israel offered this system to the Chinese for use in their Air Force. U.S. officials were concerned about the use of U.S. technology in products being sold to other foreign powers.
Radio Electronic Combat	
	An information warfare strategy which aims at disabling an enemy command and control infrastructure.
racon	radar beacon. A safety, homing, or location signal which transmits a particular signal which identifies the beacon. Beacons are intended to provide information to navigators of various types of moving vessels, such as ships, aircraft, and ground troop vehicles.
radar picket	A radar-equipped unit stationed some distance from a protected force to increase radar detection ranges.
RADINT	radar intelligence
RALT	radar altimeter
radome	a covering that protects radar equipment from weather and vandalism but which does not impede the travel of radio waves
RCS	radar cross-section. A measure of visibility on radar as based on theoretically perfect reflectance. Stealth vessels are designed to have a minimum RCS or radar 'signature.' The RCS represents the apparent radar reflectivity of an object or structure and may be larger or smaller than the actual object. The RCS can be manipulated by application of special paints or skins, nonmetallic composites, and by particular shapes that tend to scatter or have a low profile to the radar. Smooth contours or certain low profile rippled surfaces tend to lower the RCS, depending on the angle of reflection and the radar frequencies used.
RISTA	reconnaissance, intelligence, surveillance, and target acquisition
RSIP	radar system improvement kits. Technology to upgrade aging radar systems, especially on aircraft. Improvements usually include greater range and higher resolution radar imaging.
RSTA	reconnaissance, surveillance, and target acquisition
RSTER	Radar Surveillance Technology Experimental Radar. Originally a ground-based volume search radar which has since been used in the Mountaintop Program (mid-1990s) for studying technologies for airborne early warning systems (AEW). One of the goals of the Mountaintop Program has been to simulate an airborne surveillance environment. Data from the project are hosted in the CREST database.
SAGE	Semi-Automatic Ground Environment, a system for air-defense radar processing
SAR	small-aperture radar
SPADATS	Space Detection and Tracking System. A joint effort of the U.S. and Canada which tracks Earth-orbiting satellites.
SPASUR	U.S. Navy Space Surveillance system. Evolved in the early 1960s from the Vanguard Satellite Program which used a combination of intercepted and bounced signals from orbiting bodies to provide tracking and positioning information. This later was integrated into Naval Space and Command and NORAD. See NAVSPASUR.
SRIG	Surveillance, Reconnaissance, and Intelligence Group
SSR	Secondary Surveillance Radar

STARS   Standard Terminal Automation Replacement System - a joint Federal Aviation Administration (FAA) and Department of Defense (DoD) program to replace Automated Radar Terminal Systems (ARTS) and other older radar approach control facilities and towers with updated technology air traffic control services. The system, which features more powerful processing capabilities and larger color monitors to replace the small ARTS monochrome monitors, will aid in air traffic separation and sequencing, traffic alerts, weather advisories, and radar-based traffic vectoring.

TCAS   Traffic Alert Collision Avoidance System. A commercial, on-board aircraft radar system that queries the transponders of nearby craft and reports identification and location information.

TASR   terminal area surveillance radar

TMD   Theater Missile Defense. Ship-borne radars are one of the technologies used to carry out TMD missions.

TRACON   Terminal Radar Approach Control

TTR   target-tracking radar

VISTA   Very Intelligent Surveillance and Target Acquisition

WSR-88D   Weather Surveillance Radar 1988 Doppler, a component of the U.S. National Weather Service (NWS) located in the mountains of Puerto Rico since the mid-1990s. Doppler radar can be used to detect and track tornados and to help determine their velocities. Better definition and greater variety of information than previous conventional WSR-74S radar, especially for short-fuse warnings.

Electromagnetic Surveillance

7

Infrared

1. Introduction

This is the first of three chapters that combine to form *light surveillance* (sometimes called *optical surveillance*). The light spectrum includes infrared, visible, and ultraviolet spectra and is itself a subset of the *electromagnetic spectrum*. Infrared, visible, and ultraviolet devices are covered in Chapters 7, 8, and 10, respectively. Chapter 9, Aerial Surveillance, is based predominantly on the electromagnetic technologies discussed in Chapters 5 to 8.

All living and nonliving bodies in our environment radiate and reflect energy. Sometimes we can sense this energy. We have evolved eyes to specially sense the *visible spectrum*. Our skin can sense some frequencies in the *infrared* (IR) spectrum and, although we aren't highly conscious of *ultraviolet* (UV) radiation, our skin will show the effects of too much exposure to UV rays by getting sunburned. We have evolved senses that help us to detect situations that might be useful or harmful to us, in which we can locate food or mates, or in which we might be burned or otherwise injured. For the most part, we are equipped to detect levels or kinds of energy that have a direct impact on normal day-to-day life.

Many kinds of electromagnetic energy move freely through and around us all the time and most surveillance technologies depend upon remote-sensing of this electromagnetic energy. Some of this energy can be seen and heard, some of it can be felt, but much of it is 'invisible' to us and travels through our atmosphere, our walls, and our bodies, unperceived by our biological

Infrared images taken in frequencies overlapping visible frequencies, like this 1945 photo of a Japanese plane wreck near Mt. Aryat, can aid in surveillance activities, and often have a haunting aesthetic appeal, as well. Note the high contrast of the sky characteristic of visible light/infrared photos. True infrared images have a more ghostly character and do not show sharp details. [U.S. DoD historical photo by Harry Young, released.]

senses. This energy is ever-present and consistent in many respects, and thus can be harnessed and used in terms of its information-carrying capacity, even if we can't sense it directly. It is often converted from one form of energy to another in the course of its use.

Optical Technologies

Light-based phenomena can be manipulated with lenses and mirrors. This enables us to direct and convert energy to forms that are useful for surveillance. Telescopes and cameras fall within the category of *optical technologies* and are frequently used for reconnaissance and other data-gathering tasks. Devices that sense light are called *photodetectors*. Light includes energy that we can see (e.g., the colors of visible light) and energy that we can't see, including the *ultraviolet, visible,* and *infrared* spectra, extending from about 0.01 μm to about 1000 μm). These forms of radiant energy have both particle and wavelike properties. Chapters 7, 8, and 10 together provide related information on different forms of light.

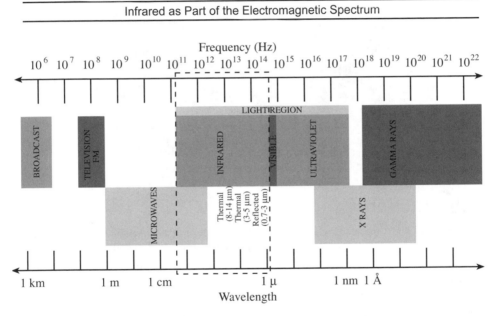

Infrared as Part of the Electromagnetic Spectrum

Infrared radiation is a region of longer wavelengths next to the visible spectrum and somewhat overlapping the shorter radio waves (microwaves). Infrared has been further subdivided into near-, middle- and far-infrared regions, though these 'boundaries' vary, depending on the application.

Representing Waves

Wavelengths are symbolically represented in two-dimensional diagrams as *sinusoidal curves*. One wavelength or repetition is called a *cycle*, or the distance from a point on a peak (or trough) to the corresponding point where the next cycle begins. Infrared wavelengths near the visible spectrum have shorter wavelengths, gradually getting longer as the frequency decreases toward the microwave region. There are no hard and fast boundaries — visible light transitions into infrared at the point where people no longer can see it, which varies from person to person — and infrared overlaps with the region of radio waves called *microwaves* because the distinctions are sometimes based upon the capabilities of the technology used to harness the energy, rather than changes in the phenomena themselves.

Two-Dimensional Representation of Wavelengths

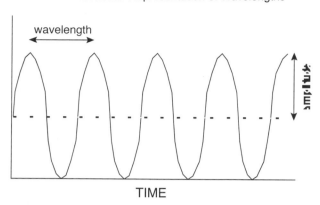

Wavelengths of radiant energy are often represented as repeating sinusoidal curves, with one wavelength shown as one cycle. Another way of representing a cycle is to imagine a second hand on an analog clock spinning around and around, with each new cycle starting when the hand passes 12. Since each new cycle would be overlaid on the previous cycle and would be hard to visually distinguish from the previous cycles, we usually 'stretch out' this graph so that each cycle appears as mountains and valleys rather than as continuous circles. Either way, the idea is to graphically represent repeating events of similar character.

Loss of Radiant Energy

Attenuation is a reduction or *diminution* of a signal from a variety of causes. Imagine a flashlight shining out into the dark. The light gets gradually dimmer at the extremes, until it no longer illuminates the objects around it. The light would continue indefinitely, except that it is scattered by reflective objects or absorbed by absorbent objects (which are usually dark or porous) and thus becomes progressively dimmer or 'weaker' as it travels away from the source.

Scatter and *absorption* are important concepts associated with attenuation.

- Scatter occurs when the radiation is reflected in a number of directions in such a way that there isn't sufficient 'signal' left to trigger a sensing device. Materials like *chaff*, strings of reflective material that deliberately scatter radar signals, are often used in countermeasures systems.

- Absorption occurs when the infrared radiation is absorbed by some object and rather than being reflected or re-emitted at the same wavelength, it may be re-emitted at some other wavelength that may not be in the range of the detector.

Units of Measure

Outside the light spectrum, the differences in the lengths of waves between very short waves in the *gamma-ray region* and very long waves in the *radio wave broadcast region* are huge, so huge they challenge the imagination. Because of these great differences, wavelengths are expressed in terms of their length in whatever units are easiest to handle mathematically and conceptually for a particular region of the spectrum.

Imagine trying to describe something very small, like a flu virus or bacterium. It isn't very practical to use feet or meters to describe microscopic sizes. Instead we use tiny units like *microns*. Now imagine trying to describe the circumference of the Earth or the distance to the Moon in microns. It's possible, but it's unwieldy and impractical—instead we use miles or kilometers. When describing wavelengths in the optical spectrum, we typically use *microns* or *micrometers* (e.g., 3 μ or 3 μm) or *nanometers*, which are 1,000 microns each (e.g., 3000 nm), and sometimes ångstroms (Å), which are discussed in the Ultraviolet chapter.

Infrared Radiation

Infrared is considered to be at the 'lower energy' end of the optical spectrum and the wavelengths associated with infrared are longer than those associated with the visible and ultraviolet regions. Some sensing devices that operate in the optical frequencies are *broadband* devices, in other words they not only detect infrared, but also visible and ultraviolet frequencies.

Infrared is invisible to human eyes, but it is not completely 'invisible' to human senses. We can feel heat from a fire without touching the flames, even with our eyes closed, because our nervous system is designed to sense infrared radiation in certain *thermal* energy regions that might do us harm. However, this sensitivity is not as specific or as well-developed as sight and hearing. In contrast to humans, some animals have very well-developed infrared sensors to aid them in locating prey or hosts, especially in low light. Several varieties of snakes have an organ in a 'pit' on each side of the head, between the nostrils and the eyes, that senses infrared to a few feet. The distance between the pits, like the distance between the eyes, aids the creature in locating a radiant source. These sensors may also help snakes to find 'warm spots' in which to sunbathe, since they are cold-blooded reptiles and must seek heat from external sources. Snakes are not the only creatures with heightened infrared sensitivity—some parasitic creatures, like the bee-mite, use infrared sensors to help them locate hosts.

Infrared radiation can be generated in many ways, through electrical activity, biochemical reactions, combustion, and nuclear interactions that are natural or manmade. Infrared comprises about half the radiation emitted by the Sun and is the primary radiation from many types of bodies. Directed infrared beams can be generated by lasers. Natural and synthetic infrared sensors enable us to detect, measure, and record this kind of radiation, which is then usually converted to electrical signals for processing and viewing. Within the infrared region, there are further categorical distinctions that will be discussed in the following sections.

Infrared radiation is all around us. Unlike radio waves, which are mostly synthetic, infrared is emitted or reflected by all bodies within our solar system. This makes it both a useful and a challenging technology. Infrared is useful because it is almost everywhere. We don't have to look hard to find it. But, infrared is challenging for the same reason—it is almost everywhere. If there's too much of it, it's harder to distinguish the relevant and useful information from all the other radiated infrared that is likely to be detected at the same time. Locating the information we want from an infrared scan can be like locating a friend in a crowded sports stadium. More than any other surveillance technology, infrared is context-sensitive, so Section 3 (Context) is more extensive in this chapter than in others. This will help the reader understand that human problem-solving strategies are often an essential aspect of infrared detection and interpretation.

Despite the technical challenges of distinguishing the signal from the noise, infrared devices are widely used in reconnaissance, navigation, imaging, astronomy, security systems, and remote-control devices for surveillance devices like cameras, video recorders, robotic sensors, etc.

Infrared devices are particularly good at detecting 'hot' objects (electrical circuits, fires, people crossing borders in the dark, etc.) and for 'seeing' in the dark, or through visual impediments such as haze. In this case 'hot' is a relative term, describing the contrast between the object being investigated to its background. To appear 'hot' to an infrared sensor, the object may need to be illuminated by a supplemental source of infrared.

Infrared can be detected by a simple thermometer, or by more sophisticated instruments such as the *thermopile, bolometer* and, more recently, *pyroelectric* devices, described in more detail later. While some photodetectors are sensitive to radiation in the ultraviolet or visible spectrums, most of them operate in the infrared wavelengths.

This chapter describes a wide range of applications using infrared technologies and explains why the use of infrared is closely linked to the specific problem being solved, and the kinds of equipment needed to solve it. From a surveillance point of view, it covers a lot of interesting technologies:

- night vision scopes and sensing devices

- remote-control and communications systems

- infrared films and cameras

- security systems, especially motion detectors and fire detectors

- aerial reconnaissance and imaging systems

- inspection, quality control, and chemical composition sensors

As can be seen from this list, infrared sensing is a versatile technology incorporated into many different types of devices, some of which are consumer-priced security products.

2. Kinds and Variations

Compared to visible light, infrared is a broad region of the spectrum. Since it borders visible light at the shorter wavelengths, it has some characteristics in common with visible light and since the longer wavelength region of infrared somewhat overlaps radio wave frequencies, it has some characteristics in common with radio waves. There are even some subregions or windows within infrared that are suitable for particular purposes. Thus, infrared is a varied, versatile, and complex region of the electromagnetic spectrum.

2.a. Infrared Categorizations

Units

To understand how the infrared spectrum is divided into subcategories, it helps to understand the units that are used to express infrared wavelengths and the relationship of a wavelength to other measures.

As has been mentioned in other chapters, *wavelength* and *frequency* are mathematically related. If you know one value, you can calculate the other, since the value for the speed of light is used as a constant (assume light is traveling unimpeded through a vacuum). Thus, the speed of light divided by the wavelength equals the frequency ($c/\lambda = f$) and, conversely, the speed of light divided by the frequency equals the wavelength ($c/f = \lambda$). This relationship is useful for calculations related to the different forms of light (infrared, visible, and ultraviolet), and for other electromagnetic radiant energies, as well.

Since we have a convenient relationship between wavelength and frequency, one of the most general and common ways to break up infrared is to divide it logically or mathematically into subregions based upon wavelength (from which the frequencies can be calculated as needed). Since infrared wavelengths are short, shorter even than most radio microwaves, they are expressed in small units. The length of the wave is usually expressed in units called micrometers or μm. These are sometimes referred to as *microns* or simply symbolized with μ. A micron is 1/1,000,00 of a meter or 1/1,000 of a centimeter. Knowing this will help you interpret the following chart, which shows three different schemes for subdividing the infrared spectrum.

Some categories are based on regions or *windows* that are known to be useful for observation and measurement (usually those that are less subject to attenuation or that have good target-background contrast ratios), still others are specified in terms of their associated detection and imaging technologies or as 'generations' of evolutionary improvement.

Infrared Categories Based on Wave's Length with Sample Applications

Designation	Abbr.	Wavelength	Sample Applications
short-wavelength infrared	SWIR	1 - 3 μm	astronomy and laser applications
medium-wavelength infrared	MWIR	~3 - 5 μm	(higher atmospheric transmission) - higher contrast, better in clear weather; space-based remote sensing
long-wavelength infrared	LWIR	~8 - 14 μm	(higher atmospheric transmission) - less background interference, better in weather with particles (fog, dust, etc.), military, industrial applications

Infrared Categories Based upon 'Distance' from Visible Light with Sample Applications

Designation	Abbr.	Wavelength	Sample Applications
near-infrared	NIR	0.7 - 0.9 μm	Adjacent to the visible spectrum, the commonly used range for photographic sensing and vidicon cameras. Most thermal energy emitted from objects up to the temperature of the Sun are in this portion. 'Hot' targets can be readily detected. Primarily photographic, photoemissive devices. Devices can be used in daylight or dark, but in dark will only detect hotter targets without supplemental illumination. Used in night-detection, photography, meteorology, and navigation.
middle-infrared	MIR	0.9 - 20 μm	Includes the peak radiation from fires and much of the energy reflected off objects illuminated by the Sun (hence, may not work as well in very sunny regions). Primarily photoconductors, photodetectors. Used in medical diagnostics, electrical and other construction inspection and quality control, and for locating hot objects obscured by smoke.
far-infrared	FIR	20 - 1,000 μm	Includes the temperatures that are cooler. Useful for high-resolution applications, line scanners, forward-looking infrared (FLIR), and other systems intended to function in light rain, smoke, etc. at normal temperatures. Used in surveillance, mapping, fire detection, and medical diagnosis.

Categories

Many of the classification schemes for infrared are not mutually exclusive but rather are ways of focusing on relevant factors that may overlap. Some classification systems are based on the character of the emissions—are they solid or gaseous? Solid emissions exhibit a continuous distribution of energy, whereas gaseous emissions have 'boundaries', that is, discrete spectral lines. These spectral lines can be extremely useful in identifying the composition of the substances that are emitting them. Note that *thermal* radiation is not a separate phenomenon from infrared, but rather a way of describing certain properties or regions of infrared light.

Two General Categories of Infrared Radiation			
Designation	**Abbr.**	**Waveleng.**	**Description**
reflected infrared	—	0.7 - 3 μm	Within this range, 2.5 μm is approximately the upper limit for reflected solar energy for remote sensing.
thermal infrared	—	3 - 5 μm	About 6 μm is the lower limit for self-emitted thermal energy [Lo, 1986].
thermal infrared	—	8 - 14 μm	Within this 'window' is a region (about 9.7 μm) which is the dominant wavelength of the Earth [Lo, 1986]. This is more easily detected on the dark side of the Earth where there is less interference to the infrared emissions from the Sun's rays.

Technological considerations, problem-solving strategies, and categorizations are discussed further in the following section (Context). It is important to remember throughout these discussions that there are rarely sharp dividing lines between categories in analog systems, that many categorical divisions are approximate, and that charts that divide middle-infrared from far-infrared, for example, indicate areas of transition rather than distinct 'walls' that separate one type of energy from another.

As can be seen in the preceding charts, the infrared region has been divided in various ways, based on the length of the waves and/or on technologically useful 'windows' (wavelengths that respond well to particular types of detectors, or that are less prone to noise from atmosphere or other confounding background radiation). Since the Earth's atmosphere reduces the contrast between the target and the background (increasing the *target-clutter* or *signal-to-noise* ratios), compensatory technologies are often applied to the raw data to try to improve the signal. At the current level of technology, the near- and middle-infrared regions have sufficiently high energy levels and reflective and emissive properties to be useful for many applications.

Near-infrared has many properties in common with the visible spectrum, such as the ability to influence film, and is thus is useful for film imaging systems.

Far-infrared regions are less useful in a general sense, in part because of their low reflectivity and interference by ambient temperatures. However, far-infrared still has emissive characteristics that may be important in situations where the target and background characteristics are sufficiently different or where their spatial characteristics are known, and can be compared with other data (such as medical imaging and fire detection).*

In general, *near- or middle-infrared* spectra exhibit vibrational motions while *far-infrared* or *microwave radiation* produce rotational motions (e.g., as in gas molecules). Distinct and consistent spectral lines can be detected in some substances (e.g., sulphur) and these characteristics

---

*Because technology is always evolving, no part of the spectrum can be discounted as permanently worthless. Some radio frequencies that were once considered unusable are now the backbone of important communications systems. In a similar way, infrared frequencies that are hard to use at the present time, might become extremely useful as a result of future discoveries.

are useful for identification and quality control using infrared devices. As the density of the substance increases, interactions at the atomic/molecular level tend to increase the level of continuity in the spectrum. Substances with temperatures above 0 Kelvin display molecular motion and hence can directly emit heat, i.e., thermal radiation [Holz, Ed., 1973].

Because so much of infrared technology depends on target-background distinctions, systems have been developed to express these quantities. Thus, *target contrast* is usually expressed in terms of ratios (e.g., 4:1) or degrees (20°). Since the information derived from infrared sensing is not always sufficient or sufficiently detailed, multispectral systems that allow the study or imaging of the same data from the point of view of different wavelengths or devices are often used together to provide a composite picture.

Devices that synthetically generate infrared radiation tend to be of two kinds:

cavity radiators - These are designed to emit, as closely as possible, *black body* radiation (a theoretical body that absorbs all incident radiation). There is no perfect human-made black body, so it is mainly used as a theoretical construct.

solid radiators - This includes any radiators which are not black body radiators.

2.b. Types of Detectors

Passive and Active Detectors

Passive detectors are those that sense radiant energy in the infrared region without supplemental illumination. Thermometers, the human skin and nervous system, and snake infrared sensing 'pits' are examples of synthetic and natural *passive infrared-sensing systems*. In contrast, *active detectors* incorporate a source of illumination to detect a target. Infrared-sensitive night-vision goggles require supplementary illumination as do infrared cameras when photographing in low-light conditions.

Purdue University, professor Jay Gore (left foreground) and research scientist Yudaya Sivathanu (right standing) have developed a sensitive detector that responds to near-infrared radiation (which may reach the detector directly or by being reflected off various surfaces). In other words, it can sound an alarm even if smoke particles don't come in direct contact with the detector. The system detects fluctuating frequencies that are characteristic of an uncontrolled flame, which helps prevent false alarms from more consistent heat sources, such as a hotplate, for example. This innovative device incorporates a fiber-optic data transmissions link to a central detection unit. In case of an alarm, a computer at the Purdue lab has been programmed to repeat safety instructions to people in the vicinity of the detected fire. The computer terminals are used to investigate air flow into a fire and fire emissions that may eventually influence building ventilation codes and aid firefighters in evaluating the presence of toxic gases. Gore's previous research includes helping NIST researchers to analyze oil well fires in Kuwait. [Purdue 1997 news photo, released.]

Infrared sensors are used in many different types of fire alarms. Traditional smoke detectors are only activated if smoke particles reach the detector, but scientists are working on infrared detectors that can respond to radiation in the infrared frequency, even if there's not enough smoke to reach the detector.

Thermal and Photon Detectors

Most photodetectors can be classified into two main categories:

thermal detectors - These are manufactured with temperature-sensitive materials to absorb incident radiation through a broad and uniform spectral sensitivity. Through thermal excitation, the kinetic energy of electrons in the higher velocity ranges is converted to potential energy, which is then used to register the presence of excitation or the level of excitation. Examples of thermal detectors include Golay cells, thermocouples, bolometers.

photon detectors - These are typically semiconductor detectors, in which excitation of absorbed photons causes the electrons to move into a *conduction band*. This response to incident radiation by electron excitation is usually expressed in the form of an electrical response that can be displayed according to its characteristics. Depending on how the response is observed, photon detectors can be

- *photoemissive* - electrons are emitted into the surrounding medium

- *solid state* - photons influence energy distribution within the material

The near-infrared region is particularly important, as it encompasses most of the thermal energy emitted from objects with ambient temperatures up to 6,000K, which is the surface temperature of our Sun.

Thermal and Pyroelectric Detectors

When bodies absorb infrared energy, there is an associated rise in temperature. *Thermal detectors* are designed to detect the resulting radiated 'heat'. Since thermal detection is based upon temperature changes rather than sensitivity to a particular wavelength, it is a broad-spectrum technology that senses not only infrared, but also visible light and ultraviolet. Bolometers based on thermal sensing are used in altimeters, satellite Earth-monitoring systems, and non-contact thermometers. Traditionally, many of these devices were ceramic-based, but they are being superseded by semiconductor devices for a wider range of applications. Semiconductor detectors generally cost less, work over a wider range of operating temperatures, and have greater resilience to large-incidence optical power (e.g., pulsed lasers).

Thermopiles are thermal detectors that generate voltages when exposed to heat and thus are good for measuring sources of continuous-wave or repetitively polled radiation. They are slower to respond than pyroelectric sensors.

Pyroelectric detectors are a more recent type of thermal detector and are more sensitive than traditional bolometers and thermophiles. They respond to temperature differences rather than generating a voltage and respond more quickly than traditional thermopiles. Pyroelectric detectors are also broadband detectors, sensitive to frequencies ranging from far-infrared to ultraviolet. By varying the absorbing surfaces of the detector, devices with different sensitivities can be designed. Chromium and specialized black paint are two substances that can be used as absorbing surfaces.

Extrinsic and Intrinsic Detectors

Intrinsic detectors usually operate at higher temperatures than extrinsic devices. They have higher quantum efficiencies and dissipate less power. One example is the quantum-well infrared photodetector (QWIP). Extrinsic detectors include the InGaAs/InP direct-bandgap detector.

Advanced Generation Systems

Infrared devices are further distinguished by their evolutionary improvements. *Second-generation* infrared-sensing systems can detect and identify targets at significantly greater distances than their predecessors. Second-generation performance is provided in a number of ways, including specialized closed-cycle cooling systems and improved detector arrays. A Standard Advanced Dewar Assembly (SADA) is an advanced type of detector module.

Scanning and Staring Arrays

Two important classifications of infrared devices, particularly those used for imaging, are *scanning* and *staring*. These are used in conjunction with *infrared detector arrays*, that is, a series of small detectors mounted such that their individual inputs can be integrated to form an image. The distinction isn't based on wavelengths or heat-sensing properties, but rather on how the image is built up. This simple example helps explain the difference between the two basic approaches:

> *scanning* - Imagine viewing a scene through a narrow tube. You can only see a little bit at a time, so you have to move the tube around to take in the various details of the scene. Despite the limited field of view, after a few movements you can build up a pretty good image of the scene in your mind. Scanning systems are similar to this. A mechanism scans the scene past a detector array to build up a composite image of the scene. Just as your brain forms a 'picture' of the scene, a video display or photographic system can be designed to display a composite picture of an infrared scene. Scanning technologies are commonly used in tactical systems.

> *staring* - Now imagine removing the narrow tube and just 'taking in' a scene all at once, the way you normally see. Staring systems are similar to this. The detector array 'stares' at the whole scene for a period of time designed to be sufficient for each detector to take in a part of the image and have it integrated and transmitted to an associated processing system. CCD cameras operate on this general principle. Systems that register the scene quickly, so that the next scene can be imaged in time to record motion, are particularly useful. Staring systems are often used for tracking and imaging.

Cooled and Uncooled Detectors

Infrared radiation is sensed in a wide range of operating temperatures, from -269°C (4K) to 40°C (313K).

While there are many types of room temperature infrared detectors, smaller, more sensitive electronics often have smaller bandgaps for detecting lower energy photons. Since the surrounding heat can influence the operation of the components, they may be cooled to control the thermal excitation. In simpler terms, cooling an infrared device helps keep the radiation associated with the device from interfering with the radiation you are trying to measure. Liquid nitrogen and liquid helium are two elements commonly used to cool infrared devices.

Substances sensitive to infrared which are commonly used in cooled infrared devices include

- lead sulphide (PbS) - above approx. 3000 cm^{-1}

- indium antimonide (InSb) - above 1400 cm^{-1}

- mercury cadmium telluride (MCT) - above approx. 700 cm^{-1}

Certain substances, such as glass or water, will impede or block infrared, and are not suitable for lenses designed to be transparent to infrared. If glass impedes infrared radiation, then how can a regular camera be used for infrared photography? It's because wavelengths in the near-infrared, adjacent to the visible spectrum, can pass through glass more readily than longer infrared wavelengths. This is one of the reasons consumer infrared-film photography only records a short distance into the infrared spectrum. Longer wavelengths can, however, be imaged with specialized systems with lens materials other than glass, materials that don't block infrared.

Precious and semiprecious gemstones (ruby, garnet) are used as components in many illumination and detection technologies, especially lasers. Illumination through sapphire, which is transparent for visible and infrared radiation to beyond 5.5 μm or even farther, can be achieved by thinning the sapphire.

Most infrared sensing is achieved with nonfiber technologies transmitting at analog rates. Typically, the wavelength (λ) and its amplitude are detected. In digital applications, however, the wavelength and the pulse amplitude are relatively fixed—reducing the usefulness of the information. Thus, the existence of a pulse and the pulse rate are sensed instead, with the amplitude being largely irrelevant [Bass, Ed., 1995].

In spectroscopy, in which the unique spectral properties of substances are studied, both the wavelength and strength of a signal are used for detection and identification.

3. Context

Context is an extremely important aspect of infrared detection and imaging. Infrared radiation is found in varying quantities throughout our world and it can be difficult to interpret data or pinpoint the specific source of infrared radiation. When infrared is used as a detection or tracking technology in automated systems, it is often backed up with evaluations and judgment by individuals or operational teams and combined with data from other surveillance technologies.

Ambient Radiation

As mentioned earlier, infrared forms part of the *optical region*. Our world is full of animate and inanimate objects radiating energy in these wavelengths. About half the energy that radiates from the Sun is in the infrared range [Holz, Ed., 1973]. Our own bodies are constantly radiating infrared energy in the form of heat. Most of the infrared in our environment occurs naturally, though it is possible to generate infrared radiation with illuminators and lasers. This is in contrast to radar, where there is little natural radiation in the radar frequencies—most of it is synthetically generated.

Some forms of infrared radiation are easy to detect and there are many consumer-priced infrared devices that take advantage of prevalent or easily distinguished infrared wavelengths (security systems, remote controls, camera autofocus (AF) systems, infrared films and lenses, night-vision scopes with illuminators). Other aspects of infrared are technologically challenging to construct or interpret, including supercooled devices and various computerized imaging systems that can be relatively sophisticated and expensive (such as aerial remote-sensing systems and astronomical telescopes).

In infrared detection, as in radar, the object that is being sought or sensed is called the *target*. The context or environment associated with an infrared target is called the *background* and objects or particles that get in the way and tend to scatter the radiation are collectively called *clutter*. Longer infrared wavelengths can move through particles in the atmosphere more readily than visible light and thus are useful for seeing through fog, dust, and other particulate matter that usually obscures the view.

Common Applications

Common infrared targets include people (customers, intruders), wildlife (game, injured livestock, research animals, insects, birds that migrate at night), astronomical bodies, electrical faults, fires, drug-growing operations, underground or building piping systems (especially heat ducts or hot water pipes), circuit boards, buildings (roofs, walls), and fires (arson, house fires, forest fires). As a rule of thumb, if it emits heat, or contrasts with the ambient temperature, it's a good candidate for thermal infrared detection. Some targets and detectors require supplemental illumination to be visible in the infrared range.

Infrared film is readily available for standard and special-purpose cameras. Most infrared film is sensitive to visible and near-infrared reflected light and so cannot be used in the dark unless a source of illumination is provided (the 'light' will not be seen by human eyes, but will reflect off objects and back to the camera to expose the film). The presence of bright sunlight will intensify photographic effects. Heat sources that are mostly reflecting in the far-infrared (about 10 to 100 μm) do not show up on infrared film, but can be seen with other types of viewing and recording systems (which often need to be cooled in order not to interfere with the incoming radiation).

Interpretation of Infrared Data

There are two interrelated aspects of infrared sensing that pose special challenges:

- detection of targets that are fast-moving, or that produce small (infrequent or quick-burst) or weak (lower energy-level, far-infrared) signals, and

- image processing/interpretation in circumstances where the target doesn't contrast strongly with the background, or in which there are infrared-opaque particles obscuring the image.

The second point is the one that brings human problem-solving skills into play. The human brain is still one of the most important 'components' of most infrared-sensing systems. Just as well-trained radar operators (e.g., air traffic controllers) become experienced in the interpretation of radar-scope images, infrared detection and imaging systems are dependent upon the ability of the user to apply good problem-solving strategies and procedures to the information received.

The problem of finding a small target in a crowd characterizes a large part of infrared technology. Scientists have to solve difficult *target-clutter* and *signal-to-noise* problems due to the ubiquitous nature of infrared radiation. That is not to say all infrared imaging problems are this challenging. If your friend is standing alone or with one or two people in a grove of trees in the dark, infrared night-vision products may help you locate his position or, at least, the position of the small group of people. Infrared products are particularly useful for this type of situation where the target is clearly distinguishable from the background, and where the characteristics of the target (such as arms and legs and walking movement) make it clear that it is a person.

Since the world is full of infrared radiation, the goal of infrared detection is to locate 'useful' information—radiation that contrasts or distinguishes the desired target, or that changes over time. Since the atmosphere and surrounding terrain emit infrared radiation that may interfere with detection (unless you are specifically sensing the atmosphere or terrain itself), there are wavelength 'windows' that are more useful than others for distinguishing a target from this background clutter.

Objects glow most noticeably in the 8 to 10 μm range, and cameras and detectors that sense within this range are useful for surveillance, security systems, and navigation systems. Thermal maps of biological organisms taken in this range are useful for biological research and medical imaging. The absorption lines of gas molecules also lie in the long-range infrared region, making it useful for atmospheric studies and monitoring.

Displaying Infrared Images

Infrared photography is a useful surveillance technology because it can be used in daylight to reveal details that do not show up on normal film ('normal' in that the colors/values approximate those familiar to the human visual system). At night it can record details invisible to human eyes. To take infrared photos at night, a high-intensity infrared light can be used for illumination. This will normally not be seen by others in the vicinity, but allows more information to be captured on film (or with a digital camera) than can be captured with visible-light systems.

Each region of infrared has its limitations for being captured on film or video. Below wavelengths of 1 μm, silicon-based CCDs (those used in video cameras) are rapidly superseding photographic plate systems for imaging infrared. For slightly longer wavelengths, from about 1 to 3 μm, the silicon-based CCDs are not quite as effective, as they decrease in quantum efficiency due to what electronics engineers call a *solid-state bandgap*. Other materials have been tested for use with longer wavelengths with the tradeoff that they are somewhat lower in resolution than the silicon imaging arrays.

Materials used to detect infrared are constantly under study and change when higher precision or less expensive options are found.*

As can be seen from this discussion, infrared sensing and imaging really represent a wide range of technologies and cut through hundreds of fields of study and application, more specific examples of which will be provided later.

4. Origins and Evolution

Most of our scientific understanding of infrared is recent, as its existence and properties were discovered less than 250 years ago. The more sophisticated devices for imaging infrared have only become common in the last few decades.

The Discovery of New Forms of Light

The discovery of the infrared region of the optical spectrum is attributed to a German-born musician/astronomer, Friedrich Wilhelm (William) Herschel (1738-1822). Herschel immigrated to England in 1757 where he became a music instructor who took time to systematically explore the sky as an amateur astronomer. In 1800, Herschel made a simple but significant observation. He reported that while using a glass mercury thermometer to measure temperatures in different regions of the visible spectrum, he noticed a change in temperature that continued past the portion perceived as the color red, suggesting the existence of wavelengths beyond those which he could see.

Herschel's discovery led to scientific debates about whether the newly discovered radiation was a type of light or a phenomenon that was different from light. The observations of Leslie in 1804 showed that different materials had quite different radiating and absorbing qualities, thus suggesting light and heat might be distinct phenomena. Herschel himself thought that the newly discovered radiant energy might be separate from visible light. Further experiments by his son led to the conclusion that the radiation that was invisible to humans was similar to visible radiation and differed not so much in character as in momentum. This radiation is now known as *infrared*, to describe it as being 'below' (infra-) the red region of the visible spectrum, in other words, at lower frequency levels than the color we perceive as red.

---

*In the near-infrared ranges from about 1 to 5 μm, InSB and HgCdTe are common. As the wavelength increases, we find Si:As BIB, Si:Sb, and Ge:Ga being used.

New Devices That Led to Further Discoveries

In the 1800s, many great minds were trying to understand physical phenomena such as light, electricity, and magnetism. At the same time, inventors were creating devices to harness their capabilities. While studying electricity and magnetism, Hans Christian Ørsted (1777-1851), a Danish scientist, and Jean B. J. Fourier (1768-1830), a French mathematician, invented the first *thermoelectric batteries*, around 1823, by welding pairs of antimony and bismuth in series.

The development of the *thermopile* (an electric thermometer) by Italian inventor Leopoldo Nobili (1784-1835), in 1829, provided a better means to detect infrared than a conventional thermometer. Bolometers were developed a number of years later and further improved by Macedonio Melloni (1798-1854), who collaborated on some projects with Nobili. Melloni developed a thermomultiplier that consisted of a cone aimed at the heat source, with an astatic galvanometer connected to the terminals of the thermomultiplier. Compared to a thermometer, Melloni's thermomultiplier was fast and sensitive. With low-intensity radiation, there was proportionality between the angular deflection of the galvanometer needle and the temperature difference between the opposite batteries. For more intense radiation, the proportion was more complex; a lookup table was often provided with the device, listing intensity in terms of degrees.

By the 1830s, studies by Thomas Seebeck with more sensitive equipment than had been available to Herschel, revealed some of the characteristics of infrared radiation, but it was not until almost 50 years after its initial discovery that it was accepted that the nature of infrared was consistent with that of visible light. In the 1870s, James Clerk-Maxwell (1831-1879) contributed some important theories about electromagnetic radiation and a series of mathematical equations that would be used to resolve theoretical and mathematical inconsistencies. Josef Stefan (1835-1893) and Ludwig Boltzmann (1844-1906) also provided important tools for modeling the characteristics and behavior of infrared radiation. The Stefan-Boltzmann Law was a theory of black body radiation first demonstrated by Stefan that was later described mathematically by Boltzmann, in 1884. Stefan was also known for research in heat conduction and kinetic theories related to heat.

In 1905, W. W. Coblentz published a classical reference book on the infrared spectra of organic and inorganic compounds, including not just solids, but gases and liquids, as well. This remarkable book, called *Investigation of Infrared Spectra*, still provides basic information for understanding spectra and their practical applications in spectroscopy.

The Emergence of Practical Applications and Accessories

By the 1930s, infrared science and technology had improved to the point where commercial and industrial applications could be manufactured. Spectrometers were improved and devices to automatically detect and/or record infrared were developed.

In 1932, Mr. Bloch at Ilford, a company known for its photographic products, developed faster emulsions on glass plates that were suitable for use in aerial photography. A deep red filter was used, instead of yellow.

In 1933, two British open-cockpit biplanes were equipped so they could survey closed Nepalese territory using infrared photos taken in the vicinity of the Himalayas. The camera was mounted in slings under the plane, at an angle, with leather pouches as shock absorbers (details of this hair-raising pioneer mission are included in the chapter on Aerial Surveillance).

Many of the significant production advancements in infrared technologies occurred during World War II, when research dollars were directed toward the improvement of surveillance devices. Infrared was of interest because it allowed field personnel to find hidden installations and to see in low light conditions. At the end of the War, improvements in the technology were incorporated into commercial systems, including faster sensors and automatic recording features.

In 1947, Kenneth Neame of the Royal Air Force took photographs in an unauthorized flight over Mt. Everest in a Spitfire XIX. Nepal was still a closed territory and had not been fully explored or mapped by westerners. What little was known about the region from the ground was determined by spies using age-old, low-tech surveillance techniques, (i.e., they disguised themselves as holy men [Greer, 1994]).

High Altitude Surveillance

In the 1950s, rocket science took off and scientists were quick to mount cameras on the airborne projectiles to record the world from new perspectives. Originally the cameras were used for recording the orientation of the rockets, but later they were also used for observation. Rockets weren't ideal as surveillance craft, as they shook, were subjected to high heat, and were only airborne for a few minutes, but they eventually became important for launching other types of technologies into space or into the upper atmosphere.

In the 1960s, improvements in aircraft and the deployment of orbiting satellites led to new applications and developments in remote sensing using infrared detectors and recording media. The interpretation of infrared data also improved. Multispectral scanners were installed in aircraft [Curran, 1985].

In the late 1960s, there was a lot of interest and research into mosaic focal planes, both non-planar and planar. However, a decade later, they were still somewhat lacking in commercial utility.

This is a Grumman OV-1C aircraft, called the Mohawk, that came into production in 1959 as a reconnaissance and photo-observation craft for the U.S. Marines and the U.S. Army. The "C" designation indicates that the aircraft was equipped with an infrared electronic reconnaissance imaging system. [NASA/Dryden Flight Research Center 1983 news photo, released.]

Infrared Photography and Imaging

The recording of infrared images on photographic film was influenced by developments in the photographic industry in general. In the 1960s there was widespread use of black and white film and experimentation with different emulsions and development processes. Infrared photography was used in many contexts by the late 1960s. As color films began to emerge and drop in price, color infrared film became available, filters were developed (for both color and black and white films), and false-color interpretation techniques were developed.

The 1970s - Commerce, Computers, and Science

Advances in technology, computerization, and miniaturization caused infrared technologies to proliferate in civilian applications in the 1970s. It became more broadly applied in atmospheric, oceanic, and agricultural studies (e.g., speculation related to world cereal production). Microcomputer-based interpretation software and display devices contributed to infrared imaging technology. Improvements occurred in computer memory capacity, especially dynamic RAM. Image display size and resolution improved as well.

Silicon chip technologies emerging in the 1970s made it possible to develop charge-transfer devices (CTDs) that could be used for infrared detector readouts. First-generation photoconductive HgCdTe (Mercury-Cadmium-Tellurium) arrays used in infrared sensing began to be available in high volume in the 1970s. HgCdTe and InSb (Indium-Silicon) are semiconductor infrared sensing materials that can be used in conjunction with a signal-processing, silicon-substrate chip to create hybrid arrays that made it possible to combine larger numbers of detectors and to image at higher resolutions. Monolithic arrays also continued to evolve.

 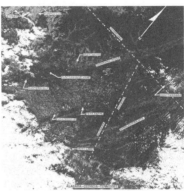

Left: This is a model for an infrared telescope intended for airborne use with the Kuiper Airborne Observatory. Right: An infrared image of the Alabama State region taken from Skylab's Earth Resources Experiment Package (EREP) using visible-light and near-infrared photography, infrared spectrography, and other sensing technologies (with labels superimposed afterward). [*NASA/Ames Research Center* 1972 news photo by Geaton Farrone; NASA/Marshall 1973 news photo, released.]

The first comprehensive global inventory to utilize new imaging technologies took place in the mid-seventies. Infrared remote-sensing was used for surveilling crops in the U.S.S.R., China, Latin America, and India. In the Large Area Crop Inventory Experiment (LACIE), Landsat sensor data were analyzed to calculate wheat production. LACIE was replaced in 1980 by AgRISTARS (Agriculture and Resources Inventory Surveys Through Aerospace Remote Sensing) [Curran, 1985].

The transition from analog to digital was gradual. Mosaic focal plane arrays were developed in the late 1970s, but they still had limitations in uniformity and dynamic range, so there was a strong interest in improving the sensitivity and resolution of detectors and imaging systems.

The 1980s - Improvements in Video and Color Technologies

Until the early 1980s, most video image displays came from 'black and white' (more correctly described as grayscale) tube systems, but these gradually were superseded by multiple-channel, false-color composite CCD-camera systems, in the early 1990s. Until about the mid-1980s, the assignment of colors to grayscale images resulted in rather unnatural, too-dull or too-bright images that were unattractive to human observers. However, by about 1987, color assignment to various tonal values of gray in specified regions had improved to the point where satellite and other aerial images of water and land, while not 'natural' in the strictest sense, nevertheless matched more closely with human preferences for terrain symbology. Multiple CCDs are now commonly combined and synchronized to provide multispectral images.

The early 1980s also saw the design and evolution of significant remote technologies: remote computer terminals, remote-controlled robots, and remote-controlled or unpiloted aerial vehicles (UAVs). UAVs are often designed with infrared sensor capabilities. These vehicles were seen as a low-cost, low-risk-to-life alternative to large-scale aerial reconnaissance planes and were commonly equipped with infrared technology and their popularity has increased in recent years.

The development and application of infrared sensors continued in the 1990s with the U.S. military's Tactical Airborne Reconnaissance Pod System (TARPS), intended to replace the RA-5C and the RF-8G. TARPS included an AN/AAD-5A infrared line scanner. The TARPS tactical photography system could be mounted under aircraft such as the U.S. Navy F-14 Tomcat. The Tomcats came into production in the early 1970s and have been used for intelligence operations for carrier battle groups, domestic disasters (earthquakes, tornadoes), and international conflicts (e.g., NATO operations).

Left: Two images of our home, the Milky Way galaxy, taken in 1983, show how dramatically our perspective can change when things are viewed with other imaging technologies. The infrared image (bottom) allows the galaxy to be seen with less of the 'haze' and dust that interferes with traditional visible light images (top), revealing the slender spiral disc shape of the galaxy. Right: This remarkable infrared image of the Earth from about 1.32 million miles away was taken during the Galileo Orbiter spacecraft mission in December 1990. The infrared image was taken using light with a wavelength of about 1μ (1 micron), which allows the rays to penetrate through the haze that is associated with Earth's envelope and thus gives a clearer picture of the surface features. South America and some of the eastern seaboard of the U.S. are pictured here. [*NASA/Ames Research Center* news photos, released.]

In the 1980s, there were significant improvements in imaging quality and resolution, along with improvements in computer imaging and display technologies.

In 1983, the Space Shuttle Columbia used the ESA Spacelab Metric Camera to take overlapping aerial color infrared photographs with a modified Zeiss RMK A 30/32 camera, including some of Mt. Everest. Much of the camera control of the large-format camera was handled remotely, from the ground.

By the late 1980s, still video was being used in military reconnaissance. Although the resolution of early systems was fairly low, it offered the advantages of filmless photography, and the potential of realtime systems (which became available within a few years). Overlapping the improvement of still-video technologies was the development of digital camera systems, which started to come down in price in the early 1990s. Digital imagery offered much higher image resolution than traditional analog video, but consumer-priced digital cameras were inferior in resolution to film images. Nevertheless, inventors continued to improve digital sensor arrays and increasingly installed them on traditional single-lens reflex cameras, and within a few years dedicated digital camera systems were being developed for military and commercial applications. The price of professional-quality digital cameras was in the $25,000-range in the late 1980s, but was destined to go much lower.

Off-the-shelf computer systems were now commonly used for reconnaissance image processing and system control, whereas in previous years, large, expensive, dedicated systems were more common.

The 1990s - Improvements and Broader Applications

In the 1990s, infrared technologies improved and widespread access to commercial infrared detecting devices, data, and images became more readily available.

By the mid-1990s, pixel arrays had improved to the point that they were similar to traditional TV screen resolutions. Output formats included a common video format called RS-170. PtSi (platinum silicon). Megapixel formats of 1040 x 1040 had been developed by 1991.

Also in the mid-1990s, as 'smart missiles' and other forms of advanced weapons and missile-tracking systems were developed, or purchased and deployed by warring nations, it became apparent that photographic reconnaissance and intelligence-gathering were not sufficiently fast or flexible to meet all the needs of war-zone activities; realtime or near-realtime intelligence systems support was needed. The U.S. Advanced Tactical Air Reconnaissance System (ATARS) represents one of the evolutionary steps from traditional systems to day or night near-realtime, data-linked, digital and infrared systems. Thus, intelligence could be gathered from reconnaissance fighter planes with less risk of danger from unfriendly ground-fire. The German Air Force IDS "Tornado", also equipped with a reconnaissance system incorporating an infrared linescanner, served a similar function.

Sensing equipment can be bulky and expensive. When the systems have to be mounted on aircraft or spacecraft with weight considerations and limited space for instruments, it is particularly important to reduce the size of systems. In some cases, microminiaturization aided in developing more efficient systems, but sometimes systems innovation solved the problem. Since visible and infrared technologies share many common attributes, inventors devised ways to image both spectra through some of the same components.

Select mirrors, that allow input to be alternately toggled among sensors (e.g., visible and infrared) enable more than one type of input to be imaged by toggling back and forth among them at very high speed. Since the human perceptual system can only resolve about 30 frames per second, multiple inputs to multiple monitors (or a single monitor in which inputs are superimposed one upon the other, or viewed in separate windows) from the same source, at anywhere from 20 to 30 or so frames per second, will appear naturalistic. Since visible data and infrared data are not received at the same rates (visible data rates may be up to 10 times faster), compensations in frame display are often made for multisensor inputs. While display systems work well at frame rates of about 30 fps, scanning rates need to be much higher, especially if the imaging system is mounted on a fast-moving aircraft.

Pushbroom technologies, imaging systems that create an image line-by-line, began to give way to *framing technologies*. Computer processing improved as well, along with motion-compensation components and algorithms.

Missile-detection systems projects continued throughout the 1990s. In one collaborative project, DARPA, the U.S. Army, and the U.S. Navy, developed the Infrared Search and Track (IRST) system. This was an advanced infrared focal-plane array sensor which, when combined with signal processing, could detect incoming test missiles without any false alarms. It was determined to be an effective countermeasure that might be particularly suitable for defending marine vessels.

Missile-guidance systems based upon infrared sensing were also being developed or requisitioned for purchase by more than a dozen nations. To counter these missile threats, Australia, Canada, the U.K., and the U.S. jointly participated in the MATES project, in the mid-1990s, developing a means of effectively breaking the track of an infrared missile-seeker, thus providing a missile countermeasure.

By the late 1990s, almost 40% of the budget of the Advanced Technology Development program was allocated for electro-optical/infrared countermeasures technologies.

Politics and Aerial Surveillance

Infrared has become an essential aspect of aerial surveillance. Almost every major type of aerial imaging platform carries at least one infrared sensor and some carry multiple sensors. These systems map and surveil the Earth from both the upper atmosphere and from space. By the 1980s, aerial surveillance had become not only a tool of science, but also a tool of business and government, for observing the planet and the activities of other nations.

The 1980s and 1990s were a time during which nations all over the world realigned themselves. New nations were created through the breakup of the U.S.S.R., Czechoslovakia, and Yugoslavia. The European common market, after decades of negotiation, was finally coming together. Political boundaries and alliances shifted in Africa. The relationship between India and Pakistan became more strained. The overrun of Tibet by China came to the attention of the western world. All this shifting of political power and political boundaries created a greater emphasis on surveillance technologies, particularly those that could help monitor political instabilities and national borders. Diminishing resources and border realignments also motivated the monitoring of local and international poaching and resource exploitation.

Further Improvements in Technology

During the 1990s, computer electronics and storage devices continued to improve in capacity and resolution and to drop in price, with the result that there were dramatic improvements in sensing and imaging systems.

Infrared sensing chips gradually became optimized for specific purposes, compared to earlier generic chip technologies. Clock and bias generation capabilities were streamlined and functionality was added in other areas. Electronic zoom capabilities were added to staring sensors.*

Throughout infrared history, there has been a search for materials that would provide greater ranges and could be used at or near room temperature. Progress was made in both these areas.

---

*Paul R. Norton provides information on photodetectors across a range of frequencies, including infrared, in Chapter 15 of Bass, Michael, Editor in Chief, Handbook of Optics, listed in the Resources section.

Through the 1980s and the 1990s, tremendous strides were made in large-bandgap, compound semiconductors. This helped implementation of quantum-well theory in a physical medium and quantum-well infrared technology became practical, with the development of commercial products with higher sensitivity levels in long-wavelength infrared.

Military Surveillance

Improvements in technology have always been of interest to the space sciences and the military, as they permitted smaller, more sensitive instruments to be used.

Infrared sensors are used in many military reconnaissance systems mounted on aircraft and these are described in more detail in the Aerial Surveillance chapter. They are also used in missile-targeting systems, described further in Section 6 (Applications).

In 1992, the U.S. Navy initiated *Project Radiant Outlaw,* a demonstration project for long-range noncooperative identification (NCID) of a variety of ground and air targets. The project included a multisensor combination of laser radar (ladar) and a shared-aperture, staring, focal-plane array, infrared sensor operating in the mid-range of 3.8 to 4.5 μm [Shen, 1994]. Thus, high-resolution, staring, passive infrared was combined with the other sensor technologies for use in various types of identification. Later, the Infrared/Electro-Optical Long-Range Photography System (IR/EO-LOROPS) project was initiated by the Navy to meet the need for long-range standoff digital infrared reconnaissance missions.

Infrared Applications in the Space Program

Many U.S. space missions have been directly involved in developing infrared technologies and some have carried infrared cameras and telescopes into space as scientific and commercial payloads. Many different kinds of sensors have evolved in step with other aspects of the space program. Ultraviolet and gamma-ray sensors were also becoming more important in the 1990s and are described in the Ultraviolet Surveillance chapter.

During the late 1990s, newer, more powerful infrared equipment was used in many industries. In the U.S. space program, a milestone was reached in early 2000 when a new thermal-infrared (TIR) imaging system carried aboard ASTER-gathered images of Earth from space that were taken at night. Visible/near-infrared systems couldn't take pictures of Earth from this distance at night for various technical reasons. When taking infrared photos on Earth's surface, a source of infrared light can be shone at a photographic subject to help it register on the film. This solution is obviously impractical in space, where great distances are involved. The TIR system overcame these limitations by sensing a region of infrared that is radiated from the Earth and is present without supplemental infrared light, enabling a new era of 'dark-side' imaging to begin.

Late 1990s - Increased Sophistication and Commercialization

The U.S. Army teamed up with *Indigo Systems Corporation* of California, in 1997, to develop the world's smallest infrared camera, resulting in the UL3 alpha that was announced in 1999. The tiny six-ounce camera is sensitive to infrared radiation in the 8 to 12 μ waveband. The camera is so small and power-friendly, it can be used as a sensor or as a component in a group of sensors, to guard minefields or other restricted areas. It is also small enough to be placed onboard some of the newer, smaller unstaffed aerial vehicles. The camera is also being evaluated for firefighting applications, since infrared can penetrate haze, and function in darker environments more readily than visible light. It may also be able to detect 'hot spots' in buildings that are not otherwise visible.

Many infrared devices are now inexpensive and easy to manufacture, and are useful for commercial and industrial applications. By the 21st century, they were becoming common in small businesses and multiple dwelling units. As prices drop, they are likely to be purchased as home consumer appliances for family and individual use. One of the more obvious commercial uses of infrared sensing is for fire detection, but security and imaging applications are popular, as well.

In the summer of 1997, the crew of the Space Shuttle Discovery completed a one-and-a-half week mission in which part of their job was to deploy the Cryogenic Infrared Spectrometers and Telescopes for the CRISTA-SPAS-2 satellite to aid in researching the Earth's middle atmosphere. [NASA/KSC news photo, released.]

At the present time, research in sensor technology continues to result in lower-cost imaging systems, high-resolution, high-sensitivity color imaging arrays, and sensor suites that can work with multiple wavelengths. Quantum-well and other new semiconductor technologies show promise. Very-high frame rates (which are especially important in equipping high-speed craft), up to at least a thousand frames per second, are now considered desirable. Research on the detection of a target within clutter, that is against a confounding background, is ongoing and will continue to be an area in which improvements will be welcome.

The *Night Vision and Electronic Sensors Directorate* of the U.S. Army and *Indigo Systems Corporation,* teamed up to create the world's smallest infrared camera, tiny enough to be used as perimeter sensors, helmet-sensors, and infrared imaging components on remote-controlled and autonomous aerial vehicles. It is also being evaluated for fire-detection applications for businesses and fire services. [U.S. Army/Indigo Systems Corporation 1999 news photos, released.]

By 2005, digital imaging began to supersede traditional film imaging. The resolution of digital cameras in the $1,000-range increased to six megapixels, and higher-priced cameras exceeded the resolution of film. The convenience of digital photography (instant images, no film-processing costs, no waste from 'bad' pictures) rang a death-knell for film processing companies and film booths started to close down. Within a few years, they may be obsolete.

5. Description and Functions

The basic types of infrared sensors have been described in earlier sections of this chapter. Some further distinctions and examples will be presented here.

5.a. General Categories and Distinctions

It is common to describe commercial equipment in terms of certain 'landmark' improvements in sensitivity or capabilities so that consumers and users can distinguish one device from another. The designations of 1st or 2nd *generation*, for example, help distinguish newer, more powerful technologies. As the generation number increases, so usually does the price, but this is not always true—breakthroughs in electronics will sometimes reduce the price of newer components that are also more powerful than the previous generation. In general, infrared and night vision products are described as follows:

generation 0 - Active night vision systems that require supplemental illumination. Electron acceleration is used for gain.

generation I - Passive systems that do not require supplemental illumination and do not employ a microchannel plate. The peak response is in the visible spectrum. Electron acceleration is used for gain.

generation II - Passive system that do not require supplemental illumination but that employ a microchannel plate* for gain.

generation III - Gallium arsenide photocathode devices that employ a microchannel plate for gain. High photosensitivity and broad-spectrum response.

Quantum-well technology is one of the more recent and interesting developments in infra-red-sensing science. Computers can do highly complex operations and calculations based upon a simple binary system. Similarly, quantum-well technologies can be stacked in such a way as to provide a structure in which the quantum energy states are limited to two states, ground and excitation—in essence, a simple but powerful binary system. Thus, an infrared detector with a sharp absorption spectrum can be designed, and particular wavelengths singled out, by tailoring the dimensions of the well.

Infrared Sensors

Infrared sensors come in many different forms, from simple thermometers to sophisticated detector arrays in which the radiation striking the detectors triggers a process to convert energy to an electrical signal that is then further processed, recorded, and displayed. For imaging systems, the trend has been to assemble denser and larger arrays to increase image resolution. Higher frame rates, more sensitive detector technologies, and better software for distinguishing target/noise distinctions are also being developed.

---

*A microchannel plate (MCP) is a perforated glass disc positioned behind the photocathode. This arrangement provides electrons to aid in gain and is characteristic of Generation II and Generation III night vision systems.

Although mirrors are used in many infrared devices, they must be made to reflect infrared. Glass tends to absorb radiation through most of the near-infrared spectrum and thus must be coated with a metal such as silver or aluminum to be effective. Transparent components are typically made of highly polished transparent minerals other than glass (e.g., quartz) to keep scattering to a minimum.

5.b. Photography and Imaging

Cameras

There are two general categories of cameras used as infrared technologies:

Film cameras that take conventional-style photos (these record reflected infrared and cannot record passive infrared sources at night, for example, without an external source of infrared illumination) and *sensor-based cameras* (such as video imaging cameras) that are equipped to do thermography. Since film cameras work with reflected infrared, a source of infrared light is needed as a source of illumination. Standard cameras with glass lenses can only be used to record infrared frequencies near the visible spectrum (near-infrared). Wavelengths greater than about 1 μm cannot be captured with standard lenses and films.

Specialized infrared cameras for wavelengths of 4 μm or longer that may use a one-dimensional array of solid-state detector elements, with one- or two-dimensional mechanical scanning, with a resolution less than that of traditional television. Traditional infrared cameras sense only to about a wavelength of 1 μm. Long-range detectors use special semiconductors (e.g., Mercury-Cadmium-Telluride) to sense up to about 4 μm or, if cooled cryogenically, sometimes can sense warm objects up to 12 μm. Each detector requires an individual amplifier to magnify the minimal output voltage, thus creating a somewhat nonuniform image from line to line [Beam, 1989].

Tube-based infrared-sensing systems tend to be more sensitive to infrared than film, but have the disadvantage of lower resolution. They detect visible and infrared wavelengths up to about 0.85 μm [Lo, 1986].

Regular cameras for shooting infrared photos typically use film sensitized to infrared in the 0.7 to 0.9 μm region of the electromagnetic spectrum or broadband film with filters to screen out most of the visible spectrum and all of the ultraviolet. Grounding may be necessary to reduce static discharges inside the camera that could affect the film.

Some off-the-shelf cameras are more infrared-film friendly than others. The Canon EOS cameras, for example, have an infrared film-loading mechanism. It has been reported by photographers that it will fog part of the outside edge along the film-tracking section (the guide holes), but the fog bleeds only marginally into the image area.

Light Meters

Regular light meters are not appropriate for use with infrared, as they are optimized to monitor wavelengths in the visible spectrum. Built-in meter systems in standard cameras can be used, sometimes with an infrared filter in place, as long as the compensations described below are used. Specialized infrared meters are available, at higher prices.

You can learn to adjust regular metering, but it is important to use a camera in which the metering can be set manually. Most of the inexpensive pocket cameras aren't manually adjustable, but many 35 mm cameras have manual overrides. The depth of focus is different for infrared than for visible light, so many swappable lenses have a little red tickmark next to the focus mark to indicate how much to adjust the focus for infrared photography (when shooting with an appropriate infrared filter).

Broadband infrared filters are opaque to visible light. In other words, our sensory equipment (our eyes) can't see through them. This makes it hard to focus and frame a shot with an infrared-blocking lens through a single-lens reflex camera. A red or yellow filter lets you see the scene, but is not sufficient for pure infrared photography, because it allows some wavelengths of visible light to penetrate the lens. However, for special effects (or special situations), a colored filter might provide the desired effect in a photo.

When photographing infrared, smaller aperture settings (higher f-stops) are preferable, approximately f/11 to f/22.

It is highly advisable to shoot at least one test roll and to record the results so that you can analyze them and compensate appropriately on future rolls. If you are planning to use more than one type of filter with more than one type of film, you may have to do several test rolls, and should probably do tests that are metered both with and without the filters in place. However, if you are planning to shoot with one filter and type of film (which is a good way to get started), then you can meter through the filter, bracket the exposures (shoot two settings over the indicated f-stop and two under it, in addition to the metered f-stop) and run a test roll. If the negatives are too light or too dark, compensate and run another test.

Metering without the filter and then adding the filter and bracketing are sometimes recommended (especially if switching filters frequently).

Filters and Focusing

A Wratten 87, 87C, or 89B filter is called a *cutoff* filter because it only admits infrared radiation, screening out the visible radiation. This provides the closest to a 'true' infrared picture. However, if some visible light is acceptable, or desired for focusing in daylight, a Wratten 25 can be used as a second choice. It won't block all the visible light (red still passes through), but it will screen out blue and ultraviolet wavelengths.

A yellow filter is used to screen out Rayleigh scatter (the particle scatter that makes the sky appear blue to us) [Lo, 1986].

The focal distance for infrared is different than that for visible light due to refractive differences. It helps to use wider-angle lenses to increase depth-of-field, and to use lenses with infrared focusing compensation (the little red tick mark on the lens). As long as automatic focusing (AF) settings on the camera can be overridden, you can usually get an initial reading through an appropriate filter and then manually adjust the focus before taking the picture. Once the film is loaded, it is best to change lenses in the dark (under a blanket, in a closet, or under a coat, if you are out in the field) and to keep it away from heat sources.

Infrared Film and Processing

Infrared film can be purchased from photo supply stores on and off the Internet. Most infrared films are available for 35-mm cameras in 36-exposure rolls, and some are available in both rolls and sheets of different sizes. Some infrared film is specialized for aerial photography, with cutdown versions for conventional cameras.

Infrared film must be handled carefully. Just as visible *light* can fog regular film, *light and heat* can fog infrared film. Film cassettes *must* be handled in complete darkness. Unlike visible light, which is blocked by the film cassette, infrared can penetrate the cassette (don't hold it in your warm hands too long or handle it near heat sources).

Some smoke detectors, motion detectors, and darkroom lights emit infrared, so take care to avoid these as well. Don't open the film cannister in the light before or after the film is exposed—this must be done in a closet, darkroom, or under a dark blanket.

Purchase the film only from a reputable supplier familiar with the storage and handling of infrared film. The film must be kept cool before and after exposure until the time it is processed. If it has been refrigerated, allow it to adjust for an hour or so to the ambient temperature before exposing the film, to reduce the chance of condensation.

Some infrared films have an *antihalation* layer. Halation is haloing, fringing or fogging, that is, the spreading of the exposure beyond the area desired. An antihalation layer offers a bit of protection against undesirable exposure 'bleed' or film fogging.

Infrared film records infrared light that is reflected off objects within certain wavelength ranges. Infrared is absorbed by such things as trees, rocks, and soil. Cool objects emit less infrared and warmer objects (people, animals, fire, electrical sources) generally reflect or radiate more infrared. The levels of contrast between the bodies that absorb or radiate infrared are recorded as different 'light' values on black and white infrared film and as different colors on color infrared film. Thus, people crossing a border at night (assuming that a source of infrared illumination is supplied) will appear as light-colored ghosts against trees that are cooler and hence will appear darker on black and white film.

Most commercial infrared films are sensitive to infrared, visible spectrum, and ultraviolet light. They tend to be a little less sensitive to green light (the dominant region of the visible spectrum). Filters can be used to exclude most or all of the visible spectrum and some or all of the ultraviolet radiation. Infrared wavelength sensitivities for film are usually expressed in nanometers (nm). 1 μm = 1,000 nm. Some common commercially available infrared films and 'nearly infrared' films follow.

Black and White Infrared Film

Most infrared images are shot in black and white. Use of multiple exposures, different filters, digital image processing, and hybrid techniques can be used to yield *false color* images, but many are left unchanged—black and white is suitable for a wide variety of applications.

Black and white infrared film is convenient because it is readily available and can be processed with standard black and white processing (with the proviso that the cannister be opened in total darkness away from heat sources). Do not trust your infrared film cannister to technicians who are unfamiliar with its handling; even if you tell them not to open the cannister, they may do so because they don't understand the technology and the risks of fogging through the film cassette. Choose an experienced lab or process it yourself.

Agfa APX 200s - This has a spectral sensitivity from below 400 nm to about 775 nm. It peaks at a level similar to the Konica Infrared film, about 725 nm.

Ilford SFX 200 - This is a red-extended-sensitivity film rather than a true infrared film used more for art photography than surveillance, but it has some interesting properties. It is panchromatic to about 800 nm. It can be used with a red filter for infrared-type pictures. Since it is a visible spectrum, somewhat-infrared film, it is easier to handle and less subject to fogging.

Kodak High Speed Infrared (HIE) - This is one of the most well-known films for infrared photography. It is a moderately high-contrast ESTAR-based film available in a variety of sizes in rolls or sheets. Kodak HIE is sensitive to about 900 nm. It must be handled carefully so as not to expose it to infrared sources (including reflective surfaces on the back of the camera) and does not have an antihalation layer. Kodak recommends an exposure index (EI) of 50, but it can be used at higher EIs to create more pronounced effects. It is a little grainier than Konica Infrared and not necessarily the best for enlargements but, on the other hand, is faster and readily available.

Konica Infrared 750 - This has two peak sensitivity ranges, at about 400 to 500 nm and about 640 to 820 nm. It peaks at about 750 nm. It includes an antihalation layer to facilitate handling. This is a slower film than the Kodak HIE and is not suitable for fast-moving objects, but is probably the most commonly used film after the Kodak HIE and is chosen when a finer grain is desired.

Color Infrared Film

Color infrared film is typically composed of color-sensitive layers in the three primary colors for light (red, green, blue - RGB). Infrared red film has an additional layer that is designed to be sensitive to infrared radiation.

Kodak Ektachrome Infrared IE - This color infrared film was developed for aerial photography. Exposures should be made with a Wratten 12 filter for preferred colors. The images will not look like regular color film as the colors match infrared energy levels rather than visible light colors as we know them, but one learns to interpret the colors with a little practice. For aerial and scientific processing, AR-5 is used. For regular slide processing, E-4 or E-6 can be used (the colors will be a little more saturated, which might serve aesthetic or publishing purposes). Check the instructions that come with the film before processing. It is sensitive up to about 900 nm.

If only infrared radiation is to be sensed, a visible light filter is placed over the lens to block the visible light.

Don't use automatic DX-coding in the camera, use a manual setting, if possible, in order to avoid the possibility of fogging from the DX scanning sensor.

6. Applications

Anything that emits heat, or patterns of heat, can be monitored using infrared surveillance systems. In some wavelengths, a secondary source of illumination may be necessary. Thus, surveillance of people in a large variety of settings is a common application. Infrared sensors can be used to monitor employee movements, possible intruders, stowaways, and border runners.

Infrared sensors can be used in many commercial applications: ensuring the safety of amusement park patrons around rides; monitoring visitors or employees around heavy equipment or potentially dangerous chemicals; standing watch in law enforcement containment areas; sensing hostile intrusions in combat areas; detecting smuggling or other covert illegal operations; medical imaging, diagnosis, and monitoring; surveillance of physical characteristics or changes, including electrical, structural or mechanical fault detection and monitoring; and chemical process monitoring.

Illumination

Just as you need a light to see objects when the sun isn't shining, you need infrared lights to see objects that don't actively emit infrared radiation.

Many kinds of infrared technologies require a source of illumination in order for the reflected light to be sensed. Infrared film photography is one example. Thus, illuminators are commercially available for wildlife observation, border patrol, maritime surveillance, search and rescue, and low- or no-light infrared photography or live surveillance.

Infrared illuminators fall into two general categories:

continuous lights - These are similar to conventional light sources that emit a continuous beam. They are used to illuminate a source being observed or photographed that does not inherently emit infrared radiation in the wavelengths necessary to activate a detector. Illumination sources are especially useful at night. Visible/infrared lights in the 40-to-50-foot range commonly cost about $150 to $200. Short-range infrared laser illuminators are in the $1,000 range. Long-range illuminators (up to about 11 miles) suitable for marine, desert, or plains applications are priced around $5,000 to $20,000.

pulsers - These illumination sources emit continuous or pulsed signals and are typically used for identification, marking, or tracking devices that are specially designed to be detected with infrared viewers. Pulsers for many typical applications are in the $60 to $800 range.

Common sources of illumination include light-emitting diodes (LEDs), filtered incandescent light sources, and laser light sources. Illuminators vary in size, and may be portable or stationary, powered by batteries or 110/220 AC wiring. High-intensity infrared searchlights may use xenon bulbs for greater range. The beam-width of an infrared illuminator may be fixed or adjustable, with adjustable varieties usually being in the higher price ranges.

Night-Vision Scopes and Goggles

Night vision equipment is one of the more popular categories of surveillance accessories. Many clandestine activities are carried out in dim light or darkness, as are nocturnal animal activities, certain bird migrations, and other areas of research and observation. Night vision scopes consist of a variety of binocular and monocular sensor-equipped optics.

Night-vision scopes are not always infrared-equipped. Some of them are basically *light-amplification systems* that boost light sources in low light so they can be more clearly seen. Those that amplify light in the visible spectrum cannot 'see in the dark' as the name might imply. Infrared-sensitivity modifications and a supplemental source of infrared illumination are needed for true 'night vision'.

True infrared and infrared-sensitive or visible-spectrum night-vision scopes are common in a variety of applications. Depending upon the type of system, they are used to detect people (poachers, border-runners, smugglers, snipers, lost hikers), animals (game, endangered species, injured livestock, lost pets), and sources of heat (electrical wiring, steam leaks, insulation gaps, structural stresses, arson, and forest fires).

The range of a scope will vary with the lens magnification (as in telescopes and binoculars) and the amount of available light. Ranges of a few hundred feet are common. However, in total darkness, supplementary infrared illumination is required and, even then, the range will be reduced to about 100 feet.

Night vision scopes based on phosphor-activating tubes do not last indefinitely and their lifetime may be reduced if they are subjected to frequent extended bright lights (which can cause burn-in just as might happen on a television or computer monitor). Similarly, and more importantly, the image intensifier is subject to aging, lasting perhaps 1,000 hours, depending upon the system. Reconditioned or used night-vision equipment should not be purchased unless you know the circumstances under which the devices were used, and for how long.

If you are planning to ship or travel with night-vision equipment, check customs and export/import regulations, as many scopes are subject to export restrictions.

Since scopes are typically portable, most of them run on batteries. Depending upon the style and size of the scope, the battery life may range from 10 to 40 hours. Most consumer scopes use AA or lithium batteries. Battery-operated handheld viewers may be used in conjunction with infrared lasers or coils. Common applications include fire detection, security, fluoroscopy, night vision, pursuit, film manufacture and processing, and mineralogy.

Equipment and Accessories

Optional lenses, filters, mounting brackets, tripods, and light sources are available for many infrared viewing products, ranging in price from $20 to $1,600. Generation I scopes are typically about $700 to $900. Generation II and III scopes are about $3,200. Some scopes have camera adaptors and some are self-contained. The higher-priced devices generally have better water-resistance, better changing-light compensation, and bright-light-cutoff safety features. They may also feature longer battery life.

The smaller pocketscopes that can be used in with video and still cameras start at around $200 (Gen. I), $2,000 to $4,000 (Gen. II), and $4,400 (Gen. III). These scopes typically have a detection range of up to 200 to 450 feet.

Mapping and Topographical Studies

Left: An aerial view of the Red Sea taken at night by ASTER, using a new thermal-infrared (TIR) imaging system. (Visible/near-infrared systems require illumination and thus cannot be used for this type of night imaging.) Lighter areas indicate regions where temperatures are higher. This is the first ever spaceborne TIR multiband sensor, a joint project of the U.S./Japan Science Team. Right: Computer scientists, like professor David Landgrebe at Purdue University, are working on computer algorithms that can automate image processing to the point where consumers can make use of technical images available from satellites and other sources of infrared. These tools, which allow particular sorts of structures, rooftops, roads, vegetation, etc. to be more easily distinguished, will add another level of information acquisition to imaging devices. Shown above is a color infrared photograph of the National Mall in Washington, D.C. that is a candidate for this type of analysis. [NASA/Goddard/ERSDAC/JPL Feb. 2000 news photo; Purdue University 1999 news photo, released.]

The mapping field is rapidly evolving and changing due to dozens of new sensing technologies that can be used to gather and process data. In addition to sensors, we have also developed new methods of transportation, higher resolution image recorders, better printers, and more sophisticated image processing methods. There have never been so many ways to access and record locations on and beyond the Earth.

Maps are no longer limited to hand-drawn renderings of streets and elevations. They now include photographic satellite images of over 80% of the Earth. They show details of countries that have traditionally been closed to foreign travel and western surveys, including buildings and farms. They chart archeological ruins and cave paintings that are invisible to conventional technologies but visible through infrared, ultraviolet, or radar probing. They are enhanced and processed with computer algorithms and image modification programs. Judging by the current rate of change, the technologies and techniques for mapping are likely to continue to evolve dramatically in the next couple of decades and infrared imaging is an important aspect of that development.

Near-infrared has many applications in infrared imaging systems, but it cannot provide night images without supplemental illumination. This makes it unsuitable for some types of imaging, including pictures of Earth taken from the upper atmosphere or from space.

Infrared Spectrometers and Reflectometers

Spectrometers are scientific/industrial instruments that are used for many purposes, from studying atoms and molecules to studying the various wavelengths that are emitted by light sources. Since different substances emit different spectral patterns, spectrometers can be used to detect and identify these patterns, which makes them useful in analytical chemistry, gemology, law enforcement, and forensics.

Spectrometers are sometimes used to determine the quantities of certain substances that are present in a sample. For example, the *Bureau of Tobacco, Alcohol, and Firearms* (ATF) makes sure that the quantity of carbon dioxide added to still wines by producers does not exceed regulated maximum limits. An infrared spectrophotometer is used to determine the amounts of CO_2 present, to make sure they comply with regulations.

 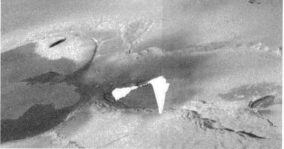

Infrared Imaging in Space. Left: A computerized infrared imaging spectrometer used in space sciences. Right: An infrared image taken from onboard the NASA Galileo spacecraft while viewing one of Jupiter's moons named Io. Images like this are sometimes taken simultaneously with two different types of imaging systems, with the pictures superimposed to give a better view of the various features. The spectrometer detects heat from sources such as this active volcano by imaging them in near-infrared. [NASA/GRC 1996 news photo by Chris Lynch; NASA/JPL 1999 News photo, released.]

Just as the visible spectrum has wavelengths that are perceived by us as different colors, the infrared spectrum has different wavelengths that can be measured for their different spectral characteristics. This is useful in many different fields.

443

Spectral analysis can be used for chemical identification in pharmaceutical applications and law enforcement (e.g., drug identification). It can be a useful tool for production line automation, quality control, and quality/composition inspection.

Spectral analysis is used by astronomers to learn more about distant planets. By analyzing the radiation emitted by planets as much as 100 light years away, it is possible to make educated guesses as to their atmospheric makeup that might include carbon dioxide, water vapor, and ozone. This aids in the search for other life in the universe and may someday help us locate other habitable planets. Since the Earth emits a considerable amount of infrared radiation, it interferes somewhat with spectral analysis from Earth-based telescopes. For this reason, space-based telescopes are also used. NASA has shown that by combining more than one telescope, it may be possible to create a space-based *interferometer* that could collect planetary infrared light with far greater effectiveness than land-based telescopes. The precision of the Space Interferometry Mission (SIM) telescope is so high, NASA claims it would be sufficient for someone on the Earth to "see a man standing on the Moon, switching a flashlight from hand to hand". This is clearly a high level of surveillant capability.

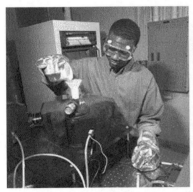

Left: Craig Washington servicing an infrared imaging spectrometer. Right: A portable version of an infrared reflectometer. Portable systems are always in demand because they can be mounted in smaller compartments and used in research and field work where it is difficult or impossible to take bulkier equipment. They also tend to consume less power. [NASA/GRC 1996 and 1999 news photos by Christopher Lynch, released.]

One of the more exciting developments in the field of infrared sensing is the invention of infrared systems that take advantage of quantum physics. In May, 2006, NASA announced a new infrared-based detector capable of sensing infrared light in a range of wavelengths (similar to the concept of color, but operating at electromagnetic frequencies that are invisible to humans).

The NASA/Goddard Quantum Well Infrared Photo Detector (QWIP) array is a gallium arsenide (GaAs) semiconductor chip with more than 100 layers of detector material. Each layer acts as a quantum well to trap electrons, releasing only those of a certain energy level to move through the array. The information derived from the chip is then processed and interpreted into a computer image. The QWIP improves upon earlier devices by sensing a range of frequencies and by making it possible to produce an image with photographic qualities that make it easier to discern shapes and contours. QWIP sensors are a flexible, generic technology that have many potential applications, including spectroscopy (e.g., crime scene analysis), fire detection, astronomy, meteorology, medical imaging, and much more.

Spectrographic sensing is an important aspect of astronomical research. In May, 2006, The National Science Foundation and philanthropists Gordon and Betty Moore announced the joint funding of an infrared, multi-object spectrograph (MOSFIRE), to upgrade the Keck I telescrope located on the summit of Mauna Kea, Hawaii. The news sensors will be capable of "simultaneously measuring up to forty infrared spectra ... of distant galaxies", thus greatly increasing the efficiency of the cosmic mapping process.

Infrared images created from QWIP quantum well photodetector data are more photographic in nature than historic true-infrared images (as opposed to infrared pictures taken in frequencies overlapping the visible spectrum) because the quantum well arrays can sense in a range of frequencies. The layered gallium arsenide sensor components are compact and reasonably-priced and thus is practical for a wide variety of spectrographic, fire detection, meterorological and astronomical applications. [NASA/Goddard News Photos, 2006, released.]

Quality Control, Assessment, and Inspection

Infrared has many industrial surveillance applications. It can be used to assess materials composition, uniformity, and to see structural anomalies that might not be apparent with other methods of inspection. It is frequently teamed up with other sensing technologies. In the early days of rail inspections, magnetic sensing was used to try to locate defects in individual pieces of track. Eventually ultrasound was found to be an effective technology for revealing problems before they could be detected visually. A whole array of sensing systems, including infrared, is now used for industrial inspection and assessment.

Infrared has been particularly valuable in electrical inspections, since there are many instruments for detecting thermal infrared. Thus, infrared detectors can be used for locating 'hot' wires or other electrical anomalies that might be hazardous. It can be used for assessing stress and wear and tear on production machinery. It is also useful in structural inspections outside of factories and is now being used to inspect bridges.

Lawrence Livermore National Laboratories (LLNL) has developed a mobile infrared-based construction inspection system. An engineer sits inside the moving vehicle of the prototype bridge-deck inspection system illustrated here. He is monitoring the data gathered from two infrared cameras that are mounted on a boom on the outside of the vehicle. Images of the road surface are transmitted to the computer inside the vehicle, which records and analyzes the images for temperature differences that might indicate cracking or rebar corrosion beneath the surface of the bridge deck. [LLNL news photo by Don Gonzalez, released.]

Fire Detection

Infrared sensors are well-suited to detecting thermal radiation in the infrared ranges that we perceive as 'heat'. This makes them valuable for monitoring industrial and scientific environments where heat-monitoring is critical for the protection of instruments, structural materials, and people. It also makes infrared sensing an excellent tool for detecting potential electrical hazards (hot wires) and for detecting fires in urban or forested areas.

Infrared forest fire detectors can be readily mounted on aircraft or all-terrain vehicles. Compact fire detectors can be mounted on walls or ceilings in buildings and can now be interconnected through computer distribution networks so that potential fire conditions can be relayed to security stations or a central console. The data could also be sent through the Internet to local fire fighters.

This well-shielded infrared imaging system, a bit larger than a basketball, is designed to mount on the side of a helicopter. Infrared sensing has been incorporated as a forward-looking infrared radar (FLIR) fire-detection system. A realtime TV monitor inside the craft enables personnel to monitor the data. The FLIR collects thermal images and temperature data in conjunction with exact location readings from a Global Positioning System (GPS) monitor inside the helicopter. This equipment aids security and safety around the *Kennedy Space Center*, especially prior to launches, and supports Florida's Division of Forestry in fighting brush fires. [NASA/KSC 1998 news photos, released.]

Electronic fire detection systems could be designed to send messages and instructions to people in specific areas of a building if danger were detected, reminding them to stay calm, to stay out of elevators, and to evacuate quickly through specified exits.

Astronomy

Infrared sensing is an integral tool of astronomers and space travelers. Various infrared wavelengths are emitted by cosmic bodies and can be analyzed to help us observe and understand the universe. Space and ground telescopes use a variety of types of cameras—by varying filters and exposure times and combining the images, spectacular three-color composites can be obtained as, for example, near-infrared images from our Milky Way taken by the *Infrared Spectrometer and Array Camera* (ISAAC) cryogenic infrared imager and spectrometer, installed on a VLT telescope in 1998.

The *Mariner 6 and 7 space probes*, launched in 1969, provided a variety of images, including TV photos and infrared and ultraviolet spectral images, as they journeyed into the solar system past Mars. Also in 1969, *Apollo 12* travelled to the Moon and brought back the TV camera that had been attached to *Surveyor 3* as it perched on the lunar surface for two and a half years. The scientists wanted, among other things, to evaluate the effects of almost three years of exposure to the lunar environment.

The *Hubble Space Telescope* is equipped with an infrared camera that provides images of the cosmos indicating how it may have looked millions and billions of years ago. Regions that were investigated with a visible-light camera were then imaged with a near-infrared camera, yielding evidence of hundreds of galaxies not seen with visible light, some of which were obscured by dust which attenuates visible light. Images from outside the visible range also showed that what previously appeared to be galaxies, might be new star regions within larger, older galaxies [Lowenthal et al., 1997].

The *Diffuse Infrared Background Experiment* (DIRBE) is a cooled reflecting telescope that is sensitive over a broad range of wavelengths encompassing 10 infrared bands from about 1.2 to 300 μm. By spinning the mounting platform, a 360° arc can be swept in each revolution, sampling about half the sky each day for about a year. Instruments for detecting different kinds of waves were installed, including a central instrument called the *Far Infrared Astronomical Spectrograph* (FIRAS).

The *Odin Satellite* is a Swedish-led multinational satellite mission involving Sweden, Canada, Finland, and France designed to conduct sensing and research into astronomy and the atmospheric sciences. The craft is equipped with the Swedish Sub-Millimeter Radiometer (SMR) and the Canadian *Optical Spectrograph and InfraRed Imager System* (OSIRIS). It was launched early in 2000 from Svobodny in Eastern Siberia into a sun-synchronous circular orbit at about 600 kilometers with an expected active lifetime of about two years.

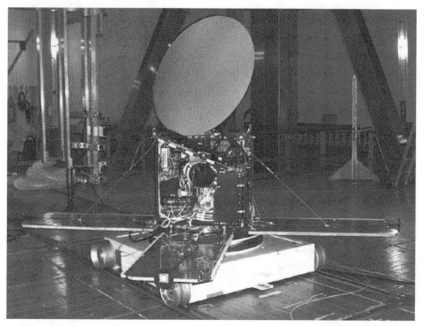

The Odin Satellite is a multinational project that includes the Canadian Optical Spectrograph and InfraRed Imager System (OSIRIS) which, among other things, will monitor ozone depletion in greater detail than was previously possible. Odin is showed in its deployed configuration in tests at the System Magnetic Tests, IABG, Munich. The OSIRIS infrared imaging system was designed and built by Routes Inc., Ontario. [Canadian Space Agency/Swedish Space Agency 1999 news photo, released.]

The Odin is equipped with a side-pointing telescope that is shielded and oriented so that it is protected from the Sun. There are also star trackers mounted on top of the upper platform, with subsystems mounted on both platforms. The system gets power from four solar panels fanned out between the shields. The optical spectrometer has four wavelength bands and sensing wavelength windows are 280 to 800 nm and 1270 nm. The infrared optical resolution is 10 nm.

In 1998, the giant *Keck II telescope* in Hawaii used a sensitive infrared camera to observe a swirling disc of dust that may indicate a new solar system being born around a star about 220 light-years from Earth. That same year, the NASA Hubble space telescope imaged a historic long-exposure infrared image of the faintest galaxies ever observed, some of which may be over 12 billion light-years away. This was accomplished with Hubble's near-infrared camera and multi-object spectrometer.

The *Next Generation Space Telescope* (NGST) was designed as a successor to the Hubble Space Telescope. The NGST will operate in a high-Earth orbit so that it will be far enough away to avoid the majority of the thermal heat emitted by the Earth. The operating temperature of the satellite is also intended to be as low as possible so as not to interfere with the astronomical emissions from other celestial bodies.

Left: A concept drawing of the proposed Next Generation Space Telescope (NGST) equipped with infrared-sensing equipment. Right: A model of the Space Infrared Telescope Facility (SITF) intended to overlap Hubble and Chandra X-Ray Observatory operations in 2001. It is being designed with state-of-the-art infrared cameras. [NASA news drawings, released.]

A number of telescopes, including the Keck Observatory telescope and the James Webb Space Telescope (JWST) are being constructed or upgraded to take advantage of developments in infrared sensing. The JWST will sense in the near and mid-infrared regions in order to explore the birth and evolution of celestial bodies.

There have also been recent discoveries of "background infrared" or "residual infrared" in the far-infrared frequencies, from analysis of data from the Diffuse Infrared Background Expoeriment aboard NASA's Cosmic Background Explorer (launched in 1989). The infrared "dust" has absorbed energy from stars, many of which are not visible to conventional telescopes, and re-radiates the light in infrared wavelengths, providing a way to trace the existence of stars that can't be viewed in other ways.

Military Surveillance

Infrared detection and imaging technologies are used in many aspects of military surveillance. Infrared sensors and recorders are installed in armored vehicles, reconnaissance planes, portable handheld units, and command centers.

The display console and control panel for a Wide Angle Surveillance Thermal Imager in use at an air base in Korea. Airman 1st Class Olivia Latham scans the air traffic control tower for infiltrators during a scheduled rear area defense field training exercise. The background sky appears as a yellow color, while the large building on the left and control tower on the right show up as pink and lavender. [U.S. DoD 1999 news photo by Val Gempis, released.]

Staffed aircraft and unstaffed aerial vehicles (UAVs) are widely used in infrared surveillance and many of them are described in other sections of this chapter and in the Aerial Surveillance chapter.

The Predator UAV is shown here in a simulated U.S. Navy aerial reconnaissance mission. The craft is launched from the carrier in the background where controllers and intelligence analysts analyze the data transmitted by the vehicle. The Predator is equipped with a variety of surveillance technologies including near-realtime infrared and color video imaging systems. [U.S. DoD 1995 news photos by Jeffrey S. Viano, released.]

Aircraft tactical reconnaissance pods carry a variety of surveillance equipment, including infrared sensors to provide imagery in darkness or bad weather conditions in low altitude assignments. Typically these pods are bullet-shaped containers weighing about 1200 pounds that are centerline-mounted under the belly of tactical aircraft. They are usually equipped with several types of cameras, recoding devices (e.g., tape recorders) and, in some of the more recent systems, realtime or near-realtime downlink technologies. Examples include TARPS and AT-RS. These tactical pods are somewhat self-contained so they can be selectively installed and potentially linked to different kinds of aircraft. The modular design also facilitates equipment upgrades and substitutions, with minimal modifications to the host aircraft.

Infrared imaging systems are frequently housed in streamlined 'pods' that can be selectively installed under the belly of various reconnaissance aircraft. Left: U.S. Navy mates remove an infrared sensor from a tactical air reconnaissance pod system (TARPS) that is attached to an F-14A Tomcat stationed in the Adriatic Sea, in 1993. Middle: A closeup of an F-14D Tomcat shows a double pod under the nose, which is equipped with a television camera system and an infrared search and tracking system. Right: The underbelly of an F-14B Tomcat equipped with TARPS. [U.S. DoD news photos by Todd Lackovitch, Dave Parsons, and Jack Liles, released.]

Missile Detection and Targeting

The detection and destruction of incoming missiles have traditionally been a difficult technological challenge. Missiles are fast and relatively small, compared to aircraft, so they are more difficult to spot and pinpoint geographically. In the United States, nonimaging infrared systems are used in a number of early warning systems. Infrared-sensor-equipped satellites, for example, have been established in ground-based radar systems to warn of impending ballistic missile attacks.

In one U.S. missile defense test scenario, satellites provide the general guidelines for the vehicle launch (a nonexplosive projectile), radar systems aid in narrowing down the target missile, and infrared sensors help in homing in during the final seconds. The intercept vehicle is controlled through a computerized communications console, which is typically ground-based. If a hit is successful, the kinetic energy from the impact destroys the missile.

The U.S. Boost Surveillance and Tracking System (BSTS) was initiated in 1984 as part of the Ballistic Missile Defense (BMD) program. BSTS was designed to replace aging systems which had been providing space-based infrared data since the early 1970s. The BSTS was the first tier of the BMD system. It was transferred to the Air Force as the Follow-on Early Warning System (FEWS) which evolved into the Alert, Locate, and Report Missiles program.

The U.S. also established a space-based infrared system in low-Earth orbit (LEO) called the Space and Missile Tracking System (SMTS), the second tier of the BMD system. This passive-sensor system performs ballistic missile boost and post-boost phase acquisition and tracking. It also performs midcourse-phase tracking and discrimination in the NMD and TMD systems. SMTS evolved out of the Space Surveillance and Tracking System and the Ground-based

Surveillance and Tracking System technology. Related experimental technologies include the Airborne Surveillance Testbed, Midcourse Space Experiment, and the Spatial Infrared Imaging Telescope series [O'Neill, 1995].

The U.S. Army midwave infrared missile seeker is mounted on a laser gyro-stabilized platform to isolate the seeker measurements from other vibrations. The seeker includes an all-reflective optical system and platinum silicide, staring, focal-plane array. The array serves as the optical sensor for the interceptor. Surveillance and fire control support are also provided by radar elements.

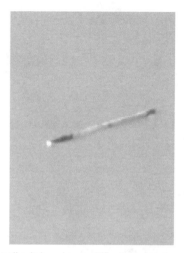 

Traditionally, it has been difficult to track and destroy incoming missiles. Infrared sensors can aid in homing in on the projectiles during the final stages of the intercept. A continuous-wave deuterium fluoride chemical laser, known as a Mid-Infrared Advanced Chemical Laser (MIRACL), is used to destroy a short-range rocket in flight (left) at the U.S. Army Space and Strategic Command's High Energy Laser Systems Test Facility, as part of the Nautilus program. The test suggests that lasers may be effective in combating hostile incoming airborne projectiles. [U.S. DoD 1996 news photos, released.]

It has been reported that at least 15 nations have anti-ship missiles with infrared guidance systems. The Electronic Warfare Systems Group has been involved in international collaborative research efforts to develop countermeasures to these infrared guidance systems. The Multiband Anti-ship Tactical Electronic System (MATES) for cruise missile defense was developed through this program. MATES was found capable of creating an optical breaklock (a jamming technique) in the missile's seeking mechanism at low power levels with more than 80% effectiveness in testing scenarios.

Remote Control and Communications Devices

Almost everyone with a TV, VCR, DVD, or stereo system has a handheld remote pushbutton device to control power, channels, and various other settings from across the room. Many of these send control signals via infrared. A remote control is a *line-of-site* device. That is, it won't work if there are infrared-blocking objects in the way, or if the controller isn't aimed in the general direction of the sensor (though sometimes you can bounce the signal off a reflecting surface). Remote-control devices work by sending pulses of invisible light that are picked up and interpreted by the sensor on the device being controlled. To visualize this more easily, imagine you and a friend are out in the dark, spaced about a block apart. Your friend uses a flashlight to send you instructions in Morse code. If a tree or house gets in the way, you don't get the message, and the light gets dimmer as you get farther apart. Infrared remote controls work much the same way, you just can't see the beam of light.

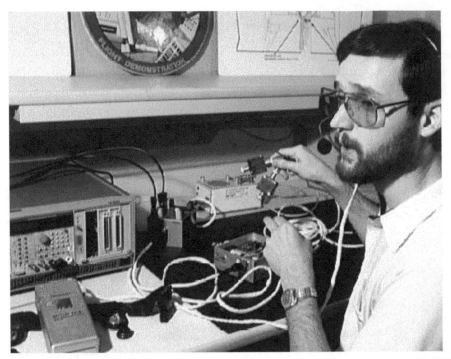

Joseph Prather is shown here with an infrared communications flight experiment (IRCFE) being readied for a flight on board the 1988 Space Shuttle Discovery STS-26 mission. The system uses light to communicate in much the same way that radio waves are commonly used. [NASA/JSC news photo, released.]

Infrared can also be used for data communications systems. Think of your friend and the flashlight again, but imagine now that you both have flashlights, so you can send messages in two directions. Since it's pretty difficult to simultaneously interpret your friend's message and send one at the same time, you arrange a signal, called a *handshake*, that indicates that you are finished sending and now you will wait while your friend signals back (like saying 'Roger, go ahead,' on a handheld radio). By coordinating the signals, messages can be alternately sent in each directions without causing confusion. Now imagine that the signals are sent with infrared data pulses rather than with flashlight Morse code pulses. This, in essence, is a wireless infrared communications device that, in turn, works very much like a radio-frequency wireless computer modem.

Motion Detectors

Security sensors commonly incorporate pyroelectric infrared (PIR) motion detectors into their systems. Typically, these are battery-operated so they can be placed in a variety of locations where electrical connections may not be available (carports, garages, worksheds, industrial yards, boat docks). These detectors are usually equipped with radio transmitters in the FM frequency that enable an alarm, recording device, or control console to be activated if the motion detector is triggered. Those which are operated on AC power may send the transmission through the electrical system, giving them a broader range, as in a large industrial complex or office building.

There are three measurements typically advertised with motion detectors:

range - the distance in which the detector is sensitive, usually about 20 to 40 feet,

breadth - the angle within which the detector is sensitive, usually about a 70° to 80° radius, and

wavelength - the sensitivity range of the detector.

Many motion detectors have light sensors that act as a simple scheduler to activate the detector only in the dark (usually at night, or when a building is closed and the lights shut off). For 24-hour surveillance, this sensor can usually be defeated by putting a piece of tape over it, fooling it into thinking it's dark so it will stay on all the time.

Many motion detectors are designed for indoor use and may not function optimally in temperature or moisture extremes. Recently there are more varieties available for indoor-outdoor use.

Since glass is opaque to infrared, it's not advisable to place a motion sensor behind a window of a door, for example, to detect someone approaching the door. Some of the infrared may reach the sensor, if it is near-infrared, but it may not be sufficient to trigger the sensor. It's best to place infrared sensors in a position where there are no impediments between the sensor and the area being sensed.

Infrared Sounders

Infrared sounders are used for atmospheric research and weather prediction. Sounders may be carried aboard satellites.

The National Oceanic and Atmospheric Administration (NOAA) uses infrared sounders to sense infrared emission sectra, including

HIRS - A High Resolution Infrared Sounder that has been in operation as a sensing instrument on a polar-orbiting satellite for two decades, providing data to the National Weather Service (NWS).

FIRS - A Fourier Transform Infrared Sounder that is a more recent than HIRS that uses a Bomem interferometer with a two-degree full-angle sensing field.

There are spectral regions in which the atmosphere can be quite opaque to infrared. While most sounders will sense in these obscured regions, scientists using the data are usually cautioned not to use data from regions known to be opaque.

Both NASA and NOAA use infrared sounders for weather prediction. It is important, in the space program, to carefully time the launch of various spacecraft to ensure safety and good operating conditions. The Atmospheric Infrared Sounder (AIRS) is a newer high-spectral-resolution infrared spectrometer. In conjunction with microwave sensors, AIRS provides weather prediction and environmental monitoring capabilities. Testing of AIRS began in the mid-1990s with plans to launch it on the EOS-PM spacecraft in 2000. AIRS has a higher spectral resolution than previous sounders and uses advanced computer processing to compensate for atmospheric and Earth emissivity and reflectivity effects.

Infrared Illumination and Thermal Imaging

Infrared sensing is useful for archaeological research and forensic investigations. Infrared and ultraviolet light can sometimes reveal features that are not obvious under ordinary light. Shining infrared light through a document, bank note, drawing, or painting can sometimes yield surprises or important clues as to the composition or history of an item. For example, infrared investigation of a historical painting of John A. MacDonald, the first Prime Minister of Canada, revealed that a portrait of a woman had previously been painted on the canvas.

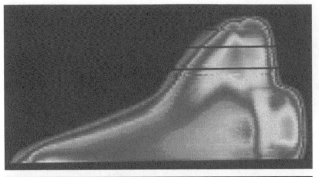

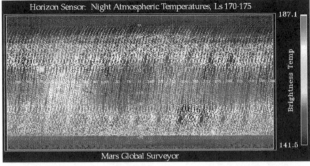

Top: This is an infrared image of a Space Shuttle on re-entry, showing the patterns of heat. The image was taken by the Kuiper Airborne Observatory (KAO) in 1982. The information on heat intensity and distribution is valuable for structural engineering and future shuttle design and also demonstrates the capabilities of the KAO infrared-sensing system. Bottom: This is an atmospheric temperature chart of Mars that shows thermal wave phenomena on the planet's surface, taken in 1999 on the Mars Global Surveyor mission. Temperature sensing and charting is a valuable implementation of infrared technology that can be applied to many fields of study. [NASA/Ames Research Center news photo, NASA/JPL news photo, released.]

Infrared-based thermal imagers are used for a number of industrial and inspection applications, including testing heat conduction and heat patterns, checking for cracks, and determining weak spots (or those subject to greater wear). They are also used for vehicle surveillance, tracking, and identification, especially in customs inspections and counterdrug operations.

Thermal imagers come in solid-mounted, portable, or headgear styles. The solid-mounted versions, if used on moving vehicles or aircraft, may include gimbal mechanisms to maintain spatial orientation.

Thermal imaging systems are used not only in space science and industrial applications, but in novel ways to assess textiles and the relative thermal values of materials intended for hostile environments such as the cold arctic and antarctic regions. The *Defence Research Establishment* in Ottawa, Canada, has created thermograms with a Dynarad fast-scan infrared camera to provide background information on experimental techniques for using thermal information to study clothing systems.

Bug Detectors

Some bug detection devices are equipped with standard or optional detection probes. In mid- and high-range detectors, there will often be an option for an infrared probe. These detectors range in price from $500 to $1,500.

Reconnaissance Aircraft (Piloted)

Reconnaissance aircraft are now routinely equipped with infrared sensors and infrared imaging systems, as are reconnaissance satellites, so there's only space for a few examples here (additional examples are listed in the Aerial Surveillance chapter).

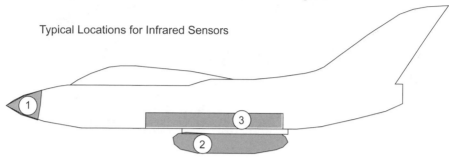

Typical Locations for Infrared Sensors

Reconnaissance-equipped aircraft most commonly carry infrared and radar equipment 1) in the nose, 2) in a fixed or detachable pod, and/or 3) internally in the belly of the craft.

The F/A-18 "Hornet" line includes a number of different models, some of which are equipped with Infrared Imaging Maverick Air-to-Ground missiles. The Hornet's night attack capabilities had been upgraded by 1989 including *navigation forward-looking infrared* (NAVFLIR) pod, night vision goggles, and special cockpit lighting for night vision devices, a digital color moving map and independent multipurpose color display. NAVFLIR is a fixed field-of-view sensor providing automatic electronic boresighting to optically align a scene on the pilot's HUD with the corresponding outside view. The second-stage upgrade was to add a FLIR pod with a built-in laser to designate targets for dropping laser-guided bombs (LGBs), cockpit modifications for night attack (night vision and lighting), and enhanced defensive countermeasures capabilities. However, in 1995, funding for upgrades were eliminated. The aircraft include targeting and navigation FLIRs.

Left: In-flight refueling between a U.S. Navy A-6E Intruder (left) and a French Navy Super Etendard. Right: Prior to decommissioning, the Navy A-6E Intruder receives its last launch signal from the deck of an aircraft carrier. The A-6Es use an imaging infrared seeker guidance system. [U.S. DoD 1996 news photos by Brent Phillips and Alan Warner, released.]

The F-14D "Tomcat", is equipped with a dual optical/electronics chin pod housing a *Television Camera System* (TCS) and the *Infrared Search and Track* (IRST) system. The aircraft has a passive infrared detection capability, targeting FLIR. Upgrades to the Tomcats include conversions from analog to digital and ATARS equipment with near-realtime traditional and infrared photos that can be transmitted to marine or ground stations.

Some of the F-14 Tomcat models have been equipped with the LANTIRN targeting systems and night vision systems. LANTIRN permits delivery of laser-guided strikes. The F-14D is equipped with the Infrared Search and Track (IRST) system. An F-18B is shown supporting the Iraqi no-fly zone in the Gulf. [U.S. DoD 1997 news photo by Bryan Fetter, released.]

The F-15E "Strike Eagle" is a long-range, all-weather, multimission strike fighter, equipped with navigation FLIR and targeting FLIR.

The F/A-18E/F "Super Hornet" was authorized by Congress in 1996, with assembly taking place in 1998. ATARS was installed in the nose of the test plane. The Super Hornet extended the Hornet's range, endurance, and survivability.

The *Low-Altitude Navigation and Targeting Infrared for Night* (LANTIRN) targeting pod system was developed in the late 1980s by Lockheed and installed in various F-15/F-16 models since 1988. As of 1995, the systems were being integrated into U.S. Navy F-14 Tomcats as part of the U.S. Navy's F-14 *Precision Strike Program*. These aircraft are intended for precision strike missions at low altitudes in darkness. The LANTIRN sensor uses a wide field-of-view FLIR system for target detection. On detection, it switches to a narrow field-of-view FLIR for target locking and weapons delivery.

Unstaffed Aerial Vehicles (UAVs)

One area of intense research and commercial interest is remotely-controlled tactical reconnaissance vehicles. They have many advantages: they can be more quickly and easily replaced than conventional craft; the cost of manufacturing and operating them is tens of thousands of dollars compared to tens of millions of dollars for conventional aircraft; and there is no loss of life if the craft is shot down.

There are two basic kinds of UAVs, those that are remotely controlled, and those that are autonomous, operating through preprogrammed algorithms or self-modifying artificial intelligence software.

Historic aerial vehicles (UAVs) had limited range and endurance, but both capabilities have been substantially improved in recent years. Many UAVs are equipped with infrared-sensing equipment. Here are just a few examples of the many types of UAVs:

BQM-147A Exdrone - This UAV was developed in the early 1980s at Johns Hopkins University. It has a delta-wing platform with an eight-foot span, weighing 40 to 80 lbs (depending upon the payload). Its top speed was 100 MPH, with a datalink range of approximately 40 miles and an endurance of 2.5 hours. It sold for under $25,000. The

Exdrone is manufactured by BAI Aerosystems and can be equipped with a variety of video and chemical sensors, and with forward-looking infrared (FLIR). At about the same time, the U.S. *Department of Defense* was initiating projects to design future ground-launched, multisensor UAVs with significantly greater range, endurance (up to 24 hours), and return capacity.

Global Hawk - A high-altitude, long-endurance aerial reconnaissance system rolled out by the *Department of Defense* early in 1997. It is capable of generating high-resolution, near-realtime imagery using synthetic-aperture radar (SAR) and electro-optical and infrared sensors.

General Atomics Aeronautical Systems Inc. (GA-ASI) - This company develops tactical endurance air vehicles, and their associated ground-station support. GA-ASI evolved their GNAT-750 (begun in the late 1980s) into the TE-UAV (Predator) which went into the test phase in the mid-1990s. These UAVs include FLIR along with video and line-of-site downlink capabilities. The GNAT-750 has an endurance of 40+ hours and can fly at altitudes of over 25,000 feet.

Predator - A Navy aerial reconnaissance craft capable of near-realtime infrared and color video sensing. Originally the Predator would beam reconnaissance information to a chase aircraft or a ground station, but it was upgraded to send signals over longer distances by relaying through satellites. These vessels have been used for training and Scud-missile spotting by the U.S. Army since the mid-1990s.

Tier III Minus DarkStar - A high-altitude, GPS-equipped endurance vehicle that can autonomously complete pre-programmed basic flight maneuvers and fly to an altitude of several thousand feet.

Hunter, Raptor, and Talon are similar vehicles used by the armed forces. Due to their size, cost, and flexibility, there is a possibility that UAVs may eventually supersede military fighter planes for many types of operations.

7. Problems and Limitations

Infrared sensors are not dangerous in the same way as X-ray emitters. Infrared radiation is everywhere and we have evolved to live in an infrared-rich environment. Most infrared sensing systems are passive systems, that is, they 'read' the infrared emanations without generating it first. Active infrared systems, like remote controls, are not typically hazardous.

Most of the limitations and problems associated with infrared sensing have to do with achieving higher frame rates and higher-resolution images and with separating the target from the clutter. Also, as the technology becomes more precise, noise and interference become areas of concern. Here is a sampling of some of the technical limitations:

- The components used to build the detector have themselves thermal radiation that can interfere with the function of more sensitive detectors.

- As infrared wavelengths become longer, it becomes more 'expensive' to create detectors. Yet longer wavelengths yield good information for some applications (e.g., astronomy), as they can penetrate dust. From 100 μm to 350 μm there are few useful windows so broadband detectors, such as bolometers, are used in these regions.

- Semiconductor photocarrier decay times are important determinants in infrared detectors [Carr 1995]. Absorption in the far-infrared is frequency-dependent for many semiconductors due to photocarrier scattering rates.

- Medium- to long-wavelength technology is still expensive and short wavelength is still somewhat limited in capabilities.

- Stealth aircraft design may make a craft less visible on radar, but may still produce significant 'infrared signatures'.

- Nuclear and space radiation create noise spikes in infrared detectors (leading to false alarms) often necessitating expensive custom adjustments in each detector.

- Emissivity (thermal emittance relative to a black body) differences can provide misleading readings. For example, if a reflective surface has a less reflective covering (paint, cloth, etc.) and part of the covering is missing or scratched, false-positive 'hot spots' may show up in the image. Visual inspection of the material can help identify the source in some situations, and imaging from a number of angles can help identify whether the hot spot is on the surface, or behind it.

- Atmospheric clutter can impede infrared sensing intended to detect chemical and biological agents and aerosols.

- Film that is sensitive to infrared can photograph only reflected radiation in the near-infrared region. It isn't sensitive enough to spot a border-runner in the dark without supplementary illumination.

- Infrared films are subject to fogging, require special handling, and shouldn't be handled by technicians unfamiliar with infrared characteristics. Special filters generally need to be used—those transparent to infrared are opaque to human eyes, making it more difficult to compose an image.

- Infrared photos of the Earth from space dramatically illustrate how infrared can penetrate 'haze' and result in clearer, brighter images for many types of applications, but there are still limitations in infrared technologies in Earth-based applications. Infrared radiation has only a limited ability to penetrate fog or precipitation at certain wavelengths. Progress is being made in this area, however, by monitoring atmospheric properties and then using data processing to compensate for the interference.

8. Restrictions and Regulations

Unlike X-ray equipment, there are very few specific restrictions and regulations for the use of infrared-sensing technologies beyond those more general considerations related to personal safety and privacy. There are health and safety regulations for the use of infrared in certain industrial environments (infrared is used for tasks such as drying printing inks), but these are typically active infrared systems, while a large proportion of surveillance technologies use passive infrared sensing, and thus are not subject to the same restrictions. Common sense is sufficient to deal with most infrared sensing technologies.

Infrared technology is subject to fewer restrictions than most other surveillance technologies. For example, radio communications are very stringently regulated by the *Federal Communications Commission* due to the fact that radio waves can overpower each other and are in very heavy demand. Infrared has many applications in wireless communications and can even be used for wireless data communications on computer networks without most of the restrictions imposed on wireless networks based upon the use of radio wave technologies. It should be remembered that infrared lies next to (and overlapping) the region designated as radio waves in the electromagnetic spectrum, and far-infrared has certain properties in common with radio microwave radiation. Infrared is a good candidate for small, local, wireless, secured computer networks.

9. Implications of Use

Overall, infrared-sensing technologies tend to be used for applications such as fire detection, remote Earth and space sensing, telescopes, electrical fault location, home and office security, missile defense, and general defense imaging. They are also used for remote-control and wireless intranet computer networks. Infrared technology is further used in medical applications, particularly for sensing circulatory conditions.

Taken as a whole, infrared technologies have numerous personal and commercial benefits. Since the benefits of the technologies, as they are currently used, seem to outweigh the negative aspects, there has been a minimum of controversy over the use of infrared for surveillance or other applications.

10. Resources

10.a. Organizations

Inclusion of the following companies does not constitute nor imply an endorsement of their products and services and, conversely, does not imply their endorsement of the contents of this text.

The Academy of Infrared Thermography (AIRT) - An independent educational institute dedicated to promoting and providing the highest standards of infrared instruction. www.infaredtraining.net/

American Institute of Aeronautics and Astronautics, Inc. (AIAA) - A nonprofit society serving the corporations and individuals in the aerospace profession to advance the arts, sciences, and technology related to aeronautics and astronautics. AIAA sponsors many technical meetings and conferences. www.aiaa.org/

CECOM - A U.S. Department of Defense mailing-list clearinghouse that automatically distributes IRIS conference proceedings to qualified subscribers. Subject to clearance by the Security Manager for the IRIA Center, this service can be requested through CECOM, Night Vision and Electronic Sensors Directorate, Security Branch, 10211 Burbeck Road, Suite 430, Ft. Belvoir, VA 22060-5806.

The Coblentz Society - A nonprofit organization founded in 1954 to promote the understanding and application of vibrational spectroscopy. It sponsors educational programs and awards in addition to providing member services. www.galactic.com/coblentz/index.htm

Electronic Development Laboratories (EDL) - Founded in 1943. Manufacturers of precision temperature measuring instruments and sensors including RTDs, thermistors, infrared, base metal and noble metal thermocouples, handheld pyrometers, solid state and infrared sensors. These products are especially useful in industrial settings for production line and quality control functions. www.edl-inc.com/ www.thermocouples.org/

FLIR Systems - A prominent commercial supplier of foward-looking infrared sensing products, including air, ground, thermographic and fire-fighting devices. FLIR sponsors the IR Info Symposium. www.flir.com/

Infrared Data Association (IrDA) - Founded in 1993 to support and promote software and hardware standards for wireless infrared communications links. www.irda.org/

Infrared Information Analysis Center (IRIA) - A service and product center sponsored by the Defense Technical Information Center (DTIC). IRIA was founded in 1954 to facilitate information exchange within the U.S. Department of Defense (DoD) to collect, process, analyze, and disseminate information on infrared technology and now also the entire electromagnetic spectrum. The IRIA Library houses over 50,000 items, including documents, proceedings, articles, and books. IRIA's classified and limited-distribution resources are available to qualified organizations meeting certain subscriber requirements. IRIA sponsors a number of special interest groups including infrared countermeasures, detectors, materials, and sensors groups. There is a broad range of information available at ERIM International Inc.'s site at www1.erim-int.com/

Infrared Space Observatory (ISO) - Launched November 1995 by the European Space Agency into an elliptical orbit to observe infrared radiation from 2.5 to 240 microns.

IRIA Target and Background Data Library (ITBDL) - Formerly the DARPA Infrared Data Library, established in 1978, this is now part of the Infrared Information Analysis Center (IRIA). Archival source of computer-compatible data tapes from DARPA-sponsored projects.

National Fire Protection Agency (NFPA) - The NFPA is an old organization, established in 1896, which promotes education and research, and scientifically based codes and standards for fire protection. www.nfpa.org/

Servo - Suppliers of thermal detection materials since 1946, particularly thermistor and pyroelectric detectors. Supplies primarily to the railroad and aerospace industries. www.servo.com/

SPIE - The International Society for Optical Engineering. SPIE has been at the forefront of optical research and development for many years and publishes many of the authoritative references on infrared technologies and their adaptation to aerial sensing and other detection technologies. www.spie.org/

Stratospheric Observatory for Infrared Astronomy (SOFIA) - A NASA/Ames Research Center project to create a nonorbiting facility with a large telescope flown in a 747 aircraft to get above most of the Earth's atmosphere. SOFIA will replace the Kuiper Airborne Observatory (KAO). This is just one of many NASA-related infrared projects. www.nasa.gov/

U.S. Air Force and FLIR Systems, Inc. - Involved in the cooperative development of a surveillance system called a Covert Adjustable Laser Illuminator (CALI) that uses a thermal imaging system to apprehend a ship's registration information in the dark. If pollution such as oil spills or covert flushing is detected or suspected, the CALI system can be used to identify vessels in the area and to record activity. Also suitable for ground surveillance and search and rescue.

10.b. Print Resources

Accetta, J. S.; Shumaker, D.L., editors, *Infrared and Electro-Optical Systems Handbook,* IRIA and SPIE, 1996. Eight volumes with contributions by eighty infrared, electro-optical specialists.

Arnold, James R., *Space Science, California Space Institute,* University of California, 1997. Chapters 9, 10, and 14 discuss infrared technologies, how they are used in space observation, and various limitations.

Avery, T. E.; Berlin, G. L., *Fundamentals of Remote Sensing and Airphoto Interpretation,* New York: MacMillan, 1992.

Bass, Michael, Editor in Chief, *Handbook of Optics Volume I: Fundamentals, Techniques, & Design,* New York: McGraw-Hill, Inc., 1995, ca. 100 pages. A multi-contributor, multi-volume comprehensive coverage of the topic, with Volume II also of interest. Extensive contributions by specialists in the field.

Beam, Walter R., *Command, Control, and Communications Systems,* New York: McGraw-Hill Publishing Company, 1989, 339 pages. Practices and techniques for designing, developing, and managing CCC (C-cubed) systems.

Caniou, Joseph, *Passive Infrared Detection: Theory and Applications,* Boston: Kluwer Academic Publishers, 1999. The basic physical principles and how they related to infrared detection technologies. Suitable as a reference text for students, technicians, physicists, and engineers.

Cindrich, Ivan; Del Grande, Nancy K., *Aerial Surveillance Sensing Including Obscured and Underground Object Detection,* SPIE, April 1994 Conference Proceedings, 1994. There are dozens of SPIE conference proceedings (too many to list here) of direct relevance to infrared technologies used in surveillance activities directed at high undergraduate/graduate-level readers. Contact your local library for a full list.

Curran, Paul J., *Principles of Remote Sensing,* London & New York: Longman Group Ltd., 1985.

Dennis, P. N. J., Photodetectors: An Introduction to Current Technology, New York: Plenum Press, 1986, 176 pages.

Dereniak, Eustace L.; Boreman, *Glenn, Infrared Detectors and Systems,* Wiley Series in Pure and Applied Optics, John Wiley & Sons, 1996, 561 pages. Written by a specialist in optical detection who has designed systems for Rockwell and Raytheon, Ereniak is a professor at the Optical Sciences Center at the University of Arizona.

Driggers, Ronald G.; Cox, Paul; Edwards, Timothy, *Introduction to Infrared and Electro-Optical Systems,* Boston: Artech House Optoelectronics Library, 1999. Comprehensive introduction to the analysis and design of infrared and electro-optical imaging systems and systems analysis, including linear shift-invariant (LSI) infrared.

Fellowes, P. F. M.; Blacker, L. V. Stewart; Etherton, P. T.; *The Marquess of Douglas and Clydesdale, First Over Everest,* John Lane the Bodley Head Limited: London, 1933, 279 pages. A 1933 team of aerial photography pioneers describes the remarkable 'state-of-the-art' adaptations and inventions that allowed them to take the first aerial photos of Mt. Everest and the closed region of the Himalayas and Nepal in both traditional and infrared modes. The authors provide illustrations and make recommendations for future similar missions.

Feynman, Richard P., *QED: The Strange Theory of Light and Matter,* Princeton University Press, 1985, 158 pages. Feynman, a Nobel laureate and Caltech instructor, created this introductory series of lectures containing some challenging but fundamental concepts in the behavior and study of photons and electrons within the framework of quantum dynamics.

Handbook of Kodak Photographic Filters, Eastman Kodak Co., Publication B-3, revised 1990. Lists information on filters suitable for infrared (and ultraviolet) photography.

Holz, Robert K., Editor, *The Surveillant Science: Remote Sensing of the Environment,* Houghton Mifflin, 1973, 390 pages. Though an older volume, this is a good selection of articles by remote-sensing specialists that starts with simpler, more basic concepts and takes the reader through the evolution to practical, applied sensing.

Hudson, Jr., Richard D., *Infrared System Engineering (Pure and Applied Optics),* New York: John Wiley & Sons, 1969, 642 pages.

Hudson, Jr., Richard D.; Wordsworth Hudson, Jacqueline, editors, *Infrared Detectors,* Stroudsburg, Pa.: Dowden, Hutchinson & Ross, 1975, 392 pages.

Iannini, Robert E., *Build Your Own Working Fiberoptic Infrared and Laser Space-Age Projects,* Tab Books, 1987, 262 pages. If you are a hobbyist who prefers to learn by hands-on methods, this book provides a project-by-project format that can help the reader get an introductory working understanding of some basic infrared technology.

Jacobs, Pieter A., *Thermal Infrared Characterization of Ground Targets and Backgrounds* (Tutorial Texts in Optical Engineering), Bellingham: SPIE, 1996, 220 pages. Introduction to heat transfer, target detection, atmospheric considerations, infrared signatures, and camouflage.

Klein, Lawrence A., *Millimeter-Wave and Infrared Multisensor Design and Signal Processing,* Artech House, 1997, 520 pages. A practical introduction to principles and applications including passive and active technologies, optics, detector theory, weapons sensors, Earth resource and weather monitoring, and traffic management.

Kodak staff, Infrared and Ultraviolet Photography, Kodak, 1961.

LaRocca, A. J.; Sattinger, I. J., Editors, *An IRIA Annotated Bibliography: Proceedings of the Infrared Information Symposia* (IRIS)(1984-1991), November 1991, ERIM Report No. 213400-118-X, 500 pages. Includes almost 2500 technical papers for the years 1984 to 1991 on a variety of topics including infrared systems for reconnaissance, surveillance, navigation, weapons guidance, countermeasures, and fire control. Available to qualified organizations.

Lloyd, J. M., *Thermal Imaging Systems,* New York: Plenum Press, 1975, 456 pages.

Lo, C. P., *Applied Remote Sensing,* Harlow, Essex, U.K.: Longman's Scientific & Technical, 1986.

McLean, Ian S., Editor, *Infrared Astronomy with Arrays: the Next Generation,* Boston: Kluwer Academic, 1994, 572 pages.

Morey, B.; Ellis, K.; Perry, D; Gleichman, K., *IRIA State of the Art Report: Infrared Signature Simulation of Military Targets,* September 1994, Unclassified, ERIM Report No. 246890-2-F.

Nielsen, H. H.; Oetjen, R. A., *Infrared Spectroscopy from W. G. Berl's Physical Methods in Chemical Analysis - Volume 1,* New York: Academic Press, 1950.

Paduano, Joseph, *The Art of Infrared Photography,* Amherst Media, 1998, 112 pages. Descriptions of the different aspects of infrared photography, darkroom information, the use of filters, exposure settings, and flash and examples of infrared images.

Proceedings of the Society of Photo-Optical Instrumentation Engineers, Esther Krikorian, Editor, Infrared Image Sensor Technology, SPIE, 1980, Volume 225, 163 pages.

Proceedings SPIE–The International Society for Optical Engineering, Airborne Reconnaissance XVIII, Bellingham: SPIE, 1994, 276 pages.

Rogalski, Atoni; Kimata, Masafumi; Kocherov, Vasily F.; Piotrowsi, Jo, *Infrared Photon Detectors,* Bellingham: SPIE, 1997, 658 pages.

Ruttledge, Hugh, *Everest 1933,* Hodder & Stoughton: London, 1934, 390 pages.

Vincent, John David, *Fundamentals of Infrared Detector Operation and Testing,* John Wiley & Sons, 1990, 504 pages. Comprehensive operation and testing reference with formulas and examples for laboratory applications. Includes detector types, radiometry, measurement and more.

White, Laurie, *Infrared Photography Handbook,* Amherst Media, 1996, 108 pages. Covers photography and spectral information, particularly with regard to Kodak HIE (black and white) film. Provides guidance in getting started in infrared photography with many examples.

Wolfe, William L.; Zissis, G. J., *The Infrared Handbook,* Washington, D. C.: IRIA, Dept. of the Navy, 1985, 1700 pages. A large, classic reference work.

Wolfe, William L., *Introduction to Infrared System Design (Tutorial Texts in Optical Engineering),* Bellingham: SPIE, 1996, 131 pages. Course notes related to detection, scanning, optics, and radiometric aspects of infrared technology.

Workman, Jerry, editor; Springsteen, Art W., *Applied Spectroscopy: A Compact Reference for Practitioners,* Academic Press, 1998, 359 pages. Practical guidance in ultraviolet, visible, and infrared reflectance spectroscopy covering various topic areas.

Articles

Aumann, H. H., Atmospheric Infrared Sounder on the Earth Observing System, Paris, France, *ASAP, Sensors, Systems and Next Generation*, available from the JPL Technical Report Server.

Bissonnette, L. R.; Roya, G.; Theriault, J-M., Lidar Monitoring of Infrared Target Detection Ranges Through Adverse Weather, *Canadian Department of National Defence R&D Report*, 1998, 15 pages.

Bruce, Willam A., Theoretical Model for Use of Infrared Detection for Host Location by the Bee-mite, Varroa Jacobsoni (Acari:Varroidae), *Agricultural Research Service*, 1996. These parasitic mites may have sensors able to detect infrared light emitted by bees.

Carr, G. L., Nanosecond Time-Resolved Far-Infrared Spectroscopy of Photocarrier Decay in GaAs, from a lecture at a meeting of the *American Physical Society*, March 1995.

Chevrette, P.; St-Germain, D.; Delisle, J.; Plante, R.; Fortin, J., Wide Area Coverage Infrared Surveillance System (Infrared Eye), *Canadian Department of National Defence R&D Report*, 1998, 36 pages. A new concept for improving infrared imaging systems and applied surveillance for search and rescue. Simulating the human eye, it allows surveillance and target acquisition in two fields of view simultaneously.

Coblentz, W. W., Early History of Infrared Spectroradiometry, *Science Monthly*, 1949, V.68, 102.

Coblentz, W. W., Investigations of Infrared Spectra, Carnegie Institute, Washington, D.C., *Publication No. 35*, 1905.

Fortin, J.; Chevrett, P., Infrared Eye: Microscanning, *Canadian Department of National Defence R&D Report*, 1998, 84 pages. The authors describe R&D on a microscanning system to enhance the resolution of a focal-plane array camera by a factor of two.

Fouche, P. S., The use of low-altitude infrared remote sensing for estimating stress conditions in tree crops, *South African Journal of Science*, 1995, V.91, pp. 500-502.

Greer, Jerry D., Airborne Reconnaissance and Mount Everest, in Fishell, Wallace G.; Henkel, Paul A.; Crane, Jr., Alfred C. (editors), Airborne Reconnaissance XVIII, *Proceedings SPIE*, 1994.

Herschel, W., Investigation of the Powers of the Prismatic Colours to heat and illuminate Objects; with Remarks that prove the different Refrangibility of radiant Heat. To which is added an Inquiry into the Method of viewing the Sun advantageously with Telescopes of large Apertures and high magnifying Powers, *Philosophical Transactions*, 1800, V.90, pp. 255-326.

Jobe, John T., Multisensor/Multimission Surveillance Aircraft, in Fishell, Wallace G.; Henkel, Paul A.; Crane, Jr., Alfred C. (editors), Airborne Reconnaissance XVIII, *Proceedings SPIE*, 1994.

Jones, R. N., The Plotting of Infrared Spectra, *Applied Spectroscopy*, 1951, V.6(1), p. 32.

Lowenthal, J. D.; Koo, D. C.; Guzman, R.; Gallego, J.; Phillips, A. C.; Faber, S. M.; Vogt, N. P.; Illingworth, G. D.; Gronwall, C., Keck Spectroscopy of Redshift Galaxies in the Hubble Deep Field, *Astrophysical Journal*, 1997.

O'Neill, Malcolm, Ballistic Missile Defense: 12 Years of Achievement, *Defense Issues*, V.10(37), 1995.

Sands, J. D.; Turner, G. S., New Development in Solid Phase Infrared Spectroscopy, *Analytical Chemistry*, 1952, V.24, p. 791.

Shen, Chyau N., "Project Radiant Outlaw", in Fishell, Wallace G.; Henkel, Paul A.; Crane, Jr., Alfred C. (editors), "Airborne Reconnaissance XVIII", *Proceedings SPIE*, 1994.

Young, B. G.; Apps, R. G.; Harwood, T. A., Infrared Reconnaissance of Sea Ice in Late Summer, *Canadian Department of National Defence R&D Report*, 1971, 17 pages. A test of the ability of airborne infrared line scanners to differentiate ice.

Journals

Astronomical Journal, American Astronomical Society professional journal in publication for over 150 years, published monthly.

Astronomy and Astrophysics: A European Journal, professional journal published on behalf of the Board of Directors of the European Southern Observatory (ESO).

The Astrophysical Journal, published by the University of Chicago Press for the American Astronomical Society, three times monthly, in publication since 1895.

The Infrared Scanner Newsletter, published seasonally by Inframetrics. Includes back-issues to summer, 1995. Describes applications of the technologies, and new products related to infrared sensing. Some of the product brochures listed in the newsletter include article reprints from industry technical journals. www.inframetrics.com/newsletter/irscanner/

Journal of Thermophysics and Heat Transfer, published by the American Institute of Aeronautics and Astronautics.

Journal of Optics A: Pure and Applied Optics, formerly *Journal of Optics,* it has been split into two publications. This journal of the European Optical Society covers modern and classical optics.

10.c. Conferences and Workshops

Many of these conferences are annual events that are held at approximately the same time each year, so even if the conference listings are outdated, they can still help you determine the frequency and sometimes the time of year of upcoming events. It is very common for international conferences to be held in a different city each year, so contact the organizers for current locations.

Most organizations describe upcoming conferences on the Web and may also archive conference proceedings from previous conferences for purchase or free download.

The following conferences are organized according to the calendar month in which they are usually held.

AIAA Aerospace Sciences Meeting and Exhibit, annual American Institute of Aeronautics and Astronautics, Inc., conference.

Military Sensing Symposium Specialty Group on Infrared Detectors, military sensing symposium sponsored by the *Infrared Information Analysis Center.* Held in conjunction with the symposia on passive sensors, camouflage, concealment, and deception and infrared materials.

AIAA Thermophysics Conference, annual conference since the mid-1960s. www.aiaa.org/

National Symposium on Sensor and Data Fusion, military sensing symposium sponsored by the Infrared Information Analysis Center,.

Symposium on Thermophysical Properties, annual symposium since the mid-1980s.

AIAA Guidance, Navigation, and Control Conference. Biennial. www.aiaa.org/

InfraRed Information Exchange, a forum sponsored by The Academy of Infrared Thermography on all aspects of infrared thermography including research and applications.

10.d. Online Sites

The Buffalo Project. An online version of a GISDATA paper on predicting urban temporal patterns by Batty and Howes. This well-illustrated site describes how remote sensing imagery can be used to predict urban change and growth. www.geog.buffalo.edu/Geo666/batty/strasbourg.html

fotoinfo.com. Has Web pages devoted to infared photography, including information on black and white negative and color slide infrared film, focusing, and processing. Surveillance uses are mentioned as well. www.fotoinfo.com/

Honeywell Infrared Projector Technology. This is a commercial site, focusing mainly on Honeywell Technology, but there are a number of well-illustrated product development descriptions in the Spotlight on Technology section that give an interesting account of some of the recent scientific and engineering inventions in the infrared field. The site is also searchable for other topics (e.g., ultraviolet). www.htc.honeywell.com/

Infra-Red Photography FAQ. A Frequently Asked Questions Web page gleaned from topics discussed on the USENET rec.photo group. This FAQ was compiled originally by Caroline Knight. www.mat.uc.pt/~rps/photos/FAQ_IR.html

Infrared, Inc. Suppliers of infrared cameras, software, and imaging systems. The site includes an introductory FAQ on infrared technology. www.infrared.com/

Infrared Spectroscopic Method. This is a good introduction to the use of Infrared Spectroscopy for identifying narcotic substances. It has a brief illustrated history, and a technical introduction to the science, followed by a good bibliography with some interesting historical references. It has been prepared by Charles Hubley of the Canadian Defence Research Chemical Laboratories and Leo Levi of the Canadian Food and Drug Laboratories.
193.81.61.210/adhoc/bulletin/1955/bulletin_1955-01-01_1 page005.html

Sierra Pacific Infrared Thermography. The site includes submissions from thermographers, papers on infrared technology, a glossary, and other resources of interest. www.x26.com/

Wireless Networks. This is good JTAP-sponsored introduction to wireless infrared networks by James Dearden, Canterbury Christ Church College. It discusses and diagrams wireless WANs and LANs, radio technology, and infrared LAN implementation. www.jtap.ac.uk/reports/htm/jtap-014-1.html

10.e. Media Resources

Through Animal Eyes, Dave Heeley, BBC-TV. 60 minutes, color. This show demonstrates how special video lenses and techniques allow viewers to imagine how various types of animals can see with senses beyond human senses. It illustrates the compound eyes of insects and crustaceans, an insect's ability to perceive ultraviolet and the infrared heat-sensing pit organs of snakes.

11. Glossary

Titles, product names, organizations, and specific military designations are capitalized; common generic and colloquial terms and phrases are not.

ABC	automatic brightness control. Since a bright light source can 'wash out' an image or in some cases, damage a viewing system's components (somewhat like a bright flashbulb aimed directly at your eyes in a dim room), ABC technology may be built into the infrared viewing system.
ACSS	ATARS common sensor suite
ADLP	ATARS DataLink Pod
AIES	Aerial Image Exploitation System. Ground-based center for processing of downlinked aerial imagery.
albedo	the reflectivity of an object; the portion of incident radiation (as off a snowfield, or the surface of the moon)
ARO	Airborne Reconnaissance Office of the U.S. Department of Defense
ATARS	Advanced Tactical Air Reconnaissance System
ATD	advanced technology demonstration
ATR	automatic target recognition. A system of sensors designed to identify various types of craft, particularly those that may not cooperate in identifying themselves.
BLP	background-limited performance. A description of limitations related to signal-to-noise ratio of the target signal to the background radiation.
blooming	Spreading of brights spots in an infrared image, usually from light or heat sources such as headlights, fires, or anything else that contrasts brightly with its background. When the blooming is seen on printed images or photos, it is sometimes also called *bleed*. Some devices are designed to have cutoff-levels or other safety devices to prevent the type of uncomfortable temporary flash-bulb-like blinding that can occur in these circumstances and to protect the sensors from damage.
CCD	camouflage, concealment, and deception; charge coupled device
COMSEC	communications security [unit]
DARO	Defense Airborne Reconnaissance Office of the U.S. Department of Defense
DASA	Deutsches Aerospace includes the Military Aircraft Division that is involved in reconnaissance craft design as the prime contractor for the German Air Force Reconnaissance System.
DL	data link, data-link [capability]
DT&E	development, test, and evaluation
ECM	electronic countermeasure
EO	electro-optical
FLIR	forward-looking infrared. Often used in aircraft surveillance pods, which can be mounted on a gimbaled turret with 360° view, with about a 3 to 5 micron wavelength band. Provides terrain view at low altitude at night. When fitted with a 50 mm lens, the field-of-view (FOV) is about 14° horizontal by about 11° vertical [Jobe, 1994]. Spotlight coverage as opposed to IRLS realtime coverage.
FMC	forward-motion compensator/compensation. Compensatory system for resolving archetypes or anomalies resulting from the forward motion of an aircraft or ground vehicle carrying active image-sensing equipment. Some systems use mirrors for FMC.
FPA	focal-plane array
GRD	ground-resolved distance
IDS	imagery distribution system
IR/EO-LOROPS	Infrared/Electro-Optical Long-Range Oblique Photography System. A standoff, dual-band (visible and infrared), digital reconnaissance system designed primarily for use on F/A-18D reconnaissance aircraft, developed under the direction of the U.S. Navy.

It is housed in a center-line station pod. Imaging is in the visible and medium-wave infrared spectral wavebands. Resolution capability in visible modes is about 2.5 to 4 feet at a range of 40 nautical miles; in infrared modes, it is about 2.9 ft at a slant range of 20 nautical miles. Imaging can be done at about 20,000 to 40,000 feet altitude (for reference, common commercial airline carriers cruise at about 30,000 feet) at speeds ranging from mach 0.4 to 1.4. About 45 of digital data can be stored on tape, or transmitted to a ground station.

IRLS	infrared line scanner. 8 to 12 micron day/night instrument which vertically scans the ground, line by line, perpendicular to the flight path of an aircraft, creating a scrolling image. The speed of the IRLS is related to the altitude, scan angle, and the speed of the aircraft. Faster systems are needed to meet the needs of fast-flying tactical aircraft. The image can be frozen to allow scroll and zoom features to be used. Very useful for monitoring natural disasters and their after-effects.
IRP	infrared projector, infrared processor
IRPU	infrared processing unit
IRST	infrared search and track system
ISU	imaging sensor unit
JSIPS	Joint Service Image Processing System
LANTIRN	Low-Altitude Navigation and Targeting Infrared for Night
LPI	low probability of intercept
MAE UAV	medium-altitude endurance unstaffed/unmanned air vehicle
MFD	multifunction display
MMSA	multisensor, multimission surveillance aircraft
NCID	non-cooperative identification
NDIR	nondispersive infrared
NDT&E	non-destructive testing and evaluation
NIRS	near-infrared reflectance spectroscopy
NOLO	no live operator - autonomous or remotely controlled vehicles and systems
PIRA	Photo and Infrared Resolution Range. Part of the U.S. Air Force Flight Test Center at the Edward's Air Force Base.
QWIP	quantum-well infrared photodetector
RC	reconnaissance-capable
RMAPS	Reconnaissance Mission Planning Software. A Fairchild commercial software product.
RMS	reconnaissance management system
RSTA	reconnaissance, surveillance, and target acquisition
TADCS	Tactical Airborne Digital Camera System. An imaging system incorporating a Kodak digital camera with MicroLITE image distribution system which is used in airborne reconnaissance craft such as the F-14.
TARPS	Tactical Airborne Reconnaissance Pod System. Flown on the U.S. Navy F-14 Tomcat is a primary supplier of tactical photography. It was originally designed as an interim solution, but upgrades have extended its life into 2000 [Hancock, 1994].
TE UAV	tactical endurance unstaffed/unmanned vehicle
TID	tactical information display. A computerized system for displaying digital imaging data which may further be downlinked to a ground station. The TID can be used to assess imagery and make adjustments to the filters, exposures, or other camera parameters. Poor images can be reshot. Information can be overlaid prior to downlink or storage on disk. Usually used in conjunction with airborne, realtime and near-realtime reconnaissance imaging systems.
TOSS	Tactical Optical Surveillance System (U.S. Navy)
UAV	unstaffed/unmanned/unpiloted aerial vehicle

Electromagnetic Surveillance

8

Visual

1. Introduction

Our eyesight contributes to our survival, the quality of our lives, and our cultural pursuits. For these reasons, much research has focused on developing technologies that correct, enhance, and supplement vision and many of these have been further adapted to aid in surveillance activities. Spectacles improve our eyesight, microscopes and telescopes enable us to see farther or more clearly, night-vision helps us to see in low-light conditions, cameras help us monitor activities in locations we cannot easily or quickly reach, and protective suits and vehicles enable us to move about and observe in hazardous areas. To supplement this 'enhanced vision', we have developed recording devices that enable us to store and review visual images.

Magnifiers, scopes, and cameras are used in space probes, Earth-orbiting satellites, unstaffed air vehicles, submarines, ATMs, personal camcorders, and more. With special suits and acces-

An electronic intrusion and visual imagery specialist, SSgt. Morgan, maintains and repairs the U.S. Air Force video imaging equipment at the Ramstein Air Base, in Germany. Video recorders are equipped with decoding cards to help secure the recorded information. [Air Force news photo by Lasky, 2006, released.]

sories we can enter radiation areas, chemical environments, and underwater habitats to carry out visual reconnaissance. Visual surveillance technologies help us organize our lives, ensure our safety, and explore the greater cosmos. Visual surveillance devices range in size from many tons to 'pinhole' video cameras so small they can be hidden in a tie clip. In fact, spy cameras have become so small and inexpensive, people have been unable to resist the temptation to use them in unethical and unsettling ways.

Visual surveillance is a very broad category of strategies and devices used by professionals and amateurs in all walks of life. Consequently, this chapter could easily be expanded to fill ten books, but it is hoped that the brief history, and representative sample of devices and applications, can provide an introduction that conveys the breadth and scope of the field.

Cameras and scopes are the most common visual surveillance products, so much of this chapter focuses on these devices, but there is also information on technologies that enable people to observe and photograph hazardous environments. Audio surveillance is built into many visual surveillance systems (e.g., video cameras) so the reader is encouraged to consult the Acoustic Surveillance chapters, as well, to learn more about the theory and application of sound.

1.a. The Prevalence of Visual Surveillance

In 1949, George Orwell published *1984*, a classic novel describing a chillingly repressive society in which a person's every move was monitored by cameras, both inside and outside of the home. Even though there were no pinhole cameras, personal computers, or Internet links in 1949, when Orwell wrote the book, his description of a surveillance society was remarkably prophetic. People in free societies recoiled when they read his outrageous and dehumanizing government control scenario. On their guard because the story had the ring of truth, many looked to the future with suspicion and anxiety.

The year 1984 came and went without Orwell's predictions coming true and there was a collective sigh of relief from many individuals who thought society was 'out of the woods' and that loss of privacy and freedom was not inevitable after all. It now appears that Orwell was not only right about many things, but that the prospect for the future may be bleaker than he envisioned. The biggest difference between Orwell's predictions and current trends is that the people wielding the cameras are not only Big Brothers (government and corporate institutions), but Little Brothers, too, citizens who are loosely regulated and quite eager to take advantage of others with no thought to the long-term consequences.

Significant violations of privacy are perpetrated by unregulated individuals who disregard fair ethical principles. Internet voyeur-site managers plant tiny cameras in extremely personal and compromising locations and sell images of unaware victims for large sums of money. 'Free enterprise' isn't supposed to include free exploitation of innocent victims and such blatant exploitation will probably result in restrictions being put into place, since these entrepreneurs have failed to practice ethical self-restraint, but legislation is slow to anticipate problems associated with new technologies.

In Orwell's *1984*, Big Brother (government) commandeered surveillance equipment to limit freedom of movement and expression. While it is true that government surveillance is increasing as the technology becomes cheaper and easier to use, and should always be monitored and regulated, government efforts have focused primarily on law enforcement and national security, rather than spying on law-abiding individuals. There are gray areas, however, and the controversy surrounding the gray areas are discussed here and in the Audio Surveillance chapter.

To illustrate trends in consumer access to technology, commercial services now offer satellite pictures of one-meter resolution (fine enough to distinguish a hot tub from a car) for less than $25 per square mile to almost anyone who wants to buy them. For the price of an Internet connection, anyone with a reasonably fast computer can access Google Earth and view aerial photographs of most of the Earth's surface without paying for individual pictures. This is an exciting development for scientists, cartographers, armchair travelers, historians, and many others, but it also makes it possible to gather intelligence on societies that have no Internet access and thus are at a significant disadvantage in economic and technological terms. Analysis of aerial photographs can yield a vasts amount of business and political intelligence.

Visual Surveillance General Applications

Government and nongovernment surveillants alike now make extensive use of visual surveillance. Major users include

- local and federal law enforcement agents (traffic control, stake-outs, patrols, prison management),

- federal military defense agents (aerial, orbital, and marine surveillance),

- private citizens and aspiring entrepreneurs (home security, nanny cams, voyeur products, prospectors),

- environmental protection groups seeking to protect the ecosystem and its inhabitants,

- public and private businesses (business security, employee monitoring, industrial safety, private detectives, salvage companies, insurance adjustors),

- search and rescue teams (military zones, natural disaster zones), and

- paparazzi and the media in general (investigative reporting, news and weather reporting, celebrity stalking).

The following examples of applications make it easier to understand the variety of visual surveillance activities and the tremendous proliferation of visual surveillance products. The first examples include many legitimate uses, whereas some of the latter examples are questionable in terms of their overall benefit and high in their potential for abuse.

- Quality assurance personnel and quality control inspectors use surveillance devices to detect and monitor production quality and maintenance and repair needs.

- Salvage operators use surveillance technologies to locate wrecks and objects of value that may be lost on land or at sea. They often use cameras and display devices in combination with sonar, metal detectors, and magnetic detectors.

- Construction contractors use surveillance to monitor progress and safety, to check structural integrity, and to limit access to hazardous areas, especially by unauthorized personnel.

- Employers monitor traffic in and out of their facilities to ensure public and employee safety, to safeguard investments, and to reduce employee theft.

- Prisons and jails utilize cameras to monitor the movement of prisoners, to protect guards, to protect prisoners from one another, and to protect prisoners from undue use of force or abuse by prison guards.

- Banks and retail stores monitor not just the inside of their business premises, but the outside as well. Some cities have as many as 2,000 cameras aimed at streets, sidewalks, and back alleys.

- Customs and immigration officials use surveillance devices to detect and locate contraband goods, illegal immigrants, stowaways, and smugglers.

- Armed forces strategists, intelligence officers, and commanders, use visual surveillance technologies to monitor borders, national security, economic trends, military activities of foreign nations, battlefields, and to engage in national search and rescue and disaster relief operations.

- Businesses use badges to track employee movements within the workplace by reading radio signals from the badges that trigger video cameras or, alternately, will install motion detectors that trigger the cameras. Some have even installed two-way mirrors and hidden cameras in washrooms and change rooms in violation of good taste, if not privacy laws.

- Video cameras are mounted in law enforcement vehicles and on street corners and, in some countries (e.g., Britain), the recordings are sold to educational and commercial buyers for distribution and resale—some can be previewed on the Web.

- Thousands of newscams are mounted on helicopters and highrises to indirectly record and broadcast the activities of citizens almost every moment of the day.

- Camera-equipped paparazzi surveil celebrities 24 hours a day (many of them living in their cars or vans) in their efforts to capture gossip or scandal images they can sell to publishers.

- Unethical entrepreneurs hire youth on skateboards to go around cities and college campuses with cameras hidden in small backpacks. These backpacks can be set down almost anywhere to tape footage that can be sold over the Internet. The images more often than not end up being balcony views of women's cleavage, or up-skirt shots taken from grates, basement staircases, and other creative vantage points. In the spring of 2000, one Internet entrepreneur, who had installed hidden 'peep' cameras throughout women's private areas (including one inside a toilet) at a California university, was broadcasting the images on the Web without the knowledge of the women in the photographs.

1.b. Potential for Abuse

Most private citizens think they are protected from illicit video recording by existing Peeping Tom laws. What they may not know is that people taping without the knowledge or consent of the surveillees may *not* be breaking existing laws (some states don't have Peeping Tom laws and others don't cover visual surveillance if the sound recording mechanisms are turned off) and, in the case of many businesses, some do not even have to inform their employees or customers that they are being observed and recorded.

There are currently hundreds of Web sites distributing personal, private, and even sexually explicit video images from hidden cameras. Some people are willing to be photographed because it is prestigious or profitable—voyeurism is big business, with many voyeur adult sites earning large sums—but there are other instances where people have consented to being taped, but are unaware that their images are being distributed to the public (until it's too late). Finally, there are thousands of people being recorded, sometimes in very compromising situations or states of undress, without their knowledge or consent.

Should we be worried about the increase in video surveillance or is it just harmless fun?

Given the prevalence of nonconsenting exploitation, there is reason for concern. Do people have the right to photograph others and distribute the images with impunity? Do surveillance videos make the streets safer or do they just drive crime underground or into other neighborhoods? There is a general perception that video surveillance is more objective and fairer than eyewitness reports, and that may be true in ideal situations, but research on cameras with remote swivel and zoom capabilities indicates that a disproportionate number of women and minorities are being observed and recorded by white males. This raises a warning flag that suggests that *a recording is only as objective as the person controlling the camera.* There have also been conflicting reports about whether cameras stop crime. Some opponents argue that the costs don't justify the results, or that crime is, in some instances, simply being displaced. These controversial research findings are also discussed later in more detail.

Not all visual surveillance is of questionable merit. Many video cameras are installed to provide safety, security, quality assurance, and other positive benefits. When cameras are installed in hazardous industrial areas or dark, vulnerable public areas, there may be clear benefits for employees and the general public.

- Subways and train stations are monitored for breakdowns and rider safety. Dark parking lots and tunnels are monitored to protect lone pedestrians.

- Bridges are monitored for traffic accidents, suicide attempts, and accidental falls.

- Amusement parks monitor machinery and the safety of patrons.

- Convenience stores videotape cashiers and cash.

- Parents monitor children and their caregivers, grown children monitor aging parents with neurological illnesses that might cause them to harm themselves if left unattended.

- Detectives track people and their activities, law enforcement agents monitor traffic and suspected criminal activities.

- Environmental scientists track land use and ecosystem damage; meteorologists track weather.

- Event planners surveil for crowd control and safety at dances, sports events, and rock concerts.

- The armed services and intelligence communities monitor the military, economic, and political activities of other nations and use surveillance devices to assess safety in areas of conflict, to locate or dismantle landmines and bombs, and to aid in search and rescue efforts.

1.c. Basics of Visual Perception

Visual surveillance technologies rely primarily on energy in the visible spectrum. *Visible* radiation is a type of electromagnetic energy in the *light spectrum,* sometimes called the *optical spectrum* (which also includes infrared and ultraviolet energy). The visible spectrum is a surprisingly narrow band sandwiched into the middle of the optical spectrum. The wavelengths associated with visible radiation are longer than ultraviolet and shorter than infrared. Sensing devices that operate in these frequencies are called *photodetectors.* Some broadband devices that detect radiation in the visible spectrum also detect a certain amount of infrared and ultraviolet.

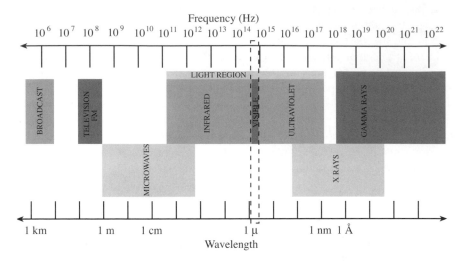

The visible spectrum is a narrow band of radiation between infrared and ultraviolet, ranging from about 390 to 770 nanometers. We perceive specific regions within the visible spectrum as different colors, with green being the dominant reflected radiation, approximately midpoint in the visible spectrum. Our visual senses, our eyes and brains working together, are well-developed to perceive and interpret the wavelengths that aid us in survival.

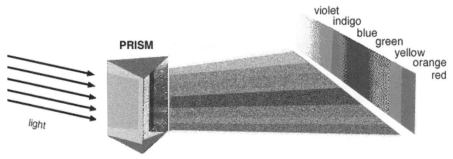

When light from our Sun is undifferentiated, we call it white light. A prism can be used to distinguish the basic colors that are combined to make white light. This is similar to the phenomenon of a rainbow, in that atmospheric moisture disperses the light into its different wavelengths. The colors are not limited to those labelled here, but blend from one to the next. The prism makes it possible to do much the same thing in terms of separating the wavelengths into its component colors.

Color imaging devices are designed to display or sense visible wavelengths and sometimes wavelengths beyond what we can see. Printers are designed to reproduce the colors as closely as possible with various pigments. Grayscale imaging devices are sensitive to values (the amount of light) in terms of brightness and contrast, but do not reproduce pigments or light as color. [Classic Concepts diagram ©2000, used with permission.]

We designate the visible spectrum as those wavelengths that are visible to *human* eyes. Some creatures may perceive some or all of this spectrum plus wavelengths outside those that are visible to humans. The visible spectrum includes wavelengths from about 390 nanometers (violet) to about 770 nanometers (red). Green is toward the middle of the spectrum from about 492 to 577 nanometers. The distinctions are not clearcut. The colors, as we perceive them, blend gradually from one to the next. The colors that we see are the ones that are reflected off the object that we are viewing. Thus, plants, that we perceive as predominantly green, are actually not green in the sense of 'being green', but rather are green in the sense that they absorb the other colors and *reflect back* the green light that stimulates our visual senses.

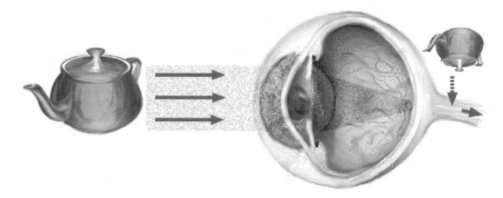

Vision is a complex interaction of 1) light energy that reaches our eyes after reflecting off objects in the path of vision; 2) physical sensing by the various structures in the eye, particularly the cornea, lens, and retina; and 3) interpretive neural processing that occurs in our brains. Since the curved shape of our eye and the way in which the light waves reach the retina on the back of our eyes cause the image to be inverted, we actually 'see' upside-down (as do many optical devices). However, our brains have adapted so well to this inversion that we are not consciously aware that we are looking at the world upside-down. [Classic Concepts ©2000, used with permission.]

Sight, the ability to perceive energy in the visible spectrum, is one of our most important and best-developed senses. With the exception of birds of prey that have better visual acuity and possibly better color vision, very few creatures on the planet see as well as people. Consequently, many common professional and consumer surveillance devices operate with visible light. Cameras are the predominant visible light technology and the primary topic of this chapter. Scopes and magnifying glasses are also discussed, as are recording devices for storing visual images.

Film, video, and digital devices that can 'capture' and record light are now so common, that there are usually several in almost every home within developed nations. 'Disposable' cameras can be purchased for under $8, film cameras typically run from $20 to $600, camcorders and digital still cameras are available for under $1,000 and even fairly sophisticated professional cameras cost less than $5,000. Visual surveillance technologies are plentiful, affordable, and versatile. These devices range in size from tiny pinhole video cameras the size of a dime and broadcast cameras that weigh 40 or 60 pounds, to telescopes that weigh several tons.

2. Kinds and Variations

Visual surveillance devices can be categorized into those that help us perceive images, those that process or encode the images, those that enable us to record them for delayed viewing, and those that aid us in visually assessing environments that are hazardous to unprotected humans. Often these functions will be combined inside one housing (e.g., a camera and a recorder combined into a *camcorder*), or closely linked by a cable or transmissions device. Often they can be subdivided into color or grayscale technologies, with some capable of functioning in both modes.

Imaging Devices - Imaging devices all have one thing in common, they are designed to enhance human vision to supplement our physiological capabilities. They help us see smaller objects, more distant objects, and enhance our vision in low-light conditions. Magnifiers enlarge and sometimes illuminate an image so we can study it in more detail. Various kinds of scopes make distant objects appear closer and more sharply defined. Cameras allow events and objects to be seen, processed, or recorded as still or moving images.

Processing Devices - These are built-in or stand-alone components that convert or modify the visual images. They include analog/digital converters, multiplexing and splitting devices, database managers, and image processors. There are many kinds of image processors; they can be used to colorize, compare, recognize, title, date-and-time-stamp, and to trigger alarms, recording devices, and playback systems.

Recording Devices - Recording devices make it possible to record and store (or transmit) and event so it can be reviewed later from another location. The record may be transient, available only for an instant, or it may be semipermanent, stored on tape or another recording medium. Common recording media include film, magnetic tapes, paper, hard drives, and memory cards. Many visual recording devices simultaneously record audio.

Probes, Robots, Remote-Controlled, and Autonomous Craft - Humans are sometimes too large, too small, too weak, too busy, or too limited in lifespan to be able to see and monitor everything that is going on. For this reason, we have developed a wide variety of mechanical and electronic aids to probe into activities and environments on our behalf. Most of these devices send back information 1) in realtime, 2) as recordings, or 3) on delayed schedules, so that they can be 'seen' vicariously at a later time.

Vehicles and Environmental Suits and Gear - There are places that humans can't go without technology. They are either beyond our physical capabilities or hazardous without protection. To aid us in surveilling from new heights, depths, or hostile environments, we have developed underwater and radiation suits, air- and spacecraft, armored vehicles, and protective accessories.

2.a. Vision Enhancement

There are a variety of devices that enable us to see better. They may magnify, alter, highlight, or brighten the visual landscape so that we can see details, motions, or colors that are otherwise difficult to see.

Magnifiers

Magnifiers make small objects appear larger so that they are easier to see. *Magnifying glasses* (handheld or worn on the face) and *microscopes* enlarge smaller objects that are nearby. They are extremely useful in forensic and other investigative sciences. They can help identify forgeries (money, signatures, photographs, stock certificates, etc.), fingerprints, blood and other tissue traces, powder burns, bullet grooves, hairs, and fibers. There is a limit to the size of objects that can be seen with optical microscopes. Generally the better the optics and the higher the magnification, the higher the price. Beyond a certain microscopic level, it is necessary to use devices other than optical microscopes. Electron microscopes, in which electrons are fired at an object and imaged as they reflect from that object, permit extremely tiny objects to be imaged. Since electrons are not color sensors in the sense that light reflects color, electron microscope images are not colored. However, false color may be applied to make the images easier to interpret and more interesting to view. Color may also be used as markers or identifiers to call attention to relevant details in an electron microscope image.

Scopes/Binoculars

Scopes are similar to magnifiers, except they are intended to make distant rather than near objects appear larger. Devices to aid in seeing over distances are some of the oldest surveillance devices. Telescopes have been used by pirates, Peeping Toms, spies, and detectives since they first became widespread during the Renaissance. When two scopes are combined side-by-side in one housing, to make them easier to handle, they are called binoculars. Specialized telescopes for seeing great distances or imaging parts of the spectrum beyond the visible range are often used by astronomers to surveil the galaxy and beyond.

Night Vision

Most scopes are dependent upon available light, which means that they become difficult to use as it gets darker. For this reason, some are equipped with image intensifiers. These are electronics that take the low light and amplify it to make it easier to see. Some night vision devices are further designed to sense infrared radiation, wavelengths that are not normally visible to humans. Most infrared night vision devices require supplemental illumination from an infrared light, though some have this built in. The less expensive night vision devices typically don't image infrared frequencies.

An image intensifier component uses photoemissive cathodes for the sensor surface. This amplifies the *incident light* image—in other words, the image brightness of the available light is increased and is typically displayed on a charge-storing target (or on film) rather than on a phosphor-coated screen. An image intensifier can also be used with the following three types of imaging systems:

vidicon - An historic tube common on closed-circuit TV that consists essentially of an electron gun and a photoconductive target for detection. Less sensitive than the orthicon, but smaller and simpler and possibly more reliable. Smearing of moving images can result from lag and the photoconductive surface is nonuniform.

image orthicon - The sensitive surface is photoemissive rather than photoconductive. The system includes amplification through internal electron multiplication. More sensitive than the vidicon, it uses a low-velocity beam and photoemissive cathodes for the sensor surface, and displays the image on a phosphor-coated mosaic.

image isocon - An historic display system that is similar to the orthicon except for the optics and the way in which only scattered portions of the signal are used.

2.b. Imaging

The most common visual imaging devices are cameras. Some of them display the images as they are recorded (within the same unit) and some must be hooked into separate recording devices to store the images. Cameras can be categorized in a number of ways, and the groups represented here have been chosen for ease of explanation and illustration. It is easier to describe technologies that use the visual spectrum than say, the infrared spectrum, because nature has equipped us with excellent visual sensing equipment that helps us to see phenomena that respond to this part of the spectrum. In contrast, with infrared technologies, what you see isn't necessarily what you get—a lot of interpretation and target-background analysis is usually necessary to understand infrared images. With visible spectrum information, what we see is pretty close to what we're used to seeing with our eyesight, and we can easily distinguish a person from a car or a building, or a person exhibiting unusual behaviors from others who are standing around.

Still-image technologies and moving-image technologies are not as distinct from one another as some might think. Most moving-image technologies consist of a series of still images played one after the other in fast succession, fooling minds into thinking we are observing natural movement. A series of still photos of a horse running, when presented at a speed of about 24 to 30 frames per second, looks like a moving picture. Our senses cannot process the information fast enough to perceive the movement as a series of still pictures presented one after the other. Through a phenomenon called *persistence of vision* our eye/brain 'blends' the still images together into 'motion' perception. There is little advantage to presenting animation frames more than about 30 or 40 frames per second, and certainly no advantage to more than 60 frames per second as far as the human vision/nervous system is concerned. Thus, movie cameras for most purposes don't have to shoot frames at very high speeds. However, if you hook a camera to a very fast plane, high frame rates are needed to compensate for the fast movement of the plane. Note that the frame rates discussed here are not the same as the *frame refresh* rates on monitors. Raster monitors that are used for television and computer view screens will refresh the screen at the rate of about 50 or 60 frames per second and there is a perceptual difference between the two speeds. However, the refresh rate is distinct from the number of animation frames and affects the clarity of the image more than perception of motion.

Film Cameras

Film Cameras are usually designed for taking individual still frames or for taking a series of frames intended to be replayed as a time-lapse sequence (which has a slower frame-rate than natural-looking animation). The film must be handled in darkness and put through a chemical process before the pictures can be viewed. The most common film sizes are 8 mm, 16 mm, and 35 mm. For high quality photography, bigger film sizes are available as sheets. Common reprint sizes for high quality photography are 4" x 5" and 8" x 10".

Handheld covert cameras are those that are easily hidden, and thus are often very small, about the size of a pack of cigarettes or smaller. There cameras the size of a thumbnail that can take tiny pictures. Very high-grain film is necessary for very small cameras as the surface area of the film is limited by the size of the camera. As the grain gets finer, usually the film requires more light exposure. Thus, there is a trade-off—very tiny cameras often don't work well in very low light conditions such as theaters, corridors, and back alleys. Digital cameras are more suitable for low-light photography.

Film cameras usually include the optics and the imaging surface (the film) in the same container. Digital and video cameras may have the optics and the imaging surface (memory cards and tape decks) in a separate container or component, linked by a communications cable. This enables the production of larger units with longer recording times while still keeping the optics small and easier to hide. Camcorders have the camera, recorder, and playback screen housed in one unit. High-resolution cameras for broadcast-quality telecasts sometimes have separate camera and film recorder units.

In some cases, the camera may be linked to the imaging device (usually a monitor or tape deck or both) through a wireless transmitter rather than a cable. Wireless connections may use radio or infrared frequencies.

In terms of how they image a scene, most camera systems fit in two basic categories:

- *frame* or *staring* camera - A 'staring' technology images the entire 'frame' or scene at one time. Most consumer and commercial photography cameras are frame-based cameras. Neither the lens nor the film is moved at the moment the image is taken. Frame cameras can be adapted for stereographic images by using two cameras slightly offset or by overlapping the individual frames. When mounted on high-flying craft, the camera is usually pointed down. When mounted on low-flying craft, there may be one camera on a pivot designed to shoot images at different angles, or there may be more than one camera.

   Frame cameras designed to take in a wider field of view result in longer, narrower images called *panoramic* images. Wider panoramas are possible by mounting two or more cameras in line with one another or shooting a success of images and combining them.

- *strip* or *scanning* cameras - A specialized 'scanning' camera advances the film while exposing it, usually as the camera moves along a scene. Synchronization of the movement of the film and the movement of the mounted camera is important to prevent or minimize image blur. This is called *motion compensation*. Strip cameras usually have a narrower field of view than frame cameras. Strip cameras are particularly useful for aerial mapping projects. Strip cameras can be used in tandem to create stereographic images.

Panoramic photography can be accomplished with frame cameras, but the width of the image is limited by the optics and distance to the film associated with the camera. Wider images are possible by mounting two or more frame cameras side-by-side, or by using a strip camera. With strip cameras, wider panoramic images can be creating by scanning back and forth while moving the camera optics or imaging surface. Different panoramic images are possible, depending upon the speed of the host craft and the direction of the scanning, whether perpendicular or at right angles to the direction of motion.

There is more information about cameras, infrared film, and aerial photography in the Infrared and Ultraviolet chapters that may be of interest.

Movie Cameras

Sometimes it is important in surveillance applications to record actions rather than still scenes. The subject of the investigation might be hiding stolen items or dumping hazardous wastes into a local watershed. These actions can be more clearly documented with moving images—moving pictures can be captured with film, video, or digital cameras.

Film cameras aren't the most suitable kind of camera for this type of surveillance—film cartridges for small, unobtrusive movie cameras (e.g., Super-8) usually only record about three minutes of action before another cartridge is needed. Video cameras, on the other hand, can typically shoot from 20 minutes to two hours of footage per cassette, depending upon the type of camera and the storage medium. Another advantage of digital cameras is that they are generally quieter than film cameras. Digital moving cameras are becoming more popular, but they still have two disadvantages for surveillance activities. Recording times for most digital movie cameras are still somewhat short, although they are longer than film cameras. This will probably improve as memory, battery, and cartridge technologies improve. The second disadvantage is that it is easy to alter the images. Digital footage can readily be changed on a computer without overt signs of tampering. This makes digital imagery questionable as courtroom evidence.*

For surveillance activities, compact, long-play, high-resolution camcorders are usually preferred. This makes Hi-8 mm and 8 mm video cameras suitable. Hi-8 has the advantages of higher image quality, more compact cassettes, and one- or two-hour recording times. Super-VHS-C is a compact camera format with good image quality, but the tapes typically don't record for as long as Hi-8 tapes, and Super-VHS-C is not as widely available as Hi-8. Regular VHS is sometimes used, but the cameras are bulky and the image quality is inferior to Hi-8. More recently, Digital Video (DV) has superseded Hi-8, and the data is flexible (easily uploaded to a computer for processing). DV cameras are very small.

Instant Cameras

For surveillance, it is sometimes important to process the pictures immediately. For these applications, instant cameras, particularly the Polaroid cameras, have been popular. Instant cameras are designed to take still pictures. They use multilayer film sheets that include the chemicals to develop the image inside the package for each image. To stabilize the image for permanent storage, a chemical fixer is included in a little tube applicator with the film pack. This allows the fixer to be rubbed across the surface of the image to protect it from fading.

Instant camera images are generally not as sharp and clear as film images, and the sheets are sometimes awkward to handle, but for many applications the speed of imaging takes precedence over the sharpness of the image.

Instant cameras that use film packs are quickly being superseded by digital cameras as the quality of digital images improves and prices continue to drop.

Digital Cameras

Digital cameras record information bit-by-bit on electronic storage media as 'data' rather than as an 'image' in analog format on film or tape. This means that digital cameras that can image quickly are neither inherently still nor moving image technologies. A series of digital frames can readily be replayed as an animation. By 1998, digital cameras were 'hot' consumer items. Now they are ubiquitous.

Digital cameras have many advantages for surveillance work, including small size, 'instant' pictures, and good sensitivity in low light conditions. No space is needed for film cassettes and the flash cards used to store the images are tiny. They are also easy to load—flipping in a memory card is easier than loading film. The pictures are almost instantly available, since there is no need to take film to a lab for processing, and many have small screens for viewing the image as soon as it is stored. Thus, the pictures can be viewed and retaken if they didn't work the first time, and they can be viewed in private without going through a public film lab. They are also easy to upload to a computer network, since no scanning or conversion from paper media is necessary.

The technology for digital photography has improved dramatically since the early 1990s. The resolution of digital cameras has improved about 16-fold since 1990. Whereas in 1995, most consumer cameras could image only about 640×320 pixels at a cost of about $800, now cameras in the $500 range can now image four megapixels and higher and the more expensive cameras rival and exceed the quality of film.

To the trained eye, digital images don't start to resemble film images until they are about

---

*There is a potential market for developers to design compression schemes or hardware tamper schemes for digital images that could reveal whether the images have been altered. This would allow surveillance professionals to use new digital technologies for the presentation of forensic evidence.

4,000 x 4,000 pixels. However, it has been found that most laypeople don't see much of a difference between digital and film images, if the digital images have a resolution of 2,000 x 2,000 pixels or higher.

For surveillance professionals, a difference in picture quality can be important. A distant picture of a driver's license or license plate can't be enlarged to read the number unless the resolution of the original image is high enough to resolve the data. No amount of computer processing can reveal details on a license plate if the original image is a fuzzy video shot at 640 x 320 pixels resolution, television shows, notwithstanding.

The chief disadvantages of digital cameras are 1) slower 'shutter' speeds, 2) limited viewfinders, and 3) limited resolutions. Low-end digital cameras typically use a liquid crystal display (LCD) to frame a shot. This can be hard to see in bright light or low light conditions. Higher-end (usually $800 and up) single-lens-reflex digital cameras have optics similar to traditional 35 mm cameras for framing a shot. This provides a clearer, crisper image, but may also be difficult to view in low light conditions.

By the end of the 1990s, very-high-density digital chips were being developed that began to rival the resolution of film. Image capture was also becoming faster, making it possible to design digital cameras that could shoot moving images.

Significant improvements in the image quality of digital cameras that occurred between 2002 and 2005 have had a dramatic impact on photography in general. It will become increasingly difficult to find film-based resources. As the demand for chemical processing of photos decreases, photography stores and photofinishing labs may not have enough clients to stay in business. Thus, the availability of consumer photofinishing services will likely decrease except for specialty enlargement, art photography, and advertising needs (e.g., billboard photography and reproductions). At the same time, the prices of chemical processing may increase.

2.c. Specialized Photography and Accessories

High-Speed Photography

In surveillance, it is sometimes necessary to record images at high speeds. The recording of fast action, such as speeding vehicles, running wildlife, fast-moving weather systems, or fleeing suspects may require specialized equipment. High-speed photography is usually defined for still images as shutter speeds of 1/1000 of a second or faster, or as moving image frame rates of hundreds of times per second or faster. High frame rates of millions of frames per second are used in scientific research and in high-speed aerial imaging, in order to capture fast-moving objects or compensate for the forward motion of the aircraft.

Forensic researchers often use high-speed photography to study bullet trajectories, blood spatter patterns, and other aspects of physics that cannot be readily seen with the unaided eye. This information in turn provides a knowledge base from which crime scene activities can be reconstructed.

Wildlife surveillance benefits from high-speed photography. There are fish that can snatch prey out of the water so fast a human can't see more than a momentary ripple. Similarly, frogs and lizards can move their legs and tongues so fast that the only way to empirically study their actions is to photographically record them. Information on wildlife behaviors can aid in scientific study and may also indicate general health and population densities that may be impacted by chemicals, poaching, or human encroachment on habitat.

Lenses

Lenses are a very important aspect of photography. Clear, crisp, well-focused images are usually desired for surveillance and better lenses tend to result in better images. Since lens-switching is often impractical, time-consuming, and sometimes even dangerous, multi-setting zoom lenses are often used. Lenses for imaging more distant objects that are compact (short in the barrel) tend to be more expensive than longer lenses with the same zoom factor. Lenses with wider aperture settings that let in more light (important for low light conditions), that is f-stops of $f10$ and lower, tend to cost more as well.

Very large lenses, including small telescopes that can be attached to a camera, often need a tripod to support their weight and keep the image steady. When distant objects are greatly magnified, the slightest tremor of the hand can shake the lens. If light conditions are low and the shutter speed slow, the image may blur.

News photographers, wildlife photographers, private investigators, and astronomers frequently use high-magnification lenses to record their subjects from a distance.

Lenses typically come in bayonet-mount and screw-mount varieties. Bayonet-mount lenses are now more common because they are quick to exchange. Screw-mount lenses tend to stay on more securely, but take a little longer to change. If you rarely change lenses, the distinction isn't important.

2.d. Image Processing

Splitters/Multiplexers

Surveillance systems will usually consist of several imaging cameras feeding into one monitoring center or processing device. For example, in aerial photography systems, where size and weight need to be minimized, infrared imaging and visible-light imaging often are captured and processed through some of the same equipment and may even be sent to the same viewing screen or recording device.

Businesses and large apartment complexes often put cameras in their hallways and parking lots for security. Their systems often include dozens of cameras feeding into a single 'command center' with viewing consoles that are staffed by one or two people. Sometimes the systems are organized so that each monitor can display *split-screen images,* that is, images from more than one camera. The typical 16" monitor usually can display about four images per screen for comfortable viewing. Sometimes six or eight images will be displayed, but beyond this, larger monitors are needed or the images are too small to see if anything unusual is happening.

Splitting and combining images requires specialized peripheral equipment called *splitters* and *multiplexers* that are hooked up between the cameras and the viewing or recording mechanisms (or sometimes built into the recording devices).

- A splitter can take the signal from a camera and send it to more than one destination. For example, a video doorway intercom might display on two different consoles, one in the kitchen and one in the garage.

- A multiplexer can take signals from several cameras, let's say four or six, and display them on a single screen in a 2×2 or 3×2 grid. A common type of multiplexer enables the user to view four images on a 'split screen' and then poll through the individual screens in full screen mode before going back to the split screen. More sophisticated systems can be computer controlled to perform these functions 1) at the touch of a key, 2) on a set schedule, or 3) if triggered by detectors near the cameras.

Date/Time Stampers/Titlers

Many still and video recording systems now have time- and date-stamping features that allow stamping right on the image at the time of the recording, or retroactive stamping of log information, including date/time and comments or names.

The date/time stamp is only as good as the person who sets it. Many systems come factory set to 00:00:00 or simply have an incorrect setting. It is important to set and occasionally double-check the date/time settings, especially if there is a possibility that the footage might be needed in court as evidence.

Image Processing

Image processing is becoming an important aspect of surveillance display technologies. Computers can be used to extract remarkable amounts of information from fuzzy video images. Since many of the less expensive time-lapse and consumer video surveillance devices provide rather low-quality images, computer processing is sometimes the only way to extract critical information such as facial features, the writing on labels and bags, license plate numbers, and other important data.

There are now software programs that can examine a frame in relation to a few preceding and succeeding frames and calculate 'missing' information in order to sharpen an image. This works because the portion of an image that is fuzzy in one frame might be clear in the next frame, even if all of the information isn't clear in any one particular frame. By combining the clearest data from a series of frames into one image, the image can be 'cleaned up'. This is useful with low-resolution systems optimized for long recordings. Image processing shouldn't be substituted for good initial images, as the cost of getting a better quality camera and/or VCR is far less than the cost of trying to clean up poor video images.

Image processing has many practical applications. For example, it can be used to track the motion of a specific person (e.g., movement through a hallway or parking lot). Software programs have been developed to recognize a human form and follow it (controlling swiveling cameras along the way) and sound an alarm if odd behavior is noted (access to restricted areas, zig-zagging in a parking lot to check cars for valuables, etc.). Other programs can isolate a person's face on a video and automatically check the features against a visual database.

Databases

Databases are powerful storage and retrieval tools, and with powerful tools come more ways to use and abuse the information that they contain. With image processors able to track movement, faces, identities, and other categorizable information, databases have become an important component of many surveillance systems. Law enforcement agencies, in particular, use many visual matching tools with online databanks, systems that are gradually superseding paper files, and books full of 'mug shots'.

Search and retrieval systems for accessing and sorting information stored in databases are becoming faster and more powerful. Similar algorithms are applied to digital processing of imagery from video cameras, and are so sophisticated that a camera can be remotely controlled by a computer to seek out, identify, and track a specific person from a database (or add each person who appears on camera) as the person moves through a hallway, parking lot, or office complex. Side views, hats, and beards are no longer a deterrent to computer algorithms. Celebrities, ex-employees, suspected criminals, disgruntled spouses, and other people can be singled out and tracked without their knowledge or consent.

2.e. Recording Devices

Some of the technologies mentioned so far, such as cameras, are designed to image and record at the same time. Others image on one unit and record on another, with a wired or wireless link to connect them.

Recording mechanisms are sometimes built into imaging devices, rather than housed as separate components. Film cameras and video camcorders are the two most common types of imaging and recording systems. Video cameras are sometimes separate from the recording devices, linked through a cable or radio transmitter.

There are many kinds of recording media: tapes, diskettes, CD-ROMs, DVDs, PCMCIA cards, proprietary memory cards, hard drives, etc. All of these are used in various types of surveillance devices and each has advantages and disadvantages. One of the dilemmas facing surveillance professionals, especially those doing forensic work, is that courtroom evidence must be considered to be free of tampering. Images captured with digital devices can readily be changed with computer software. Until tamperproof products are developed for forensics work, it may be necessary to capture some kinds of evidence with traditional photographic film.

Since recording devices are important to surveillance activities, and covert cameras often have to be small, with the recording device located away from the camera, a number of kinds of remote recording systems are used:

continuous recorders - Short-duration recording devices usually perform continuous recording at about 20 to 30 frames per second (and much faster for slow motion photography and high-speed photography such as aerial photography). Most analog recorders record to tape or film, but it is impractical to have tape or film that is longer than a few hundred feet. Thus, many continuous recorders are not practical for 24-hour surveillance. Most analog continuous recorders can record from 20 minutes to about 6 hours. In the late 1990s, digital continuous recorders and portable models usually had recording limits of a few minutes up to about 20 minutes. However, this is rapidly increasing as the capacity of digital memory increases and the cost decreases. Digital recorders may record for hours or days if they are linked to very large hard drives with good compression algorithms.

time-lapse recorders - A long-duration recording device usually uses a time-lapse system that records a 'frame' every few seconds or minutes. If you've seen video images of a flower blooming in a few seconds or clouds rapidly scudding across the sky, you were probably watching time-lapse photography. The biggest advantage of time-lapse recording is that long durations can be captured on limited-length recording media. The biggest disadvantage is that an important event might occur during one of the 'blank' moments between individual frames.

2.f. Environmental Vehicles, Suits, and Gear

Environmental Suits

Environmental suits enable people to enter hazardous environments to carry out visual surveillance activities. Most people are familiar with *scuba* gear (self-contained underwater breathing apparatus). Divers use scuba systems to stay underwater for extended periods of time, sometimes in the very cold temperatures found at higher latitudes or greater depths. Wreck divers use scuba gear to find and salvage airline or marine wrecks and search and rescue divers to find debris and human remains. It is not uncommon for divers to take along handheld, chest-worn, or helmet-worn video and still cameras to record debris, wrecks, and their activities while diving.

Water will harm most precision instruments and detection devices. For this reason, divers have to use special underwater cameras and recording devices or waterproof housings to protect conventional cameras. Underwater video cameras may be self-contained or may connect, via an 'umbilical cord' to a recording console at the surface.

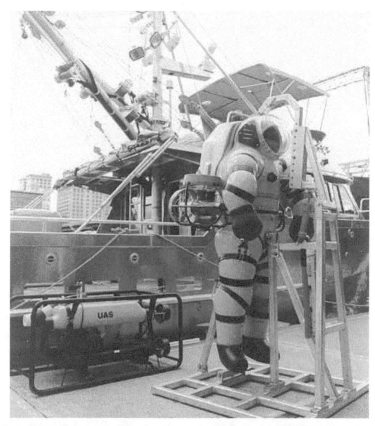

This unique hardsuit from *Underwater Atmospheric Systems* (UAS) enables divers to dive deeper and longer than in scuba suits, due the built-in pressurized 'atmospheric' life support systems (that prevent 'the bends'). The suit allows divers to enter depths of 1250 feet for up to 48 hours, with typical operating times of 6 to 8 hours. The umbilical provides electrical connections, and communications are hardwired with acoustic through-water backup system. Thruster packs can be attached to the suit to increase mobility. The suit is constructed mainly of T6061 cast aluminum, designed and developed by the Ceanic Corporation. In the background is a UAS boat equipped with several surveillance systems, including sonar and radar. The hardsuit can be used for underwater observation, salvage, insurance and maintenance inspections, appraisals, damage assessment, and video recording. [Photo © 2000 UAS by Robert Mester, used with permission.]

Diving bells or remote-controlled undersea vessels are often used to reach greater depths or to supplement the work of divers. *Hardsuits*, specialized diving suits that contain their own pressure systems, are useful for activities at depths that are too great for a soft-suited diver to access. Thus, they can be used for platform and vessel inspection (quality assurance, insurance claims, appraisals), maintenance, repair, salvage, rigging, testing, and other deep-water applications.

Goggles, masks, helmets, gloves, boots, and other accessories are typically incorporated into environmental suits and are, at times, used individually. Communications links are sometimes established through tethers and cables. The tethers may be attached to the host vessel on the water's surface or may attach to a communications buoy that uses radio waves to transmit to and from the host.

Left: When TWA Flight 800 crashed off Long Island, New York, U.S. Naval divers assisted in search and recovery operations. Brad Fleming dons a Mark-21 diving helmet. Right: A U.S. Navy Commander uses a camera 240 feet below the surface of the ocean as he inspects and photographs an artifact on a sunken U.S. Civil War ship. The vessel was lost in a battle in March 1862 in the first battle fought by iron armored vessels. [U.S. DoD 1996 news photo by Charles L. Withrow; 1999 news photo by Eric J. Tilford, released.]

Gas masks and radiation suits have been regularly issued by armed forces personnel to surveil areas in which there have been chemical spills or radiation leaks, or where it is suspected that chemicals or radiation might be used against them by other forces. With the development of new synthetic materials, the technology for these has considerably improved. Protective suits are lighter and more flexible than those that were in use during and immediately after World War II.

Left: Sgt. Putnam in a gas mask and permeable cloth helmet during a chemical warfare decontamination demonstration in 1944. Right: Airborne Corps historian wearing M-17A1 protective mask during an Iraqi scud attack in January 1991. Protective gear enables people to work, scout, and patrol in hazardous environments. Cameras will sometimes also be equipped with housings to protect from chemical, water, and radiation exposure. [U.S. Army 1944 Signal Corps photo by Sgt. Bradt and U.S. Army 1991 Airborne Corps archive by John F. Freund, released.]

Protective Vehicles and Stations

Sometimes a protective suit is not enough. Humans are still vulnerable to explosives, radiation, and gunfire. Vehicles that make it possible to surveil in environmentally or socially hostile environments range widely in form and function and include aircraft, spacecraft, diving bells, submarines, amphibious vehicles, and large armored tanks. They allow people to enter, view, and record in environments that are too hot or cold, air deficient, contaminated with nuclear fallout, or subject to hostile fire. They may also enable workers to access more territory in a shorter period of time than is possible on foot. One example of vehicle-facilitated surveillance is the lunar rover used by the Apollo 15 astronauts to investigate and photograph about 17 miles of the Moon's surface in the early 1970s.

Left: A Light Armored Reconnaissance Vehicle that is part of the Strategic Reserve Force of the Stabilization Force of the U.S. Marines, in a 1998 training exercise. Right: In 1999, the vehicles were used for urban patrol in Kosovo (right). [U.S. DoD 1998 news photos by Steve Briggs and C. J. Shell, released.]

In the early 1970s, Skylab, an 89-foot Earth-orbit workshop, provided a means for people to work and observe from above the Earth's atmosphere. Skylab made possible a new level of astronomical observations and Earth resources observations.

Left: A U.S. Marine Assault Amphibious Vehicle coming ashore in Croatia during a training exercise. Right: Amphibious Vehicle coming ashore during maritime combined operational training exercises among NATO nations to improve skills for NATO-led peacekeeping support operations. [U.S. DoD 1998 news photo by Steve Briggs, U.S. Navy and Timothy A. Pope, released.]

2.g. Probes, Robots, Remote-Controlled and Autonomous Craft

As was mentioned in the previous section, many technologies have been developed to enable people to enter and surveil environments that could be hazardous or lethal to humans, including the deep sea, the upper atmosphere, space, feep underground caverns, radiation-contaminated areas, and small or confined structures. Sometimes, however, it is safer or more efficient to send in a robotic vessel rather than a human in a protective vehicle or suit.

The Mini-Remote Vehicle 1 (MR1) (left) and MR2 (right) are used by the U.S. Navy to probe hazardous deep sea environments. They are self-propelled and equipped with lights, still and live video cameras, and rotating-head sonar to survey marine environments to depths of about 1000 feet. The Navy used this technology in 1996 for search and recovery of data flight recorders from the TWA 800 airliner that crashed off Long Island, New York. [U.S. DoD 1996 news photo, released.]

To help us maneuver, or peer into unfriendly regions, scientists have developed not only special suits and a variety of staffed and unstaffed vessels and probes, but also special radiation-proof or waterproof camera housings, and an extraordinary variety of robots that can search, detect, respond to stimuli, or send information back to a communications station. Often specialized cameras and other surveillance devices are used in conjunction with these environmental suits, vehicles, and robots.

Most autonomous vehicles, probes, and robots have video vision systems. Many remote-controlled vehicles are also equipped with cameras. The vision systems are sometimes used to facilitate navigation. At other times, they make it possible for a pilot to control the device from a distance. Many are equipped with more than one vision system to serve more than one purpose, including navigation, surveillance, and environmental sensing or measurement tasks.

Robots are even more diverse than the people who invent them. They range from huge industrial production robots to miniature robots the size of an insect. There are also ideas for microscopic robots, called *nanites* described in science fiction literature that may one day be a reality. There is only room here for a few robot examples, but at least it can provide a taste of the utility and diversity of robots with surveillant capabilities.

Robots use a variety of locomotion systems. Some roll, others bounce, still others fly, wiggle, or slither. Each style of movement lends itself to navigating a different kind of terrain.

A 'slithering' robot that mimics the movement patterns of snakes has been developed by a California animator, Gavin Miller. The shape and movement of the robot lets it enter and traverse environments that might be difficult for other robots. Slithering robots are not common. It is a challenge to control and articulate a large number of 'limbs' even if those limbs don't have any legs.

Left: Ralph Etienne-Cummings and a robot that can follow a line on the ground and avoid obstacles. Right: Two vision sensors are mounted as 'eyes' on the front of the vehicle-shaped robot. These microchips give the robot the vision capabilities it needs to navigate and carry out its tasks. The 'bot also has learning algorithms that enable it to remember how it avoided previous obstacles so it can apply the information to later situations. [Johns Hopkins 1999 news photo by Mike McGovern, released.]

At Johns Hopkins University, Ralph Etienne-Cummings, an electrical engineer, has developed a robotic vision system on a microchip for navigating and avoiding obstacles. The single chip design, also known as a *computation sensor,* handles analog and digital processing functions, extracts relevant visual information, and then processes the data and responds accordingly. If it 'sees' an obstacle, it will move around it. If used in a surveillance system, it can follow a moving target such as a person. This concept could also be adapted to hands-free videoconferencing.

The advantages of having the basic functions on a single electronic chip are speed, portability, and lower power consumption. Smaller systems typically require less power. Thus, chip-based robots could be used on boats, airplanes, and autonomous vehicles.

Sometimes robots must deal with stairs or holes. They may even need to jump and move to evade projectiles.

Jumping robots, equipped with tiny pan and tilt cameras, have been developed by Nikolaos Papanikolopoulos at the University of Minnesota in Minneapolis. These nine-ounce, sugar-tin-sized robots can be sent as a 'scouting party' in situations that might be hazardous to humans. Equipped with a spring-controlled 'bounce reflex' they can hop up and down stairs and over obstacles to survey the area and radio back visual and auditory data, as well as other information such as the presence of dangerous chemicals or gases. Support scouts with sound/vibration sensors, rather than video cameras, have also been developed. Both varieties can transmit information to a mobile radio repeater that relays it to a surveillance console at a safe distance from any hostile activities. The robots could be used for rescue operations, military activities, minesweeping, hazardous environment inspection, and investigative reporting.

At the U.S. Space and Naval Warfare Systems, robots are researched and developed in two divisions: Advanced Systems and Ocean Systems. One of the more recent projects, the ROBART III robot, is equipped with a wide variety of sensors and autonomous systems for intrusion detection and response. It has further been developed so that it can be used with a system of 'slave' robots (essentially a distributed network system) that accompanies ROBART III for deployment onsite.

ROBART III's head-mounted sensors include two Polaroid sonar transducers, a Banner near-infrared proximity sensor, an AM sensor microwave motion detector, and a video surveillance camera. The output of the CCD camera is broadcast to the operator over an analog radio-frequency (RF) link and simultaneously fed to an onboard video motion detector that provides azimuthal data, allowing the head pan-axis controller to automatically track a moving target. Azimuthal and elevation information from the motion detector will be similarly fed to the pan-and-tilt controller for the six-barrel pneumatically fired dart gun for automated weapon positioning.

Additional Polaroid sensors and near-infrared proximity detectors are strategically located on the system to provide full collision-avoidance coverage to support the advanced teleoperation features.

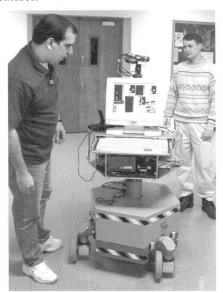 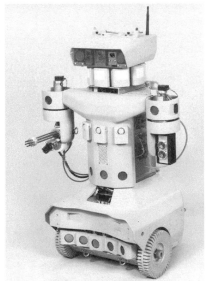

Left: Purdue Doctoral students DeSouza and Jones work with "Peter" a mobile robot equipped with a number of different sensors. Peter navigates by converting images to 3D measurement grids and comparing the surroundings with a pre-programmed destination. The robot has cameras and ultrasonic detectors to help it carry out its tasks. Right: ROBART III, evolved from Bart Everett's original Robart I, was developed in 1992 as part of the U.S. Naval Mobile Detection, Assessment, and Response System (MDARS) program. [Purdue 2000 news photo; U.S. Space and Naval Warfare Systems Center news photo, released.]

At NASA and its various associated space centers, robotics and vision systems that can be used in space missions are the subject of continuing research and development. Autonomous robots that can explore planetary environments, robot simulators that can be used to train astronauts, and remotely-controlled robots that can be operated by astronauts from a distance, have all been used in one way or another to enhance space exploration and the collection of samples to bring back to Earth. Many of these robotic systems are enhanced with one or more vision systems or visual recording devices.

Left: An artist's rendering of Robonaut at work while being guided by an astronaut inside the space vessel. Robonaut is sufficiently human in form and movement that you could almost call it an android (humanoid robot), but it is actually a robotic 'surrogate'. Developed by NASA at the Johnson Space Center, this robot has an unusual control system by which it can mimic the movements of the astronaut wearing a controlling suit at a distance. Right: Astronaut Jeffrey Hoffman is shown, in 1993, wearing a special helmet and gloves that allow him to remotely view and position a robot arm, the Remote Manipulator Systems (RMS), that can be used to manipulate equipment from a distance (e.g., parts of a telescope). [NASA/JSC news photos, released.]

Developed by NASA at the Johnson Space Center, the Robonaut android-like mechanism has an unusual control system. Most robots are either autonomous, following programming algorithms or artificial intelligence heuristic problem-solving methods, or remote-controlled, relying on commands from the operator. Robonaut stands out for its ability to mimic the movements of a nearby human wearing a motion-sensitive body suit. The person controlling the robot can remain safe behind a protective barrier, as in a spacecraft, laboratory, or nuclear installation, watching what the robot sees with its vision system on a display system. Robonaut is remarkably dexterous, with over 100 sensors and controllers in the arm alone. Special materials protect the robot from damage from projectiles and fire.

Hawk Protections Systems, has developed a heavy-duty remotely-controlled robot resembling a small tank, that has options for detecting gas, explosives, radiation, or chemicals. The system has a built-in lighting system and a remote-controlled video camera. The system could be applied to law enforcement, security, private industry, and military users. [Hawk Protection Systems, Inc., news photo, 2005.]

The concept behind Robonaut has been applied to other types of devices that mimic portions of the human body rather than the whole body. Devices that mimic the functions of the head and hands are especially useful as they can 'see', 'hear', and collect things. Gripper arms on diving vehicles, for example, can be made to follow the movements of an operator in a nearby marine vessel. Thus, the operator can remotely move, inspect, and pick up objects for collection from a safe distance. This is particularly useful for deep sea exploration and explosives handling.

Robotics are versatile systems that can be equipped with a number of different sensors. Recognizing this, some commercial robotics vendors have developed basic robots that can be purchased with optional sensors depending upon the user's needs.

3. Context

Visual surveillance technologies are used in many different settings: in laboratories, businesses, toll booths, city streets, submarines, aircraft, cars, bedrooms, day care facilities, schools, shopping malls, subways, ferries, change rooms, space probes, military installations, wilderness areas, wildlife preserves, forensic laboratories, and observatories. It's difficult to think of a context in which a camera, telescope, or microscope isn't useful in some way.

Cameras are the most familiar type of surveillance device. They are used by photographers, detectives, and automated satellites and robots to document crimes, news, people's activities, and workplace environments. In summary, the most common settings and applications for film and video photography include

- *safety and security systems* installed in homes, offices, parking lots, industrial complexes, bridges, research laboratories, schools, detention facilities, trains, malls, prisons, and financial institutions,

- *forensic photography* for recording 'mug shots' of convicted criminals, documenting crime scenes, establishing evidence, examining details that might be difficult to see under normal circumstances, and recording the activities of persons under suspicion in the process of building a case for arrest and prosecution,

- *investigative surveillance* routinely practiced with cameras by detectives, journalists, news crews, divers, spies, wildlife conservationists, environmental engineering and cleanup crews, customs and immigration officials, reconnaissance specialists, and search and rescue teams, and

- *routine visual recording* of activities by professions in which accountability is important, including medicine, law enforcement, large-scale construction and engineering, archaeology, political and environmental advocacy, and armed forces operations and exercises.

4. Origins and Evolution

4.a. Human Aspects

The evolution of visual surveillance devices is intrinsically linked to human needs and impulses. Humans are highly motivated by curiosity and a desire to feel safe. We develop and commandeer various types of technology to gather information to foster a feeling of security and control in our lives. Visual surveillance technologies are tied to the evolution of optics, vacuum tubes, and recording devices. In less than 200 years, these devices have changed from elite experimental systems to everyday consumer items.

The Impulse to Observe

Humans are naturally curious. The impulse to observe is strongly tied to our biological origins as predators and gatherers. A predator needs to be curious about movement, form, or smell, and to be able to locate and recognize the source of the movement in order to procure food. The same observational skills can be important in recognizing and gathering food, sensing dangers, procreating, and caring for offspring. Curiosity, jealousy, and competition for resources and mates may account for the earliest incidences of spying in human history. There are many stories of spying and the covert transmission of information that predate visual surveillance technologies as we know them, so it is not surprising that new technologies are regularly exploited for their surveillance potential, even if they were originally invented for some other purpose.

The impulse to observe is counterbalanced with a human need for privacy. By the 1500s, the citizenry in Europe had implemented some of the early "Peeping Tom" laws to protect people from being spied upon through windows and doorways.

In the United States, the Constitution and Bill of Rights of 1789 and 1791 formalized many basic aspects of freedom and privacy that underlie our present social framework and system of government. This foundation and the subsequent legislative amendments that have followed were designed to safeguard our most basic needs, some of which are the needs for trust, security, and the opportunity to be alone and uncensored.

In 1890, Warren and Brandeis wrote an article that was published in the Harvard Law Review that referred to Cooley's definition of privacy as the "right to be left alone" and argued that this was "the most comprehensive of rights, and the right most valued by a free people". This implied a need to establish a boundary between the desire to observe and the need to not be observed.

The Impulse to Exchange Information

People relate strongly to their visual senses. They enjoy seeing, recording what they see, and conveying their visual experiences to others. People readily share images of special moments—things they want to remember. They exchange pictures of family members that document important events, animals, and beautiful scenery. Businesses exchange information for developing projects, communicating goals, marketing, and otherwise carrying on business. Governments share images to carry out their responsibilities on behalf of their constituents. They use pictures to monitor trade, to carry out defensive actions, to curtail criminal actions, and to keep abreast of the activities of other nations.

The Impulse to Record

The most ancient and primitive 'recording devices' were fingers or sticks that people used to make draw in sand or on rocks (e.g., cave paintings). This desire to record images and events later evolved into more portable forms of pictures and pictograms inscribed on papyrus. The history of art includes many examples of historical records of events. Artists were once employed to visually record important transactions and occasions such as weddings, the rise of a new monarch, the outcome of a significant battle, or the death of a hero. Accuracy was dependent upon the skill and point of view of the artist creating the record. This desire to record and preserve information more quickly and accurately than an artist could render with paint and canvas was at the heart of the development of photography.

4.b. Technological Aspects

The Evolution of Optics

The field of optics probably began with water rather than glass. Water can reflect images and a piece of contoured ice or droplet of water can 'display' or 'project' an image of objects behind it on its upper surface. You can observe this basic phenomenon by placing a large drop of water on a laminated document, like a driver's license, and looking at it from the side. The document will be clearly visible on the curved upper surface of the water drop. Aristophanes (ca. 450 B.C.–ca. 388 B.C.) apparently demonstrated various refractive characteristics of water (and possibly also glass) in ancient Greece. About 400 years later, Lucius Annæus Seneca (ca. 4 B.C.–65 A.D.) recorded his observations on the magnifying properties of liquids.

Almost all visual surveillance devices exploit the reflective and refractive properties of transparent materials in the form of lenses. Lenses let you control the direction of light and are extensively used in cameras, scopes, and illuminators. Lenses and the properties of light have been of interest for at least 2,000 years; Claudius Ptolemæus (85-165) apparently authored five volumes on optics, though only one survives. Another great scholar, Abu Ali Hasan (sometimes written Al-Hazen) Ibn al-Haitham (965-ca. 1040), created hundreds of scientific documents, including critical reviews of Ptolemy's writings, along with his own theories about optics, the nature of light, light passing through water-filled vessels, and the reflection of light from variously shaped surfaces (*Kitab-al-Manadhir*). His theories strongly influenced western thought through Latin translations of his documents about 200 years later, including *Opticae Thesaurus*.

The Development of Lenses

Water may have made humans aware of the magnifying properties of curved surfaces, but it wasn't a very practical material for making lenses. To create solid lenses that embodied the optical properties of water, it was necessary to develop skills in handling appropriate transparent materials, especially glass and quartz. Anecdotal histories by writers like Pliny (*Naturalis Historica,* 77 A.D.) suggest that the ancient Phoenicians discovered glass several thousand years before biblical times. Egyptian glass beads and Babylonian glass rods from about 2500 B.C. and earlier have been found and artifacts that may be glass lenses date back as far as 2000 B.C.

Left: From a basic smoothly-ground lens, inventors developed spectacles, spyglasses, microscopes and many other essential tools that enabled them to more easily surveil their environment. Right: Lenses have now been supplemented with 'vision-sensing' fiber-optic mesh systems such as that developed at MIT. [Classic Concepts photo, 2000, and Classic Concepts drawing, 2006, used with permission.]

Many historic lenses were made from blown glass. Glass-blowing houses appear to have been established in Greece before 5 B.C. However, it was another 200 years before pioneer artisans and scientists systematically harnessed the light-influencing principles of the curved water droplet in glass and gems.

Optical glass and gems probably existed in Asia in antiquity, though we do not have many historic details in the English language. Occasional references to the magnifying properties of curved glass can be found in European documents from about the third century onward, but melting, shaping, and grinding skills and tools were not widespread until around the 1200s in Europe, when the commercial manufacture of window glass was established. Quartz and beryl were used in early lenses and are still important in optics manufacturing, especially for scientific applications or specialized photography.

Practical Viewing Devices

Lenses with various concave, convex, and flat surfaces are widely used in microscopes, telescopes, cameras, and eyeglasses. Historically, the development of lenses for one application often improved technology for related applications. Mirrors are used with many viewing devices, sometimes in combination with one or more lenses, and the improved manufacture of mirrors during and after the Renaissance aided in the development of many surveillance technologies.

Ground crystalline substances known as 'reading stones' were used by monks in the Middle Ages, around 1000. Over the next several centuries, as reading spread from the cloisters to the general population, the demand for reading stones increased. The problem with reading stones was that the reader couldn't comfortably hold a book with both hands and use a stone at the same time. It had to be continuously moved across the text.

Roger Bacon explored better ways to improve vision. Sometime around 1250, he demonstrated that ground lenses could improve eyesight and recorded this use of crystal or glass in his *Opus Majus* in 1268. As the craft of writing and publishing evolved, so did the craft of developing stones, spectacles, and other instruments to make writing clearer. Eyeglasses that could rest on the head, thus leaving the hands free, soon became popular.

Alessandro di Spina is credited with inventing the first eyeglasses, in 1280, though it is possible that monocles may have been used in ancient Greece and there are some written references to eyeglasses a few years earlier than the attribution to di Spina. Sofronius Eusebius Hieronymus (340-420) is depicted in portraits with eyeglasses. Eyeglasses with colored glass also existed in earlier times in Asia, though their design suggests they may have been used for religious or aesthetic rather than utilitarian purposes. In 1289, di Popozo referred to spectacles for older, failing eyes in his manuscript *Traite de con uite de la famille*.

A glass industry emerged in Constantinople and the refinement of lenses and their combination led to important magnifying tools like telescopes and microscopes.

Leonardo da Vinci may have been working on the invention of magnifiers or telescopes in the early 1500s, but no actual models survive. He is said to have had disagreements with German glassmakers and to be working in secrecy. Perhaps he was developing telescopes as a tool of warfare, an activity that may have demanded a certain amount of secrecy at the time.

Going Beyond the Basics

The evolution from spectacles to more sophisticated imaging devices was a natural development. By combining two lenses, you could create a basic refracting telescope or compound microscope. German-born Hans Lippershey (ca. 1570-1619) of The Netherlands applied for a patent for a dual-lens telescope in the early 1600s (which was denied) and developed several telescopes for the government.

In about 1590, in The Netherlands, Hans Janssen, apparently aided by his son Zacharias Janssen, a youth at the time, created drawings for lens configurations that could be considered rudimentary compound microscopes. Zacharias built a telescope-like three-tube microscope around 1595 that could magnify from 3 to 10 times by extending the tubes. The eyepiece was bi-convex and the objective lens plano-convex [Davidson and Abramowitz, 1999]. Practical microscopes emerged about half a century later.

Though it was apparently not the first such instrument, Galileo Galilei (1564-ca. 1642) had developed a telescope by 1609. He then improved upon it and helped to popularize the invention. The use of telescopes to observe the heavens was somewhat suppressed in the west by religious superstitions and taboos. In many western religions, it was considered audacious and sinful to examine the heavens, as it implied peering into the affairs of God. This is perhaps one of the reasons why optics developed earlier in the east than in the west up to the time of the Renaissance. In some eastern areas, it was considered socially acceptable to determine the exact location of God, in order to find the right orientation in which to pray. In the west, however, early telescopes were primarily used for navigation and the surveillance of animals and people. In spite of various restrictions and religious doctrines that hampered western science, Galileo dared to stare into the heavens and chart the skies with his new instruments, publishing his astronomical observations in *Siderius Nuncius* in 1610.

It is not uncommon for practical inventions to come into use before the theory of how they work is well understood. Johannes Kepler (1571-1630) made important theoretical observations about light transmission and applied his ideas as practical improvements to the telescope by describing different combinations of convex and concave lenses. The technology was further developed by Robert Hooke who published *Micrographia, or some Physiological Descriptions of Minute Bodies made by Magnifying Glasses with Observations and Inquiries thereupon,* in 1664.

Around the same time as the theory and practical application of microscopes became important, the construction of these devices began to develop in two general directions: English designers created tripod microscopes of turned wood, pasteboard, and leather, while Italian designers favored turned wood and brass.

New Forms of Optical Enhancements

As was mentioned earlier, reading stones were somewhat awkward to use, as they required a hand to be constantly moving over the text. Spectacles were better, but even they had some disadvantages, since prescription lenses had not yet evolved into a science. The idea of enhancing or altering human vision by direct contact with the eye, as in contact lenses, may have been foreshadowed by some sketches of the cornea drawn by Leonardo da Vinci in the early 1500s (Codex D), but he doesn't appear to have been designing a way to correct vision or to have produced any working models.

By the 1600s, however, inventors were trying to find a way to directly enhance vision, without stems or handheld impediments. A *contact tube* was described by René Descartes (1596-1650) in *La Dioptrique,* prepared as an essay for publication, in 1636, and by Philippe de la Hire (1640-1718) half a century later. Thomas Young (1773-1829), a British physician, tried placing microscope lenses directly on his eyes in 1801.

For many centuries, gems and minerals have been important components in the manufacture of lenses. Sunglasses in the mid-1700s were made with slightly opaque quartz lenses. Quartz is still used in many scientific optics and for infrared and ultraviolet photography.

Usable contacts were developed in the late 1800s by F. E. Muller, Adolf Eugene Flick, Eugene Kalt, and others, but could not be worn by many people because they irritated the eyes.

Most early contact lenses were made of blown glass.

Mirrors are integral to many optical devices. Metal mirrors were available in the Renaissance, but modern glass mirrors weren't used regularly until the 1800s. Isaac Newton (1642-1727) solved some of the problems of color aberration associated with refracting lenses by using a polished metal mirror instead of curved glass as part of a telescope assembly, thus creating a reflecting component. James Gregory (1638-1675), in 1663, and N. Cassegrain, in 1672, developed variations on this mirrored telescope. Mirrors constructed of polished metal tended to scratch and corrode. Léon Foucault (1819-1868) created modern mirrors by binding silver to glass, producing clearer images and longer-lasting mirrors.

Evolutionary Improvements

In the 1600s, many discoveries were made and discussed among scientists and inventors. The pace of evolution in optics increased at this time.

Very long telescopes were developed by Christiaan (1629-1695) and Constantijn Huygens. By 1655, Christiaan had made some important astronomical studies of Saturn with a telescope of his own design. He is also credited with discovering the phenomenon of polarization. Jean Dominique Cassini (1625-1712) used longer and longer telescopes to probe the skies, discovering many attributes of the planets in our solar system.

Robert Hooke (1635-1792) was an important inventor of telescopic and microscopic devices, as well as thin films. One of his contemporaries, Antonie van Leeuwenhoek, (1632-1723) made advancements in lens grinding and microscopy. He invented and manufactured many devices to improve upon the compound microscopes of the time.

Eyeglasses and contact lenses improved and were produced in increasing numbers. John F. William Herschel (1738-1822) described a contact lens in *"Encyclopedia Metropolitana"*. In 1784, Benjamin Franklin (1706-1790), scientist and statesman, combined two types of lenses into a single frame to invent bifocals. Joseph Jackson Lister (1786-1869) developed a system of lenses that were spaced to correct for chromatic aberrations and to reduce spherical aberrations. Joseph Fraunhofer (1787-1826) began experimenting with telescopes to detect radiation outside the visible spectrum.

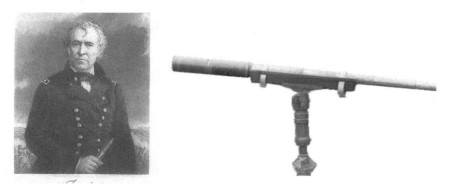

Telescopes were a prevalent visual surveillance tools used in seafaring and land-based warfare and exploration. Left: Zachary Taylor (1784-1850) is shown in military uniform holding a telescope. Right: A European 19th century wooden mariner's telescope. [Library of Congress By Popular Demand collection, copyright expired by date; Classic Concepts photo ©1999, used with permission.]

For modern optics to emerge, it was necessary to understand the mathematics of optics. Johann Karl Friederich Gauss (1777-1855) developed a theory of lenses (Gaussian optics) that provided a mathematical framework for optical imaging theory.

Modern Optical Manufacture

Improvements in glass manufacture and polishing and the positioning of lenses to reduce color aberrations contributed to the evolutionary development of lenses in the late 1700s and early 1800s. The world's largest refracting telescope was established at the Dorpat Observatory in Estonia, making it possible to systematically chart the astronomical bodies of the Northern Hemisphere. Not long after, lenses became important components in photographic devices, evolving into still cameras, moving film cameras, and video imaging devices. Simple devices for enhancing human vision led to sophisticated devices to record visual events.

By the mid-1800s, some significant optical companies were founded, including Karl Zeiss Optical.

Miniaturization

Portability is often an important aspect of surveillance technologies. Scopes, lenses, microscopes, and other viewing devices that could be built in small packages, without sacrificing clarity or precision, are favored for travel, scientific fieldwork, onsite forensics, and covert observation.

By the mid-1800s, designers were better at designing optics that would fit into small casings. The Nachet pocket microscope was a portable brass instrument with good quality optics and interchangeable lenses. The platform was designed to reverse so the scope could be transported. The Nachet microscope was available for purchase up to about 1872.

A pair of scleral contact lenses from around 1948, from the collection of Thomas A. Farrell, M.D., show that contact lens technology had continued to develop as well.

The development of synthetic lenses had an important impact on the optics industry. Plastic is now used to create impact-resistant lenses for a variety of uses. Plastic contact lenses were developed after World War II, but did not become widespread until the 1970s.

Visual Recording Technologies

In China, the concept of the pinhole camera had been recorded by 500 B.C., but the capability to accurately record an event and easily review a visual transcript at a later time is less than 200 years old.

By the Middle Ages, pinhole images were known in Europe along with the idea of controlling the image size and quality by manipulating the size and distance of the aperture. By the mid-1500s, Daniello Bararo had experimented in channeling the light going through the pinhole by using a lens.

At this point, inventors sought ways of 'registering' the image to create a more permanent record. Semi-permanent recordings, such as 'sun prints' were developed by 1800 and, over the next three decades, experimental components evolved into heliographs and daguerreotypes. It was now possible to create a picture that could be viewed over and over again over time. In fact, daguerreotypes are so robust, many still exist.

The first photographs on metal created by Joseph Nicéphore Niepce were called *heliographs*. Louis Jacques Mandé Daguerre (1789-1851), another pioneer of photography, went into partnership with Niepce in 1829. He began experimenting with silver salts on copper plates and found that light could darken the salts. Daguerreotypes exhibited good permanency but were relatively awkward and expensive to create compared to later paper prints. The copper base was heavy (though tin was used later) and the images were somewhat faint and ethereal depending on how the plate was held in the light for viewing.

Left: A daguerreotype photograph taken around 1840 to 1860. The subject of the portrait, Mary Manuel Lisa, is wearing a pair of glasses and holding a second pair in her hand. One wonders if the possession of two pairs of glasses was a sign of status or pride at the time, since they are so prominently displayed in the portrait. Right: Another early daguerreotype, taken around 1856 to 1860. The subject, John Hanson, a Senator from Bassa County, is wearing spectacles in the style of the time. [Library of Congress America's First Look into the Camera collection photos, copyrights expired by date.]

In 1833, William Henry Fox Talbot made significant strides toward industrializing photography when he found a way to fix silver onto coated paper. He created early examples of contact prints, a development that evolved into prints on paper, which then became a primary means of recording photographs.

By 1847, Calude Felix Abel Niepce had found a way to set an emulsion on glass to create a negative, which could then be duplicated many times as a positive. Eventually this was done on plastic rather than glass. Thus, Daguerre, Talbot, and Niepce created a new industry that was to have far-reaching effects on human social structures and technological evolution, particularly in the fields of communications, entertainment, and surveillance.

Capturing a Moment in Time

Once the technology became available, photographers started recording everything in sight: people, plants, animals, instruments, vehicles, oceans, deserts, and constellations. It was discovered that film was more sensitive to the light from celestial bodies than a human eye peering through a telescope. Thus, the development of telescope optics aided in the creation of cameras, and the creation of cameras aided in surveillance of the heavens with telescopes. More stars could be observed by recording them on film, movement could be recorded by taking successive frames, and changes over time could be compared with photos taken at longer intervals. In other words, photography didn't just record what humans could see, sometimes it recorded what humans couldn't see. The same principle applies to surveillance in general, the recording of information allows it to be shared, compared, and analyzed in ways that are not possible otherwise.

Left: Lieutenant Colonel Huntington and Sergeant Goode using a spyglass to spot Spaniards at Camp M'Calla in June 1898. Right: A photograph of a painting taken around 1900 to 1920. The painting depicts a woman using a telescope to search the distance. [Library of Congress Touring Turn-of-the-Century America photo by Edward Hart; Detroit Publishing Co. photo of painting possibly by Charles Reinhart, copyrights expired by date.]

A scientist using a microscope and other devices in a workshop around the 1910s. [Library of Congress, Fred Hulstrand History in Pictures collection, NDIRS-NDSU, Fargo, copyrights expired by date.]

Surveillance Craft and Gear

A large part of visual surveillance involves the use of vehicles and specialized gear to carry a human observer into locations where he or she would otherwise be vulnerable to the environment or to detection. Thus, diving suits, space suits, diving bells, submarines, aircraft, and a great variety of other technologies have been created over the last couple of hundred years to allow a human to observe and record more safely in unfriendly environments. Sonar is one of the primary surveillance technologies used in submarines but the earliest submarines were in use almost half a century before the development of modern sonar and the primary means of surveillance on early subs and diving bells was through direct observation or enhanced observation with telescopes and periscopes.

Submarines in the 1800s were manually powered, with almost all the crew members charged with turning a hand crank while another steered and observed the surrounding waters. Submarines were put into service as patrol vessels and torpedo-launchers during the American Civil War. From there they evolved dramatically. By World War II, not only did subs incorporate sonar devices, but the Germans had created an experimental synthetic rubber skin designed specifically to counter sonar probes.

 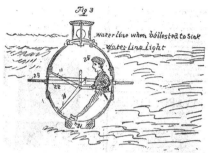

The Confederate submarineH. L. Hunley was a historic stealth vessel that patrolled the oceans and, during the Civil War, became the first to sink a battleship. It was simple technology, one crew member observed and steered while eight others powered it with a hand-turned crank. [U.S. Naval Historical Center archive photos of painting by R. G. Skerrett, 1902 and drawing by W. A. Alexander, ca. 1863, copyrights expired by date.]

Left: Documentary photographers having lunch in the Bull Run area before the second battle in 1862. Right: Photographic evidence of the destruction of a stone bridge at Bull Run, Virginia, in March 1862. [National Archives and Records Administration Photos by George Barnard and James Gibson, copyrights expired by date.]

As soon as photographic equipment and processes became portable and practical, photographers set out to document the events of their times and to offer their services to businesses and governments that wanted a record of operations or covert images to aid in reconnaissance operations. A number of photographers became noted for their portraits of Civil War dignitaries and for reconnaissance photos that assisted military generals during the War. They also created historical records of their travels and the plight of the less fortunate who perished in the War.

Left: This haunting and moving documentary photograph by Capt. Andrew J. Russell shows the dead Confederate soldiers behind the stone wall of Marye's Heights in Fredericksburg, Virginia. The soldiers were killed during the Battle of Chancellorsville, May 1863. Right: A grim, chilling, but perhaps necessary reminder of the realities of war in which Timothy H. O'Sullivan photographically captured both Union and Confederate dead at the battlefield of Gettysburg, Pennsylvania. When prepared for burial, secret documents were sometimes discovered in clothing and in accessories carried by the soldiers. [National Archives and Records Administration Photos by Andrew Russell, copyrights expired by date.]

Left: George N. Barnard's photographic equipment set up southeast of Atlanta, Georgia, in 1864. A row of tripods is in the background. Right: A documentary photograph by Barnard of the fortifications near the Potter House in Atlanta, Georgia in 1864. [National Archives and Records Administration Photos by George Barnard, copyrights expired by date.]

Surveillance from New Heights

Businesses and armed forces were highly motivated to develop craft that would allow people to see the world from new heights. Entrepreneurs offered sightseeing flights and hunting expeditions, scientists studied new lands, and topological, geological, and archaeological structures not visible from the ground, military agents scouted new lands and latitudes with airplanes, sometimes while dodging the bullets of unfriendly nations.

Many important milestones in aviation surveillance occurred in the early 1900s. National defense was a priority and had traditionally been carried out by the navy, but the world was speeding up and ships were relatively slow-moving vessels. Thus, the possibility of deploying fast-moving aircraft from short runways on heaving sea vessels led to the development of some of the more sophisticated visual/aerial surveillance technologies.

After seeing public demonstrations by the Wright brothers in 1908, naval agents evaluated the potential of aircraft as fleet scouts. In December 1908, an aviation report by Lt. George C. Sweet was presented to the Secretary of the Navy, specifying an aircraft capable of performing scouting and observation missions from naval vessels, essentially the first proposal for an aircraft carrier fleet. Apparently the technology was not considered sufficiently advanced, for in August 1909, the Acting Secretary rejected a request for two 'heavier than air flying machines'. Dirigibles and balloons were then receiving a lot of attention and were probably given higher priority.

As often happens, international and commercial competition motivated the Navy to change its mind. In 1909, F. L. Chapin, the U.S. Naval Attaché in Paris, reported on the Rheims Aviation Meet in Europe, promoting the usefulness of aircraft in naval operations. The viability of marine 'airstrips' was further demonstrated when a civilian pilot took off from an anchored ship, in November 1910, using a wooden platform on the bow.

Meanwhile, in the United States, Captain W. I. Chambers continued to advocate the potential of aircraft. By 1911, both a successful shipboard landing and a hydroaeroplane landing on water had been demonstrated, so the first naval aircraft were purchased and naval pilot training commenced. The Navy also began to experiment with another essential surveillance technology—wireless communications. Radio sets could enable pilots to coordinate their scouting missions and transmit their observations to remote command stations. From this point on, the evolution of naval aircraft, and the means of launching and catching the planes on moving ships, were continuously studied and refined.

By December of 1912, naval reconnaissance planes were ready for peacetime patrols and testing of specific tasks, including submarine spotting, mine and target detection and location, and aerial photography. Within a year, experimental craft, including an amphibious flying boat called the OWL, for Over Water Land (an A-2 redesignated as E-1), was being tested. In early 1914, a wind tunnel for aviation experiments was constructed at the Washington Navy Yard.

Thus, with the outbreak of the Mexican conflict and, later, World War I, the basic technologies related to aerial surveillance had been established and the Armed Forces looked for ways to put them into service.

By spring 1914, the U.S. Navy was using AH-3 hydroaeroplanes for reconnaissance over Mexico, in spite of the fact that naval aviation was not yet formally recognized. The Navy subsequently commissioned swept-wing Burgess-Dunne hydroaeroplanes for the AH line and, by July 1914, naval aviation was finally acknowledged when the U.S. *Office of Naval Aeronautics* was established.

During this period of aviation development, cameras were improving in portability and utility and the Navy had an interest in recording from the air. In May 1916, the *Naval Observatory* sent a Hess-Ives color camera to the *Naval Aeronautic Station* at Pensacola for testing in aeronautics. Outfitting of ships to serve as aviation hosts was also underway and, in July, the North Carolina (ACR 12) became the first U.S. Navy ship equipped to carry and operate aircraft. By December, it was reported that an Eastman Aero camera had been tested in aircraft at altitudes of 600 to 5,100 feet and produced very satisfactory photographs for naval purposes. The following month, 20 Aeros were requisitioned. Visual surveillance from the air had evolved in just a decade to aerial surveillance, and over the coming years other types of photography and imaging were developed.

Around 1918, naval forces began painting many of their reconnaissance vessels to reduce their wartime visibility (a foreshadowing of the development of low-sonar and low-radar-signature vessels two decades later). In the spring, flying boats were used for long-range reconnaissance off the German coast and the H-16 aircraft was engaged in submarine-spotting from U.S. and European stations. Kite balloons with weather-sensing instruments were launched, as well.

Visual surveillance sometimes has its comic side. In one case, a kidnapper was arrested after the New York Police Department (NYPD) tailed his carrier pigeon. In 1929, the NYPD began its first Aviation Unit and was conducting regular air patrols by 1930.

In Germany, at this time, aircraft and Graf Zeppelin dirigibles were being used in reconnaissance and bombing raids against the allied forces, and Britain was employing a variety of airplanes and airships. Visual surveillance from the air was now firmly established. After the war, these technologies continued to evolve and were becoming widespread in commercial aviation by about 1937 and were again put into the service when the European War (World War II) broke out.

In 1947, the NYPD acquired the first helicopter and helicopters eventually superseded the fixed-wing planes in the Aviation Unit. The NYPD now conducts patrol and surveillance activities, and responds to emergencies and medical evacuation needs.

Following World War II, the invention of the transistor, in 1947, greatly enhanced the capability of engineers to install small, fast surveillance technologies on fast-moving aircraft, and their inventions are described further in various sections in this book (see the chapters on Aerial Surveillance and Radar Surveillance, which include not only information on various imaging systems but also further information on the history of balloons and dirigibles as support vehicles for visual surveillance.)

The Development of Image Transmission

Consumer video cameras and recording devices evolved hand-in-hand with broadcast television technology.

The Nipkow disc, pioneered by Paul Nipkow, directed light though a rotating disc perforated with a spiral pattern of holes. Photosensitive selenium behind the disc registered dark and light areas as the image was sequentially scanned. The device is from the collection of the American Antique Radio Museum. [Classic Concepts ©1998, used with permission.]

One of the most significant events in television history was the discovery of the light-sensitive properties of selenium. In France, in 1878, M. Senlacq suggested that selenium could register dark and light shapes and hence might be used for recording documents. By 1881, Shelford Bidwell in Britain had succeeded in transmitting silhouettes, but the effect was primitive and, for a while, no big advances were made.

The idea of transmitting images over distance caught the imagination of many inventors around this time, and there was a lot of unspoken competition to develop the first practical system. Paul Gottlieb Nipkow was one of the first inventors to receive a patent for a historic television, in 1884, and patents by other inventors soon followed. Most of these early devices were low resolution, somewhat unstable systems but they represented *proof of concept*. There were a number of basic aspects that needed to improve to create practical systems:

Signal Amplification. The early television pioneers needed to amplify weak signals before they could develop the technology into practical devices, but vacuum tubes had not yet been invented. In 1904, John Ambrose Fleming (1849-1945) developed the Fleming valve, the first vacuum tube that could function as a simple rectifier. In 1906, Lee de Forest, an ambitious experimenter, added a crucial third element to a Fleming valve, a grid that was able to control the flow of electrons from the cathode to the anode, providing a means of amplification. While de Forest himself didn't understand how the tube worked, his invention, called the *Audion*, represents a significant milestone in the history of electronics.

Power. These early inventions were developed in the days when houses were lit by candle-light and heated by wood, coal, or sod fires. The production and transmission of power were important aspects of the development of image transmission technologies. Charles Proteus Steinmetz (1865-1923) was one of the pioneers to investigate the alternating current AC phenomenon. This important groundwork led to the development of better motors, generators, and power distribution systems, all of which are crucial to broadcast technologies. Nikola Tesla (1856-1943) was an eccentric genius who had regular communications (not all of which were friendly) with Thomas Edison. At one point, the Nobel Prize was offered jointly to the two men, but Tesla eschewed the offer and it went instead to a Swedish scientist. Tesla championed alternating current (AC) at a time when other renowned scientists were promoting direct current (DC), and made some important discoveries that have since been incorporated into commercial systems, including some implemented by George Westinghouse (1846-1914).

Image Display. Breaking up an image and amplifying an image are not enough to produce a television set. There needs to be a way to display this image at a reasonable brightness and size. A specialized electron tube developed by Vladimir Kosma Zworykin (1889-1982) provided this means. Zworykin patented the cathode-ray tube (CRT) in 1928 and further developed the *iconoscope* based on his CRT invention. The cathode-ray tube has since become an indispensable aspect of many surveillance technologies, from televisions to radar screens and computer monitors, cathode-ray tubes, in spite of being awkward, large and bulky, have been a primary display technology for almost 70 years. Zworykin went on to become the director of research for the Radio Corporation of America (RCA).

The first primitive vacuum-tube-based television systems began to appear in the mid-1920s. Scottish-born John Logie Baird (1888-1946), Japanese inventor Kenjito Takayanagi, and American Philo T. Farnsworth, independently demonstrated working televisions systems within the span of about a year. Most historic televisions were based upon mechanical disc technology but Zworykin soon patented an electronic scanning system, in 1928.

Baird is less known for inventing another type of visual technology. In 1928, he introduced a *video album*, a means to store video frames on a 78-rpm recording disc that could be replayed at 12.5 frames per second. Video tape recording media emerged in the late 1940s but initially suffered from a number of limitations that were worked out over the next few years. Video tape recorders as we know them began to become practical in the early 1950s.

Video tape recorders would dramatically change television broadcasting which, up to now, had entirely live performances. They would also provide a practical medium on which to store surveillance footage so that a human operator didn't have to watch a monitor 24 hours a day.

In 1956, the first television program that was recorded to tape was broadcast by replaying the tape. Prerecorded programming and reruns quickly became a staple of television programming. The price of the first blank commercial black and white reel-to-reel video recording tapes was astronomical compared to today's $2 consumer tapes. The first professional tapes sold for about $300 each, about two month's wages for the average person at the time. But the television network could replay a tape many times and use a prerecorded tape to increase programming options and expand time slots, so the price of a tape was for them a very worthwhile investment. For the average consumer, however, the technology didn't become available until the mid-1960s and wasn't affordable until the mid-1970s when JVC introduced the VHS format.

Image Broadcasting

As soon as practical television systems appeared, broadcast stations sprang up all around the country. Radio broadcasts could be profitable and many entrepreneurs were convinced that television broadcasts would be even more profitable. The first broadcast stations were established in the late 1920s, and by the mid-1930s, the public broadcasting industry emerged. Still, it took another 20 years before commercial television sets were mass-produced. The black and white sets that were eagerly bought by consumers in the late 1940s and early 1950s were characterized by fuzzy pictures, test patterns, and lots of fiddling with the horizontal- and vertical-hold knobs to stabilize the images.

Closed-circuit television systems were developed along with public broadcast systems, but there were no subsidies or advertising revenues for closed-circuit systems, so they remained expensive until the late 1960s, when black and white closed-circuit systems became practical and affordable for security surveillance in the business world. The biggest disadvantage of these early systems was that they required human eyes to watch the system at all times. Video recording was cumbersome and expensive until the introduction of the VHS and Beta video-tape-recording formats.

There were a number of historic developments between 1962 and 1972 that set the stage for future video surveillance systems technology:

- In 1962, satellite broadcasting began with the Telstar 2.

- Up until this time, the only news about warfare that had been available to the public was through newspaper and radio broadcasts. In 1968, television news broadcasts from Vietnam were shown to the public.

- In July 1969, American astronauts touched down on the Moon and sent back live images of their historic first steps while millions watched on televisions from around the world.

- Digital television began to be used in commercial broadcasting in 1971, when the British Broadcasting Service (BBC) combined the audio signal with the video signal in their transmissions.

- Commercial television broadcasting was first available through the geostationary ANIK 1 satellite that was pioneered by the Canadian Broadcasting Corporation (CBC), in 1972.

- In 1968, Intelsat was establishing a global satellite communications system that was launched in 1972.

Thus, the essential building blocks for television broadcasting and the capability of global transmissions were in place by the early 1970s. Institutions that could afford closed-circuit television started almost immediately to use the new technologies for security surveillance but, for the average user, television was still too limited and expensive to use for security purposes. In 1976, Ampex began creating video-recording technologies that would greatly improve the viability of television broadcasting as a surveillance medium, making it possible to record video images in slow motion. Extended recording formats and time-lapse systems from a number of developers would follow over the next several years.

Video Viewing Technologies

As solid-state technology evolved, smaller and smaller components were designed and manufactured. Most people watched television in their living rooms, but there were some who wanted portable TV. The first transportable televisions were luggables that needed the same AC power source as stationary sets. Then basketball-sized transportable sets were developed and, eventually, clock-sized sets that could run on batteries. At the same time that TV sets were becoming smaller, the technology for image resolution was improving. The increasing popularity of computers in the 1980s improved viewing technologies in general. Because readable text on a computer screen required better display technologies than the average television program, there was a demand for improved monitors which, in turn, spilled over into television technologies.

Video technology-based surveillance systems began spreading in the 1980s and closed-circuit television for consumer applications became a practical reality in the early 1990s.

Visual Surveillance in Homes and Offices

Standardized video cameras that could be wired into a central switching console and displayed on television sets or computer monitors were prevalent by the early 1990s. Homes and small offices could now afford to wire their premises for video and retail stores eagerly installed the systems to monitor shoplifting and employee theft.

Small cameras, camcorders, time-lapse recording decks, and inexpensive video/audio transmitters all contributed to the explosion of fairly sophisticated video surveillance technologies in the mid-1990s. Broadcast quality images were in professional hands until about the late 1980s, when Hi-8 mm video formats began to catch on. By the mid-1990s, handheld Hi-8 camcorders were under $1,600 and dropped to less than $700 each by 1999. Hi-8 recording devices were small enough to fit in a purse or a briefcase. Coupled with a tiny pinhole camera hidden in a pair of sunglasses or a tie clip, it was now possible to wear a 'video wire,' a video version of the covert cassette recorder that had been a surveillance staple for the last couple of decades.

In the 1970s and 1980s, some parents installed intercom-like 'baby monitors' in their infants' and children's rooms. By the late 1980s and early 1990s, small video monitors were designed especially for this purpose, becoming affordable by the mid-1990s. About this time, television news teams revealed that some of the cameras had picked up gross examples of abuse by child caregivers while parents were away. The shocked public began to buy more cameras, especially pinhole cameras, to ensure their children were getting the best care. This demand fueled development of a new class of products called *nannycams*. Nannycams were small, discrete cameras and transmitters or recorders that could broadcast video and audio signals for a short distance, usually up to about 300 feet. By the year 2000, nannycams were becoming more sophisticated, with FCC-approved distances, from reputable vendors, up to one or even three miles. With an appropriate antenna, this range could be extended even further. This made it practical to monitor the caregiver from the workplace or a friend's place across town. It also made it possible for other people within range of the transmissions to pick up the broadcast and look in on the home that was being monitored.

Surveillance and the Legal System

By the mid-1960s people were beginning to introduce video surveillance tape as courtroom evidence and the courts were working out the issues of the admissibility and interpretation of this kind of evidence.

In the early and mid-1990s, various court cases related to nonconsensual video taping in various workplaces, and public areas were beginning to be brought to court. In general, monitoring and recording in public places were not judged to be violations of a person's rights, whereas monitoring in dressing rooms and washrooms was, in some cases, found to be a violation of a person's privacy. These cases are still being debated and no firm pattern has yet been established.

By the late 1990s, it was estimated that there were at least a million video cameras installed in the United States for the purpose of promoting public safety and security, not counting those covertly installed for the purpose of spying.

By 1999, security advocates and privacy advocates were meeting at various conferences and summits to debate the pros and cons of video surveillance monitoring and recording of public areas for safety and security purposes. Security vendors wanted to freely promote their commercial products to prospective buyers, and privacy advocates wanted to carefully evaluate the ramifications of the use of these products and possibly curtail the indiscriminate commercialism of technologies that could have negative consequences if used in unethical ways. It was pointed out that a clandestine or covert video tape was, in many ways, more intrusive than an audio wiretap.

Digital Technologies

In the early 1980s, television technologies began to be interfaced with personal computers or with specialized microcomputer electronics. Chyron, a leader in video titling equipment, introduced a character generating system in 1982 that could be controlled by a personal computer. Two years later, a small group of visionary computer designers created the Amiga computer, which debuted late in the summer of 1985. Unlike other systems, the Amiga, with a few hardware additions, could be readily interfaced with video equipment to provide control, titling, and animation capabilities that previously had only been available on professional video systems costing tens of thousands of dollars. The Amiga precipitated a desktop video revolution that dramatically reduced the price of computerized video technologies. It also presented new ways of generating and manipulating video images that could be used for courtroom simulations and forensic re-creations of relevant events.

As digital video technology evolved, so did digital still technology. Digital still cameras were available to consumers in the 1980s, but the prices were initially high, sometimes as much as $25,000 for a single-lens reflex camera that could only shoot medium-resolution images. By the late 1990s there were cameras with faster shutter speeds, higher storage capacities, and similar imaging capabilities for about $1,100. Digital cameras thrived commercially partly because of the convenience of instant pictures and also due to the ease of uploading the images to Websites. Internet auction and store sales were often improved by images, providing a strong motivation for online vendors to purchase digital cameras.

Encrypted digital video was a mixed blessing for law enforcement. Digital video is convenient, but easily altered and thus may not meet all criteria for use as courtroom evidence. If various encryption or tampering mechanisms can be built into digital video, then its use might become practical in the courtroom, but until then, the authenticity of the images can be challenged.

Databanks and Visual Surveillance

In the 1980s and 1990s, there were significant developments in visual surveillance. Within two short decades, programmers had developed ways to read, store, and manipulate images, and artificial intelligence researchers had developed sophisticated algorithms for pattern recognition and artificial vision. Taken together and interfaced with video data, these made it possible to automatically detect, recognize, track, and analyze video images from security cameras and covert surveillance devices with a minimum of human intervention.

While the technical tools for handling the images were evolving, the Internet and the World Wide Web were also evolving and gradually revolutionizing global communications. Now it became possible for image processing to be monitored and transmitted from and to any location with an Internet connection. A company with a head office in California could visually monitor the work of employees in a branch office in Sweden without making long-distance calls. A police officer in New York could check a suspect's record against national databases and discover that he or she was wanted for a violent crime in Utah three months earlier. A shopping mall could store a database of known troublemakers in a central security facility and configure the system to ring an alert when any person found in the database walked in the door, even if he or she wore a hat or glasses. Anonymity in public places was quickly disappearing and personal privacy was slowly disappearing.

International Compliance

Many industries must conform to strict industrial and emissions guidelines. There are also many international treaties governing weapons production and the detonation of nuclear armaments. Starting around the 1990s, video technologies were gradually being integrated into compliance and safety regimes.

For example, in the fall of 1995, the *International Atomic Energy Agency* announced that it was planning to install Gemini, a fully authenticated digital video surveillance system, at global nuclear sites, to enforce the Non-Proliferation Treaty, a multilateral agreement intended to prevent the proliferation of nuclear weapons. The Gemini system is housed in a tamper-proof enclosure with two independent surveillance units, each with a camera to capture and authenticate images. The system is further designed with Ethernet linking capability to enable it to interface with computer workstations optimized for high-bandwidth video data display.

Thus, by the year 2000, video cameras could be seen in almost any environment within developed nations, and sales projections indicated that this was only the beginning. The numbers were expected to triple over the next couple of years.

5. Description and Functions

Visual surveillance is probably the most important surveillance technology because humans are highly visual and devices that aid in visualization are plentiful and often inexpensive and readily available. Whereas radar and infrared typically involve expensive technologies or technical knowledge to build them, most people can readily use a magnifying glass, pocket scope, or portable camcorder. Thus, the number and variety of vision-enhancing and recording devices is large and expected to continue to increase.

5.a. Basic Equipment

Magnifying Glasses

Magnifiers can make nearby objects appear larger. Since a handheld magnifying glass is relatively portable, it may be carried in a pocket or briefcase, and magnifying eyeglasses are worn like prescription eyeglasses. Magnifying glasses are often used in laboratories and classrooms and at crime scenes. Good lighting is important and some magnifiers are equipped with supplemental lighting.

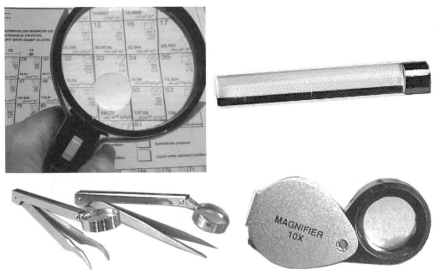

Handheld magnifying glasses (top left) are widely used in laboratories and at crime scenes. Bar magnifiers (top right) aid in reading small print, revealing the presence of code in small markings (e.g., microdots), and evaluating security features such as watermarks, impregnated substances, and special fibers. Tweezer magnifiers (bottom left) help in handling very fine evidence such as hairs and fibers. Pocket magnifiers (bottom right) are handy for many jobs in the field and the cover helps keep debris off the surface of the lens. Sometimes pocket knives have built-in magnifiers. Magnifiers are widely used in many fields. They aid in studying fiber evidence, stains, bruises, fingerprints, documents, security markings on negotiable items, and many other items of interest to police agents, scientists, and detectives. [Classic Concepts photos ©1999, used with permission.]

Scopes

Scopes make distant objects appear larger and, thus, easier to view. Scopes come in a wide variety of sizes and magnifications, from small handheld hiking and night-vision scopes to observatory-sized astronomical telescopes. Spyglasses are simple scopes, like those pictured in paintings of historic mariners, especially pirates and explorers. Binoculars are two scopes combined side-by-side to provide magnification for both eyes. Periscopes are viewing scopes used in submarines. Microscopes, which magnify tiny objects, are important laboratory tools, especially in the forensic sciences.

Scopes range in cost as much as they range in size, with higher magnifications generally associated with higher prices.

Just the act of using a scope may alert the target to the fact that s/he is being surveilled. Since most scopes use glass lenses that are reflective at certain angles, the glint off of the scope may attract attention. For this reason, there are scopes made from different materials, some with baffles, coatings, or other reflection-reducing components and chemicals.

A tripod might be necessary for steadying a scope, since high levels of magnification will cause the image to shiver from even the slightest vibration.

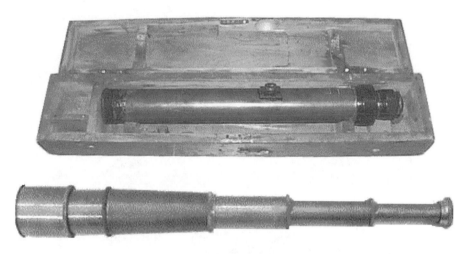

A variety of telescopes are shown here, from a World War I military scope (top) and mariner's spyglass (middle), to modern consumer Meade telescopes (bottom left) and astronomical observatory telescopes stationed in Hawaii. [Classic Concepts photos ©1999, used with permission; bottom right, photo public domain.]

Top Row: Historic scopes from the late 1800s and early 1900s are essentially 'spyglasses'. Second row: A variety of military binoculars and scopes. Bottom left: A high-power microscope suitable for medical, laboratory, and forensic applications. Bottom right: A true 'spy' scope, a telescope disguised as a walking stick. [Top Photos: Library of Congress, copyrights, expired by date; Second Row and Bottom Left: U.S. DoD news photos, released; Bottom Right: Classic Concepts ©1999, used with permission.]

Left: An M-137A1 Panoramic Telescope is used in a tactical employment demonstration by members of a peacekeeping mission in Bosnia and Herzegovina. Sgt. Thomas McEwen, U.S. Army, explains the use of the telescope to members of the Russian Separate Airborne Brigade. Right: Night-vision goggles are used for ground and air surveillance (and sometimes for night-time parachute jumps). This hands-free set is used on helicopter flight helmets by the U.S. Army. [U.S. DoD 1999 news photo by James V. Downen, Jr., released; 1997 news photo by G. A. Bryant, released.]

Goggles

Better night vision can be provided by magnifying low light-level images or by using infrared-sensing technologies. Those that use light intensifiers have safety 'shutoff' mechanisms to prevent a sudden blinding light that could injure the yes of the user. Those that work with infrared frequencies sometimes require a separate source of infrared illumination.

5.b. Image Recording Devices and Accessories

Film

Since most reconnaissance activities involve field work in less-than-ideal working environments, there are some concerns particular to the use of film in surveillance.

Film Roll Lengths

Most of the film used in surveillance activities tends to come in longer rolls, usually 36 exposures for traditional still photography, and rolls of 100 or 250 feet or more for aerial photography. Longer rolls reduce the likelihood of having to stop to replace film (which might attract attention) and there is less chance of missing a shot. Aerial photography often requires a lot of territory to be scanned and imaged in a short period of time, necessitating long rolls of film. As with most technologies, there are tradeoffs. Longer rolls are heavier and associated with more inertia, making it difficult to advance the film quickly and smoothly. Special mechanisms are needed to facilitate film advancement. Long rolls are also larger and heavier, requiring stronger reel mechanisms, larger canisters, and sturdier mounting systems. Estar films are favored for many aerial film needs because the stronger base allows the film to be thinner and hence lighter. A 100-foot roll may weigh about six pounds.

Curl

Most photographers are familiar with the fact that film tends to curl. This is partly because film is rolled and partly because film is composed of more than one layer sandwiched tightly together, and the layers may have different compositions. To minimize curl, many specialized aerial films have a gelatin coating on both sides of the base layer. An antihalation coating may also be included (this is discussed in more detail in the Infrared chapter) to reduce film fogging.

Environmental Extremes

While conventional film is fairly robust in various extremes of temperature and humidity, specialized films, like coated aerial films, infrared films, etc., may be sensitive to heat and humidity. Extreme cold will make film brittle, while extreme heat or humidity will blister or fog the film (especially infrared-sensitive film).

Image and Data Processing

Computer processing adds an entirely new dimension to the handling of visual information. Some of the more sophisticated digital imaging systems can recognize a human and track the person's movements through a building complex, even identifying the person from matching items in a computer database. Computer networks make it possible to view the data from anywhere in the world through the Internet.

When surveillance video cameras are combined with intelligent computer visual recognition software, they create powerful surveillance systems. It is possible to process the images to detect movement, home in on an approaching face, switch to another camera as the person moves through the complex, consult a database to identify the individual person, and play spoken messages to that person, if desired. This example is simulated, but computerized systems that perform these functions are already in use. They have been interfaced with databanks of identified criminals and installed in locations such as public malls. [Classic Concepts ©2000, used with permission.]

Computer processing makes it possible to manipulate and transmit data in many ways. Here are some of the most common ways that are relevant to surveillance:

- The tedious job of retrieving and analyzing visual surveillance data can be partly automated. Computer software can preview hours of footage and 'flag' areas of video where movement or anomalies may have occurred.

- Computers can be programmed to recognize certain people or events that can then serve as a trigger for automatically starting up recording machines, checking databases, or sounding alerts to a human operator.

- Image processing can be used to take data from a number of successive frames and incorporate them into a clearer picture so that a specific person, or the make and model of a car (and sometimes even the license plate), can be recognized. However, this is expensive, difficult work that may only be partly successful and should only be done when necessary. (Note: It is usually less expensive to buy better resolution equipment than it is to try to clean up poor video footage.)

- Data compression can be used to store large amounts of information in smaller amounts of space.

- Data encryption can protect the data.

- Computer networks can be used to access or transmit the information around the office or around the globe.

5.c. Management and Storage of Recording Materials

When sales representatives promote visual surveillance systems to various organizations, they don't always spell out the costs of maintaining and repairing the systems and archiving and retrieving the recorded data. These are administrative aspects that are important to consider before purchasing multiple-camera surveillance systems.

The costs of monitoring will depend upon the system. Some systems are monitored by live operators and the events are not recorded, while some are monitored by live operators and also recorded. Some systems are automated and live operators are only summoned if an alert is triggered. Other systems record continuously and the recordings are only reviewed if an unusual event happens (e.g., a convenience store robbery). Whichever method is used, there are labor and maintenance costs associated with monitoring, taping, storing, and reviewing events.

If a location that is being surveilled is *not* recorded, then an employee has to be available at all times to watch the monitors. If surveillance is 24 hours a day, as in a subway or busy hotel, then three 8-hour shifts are typically set up. Employees must either watch the monitors constantly, or watch them when motion is detected and triggers an alarm. In cases of dangerous environments (e.g., industrial settings) or environments where several cameras are used in various locations, or in applications where it is important to create a record, live monitoring may be supplemented with recordings (e.g., video replays of an industrial accident or refereed sports event).

It is impossible to know in advance if a recorder will record a significant event. The famous broadcasts of Princess Diana and Dodie al Fayed exiting the hotel before their untimely deaths in an automobile accident, in 1997, demonstrate how a seemingly unimportant piece of video footage can later become crucial to an investigation. After the Princess and her companions lost their lives, the hotel tapes were aired to the public and scrutinized in great detail to see whether the driver might be impaired, whether the party may have been followed, or whether any other clues could be gleaned from their activities. Administrators seeking to install visual surveillance systems need to establish a process for organizing and maintaining recording archives that can be efficiently retrieved and viewed after the fact. They must also decide how long tapes should be retained, since few organizations have the resources to keep tapes indefinitely. In some cases, regulations require that recordings be destroyed after a certain amount of time has passed or if certain events occur (e.g., an employee leaving a company).

Example Costs of Archiving

As a basic example, imagine that a community is surveilling a park or a hospitality business or a small casino is surveilling its public areas. Assume there are 20 cameras in operation 24 hours a day saving to realtime or near-realtime recording machines (VHS, S-VHS, 8 mm, Hi-8 mm, and computer digital formats are most commonly used). For this example, let's say the video format is S-VHS (a high-resolution format that is used in many law enforcement activities). While time-lapse systems often increase the recording duration of a tape, keep in mind that it can take as little as 7 seconds to enter and exit a room to steal money, jewelry, or a purse. Thus, time-lapse systems with a frame-rate of less than 10 frames per minute are not suitable for monitoring all kinds of human activities. For this example, assume that non-time-lapse recording systems optimized for 12 hours of continuous recording are used (there are machines that record for longer, but there is usually a large trade-off in image quality or cost). Finally, assume that the tapes in this example are kept for three weeks each.

Given this basic scenario, at least two full-time employees are needed to swap tapes and to carefully label, box, and archive 40 tapes a day, at a cost of about $300 per day for materials, floor space, and labor (not including the cost of cameras, monitors, VCRs, repairs, or the cost of later viewing the tapes).

You can't just hire someone to come in twice a day to replace tapes. The tape replacement has to be staggered so that the machines are timed to end at different times. Otherwise there will be 'blank spots,' security *dead zones* in which nothing is recording because 19 machines are idle while waiting for new tapes while the first one is being replaced. Thus, machines should be staggered to end about 20 minutes apart, giving the operator time to prepare tapes for each machine and to swap them as needed. With 20 recorders, this process alone takes over six hours and must be performed twice a day.

After a period of three weeks, the archive will hold 840 tapes, requiring about 15 cubic feet of shelf space. Tapes can be reused about three to five times before the quality starts to significantly degrade and then it's best to replace them. Thus, the total cost per month of managing the machines and tapes can easily reach $12,000 ($144,000/yr). To this must be added the cost of maintaining and repairing the cameras, monitors, and tape decks.

Computerized recording systems with sufficient storage space (e.g., large hard drives) can sometimes be used to provide longer recording times, so tapes don't have to be constantly swapped, but the logistics of simultaneously recording two or 30 camera feeds is beyond the scope of most personal computers and off-the-shelf software (at least at the present time). Computer workstations with custom software and hardware adaptations are available at corporate prices, however, which means that consumer versions may be available soon.

In both tape and computer systems, some optimization of storage can be achieved with motion detectors placed in low traffic areas that trigger the recording devices and record for a preset amount of time after motion is no longer detected.

Assuming a computer security system has 20 video inputs and can store data from all 20 at the same time (or perhaps there are five computers, each storing input from four cameras), there seem to be some advantages, but is it really cheaper? Most video tape decks store images in higher resolution than most computerized systems digitizing input through multiple videocams. Computer data need to be backed up. All computer data are vulnerable to corruption, accidental deletion, and loss due to hard-drive failures. That is not to say that computer hardware is inherently less reliable than tape recording machines, but rather that operator error and programming bugs are more likely to impact the integrity of data in a computer system than in a traditional VCR. Therefore, computer backups to high-capacity tapes or cartridges need to be made from the original computer data. In other words, you end up with the same storage, labeling, and archiving concerns for computer backup tapes or cartridges as you would with traditional video tapes. It's important to assess the advantages and disadvantages of each system before investing in visual recording systems.

6. Applications

Since this volume describes surveillance technologies as they are used, rather than how they should be used, this section includes descriptions of both legal and illegal uses of visual surveillance. Consult Section 8 (Restrictions and Regulations) for more information on lawful uses and restrictions on the use of visual surveillance devices. Note that some uses and technologies may be legal within home or educational contexts, but not in other applications.

There are thousands of different applications of visual surveillance technologies, so it's not possible to list them all, but there are enough examples here to represent common uses.

6.a. Community Safety and Security

Art Theft

Investors and cultural historians have long considered fine art to be one of our greatest legacies. Unfortunately, the value of artworks makes them vulnerable to theft and sometimes to vandalism. Many museums have installed visual surveillance systems to protect works of art and to monitor patron safety and compliance with preservation guidelines (e.g., no flash pictures).

Surveillance systems can be difficult to install on the limited budgets of organizations that safeguard public works. Even centers with surveillance measures in place occasionally experience thefts. In January 1998, a Greek stele with 4th century B.C. inscriptions was stolen from the Louvre, in spite of photo surveillance. In May 1998, a Jean-Baptiste Camille Corot painting in the Louvre was stolen from its frame from a room that lacked camera surveillance. The police searched for fingerprints on the frame and glass but had to investigate the theft without any identifying visual information.

The same month, Italy's only two van Gogh paintings, and a work by Paul Cézanne, were stolen from the Rome Museum. This tragic theft occurred in spite of the existence of a camera surveillance system. The armed and masked thieves bound and gagged the guards, stole the works, and removed a surveillance tape from the closed-circuit recording system.

In fact, many thieves now look for surveillance tapes to try to remove the evidence. The recording devices should be locked and hidden. Some institutions have two recording systems, one that is a decoy and one that is hidden as a backup.

Bullet Forensics

Magnifying technologies are used in almost every aspect of forensics. Visual technologies including microscopes and magnifying classes are used to investigate bullets that have been recovered from crimes. Bullets can be visually examined to determine whether they have been fired, and what type of gun fired them. The microscopic patterns engraved into a bullet when it spirals down a gun barrel are unique to that barrel and can be matched to a specific gun. National databases and gun and ammunition collections further aid in visual identification and categorization.

Building Inspection

Visual surveillance techniques are used in many aspects of building inspection and industrial fault detection. Infrared and ultraviolet cameras and photographic records are common in this field of application, as well.

Crime Scene and Accident Records

Photographs to indicate the state of a scene at the time of a crime, accident, or discovery of unusual activities are often extremely important for solving a crime, linking an event with other similar situations, or justifying damages for insurance or job-retention purposes.

You cannot always know in advance what aspects of a crime scene might be important. It is vital not to disturb or trample things in the process of photographing them. Footprints and tire tracks should always be photographed before people walk around the crime scene, with plaster casts taken, if possible. Anything that appears to have been disturbed should be photographed because it isn't usually known in advance what might be relevant to the crime later.

In auto accidents that involve insurance claims and liability suits, it is important not only to photograph the vehicles and people involved, but also the conditions (e.g., weather) at the time of the accident (as these may change dramatically by the time the claim is adjudicated or the suit heard in court), along with blind spots, signs, tire tracks, and sometimes even witnesses. It may also be important to photograph injuries, including bruises, broken limbs, etc. The photos should be carefully labeled with the event and the time and date of the accident. Any unusual circumstances that can't be photographed should be noted (oil spills, a hail storm, construction, a deer crossing the road, etc.). No matter how clear these details may seem at the time of the event, the clarity of the memory will fade in time. If the cause of an accident is a belligerent or drunk driver, the other person may not be cooperative and the photographic evidence may be the only way to assert the truth.

Explosives Detection and Disposal

The detection and handling of explosives is one of the most difficult areas of police and military work. Unexploded 'bomblets' and mines and deliberately planted terrorist bombs must always be handled with care, even by seasoned experts. Surveillance technologies can help to locate, neutralize, and otherwise dispose of dangerous explosives. Technologies that help locate explosives before people stumble across them or that allow experts to neutralize the explosives or intentionally detonate them from a safe distance, are welcomed by those concerned with public safety. Traditionally, bomb experts have used heavy Kevlar bomb suits to protect them from a possible explosion, but new options are becoming available through remote-controlled cameras and robots.

Vision-equipped robots can be used to get close to hazardous materials to locate and disable explosives that might pose a danger. This Remotec Andros 5A, developed at the Sandia Labs, is equipped with three cameras, a gripper, and a bomb disabling gun. The underlying software is designed so that various capabilities can be 'mixed and matched' to create a variety of robots specialized for specific operations such as waste cleanup or accident remediation. [Sandia National Laboratories 2000 news photo by Randy Montoya, released.]

Many surveillance devices are now available for locating explosives and are described in several of the other chapters. In terms of visual surveillance, some of the more interesting technologies include vision-equipped robots that not only detect bombs and other explosives, but may have the capability to safely dispose of them as well.

As an example, Sandia National Laboratories has developed a robot vehicle with three cameras designed to aid in bomb disposal. The Remotec Andros 5A has a gripper and a bomb-disabling gun in addition to the three vision systems. The robot is now being integrated into the Albuquerque Police Department's bomb disposal unit.

To operate its various robots, Sandia has developed a Sandia Modular Architecture for Robotics and Teleoperation (SMART) which is a 'stackable' software base consisting of inter-changeable modules for each robotic function. Using this system, the desired robot capabilities could be mixed and matched 'off-the-shelf' and incorporated into custom systems. Thus, the robotics system needn't be constrained to just disabling explosives, it could also be configured for cleaning up wastes, hazardous materials spills, or responding to accidents. Since this type of flexible technology can be applied to many law enforcement situations, the FBI provided advisement to Sandia on the project.

Public Areas Monitoring

During the 1980s, security cameras were increasingly installed in subway systems, parking lots, trains, and corridors, but they were not yet prevalent in town squares or public sidewalks.

As of 1998, the number of cameras installed in public areas dramatically increased to the point where some cities have so many, it's difficult to walk down a public sidewalk without seeing cameras every block or two. They monitor ATMs, banks, intersections, retail outlets, gas stations, convenience stores, campuses, and gated or secured neighborhoods. Some say they prevent crime. Others say they displace crime or prompt different kinds of crime. These assertions are difficult to study empirically. It appears fairly certain that video monitoring aids in identifying subjects and in convicting suspects, but it is not certain beyond a doubt that crime overall is reduced by video cameras. Given the cost of installing, maintaining, and monitoring video surveillance cameras and recordings and the significant impact on personal privacy, it is important to be aware of all the issues related to video surveillance of public areas.

Here are just a few ways in which video cameras are used in security monitoring:

- In the Bronx, cameras monitor schoolyards.

- The *Housing Authority* is installing bulletproof cameras in corridors of city housing projects. In Harlem grant houses, almost half a million dollars were invested in surveillance camera systems, with a reported drop in crime.

- Valhalla and Dutchess Counties have added cameras to guard helmets, visiting rooms, and some prison cells.

- Los Angeles, Oakland, Tacoma, Seattle, New York, Baltimore, and Charleston are just a few of the cities that have installed video surveillance systems in public areas, most often parks, streets, parking lots, and bridges. In some cases the recommendations were defeated before the cameras were installed. In some, the cameras were put in place without public hearings and later removed as a result of public concerns. In some, the cameras were supported and implemented and later removed. And, in a few cases, the cameras were supported, implemented, and continue to be used.

- In Gulfport, cameras have been proposed to help police monitor high crime areas and to act as a deterrent. These are intended to be mounted on telephone and light poles.

- Throughout the U.S., Canada, and the European Union, video surveillance cameras are promoted as a way to reduce crime and promote public safety. There is as yet no hard evidence that this is so, partly because the technology is new, and partly because it is hard to tell when reduction in crime is strictly due to one change or a number of changes, or is related to displacement, that is, the movement of criminals from a surveilled area to an unsurveilled area.

- A small community in interior British Columbia was experiencing repeated thefts and vandalism and the residents were frightened that the activities might escalate. The community banded together to set up human surveillance shifts around the clock and used handheld cameras and clipboards to record the comings and goings of everyone into and out of the community. The concerted effort resulted in a halt to the series of thefts.

- In Newham, East London, a video system in a shopping center was installed that could create a facial code from digital video and compare it to a database of local criminals. A match sounds an alarm in a control room and the security person on shift assesses the computer match and alerts local authorities Similar systems are being evaluated for airports and border crossings.

Traffic Monitoring

Speeding, running red lights, and showing lack of care and attention when approaching pedestrian areas, results in thousands of deaths each year at the hands of irresponsible motorists. Some cities have instituted speeding violation surveillance systems and intersection cameras to detect those who run the lights. Some of the more sophisticated systems can sense the sounds of a crash and store the previous several minutes of video in order to evaluate what happened to cause the accident.

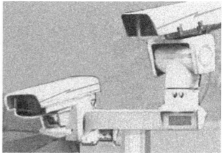

Powerful, zoom-enabled, swivel-mounted CCTV-based surveillance camera systems have been installed on many public sidewalks and intersections in the U.S. and the U.K. These are intended to monitor accidents and to reduce street crime. Traffic advisory cameras have also been installed on interstate highways. [Classic Concepts ©2000, used with permission.]

Border Patrol

The U.S. Customs and Immigration services were some of the first agencies to make regular use of visual surveillance technologies to patrol borders and inspect cargo and luggage to make sure they complied with import restrictions and safety guidelines. Over the years, they have also equipped aircraft and marine vessels with a number of visual, infrared, and ultraviolet sensors to aid in patrolling coastal areas in various types of weather and conditions of light or dark.

Many border checkpoints are equipped with cameras that monitor vehicles as they pass through Customs and Immigration checkpoints. Inside the booths, computers are used by officials to collect and verify border traffic statistics. [Classic Concepts ©1999, used with permission.]

Customs has installed a number of visual surveillance systems that can read the license plates of vehicles passing a checkpoint and automatically relay law enforcement information that may be relevant to the computer terminals of inspectors in nearby booths or buildings.

Customs officials sometimes use a high-tech 'optical viewing instrument called a *fiber-optic scope*. This is similar to instruments in medical diagnostic equipment to probe tiny areas that are difficult to reach any other way.

Optical fibers are tiny filaments that can readily transmit light and bend around corners.

They are well suited for instruments that must slide into tiny openings and otherwise inaccessible pockets or compartments. In customs work, they are used to probe inside vehicle walls, gas tanks, and compartments. In other aspects of surveillance, optical fibers are also used to transmit computer data or broadcast programming (e.g., digital cable).

Left: Visual inspection is frequently used by customs agents to check cargo or may be used to confirm the detection of anomalies flagged by high-tech devices. Right: The inspection technologies used by customs agencies are so varied, some of them are difficult to categorize. This image shows an instrument used by U.S. Customs to measure density variations. Since many smugglers create false walls or floors in their vehicles, a density monitor with a visual readout can help detect abnormal variations that might indicate hidden articles. [U.S. Customs news photos by James R. Tourtellotte, released.]

6.b. Home and Family Safety

Domestic Help Monitoring

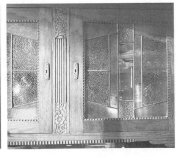

Some of the common household objects in which people hide pinhole cameras (not all of which are pictured here) include potted plants, books, stuffed animals, smoke detectors, thermostats, cabinets, keyholes, stereo systems, lamps, picture frames, speakers, pencil sharpeners, and clock radios. [Classic Concepts photos ©2000, used with permission.]

With costs dropping, small cameras are now being purchased by home consumers, usually to watch their infants and children and sometimes to watch aging parents who need extra care. They have also been used on occasion to monitor contractors and house cleaners.

Cameras installed in private homes have uncovered the disturbing fact that some contractors and housekeepers steal from their clients. Money from drawers, piggy-banks, jars, and closets, and small-but-valuable items such as jewelry are most often taken. Even when confronted with their actions, most of the thieves deny the thefts. They don't usually admit their actions until they are shown the tapes or the money is found in their possession.

Pinhole nannycams have disclosed that a small percentage of caregivers are appallingly abusive. Video recorders have shown that there are nannies who regularly hit children, neglect them, or drag them without considering their safety and well-being. While the incidence of this type of abuse is low, it shouldn't happen at all and parents have been motivated to record the conduct of these workers to make sure their children are safe. Surprisingly, many of the abusive caregivers caught on tape had been informed in advance that they might be recorded and still engaged in this astonishing behavior.

Entry monitoring

Doorways and entries to amusement park rides and sports stadiums are often monitored to see who is entering and exiting and to make sure there are no unexpected obstructions to fire exits or entry to hazardous areas without correct attire or authorization. Entry monitoring is also an important part of surveillance on transportation systems, cruises, air flights, and train trips (see the X-Ray and Magnetic Surveillance chapters).

Employee Monitoring

Surveillance cameras are being installed in the workplace in increasing numbers. Some of these installations are to monitor hazardous areas where there might be heavy machinery, chemicals, or the possibility of radioactive contamination. However, some employers are also monitoring office areas, staff rooms, dressing rooms, and washrooms. These activities are questionable, particularly when the management isn't similarly scrutinized. Any employee being monitored should be explicitly notified and the method and schedule for archiving made known to those who are being taped. Tight security should be maintained on archive tapes, so they don't get copied or put into the wrong hands. There should be restrictions on how long an employer may keep the tapes as well as regulations that prevent an employer from duplicating or selling the tapes to outside parties such as insurance companies or marketing agents. There should also be guidelines on how long recordings can be kept after an employee has left the company.

Hazardous Area Monitoring

People seem compelled to break rules. If a popular hiking area is closed due to avalanche danger, a dozen people a day will hop the fence anyway. When dangerous cables or electrical systems or bridges are posted with No Trespassing signs, children and teenagers will still play in and under them. Unfortunately, search and rescue operations are often financed by tax dollars, pranksters may be injured or killed, and people breaking the rules will sometimes surprise everyone by suing for their injuries even though they clearly broke the rules. For these reasons, surveillance cameras are often aimed at hazards, to avert possible tragedies.

Unfortunately, surveillance cameras are only as good as the people who maintain them. In a 1999 accident on a popular tourist bridge near Vancouver, B.C., it was reported that two surveillance cameras were monitoring the bridge when a mother unexpectedly dropped her infant daughter almost 200 feet into the edge of the riverbed below. The baby's life was saved only because the fall was broken by trees and foliage below the bridge. Law enforcement officials were having difficulty determining if the mother's actions were deliberate or accidental. It was discovered that the surveillance cameras were apparently not functioning at the time of the fall, and the only relevant footage was shot by a tourist moments before the accident. In this case, surveillance photos might have helped authorities determine the facts of the case.

Munchausen Syndrome by Proxy

Munchausen Syndrome by Proxy is an illness in which a parent or caregiver abuses a sick or healthy child by endangering his or her health. The person may even be trying to murder the child (most often through poisoning or smothering) but may deny the fact to others and sometimes even to him- or herself. The parent with the syndrome typically claims to be concerned about the child's welfare and may appear to be caring for the child. The parent may exhibit other types of mental illness and may engage in frequent fabrications. The child may be in a home, a social agency, or a health care facility. Since the person with the syndrome is often adept at hiding the abuse, it is sometimes difficult to establish whether indeed Munchausen Syndrome is a factor.

In most of the news coverage and studies on Munchausen Syndrome, the focus has been on mothers; however, it has been reported that fathers also carry out the abuse [Meadow, 1998].

In the past, the primary means of confirming the syndrome has been to remove the child from the care of the parent and see if the child's health improved. Obviously, there are many situations where this may not be sufficient evidence to establish the syndrome and it may be difficult to find the initial justification to remove the child from the parent. Since proving the syndrome can be difficult, social agencies and health care facilities are now installing video surveillance systems to determine whether the parent is a danger to the child. Sometimes two cameras with different vantage points are necessary, since the parent may have a back turned to the camera.

6.c. Space Surveillance

Left: An aerial view of 'Science City', which includes the Maui Space Surveillance Complex atop a dormant volcano on Maui in Hawaii. Space telescopes are generally located away from cities in locations where there is less interference from light and air pollution. Right: The Maui Space Surveillance Complex, Detachment 3. Note the numerous telescopes of different sizes. [U.S. Air Force news photos, released.]

Amateur, commercial, and military surveillance of outer space has yielded an astonishing amount of information about our origins and the billions of galaxies and solar systems that were once beyond the reach of our technology and almost beyond the reach of our imaginations. The continued evolution of telescopes and the launching of the early space vehicles prompted the development of both ground and space-based space surveillance strategies and complexes.

The 18th Space Surveillance Squadron operates the *Maui Space Surveillance Complex* on top of a dormant volcano on the island of Maui. The contracted staff members detect, track, and identify space objects with an electro-optical deep space surveillance system (GEODSS) and the Maui Space Surveillance System (MSSS). The Space Object Identification data are supplied to several agencies including the Space Control Center, the 1st Command and Control Squadron Control Center, and the Combined Intelligence Center in Colorado.

Closeup views of telescopes within the Maui Space Surveillance Complex. On the left is a 1.0-meter telescope. Specialized optics and highly polished precision mirrors are used in many aspects of telescope design. A new infrared capability is being built into the Keck telescope. [U.S. Air Force news photos, released.]

6.d. Media Applications

Journalism

Gossip journalists track many people who don't want to be photographed using long lenses and covert surveillance technologies. There has been a lot of controversy over this, particularly when the journalists shoot compromising private photos of people and their families in and around their homes or vacation getaways.

Investigative journalists often prepare news reports on people who don't want to be photographed. In some cases the unwilling subjects don't want their privacy and safety compromised. In other instances, journalists have been known to uncover evidence of criminal activities and have made law enforcement agents aware of the situation. Thus, journalists make use of visual technologies not only for regular community features and articles, but also for investigative reporting and covert investigations of criminal suspects.

6.e. Scientific Research

Atmospheric Studies

Atmospheric physicists use specialized cameras to study phenomena called *sprites,* fleeting light phenomena that tend to appear in the mesosphere during lightning storms, with night-vision (image-intensified) cameras that can shoot from 1,000 to 6,000 frames per second, compared to a conventional 16 fps camera [Baker, 2000]. By using the cameras in conjunction with signals from Global Positioning System (GPS) satellites, the phenomena being observed can be located. Clearly these specialized high-speed, low-light cameras would be suitable for many different types of surveillance activities.

Marine Studies

Underwater cameras are widely used in diving, and marine studies of creatures and plants and their habitat. Since water and delicate instruments rarely mix well, special waterproof cameras or waterproof camera housings are typically used. Earlier in the chapter, a number of camera-equipped diving vessels and suits were illustrated.

While not every part of the Earth's terrain has been viewed on foot, scientists have certainly endeavored to inspect every aspect of the Earth's terrain with cameras installed in airplanes, helicopters, and satellites.

Terrain Studies

Environmentalists, geographers, paleontologists, cartographers, agronomists, mineral and oil prospectors, miners, researchers, and military strategists are interested not only in Earth's terrain, but in the terrain of other planets as well.

Purdue University professor Jon Harbor and two students (Luke Copland is shown here) traveled to Switzerland to gather video images from beneath a glacier. The special, waterproof video camera provided a detailed view of network channels, streams, and changing ice structures. A high-pressure hot-water drilling system was used to drill vertical boreholes through the Arolla Glacier near the Matterhorn. The holes ranged in depth from 100 feet to over 450 feet. [Purdue 1996 news photo, released.]

6.f. National Security and International Peace-Keeping

Remote Administration

Videoconferencing technologies are now used in many aspects of military communication. They provide a visual/auditory link for meetings, tactical discussions, personal calls for service members deployed abroad, and long-distance enlistment.

This video camera and monitor form part of a videoconferencing system aboard a U.S. aircraft carrier. Here it is being used for an administrative ceremony, in which P.O. 2nd Class David Lee, Jr. is re-enlisted while onboard the carrier in the Persian Gulf. The ceremony itself was conducted remotely at the *Navy Command Center* in the Pentagon. [U.S DoD 1997 News Photo by Brian Fleske, released.]

Operations Monitoring

Cameras are used extensively by the U.S. military to document operations and exercises. Here, a U.S. Navy Photographer's Mate uses a Hi-8 mm video camera with an extended lens to record activities on an aircraft carrier deployed in the Persian Gulf. [U.S. DoD 1997 news photo by James Watson, released.]

Visual surveillance technologies are now routinely used for recording military operations, equipment, and personnel. This facilitates the review and revision of exercises and operations, provides accountability, and a historical record of events.

Surveillance photos can help assess the effectiveness of commercial or military activities after some action has been taken. They can record mine explosions designed to release more ore, controlled burnings intended to prepare land for crops, or reconnaissance imaging to assess the effect of bombing missions. These U.S. *Department of Defense* 1998 aerial photos are reported by the DoD as being bomb assessment images. They include the *Tikrit Radio Jamming Station* (left) and the *Al Basrah Military Cable Repeater Station*, both in Iraq. Buildings reported by the DoD as being destroyed in Dakovica, Kosovo are shown here before and after the bombing mission. [U.S. DoD 1998 news photos, released.]

Aerial Surveillance

See the chapter on aerial surveillance for a more extensive treatment of this application. (A large part of aerial surveillance is carried out with infrared technology; see the Infrared Surveillance chapter for additional information).

Aerial surveillance from and of aircraft are accomplished with a variety of devices, from basic scopes to sophisticated digital computer imaging systems. Left: Cpl. O. Villarreal of the 2nd Low Altitude Air Defense Squadron uses binoculars to spot aircraft in a multinational tactical air defense exercise at the NASA White Sands Test Site. Right: On board an aircraft carrier Intelligence Center, Intelligence Specialist Rasch (IS2) performs imagery interpretation, photometrics, and analysis of Tactical Air Reconnaissance Pod System (TARPS) film. [U.S. DoD 1996 news photos by Benjamin Andera and Daisy E. Ferry, released.]

Aircraft tactical reconnaissance pods are designed to house a variety of surveillance equipment. Visual systems are often supplemented with infrared sensors to provide imagery in darkness or bad weather conditions in low altitude assignments. Typically these pods are bullet-shaped units of about 1200 pounds that are centerline-mounted under the belly of tactical aircraft. They typically include several kinds of cameras, recording devices (e.g., tape recorders) and, in some of the more recent systems, realtime or near-realtime radio-frequency downlink technologies.

Hostile Territory or Political Climate

Left: A U.S. Army private of the 4th Chemical Company dons chemical gloves during a nuclear, chemical, and biological skills challenge held in the Republic of Korea. Right: The gas mask and goggles are to help protect against dangerous nerve agents (pills are sometimes also given to help protect against nerve agents). Environmental suits allow people to enter and observe in hazardous areas. [U.S. DoD 1998 news photo by Steve Faulisi, released.]

Virtually all visual surveillance technologies and associated gear, environmental suits, and strategies are used for monitoring hostile territories or nations at one time or another. Many of these technologies, particularly environmental suits, are not so much surveillance devices in themselves as they are a means for human beings, with their excellent visual acuity and problem-solving skills, to enter unusual or unfriendly environments.

Compliance Monitoring and Enforcement

Left: Documentary photos of animals that have been injured or killed by industrial accidents have prompted tighter regulations. This unrecognizable animal is a sea otter that died as a result of oil from the Exxon Valdez spill. Since that time ship hull designs have been improved and additional safety procedures implemented. Middle and Right: Photos can help document illegal products created from endangered species that are killed by poachers. This chilling photographic evidence of glue made from the bones of endangered tigers illustrates the lengths to which resellers will go to make a profit from precious and fast-disappearing resources. There are less than 5,000 remaining wild tigers on the entire planet. In human terms, that's barely enough to populate one small village. [U.S. Fish and Wildlife Service news photos, released.]

There are many ways in which surveillance is used to safeguard public safety and enforce laws. Wildlife management requires constant monitoring to ensure compliance with licensing, limits, and endangered species laws. Sensitive areas need to be protected from construction, mining, and trampling by hikers or tourists. Treaty agreements are monitored and coastal offsets are patroled on a regular basis. Large industries with the potential to pollute are regulated and monitored by law enforcement officials and environmental organizations. Visual surveillance strategies and technologies form an integral part of these processes. Gear for harsh environments or hazardous areas sometimes are used with the cameras to inspect places that might otherwise be inaccessible.

6.g. Emergency Services

There are two main trends in the emergency services industry: 1) 'visual-911' systems, that is, emergency call boxes equipped with video surveillance cameras and 2) CCTV systems to record the activities of emergency deployment teams. Emergency teams are sometimes recorded by third-party news agencies or production companies, with the footage sometimes aired on television or the Internet.

Visual surveillance technologies are intrinsic to search and rescue operations. Scopes, night-vision goggles, illuminators, aircraft, and imaging devices are all important in assessing terrain, movement, the presence or absence of people, damage, chemical leakages, and structural integrity. These tools are often used in conjunction with other surveillance technologies including infrared, X-ray, radar, robots, and tracking dogs.

Search and rescue operations take place in many kinds of marine, forest, mountain, and urban environments. In cases where a missing or injured person is suspected of being a victim of foul play, law enforcement and intelligence agents may work in cooperation with search and rescue personnel. This is important, since the very act of seeking and retrieving individuals may disturb a crime scene in ways that could interfere with an investigation. This is particularly true in homicides, kidnappings, and bombings.

Left: FBI agents, fire fighters, and FEMA Urban Rescue Task Force members work together in 1995 to carry out a visual search to find survivors and clues in the Oklahoma City bombing. Right: Protective gear, including hardhats and dustmasks, protect workers looking through a collapsed structure in Puerto Rico. Disaster situations must be carefully administrated so rescue workers who help victims don't inadvertently damage evidence that might reveal clues about the cause of the disaster. [Federal Emergency Management Agency news photo, released.]

Interagency cooperation often happens in cases of searches for missing persons (who may be ill or injured) and runaways (who may cross state or national borders). Border patrol agents have been known to supply equipment, air surveillance, and tracking expertise in locating missing persons.

6.h. Commercial Products

Pinhole Stationary Video Cameras

Video surveillance technologies considered to be a growth industry. The sale of security cameras is billions of dollars annually.

Cameras the size of a quarter can be hidden in many common consumer products. They can be attached to a cable that interfaces with a VCR or other recording device, or they can be attached to a wireless transmitter (usually broadcast FM, 900 MHz, or 2.4 GHz frequencies) that sends the signal to a remote recording device. Most consumer wireless pinhole cameras have a range from about 100 to 300 feet with a few that will transmit to about 1,000 feet. It should be remembered that anyone in the vicinity, like next-door neighbors, with a receiver tuned to the same frequency can also intercept wireless broadcasts.

Pinhole cameras can be purchased in three basic configurations:

> *board level* - This is a circuit board integrated with the camera that comes without a housing. It can be self-installed by individuals with sufficient technical knowledge. Often the boards have not been FCC approved for resale or use outside of hobby or educational settings.

> *encased* - This is a camera installed inside an indoor or outdoor casing. The long narrow casings are known as *bullet* housings. These consumer products are usually, although not always, FCC approved.

> *hidden* - This is a 'consumer-ready' camera that is already built into a consumer electronics device, home furnishing, or other host. Common examples include VCRs, clock radios, clocks, teddy bears, paintings, pencil sharpeners, speakers, pagers, smoke detectors, telephones and other common items that are typically in a position to survey a room without attracting attention. If you don't know where the device is, it can be very difficult to detect by sight alone. Some of the hidden cameras are in dummy devices (ones that don't work) in which the covering is just a shell for the camera. Others are hidden in working devices (clock radios, clocks, etc.).

Pinhole Portable Video Cameras

Portable video cameras have the same basic capabilities as pinhole stationary video cameras. Their resolution and features are similar. The main difference is that they are battery-operated to allow them to be carried around. They are usually built into wearable items such as tie tacks, watches, pens and belts and carryable items such as pens, pagers, cell phones, etc.

Most portable pinhole cameras are used with Hi-8 mm or DV recording decks since the decks are compact enough to fit in a briefcase or purse. Some of the tiniest recording decks can be body-worn, but are more expensive.

Pinhole cameras have also been specially installed in sunglasses so they can be unobtrusivey worn while recording. They typically link to portable recording units that are worn elsewhere on the body or connect with body-worn transmitters worn on the hip or in a pocket. The ones with transmitters can send the video signal to a monitor in another room or a recording system offsite in a van or another building nearby. These range in price from $700 to $3,700.

Body-worn cameras can be detected easily by metal detectors or pat-down searches and thus are vulnerable to discovery. The battery life is limited and the ones with transmitters can be picked up by anyone with an appropriate video scanner. Nevertheless, there are situations in which they are effective surveillance devices.

Two Varieties of Wireless Miniature Cameras

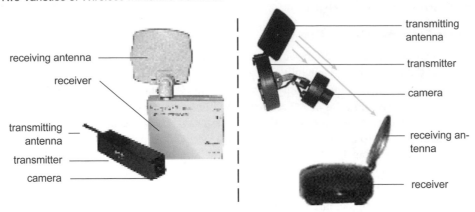

receiving antenna

receiver

transmitting antenna

transmitter

camera

transmitting antenna

transmitter

camera

receiving antenna

receiver

There are a variety of designs for wireless pinhole cameras. On the left is a 2.4 GHz wireless NTSC 0.5 lux color camera that weighs less than 50 grams and includes a wireless transmitter inside the dark casing that is about the size of a screwdriver handle. It can transmit with a range of about 100 feet to the 100 gram Navigator® receiver. The suggested retail price is $400. On the right is a different design approach. The XCam2™ NTSC color 'bullet' camera with audio has a directional 2.4 GHz wireless 100-foot range transmitter attached to the base of the camera. It is rated for outdoor use with a suggest retail value of about $300. [Classic Concepts photos and illustration ©2000, released.]

Miniature still cameras, often called *spy cameras,* are frequently associated with secret agent movies and, indeed, they have been used for decades for clandestine photography. The smallest of the spy cameras are the *subminiatures* which are remarkably small, about 3" x 1", not much bigger than a matchcase.

Public Security Cameras

Optech Integrated Systems has installed more than 100 of their Passenger Help Point Systems in rail link centers in London and selected London Underground stations since 1987. These are, in essence, visual 911 centers. Passengers needing information or emergency assistance can activate an alarm button on the console that switches on a video camera that is aimed at the caller. The emergency or dispatch staff at a central control center can then see the caller and communicate with him or her while viewing a map of the area on a computer terminal. Depending upon the nature of the call, they can send out whatever assistance might be appropriate. The communication is automatically recorded in much the same way that 911 phone calls are currently recorded to tape.

Specialized Cameras

Many aspects of surveillance require special cameras. Surveillance often occurs in low light, or may require special filters for sensing infrared or ultraviolet. Sometimes high-speed photography is used to capture fast events or events that need to be viewed in slow motion.

The U.S. Department of Energy and Sandia National Laboratories *Albuquerque Full-Scale Experimental Complex* (AFSEC) provides a number of specialized photographic technologies including high-speed and ultra-high-speed photography, image-motion photography, still, time-lapse, and Schlieren photography. These are especially useful for scientific research, reconnaissance, and forensic investigations.

Very fast frame rates can be used to photograph explosions, projectiles, fast-moving aircraft, and natural phenomena for later analysis. Schlieren photography can be used to record shockwaves associated with supersonic events.

AFSEC Photometrics products include high-speed and ultra-high speed cameras that can produce between 40,000 and 26 million frames per second. This selection from a series of images of a bullet impacting an armor plate was captured as the bullet travelled 2,750 feet per second. [U.S. DoE and Sandia National Laboratories news photos, released.]

Video Intercoms

Video Intercoms are essentially the same concept as the more familiar audio intercoms. A small intercom camera unit can be located by a door, using power from the existing doorbell outlet and will transmit images to a receiver elsewhere in the house either by a wired connection, or a wireless connection. The video image area is quite small. They sell for around $400.

Vehicle Environment Camera/Recorder Systems

Trucks and cars can now be equipped with cameras that allow drivers to view their blind spots, especially behind the vehicle or the back right-hand side. Others are designed to monitor and record events happening all around the vehicle so that if some important event or accident occurs, there will be a record of the event. Docudrive has such a system for $5,000. The system records the immediate environment of the vehicle and stamps the video images with the date, time, and other information.

6.i. Display Devices and Accessories

Monitors, Monitor/Receivers

Most video surveillance signals are either displayed in realtime or recorded and displayed in continuous mode or time-lapse mode. Display monitors are somewhat generic, so it is often possible to display a video signal on a variety of devices ranging from consumer TVs to video monitors, to computer monitors. Monitors with built-in receivers operate in a variety of frequencies, including UHF and 900 MHz.

Rows of security monitors enable National Guard Pfc. Ellington monitor the landscape at the U.S. Border Patrol Communications Center near Yuma, Arizona. [U.S. Army news photo by Greenhill, 2006, released.]

Multiplexing or Splitting Units

In surveillance, it is common for more than one video source to be monitored at the same time and it is convenient to display a large number of camera inputs on a smaller number of display monitors, especially if the viewing room is small. A multiplexer allows multiple inputs to be displayed on a single monitor. A splitter allows a particular input to be sent to more than one monitor. Multiplexers are commonly used for surveillance. Unless you have large, high-resolution monitors that can display up to about 6 or 8 scenes, it is usually advisable to limit a multiplexed display to four views. In other words, four camera sources can be multiplexed onto one screen and, with a good multiplexing system, individual inputs can be displayed full-screen as well, if desired, at the touch of a switch or as set by a timer.

Time-Stamp Generators

A time-stamp is often an important aspect of surveillance videos, especially those used for forensics. Most camcorders allow the time and date to be preset and then 'stamped' on the image as it is being filmed. Live camera images that are sent through a cable or a video transmitter to a VCR can sometimes be time-stamped on the VCR (depending on the system), or can be passed through a specialized time-stamp generator, a component that resembles a switcher. Some time-stamp generators are also equipped to superimpose the camera position onto the video image since there are now cameras that can swivel and pan to capture video from many directions. Since time-stamp-generating components and VCRs are usually specialized for surveillance purposes, they may also have autostart and timed-recording mechanisms.

Time-stamp, autostart VCRs range in price from $200 to $1600 and peripheral time-stamp components range from about $50 to $120.

6.j. Recording Systems and Devices

Event Timers, Time-Lapse VCRs

Recording for long periods of time is important for many types of surveillance activities. Retail stores don't want to have to change tapes in dozens of machines every six hours, as is typical for consumer VCRs on extended play modes. Time-lapse VCRs commonly can record from 24 to 960 hours for about $500 to $1800, depending upon the length of the recording, the frequency of the frames and the quality of the image. Black and white time-lapse VCRs typically feature higher resolutions and lower prices than color systems.

Portable/Covert VCRs

Body-worn spy cameras that are incorporated into sunglasses and tie-tacks require either a wire or a tiny transmitter to send the image from the source to a recording or display device. Thus, video recorders that are intended to be used with a wired body-worn cameras are designed to be small enough to fit in a large purse or small briefcase. Hi-8 mm is a very high resolution video format in a small package, and thus is very convenient for use with portable recorders. Some of these VCRs have time-lapse capabilities, though most are continuous-recording devices. They range in price from about $1,100 to $2,000 and weigh about three pounds with the tape and battery.

A normal three- or four-pound camcorder can sometimes be used with wearable cameras. By leaving the lens cap on the camcorder (primarily so the lens won't get scratched) and running RCA cables from the camera to the camcorder and then setting the camera, if necessary, so that it accepts input from the line rather than from the lens, it can be used as a recording device with a pinhole camera.

The biggest problem with using consumer camcorders for surveillance is that many of them have a *power-saver* mechanism that shuts off the system to conserve power if the camcorder is not used for a few minutes. In some cases, this feature can be disabled, but not always. Since it is hard, in surveillance activities, to anticipate what needs to be recorded and when it needs to be recorded, it may be impractical to check the camcorder to see if it has powered down automatically or is ready to record. However, there may be situations where some kind of recording is better than none, with a camcorder filling the gap.

Intranet and Internet Remote Video Surveillance

Organizations that have a number of branch offices or a number of people who work off-site sometimes use remote video surveillance to monitor their employees and contractors. A head office in Chicago, for example, could monitor branch offices in Texas, Germany, Florida, and Washington in realtime or near-realtime over the Internet. Systems designed with these capabilities typically require certain basic components, including

- a digital or analog video camera at each station,

- an interface box that handles the input from several cameras or a personal computer or interface box associated with each camera,

- a fast Internet link that can be accessed at prearranged times or that is live all the time,

- interpretation software at the receiving end, and

- some type of storage and display system (usually a computer).

Many of these network video systems work on a time-lapse basis, transmitting an image every few seconds rather than continuously. This saves computer processing power and transmission time.

Fast connections work best for these applications. The connections can be through a variety of services, including ISDN, DSL, satellite, or cable modem. Slower, but still useful connections, can be achieved through 33K or 56K fast modems through normal phone lines. Some systems operate with dedicated software, others use Web browsers to display the camera images which may or may not be enhanced with Sun Microsystem's Java programming routines.

Some Internet video connections use software or hardware motion detectors to trigger alerts if an event happens at one of the remote camera stations. Currently, hardware motion sensors work better. Computer software motion sensors that process the data from the video cameras tend to have false alarms from changes in light or other events that aren't relevant.

7. Problems and Limitations

Visual surveillance is such a large and diverse field that it's difficult to make generalizations about limitations of the various types of systems. Price used to be a limitation of visual surveillance devices; digital cameras and high quality video cameras and recording devices cost thousands of dollars each a few years ago, but this is no longer true. There are now many good quality options for under $1,000. However, it is still true to some extent that the more expensive systems provide better quality images, longer recording times, and better processing capabilities.

Resolution and Image Clarity

Many older surveillance systems suffered from low image quality, but newer systems are remarkably good and eventually the older systems will be replaced by crisper, smaller, more efficient systems. Hi-8 and digitial video (DV) formats are preferred over VHS for most surveillance applications, though many retail outlets still use VHS. Super-VHS offers superior image quality but requires larger cameras and decks than Hi-8. Compact VHS (VHS-C) offers an option for smaller, portable camcorders, but doesn't provide the high image quality or long recording times of Hi-8.

Digital Versus Analog Technologies

The new digital video and still cameras are convenient in many ways. They are small, relatively easy to use, and the memory cards can be swapped in and out more easily than film can be changed. They work in reasonably low light conditions. The still cameras usually store more images than traditional film cameras (50 to 200 images versus 20 or 36 on film).

The most convenient aspect of digital cameras is that there's no film to take to the processing center and no wait to see the images. It takes less time to link a digital camera to download images to a computer than it does to process and scan photos or film negatives.

However, digital technologies also have some disadvantages. Except for the most expensive cameras, the resolution of digital images is not as high as film images. Most digital movie cameras in the consumer price range do not offer the options of interchangeable lenses that are currently available for film cameras (this will probably change as prices drop and digital camera options improve). For forensics, one of the biggest problems with digital images is that they are not 'tamper-proof'. Courtrooms are reluctant to accept evidence that can be changed without any sign of tampering, as is possible with digital photography. There are companies that have designed video encryption systems for international monitoring and it is possible that this technology will eventually be built into digital cameras for forensics work so that digital images can be effectively used as evidence in courtroom trials.

Recording Times

Probably the biggest limitation of video surveillance is the limited recording times for tapes or memory for digital images. In general, consumer tapes hold about two hours of realtime video in regular play modes and up to about six hours of video in extended play modes. For 24-hour, seven-day-a-week recording applications, this is obviously inadequate as someone would have to be on hand to constantly change the tapes. If 20 cameras are in operation, this would mean changing tapes 80 or more times per day. Time-lapse recording systems are designed to greatly extend recording times, to as much as 20 hours per tape and sometimes even up to 960 hours on high-end systems. There is a trade-off in terms of image quality and there may be small events that go unnoticed because they happen between the moments when the system is recording but, generally, time-lapse video is a good compromise in terms of price and convenience. Newer digital video systems may use gigabyte and terabyte hard drives and image compression schemes to record a large amount of video, more than can be saved on some time-lapse systems. As the price of computer hard drive storage continues to drop, huge hard drives will probably be favored over tape systems. At the present time, many systems still rely on tape, for reasons that have been discussed in earlier sections.

8. Restrictions and Regulations

When a new product or technology is introduced, it will often be distributed for a while without legal restrictions. If responsible use of the product shows that it poses no danger to health or safety, then unrestricted distribution may continue. However, sometimes new products are used by individuals in ways that impact the lives of others and people will want to establish restrictions or penalties for their improper use. Drivers' licenses weren't established until inexperienced drivers had been involved in many accidents. Motorcycle helmets and seatbelts weren't required until accident statistics showed that they reduced fatalities. Automotive radar detectors weren't illegal when they were first introduced. For each new technology or product, there is a period during which the public and the lawmakers gradually become aware of the product and its impact on their lives. During this 'ramp-up' period, citizens have a choice as to whether they will use the technologies in responsible or irresponsible ways. If they are used in irresponsible ways, innocent victims who may be affected by their conduct will often lobby Congress to regulate the technologies.

When film and video cameras were first introduced, people used them to photograph special events and friends and families. Employers used them to monitor construction yards and hazardous environments. Abuse was not prevalent and the use of the cameras went unregulated for several decades. However, now that tiny video cameras are inexpensive and recording devices widely available, people are starting to use video cameras in questionable and sometimes highly irresponsible ways. Some blatant abuses of privacy and decency have occurred over the last three or four years. Employers have been putting hidden cameras in washrooms, without informing their employees, and Internet entrepreneurs have taken covert nude shots and uploaded them to the Internet, charging money for them without the subjects' knowledge or consent. These types of actions have provoked concern among citizens and privacy advocates, who are now asking lawmakers to step in and provide protections.

One of the first laws to handle an intrusion into a person's privacy was put in place in 1903. It became known as personality rights, the right to control the use of one's name and image for commercial purposes. Essentially it established property rights to one's persona. There have been many debates over this protection, particularly since the news media want the power to report on individuals and their activities. At the present time, the use of images of public personalities in documentary news reporting is permissible.

In the 1960s, searches and seizures were more stringently regulated to protect a person from arbitrary and unwarranted intrusions. Prisoners were considered to be exempt from these regulations, however.

With the increase in travel, communication, and computer networks in the early 1980s, the international community began discussing privacy regulations. The *Protection of Privacy and Transborder Flows of Personal Data of 1980* is a set of international guidelines governing the flow of personal data between countries.

Now that interactive TV is becoming a reality, it is technologically possible for broadcasters to monitor their subscribers and they may eventually want to supply small TV-top video cameras with their cable subscriptions that allow them to personally communicate with and visually monitor television viewers. However, some people are concerned about the consequences of using the technology in this way. They are particularly concerned with the safety of their children and the security of their homes. As a result, the *Cable Communications Policy Act of 1984* was put forward to prevent interactive cable operators or third parties from monitoring the cable consumers' viewing or buying habits. It also prohibits the collection of personally identifiable information without the consumer's proper consent, except as may be needed to render a service to the consumer as provided by the operator. The cable operator is required to provide notice, in the form of a separate, written statement to such subscriber, that clearly and conspicuously informs the subscriber of the use of any information that may have been collected.

Current Regulations

Peeping Tom laws are intended to protect us from prying eyes, so most people assume they protect us from prying video devices. However, many states don't have Peeping Tom laws. If someone watches you undressing in the shower through a bathroom peephole, they might be charged in a region that has Peeping Tom laws. However, if the same person videotaped you undressing in a public dressing room, it may not be possible to lay charges against them due to the wording of the law, or as we call it the *letter of the law*. Private citizens often assume laws are based on logic, when in fact, much of the interpretation of law is based on definitions and precedents. Here is an example:

> *In the 1970s, a male prostitute was apprehended on charges of prostitution with ample evidence to make a conviction. The legal definition of a prostitute in that jurisdiction stated explicitly that a* prostitute *was a* woman. *Since he was a man, he could not be considered a prostitute according to the law and he was freed.*

There are many laws like this that are written to reflect a specific social attitude or technology at a particular point in time. When new situations come up that reveal the limitations of laws, they may be revised, but it can sometimes take years to change a law. Many current laws intended to safeguard privacy don't protect us from new and evolving technologies like video surveillance cameras. Often, when a conviction *is* made, it is because other aspects of law are brought into the case. Here is an example of a situation in which the primary offense could not be prosecuted, but a secondary aspect was used instead:

> *A landlord installed a two-way mirror in the bedroom of an adjoining suite and rented the room to two female college students. He videotaped them in compromising situations. The tape was discovered and the man was charged. The prosecuting attorneys were unable to get a conviction based on the video tape evidence. They were, however, able to prosecute the landlord under laws related to wiretapping, based on the fact that he recorded sound on the audio portion of the tape.*

This specificity of laws creates a difficult situation for citizens, law enforcement agents, attorneys, and judges concerned with protecting community welfare in a society where technology is evolving and changing rapidly. It is complicated further when the Peeping Tom is aware of the 'letter of the law' and blocks the microphone, thus eliminating the audio portion of the tape in order to avoid prosecution.

It may surprise the reader to discover that in the workplace, the rules are even more lenient. Employers have very broad powers to protect their investments. Thus, it may be completely lawful in some regions to videotape employees anywhere, even in dressing rooms or bathrooms. Given that most employers currently have no legal obligation to let employees know they are being surveyed (this is gradually changing) and given that employees have no equivalent opportunity to visually surveil managers and upper-level executives, it is not only privacy that is at stake, the potential for double standards or discriminatory practices are great as well.

In general, employers currently may put cameras and other monitoring devices almost anywhere. In *Vega-Rodriguez v. Puerto Rico Telephone Company*, 1997, the company could monitor employees in open work spaces. However, there have been some cases in which videotaping of employees was found to infringe on the employees' rights, as in *Anderson v. Monongahela Power Co.*, January 1995, in which employees were videotaped in locker rooms by covert security cameras.

Advanced Swivel-and-Zoom Capabilities and Voyeuristic Temptation

Hidden or remotely-controlled cameras are becoming highly sophisticated. It is now possible to swivel and aim them from remote-control command center booths. Some camera systems are now powerful enough to zoom in on newspaper text held in the hand of a person across the street. If you doubt this, keep in mind that orbiting commercial satellites far above the Earth can resolve an object on the ground the size of a doghouse (military surveillance systems are even more powerful). Compared to that, reading a headline a few meters away no longer seems surprising.

Recent research has found that operators controlling swivel-and-zoom cameras designed to protect public streets and transportation systems have been taking liberties with the technology. Cameras installed on traffic lights, for example, are intended for monitoring intersections for traffic violations, but they are sometimes aimed down the blouses of pedestrians who are crossing the street.

Retailers, too, have made intrusions on privacy by installing cameras and two-way mirrors in dressing rooms, explaining that this is where most shoplifting occurs. That may be true, but it has also been shown that men are monitoring women's dressing rooms whether or not there is just cause to suspect the shopper might be shoplifting.

Privacy violations are not limited to men. Anyone staffing a security system has the opportunity to step outside the bounds of decency and professionality and observe people in inappropriate ways. However, statistically, there are more men than women monitoring security systems, and there are more women than men being watched through the systems with voyeuristic intentions. The one exception is with cameras monitoring high crime areas. In these instances, men are targeted more often than women, on the assumption that men are more likely to commit crimes than women. The higher percentage of men arrested or incarcerated in prisons is sometimes used as a justification, but it is not a good one, as it becomes a self-fulfilling prophecy. Discrimination at the basic level of observation for which the cameras are intended should be discouraged.

Due to these various problems, there have been calls for laws analogous to the *Electronic Communications Privacy Act* to protect citizens from video surveillance.

With the exception of certain FCC emissions requirements, the designing, constructing, and selling of video surveillance devices such as pinhole cameras and tiny recording decks are loosely regulated. The onus is on the user to use this equipment lawfully. There are many legitimate personal, commercial, and educational uses of cameras and restricting them completely is not practical or desirable in a free society. Just as it is incumbent on the public not to use a hammer to bludgeon a next-door neighbor, it is incumbent on the public not to use a video camera to invade another person's privacy.

Examples of Privacy-Related Bills

In 1890, a Warren and Brandeis article in the Harvard Law Review referred to Cooley's definition of privacy as the "right to be left alone" and argued that this was "the most comprehensive of rights, and the right most valued by a free people". This established, in a social and legal context, a boundary between the desire to observe and the right to not be observed. (Discussion of the debate over privacy rights can be cross-referenced in the *Introduction Chapter* and the *Audio Surveillance Chapter*.)

The following sampling of privacy-related bills from the state of Maryland helps illustrate the kinds of concerns lawmakers have about intrusion into privacy from visual surveillance devices, and also gives some insight as to how slow and reactive (rather than proactive) the judicial process can be compared to changes in technology.

House Bill 273 "Crimes - Visual Surveillance", Criminal Law - Substantive Crimes. Delegate Dembrow.

Broadening the application of provisions prohibiting the visual surveillance of a person in certain places; allowing certain damages to be awarded; and making it a felony to break and enter, enter under false pretenses, or trespass on any premises with the intent to place, adjust, or remove visual surveillance equipment without a court order.

Status: Mar. 1996 passed in the House; Mar. 1996 at hearing stage in the Senate. See House Bill 780.

House Bill 779 "Crimes - Visual Surveillance - Private Residences", Criminal Law - Substantive Crimes. Delegate Dembrow.

Prohibiting a person from placing or otherwise bringing, or procuring another to place or otherwise bring, a camera onto real property for purposes of conducting visual surveillance of persons inside a private residence on the property; providing that ownership is not a defense if the owner is not an adult resident or the resident's legal guardian; providing a defense for good faith reliance on a court order; providing penalties and exceptions; providing for a civil cause of action; defining terms; etc.

Status: Mar. 1997 passed in the House; see House Bill 170 for continuance.

House Bill 780 "Crimes - Visual Surveillance", Criminal Law - Substantive Crimes. Delegate Dembrow.

Prohibiting persons with prurient intent from conducting or procuring another to conduct visual surveillance within specified private places; providing penalties; providing for a civil cause of action; defining terms; providing specified exceptions; providing that the Act does not abrogate or limit specified other remedies; etc.

Status: Mar. 1997 passed in the House; Mar. 1997 Senate hearing. See House Bill 170.

House Bill 170 "Crimes - Use of Cameras and Visual Surveillance", Criminal Law -Substantive Crimes. Delegate Dembrow.

Prohibiting a person from placing or procuring another to place a camera on real property to conduct deliberate surreptitious observation of persons inside a private residence; prohibiting the conducting of visual surveillance with prurient intent in dressing rooms, bedrooms, and rest rooms in specified places; providing that ownership is not a defense; providing a defense; providing penalties; providing for a civil cause of action; etc.

Status: Feb. 1998 passed in the House; Apr. 1998 Senate hearing. See House Bill 95.

House Bill 95 "Crimes - Use of Cameras and Visual Surveillance", Criminal Law - Substantive Crimes. Delegate Dembrow.

Prohibiting a person from placing or procuring another to place a camera on real property to conduct deliberate surreptitious observation of persons inside a private residence; prohibiting the conducting of visual surveillance with prurient intent in specified private places; providing that ownership of the private residence is not a defense; providing that a good faith reliance on a court order is a complete defense; providing penalties; providing for a civil cause of action; etc.

Status: Feb. 1999 passed in the House; Apr. 1999, passed in the Senate. May 1999, vetoed by the Governor as duplicative (see Senate Bill 689).

Senate Bill 689 "Crimes - Use of Cameras and Visual Surveillance", Criminal Law - Substantive Crimes. Senator Forehand.

Prohibiting a person from placing or procuring another to place a camera on real property to conduct deliberate surreptitious observation of persons inside a private residence; prohibiting the conducting of visual surveillance with prurient intent in dressing rooms, bedrooms, and rest rooms in specified places; providing that ownership is not a defense; providing a defense; providing penalties; providing for a civil cause of action; etc.

Status: April 1999 passed in the Senate; May 1999 signed by the Governor.

(See the Audio Surveillance and Aerial Surveillance chapters for other privacy and surveillance regulations. The introductory chapter also has general information on privacy regulations.)

9. Implications of Use

9.a. Social Implications

As mentioned in the introduction to this chapter, the prospect of a world based on Big Brother surveillance was depicted George Orwell's famous novel *1984*, released in the aftermath of World War II. Orwell's classic is still required reading in many educational programs. However, it lost part of its clairvoyant mystique when 1984 came and went without the dystopian scenario coming to pass—people who valued personal freedoms and feared excessive government monitoring sighed with relief. Many concluded that Orwell was overly pessimistic.

Given the proliferation of technology since 1984, it is time to reassess. Orwell's primary theme was audio/visual surveillance of people's personal and business activities through systems that were installed with tax dollars by an overly controlling government. Surveillance technologies didn't become highly prevalent until 1994, but now they are increasing rapidly, not just in government applications, but in businesses and homes. We now spy on children, nannies, housekeepers, and employees on a regular basis. Satellite and aircraft surveillance of the entire planet is routine; government and community monitoring of criminals has increased.

Surveillance is no longer limited to audio/visual technologies. Vehicle traffic is monitored with radar; military personnel and criminals are subject to mandatory DNA testing. Violent offender lists are circulated through communities when convicts are released. Video cameras automatically match faces to individuals listed in databases in shopping malls. It could be argued that in some ways we are already surveilling people more closely than even Orwell anticipated.

So why did the Orwellian scenario frighten and enrage citizens twenty years ago, while only a small percentage of protesters now oppose these activities? What happened to the justified concern about the reduction of our privacy and liberty? There are several reasons why surveillance is becoming prevalent:

1. *When people are frightened, they more willingly give up their liberties.* Most people would oppose surveillance cameras in public washrooms, for obvious reasons, but if a series of assaults or child kidnappings were to occur in washrooms in airports or shopping malls, and cameras were installed to safeguard the safety of citizens, most

people would grudgingly or even willingly accept the technology rather than risk disfigurement, murder, or potential harm to their children.

2. *When private citizens have access to the same technologies as governments and corporations, they appear to be less threatened by the technologies.* People have a tendency to accept the familiar, even if it is potentially harmful. Because video cameras are in the hands of many private citizens, they may not immediately recognize the long-term consequences of surveillance. Surveillance by Little Brothers may be a bigger threat than surveillance by Big Brother but may not be perceived as readily as a threat.

The benevolence of 'equality of technology' is assumed rather than proven. In other words, private citizens may have access to GPS receivers and thus not feel threatened when governments are using them, too, but may overlook the fact that larger social organizations sometimes have wider legislative leeway than private citizens (e.g., higher resolution transmissions). As an example, citizens may have infrared cameras for ground-level photography, but the government usually has cameras with greater capabilities, such as aerial-mounted infrared cameras or regular video cameras on every street corner. This initial impetus for having street cameras might be to report traffic congestion or icy road conditions, information-gathering intended to benefit all, but the records might later be analyzed for other purposes, such as determining whether a car-jacking took place at the corner or whether a particular person passes by the corner at the same time every day.

With better technologies and higher-resolution cameras, there might come a time when a competent lip-reader could look at the images and interpret a private business conversation. If a lip reader can do that, then a computer program could potentially be designed to do the same thing. Currently there are few legal impediments to commandeering information gathered for one purpose to serve another. This is not to say that governments are the primary abusers of power and technology. The incidents of entrepreneurial voyeur sites that publish intimate images of people without their consent indicate that immoral and unscrupulous individuals who are regulated less than law enforcement officials are potential abusers of technology. The incidence of employee surveillance in corporations, without a balancing ability on the employees to surveil their bosses, is another area in which there is considerable potential for abuse.

3. *When change occurs gradually, a threat is not always recognized.* High school teachers often tell the parable about dropping a frog on a hot frying pan. When it hits the pan, the frog will jump out and try to get away, whereas if you put it in a pot of cool water and raise the temperature very gradually, it will overlook the danger and eventually boil to death. There are certainly precedents in human history that support the hypothesis that changes that occur slowly are less likely to provoke public protest. Visual surveillance, like the slowly rising temperature of the water, has insinuated itself into our society gradually. First there were hidden cameras in banks and department stores where shoplifting was prevalent. Now there are cameras in banks, many retail stores, homes, businesses, industrial complexes, private clubs, ATMs, schools, and amusement parks. Many people have videocams in their homes and offices that broadcast directly to the Internet. There are even people who have cameras in every room in the house.

9.b. Balancing Privacy, News Coverage, and Safety

To give an idea of how far some employers will go in visually monitoring their employees, some employers have sent detectives to videotape employees taking sick leave to see if they are actually sick. Many have put video cameras in private areas in their businesses where employees go on their breaks. It has been argued that employees have a right to privacy during break times and also that the employees' work can suffer if they are constantly being monitored and don't feel that they are trusted. Currently, it is legal in all but a few states for employers to place hidden cameras in locker rooms and even bathrooms.

There are some aspects of law enforcement in which video cameras do provide benefits. It has been found that response times to crimes increase in regions where patrol rates are low and where video cameras supplement patrols. It has been found that the rate of confessions increases when suspects see that their actions have been caught on tape. It has also been found that convictions and prosecutions that are supported with video evidence generally go more quickly and smoothly. However, not every kind of law enforcement video surveillance system has clearcut benefits. Some systems have been opposed and others have been removed after it was found that they were not as effective as was initially hoped, or were too expensive to maintain. Here are some examples that help illustrate the complexity of the issues:

- After twenty-two months of monitoring surveillance cameras in Times Square, it was found that the systems resulted in only 10 arrests. The cameras were later removed.

- A CCTV-based surveillance system was proposed by a Police Department in 1996. This was a high-resolution, swivel-mounted, zoom system that could capture movement and details at a distance of up to about a mile. After some debate about privacy issues and efficacy in reducing crime, the Police withdrew the recommendation for the camera systems to be installed.

- The city of Seattle established a privacy policy because of law enforcement surveillance in the 1960s and 1970s. This ordinance prohibits targeted surveillance of individuals or groups solely because of their political views. Those shooting footage of protesters were asked to destroy their tapes and those investigating the incidents were appalled that the videographers had not identified themselves while shooting the footage.

- A New York Civil Liberties Union volunteer canvas of surveillance cameras focused on public areas in December 1998 located 2,380 apparent cameras in the New York area, and it is expected that there are many more hidden cameras. Most of the systems are privately operated, secured to rooftops, building entrances, and lamp posts. Some are swiveling globe-covered cameras. Recommendations derived from this informal tally include the registration of cameras that are aimed at public places and limiting the duration of taping that may be stored. The NYCLU has published a map on the Web that shows the location of the identified surveillance cameras.

- In spring 1999, southern California officials were seeking to install street-based surveillance cameras aimed primarily at sidewalks and parking lots. It was intended that signs be posted to let people know they are being monitored. Cameras would be monitored by law enforcement agencies. After a trial period, if the program was successful, the goal was to expand the installation of the cameras. This was after a similar plan was rejected two years earlier.

- On Sullivan's Island in Charleston, an infrared camera points offshore to monitor boats that may be lingering for too long near the wreck of the Confederate submarine Hunley. Suspect boats are visited by the Coast Guard.

James Ditton, of the Scottish Centre for Criminology, who monitored cameras in Glasgow for four years, has cast doubts on the benefits of video surveillance. He reported that crime had indeed fallen in the surveilled area, but that the data had to be considered in conjunction with data from the non-surveilled areas, where crime fell even more. This raises not only questions about the cost-effectiveness of the video surveillance programs but also about how the effectiveness of assessing camera systems. If surrounding areas that don't have the cameras are not included in the research statistics, there is reason to doubt the validity of any reports of increase or decrease in crime.

In some cases, judges have been offering the opinion that public spaces such as parks afford no expectation of privacy, yet people have often used publicly accessible spaces for precisely that, as places to conduct private business or personal conversations, for lovers to spend personal time together, and for grieving people to find time to be alone.

News Surveillance

Over the last several decades, news photographers have become very aggressive in shooting video footage of private moments. The news media have a great deal of leeway in reporting the news and in exercising its rights of freedom of speech. However, some news photographers have stepped over the line by climbing over private fences and aiming lenses in the bathroom windows and backyards of celebrities and sometimes photographing their children and friends. Some of them live in their cars, following celebrities 24 hours a day, waiting outside doorways, hotels, and recreational establishments to catch images of the people they are 'staking out'.

> The American rule is that photographers may shoot whatever they can see while standing in a place to which they have the right of access. Nonetheless, courts have put limits on photographers who, at least in the courts' eyes, behaved badly in pursuit of individuals. This has happened with or without a finding that the behavior of the news people could have been enough to break a specific law, and even when the photographer stood in a public place.

> [Alice Neff Lucan, Existing Limitations on News Photographers in Pursuit of Individuals, Web article, 1998.]

In the late 1990s, TV news crews came under criticism from the public for accompanying police departments on their patrols. These news crews will sometimes wear police caps and vests, making it appear to the public that they are directly associated with the police and have the same authority. This causes a great deal of confusion for individuals dealing with the police. The rules for interacting with the police are not the same as those for news crews. If a police officer stops you while driving and tells you to roll down your window and step out of a car, you are legally obliged to comply. If a news agent tries to stop you while driving or order you out of your vehicle, you have no obligation whatsoever to comply. Similarly, if a news agent steps into a house with a video camera and begins taping while police officers are issuing a search warrant, the average citizen has no idea whatsoever what his or her rights are with regard to the video technicians and, in such a stressful situation, can't distinguish between a news crew member or a police officer if they are all wearing clothing labelled 'Police'.

This is not a hypothetical problem. This scenario has occurred in several regions in which news crews accompanied police on patrols and searches. Given that a person is considered innocent before being found guilty, there are also questions as to whether the news crews have the right to air privileged footage that is only available through the execution of the police search warrant, without actual proof of guilt. It could be argued that this contributes to defamation of character. The police have generally supported these news tapings to help the police educate the community about their activities and there are positive benefits from increased public awareness of police activities, but three things need to change: 1) news crews should be clearly distinguishable from the police, 2) video footage should not be shown until after a person has been found guilty or innocent of an offense, and 3) if the person is innocent, should not be be shown without informed consent from the person taped, as it could contribute to defamation of character and other negative consequences.

In other instances, news crews have taped hostile conflicts overseas, or situations involving domestic violence or hostage taking. Armed forces personnel and police have expressed concerns that live news coverage could endanger the lives of agents trying to handle volatile situations, and may actually affect the outcome of a situation. Unstable individuals can be provoked by the presence of cameras or, in some cases, may grandstand for the camera, carrying out acts of violence for publicity purposes that they may not otherwise have attempted. These factors must be considered by news personnel when covering sensitive stories and events.

In some cities, the news media have entered into voluntary agreements with police departments to restrict live coverage of terrorist and hostage situations and incidences in which unbalanced or distraught individuals have threatened violence or suicide.

9.c. Prejudice and Monitoring

We like to think that surveillance systems provide objective records of human behavior, but it has been found that the placement and use of surveillance cameras sometimes magnify the prejudices of those installing or operating the systems, or those selecting or evaluating the records. Several research studies have shown that people of color, young people, and males were disproportionately targeted as suspects of shoplifting or other criminal behaviors, and that a high proportion of white male camera operators have been surveilling females for voyeuristic rather than security reasons. Homeless people were also disproportionately targeted, as were people who challenged the right of the camera operators to monitor them.

The aim of one study by Norris and Armstrong was to evaluate who was being watched by public CCTV surveillance systems. The researchers studied operators of 148 cameras in three major areas. They reported in their results that 40% of people were targeted for no obvious reason related to criminal behavior and were apparently selected for belonging to a subculture (black, male, or homeless). Those wearing uniforms appeared to be exempt from targeting. In the conclusion, the authors reported:

> The gaze of the cameras does not fall equally on all users of the street but on those who are stereotypical [sic] defined as potentially deviant ... singled out by operators as unrespectable. In this way youth, particularly those already socially and economically marginal, may be subject to even greater levels of authoritative intervention and official stigmatisation, and rather than contributing to social justice through the reduction of victimisation, CCTV will merely become a tool of injustice through the amplification of differential and discriminatory policing.

9.d. Privacy and Identity Protection

Commercial businesses have always been interested in identifying shoplifters, but now they are also collecting data on regular customers, previous clients who haven't paid their bills, and other commercially significant individuals. We are already seeing examples of this type of *identity profiling* on the Internet, with companies using Web browser 'cookies' to see who is logging onto their systems and making purchases. Currently these cookies are used to streamline shopping, especially shopping with credit cards, so the customer doesn't have to fill out a form every time he or she wants to buy something online. However, it's only a matter of time before everyone using a computer has a small videocam attached to the computer that provides videoconferencing capabilities to the online database and the seller could potentially collect demographics on the unwary shopper from the camera images, including gender, approximate age, ethnic group, etc., and match them with purchase preferences and income levels. All it would take to make this a reality would be to build the videoconferencing capabilities into the Web browsing software. Some proprietary Web browsers already do this.

A great deal of unregulated targetable information would then be in the hands of retailers. Since a majority of trusting consumers have historically been vulnerable to marketing hype, propaganda, and commercial coercion, this brings up questions about how they can be protected and educated about identity profiling and lifestyle surveillance on the part of retailers who may be gathering personal information without their knowledge or consent.

There are also indications that the databases from many different agencies are being linked together through the Internet to share data, for a fee. This is similar to junk mail lists. If you buy a magazine from a mail order vendor, that vendor often makes more money selling your name and address to other vendors than from selling you the magazine. Currently, there is not an effective process for having yourself removed from these mailing lists (in spite of efforts to set up exclusion lists). There isn't even an effective way to correct errors. Vendors are supposed to offer the consumer the option of privacy but they often require the consumer to take extra steps, like writing a letter or filling in a little box buried in the fine print. In most cases, without the consumer's explicit written request to be kept off mailing lists, vendors are free to sell their names to other agencies. These same principles can apply to images and information acquired off the Internet when the person makes a purchase online. If videocams are integrated into Web browsers, these detailed databases may include a picture of the face of the computer user and even the titles on the books on the shelf behind him or her.

It may seem obvious that all you have to do is turn off the videocam while visiting sites that potentially want to market your private information to others, but marketing reps have many ways to convince people to give up their freedoms. Supermarkets currently run deep discount specials on desirable shopping items to 'members only' to get people to sign up for electronic member cards that store their shopping preferences in a database. Another very effective method of acquiring information is to run high-stakes contests and sweepstakes. By offering million-dollar prize opportunities to consumers who put their names and faces in a visual database, most people can be convinced to give up their identities. Since these marketing strategies are already being used to assess shopping habits in supermarkets and to collect names and addresses of Internet users, it's likely it will eventually be used to collect visual data when the technology is put in place.

9.e. Evidence Issues

From a surveillance point of view the distinction between analog and digital data is sometimes important. Analog technologies are more difficult to 'forge', that is, it is harder to insert or delete images on an analog video tape (or a film photo) without evidence of tampering. Thus, some analog technologies may serve as stronger evidence in court. The disadvantage to analog technologies is that it may be difficult to get good still images for closer analysis. If you have paused the tape on a VCR, you have probably noticed the degradation in signal quality and the relative coarseness of the image that occurs when viewing still frames compared to images on DVDs. Surveillance tapes of convenience store robberies are an excellent example. They may show the general features of the robber, but not details like the color of the eyes, the brand of watch, small moles, or other identifying features.

Digital technologies, on the other hand, provide better still frames, but the images are easy to alter with computer software. The potential for tampering or forgery is very high. Thus, a digital photograph may not serve as strong evidence in court and may not even be admissible in some circumstances. If the imagery is used in law enforcement and crime investigations, it is important to consider the end-goal of the images before choosing a recording device.

Keep in mind, however, that even traditional photos can now be digitally altered. The traditional photo can be scanned with a very high-resolution digital scanner, altered with an image-processing program and printed to a high-resolution dye-sublimation or ink-jet device that creates images that resemble photos. At one time a photo was considered a 'true' record of events, and photos are widely used in the justice system, but there are now serious difficulties with authenticating any type of photographic evidence.

For genealogists and historians, the 1990s represents a significant milestone in the sense that families and events as represented by photos are now altered on a regular basis. There are commercial image processors who will charge a fee to change your business photos and family photos in any way you like. There are examples of divorced couples having all their family photos of their life together altered to remove their ex-spouses. In other words, instead of pictures of both parents with the house and children, only one parent is seen in the picture, as though the other never existed. This revisionism will make it impossible for future generations to look at past visual records and know what is 'true' and what is not. It also makes it easier for clandestine activities to be covered up and for fraudulent 'news' agencies to make up the news. The truth may be out there, but it may no longer be possible to distinguish the fact from fiction in visual images.

10. Resources

10.a. Organizations

Inclusion of the following companies does not constitute nor imply an endorsement of their products and services and, conversely, does not imply their endorsement of the contents of this text.

American Society for Industrial Security (ASIS) - The world's largest and oldest association for security professionals with more than 30,000 members worldwide. Members are mainly involved with internal commercial security and make wide use of access-control systems and security cameras. An annual conference is sponsored in the fall. www.asisonline.org/

Analog VLSI and Robotics Laboratory - Part of the Indiana University's Department of Computer Science, there are projects on robotics, analog VLSI, and emerging computation. The Stiquito project is an interesting effort to build a colony of tiny robots in order to study cooperative behavior of autonomous agents. www.cs.indiana.edu/robotics/avlsi.robotics.html

CECOM - A U.S. Department of Defense mailing-list clearinghouse that automatically distributes IRIS conference proceedings to qualified subscibers. Subject to clearance by the Security Manager for the IRIA.

Consumer Project on Privacy - Organized by Ralph Nader in 1995 to focus on telecommunications regulations and pricing, fair use issues, and the impact of technology on privacy. www.cptech.org/privacy/

Laboratory for Integrated Advanced Robotics (LIRA). Located at the University of Genoa, Italy, this lab focuses on artificial vision and sensory-motor coordination from a computational neuroscience perspective. www.lira.dist.unige.it/Introduction/intro.html

Neuromorphic Vision and Robotics - At the Higgins Lab at the University of Arizona, this research lab focuses on neuromorphic engineering, with a focus on vision and robotic systems. neuromorph.ece.arizona.edu/

Office of the Information & Privacy Commissioner for British Columbia - The Commissioner exercises duties and authorities under the Canadian Freedom of Information and Protection of Privacy Act. The Web site includes news releases and information on publications, investigations, court decisions, and legislation.

Privacy Commissioner of Canada - Information on the role of the Privacy Commission, the Privacy Act (put into effect in 1983), and various reports associated with the office. www.privcom.gc.ca/

Privacy Commissioner for New Zealand - This site includes information regarding the office of the Privacy Commissioner, the Privacy Act 1993, and Privacy Act Reviews with summaries and charts. privacy.org.nz/top.html

Privacy International - A London-based civil rights group. Hosts an annual "Big Brother" ceremony (since fall 1998) in which awards are given to organizations judged to have contributed to the destruction of personal privacy and those who have contributed to the protection of privacy in Britain. Judging is carried out by selected academic, media, and legal professionals. www.privacyinternational.org/

Robotics and Intelligent Machines Coordinating Committee (RIMCC) - Since 1993, RIMCC has been dedicated to helping accelerate the advancement of key robotics technologies and providing a link between government- and university-based research and the community of users at-large.

Robotics Manufacturing Science and Engineering Laboratory - Located at Sandia National Laboratories where there are many research projects devoted to robotics applications. www.sandia.gov/

The Surveillance Camera Players - This group of actors/activists stages live plays/protests in front of surveillance cameras, accompanied by an attorney to protect their civil rights. The actions of the players have resulted in a number of arrests in spring 1999, some of which were said to be unconstitutional.

Visual Computing Lab - The University of California, San Diego, Visual Computing Lab has research projects on video surveillance and monitoring (VSAM). Projects include Multiple-Perspective Interactive Video (MPI-Video) technology-development. vision.ucsd.edu/Vsam/

10.b. Print

Biberman, Lucien M.; Nudelman, Sol, Ed., *Photoelectronic Imaging Devices,* two volumes, New York: Plenum Press, 1971.

Brin, David, *The Transparent Society: Will Technology Force Us to Choose Between Privacy and Freedom?* New York: Addison-Wesley, 1998. This noted nonfiction and science fiction author discusses the electronic age and its impact on society.

Brugioni, Dino A., *Photo Fakery: The History and Techniques of Photographic Deception and Manipulation,* London: Brassey's, Inc., 1999, 256 pages. The author is a former CIA photo interpreter and founding member of the National Photographic Interpretation Center who describes how deception is accomplished and how to spot it. Includes examples.

Cohen, S., *Visions of Social Control,* Cambridge: Polity Press, 1985.

Davies, Simon, *Big Brother: Britain's Web of Surveillance and the New Technological Order,* London: Pan Books, 1996.

Edwards, Charleen K., *A Survey of Glassmaking - From Ancient Egypt to the Present,* Chicago: The University of Chicago Press, 1974.

Geraghty, Tony, *The Irish War: A Military History of a Domestic Conflict,* London: Harper Collins, 1998, 404 pages. The Irish-born British journalist describes how surveillance technologies that were developed for monitoring the IRA in Northern Ireland are being adapted for surveillance of the general population. British subjects do not have the same constitutional protections as Americans. Geraghty is also the author of Brixmis, an account of western intelligence missions inside the Iron Curtain. It was reported in May 1999 that Gerahty had been charged with offenses under the British Official Secrets Act.

Jensen, Niels, *Optical and Photographic Reconnaissance Systems,* New York: John Wiley and Sons, Inc., 1968, 211 pages. From the Wiley Series on Photographic Science and Technology and the Graphic Arts. The author was affiliated with the Reconnaissance & Intelligence Laboratory at Litton Data Systems Division and associated with Hughes Aircraft on a project to photograph the moon. The volume covers fundamentals of reconnaissance imaging systems and atmospheric optics including some of the basic math.

King, Henry C., *The History of the Telescope,* New York: Dover Publications, Inc., 1955.

Land, Barbara, *The Telescope Makers: From Galileo to the Space Age,* New York: Thomas Y. Crowell Company, 1968.

Lyon, D., *The Electronic Eye,* Cambridge: Polity Press, 1994.

Marchand, Donald A., *The Politics of Privacy, Computers, and Criminal Justice Records: Controlling the Social Costs of Technological Change,* Virginia: Information Resource Press, 1980.

Marx, Gary T., *Undercover: Police Surveillance in American Society,* Berkeley: University of California Press, 1988.

Pritchard, Michael; St. Denny, Douglas, *Spy Camera: A Century of Detective and Subminiature Cameras,* London: Classic Collections, 1993. Documents several hundred subminiature and 'spy' cameras sold at Christie's London Auction in 1991 from the collection of David Lawrence.

Ross, Douglas A., *Optoelectronic Devices and Optical Imaging Techniques,* London: Queen Mary College, 1979.

Rothfedder, Jeffrey, *Privacy For Sale: How Computerization Has Made Everyone's Private Life An Open Secret,* New York: Simon and Schuster, 1992.

Rule, J., *Private Lives, Public Surveillance,* London: Allen-Lane, 1973.

U.S. Department of Energy, *Robotics and Intelligent Machines Critical Technology Roadmap,* Washington, D.C., October 1998. Describes the future robotics technology needs of the DoE and provides a structure for meeting those needs. This came about as a result of the 1997 Congressional Exposition on Robotics and Intelligent machines, sponsored by the DoE, RIMCC, and Sandia National Laboratories.

White, William, *The Microdot: History and Application,* Williamstown, N.J.: Phillips Publications, 1992. Documents techniques of reducing text and images using photographic and electrographic techniques. Includes magnified examples of microdots and related items.

Articles

Baker, Oliver, The Importance of Being Electric, *Science News*, 2000, V.157(3), pp. 45-47.

Boal, Mark, Spycam City: The Surveillance Society, *The Village Voice* online newspaper. www.villagevoice.com/

Davidson, Michael W.; Abramowitz, Mortimer, Museum of Microscopy, sponsored on the Web by Olympus America and Florida State University, 1999.

Davies, Simon, Privacy in the Workplace Survey Report, *Society for Human Resources Management, 1991 SHRM*, Virginia, 1992, p. 192.

Dixon, Tim, Invisible Eyes: Report on Video Surveillance in the Workplace, *Committee of New South Wales paper no. 67*, August 1995.

Etienne-Cummings, Ralph, A Visual Smooth Pursuit Tracking Chip, *Advances in Neural Information Processing Systems,* 1996, V.8, pp. 706-712.

Etienne-Cummings, Ralph, Intelligent Visual Sensors: Will Robotics Benefit? *Proceedings Workshop on Biomorphic Robots,* IROS98, Victoria, B.C., 1998.

Everett, H. R. Breaking Down the Barriers, *Unmanned Vehicles,* V.3(1), 1998, pp. 18-20. A brief history of robotic security developments.

Everett, H. R.; Gage, D. W., From Laboratory to Warehouse: Security Robots Meet the Real World, *International Journal of Robotics Research*, Special Issue on Field and Service Robotics, V.18,(7), pp. 760-768.

Flaherty, David H., Controlling Surveillance: Can Privacy Protection Be Made Effective? *Technology and Privacy: The New Landscape*, Philip E. Agre and Marc Rotenberg, Editors, Boston: MIT Press, 1997, p. 170.

Flaherty, David H., Workplace Surveillance: The Emerging Reality, *Labour Arbitration Yearbook 1992*, William Kaplan et al., Editors, Butterworths, Lancaster House, 1992, p. 189.

Gage, D.W., Security Considerations for Autonomous Robots, *Computer Security Journal*, 1990, V.vi(1), pp. 95-99.

Goldberg, Matt, Machine Age: Is the CEO Reading Your E-Mail? *Village Voice*, November 1998. The situation of employee dataveillance and concerns about employee/employer rights and privacy.

Heath-Pastore, T.; Everett, H. R.; Bonner, K., Mobile Robots for Outdoor Security Applications, *American Nuclear Society 8th International Topical Meeting on Robotics and Remote Systems (ANS'99)*, Pittsburgh, Pa., Apr. 1999.

Honess T.; Charman, E., Closed Circuit Television in Public Places, *Crime Prevention Unit paper no. 35*, London HMSO, excerpt from Privacy International, CCTV FAQ.

Kelly, Sean; Blankenhorn, Dana, Capturing the night; New technology aids war coverage, *Electronic Media,* 4 Mar. 1991, p. 14.

Kirven, Tristan H., Beyond the Negative: A Criticism of Police Pictures: The Photograph As Evidence, *Journal of Criminal Justice and Popular Culture,* V.6(1), 1998, pp. 10-14. An editorial art criticism of S.S. Phillips' historical survey of forensic photography shown July 1998 at the Grey Art Gallery at New York University.

Krupp, E.P., Echoes of the Ancient Skies: The Astronomy of Lost Civilizations, Meadow, Roy, Men capable of Munchausen syndrome by proxy abuse, *Archives of Disease in Childhood*, March 1998.

Meadow, Roy, Men Capable of Munchausen Syndrom by Proxy Abuse, *Archives of Disease in Childhood,* March 1998.

Norris, Clive; Armstrong, Gary, The Unforgiving Eye: CCTV Surveillance in Public Space, Hull: Centre for Criminology and Criminal Justice, Hull University study.

Rochelle, Carl, Public cameras draw ire of privacy experts, *CNN Interactive U.S. News*, March 1996.

Security Industry Association and International Association of Chiefs of Police Information Brief for Guildine On Closed Circuit Television (CCTV) for Public Safety and Community Policing, 31 October 1999. Describes intentions to integrate the best practices of public and private policing using state-of-the-art technologies. www.siaonline.org/cctvinfobrief.html

Waters, Nigel, Street Surveillance and Privacy, *Privacy Issues Forum,* New Zealand, 1996, p. 3.

Whalen, John, You're Not Paranoid: They Really Are Watching You, *Wired,* March 1995. The author crashes a security conference and discusses his impression of the business security industry and employee monitoring.

Journals

Computer Vision and Image Understanding, Academic Press publication covering computer analysis of pictorial information.

Imaging and Vision Computing, Elsevier publication on theory and applied research in image processing and computational vision.

Imaging Science and Technology, a publication of the Society for Imaging Science and Technology covering a broad range of research and applications.

The International Journal of Computer Vision.

Journal of Electronic Imaging, a publication of the Society for Imaging Science and Technology covering design, engineering, and applications.

Machine Graphics & Vision, a juried international journal published quarterly by the Polish Academy of Sciences.

Privacy Times, a Washington newsletter published by Evan Hendricks.

Security Newsbriefs, Executive briefing from a survey of over 1000 publications for corporate, industrial and professional security managers. Each day's issue includes about a dozen summaries on topics that include shoplifting, employee theft, computer crime, industrial espionage, CCTV, alarms, and lighting. Available to ASISNET subscribers. www.asisonline.org/aboutnews.html

10.c. Conferences and Workshops

Many of these conferences are annual events that are held at approximately the same time each year, so even if the conference listings are outdated, they can still help you determine the frequency and sometimes the time of year of upcoming events. It is very common for international conferences to be held in a different city each year, so contact the organizers for current locations.

Many of these organizations describe the upcoming conferences on the Web and may also archive conference proceedings for purchase or free download.

The following conferences are organized according to the calendar month in which they are usually held.

ACCV2002, annual Asian conference on computer vision.

IEEE Workshop on Visual Surveillance. A one-day workshop international workshop.

AVBPA. Audo- and Video-based Biometric Person Authentication, annual conference since 1998. Techniques for the representation and recognition of humans.

ABA TechShow sponsored by the American Bar Association. Topics included technology planning and high tech trials.

Audio and Video Evidence: Tuning the Tapes to the Defense Frequency, National Defender Investigator Association's National Conference.

Effect and Impact of the Courtroom of the Future, Claude W. Pettit School of Law - Law Review Symposium,.

Scientific and Demonstrative Evidence: Is Seeing Believing? National conference on science and law.

Automatic Face & Gesture Recognition, annual international conference since 1995. Conference proceedings for previous conferences are available through IEEE.

PETS, 1st IEEE international workshop on performance evaluation of tracking and surveillance.

ISR, International Symposium on Robotics.

International Robots & Vision Show. A biennial conference with practical solutions for improving product quality and increasing production.

IEEE International Workshop on Visual Surveillance. Held in conjunction with CVPR since 1998. Theoretical and practical aspects of visual surveillance, including tracking, scene interpretation, object detection, recognition, and other topics.

ASIS International. Industrial security issues and products.
www.asisonline.org/seminar/seminar.html

Advanced Surveillance Technologies, an international conference sponsored by Privacy International, Electronic Privacy Information Center. An overview of the program is available at Privacy International's Web site. www.privacy.org/pi/conference/copenhagen/final.txt

Video Surveillance and Monitoring Demo II and Workshop. The VSAM project is developing automated video processing technologies for use in urban and battlefield surveillance applications for situations in which human observation is too dangerous or costly. Examples include security for restricted areas, buildings, airports, and their environs. VSAM is sponsored by the Defense Advanced Research Projects Agency, Information Systems Office (DARPA ISO) as part of the *Image Understanding for Battlefield Awareness* program.

International Conference on Information, Intelligence, and Systems (ICIIS), combined technical conference for interaction between scientists and practitioners from diverse fields dealing with complex problems in neuroscience, biology, robotics, image, speech and natural languages, and their integration.

10.d. Online Sites

The following are interesting Web sites relevant to this chapter. The author has tried to limit the listings to links that are stable and likely to remain so for a while. However, since Web sites sometimes change, keywords in the descriptions below can help you relocate them with a search engine. Sites are moved more often than deleted.

Another suggestion, if the site has disappeared, is to go to the upper level of the domain name. Sometimes the site manager has changed the name of the file of interest. For example, if you cannot locate www.goodsite.com/science/uv.html *try going to* www.goodsite.com/science/ *or* www.goodsite.com/ *to see if there is a new link to the page. It could be that the filename* uv.html *was changed to* ultraviolet.html, *for example.*

A Complete History of the U.S.S. Dyess. This is an anecdotal history of an American destroyer that conducted a variety of types of visual, audio, and radar surveillance. The historical journal is provided by Chief Petty Officer Ralph J. Brown, Sr. who served on the vessel from 1955 to 1960. www.extremezone.com/~pomeroy/dyess/history.html

American Civil Liberties Union (ACLU). A prominent civil liberties organization that provides nonpartisan, nonprofit education and advocacy on behalf of over 1/4 million members. Its primary goal is assuring that the American Bill of Rights is upheld and continued for future generations. There is quite a bit of searchable information on this site related to video surveillance and political issues associated with the use of the technology. www.aclu.org/

An Appraisal of the Technologies of Political Control: An Omega Foundation Summary & Options Report For the European Parliament. A summarized document (approx. 20 pages) of an Interim Report that discusses recent developments in surveillance technology, tracking systems, face recognition, vehicle recognition, data matching, and the proliferation of this technology in light of its practical applications and potential for use and abuse within our political structures. edd.www.cistron.nl/stoa2.htm

AsherMeadow Organization. This site includes resources for the Munchausen Syndrome by Proxy community including case studies that describe the use of video surveillance in hospital rooms to detect child abuse related to this syndrome. www.bcpl.net/~agravels/truestories.htm

AVS-PV Advanced Video Surveillance - Prevention of Vandalism in the Metro. This Italian site illustrates vandalism in public transit in Europe, with a number of metro stations as trial sites. The focus of the project is on the economics and prevention of vandalism (graffiti, crassity, defacement, etc.) through digital processing of visual images and preventive actions undertaken by surveillance operators in the metro. Numerous photos and diagrams are included. dibe.unige.it/department/imm/avspv3.html

Building to Reduce Crime: Guidelines for Crime Prevention Through Environmental Design. This site includes many drawings and photographs demonstrating how environmental planning impacts human behavior and how visual surveillance for safety can be achieved through the design of buildings and related structures. Mixed-use neighborhoods, open areas, lighting, street-facing porches, clear sightlines, and other strategies are described and illustrated. References and information about other crime-prevention initiatives are included. Prepared by Melanie D. Tennant for the Development Services Department, City of North Vancouver. trinity.cnv.org/CrimePreventionThroughEnvironmentalDesign.htm

CCTV Archives. A site that asks the question of whether the more than one million CCTV cameras installed throughout Britain affect our daily lives. Video clips are displayed, archived, highlighted, and sold on this site. One of the selections is Police Stop! a video archive of the tapes that are collected from video cameras mounted in police vehicles. Another is Really Caught in the Act which shows crime in many retail establishments and workplaces. There are also monthly featured clips. A site that is interesting from an educational point of view, but also has a voyeuristic draw and the unsettling reality that people's activities are being disseminated worldwide. www.cctvarchive.com/

History of the 225th Surveillance Airplane Company. Prepared by Major Gary L. Petesch, Unit Historical Office, site maintained by Howard Ohlson. This site describes activities for the year 1968 during which ground-sensing data, traditional photographs, side-looking radar (SLAR) images, and visual surveillance were gathered for a Vietnam combat surveillance and target acquisition capabilities assessment mission. ov-1.com/225th_AVN/225th-history68.html

Human Rights Watch. This New York City-based organization has an extensive selection of online reports on many rights topics, including landmine monitoring, armed conflicts, the disabled, child labor and abuse against women, including a number of reports on visual surveillance, for example, allegations of excessive and unwarranted observation and videotaping of women in various states of undress in the prison system. www.igc.apc.org/hrw/

In Plain and Open View: Geographic Information Systems and the Problem of Privacy. Michael R. Curry, Department of Geography, University of California, LA. This Web page describes the important issues of privacy related to the constant geographical mapping of our planet, which includes our cities and individual homes. Much of this geographic imagery is paid for by tax dollars and is available to anyone who can download it through a computer. The Web page lists a large number of surveillance/privacy-related documents. www.spatial.maine.edu/tempe/curry.html

Museum of Microscopy. This extensive scientific site provides excellent illustrations and descriptions of microscope design and history including photographs and rendered recreations of significant scopes. It is well-organized and managed by M. W. Davidson, M. Abramowitz, through Olympus America Inc., and the Florida State University, in collaboration with the National High Magnetic Field Laboratory. micro.magnet.fsu.edu/primer/

The NYC Surveillance Camera Project. This site was compiled by citizens concerned with the proliferation of surveillance cameras, most of them privately owned, which are aimed at public places. To date there is little legislation governing the use of these cameras or materials recorded. The site includes maps of camera locations, information about the project, online submission of information about cameras not currently listed, and general news about video surveillance. www.mediaeater. com/cameras/index.html

Privacy International Statement on Closed Circuit Television (CCTV) Surveillance Devices. A statement that reflects the policies and concerns related to the erosion of privacy and the extraordinary growth of the electronic visual surveillance industry, with entreaties for legal safeguards. www.nonline.com/procon/html/conCCTV.htm

Safe Schools Design Guidelines. This interesting site from the University of South Florida (USF) Center for Community Design and Research includes a series of pages showing Safe Schools Design Guidelines (1993) that provide rationales and planning suggestions for the physical shapes and orientations of buildings and play areas. Includes numerous diagrams and explanations of how facilities planning can enhance the ability of the staff and community to monitor the activities and safety of students to reduce vandalism, crime, and accidents. www.fccdr.usf.edu/projects.htm

United States Postal Inspection Service. This site describes the functions and departments of the U.S. Postal Inspection Service Crime Lab Forensic & Technical Services Division (FTS&D). The service manages a national forensic laboratory and four regional labs, plus five technical services field offices. www.usps.gov/postalinspectors/crimelab.htm

U.S. Navy ROBART Series. An interesting illustrated site showing a series of robots equipped with a variety of types of surveillance technologies, developed since the early 1980s. The pioneer systems could only detect potential intruders, while later systems can now detect and assess and, in some cases, respond to the intrusion with defensive or offensive actions. A variety of detection systems have been incorporated into these robots, including infrared, optics, ultrasonics, microwave, vibrational sonics, and video motion detectors. www.spawar.navy.mil/robots/land/robart/robart.html

Video Movement Tracking. The Gerhard-Mercator-University Duisburg in Germany has three animated GIF files that demonstrate computerized image recognition and motion tracking capabilities, e.g., a human enters a door in a hallway and his motion is tracked digitally, using a combination of Pseudo-2D Hidden Markov models and a Kalman filter, as he progresses along the hallway. References for the techniques are also included. www.fb9-ti.uni-duisburg.de/demos/tracking.html

Video surveillance by public bodies: a discussion. Investigation Report P98-012 submitted by David H. Flaherty, Information and Privacy Commissioner for British Columbia, March 1998. This investigative report was based on research and site visits to public bodies that engage in video surveillance. It discusses the installation and pervasiveness of various types of visual surveillance, associated costs/benefits, and assessment of adverse consequences for personal privacy. It further surveys the research on the prevalence of video surveillance and reports on the efficacy of the technology. This is a valuable document for anyone planning to install video surveillance, or researching the impact of video surveillance on the public who seeks to be aware of the various pros and cons. www.oipcbc.org/investigations/reports/invrpt12.html

Vision Chips, or Seeing Silicon. This site has LOTS of information on vision chips and smart sensors, along with links to conferences, research labs, etc. Visit the World of Vision Chips link to get an idea. www.eleceng.adelaide.edu.au/Groups/GAAS/Bugeye/visionchips/

Visual and Acoustic Surveillance and Monitoring (VSAM). A project at the Computer Vision Laboratory at the University of Maryland. This site includes several pages, some with illustrations, of the goals and progress of the project to research visual and infrared spectrum and acoustic computerized surveillance mechanisms for a variety of practical applications including military, law enforcement, traffic management, and urban security. Demonstrated systems include personal computer-based real-time tracking of people and their body parts (heads, feet, etc.). www.umiacs.umd.edu/users/lsd/vsam.html

Watching Them, Watching Us - UK CCTV Surveillance Regulation Campaign. This site provides public information about political developments, funding, installation, and use of CCTV surveillance equipment. It engages in important debate about the efficacy of cameras, licensing issues, the integrity of evidence, the proliferation of Webcams, and more. www.spy.org.uk/

What Man Devised That He Might See. An attractively illustrated history of spectacles by Dr. Richard D. Drewry, which takes the viewer through various technological improvements over the centuries from ancient times to the present. www.eye.utmem.edu/history/glass.html

10.e. Media Resources

Caught in the Act, TV series featuring video footage of various individuals caught in the workplace and other settings engaged in criminal, unethical, dangerous, or unsavory activities. Distributed by NTV Entertainment.

Cheaters, TV series featuring video footage of surveillance of suspected cheating couples.

Cops, series in which videotaped patrol activities are televised. www.tvcops.com/

To Serve and Protect, television series of videotaped patrol activities of the Vancouver Police Department.

Sliver, feature film starring Sharon Stone, Paramount Pictures, 1993. Video cameras are featured in an urban apartment complex drama. Rated R.

The Truman Show, Motion picture starring Jim Carrey, directed by Peter Weir, Paramount Pictures, June 1998. A small-town insurance salesman encounters video surveillance in a surprising way. The screenplay is by Andrew Niccol, author of the movie *Gattaca,* which is listed in the DNA chapter.

Museum of Opthalmology. A foundation of the American Academy of Ophthalmology in San Francisco. The extensive holdings of the museum include historical lenses with about 100 items on display at any one time. Located at 655 Beach Street, San Francisco.

11. Glossary

Titles, product names, organizations, and specific military designations are capitalized; common generic and colloquial terms and phrases are not.

Cartesian surface	a reflecting or refracting surface that forms an image with 'perfect' fidelity to the original
CCD	charged-couple device, a system utilizing light-sensitive photo diode elements to register light intensities that are translated into digital signals. The resolution of a CCD device is determined in part by the type, placement, and quantity of CCD elements, which can be arranged in lines or grids. Many digital scanning and image devices are CCD-based.
CCTV	closed-circuit television
CCU	camera control unit
CEPTED	crime prevention through environmental design
CFR	computerized face recognition
displacement effect	an adjustment that occurs when surveillance of any area results in criminals moving their activities elsewhere in order not to be watched
FOV	field of view, usually described in degrees. Used to describe the angle of imaging on lenses and other imaging devices.
MTF	modulation transfer function
papparazzo	a freelance photographer who persistently watches and aggressively pursues celebrities for the purpose of capturing candid photographs for publication
photoelectric cell	a light-sensitive sensing device commonly used in security systems, automatic lighting systems, automatic doors, etc. A basic photoelectric cell can be created with a vacuum tube by coating one of the electrodes with cesium.
photovoltaic device	a type of specialized semiconductor for converting light into electrical energy. Photovoltaic slices can be combined into an array, also known as a *panel*, and used, for example, as solar collection panels. Photovoltaic panels are used for solar powering many types of solar devices, from watches, pocket lights, and security lights, to satellite transceivers.
Picturephone	one of the earliest teleconferencing systems, designed by the AT&T Bell Laboratories in the mid-twentieth century, but not introduced to the public until 1970. It was far ahead of its time. Videoconferencing didn't became a mass-market product until about 1998. Picturephone-type technologies are now used not only for remote conferencing, but for monitoring of remote video security systems.
PSTN	public switched telephone network. The national public telephone infrastructure that provides universal access as originally put forth in the 1934 Communications Act.
quartz	Silicon dioxide (SiO2), a mineral that ranges from transparent to white and may include colored impurities. It is common, a principle component in sand and sandstone, also

	found in granite. Quartz is used in many types of lenses and as lens coatings.
scopophilia	the love of looking/viewing/observing. Voyeurism is a subset of scopophilism in which the observer preferentially watches scandalous or sexual activities.
VHS	Video Home Systems. This is a widely distributed sound/image recording and play-back format originally developed by JVC. S-VHS (Super-VHS) is a higher-resolution version of this format which is downwardly compatible. VHS is gradually being superseded by Hi-8mm tape and DVD digital formats.
video tape	An analog video (and sound) recording medium developed in the late 1920s and introduced commercially in the 1950s. Video tape revolutionized not only home movies, but the entire television broadcast industry. No records remain of many of the earliest television broadcasts because the shows were aired live and never taped. However, reruns of popular shows from about the mid-1950s on still garner strong fan and sponsor support through reruns. Live broadcasting is now rare, mostly saved for special events, awards ceremonies, and surveillance activities in which no records are required. However, the majority of surveillance activities are taped in order to provide evidence or a record of events that can be analyzed at a later time.
vidicon	a television with a photoconducting pickup sensor
voyeur	One who watches sordid, scandalous, or sexually explicit activities, often for sexual gratification. Voyeurs frequently carry out their observations surreptitiously.

Electromagnetic Surveillance

9

Aerial

1. Introduction

Aerial surveillance is not technically an electromagnetic technology, but a broad category of technologies used to view things from above. About 90% of aerial surveillance devices rely on electromagnetic phenomena, however, so the chapter has been included in the Electromagnetic Surveillance section. Note that aerial surveillance will be better understood if you read or scan the Infrared and Visual Surveillance chapters first.

Aerial surveillance began when humans climbed trees and cliffs to get a better view for spotting food, shelter, fuel sources, predators, and the activities of other people. When great fleets of ships began to sail the oceans, telescopes were carried up to crow's nests to observe lands and other ships from a distance. Not long after humans first flew in hot air balloons, they realized they could use telescopes, binoculars, and photographic equipment from better vantage points if they combined them with aerial technologies. Soon airplanes enable them to see from unprecedented heights and the development devices specifically designed for surveilling from planes, helicopters, and spacecraft began.

Aerial surveillance is one of the most important political and scientific developments of the twentieth century. Viewing the Earth from above, and the solar system from beyond the Earth

---

When U-2 pilot Francis Gary Powers was shot down on a spying mission over Soviet territory in 1960, the United States hastily applied NASA markings to a U-2 aircraft (enlargement), issued coverup news releases that the plane was a stray weather research craft, and invited the press to view and photograph the plane. The Soviet Premier later completely disproved the story by presenting the pilot, the recovered wreckage, and the incriminating surveillance tape. The history of spy planes is described further on page 9-20. [NASA/Goddard 6 May 1960 news photo, released.]

has changed our perception of ourselves and has provided dramatic new ways to explore the cosmos. In the context of aerial surveillance, flight has stimulated the development of faster, higher-resolution imaging technologies and, in the last few decades, aerial photography has became almost routine.

Most aerial surveillance devices are designed to image surface features and vessels, but aircraft are sometimes used to visually surveil marine environments and to help in the deployment of acoustic devices such as towed sonar. Aerial surveillance is not a specific technology, but rather a general means of deploying a variety of technologies with applications sufficiently diverse and important to warrant a chapter on its own.

Capturing still or moving images from great heights or at great speeds is a difficult technological challenge that scientists have tackled with a remarkable degree of success. Many aerial imaging devices have been adapted for space surveillance as well. Orbiting satellites and space probes provide not only a view of near and distant parts of the universe, but also information on how the universe began.

The Shift in Power

People engage in aerial surveillance for many reasons, including curiosity, scientific research, military security, and the gathering of business intelligence. The capability to unobtrusively observe other people's business, resources, and activities is valuable information. With that power comes responsibility, whether or not it is mandated by law. Until the mid-1990s, the power of surveillance was mainly in the hands of local and federal government agents and, to a lesser extent, private detectives. This is no longer true. It is now possible to observe the planet through Google Earth and to purchase an aerial photo of your neighbor's backyard that is sufficiently detailed for you to distinguish between a large dog house and a small hot tub. You can purchase an equally detailed picture of government buildings in foreign nations, a refugee tanker, a controversial logging site, or the production yard of your chief business competitor on the other side of town. Sources of these images are discussed later in this chapter.

Technology changes faster than laws can be drafted to protect the vulnerable. Currently, private citizens don't need special permission to purchase or own most types of images—you can download them off the Internet. Satellites and network distribution channels have put access to information in the hands of ordinary people in most of the democratic developed nations. More than half the American populace now has access to the Net. While military entities are still the only ones who can get the highest resolution imagery, sometimes accessible in realtime, the gap between information available to civilians and those in traditional positions of power (e.g., national defense) has narrowed dramatically. This shift of access puts a great deal of power in the hands of a largely unregulated public which, in free societies, is bound to result in some unethical and unscrupulous exploitation of the technology on the part of individuals who are willing or eager to take advantage of others.

To resolve these issues we must either evolve as a society to take more personal responsibility to respect the rights of others, or we must give up a portion of our freedoms and more stringently regulate access to and use of new technologies. Either way, society must change and adapt, because satellites and unstaffed aerial vehicles can recognize and track not just backyard hot tubs, but individual people, without their knowledge. It won't be long before every move you make can be recorded in realtime by satellites and intelligent software (see the chapter on Visual Surveillance for information on software that can identify individual faces on a crowd and track their movements automatically).

Aerial Photography in the Context of Progress

The technology to surveil from higher vantage points has evolved dramatically since our ancestral days, but before we call it progress, it is prudent to note that human motives haven't changed much over the centuries. Most aerial surveillance is used to monitor predators (primarily other humans), food (esp. crops and herds), shelter and fuel (esp. oil and forests), and the activities of other people—essentially the same reasons we spied from cliff- and treetops 100,000 years ago.

This chapter describes a number of representative airborne and spaceborne aerial technologies, including highlights in the history of the technology, applications, and device deployment. The basic physics of the specific spectra used in aerial surveillance are described in more detail in other chapters.

2. Kinds and Variations

Aircraft and spacecraft can be loosely categorized as those that fly through the Earth's atmosphere and those that fly primarily in space (first passing through the Earth's atmosphere). Some, like the NASA space shuttles, are designed to travel in both air and space. Some craft are dependent upon wind or 'loft' to stay airborne (parachutes, gliders, kites), some utilitize lighter-than-air gases like hydrogen and helium, and some are self-powered (planes, ultralights, rockets). Most spacecraft are launched through the atmosphere to a distance above Earth where they can achieve a more-or-less stable orbit for a period of time (usually a few years), while some are launched far into space, where they can travel for long distances until they come within the gravitational influence of other bodies.

In the early days of the air travel industry, balloons and dirigibles were known as airships and fixed wing craft were commonly known as 'flying machines'. For a few decades this terminology was sufficient to describe most of the aerial vessels that were developed and used for civilian and military purposes. In the last 20 years or so, it has become more difficult to describe the great variety of airborne craft that are used to view the world. Aircraft and the more recent spacecraft, come in many shapes and sizes. Recently, we've added rocket ships, shuttle craft, autonomous and remote-controlled air vessels, and orbiting space stations and satellites. All of these craft, small and large, fast and slow, can be used in one way or another for surveillance. Their prices range from about $500 to hundreds of millions of dollars. Most of them can be fitted with a variety of passive and active remote-sensing technologies, including radar, infrared, ultraviolet, radio, photographic, and tape recording devices.

2.a. General Categories of Aerial Surveillance Craft

The kinds of sensing technologies that can be used with a particular type of airborne craft will depend on the size and weight of the craft, its visibility, the flying or orbiting altitude, cost, maneuverability, and ability to carry equipment. Some of the basic categories of air- and spaceborne vessels that can be used to observe and record include wind-lofted, light-than-air, and self-powered craft.

Wind-Lofted Craft

Typical wind-powered aerial craft include parachutes, kites, and gliders. These craft are dependent on differences in air pressure between their upper and lower surfaces and, in the case of parachutes and some gliders, may need to be towed, dropped, or launched.

Parachutes, kites, and gliders have all been used at various times to carry surveillance devices. Sometimes radio gear is dropped by parachute for scientific and military surveillance missions. Here, a U.S. Marine Corps captain pulls in a parachute in a Kuwait drop zone as part of Exercise Eager Mace 98. Capt. Rick Uribe dropped 1,250 feet from a cargo plane in 1997 as part of a joint training exercise with U.S. and Kuwaiti armed forces. [U.S. DoD 1998 news photo by Mike Wentzel, released.]

The practical weight limit for most wind-dependent craft is a few hundred pounds, and their steering capabilities are often limited, as well. Yet they still have characteristics that are important in surveillance. They are quiet and easier to transport than larger, heavier craft and can be quite inexpensive. Helmets for use with parachutes and gliders can now be purchased with mounting brackets for cameras (skydivers often use these) and some even have small built-in video cameras. There are also harnesses that allow a camera to be securely fastened to a person's chest. Smaller autonomous or remote-controlled versions of wind-dependent craft can be equipped with remote-controlled cameras designed to record video images or take pictures at predetermined intervals

Lighter-Than-Air Craft

Air is made of a variety of elements and the lighter ones tend to rise to the top (toward weaker gravitational forces). Hence, the air on mountain tops is 'thinner' than at lower altitudes—in other words, there is less oxygen as you move away from the Earth. Mountain climbers, high-altitude flyers, and skydivers need oxygen supplementation or they risk confusion, brain damage, or cerebral edema. The characteristic of lighter elements rising makes it possible to launch hot-air balloons and dirigibles ('steerables') that contain lighter-than-air hydrogen or, more commonly, helium (which is less flammable). There are both rigid and nonrigid dirigibles.

Balloons and dirigibles have been used for surveillance since the earliest flights and even now, some are being constructed to carry cameras and radar systems. Their carrying capacity is related to the size of the vessel but can be many hundreds of pounds. Indeed, in the past, dirigibles were used for commercial travel.

Self-Powered Craft

There are many types of self-powered aircraft, ranging from a few ounces to hundreds of tons, so it is difficult to make generalizations about their carrying capacity or suitability for specific types of surveillance. These versatile devices, which comprise the majority used for aerial surveillance, include

- remote-controlled planes, helicopters, and air vessels,
- autonomous, unpiloted air vessels (autonomous UAVs), and
- piloted aircraft, helicopters, and space shuttles (return trip).

In addition to the basic categories of Earth-based aerial craft specifically designed for surveillance, there are also a number of spacecraft and projectiles that are retroactively fitted with surveillance devices or that are outfitted with components for testing new surveillance technologies.

Launched Probes and Orbital Craft

Looking beyond our current frontiers is an important aspect of surveillance. During the years when three- and four-masted ships were important trade and military vessels, one would climb a mast to the Crow's Nest and scope the surrounding waters with a telescope. However primitive it might be, the telescope was essential to survival; finding land after a two-month journey, with food and fresh water depleted, could mean the difference between life and death. It could also reveal frontiers that yielded great wealth from the exploitation of new and abundant lands. Now that all the land on the planet has been mapped and staked out, space is seen as the next frontier. Astronomers look to the heavens for new understanding of our universe, but also for places to colonize, and what we learn in the process may mean the difference between life and death for the entire human race in a future with increasing populations consuming the planet's finite resources. If we can find water, land, medicine, and other resources outside of Earth's boundaries, it may contribute to our survival as well as our knowledge of the cosmos. Even in the 1960s, when the population was only half of what it is now, space pioneers felt a sense of competitive urgency to get out into space when the Russian Federation put a satellite into orbit. The U.S. immediately responded in kind, and the resulting developments came to be known as the "Space Race".

Launched space probes or orbital craft include space shuttles (partially self-powered), rockets, satellites, and specialized planetary and interplanetary space probes. Air and spaceborne telescopes also contribute to aerial surveillance.

Projectiles

Almost any projectile that travels a reasonable distance in the air can be equipped with a small camera. Wireless pinhole cameras are now so tiny, you could attach one to the shaft of an arrow. Other projectiles that have been outfitted with surveillance devices include missiles, rockets, and fireworks (which are essentially a type of rocket designed to 'self-destruct').

2.b. Reconnaissance and Surveillance

Aerial surveillance came of age when humans took to the air in balloons and airplanes. Environmentalists often track plant and animal resources and monitor the legal and responsible use of these resources using aerial craft and photos. Firefighters use aerial surveys to locate and track forest fires and to administrate firefighting activities. Land speculators use aerial surveillance to locate and assess potential property holdings.

Military Reconnaissance

Reconnaissance is a preliminary or exploratory survey that accounts for a large proportion of military aerial surveillance. Assessing a situation before sending in troops or weapons is an important aspect of military preparedness and combat strategy. Military operations utilize a great number of different aerial craft, including planes, satellites, dirigibles, gliders, and unpiloted aerial vehicles.

Aerial reconnaissance played an important part in World War I, but the technology itself was crude and, as yet, undeveloped. For the most part, military aerial reconnaissance in the modern sense was developed and institutionalized in the 1940s.

During World War II, U.S. aviation reconnaissance was divided into two general categories:

strategic   long-range operations, and

tactical   visual and photographic missions (installation recording, target imaging, and mapping, etc.) for the Tactical Air Force, the Tactical Air Division, and associated ground forces.

Military reconnaissance has a prominent role in present society. Considerable funding and personnel are involved in maintaining early warning systems and global monitoring activities. Military applications of technologies such as satellite sensing and Global Positioning Systems (GPSs) are often funded by military interests before they came into use as civilian commercial products. As a consequence, many of the examples in this chapter are military aerial reconnaissance systems and activities.

Commercial and Law Enforcement Surveillance

A variety of aerial craft are used in civilian and law enforcement surveillance, including light planes, helicopters, and remote-controlled aircraft. Producers looking for 'reality programming' footage for television broadcasts often travel with law enforcement agents and news crews, sometimes supplying the aircraft and sometimes using those available.

Local law enforcement agents now also make regular use of aircraft to monitor fugitives, accidents, natural disasters, national borders, coastlines, and crime scenes. Individuals who run from crime scenes or speed to avoid arrest have a far more difficult time evading arrest when a helicopter or other aerial surveillance craft is radioing his or her position to police on the ground. Marine vessels suspected of transporting drugs or illegal aliens may be spotted by boats or by helicopters. Police departments in some of the larger cities use helicopters to deter or investigate crime.

3. Context

High-resolution aerial imaging as a technology is less than forty years old. Historic images were crude, costly, and often classified. When aerial images were expensive and low-resolution, they were primarily sold to researchers, large corporations, and military analysts. In the mid-1990s, it was not unusual for an image to cost over $4,000. Now that high-resolution satellite images can be purchased for under $25 per square mile (or viewed for free on Google Earth), a dramatic shift is occurring in the applications for which these images are used and the people who use them.

Scientific Inquiry

Aerial surveillance technologies, especially those related to imagery, are useful for almost every type of scientific inquiry including ecosystems monitoring (weather, pollution, climate change), geology, archaeology, wildlife monitoring, etc.

Left: A portion of an image of New Mexico from the Department of Energy's R&D Multispectral Thermal Imager (MTI) satellite launched March 2000. The project was led by Sandia Labs. It senses three visible and 12 infrared spectral bands, collects pictures of volunteer ground sites, and comp56ares the images with other data. Right: A sample fish map generated from satellite data. [Sandia news photo, released; OrbImage/GeoEye ©2000 news photo, www.geoeye.com, used as per copyright guidelines.]

The data from multispectral imaging satellites can potentially be used for hazardous waste site characterization, climate research, and treaty monitoring applications. Fish maps can be used for charting fish stocks and migration patterns and managing marine resources.

Government Applications

Aerial surveillance is used in many government and commercial activities, including national defense, local law enforcement, disaster assessment and relief, community planning, building permit issuance and compliance monitoring, resource exploration, wildlife monitoring, property tax assessment, border patrol, search and rescue, camouflage detection, and treaty negotiation and verification.

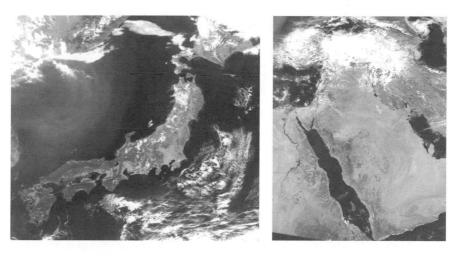

Aerial Images of Foreign Nations. Left: The islands of Japan, 9 May 1998. Right: Egypt, Saudi Arabia and the Middle East, 3 April 1998. [OrbImage ©1998, www.orbimage.com, used as per copyright instructions.]

Commercial Applications

There are thousands of ways in which aerial surveillance, particularly satellite images, are useful in commercial applications.

Left: A NASA aerial photograph of the *Stennis Visitors Center and Administrative Complex*. Stennis is a lead center for commercial remote sensing. Right: A composite aerial perspective radar photo of Pasadena, Ca., with a Landsat image overlay. The detail is supplied by USGS digital photography. [NASA 1990 news photo, NASA/JPL 2000 news photo.]

Commercial applications include, but are not limited to, weather reporting, property value and damage assessment, marketing and promotion, travel planning, contractor planning, city planning, architectural planning, investigative journalism, ranching and livestock monitoring, surveying, fish finding, crop yield assessment, forestry, firefighting, and managing livestock on large ranches.

Detailed weather images are useful not only to weather forecasters, but are of commercial interest to newscasters, boaters, commercial fishers, firefighters, production companies, homeowners and businesses in the path of a hurricane, insurance adjusters, construction companies, and others.

Nonprofit and Public Welfare Applications

Charitable, nonprofit, and other public welfare organizations use aerial imagery for monitoring the environment (habitat, pollution, endangered species, waterways, etc.), human rightss, refugee, public safety, and public resources.

Personal Applications

Satellite imagery is now inexpensive enough that consumers can afford to purchase photos of home and hearth, hobby farms, and land investments for small business use or even for hanging above the fireplace or sending to Grandma. They may also be useful for trip planning, genealogical research, and home-schooling applications.

4. Origins and Evolution

While people have undoubtedly been spying from trees and cliffs for tens of thousands of years, the history of aerial surveillance technology is very recent, less than 300 years old. Most of the significant inventions, air vehicles, portable cameras, and practical telescopes, didn't emerge until the late 1700s and early 1800s.* Dramatic improvements did not come about until the mid-20th century, and satellite imagery is less than 40 years old. This section charts some of the significant inventions and inter-relations between flight and photography, the two most important technological aspects of modern aerial surveillance.

---

*Earlier examples of *camera obscura* (pinhole cameras) lacked the ability to record or 'fix' the images and so are not covered here.

Aerial surveillance involves the evolution of two main aspects of technology, devices to help us fly and devices to help us record images.

The Late 1700s - Pioneer Aviation and Emerging Photography

The first attempts to put cargo into the air probably started with birds and kites thousands of years ago. The first successful attempts to put people in the air were with kites and balloons. The first significant hot air balloon flight was launched in France on 5 June 1783. Later that year, on 21st November 1783, humans traveled over Paris in a balloon constructed by the brothers Joseph and Etienne Montgolfier. It reached a height of almost 6,000 feet and traveled nearly a mile. This achievement was sufficient to inspire inventors worldwide to seek ways to put people in the air. As a result, over the next several decades, many experimental flying technologies were developed, most of which still exist in one form or another today.

The other key aspect of aerial surveillance, photography, was also emerging. Pinhole cameras (*camera obscura*) and sun paintings (heliotropes) were being developed around this time.*

The Early 1800s - Pioneer Photography and Aerial Photography

Aerial photography appears to have begun almost at the same time as photography itself. In fact, the first acknowledged photograph by Joseph Nicéphore Niépce (1765-1833), taken in 1826, was shot from a height, looking down over the roof of an outbuilding.

This historic photograph, by J. N. Niépce, was shot from an upper story window, providing an elevated vantage point of his estate. One of the most significant limitations of early photography was that exposures like this took many hours, so it was impossible to take a photograph from a moving conveyance like a kite or balloon. [Copy print of the original 1826 heliograph, from the Permanent Exhibitions at the Harry Ransom Center at the University of Texas, Austin, copyright expired by date.]

As soon as cameras were available outside science labs, inventors and entrepreneurs began rigging them to trees, cliffs, birds, masts, kites, and balloons. Many of these early attempts at aerial photography were unsuccessful and occasionally disastrous to equipment or bystanders, yet they suggest that the desire to obtain images from the air is a basic impulse, waiting only for the technology to provide a better way to take advantage of opportunities.

The development of equipment for aerial photography fueled the invention of timers, image and camera stabilizers, automatic shutters, smaller cameras, and automatic film advance mechanisms. The development of photography is intrinsically linked with the human impulse to witness and record unusual scenes, events, or covert activities. Many of the most sophisticated cameras are those that have been developed for use on reconnaissance planes that fly very high and very fast, with the advanced technology that makes these feats possible, gradually filtering down into consumer products.

The first historic photographs weren't recorded on paper. Photographic negatives were imaged on materials that were difficult to handle and had to be manually placed and removed. Many historic pictures were temporary, and the permanent ones were usually printed on a ceramic or metal base, which made them durable but also somewhat awkward and expensive to process. These early systems were too cumbersome for most surveillance tasks.

Louis Jacques Mandé Daguerre (1789-1851) made a significant contribution to photography by developing a process for imaging on a copper plate so that reproductions could be made commercially, resulting in the *daguerreotype*. This was an important milestone in photography, but there were still some hurdles to overcome before photography became an intrinsic tool of surveillance. In Daguerre's day, exposures could last almost an hour. It was faster than Niépce's process, but still too long for recording normal human activities or anything that moved.

In spite of its limitations, the world was quick to embrace the new technology. In the mid-1800s, daguerreotypes were introduced to Japan through trade with The Netherlands.

Gradually, film negatives and paper prints were developed and automated or semi-automated film advance, shutters, and focusing were added. The feature film industry, which depended upon long reels to store thousands of individuals, also provided key technology for aerial photography. Planes fly at high speeds, necessitating fast frame rates and long reels of film, and the special problems of stability, registration, and drag on the reels had to be solved. At the same time, inventors were experimenting with different kinds of film, including those that were more sensitive in the infrared spectrum. (For further details on the history of photography, see the history sections of the Infrared and Visual Surveillance chapters, as well.)

The Mid-1800s - Rise of Photography and Airships

The Crystal Palace, built for the 1851 Great Exhibition, was photographed in detail by pioneer photographer Philip Henry Delamotte, thus initiating a new age of documentary photography. Aerial surveillance photography has become an important aspect of modern documentary news photography and is increasingly used to monitor vehicular traffic. [Photo by Philip Henry Delamotte, public domain by date.]

In the mid-1850s, the commercial value of photography was acknowledged through some of the first photographic exhibitions, and photography for documentary use was pioneered by Philip Henry Delamotte (1820-1889), who meticulously recorded the Crystal Palace in London, England. Newscasters, investigative journalists, and political documentors now rely heavily on aerial photography.

In the mid-1850s, another important aspect of aerial photography was introduced. *Aerostats*, which included balloons and airships, were getting a lot of attention from inventors and the public. Henri Giffard (1825-1882) created a coal-gas-powered airship that achieved the first powered, controlled flight in 1852. *Dirigibles* (literally 'steerables') were distinguished from balloons by their propulsion and control systems, allowing them to be navigated horizontally without dependence upon the direction of the wind.

In the late 1850s, the idea of transatlantic flight fired the imaginations of a number of pioneer balloonists, including John Wise, who oversaw the construction of a huge balloon with a passenger gondola that he optimistically christened the Atlantic. Unfortunately, on an inaugural flight in 1859, the balloon encountered a storm and the balloonists were treated to the discomfort and indignity of landing in an elm grove—much to their disappointment. They did succeed, however, in establishing a long-distance record of 809 miles that remained unbroken for half a century.

An Official Balloon Reconnaissance Corps

In April 1861, Thaddeus S. C. Lowe (1832-1913) touched down in South Carolina after traveling more than 500 miles in a balloon. Like Wise, Lowe's original intention was to establish transatlantic balloon travel, but political unrest caused him to change his priorities and he began to promote the idea of an aerial reconnaissance corps to support the northern armies. In June, he demonstrated to President Lincoln that a balloon could be used for aerial reconnaissance by going aloft and sending a message through a War Department telegraph system connected to the ground through a tethered cable. He also took along a camera and shot historic aerial photographs. The *Balloon Corps* was established by Lincoln as a civilian corps under the Union Bureau of Topographical Engineers and Lowe served as the head of the Union Army aeronautics service during the American Civil War.

Lowe wasn't the only ballooning pioneer to envision a military air corps. John LaMountain, who had accompanied John Wise on his 1859 trip, was also promoting the idea of using balloons to surveil enemy troops from the air. While Lowe got the early favor of politicians, LaMountain nevertheless made reconnaissance trips later in 1861, surveying Confederate troops near Newmarket Bridge, Virginia—thus becoming the first to officially deliver enemy intelligence obtained from an airborne balloon.

 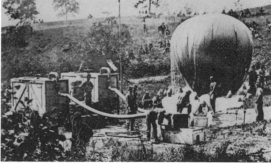

Left: Thaddeus Lowe viewing the battle from an aerial balloon, the "Intrepid", on 1 May 1862, during the Civil War Peninsular Campaign. Right: The Federal observation balloon being inflated during the period of the Battle of Fair Oaks, Virginia, 1862. [Library of Congress compiled by Milhollen and Mugridge, photos public domain by date.]

Concerned about the northern forces' Balloon Corps, Capt. John Randolph Bryan volunteered to supervise the construction and deployment of a surveillance balloon for the Confederate army. In April 1862, he sailed over Yorktown, Virginia, in a hot air balloon, as part of the Peninsular Campaign. He maintained communication with ground forces through semaphore. When safely out of shooting range of Union troops, he created a map of Union positions. The same month, General Fitz John Porter was borne into enemy territory in a balloon that strayed, but managed to return to base. The die had been cast and further balloons followed, with the popular story being that southern belles donated their finest silk dresses for the making of balloons for the War effort. In fact, it is more likely that the silk balloons were constructed out of silk dress fabric than of actual silk dresses.

Commercialization

Many balloonists offering their services to the Civil War effort originally intended to establish commercial balloon travel services, particularly trips across the Atlantic. Once the War ended, commercialization efforts were renewed. By 1874, passenger balloon flights were being offered over San Francisco and, within thirty years, the Thomas A. Edison, Inc. company was creating animated aerial moving pictures from balloons.

The Late-1800s - New Technologies and Chemistries

Toward the end of the century, helium was discovered. It was lighter than air and without the explosive nature of hydrogen, and thus displaced hydrogen as the preferred gas for filling aerial balloons (especially following the tragic explosion of the Hindenburg in 1937).

In 1894, Octave Chanute (1832-1910), an American immigrant from France, improved public opinion of flying as a field by publishing *Progress in Flying Machines*—a history of attempts at flight up to that time. Over the next several years, Chanute applied his engineering knowledge to the design of gliders, and later served as a flight consultant to the famous Wright brothers.

In 1899 in America, Orville and Wilbur Wright were experimenting with kites, with the goal of putting a man in the air, while overseas, in Europe, a different type of craft was developed by German entrepreneur and intelligence agent Graf (Count) von Zeppelin.

The 1900s - The New Age of Flight

Zeppelin Historic Commercial and Military Airships. The LZ1 Zeppelin at Lake Constance, created and constructed by Graf von Zeppelin and Theodor Kober. [Zeppelin Museum Friedrichshafen photos, public domain by date.]

The invention of dirigibles and balloons provided new ways for people to view the world from aloft, even if it wasn't for any significant distance or period of time.

As the 20th century approached, Ferdinand Adolf August Heinrich Graf von Zeppelin (1838-1917) was perfecting his Zeppelin airship. He was seeking ways to make flying more efficient and comfortable. The first flight of the Graf Zeppelin in south Germany on 2 July 1900 extended almost four miles. In 1908, von Zeppelin established the Zeppelin Foundation for the development of aerial navigation and airships and the Zeppelin airship was put into service as a commercial craft in 1910.

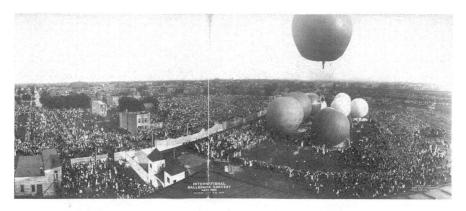

Ballooning captured the popular imagination in the early 1900s and soon competitions were being held. This picture shows the International Ballooning Contest held at Aero Park, Chicago, on 4 July 1908, with thousands of spectators enjoying the sight. Since there were no highrises from which this photo could have been taken and the terrain in the foreground is reasonably flat, it's reasonable to assume that it may be an aerial photograph taken from a nearby balloon. [Library of Congress, George R. Lawrence Company, public domain by date.]

The Wright brothers graduated from building kites to creating a new form of fixed-wing flying machine. The main stumbling block to aerial photography at this time was not so much the ability to get into the air, which was improving by the day, but rather the bulky nature of cameras at the time. They were not awkward to handle and limited in their utility by long exposure times that required the camera (and the subject) to remain completely still. As cameras improved, however, so did aeronautical science.

The historic first piloted airplane to sustain flight. It was flown by Orville Wright and later that day in a longer flight by his brother, Wilbur Wright, 17 Dec. 1903. The plane had a wingspan of just over 12 meters, powered by a 12-horsepower gasoline engine. [The *Smithsonian National Air and Space Museum* collection. The original negative was imaged on glass, public domain by date.]

On 17 December 1903, Orville Wright piloted the first sustained powered airplane flight from Kitty Hawk, North Carolina (at least, it was the first to receive widespread attention in the press). Within two years the brothers had improved on the technology enough to develop a practical model. Meanwhile others continued to research balloon flight in parallel with the development of planes.

The armed forces were now giving the new aerostats and flying machines some serious consideration as tactical military vessels. Thomas Scott Baldwin, an associate of the Wright brothers and Alexander Graham Bell, was appointed by the U.S. Army to oversee the construction of balloons and dirigibles for military use, beginning with an airship similar to the Zeppelin, in 1908.

R. F. Collier, of *Collier's Magazine*, purchased a Wright brothers plane in 1910, which he turned around and offered to the U.S. Army. The plane was demonstrated for possible use as a courier craft for travel among military stations.

A pioneer aerial image, this photo of Washington, D.C. (left), was apparently taken from a hydrogen balloon on 4 June 1907, possibly from the U.S. Army balloon, or one like it, that is depicted on the right. [U.S. Army historic archives, public domain.]

U.S. Signal Corps Dirigible No. 1 in the hangar (left) and being launched (right). Originally the Baldwin dirigible, it was first tested in the summer of 1908 and formally accepted into the Signal Corps fleet in August. [U.S. Air Force historic archives, public domain.]

The 1910s - Flying Machines and Consumer Cameras

Many covert photographic assignments require 1) smaller cameras, so they can be packed, hidden, or attached to balloons, remote-controlled planes, or other airborne craft and 2) often require wider aperture capabilities, so pictures can be taken in lower light conditions. Both these problems were addressed by Oscar Barnack, a German designer with the Leitz optical company, who developed a prototype of the Leica camera by about 1913. However, World War I interfered somewhat with Barnack's realization of his finished model. Nevertheless, when the new 35 mm available-light Leica was introduced, mainstream news photographers and celebrity newshounds quickly began to make use of the new capabilities of the camera and the photographic medium for surveillance purposes.

The development of the Zeppelin airship in Germany spurred the British Admiralty to construct a rigid airship of similar design. By 1910, both Germany and Britain had acknowledged the surveillance and weapons-carrying capacity of airships. Several prototypes were submitted and some were developed, but the first to actually fly was the "HMA. No. 9", originally designed in 1913, but delayed in construction and flight until 1916, due to an interim emphasis on non-rigid airships.

The development of the Zeppelin in Germany spurred the creation of an airship fleet in Britain. Top: The British HMA No. 9, the first rigid British airship to fly successfully. Bottom: A British 'submarine scout' (designated "S.S.") non-rigid reconnaissance airship. [Airship Heritage Trust historical photos, copyrights expired by date.]

When the Mayfly, the first of the rigid British airships, was destroyed, the Royal Navy chose to develop a fleet of over 200 non-rigid airships for submarine, coast, and ocean surveillance, including the NS No. 10 North Sea class vessel shown here. NS vessels were equipped with separate command cabins and engine rooms slung under the bellies of the airships, as is shown. These were often joined by a walkway for braver souls. The ships could stay aloft for about 24 hours. [Airship Heritage Trust historical photo, copyright expired by date.]

World War I was the first significant war fought both in the air and on the ground. The Germans, the British and, to a lesser extent, other Europeans and Americans, used rigid and non-rigid airships and equipped their crews with binoculars and telescopes for various surveillance activities, including submarine-spotting, coastal patrol, and target assessment.

In America, the U.S. Signal Corps officially formed a new Aviation Section on 18 July 1914, thus recognizing the role of aviation in military activities. The Section was small at first, consisting of just a single squadron, and the Aviation School.

In 1916, Brigadier General John Pershing began to make extensive use of innovative technologies, including mobile Radio Tractor units and aerial reconnaissance systems. Captain Benjamin Foulois commanded the 1st Aero Squadron, using several Curtiss JN-3 aircraft, when Pershing ordered troops into New Mexico to respond to the raid of Pancho Villa. Pancho Villa wasn't found, but the events demonstrated the value of aerial reconnaissance and radio devices in military reconnaissance.

Lothar von Wallingsfurt (the Sepia Baron) had a fertile imagination and a fervor for air transportation and is perhaps best known for an unsuccessful attempt to fit out Parseval-Sigsfeld observation balloons for use as fighter interceptors, in 1915. He later persuaded the German Naval Airship Service to use Zeppelins in warfare. On its first raid in 1917, the L.49 airship crashed over London. Wallingsfurt then served with the German Secret Service in Zürich where he continued to advocate the use of airships for warfare and transport. [Janus Museum photo and poster, copyrights expired by date.]

On a historic day in July 1919, half a year after the declaration of the Armistice, the British R34 became the first airship to cross the Atlantic. The vessel, which had been constructed in Glasgow, Scotland, negotiated the winds of the North Atlantic for thousands of miles from East Lothian to Long Island, New York.

The 20s - Transatlantic Flight and the Lure of Outer Space

The successful flight across the Atlantic by the R34 probably was as inspirational as the invention of the airplane. Human imagination quickly dreamed up rockets that could fly into space, after the Wright brothers demonstrated that piloted heavier-than-air craft were not only possible but practical. In 1926, Robert H. Goddard (1882-1945), a prolific inventor, launched a rocket on a flight that achieved a height of 41 feet using a liquid propellant. Forty-one feet wasn't exactly outer space, but it was sufficient to convince him and other inventors, who were discovering the wonders of the heavens through new, powerful telescopes, that such a thing was possible.

It was an era of adventure and firsts in aviation history. A year after the Goddard rocket flight, Charles A. Lindbergh (1902-1974) achieved enduring international fame by completing a solo, nonstop, monoplane flight across the Atlantic in the Ryan NYP *Spirit of St. Louis* in a time of 33.5 hours, at an average air speed of a little more than 100 mph. Lindbergh continued to be a goodwill ambassador and champion of commercial flight for many years, stimulating air transport design and development.

The '30s - Paparazzi, Pioneers, and Global Unrest

Erich Salomon (1886-1944) was one of the first photographers to persistently take pictures in places where cameras were not permitted, often concealing the camera in an attaché case—the archetypal 'spy' accessory. Salomon gained notoriety when he published covert photographs of a murder trial. In 1931, he published *Celebrated Contemporaries in Unguarded Moments*, which included candid shots of about 150 dignitaries and celebrities of the time.

Aerial explorations of cultures, with benefits for archaeology and anthropology, also became practical. Atlantis Verlag produced a number of books in the 1930s to 1950s that showed the culture and topography of foreign nations that were otherwise rarely seen by westerners. Wulf Diether Castell took some remarkable and beautiful photos of the Gansu, China, with a Leica camera. Over 100 of these black and white photos were published in *Chinaflug* (Flight over China), in 1938. The book, unfortunately, is out of print and difficult to find, but represents some of the earliest high quality aerial photos of places in the world that were hard to reach.

By 1933, exciting and downright dangerous efforts spurred interest in aerial and infrared photography. In an ambitious pioneer survey, cameras were mounted on two British open-cockpit biplanes, which had been converted from bombers to reconnaissance planes for the purpose of aerial surveillance. These planes had a maximum cruising speed of about 135 miles an hour. In addition to traditional black and white photos, the adventurers shot black and white Ilford infrared images [Greer, 1994]. They used gimbal-mounted Eagle III Williamson aircraft cameras that could shoot up to 125 five-inch photos. Nepal was closed at the time, so permission was needed to do a flyover, and the plane was not permitted to land in Nepalese territory. The surveys were done on a trip over Everest—traditional and infrared photographs were taken in the vicinity of the Himalayas.

Almost everything associated with the historic aerial trip was improvised. The camera was constructed of plywood by the art editor of the London Times newspaper. The designers had to find an effective way to attach the fully manual camera so it was aimed at the terrain. They mounted it in slings under the plane, facing forward, and tipped it down about 6°, with stuffed leather pouches as shock absorbers. The system was primitive by modern standards—the camera was aimed by aiming the plane. Double plates had to be individually mounted and removed and the shutter manually wound and released. The signal for camera readiness was a jerk on a string to the pilot. Despite the difficulties, the mission was a success and set the stage for future reconnaissance planes and aerial surveys.

A number of amateur and professional photography associations were established around this time and some of the prominent camera companies were formed. In 1934, the American Society for Photogrammetry and Remote Sensing (ASPRS) was founded. The Japanese company now familiar as Minolta began lens production in 1937 as hostilities were gaining momentum in Europe. The global unrest resulted in many photographic technologies being put into service for wartime activities and in companies modifying consumer products to market them in the service of war. Konishiroku Co., Ltd., began constructing aerial cameras and X-ray photographic systems in 1937. Minolta created a *Rokkor* aerial camera lens in 1940, and the company was put into service by the Imperial Japanese Navy in 1942 to manufacture optical glass. Fuji Photo Film Company, which was gradually expanding into other photographic products besides film, manufactured aerial cameras and lenses during World War II, as well [Ono, 1996].

War and Post-War

Not all balloons were designed for carrying people or cameras. Some were intended as defensive obstacles, for impeding the flight of foreign reconnaissance or fighting aircraft. Tethered *barrage balloons* were in use at least by the mid-1930s and occasionally they would explode in flames over the areas they were intended to protect, an event that was captured in a London newsreel in 1934. Before and during World War II, the balloons were installed to try to prevent air attacks on vulnerable areas such as major cities and important shipping lanes. The heavy tether cables could slice the wing off a low-flying aircraft that tried to fly through the barrage.

CounterSurveillance. Left: U.S. Marine Corps with a barrage balloon on Parris Island, South Carolina in May 1942. The gas-filled barrage balloons and their cables were used as defensive obstacles to deter enemy reconnaissance and fighter aircraft. Right: Unfortunately, it's difficult to see them in this small photo, but there appears to be a combination of at least a dozen airships and barrage balloons in this moving image of "D-Day" during World War II on the French beachhead in Normandy. [Library of Congress FSA/OWI photo by Alfred Palmer, public domain; U.S. Army 6 June 1944 historical archive photo by Steck, released.]

The war years were a strong stimulant to aircraft development and piloting technology. Existing navigation devices were put into the service of the war effort, and research dollars were channeled toward these technologies. It is therefore not surprising that many advances in aircraft design and surveillance photography were improved in the 1940s.

Jet engine technology began to appear in the early 1940s and the new designs were quickly put into military service. The higher speeds and higher altitudes at which planes could now fly made it more difficult to hang out a window with a camera to take aerial pictures, as had been done in earlier decades. It also made it necessary for the pilots and crew to wear protective articles like arctic gear and oxygen masks.

At this time, allied military reconnaissance was conducted in part by camera-equipped, modified four-engine B-17 and B-24 aircraft and modified two-engine P-38 and F-5 fighter planes, along with the British Mosquito. The P-51 Mustang, equipped with an oblique camera, was also used for reconnaissance.

Airships played an important part in surveillance in World War I and, despite the newer, faster airplanes that were developed for use in commercial and military aviation, airships continued to be used for a variety of activities during and following the war, including submarine spotting.

Airships travelled more slowly than fixed-wing aircraft, were generally more fuel-efficient, could stay airborne for longer periods of time, and were faster than most marine vessels. These characteristics made them useful for many types of visual and telescopic surveillance (in recent years they are coming into use as radar-carrying platforms, especially for large-aperture radars). For their size, airships had relatively low radar signatures due, in part, to their streamlined shape.

L. Fletcher Prouty offers this personal perspective on his war experiences:

> We also learned while we were there that Frankfurt was the European base for the border flying and other aerial surveillance activities. This was before the U-2 started operating; it later became the European base for U-2's. We had aircraft flying the borders, doing surveillance with either radar or photography in that period. They were quite effective. We also had an enormous balloon program. We would launch large balloons, loaded with leaflets or loaded with instrumentation, that would provide various propaganda information throughout Eastern Europe....
>
> It was an interesting program. You'd think that just random balloons wouldn't accomplish much, but they apparently did.... There was a base at Wiesbaden which was entirely operated under what we called "Air Force cover", but was for the operation of CIA aircraft. And they were very active all over Europe.
>
> [David T. Ratcliffe electronic edition of *Understanding Special Operations And Their Impact on the Vietnam War Era: 1989 Interview with L. Fletcher Prouty, Colonel USAF (Retired): Military Experiences 1941-1963*, rat haus reality press.]

The 1950s - The Postwar Boom and the Cold War Years

On Soviet Aviation Day in 1955, foreign observers saw, during an exhibition in Moscow, a surprisingly large number of new heavy bombers fly by, leaving the impression that the Soviets were building a powerful bomber force. However, it was learned later that the same bomber squadron had been flying around in circles. Since 1956, U.S. U-2 operations supplied evidence that the Soviets, in fact, are concentrating on rockets and not on the production of bombers.

[Dr. George Walter, Intelligence Services, *Military Review*, August 1964.]

The fear and distrust that had flared up between the U.S. and the Soviet Union in World War II wasn't alleviated by later events. Each nation spied on the other and western concerns were fueled by some of the expansionist and repressive policies of the Soviet government. In turn, the Soviet government perceived the Americans as militarily aggressive and unsympathetic to communist doctrine. All in all, satisfactory political negotiations were not prevalent at the time.

President Eisenhower wanted more military intelligence on the capabilities of the Soviets than they were voluntarily revealing and, in December 1955, authorized project GENETRIX, to include of high-altitude balloons carrying photographic equipment. The balloons drifted over Soviet territory in early 1956, collecting images along the way, and were subsequently recovered in the air over the Pacific Ocean. Not surprisingly, the Soviets objected strongly to this, as no doubt the U.S. would have done if unauthorized Soviet balloons had flown over the Midwest, and the program was discontinued, only to resurface a few months later in the form of the first significant U.S. spy plane.

The Beginnings of Space Travel

In 1957, the Russian Federation announced in print that they were going to send up a satellite and published the frequencies so that radio operators worldwide could monitor its progress. They made good on their announcement and launched Sputnik I in October 1957. Considering the advance notice, it is somewhat surprising that the U.S. reacted with so much surprise. A month later, the Federation sent up Sputnik II with a dog named Laika. Spurred into action, the U.S. stepped up its own rocket program and, in December, before a television audience, attempted to launch the Vanguard satellite—the Vanguard blew up on the launch pad.

People were, in some ways, more trusting of the government in the 1950s. Television was still rare and only scattered news came through the radio. Since it was difficult to know just what the government was doing at any given time, compared to today, there was less of an expectation of being able to have a voice in political affairs. As a result, there was less pressure on public officials to reveal the internal workings of their defense and spying activities. Occasionally accidents on foreign soil would reveal clues as to what kinds of surveillance activities were being undertaken, but often operations were clothed in secrecy.

The First Spy Planes

Planes were used to spy almost from the beginning, but it was not until the 1950s that very specialized stealth planes were designed and manufactured.

Construction of the single-engine *CL-282 spy plane* began at the Lockheed Experimental Department in the mid-1950s, with the aircraft becoming operational in June 1956. Its first use was during the Suez Canal Crisis. The CL-282 was specifically designed to fly higher and longer than previous planes, in order to carry out foreign reconnaissance, particularly over the Soviet Union. In other words, it was effectively a replacement for the GENETRIX balloon program. In spite of its covert intentions, there were internal reports that the craft, now better known as the U-2, was not particularly difficult to detect or track. In spite of this vulnerability, it managed to fly many missions before it was shot down over another nation's airspace. During the course of its service, scientists and engineers discovered that it was more difficult to use radar to track planes that were flying at supersonic speeds. By 1959, approval for improved stealth aircraft was given by the President.

In early 1960, the general public knew nothing about the U-2 aircraft. The Soviets had been protesting American violation of Soviet airspace, while the Americans denied any such activities. Then news reached President Eisenhower that a U-2 spy plane, dispatched on a surveillance mission over Soviet territory, had disappeared on 1 May 1960. Assuming that the pilot either couldn't have survived the crash or would have sacrificed himself and the plane rather than being captured, the Eisenhower administration announced that it was nothing more than a U.S. weather plane, studying the upper atmosphere; it must have gotten lost over Turkey, perhaps even crashed. To support the story, a U-2 was given NASA decals and presented to the press. A few days later, the Soviets revealed the falsity of the American assertions by announcing that they had recovered the live pilot, the wreckage, and the exposed surveillance film. This caused significant embarrassment to the President. More importantly, however, it resulted in the cancellation of planned Presidential trips to Paris to attend an important multinational summit meeting, and to Moscow to attend a June diplomatic meeting with the Soviets.

The U-2 continued to fly surveillance missions along the Soviet borders, sometimes straying across those borders, with continued protests from the Soviets and various apologies from the U.S. Later, the U-2 was also used to provide aerial intelligence in the region between Florida and Cuba during the Cuban Missile Crisis, in 1962—a serious diplomatic situation that almost led to a major war. Plans for adapting the U-2 craft so it could be launched from aircraft carriers were soon underway, with testing of modified U-2s beginning in August 1963. It was found that the planes could take off from carriers, but needed more modifications to be able to land effectively.

In science and technology, one of the most important areas of research and engineering in the early 1960s, was the development of titanium alloys. These strong industrial materials enabled construction of new generations of aircraft and deep-ocean-going craft. Research into materials, designs, and strategies that could reduce radar signatures was also continued and applied to the U-2 planes.

Thus, general and applied research into stealth technology and airplane design continued and the successor to the U-2, the A-12 (part of a project codenamed OXCART), flew in 1962. The A-12 could travel at three times the speed of sound and had a range of more than 3,000 miles. It would cruise at altitudes three times as high as large commercial airline services. The A-12 was less vulnerable than the U-2 and incorporated better photo-reconnaissance equipment. It also incorporated special quartz windows that could admit infrared light. Perkin-Elmer cameras were selected with Eastman Kodak as a backup supplier. The A-12 project didn't continue past 1968, but it provided research information that resulted in a fleet of two-seater A-11s, now known as SR-71s.

The Beginnings of Space Surveillance

While atmospheric surveillance aircraft improved, the U.S. space program was getting underway, and many aspects of it were top secret.

The X-15 was a rocket-powered, missile-shaped aircraft designed for high-altitude testing and the deployment of scientific sensing and data-collection instruments. Like the later Space Shuttle, the X-15 was launched from another craft at high altitude—in this case, a B-52 aircraft (right). It would then glide in powerless flight and execute a glide landing. There is an X-15 rocket plane exhibited in the National Air and Space Museum in Washington, D. C. [NASA/ Dryden Flight Research Center 1960 and 1968 news photos, released.]

There followed a great deal of overlap in the research and design of high-altitude, high-endurance aircraft and the design of rockets and space capsules as the U.S. space program got underway. In 1959, the X-15 research aircraft was developed to provide in-flight information on aerodynamics, flight control, and the effect on pilots of high-speed, high-altitude flight. The aircraft was also used to carry sensors and scientific payloads beyond the Earth's atmosphere, remaining in regular service for the next decade.

CORONA

In the early days of space missions, public sources of information on the U.S. surveillance satellite programs were scarce. Occasional news reports appeared in the media, as when John

Finney reported a satellite capsule-recovery in an August 1960 issue of the *U.S. Times*, but that was only the tip of the iceberg. Three decades later, oral histories and declassified information began to show the extent of the programs and fill in some of the gaps.

Left: Mobile launcher I was constructed during the period from 1966 to 1968 by NASA at the Kennedy Space Center. It was an integral part of the Apollo program. The structure was altered in the mid-1970s and finally disassembled in 1983 soon after this picture was taken. Right: Earth from 250,000 miles away as photographed by the astronauts of the Apollo 10 space mission in May 1969. [Library of Congress HABS/HAER collection; NASA/Ames Research Center news photo, released.]

The CORONA space reconnaissance satellite was conceived and contracted out in the 1950s as part of the Air Force Weapon System 117L (WS-117L). The CIA was given responsibility for the surveillance aspects of the project. In 1957, the project came to be known by the codename SENTRY and thereafter as Samos. By 1960, Samos consisted of a number of basic aerial imaging technologies that characterize much of aerial surveillance today, including

Project 101A - high-magnification visual surveillance images readout,

Project 101B - high-magnification visual reconnaissance film recovery,

Program 201 - high magnification visual reconnaissance film recovery,

MIDAS - Missile Defense Alarm System, an infrared surveillance sensor intended to detect the hot gases from launched weapons, and

Discoverer (CORONA) - visual reconnaissance film recovery.

Beginning in August 1960, Lockheed's CORONA satellite, the first photo reconnaissance system, collected images of the Earth sized at about 12 meters resolution (the width of a small city lot), using an Eastman Kodak Company camera. By the early 1970s, the resolution had been improved to about three meters, the length of a small bedroom, and the volume of film that could be housed had been increased as well. While President Johnson alluded to the use of satellites for surveillance activities in the mid-1960s, classified documents regarding the CORONA system were not released to the public until 1995. Some of the details will never be known—much of the information was conveyed by mouth and never recorded in print. During its tenure, the CORONA project imaged Soviet submarines, bomber and fighter jets, atomic weapons storage installations, ICBM complexes, the Missile Test Range north of Moscow, and ocean-going Soviet vessels. The MIDAS sensing system eventually evolved into early warning satellite systems.

Continued Aircraft Surveillance

As was mentioned earlier, the A-12, A-11 and, eventually, the SR-71 aircraft, evolved out of the original U-2 surveillance planes, and project OXCART included the readying of planes for the possibility, if not the fact, of flying over Communist China. This resulted in Operation Black Shield, which was called up in 1965 to address the needs of Far East operations, based out of Kadena Air Force Base in Okinawa.

In May 1967, the first Black Shield missions were flown over North Vietnam and the Demilitarized Zone. Reconnaissance photos of 190 SAM sites were taken from the aircraft and there were no indications that they had been detected by foreign radar on the initial flights. Later flights that detected radar tracking signals were not greeted with hostile actions and the sorties determined that North Vietnam did not have surface-to-surface missiles. The planes were fast. A single pass over North Vietnam in the A-12 took less than 13 minutes. On landing, the photographic film was removed and sent for processing.

On one U.S. reconnaissance flight over North Vietnam, the American surveillance craft was detected and a missile fired, all of which was documented photographically by the reconnaissance camera. It was the first time the electronic countermeasures equipment associated with the A-12 had been used in real action and it apparently functioned effectively.

By the late 1960s, the public was becoming better informed about satellite surveillance, as less-speculative articles about military satellites were published in popular journals.

The '70s - Space Stations and Civilian Satellites

Commercialization of the satellite imagery market was not a smooth process, but it has been an important one. Access to detailed information about the Earth not only influences the balance of power between governments and civilians, but also greatly affects how surveillance professionals, environmental groups, businesses, and individuals can access information. The quality and price of images also strongly affects how the information is used and by whom.

Landsat, the first civilian remote-sensing satellite was developed by the U.S. *National Aeronautics and Space Administration* (NASA). Launched in 1972, it was designed to capture multispectral images at a resolution of 80 meters (about the size of a playing field). Due to the limitations of cost and resolution, these images were mostly of interest to scientists, educators, and government agencies, and didn't yet meet the requirements of most commercial entities. With Landsat 4 and 5, which had an improved 30-meter resolution (about the length of a city lot), commercial marketing of the technology became feasible. Thus, commercialization was initiated with the Carter administration *1979 Presidential Decision Directive* that transferred operation of Landsat to the *National Oceanic and Atmospheric Administration* (NOAA) and directed NOAA to increase private sector access to the technology. The commercialization was to be initiated gradually as the market developed.

At the same time that new satellites were being launched, ground-based space surveillance complexes were being developed. With so many more vessels launched into space, there was a strong motivation for developing telescope-based centers that could detect and monitor all the new space objects. The interest in deep space and the celestial environment outside our solar system was increasing as well. The *Maui Space Surveillance Complex* was one of the centers established at this time to surveil a larger part of our universe.

The 1980s - Politics, Peace, and the Pressure to Commercialize

In 1982 and 1983, the U.S. government commissioned some studies to determine the feasibility of commercializing the satellite-imaging field and found that the commercial market for Landsat imagery, at the time, was underdeveloped and that commercialization "should be done gradually". Three subsequent studies supported this conclusion and some analysts warned that subsidies would be necessary if commercialization proceeded too fast. In spite of this, the Reagan administration chose to accelerate commercialization.

The U.S. Congress responded to the President's commercialization directive by signing into law the *Land Remote Sensing Commercialization Act* (P. L. 98-365), in July 1984, the year Landsat 5 was launched. The intent of the LRSCA was to increase commercial access on a nondiscriminatory basis and to establish a licensing process. The commercial sector still didn't respond enough to support commercial ventures, so the U.S. government agreed to approximately $300 million in subsidies through the late 1980s, with EOSAT to assume all operational costs upon launch of Landsat 6. In a reversal of support in the 1987 Reagan budget proposal, nearly half the promised subsidy ($125 million) was deleted and EOSAT instituted layoffs and a dramatic downsize in marketing.

While U.S. politicians were busy trying to work out the timetable and terms of commercialization, foreign nations were stepping up their own satellite programs, with France launching Spot-1 in February 1986.

After more studies and proposals, by spring 1988, the U.S. government agreed to subsidize EOSAT in the development of Landsat 6. Any project cost overruns were to be covered by EOSAT.

The U.S. IRS-1A satellite, with visible spectrum and several ranges of infrared imaging, was launched in 1988. It had a spatial resolution of about 72 meters with a 148 km swath.

Over the next several years of the George H. Bush administration, subsidies to continue operations of Landsat 4 and 5 beyond the original contracts were approved.

Unfortunately, the premature commercialization of the satellite imaging market increased the price of image scenes to over $4,000 each. As costs rose, demand fell. Between the mid-1980s and 1990, purchases dropped to a quarter of their previous volume. Even research institutions were finding it difficult to justify the increased costs.

In spite of various setbacks and disappointments in commercialization of satellite imagery, a number of private and public companies began to plan ways to market imagery. One of these was Aerial Images, Inc. (founded in 1988), partnering with Microsoft and Kodak to create the Terra-Server online image server, in the 1990s. A similar venture was founded in 1995 as Space Imaging, also in association with the Eastman Kodak Company. Back in the late 1980s, however, it was difficult to predict how successful these ventures would be, and how long it would take to launch commercially viable private satellites.

The 1990s - Global Conflicts and the Commercial Sector

In the 1990s, both surveillance and commercial satellites were beginning to share the orbital space around the Earth and, like good land, good orbits began to be much in demand.

In 1990, France launched Spot-2, and while it did not meet the spectral and swath specifications of the U.S. Landsat systems, it could compete with higher resolution images, thus potentially challenging U.S. dominance of the satellite imaging market.

The 1990-1991 Persian Gulf War was a technological war on many fronts. Not only were new land- and air-based missile-sensing and targeting systems put into use, but the Defense Intelligence Agency (DIA) reported that space satellite images from the Landsat systems were used for military surveillance, including terrain analysis, and planning and execution of some of the tactical maneuvers in Operation Desert Storm.

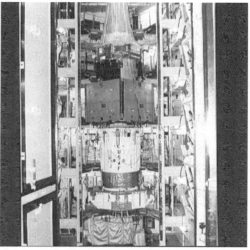

The STS-44 is a Defense Support Program (DSP) surveillance satellite that was designed to detect missile and space launches and nuclear detonations. An infrared sensor was installed in the top of the craft and it was scheduled for launch into a geostationary orbit in 1991. [NASA/JSC 1991 news photos, released.]

The IRS-1B satellite, with visible spectrum and near-infrared imaging, was launched in 1991. The craft's spatial resolution for imaging was about 36 meters with a 146 km swath.

In May 1992, a Presidential Directive from the Bush administration enabled more people, including environmental scientists, to make use of aerial surveillance data for monitoring the globe and its environment.

In the private sector, entrepreneurs were now interested in satellite imaging profits. In July 1992, WorldView, Inc., applied to NOAA for a license to operate a commercial satellite with three-meter (about the size of a bedroom) panchromatic resolution.

The Boeing EX, shown here in model form in a Transonic test tunnel, is an advanced surveillance aircraft concept proposed by Boeing to replace the Grumman E-2C "Hawkeye". Each wing is designed to hold active-aperture radar arrays to make the craft more aerodynamic than previous aircraft. [NASA/Langley Research Center 1993 news photo, released.]

In response to the increased use of satellite images in national defense and other activities, the U.S. Congress passed the *Land Remote Sensing Policy Act* (P.L. 102-555), signed into law by President Clinton in 1992. The act recognized the importance of satellite imagery in studying the environment and in carrying out activities related to national security. The act further acknowledged some of the problems associated with previous attempts at full commercialization and the loss of global leadership in the U.S. satellite imaging market. It consequently transferred control of the program to the *National Aeronautic and Space Administration* (NASA) and the *Department of Defense* (DoD). The authority for licensing private satellites remained with the U.S. Secretary of Commerce.

In January 1993, The WorldView, Inc., commercial license was granted, opening the doors to private-sector commercial satellite imaging ventures in competition with the Landsat systems. A few months later Commercial Remote Sensing System (CRSS) applied for a license through Lockheed Martin for its IKONOS 1 system.

Satellite and aircraft imaging in the early days was primarily used for scientific research, early warning systems, and foreign spying. Now these systems were being used in commercial applications, newscasting, and federal and local law enforcement on a regular basis. As an example, in April 1993, the FBI used aircraft-mounted forward-looking infrared radar (FLIR) to surveil the Branch Davidian compound in WACO, Texas. The videotaped transcript also included some audio of radio messages. Two tapes containing about 3.5 hours of video footage were later released to the public, in September 1999.

Improving Aerial Photography and Resolution

It was clear, after control of the Landsat program was transferred to NASA and the DoD, that scientists and administrators recognized the need for higher resolution images. For future systems, NASA favored 15-meter resolution (about the size of a house), whereas the DoD favored 5-meter resolution (about the size of a living room). As with most technologies, there was a trade-off between price and resolution—higher-resolution images would be more costly. While NASA and the DoD were airing their differences, the unexpected happened. In October 1993, the realities of launching expensive equipment into space became apparent when Landsat 6, which had cost almost $300 million, plunged into the ocean only a few minutes after launch. The DoD subsequently left the Landsat program.

The following year, a *Presidential Decision Directive* (PDD-2) announced that NASA would be responsible for the development and launch of Landsat 7, with NOAA operating the satellite and ground systems. Data would be archived and distributed by the *Department of the Interior*. The Directive also eased up on restrictions regarding the sale of satellite imagery to foreign agencies.

With declassification of U.S. documents on CORONA, ARGON, and LANYARD early in 1995, and release of CORONA photographs over the next several months, information about the CORONA surveillance satellite project started reaching the popular media. Nearly a million images of the Earth's surface, collected between 1960 and 1972, were declassified. Multinational cooperation in space and satellite programs increased at this time.

In 1995, it was reported that the KH-11 reconnaissance satellite was being updated to increase the downlink rate in order to bring it closer to realtime coverage. Some SR-71 reconnaissance aircraft, descendents of the U-2, were reactivated for tactical purposes.

It became apparent that there was a trend to change the structure of the U.S. surveillance satellite program from a few large-scale satellites to more small-scale satellites and to attempt to add stealth components to the vessels similar to those used in stealth aircraft.

Commercial Competition and Image Servers

By March 1994, the U.S. Department of Commerce began issuing licenses to commercial satellite companies. One of the first to be granted a Federal Communications Commission (FCC) license was Space Imaging, chartered by Lockheed Martin and Raytheon, with its IKONOS system eventually becoming a major contractor. In February 1995, Eastman Kodak Company and Space Imaging, Inc. announced a strategic business alliance to market satellite imaging products and services. This alliance would lead to consumer access to imaging products from a variety of satellites.

In April 1995, the commercial OrbView-1 satellite was successfully launched by Orbital Sciences Corporation, equipped with atmospheric instruments to provide weather-related data to U.S. government agencies.

In 1996, Lockheed's Space Imaging purchased EOSAT and its Mapping Alliance Program in a bid to become the commercial leader. Meanwhile other companies and nations were seeking to enter the potentially lucrative market.

In May 1997, Aerial Images, Inc., Microsoft Corporation, and Digital Equipment Corporation announced the Terra-Server project, a joint project to provide a global online atlas of two million square miles of two-meter resolution satellite images derived from the Russian Space Agency's SPIN-2™ satellite. On 18 February 1998 the satellite was successfully launched into orbit from Kazakhstan. Images from SPIN-2 are typically 500 MB, but the Terra-Server is designed to 'slice' these images to suit the user. The images can be downloaded as tiles, previewed, and purchased for about $30 per square mile if they meet the consumer's needs. Thus, the Terra-Server commercial system is essentially a map-on-demand system served 24 hours a day over the Web providing posters, custom images, and digital data on CD-ROM. This project is significant not only in its scope, but in that it represents a cooperative commercial effort between the U.S. and Russia.

The Terra-Server and other Web-based commercial satellite image servers such as Space Imaging, promise to fundamentally change the relationships between government agencies, intelligence organizations, and individuals. Information that was previously impossible to get, even by intelligence agencies, can now be downloaded in a few minutes through the Web.

Following the launch of OrbView-1 in 1995, OrbView-2, a color imaging satellite providing daily updates, was launched in August 1997. A Web image server was developed over the next several years, offering Web-based browsing and purchasing, with images delivered via Internet, computer tape, or CD-ROM. The OrbView-2 system was equipped with the Sea-Viewing Wide Field-of-View Sensor (SeaWiFS) which included six visible and two infrared imaging channels. OrbView-3 and Orbview-4 were added to the launch schedule for the first half of 2001 to provide high resolution options. OrbView imagery was made available through OrbImage and other commercial imaging partners of Orbital Sciences Corporation.

The Old and the New

During the 1990s, old technologies were brought back to be combined with new innovations and entirely new kinds of craft were built to serve surveillance objectives. Some of the most interesting of these were remote-controlled and autonomous aircraft.

Very few people in the 1970s or the 1980s would have predicted the return of airships in the late 1990s. After laying dormant for almost half a century, the historic Zeppelin airships were put back in service, showing their continued viability as a technology. By the mid-1990s, the U.S. Navy was building an airship sufficiently large to carry a phased-array radar system and entrepreneurs were attaching still and video cameras to a variety of balloons and airships for taking aerial photos.

This is the 71M Aerostat Test Balloon equipped with a radar system in a streamlined protector on the belly of the balloon. The system can be used to track aircraft and cruise missiles up to a range of about 150 nautical miles. It is shown ready for liftoff from the Aerostat Test Bed in a multinational defense exercise during Operation Roving Sands. [U.S. DoD 1996 news photo by Benjamin Andera, released.]

In the mid-1990s, the U.S. Navy began flying Predator aircraft off their carriers in simulated aerial reconnaissance operations. The Predator was a new unstaffed air vehicle (UAV) equipped with near-realtime infrared sensing and color video capabilities. The sensor data can be transmitted to controllers in ships and ground stations.

UAVs were used in NATO scouting missions over Kosovo for imaging troops and installations, relaying the signals to overhead aircraft.

The Altus 1 was built for NASAs Environmental Research Aircraft and Sensor Technology program and was followed by the Altus 2. Both are variants on the Predator surveillance drone built by General Atomics/Aeronautical Systems Inc. The Altus remotely piloted drones are designed for high-altitude, long-duration, scientific sampling missions.

The Altus 2, a variant of the Predator unpiloted air vehicle (UAV), is designed for scientific sensing/sampling. On the right it is shown retracting its landing gear. [NASA/Dryden Flight Research Center 1997 news photo, released.]

Aerial and Satellite Technologies - Successes

In April 1999, Landsat 7 was launched, with a resolution of 15 meters for panchromatic images and 30 meters for multispectral images of the Earth. The images were to be made available to consumers at cost. Thus, prices dropped tenfold in a few years, averaging about $550, depending on the degree of processing and correction applied to the data. With these services, the U.S. was endeavoring to recapture the global satellite imaging market. Meanwhile commercial companies were improving their technology, as were other countries.

Before the final testing and calibration of IKONOS, Space Imaging, which offers images from IKONOS 1, the Indian Remote-sensing systems, U.S. Landsat, Canadian RADARSAT, Japanese JERS, and the European Space Agency's ERS systems announced, in Nov. 1999, the imminent availability of high resolution images. Through its CARTERRA™ product line, Space Imaging specifically states that the imagery can be used for "monitoring specific natural and human events" leaving little doubt about its surveillance potential. Space Imaging lists utility and transportation companies, resource managers, real estate developers/brokers, and intelligence agents as potential buyers.

Much of the satellite contracting business had been handled by Lockheed Martin since the early 1970s, but in fall 1999, the Seattle-based Boeing Corporation received approval to build a new generation of imaging satellites through a NRO Future Imagery Architecture (FIA) funding award.

Aerial and Satellite Technologies - Problems

There is always a tug-of-war between what the public wants to know and what the government wants to reveal. As of about April 1999, the *Department of Defense* no longer provided tracking data on military satellites for posting on NASA's Orbital Information Group Website. Until that time, the site had been providing object orbit location and apogee/perigee information.

Commercial aerial imaging had evolved to the point that the U.S. Government began purchasing commercial images to supplement the information received from government systems. They were especially interested in Ikonos 1, Orb View 3, and QuickBird-1 high-resolution images. These sources have the benefits of easy access and lower classification levels and thus less red tape. One of the disadvantages of using these sources for government purposes is the general vulnerability of commercial systems, which are built for profitability rather than security. They lack stealth technology and extra shielding and thus may be vulnerable to attack.

The positioning of satellites is an art in itself and is not always successful. In April 1999, Lockheed Martin and its associates attempted to place the IKONOS 1 satellite at 423 miles above Earth, but the launch vessel was unable to accomplish the task. On 24 September 1999, Lockheed Martin, in association with Raytheon, Mitsubishi Corp., Eastman Kodak Company, et al., successfully launched its IKONOS imaging satellite aboard the Athena II launch rocket and successfully placed it in a sun-synchronous, circular, low-Earth orbit. IKONOS was designed to deliver satellite images with one-meter-resolution monochrome (the size of a desk) and four-meter-resolution color (the size of a room).

While new commercial satellites were being launched, regular military atmospheric reconnaissance continued closer to home. Aerial reconnaissance in unfamiliar territory has many hazards—in July 1999, crew members of a U.S. Army Dehavilland RC7 reconnaissance plane were reported missing in southern Columbia, South America, while flying a counter-narcotics surveillance operation. Unfortunately, the occupants of the plane didn't survive the crash that occurred in a remote, mountainous region of forest.

New Multinational and Humanitarian Projects

Fast planes, satellite images, the Internet, and human mobility have created a stronger sense that we are a global community and the effects can be seen in greater cooperation in scientific and commercial ventures. The late 1990s and early 2000 were a time when reconnaissance and surveillance technologies and personnel were used to aid victims of many disasters, including floods, hurricanes, earthquakes, and volcanic eruptions. Aerial surveillance information can be extremely valuable for relief efforts.

Left: A U.S. Air Force Combat Shadow (MC-130P) flies over South Africa on a reconnaissance mission to surveil damaged roads in Central Mozambique. Relief efforts aided the Africans following torrential rains and flooding. Right: P.O. 3rd Class Anita Lillibridge rotates the propeller on a P-3C Orion aircraft at Entebbe, Uganda. The Orion was used for reconnaissance missions over Zaire to support United Nations refugee relief efforts. [DoD March 2000 and Nov. 1996 news photos by Cary Humphries and Barbara Burfeind, released.]

In early 1999, the U.S. *Department of Defense* hosted Japanese specialists from the *Japanese Defense Agency* to train as space imagery interpreters and analysts in preparation for a Japanese satellite reconnaissance program scheduled to be put in place over the next several years. The satellites were developed by *Japan's National Space Development Agency*, using some U.S. technology. These actions were in part initiated because of political unrest between Japan and North Korea.

In June 2000, a more surprising alliance occurred when President Clinton and the new Russian leader negotiated the beginning of a U.S./Russia early warning system—a milestone in collaborative international surveillance.

Clearly, aerial surveillance is, and will continue to be, an important tool for observing our world and monitoring resources, people, and activities.

5. Description and Functions

This is a supplemental rather than a primary chapter, because aerial surveillance uses a range of technologies rather than being a technology in its own right. The basic technologies used in aerial imaging are discussed in the Infrared, Visual, Ultraviolet, Radio, and Radar Surveillance chapters and should be cross-referenced. To some extent, sonar is also dropped or towed by helicopters, planes, and remote-controlled craft, and submarine-spotting is an integral aspect of military aerial reconnaissance, so the Sonar Surveillance chapter is relevant, as well.

5.a. Cameras

There are hundreds of different cameras and camera-like imaging systems used for aerial surveillance and so just two examples associated with spaceborne systems are given here. You are encouraged to cross-reference information in the Radar, Infrared, Ultraviolet, and Visual Surveillance chapters, as there is considerable diversity and overlap in various camera-related imaging systems. The Infrared and Ultraviolet Surveillance chapters have more details on the specifics of taking pictures with special IR and UV films and filters.

TK-350 - A Russian high-resolution topographic satellite-based imaging camera. Accurate, detailed, with wide swatch capabilities, this camera is suitable for photogrammetry and relief definition mapping applications. Topographic maps of 1:50,000 scale can be created. It has a rectangular frame format with the long edge coincident to the flight direction. The focal length is 350 mm, with a relative aperture of 1:5.6 and stereoscopic overlap of 60% or 80%. Press rollers are used to maintain film flatness. It has an accuracy of 10-meter ground resolution, which can be enhanced with concurrent two-meter data from the KVR-1000.

KVR-1000 - A Russian high-resolution panoramic satellite imaging camera. This camera is accurate, detailed, with wide swatch capabilities. It is suitable for ground identification applications with an accuracy of two-meter ground resolution at an altitude of 220 km. The focal length is 1000 mm, with a relative aperture of 1:5 and a longitudinal overlap of 6% to 12%. It is used in conjunction with the TK-350 in the creation of satellite cartographic systems.

5.b. Remote-Controlled and Autonomous Vehicles

Many of the newer aerial surveillance technologies are based upon radio remote-control or computer-programmed autonomous flight.

Remote-controlled vehicles are linked to a control console by wired or wireless links. Wireless links are usually preferred because they offer greater flexibility of movement. Wired links are usually used when not much movement is required (sometimes just pan and tilt) and where very long ranges or tighter security might be needed. Some use a combination of wired and wireless links, typically set up in relays. It is likely that wired and wireless links will continue to coexist, rather than one superseding the other, due to the different strengths and benefits of each method, and because they can be used in combination.

The Pioneer I is a remotely piloted vehicle, shown here in a 1986 test on a U.S. Navy battleship. The Pioneer is equipped with a stabilized television camera and a laser designator designed for over-the-horizon targeting and reconnaissance. The system can be controlled from the host console to a range of about 110 miles with an operating endurance up to about eight hours. [U.S. DoD news photos by Jeff Hilton, released.]

Autonomous air vehicles are those that fly by themselves once they are set to go. Computer programming and intelligent algorithms have made autonomous vehicles possible and the more sophisticated ones can take off, fly a programmed mission, return, and land without further instruction. Hybrid vehicles can also be designed to be capable of autonomous flight, but may also accept instructions or programming changes en route. Autonomous vehicles are useful in commercial and military applications where larger craft are not cost effective or where there is risk of injury to a pilot and crew.

This is a U.S. Air Force Tier III Minus, or "DarkStar", which doesn't require a pilot or remote-control mechanism to fly missions. The DarkStar can autonomously execute preprogrammed flight maneuvers with help from differential Global Positioning System (d–GPS) data. It is a high-altitude, high-endurance air vehicle optimized for reconnaissance up to a distance of about 500 nautical miles in highly defended areas. Sensors can optionally be electro-optical or synthetic-aperture radar (SAR). Darkstar's initial flight was in March 1996. [U.S. DoD 1996 news photo, released.]

This unpiloted air vehicle is a U.S. *Department of Defense* Global Hawk, a high-altitude, long-endurance aerial reconnaissance system designed to provide high-resolution, near-realtime imagery of large areas. It includes a variety of surveillance technologies, including synthetic-aperture radar (SAR), electro-optical, and infrared sensors. It can survey an area of about 40,000 square nautical miles in a day to a resolution of about one meter. [U.S. DoD 1997 news photo by David Gossett, courtesy of Teledyne Ryan Aeronautical, released.]

5.c. Commercialization

Now that it has become less expensive to create aircraft and satellite images and since federal regulations have been relaxed somewhat to promote commercialization, there are a number of sources of aerial imagery available to members of the general public. These can be purchased to aid in land management, law enforcement, construction, resource reclamation, wildlife habitat assessment, and even to provide pictures of a person's home from the air.

Satellite photos can be purchased for $20 or less for about a half block square from sources on the Web that are listed in the resources section at the end of this chapter. It is important to have a good idea of the specific part of the world you wish to image and to remember that if you live in an area with constant cloud cover, it may not be easy to obtain a clear picture. Keep in mind also that the images on hand in the databanks aren't necessarily recent. You may have to use remote-controlled airplanes or balloons to get timely photos.

Licensed Commercial Satellite Systems (1984 to 2004)				
Company	**Applied**	**Approved**	**System**	**Web Site**
WorldView Inc./Earth Watch	15-Jul-92	04-Jan-93	EarlyBird	www.digitalglobe.com/ewhome.html
EOSAT	06-Oct-92	17-Jun-93	Landsat 6	www.spaceimaging.com
Lockheed/Space Imaging	10-Jun-93	22-Apr-94	IKONOS-1	www.spaceimaging.com
OrbImage	14-Dec-93	05-May-94	OrbView-1	www.orbimage.com
OrbImage	14-Dec-93	01-Jul-94	OrbView-2	www.orbimage.com
Astrovision	26-Mar-94	25-Jan-95	N/A	
Space Imaging		Oct-1995	CRSS	www.spaceimaging.com
EarthWatch/Ball	18-May-94	02-Sep-94	QuickBird	www.digitalglobc.com/ewhome.html
GDE Sys. Imag./Marconi N.A.	02-Mar-95	14-Jul-95	N/A	www.marconi-is.com
Motorola	31-Mar-95	14-Jul-95	(license terminated)	
Boeing Commercial Space	19-Jan-96	16-May-96	N/A	www.boeing.com/defense-space/space/
CTA Corporation	06-Sep-96	09-Jan-97	(license terminated)	
RDL Space Corporation	01-Mar-97	16-Jun-98	RADAR-1	(license terminated)
Space Technology Dev. Corp.	11-May-98	26-Mar-99	NEMO	www.spacetechnology.com
Ball Aerospace/Technologies		21-Nov-00		www.ball.com/aerospace
DigitalGlobe		06-Dec-00	QuickBird	www.digitalglobe.com
Space Imaging		06-Dec-00	IKONOS	www.spaceimaging.com
DigitalGlobe		14-Dec-00		www.digitalglobe.com
TransOrbital		06-Mar-02	TrailBlazer	www.transorbital.net
DigitalGlobe		29-Sept-03		www.digitalglobe.com
Space Imaging		14-Oct-03		www.spaceimaging.com
Northrop Grumman		20-Feb-04		www.northropgrumman.com

6. Applications

Aerial photography can be used in almost every field of scientific study and thousands of commercial and industrial projects. It should also be remembered that aerial photographs are not just taken from planes, helicopters, and remote-controlled balloons, sometimes they are taken while the photographers are skydiving, leaning over tall buildings, climbing towers or bridges, or perching on the top of trains, as some of the following examples show.

Agriculture, Wildlife and Park Management

Aerial surveys make it possible to evaluate property before it is purchased and to manage it for specific uses, including crops, livestock, wildlife, and parks. They also make it easier to assess damage from storms, pests, or disease and can aid in limiting the spread of the damage.

Many of the resource management concepts that apply to agricultural lands are also used in the development and management of parks, campgrounds, and wilderness areas—including construction and disaster response (and sometimes also search and rescue).

Top Left: The Sagami River rice harvest in Japan, 1997. Top Right: The Genava State Park small harbor and breakwaters project which combined federal and local funding. Bottom left: The Arlington National Cemetery amphitheater, 1997. A dramatic photo of the Mt. St. Helen's volcanic eruption. The eruption necessitated search and rescue, disaster relief, and restoration efforts from a number of agencies. The U.S. Army Corps of Engineers is involved in many federal projects. [First three photos U.S. Army Corps of Engineers news photos 1997 by Doyal Dunn, 1992 by Ken Winters, and 1997 by Susanne Bledsoe, released. Mt. St. Helens photo May 1980 U.S. Geological Service (USGS), released.]

Archaeology

Aerial photography has many benefits in archaeology, from locating suitable dig sites to finding patterns in the topography that have been left by ancient civilizations that are not visible from the ground. By adding infrared photographs, impressions of old villages may be visible, as well.

Sometimes dam, bridge, and flood construction or damage restoration projects will unearth new archaeological sites or, conversely, will be planned for areas that are already established as archaeological sites. Aerial photos can help determine the importance and range of the sites and record the process of locating artifacts and fossils. Left: An archaeological dig in the State Road Coulee flood control area, Wisconsin, showing the 'grid' method. Right: A Gray's Landing Lock and Dam dig. [U.S. Army Corps of Engineers news photos, 1991 by Ken Gardner and 1988 by Bill Lukitsch, released.]

Engineering and Construction

There are almost unlimited opportunities for using aerial photographs for monitoring engineering and construction projects. From bridge and dam construction to flood control channels and hurricane barriers, aerial photos help evaluate the landscape and settlements before doing any work, can monitor projects in progress, and record finished projects and any problems or changes that occur over time.

Aerial photos also provide a way to record and evaluate demolition projects and to assess damage from hurricanes, floods, and earthquakes. These are useful for insurance claims and reconsruction.

Left: The Alcan Highway White Pass Line in the Yukon, in 1942. This very scenic and sometimes hair-raising route is now the main artery to the new Alcan International Highway. Right: The Blue River Dam project in Oregon, a flood control and recreational project. [USACE Office of History photo; Army Corps of Engineers news photo by Bob Heims, released.]

Top left: An aerial view of the damage caused by the Kansas City District flood in 1993. Top Right: A new levee installed for Westwego/Harvy hurricane protection in 1998. Bottom Left: A Pentagon renovation project. Bottom Right: San Francisco and the Golden Gate Bridge. [U.S. Army Corps of Engineers news photos 1997 by Susanne Bledsoe and Robert Campbell.]

Environmental

There are many environmental applications for aerial photographs.

Left: Operation Fish Run, in which juvenile salmon were shipped downstream on a barge for release below the dam, 1984. Right: A General Motors chemical waste site in which silt curtains are being installed prior to a planned dredging cleanup in this section of the St. Lawrence River. [U.S. Army Corps of Engineers news photos, released.]

Aircraft and satellite environmental imaging applications include geological and marine ecology assessment, pollution and toxic wastes monitoring, habitat restoration, dike and levee construction, and wildlife tracking and conservation. Aerial photographs have been used to survey reefs and active volcanoes, to track whales and bears, and to aid in selecting habitat for parks and preserves. They have also been used to locate industrial wastes and to provide valuable data for waste reduction and removal.

Top Left: Colonel Crear arriving for the Mississippi Delta Wildlife and Wetland Tour in the Vicksburg District in 1987. Top Right: Alligator mom in the Mississippi wetlands, just one of the many species observed during the Wetland Tour. Bottom: Erosion protection measures are illustrated where the mouth of the Mississippi River flows into the Gulf of Mexico, including banks and rock dikes, 1994. [U.S. Army Corps of Engineers news photos, released.]

Military Reconnaissance

Aerial images are used in almost every aspect of military reconnaissance and many of the technologies developed for use on fast planes are highly sophisticated—with high resolution, fast frame rates, and computer processing of the resulting images.

Since before World War I, images from high vantage points were used for tactical and strategic missions. As the sophistication of the technologies improved, the number of applications for which they were used have steadily increased. At the present time, aerial imaging is primarily used to patrol borders and to assess foreign military activities weapons storage installations and armament purchases and manufacture, including nuclear and chemical production facilities. The military makes regular use of both air and space images from a variety of craft and communications systems. The images come not only from classified photo systems and surveillance craft, but also from commercial sources.

Left: U.S. surveillance photo of the Shifa Pharmaceutical Plant in the Sudan. Right: U.S. aerial photo of the Zhawar Kili Support Complex, Afghanistan. These images were used by the U.S. Secretary of Defense and General H. Shelton of the U.S. Army to brief reporters in the Pentagon on the U.S. military strike on a chemical weapons plant in the Sudan and training camps in Afghanistan. [U.S. DoD 1998 news photo, released.]

A P-3C Orion patrol aircraft flies over a U.S. and a Korean submarine during RIMPAC '98 exercises. Historically, many types of aircraft have been used for submarine spotting including fixed-wing planes, helicopters, rigid and nonrigid airships, and balloons. Now remotely operated and autonomous air vehicles are being added to the arsenal. [U.S. DoD 1998 news photo by August Sigur, released.]

Top: These U.S. Navy F-14B Tomcats were used to patrol over a Persian Gulf no-fly zone, operating from a nearby aircraft carrier. Bottom: U.S. Air Force F-15C Eagles flying on a patrol mission over Southern Iraq. Imaging 'pods' are often attached under the belly of aircraft. [U.S. DoD 1998 news photos by Bryan Fetter and Greg L. Davis, released.]

Left: A U.S. *Department of Defense* image of the Khamisiyah Ammunition Storage Complex in Southern Iraq, taken 10 Feb. 1991 and released to the public as a news photo in 1996. The smaller rectangle in the upper left is a region called 'The Pit'. The photo on the right is described by the DoD as a U.S. Navy photo of a submarine purchased from Russia by Iran that was being towed through the Mediterranean Sea toward Egypt. [U.S. DoD 1996 news photos, released.]

Military Communications

Wireless links through satellites are now an essential aspect of military communications.

This mobile 20-foot satellite antenna designed for military communications was set up by U.S. Air Force personnel from several Combat Communications Squadrons. Called the Quick Reaction Satellite Antenna, it was used in southwest Asia in Feb. 1998 in support of Operation Southern Watch. [U.S. DoD 1998 news photo by Efrain Gonzalez, U.S. Air Force.]

Space Surveillance

Amateur, commercial, and military surveillance of the far reaches of space has yielded an astonishing amount of information about our origins and the billions of galaxies and solar systems that were once beyond the reach of our technology and almost beyond the reach of our imaginations. The continued evolution of telescopes and the launching of the early space vehicles prompted the development of both ground and space-based space surveillance strategies and complexes. Space surveillance is discussed in the Infrared, Visual, and Ultraviolet Surveillance chapters.

Marine Patrol

There are many activities that are monitoried in marine environments.

Coast Guard Surveillance. Left: The HC-130 Hercules is used by the U.S. Coast Guard as a long-range surveillance and transport aircraft for search and rescue, law enforcement, marine environment protection, military readiness, and refugee monitoring. Right: The HH-65A helicopter serves similar functions as the Hercules and can take off from land or from medium-endurance Coast Guard cutters. [U.S. Coast Guard news photos by PAC Tod Lyons and USCG, released.]

Aerial surveillance vessels and equipment are used regularly in the enforcement of coastal territories for policing fishing rights, smuggling detection and prevention, search and rescue, salvage operations, and wildlife conservation.

The most common surveillance technologies used in marine patrol are sonar, radar, and aircraft-mounted optical devices (mostly visible and infrared).

Refugee Monitoring

Most nations are concerned about refugee movements, not only because they can represent an influx of illegal immigrants, but because refugees may be suffering from significant economic hardships and health problems that could pose risks to themselves and others.

Refugees traveling on foot on a road near Goma in Zaire (left) and in a refugee tent camp in Kilambo (right) as photographed by the U.S. Navy. [U.S. DoD 1996 news photos, released.]

Commercial Imaging Products

These are just two examples of the many thousands of types of imagery that are now available from satellite databanks, but they are sufficient to give an idea of the resolution that is now possible for commercial and personal applications. When reproduced full size, they are sufficiently detailed to distinguish bushes, architectural structures, and individual cars.

Left: Monochrome image of Tiananmen Square and the Temple of Heaven in Beijing, China, imaged by the IKONOS satellite, 22 October 1999. Right: Monochrome image of residential area from the CARTERRA™ Precision database. Such photos can be used for census confirmation, property assessment, community development, insurance adjustments (e.g., after natural disasters), licensing, permits, and other applications. [© 1999 Space Imaging news photos, used as per copyright requirements. www.spaceimaging.com/]

7. Problems and Limitations

Technical Limitations

Aerial surveillance encompasses such a diverse spectrum of technologies that it is difficult to make generalizations about the limitations of aerial surveillance devices. Limitations of specific categories of devices are discussed in other chapters. In general, resolution, storage capacity, image stabilization, and frame speed have been the main limitations, but even these have been overcome to a remarkable extent since 1995. The price of aerial images is no longer even a significant limitation, as digital storage has dropped dramatically in price, as has the cost of commercial satellite imagery. Online viewing of the Earth's topography from a "bird's-eye" view on Google Earth is available for the price of an Internet connection and a moderately fast personal computer.

Satellites are not suitable for all kinds of observation. Aircraft, dirigibles, and the newer autonomous and remote-controlled air vehicles will continue to be used for many applications, especially those that require realtime information associated with specific locations, including scientific observation, wildlife and border patrol, news broadcasting, and tactical reconnaissance activities.

Political Considerations

One of the problems with global economics is that those who have more continue to have the resources to widen the gap. Thus, discrepancies between the economic resources of developed and undeveloped nations continue to widen as technology advances. This is also true in the commercial use of aerial imaging. Countries that can get the information from aerial images have clear political and economical advantages over those that can't. Many entrepreneurial opportunities provided by the new imaging technologies favor those with state-of-the-art resources.

Because it is a relatively new field, many trade secret protections do not apply to satellite imagery. If you want to get a picture of the production yard of a competing contractor in your own country, or in another, there are currently few restrictions on doing so. However, the long-term trade imbalances that could result from open access to images may come under scrutiny in the near future.

8. Restrictions and Regulations

International law is still being developed with regard to aerial surveillance and aerial images. Presently, it favors the open use of imaging from outer space, since the benefits of international open access are many, but that could change.

In *Presidential Decision Directive (PDD) 54*, President Carter transferred operation of Landsat systems to the *National Oceanic and Atmospheric Administration* (NOAA) within the *Department of Commerce*. NOAA was directed to further private sector access to civil remote-sensing to move the technology into the hands of the private sector, to encourage growth of the industry. In recent years, commercialization has steadily grown.

Two United Nations documents of relevance to aerial surveillance include

The Outer Space Treaty of 1967 - States that outer space cannot be claimed as national territory, consequently, satellites can travel over any territory, and

U.N. General Assembly 1986 - The Assembly adopted a set of legal principles on civilian remote sensing that do not require prior consent. It declares that the sensed State shall have access to them on a nondiscriminatory basis and on reasonable cost terms.

In 1992, the U.S. Congress passed the *Land Remote Sensing Policy Act* (P.L. 102-555), which was signed into law by President Clinton. The Act recognized the importance of satellite imagery in studying the environment and in carrying out activities related to national security. It further acknowledged some of the problems with previous attempts at full commercialization and the loss of global leadership of the U.S. satellite imaging market. It consequently transferred control of the program to the *National Aeronautic and Space Administration* (NASA) and the *Department of Defense* (DoD). The authority for licensing private satellites remained with the Secretary of Commerce.

Aerial surveillance often involves transmitting radio-wave data to Earth. Thus, *Federal Communications Commission* (FCC) regulations apply to many aerial surveillance transmission. The regulation of bandwidth has not been an easy task. Radio wavelengths are limited and are tightly controlled. Amateurs, in particular, tend to lose operating frequencies to larger commercial interests through political pressure on the FCC. This applies to other surveillance technologies besides just aerial surveillance. For example, in 1998, the FCC temporarily restricted amateur access to 76 to 77 GHz to provide increased bandwidth for commercial development and use of frequencies for applications such as vehicle radar collision-avoidance systems in spite of reports that incompatibility would not be a problem. Sometimes trade-offs are negotiated, as when the FCC upgraded amateur and amateur-satellite allocation of 77.5 to 78 GHz from secondary to coprimary.

9. Implications of Use

The Emerging Surveillance Society

The availability of surveillance imagery to the general public opens the door to broader 'policing' of commercial, military, and civilian activities and the more stringent monitoring of abuses of our global populace and global resources. Nonprofit organizations and watchdog groups are taking advantage of the wider dissemination of information to enlist volunteer help in reducing weapons proliferation, human rights abuses, and environmental exploitation and destruction.

Amnesty International, concerned about allegations of serious human rights violations in Kosovo, made an urgent plea to those with good reconnaissance and intelligence capabilities to monitor humanitarian concerns and to divulge the results to the public, where appropriate.

There are few restrictions on how a generic technology like an aerial photo can be used. Thus, they are at great risk of being abused. For example, a nation or terrorist group planning a deadly strike, could purchase aerial images from a commercial entity, without having to state the reason for the purchase.

Future Technologies

Unstaffed aerial vehicles (UAVs) and DARPA micro aerial vehicles (MAVs) as small as six inches have become commercially practical. Even smaller devices that fly like hummingbirds or bees have been considered. If these technologies are successful, there may one day be tiny spies everywhere.

In the 21st century, the emphasis on aerial surveillance is shifting from aircraft to spacecraft. An orbiting satellite in space cannot be shot down as easily as a plane flying over foreign territory. If an orbiting satellite can unobtrusively record and transmit images of one meter or better, there is less reason to hire planes and helicopters to chart the globe. If an orbiting vessel can function for years off of a few solar panels, it's economical to operate compared to a piloted jet plane consuming hundreds of gallons of fuel. There are many international restrictions on aircraft that do not apply to satellites—once launched, the cost of keeping a satellite in orbit is negligible compared to the cost of keeping a plane or helicopter in the air for 24 hours a day.

Getting a satellite into space is still a difficult challenge—many projects have died on the launchpad. In September 2001, for example, Orbimage's first high-resolution imaging satellite, the OrbView-4, experienced a launch failure. Finally, in June 2003, OrbView-3, a system able to image at resolutions as high as one meter, successfully reached orbit.

Left: The NASA Helios "Flying Wing" was constructed as part of the Environmental Research Aircraft and Sensor Technology (ERAST) program. The craft has a wingspan of 247 feet (about twice the length of a city lot) and is controlled by remote control. Solar cells along the upper surface provide power during daylight hours. Unfortunately, the aerial vehicle, designed to stay aloft for weeks at a time, crashed for undetermined areasons on 27 June 2003, near Kauai. [NASA news photos, July 2001 by Nick Galante and June 2003 by Carla Thomas, releases.]

Newly developed atmospheric craft have had problems, as well. The same month that OrbView-3 reached orbit, NASA's Helios was lost due to a crash. A successor to the Centurion "Flying Wing", the Helios was equipped with GPS to aid navigation and was suitable for environmental monitoring and telecommunications relays. It was built to stay aloft for weeks at a time at high altitudes and, before crashing, set a new altitude record for propeller-driven aircraft, flying to a height of almost 97,000 feet.

Government Policies on Remote-Sensing Systems

In 2002, the U.S. Government began a review of U.S. space policies, including the relationship of government to commercial remote-sensing interests. Prior to this, no major revisions in policy had occurred since President Clinton signed a directive in March 1994. In May, 2003, the White House released a fact sheet regarding the U.S. Commercial Remote Sensing Space Policy, with an emphasis on the competitiveness of remote-sensing systems and their utility for research, data products, climate, weather, and agricultural monitoring, hazard response, transportation, and infrastructure planning. The statement emphasized, however, that the "fundamental goal of U.S. commercial remote sensing space policy is to advance and protect U.S. national security and foreign policy interests by maintaining ... leadership in remote sensing space activities ..."

10. Resources

Inclusion of the following companies does not constitute nor imply an endorsement of their products and services and, conversely, does not imply their endorsement of the contents of this text.

10.a. Organizations

American Society for Photogrammetry and Remote Sensing (ASPRS) - A scientific association founded in 1934 with over 7,000 international members. ASPRS seeks to understand and promote the responsible use of photogrammetry, remote-sensing, and geographic information systems (GIS). ASPRS holds an annual conference and publishes a peer-reviewed journal. www.asprs.org/

Commercial Satellite Imagery Library (CSIL) - Hosted by the Defense Intelligence Agency, this library is available on the Intel Link Network.

National Aeronautics and Space Administration (NASA) - NASA handles a vast research, development, and applications structure devoted to space science and related spinoff technologies. NASA cooperates with many agencies and contractors and disseminates a great quantity of news and educational information related to its work. Of interest are the many satellite and other aerial sensing systems that have been used and are continually being developed. www.nasa.gov/

National Imagery and Mapping Agency (NIMA) - A resource and repository for geospatial and imagery information to support U.S. national security objectives. Many of the NIMA aeronautic, nautical, hydrographic charts, and publications are available for purchase to the public through NOAA Distribution. NIMA topographic maps and gazettes are available through the USGS. www.nima.mil/

National Photographic Interpretation Center (NPIC) - The Center is managed by the Directorate of Science and Technology and functions with the National Reconnaissance Office (NRO), which designs and operates the nation's reconnaissance satellites and supplies them to patrons such as the CIA and the Department of Defense. The NRO is part of the U.S. Intelligence Community. www.nro.odci.gov/

Nigel J. Clarke Publications - This is a commercial U.K. organization selling Luftwaffe Aerial Reconnaissance photography taken from 1939 to 1942. Photos include grain silos, barrage balloon depots, barracks, wireless stations, etc. around Bristol, Cardiff, Dover, and Dublin, etc. They are also available in book form in two volumes as "Adolf Hitler's Holiday Snaps".

ORBIMAGE - An affiliate of Orbital Sciences Corporation which provides imagery and services from a global system of advanced imaging satellites, ground stations and Internet-based distribution channels. Their first OrbView system was launched in 1995, the second in 1997, with more planned. OrbView-2 provides daily color images of Earth's topography. www.orbimage.com/

Skyscan Aerial Photography - A commercial U.K. company offering resources for balloon-based aerial photography, an extensive aerial photo library and information on infrared photography. www.skyscan.co.uk/

Smithsonian National Air and Space Museum - The largest collection of historic air and spacecraft in the world. It is a center for research into space flight, located in the National Mall in Washington, D.C. ww.nasm.si.edu/

Space Imaging - Commercial satellite images, including the CARTERRA™ series of images. www.spaceimaging.com/

TopoZone - Resources for outdoor enthusiasts and other users of topographic maps available through cooperation with the USGS. TopoZone provides detailed interactive maps of the entire United States. www.topozone.com/

United States Air Force Museum - Among the exhibits is a replica of the 1897 glider designed by Octave Chanute.

University of Texas History of Aviation Collection - This extensive collection includes *over two million items related to aviation history* comprising more than 200 donations of private collections. It includes documents and photos related to lighter-than-air vessels, general aviation, World Wars I and II, the Air America Association, and much more. Researchers and potential donors are invited to contact the Eugene McDermott Library Special Collections section. www.utdallas.edu/library/special/aviation/

U.S. Geological Survey (USGS) - The USGS has extensive data repositories of satellite and other images gathered, processed, and archived since the 1960s beginning with the CORONA era of satellite-imagery. They are used by many levels of government and by scientists and commercial industry. They include data such as demographics, environmental wildlife and ecosystem trends, hazards, resources, and much more. Many of these various types of images and topographic maps are available for download or print/negative/positive purchase, usually at cost. www.usgs.gov/

10.b Print Resources

These annotated listings include both current and out-of-print books. Those which are not currently in print are sometimes available in local libraries and second-hand book stores, or through interlibrary loan systems.

Arnold, Robert H., *Interpretation of Airphotos and Remotely Sensed Imagery,* Prentice-Hall, 1996, 262 pages.

Arthus-Bertrand, Yann; Bessis, Sophie; Baker, David, *Earth From Above,* London: Thames & Hudson, 1999, 424 pages. Not a technical reference, but one of the most aesthetic political essays of the Earth as told with over 200 aerial photographs captioned by scientists and sociologists—it appeals to all age groups.

Avery, Thomas Eugene; Berlin, Graydon Lennis, *Fundamentals of Remote Sensing and Airphoto Interpretation,* New York: MacMillan Pub. Co., 1992, ca. 496 pages.

Ball, Desmond, *Soviet Signals Intelligence (SIGINT)–Intercepting Satellite Communications,* 1989, out of print.

Ball, Desmond, *A Base for Debate: The U.S. Satellite Station at Nurrungar,* Sydney: Allen & Unwin, 1987, 122 pages.

Beschloss, Michael R., *Mayday: Eisenhower, Khruschev and the U-2 Affair,* New York: Harper & Row, 1986. The events, politics, and people involved in the shooting down of the U-2 spy plane.

Bewley, Robert; Donoghue, Danny; Gaffney, Vince; van Leusen, Martijn; Wide, Alicia, Editors, *Archiving Aerial Photography and Remote Sensing Data: A Guide to Good Practice,* Oxbow Books Ltd., 1999, 46 pages. A peer-reviewed reference on the issues of standardization and archiving, preservation, and retrieval of archaeological survey data.

Bowden, Mark, *Black Hawk Down: A Story of Modern War*, New York: Atlantic Monthly Press, 1999, 386 pages. An account of a 1993 firefight in Somalia which in part utilized satellite imagery, SIGINT, camera-equipped helicopters, and other surveillance technologies. The interesting account is very relevant to aerial surveillance and application of current technologies.

Burrows, William E., *Deep Black: Space Espionage and National Security,* New York: Random House, 1986.

Burrows, William E., *Deep Black: The Startling Truth Behind America's Top-Secret Spy Satellites,* Berkeley Publishing Group, 1988. Aerial surveillance from the World War to the present and how politics have been influenced by information from spy satellites.

Cameron, Robert, *Above London, Above Paris, Above Los Angeles*, et al. As a photographer, Cameron has coauthored, over the last two decades, an extensive series of photographs of major cities from the air, available from a variety of publishers. There is a similar series by David King Gleason which includes *Over Boston, Over Miami,* and others.

Campbell, Melville, *City Planning and Aerial Information,* out of print.

Cindrich, Ivan; Del Grande, Nancy K., *Aerial Surveillance Sensing Including Obscured and Underground Object Detection: 4,6 April 1994* Orlando, Florida, Bellingham, Wa.: SPIE, Volume 2217. This is only one of many dozens of technical publications and conference proceedings by the Society of Photo-Optical Engineers that relate to remote sensing and aerial imaging.

Collins, Mary Rose, *The Aerial Photo Sourcebook,* Scarecrow Press, 1998, 224 pages. A beginner's illustrated reference bibliography of over 800 books and articles related to aerial photography. Includes government and commercial sources and collections.

Conway, Eric D.; Maryland Space Grant Consortium, *An Introduction to Satellite Image Interpretation,* Baltimore, Md.: Johns Hopkins University Press, 1997, ca. 256 pages.

Darvil, Timothy, *Prehistoric Britain from the Air: A Study of Space, Time and Society* (Cambridge Air Surveys), Cambridge: Cambridge University Press, 1996.

Day, Dwayne,; Logsdon, John M.; Latell, Brian, Editors, *Eye in the Sky: The Story of the Corona Spy Satellites,* Washington, D. C.: Smithsonian Institution Press, 1998. Essays dealing with the technical and political aspects of the surveillance satellites.

Donald, David, *Spyplane: The Secret World of Aerial Intelligence-Gathering,* Middlesex, U.K.: Temple Press, 1987, 127 pages. Out of print.

Erickson, Jon, *Exploring Earth from Space,* Blue Ridge Summit: Tab Books Inc., 1989, 192 pages. Covers early space travel, planetary probes, spy and communications satellites, image interpretation, different spectra used for aerial observation, and sample applications. The book is well illustrated with line drawings and photographs and written at a clear, succinct college/high school reading level.

Gerken, L., *Airships: History and Technology,* American Scientific Corporation, 1990.

Goddard, Brigadier General George W. (USAF Retired); Copp, DeWitt S., *Overview: A Life-long Adventure in Aerial Photography,* New York: Doubleday & Co., 1969. Out of print.

Holz, Robert K., *Surveillant Science: Remote Sensing of the Environment,* John Wiley and Sons, 1985, 413 pages. Updated version of Houghton Mifflin 1973 publication. Out of print. While an older text, it is worthwhile in that it brings together basic technical discussions from many different sources and experts in the various fields, including aerial surveillance, in a graded manner from basic physics concepts to more technical aspects of implementation.

Hough, Harold, *Satellite Surveillance,* Port Townsend, Wa.: Loompanics, 1991, 196 pages. Loompanics tends to publish information on gray area subjects that traditional publishers may choose not to print. Their books are often intended to point out issues of concern to the general public; their authors vary somewhat in expertise, but most appear credible. The book describes how government tools are now within the purview of the public and how private citizens can gain access to aerial 'spy' services and images (this is now somewhat dated).

Kagan, Boris M., *Soviet ABM Early Warning System: Satellite-based Project,* out of print.

Lashmar, Paul, Spy Flights of the Cold War, Great Britain: Sutton Publishing Ltd., 1998 (new edition). Aerial surveillance between the Soviets and the western nations.

Layman, R.D., *Naval Aviation in the First World War: Its Impact and Influence,* London: Chatham Publishing and Annapolis, Md.: The Naval Institute Press, 1996, 224 pages. A well-researched, illustrated overview of many aspects of naval aviation, strategy, and operations.

Lillesand, Thomas M.; Kiefer, Ralph W., *Remote Sensing and Image Interpretation,* New York: John Wiley & Sons: 1999 (4th edition), 736 pages. Illustrated, plus color plates. An introduction to remote sensing for students studying upper level resource management who are already familiar with the remote-sensing nomenclature. Includes photointerpretation, hyperspectral scanning, satellite systems, and classification topics. More theoretical than how-to.

Lind, Marilyn, *Using Maps and Aerial Photography in Your Genealogical Research,* Linden Tree, 1985. Note, there is a 1987 supplement to this book.

Lloyd, Harvey, *Aerial Photography: Professional Techniques and Commercial Applications,* New York: Amphoto Books, 1990, 144 pages. This book is a beautiful showcase of aerial photographs and discussion of some of the basic equipment, but note that it is not an in-depth technical book on how to recreate images, as the title might suggest. The Kodak Guide to Aerial Photography, by Rokeach, 1998, is similar in that there are many excellent examples, but less technical detail on improving professional technique.

Parkinson, Claire L., *Earth from Above: Using Color-coded Satellite Images to Examine the Global Environment,* University Science Books, 1997, 175 pages.

Pedlow, Gregory W.; Welzenbach, Donald E., *The Central Intelligence Agency and Overhead Reconnaissance: The U-2 and OXCART Programs, 1954-1974,* CIA, 1992. Available now in Adobe Acrobat format as The CIA and the U2 Program, 1998.

Peebles, Curtis, *The CORONA Project: America's First Spy Satellites,* Annapolis, Md.: Naval Institute Press, 1997, 352 pages.

Pocock, Chris, *Dragon Lady: The History of the U-2 Spy Plane,* England: Airlife, 1989, 128 pages. The development and mission of the U-2.

Rich, Ben R. with Janos, Leo, *Skunk Works: A Personal Memoir of My Years at Lockheed,* New York: Little, Brown and Co., 1994, 372 pages. A history of spy and stealth planes and the secret division at Lockheed.

Richelson, Jeffrey T., *America's Secret Eyes in Space: The U.S. Keyhole Satellite Program,* HarperCollins, 1990, out of print.

Richelson, Jeffrey T., *America's Secret Eyes in Space: The U.S. Spy Satellite Program,* New York: Harper and Row, 1990.

Richelson, Jeffrey T., *America's Space Sentinels: DSP Satellites and National Security,* University Press of Kansas, 2001 (reprint).

Richelson, Jeffrey T., *America's Space Sentinels: DSP Satellites and National Security,* University Press of Kansas, 1999, 330 pages. Richelson is associated with the National Security Archive and has written numerous books on the topics of intelligence and national security. This text offers a comprehensive look at satellite technologies, infrared surveillance, and early warning systems, beginning in the 1960s up to the SBIRS of the current century.

Ruffner, Kevin C., Editor, *CORONA: America's First Satellite Program,* Washington, D. C.: CIA History Staff, 1995. Declassified images and documents related to the CORONA project.

Skrove, Johnny, *The Kola Satellite Image Atlas: Perspectives on Arms Control and Environmental Protection,* out of print.

Skyscan Photography; Burton, Neil, *English Heritage from the Air,* Sidgwick & Jackson, 1994. Picture book of about 400 aerial photos of British heritage sites.

Steinberg, Gerald M., *Satellite Reconnaissance: The Role of Informal Bargaining,* Praeger Pub. Text, 1983, 208 pages.

Thompson, Don W., *Skyview Canada: A Story of Aerial Photography in Canada,* Ottawa: Information Canada, 1975, 270 pages. The author describes the history, equipment, and persons involved in aerial surveying in Canada. Illustrated.

van Zuidam, R.A., *Aerial Photo-Interpretation in Terrain Analysis and Geomorphologic Mapping,* out of print.

Yost, Graham, *Spies in the Sky (World Espionage Series),* New York: Facts on File, Inc., 1990, 140 pages. A light but interesting overview of aerial intelligence written at a young adult level that can serve as an quick intro for anyone new to the field of aerial surveillance. Describes the U-2 missions, spy satellites, and related topics.

Articles

Air Intelligence Agency, History of the Air Intelligence Agency: RCS: HAF-HO(A&SA)7101, declassified U.S. Government document, 15 December 1995.

Brown, Stuart F., America's First Eyes in Space, *Popular Science*, Feb. 1996, pp. 42-47. Surveys and illustrates the development of U.S. satellite surveillance.

Bulloch, Chris, View from the Top–Intelligence Gathering from Aircraft and Spacecraft, *Interavia,* V.39, Jan. 1984, pp. 543-548.

Cairns, Donald W., UAVs–Where We Have Been, *Military Intelligence*, Mar. 1987, pp. 18-20. A brief history of UAVs in the U.S. military.

Day, Dwayne A., CORONA: A View Through the KEYHOLE, *Intelligence Watch Report Quarterly*, V.2 (1) 1995, pp. 17-21.

Evans, Charles M., Air War Over Virginia, *Civil War Times*. The author is the founding curator of the Hiller Air Museum in Redwood City, California. This article is a fascinating account of the early air balloons that were used in the civil war and the rivalry that existed not just between warring sides, but between individual balloon promoters on the Union side.

Falk, Richard A., Space Espionage and the World Order: A Consideration of the Samos-Midas Program, *Essays on Espionage and International Law,* 1962, pp. 45-82.

Greer, Jerry D., Airborne Reconnaissance and Mount Everest, in Fishell, Wallace G.; Henkel, Paul A.; Crane, Jr., Alfred C., Editors, Airborne Reconnaissance XVIII, *Proceedings SPIE,* 1994.

Klass, Philip J., Military Satellites Gain Valuable Data, *Aviation Week & Space Technology*, 15 Sep. 1969, pp. 55-61.

Krepon, Michael, Spying from Space, *Foreign Policy*, V.75, Summer 1989, pp. 92-108. The author proposes that a three-tiered system is developing with regard to space surveillance.

McDonald, Robert A., Editor, Between the Sun and the Earth: The First NRO Reconnaissance Eye in Space, *American Society for Photogrammetry and Remote Sensing Monograph,* 1997. The story of the first U.S. reconnaissance satellite.

Ono, Philbert, PhotoHistory, *PhotoGuide Japan*, 1996. photojpn.org/

Orlov, Alexander, The U-2 Program: A Russian Officer Remembers, Center for the Study of Intelligence. An interesting historical perspective on the U-2 spy missions over Soviet territory.

Ruffner, Kevin C., Editor, CORONA and the Intelligence Community: Declassification's Great Leap Forward, *Studies in Intelligence 39*, No. 5, 1996, pp. 61-69. Preparation of CORONA materials released to the public in May 1995.

Sweetman, Bill, Spies in the Sky, *Popular Science*, Apr. 1997, pp. 42-48. A good summary overview of a variety of U.S. surveillance satellites.

Wheelon, Albert D., CORONA: The First Reconnaissance Satellites, *Physics Today*, Feb. 1997, pp. 24-30. The author was Deputy Director for the CIA Science and Technology department in the early 1960s. He provides an overview of the development and operation of the CORONA surveillance satellite.

Journals

Aerospace Technology Innovation, is a NASA Commercial Technology Division publication. The publication can be downloaded in Adobe PDF format. Previous issues dating back to 1993 are available online. ctd.hq.nasa.gov/innovation/index.html

Air & Space Magazine, published by the Smithsonian Institution.

Aviation Week and Space Technology, published by McGraw-Hill.

Earth Imaging Journal, Earthwide Communications, bimonthly publication devoted to remote sensing images and technology. www.eijournal.com/data/subscription.asp

Imaging NOTES, Earthwide Communications, Colorado. Consumer/professional magazine with articles on the applications of satellite imagery in a broad range of fields, including the environment, mapping, data integration, utilities, telecommunications, etc. Six issues per year. www.imagingnotes.com/

Journal of Optical Technology, a monthly English translation of *Opticheskii Zhurnal* from the S.I. Vavilov Optical Institute in Russia. Topics are of interest to optical scientists and include remote sensing and aerial imaging technologies, as well as other aspects of optics.

The International Journal of Aerial and Space Imaging, Remote Sensing, and Integrated Geographical Systems.

10.c. Conferences and Workshops

Many of these conferences are annual events that are held at approximately the same time each year, so even if the conference listings are outdated, they can still help you determine the frequency and sometimes the time of year of upcoming events. It is very common for international conferences to be held in a different city each year, so contact the organizers for current locations.

Many of these organizations describe the upcoming conferences on the Web and may also archive conference proceedings for purchase or free download.

The following conferences are organized according to the calendar month in which they are usually held.

Satellite Conference, Access Intelligence-sponsored annual conference and trade show.

Aerial Imaging, a Kodak-sponsored aerial imaging products trade show that is held several times a year in Canada and the U.S.

GIS, annual conference. Topics in 2000 include forestry, natural resources, business geographics, Internet GIS, Telco and more. Applications and solutions-oriented program.

GITA Annual Conference XXIII/GITA, sponsored by the Geospatial Information & Technology Association to provide education and information exchange on the use and benefits of geospatial information and technology in telecommunications, infrastructure, and government utility applications. www.gita.org/

National Space Symposium, www.spacefoundation.org

ASPRS. Start the 21st Century: Launching the Geospatial Information Age American Society for Photogrammetry & Remote Sensing. ASPRS holds an annual conference and publishes a peer-reviewed journal devoted to Earth and space remote-sensing technologies and applications. www.asprs.org/

No More Secrets? Policy Implications of Commercial Remote Sensing Satellites, Carnegie Endowment. International conference sponsored by the Carnegie Endowment Project on Transparency focusing on the technical and policy issues associated with commercial high-resolution satellites. www.ceip.org/

AFCEA Association for Communications, Electronics, Intelligence & Information Professionals.

Annual ESRI International User Conference, geographic Information Systems (GIS) practical applications. Session topics with over 800 papers include the fields of agriculture, business, cartography, databasing, archaeology, forestry, law enforcement, telecommunications, transportation, public works, and more. www.esri.com/eventus/uc

Interagency Workshop on Requirements for Measuring/Monitoring Fires from Space, sponsored by the NASA Langley Research Center and the U.S. Forest Service. Includes imaging of fires from space for detection, monitoring, and surveillance.

NSGIC, sponsored by the National States Geographic Information Council. www.nsigic.org/

FutureView, a Pictometry International Corporation conference in digital, oblique aerial imaging.

10.d. Online Sites

Airship Heritage Trust. A charitable organization in Britain which provides an illustrated history of the development of British airships and ways in which they were used in military and commercial applications. www.aht.ndirect.co.uk/

Aviation Museums. Sponsored by the Aeroclub of Chania, this site lists aviation museums worldwide (U.S., Canada, Australia, etc.) and some relevant jourals. www.otenet.gr/aeroclub/link_museums.htm

Google Earth. A freely-distributed software application that enables users to access three-dimensional images of earth consolidated from a number of imaging systems.

Space Imaging Launch Information. Interesting illustrated educational and press information about commercial imaging satellite technology in text, video, and audio formats is available on this site. Topics include the launch of the IKONOS 1 (including a video of the launch), the satellite, the camera, and the ground station. Press releases detailing the history of the venture and video of the post-launch news conference are also archived. www.connectlive.com/events/spaceimaging/

SPIN-2 Space Survey Photocameras for Cartographic Purposes. A white paper by Viktor N. Lavrov, Sovinformsputnik, which provides details and charts of the specifications of two Russian high-resolution space cameras which, used together, form a high-resolution (two-meter) wide-swath cartographic satellite imaging system. This site also describes Web server provisions of remotely sensed images, including those from the SPIN-2. www.spin-2.com/

U.S. Air Force Military History. An illustrated history with many historic events, people, and technologies related to the military from the early part of the 20th century. www.wpafb.af.mil/museum/history/

10.e. Media Resources

Code Name: Project Orion, part of the History Channel *History Undercover* series. Project Orion is about declassified documents that give a glimpse into the secret dreams at the dawn of the space age. Physicist and author Freeman Dyson, who worked on deep-space projects in the 1950s, discusses their goals and projects. VHS, 50 minutes. May not be shipped outside the U.S. and Canada.

Francis Gary Powers: The True Story of the U-2 Spy Incident, a feature film starring Lee Majors and James Gregory, directed by Delbert Mann. A 1976 dramatization of the capture of Gary Powers and his subsequent imprisonment. This received only lukewarm reviews, but may be of interest. VHS, 1 hr. 40 minutes.

Pioneers in Space, from the History Channel *Pioneers in Space* series. This is a chronicle of the birth of the American space program in the early 1960s. It includes a closer look at the Mercury Project. VHS, 50 minutes. May not be shipped outside the U.S. and Canada.

Project Manhigh, from the History Channel *History Undercover* series. This project began even before the space program, in the 1950s, when the Air Force had grand, seemingly impossible, goals of putting a man 20 miles above the Earth in a capsule attached to a helium balloon. VHS, 50 minutes. May not be shipped outside the U.S. and Canada.

11. Glossary

Airborne and spaceborne vehicles and craft employ a wide variety of surveillance systems, including infrared, visible spectrum, and radar technologies. Here are some common acronyms associated with air and space imaging to extend and supplement those included in the Infrared, Optical, and Radar Surveillance chapters.

Titles, product names, organizations, and specific military designations are capitalized; common generic and colloquial terms and phrases are not.

AASR	advanced airborne surveillance radar
AM/FM	automated mapping and facilities management
ARS	airborne radar system, aerial reconnaissance and surveillance
ASARS	advanced synthetic aperture radar system
ASR	airport surveillance radar
ATAR	advanced tactical radar, automatic target recognition
BWER	bounded weak echo region. A region of weak radar reflectivity which is bounded by a differently reflecting environment, as might occur in a storm with variations in updraft and precipitation.
CANOPUS	Canadian satellite observing systems incorporating advance imaging radar, optical, and sensing instruments.
CANTASS	Canadian Surveillance Towed Array Sonar System. A long array of small hydrophones which is towed. Used in submarine sensing and warfare.
CARS	Contingency Airborne Reconnaissance System
DAIS-1	Digital Airborne Imagery Systems. A commercial, multispectral, digital image-capture system designed and built to Space Imaging's specifications for producing radiometrically calibrated aerial imagery.
DAR	defense acquisition radar
DEM	digital elevation model. A geographic information systems (GIS) aerial map showing vertical dimensions of a region. In GIS modeling, this often comprises the topographic base map.

ESM   electronic surveillance measures

ETRAC   enhanced tactical radar correlator

FAAR   forward area alerting radar

FFR   fire finder/finding radar

FIRESTORM   Federation of Intelligence, Reconnaissance, Surveillance, and Targeting, Operations and Research Models

FLIR   forward-looking infrared radar

FLTTD   Field Ladar Technical Transition Demonstration. A laser radar program of the U.S. Army.

FOPEN   foliage penetration

FOSIC   Fleet Ocean Surveillance Information Center

FSS   frequency surveillance system

GBR, GSR   ground-based radar, ground surveillance radar

GCA   ground control approach. Aircraft landing systems which typically include radar systems monitored by air traffic controllers.

GIS   geographic information systems

GPR   ground-penetrating radar

GSTS   ground-based surveillance and tracking system

HIPAR   high power acquisition radar

IFSAR   interferometric synthetic aperture radar

ISTA   intelligence, sureiillance, and target acquisition

Joint STARS, J-STARS

                   Joint Surveillance and Target Attack Radar System. An air-to-ground E-8C aircraft surveillance system for locating, classifying, and tracking ground targets. See

Electromagnetic Surveillance 10

Ultraviolet

1. Introduction

Ultraviolet radiation is invisible to humans. It extends just beyond the part of the visible spectrum that we perceive as the color violet. Ultraviolet, infrared, and visible radiation together comprise the range of frequencies we call the *light spectrum*. Most people are familiar with 'black lights' that cause some colors to emanate an eerie blue-white glow. This glow is an example of ultraviolet *fluorescence* in which the ultraviolet radiation excites chemical substances to emit light that we can see in the visible spectrum.

Ultraviolet wavelengths are shorter than visible lightwaves which, in turn, are shorter than infrared. Sensing devices that operate in the optical frequencies are called *photodetectors* and some broadband devices that detect radiation in the visible spectrum also detect infrared and ultraviolet frequencies. Advanced multispectral imaging systems often sense a wide portion of the optical spectrum, including UV frequencies.

While the ultraviolet spectrum consists of wavelengths that are invisible to human eyes, other creatures may perceive all or part of this ultraviolet spectrum. Birds are known to respond

Because humans can't see ultraviolet light, it is difficult to gauge our exposure to Ultraviolet-B. UV-B can have adverse effects on the human body, the most obvious of which is sunburn. Not all damage from ultraviolet can be readily seen. Purdue researcher Richard Grant and Forest Service worker Gordon Heisler explain that UV-B doesn't shine as directly as visible light, but rather bounces around in the atmosphere. They took fish-eye photos like the one above to get a complete picture of the sky and how much was actually obstructed, employing a novel strategy for assessing UV-B exposure. [Purdue 1997 news photo, released.]

to ultraviolet reflected from specialized feathers (e.g., on some birds' crests) and some insects appear to locate food sources through ultraviolet colors on flowers that make the pollen or nectar areas stand out.

Radiation in the visible spectrum is not generally harmful to humans and the infrared spectrum, which can sometimes be sensed as heat, can be avoided, but the ultraviolet spectrum may be difficult to detect and may cause harm from prolonged exposure. Ultraviolet rays can damage skin before the person is aware that it has happened. Tanning lamps, sunlight, and specialized UV lights emit ultraviolet radiation. Fortunately, the Earth's atmosphere filters out some of the harmful rays from the Sun, making the planet safer and more habitable—ozone contributes substantially to shielding us from ultraviolet. Frequent or prolonged exposure may predispose a person to skin cancer and health care professionals warn us to wear hats, glasses, and sunscreen when exposed to ultraviolet from natural or synthetic sources. Due to the health hazards associated with ultraviolet exposure, UV lamps are not usually directed at people unless goggles are worn.

Ultraviolet illuminators are used a variety of surveillance tasks. They can be helpful for inspecting artworks, documents, entry access stamps on skin, and ancient murals and paintings on the walls of archaeological finds. Astronomers use UV rays to glean information about the cosmos and UV technologies are incorporated into space-based spectral imaging systems. Superconducting tunnel junctions (STJs) are being used to develop devices for sensing in the extreme ultraviolet frequencies.

Electromagnetic Spectra with Ultraviolet Between Visible Light and X-Rays

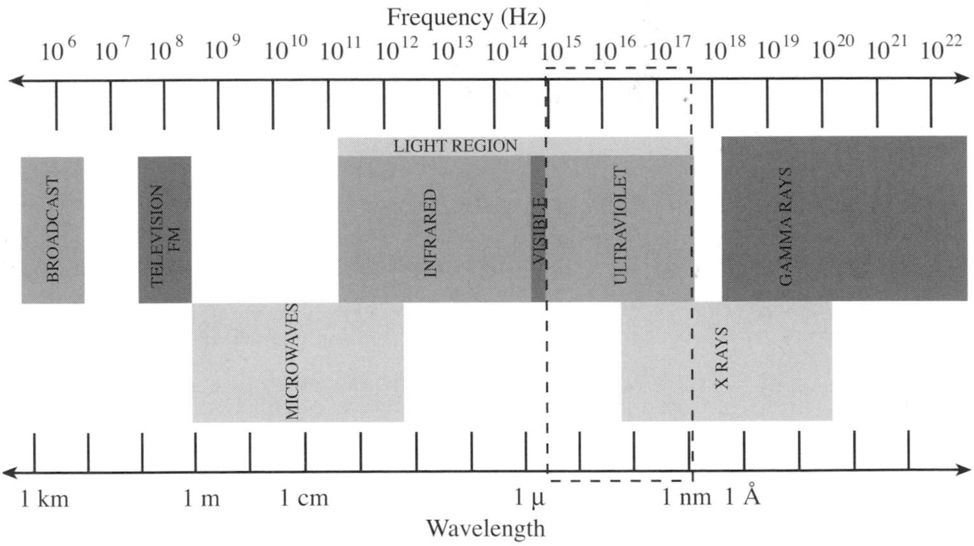

The ultraviolet spectrum is a band of short-wavelength radiation beyond the visible spectrum, ranging approximately from 190 to 400 nanometers. As the scale across the bottom of the chart shows, this can also be expressed as 1900 to 4000 ångströms or 0.19 to 0.40 microns. [Classic Concepts ©2000, used with permission.]

Ultraviolet is interesting and useful. Some substances react to ultraviolet radiation by re-emitting longer wavelengths that can be seen by the unaided eye. This process is known as *fluorescence*. By applying substances and chemicals that are known to fluoresce, it is possible to discretely 'mark' items for tracking or identification.

This chapter describes a variety of ultraviolet surveillance technologies, including entrance control, security, identification, aerial imaging, forensics, astronomy, and archaeological site investigations.

2. Kinds and Variations

Ultraviolet is part of the optical spectrum, so much of the terminology is the same as that associated with infrared and visual spectra. Depending upon how you view it, light sometimes appears to behave as a wave and sometimes as a particle, but for the purposes of discussion, an individual unit or 'packet' of ultraviolet radiation is called a *photon*.

Units

Ultraviolet wavelengths are very short, so they are typically described in small units such as *nanometers* or *ångströms* (Å). A nanometer is one billionth of a meter. An ångström is even smaller, one ten-billionth of a meter. One nanometer = 10 ångströms, thus, 320 nm = 3200 Å. The ångström (often written *angstrom* in English) is named after Swedish physicist Anders J. Ångström.

Categories

Ultraviolet is produced naturally by the Sun and other heat-emitting celestial bodies and can be generated synthetically by heating substances to high temperatures.

Ultraviolet radiation is generally grouped into three categories, based upon wavelength. There is no actual dividing line between the categories, but rather a gradual transition of effects from one to the next. Nevertheless, these categories are useful for discussing ultraviolet characteristics and effects:

UV-A - between about 320 and 400 nm. This is the lowest energy (longest wavelength) form of ultraviolet radiation that is emitted by our Sun. Most of the Sun's UV-A radiation reaches the Earth, as it is only partially absorbed by the ozone layer. This is the region emitted by the lights in tanning beds. Over time, UV-A can break down collagen and elastin tissues in biological organisms.

UV-B - between about 290 and 320 nm. In the mid-range, at slightly higher energy levels, some of the ultraviolet rays are absorbed by the ozone layer, but some also reach the Earth's surface. Ultraviolet can influence a wide variety of biological processes on Earth and cause sunburn. Variations in ultraviolet radiation which are, in part, influenced by the amount of ozone in the atmosphere, can cause changes in the Earth's environmental balance that may significantly affect our lives over time. Smog may be more prevalent in ozone-depleted regions.

UV-C - between about 190 and 290 nm. In the higher-energy-level ranges, near the region called X-rays, ultraviolet light can have significant harmful effects. We are protected from these effects by the Earth's atmosphere, which filters out UV-C before it reaches the Earth's surface.

For astronomical research, extreme-ultraviolet radiation (EUV) has been defined as wavelengths between about 100 Å and 1,000 Å (10 nm to 100 nm).

Chemical Interactions

Ultraviolet is useful for surveillance because of the way it affects various substances. Some substances fluoresce naturally when exposed to ultraviolet illumination and others will fluoresce if chemically treated. Various UV reactions enable inventors to create technologies that exploit these properties and has helped practitioners use them for identification purposes.

Sunscreens are products developed to shield the body from the harmful effects of ultraviolet radiation on the skin. They range from products that absorb ultraviolet to those that absorb and reflect. Titanium dioxide and zinc oxide are two substances that are used to block out ultraviolet. Sunscreens and sunblocks are used by sunbathers and those who frequently work outdoors, but they are also used by those exposed to ultraviolet radiation in a laboratory setting. An ultraviolet-radiation numerical index scale has been developed to provide information on the danger of skin damage. Sunblocks can be useful for field or laboratory experiments in which selective blocking of ultraviolet radiation or reflection is desired by applying them to surfaces.

3. Context

Ultraviolet radiation is naturally occurring, so it helps to understand the sources of ultraviolet in order to use the phenomenon to best advantage. Synthetic ultraviolet surveillance devices must be used in ways that minimize interference from natural sources. Natural ultraviolet devices need to be used in ways that best exploit their properties.

Like infrared and visible radiation, ultraviolet is particularly useful for imaging or detecting marks that are invisible to the unaided eye. For security and surveillance tasks, it's very convenient to have the capability to mark something invisibly in a way that the mark can be revealed at will.

Sources and Intensity of Ultraviolet Radiation

Most of the ultraviolet radiation in our environment comes from the Sun. The amount of ultraviolet radiation reaching the surface of the Earth depends upon many factors, including how much is emitted by the Sun, its frequency, how much is scattered or absorbed by the Earth's atmosphere, and its angle of travel in relation to the Earth.

The Sun's rays hit the Earth at different angles at different times of the day and year. Ultraviolet reaches the Earth at a more direct angle over the equator and during the summer. When the rays travel a shorter distance, they encounter fewer obstacles and are subject to less attenuation. In other words, the UV radiation hitting the equator (and other latitudes during the summer months) is more intense than at the poles. Around noon, when the Sun is at its 'highest' (most direct) angle in relation to the Earth, the ultraviolet rays are more intense. (This is why medical practitioners advise bathers to limit their exposure to the Sun during the noon hours when they are more vulnerable to burning.)

Similarly, ultraviolet radiation travels a shorter distance when it encounters mountain tops than when it reaches sea level. Not only is the atmosphere thinner at higher elevations, thus absorbing less of the radiation, but it is also a 'nearer' destination, resulting in slightly less attenuation than if the rays had to travel farther through the more numerous particles and gases associated with the atmosphere near sea level. In general, ultraviolet radiation is more intense on Earth's mountain tops than it is at sea level, in addition to which UV rays reflect off the snow and hit nearby objects from all angles.

The Earth's *ionosphere* is a region of free electrically charged particles extending from approximately 50 to 500 kilometers above the Earth's surface. There are subregions within the ionosphere that have varying properties, depending upon the Sun's activities (e.g., Sun spots) and the time of day. Ion sources include the solar winds as well as gases that have been ionized by ultraviolet radiation. Thus, ionization from ultraviolet radiation makes an important contribution to surveillance technologies by providing an ionospheric 'surface' off of which to bounce radio waves to achieve greater distances.

Ozone is a gas formed through a photochemical reaction when oxygen is ionized (e.g., by ultraviolet radiation). Ozone is a pungent, irritating gas that is harmful to humans, but extremely beneficial up in the stratosphere around the Earth, where it blocks ultraviolet rays, especially UV-C, thus protecting us from the most dangerous ultraviolet frequencies.

Absorption, Reflection, and Scattering

When the atmosphere is 'thicker', in other words, when it is cloudy from increased water vapor in the air, ultraviolet radiation is absorbed further, and less of it reaches the Earth. Thick clouds will tend to absorb more ultraviolet than thin clouds, as will precipitation that both absorbs and scatters the radiation. Particulate matter such as dust from storms, or pollution, will also to cut down on the amount of UV that reaches the Earth's surface.

Ultraviolet doesn't just stop moving when it reaches the Earth or the objects on its surface. Some of the radiation is reflected back from surfaces in varying quantities. Water tends to absorb most ultraviolet radiation, reflecting back only about 5%, whereas lightly colored, smooth, reflective surfaces like some buildings, parabolic antennas, light-colored automobiles, and snow, may reflect up to 85% of the radiation, bouncing it around to other reflective and nonreflective surfaces until it eventually dissipates or is absorbed.

Because ultraviolet is emitted by the Sun, there can be interference from sunlight when using ultraviolet for detection purposes. Unlike infrared, however, ultraviolet is quite easy to screen out. A dark room, dark box, or even a hand casting a shadow will often sufficiently screen out short ultraviolet rays so that a synthetic source can be used to examine objects with ultraviolet signatures or imprints. This is one of the characteristics that makes UV useful for surveillance tasks.

Some surveillance systems need to be protected from the Sun's ultraviolet radiation—especially those launched into space where the Earth's atmosphere cannot protect them. Satellites, high-flying aircraft, and space probes all may be exposed to significant ultraviolet doses that can break down paints, plastics, and delicate structures. Vehicles, boxes, and compartments for storing or deploying sensitive electronic equipment may also be coated, to resist damage from ultraviolet. There are a variety of commercial products designed to reduce or block ultraviolet radiation, including gel coatings and specialized paints.

Some substances will absorb ultraviolet radiation and re-emit the energy at another wavelength. Substances that *fluoresce* are absorbing radiation and re-emitting it in the visible spectrum. Out in space, regions of cosmic dust have been found to absorb ultraviolet radiation and re-emit it as mid- and far-infrared, allowing us to sense otherwise difficult-to-detect starlight with infrared telescopes and probes.

Astronomers are interested in the different kinds of radiation emanating from bodies in and beyond our solar system. Since 'hot' bodies typically emit ultraviolet, there is much to be learned by developing detection systems and telescopes that can image ultraviolet. However, hydrogen and helium are gases that are abundant in the universe absorb extreme ultraviolet, making it a technical challenge to observe ultraviolet radiation emanating from celestial bodies. Conventional telescopes are not well-suited to ultraviolet observation, as the radiation is influenced by conventional coated surfaces (e.g., mirrors). Adjustments in mirror angles and surfaces are necessary to reduce scattering and absorption.

Chemical Marking and Illumination

Ultraviolet-reflecting chemicals are very handy for marking papers, cartons, clothing, currency, valuables, and skin. They are also used as markers or 'tags' for microscopic examinations and for marking birds for experimental observation or conservation. Ultraviolet chemical stamps are frequently used as entrance controls at public events.

In surveillance devices, ultraviolet is typically used with ultraviolet illuminators and ultra-violet-reflecting chemicals. As an example, a person's hand can be 'stamped' with an ultraviolet-sensitive chemical that may itself be either visible or invisible. When an ultraviolet-emitting light source illuminates on the mark, it will 'glow' faintly (fluoresce). Ultraviolet-detection and analysis technologies are also used in astronomical research—many satellites are equipped with ultraviolet sensors.

Some substances fluoresce naturally when exposed to ultraviolet light. Others fluoresce if they have been chemically treated. In some instances, a simple handheld UV illuminator is used, in others, the UV light is carefully processed through a filter to isolate the desired wavelengths. Consumer devices tend to work with simple commercial illuminators, while scientific devices may require precise filtering or staining to achieve the desired result. In fact, the science is now sufficiently advanced that it is possible to detect as few as 50 fluorescing molecules per cubic micron through a microscope.

4. Origins and Evolution

When we represent infrared and ultraviolet symbolically, they are shown as regions on either side of the visible spectrum, with infrared at the longer wavelength (red) and ultraviolet at the shorter wavelength (violet). Infrared and ultraviolet share characteristics with the visible spectrum that led to their discovery at about the same time color was being more closely studied. The nature of light itself was of great interest, starting around the 17th century, and theories and experiments to discover the wave/particle aspects of light provided an important basis for understanding ultraviolet radiation once it was known to exist. It is helpful to cross-reference the histories of Visual Surveillance and Infrared Surveillance to get an overall understanding of the development of light technologies.

Early Observations

The most important historic contributions to the discovery and development of ultraviolet technologies include the understanding of lightwaves and basic observations on ultraviolet radiation and its place in the optical spectrum.

During the 1600s, scientists generally considered light to be a wave phenomenon. By the 1700s there was important speculation and experimentation by Isaac Newton (1642-1727) leading to the theory that light might, in fact, be a particle phenomenon. He described his 'cor-puscular' theories of light in *Opticks* in 1704. This was almost a century before the discovery of invisible radiation within the optical spectrum, so Newton didn't live to see that wave and particle theories of light would eventually be applied to wavelengths that are invisible as well as those that are visible.

Many great discoveries in science are based upon simple observations and subsequent theories to confirm and predict simple effects. Karl Wilhelm Scheele (1742-1786) was aware, in 1777, that silver chloride darkened when exposed to sunlight, but he made a further impor-tant observation that it darkened most rapidly at the violet region of the visible spectrum. This observation of different effects in different parts of the spectrum was an important step toward the discovery of ultraviolet and other forms of radiation.

In 1792, Thomas Wedgwood was working with bodies heated to high temperatures in kilns and noted that all objects turned certain colors at certain temperatures, regardless of their com-position. Thus, metal or clay heated to several hundred (or thousand) degrees, would glow with the same colors, in spite of the fact that they had markedly different compositions. Gustav Robert Kirchhoff (1824-1887) further refined Wedgwood's observations and proposed a mathematical relationship between the emitted and absorbed power of a surface in thermal equilibrium.

In 1800, Thomas Young (1773-1829) challenged Newton's theory of light 'corpuscles', or particles, and proposed a wave model based on longitudinal waves. Meanwhile, F. Wilhelm (William) Herschel (1738-1822) was studying the colors of light through a prism with a thermometer. He noticed that changes in temperature that were apparent in the different colors of the spectrum continued beyond the red region even though he couldn't see any more colors in that region. From this, he correctly concluded that there were wavelengths beyond the color red that were invisible to humans, but of a similar character to visible light. Thus, Herschel discovered *infrared*.

Prism experiments with light were important at this time, but in 1801, Johann Wilhelm Ritter (1776-1810) used chemicals to further investigate the effects of individual colors. Just as Herschel had inferred the presence of wavelengths beyond the visible spectrum from the change in temperature in the infrared region, Ritter inferred the presence of ultraviolet by the interaction of each portion of the spectrum with chloride. In the visible violet region, the light darkened the chloride—in the invisible ultraviolet region, it darkened it even more. Ritter had taken Scheele's observation and Herschel's experiment one step further and discovered *ultraviolet*.

In 1842, Edmond Becquerel (1820-1891) photographed the solar spectrum and, since the film was sensitive to ultraviolet, his photographs included wavelengths extending into the ultraviolet.

Theoretical/Mathematical Frameworks

In the 1850s, James Clerk-Maxwell (1831-1879) made important observations and developed theories related to electricity and magnetism. He developed the framework within which we now understand that ultraviolet is a type of *electromagnetic* energy along with radio waves, infrared, visible light, X-rays, and gamma rays. Clerk-Maxwell's predictions and calculations, known as *Maxwell's equations*, were remarkable considering that most of these forms of electromagnetic energy would not be discovered for another four decades.

Fluorescence has been observed for thousands of years, as it occurs naturally, but the phenomenon of fluorescence was not really understood scientifically until about the middle of the 1800s. George Stokes (1819-1903) observed that fluorspar would 'glow' when it was illuminated with ultraviolet light and coined the term *fluorescence*. He further observed that the fluorescent light had longer wavelengths than the light causing the excitation. Since that time, we have learned more about materials that will fluoresce in their natural form or when chemically treated—knowledge that we now use in commercial applications.

In the 1870s, scientists studied the influence of ultraviolet on biological organisms. They observed that bacteria and certain pathogens are influenced by specific wavelength regions of sunlight and that UV-C could slow or stop their growth. This observation would eventually lead to new methods of environmental control of bacteria, molds, and other organisms. The discovery also led to practical applications a century later in chemical surveillance, 'clean' rooms, hospitals, and waste-destruction facilities.

In 1886 and 1887, Heinrich (Rudolf) Hertz (1857-1894) performed some important experiments that helped confirm Clerk-Maxwell's theories about electromagnetic energy by applying voltages to metal spheres with a narrow gap between them. He further discovered that ultraviolet light can prompt a metallic surface to emit electrons. Thus, it was determined that it was the *frequency* of the light that was important to this effect and that the light's intensity (which we perceive as brightness) influenced the number of electrons ejected, but did not influence their kinetic energies (Einstein later studied this *photoelectric* effect in more detail). As a picture of the electromagnetic spectrum and its characteristics emerged, visible light, which is so significant to humans, was found to be only a tiny portion of the electromagnetic spectrum.

In the early years of the 20th century, researchers looked beyond the Earth to the primary source of our electromagnetic energy—the Sun. They studied sunspots and speculated as to their character. By study and experimentation over a period of about 30 years, they found an electrified region of the Earth's atmosphere that was reflective and decided that most of the ionization in the atmosphere resulted from ultraviolet radiation from the Sun. They further realized that long-distance radio communications were intrinsically linked with this ionization process.

The Emergence of Quantum Physics

A revolution in physics was occurring in the early 1900s. Physicist Max Planck (1858-1947) was endeavoring to explain *blackbody* radiation (a black body is an idealized model of an object defined as absorbing and radiating all frequencies). Until this time, our understanding of thermodynamics and electromagnetism was not complete enough to explain a few important concepts. For example, scientists mathematically modeling the optical spectrum predicted that the intensity of the radiation would continue to increase indefinitely, from infrared to ultraviolet, implying that any body would emit an infinite amount of energy. Practical models didn't seem to support this theoretical prediction and the discrepancy was termed the *ultraviolet catastrophe*. Planck pointed out that a particle explanation, rather than a wave explanation, could resolve the theoretical and experimental discrepancy.

Einstein's Contributions to Physics. Left: A portion of a letter written to G. E. Hale by Einstein in 1913, with an illustration of how gravity should affect light passing near the Sun, causing deviations in the visually recorded position of stars, depending upon the position of the Sun—thus implying the warping of space. Right: Albert Einstein with his wife and son around the time he was formulating his Special Theory of Relativity, approximately 1904. [Detail: The Albert Einstein Archives, The Jewish National & University Library, The Hebrew University of Jerusalem, Israel, ca. 1904. The Observatories of the Carnegie Institution of Washington, 1913, copyrights expired by date.]

At this time, Albert Einstein (1879-1955) was developing his Special and General Theories of Relativity and was probing another problematic area of physics called the *photoelectric effect*, discovered two decades earlier by Hertz. Einstein improved our understanding of the nature of light by explaining the photoelectric effect in terms of particles or *photons*, thus resolving some of the inconsistencies encountered by previous scientists.

Other scientists were providing important significant insights to our understanding of the universe. *Quantum physics,* which called into question all previous ideas about physics at the atomic level, was emerging. As a result, our notions of optical radiation as being either waves or particles were challenged, with the bold assertion that it depended on how you looked at it. Since we cannot directly observe atoms, we must study them indirectly, with precision instruments that, in themselves, make assumptions about what they are testing and observing. It was found that under some circumstances, light appeared to behave as particles, under others, it appeared to behave as waves. The more precisely you tried to design your experiments, the more elusive the answer became. Yet, over the next several decades, the field of quantum physics gained an uneasy credibility alongside classical Newtonian physics. Now, it is one of the most important and exciting areas of research and is better understood from the perspective of probabilities than of hard and fast rules more characteristics of classical models.

Experimental and Practical Applications

In spite of economic difficulties, a great deal of scientific discovery occurred during the depression years, in both America and Europe, that related directly to surveillance technologies. Radio-ranging technologies for navigation and object-detection, cathode-ray tubes for viewing sensor data, radio communications, and ultraviolet technologies, all developed in important ways, in the early and mid-1930s. The World War II years, in which research funding was made available for defense and warfare technologies, were also significant for practical applications of ultraviolet, as were postwar commercial developments.

Karl Von Frisch (1886-1982), a Viennese zoologist, studied communication in animals and insects. He is best remembered for his work with bees. In the course of his research, he paired colored cards with bowls of sugar water and discovered that bees were attracted to the bowls on the basis of the color on the cards. He noted that they could see particularly well at the violet and ultraviolet end of the spectrum, beyond human vision. This contradicted earlier assertions that bees were color blind and sparked interest in ultraviolet vision systems.

Astronomers had known for decades that the planet Venus was shrouded with a mysterious white cloud, but they could only speculate as to what it was and what it might hide. Frank E. Ross (1874-1966), an American astronomer, made some of the first ultraviolet photographs and some of the pioneering planetary photographs, as well. Using plates and ultraviolet-sensitive film, he photographed the cloud formations around Venus, in 1928, providing new insight into the composition of the particles that cloaked her surface.

In the 1930s, Haitinger and others developed secondary fluorescence that allowed chemical treatments to 'stain' various materials. This has been useful in creating ultraviolet 'markers' and in staining substances for examination with specialized microscopes. It has gradually come to be an important tool of forensic chemistry.

In the 1950s, Coons and Kaplan demonstrated important biological localization techniques using tissues stained with fluorescein-tagged antibodies. This was an important contribution to microscopy.

With new technologies, the interest in studying our solar system and the rest of the galaxy increased. Scientists began to detect extreme-ultraviolet by the late 1950s, when they found shorter wavelengths emanated from the Sun. Because the extreme-UV is easily absorbed or scattered, it is difficult to detect from conventional ground-based telescopes.

The 1970s was a time when ultraviolet began to be used in a wide variety of applications, including astronomical observation, birding, and lasing.

Until the late 1950s and 1960s, astronomers considered it extremely difficult to observe radiation in the extreme ultraviolet, yet there was potentially a great deal of information about the universe that could be gained by finding solutions. Experimental ultraviolet astronomical photography began to show results by about the mid-1960s.

Left: The S-13 experimental Ultraviolet Astronomical Camera tested on the Gemini II space flight in 1966. The purpose of the experiment was to image the ultraviolet radiation from hot stars and to develop new ultraviolet photography techniques and devices. Middle and Right: Two frames of film from one of the early ultraviolet-sensing photography experiments carried out on the Gemini flight, in August 1966. Black and white 70 mm film was used. [NASA/JSC news photo, released.]

Research into space flight, telescope modifications, and mirror optics in the mid-1970s began to make it possible to develop extreme-ultraviolet (EUV) sensing systems and, eventually, EUV telescopes and space missions.

Left: Ultraviolet instruments require special attention to fabricating the smoothness, coatings, and angle of installation of the mirrors, as ultraviolet is easily scattered or absorbed. This image shows a prototype mirror for the Far-Ultraviolet Sensing Explorer (FUSE) satellite about to be coated with a UV-reflective material. Two of the FUSE mirrors were coated with lithium fluoride (LiF) and two with silicon carbide (SiC). The coating is carried out in a 'clean room' in a vacuum tank, shown here at the NASA Goddard Space Flight Center. Right: Each of the LiF mirrors in the FUSE satellite is equipped with a Fine Error Sensor built by Com Dev, Ltd., of Cambridge, Ontario, supplied to the FUSE project by the *Canadian Space Agency*. Shown here is a full-scale model with the CCD camera, radiator, and controller box. [NASA 1997 news photos, released.]

UV Photos from Space

Taking UV photographs from a spaceship is both a challenge and an opportunity. The spacecraft has to be carefully oriented and care has to be taken to protect the crew from ultraviolet exposure, but the amount of ultraviolet radiation in space is considerable and can yield information that can't be obtained from within the protection of Earth's ozone layer.

In July 1971, Apollo 15 was launched into space on what was considered primarily a science mission that would include ultraviolet photography. Because ultraviolet passes more readily through quartz than through glass, Window 5 in the Command Module had been specially fitted with quartz window panes. The high quality Hasselblad cameras carried along on the mission could then also be used to take ultraviolet-sensitive photos of the Earth and the Moon. When used for UV photos, a Hasselblad was fitted with a 105 mm UV-transmitting lens and four exchangeable filters. Since there was no ozone layer to protect the astronauts from dangerous UV radiation coming through the window, a special transparent polycarbonate shade was designed to cover the window between photographic sessions.

In order to shoot UV photos from space, the Apollo spacecraft had to be oriented so that Window 5 was facing the Earth or the Moon. While in position, about eight images could be taken, with two images for each of the four different filters.

The use of ultraviolet photography continued in 1972 aboard the Apollo 16 mission, with a specialized far-ultraviolet camera.

Left: After the success of UV photography on Apollo 15, a far-ultraviolet camera was prepared for the Apollo 16 mission, in 1972. Here it is being examined by astronauts Charles Duke, Jr., and John Young. Right: This far-ultraviolet (1304 Å) image of the Earth was taken from space by John Young, Tomander of the Apollo 16 mission, on 21 April 1972. [NASA/JSC news photo, NASA/Kennedy Space Center photo, released.]

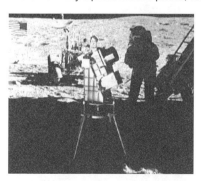 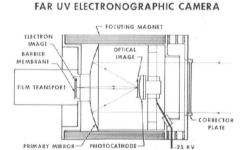

Left: In the foreground, in front of Astronaut John Young, is an ultraviolet camera with a protective coating of gold to help maintain a safe operating temperature. The camera is aimed to record the large Magellanic Cloud from the surface of the Moon, on 22 April 1972. Right: A November 1973 engineering drawing for the design of a far-ultraviolet electronographic camera to be used on Skylab 4 to photograph aspects of the comet that can't be seen from the surface of the Earth. [NASA/JSC news photos, released.]

Photos during Interplanetary Fly-bys

In November 1973, the Mariner 10 space probe used Venus' gravity to help it reach Mercury in 1974. It provided the first close-up ultraviolet images of Venus' atmosphere and the scope and behavior of its cloud cover. The craft was able to accomplish three fly-bys of Mercury before it ran out of the gas used to control is attitude, whereupon it moved into a solar orbit. In all, it was able to transmit more than 12,000 images of Venus and Mercury.

Left: Assembly of the Mariner 10 space probe. Middle: The launch in November 1973 on the Atlas Centaur launch rocket. Right: Artist's rendering of the Mariner 10 as it would have appeared during flight. The craft gathered data that indicated a thin atmosphere and magnetic field associated with the planet Mercury and earlier provided ultraviolet images of the atmosphere of Venus. [NASA 1973 news photos, released.]

In 1975, the Apollo-Soyuz Test Project, a joint project of American astronauts and Soviet cosmonauts, provided some exciting new information when NASA crew members used a command and service module to aim an extreme-ultraviolet telescope at a number of preselected targets. The successful detection of several of these targets established the feasibility of EUV surveillance of space.

In January 1978, the International Ultraviolet Explorer (IUE) satellite was launched as a cooperative effort among three agencies: the Smithsonian Environmental Research Center (SERC), The European Space Agency (ESA), and the National Aeronautics and Space Administration (NASA). It was placed in a 24-hour elliptical orbit around the Earth to communicate with the Goddard Space Flight Center and the Vispa ground-station near Madrid, Spain. The IUE showed a hot gaseous region around the Milky Way galaxy, among other discoveries.

Left: An ultraviolet photo of Io, one of the moons of the planet Jupiter, taken on a Voyager 2 spacecraft mission, the evening of 4 July 1979. The dramatic bright blotch on the right is the plume of a volcanic eruption, more than 200 kilometers high. Right: Callisto, a moon of the planet Jupiter, taken on 8 July 1979. This is a composite false color image, with ultraviolet light used to add a pseudo-blue component. Ultraviolet helps provide contrast to the surface regions to make the image easier to interpret. [NASA news photos, released.]

In the early 1970s, scientists began to study ultraviolet-sensing capabilities in birds and, by 1978, it was known that pigeons could sense ultraviolet light. There is speculation that this helps them find ripe foods and might aid in navigation. Since then, it has been discovered that many migratory birds are sensitive to ultraviolet (and possibly also infrared). This means that leg bands or dyes for bird surveillance should be carefully chosen so they don't interfere with the birds' mating or foraging behaviors when they are marked and studied. Fluorescent and UV markers or aerosols have traditionally been used for bird research and may be suitable in some circumstances but not others.

Scientists at the IBM Research Center noticed that ultraviolet pulsed lasers interacted with organic matter in unique ways—they would decompose in a particular way with as little as one laser pulse. This knowledge could be used to design advanced laser 'etching' tools.

Ultraviolet Effects

We began using ultraviolet in commercial applications about a decade before we really understood its influence on our ecosystem. The 1980s was a time when scientists more carefully studied extraterrestrial sources and uses of UV radiation and further refined the technologies within a broader framework of understanding. Scientists began putting ultraviolet sensors into space on board satellites even though, at this time, they were still a fairly new technology.

The Viking satellite, Sweden's first satellite, was launched in 1986 with a variety of sensing instrument payloads, including an ultraviolet imaging experiment. The University of Calgary in Canada supplied two ultraviolet cameras that captured auroral images to give a new view and set of data of the northern lights. The UV imaging experiment also provided data that could be used in conjunction with or compared to data from the other sensors (magnetometer, electric field sensors, and others).

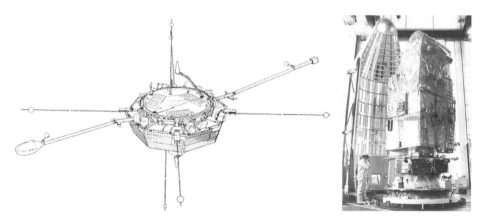

The Viking satellite, after some delays, was launched from Korou on the Ariane 1 rocket in February 1986. It was initially placed in an 822 km circular orbit, that was subsequently raised to 13,530 km. It was controlled from a ground-station at Esrange until a short-circuit in a power modulator on the satellite caused communications contact to be lost about a year later. [Swedish Space Corporation 1986 news photos, released.]

In the 1980s, we began to understand the deleterious effects of pollution on the ozone layer and to notice the thinning of this layer. Ozone was found to be essential to our survival, as it screened out deadly solar UV rays.

Research and Applications

In the 1980s and 1990s, in the natural world, fish and birds were studied for their ultraviolet-sensing abilities. In research labs, there was commercialization of UV technologies and further investigation into synthetic and natural ultraviolet sensors. For example, forensic scientists used ultraviolet for investigations and weathering tests. Above the Earth, satellites sensed the extreme-ultraviolet regions and brought new insights into the cosmos.

Science has to prove itself fairly definitively before it is allowed into the courtroom. Fingerprint technology has been gradually accepted as a routine identification technique, but other applied sciences, including lie detector tests and DNA profiles, have been slower to gain acceptance and are still not admissible in courtrooms in some areas. However, the application of science to investigations and courtroom proceedings is gradually becoming accepted as these sciences become more precise and able to identify violent criminals who might otherwise continue to endanger other people.

Since the 1970s, researchers have increasingly demonstrated the application of ultraviolet research to various branches of forensics. It was found that ultraviolet weathering could help determine the age of materials, that ultraviolet lights could help reveal stains and marks that would otherwise go unnoticed, and that ultraviolet markings could be used to track materials that might have been disturbed or stolen.

Astronomical Research Using Ultraviolet

Ultraviolet sensing evolved more rapidly in the later 1980s and 1990s. Space Shuttle missions, space probes, aerial UV-sensing aircraft, and many other ultraviolet technologies were developed and tested at this time.

Once the basic logistics of taking pictures with special lenses, filters, cameras, and ultraviolet-sensitive film had been worked out, scientists began combining images taken with different filters and creating *pseudocolor* or *false color* images that were easier to interpret and more enjoyable to view.

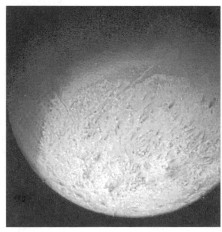 

Left: An image of Triton, the planet Neptune's largest moon, taken on the Voyager II space mission, in August 1989. Pictures were taken through green, violet, and ultraviolet filters and combined to create this image. Right: This is the Astro-1 on the Space Shuttle Columbia mission in December 1990. In the foreground are a variety of X-ray and ultraviolet sensors, including the Hopkins Ultraviolet Telescope (HUT), the Ultraviolet Imaging Telescope (UIT), and the Wisconsin Ultraviolet Photopolarimetry Experiment (WUPPE). [NASA/Marshall news photos, released.]

The Extreme Ultraviolet Explorer (EUVE) was launched from a Delta II rocket on June 1992 to engage in all-sky (50 to 740 Å) and deep ecliptic (65 to 360 Å) studies. This was a milestone in space surveillance, as these bands had not been explored to any great extent up to this time. The EUVE included three grazing-incidence scanning telescopes and an extreme ultraviolet (EUV) spectrometer/deep survey instrument. The first portion of the satellite mission was to survey and map the EUV sky with the scanning telescopes, the second portion to act as a space observer.

In April 1993, the ALEXIS (Array of Low-Energy X-Ray Imaging Sensors) satellite, funded by the U.S. Department of Energy, was launched into Earth orbit. ALEXIS was equipped to make celestial observations in the extreme-ultraviolet with six remarkably compact telescopes. The satellite also carried scientific payloads that sensed other types of electromagnetic radiation, including the Blackbeard radio experiment and the ALEXIS telescopes.

Other Branches of Science

In the biological sciences, particular types of fish were found to possess photoreceptors sensitive to ultraviolet and salmon and rainbow trout became the subjects of numerous studies. It was discovered that ultraviolet-detecting structures in the retinas of young salmon gradually disappear until they are essentially gone from the main retina at about seven months of age [Kunz et al., 1995]. Trout have a similar pattern, although the mechanism can be influenced by the experimental introduction of hormones [Browman and Hawryshyn, 1994].

Lasers have been an important aspect of many technologies, including printers, measuring instruments, cutting tools, sighting mechanisms, and much more. Different kinds of lasers and different potential applications continued to be developed through the 1990s. Until the early 1980s, lasers almost exclusively used visible or infrared light. By the mid-1990s, however, scientists had developed practical ultraviolet lasers that could precisely etch biological and synthetic materials without further chemical processing.

Airborne Ultraviolet Sensing

Aerial sensing in all regions of the electromagnetic spectrum became important in the mid-1990s. Infrared, visual, and ultraviolet detectors, and imaging technologies were fitted into a wide variety of scientific, military, and commercial aircraft.

Experimental, UV-Equipped Aircraft. Left: An experimental SR-71 Blackbird was equipped with a UV video camera in the nose, in March 1993, to test the feasibility of the equipment and the effects of turbulence. Here, a Blackbird is shown over the mountains, in December 1994. The Blackbird is capable of speeds of over 2000 mph (Mach 3+). Right: The eerie Blackbird stealth plane, shown here in 1997, would disappear if it were photographed against a dark background. It was designed three decades earlier as a reconnaissance plane for the U.S. Air Force. [NASA/Dryden Flight Research Center news photo, released.]

In March 1993, NASA sent an experimental SR-71 Blackbird aircraft into the descending night from the Dryden Flight Research Center. The SR-71 was equipped with an ultraviolet video camera mounted on the nose so that it was aimed at the sky. It was intended to capture UV images of stars, comets, and asteroids. Part of the reason for the flight was to test the camera's reaction and performance in turbulent conditions at high speeds. Another reason for equipping the Blackbird with UV-sensing was its high-altitude capacity. The aircraft could fly to over 85,000 feet, a photographic vantage point that could not be attained by the telescopes at the Earth's surface.

The Continued Evolution of Space Science

In the mid-1990s, research into the astronomical use of ultraviolet sensing became more sophisticated and was now a regular part of space science missions.

In April 1996, the Midcourse Space Experiment (MSX) was launched into a circular orbit around Earth at an altitude of about 900 kilometers. It is used by the U.S. Ballistic Missile Defense Organization for the purpose of characterizing ballistic missile signatures during midcourse flight, that is, the period between booster burnout and re-entry. In the past, detection and targeting of incoming ballistic missiles were difficult technical challenges. The MSX was designed to observe a wide spectrum of wavelengths, including far ultraviolet. It included five primary instruments with eleven optical sensors aligned for simultaneous observations to more precisely track moving targets. The system had other sensing applications, including the capability to monitor global environmental changes, especially ozone and carbon dioxide, and the ability to surveil terrestrial and celestial targets. The estimated mission life at the time of launch was about four years.

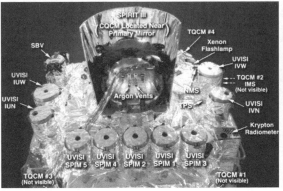 

Ultraviolet and Infrared Sensors. Left: The numerous sensors incorporated into the instrument section of the MSX are shown here. Note the UVISI sensors across the bottom. The MSX also includes an infrared telescope. Right: The MSX, with far-ultraviolet sensing, is launched by a Delta rocket from the Vandenburg Air Force Base in April 1996. [U.S. DoD BMDO 1996 news photos, released.]

In September 1996, after 18 years of successful service, the International Ultraviolet Explorer (IUE) satellite was turned off. It was the main ultraviolet-sensing instrument for the international astronomical community during its years of operation.

In February 1997, the Space Shuttle Discovery made a rendezvous with the orbiting Hubble Space Telescope (HST). On this mission, among other changes, the Faint Object Spectrograph (FOS) was removed and replaced by the NICMOS infrared camera/spectrograph. The FOS, which was installed in Hubble's axial bay, generated data in the ultraviolet spectrum.

In June 1999, NASA launched the Far-Ultraviolet Spectroscopic Explorer (FUSE) satellite. Orbiting at 768 km, it provided new ways to surveil the composition and origins of the universe and preliminary results have been promising. FUSE is equipped with four telescopes that UV light and channel it to a high-tech instrument that breaks down the UV into component wavelength regions spectrographically. The far-ultraviolet portion of the light spectrum sensed by FUSE is not visible to many of the other space-based sensing platforms, including the Hubble Space Telescope. The FUSE spectrogaph is serving as a model for updating older systems and led to the Cosmic Origins Spectrograph, an instrument scheduled for installation in the Hubble Space Telescope platform that has met with some delays.

Left: The FUSE satellite during final tests in the 'clean room' at the Orbital Sciences Corporation. The telescopic instruments are attached to the top of the structure. Middle Left: The FUSE ready to go with solar panels attached prior to transport to the launch pad. Middle Right and Right: The Delta II rocket second-stage assembly being hoisted into position on the launch pad in preparation for the launch of the FUSE satellite and the launch on 24 June 1999. [NASA 1999 news photo by NASA; 1998 news photo by Orbital Sciences Corp.; 1999 news photo by NASA; 1999 news photo by NASA/KSC, released.]

The Increasing Importance of UV Sensing in Space

The payload of the 1998 Space Shuttle STS-95 mission was significant for at least two reasons. Not only were many of the payload instruments designed for ultraviolet sensing, but the Payload Specialist was the famed astronaut John Glenn, then in his 70s, taking a return flight almost 37 years after becoming the first American astronaut to orbit the Earth.

One of the things that space exploration has taught us is that the world beyond our planet is rich in variety and phenomena we haven't even begun to map or understand. Ultraviolet emissions form an important part of this diverse environment. The STS-95 Shuttle Mission was primarily a scientific mission.* Other highlights related to astronomical surveillance focused heavily on ultraviolet studies and included

UVSTAR - an extreme-ultraviolet spectrographic telescope that can sense and resolve images of sources of plasma in space, including the plasma from hot stars and Io, one of Jupiter's moons. It can be used to study Earth's ultraviolet emissions.

EUVI - an extreme-ultraviolet imaging system aboard the UVSTAR for taking extreme-UV measurements of Earth's atmosphere. Two imagers are used to scan the Earth's 'shadow line', to map the intensity of helium and oxygen ions in the atmosphere. It permits precise measurements of the Earth's plasmosphere and ionosphere.

STARLITE - a telescope and imaging spectrograph for studying astronomical targets identified as ultraviolet targets, making it possible to gather data on sky background

---

*One of the studies included checking on the health, progress, and physiological changes in John Glenn, the oldest astronaut ever to go into space.

emissions, supernova remnants, nebulae, star-forming regions and volcanic emissions from the moon Io.

SEH - instruments designed to gather data on absolute extreme-ultraviolet/far-ultraviolet fluxes (energy outputs). SEH interprets EUV/FUV emissions from objects in the solar system, plasmosphere, and magnetosphere, as well as solar system objects. Changes in the Earth's atmosphere resulting from solar extremes during the daytime can also be assessed.

Some of the UV instruments described here have been improved and updated since the STS-94 flight as research in UV sensing continues to result in better sensing technologies.

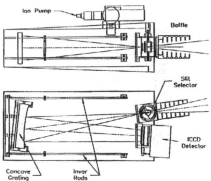

Left: The UVSTAR spectrographic telescope shown here operates in the 500 to 1250 Å waveband. Right: The UVSTAR consists of a pair of telescopes and concave-grating spectrographs to cover overlapping spectral ranges. The first telescope operates in the far-ultraviolet range, the second in the extreme-ultraviolet spectral range. [NASA and Italian Space Agency news photos, released.]

XUV Research

While many exciting developments in UV sensing in the late 1990s were applied to astronomy, there were also breakthroughs in the use of ultraviolet in optical physics and microelectronics. Ultraviolet is at the high-energy end of the light spectrum, approaching X-rays and, thus, extreme-ultraviolet (XUV) shares some common properties with X-ray technologies that enable some of the research on X-radiation to be applied to ultraviolet radiation. In microelectronics, ultraviolet has been studied for its potential for lithography for the production of integrated circuits (ICs) and for the fabrication of high-precision mirrors. Thus, ultraviolet may eventually evolve into an important industrial tool for the production of optical and microelectronic surveillance devices.

Not all UV sources are hazardous. UV light at the lower end of the UV spectrum (near-ultraviolet, bordering the visible spectrum rather than X-radiation), when used in certain environments or with protective gear or chemical blocks, has a multitude of applications. Sandia Labs and Brown University researchers jointly have combined UV radiation research with laser technology to create a solid-state microcavity laser that stimulates photons to emit UV light. These vertical-cavity surface-emitting lasers (VCSELS) can be coated with phosphors to generate white light suitable for commercial lighting applications that would last longer than conventional fluorescent bulbs. Since they are very small and can be arranged in arrays, they have a huge advantage in terms of flexible arrangements of the components, and of portability.

More recently, Sandia has demonstrated deep-UV semiconductor components, based upon a sapphire substrate with conductive layers of aluminum gallium nitride, that emit wavelengths of 290 nm and 275 nm. These can be used in some of the same applications as near-UV components (e.g., VCSELs) for biochemical sensing (e.g., the detection of anthrax), but also could be used for sterilization/decontamination applications (e.g., purifying water) due to their high-energy levels, and for chemical processing—which may be useful in the manufacture of other sensing technologies.

Left: Jung Han holds up a VCSEL-based laser component. These are compact long-lasting sources of near-UV light and thus have great potential as sensing components for detecting dangerous bacteria and weapons-grade fissionable materials for health monitoring (natural pandemics) and military surveillance (biochemical weapons). Right: Andy Allerman demonstrates deep-UV semiconductors, which are suitable for biochemical sensing, as well as for purification/decontamination applications, and chemical processing. [Sandia Labs news photo by R. Montoya, 2000 and 2003, released.]

Forensic Sciences

In the 1990s, the use of ultraviolet in forensics increased with the availability of fluorescing fingerprint powders, ultraviolet illuminators, specialized UV cameras, and microscopes that could be used to minutely analyze fibers or documents in visible and ultraviolet spectra.

Ultraviolet experiments in astronomy and other sciences, and the practical application of ultraviolet sensing to forensics and archaeology, continue.

Robotics and Industrial Applications

Robots can be equipped with a variety of sensors for detecting chemicals, light, movement, and other environmental stimuli. Since deep-UV can be dangerous to biological systems, it is sometimes practical to mount a UV light or UV-based sensing component to a robot rather than having a human operate the device. The UV sensor, in turn, can be used to sense biological agents and chemicals.

In industrial fabrication, ultraviolet light can be used to sterilize components and for photolithography.

5. Description

Photographs are an essential aid to scientific, historic, and forensic documentation. Sometimes, ultraviolet light is used to illuminate old paintings, frescoes, cave drawings, crime scene stains, and historic documents. This means of viewing them is invaluable, but is even better if the discoveries can also be photographically documented.

One of the most valuable aspects of ultraviolet in surveillance is the fact that it can detect faint or otherwise nonvisible phenomena. Thus, one of the areas that is of interest to surveillance professionals is the photographic recording of ultraviolet light. Here are some introductory guidelines for basic ultraviolet photography.

Film and Lenses

Black and white film is suitable for most medium-ultraviolet documentary applications as ultraviolet is not perceived by us as color, and most color films have a layer that blocks UV light. Many people also have inexpensive UV-blocking filters to slightly reduce haze in their pictures and to protect their expensive camera lenses. Remember to remove UV filters when taking pictures intended to emphasize ultraviolet. The lens and associated filters must let ultraviolet light pass through to the film. Unfortunately, with some coated lenses, the coating itself in part screens out UV (to reduce 'haze'), which interferes with UV photography. Using older uncoated lenses can sometimes reduce the possibility of this happening.

Quartz lenses allow ultraviolet light to readily pass, but may be out of the user's price range. If a spectrophotometer is available, you can check if the desired glass lens allows UV light to pass.

UV photography generally uses the longer UV frequency ranges from about 300 to 400 nanometers. Some filters allow both visible and ultraviolet light to pass through the lens and others screen out the visible light and admit only the ultraviolet—which is best depends on the application. For most forensics work, it is helpful to have the visible information as a guideline (e.g., detecting a forged signature over another that has been erased on a bank check), but even here there are exceptions.

TMAX film with 400 or 3200 ASA is suitable for UV photography. For professional quantities of films that are sensitive to ultraviolet, one option is the Fine Grain Positive Release bulk roll film (#5302) available from Kodak.

Filters

An *exciter filter* can be used to selectively pass fluorescencing wavelengths. The most common filter for this is the Wratten 18A. Like ultraviolet, infrared rays focus just slightly farther from the lens than visible light rays, due to refractive differences, and adjustments need to be made. Many lenses have a little red tick mark next to a yellow tick mark that indicates UV adjustment. If a fluorescing screen is used, the sharpness of the screen image will indicate if the scene is properly focused.

While black and white films are preferred for many UV applications, color negative and slide film can also be used with an exciter filter. A *barrier filter* that allows only fluorescence to pass through the lens is pale yellow, such as the Wratten 2A, 2B, or 2E. With yellow filters used to record fluorescence, focusing is done in the normal way and most current meters are sensitive enough to automatically calculate the exposure.

Some photographic filters will screen out most visible light (around 400 to 750 nanometers), while permitting light energy toward the infrared and ultraviolet regions to be transmitted to the film. Since infrared and ultraviolet have different focal points, either one or the other will be in focus—but not both. The Schott UG1 filters out visible light and allows ultraviolet radiation from about 280 to 420 nanometers to be admitted, peaking at about 360 nanometers, and infrared radiation from about 690 to 1100+ nanometers to transmit, peaking at the transition to near-infrared, which is about 750 nanometers.

Hoya has a series of specialized filters for ultraviolet photography:

> *ultraviolet and visible filters* - These filters include the UV-22 which has a fairly broad range of transmission and UV-28, UV-30, UV-32, UV-36, and UV-38, which absorb

the shorter wavelengths and transmit visible light. The numbers correspond roughly to the lower end of the range of wavelengths that can pass through the filter. For example, UV-28 transmits UV at about 280 nanometers and above, and UV-32 transmits light at about 320 nanometers and above. These frequencies correspond approximately to the line spectra of mercury lamps. The filters themselves appear almost black.

ultraviolet filters - There is a series of Hoya filters for transmitting ultraviolet and screening out (absorbing) visible light (they also allow a bit of infrared to be admitted). These include the U-330, U-340, U-350, and U-360. As the numbers go higher the breadth of the spectral admission tends to decrease. For example, U-330 allows wavelengths from about 200 to about 450 nanometers to be transmitted through the lens, a range of about 250 nanometers, peaking at about 330 nanometers. The U-350 transmits from about 300 to about 480, with a range of about 180 nanometers, peaking at about 350 nanometers.

Flash and Illumination

When ultraviolet is photographed, the imaging system uses either *reflected ultraviolet* from a secondary source of illumination or the *radiated illumination* from a substance which fluoresces when stimulated. Reflected ultraviolet is mainly used in forensic and medical photography. Fluorescence is used in biochemical sciences, archival applications, archaeology, and sometimes in forensic and medical applications.

If extra illumination is needed, it is best to use a professional mercury arc lamp with certified eye protection, but in some instances a commercial "black light" or flash that does not have a coating to absorb ultraviolet light can be used. A "black light" is a mercury-vapor tube coated with phosphors on the inside (somewhat like the inside surface of a computer monitor screen). When stimulated by the mercury vapors in the tube, the phosphors emit longer-wave UV near the violet portion of the visible spectrum. Since certain UV frequencies can kill bacteria, the technology is used in medical sterilization procedures and there are medical lamps that emit shorter-length UV rays, but be sure to use protection.

As lamps emit shorter UV waves, the price generally increases because quartz has to be used rather than glass (which hinders short-UV).

If artificial UV illumination is used, depending on the filter, it may be necessary to block the windows, turn off the room lights, and use a high-ASA (fast) film and long exposure times. Take several pictures at different aperture settings (this is called *bracketing* the exposures). Photographic and scientific suppliers can aid in selecting UV illuminators and protective eyewear.

When photographing short-wave ultraviolet, the photography can be done in the dark without the filter, but special lenses are usually necessary, as regular optical glass tends to absorb the desired wavelengths. Pinhole cameras, quartz lenses, and special plastic lenses may be used. Substances that fluoresce will be recorded as a combination of the reflected ultraviolet and some of the visible light that is emitted in the fluorescing process. For very-shortwave ultraviolet, normal film emulsions may absorb the radiation and special emulsions such as those coated with silver halides must be used.

Digital Processing

With digital processing software, even more possibilities are open for ultraviolet image analysis. By shooting images with and without different UV filters, scanning the results at high resolution, and superimposing the images, it becomes possible to reveal and study features from new perspectives.

6. Applications

Ultraviolet-reflecting substances are useful for many types of marking and identification, including crowd control, valuables marking, fingerprint detection, mineral identification, and stamp and currency security encoding. Ultraviolet itself is useful for forensic weathering tests and astronomical research.

6.a. Archival Investigations and Preservation

When ultraviolet light is shone on paper, ink, or paint, a surprising new world sometimes opens up, where details that couldn't previously be seen can be studied or read. This makes it a valuable surveillance tool for historians studying old documents, archivists developing preservation techniques, or investigators trying to locate subtle clues or to determine if something is a forgery.

Ultraviolet has aided archivists in examining, studying, and preserving historical documents. Ultraviolet has even made it possible, in some instances, to detect hidden writing or older writing (or painting) under the current surface layer. It can help reveal details of the ink and paper and changes in the materials over time, which provides valuable information for the development of preservation techniques. It can also help reveal possible forgeries or aid in determining the age of a specimen.

Ultraviolet fluorescence and ultraviolet reflectance are used in assessing works of art, in combination with infrared and laser-illumination technologies.

6.b. Scientific Investigations

Archaeology and Anthropology

Just as ultraviolet can be used to examine documents and paintings, it can also light up cave paintings and marks on old pottery, papyrus, and walls. In catacombs in the Mediterranean region, ultraviolet was found to reveal intricate, ancient embellishments that had never been noticed before in previous investigations with candles and flashlights. This provides a new glimpse into history and adds to our understanding of ancient cultures.

Ultraviolet can also reveal clues as to age, cracks, moisture content, and other important data that help determine the condition and history of historical artifacts.

Because of the importance of recording archaeological discoveries and the process of recovery and restoration, longwave (near-) ultraviolet, which is the easiest to record photographically, is generally used for documentation in this field. Shortwave (far-) ultraviolet does not pass through glass, which is why most cameras are not suitable for shortwave UV photography.

Remains Identification

Sometimes tools common to archaeological research are used in forensics to identify the bodies of victims of wars or accidents that are many years old. In one interesting case, the U.S. Army *Central Identification Laboratory* (CIL) helped identify the crew members of a C-87 cargo plane that crashed in the Himalayas in 1944.

The wreckage of the plane was discovered in 1993 on a glacial slope. The remains of three of the bodies were recovered and returned to the U.S. DNA identification was still a young science at the time and sometimes other means of identification are used in conjunction with DNA technology, as in this case. CIL members hiked to the wreckage and noted the serial number of the plane. Meal chits were discovered that held the signatures of the crew members.

Using ultraviolet and infrared illumination, the CIL staff brought out the information on the faded cards, revealing three out of five names. This made it possible for the relatives to bury their loved ones and receive closure after half a century of waiting and wondering what had happened to their family members [Rodricks, 1997].

In another, more celebrated case, ultraviolet photography has been used to try to help confirm or deny the authenticity of the Shroud of Turin, a shroud in which Jesus was said to have been buried. This and various other blood and chemical evidence shows, however, that the shroud probably originated at about the same time that it first came to public attention, in the early part of the Italian Renaissance. One of the most interesting observations scientists have made about the shroud is that the shape of the face doesn't match the fat, stretched-out shape that always occurs when a shroud is wrapped around a rounded object, like a head (you can try this for yourself). The natural proportions of the face on the shroud would likely only arise if it had been somehow painted on the fabric or impressed on the fabric from a shallow bas-relief (which were common adornments on buildings at that point in history). Some have explained this by saying the body "passed through" the shroud when it made the impression.

Industrial and Resource Sciences

In industrial research, the chemical and geological sciences are often combined to assess natural resources and carry out gas and mineral exploration. Spectrometers and chromatographs (discussed further in Chemical Surveillance) are used in this research, including ultraviolet-visible spectrophotometers. Spectrophotometers are used in planetary sciences, fiber forensics, and for identifying narcotics and various other chemical samples.

A single-beam, microprocessor-controlled, ultraviolet-visible spectrophotometer with a diode array detector can help determine the transmission and absorbing properties of materials in the frequency ranges from about 200 to 800 nanometers. Different substances transmit and absorb in different patterns and proportions, yielding profiles that can aid in identifying a substance. This allows a spectrophotometer to be used for cataloging and comparing fibers and various other materials. [NASA/GSC news photo, released.]

Ultraviolet-visible spectrophotometers are used in the Johnson Space Center (JSC) *Environmental Health Laboratory* along with other instrumentation to monitor work environments on Earth and in space, including the assessment of air, water, and hazardous waste materials.

Sometimes it is important to detect certain gases in the air, e.g., the presence of hydrocarbon vapors. Ultraviolet and infrared illuminators can be used to radiate the area and, when intercepted by a receiver and compared against reference values, indicate the presence of certain gases [Dankner et al., 1995].

6.c. Crime Scene and Accident Investigations

Ultraviolet is used in a number of ways to gather crime evidence. When a victim of homicide is discovered at a crime scene, for example, the body is sometimes days, weeks, or even months old. To help establish the time of the crime, the remains are examined for insects, decay, weathering, and other clues. Sometimes, however, it's not a murder case that is under investigation. Crimes of vandalism, theft, kidnap, or rape require other types of clues to reveal details about the events and when they took place.

Accelerated Weathering

Sometimes when a crime scene is very old, it becomes difficult to reconstruct the date that it occurred. However, there may be objects at the scene that have been subjected to sun exposure and weathering, such as photos, books, a baseball mitt, watch, or purse. Sometimes it may be possible to find similar materials and subject them to artificial weathering to gain the same effect and then compare the weathered materials to the materials found at the crime. This, in turn, can sometimes provide clues as to the length of time that has passed since the objects were first placed at the scene.

Accelerated weathering is a way of exposing a substance to ultraviolet and/or chemicals that have been predetermined to simulate the weathering effects of sunlight, condensation and the level of precipitation that might be common to the area. Both UV-A and UV-B lamps may be used in this process, along with moisture to simulate the condensation and rain.

Stain Detection

Stains on clothing, sheets, mattresses, walls, documents, phones, weapons, and vehicles can all provide clues at a crime or accident scene. Mercury-xenon lamps are sometimes modified with filters to provide an ultraviolet illumination source. Using protective goggles, an investigator can shine the light on various areas suspected of having stains and locate them more readily. Bodily fluids tend to fluoresce at a UV level of about 440 nanometers. The frequency can be controlled to some extent with different filters, which usually range around 420 ± 30 nanometers. Once a stain is located, it can be sampled and studied further in a lab, which will probably include microscopic examination in visible and ultraviolet frequencies and may also include ultraviolet-visible spectrophotometry.

Note that if stains, prints, and other clues are being photographed with ultraviolet light as part of an investigation, it is important to take the photos before covering the evidence with protective plastics, tapes, or varnishes. Part of the way in which these materials preserve the evidence is by screening out ultraviolet radiation, in which case they would also prevent the ultraviolet rays from exposing the photographic film.

Fiber and Natural Substance Detection

Different microscopes are used in forensics to evaluate hair, fibers, and other substances that might be found on a victim's body or in the suspect's clothing, shoes, house, or vehicle. Cat hairs, tobacco leaves, and car mat fibers are all examples of materials that have helped investigators link a suspect to the scene of a crime. Ultraviolet-visible spectrophotometers may be used in this type of investigation.

Fraud Detection

Heat-sensitive recording materials, such as thermal papers, can be impregnated with chemicals that respond to ultraviolet or infrared stimulation. This permits the fabrication of *security papers* and *security inks*, materials that can be used for certificates, bank notes, and other negotiable instruments or confidential documents and *security fabrics* that can be used for special purposes. Ultraviolet-sensitive powders can be used to coat materials to detect tampering or theft.

Fluorescing inks can be made with chemicals such as fluorescein and certain dyes. Since inks can be impregnated with ultraviolet-sensitive chemicals, it follows that typewriter and computer printer ribbons could also be designed this way, to add an extra measure of protection to documents that are not mass-produced.

Ultraviolet products are commonly used for monitoring entrances and exits at music, religious, political, and sports events. These applications are likely to continue as they are inexpensive and relatively easy to use.

Narcotics Detection

Stain detection and narcotics detection use some of the same laboratory tools and techniques. For example, ultraviolet-visible spectrophotometry can be used to investigate a chemical solution with ultraviolet and visible light ranging between approximately 300 and 800 nanometers. Depending upon its composition, the substance will absorb some of the light, causing the energy level of the light to be shifted. The detector then assesses the amount of light that transmits through the substance. By calculating the relationship of the absorbed light to the transmitted light a chemical 'signature' can be generated that aids in identification. Each substance has its own signature, which includes peak absorption levels at certain wavelengths. When these patterns are plotted on a graph, a 'picture' of the chemical can be created that can be compared to reference chemicals to attempt to find a match.

Injury Detection and Assessment

When examining a body for forensic purposes, it is important to document any injuries that may be present. These may not be easy to see with the unaided eye. Ultraviolet light can be used to make surface bruises, bite marks, scars, cuts, and scratches more visible and ultraviolet photography can aid in documenting these marks.

6.d. Scientific Research

Fluorescent Microscopy

The level of light that reaches the eye when ultraviolet is used to fluoresce an autofluorescing or chemically treated substance is very small. Most of the photons are absorbed, rather than reflected back to the eye. For this reason, powerful illuminators, usually arc lamps, are used with fluorescent microscopes to increase the level of excitation. Xenon and mercury-burner lamps are commonly used.

Astronomy and Cosmology - Ultraviolet-Sensing Satellites

As described in the historical introduction in Section 4, ultraviolet sensing has fairly recently become an essential aspect of astronomical research. Ultraviolet-sensing satellites are designed to allow us to surveil Earth's envelope and characteristics of other bodies in the universe. 'Hot' celestial bodies emit ultraviolet, giving us a means to detect and map their characteristics and positions. White dwarfs and binary stars are just two types of celestial bodies that emit ultraviolet radiation in sufficient quantities to allow us to detect them from greater distances than is possible with visible light sensors.

Ultraviolet telescopes are designed to overcome some of the UV-scattering characteristics of conventional telescopes, with special adjustments and ultra-smooth mirrors. UV-detectors often need to be highly selective when they are operating in Earth-orbits where there is interference from the Sun's UV radiation.

Earlier missions that sensed ultraviolet radiation include the Astronomy Netherlands Satellite and NASA's Orbiting Astronomical Observatories. Examples of more specialized UV-sensing astronomical missions include

The International Ultraviolet Explorer - launched in January 1978, the IUE was designed to sense ultraviolet radiation in the following wavelengths:

- 115 to 200 nanometers with a short-wavelength Prime camera (SWP). and

- 190 to 320 nanometers with long-wavelength Prime (LWP) and Redundant (LWR) cameras.

During observations, the spectrum is displayed as an image through a television camera with an optimum signal-to-noise ratio of about 20:1. Since camera and film technologies vary in sensitivity in different wavelengths, the calibration and design of these instruments can be quite challenging.

The Extreme-Ultraviolet Explorer - the EUVE satellite, launched in June 1992 is designed to sense extreme-ultraviolet radiation. It was placed into low-Earth orbit (LEO) below the Earth's geocorona, so that it could be serviced by space shuttle missions. (Space shuttle astronauts have carried out a number of repairs and adjustments to various orbiting platforms, including the Hubble Space Telescope and the MIR space station.) EUVE sensing ranges include

- all-sky (approx. 50 to 740 Å), and

- deep ecliptic (approx. 65 to 360 Å).

Other space agencies have been adding to the knowledge base that makes space exploration possible. ASTRID is a small, spin-stabilized, scientific satellite launched by the *Swedish Space Agency* in January 1995. ASTRID carries a number of imaging sensors, including an *energetic neutral atom analyzer*, an *electron spectrometer* and two *ultraviolet imagers* for imaging the aurora. The platform was designed and developed by the *Swedish Space Corporation's* Science Systems Division in Solna, Sweden, while the payload was developed by the *Swedish Institute of Space Physics* in Kiruna. The control and command station is in Kiruna, a town in northern Sweden where a large parabolic antenna complex has been established.

The ASTRID, a small, scientific satellite, is equipped with ultraviolet Miniature Imaging Optics (MIO) - stainless-steel-mounted photometers installed in the satellite spin plane. One sensor observes the Lyman alpha emission from the Earth's geocorona, while the other observes auroral emissions. The satellite was launched along with the Russian navigation satellite Tsikada on a Kosmos-3M rocket supplied by the Design Bureau Polyot, Omsk. [Swedish Space Agency 1995 news photos, released.]

ASTRID-2 is the successor to the Swedish ASTRID satellite. This spin-stabilized, scientific satellite was launched from Plesestsk, in December 1998, aboard a Kosmos 3M rocket. Contact was lost in July 1999, but a large amount of data was secured during the satellite's operation. One of the missions of the satellite was to collect ultraviolet auroral images and atmospheric UV-absorption measurements.

Left: Tests at Plesetsk necessitate the disassembly of the main avionics unit to fix a problem on the main processor board. Middle: The Astrid-2 Kosmos 3M rocket ready to launch in Plesetsk. Right: The Astrid-2 control center at the Swedish Space Agency headquarters at Solna, Sweden. [Swedish Space Agency 1998 and 1999 news photos, released.]

The Ultra-Violet Auroral Imager

The Canadian Space Agency's Ultra-Violet Auroral Imager (UVAI) was launched onboard Russia's Interball-2 satellite. Images from sensing systems like these are sometimes hard to interpret by themselves, but they can be combined with photographs and computer-generated pictures to provide reference points and a view of the relationship of the recorded phenomena to the Earth.

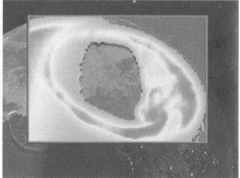

This composite image shows the position of the Earth from space, featuring a 'swirl' over northern Canada from a UV Auroral Imager (UVAI) image of a magnetic space storm that was superimposed over the image of the Earth for reference. The UVAI image was taken 1 March 1997 and has been color-coded to symbolically represent the intensity of the aurora. [Canadian Space Agency 1997 news photo, released.]

The Far-Ultraviolet Spectroscopic Explorer

The Far Ultraviolet Spectroscopic Explorer (FUSE), launched June 1999, provides several telescopes equipped to image radiation from distant galaxies, providing information on the character and origin of the universe.

The design, development, and launch of the FUSE satellite have involved the collaborative efforts of many corporations and educational institutions. The illustrations below show three instrumentation components that make up part of the 'payload' of the FUSE system, each assembled at a different location.

Left: One of four telescope mirror assemblies, assembled and tested in the 'clean room' at the Johns Hopkins University Homewood Campus. The light from the four mirrors is aligned and channeled into the spectrograph. Middle: One of the two detector assemblies, both of which serve as electronic 'retinas' to sense ultraviolet light and convert the energy into electrical signals that form the digital data that are communicated back to the Earth-based station. The sensors were constructed and tested at the University of California, Berkeley. Right: The FUSE spectrograph which contains the gratings (top) and detectors (bottom center) used to detect and analyze ultraviolet light. The spectrographic analyzer was assembled at the University of Colorado. [NASA 1998 news photos by Johns Hopkins University/APL; UC Berkeley; and University of Colorado, released.]

Left: A view down through one of the four telescopes in the FUSE satellite, called Lithium Fluoride #1. Right: The FUSE Satellite Control Center at Johns Hopkins University Homewood Campus. Satellite communications and data are sent and received at this center. [NASA 1999 news photo by the CCAS Mechanical Team; 1998 news photo by Johns Hopkins University, released.]

Ultraviolet-sensing satellites often have other types of sensors in additional to UV sensors. X-ray sensing is an important aspect of astronomical research that often goes hand-in-hand with UV sensing.

The FUSE ground station antenna at the University of Puerto Rico, Mayaguez, during installation in 1998 and from another angle, after subsequent installation of a protective radome. [NASA 1998 and 1999 news photos by UPRM, released.]

The Galaxy Evolution Explorer

In April 2003, the Galaxy Evolution Explorer was launched by NASA and Caltech into a high-Earth orbit to image galaxies in the ultraviolet range to aid in the study of stars and their history. The Explorer imaged using state-of-the-art ultraviolet detectors that give a very different image of galaxies from visible light images. The system images "sections" of the sky which are then combined into a composite image, with about nine image tiles to make up a picture of a galaxy.

Andromeda, as seen in the UV range and the visible range. Left: The Andromeda galaxy, a near neighbor of our Milky Way galaxy, as imaged by the Galaxy Evolution Explorer with ultraviolet sensors. Note that the UV image does not have the "fuzziness" characteristic of visible light images (right)—the solar system "clusters" and bands of space between them are more clearly visible. [NASA/JPL-Caltech news photos by GALEX and John Gleason.]

6.e. Spill Detection

Ultraviolet remote-sensing techniques are used in conjunction with other detection and imaging technologies for various environmental studies, including coastal monitoring, the detection of spills, and the determination of spill thickness, spread, and scope. Ultraviolet is also useful for measuring the effectiveness of various dispersants. Many oil-spill aerial-detection systems combine one or more visual imaging systems in combination with an ultraviolet camera and an infrared camera.

6.f. Commercial Products

Detection Substances

Ultraviolet detection products (known as 'spy dust') are commonly available for monitoring theft and tampering and to provide positive identification. They are used for marking currency, valuables, doorknobs, safes, and items that need to be identified at some later date or at a different location. The powder, gel, or UV 'ink' is usually dispensed from a small vat, tube, or marking pen and when applied, cannot be seen with the unaided eye. Then, if there is tampering or theft, the area or the recovered object can be studied by exposing it to appropriate frequencies of ultraviolet light. It can be noted whether they have been touched or transferred to other objects or someone's hands or clothing and sometimes even fingerprints can be detected.

UV detection powders typically sell for about $15 per ounce, with lower prices for bulk purchases. An ultraviolet illuminator is also required to stimulate the particles to fluoresce so they can be seen if they are later inspected.

Flame Detectors

Ultraviolet flame detectors can sense the UV radiation from hydrocarbon flames and indicate *flame* or *no flame* conditions for fire surveillance. They are used in a variety of industrial settings where generators or burners might be at risk of combustion.

Honeywell products include a gallium nitride solid-state sensor and two types of Geiger Mueller detectors that are sensitive to ultraviolet at 1800 to 2600 Å. At this spectral-response level, the detector does not respond to ground-level solar radiations nor blackbody emissions up to 2600°F.

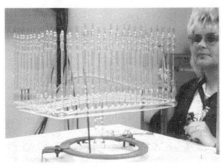 

Ultraviolet flame detectors from Honeywell include tube (minitube and power tube) and solid-state varieties. Left: Minitube technology is more robust for environments with heat and vibration. Right: Power tube technology has the same sensitivity but is more suitable for lower heat/vibration space- and aircraft applications. [Honeywell 1999 news/product photos, released.]

The Honeywell flame-monitor system has been utilized in fire surveillance applications on the NASA Skylab.

Maritime Surveillance Products

Ultraviolet sensing devices are often used in conjunction with visible and infrared sources for many types of aerial surveillance activities, such as aerial surveillance of marine craft and surface and underwater marine environments.

Left: The Daedalus AAD1221 Maritime Surveillance Infrared/Ultraviolet (IR/UV) Scanner operates in the 8.5 to 12.5 μm region and in the 0.32 to 0.38 μm region (ultraviolet) as part of the Maritime Surveillance System supplied by the *Swedish Space Corporation*. The Swedish Coast Guard has been using this system on Cessna 402C airplanes. [Swedish Space Corporation news/product photos, released.]

The *Swedish Space Corporation* (SSC) offers an infrared/ultraviolet scanner, the Daedalus AAD1221 Maritime Surveillance Scanner. This is a subsystem of the Maritime Surveillance System (MSS) originally developed for the Swedish Coast Guard, in operation for over 15 years. This equipment is suitable for use at low altitudes for imaging oil spills, for example. The system obtains ultraviolet data during the day and maps the extent of a slick, irrespective of thickness. False-color images are displayed in realtime. Other components of the MSS include side-looking airborne radar, a microwave radiometer, and handheld cameras.

Experimental Sensors

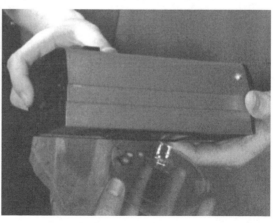

Purdue chemistry student Matthew Allen shines ultraviolet light on a dish of porous silicon held by chemist Jillian Buriak. Buriak has found a way to stabilize the surface of porous silicon so its light-emitting properties can be used to develop new types of sensors and opti-electronic devices. The silicon in the dish responds to the UV stimulation by emitting a bright orange light. Buriak has succeeded in overcoming some of the conditions that limit the photoluminescent properties of porous silicon, and it may someday be practical for developing fine-tuned sensors to perform realtime measurements or new flat-screen computer displays. [Purdue 1998 news photo, released.]

Ultraviolet Light Meters

Light meters that respond to ultraviolet light are used in a number of chromatography, sterilization, and forensic investigation tasks. They are also known as radiometers. UV light meters use photoelectric cell sensors and filters to meter only the UV light. For professional purposes, they are priced around $1,000 and it is best to get meters that have been certified (e.g., U.S. NIST).

7. Problems and Limitations

Because of their potential for damage to humans, ultraviolet-generating products should be handled with caution. Passive systems, that sense ultraviolet without creating it, are usually not harmful, but the rays influencing them may be.

Visibility

One of the limitations of ultraviolet-sensing technologies is that ultraviolet cannot be seen by humans. Thus, we must rely on data- or light-conversion techniques to detect or image ultraviolet sources. Making substances fluoresce is one way of taking advantage of their chemical properties to reveal the presence of UV in the visible spectrum. Using ultraviolet-sensitive film and photographic techniques is another way to make ultraviolet emissions 'visible'.

Hazards

Ultraviolet devices must be used with reasonable care. Remember that the far-ultraviolet spectrum is adjacent to X-rays, which can inflict significant harm on humans and other living things.

Many ultraviolet products are not shone directly at humans or, if they are, only for a short time. Some ultraviolet products illuminate in the portion of the ultraviolet spectrum close to the visible spectrum (violet), which is less dangerous to humans. Nevertheless, it is prudent to remember that ultraviolet exposure can permanently damage skin and eye tissues and that people taking antibiotics, tranquilizers, or birth control pills may be especially sensitive to their effects. Sunblock products (SPF 30+), can also be used to protect hands or other body parts that may be exposed to ultraviolet illuminators in labs or other environments.

The *UV Index* is expressed on a scale from zero to 10 that is used by the *Environmental Protection Agency* and the *National Weather Service* to report on the amount of UV radiation reaching an area. Levels of five or above are considered hazardous and protection should be worn.

Surveillance devices that emit UV radiation should not be left on for extended periods of time and should not be shone directly at people's eyes. UV-blocking goggles should be worn, especially if the ultraviolet is in the higher frequency ranges. For basic goggles to protect from the blue 'haze' that emanates from UV wavelengths, ANSI 287.1-1989 and OSHA 1910.133 goggles can be worn. Goggles can also help increase contrast when viewing UV-illuminated evidence or documents.

Fortunately, most small commercial UV illuminators are designed to work in wavelengths that are close to the violet region of the visible spectrum, which are less harmful, but it's still wise to be careful. Used with common sense, UV surveillance tools are especially useful for crowd control at large events, for detecting tampering or theft, and for studying archaeological finds.

8. Restrictions and Regulations

With the exception of industrial standards for UV-sterilization products, safety gear, UV-resistent paints and gels, and ultraviolet lights (such as tanning lights and UV illuminators), there are not a lot of restrictions to using ultraviolet-sensing technologies for a variety of commercial, scientific, educational, and personal applications. Thus, ultraviolet has become a valuable aspect of surveillance, particularly in archaeology, space science, geology, biochemistry, and forensics.

9. Implications

Many of our discoveries about ultraviolet are new. Ultraviolet-sensing studies in astronomy and wildlife biology are very recent, barely three decades in many cases, and less than fifteen years in some. Ultraviolet lasers are also recent inventions. The author feels certain there is more to learn about ultraviolet and that there are probably many practical applications for UV-sensing that will be developed in the future.

The study of ultraviolet-sensing in birds and other creatures is not just important for scientific understanding and the possible medical benefits available through technology; it is also important because advancements in biology and chemistry allow us to genetically engineer cells to contain characteristics not spontaneously found in nature. This opens up a Pandora's Box of possibilities.

Many species of birds have better vision than humans. In fact, as a group, birds tend to have better vision than mammals. Many mammals have poor visual acuity and poor color-sensing abilities. Even the octopus, a cephalopod mollusk, has better vision than most mammals. Humans are unusual in their ability to see well and to see colors. Dogs and cats (with the exception of Siamese cats) apparently see the world in monochrome. In contrast, many birds, especially birds of prey, have remarkable acuity and color vision. Some of them can see five to 10 times better than humans and they apparently see not just the colors humans see, but can sense into the ultraviolet and infrared spectra, as well. Bird plumage probably appears more 'colorful' to birds than it appears to us, with ultraviolet variations that add to each bird's individuality.

Just as dolphin research has helped us understand and design sonar devices, bird research has taught us new things about ultraviolet. This knowledge may lead to new types of chemical sensors or even to birds trained to seek out ultraviolet and sound a signal. We have also learned that the extra ultraviolet-sensing abilities of birds are at least partly chemical and there is undoubtedly some hormonal and perceptual processing that works in conjunction with the chemical structures in the eye itself. Despite the difficulties inherent in genetic procedures, scientists and philosophers have already proposed genetically inserting ultraviolet-sensing genes into humans in the embryonic stage in order to see if the human visual range can be extended into the ultraviolet [Sandberg et al., 1997]. The idea may not be farfetched.

Sunblock creams can screen or block ltraviolet light, and can be used in scientific studies of birds and other animals to note behavioral changes related to reduction in ultraviolet 'colors'. But in terms of surveillance, they could also theoretically be used by someone wanting to hide ultraviolet markings, perhaps for clandestine or illegal purposes.

10. Resources

Inclusion of the following companies does not constitute nor imply an endorsement of their products and services and, conversely, does not imply their endorsement of the contents of this text.

10.a. Organizations

American Astronomical Society (AAS) - A Washington, D.C.-based professional society for astronomers providing support, publications, grants, and educational programs. www.aas.org/

Astro Space Center (ASC) - A branch of the Lebedev Institute of Astrophysics in Moscow, Russia. The ASC engages in upper atmospheric, solar, and astronomical research. Research includes spectrographic studies and astronomics ranging from ultraviolet to cosmic rays. The facility is equipped to design, construct, and calibrate space experiments. www.asc.rssi.ru/

Canadian Space Agency (CSA) - A Canadian Agency, founded in 1989 from scientific developments dating back to the 1800s, that is involved in atmospheric sciences, space exploration, and space astronomy. CSA plans and implements Canada's Space Science Program. CSA also supplies scientific instruments to a variety of NASA projects including the FUSE satellite. www.science.sp-agency.ca/

Center for EUV Astrophysics (CEA) - CEA was opened in September 1990 to support research into extreme-ultraviolet astronomical sensing. In 1997 CEA assumed command and control operations for the EUVE satellite system. CEA publishes an EUVE Observation Log, image and information archives, and supports a Guest Observer (GO) program online. It is located at the University of California, Berkeley. www.cea.berkeley.edu

Center for Science Education - U.C. Berkeley Space Sciences Laboratory educational site which includes resources for educators and scientists, including lesson plans, study units, and information about educational programs. The Solar Max 2000 section includes information on space flight and solar cycles and the Center for Science Education has a Light Tour guide to the optical spectrum which allows the user to input wavelengths of interest. cse.ssl.berkeley.edu

Fluorescent Mineral Society (FMS) - This is an international nonprofit society of professional and amateur mineralogists and gemologists founded in 1971. The organization shares knowledge and examples of luminescent minerals and organizes seminars and research projects. www.uvminerals.org

Forensic Consulting Associates along with **New England Forensics** - Provides seminars on specific aspects of forensic investigations, including specialized and advanced photography courses, some of which utilize ultraviolet photography. www.forensicconsulting.com

Forensic Document Services (FDS) - This Australian firm, with branches worldwide, provides scientific examination services to government and commercial firms for investigating handwriting and documents. Technologies include spectral comparison, microscopy, and ultraviolet and infrared imaging. The company also provides courses and seminars. www.asqde.org/fds

International Association of Financial Crimes Investigators (IAFCI) - A nonprofit, international organization of law enforcement officers and special agents founded in 1968 to provide professional fraud prevention consulting and fraud investigation services. www.iafci.org

International Astronomical Union (IAU) - A professional (Ph.D. and beyond) organization founded in 1919 to promote the science of astronomy through international cooperation. www.iau.org

International Ultraviolet Association (IUVA) - A professional association interested in the development and research of ultraviolet sciences. Its members are involved in general and specialized areas of UV research, including sterilization, water purification, etc.

National Aeronautics and Space Administration (NASA) - NASA handles a vast research, development, and applications structure devoted to space science and related spinoff technologies. NASA cooperates with many agencies and contractors and disseminates a great quantity of news and educational information related to its work. Of interest to ultraviolet sensing are the many UV telescopes, photographic systems, and aerial sensing systems that have been used and are continually being developed. www.nasa.gov/

10.b. Print Resources

Green, A. E. S., Editor, *The Middle Ultraviolet: Its Science and Technology,* New York: Wiley, 1966, 390 pages.

McEvoy, R. T., *Reflective UV Photography, IX Log 911, Rochester,* New York: Eastman Kodak, 1987.

Rabalais, J. Wayne, *Principles of Ultraviolet Photoelectron Spectroscopy,* New York: Wiley, 1977, 454 pages.

Radley, J. A.; Grant, Julius, *Fluorescence Analysis in Ultra-Violet Light,* New York: Van Nostrand, 1954

Redsicker, David R., *The Practical Methodology of Forensic Photography,* New York: New York, 1991.

Articles

Barsley, Robert E.; West, Michael H.; Fair, John A., Forensic Photography - Ultraviolet Imaging of Wounds on Skin, *American Journal of Forensic Medicine and Pathology*, 1990, V.11(4), pp. 300-308.

Bennett, A.; Cuthill, I, Ultraviolet Vision in Birds: What is its Function? *Vision Research*, 1994, V.34, pp. 1471-1478. Reviews evidence for UV vision in birds, discusses the properties of UV light, and discusses the functions of UV vision in birds.

Bennett; Cuthill; Partridge; Maier, Ultraviolet vision and mate choice in zebra finches, *Nature*, 1996, V.380 pp. 433-435.

Dankner, Yair; Jacobson, E.; Goldenberg, E.; Pashin, Sergey, Optical-Based UV-IR Gas Detector for Environmental Monitoring of Flammable Hydrocarbons and Toxic Gases, *SPIE Proceedings*, V. 2504, 1995, pp. 35-38.

Elias, Eliadis, UV Photography, available on the Web. The author has experience on archaeological photography. The author describes color film and slide photography, UV light sources and many practical tips for getting the best UV mineral pictures. A very worthwhile, illustrated article that can be read on the Fluorescent Mineral Society Web site.

Emmermann, Axel, Photographing Fluorescent Minerals, Antwerp Mineral Club, Belgium.

Holberg, J. B.; Ali, B.; Carone, T. E.; Polidan, R. S. Absolute Far-Ultraviolet Spectrophotometry of Hot Subluminous Stars from Voyager, *Astrophysics Journal*, V.375, pp. 716, 1991.

Holberg, J. B. EUV Results from Voyager, *Extreme Ultraviolet Astronomy*, R.F. Malina and S. Bowyer, Editors, Pergamon Press, 1989.

Krauss, T. C., Forensic Evidence Documentation Using Reflective Ultraviolet Photography, *Photo Electronic Imaging*, February 1993, pp. 18-23.

Krishnankutty, Subash; Yang, W.D.; Nohava, T.E., UV Detectors at Honeywell, *Compound Semiconductor Magazine*, 1998, Summer, pp. 4-5.

Kunz, Y. W.; Wildenburg, G.; Goodrich, L.; Callaghan, E., The Fate of Ultraviolet Receptors in the Retina of the Atlantic Salmon, *Opthalmic Literature,* 1995, V.48(2), p.116.

Linick, S.; Holberg, J. B., The Voyager Ultraviolet Spectrometers - Astrophysical Observations from the Outer Solar System, 1991, *J.B.I.S.,* V.44, p. 513.

Marshall, Justin; Oberwinkler, Johannes, Ultraviolet vision: The colourful world of the mantis shrimp, *Nature,* 1999, V.401(6756), pp. 873-874. Describes four types of UV photoreceptors found in the shrimp.

Pex, James O., Domestic Violence Photography, Oregon State Police, Coos Bay Forensic Laboratory. This describes the basics of photography and provides some practical tips on infrared and ultraviolet photography.

Rodricks, Dan, After Nearly 54 Years, Pilot Comes Home, *SunSpot,* 26 Dec. 1997.

Sandberg, B.; Tuominen, H.; Nilsson, O.; Moritz, T.; Little, C. H. A.; Sandberg, G.; Olsson, O., Growth and Development Alteration in Transgenic Populus, Fort Collins, Co.: U.S. Forest Service, Rocky Mountain Research Station, 1997, pp. 74-83.

Schneider, Russell E.; Cimrmancic, Mary Ann; West, Michael H.; Barsley, Robert E.; Hayne, Steve, Narrow Band Imaging and fluorescence and its role in wound pattern documentation, *Journal of Biological Photography,* 1996, V.64 (3), pp. 67-75.

Starrs, James E., New techniques: Ultra-violet imaging - Don't go west, young man, Mississippi Court says; The West Phenomenon seen as less blue light than blue smoke and mirrors, *Scientific Sleuthing Review,* 1993, V.17(1), pp. 13-14.

West, M. H.; Barsley, R. E.; Hall, J. E.; Hayne, S.; Cimrmancic, M., The Detection and Documentation of Trace Wound Patterns by the Use of an Alternative Light Source, *Journal of Forensic Sciences,* November 1992, V.37(6), pp. 1480-1488.

West, Michael H.; Billings, Jeffrey D.; Frair, John, Ultraviolet Photography:Bite Marks on Human Skin and Suggested Technique for the Exposure and Development of Reflective Ultraviolet Photography, Sept. 1987, *Journal of Forensic Sciences,* V.32(5), pp. 1204-1213.

West, Michael H.; Frair, John A.; Seal, Michael D., Ultraviolet Photography of Wounds on Human Skin, *Journal of Forensic Identification,* 1989, V.39(2), pp. 87-96.

Journals

Astronomical Journal, American Astronomical Society professional journal in publication for over 150 years, published monthly.

Astronomy and Astrophysics: A European Journal, professional journal published on behalf of the Board of Directors of the European Southern Observatory (ESO).

The Astrophysical Journal, published by the University of Chicago Press for the American Astronomical Society, three times monthly, in publication since 1895.

Journal of Forensic Identification, a scientific publication of the International Association for Identification. www.theiai.org/publicationjfi.htm

Journal of Forensic Sciences, a publication of the American Academy of Forensic Sciences. www.aafs.org/Journal.htm

Journal of Optics A: Pure and Applied Optics, formerly *Journal of Optics,* it has been split into two publications. This journal of the European Optical Society covers modern and classical optics.

Physics in Medicine and Biology, applications of theoretical and practical physics in medicine, physiology, and biology.

Security Journal, international journal with contributions from many scientific disciplines related to protecting assets from loss, published in association with the American Society for Industrial Security Foundation.

10.c. Conferences and Workshops

Many of these conferences are annual events that are held at approximately the same time each year, so even if the conference listings are outdated, they can still help you determine the frequency and sometimes the time of year of upcoming events. It is very common for international conferences to be held in a different city each year, so contact the organizers for current locations.

Many of these organizations describe the upcoming conferences on the Web and may also archive conference proceedings for purchase or free download.

The following conferences are organized according to the calendar month in which they are usually held.

Meeting of the American Astronomical Society, a regular meeting that has been held over 200 times.

From X-rays to X-band: Space Astrophysics Detectors and Detector Technologies, proceedings are available for download. www.stsci.edu/stsci/meetings/space_detectors/

The United Nations Conference on the Exploration and Peaceful Uses of Outer Space (UNISPACE), first held in 1968 and every decade-and-a-half since then. Committee meeting are held more frequently and information is available on the U.N. site.
http://www.unis.unvienna.org/unis/pressrels/2005/unisos307.html

International Symposium: Adaptive Optics: From Telescopes to the Human Eye, this topic was discussed at the 2000 symposium in Spain.

International Conference on Space Optics-ICSO, annual international conference.

10.d. Online Sites

The following are interesting Web sites relevant to this chapter. The author has tried to limit the listings to links that are stable and likely to remain so for a while. However, since Web sites sometimes change, keywords in the descriptions below can help you relocate them with a search engine. Sites are moved more often than deleted.

Another suggestion, if the site has disappeared, is to go to the upper level of the domain name. Sometimes the site manager has changed the name of the file of interest. For example, if you cannot locate www.goodsite.com/science/uv.html *try going to* www.goodsite.com/science/ *or* www.goodsite..com *to see if there is a new link to the page. It could be that the filename* uv.html *was changed to* ultraviolet.html, *for example.*

Center for Science Education - U.C. Berkeley Space Sciences Laboratory eduational site which includes resources for educators and scientists, including lesson plans, study units, and information about educational programs. The Solar Max 2000 section includes information on space flight and solar cycles and the Center for Science Education has a Light Tour guide to the optical spectrum which allows the user to input wavelengths of interest. cse.ssl.berkeley.edu/

Electronic Newsletter of the EUVE Observatory - A quarterly publication of the EUVE Science Archive group issued by the Center for Extreme Ultraviolet Astrophysics at Berkeley, which describes the Extreme Ultra-Violet Explorer program and scientific discoveries made since the launch of the

EUVE in 1992. Issues dating back to 1991 are available in text form, current issues by email subscription request. The site also includes some prelaunch information from a NASA brochure which explains the unexplored ultraviolet radiation surveillance windows and their potential for scientific discovery. www.cea.berkeley.edu/~science/html/Resources_pubs_newsletter.html

Far-Ultraviolet Spectroscopic Explorer (FUSE) - This site provides background information, mission status, technical details, news, and FAQs related to the NASA-supported astronomical explorer launch in June 1999. fuse.pha.jhu.edu/

IUE Analysis - A Tutorial. This is a step-by-step multipage technical site providing information on the *International Ultraviolet Explorer Satellite*. It explains how to extract data from an IUE tape and conduct a simple data analysis. It was created by R.W. Tweedy and Martin Clayton in 1996 in association with CCLRC/Rutherford Appleton Laboratory, Particle Physics & Astronomy Research Council. A hard copy is also available for download. www-star.st-and.ac.uk/starlink/stardocs/sg7.htx/sg7.html

Microscopy Primer - Fluorescence Microscopy. This extensive scientific site provides an introductory history and explanation of ultraviolet as it relates to fluorescence and microscopic research. It includes fundamentals, light sources, transmitted and reflected light, optimization and fluorochrome data tables. It is well-organized and -illustrated and includes Java demos. The site is managed by M. W. Davidson, M. Abramowitz, through Olympus America Inc., and the Florida State University, in collaboration with the National High Magnetic Field Laboratory. micro.magnet.fsu.edu/primer/techniques/fluorescence/fluorhome.html

School of Photographic Arts and Sciences at RIT. This excellent site recounts several dozen examples of experiments with infrared, ultraviolet, and conventional photography. Professor Andrew Davidhazy has compiled these works which include suggestions for exposures, films, filters, and lenses. The author even describes how to shoot high speed images of moving bullets with an Agfa 1280 digital camera and demonstrates how ultraviolet and infrared films can be used to reveal 'hidden' writing and art forgeries or altered works. www.rit.edu/~andpph/articles.html

11. Glossary

Titles, product names, organizations, and specific military designations are capitalized; common generic and colloquial terms and phrases are not.

black light	colloquial term for a relatively inexpensive phosphor-coated mercury-tube lamp which emits light in the near-ultraviolet and causes certain things (especially whites) to appear to humans to glow
diffraction grating	a device designed to spread light into its component wavelengths, often with many finely etched parallel (or nearly parallel) lines or grooves. Diffraction gratings are used in precision instruments like spectrographs.
FARUV	far ultraviolet
FES	fine-error sensors. Instruments used in conjunction with precision instruments to aid in such tasks as navigation and aiming (as in a telescope).
fluoresce	to react to stimulation by re-emitting wavelengths that are 'shifted' in frequency such as ultraviolet energy being absorbed and then re-emitted in the visual spectrum
FUSE	Far Ultraviolet Spectroscopic Explorer
geocorona	a huge region of gases encircling the Earth that becomes thinner at the outer boundaries as it transitions into space. Ionization occurs in this region and the ion particles are held by the Earth's gravity. The geocorona causes scattering of extreme ultraviolet radiation, a phenomenon that is less pronounced at night.

stains - fluorochrome

secondary fluorescence stains that are used in microscopy to 'tag' specific objects or substances so that they will react to ultraviolet light if they do not automatically fluoresce.

STJ
super-conducting tunnel junction. A form of cryogenic detection component that can detect radiation from infrared through to X-ray frequencies. It is now possible to combine them into arrays.

SUSIM
Solar Ultraviolet Spectral Irradiance Monitor. One of the instruments used upon the UARS system.

UARS
Upper Atmosphere Research Satellite. An environmental satellite launched in September 1991. One of its many tasks is to provide data on the ozone layer which shields us from ultraviolet radiation.

UV
ultraviolet

UVAI
Ultra-Violet Auroral Imager. A Canadian Space Agency imager on the Russian Interball-2 satellite.

UVIR
ultraviolet infrared

UVPI
ultraviolet plume imager/instrument

UVSP
Ultra-Violet Surveillance Program

VLIM
very local stellar medium. A natural space phenomenon consisting of atoms of interstellar gas that move through the solar system as we move through our galaxy. These electrically neutral interstellar 'winds' scatter extreme ultraviolet radiation, creating an apparently unavoidable interference or 'noise' to UV detectors and telescopes.

WUPPE
Wisconsin Ultraviolet Photo-Polarimetric Experiment. A pioneer space-based polarization and photometry experiment in the UV spectrum jointly carried out with NASA.

Electromagnetic Surveillance

X-Ray

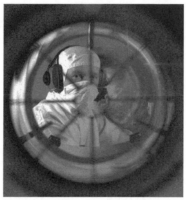

1. Introduction

X-rays are invisible penetrating rays that have a seemingly magical ability to travel through many substances. Since some substances will impede X-rays more than others, it is possible to generate images that show the relative density of materials—making it possible to look inside solid objects. This adaptation of X-rays is often colloquially called 'X-ray vision'. X-ray vision is symbolic for many surveillance activities, from medical imaging and extraplanetary observation, to clairvoyance and voyeurism. In the context of this chapter, X-ray technology is described as it pertains to common professional applications that use X-radiation for sensing applications.

X-rays can be harmful to humans because they disturb our basic atomic structure. The Earth's ozone and atmospheric envelope protects us from natural X-radiation. Because X-rays are not prevalent on our planet and because they have particular high-frequency characteristics, humans have not evolved any natural 'immunity' or protection against damage that might be caused by exposure to X-rays. When X-rays are synthetically generated, they are used in carefully controlled situations to prevent human exposure.

A nickel prototype mirror being optically tested at the *Advanced X-Ray Astrophysics Facility*-S (AXAF-S). X-ray telescopes are providing new ways to study black holes, cosmic clouds and other phenomena in the universe that we have never before been able to see or record. [NASA/Marshall news photo, released.]

X-radiation is a very useful phenomenon. It behaves in predictable ways, making it a practical tool for studying substances down to the atomic level, and for imaging materials that are sensitive to its effects. X-ray technologies are often used with other surveillance technologies, including chemical and visual surveillance devices. X-rays enable us to look inside containers and living beings without direct handling or surgical intervention. This makes it possible to detect things like smuggled items, manufacturing defects, contamination, and various diseases or physical abnormalities. X-ray technologies are also used in astronomy for surveilling cosmic emanations and studying the origins of life.

X-rays penetrate different materials to different extents, depending upon their density. Since bone is denser than skin, X-rays passing through a body onto a photosensitive surface like an X-ray film or digital X-ray detection array, will affect the imaging surface more or less, depending on whether the rays passed through bones or skin. The rays don't pass as readily through bone, so they don't stimulate the imaging system as much as the rays that passed through the skin. Similarly, in airport security systems, radiant energy travels more readily through clothes than it does through a gun or laptop computer. This creates an image with lighter or darker areas, depending upon the characteristics of the imaging surface. With newer, digital imaging systems, the values can be inversed or processed with colors to make the picture more readable. The characteristics of medical X-rays can be readily applied to industrial and other applications, as well.

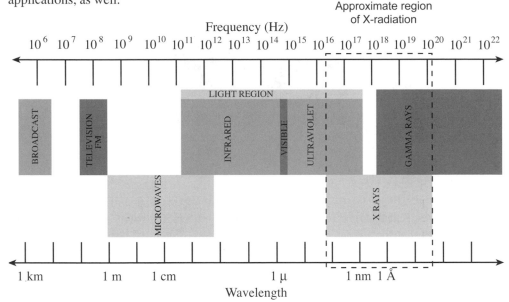

2. Kinds and Variations

X-ray machines are used for many different kinds of surveillance and are commercially manufactured in different sizes, with a variety of levels of portability, power, and cost. This text does not go into depth about X-rays that are used for medical purposes—radiological and diagnostic medical texts should be consulted for this specialized information. This text focuses more on X-rays used for covert, military, or security purposes.

The size of an X-ray machine used to be a good indicator of the power and price of the system, but with computer technologies, this is no longer true. It is now possible to pack powerful capabilities into smaller systems. Microelectronics even make it possible to equip vans with high-power X-rays for special events.

X-ray technology is basically passive or active:

passive X-ray detector system - one that reacts to X-rays without generating X-rays. Passive systems are used in detection devices such as X-ray space telescopes, that sense cosmic rays, or in situations where X-rays are suspected of being used as a weapon. Passive systems are also used to detect and measure leakage from active X-ray machines.

active X-ray system - one that generates X-rays and which may or may not have a detector or imaging system to 'capture' the emitted rays. An *active sending system* simply generates X-rays, as in a scientific experiment to observe their effects on bacteria. An *active sending-and-sensing system* is the one we commonly see in airports or hospital radiology labs—the system sends out X-radiation and detects and displays the information that is generated by the X-ray probe.

X-ray images from passive sensing systems or active sending-and-sensing systems are displayed in a number of ways:

traditional grayscale photographic images on film - these resemble large negatives and are the types of images used by radiologists and some manufacturers or quality assurance testers who are recording materials fabrication processes or documenting possible defects in production components.

realtime 'light-box' style illuminators - images that are projected onto realtime or near-real-time viewers in airport- and crowd-security systems, especially the small conveyor-belt systems used in airports or small-scale industrial production lines.

realtime or delayed computerized display - digitized images that may be realtime, near-real-time, or on-demand database entries that can be displayed and sometimes manipulated on a computer screen. They are used for more specialized applications, usually for experiments or observations in scientific labs.

When X-rays are sensed, they are most often viewed as grayscale images, since X-rays are not part of the visible spectrum that we perceive as color. However, sometimes assigned colors can aid in interpreting an image—'false color' or 'pseudocolor' systems are becoming more common. These color assignments are also found in infrared and ultraviolet imaging systems. Some pseudocolor systems can toggle between the grayscale and color interpretations so that the 'raw' data can be cross-referenced with the image that has been colored by the computer. Some airport security systems now have several viewing options that display the contents of a suitcase in a variety of ways so that the identity of objects may be more easily determined without opening the suitcase.

Basic System Design

Baggage inspection machines are one of the more common X-ray surveillance technologies. This text doesn't go into detail on the mechanics and electronics of the different types of X-ray machines, but a simple baggage X-ray machine will be described to give an idea of how X-ray baggage-inspection systems work.

- X-rays are generated by using an electron gun that emits electrons, a system that is similar to the electron guns in a cathode-ray tube (CRT) to display TV or computer images. High voltages are then used to accelerate the electrons until they collide with a barrier (usually metal). The speeding electrons are abruptly halted and scattered by the barrier, resulting in the generation of radiation in the X-ray spectrum. Sometimes more than one energy level of X-rays is produced, depending upon the purpose of the system.

- The generator that produces the X-radiation is usually positioned above or below the baggage conveyor belt. When the machine is activated, the beams are directed up or down through the belt and through any baggage that is passing by on the belt.

- Between the X-ray generator and the conveyor belt are devices called *collimators,* that control the direction of the beam, usually processing the rays so that they are parallel and spaced as desired. X-rays normally spread out in a cone shape, like a flashlight beam, and it is usually necessary to narrow and focus the beam for practical use.

  - On the other side of the baggage, away from the X-ray emitter, are *detector bars* that are stimulated when they are hit by the X-radiation. The information from the detectors is translated into data that may be stored or processed and which is sent to a display system. On baggage systems, this display looks like a light box with ghostly grayscale or pseudocolor images of the various objects displayed against a light background.

Dual-energy beams that use two different voltages have been devised to aid in discriminating between different substances, such as between organic and inorganic materials.

Improvements in Technology

Since digital technologies are becoming more prevalent and microelectronics finer and more precise, it is now possible to design X-ray sensor arrays that are very high resolution. As the number of elements in the array increases, generally the resolution increases (as does the price).

Note, this book doesn't have a separate chapter on gamma-ray surveillance, because the science is recent and is mainly used in theoretical and space sciences, but there are some interesting developments in which low-dose gamma rays are being designed into cargo-inspection systems, without some of the protective enclosure limitations of certain X-ray technologies.

Since gamma-rays lie at the extreme high-frequency end of the electromagnetic spectrum, roughly adjacent to X-rays, many of the more general concepts relating to X-rays also apply to gamma rays, especially in space telescope applications. You are advised to consult physics texts and industrial suppliers to learn more about gamma ray applications.

Product Specifications

X-ray systems are usually specified in terms of their overall size, the size of the opening for the X-rayed objects, their weight, power consumption, beam orientation, resolution, storage capacity (digital systems), color (grayscale or pseudocolor), dosage per inspection, radiation leakage, and penetration. If the X-ray system includes a conveyor system for moving objects through the path of the X-ray beams, the speed of the belt will also be specified.

Beam orientation is important in systems designed to X-ray larger containers, such as cargo pallets. It is necessary to match the system and the direction in which it points to the needs of the task. For small baggage inspection, a downward-pointing beam is fine, but this wouldn't work for scanning a cargo truck or shipping container from the side.

3. Context

While X-ray surveillance is not inherently tied to any particular setting or application, because of its risks and cost, X-ray technology tends to be used almost exclusively in professional contexts. Trained personnel and special housings and facilities are required to contain the X-radiation. Traditionally, X-ray equipment has been large and somewhat immobile, though new computer technologies are succeeding in putting certain X-ray capabilities on electronic chips, which may spawn a new generation of portable equipment. In general, though, X-ray technology is not a casual consumer product and X-ray equipment is not commonly used outside of professional settings (and is generally used by personnel who have been through safety training or certification programs).

The contexts in which X-ray surveillance are most often used include:

medicine and preventive health care - X-rays are used for anatomical studies, medical monitoring and diagnostic procedures, forensic investigations (e.g., autopsies of injuries leading to death) and, in some cases, assessment of prenatal conditions.

cargo and luggage inspection - Many organizations use X-ray machines for luggage inspection, including public and military transportation providers, customs and border patrol officials, private and public mail and courier services, and public events crews.

archaeological, artifact, and art inspection - X-rays enable artifacts, art, bones, caves, ancient buildings, and other objects to be inspected for characteristics that cannot otherwise be seen, to aid in their identification, preservation, and authentication.

fraud detection - X-rays are sometimes used to examine documents or items that have been subjected to vandalism or tampering, though other less hazardous technologies are usually chosen first.

industrial testing, inspection, and quality control - X-rays can to used to check for quality, consistency, defects, contamination, internal characteristics, and other production fabrication qualities.

military - X-rays have been studied for specialized military applications such as weapons and detection devices and for assessing the health of personnel.

astronomy - X-ray telescopes are one of the more exciting technologies that are used to probe the cosmos and several have been put into space in the last several years, often in conjunction with infrared and ultraviolet sensing devices.

It takes skill and experience to interpret X-rays, particularly medical X-rays. It takes a different type of skill to interpret the X-ray view screens on conveyor-belt industrial product control or airport security systems. On conveyor-belt systems, a constant parade of objects is moving by at medium or high speed. It requires constant attention to assess each and every image that goes by hour after hour, especially if irate travellers are complaining about delays and missed connections. To maximize efficiency and decrease the chance of errors, those viewing the screens should be put on rotating shifts that allow them to do some other type of activity at least every two hours, to give their eyes and brains a break.

Most X-ray systems require visual interpretation, but some of the computerized systems may also have alarms that can sound if certain triggers are detected. Wholly automated systems may also exist in some circumstances where the objects being inspected are all alike or follow similar patterns, as in industrial production line or quality inspection systems. Sometimes X-ray surveillance systems are used in conjunction with metal detectors and chemical sensors. In a baggage inspection system, for example, if an object cannot be definitely identified or is anomalous in some way, it's characteristics may be tested further or affirmed by sensing it with a metal detecting wand or with a handheld chemical 'sniffing' device. Systems like this are now used in airports.

Displays

The kinds of monitors that are used to display X-ray imagines are dependent upon the context. In medicine, where accuracy and high resolution are essential to the interpretation of sensitive medical information, a system which takes longer to display (or chemically develop) the image is usually better than a fast system with lower resolution. However, digital high-resolution systems with the potential to revolutionize X-ray technology in medicine were becoming available in the late 1990s.

In transportation security systems, realtime or near-realtime display is important, since it isn't practical to keep passengers waiting in long lines while X-rays are being developed and processed. Hence, these systems tend to use low or medium resolution images displayed on a small screen that resembles a 'light box' which is brightly illuminated, and shows the contents of baggage. Some of the more automated systems will sound an alarm or highlight an area if they detect objects of a suspicious nature.

For industrial purposes or cargo surveillance, sometimes multiple-monitor systems are used. Two screens make it possible to display cargo from different vantage points, while a third might be used to display information from a barcode or database that gives information about the cargo, such as the cargo manifest and shipping history.

Spin-Offs from X-Ray Technology

Holography is an interesting technology that evolved from X-ray research. It now has many applications in security and surveillance.

Holography has contributed to the design of counterfeit-deterrence features, UPC scanners, non-destructive quality and defect testing, and pattern-recognitions systems in computer software and robots. Holographic storage concepts, which are related to holographic recording concepts, are important to microcomputer research as they hold promise for improved mass storage devices. Inexpensive mass storage is desirable for the storage of surveillance information, particularly 'high-bandwidth' applications like sound and video.

4. Origins and Evolution

The existence of X-radiation wasn't known until less than 200 years ago, making it a very recent discovery and an even more recent surveillance technology. Unfortunately, the pioneers of X-ray science didn't realize how very dangerous high-level X-ray energy could be or how it could damage human tissue. The history of X-ray technology is not just a history of the generation of X-rays and improvement of the technology, but is also a history of how to protect ourselves from its dangerous effects and side effects.

The Discovery of X-Rays

William Crookes (1832-1919) was an English physicist with a strong interest in spectroscopy, radiation, and the phenomenon of luminescence. In the course of his research, Crookes noted that cathode rays traveled in straight lines and had sufficient energy to turn a small wheel. There was discussion at the time over whether cathode rays were a wave phenomenon or a particle phenomenon. While doing experiments with a Crookes tube, Crookes observed some of the effects of X-rays but didn't recognize the significance of the observations until after X-rays had been more systematically studied by Wilhelm Röntgen.

Wilhelm Konrad Röntgen (1845-1923) was a German physicist working at the University of Würzburg, Bavaria. He had been experimenting with Crookes tubes and generating cathode rays, the technology that eventually led to cathode-ray tubes (CRTs) several decades later, when he observed that some chemicals had a particular tendency to luminesce (to light up when stimulated). To better see the effect, he darkened the room and applied cardboard light-blockers and discovered that materials at a distance luminesced when the cathode rays were activated. The rays had passed right through the cardboard! He found that it would work even through walls. He had discovered 'invisible' rays that could pass through solid matter. Since he wasn't able to immediately identify the nature of the rays that were apparently passing through solid substances, he called them X-rays, after the convention of mathematicians to use the symbol x to indicate an unknown value or quantity. These rays were called Röntgen rays (in English it is often spelled Roentgen), in honor of Röntgen's discovery, right up until the 1950s, but the shorter term *X-rays* is now more prevalent. X-ray doses are still expressed in units called *roentgens*.

A little more than a month after his important discovery, after numerous experiments, Röntgen published a paper on X-rays, in December 1895. At a lecture demonstration about a month later, Röntgen X-rayed the hand of an octogenarian volunteer and spurred a fever of research and activity over the new discovery.

Soon after Röntgen published his important paper on X-rays in Europe, studies of X-rays were beginning on the other side of the world, in Japan. Kenjiro Yamakawa and Han-ichi Muraoka, who had studied in Röntgen's lab at Strasburg University, took an X-ray photograph, aided by the founders of the Shimazu Company. These early X-rays clearly showed a pair of glasses, a key in a lock, and the bones of a hand wearing a wedding band.

Discovering the Properties of X-Rays

Subsequent experiments revealed that X-rays were increasingly impeded as materials became denser. This made it practical for taking 'pictures' of human tissue that showed the bones and cartilage in contrast to the soft tissues, effectively making it possible to 'look inside' human beings. No one yet understood the cumulative effect X-rays would have at the basic cellular level.

It took a while to determine the type of energy X-rays represented. Some scientists thought it might be a longitudinal wave phenomenon, somewhat like sound. Others asserted that it was a particle phenomenon, like cathode rays. Still others felt it was electromagnetic, like light (which had itself not yet been fully explained).

Charles Glover Barkla (1877-1944), an English physicist, began investigating X-rays while at Cambridge and noticed that X-rays could reveal information about gases by the ways in which the rays were scattered when directed at the gases. The pattern of scattering was related to the density of the particular gas and thus could reveal its molecular weight. It also revealed, in 1904, that X-rays apparently were a particular kind of wave that was unlike longitudinal sound waves. Barkla further observed that the degree to which the X-rays could penetrate matter was related to which element had been used to scatter the rays.

A German physicist, Max T. F. von Laue (1879-1960), who had been an apprentice to the great quantum physicist Max Planck, began to experiment with crystals as a natural type of 'atomic grating' that could be used to control the passage of rays such as X-rays. Just as Barkla had discovered that gases could scatter X-rays to reveal information about those gases, von Laue discovered that crystals could influence X-rays to reveal information about the crystals and the X-rays themselves. Crystals are used in science because they have a very regular and somewhat predictable atomic structure that makes them useful for precision applications that cannot be easily achieved by other means. Many crystals also have particular vibratory qualities that are valuable for timing applications in electronics. Von Laue found that crystals could also reveal more about the nature of various radiant energies. By aiming X-rays at particular crystals and observing the resulting effect or direction of the scattering of the rays (depending upon what was being studied), clues as to the nature of the rays could be discerned. Von Laue took advantage of the nature of crystals to study X-rays and not only added evidence to the argument that X-rays were a form of electromagnetic radiation, but also created a new way to study the crystals themselves. Thus, the science of X-ray crystallography emerged.

Research on X-rays continued in both Japan and Europe. A father and son team, William Henry Bragg and William Lawrence Bragg, picked up on the work of von Laue and continued to study the effects of passing X-rays through different crystals. By doing so, they learned more about the energy frequencies emitted by X-rays and where they stood in the electromagnetic spectrum in relation to other phenomena such as visible light. This work earned them the 1915 Nobel Prize in physics. A Japanese researcher, Torahiko Terada, also did pioneer work on X-ray diffraction. Peter J. W. Debye subsequently demonstrated somewhat surprisingly that powdered crystals or mixtures of powdered crystals could be used as well as solid chunks of crystal for X-ray analysis. This opened up many possibilities for variations on the technology and possible industrial applications.

While the relationship may not be immediately apparent, X-ray technology enabled scientists to fill in many gaps in the Periodic Table, a cataloguing of atomic elements that was as yet somewhat rudimentary and incomplete. Henry Moseley (1887-1915), an English physicist, made use of the potential of William H. and William L. Bragg's crystallographic research for comparing X-ray wavelengths with atomic weights. The Periodic Table was reshuffled a bit into a more workable pattern through the work of Moseley, who formulated the concept of the atomic number in 1914. Thus, X-rays became an important aspect of analytical chemistry and Moseley developed the early science of X-ray spectrometry.

It was about this time that the first basic X-ray systems were developed in both Europe and Japan.

Karl Manne Georg Siegbahn (1886-1978), a Swedish physicist, built on the work of earlier scientists and refined the study and production of X-rays. Just as visible light is divided into different spectra that we perceive as different colors, X-rays had spectral qualities that were somewhat similar to light. In 1924, Siegbahn was awarded a Nobel Prize in physics for the development of X-ray spectroscopy.

The 20th Century - Practical Applications

The obvious application for X-rays was in medicine and physiology. Radiology books and journals became common at the beginning of the 1900s, with many picture examples of broken or abnormal bones and cartilage.

The prospect of looking inside things, especially inside humans, was so irresistible, that X-ray machines were showing up everywhere. Some of the major shoe manufacturers were even putting X-ray machines in shoe stores, where the sales reps would X-ray a person's foot in order to equip him or her with the 'perfect fit'. Remarkably, after many scientific experiments had been done with X-rays, people still didn't fully realize its dangers. X-ray enthusiasts created X-ray movies of people eating and swallowing. These are delightful to watch, but may have had serious health consequences for the subjects of the movie. Some people directed X-rays at their heads, trying to get a 'buzz'. Not only were X-rays dangerous, but many of the early machines delivered high doses of radiation. They a danger to the people near the machines, and would travel through the walls and across the street to expose the people working many meters away.

The idea of using X-radiation for industrial inspection may have been born around the same time that experimenters were developing devices that used acoustics or magnetism to inspect tracks, girders, and large industrial vessels such as military ships and tanks. X-rays provided a new way to look inside certain kinds of materials, particularly wood, thin metals, fabrics, and liquids. Lead was found to be impermeable to X-rays, which made it suitable material for masking off areas or providing protection from hazardous radiation.

In 1947, Dennis Gabor (1900-1979), a Hungarian scientist living in Britain, was seeking a means to greatly magnify X-ray images and essentially invented holography in the process. He was awarded a Nobel Prize for physics in 1971.

By the 1950s, X-ray diffraction techniques were used to discern more about a structure called DNA, found in living tissues—a fundamental 'blueprint' for the development of living cells (see the Genetics Surveillance chapter).

In the 1950s, many older dentists using traditional methods of diagnosis weren't yet using X-rays, but younger dentists, who had more recently graduated from dental school, were installing X-ray equipment to aid in detecting tooth decay and abnormalities. X-rays in those days were sometimes used overzealously. There are dentists from that time who are reported to have cellular damage to their fingers from exposure to X-rays and the machines still delivered fairly high doses. There was concern that X-ray exposure could lead to cancer, particularly leukemia.

Space Science

Like ultraviolet astronomy, X-ray astronomy is a relatively new discipline. Inside the Earth's protective ozone envelope, it's difficult to detect or measure any of the weak sources of X-rays that come from other parts of the universe. It's also difficult to separate weak X-ray sources from the stronger X-rays emanating from our Sun. Space travel really needed to get off the ground before extraplanetary X-ray surveillance could come into its own as a science.

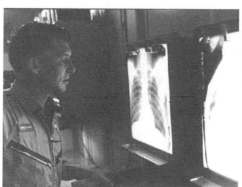 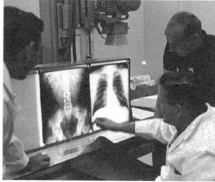

X-ray technology is versatile. In the space program, X-rays were used not only to ascertain the health of astronauts before subjecting them to the rigors of space, but also as reference documents, for assessing any changes that might be discerned after the astronauts returning from their flights. At the same time, X-ray telescopes were launched into space to study the universe beyond our planet and our solar system. Left: Astronaut Walter Schirra views his X-rays after his 1962 space flight. Right: Astronaut Charles Conrad (in the dark shirt) was the pilot for the Gemini 5 space flight. He is shown discussing X-rays with the medical team at Cape Kennedy in August 1965. [NASA/JSC news photo, released.]

After the launching of the Russian Federation Sputnik satellite in 1957, the United States embarked on its own space program and launched a number of rockets, during the late 1950s and early 1960s. Telescopes and other instruments were sometimes sent along with the rockets as the *payload,* to provide scientific data about space exploration and the environment of space. John Lindsay and other members of the *NASA Goddard Space Flight Center* took the first X-ray image of the Sun from a rocket, in 1963.

As space travel improved, the first of a number of scientific flights and laboratories were established in space. The Skylab mission was a way to study the characteristics and rigors of space more closely and provided a way to deploy telescopes outside the Earth's atmosphere and ozone layer. This made it possible to study emanations that would never reach the telescopes on Earth. One of these was an X-ray telescope that recorded many thousands of images over a period of nine months.

Scientists began to look for ways to take X-ray pictures of cosmic phenomena outside our solar system. In 1975, Paul Gorenstein, with the *Smithsonian Astrophysical Observatory*, and his colleagues, used a mirror and X-rays to photograph the constellation Virgo.

X-Ray Technologies in Other Disciplines

By the 1960s, research into holography was improving the science and by the early 1970s, artists and scientists had produced some stunning images, some as large as 1×2 feet, that were exhibited around North America in a traveling show that inspired commercial, industrial, and artistic development of the technology, some of which is now used in computer and surveillance applications.

X-rays were increasingly used to check luggage and cargo containers, to aid in quality assurance in industrial and commercial settings, and in many aspects of medicine.

X-rays were also used to monitor the progress of a fetus in the womb, but as concern about the dangers of X-rays to the developing embryo increased, this practice was largely discontinued by the 1970s, except for instances of serious problems that could not be assessed any other way. Gradually, the use of fetal X-rays gave way to the use of ultrasound which, when used correctly, is much less hazardous to human tissue than X-rays.

X-Ray Imaging of Luggage and Hand Baggage

As X-ray machines on public transportation systems continued to develop and concerns over terrorist threats mounted, public concerns about damage to photographic film increased, as well. In March 1978, the *Federal Aviation Administration* (FAA) required that signs be posted informing passengers that they had the right to request hand inspection of their photographic film in order to avoid X-ray exposure of the film.

In 1988, Pan Am Flight 103 exploded over Lockerbie, Scotland, from plastic explosives hidden in a suitcase. This tragedy focused greater attention on airline security and resulted in the mobilization of government intelligence agents and the local installation of equipment and personnel to help prevent such disasters in the future.

The 1990s - Transition to Digital

The entire electronics industry was substantially affected by the microcomputer revolution and, like many other technologies, X-ray devices gradually began to incorporate digital electronics. In earlier systems, however, mass storage and resolution were limitations that made it difficult to match the clarity of traditional film images. This changed in the mid-1990s, however, when multigigabyte computer drives dropped dramatically in price and resolution increased. In November 1999, PerkinElmer Optoelectronics announced the introduction of a digital X-ray camera with a monolithic active detector that was equal in size to a traditional sheet of X-ray film.

Digital technology had many advantages for X-ray images. A computerized system could provide features like image pan and zoom. It also made it possible to email the image or discuss it with other professionals through Internet or intranet 'whiteboarding' (simultaneous viewing), to compare two X-rays side-by-side, or to display them as thumbnails of images taken over a period of time. The speed of viewing of the image was greatly increased as well, as no photographic chemical development was necessary with a digital system, and X-rays can be retaken immediately if there were a problem with the first image—without having to reschedule followup appointments. These advantages benefited medical radiologists, industrial inspectors, and production-line quality assurance professionals using X-rays.

1980s and 1990s Space Science

Space science had come a long way from historic experiments in the 1960s. A great deal of new information about the universe was gathered in the 1980s and 1990s through computer and imaging technologies, including infrared, ultraviolet, and X-ray photography from space and on other planets (e.g., Mars).

Space exploration involves putting equipment and people in unfamiliar environments and subjecting them to stresses that aren't a part of normal Earth travel. This requires a commitment to studying the effect of space travel on various materials and instruments and the development of many new technologies. In 1994, the *Light Alloy Laboratory* was opened at the *NASA Langley Research Center*, to provide state-of-the-art analytical instruments as well as materials testing and processing equipment. Along with various microscopes and thermal analysis devices, X-ray diffraction equipment could be used for various identification and analysis tasks. Not all the technology is used in space—special alloys, coatings, and paints, also have applications for aeronautics research for Earth-based, high-speed and high-altitude surveillance aircraft.

Left: An artist's concept of the *Advanced X-Ray Astrophysic Facility* deployed in space. Right: Engineers at the *X-Ray Calibration Facility* (XRCF) at work, integrating the High Resolution Mirror Assembly into the *Advanced X-Ray Astrophysics Facility-1* to create the most powerful X-ray telescope ever built. [NASA/Marshall news photos, 1995 and 1996, released.]

In 1997 the Mars Pathfinder project included a number of interesting technologies, including an X-ray spectrometer that was used on the Mars rover—a terrain robot was used to explore Mars.

Surveillance technologies have resulted in some remarkable discoveries. Black holes were once just theoretical ideas, but in 1998, scientists were able to indirectly observe matter spiraling into a black hole and gaseous jets appearing to escape from the hole. X-rays were used to detect the matter disappearing into the black hole and infrared and radio waves to detect the jets of gas appearing to emanate from the black hole.

Left: An X-ray defractometer is extended out from the arm of a Marsokhod Rover, a planetary exploring robot. The defractometer was tested by sensing a rock, in this simulated Mars-like terrain, in 1995. Right: This is the rover robot that was carried on the Mars Sojourner interplanetary mission to study the planet Mars. The rover was photographed on Mars in 1997 while using the Alpha Proton X-Ray Spectrometer (APXS) to study a rock the scientists named "Moe". The Mars Pathfinder mission was developed and managed by the *Jet Propulsion Laboratory* for NASA's *Office of Space Science*. [NASA/Ames and NASA/JPL news photos, released.]

The *Advanced X-Ray Astrophysics Facility*, now called the *Chandra X-Ray Observatory* consists of a spacecraft, a scientific instrument module (SIM), and a highly sensitive X-ray telescope. The Observatory is designed to help astronomers detect black holes and high-temperature gas clouds, yielding information about our origins and the universe around us. Chandra was designed to be launched on board the Space Shuttle Columbia. Left: The 50,162-pound *Chandra X-Ray Observatory* being prepared for its mission in March 1999. Right: Astronauts took this photo of the Chandra tucked against the Space Shuttle while in Earth orbit. Here it is being tilted just prior to release from the Shuttle Columbia. [NASA/KSC news photos, released.]

Border Surveillance

During the 1990s, U.S. Customs invested in additional staff and technologies, including X-ray devices, to aid in the detection of smuggling of illegal goods and substances into the U.S. Additions and upgrades in equipment during the 1990s included baggage, pallet, and mobile X-ray units for surveilling trucks and containerized cargo. These were often used with other detection measures, including substance-sniffing dogs and fiber-optic scopes to examine gas tanks and other enclosed areas. For secondary inspections, a mobile X-ray van would sometimes be used for large or heavy items. A truck X-ray system was in use at the Otay Mesa crossing, along the southern border, for carrying out random and referred X-ray inspections of large cargo vehicles. For visual inspection large trucks and containers without the use of X-rays, a 'cherry picker' (a hoist like those used by above-ground utility-line technicians) would sometimes be used.

As part of the 1998 National Drug Control Strategy, the U.S. Government planned the implementation of a *Port and Border Security Initiative,* to include new Border Patrol agents and new technologies such as advanced X-ray devices and remote video surveillance along the southwest U.S. border. Nonintrusive inspection/surveillance technologies were also included.

X-ray Large-Scale Cargo Inspection. Left: Large items such as semi-trailers and cargo containers can be inspected in this large-scale New Mobile Sea container system. Note that an X-ray hazard sign on the top of the structure warns against inserting any part of the body when the system is energized and a light at the right indicates when the X-ray is on. Right: A train is positioned in preparation for X-ray inspection. [Customs and Border Protection news photos by G. Nino, 2005, released.]

Innovations

The more important innovations in X-ray technologies emerging in the 1990s included the development of X-ray telescopes, pseudocolor display systems, digital X-ray imaging systems, and 3D imaging arrays. Stereoscopic X-ray systems which make it easier to visualize the spatial orientation of objects being scanned are particularly valuable for luggage and cargo inspection, as they help the inspectors interpret the display more easily.

Illegal Entry and Smuggling Concerns

In the early part of 2000, the U.S. Congress debated the proposed rehabilitation of the San Diego and Arizona Eastern (SD&AE) Railroad, a service that would cross the southern border in two areas. Opposition to the reopening of the railroad was based, in part, on a fear that crime would increase along the U.S.-Mexico border. Opponents felt that the economic benefits would be overridden by substantial needs for increases in surveillance, enforcement, and equipment to safeguard the operation of the line. Input from the U.S. *Immigration and Naturalization Service* (INS), the U.S. *Customs Service*, the U.S. *Border Patrol*, and the *Office of National Drug Control Policy* (ONDCP) were sought, in order to assess the increased security needs that would be involved in this project. Currently, the border has problems with theft, illegal entry, and smuggling. Railways are one means by which these criminal activities are carried out and thus must be anticipated and handled. Over 30,000 illegal entrants are reported to have been found in Union Pacific rail cars in 1999. The risk of injury and sometimes death is present when aliens jump trains or hide for hours without food or water in hot rail cars. In congressional testimony on transportation, Duncan Hunter reported the following:

> ... I requested the General Accounting Office (GAO) to conduct a comprehensive study on what resources would be required by border enforcement agencies to properly maintain control of the border should the SD&AE Railroad be reopened. Stating it would be difficult to estimate resource needs for modern freight service, U.S. Customs and Border Patrol officials made their recommendations for basic service, of non-cargo freight only. Among the equipment listed as needed were radios, railcar X-rays, generators, video surveillance and security devices, inspection and office space, computers and as many as 31-35 additional agents....

> The amount of time needed to adequately inspect railroad cars at border crossings must also be addressed. As a result of both U.S. and Mexican Customs officials examining each railway car, the time to properly inspect these cars and cargo can often take up to twenty-four hours. Consequently, a great strain is realized, making this project practically unfeasible with current resources...

> [Duncan Hunter, Testimony of Congressman Duncan Hunter (CA-52): House Appropriations Subcommittee on Transportation, 10 February 2000.]

X-ray surveillance is sometimes used to gather long-term health data on workers exposed to hazardous materials in public gathering places, mining operations, and industrial facilities. X-rays can monitor people for tumors, lung diseases, and other problems associated with exposure. Asbestosis is a scarring of lung tissue that occurs through inhalation of asbestos fibers. Coal workers' pneumoconiosis is a lung disease caused by exposure to coal dust. Emphysema and lung cancer are risks for employees (both smoking and nonsmoking) who work in smoky bars and breath large quantities of sidestream smoke.

The *National Institute for Occupational Safety and Health* (NIOSH) reported in 2002 that about one in 20 miners surveyed in their program showed X-ray evidence of pneumoconiosis. If exposure continues, the disease can progress to massive fibrosis, known colloquially as "black lung" disease. As a result of legislative acts in 1969 and 1977, two X-ray programs were established so that miners would be offered at chest X-ray at the beginning of their work and could be monitored periodically thereafter.

After the terrorist attack on the World Trade Center in 2001, security was increased on major transportation systems, especially airlines. Thus, more powerful X-ray scanning equipment was added and, in 2003, the *Federal Aviation Administration* (FAA) issued an advisory that the stronger scanning equipment would fog any unprocessed film (not just high-speed film). Fortunately, by this time, digital cameras were beginning to supersede film cameras, thus reducing the problem for at least some photographers and tourists. However, many camcorders use magnetic tape to store motion images, and the tape might be affected by X-rays or by higher voltage fields generated in and around the machines. X-ray exposure usually shows up as graininess and a light fogging of part or all of the image.

X-rays are high-energy radiation, compared to visible light frequencies. They must be used with caution. Recently, there have been companies promoting X-radiation-based 'backscatter' body imaging systems for surveillance. This is highly controversial. Most human rights advocates and even private individuals would assert that X-ray surveillance (e.g., for airport security) is not only potentially dangerous to health, but also dehumanizing. The images quite clearly show a person's physique—every lump and sag, every appendage, even conditions like bowlegs or surgical alterations. Denser materials like corsets, jewelry, and trusses are also visible. The surveillee may have no control over who has access to the images (which can easily be stored in a database for future reference) or how long they are stored. You would think that such a system would be dismissed out of hand, but the Transportation Security Administration (TSA) announced a proposal, in 2005, to purchase this backscatter X-ray systems for searching travellers at selected airports. See Problems and Limitations for further discussion.

5. Description and Functions

X-ray detection is a versatile technology used in many industries. The list of applications is long. Mail and courier services use it to detect hazardous cargo; customs agents to enforce public safety, and import and export laws; industrial production lines to assess quality and to detect flaws; astronomers to detect emanations from black holes and high-energy cosmic events; archaeologists and art historians to inspect artifacts; geologists to study crystals; physicists to investigate atomic properties; and forensic investigators to detect fraud or injuries to victims of crime. The next section on applications provides more specific descriptions and illustrations of X-ray technology in practical use.

6. Applications

Cargo Surveillance

X-ray Cargo Inspection. CBP officers unload palettes of cargo in preparation for X-ray inspection. Right: A CBP officer scans a seaport container using a mobile truck X-ray device and watches the monitor while the scan is in progress. [Customs and Border Protection news photos by J. Tourtellotte, 2003, released.]

Luggage and hand baggage carried by passengers on aircraft are now routinely inspected with X-rays to check for explosives, weapons, and sometimes smuggled items. The primary reason given for use of these machines is staff and passenger safety. The radiation intensities used for luggage are usually stronger than those used for hand baggage. Large-format air cargo that is taken on cargo transports is also screened with more powerful X-ray equipment, since it must penetrate large pallets and containers. It is similar to systems for surveilling transport trucks. Gamma rays, which share many of the characteristics of X-rays, are also used for cargo inspection at border crossings.

U.S. *Customs and Border Protection* (CBP) uses a variety of strategies to prevent smuggling, human cargo, and weapons from passing illegally into the country. More than 160 X- and gamma-ray imaging systems have been installed at official points of entry for examining trains, cargo containers, trucks, and cars.

X-ray Image Interpretation. Creating an X-ray scan is only part of the job. Once the interior of a cargo shipment has been X-rayed, it is still necessary for human operators to analyze and evaluate the images to see if there are any anomalous contents that might need further investigation. [Customs and Border Protection news photos by J. Tourtellotte, ca 2004, released.]

Microscopy

X-ray microscopes make it possible to view materials in a different way. Very high resolution images are possible with X-ray technology.

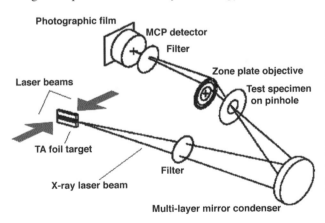

Left: This is a multilayer water-window imaging X-ray microscope demonstrated by its inventor. Right: A basic schematic of the main parts of an X-ray microscope. [NASA/Marshall news photo, 1992; Lawrence Livermore National Labs tutorial photo, released.]

Conventional microscopes use light to illuminate a subject. An X-ray microscope, on the other hand, uses X-rays to selectively penetrate the subject and, instead of the operator looking at a slide on a lit platform, the image is sent to a camera or digital system for display.

X-ray microscopes make it possible to study substances in new ways. To view most specimens under a conventional microscope, it is often necessary to slice the specimens into very fine pieces. This isn't practical for all specimens and impossible for many. The X-ray microscope enables thicker specimens to be viewed because the X-rays can penetrate more deeply into the tissue or object than the light from an optical microscope.

X-rays are absorbed by water such that organic substances contrast quite strongly with water in an X-ray image. This makes it a good technology for studying organic materials and thus a useful technology for surveillance, with particular relevance for agricultural testing, quality control, and the detection and identification of possible contaminants in a variety of agricultural and industrial environments.

Crystallography

X-ray crystallography is used in many fields, but it is especially important to materials science and engineering. It is widely used in mineralogy, because it allows us to determine crystalline structures. This, in turn, provides information that is useful in classifying minerals and studying the properties of gemstones. X-ray crystallography can help us learn more about minerals on Earth and the mineral structures of samples gathered from other planets.

Essentially, X-ray crystallography is based on aiming intense X-ray beams at crystalline substances and collecting and studying the diffracted X-rays. By looking at the different patterns and comparing them to the sources, very fine discriminations can be made and information about their basic atomic structure inferred.

Astronaut Dr. Bonnie Dunbar gets a briefing about the *X-Ray Crystallography Facility* from engineer Lance Weiss (left) and engineer Stacy Giles (right) while the facility was under development at the *Center for Macromolecular Crystallography* of the University of Alabama in Birmingham. This equipment expedites the collection of information from crystals grown on the *International Space Station*. [NASA/Marshall news photos, 1999, released.]

At *Lawrence Livermore National Labs* (LLNL), there is an X-ray crystallography group with the Biology and Biotechnology Program. The lab includes computer-controlled crystallization robots, microscopes, and temperature-controlled rooms for studying many aspects of molecular biology.

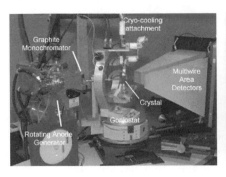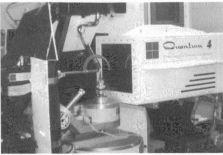

Lawrence Livermore National Labs X-ray diffractometers. X-ray diffractometers are traditionally used in high energy physics. Left: There are two detectors in this ADSC dual multiwire
system, located 50 to 150 cm from the crystal. The instrument is kept running continuously.
Right: The LLNL four-module X-ray detector used in cryo-crystallography. This detector uses
CCD technology, focusing mirrors, and a cryo-cooler. When X-rays hit the system, the imaging
phosphor faceplate emits photons. [LLNL news photos, released.]

Medicine

X-ray imaging of human and animal tissues is a regular part of human and veterinary
medicine. X-rays are used to determine growth, damage, abnormalities, dental decay, and many
other aspects important to diagnosing disease and maintaining health. They are also useful for
medical autopsies.

Teleradiology

Teleradiology is a term for digital X-rays that are sent over computer networks or other
communications pathways to be displayed on a monitor at a remote location. This technology
was fairly new in the late 1990s, and not yet widespread, but will be particularly valuable for
agents or military personnel stationed overseas or on ocean-going vessels. Missionaries, relief
workers, foreign diplomats, travelers, and military service members on aircraft carriers and
submarines can be away for months at a time, and many of them experience illnesses of one
sort or another. Many foreign countries do not have easy access to good medical resources or
diagnostic information. The capability of sending information to medical specialists thousands
of miles away in a few seconds greatly extends the effectiveness and reach of people working
far from home.

Remote X-rays can also be used for engineering professionals and others to collaborate on
the production of large-scale industrial projects, including oil rigs, highrises, ships, and bridges
in remote locations.

In terms of covert surveillance, teleradiology can be used to forward information to agents
stationed abroad, who need medical, industrial, or other information that might originate from
X-ray images generated at a remote station.

Industrial Fabrication and Testing

X-rays can be used to test for defects, conformity, contaminants, and other problems that
may show up in industrial production. The Total X-Ray Reflection Fluorescence (TXRF) system,
for example, is a nondestructive X-ray technology used to assess contaminant concentration,
using conventional and synchrotron X-radiation. It may detect contamination from a variety
of organic and inorganic sources, including chemicals, gases, dirt, and ion effects. It can yield
not only structural information, but also information about chemical composition.

Environmental Research and Monitoring

When we think of the environment, we don't usually think of X-ray technologies, but they can help facilitate chemical analysis of a variety of substances, making the technology of *X-ray fluorescence* applicable to many areas.

An X-ray fluorescence analysis system developed by the *Pacific Northwest National Laboratory.* This system can chemically analyze a variety of synthetic and natural materials, including biological and geological substances. It serves in environmental research and monitoring applications. [Courtesy of the Pacific Northwest National Laboratory, 1996 news photo.]

Forensic Investigations

X-ray technology is more expensive than many other sensing technologies. It is also more hazardous and less portable than ultraviolet and infrared devices, and thus not used as extensively at crime-scene sites as other surveillance technologies. However, it is useful for studying objects that can be moved from a crime scene or archaeological site to an X-ray facility, and is a useful tool for medical autopsies to determine injuries or cause of death.

X-ray examinations have recently been used in novel ways to examine historical artifacts. The Archimedes palimpsest, a priceless goatskin parchment dating back 1,000 years to a scribe who painstakingly copied down the work of the mathematical seer, has been erased and overwritten by subsequent generations. Thus, the original writings have been obscured from visual inspection and the parchment refused to reveal all of its secrets, even after scrutiny by a variety of sensing technologies. Then, in May, 2005, a synchrotron X-ray beam at the *Stanford Linear Accelerator Center* probed the document and revealed new details not previously visible. The idea for X-ray imaging came from an insight by Uwe Bergmann that some of the older texts contained iron pigments. Thus, the synchrotron X-ray beam would cause the traces of iron ink to fluoresce, "illuminating" portions of the manuscript that could not be seen as easily with other methods. Distinguishing information on different layers and deciphering the old Greek text are still significant challenges, but the X-ray images, along with other information, have provided a unique glimpse of history not accessible in other ways.

Another useful tool for forensic investigation is a soft-X-ray (SXR) microscope. These devices have higher resolution than visible light microscopes and, in some cases, don't require as much sample preparation as samples that are to be viewed with electron microscopes. The images have good contrast and provide useful spectrographic information. Soft X-ray microscopes are based on synchrotron emissions that are 1) guided along a beam-line and the rays translated to a CCD image or 2) scanned and imaged according to absorption-versus-position data.

X-ray imaging is a field in which many discoveries and improvements are occurring. There are about fifty synchrotron centers worldwide that can generate X-radiation across a broad spectrum of frequencies and facilities like the LLBL *Center for X-Ray Optics* carry out ongoing research on X-ray technologies like soft X-rays.

Synchroton technologies and soft X-ray microscopes can be used to investigate fingerprints, chemicals (e.g., explosives), toxins, tissues, human remains, and other materials that might relate to scientific research, law enforcement, or national security.

7. Problems and Limitations

Health Hazards

The most significant limitation of using X-rays for surveillance is that the radiation is hazardous to living tissue and thus must be used with effective shielding and only by trained technicians in carefully monitored settings.

People who take the risk of stowing away in trucks and cargo ships to illegally enter another nation run the risk of being exposed to the higher radiation levels from X-ray machines that are used to check the presence, composition, and integrity of cargo at border crossings. These machines are designed for imaging through metal and packages and are too strong for human tissue.

Damage to Film from X-Ray Machines

The X-ray machines used for security purposes are not all alike. Sometimes more powerful systems are used for luggage, as opposed to hand baggage. However, since the effects of X-ray exposure are cumulative, several trips through X-ray machines, as when changing planes, can fog film.

There were reports that early X-ray machines in public transportation systems might fog high-speed films, but most people were told that there usually wasn't a problem. This is not entirely true, as high-intensity scanning machines have been developed that can affect film. X-ray machines in airports can fog film, according to tests conducted by the Photographic and Imaging Manufacturers Association. The widely distributed CTX-5000 X-ray machine, which is used for checked luggage, can fog all unprocessed film.

Lower-energy hand-luggage surveillance systems may not cause as many problems unless the film is very fast or of an especially sensitive nature (e.g., medical films) but newer systems for large luggage have been shown to fog all film. In the United States, travellers may request visual inspection of sensitive photographic films as per the *Federal Aviation Administration* (Reg. 108.17). Machines installed in airports since 2003 will definitely fog films, even from only one exposure. Remove film, cameras, and camcorders (and recorded tapes) from luggage and ask for them to be visually inspected or put them in the trays supplied at most security checkpoints rather than having them sent through the X-ray machine. Lead-lined pouches don't usually help because pouches look suspicious on the X-ray viewing screen and the contents will probably be taken out of the pouch and sent through again before the traveller can stop the baggage inspector.

Vigilance

X-ray devices to detect weapons, explosives, and contraband are only as good as the people who are using them. Training, experience, vigilance, and the ability to concentrate for extended periods are necessary for the reliable identification of materials that may be hazardous. A concerned attitude and rotating shifts, to reduce boredom and eyestrain, are as essential as training and good equipment. These guidelines are particularly important in situations where X-ray detection is used to detect terrorist weapons.

Vigilance has a psychological aspect as well. Some people are too young, or too yielding, to withstand the barrage of complaints from people who are in a rush to catch their train or plane or who are trying to get across a border. X-ray operators must be able to concentrate on their task and slough off the persistent complaints rather than allowing them to distract them from their work.

Expense and Inconvenience

Unfortunately, the use of X-ray surveillance for surveilling luggage in public transportation stations, border crossings, museums, sports events, and various workplaces creates greater expense (in equipment, personnel, and maintenance) and greater inconvenience for everyone. 100% of travellers are impacted by the 2–10% or so who might be carrying weapons, controlled substances, or smuggled items.

8. Restrictions and Regulations

Because X-radiation may harm living tissues, there are a number of federal and international standards for the size and composition of X-ray cabinets, for the strength and direction of the beam, the amount of permitted leakage, and the specific applications for which X-rays may be used (e.g., Federal Standard 21-CFR 1020.40). Because X-rays can pass through materials, they keep going right past the person or object X-rayed, through walls, even across the street.

X-Rays as Forensic Evidence

X-ray images may be used as evidence in court cases involving personal injury, smuggling, and illegal entry (stowaways). Authentication of these X-rays before presentation in court is important. The interpretation of X-rays is a specialized skill and lay jurors are not permitted to interpret X-rays without advisement by an expert. Unfortunately, some X-rays may not be permissible. As digital X-ray systems supersede X-ray film images, it becomes easier to tamper with the images and more difficult to detect tampering.

Mine Workers Surveillance

The *Federal Coal Mine Health and Safety Act of 1969* was established to set respirable dust standards for coal mines in order to reduce diseases such as pneumoconiosis (black lung) and silicosis. This led to great improvements, but the *Department of Labor Mine Safety and Health Administration* also set forth actions for an X-ray surveillance program for surface miners to further detect and prevent lung problems associated with the work. As was mentioned earlier, mine workers may request X-rays at the start of their employment and periodically thereafter to check for heart or lung problems that may be job-related.

Medical X-Rays

X-radiation is high energy—it can knock electrons, one of the building blocks of life "out of orbit", in a manner of speaking. In other words, it can harm living tissue (and damage nonliving components). X-rays must be considered dangerous and should be avoided unless absolutely necessary. The *American College of Radiology* has stated that "Chest radiographs should not be required solely because of hospital admission. All 'routine' diagnostic studies for the sake of 'routine' should be avoided". There are training, certification, and use guidelines in place in nearly all aspects of medicine that govern procedures and careful use of X-ray equipment.

Agricultural X-Rays

There are a number of health and agricultural guidelines governing the use of X-rays in products that are for human and/or animal consumption. The World Health Organization, for example, publishes guidelines for the X-raying of foodstuffs. The U.S. Food and Drug Administration (FDA) also publishes guidelines.

Transportation Security

Transportation systems are vulnerable to terrorist activities, as are large events where people congregate. There are a number of guidelines that are designed to aid in the protection of conveyances and spaces that are often occupied by many people. One example is the 1990 *Aviation Security Act*.

In 1997, the *Federal Aviation Administration* (FAA) announced that it was purchasing over $7 million worth of explosives-detecting equipment for screening airline baggage. The automated systems would take some of the strain off operators by automatically alerting them to suspicious objects. These systems were planned to be in place by 1998.

Systems used for security by the FAA, include dual-energy automated X-ray machines and vertical-beam, dual-energy devices.

In 1998, the FAA also addressed the need for properly trained X-ray operators to ensure that operators were sufficiently skilled to operate X-ray machines effectively.

In 2004, the *Intelligence Reform and Terrorism Prevention Act* was enacted and included a means for travellers to correct inaccurate information if they found they had been added to terrorist watch lists that prevented them from using transportation systems. This may seem adequate protection for innocent travellers in theory, but has been found very difficult to accomplish in fact. Many people who have tried to correct erroneous data have had little or no success in overcoming the bureaucratic hurdles. This makes them more vulnerable to more invasive searches. If their name comes up in a database, they may be asked to accede to a strip search or backscatter X-ray search.

In 2005, the *Transportation Security Administration* (TSA) announced a proposal to purchase X-ray backscatter systems for imaging people at airport security checkpoints. The system scans a body with X-rays to create what looks very much like a photo of the person standing in the nude and may reveal weapons hidden under clothing, but doesn't reveal objects inside body cavities and thus is not necessarily more effective than other kinds of searches.

You would think that such a blatantly invasive system would be dismissed out of hand, but the argument in favor of the machines is that they are less invasive than a strip or pat-down search. That may be true, but giving a person the option of an X-ray rather than a strip search borders on coersion. The ease with which the machines can be used may cause them to be used with unnecessary frequency and may blind the operators to the dangers inherent in X-ray technology and to the potential and incremental loss of privacy they represent as a whole. In some ways a backscatter X-ray image is worse than a strip search, because the data could be viewed by many people, shared (perhaps even on the Internet by unscrupulous employees selling image files on the black market), sold to tabloid newspapers, and stored indefinitely. People imaged this way may not even get to see the images themselves or be told that the images may be stored for later viewing by others. The potential for personal humiliation from this kind of technology is very high. The temptation for operators to accept bribes to sell copies of the images is also high. The fact that we are even considering the use of such machines is indicative of problems in our society that should probably be solved in better ways.

9. Implications of Use

The four main areas in which X-ray detection devices are used are cargo surveillance, industrial inspection, medical surveillance, and astronomical surveillance.

Cargo-checking for dangerous or smuggled items is done on a daily basis at thousands of ports, mail depots, and border crossings around the world and is generally considered to be a deterrent to terrorist activities. The one area of concern is possible harm to illegal aliens stowing away in cargo containers. Since fairly high levels of X-rays are used to probe large trucks and shipping containers, anyone hiding inside them is exposed to levels much higher than from other sources, such as medical X-rays.

Extraplanetary surveillance through X-ray telescopes and space probes is yielding a great deal of new information about the cosmos. Even though the surveilling of stars and black holes may not seem important to daily life, the knowledge and technology that come from these activities and experiments lead to new fabrics, electronics, coatings, medical procedures, computer algorithms, and new ideas. Satellites that were originally launched to explore space are now used as communications relays. Coatings that were developed to launch a telescope into space on a rocket are used in industrial environments.

As long as X-radiation is used within strict safety guidelines, it is generally considered to be beneficial. However, backscatter systems for transportation security surveillance of people is highly controversial and of dubious benefit.

10. Resources

Inclusion of the following companies does not constitute nor imply an endorsement of their products and services and, conversely, does not imply their endorsement of the contents of this text.

10.a. Organizations

Australian X-ray Analytical Association (AXAA) - Formed in 1968 to promote the exchange of scientific and technological information in the fields of X-ray diffraction, fluorescence, surface analysis, and crystalline structures. www.latrobe.edu.au/www/axaa/

Federal Aviation Administration (FAA) - The primary U.S. aviation regulatory body, the FAA publishes standards, overseas certification of pilots, and serves as a handler for aviation-related travel and mechanical reports. One of the jobs of the FAA is overseeing security in aircraft and airports. www.faa.gov/

Hard Facts Investigative Engineering - Art graphic presentation including photographic documentation, X-rays, computer graphics, aerial photography, video surveillance, and courtroom displays. www.hardfacts.net/grafpres.htm

International Union of Crystallography (IUC) - A member of the International Council for Science which provides professional services for crystallographers. www.iucr.org/

MicroWorlds - This is an online science education publication from Lawrence Berkeley National Labs which includes quite a bit of information on X-ray microscopy and images of the equipment and X-rayed samples. Lawrence Berkeley Labs has a Center for X-Ray Optics. www.lbl.gov/MicroWorlds/

National Synchrotron Radiation Laboratory (NSRL) - This facility is in China, at the University of Science and Technology. It is China's first dedicated synchrotron radiation facility, built between 1984 and 1989. Among many other applications, synchrotron technology has been used to research microscopy, X-ray lithography, holography, and spectroscopy.

PerkinElmer Instruments, Inc. - A prominent supplier of X-ray and metal detecting devices for surveillance applications. The Web site includes specifications for a wide variety of systems. instruments.perkinelmer.com/

World Health Organization (WHO) - WHO promotes health and well-being and maintains guidelines for the use of X-rays and the X-raying of foodstuffs. www.who.int/

10.b. Print Resources

Some of the following publications may be out of print. If so, it is sometimes possible to find them in second-hand book stores or to borrow them through interlibrary loans.

Aprile, Elena, Editor, *Gamma-Ray Detectors,* Bellingham, Wa.: SPIE, 1992, 311 pages. Illustrated proceedings of the 1992 San Diego conference. Of interest for gamma-ray spectroscopy.

Bertin, Eugene, *Principles and Practice of X-Ray Spectrometric Analysis,* New York: Plenum Press, 1970.

Carpenter, John M., *Neutrons, X Rays, and Gamma Rays - Imaging,* Bellingham: SPIE, 1992, 369 pages. Conference proceedings.

Fabry, David, *Explosives Detection Devices Developmental: Final Report,* Atlantic City, N. J.: Atlantic City International Airport, FAA Technical Center, 1993.

Hoover, Richard B., *X-ray Detector Physics and Applications*, Bellingham, Wa.: SPIE, 1992, 250 pages. Related to *X-ray Spectroscopy,* proceedings of a 1992 conference.

Jenkins, Ron, *X-ray Fluorescence Spectrometry,* New York: John Wiley and Sons, 1988.

Saraceni, Pete, Jr., *Quick Reaction Automated Explosive Detection,* FAA Technical Center, 1983. For FAA use only.

Workman, S. T., *Modular Automated X-ray Experimental Airline,* Washington, D.C.: Department of Transportation, FAA, 1980, 55 pages.

Articles

Bazalon, Judge David, Civil Liberties - Protecting Old Values in the New Century, *New York University Law Review*, 1976, V.505, p. 511. An impassioned opinion on the emerging surveillance technologies and the difficulties of enforcement given their 'invisible' nature.

Beardon, J. A., X-Ray Wavelengths, *Review of Modern Physics*, 1967, V.39, p. 78. X-ray emission-line energies.

Evans, J. P. O.; Robinson, M., The development of 3D X-ray systems for airport security applications, Boston, Mass.: SPIE V.1824, *Applications of Signal and Image Processing in Explosives Detection Systems*, pp. 171-182. Evans and Robinson have written an extensive series of articles on 3D X-ray imaging technologies and techniques.

Fuller, Matthew, Talk for New Visions, 1994. While this does not directly relate to surveillance, it does discuss the use of technologies such as X-rays for purposes outside the mainstream of thinking and thus provides a fresh perspective, which is sometimes a welcome respite and a good way to stimulate lateral thinking. www.altx.com/interzones/london/new.visions.html

Robinson, I. K.; Tweet, D. J., Surface X-ray diffraction, *Reports on Progress in Physics*, 1992, V.55(5), pp. 599-651. Introduction to X-ray diffraction and how it can be applied to surface and interface study.

Sincerbox, Glenn T., Holographic storage: are we there yet? University of Arizona. The author is the Director of the Optical Data Storage Center at U of A. The basics of holographic recording and storage are discussed along with information on mass storage devices.

Sincerbox, Glenn T., Editor, Selected Papers on Holographic Storage, Bellingham, Wa.: SPIE Optical Engineering Press, 1994.

Templeman, Bob, Generating X-Rays with Receiving Tubes, *The Bell Jar*, October 1994, electronic issue No. 2. The use of old TV tubes as cold cathode X-ray emitters.

The X-ray Century, this was published in the 1890s and contains illustrated articles of the early discoveries in X-ray science. It has been republished by the Medical Physics Publishing Corp. Information is available online from the Department of Radiology at Emory University. www.cc.emory.edu/X-RAYS/century.htm

Journals

American X-Ray Journal and American Electro-Therapeutic and X-Ray Era, historic journals from late 1800s to early 1900s which are usually shelved in rare materials sections in medical libraries. Interesting historical perspective on the development of X-ray science and technology.

Synchrotron Radiation Online, a service for subscribers to the *Journal of Synchroton Radiation.* Includes information on source technology, instruments, and techniques over all spectral ranges in synchrotron radiation research, published by the International Union of Crystallography.

10.c. Conferences and Workshops

Many of these conferences are annual events that are held at approximately the same time each year, so even if the conference listings are outdated, they can still help you determine the frequency and sometimes the time of year of upcoming events. It is very common for international conferences to be held in a different city each year, so contact the organizers for current locations.

Many of these organizations describe the upcoming conferences on the Web and may also archive conference proceedings for purchase or free download.

The following conferences are organized according to the calendar month in which they are usually held.

AXAA National Conference, sponsored by the Australian X-Ray Analytical Association.

X-99, Conference on X-ray and Inner-Shell Processes, annual international conference since the early 1980s.

Denver X-ray Conference, annual conference held since the 1940s. www.dxcicdd.com

XAFS Conference on X-Ray Absorption Fine Structure, sponsored by the International XAFS Society for physicists working with atomic excitation using X-rays and electrons.

The X-Ray Conference, annual conference since the early 1950s, sponsored by the International Centre for Diffraction.

10.d. Online Sites

The following are interesting Web sites relevant to this chapter. The author has tried to limit the listings to links that are stable and likely to remain so for a while. However, since Web sites do sometimes change, keywords in the descriptions below can help you relocate them with a search engine. Sites are moved more often than they are deleted.

Another suggestion, if the site has disappeared, is to go to the upper level of the domain name. Sometimes the site manager has simply changed the name of the file of interest. For example, if you cannot locate www.goodsite.com/science/uv.html *try going to* www.goodsite. com/science/ *or* www.goodsite.com/ *to see if there is a new link to the page. It could be that the filename* uv.html *was changed to* ultraviolet.html, *for example.*

Airport Security: A Seven-Part Series. This illustrated series of articles on the Newsday site discusses airline bombing disasters, FAA regulations, passenger and baggage checking, and cargo containers. www.newsday.com/jet/airdex.htm

Archimedes Palimpsest. This site includes downloadable high-resolution X-ray images of the priceless Archimedes palimpsest which, underlying other writings and images is information in Archimedes own hand. X-radiation made it possible to see details that were not viewable with other forensic sensing technologies. www.archimedespalimpsest.org

LLNL X-Ray Laser Sources. There is a very good technical information site on X-ray lasers and X-ray microscopes, with lots of illustrations on the Lawrence Livermore National Laboratories site. www.llnl.gov/science_on_lasers/11RSources/RS-C_XRL.html#FigVI-21

Ricoh X-Ray Science in Japan. The Ricoh site sponsors a brief illustrated history of the course of X-ray science and technology as it progressed in Japan from the 1800s to the present. In the early days, Japan was almost neck and neck with the Europeans and there were cross-communications between scientists in those parts of the world. The content is provided by Masatoshi Kobayashi of Tokai University. www.ricoh.co.jp/net-messena/NDTWW/JSNDI/XRAY.html

X-Ray for kids. This is a great site for any age. There are basic introductory image sample X-rays of plants, animals, insects, objects, and shells. This gives an introduction to the ways in which certain tissues and structures will admit or impede X-radiation. www.yhrad.com/kids.html

X-Ray Systems and Research. There is a good Web site at the Nottingham Trent University, in collaboration with the Police Scientific Development Branch (PSDB) in the U.K., which has illustrated pages showing the basic components of an X-ray system and sample images of a variety of types of X-rayed objects. eee.ntu.ac.uk/research/vision/asobania/index.html

X-Ray WWW Server. Since 1994, this server from the Dept. of Physics at Uppsala University, Sweden is a COREX biibliography and database repository of the Henke atomic scattering factors and other information pertinent to X-ray spectroscopy, including research, conferences, and other Web sites. xray.uu.se/

10.e. Media Resources

Barbara McGill Balfour - SoftSpots, an art exhibit of abstract printmaking related to medical imagery, organized by the Southern Alberta Art Gallery and the Canada Council, 17 Oct. - 22 Nov. 1998. A quote about this exhibit is probably the best way to illustrate why it has been included in this book

about surveillance technologies, "The medium is an appropriate metaphor, as printmaking involves the emergence of an image or marks on a surface, and some of the terminology involved - stretching, or bleeding for example - can carry a double meaning in this context. Her work also deals with the pervasiveness of medicine in our contemporary lives, and the constant surveillance - through X-rays, blood tests, and biopsies - that we have set up as a line of defence against disease".
earth.online.uleth.ca/~saag/exhibits/balfour/index.htm

Medical Imaging, is a History Channel *Modern Marvels* series show on ultrasound and X-ray history and current uses in medical imaging. It also chronicles the case of an executed murderer who left his body to science. VHS, 50 minutes. May not be shipped outside the U.S. and Canada.

11. Glossary

APXS	Alpha Proton X-Ray Spectrometer - an instrument used on the Mars Sojourner
AXAF	Advanced X-Ray Astrophysics Facility
burst	brief, intense emission
EMA	electron microprobe analysis
HXR	hard X-ray. High-energy X-ray photon. When HXRs are generated by solar bodies like our Sun, they are known as solar flares.
HXT	hard X-ray telescope
scintillator	a crystal which converts gamma-ray energy into detectable light such that it can be detected
synchrotron emission	radiation emitted by charged particles that have been accelerated against a barrier by a magnetic field. A synchrotron machine accelerates charged particles and maintains them (through use of the magnetic field) in a "circular" (fixed radius) orbit. The radiation generated ranges through the X-ray and infrared frequencies.
SXR	soft X-ray, X-rays that are less harmful to biological systems and thus more commonly used for medical imaging or sensing of more delicate/fragile structures (e.g., microscopic specimens)
TRXRF	total reflectance X-ray fluorescence
XBP	X-ray bright point—a brief, 'brighter' spot on an X-ray image that indicates a small magnetic bipole. In larger-scale X-ray sensing, the Sun shows fluctuating XBP activity. XBPs were first observed through X-ray telescopes in the late 1960s.
XRF	X-ray fluoresence spectroscopy
XTE	X-Ray Timing Explorer

Surveillance Technologies

Biochemical

Biochemical Surveillance

Chemical & Biological

1. Introduction

Chemical and biological surveillance techniques are not as well-known outside of scientific circles as consumer technologies such as video cameras, and yet humans engage in significant amounts of chemical/biological surveillance. Smell and taste are chemical detecting senses that provide us with remarkably accurate information about various substances. Many animals have better chemical sensors than humans—deer and dogs have keen noses that enable them to sense predators or hunt for food and mates.

Insects like the Cecropia moth can locate mates and food sources at distances of up to seven

Eugene Mizusawa, a postdoctoral fellow at Lawrence Livermore National Laboratory, peers into a beaker to calibrate a new type of fiber-optic sensor for monitoring chemical contamination of groundwater. The intensity of the yellow fluorescence indicates the amount of carbon tetrachloride present in the solution. LLNL scientists are developing other sensors for identifying and measuring a variety of chemical contaminants and solvents. The system can be used in the field, reducing the need to send samples to the lab. [Lawrence Livermore National Labs news photo by Jim Stoots, ca 2000, released.]

miles away with highly sensitive chemical detecting antennae. We have learned a great deal from studying and trying to reproduce natural chemical detection systems.

Chemical and biological surveillance are technical fields—much of the research involves laboratory work on a microscopic scale, sometimes in sterile or carefully monitored environments. Gradually, portable 'minilabs' are being developed for work in the field, but the results of onsite analyses may have to be double-checked or subjected to further analysis by a specialist in a lab.

Chemicals are useful for detecting, marking, tracking, and influencing chemical reactions or behavior. Sometimes they are even used as decoys. Chemical/biological surveillance includes the detection and analysis of evidence at a crime scene, vapors at an industrial accident, hormones on a chicken farm, or the presence of chemical weapons in hostile territory or at border crossings. Chemical/biological devices are used fordrug detection, sample preservation, industrial quality control and inspection, safety and hazards monitoring, search and rescue, firefighting, arson investigation, and military reconnaissance. (Note that DNA profiling is covered in the Genetics Surveillance chapter. Animals that are used for biological/chemical surveillance tasks such as sniffing out bombs or people are also covered separately in the Animal Surveillance chapter).

This chapter focuses on chemical and biological technologies that are used in forensics, archaeology, environmental studies, quality assurance, and national defense. Spectrographic analysis to determine chemical composition is also widely used in astronomy, but the reader is advised to consult astronomy books for more information on this interesting subject.

2. Kinds and Variations

2.a. Basic Terms and Concepts

'Chemical surveillance' is an ambiguous phrase that is easier to explain with a few examples than with a definition.

1) Chemicals can be used to surveil other things (which may include other chemicals):

- Chemical stains can reveal faults in components moving through a production line.
- Marking powders can indicate tampering on a doorknob.
- Exploding indelible-dye packs can be placed in money sacks to mark bank robbers.
- Luminol™ can be sprayed on walls or textiles to reveal otherwise invisible blood stains.
- Quick-hardening plaster can preserve a physical impression like a shoe or tire print.

2) Other technologies can be used to detect or track chemicals:

- Ultraviolet can reveal marking ink, faded rock paintings, or underlying paint or ink traces.
- Infrared light can detect spills or people crossing a border at night.
- MRI scans can reveal inflammation and other chemical anomalies.
- A sniffing dog can detect explosives or drugs or find someone who has wandered away or been kidnapped.

- A microscope can be used to match lipstick marks on a cigarette butt with a mouth print.

Both *chemical surveillance* and *surveillance of chemicals* are introduced in this chapter.

The terms *biological*, *chemical*, and *biochemical* are used somewhat loosely in this text because there is considerable crossover in the technologies. Many devices sense or utilize chemicals that have a direct impact on biological systems and biological cells are very much like little semiporous balloons filled with chemical factories, so the distinctions blur when describing technologies that have both biological and chemical sources or consequences. The casual use of terminology is not intended to mislead, but to simplify statements that refer to biological/chemical technologies that are interchangeable or not easily categorized.

A large part of chemical surveillance involves detecting and identifying substances and diseases that are hazardous to humans, so it is helpful to know some of the basic terms related to poisons and human illnesses.

germ - a generic, colloquial term for any microorganism that may cause disease. It encompasses a variety of viruses and bacteria that are too small to see with the unaided eye and thus difficult to distinguish as causal agents.

infectious diseases - diseases that are transmitted by direct contact.

contagious diseases - diseases that can be transmitted by proximity, usually through air or water vapor. Leprosy is infectious (requires contact) and is not easily spread. Most sexually transmitted diseases are infectious—they require interaction of bodily fluids and are not transmitted by mere proximity or casual interaction. On the other hand, common cold germs are contagious, as they can be carried through the air on saliva or picked up from contact with objects that are infected with the germs.

poison - a generic term including a range of substances that can make a person ill. Poisons may be organic or inorganic and usually require a certain threshold before they are harmful. For example, nutmeg in small quantities is a popular baking spice. Nutmeg in large quantities can produce hallucinations and toxic reactions. A *toxin* is produced through metabolic activity in living organisms (e.g., snake venom) and usually promotes antibody production. The word *toxic* really should refer to an effect specific to *toxins* but in popular literature it is broadly used to describe the affect of a toxin *or* poison.

pathogen - a specific disease-causing agent (e.g., smallpox virus) while an *antigen* is a substance capable of triggering an immune response.

antibodies - biological immunoglobulins produced by a body to try to counter the effects of antigens.

antidote - a remedy intended to counteract the effects of toxins (e.g., antivenon for snake bites) or poisons.

Pathogens, antigens, and toxins can be variously contracted through foods, beverages, injections, bites, and mists.

Chemical surveillance includes a lot of scientific *jargon* outside the scope of this book, including many chemical compound names, pharmaceutical brands, and medical terms. Unlike the glossaries in other chapters, which are very brief but somewhat complete in themselves (from *a* to *z*), it wasn't possible to create a similar glossary for chemical surveillance within the space allotted. A short sample glossary is included, but the reader is encouraged to consult a specialized chemical dictionary for more in-depth information.

2.b. Detectors

Chemical detectors (that is, detectors that scout out biological and chemical traces) can be natural, synthetic (biological), or electronic. Humans are born with chemical sensors (including taste and smell) as are other creatures. Synthetic detectors include lab-developed biological substances and compounds, and electronic sensors. Electronic chemical detectors that work somewhat like a dog's nose are colloquially called 'sniffers', and they are becoming so sophisticated that they may eventually supersede sniffing dogs for certain applications.

Some detectors only detect the presence of chemicals, others may provide additional information about their quantity, identity, composition, and sometimes even their direction of origin.

Dusting for fingerprints and coating valuables with ultraviolet inks or dyes are common ways to use chemicals for detecting and marking, using chemical substances. These aspects of chemical/biological surveillance are covered in more detail in the Biometrics Surveillance and Ultraviolet Surveillance chapters.

2.c. Categorization

Chemical surveillance is more complex and varied than other surveillance technologies and more difficult to group into simple categories for the purposes of discussion for the following reasons:

- Chemicals are basic biological building blocks and thus are used in all industries—a chemical used in an industrial refinery may also be used for testing foodstuffs or decorating a baby's crib.

- Chemical/biological agents may have different effects at different doses—what is therapeutic in small doses may be toxic in larger doses.

- Chemicals/biologicals may have unpredictable interactions in biological systems (e.g., genetic variations may add another level of complexity). A bacterial agent that makes one person slightly ill might kill another, and have no affect on a third.

- Many biological agents are living organisms (e.g., smallpox), they can reproduce and mutate with unexpected results.

- Organisms like viruses and bacteria are too small to see without special equipment and thus may go undetected until they begin to influence their host. We are constantly struggling to determine the presence of these organisms and to understand and adapt to their changes.

For the sake of simplifying a broad topic into something manageable that can be handled in one chapter, four general areas have been emphasized as follows:

archaeology - Chemical surveillance is used to investigate, reclaim, and restore historical art, tools, cultures, records, structures, and fossils.

environmental studies - Chemical investigation can help assess the age, composition, character, and status of every aspect of our environment, and the creatures that inhabit the ecosystem. This includes research and assessment of weather, geology, marine ecology, forestry, agriculture, and others.

product and cargo inspections - Chemical surveillance is used in agriculture and industry to automate production lines, to assure quality and compliance, and to inspect products during or after growth or manufacture. It is useful in customs and local law enforcement for detecting contraband and controlled substances, particularly narcotics, regulated agricultural products, explosives, and weapons.

> **search and rescue, forensics, and national defense** - Various law enforcement and national defense agencies use chemical surveillance technologies to find and identify lost persons, runaway or kidnapped children, wandering elderly, lost hikers, and victims of bombings or natural disasters. They are further used in investigations of industrial spills, contaminations, arson, vandalism, kidnappings, rapes, and homicides. Chemical surveillance is used to identify, assess, and neutralize poisonous agents that can occur as a result of natural disasters, spills, bombings, and hostile conflicts.

3. Context

Prevalence

Chemicals and biological agents are used in homes, businesses and labs. Biochemical substances are so intrinsic to our lives, we take them for granted. Few of us dwell on chemistry, yet we use chemicals every day to cook and clean, to paint and glue, and to perfume our hair and skin.

Chemicals are not just external substances, they are also part of our bodily makeup. They influence our thoughts, our memories, and our emotions. Of interest to surveillance professionals is the fact that we leave a biochemical trail everywhere we go, as we slough off skin, hair, lotions, and fluids. These residues form a 'trail' that investigators can follow through chemical surveillance.

Chemicals are used in agriculture to grow, transport, clean, ripen, preserve, and process foods for the marketplace. They are used to prospect, mine, refine, fabricate, color, coat, and preserve. They are further used by scientists to test, explore, investigate, and reveal. They are indispensable in every aspect of lab work, not just as a focus of scrutiny, but as tools to stain, separate, preserve, and selectively destroy. They aid us in surveilling chemical spills, pollution, environmental trends and changes, weather, health, murders, and genetic relationships.

Use and Abuse

Chemicals can be helpful or harmful. The use of chemicals/biologicals to enhance our lives is an ongoing area of research and chemical surveillance and is an important aspect of several fields including health and safety, compliance monitoring, and quality control.

Sometimes chemicals are used by naive individuals to self-medicate or by unscrupulous individuals to cheat (e.g., steroid use by competitive athletes). Athletics administrators are constantly being called upon to regulate and detect illegal performance-enhancing drugs. These drugs are regulated and monitored for two reasons: to protect the health of the athlete who might be harmed by taking drugs (or adulterated imitations of the drugs), and to protect the competitive rights of the athletes who are endeavoring to excel without breaking the rules.

In agriculture, growers and breeders may illicitly use chemicals in ways that contravene health and safety guidelines (e.g., incorrect or excessive use of antibiotics).

Because of evidence of abuse in various industries, chemical use is monitored and legislated, and chemical surveillance is regularly practiced by watchdog agencies, inspectors, insurance assessors, customs and law enforcement officials, and military reconnaissance teams.

Chemical Weapons Threats

When chemicals are used in the context of war to inflict harm, they are known as *chemical warfare agents*. Chemical warfare is an emotional topic, and information about its use and effects is often kept away from the general public. It is feared that reports of epidemics, mass murder, or uncontrolled release of chemical agents or harmful bacteria can create social instability and hysteria. Fears of chemical contamination or uncontrollable consequences are not entirely unjustified. Nature is too adaptable and sophisticated for humans to be able to control every aspect of biological/chemical use and our ability to locate and create new chemicals far exceeds our ability to anticipate all the possible consequences of their use.

Most chemical threats come from two quarters:

- the deliberate release of biological/chemical agents to do harm, and
- the accidental release of these agents from insecure stockpiles or through incorrect or fraudulent use.

Chemical/biological threats are much like bomb threats—because of their potential for harm, authorities generally prefer to assume they are real until they can establish otherwise.

There is a general feeling on the part of health professionals that more could be done to protect the public from chemical warfare agents, and many labs and commercial firms have been seeking ways to develop better detectors and neutralization strategies and technologies. However, on the negative side, there is also a feeling that more sensitive detection instruments and computer data logs are contributing to an erosion of personal privacy, which is true. Yet, given the complexity of chemicals and their ubiquitous nature, in general, they have served us well and will hopefully continue to be used in ways that enhance the quality of our lives.

4. Origins and Evolution

The Evolution of Chemical Use and Abuse

Some chemicals survive for many thousands of years and others quickly vanish without a trace. Those that remain, help scientists learn about ancient civilizations and much of that knowledge can be applied to understanding the customs of our ancestors and how we arrived at the current state of our civilization.

Chemicals reveal clues in many ways. Detectives, pathologists, forensic investigators, doctors, paleontologists, astronomers, and environmentalists all use chemical surveillance techniques and technologies to analyze the present, the recent past, and the very distant past. They have learned that humans have been using chemicals for thousands of years to adorn their bodies, cook and preserve their foods, tan their hides, preserve their dead, and record their activities and thoughts. Well-preserved ancient human remains reveal the use of pigments, powders, embalming fluids, and tanning chemicals. We have learned that past civilizations used poisoned darts and arrows to hunt prey, and ochre and charcoal for makeup and paint. They used arsenic and other poisons to do away with unpopular rulers, and noxious materials, disease-riddled clothing and blankets, and fumes, to overcome enemies.

Early Regulations

Chemical substances have been subject to regulation for many centuries. Pigments, spices, medicines, liquor, and chemicals considered to be of value for cosmetics or other uses have long been subject to restrictions or tariffs. Historic river and sea routes were often patrolled to control the movement of goods, and to exact tariffs to fill royal treasuries. By the 1700s, Europe had set up customs warehouses wherein goods for import or export could be stored and inspected. By the early 1800s, the "Waterguard" or "Coastguard" in Britain served the function of customs and excise officials, regularly boarding boats and other vessels to inspect cargo and luggage.

Disease and Warfare

While medical *science* is a fairly recent development, medical *techniques*, whether effective or not, have been in use for tens of thousands of years. Poultices, tinctures, and herbal medicines have evolved through trial and error and have been passed down through oral and written histories. Germs weren't really understood until lenses and microscopes were developed, but people in the Middle Ages and earlier must have known that some conditions are contagious, because they took steps to quarantine people with leprosy or plague. They are also reported to have tried to vanquish enemies in besieged cities by catapulting contaminated corpses over their battlements. In letters to his father during the Renaissance, Michelangelo indicated that doctors in Florence understood the source of the Black Plague. Two hundred years later, Russian troops are reported to have used plague-infected corpses against the Swedish. In 1763, British soldiers in North America gave blankets to Native Americans, knowing that they were infected with smallpox. These deliberate acts of 'chemical warfare' were probably responsible for the smallpox outbreak that decimated bands in the Ohio River region.

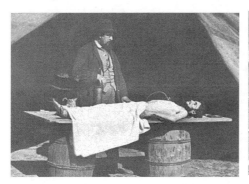 

Left: A Civil War surgeon needed a working knowledge of chemistry in order to do embalming work on a soldier's body, as is shown here. Right: The body of a slain Confederate soldier, in 1864. When they died and were searched, it was discovered that some of the soldiers were spies, as revealed by secret messages that were found hidden in their boots, buttons, and seams. [Library of Congress Civil War collection photos, photo on right by Timothy O'Sullivan, copyrights expired by date.]

As much as conflict is a part of human lives, there is a part of the human makeup that recoils at the thought of biological/chemical warfare. People sense something unsporting and truly evil about the practice, which has so far prevented us from annihilating ourselves, despite our ability to do so. Yet there is also a segment of society willing to engage in the study and manufacture of biological/chemical weapons and to consider their use.

To counter this very small but dangerous minority, society has condemned the use of chemical warfare for at least the last century and a half. In April 1863, the U.S. *War Department* issued *General Order 100,* proclaiming, "The use of poison in any manner, be it to poison wells, or foods, or arms, is wholly excluded from modern warfare". Obviously this statement would not have been made if chemical warfare had not been considered or used in earlier times.

As prohibitions on chemical warfare were being formalized, so were customs practices throughout Europe and America. The import and export of foods, cosmetic pigments, powders, and drugs had been regulated in one way or another for many hundreds of years in Europe and Asia, with inspections and tariffs not unlike what we have now. By 1789, the U.S. had a tariff act and, by 1909, the *Customs and Excise* body in Britain was consolidated and formalized. Many of the current revenue and inspection organizations, including the U.S. *Coast Guard* and the *National Bureau of Standards*, originated through the execution of customs responsibilities.

Chemical Warfare

In 1915, the German army used poisonous gas during the battle of Ypres and the British countered by using it in the battle at Loos. Neither side used the gas very effectively. It was discovered that chemical agents were not as controllable as bullets. Bullets went where they were aimed. Gas, on the other hand, was influenced by the wind, which might be blowing right back in the faces of the people who unleashed it. At times the enemy would set the gas cloud ablaze, or the gas would give away the position of the aggressors who would then discover that the other side was better equipped to withstand the effects of the gas.

Many different gases were tried during the war, including tear gas, mustard gas, chlorine gas, and others. Mustard gas had the most horrific consequences, causing the most deaths and leaving others with severe respiratory problems and ulcerations of the skin and eyes.

Left and Middle: A clipping from the Dayton Forum in October 1880, describes some of the horrible consequences of exposure to mustard gas. Right: A nonlethal tear gas alternative is announced for riot control in this 1919 Pittsburg news item that also states that France was 'perfecting various gases for the government' at the time. [Library of Congress, African-American Experience in Ohio Collection, copyrights expired by date.]

Prohibitions on Chemical Use

In June 1925, largely due to the use of poison gas in the war, the Geneva *Protocol for the Prohibition of the Use in War of Asphyxiating, Poisonous, or Other Gases, and of Bacteriological Methods of Warfare* was instituted.

While international agreements were being worked out, nations continued to study chemical agents. In the 1930s, Russia ran a *Central Military-Chemical Polygon*, a chemical warfare research center. In Germany, scientists were developing sarin gas, in the late 1930s. Allied nations were known to have produced sarin during World War II, an organophosphorus compound that can cause nausea, headache, weakness, serious cardiopulmonary problems, and cardiac arrest.

Six men wearing gas masks while standing either in an industrial facility or possibly on a government vessel, in June 1920. Masks were apparently ineffective against mustard gas. [Library of Congress Historical Society of Colorado photo, copyright expired by date.]

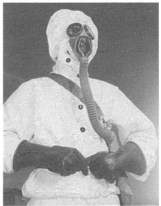

Left: A chemical laboratory at the House of David, in 1940. Right: A U.S. Naval Air Base sailor dressed in new protective clothing and gas mask designed for use in case of chemical warfare. Historic notes describe this 1940s suit as being lighter than chemical-protecting suits from earlier times. [Library of Congress FSA/OWI Collection photos by John Vachon and Howard Hollem, public domain.]

Japan also conducted biochemical research and warfare during this period, using a variety of biological and toxic agents against the Chinese.

During the mid-1930s Disarmament Conference, international agreements were proposed to prohibit the production and stockpiling of biological and chemical weapons, but progress on prohibitions was slow.

By the time World War II broke out, some progress had been made in limiting the use of chemical agents. Even though more lethal biological and chemical weapons were developed during World War II, they were not put into use as they had been in World War I.

During the Vietnam War, the U.S. forces used herbicides to spray crops in Vietnam. These herbicides worked in a number of ways, interfering with plant metabolism or moisture retention, and were sprayed at levels 30 times higher than normal agricultural practices. Within weeks of spraying, leaves, flowers, and fruits were lost and many trees died. Dioxin was one of the chemical sprays that was used, known colloquially as *Agent Orange*. Many deleterious health effects have been attributed to exposure to these herbicides during the War, including a high rate of susceptibility to cancer and a variety of birth defects, including spina bifida. Since the War, residual chemicals have been moving up the food chain, becoming apparent not only in plants, but in fish and humans, as well.

Postwar Disease Control and Forensic Chemistry

Many aspects of chemical and biological surveillance have been developed during wartime or after incidences of accidental death or homicide. The human need for closure and ritual in situations related to death has resulted in whole branches of science being devoted to the location, handling, investigation, burial, and exhumation of human bodies.

Another outcome of World War II was the study of various diseases contracted by military personnel serving abroad and their potential to spread to populations at home after returning from their tours of duty. Even during peacetime, service members were traveling to distant countries and contracting illnesses that required attention and sometimes quarantine in order to prevent their spread. The forerunner to the *Centers for Disease Control and Prevention* (CDC) originated in 1946 as the *Communicable Disease Center* that opened in the offices that handled malaria control during the war.

Epidemics, such as the polio epidemics of the 1950s, spurred research into the causes, spread, and prevention of disease. Medical surveillance increased so that health warnings could be issued, preventive measures taken, and broadscale vaccination programs implemented. In 1951, the *Epidemic Intelligence Service* (EIS) was established, and international and domestic disease surveillance programs put in place. In 1955, the *Polio Surveillance Unit* was established.

The Control and Surveillance of Imports and Exports

In the 1950s and 1960s, commercial air travel, automobiles and trucks, and the booming postwar population all increased steadily. The bigger population and increased mobility put greater strains on customs bodies throughout the world to maintain order at borders and to continue to manage customs inspections and tariffs.

Until the 1960s, the vast majority of customs, border patrol, and immigration surveillance techniques were hands-on visual inspections by human agents. Gradually, however, mechanical and electronic means to inspect cargo and luggage were added to stem the tide of illegal aliens and monitor controlled substances (foodstuffs, pharmaceuticals, narcotics, and various dutiable items).

Customs inspections and administration couldn't be handled entirely on a country-by-country basis. Clearly, international cooperation would be needed to manage the increased movement of goods, and global conventions were held. In 1961, for example, the United Nations held a conference to bring international drug control under one convention. This was adopted as the *Single Convention on Narcotic Drugs* which defined various narcotic substances and established the *International Narcotics Control Board*. The role of the Board was to promote government compliance with the terms of the convention and to assist them in their efforts.

Increased mobility had another important consequence. It was now more common for illnesses to be spread from one country to another, through airplane and train travel. Thus, a global perspective had to be taken to control the spread of disease pathogens, which are indifferent to human political borders. In 1966, the *Communicable Disease Center* established the Smallpox Eradication Program to eradicate smallpox and to control measles in almost two dozen African countries.

Negotiation on various arms treaties continued, impeded somewhat by suspicious and sometimes hostile relations between the U.S. and the Soviet Union. In spite of the various treaties and attempts to eliminate chemical and biological warfare, there were reports that the U.S. armed forces shipped a 'small amount' of sarin nerve gas to Vietnam in 1967, though it was claimed that it was never used.

The 1970s - International Control of Disease, Arms, and Drugs

The emphasis on global management of diseases, weapons of warfare, and narcotic substances continued into the 1970s. Hence, surveillance of compliance to these regulations was of importance, as well.

In 1970, the *Communicable Disease Center* was making progress in eradicating smallpox worldwide. The CDC's name was changed to the *Center for Disease Control* to reflect a broader health-centered focus.

The Pentagon reported (in 1998) that chemical agents stored in Okinawa were removed in 1971.

The General Assembly and the Conference of the Committee on Disarmament (CCD) did not result in agreement, but in 1971, the Soviets and their allies introduced a revision limited to biological weapons and toxins and an agreement was worked out with western nations. In December, the General Assembly unanimously adopted prohibitions. France abstained but, the following year, enacted internal legislation prohibiting biological weapons.

In 1971, the United Nations held the *Convention on Psychotropic Substances* in response to the various new drugs that were being synthesized in labs at the time and which had been reported widely in the media. The articles of the convention sought cooperation with the *World Health Organization* in assessing the degree of impact on physical or psychological health that might be associated with various substances. The information would aid in determining substance controls: which ones should be dispensed with prescriptions? how they should be manufactured? how should they might be imported and exported? should they be prohibited or permitted for medical and scientific uses? etc., in order to develop policy and controls. In 1972, the *Single Convention on Narcotic Drugs* was amended to reflect changes and additions.

The international community was busy that year working on amendments for global narcotics control. In April 1972, more than 100 countries signed the *Convention on the Prohibition of the Development, Production, and Stockpiling of Bacteriological (Biological) and Toxin Weapons and on their Destruction*. Existing stocks of biological weapons were claimed to be destroyed and only defensive research with these materials could be continued according to the terms of the prohibition convention.

In spite of international efforts to banish chemical and biological warfare agents, some of the middle eastern countries are suspected of having biological weapons and it was found that some western nations hadn't entirely eliminated their stockpiles.

In the late 1970s, the *Center for Disease Control* made progress in its programs to eradicate communicable diseases, including the apparent eradication of smallpox and wild polio (wild strains tend to mutate and may differ from vaccine-related strains). In 1978, the CDC opened a special maximum-containment laboratory for the safe study and handling of dangerous viruses. In 1980, the CDC, which had grown to many branch offices at this time, was renamed *Centers for Disease Control*.

1980s and 1990s - Biochemical Warfare and Health Concerns

In 1988, the United Nations made adjustments to previous conventions and took further steps to promote cooperation among international parties to effectively administrate aspects of various drugs with the *Convention Against Illicit Traffic in Narcotic Drugs and Psychotropic Substances.*

In the early 1990s, reports of persistent illness from American service members who had served in the Gulf War began to filter out. It was not known whether these illnesses were caused by emotional trauma, exposure to chemical agents, or side effects from antibiological/chemical warfare medications given to service members to protect them from possible harm.

The Mid-1990s - Treaties, Training, and Terrorism

In terms of the politics and science of biochemical surveillance, the mid-1990s were an active time. Various international agreements were being worked out, response teams were set up, certification programs were implemented and, in a shocking setback, terrorist chemical weapons were used in Japan.

The Army Corps of Engineers was involved in the renovation of a *Central Chemical Weapons Destruction Analytical Laboratory* as part of the Cooperative Threat Reduction Russian Chemical Weapons Destruction program. This lab is located in Moscow, with the project administrated through the *Transatlantic Programs Center.* Left: A new entrance being constructed for the facility. Right: Putting up a new wall for the lab. [U.S. Army Corps of Engineers news photo, released.]

In January 1993, it looked like progress was being made in international agreements to ban chemical weapons. The *Chemical and Biological Weapons Nonproliferation* project was launched to examine the issues that had been debated since the 1920s. The project was pursued in part to support the implementation of the *Chemical Weapons Convention* and to further strengthen the 1972 *Biological and Toxin Weapons Convention.* More than 120 countries were signatories to the *Convention of the Prohibition of the Development, Production, Stockpiling, and Use of Chemical Weapons and on Their Destruction.*

In the mid-1990s, a U.S. presidential initiative funded a prototype for Metropolitan Medical Strike Teams consisting of trained personnel organized to assist community response to potentially hazardous events. This was further developed to become the Metropolitan Medical Response System (MMRS) to link the resources of first-response, health care, and public health systems.

In June 1994, residents in Matsumoto, Japan, sought medical attention and were found to have reduced cholinesterase levels. They were treated for poisoning, but several died and hundreds became ill. It was later determined that they had been exposed to sarin nerve gas.

The toxin was identified through gas chromatography and a document called *The Matsumoto Incident: Sarin Poisoning in a Japanese Residential Community* was prepared in the fall of 1994 to discuss the event and relevant issues [available online at www.cbaci.org].

In March 1995, the world was stunned to hear that a sarin gas attack on Tokyo subway travelers had left hundreds of people sick and about a dozen dead. It thus appeared, in hindsight, that the June 1994 incident may have been a 'test' for the 1995 attack.

Certification and Countermeasures Programs

In 1995, the American Board of Criminalistics was funded to develop written certification tests for a number of specialties in the field of forensic science, including forensic biology, fire debris analysis, hair and fiber analysis, narcotics identification, and paint and polymer analysis. This project was implemented so that criminal science specialists involved with forensic detection and analysis would be certified to have a level of knowledge and experience appropriate for making recommendations to investigators and attorneys.

This ink drawing is an artist's rendering of the Tarhunah underground chemical plant said to be located in Libya. [U.S. DoD 1996 news photo, released.]

In 1996, DARPA launched the Biological Warfare Defense Program to develop technologies that could aid in thwarting the use of biological warfare agents by military opponents and terrorists.

Also in 1996, the Nunn-Lugar-Domenici Amendment was made to the *FY 97 Defense Authorization Act* "Defense Against Weapons of Mass Destruction, Subtitle A: Domestic Preparedness".

Progress on Treaties

In 1997, the *Chemical and Biological Weapons Threat Reduction Act* was proposed in the U.S. Senate and passed as amended. The intent of the act was to make it U.S. policy to take preventive measures against and to discourage and prohibit chemical and biological weapons.

The status of global weaponry was scrutinized with global human intelligence and technological surveillance devices that became available in the 1990s, with the result that the Pentagon reported that more than 25 nations had developed or might be developing nuclear, biological, and chemical weapons. In 1997, China pledged not to engage in any new nuclear cooperation with Iran and, in November, Russia ratified the *Convention on Chemical Prohibitions of 1993*.

When the *Chemical Weapons Convention,* an international effort at international chemical weapons disarmament, came into force in 1997, it mandated the creation of the *Organisation for the Prohibition of Chemical Weapons* (OPCW) to eliminate and verify the destruction of declared chemical weapons stockpiles. Considering all the previous agreements concerning the elimination or reduction of biochemical weapons and promises of various nations to do away with these materials, it is sobering to know that the OPCW member parties announced that more than seventy-one thousand metric tons of extremely toxic chemical agents would need to be destroyed (and their destruction verified). That's about the same tonnage as over 220 of the largest fully loaded commercial planes (including the weight of the plane itself). To give an idea of the scope of this quantity, the OPCW points out that a single drop of nerve agent smaller than a pin head can kill a human in minutes.

Russian experts slated a number of chemical-weapons production sites for destruction late in 1999. The sites apparently had been used to produce toxic chemicals including mustard, sarin, and soman gas, VX, and lewisite, and munitions laced with hydrogen cyanide and phosgene. The removal process would take several years due to the complexity of safely reactivating and removing chemical agents and decontaminating surrounding support structures. Financially stressed in the 1990s, Russia sought funding assistance from other nations.

Left: In spring 1997, the U.S., Canada, Germany, and The Netherlands participated in exercises to refine operations skills under simulated high-threat environments. In this case, participants were garbed in chemical gear, chemical attacks were executed by helicopter, and service members briefed after the exercise. Crew members headed for the Persian Gulf were trained, prior to the trip, for chemical and biological warfare. Right: A room full of chemical protective suits are inspected by fireman Russell Legett on the USS George Washington (CVN 73) as it heads for the Persian Gulf in November 1997. [U.S. DoD news photos by James Mossman and Joseph Hendricks, released.]

Preventive Health Measures

Late in 1997, a decision was announced by the U.S. Defense Secretary to inoculate about 1.5 million service members against anthrax, considered to be a highly lethal candidate for use as a biological weapon. As a first step, in 1998, personnel deployed to Korea and Southwest Asia were inoculated. Trials of a newer, cleaner smallpox vaccine, grown in a sterile lab, were carried out on U.S. Army participants.

 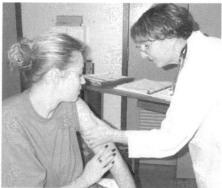

Left: P 1st Class Suderman draws blood from P 1st Class Lavenhous, one of 150 volunteers in clinical trials for a new smallpox vaccine at the *Army Medical Research Institute* for Infectious Diseases. Right: Dr. Coster, Chief of Clinical Studies examines Army Spc. Winnona Yanson, another volunteer. [U.S. DoD news photos by Douglas J. Gillert, released.]

International Affairs in 1998

There were many events in 1998 related to the development of international treaties and worldwide control of chemical weapons production and stockpiling.

The United Nations Special Commission, a body charged with eliminating Iraq's weapons of mass destruction, claimed that the Iraqi government was trying to conceal some of its programs.

Left: An aerial reconnaissance photo announced by the U.S. as being the Shifa Pharmaceutical Plant in the Sudan. The U.S. military had launched a strike on an apparent chemical weapons plant in the Sudan and used this as an exhibit in press briefings. Right: A simulated casualty from a chemical attack is loaded by service members in chemical suits. This exercise was executed jointly with the U.S. and the Republic of Korea as Exercise Foal Eagle, in October 1998. [U.S. DoD news photo on right by Jeffrey Allen, released.]

In 1998, the U.S. presented new devices intended for the detection of weapons and chemicals.

In June 1998, the U.S. claimed that Syria had installed sarin nerve gas in missiles, artillery, and on board aircraft. In July, a U.S. Pentagon review denied any evidence to support allegations that U.S. troops used sarin nerve gas in 1970 to hunt down American defectors. Iraq was accused by the U.S. of having laced a missile warhead with VX nerve agent, resulting in continued sanctions against Saddam Hussein, the Iraqi leader. At one point it was claimed that the material was used in equipment calibration and Iraqi supporters demanded analysis to confirm the presence of the nerve agent. It was also claimed that U.N. inspectors had planted the V nerve agent in the missile warheads. Experts later destroyed tiny quantities of the VX and Iraq refused to comment further on the matter.

In the fall, the U.S. and the Republic of Korea participated in joint training exercises to test rear area protection in which chemical attacks were simulated.

Customs and Border Control

Throughout the years, the U.S. Customs department had been updating and improving inspection practices and incorporating new technologies as they became commercially viable. In the late 1990s, new equipment and sensors were added to a number of border stations and canine sniffing programs were continued or expanded. Operation Brass Ring was instituted in 1998, a program in which random and unpredictable strategies were used to make it difficult to anticipate the inspection actions of customs personnel. At the West Texas and New Mexico border region, in one year alone, agents seized more than 200,000 pounds of narcotics that individuals attempted to smuggle across the border undetected.

In February 2002, the U.S. Customs Service announced that it was recruiting 1,300 customs inspectors, in support of anti-terrorism efforts to limit smuggling of biochemical or nuclear material into the U.S.

In 2002 and 2003, in cooperation with a number of Canadian, European, and Asian cities, the U.S. began working in reciprocal programs with foreign customs officials to survey cargo with high terrorist risk. For example, in the United Kingdom, under the *Container Security Initiative* (CSI), CBP officers began working with U.K. officials at Felixstowe to surveil cargo destined for the U.S.

Cargo Chemical Surveillance. Left: A Customs and Border Control officer uses surveillance equipment to check a seaport container for potentially dangerous prohibited chemicals. Right: A CBP officer and trained canine inspect trucks in Detroit, Michigan, for bombs. [CBP news photo by J. Tourtellotte, ca 2005, G. Nino, 2006, released.]

Not all border chemical surveillance is about detecting terrorist weapons or hallucinogenic drugs. A great deal of routine surveillance is to uncover taxable items and prohibited agricultural products. Officials try to catch drug dealers/carriers who cross borders to sell Ketamine ("Special K"), a controlled substance developed as an animal anaesthetic that is used to anaesthetize victims to rob or rape them. Sometimes it is further processed and sold as a "party" drug for producing feelings of euphoria.

Terrorist Acts

There have always been terrorist acts, but since the advent of television, these events receive media attention which may, in some cases, increase public reaction and fear. Deaths from terrorist acts are far less frequent than those from auto accidents, preventable illnesses (e.g., diet- and smoking-related illnesses), and murder, but acts of terrorism carry with them the discomfort of the "unseen enemy" and are no less tragic for the victims and their families. Terrorist acts like the attack on the World Trade Center in 2001 resulted in changes in legislation, the structure of the intelligence community, and prompted increased spending on security and surveillance.

Annual Deaths in the U.S. from a Number of Causes			
Cause	Approx. Deaths	Source	Year(s)
Auto Accidents	44,800/year	National Safety Council	2003
Cancer/Heart Disease from Secondhand Smoke	38,112/year	CDC	ca 2000
Suicide	31,000/year	NCICP	2001
Murder	21,000/year	NCICP	2001
Falls	16,200/year	National Safety Council	2003
Poisoning	13,900/year	National Safety Council	2003
Deaths in Fires	3,675/year	National Fire Protection Assoc.	2006
Drowning	2,900/year	National Safety Council	2003
Suffocation	2,600/year	National Safety Council	2003
Infant Death from Smoking in Pregnancy	910/year	CDC	1997-2001
All Above Causes	174,187/year		
Terrorist Acts	94/year	Terrorism Knowledge Base	1968–2006

On 7 August 1998, U.S. embassy buildings in Kenya and Tanzania were devastated by terrorist bombs, with thousands of people injured and more than two hundred lives lost. This act raised concerns that terrorists who were willing to kill so many people with bombs might also be willing to kill people with biological or chemical weapons.

On 20 August, the U.S. government launched missile attacks on installations in Afghanistan and the al-Shifa pharmaceutical factory, claiming it was being used for making chemical weapons, a claim that stirred international controversy. Sudan requested that the *United Nations Security Council* pursue a fact-finding mission to verify American claims. Such a mission would include chemical surveillance to locate evidence of alleged chemicals in the soil and debris at the bombing site.

The *U.N. Security Council* postponed a decision on the mission. The Sudanese claimed that if al-Shifa had, in fact, been a chemical weapons site, then bombing it would risk releasing the chemicals into the immediate vicinity, endangering thousands of civilians. The counter-argument was that the precursors used to developing toxic weapons such as nerve gas, are not of the same toxicity as the final product. Westerners who had been at the factory claimed they had seen no sign of chemical weapons. Some still feel, at this point in time, that the incident was a false alarm and the actions were taken in haste. The U.S. continues to defend its actions.

Left: A victim of the tragic bombing of the U.S. Embassy in Nairobi, Kenya, is honored in a ceremony at the Andrews Air Force Base, Md. in August 1998. [U.S. DoD news photo by Mark Suban, released.]

Chemical and Crime

Because biochemical warfare is such a frightening prospect, a good deal of surveillance is focused on anticipating and preventing its use. However, biochemical warfare is not the only aspect of surveillance that involves chemical agents. Chemicals are sometimes used in more specific and direct ways to inflict harm. Cyanide, arsenic, and strychnine are poisons that are sometimes used to kill people or pets and thus are of concern to forensic pathologists. When older people are poisoned, there may not be an autopsy performed, due to the age of the victim. This makes it harder to know whether chemicals were related to the death. In recent years, however, improvements in spectrophotometry, mass spectrometry, and X-ray analysis techniques have made it easier to determine if there are toxic chemicals in a person's body tissues that could indicate foul play. By the late 1990s, lab techniques to detect poisoning had greatly improved.

Chemicals are not always used in direct ways to commit crimes. Sometimes they are an indirect means to obscure other objects, activities, or goods intended to be hidden, installed, or detonated. In 1999, in one customs seizure of goods, it was discovered that smugglers had chemically altered cocaine to darken it and make it look like metal.

Industrial Spills

Not much is said about industrial chemical spills, but these have been a problem for decades, with factories either not being sufficiently regulated or illegally dumping effluent and toxins into watersheds, oceans, lakes, and landfills. Chemical surveillance is important in the management of industrial effluent and the detection of illegal dumping.

A few examples from 1999 and 2000 help to provide a picture of the prevalence and variety of types of industrial spills that occur. It also underlines the challenges facing investigators and experts who often must determine the chemical composition of the spills, their potential danger to humans, and the best strategies and materials for cleanup. Chemical surveillance can provide early warning of spills and evidence of chemical composition for prosecution in instances of negligence or where responsibility for cleanup funding must be established.

April 1999 - a chemical spill at a milk-processing plant in Wisconsin sent a yellow plume high into the sky. Hundreds of residents and workers were evacuated.

June 1999 - Washington State, a pipeline leak into a creek ignited a series of explosions near a chlorine water treatment plant. Residents had no way of knowing if the thick billows of smoke were toxic. Early warning signs included a strong smell of 'gasoline' near the creek. One report claimed that computers correctly recognized there was a problem and shut off the system, but that personnel turned it back on again prior to the explosions. Since several boys were killed, the proof or denial of allegations is ongoing, as are cleanup efforts.

July 1999 - a corrosive chemical spill during unloading activities at a toiletries factory in Iowa necessitated the evacuation of about five thousand people.

August 1999 - a barge collision on the Ohio River caused a gasoline spill to contaminate the water.

June 2000 - a test of a well in British Columbia revealed mercury levels in the ground near a landfill to be thousands of times higher than established safety limits, raising concerns about leakage into the water table and nearby residential areas.

Chemical Surveillance Measures

Sometimes other kinds of surveillance procedures are combined with chemical 'sniffers". Since many chemicals are delivered in liquid form, sometimes devices that detect changes in moisture or electrical conductance (e.g., resistance) can help signal the presence of chemicals. Electromagnetic survey (EM) devices and ground-penetrating radar can aid in surveying an area suspected of contamination (ground-penetrating radar is discussed in the Radar Surveillance chapter). A flame ionization detector (FID) may be used to locate hydrocarbon spills, and a gas chromatograph may be used to more directly 'sniff' chemical vapors. Optical technologies are sometimes also used.

In the early 1990s, the U.S. *Department of Energy* began working on light detection and ranging (LIDAR) equipment to aid in the detection of air and surface chemical agents. This was intended at least partly as a preventive tool, but has been found to be more effective for identification, tracking, and recovery missions [N. Williams, ca 2005].

In March 1999, federal researchers were encouraged to step up development of better chemical and biological agent detection systems so they could be put into place by the next Winter Olympics. The Chief United Nations weapons inspector reported that Iraq had converted a livestock vaccine facility into a production facility for biological warfare agents.

In July 1999, chemical experts in Baghdad were ordered to destroy VX nerve agent in a United Nations laboratory after Iraq's allies failed to convince the Security Council that they should be held for further analysis.

Left: A chemical suit stands in mute testimony on the left while Secretary of Defense William Cohen holds up a copy of *Proliferation: Threat and Response* which discusses nuclear, chemical, and biological threats and DoD countermeasures. Right: A treaty signing in the Pentagon committed the U.S. and Azerbaijan to cooperation in the counter-proliferation of nuclear, chemical, and biological weapons and related materials. The signing between the U.S. Deputy Secretary of Defense and the Minister of Foreign Affairs took place in September 1999. [U.S. DoD news photo by Helene C. Stikkel and Jerome Howard, released.]

In 1999, the U.S. Director of the CIA delivered a report on weapons acquisition in various nations. He reported that Iraq had stockpiled chemical weapons, "including blister, blood, and choking agents and the bombs and artillery shells for delivering them". He further reported that North Korea was producing "a wide variety of chemical and possibly biological agents, as well as their delivery means" and that Syria had "a stockpile of the nerve agent sarin and apparently is trying to develop more toxic and persistent nerve agents".

In October 1999, Hurricane Floyd battered the east coast, causing concerns that bacteria and chemicals may have contaminated the North Carolina waterways.

Into the 21st Century

The year 2000 was characterized by continued international efforts to reduce the proliferation of biological and chemical weapons and by environmental concerns about contamination from industrial effluents, spills, PCB storage, and nuclear wastes.

In January 2000, in armed conflicts between Russia and rebels in Chechnya, the Russians reported that Chechens had detonated bombs containing poisonous chlorine and ammonia.

The U.S. Government sought funding for developing blood tests to measure toxic exposure in humans. Agents within the FBI were assigned to counter-terrorism, which included biological/chemical investigation and countermeasures.

In February, 2000, equipment repair personnel at a chemical weapons incinerator in Utah experienced accidental exposure from a leak into the room in which they were working. The same month, San Martin Lake, in Texas, was fouled by a toxic chemical spill that killed about six million fish and large numbers of birds.

In a development that probably surprised Westerners more than Asians, the cult accused of attacking subway commuters with sarin gas in Tokyo in 1995 admitted responsibility and offered to pay compensation to the victims of the attack.

National Biodefense

Concerns about biochemical warfare and general surveillance of health risks is ongoing. Fears of West Nile disease, bird flu, and other infectious and contagious agents spread by natural means or by aggressors is constantly monitored. As human populations increase, so do concerns about America's ability to cope with a possible pandemic.

In 2003, the *Project Bioshield Act*, a bill to accelerate research and development and effective countermeasures against bioterror agents, was jointly submitted by representatives of Health & Human Services (H&HS) and the Department of Homeland Security. The bill was signed into law in July, 2004, and is described on the H&HS Website "as part of a broader strategy to defend America against the threat of weapons of mass destruction". This emphasis on weapons of mass destruction seems more narrow than most bills of this nature and might be interpreted by some as a knee-jerk reaction to the events of 11 September 2001 but, in more general terms, includes provisions to accelerate research and development and improve the availability of effective medical countermeasures. Responsibility for implementing the bill are in the hands of H&HS, in collaboration with the *Department of Defense* and the *Department of Homeland Security*.

In 2005, the *National Biodefense and Pandemic Preparedness Act* was introduced in the U.S. Senate to amend the *Public Health Service Act* to address biosurveillance capabilities and responses. It covers the restructuring of the *National Biodefense Initiative* (including the establishment of a *National BioVenture Trust*), vaccine manufacture, improvement of the "Bioshield", incentives for countermeasure development, injury compensation, and strengthening public health readiness for pandemics.

In 2006, the *Organisation for the Prohibition of Chemical Weapons* (OPCW) reported that over the last six years, about 20 percent of the world's declared chemical weapons stockpile had been destroyed, leaving about fifty-eight thousand metric tons still extant.

5. Description and Functions

Biological and chemical surveillance activities are primarily concerned with the detection of synthetic and biological chemicals associated with crimes. The analysis of these chemicals may include not only identification, but the determination of their origin and composition. The neutralization, cleanup, and decontamination of chemicals are closely related to chemical surveillance and so are briefly discussed here.

Surveillance measures are further used to determine compliance with international and domestic biological/chemical nonproliferation treaties and bans, and to detect the presence of biochemical weapons.

5.a. General Guidelines

Detection of biological/chemical and, in some cases, nuclear contamination, requires a strategy and one or more synthetic or biological devices or detection systems. Biological agents may be detected with various electromagnetic (EM) devices to get a general idea of whether there is contamination. EM devices can sometimes also provide information on the location and general extent of any biochemical agents. However, to get more precise data, chemical kits, and laboratory follow-ups (if possible) are often used. In food and water supplies, it is important to test before adding any other chemical agents (e.g., chlorine) that might obscure the testing results. In the case of chemical spills, ground-penetrating radar, lidar, flame ionization, and vapor detectors may also provide valuable information to locate and determine the character of spilled chemicals.

Protective Environments and Gear

When dealing with chemicals, it is always important to use appropriate safety gear, including suits, masks, gloves, and goggles. In some cases, as in the military and medical fields, preventive steps such as vaccinations and X-rays are given to personnel who might be working with chemicals and biological organisms on a regular basis or might be exposed to them in the course of their duties.

Decontamination tents, trailers, and quarantine/isolation rooms are used in some circumstances, as in the space program and in certain military and medical stations.

Decontamination and Sterilization Products

When chemical agents are detected, they can sometimes be contained or moved, but sometimes it is necessary to neutralize or destroy them, especially in the case of those that are potentially hazardous. This is usually done by decontamination or sterilization. Medical wastes, bacteria, viruses, toxins, and poisons are often routinely treated or sterilized to render them less harmful to humans and the environment.

There is significant overlap in the chemicals and processes used to sterilize and those used to decontaminate, but given the complexity of chemicals and the differences between decontaminating a person and sterilizing an object, there is not 100% overlap. Sometimes the processes are used together.

Because the focus of this book is surveillance rather than remediation, decontamination and sterilization are not discussed further. It is sufficient to mention that heat, electricity, and chemical agents are the primary means for decontaminating and sterilizing. Anything that breaks or reshapes chemical bonds is a potential candidate for these purposes. Methods that cause the least harm to the object or living organism being decontaminated or sterilized are usually favored though. In a few instances, economic factors take precedence and there are cases where preservation of the original contaminated object or substance is not required and it may be incinerated or otherwise destroyed.

Preservation

The preservation of evidence through chemical means is an important aspect of law enforcement. Getting good samples for DNA analysis is discussed in the Genetics Surveillance chapter. Preserving latent prints and other bodily remnants is described in the Biometrics Surveillance chapter. A few general methods are described here.

Tapes and Samples

Specialized clear tapes are used to 'lift' and preserve many types of evidence including blood smears, paints, spills, and fingerprints. Once a print has been made more visible with a powder, adhesive tape or precut strips can be placed over the print, pressed gently, and placed over a non-acidic card or paper to protect the image on the tape. This takes practice. The tape must be large enough to cover the area to be lifted and must be laid down gently and evenly so as not to create creases in the tape or distort the image by adhering part of it and then accidentally pulling on it. Then the correct amount of pressure must be applied. Too much pressure can 'squash' a print. Too little can cause some of it to remain on the original surface and the odds of correctly lining up the tape a second time to get the rest of the print are almost zero.

Light is sometimes used to reveal very faint prints that cannot be brought up by powders. In these cases, photographic evidence must be taken as there is no way to use a tape or other medium to physically transfer the prints. The same is sometimes true of other bodily residues.

Since tape glue is itself a chemical substance that might interfere with the analysis of trace materials, it is not always appropriate to use evidence tape. Some samples should be put in dry, clean, acid-free envelopes or bags instead of using tape. Tweezers and small jars can be used for fragmentary evidence such as carpet and clothing fibers that might be used to connect a victim or suspect with a crime scene.

When hairs are sampled from a living person, as in a poisoning investigation, or when a hair is compared to a sample from a crime scene, the entire hair should be taken and pulled (if the person can tolerate it) rather than cut. This serves two purposes. Since the 'history' of chemical use is contained in various parts of the hair, it preserves the entire timeline. The second reason is that cutting the hair might make it unclear which end of the hair is most recent. If checking a poisoning timetable, it can be important to know if arsenic, for example, was administered six months previously or immediately before the report of the crime (or admission to a hospital). A cut hair makes it possible to get the timeline backward, confusing the investigation and leading to the wrong conclusions. If the hair roots cannot be pulled, then the hairs should be oriented in the same direction and clearly labeled as to which end was near the scalp. For DNA tests, it is important to wear gloves and to get the root as well as the hair and, further, to transport it in a sterile container so the hair doesn't get contaminated by other sources of DNA.

Preservatives

Chemical preservatives such as formaldehyde may be used to preserve certain specimens such as animal/human tissues. If the tissues are to be used for DNA sampling, Luminol™ and formaldehyde should not be used, as they will chemically alter the sample. It is better to freeze or air-dry and bag. Chemical preservatives should be handled carefully (or not handled at all) as some are poisonous and some may provoke allergic reactions.

5.b. Biochemical Markers

Chemicals are used to mark targets or vehicles so they can be easily located by lights or special vision enhancers. This may help find a person in a crowd or a vehicle in a parking lot. Chemicals are often chosen because they show up in special lighting or with special infrared or ultraviolet viewers or imaging systems. Chemical markers are sometimes used along with radio tracking beacons.

Chemical dyes are sometimes placed in bags that hold banknotes or other important negotiable items. If the bag is not opened correctly, the dyes may disperse or 'explode' onto the hands and clothing of the unauthorized person holding the bag. Dyes that cannot be washed off by any normal means are used. It is important to make the dyes work in such a way that they do not spray in the eyes, as indelible inks can cause temporary or permanent blindness.

5.c. Detection of Chemical Prints

Fingerprinting

Fingerprinting is an important form of biochemical surveillance. Fingerprints themselves are chemical traces of perspiration, body oils, and soaps, and chemical means are used to locate and preserve the prints (fingerprinting is discussed in the Biometrics Surveillance chapter).

Chemical 'Fingerprinting'

Chemical forensics continues to improve, with many new techniques having been added in the 1990s. Chemical forensics is also used in conjunction with electronic chemical detectors and accessories. One innovative development in chemical surveillance is *molecular fingerprinting* in which the molecules of synthetic products such as plastics are manipulated to create a coded layer that can be applied to many products. Horcom, based in Ireland, has succeeded in creating a coded liquid suspension that is invisible to unaided eyes, but that can be detected with a spectrometer.

Identifying Remains

One of the more important aspects of biochemical surveillance is the identification of human (and sometimes animal) remains resulting from accidental death, suicide, homicide, or warfare. This is principally carried out by *forensic anthropologists,* though many other branches of science are involved. Identification aids in convicting criminals, in establishing evidence for war crimes, in detecting and preventing the spread of diseases, and in notifying distraught relatives and providing them with closure so they can move on with their lives with a few sensitive questions answered and put to rest.

Entomology is the study of insects and insects will inhabit human and other remains. When forensics is used to determine information about a corpse, how and when a person died, and sometimes even where, it is sometimes necessary to have a good understanding of insect habits and lifespans in order to make good scientific observations. This is the field of *forensic entomology.* It's a rather gruesome area of study, but insects and bacteria are so quick to move in on a 'fresh' meal that it would be irresponsible for scientists and technicians to ignore the information that can be gleaned from observing their habitation of a dead body. In fact, training programs in which corpses are buried or left to decay and then studied at each stop of the process, are an important aspect of law enforcement. Forensic entomology can aid in establishing time of death, the possible origin or location of a corpse, if it has been moved, and sometimes the cause of death.

Changes in body temperature can help establish time of death in criminal investigations, but if the body has been dead for more than about three days, other methods must be used. Gail Anderson, a Canadian forensic entomologist with *Simon Fraser University's* criminology school, has created a national database of bug-colonization patterns to aid investigators in determining various geographic and climatic conditions that can help establish time of death if an extended period of time has passed. Anderson began the project in 1992, and has since catalogued millions of insects. These databases are important because of the vast differences in insect species and behaviors over different climates and geographic ranges.

Insect Data for Forensic Investigation and Evidence Comparison. Left: Criminology professor Gail Anderson smiles between cages of buzzing insects. Anderson, who has spent eight years building a forensic entomology database, has received a grant for equipping the first forensic entomology lab in Canada Middle: Akbar Syed is a forensic entomologist who manages one of the best insectaries in North America. Syed has developed techniques for using insects to estimate time of death and has assisted the RCMP and the coroner's office. Right: Niki MacDonell, a graduate student with plans to be a coroner, made routine visits to a series of clothed and submerged pig carcasses to study the life cycles of insects that colonize the dead bodies. [Simon Fraser University 1999, 1998, and 1997 news photos, released.]

The identification of remains is discussed further in the Biometrics and Genetics Surveillance chapters.

5.d. Chemical Microscopy and Analysis

Paint Identification

Vehicles are frequently involved in crimes. Sometimes the vehicle is the subject of the crime itself, as in car theft or smuggling. In other cases, cars and trucks are used as getaway vehicles, and may be associated with violent crimes such as homicides and kidnappings. Sometimes it is important to be able to tie a vehicle's location at a particular time to the scene of a crime.

It is often important to chemically assess a vehicle's paint to see where it has been or where it originated. Other types of vehicles or industrial equipment may be subject to the same types of investigation. The carpet fibers, dirt in the tires, vehicle contents, and paint are important targets of investigation.

Paint is generally a combination of pigments, additives, waxes, solvents, and polymers. Microscopes and chemical processes can be used to evaluate the layers, their composition, color, and age. Sometimes it's even possible to determine if the vehicle has been subjected to unusual temperatures, storms, or other environmental conditions that might indicate where it has been and for how long.

Weathering tests can be conducted to see if a particular paint responds to environmental influences in a certain way. In some cases (and with some luck), a paint chip can be traced not only to the make and model of the vehicle, but sometimes even to the time of year the vehicle was first released, as paint mixtures are sometimes changed by manufacturers during the course of the year. More often, however, a forensic scientist will be asked whether paint chips match a particular vehicle, an easier question to answer than finding the origin of a particular chip.

Paint evidence should be put in noncontaminating (acid-free) containers or bags whenever possible, as the chemicals in adhesive tape could interfere with analysis of the paint.

Paint can vary from one part of a vehicle to another (e.g., the trunk may have been repainted following a rear-end collision), so it is best to collect samples as near as possible to where the chips may have originated. It is also a good idea to collect samples from various parts of the vehicle, labeling them carefully, because information about different paint on different parts of a vehicle can sometimes be tied to the repair and body work history of the vehicle. For example, a hit-and-run suspect may have had the damaged hood of a car replaced.

When collecting samples, the various layers down to the bare metal should be collected in one piece, if possible, as the composition and order of the layers may aid in identification.

The same general principles apply when collecting paint from furniture or walls. When possible, collect below the lowest layer of paint.

Infrared microscopy is one of the tools used in paint identification and matching.

5.e. Chemical Detection and Identification

Chemical detection is used to detect illegal smuggling or use of foodstuffs, accelerants, explosives, and controlled substances. It is also used in many aspects of forensics related to finding out whether an individual committed suicide or was murdered by poisoning.

Arson and Explosives

An important aspect of chemical detection is the investigation of arson or the detection or investigation of explosives. There are thousands of incidences of arson every year and about 2,500 attempted and actual bombings. Bomb threats also result in a variety of surveillance technologies, including chemical surveillance used to detect the bomb, to determine what type of bomb it might be in order to disarm it, and for use in detonating or disarming it.

When arson is suspected, the presence of accelerants or suspicious electrical anomalies can give clues as to whether a fire was accidental or deliberate. Fire follows patterns of behavior that are known to arson specialists and chemical residues from accelerants are not necessarily destroyed by the subsequent fire. Dogs are used to locate specific accelerants (and sometimes victims and survivors) and also have been trained to recognize a variety of types of explosives.

Foodstuffs

When foodstuffs that threaten local agriculture or business economics are suspected of being smuggled, various surveillance techniques are used to detect and locate the materials, including visual inspection, substance-sniffing dogs, X-ray detectors, and even gamma-ray detectors for checking large shipments of cargo.

Food Diagnostics

It is estimated that there are as many as 33 million incidences of food-induced illnesses each year in the U.S. alone. Biological/chemical surveillance is used by growers, manufacturers, and distributors to ensure safe products.

> *BSD sensors* - A series of biosensor products designed to detect pathogens and toxic substances in realtime. The sensors can be configured for species-specific microorganisms for use in identifying specific food pathogens. Biosensor Systems Design, Inc.

Narcotics

Narcotics-detection is often done with the aid of dogs, as described in the Animal Surveillance chapter. However, there are steadily improving electronic 'sniffers' that may someday supersede canine corps for certain types of detection and identification tasks. These electronic devices can be set to sniff out a variety of substances and some can even indicate the direction from which the chemicals are coming, which makes them potentially useful for other tasks, such as locating chemical contaminants and chemical warfare agents.

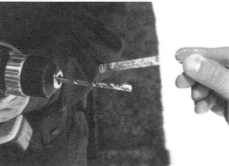

The U.S. Customs service (now Customs and Border Patrol) is responsible for the detection of illegal foodstuffs, undeclared goods, restricted pharmaceuticals, narcotics, and explosives. Left: Canines are trained to detect a number of specific substances. Beagles aid in detecting foods, and Yellow Labs and Golden Retrievers are used in a variety of explosives and drug detection activities. Here, Big Nick zeroes in on drugs underneath a car while in training. Right: After a canine 'hit' on a boat hull, the hull was drilled and the hit confirmed with a chemical detection kit. The chemical turned blue, indicating cocaine, which was subsequently found stashed in the hull. [U.S. Customs news photos by James R. Tourtellotte, released.]

These are examples of some chemical detectors (they are included as examples only, and their inclusion does not imply endorsement):

AP2C - A handheld chemical warfare agent detector, based upon flame photospectrometer technology, that detects nerve and mustard gas agents and identifies liquid and vapor Vx.

> *EVD-3000* - A handheld, rapid-response, high-sensitivity explosives detector available from American Security.

> *Ionscan 400* - A hybrid drug and explosives detector designed to detect and identify up to 30 different substances. Automated operation, self-calibrating. Suitable for use by law enforcement agents, airport security agents, and investigative scientists. Available from American Security.

> *M256A1* - A portable carry-case detection kit with individual pouches containing materials designed to detect nerve, blood, and blister agents. It is intended to confirm chemical presence and its concentrations. This kit was used by the U.S. military in the Gulf and Iraqi conflicts.

> *NDS-2000* - A portable, handheld, battery-operated narcotics detector for detecting cocaine, cannabis, crack, heroin, methamphetamines, and others. Available from American Security.

Smiths Detection LCD (lightweight chemical detector). This is a handheld kit for individual detection of toxic chemicals and chemical warfare agents that may be clipped to a belt or harness. It is used by emergency response organizations and armed forces. Smiths also markets a handheld chemical warfare agent monitor (CAM) for military, law enforcement, and industrial chemical detection. It confirms presence of chemicals and their concentrations.

Drugs and Poisons

It is sometimes difficult to determine whether someone has died of poisoning or an overdose of drugs and whether it was intentional, accidental, or a case of homicide.

Forensic toxicologists regularly examine chemical indicators found during autopsies, or collected at the scene of a suspicious death.

Air Quality

Various detectors and measuring instruments have been developed to surveil air quality, particularly in large cities and industrial centers. These monitors are used by companies to ensure compliance with regulations and are by newscasters and government monitoring agencies to check and report air quality. People with illnesses or compromised immune systems can be particularly susceptible to air pollution.

Dr. John Kauer, and chemist David Walt, of Tufts University, teamed up to create a fiber-optic sniffer that can identify various substances by their smells. It has many applications in surveillance, including the capability to monitor internal and external air quality, changes in blood chemistry in the human circulatory system, or chemical leaks in a tanker or factory.

Explosives Detectors

Explosives detectors work on a number of different principles, from X-ray machines that might indicate a bomb in a briefcase, to dogs trained to sniff out certain chemicals, explosives detection is used in border security, domestic terrorism, and military operations. A new type of machine was certified by the FAA in late 1998. Also, in 1998, it was reported by the U.S. *Dept. of Transportation* that bomb detection machines for screening luggage were not working as well as was hoped, partly due to slow implementation by airlines.

Firearms Matching and Gun Residues

Firearms identification and matching is accomplished through a variety of surveillance technologies, including visual, infrared, and physical, but since a portion of it involves residues analysis and chemical and microscopic techniques, firearms are mentioned in this chapter.

Identifying or matching a firearm is an interesting aspect of forensic science. It involves inspecting, comparing, loading, firing, tracing, and sometimes chemically treating the firearm in order to determine its characteristics. Firearms used in crimes are sometimes modified to hide their identity—sometimes they will have the serial number filed off. Thermal, magnetic, and chemical processes can sometimes be used to retrieve the information even when the number is no longer visible. A computer floppy diskette that has been erased still has minute traces of patterns on its surfaces that can be discerned with the proper tools and techniques and the same goes for gun serial numbers. The very act of stamping the serial number may have disturbed the metal enough that the information can be revealed. (These techniques can sometimes also be used to retrieve numbers from stolen vehicles or tools.)

Each gun has a unique 'signature' in the sense that the marks in the barrel and the marks in the projectile will have certain characteristics in common. In a more general sense, firearms will share characteristics with those of the same type or from the same manufacturer or plant. This information can also be helpful if the weapon itself is not available, with only evidence of its having been fired.

The FBI and related organizations maintain a number of databases and collections of firearms to aid forensic scientists in learning the characteristics of individual types of guns and using them in comparison tests. They are also used in experiments to determine their range and the types of effects they inflict on different sorts of materials. Samples can help answer many questions including: How far do the bullets penetrate? How do angles affect their influence, or temperature, or precipitation? How strong do you have to be to pull the trigger? How far away can you hear the shot? How widely does the shot scatter? How much does the bullet deform on impact? (See the Introduction & Overview chapter for more information on the FBI database.)

Reference Fired Specimen File - FBI Firearms-Toolmarks Unit that includes examples of ammunition illustrating what happens to them after they are fired.

Reference Firearms Collection (RFC) - a physical repository of firearms with more than 5,000 items. This aids in classification and parts identification in cases of evidence derived from a crime scene or from surveillance images.

Standard Ammunition File (SAF) - a physical collection of more than 15,000 commercial and military specimens of foreign and domestic ammunition. Both whole and disassembled versions are available as well as pellets, bullets, cases, etc. The information has been transferred to a computer database for faster search and retrieval for initial investigations.

DRUGFIRE - an FBI database providing examiners with a means to compare and link evidence in serial shooting investigations. After a weapon has been test-fired, images of the results can be digitally stored to add to the system for comparison with thousands of other categorized images. Since this can be accessed over a network, crimes committed in one area can be linked with crimes in another. This has gradually been absorbed into the National Integrated Ballistics Information Network (NIBIN).

National Crime Information Center (NCIC) - NCIC maintains a database of serial numbers of firearms. Weapons reported stolen are included in the list.

5.f. Industrial Prospecting and Production Monitoring

Chemical surveillance is an important aspect of resource- and manufacturing-based industries. Mining, refining, food production, pharmaceutical research and production, and cosmetics are just a few of the areas that use chemical surveillance to ensure uniformity, compliance, and quality in their products. What was once accomplished by following certain protocols and procedures is increasingly accomplished with sensors and detection methods that provide more objective measures of performance that can be automated when integrated with computer algorithms.

SeptiStat, from Biosensor Systems Design, Inc., is an example of a pathogen-specific sensor system for quick assessment of contamination levels in food and industrial environments. Sensors can be used individually, or can be multiplexed for pathogen specificity.

Production Analysis

Chemical surveillance is widely used in the manufacture of consumer products to determine their composition and ensure uniformity. It is also sometimes used by companies to analyze the components of their competitors' products. It is further used by production managers and inspectors to make sure products comply with health and safety guidelines.

Minispec - A sensitive nondestructive analyzer that has a number of options that allow further types of analyses, including fat composition (e.g., margarine production), fluorine content (e.g., toothpaste products), hydrogen. It is suitable for food, polymer, petroleum, and medical industries. Available from Bruker.

Process Analysis and Control

There are many industries in which chemical reactions and interactions determine when the process is ready for the next step. Wine-making, paper-making, dying, cheese-making, and metals refining and fabrication are all examples of processes that go through a number of stages that are, at some points, critically linked to the previous stage or a chemical state of readiness. Chemical surveillance products can assist in determining when to stop a process or when the right stage of 'readiness' has been reached for continuing to the next process and further can allow quality control monitoring during various stages of production.

Process Control Sensors - Sensors which can be set and calibrated and operated online or offline to alert personnel or machines as to the status or state of readiness of a batch being set up for the next step.

5.g. Biochemical Warfare Agents and Medical Diagnostics

Many biological/chemical agents have been used to try to deliberately poison or infect others, including instances of assassination, warfare, and homicide. Archaeologists have discovered, through the study of disinterred remains, that indicate that arsenic poisoning has been around for a long time.

Biological and chemical agents that can be used to inflict harm include anthrax, the plague, ebola, pox viruses (e.g., smallpox), botulism (which is produced in anaerobic environments), tularemia, venoms, and various nerve gases. Unlike most weapons, many biochemical agents cannot be controlled once they are 'fired'. The chances of injuring the person who unleashed them may be quite high, in the case of disease pathogens, unless they have been specifically immunized or are wearing special protective gear. Fortunately, many of these agents of warfare are not as easy to manufacture as is generally thought, and humans have built-in protections against many biological agents. There are also a large number of organizations working tirelessly for the destruction of biochemical weapons and for laws to eliminate their manufacture and they are achieving some degree of success.

Purdue researcher Fred Regnier (right), and doctoral student Bing He, examine a silicon wafer that contains a scaled-down version of a liquid chromatograph, a system used to separate chemical compounds. This miniature laboratory can carry out many of the chemical functions of the larger chromatograph on the desk. Regnier has placed multiple mini-labs on a computer chip to create a micro-chromatograph with no moving parts. This trend to miniaturization of complex lab procedures and equipment is expected to continue for some time and is a boon to field scientists and forensic experts who need hands-on methods to analyze data at a scene before the site becomes contaminated or the trail 'gets cold.' [Purdue news photo, released.]

DNA analysis and profiling can provide information on biological agents, which sometimes provides great specificity in their type and origins.

OptiSense Technology™ - This represents a series of biosensor products. The medical diagnostic products are designed to detect pathogens or critical molecules in body fluids. Ex vivo and multiple blood and urine analyses can be performed in less than a minute. A handheld analyzer works in conjunction with disposable cartridges. Bacterial- and viral-screening sensors are also in development. Available through Biosensor Systems Design, Inc.

There are a number of experimental technologies in development.

DARPA is considering a nitric oxide detector, since infected people may exhibit higher levels than those who are not, before any overt symptoms may be noticed.

IatroQuest Corporation in Ottawa, Canada is working on a pocket-sized chemical and biological warfare detector designed to replace larger machines used in military operations. It could also be used in vulnerable urban settings. It combines biological molecules with inorganic material such that it behaves like a 'smart antibody', reacting to chemicals somewhat similar to the human immune system reacting to foreign agents. It then performs a spectral analysis and alerts the user, specifying the direction from which the foreign matter was detected, and optionally displaying instructions.

There are many motivations for developing smaller, more sensitive detection kits. Besides being more cost-effective and potentially 'expendable' (usable in high-risk situations), small kits can potentially be catapulted or parachuted into suspected hazardous areas ahead of investigators and can be packed and hidden more readily than larger units.

Chemically separating mixtures into pure components, called *capillary chromatography capillary electrophoresis*, is a method used to analyze blood and tissue samples in medicine and forensics for drug discovery and, in medical research (especially pharmaceuticals). A *chromatograph* is an instrument used in both purification and analysis processes.

6. Applications

6.a. Chemical Contaminants Detection

Chemical contamination occurs in many situations, from urban and industrial pollutions and spills, to deliberate contamination by terrorists and hostile agents. Food and water supplies are monitored for pesticides, heavy metals, and contaminants. Chemical surveillance tools allow scientists to assess many different substances and to help determine whether contamination has occurred and what contaminants might be present. This, in turn, can aid in preventing further contamination and cleanup measures.

In the U.K., the Food Advisory Committee monitors and report on food standards, labeling, and food chemical surveillance. The Committee advises government Ministers on general matters related to food safety and specific matters related to the *Food Safety Act* of 1990. Their recommendations form the basis of much of the U.K.'s food legislation.

An important aspect of chemical detection is identifying the kinds of chemicals that may be present. This is important in pollution-monitoring, chemical manufacturing processes, and for detecting chemical contamination.

At Purdue University, missile technology has been used to develop an instrument that can search rapidly for chemical catalysts, which could then be incorporated into an extensive database for reference for future projects. This new nondestructive technique can apparently create and test thousands of chemical samples in the time it used to take to sample just one. The system uses infrared technology developed for heat-seeking missiles which has recently been declassified.

Another chemical analysis system at Purdue is a near-infrared Raman Imaging Microscope (NIRIM) that uses laser light to analyze composite materials thousands of times faster than older methods. When a laser stimulates chemical, it reacts with molecular vibrations that produce color changes in a process called Raman scattering, thus producing a reaction unique to that chemical. This can potentially be used in industrial and robotic vision and detection systems, in industrial separators, and for medical diagnostics.

The U.S. Army Corps of Engineers has worked on a number of cleanup projects, including industrial spills and storage projects. Top Left: An example of a HTRW Level B protective chemical suit used in the Detroit District for cleanup. Top Right: Drums have been compacted in preparation for removal from Barter Island, Alaska in 1998. Bottom Left: Engineer Mark Wheeler in protective gear inspects the oily lagoon at the Bridgeport BROS Superfund site in New Jersey in the process of remediating the hazardous environment. Bottom Right: Soil contaminated with low levels of nuclear waste from the former Fort Greely Nuclear Power Plant are systematically removed in 1998. [U.S. Army Corps of Engineers news photos, ca 1999, released.]

Mass spectrometry is another laboratory method used to identify the chemical nature of a substance. Researchers from the *Scripps Research Institute* and *Purdue University* have been teaming up to combine the properties of porous silicon with mass spectrometry to streamline and automate the analysis of biological molecules. The pores in the silicon give the material photoluminescent properties—the substance emits light when stimulated with an electric current. When exposed to ultraviolet, the pores absorb and emit light.

The researchers found that porous silicon could be used in place of traditional materials for a wide range of biomolecules, including sugars, drug molecules, and peptides. This led Buriak, Wei, and Siuzdak to develop a new technique, called *desorption ionization on silicon* (DIOS) that could overcome some of the limitations of distinguishing drugs with a relatively low mass, which are difficult to detect with traditional laser spectrometry.

In projects funded in part by the *National Science Foundation*, two new rapid chemical analysis systems have been produced at Purdue. Left: Chemical engineer Jochen Lauterbach and grad student Gudbjorg Oskarsdottir use a rapid-scan Fourier transform infrared imaging system to rapid-scan chemical catalysts. Infrared reveals a unique signature corresponding to each chemical, in essence, a molecular 'fingerprint'. Right: Chemist Dor Ben-Amotz uses a Raman-scattering spectroscopic microscope to analyze chemical components in a plant cell. The system uses laser light to analyze and identify materials in realtime. Both systems enable faster, more automated detection and identification of chemicals. [Purdue 2000 news photos by David Umberger, released.]

6.b. Biological Hazards

Biological hazards that exist in nature can be anticipated to some extent, allowing us to protect ourselves against them, but there are still situations where it is useful to have technology to help assess an environment's safety. In regions where there have been natural disasters resulting in the deaths of people and animals, detection systems for disease and biological pathogens can help assess what medical assistance and relief supplies might be necessary. Lately, monitoring for bird flu (H52N) and West Nile disease have been priorities.

The *Centers for Disease Control* monitors health on a global scale, since the health of other nations affects all of us, particularly those who travel or are stationed abroad. Information from the CDC is available on epidemics, medical assistance, inoculation programs, and much more. The U.S. *Department of Agriculture* also has information on biological hazards that may be contracted through agricultural products.

6.c. Terrorism

Domestic and foreign acts of terrorism are not new. In the 1970s, someone apparently poisoned store supplies of pain-killers. Since then, most over-the-counter drugs come with plastic wraps and other 'tamper-proof' safeguards to make it easier to tell if a bottle has been opened. In the 1980s, cult members poisoned restaurant food with salmonella. A more serious attack occurred in March 1995, when a Japanese religious cult released toxin sarin gas into Tokyo subway stations, injuring thousands of commuters. Sarin exposure can result in pupil dilation, numbness, nausea, coughing, and sometimes more serious convulsions. Depending upon the strength of the solution and the amount of exposure, sarin gas exposure can be fatal.

When terrorists release harmful biological agents, biochemical surveillance is needed to determine the type of agent and its concentration. This information is used to mobilize medical, defensive, and cleanup operations.

Toxic agent attacks in urban environments can have serious secondary consequences as it takes time for medical professionals to determine the cause of the illness and how to contain it.

In warfare, biological agents can be released to confuse, impede, or disable a hostile force. Releasing an agent for which the delivering side has been immunized might provide an immediate or longer-term advantage.

6.d. Biological/Chemical Warfare

Bacteria and viruses are microscopic and difficult to detect—we often don't recognize them until a large number of people are observed to have similar symptoms. Gases may also be difficult to see or smell and thus are strong potential warfare agents that may be deployed in hostile situations.

Left: Danielle Williams of the U.S. Army dons protective goggles and gloves in a nuclear, chemical, and biological skills challenge held in the Republic of Korea, in 1998. Right: A U.S. Air Force member talking to a radio link through a chemical/biological protective mask in an exercise to accustom service members to working with protective gear, in 1999. [U.S. DoD news photos by Steve Faulisi and Lance Cheung, released.]

Biological and chemical weapons are unsettling for many reasons. Besides their relative invisibility, they can be inserted in the head of a missile and launched into the center of an opposing force, scattering gas, bacteria, and viruses in a cloud far from the source of the missile. Countermeasures in the form of sensitive detectors, evacuation strategies, antidotes, an special chemical suits and accessories are some of the ways in which the military responds to this kind of weapon. Chemical plants throughout the world are monitored to prevent this type of scenario through chemical weapons bans and international inspections help stem the proliferation of biological/chemical warfare.

One of the better-understood methods of counteracting biological warfare is to immunize the population or selected populations that might be exposed to infections and contagions. Anthrax and pox viruses are two threats for which immunization has been carried out by the U.S. Army.

Left: Gas masks are adjusted as part of a simulated chemical attack exercise carried out in the Republic of Korea in October 1998. Right: A pressure washing system is used to decontaminate an M-3 vehicle as part of a chemical training exercise conducted by the U.S. Army in 1998. Spc. D. Shewfelt is dressed in chemical goggles and gloves for carrying out the task. [U.S. DoD news photo by Steve Faulisi, released.]

Drugs that provide temporary immunity or at least some measure of protection are also used, such as drugs to help prevent malaria. The antinerve-agent drug pyridostigmine bromide was given to service members in the Gulf War in 1991 to help protect against soman, a toxic nerve agent. Unfortunately, some of these medications can have mild to severe side effects and there is always a small percentage of individuals who do not tolerate them well or suffer allergic reactions.

Water spray from a fire truck is used as part of a decontamination procedure in a simulated chemical agent exposure held at the Pentagon in May 1998. The exercise was hosted by the *Defense Protective Service* along with the *Office of the Secretary of Defense*. The exercise was carried out to test civilian emergency response procedures. [U.S. DoD news photos by Jeffrey Allen and Renée Sitler, released.]

Protective clothing in the form of masks, suits, gloves, and hoods are intended to provide some protection in contaminated areas or while evacuating from a region suspected of containing hazardous substances.

Chemical sniffers and vapor detectors can provide a certain amount of advance warning of some types of agents, and more sophisticated detection systems are under development all the time.

The U.S. Army has developed a Long Range Biological Standoff Detection System that uses a number of different surveillance technologies to detect biological hazards. It is specifically designed to detect, track, and map large aerosol clouds. The system uses an eye-safe laser transmitter, a 24-inch receiving telescope, and a transferred electron-intensified photodiode detector.

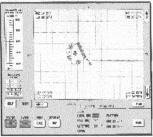

This system is designed to detect biological aerosol clouds at ranges up to about 30-50 km (depending on the system). The cloud configuration is then digitally mapped. [U.S. Army news photos, released.]

Electronic chemical sniffers aren't limited to any one area of surveillance. They have a broad range of applications that include food testing, medical diagnosis, pharmaceuticals monitoring, detection for customs or law enforcement, air quality monitoring, and the development of fragrances and scented products. This system at the *Pacific Northwest National Laboratory* is integrated with a computer that has an algorithmic neural network, a type of artificial intelligence programming. [Courtesy of the Pacific Northwest National Laboratory, 1995 news photo.]

Public safety organizations should be particularly concerned that genetics-targeted biological/chemical weapons could become a reality. DNA-specific agents could be used to single out and harm people of a certain gender, color, or with identifiable genetic abnormalities.

A Nuclear-Biological-Chemical (NBC) reconnaissance vehicle (XM-93) from the 91st Chemical Company on display in October 1990. [U.S. Army 1990 Airborne Corps History Office photo by Randall M. Yackiel, released.]

6.e. Natural Resources Management and Protection

The protection and management of natural resources, whether or not they are food sources, are an important aspect of a nation's cultural and economic survival. Surveillance strategies and technologies are used to monitor commercial harvesting, wildlife ecology, and poaching activities throughout the world. Without these protections, our resources might soon be completely depleted, as has happened in the past in unregulated areas. DNA-monitoring, radio collar tracking, sonar, and optical surveillance are examples of technologies that are used regularly to monitor natural resources and those who seek to abuse them.

6.f. Commercial Products

Bags and Vials

Clean, plain bags and envelopes and glass vials are regularly used for holding evidence gathered at a crime scene. Gloves are worn when collecting the evidence and clean (preferably sterile) tweezers are used to pick up small objects. Bags and envelopes that are preprinted so that they can be consistently labeled are commercially available. Vials usually have caps and come in a variety of sizes.

Evidence Tapes

Evidence tapes are frequently used to 'lift' small items without touching them. Delicate evidence, such as dusted latent fingerprints, are often lifted with tape after being photographed. Clear tape, poly tape, Handi-Lifts™, Lightening Lifts™ pre-cut strips are common commercial products. The best tapes are those that are clear, don't wrinkle easily, and have the least intrusive chemicals in the adhesive itself. If the evidence is to be stored for any length of time, it is best to use acid-free products.

Backing Cards

Blank and printed cards are useful for marking or storing evidence. Lifted prints, stains, and other evidence are often lifted with specialized tapes that may then be adhered to a card and labeled. Since chemical analysis of the stains is often carried out, it is important to use cards that do not interfere with the chemicals and last a long time. Acid-free *archival* cards are preferred. Preprinted cards sometimes simplify the task of labeling and may aid in reducing the number of inadvertent omissions. They may also be numbered for file references to other documents. Numbers may further reduce the chance of misplacing or mislabeling evidence.

Computerized Analysis and Databanking Products

CAL-ID - computerized fingerprint-processing system used to store and retrieve records.

Automated Fingerprint Identification System (AFIS) - search and retrieval from inked fingerprint cards.

Automated Latent Print System (ALPS) - latent print search for matches that generates a candidate list for a qualified examiner to investigate.

7. Problems and Limitations

The main problems associated with the tools of chemical surveillance and chemicals that are used in surveillance are related to health, safety, and contamination of evidence. With the exception of forensic fingerprint dust and a few other common items, most chemicals are used by trained lab technicians and specialists. They should be used carefully, according to instructions, and disposed of in environmentally responsible ways.

Individual chemical and biological hazards are too diverse to discuss in any detail in this section. The surveillance of biological and chemical hazards is undertaken by hundreds of agencies who provide specialized information on problems and limitations. Some of these are listed in the *Resources* section and you are encouraged to contact the many government bodies that provide free printed pamphlets and information on their Websites.

8. Restrictions and Regulations

Because of their potential for harm to biological organisms, the use and distribution of chemicals are heavily regulated. However, due to their diversity and differences in quantities used by individual industries, regulations on specific chemicals cannot be listed here. Some of the more general regulations regarding chemical and biological warfare include

1925 Geneva Protocol - This Protocol expressly prohibited the use of poisonous gas and bacteriological warfare.

1962 Eighteen-Nation Disarmament Committee (ENDC) - This group's recommendations included strategies for disarmament that included provisions for eliminating chemical/biological weapons.

Chemical Weapons Convention - This international convention aided in implementation of the *Convention of the Prohibition of the Development, Production, Stockpiling, and Use of Chemical Weapons and on Their Destruction* and strengthened the 1972 *Biological and Toxin Weapons Convention*. In the 1990s, over a hundred countries signed the CPDPSUCWD.

Armed Forces 1994 Joint Service Agreement - This armed forces agreement established an internal structure and process for developing and validating operational requirements to support chemical/biological defense needs.

Chemical and Biological Weapons Threat Reduction Act - This is a U.S. policy intended to take preventive steps to reducing biochemical threats.

Public Law 102-585 - This established the *Veterans Association's Persian Gulf Registry,* in August 1992, in which veterans are eligible for physical examinations with basic lab studies. This was a response to various persistent health complaints collectively called 'Gulf War Sickness'. or 'Gulf War Syndrome'.

American Board of Criminalistics - This organization has developed a number of written certification tests for forensics-related activities, some of which include contact with chemicals in the course of investigative duties.

Federal Bureau of Investigation (FBI) - The FBI publishes a number of evidence standards that are relevant to chemical/biological surveillance pertaining to crime scene investigation and forensics. The FBI publishes some of this information on its Website and in printed bulletins. www.fbi.gov

9. Implications of Use

Chemicals have many beneficial uses and are inseparable from life itself. They can be used to help solve crimes and to detect and combat disease. Used well, they contribute to the quality of our lives and, in some cases, to our longevity.

When biological or chemical agents are used to inflict harm, through terrorist bombs, poisonings, or chemical warfare, it is important to detect and ideally to prevent the use of chemicals as weapons of destruction. Considerable effort has been made on an international scale to try to reduce and prevent chemical warfare. There is still work that needs to be done and technologies that can aid in controlling their proliferation, but some progress has been made in these areas.

The various U.S. governmental and health organizations have admitted that more could be done to prepare and protect from chemical contamination and biological warfare, but at least the understanding that we can improve our systems is there and hopefully a greater awareness of the relevant issues will improve public support for future efforts.

10. Resources

Inclusion of the following companies does not constitute nor imply an endorsement of their products and services and, conversely, does not imply their endorsement of the contents of this text.

10.a. Organizations

American Academy of Forensic Sciences (AAFS) - A professional organization of physicians, criminalists, toxicologists, anthropologists, engineers and others involved in the application of science to law. www.aafs.org

American Board of Forensic Anthropology, Inc. (ABFA) - This nonprofit board was established in 1977 to provide a program of certification in forensic anthropology that can be used to identify forensic scientists qualified to provide professional services to judicial and executive government bodies. www.csuchico.edu/anth/ABFA

American Chemical Society (ACS) - Provides members with technical and educational information resources, professional development assistance, industry advocacy, awards, and insurance programs. Supports over 30 specialty divisions, including Analytical Chemistry (ANYL) and Chemical Toxicology (TOXI) which have their own publications in addition to the ACS publications. www.acs.org

American Society for Investigative Pathology (ASIP) - A society for biomedical scientists who investigate the mechanisms of disease. The discipline uses a variety of structural, functional, and genetic techniques, applying the research results to the diagnosis and treatment of disease. ASIP supports professional development and education of its members. asip.uthscsa.edu

American Society of Clinical Pathologists (ACSP) - The ACSP is a not-for-profit medical society engaged in educational, scientific, and charitable activities which promote public health and safety through the appropriate application of pathology and laboratory medicine. www.ascp.org

Bonn International Center for Conversion (BICC) - Deals with issues and policies related to dismantling weapons remaining from conflicts or resulting from arms reductions actions and the conversion of military resources for civilian purposes. www.bicc.de/weapons

Center for International Security and Cooperation (CISAC) - A multidisciplinary community within Stanford University's Institute for International Studies dedicated to research and training in issues of national security. www.stanford.edu/group/CISAC

Center for Research on Occupational and Environmental Toxicology (CROET) - CROET is an organization of scientists, educators, and information specialists within the Oregon Health Sciences University in Portland, Oregon who conduct applied research with labor, industry, government, and community members. CROET maintains the *Oregon Chemical Surveillance Project* on the Web, which charts the distribution, use, trends, and possible adverse effects of potentially hazardous substances distributed around the state of Oregon. www.ohsu.edu/croet

Centers for Disease Control and Prevention (CDC) - An agency of the U.S. Department of Health and Human Services which promotes health and quality of life by preventing and controlling disease, injury, and disability. The CDC has almost a dozen specialist centers and provides data and statistics, publications, and funding opportunities. It also performs many administrative functions for its sister agency, the Agency for Toxic Substances and Disease Registry. The site is searchable and includes more than 100 references to documents about chemical warfare, bioterrorism, and the safe disposal of chemical arsenals. www.cdc.gov

Chemical and Biological Arms Control Institute (CBACI) - A nonprofit corporation established to promote arms control and nonproliferation, especially of chemical and biological weapons. CBACI stresses assistance to global industries in implementing the Chemical Weapons Convention. www.cbaci.org

Chemical Science and Technology Laboratory (CSTL) - Within the U.S. Department of Commerce, the CSTL is one of seven NIST measurement and standards laboratories. It works to promote U.S. economic growth by working with industry to develop and apply technology, measurements, and standards. It includes five divisions: analytical chemistry, biotechnology, physical and chemical properties, process measurements, and surface and microanalysis science. www.cstl.nist.gov

Chemical Warfare/Chemical Biological Defense (CW/CBD) Information Analysis Center (CBIAC) - CBIAC collects, reviews, analyzes, appraises, and analyzes information related to CW/CBD and provides a searchable database and some well-chosen links to other sites on a number of topic areas. www.cbiac.apgea.army.mil

Department of Peace Studies, University of Bradford - This U.K. university center provides research and degree programs in peace studies. It includes the Centre for Conflict Resolution and a number of projects including the ongoing improvement of the *Biological and Toxin Weapons Convention* database. www.brad.ac.uk/acad/peace

Federation of American Scientists (FAS) - Since 1990, the Biological and Toxin Weapons Verification Group has endeavored to explore access to biotechnologies in the pursuit of peace and to develop technical and political confidence-building measures to encourage the adoption of strong verification protocols. The group archives a number of working/briefing papers online. www.fas.org/bwc

Harvard Sussex Program (HSP) - A collaboration with the Belfer Center for Science and International Affairs at Harvard University. Research is conducted on chemical- and biological warfare-related technologies and their policy implications. HSP publishes the CBW Conventions Bulletin.

Henry L. Stimson Center - The Center hosts the *Chemical and Biological Weapons Nonproliferation Project* which examines issues associated with biochemical weapons. The project was launched in 1993 and includes problem-solving and information-dissemination on topics such as the implementation of chemical weapons conventions, weapons destruction technologies, and export controls. The project is funded primarily by the Carnegie Corporation and a number of foundations. www.stimson.org/cwc/index.html

International Association for Identification (IAI) - A nonprofit, professional organization for professionals engaged in forensic identification and scientific examination of physical evidence. The IAI provides a range of education and certification programs including latent fingerprint examination, crime scene certification, forensic artist, etc. Descended from the International Association for Criminal Identification, founded in 1915. www.theiai.org

International Association of Forensic Toxicologists (TIAFT) - Established in the early 1960s, this association has over 1400 members worldwide who are engaged in analytical toxicology and related areas. TIAFT promotes and encourages research in forensic toxicology, hosts meetings, and provides case notes, online reviews, and other member resources. www.tiaft.org

International Narcotics Control Board (INCB) - An independent, quasi-judicial control organization for implementation of United Nations Drug Conventions established in 1968. Members are elected by the United Nations Economic and Social Council (ECOSOC). International Drug Conventions are archived on the site along with technical reports and information on the General Assembly. www.incb.org

Lightning Powder Company, Inc. Supplies chemicals, including powders, to crime scene investigators and provides informational articles on their Web site on fingerprint technology. www.redwop.com/

National Climatic Data Center (NCDC) - A national environmental data center operated by the National Oceanic and Atmospheric Administration (NOAA) of the U.S. Department of Commerce. The NCDC maintains the world's largest active archive of weather data. www.ncdc.noaa.gov

National Disaster Medical System (NDMS) - Created by the Federal Government to serve as a national medical system for responding to major mass casualty situations resulting from civilian disasters or overseas conflicts.

National Geophysical Data Center (NGDC) - A national environmental data center operated by the National Oceanic and Atmospheric Administration (NOAA) of the U.S. Department of Commerce. The NGDC is a national repository for geophysical data, providing science data services and information. www.ngdc.noaa.gov

National Ground Intelligence Center (NGIC) - One of the services of the NGIC is a Web-based network that provides information on the estimated consequences of biological, chemical, nuclear, or radiological materials to assist decision-makers in formulating emergency responses.

National Institute of Standards and Technology (NIST) - NIST is an agency of the U.S. Department of Commerce's Technology Administration, established in 1901 as the National Bureau of Standards and renamed in 1988. It aids industry in developing and applying technology, measurements, and standards through four major programs. The Chemical Science and Technology Laboratory is one of seven NIST measurement and standards laboratories. www.nist.gov

National Law Enforcement and Corrections Technology Center (NLECTC) - A program of the National Institute of Justice which includes timely news and Internet information resources, many of which are relevant to forensics and various aspects of surveillance. nlectc.org/inthenews

National Oceanographic Data Center (NODC) - A national environmental data center operated by the National Oceanic and Atmospheric Administration (NOAA) of the U.S. Department of Commerce. The NODC was established in 1961 and serves to acquire, process, preserve, and disseminate oceanographic data. This includes physical, chemical, and biological data, some of which are related to the presence and composition of chemical substances and pollutants. Selected data are available on magnetic media or CD-ROM and, in some cases, over ftp links. www.nodc.noaa.gov

Office of Emergency Preparedness (OEP) - A department of the U.S. Public Health Service.

Organisation for the Prohibition of Chemical Weapons (OPCW) - An international organization dedicated to the destruction and verification of destruction of the world's chemical weapons stockpiles within the mandate of the *Convention on the Prohibition of the Development, Production, Stockpiling and Use of Chemical Weapons and on Their Destruction* ("Chemical Weapons Convention"). www.opcw.org

Southern California Association of Fingerprint Officers (SCAFO) - A nonprofit organization founded in 1837 to support professional identifiers. It now includes members in more than 50 law enforcement agencies. The Web site has an excellent fingerprint bibliography. www.scafo.org

Stockholm International Peace Research Institute (SIPRI) - SIPRI was established in 1964 to contribute to the understanding of the preconditions for a stable peace and for peaceful solutions to international conflicts. As an independent foundation, SIPRI researches from open sources, and makes available, information on weapons development, arms transfers and conduction, military expenditures, and arms limitations and disarmament activities. SIPRI research includes the Chemical and Biological Warfare Project. www.sipri.se

SWGFAST - The Scientific Working Group on Friction Ridge Analysis, Study, and Technology was founded as the result of a 1995 FBI meeting of latent print examiners. SWGFAST (formerly TWGFAST) provides guidelines, discusses analysis methods and protocols, and provides support for the latent print professional community. Information is available through the FBI Laboratory. The Scientific Working Gorup on Materials Analysis (SWGMAT) group is similar. www.fbi.gov

U.S. Armed Forces Joint Nuclear, Biological, and Chemical Defense Board - This board and its subgroups help articulate, coordinate, and expedite tactics, doctrine, training, and equipment. www.chembiodef.navy.mil

See also the U.S. Navy Chemical-Biological Defense. www.cbd.navy.mil

U.S. Army Chemical Defense - The U.S. Army has a number of departments that deal with Biological/Chemical detection as it pertains to defense and warfare, including

> The Edgewood Chemical Biological Center (ECBC), the principal R&D center for chemical/biological defense technology, engineering, and service.

> The Office of the Program Director for Biological Defense Systems (PD Bio), responsible for the development, production, fielding, and logistics support of assigned biological defense systems in the area of detection. This office includes

>> The Biological Integrated Detection System (BIDS)

>> The Long Range Biological Standoff Detection System (LR-BSDS)

>> The Short Range Biological Standoff Detection System (SR-BSDS)

>> The Integrated Biodetection Advanced Technology Demonstration (ATD)

>> The Chemical Biological Mass Spectrometer (CBMS)

U.S. Army Medical Research Institute of Infectious Diseases (USAMRIID) - USAMRIID is the lead biological warfare defense laboratory for the U.S. Department of Defense. It researches drug development and diagnostics for lab and field use. It also formulates strategies, information, procedures, and training programs for medical defense against biological threats. USAMRIID collaborates with the Centers for Disease Control and Prevention, the World Health Organization and a number of academic centers. www.usamriid.army.mil

U.S. Army Soldier & Biological Chemical Command (SBCCOM) - This command provides support in defense research, development, and acquisition; emergency preparedness and response; and safe, secure chemical weapons storage, remediation, and demilitarization. www.sbccom.army.mil

10.b Print Resources

The author has endeavored to read and review as many mentioned resources as possible or to seek the recommendations of colleagues. In a few cases, it was necessary to rely on publishers' descriptions on books that were very recent, or difficult to acquire. It is hoped that the annotations will assist the reader in selecting additional reading.

These annotated listings may include both current and out-of-print books and journals. Those not currently in print are sometimes available in local libraries and second-hand book stores, or through interlibrary loan systems.

Alibeck, Ken; Handelman, Stephen, *Biohazard: The Chilling True Story of the Largest Covert Biological Weapons Program in the World,* Random House, 1999. A Russian Colonel who left in 1992 was involved in the U.S.S.R.'s biological warfare program. He tells his story and reinforces the importance of keeping these weapons out of the hands of terrorists and suggests responses to biological warfare.

Dando, Malcolm, *Biological Warfare in the 21st Century,* London, New York: Brasseys, 1994. The author, a professor of International Security in the Department of Peace Studies at the University of Bradford, describes biochemical weapons developed by several nations, the implications of their existence and use, and how proliferation might be contained and controlled.

A FOA Briefing Book on Chemical Weapons, FOA, S-172 90, Stockholm, Sweden.

Geberth, Vernon, *Practical Homicide Investigation,* Boca Raton, Fl.: CRC Press, 1998.

American Journal of Pathology: Cellular and Molecular Biology of Disease, American Society for Investigative Pathology. This is now available online with searchable back issues. www.amjpathol. org/

Office of Technology Assessment, U.S. Congress, *Proliferation of Weapons of Mass Destruction: Assessing the Risks.* Publication OTA-ISC-559, Washington, D. C., U.S. Government Printing Office, 1993.

Price, Richard M., *The Chemical Weapons Taboo,* Ithaca: Cornell University Press, 1997, 233 pages.

Preston, Richard, *The Cobra Event,* Ballantine Books reprint, 1998, 432 pages. A novel about a biological terror attack on the city of New York.

Report to Congress: *Response to Threats of Terrorist Use of Weapons of Mass Destruction,* Washington, D.C.: U.S. Dept. of Defense, 31 January 1997.

Tenet, George J., *Statement by the Director of Central Intelligence Before the Senate Select Committee on Intelligence on the Worldwide Threat in 2000: Global Realities of Our National Security,* www.cia.gov, 2 February 2000. This statement describes a variety of threats, nation by nation, some of which include chemical/biological warfare. Also related to the subject in this report is another from the CIA Nonproliferation Center, *Unclassified Report to Congress on the Acquisition of Technology Relating to Weapons of Mass Destruction and Advanced Conventional Munitions, January through June of 1999.*

Zatchuk, R., Editor, *Textbook of Military Medicine,* Washington, D.C.: U.S. Dept. of Army, Surgeon General, and the Borden Institute, 1997.

Articles

Carnes, S.; Watson, A., Disposing of the U.S. chemical weapons stockpile, *Journal of the American Medical Association,* 1989, V.262, pp. 653-659.

Centers for Disease Control and Prevention, CDC Recommendations for Civilian Communities Near Chemical Weapons Depots, *60 Federal Register,* 1995, pp. 33307-33318.

Franz, D.; Jahrling, P.; Friedlander, A. et al., Clinical Recognition and Management of Patients Exposed to Biological Warfare Agents, *Journal of the American Medical Association,* 1997, V.278, pp. 399-411.

Inglesby, T.; Henderson, D.; Bartlett, J. et al., Anthrax as a biological weapon, *Journal of the American Medical Association,* 1999, V.281, pp. 1735-1745.

Ibid., Smallpox as a biological weapon, pp. 2127-2137.

Kent, Terry, Recent Research on Superflue, Vacuum Metal Deposition and Fluorescence Examination, *PSDB,* July 1990, pp. 1-4.

Kern M., Kaufmann, A., Principles for emergency response to bioterrorism, *Ann. Emergency Medicine,* 1999, V.34, pp. 177-182.

Lee, Henry C.; Gaensslen, R. E., Cyanoacrylate Fuming, *Identification News,* June 1984, pp. 8-14.

Macintyre, Anthony G. et al., Weapons of Mass Destruction Events With Contaminated Casualties: Effective Planning for Health Care Facilities, *Journal of the American Medical Association,* Special Communication, 12 January 2000. Includes background information, strategies, and references.

National Institute of Justice, Certification of DNA and Other Forensic Specialists, *National Institute of Justice Update,* September 1995, 3 pages.

Noll, G.; Hildebrand, M.; Yvorra, J., Personal Protective Clothing and Equipment, in Hazardous Materials, Stillwater: Fire Protection Publications, Oklahoma State University, 1995, pp. 285-322.

Nozaki, H.; Aikawa, N.; Shinozawa, Y. et al., Sarin gas poisoning in the Tokyo subway, *Lancet,* 1995, V.345, pp. 980-981.

Nozaki, H.; Aikawa, N.; Shinozawa, Y., Secondary exposure of medical staff to sarin vapor in the emergency room, *Intensive Care Medicine,* 1995, V.21, pp. 1032-1035.

Sidell F., What to do in case of the unthinkable chemical warfare attack or accident, *Postgraduate Medicine,* 1990, V.88, pp. 70-84.

Sokes, J.; Banderet, L., Psychological aspects of chemical defense and warfare, *Military Psychology,* 1997, V.9, pp. 395-415.

Sullivan, J.; Krieger, G., Hazardous Materials Toxicology, Baltimore, Technical Working Group on Crime Scene Investigation (TWGCSI), Crime Scene Investigation: A Guide for Law Enforcement, *National Institute of Justice,* Research Report series, January 2000, 49 pages.

Thomas, Evan; Hirsh, Michael, The Future of Terror, *www.newsweek.com,* 10 January 2000. A discussion of the some of the possible organizational biochemical threats to America.

Torok, T. J.; Tauxe, R.V.; Wise, R.P. et al., Large community outbreak of salmonellosis caused by intentional contamination of restaurant salad bars, *Journal of the American Medical Association,* 1997, V.278, pp. 389-395.

Walsh, Anne-Marie, Irish Invention Spots Fakes by 'Fingerprints', *London Sunday Times Online,* 15 Nov. 1998.

Wheeler, Michael D., Evidence Analysis: Photonics Cracks Cases from Tiny Clues, *Photonics Spectra,* 1998, V.32(11), p. 100.

Journals

CBW Conventions Bulletin, formerly the *Chemical Weapons Convention Bulletin,* quarterly journal of the Harvard Sussex Program. Provides news, background, and opinion on chemical/biological warfare issues.

Drug Discovery Today, a professional journal of current news and reviews for the drug discovery community.

Forensic Science Communications, a quarterly journal from the FBI Laboratory personnel (superseding the Crime Laboratory Digest). Previous issues dating back to April 1999 are now available online. Articles are searchable.

Modern Drug Discovery, a practical professional journal for chemists and scientists in drug discovery with articles on medicinal chemistry and molecular biology. Published by the American Chemical Society.

Peace Studies News, published three times a year by the Department of Peace Studies, University of Bradford, along with a number of departmental and staff publications.

SIPRI Chemical & Biological Warfare Studies, a series edited and published by the Stockholm International Peace Research Institute.

Strategies to Protect the Health of Deployed U.S. Forces, a series of reports by the National Academies' Institute of Medicine and National Research Council, published by the National Academy Press in 1999 and 2000. They discuss the technical aspects of identifying and assessing hazards, improving surveillance activities, and reducing the risk of exposure. The reports are intended to aid in developing long-term strategies and policies.

Treatment of Chemical Agent Casualties and Conventional Military Chemical Injuries, Washington, D.C., U.S. Government Printing Office, 1990.

Xenobiotica, a professional journal which covers general xenobiochemistry, including the metabolism and disposition of drugs and environmental chemicals in animals, plants, and microorganisms and related methodologies and toxicology.

10.c. Conferences and Workshops

Many of these conferences are annual events that are held at approximately the same time each year, so even if the conference listings are outdated, they can still help you determine the frequency and sometimes the time of year of upcoming events. It is very common for international conferences to be held in a different city each year, so contact the organizers for current locations.

Many of these organizations describe the upcoming conferences on the Web and may also archive conference proceedings for purchase or free download.

The following conferences are organized according to the calendar month in which they are usually held.

Bioterrorism in the United States: Calibrating the Threat, CBACI conference held in conjunction with the Centers for Disease Control and Prevention.

Annual Western Spectroscopy Association Conference, annual conference since the early 1950s.

DoD Medical Initiatives Conference and Exhibition: Weapons of Mass Destruction, Department of Defense conference.

ASCP/CAP Spring Meeting, the national meeting of the American Society of Clinical Pathologists.

Drug Discovery Technologies, annual European conference for pharmaceutical and biotechnology drug discovery researchers.

Experimental Biology, annual meeting of the American Society for Investigative Pathology (ASIP).

Analytica, The International Trade Fair and Conference for Analysis, Biotechnology, Diagnostics, and Laboratory Technology, annual since the early 1980s.

California State Division International Association for Identification, Annual Training Seminar since about 1926.

California Conference of Arson Investigators, a large national association the seeks to transcend the gap between public and private sectors. The CCAI condust semi-annual training seminars, regional roundtable meetings, and other educational activities. arson.org/

NDA Waste Characterization Conference. A forum for radiological and hazardous waste characterization issues.

Annual Worldwide Chemical Conference, Armed Forces conference.

International Association for Identification, annual professional education conference on forensic identification and investigation.

Advanced Death Investigation Conference.

Drug Discovery Technology, Annual Exposition and Symposium of pharmaceutical and biotechnology drug discovery researchers, since mid-1990s. Topics include automation and robotics, assay development and screening, miniaturization and chip-based technologies, integrating chemical, biological and genomic data, and more. www.drugdisc.com/

International Mass Spectrometry Conference, international conferences that are held every three years.

Forensic Science - Challenges for the New Millennium, sponsored by the European Academy of Forensic Science (EAFS)

Bloodstain Pattern Analysis Workshop, Metropolitan Police Institute.

Investidation and Forensic Science Technologies, a Photonics East conference. Proceedings are archived by S.P.I.E. S.P.I.E. sponsors a number of optics, remote sensing, and forensic science events. spie.org/

10.d. Online Sites

The following are interesting Web sites relevant to this chapter. The author has tried to limit the listings to links that are stable and likely to remain so for a while. However, since Web sites do sometimes change, keywords in the descriptions below can help you relocate them with a search engine. Sites are moved more often than they are deleted.

Another suggestion, if the site has disappeared, is to go to the upper level of the domain name. Sometimes the site manager has simply changed the name of the file of interest. For example, if you cannot locate www.goodsite.com/science/uv.html *try going to* www.goodsite. com/science/ *or* www.goodsite.com/ *to see if there is a new link to the page. It could be that the filename* uv.html *was changed to* ultraviolet.html, *for example.*

Bradford-SIPRI Chemical and Biological Warfare Project - As an international foundation, SIPRI carries out research and disseminates information related to disarmament and the foundation of peace. The CBW site includes research and documentation, educational modules, publications, and general information about the project. It is being strengthened with support from Bradford University which further hosts the Biological and Toxin Weapons Convention (BTWC) database. projects.sipri.se/cbw/cbw-mainpage.html www.brad.ac.uk/acad/sbtwc

Bureau of Nonproliferation - This is a bureau of the U.S. Department of State that has Web pages devoted to information on Weapons of Mass Destruction, including biological and chemical weapons. It provides links to congressional testimony and briefings, fact sheets, and treaties. www.state.gov/www/global/arms/bureau_np/wmd_np.html

Oregon Chemical Surveillance Project. The Center for Research on Occupational and Environmental Toxicology (CROET) has established a site based on eight databases that describe hazardous chemicals used and stored in Oregon, the extent of releases, and associated human exposures. It also includes information on adverse health effects and displays the spatial distribution of the chemicals on a map. www.ohsu.edu/croet/database.html

Organization for the Prohibition of Chemical Weapons (OPCW) - OPCW maintains the Chemical Weapons Convention Website with information about member states, activities, courses, and the text of the Chemical Weapons Convention in several data formats. The CWC seeks progress toward disarmament, including the elimination of weapons of mass destruction, including asphyxiating, poisonous or other gases, and bacteriological methods of warfare, according to the Geneva protocol of 1925. The site also includes a fact file on chemical weapons. www.opcw.nl

United States Institute of Peace (USIP) - USIP is an independent, nonpartisan federal institution created in 1984 and funded by the U.S. Congress to strengthen the nation's capacity to promote the peaceful resolution of international conflict. The organization sponsors conferences, library services, publications, educational activities, and grants. Publications can be searched on the Web (e.g., discussions of biological, chemical, and nuclear weapons in the Iraq Crisis). Chemical weapons are discussed in White Papers, funded project reports, and Special Commission reports. www.usip.org/

10.e. Media Resources

Note that television programs are often available on VHS tape after broadcast. Contact the broadcaster for information.

Clouds of Death, This History Channel *History Undercover* program traces the history of biochemical warfare from medieval times to the use of sarin gas in World War I and the Tokyo subway incident. Includes interviews and documentary footage. Restricted to the U.S. and Canadian markets. It is available on VHS, 50 minutes.

Coming Home: Agent Orange/Gulf War, is part of the *20th Century with Mike Wallace* series on the History Channel. It looks at the aftermath of the after effects of servicemember's possible exposure to toxic agents in the execution of their duties. The case of the Zumwalt family is presented. Restricted to the U.S. and Canadian markets. It is available on VHS, 50 minutes.

Declassified: Human Experimentation, a History Channel *History Undercover* series program traces the history of top-secret experiments conducted by the U.S. military on human subjects, subjecting them to radiation and poisons without their direct knowledge or consent. Restricted to the U.S. and Canadian markets. It is available on VHS, 50 minutes.

Insidious Killers: Chemical and Biological Weapons, is part of the *20th Century with Mike Wallace* series on the History Channel. Traces the development of the proliferation of weapons from World War I to the terrorist attack in Tokyo's subways with sarin gas. Restricted to the U.S. and Canadian markets. It is available on VHS, 50 minutes.

Into the Fire: Arson: Clues in the Ashes, explores fires in New York City, nearly half of which appear to be deliberately set. Shows fire marshalls looking for clues in the ashes. A Discovery Channel TV program aired in spring 2000.

Into the Fire: From the Ashes, explores insurance torchings, arson murders, and other crimes in which arson appears to have been used as a murder weapon and/or to hide clues of a crime. Shows arson investigation and undercover work. A Discovery Channel TV program aired in winter 1999/2000.

11. Glossary

It is more difficult to compile a succinct glossary for this chapter than for most of the others in this text, as it would have to include not only a lot of chemical and surveillance terms, but also many medical and industrial terms. This short selection gives a flavor of some of the types of terms and concepts that are common to chemical surveillance activities and chemical agents. Terms for Biometrics and Genetics Surveillance, aspects of biochemical surveillance, are listed in the chapters specific to those topics.

Titles, product names, organizations, and specific military designations are capitalized; common generic and colloquial terms and phrases are not.

ABC	atomic, biological, and chemical
ABG	arterial blood gas
AC	hydrogen cyanide
ACAA	automatic chemical-agent alarm
ACADA	automatic chemical-agent detector
accessible form	an undiluted agent that has not been decontaminated or neutralized, but which might be removed for unauthorized purposes.
ACPG	advanced chemical-protective garment
ACPLA	agent-containing particle, per liter of air

ACPM	aircrew protective mask
Action Level	a concentration designated in Title 29, Code of Federal Regulations for a specific substance, calculated as an 8-hour time-weighted average, which initiates certain required activities such as exposure monitoring and medical surveillance.
acute toxicity	immediate toxicity associated with mortality. It should not be confused with acute exposure.
ADI	acceptable daily intake - the estimate of a dose due to exposure to a toxicant that is unlikely to be harmful if continued exposure occurs over a lifetime
ADS	area detection system
adverse effect	a biochemical change, functional impairment, or pathological lesion that impairs performance and reduces the ability of the organism to respond to challenges
adverse effect level	exposure level at which there are statistically or biologically significant increases in frequency or severity of deleterious effects between the exposed population and its appropriate control group
aerosol	airborne solid or liquids such as dusts, fumes, mists, smokes, or fogs
AERP	aircrew eye/respiratory protection
AIC	acceptable intake for chronic exposure - the estimate of health effects from chronic exposure to a chemical. It is similar to a Reference Dose (RfD).
AIS	acceptable intake for subchronic exposure - the estimate of health effects from sub-chronic exposure to a chemical. It is similar to a Subchronic Reference Dose.
ALAD	automatic liquid-agent detector
AMAD	automatic mustard-agent detector
antidote	a substance or agent that inhibits or counteracts the deleterious effects of a poison
assay	to analyze for one or more specific components
BW	biological warfare
CWC	Chemical Weapons Convention
FDI	forensic digital imaging. Digital scanning, enhancement, and processing of forensic materials to contribute characteristics information and knowledge not obtainable by other means.
GB	sarin, a nerve agent
GD	soman, a nerve agent
GLC	gas/liquid chromatography
GPFU	gas-particulate filter unit
HAZMATs	hazardous materials
incendiary	constructed or consisting of materials intended or predisposed to ignite and/or burn
PPE	personal protective equipment
vesicant	a blistering agent such as phosgene oxime and mustard compounds, that may blister or burn sensitive tissues such as eyes and lungs.
WMD	weapons of mass destruction
zootoxin	the toxin or poison of an animal, such as venom from a spider, wasp, or snake

Biochemical Surveillance

13

Biometrics

1. Introduction

Humans have many unique attributes that provide biometric measures. Some, like finger-prints, toeprints, and basic voice characteristics, remain the same throughout our lives. Others may change with surgery, dyes, weight gain or loss, or aging. Those that remain constant are of particular interest for use as biometric identifiers. Voice characteristics can be charted and analyzed through computer processing, the retinas in the back of our eyes can be scanned for their unique patterns, as can the irises at the front of our eyes. As we mature, our faces take on recognizable features and proportions that remain reasonably constant. Each of these sets of features can be measured, stored, compared, and retrieved using computer technologies. With improvements in image processing, databases, and transmission through computer networks, it becomes possible to exploit these characteristics in surveillance applications.

Biometric surveillance is the detection, identification, and tracking of individuals based upon unique physical characteristics or attributes such as proportions, size, color, fingerprints, iris

A Royal Canadian Mounted Police (RCMP) officer examines a gun for fingerprints while a large handprint adorns the wall behind his desk. [Canadian National Archives, copyright expired.]

patterns, or voice patterns. Biometrics is an emerging technology that is rapidly commercializing because of advancements in sensing technologies and computer processing.

Biometric surveillance is closely related to chemical surveillance. When investigators dust for fingerprints, they take advantage of the chemical qualities of the oils and sweat in the fingerprints to hold onto the dust. The dust pattern then reveals the latent chemical traces or "prints". Chemical approaches to biometric identification, are gradually being extended and, in some cases, superseded by visual surveillance and electronic scanning and processing systems—it's sometimes easier to use a digital sensor to take an image of a print than to dust or stain, but chemicals will continue to be important even as the digital biometrics industry expands.

As an example of digital biometrics procedures that are changing with the technology, fingerprint identification cards used to be created by rolling a finger in ink and pushing the finger against a card to set the ink on the paper. Digital systems, in which a camera or graphics scanner takes a scan of the fingerprint and processes it directly into a database, are now available. Such systems scan a whole hand in the time it takes to create a chemical print of one finger. Eventually a special sensor, like an ultraviolet scanner, may locate and process prints in realtime, without dust, sending the data immediately to a central law enforcement database and transmitting back a picture of the person who left the prints.

This chapter describes some of the most common biometrics technologies used in surveillance, and advances in computer technology in recent years that have enhanced their effectiveness.

2. Kinds and Variations

The most common biometric technologies for detection and identification include

finger, thumb, and hand prints - Widely used in law enforcement and now for some kinds of banking transactions, as verification systems. They are also used in certain federal ID cards, including Permanent Residence cards (more commonly known as "green cards"), and entry and exit systems, in some prisons.

toe- and footprints - Used in some hospitals to identify and record newborn babies and for some law enforcement purposes. Toeprints are sometimes substituted for fingerprints for subjects who may not have hands as a result of accidents or birth defects.

iris scans - Visual scans of the colored portion of a person's eye. By 2000, iris scanners could recognize an iris at a distance of up to two or three feet from the scanning imager. A person's iris pattern remains constant, for the most part, from infancy to old age. Iris scans are sometimes used for entry access systems.

retina scans - Visual scans of the structures at the back of the eye generally require that the subject stand near the camera, usually with the eye resting against an eye cup. These are not as convenient as iris scans but are sometimes used for entry access systems.

voice prints - Assessments of the complex patterns and characteristics unique to each person's voice can identify a person. While voices create unique patterns, the variations that are possible within an individual's modes of speaking may change slightly. Excitement tends to make voices higher, illnesses may make voices scratchier or softer. While recent improvements in voice print processing have made them more practical and reliable, voice print data is sometimes combined with other biometrics or visual identification systems for confirmation. Voice analysis may be used for identification and also for evaluation of a person's emotional characteristics and is incorporated into some lie detector systems.

face prints - Assessments of the complex proportions, colors, and individual features can visually identify a face. Systems that rely on thermography — scanning to provide infrared 'signatures' rather than a visible-light signature will map the face according to its warm and cool spots rather than it's visual features. Thermographic data can be combined with visual data to more specifically distinguish two similar faces. Face prints are used for entry access systems and are beginning to be used for video tracking systems and transactional verification systems. (Thermographic systems tend to be more expensive than visible light scanning systems. See the Infrared Surveillance chapter for more information on infrared imaging.)

DNA profiles - Unique genetic profiles of a person's biological makeup contained in live cells can indicate gender, familial relationships, and race, and can sometimes provide medical estimates and other intimate information. (This is a broad and important topic that it is discussed in more detail in the Genetics Surveillance chapter rather than being discussed in details in this chapter.)

other measures - Other physiological measures are sometimes used for detection, such as height, weight, patterns of movement, and general physiological proportions. Heart rate, respiration rate, and perspiration measures are included, as is handwriting analysis. Physiological measurements are more difficult to acquire than fingerprints and eye scans, so they tend to be used in more specialized scientific applications, although some retail outlets now use body mapping to tailor clothes to the proportions of an individual, using laser body contour scans. These measures are sometimes also used to study movement in dancers and athletes and to map a person's form for the fitting of prosthetic devices (or their teeth for creating custom crowns).

A biometric print is usually created either by 1) contact, or 2) scanning (from a few inches or few feet away). Scanning may include visual or infrared scans, X-ray scans, or measures of electrical or magnetic conductivity, temperature or moisture (e.g., perspiration). The most common biometric technologies used in surveillance include

contact prints - Traditionally, fingerprints have been acquired by contact. The finger is inked and pressed against a surface. The same method can be used with thumb, hand, toe, and footprints. It can even be done, to some extent, with facial prints, although the process would be objectionable for most practical applications.

image scanning - This method is becoming more practical and reliable due to advances in computer technology. By scanning a retina, iris, face, finger, toe, or body contours, it is possible to electronically measure and analyze the components that make up the person's unique features and compare the data with known parameters or individual listings in a database. The majority of new products use some type of image scanning technology.

physiological indicators - This encompasses a wide variety of biological and chemical indicators that occur as a result of a person's individual physiological characteristics and responses. Lie detector tests, for example, measure a person's stress levels by assessing heart rate, breathing, galvanic skin response, and sometimes other measures. Since physiological indicators are highly variable with a person's mood and health and since their measurement requires special instruments and trained experts to interpret the results, they are not widely used for basic identification systems. However, traditional lie detector tests are used at times in law enforcement to confirm confessions or gather evidence. More recently developed *voice stress analyzer* machines can be used to assess emotional states and have been built into some software programs to provide indicators as to whether a person might be lying. While all lie detector technologies are only somewhat reliable and require baseline data and expert interpretation, business negotiators have expressed interest in using computerized lie detector products for corporate surveillance.

3. Context

The most common contexts in which biometrics are used for surveillance are for criminal and suspect identification (fingerprints and mug shots), entry access, and transaction verification. Thus, biometrics are primarily used in law enforcement, in the workplace, and in retail and financial centers (they are applied in more specialized laboratory research as well, as in anthropology, kinesthetics, and medicine). Biometrics are used in the following contexts:

law enforcement - Biometrics can help identify suspects or individuals who may have a warrant out for their arrest. They may help determine if a person was present at a crime scene or had contact with key items such as guns, vehicles, or valuables. They may help locate and identify kidnap victims, hostages, runaways, and victims of suicide or homicide. Forensics professionals use biometrics to reconstruct crime scenes and examine documents for fraud and forgery.

national security - Biometrics are used by customs and immigration officials and intelligence agents to identify refugees, illegal aliens, foreign spies, and potential terrorists. They are also used to facilitate border crossings and to verify authorized entrants. Photos are used on passports as a basic form of visual biometric information. A fingerprint on a Permanent Resident card further helps identify a card-holder.

entry security - Biometrics are increasingly used to control access to restricted or hazardous areas, to special high security areas (e.g., for athletes at major sports events), to members-only areas, and to workplace buildings. They are also used by entertainment services to expedite access for season ticket holders.

transaction verification - Biometrics may be used to double-check the identity of a person before granting permission for them to make financial transactions, such as bank machine withdrawals, or purchases at a retail store with a check or credit card. This may infringe on privacy because an individual has no control over what the company may try to do with the personal information, especially if it is stored in computer memory.

4. Origins and Evolution

Fingerprints and mug shots have been used for more than 100 years in organized law enforcement. In fact, fingerprinting may have been used for hundreds of years in Asia and was perhaps even used in ancient Egypt. It is still one of the most important verification tools in the justice system today. Mug shots have been in regular use since about the time of the Civil War.

Mug Shots

Since photography is a relatively new technology, the practice of taking mug shots is not as old as the practice of taking fingerprints. However, mug shots appear to have been introduced to western law enforcement as a regular practice a little earlier than fingerprints and 'wanted posters' with hand-drawn mug shots were being created by portrait artists long before photographic mug shots were available.

As has been described in the introductory chapter to this book, Allan Pinkerton, the founder of *Pinkerton's Detective Agency*, was the first to develop an extensive collection of mug shots in America, based upon drawings, descriptions, posters, newspaper clippings, and any other sources the agency could find that would aid in identifying subjects of investigations. Pinkerton was active in President Lincoln's secret service during the Civil War and returned to his private practice after the War, continuing to extend and expand his manual database of criminal suspects. Soon, many law enforcement agents were actively adopting the practice of using mug shots and, by the middle of the twentieth century, the practice was widespread and routine.

The Origins of Prints as Identification Marks

In 1880, an influential article appeared in *Nature* magazine, a reputable scientific journal. The article titled "On the Skin-Furrows of the Hand", written by Henry Faulds, describes how he noted fingertip impressions in pottery that led him to examine the fingers of humans and monkeys. Faulds reported on how he began collecting fingerprints from people of different nationalities and of both genders, looking for contrasts and similarities. The article also described how to take a basic print. When he was in China, he heard that criminals there had "from early times" been made to provide impressions of their fingers "just as we make ours yield their photographs". He reported that "Egyptians caused their criminals to seal their confessions with their thumbnails, just as the Japanese do now...". He describes his initial observations as a "rich anthropological mine for patient observers". Faulds article was a catalyst for further discovery and many modern practices.

Faulds was not the only observer intrigued by fingerprints. W. J. Herschel responded to Faulds' article by publishing a response in the 25 November 1880 issue of Nature magazine describing how he had been taking 'sign-manuals' comprising 'finger-marks' for over 20 years and introduced them "for practical purposes in several ways in India with marked benefit". Herschel reports having introduced fingerprints as a subsitute for signatures for pensioners lacking in vitality and for prisoners entering jail. Reported Herschel, "The ease with which the signature is taken and the hopelessness of either personation or repudiation are so great that I sincerely believe that the adoption of the practice in places and professions where such kinds of fraud are rife is a substantial benefit to morality". Herschel further pointed out that fingerprints remained unchanged over time.

It took a while for fingerprinting concepts and methods to become firmly established in law enforcement, but by the 1920s, the *Federal Bureau of Investigation* was collecting fingerprints, and by the 1930s there were good references describing the art and science of taking prints, as the practice rapidly spread.

Polygraphs

Encouraging people to divulge secrets, or coercing people into telling the truth have long been traditions in law enforcement and warfare. In earlier times, beatings and torture were sometimes used to extract information. By the early 1900s, inventors were looking for more humane and effective ways to harness new technologies to determine the truthfulness of people's assertions.

People cannot consciously control all their subconscious biological functions. Thus, different aspects of a person's physiology have been studied to see what they reveal separately from a person's words, including breathing, heartbeat, and perspiration. Sensors have been developed to detect and measure these phenomena. The polygraph machine, which uses physiological sensors to measure biological changes, and creates a visual readout in the form of a graph was first developed by John Larson and Leonard Keeler.

Using the readouts in a structured situation, with "baseline" questions asked at the beginning to establish a standard against which to compare further readings, a person's reaction to the target questions can be evaluated by a qualified examiner. These can then be viewed on a traditional paper-based polygraph or on a digitizing system that displays (and sometimes interprets) the data on a computer screen.

Modern Fingerprint Methods

Fingerprinting is one of the most common biometric identification systems. For optimal effectiveness, dark, clear prints are desired and a variety of inks (sometimes using conventional stamp pads) are used. Ground carbon particles in a fluid medium can produce a clear print, but are somewhat objectionable to the person providing the print.

Fingerprints may then be stored in a physical card catalog or scanned into a digital database for future reference.

Dusting for fingerprints is an art that requires practice, patience, special fine dusts, high quality brushes, and skill. Inventors have looked for ways to simplify the process and to record prints without dust and, gradually, as technology improves, have had some success. By the late 1970s, scientists were exploring lasers as a means of finding latent fingerprints and, by the 1980s, had discovered ways to reveal 'difficult' prints—prints on paper and other materials that could not be readily revealed with older methods.

Many people object to the mess and inconvenience of having ink on their hands when providing finger or handprints. Finding a way to get clear fingerprints without the staining and odor associated with traditional carbon-based inks was considered desirable. Some low-carbon systems had been developed over time, but these were not yet truly 'inkless' systems. However, new methods of using light scattering were developed in the late 1970s and early 1980s, and begin to supersede traditional methods in the late 1980s and 1990s.

As video technologies improved, the quality and resolution of charge-coupled devices (CCDs) made it practical to devise fingerprint scanners based on the reflectance and/or scattering of light from a finger pressed against an image sensor. Some systems even went a step further than traditional prints by incorporating the unique pressure patterns of a person's finger in addition to the ridges and groove patterns.

The 1980s - The Development and Testing of New Technologies

In the late 1970s and early 1980s, new miniature electronics and microcomputer technologies enabled many new inventions to be developed and some of these technologies were applied to the development of innovative biometric detection devices and database systems.

Microminiaturization and solid state electronics made many new technologies possible in the late 1970s and some were applied to biometric sensing.

- In 1977, Robert Hill of Washington State, submitted a patent for identifying individuals by the vascular patterns inside their retinas. The system was based on the idea of scanning the eye in a selected pattern and analyzing the light that was reflected off the retina, thus revealing the patterns of blood vessels at the back of the eye, which are unique to each person and to each eye.

- In 1981, Feix and Ruell of *Siemens Corporation* submitted a patent for a combination voice signature and facial recognition system in which a key word spoken into a microphone could be compared to a pattern matcher in a database to form a 'familiarity score' aiding in identification.

- In the early 1980s, the *Japanese National Police Agency* established a computerized retrieval system for fingerprint files that has been used as a model by a number of American police agencies. Various state agencies in the United States were implementing similar systems.

- By the mid-1980s, some law enforcement agencies were beginning to switch over from manual fingerprint systems to automated fingerprint systems that scanned or otherwise electronically imaged or recognized a fingerprint. Palmprint systems started to become practical around the same time. Computerized systems could be linked to a central electronic database.

- In 1985, Leonard Flom and Aran Safir submitted a patent application for an iris identification system in which an image of the iris and the pupil could be compared against stored image information. This was said to have several advantages over retinal identification—not only was the iris pattern not as dependent on pupil size (a constricted pupil could obscure the retina), but it was not necessary to put the eye as close to the scanner as in retinal scans.

- In 1988, Laurence Lambert, in association with the U.S. Secretary of the Air (Washington, D.C.) submitted a patent for an autonomous face recognition machine. This was a system capable of locating faces in video scenes with mixed/random content and of identifying those faces. This automated system could function without human input or intervention and was not disrupted by scale, brightness, or general discrepancies in focus. The system could be interfaced with motion detection features to improve recognition time to within about a minute.

In the early 1980s through the 1990s, increasingly sophisticated ways of processing latent fingerprints have been described in the scientific literature. In the past, prints on paper or other rough or absorbent materials were difficult or impossible to see clearly. Now it was possible to reveal prints in new ways as magnetic and fluorescing powders were gradually developed.

The 1990s - Proliferation of Biometrics Systems

Biometrics systems became practical and less expensive in the 1990s and companies began heavily promoting the products to government agencies, financial institutions, corporations, and retailers.

The 1990s is characterized by the improvement of many of basic technologies developed in the previous two decades and by commercialization. It was a time when organizations such as the FBI, the NIST, and others, began to develop standards for fingerprint digitization and compression. Some examples of improvements and practical applications include the following:

- In 1991, John Daugman of *Iri Scan Incorporated* submitted a patent for an iris analysis system based on using the eye as an 'optical fingerprint', thus introducing biometric identification systems that were more resistant to forgery than traditional systems. This system improved on previous systems and provided greater automation.

- In 1993, the *Immigration and Naturalization Services* (INS) began establishing a *Passenger Accelerated Service System* (INSPASS). This program used biometric technology to inspect frequent travelers who carry a special card after enrolling in the program.

- In the mid-1990s, traditional and electronic identification programs were upgraded or enhanced by the U.S. *Department of Justice* (DoJ). These included the INS *Automated Fingerprint Identification System* and the *Computer-Aided Detection and Reporting*

Enhancement System, initially installed in Texas and California. These upgrades were put into effect to improve remote-sensing capabilities and electronic surveillance of the southern border.

- In the mid-1990s, a number of entertainment complexes started using hand scanners to expedite admissions and to monitor employees and season pass holders.

- At a hospital in Temple, Texas, a new system was put in place so that physicians could access the hospital's electronic medical record system and other computer applications by using a biometric fingerprint scanner. Users indicated that the biometric access was more convenient than remembering a password. The system also provided tighter security and accountability, since the identity of the person typing a password could not be traced as readily to a specific individual as a biometric identifier.

- Iris scanners were also used at high-level sports events, like the Nagano, Japan, Winter Olympics, where they controlled access to the rifles used in the biathlon competition.

- Throughout the 1990s, U.S. *Customs and Immigration Services* continued to explore the use of a number of biometrics systems to expedite border access, including voice recognition and hand scanning. Several systems have been tested and some implemented for these purposes. In some cases, systems are used together, such as a hand-scan verified by a voice print.

- In 1996, Penine Katz from *MCI Communications* filed a patent for a telephone-based personnel tracking system that could detect personal identification codes and automatic number identification on incoming telephone calls. The system is designed to create a report of incoming calls that includes the location of the calling telephone and the person making the call. One of the stated objects of the invention was to use voice recognition to recognize a caller's voice and spoken commands and to match the voice to the caller's identity through voice print matches. The intention of the system was to monitor and record the arrival and departure of field-based employees, but clearly the technology is generic enough to be used in many telemarketing and technical support applications.

- In the U.K., some automated teller machines (ATMs) have been equipped with iris scanners. These are generally less objectionable to users than retinal scans, as it is not necessary to put the eye against an eye cup in some iris systems. Some of the newer systems can register an iris up to a couple of feet. However, providing such personal information to financial institutions, without having any control over how long they keep it or how they may use it raises some serious issues about privacy.

- In 1998, the U.S. *Immigration and Naturalization Service* (INS) installed hand scanners in a number of airports in the United States and Canada. These systems scan a hand and match it with entries in an INS database, with the intention that it help expedite passage through the checkpoint.

- In June 1998, the *National Highway Traffic Safety Administration* (NHTSA) issued proposed regulations to require all states to include social security numbers and security features like biometric identification on drivers' licenses by a specified deadline.

The Motivation for Biometrics Surveillance

Fraud was one of the biggest reasons why the demand for biometrics identification systems increased. The fact that the technology was becoming more powerful, more reliable, and less expensive was also a motivating factor.

Check fraud losses to retailers and financial institutions were estimated to be around $10 billion a year by the mid-1990s. Welfare and insurance fraud was also prevalent, particularly medical insurance fraud. One of the ways in which people were defrauding welfare and insurance providers was by using multiple identities. The use of advanced copier machines to counterfeit negotiable notes (checks, food stamps, coupons, etc.) was also increasing. Discovery and conviction rates for these crimes had traditionally been low.

Signature verification has traditionally been used to verify contracts, documents, and negotiable items, but signature verification and handwriting analysis are hit-and-miss. Not everyone is good at interpreting or recognizing signatures, not everyone writes his or her signature consistently, and many cashiers simply fail to check the signatures on checks and credit cards.

Biometrics has been seen as a way to deter fraud, as well as a way to detect it once it has occurred, and as a means to apprehend the perpetrators and prove fraud in a courtroom.

By the late 1990s, a number of banks began using fingerprint identification systems for noncustomers depositing or cashing checks. Again, this brings up questions about the erosion of privacy.

A further reason for promoting biometrics identification was the increasing prevalence of electronic transactions. Through computer networks, it was possible to create contracts and enact financial transactions, but it was difficult to authenticate the source and signer of these documents. Because of the potential for profit that is inherent in conducting transactions very quickly, business agents wanted ways to bypass signatures that were sent by courier or postal service. Developers and e-business proponents saw biometrics as a solution to these problems.

One final motivation for the promotion of biometrics technologies was simple entrepreneurial opportunity. If new products became available and sales representatives could convince companies to buy them, there might be opportunities to make a profit.

Concerns About Privacy and the Potential Abuse of Data

The five biggest concerns of privacy advocates are 1) loss of freedom, 2) loss of anonymity, 3) commercial exploitation and storage of intimate information without a person's consent or knowledge, 4) coercion on the part of those in control of sensitive biometric data, and 5) changes in the purpose for which data were originally collected. In the 1990s, there were many, many examples of abuse and overstepping of the original mandate for collecting biometric information, giving genuine cause for concern.

There isn't space to detail the many examples of privacy rights violations that have been associated with the abuse of personal biometric information over the last few years. Some of these have been mentioned in other chapters and others can be referenced from the resources listed at the end of this chapter. For the purposes of illustration, just one example is given here, and a few are mentioned in Section 9 (Implications).

In June 1997, the Secretariat for *Criminal Justice Coalition* in Australia sent a strongly worded complaint to the New South Wales (NSW) Ombudsman Office entitled *Against Department of Corrective Services Biometric Scanning of Visitors, Workers & Children*. The complaint was specific to the Implementation of Biometric Fingerprint Scanning in NSW Maximum Security Prisons. NSW had implemented the biometrics security system in all maximum security prisons after reviewing voluntary visual identification systems while engaging with discussions with the Privacy Committee in 1995.

Without further substantial discussion or endorsement from the Privacy Committee, however, biometrics scanning technologies were introduced into the prison system in 1996. This incident is illustrative of many political and institutional developments in which systems are installed with a minimum of public interaction or approval, or in which the stated objections and the discussion process are bypassed or overridden in favor of pushing through the technologies. Part of the controversy surrounding this case is because the system was initially stated to be a voluntary means of facilitating the visitation process, but it was found that once the units were installed, the emphasis shifted to using it as a means of preventing escape.

This is one of the biggest problems with new surveillance systems—there have been many situations in which the original intention and mandate for installing a system seemed benevolent, and was approved, but which changed, without public discussion, *after* installation. Thus, privacy advocates are not so much opposed to the technologies themselves as they are to the ways in which they are used—particularly when people who implement them stray from the original stated purpose.

A secondary problem is the temptation for employees to steal the data and sell it on the black market. There was a fairly high-profile news story in which it was found that an employee had accessed personal information on millions of *America Online* (AOL) users in the early 2000s. This is not an isolated incident. Many employees have access to databases and it's often a simple matter to copy data onto another storage medium and sell it to tabloids, insurance companies, retailers, marketing companies, disgruntled spouses, detectives, spies, identity thieves, and others who might desire private information on a specific person or group of individuals. Some people have already made considerable sums for selling identity information on the black market.

Another problem arises when biometric data is combined with tracking devices. With the trend to add more and more biometric data to identification cards, it becomes more difficult to go about one's business without being followed by computer tracking systems. As new ID systems come into existence, companies invariably find ways to utilitize the information. Social security numbers are one example—many companies will not process a mortgage or an insurance application without a SSN. In the future, they may refuse to grant mortgages, credit cards, or insurance without biometrically coded ID cards. When tracking/tailing technologies, such as RF/ID chips (radio transceiving chips) are added to biometric IDs, it is not only possible to uniquely identify a person, but also to track their movements whenever they are within range of an RF/ID sensor—even if that person has not violated any civil statutes or committed any crime. This is not speculative technology. The federal government was already taking steps toward adding RF/ID chips to U.S. passports in 2006. Passports and Permanent Resident cards have biometric data already (photos and, in some cases, fingerprints). Thus, once an RF/ID had been matched up with the info on the card, a traveller could be detected and identified the minute he or she set foot in an airport, train station, or border crossing, without the individual knowing he was being tracked.

5. Description and Functions

Fingerprints as a Preferred Identity Tool

Fingerprint technologies form the greatest proportion of biometric identification and surveillance devices. Fingerprints are considered to be unique and essentially unchanging throughout a person's life (except for changes in size during the growth period of childhood) and stable, except for unusual accidents or dismemberment. It is difficult to remove or fabricate fingerprints and those who make the attempt usually fail or cause themselves unnecessary pain or disfigurement. For these reasons, fingerprints are favored by many organizations as an identity marker.

Fingerprint identification, if stored in databases, can provide information about a person's whereabouts, activities, and identity. This information has been used to verify identity for drug testing, financial transactions, insurance applications, scholarship exams, and job applications. It has also been used for locating and convicting criminals, finding missing persons (especially young children or individuals with amnesia or mental illnesses), or detecting and verifying tampering or fraud.

The primary means of detecting fingerprints at crime scenes is with specialized dust. The primary means of using fingerprints for verification or access is with inked systems, but inkless systems and computer image scanning systems are beginning to supersede ink systems.

Fingerprint Acquisition and Identification

Fingerprinting is an important investigative tool and means of documenting a person's presence at a particular location. Because fingerprints are widely used in security and crime investigation, there are many products for processing prints and even some that increase the likelihood of prints being left behind.

Human skin is constantly excreting fluids and moisturizing itself with body oils (oils may also adhere from touching hair) and these fluids and any dirt or residues from the hands (or feet) may stick to surfaces when they are touched. Prints are often found on doorknobs, windows, flashlights, electronics devices, and steering wheels. Glass and metal surfaces tend to hold prints better than coarser, more porous surfaces like fabric or paper, but even latent prints are becoming easier to detect.

There are five general aspects of fingerprinting: 1) acquisition, 2) detection, 3) matching/ identification, 4) preservation, and 5) database indexing (for storage and retrieval). Chemicals are particularly important in acquisition and detection, and somewhat important in preservation (usually in cases where manual rather than computerized systems are used).

acquisition - Fingerprints are commonly acquired for routine records, for ID cards, for preventive identification records, and for investigations. Examples include bonded and security personnel databases; passports, visas, or resident identification cards; police booking procedures; tampering evidence; those at-risk for child kidnapping or elderly wandering. For acquisition, the emphasis is on getting good quality prints that are clean and clear, that last long enough for the application at hand. Specially designed chemical fingerprint pads, inks, and papers are available for this. Computerized 'inkless' systems are now available as well.

detection - Fingerprint detection is important when investigating tampering and crime scenes and, to a lesser extent, when checking ID against the person carrying it. Fuming may also reveal latent fingerprints. In the case of tampering or crime investigation, ultraviolet lights, detection powders, photographs, and fingerprint dusting are commonly used with appropriate chemicals and chemical sensors.

matching/identification - A fingerprint isn't useful unless it can be matched to the person who left the print. Matching and identification are usually carried out visually or with computerized surveillance aids. Chemicals are less important than human analysis or database matching algorithms for this step in the process.

preservation - The preservation of fingerprints, provided voluntarily or taken during bookings, ideally requires good quality, permanent, indelible inks stamped on acidfree archival papers or cards kept in storage facilities that are temperature-regulated and safe from fire and moisture. In the case of ID cards that are renewed every few years, lamination is usually sufficient to preserve the prints. The preservation of prints found in an investigation may involve photographing the source of the prints, dusting them for better visibility, lifting them off surfaces with tape, bagging specific objects that hold prints (guns, phones, knives), sealing the prints with a plastic coating, and photographing the prints themselves or scanning them into a computer matching system/database. Chemicals are most often used in coating and storing the sources of the prints if they are to be held for some time and, of course, are used in photographic processes associated with recording the prints, if a film camera is used.

indexing, search, and retrieval - Traditionally, fingerprints were kept on cards or in files containing objects and envelopes, in banks of filing cabinets. One national repository alone includes over 34 million sets of prints stored this way. Increasingly, however, prints are scanned into computers and retrieved electronically. Sometimes computer matching and identification software is included, to aid in narrowing down searches or developing candidate lists. Chemicals are not a direct aspect of computer databanking, but the quality of the prints created with chemical inks and latent print processes will influence the quality of the images available in the computer database and should be processed with the utmost attention to quality. It is also important, whenever possible, to preserve the original evidence in case computer data are compromised or destroyed.

Patent prints are those that are somewhat obvious to the unaided eye and *latent* prints are those that are difficult to see or detect without technological aids such as chemicals or image-processing systems.

Because fingerprint analysis is central to documenting and solving many crimes, criminals have attempted to hide or remove prints. The most common and obvious way is to wear gloves. The second is to wipe down surfaces or burn evidence to remove prints, hair, and skin. Some have attempted to hide their prints by attaching 'fake prints' to the end of their fingers, made from thin, flexible molded materials. Others have worn 'chemical gloves' to reduce the chance of leaving prints. Some have even taken extreme measures by surgically or chemically removing the skin on the tips of their fingers. This doesn't always work. The cracks and indentations peculiar to scar tissue, in cases where surgery didn't work well, can be just as unique as an actual print.

Inked and Inkless Systems

Fingerprinting has traditionally been carried out with carbon-based inks, usually supplied as specialized stamp pads. The disadvantage to this system is the objectionable soiling that occurs on the body part that was printed.

Other types of 'inks' in the form of leuco dyes have been tried—they prevent staining, but are not as permanent as traditional black inks.

Inkless systems now exist. Most of these require two steps to process the prints. In the first, the fingers (or other body parts) are coated with a chemical reagent, then a developing reagent is applied to the medium on which the prints have been applied to darken the surface. This involves more work for the person collecting the prints, but is still favored where good taste or diplomacy in the matter of soiled limbs is considered important. Unlike inked prints, the person being printed must wait until the processing is complete to make sure the prints are good. If not, the process will have to be repeated. There is a higher chance of allergic reactions by sensitive individuals taking the reagent prints than inked prints.

Inkless systems that use a thermal rather than a chemical process have also been developed.

Thermal systems have some drawbacks. Specialized paper must be used and, to some it has an irritating texture or odor. It also tends to curl and fade, depending upon the brand. In September 1998, D. Arndt filed for a patent on behalf of Identicator, Inc. for an inkless system based on alcohol-soluble dyes dissolved in fatty acid esters, which could be produced in a variety of colors. This very-low-stain fingerprinting ink can be made resistant to fading.

Visual scanners are similar to the image scanners used for computer graphics applications. Software to detect and analyze the prints is still in development, and there are many different approaches to print analysis, but the potential for automation is very good, as some some systems can analyze up to as many as three dozen characteristics of an individual print and can automatically select create and create a list of possible matches out of prints stored in a database.

Computer Analysis and Identification Systems

Computerization of fingerprint databanking and analysis is an important evolution in the technology of fingerprint science.

There have been a number of computerized systems suggested and developed to speed up the process of finding matches between a reference fingerprint and prints stored in an electronic database. With databases increasingly being linked across the country, the number of possible matches greatly increases and it is not humanly possible to individually search millions of prints, even when they are broken down into categories. Given that there may eventually be a national electronic database of prints that exceeds 50 million entries, obviously some type of automation is desirable.

Computer imaging and the cataloging of visual information have a lot of things in common with robotic vision systems. Image recognition is based on different ways of analyzing the data, including an assessment of attributes such as:

- the proportion of light and dark,
- the complexity of the overall image or parts thereof,
- the direction of the lines, and
- the proportion of verticals to horizontals (or diagonals).

Some computer systems even measure the amount of pressure exerted from the finger being printed.

The process of recognizing a print can be mathematically symbolized with numbers that are based on these attributes. If the numbers are the same, or nearly the same, a list of close matches can be generated. The process is similar to using a search engine. If the user is getting too few near-matches and is not finding the desired data, the search can be expanded to generate more 'hits'.

Similar computer algorithms are used in fingerprint (or handprint) identification systems for entry/exit applications, in which the person gains authorized access by placing part or all of the hand in a computerized scanning device that checks the print against a list of authorized users, before allowing a doorway to be entered or exited.

There are two basic ways to store visual information as computer image files, *raster* and *vector*. There are electronic fingerprint systems based on both formats. Some systems will delete the image file, for security reasons, after it has been mathematically coded.

A raster image is typically displayed as dots, pixels, or discrete units of information. In raster storage formats, there is no data relationship assumed between a pixel and those nearby (except for purposes of compression). However, the relationship of dark and light pixels *is* of interest for matching purposes because it aids in determining a pattern or similarity. Raster images of sufficient resolution to provide a really good image of a print can vary from 500 kilobytes to 2-3 Mbytes of data storage (or more for a handprint).

Lossless compression formats (those which don't average or extrapolate the data when being compressed or decompressed) are preferred for data storage of images that must be exactly the same each time they are retrieved or analyzed. Tag Image File Format (TIFF) is a robust, well-supported, lossless professional graphics format that is preferred for many types of raster image storage and display applications. TIFF files can be reduced in size using LZH compression without loss of visual information. Lossy compression formats (those that lose information) are sometimes used. Those that employ wavelet algorithms appear to provide some of the best 'recovery' from lossy data, but some data will still be lost. Raster formats do not magnify well due to their structure—staircasing or 'aliasing' occurs when their resolution is increased. Grayscale images are usually preferred over black and white (line) images because they improve the 'perceptual' resolution of the image. Scanners and printers typically use raster systems.

A vector image is composed of lines that are mathematically defined rather than discrete dots. If you take a circle defined in a raster file and magnify it, it will look coarse and blocky, because there is no information in the file to explain how to fill in the extra pixels when the circle is enlarged—the software doesn't know the pattern of dots was a circle. In contrast, if you take a circle defined mathematically and display it at 600 dpi (the resolution of most desktop printers) it looks pretty good. If you double the size, it looks even better. If you blow it up to the size of a room (assuming you have a large enough plotter or printer), it still looks smooth and round. This is because the program is generating the circle 'on the fly' from a mathematical formula for a circle at whatever is the best resolution of the output device, whether it is a handheld electronic notepad or a 20-foot billboard.

Mathematical comparison of vector-format prints can sometimes be more efficient than raster-format prints for search and retrieval purposes. It depends on the system and the nature of the data. Adobe PostScript is an example of a vector-based language that is used to render images and fonts. CAD programs for design and drafting also use vector formats. Plotters typically use vector shape-description systems such as PostScript or HPGL.

Both raster and vector systems exist for computerized fingerprint identification and matching systems. Each method has its advantages. The jury is still out on which will eventually be found to be most practical and powerful for the job, but it's possible that vector systems may have a slight edge in the long run for actual search and retrieval, and data storage, and raster systems may continue to be used for printing or displaying the images.

6. Applications

6.a. Examples

A few examples of biometrics systems were described earlier in the chapter, in Section 4 (Origins and Evolution). This section describes some that came into use in the 1990s and 2000s.

- In Texas and Arizona, public assistance and food stamp applicants are fingerprinted using a computerized electronic finger imaging system. The scanner records the finger image of the index finger of each hand and compares it to images stored in the client database. Portable versions exist to service homebound clients. The program was instituted to deter multiple enrollments.

- In Canada, Citibank and Metro approved a finger-scanning system called the Client Identification and Benefits System (CIBS) for business transactions and secure access to benefits. This system takes a finger scan, converts it to digital format, encrypts the data, and stores the biometric record, destroying the original image scan to secure the data from visual 'snooping' or duplication.

- Since 1990, the Cook County, Illinois, Sheriff's Department has been using retinal scans to scan prisoners for identification and monitoring purposes. The scanning system is linked with a database of more than a quarter million reference patterns.

- In 1997, initiatives for developing computer-based generic biometric applications programming interfaces (APIs) resulted in the announcement of the Human Authentication Application Program Interface (HA-API), developed by the *National Registry* and the U.S. *Department of Defense*.

- In 1999, Fairfax County, Virginia, police and sheriff's officers began using technology that combined electronic fingerprint identification with digital facial image. The system is intended in part to help identify suspects who use numerous aliases and also to cut down on lengthy searches through manual 'mug shot' books.

The FBI has been working with West Virginia University to offer degree programs in forensic identification, in areas such as fingerprint and biometric technologies. The *International Association for Identification* will require that persons employed in forensic identification have a bachelor's degree, to take effect by the year 2005.

6.b. Equipment

There are hundreds of commercial products for providing biometric access to buildings, machines, and services. Here is just a small sampling to give an idea of the depth and range of commercial products (inclusion does not imply endorsement of these products—they are listed as examples only).

Finger Imaging Recognition

APrint™ HoloPass™ - Advanced Precision Technology, Inc., uses patented holographic optics to produce images for one-to-one verification and one-to-many matching.

Fingerprint Identification Unit™ (FIU) - Sony software with 56 and 128-bit encryption. Third-party vendors have used the system for developing computer logon systems, smart card terminals, and physical access controllers.

IriScan and GTE are creating biometrics-secured electronic commerce transactions. They are working on prototypes of the Iris Certificate Security (ICS) system.

Puppy™ Secure Biometric Logon System - A Windows NT™-based biometric fingerprint logon sytem.

To effectively capture the fine details of a fingerprint, prints are usually scanned at resolution of at least 500 dots-per-inch and saved in compressed formats to minimize file size for storage.

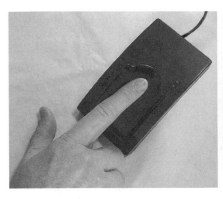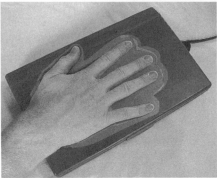

Fingerprint scanners use a variety of software algorithms to analyze and identify the users. Some even measure the pressure of the finger against the scanning mechanism in addition to capturing the skin indentations. Hand scanners (shown here as a conceptual drawing) are less common than fingerprint scanners. [Classic Concepts photos ©2000, used with permission.]

In 2001, Gerald Black was awarded a U.S. patent for a fingerprint recognition that operates on the basis of imaging through a pen. With the increase of pen-based computing devices, it was seen as a way to identify a person without the use of a card. When the user picked up the pen, the system would check the prints of thumb and forefinger against a database and authorize the person to use the instrument or to sign a document (e.g., a contract or financial transaction). Practical application of such a system involves having fingerprints on file for all the people who might be using the system. In the workplace, this might be practical. Outside the workplace, considerations of privacy for the person using the pen must be considered, especially if the fingerprint information is to be stored or used for purposes other than the original application.

Voice Recognition

SpeakEZ Voice Print^{SM} - A voice verification technology that can be used with financial systems, building-access devices, and wireless and wired telephone services.

biometric border control system - Installed at Scobey, Montana, on the Canadian border, this system uses a magnetic stripe card to identify people, combined with voice verification through a telephone handset.

The *Rensselær Polytechnic Institute* has developed a voice-activated lock for doors for high-security areas.

Face Recognition

FaceIt™ - by *Visionics Corporation*. This is facial recognition software that can be built into a variety of applications to rapidly detect and recognize faces. It is a software development tool designed to be scalable and adaptable to a range of software products. It is capable of recognizing single or multiple faces and functions in one-to-one matching or one-to-many matching modes. It is also capable of following faces and isolating them for tracking purposes. FaceIt creates a digital template unique to an individual and compresses the data down to about 84 bytes, for storage and retrieval, and for determining degree-of-similarity to stored images.

TrueFace™ Access - by Miros, a division of eTrue.com This is facial recognition software certified by the *International Computer Security Association*. TrueFace ID identifies a person's face from image files or surveillance video when compared to images in a database. TrueFace Network is client-server software that provides secure access to server data using face and/or finger verification.

The *Central Intelligence Agency* (CIA) has been involved in the development of a number of biometric identification systems for use by federal agencies. Much of the information on these technologies is classified, as is biometrics technology developed by the U.S. *Department of Defense*.

Eye Recognition

The human iris, the colored portion of the eye surrounding the pupil, is unique to each individual and to each eye, and remains stable from infancy to old age, thus making it suitable as a biometric identification tool.

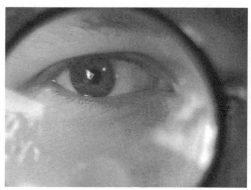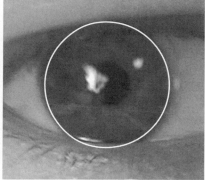

The human iris is suitable as a biometric identification tool. Looking into an iris-scanning device is more convenient than looking into a retina-scanning device, because the subject doesn't have to be as close to the scanning mechanism for the scanner to image the iris (which is at the front of the eye). Newer systems that use narrower wavelength light beams are less likely to be disrupted by reflections caused by glasses or contact lenses. [Classic Concepts ©2000 photo, used with permission.]

eyescanning ATM technology - Sensar eyescanning was demonstrated at the 1997 Banking Administration Institute Conference. When a banking customer inserts a card into an ATM , the stereo camera locates the face, finds the eye, takes a digital image of the iris at distances up to three feet, compares the captured image to one on file with the bank, and rejects access if there isn't a match. It works with glasses, contact lenses, and at night. No beam is shone in the clients' eyes, passive technology in the form of a zooming camera is used.

Motion Recognition

IBM - Clients sign with a pen that measures speed, pressure, and direction of movement, which goes beyond just the shape of the signature (which can be forged) to measure several attributes, making forgery virtually impossible.

There are also systems that will monitor eyeblinks to determine if a subject may be lying or evading, or to determine if an employee is fatigued and should perhaps be rotated on a production line, or removed if working with dangerous equipment.

Multiple Features Recognition

Digital Justice Solution™ - By Printrak International Inc. combines realtime automated fingerprint identification, computerized criminal history, a mug shot, and document storage and retrieval capabilities. The system can also be identify latents prints lifted from crime scenes, comparing them against over a million records on file in the Louisiana system.

e*PICS™ - A system from *Electronic Identification, Inc.,* that uses a digital camera and ultrasonic fingerprint integrated into a Web-enabled workstation. The data are converted into a digital signature using proprietary, patented software compression technology, and burned into a chip to create an ID card with the digital profile printed on the face of the card. The card works with e*PICS authentication readers.

layered biometric ATM - Diebold, Keyware Technologies, and Visionics - Uses speaker verification and face recognition to process transactions without the use of a PIN or password.

SPIKE™ *Suspect and Prisoner Identification Key Evaluation System* - Combines face recognition, voice verification, and AFIS (Automated Fingerprint Identification System) fingerprint verification. Developed for use by law enforcement agencies, the system incorporates a computer, high-resolution digital camera, directional microphone, and fingerprint scanner, integrated with SPIKENET™ Internet connectivity.

Traditional lie detectors monitor a number of features including galvanic skin response, heart rate, etc. Digital lie detector systems may also evaluate voice patterns.

Fingerprint Detection

Fingerprint kits vary in portability, quality, and ease of use. They are also somewhat dependent upon the skill of the person taking the prints. Consequently, practice cards and training videos are supplied by many of the major vendors. Kits may print a single finger (these are often used by banks), or may be designed for printing all digits or even palm or foot prints. When printing larger areas, it is sometimes necessary to have a thin foam layer under the printing card for the surface area to evenly contact the card. This is supplied with many kits for palm or foot prints. The cards used for storing the prints are designed accordingly.

Fingerprint kits are commonly of inked (stamp-pad style) or inkless. Inkless kits are convenient to the fingerprintees since most people find ink on their fingers objectionable, but have a higher incidence of allergic sensitivity. They also require a two-step process to 'develop' the image of the print. Many systems require cartridges or inking pads to be periodically replaced, usually about every 500 to 1500 inkings or so.

Fingerprint Collection

Basic stamp pad/index card system are inexpensive, and may be suitable for ID cards or temporary records, but might not meet standards for crime investigation evidence.

Specialized inks, designed to provide a finely textured, detailed, dark print, are sold in containers of various sizes and shapes. Federal bureaus use them for ID cards and banks sometimes use the small round ones for signature ID prints on negotiated checks, usually third-party checks or checks cashed by customers without accounts at the cashing agency (banks, casinos, money depots). These vary in price from a few dollars to $80 or so, depending upon the quality of the ink, the number of impressions, and the kinds of ink.

Compact inkless fingerprint pads are about $600 with replacement cartridges about $80 to $200 each, depending upon the number of impressions.

Identification kits, sometimes called 'portable fingerprinting stations', containing cards, ink or inkless chemicals, sealers, and other accessories can be purchased for about $120 to $300.

Special-purpose kits, designed to take prints from people with arthritic or differently shaped fingers, or from cadavers, are also available in about the same price ranges as regular kits. The main difference is that there is usually a spatula or spoon-like implement included for rubbing against a curved surface to get the print.

Shoe kits to record shoe tread patterns (or telltale shapes in other objects) are sold for about $200 and are valuable for scanning into a computer or saving in a file so that the original shoe doesn't have to be pulled out each time, and so that unimpounded shoes (e.g., other suspects) can be referenced later, if needed.

Latent Prints

There is a lot of chemical science involved in the acquisition and processing of prints and they can't be covered in detail, but a sampling of chemicals and techniques is described here. (If you want more information on this important and interesting aspect of biochemical surveillance, there are numerous bibliographic references at the end of the chapter and a good Website at SCAFO mentioned in Section 10.e. Media Resources.)

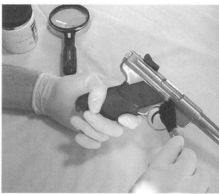

A number of different kinds of fingerprint dusts are used to reveal latent prints on windows, jewelry, firearms, the handles of cars and other objects, including magnetic dusts, colored dusts, and dusts that fluoresce when exposed to ultraviolet light. Sometimes chemical fuming helps to bring up prints to make them more visible. Some more recent innovations include lasers that reveal latent prints without dust. [Classic Concepts ©2000 photos, used with permission.]

'Dusting for prints' is familiar to the general public through TV shows and newscasts.

Latent fingerprints may not be visible to the unaided eye. The use of dusts, magnifying glasses, and fuming agents may aid in revealing or further developing prints so they can be seen and processed.

Dusting Powders

Fingerprint detection and collection are intrinsic aspects of crime scene investigation. Many criminal convictions have been based upon fingerprint evidence.

Fingerprint dusting powders come in nonmagnetic and magnetic varieties. When they are being photographed, prints on a dark background show up better with a light dust and prints on a light background show up better with a dark dust. To accommodate differing needs, powders are available in a variety of colors, commonly black, white, and metallic shades. For prints that are going to be lifted with tape and placed on a card for examination and filing, dark powders are favored. Bichromatic™ is a commercial product that combines black and sliver-gray powder, which tends to look lighter on dark surfaces and vice versa, and is a practical solution for some situations.

Powders are typically applied with soft, wide, hair or fiberglass brushes that resemble makeup brushes. Powders should be fine and loose, designed for the purpose. Dessicant packets or rice grains can help prevent moisture from clumping the powder. Small ball bearings are sometimes used to help break up the clumps. If there is too much powder, it can damage the print by overloading it and filling in the channels, so the powder is built up gradually, in fine layers. It is important not to spread too much powder around the area being dusted, as it might contaminate other parts of a crime scene.

Magnetic powders are less common, but are suitable for use on shiny glossy coatings, plastic containers, and bags, etc. A special applicator is used with magnetic powders. The applicator is lightly moved over the surface with the prints without touching the surface being dusted. Excess particles can be returned to the container.

Tapes

Specialized clear tapes are used to lift and preserve many kinds of evidence, including blood smears, paints, spills, and fingerprints. Once a print has been made more visible with a powder, adhesive tape or precut strips can be placed over the print, pressed gently, and placed over a non-acidic card or paper to protect the image on the tape. This takes practice. The tape must be large enough to cover the area to be lifted and must be laid down gently and evenly so as not to create creases in the tape or distort the image by adhering part of it and then accidentally pulling on it. Then the correct amount of pressure must be applied. Too much pressure can 'squash' a print. Too little can cause some of it to remain on the original surface and the odds of correctly lining up the tape a second time to get the rest of the print are almost zero.

Graphic artists who have done a few years of 'table work' or artifact restoration, that is, a lot of drawing, ruling, inking, and film handling, usually have acquired the hand-eye expertise to dust and lift prints. It requires patience and a steady hand and, considering the importance of this type of evidence and the risk of damaging prints, should be handled by an expert.

Print-Gathering Supplies

Fingerprint Dusts and Brushes

Camel-hair and fiberglass brushes are preferred by most professionals. It's important to avoid poorly bound brushes that lose bristles. Wide, soft brushes are preferred. The bristles of the brush should not be handled, to avoid leaving oily finger residues. Excess powder should be tapped into a dish or the powder container before applying to the surface and layers should be built up gently, taking care not to apply too much (it's difficult to remove without harming the print). Photograph the print as soon as it is visible. Don't try to overwork the dust to improve the print. Tap the brush gently to shake out excess residue and keep it dry. Use separate brushes for dark and light powders.

Latent Print Developers

The term 'latent' refers to objects that are present but not visible. Latent prints are not easily seen or otherwise detected and chemicals are often needed to reveal or further 'develop' the prints to make them visible for the use of investigators, attorneys, and juries. There are many ways to develop latent prints and the research is ongoing. Lasers, UV light, image processing, fuming, and chemical development are all options, with no clear superiority of any one approach.

Since chemicals must be carefully stored and are used by people, nontoxic, nonflammable brands are preferred. However, some chemicals with useful qualities have toxic or irritating effects (and there may be import/export restrictions on some), in which case appropriate ventilation and precautions must be observed. These are some common techniques for revealing latent prints:

- *Ninhydrin* mixed with a number of different fluids can give results ranging from adequate to excellent in revealing latent prints.

- *Gentian violet* is generally purchased as a powder. For use, it is dissolved in distilled water with pH adjustments with a very small quantity of ammonia. Tape with a latent print is soaked in the solution, rinsed, and examined under the microscope where stained cells that make up the ridges can be seen as a distinct violet color. Sometimes prints are transferred from the tape to photographic or other papers before processing with gentian violet. Whenever possible, make as few transfers as possible. Like a photocopy of a photocopy, information and clarity are lost each time particles are transferred. If a print must be transferred, it is usually best to use a chemical reagent or heat.

- *Powder-detergent suspensions in water* may help reveal prints. A thick, iridescent dark or light mixture (depending on what is needed for contrast) is created, which is applied to an adhesive surface. The tape is rinsed, pulled through the mixture, or the mixture applied carefully with an implement, and then rinsed again.

Sometimes fingerprints include contaminating traces of blood. Protein dyes that respond to the proteins in the blood stain can be used to stain the blood without affecting the latent print. Protective gear is recommended when using these kinds of products.

Backing Cards

Blank and printed cards are used for marking or storing evidence. Lifted prints, stains, and other evidence are often lifted with specialized tapes, which may then be attached to a card and labeled. Since chemical analysis of the stains is often carried out, it is important to use cards that do not interfere with the chemicals and which last a long time. Acid-free *archival* cards are preferred. Preprinted cards sometimes simplify the task of labelling and may aid in reducing the number of inadvertent omissions. They may also be numbered to reference other documents. Numbers may further reduce the chance of falsified or mislabeled evidence.

Evidence Tapes

Clear tape, poly tape, Handi-Lifts™, Lightening Lifts™ pre-cut strips are all used to lift prints. The best tapes, in most cases, are those that are clear, that don't wrinkle easily, and have the least intrusive chemicals in the adhesive itself.

7. Problems and Limitations

Forensic Identification of Fingerprints

Some of the difficulties in using computerized matching systems relate to differences between traditional reference fingerprints and the fingerprints left at a scene. In most cases, reference fingerprints are made by the print-taker grasping the individual's finger, exerting light pressure and rolling the finger back and forth to get a print that clearly shows the broadest possible area. The prints left at scenes, however, usually only show part of the finger that protrudes the most, or those parts that come in contact with an object when it is being grasped. The computer software must be able to analyze the prints in such a way that differences in the outer edges stemming from the means of procuring the prints do not interfere with the possibility of a match. This can be done, mathematically, but may require methods different from those that are used in access surveillance applications (e.g., interpreting an image from the center and outward).

Optimizing Iris and Retina Scans

Iris and retinal scan systems tend to require clear, well-focused images that may be hindered, in some instances, by contact lenses or eyeglasses. Earlier systems that shine a light source at the subject's eye tend to have greater problems with incidental reflections than passive light scanning systems. Removal of eyeglasses, and particularly of contacts, is considered inconvenient or intrusive.

Retinal images usually require that the subject be very close to the scanner. Sometimes an eye cup is used to position the eye, which may be inconvenient for the user and brings up some issues of sanitation, since eyes are a source of mucous and bacteria that can be passed from one person to the next.

In recent years, narrow bandwidth light sources have been used to overcome some of the problems of glasses and contacts and the reflectance associated with these accessories.

In a general sense, related to all biometric technologies, injuries and illnesses sometimes cause physical changes that can prevent access or recognition. Some individuals are lacking limbs or eyes due to accidents or birth defects and thus might be denied rightful access to workplaces or transaction machines unless special provisions are made.

Reliability

Electronic biometrics devices were, in many cases, somewhat unreliable until the 1990s at which time many of them began to improve and become commercially practical. However, reliability is not just a factor of electronics and software. Some systems tend to be more reliable due to physical characteristics and changes of a person's body over time. Some kinds of processing are more difficult than others and some physical aspects are more variable from day to day (if you've ever had laryngitis you're aware of how much a voice can change under certain circumstances). Voice systems tend to fail more often than fingerprinting or facial recognition systems.

In general, reliability is improving as systems and software improve and multiple-input systems can sometimes overcome problems.

Identical twins can fool some biometric systems. If the twins are not mirror twins (with opposite features) and close relatives of the twins can't tell them apart, it's possible that the biometric systems can't tell them apart either, on the basis of facial or voice recognition. However, other aspects, such as handwriting style and iris patterns may differ. When multiple sensors are used, it becomes slightly easier to distinguish identical twins. Mirror twins are easier to identify since faces tend to be asymmetric, and imperfections and irregularities will be on opposite sides—thus aiding identification.

8. Restrictions and Regulations

Protection of Personal Information

Biometric surveillance is a relatively new field. Some aspects of it, such as fingerprint files and mug shots, have become standard law enforcement tools, but new technologies that enable any organization or retailer to use biometrics to monitor employees or customer profiles, are now widely available and there are few restrictions on their use.

In some aspects of law, biometric information is regarded as 'personal information'. In a number of codes, personal information is defined as any information that:

> ... describes, locates, or indexes anything about an individual or that affords a basis for inferring personal characteristics about an individual including, but no limited to, his education, financial transactions, medical history, criminal or employment records, finger and voice prints, photographs, or his presence, registration, or membership in an organization or activity or admission to an institution.

Legal Force of Biometric Signatures

On 16 June 2000, the *Electronic Signatures in Global and National Commerce Act* was passed unanimously by the U.S. Senate. This made electronic signatures as legal and binding as traditional written signatures, providing a means to authenticate electronic transactions. The definition of *electronic signature* is broad enough to include biometrics:

> The term 'electronic signature' means an electronic sound, symbol, or process, attached to or logically associated with a contract or other record and executed or adopted by a person with the intent to sign the record.

FBI Support Materials

The Federal Bureau of Investigation publishes a number of standards related to evidence that are relevant to chemical/biological surveillance, as they relate to crime scene investigation and forensics such as the *Electronic Fingerprint Transmission Specification* (24 August 1995).

General Orders

On a broader scale, *Executive Order 13083* (called Federalism) was signed by President Clinton on or about 14 May 1998 (revokes Order 12612). Its passage was quiet, but it has since become a controversial document that some claim has reduced rather than expanded the divisions of power within the U.S. constitutional system by slanting primary lawmaking functions toward the President and away from Congress.

Congressional Trends

In compliance with the *Illegal Immigration Reform and Immigrant Responsibility Act of 1996*, Border Crossing Cards (BCCs) issued after 1 April 1998, were required to include a biometric identifier, such as a fingerprint.

In 2002, the DoD *Biometrics Fusion Center* (BFC) sponsored a Biometrics Test Database biometrics collection gathering to support Homeland Security in collecting biometrics data for testing and evaluation of commercial biometric technologies for use by the DoD. The *Biometrics Management Office* (BMO) was specifically investigating fingerprint, hand geometry, iris scan, face and voice recognition, and signature verification systems.

In 2004, the *Intelligence Reform and Terrorism Prevention Act* included biometric border entry/exit provisions and the report of the 9/11 Commission supported the concept of funding and completing biometric entry/exit systems as essential to national security. The Commission reported that fingerprints and face recognition systems were the most commonly used biometric technologies, with iris scans described as "promising for future applications".

Also, in 2004, the House of Representatives and the U.S. Senate passed H.R.-4417 to extend deadlines in the *Enhanced Border Security and Visa Entry Reform Act of 2002*, pertaining to the implementation of machine-readable, tamper-resistant entry/exit documents (e.g., passports) that contain "biometric and document identifies comporting with specific standards". In essence, it gave those issuing visa waivers until 26 Oct. 2005 to include biometric data, a system intended to work in conjunction with enrollment of travellers through the U.S. *Visitor and Immigrant Status Indicator Technology* (US-VISIT) program administrated by Homeland Security. The system requires index finger scans and a photograph for identity verification. In addition to this, the *Department of State* was developing U.S. biometric passport standards.

9. Implications of Use

Many concerns over biometrics identification systems have been raised, most of which relate to administrative changes and use of the technologies rather than the technologies themselves. The most prevalent concerns include reduction of boundaries between restricted and free citizens, changes in administration and use, lies, profit motives, and centralization due to economics.

Reduction of Boundaries and Database Errors

The trend for standardized, centralized databases is increasing. In these databases, a criminal who is incarcerated for murder looks much the same as a law-abiding private citizen. The computer puts the picture in one part of the screen, the name and address in another, and the personal information in a third. The electronic distinction between a criminal, someone on a watchlist, and an honest person may be as small as a single checkbox in the bottom right corner which, given human error, could easily be checked accidentally.

This may not seem like a problem until you examine some of the dynamics of these computerized systems and how they are developed and sold. Databases are structured to demand certain information in certain ways. Anyone who has done data entry is aware that certain data entry fields require inputs before the program will let you continue or before it will perform certain types of processing tasks. When systems are designed to record fingerprints, mug shots, and other personal identification measures, they are often marketed initially to law enforcement agencies and other bodies that have the resources to purchase them. Then, with a few modifications (or no modifications), the companies that design these systems seek to broaden their markets, increase their profits, and offset their development costs by selling them to other agencies. These could be government agencies (e.g., authorities issuing drivers' licenses), private companies, or retail establishments. In many cases, the database structure is essentially the same. In fact, with the possibility of connecting through the Internet and pressure on developers to standardize systems, the potential for cross-system data sharing increases. Once the institutions adopt the software for use for applications involving the general public, there is new pressure on the public to provide the same kind of information that a criminal must supply while being incarcerated.

This homogenization of biometric identities blurs the line between criminals and honest citizens and has already occurred in some prison and government database systems. Since it has been found that database entry operators make errors from 10% to 30% of the time, depending on how carefully they are supervised and crosschecked, the potential for misinformation to spread rapidly through computer systems is great.

Changes in Administration

One of the biggest concerns in the protection of rights and privacy comes with changeovers in executive or governmental administrations. There are thousands of historical examples that show that data collected for one purpose are often used later for another purpose, without the permission or knowledge of the person originally supplying the data. This may be by the administration collecting the data or by a subsequent administration with very different political views and motives. Since computer information is easy to duplicate and transmit, there is no guarantee, if one file is deleted, that it does not exist in multiple copies elsewhere.

Profit Motives

Many retail stores and supermarkets are now issuing 'member cards' or 'loyalty cards' that reward frequent and regular customers with discounts and other percs. They do so to collect data on customers. Some of these companies sell your personal information to other companies and once it is out of their hands, they have no control over who has it or how they use it. As biometric data are added to identification and member cards, and as computer networks make it possible to amass substantial information on individuals through data mining, very intimate information profiles are being developed on individuals.

Since many of these data lists are sold on both the open market and the black market, they can be purchased by secret service agencies, foreigners, and marketing professionals.

It's not difficult to configure a computer to analyze the data and figure out how many children a person has, what schools they attend, how much money they make, and what kinds of products they buy. This kind of intimate information can be used by unscrupulous employers or gray market business owners to stalk, coerce, or defraud.

Even government agencies have sold private information to commercial companies. In some states, drivers' license information has been sold. In some states the sales were blocked, but in another, the court ruled that drivers' license information was not private and that the sale was legal, in spite of public protests.

Informed Consent

One of the biggest ethical problems related to biometric profiling is the issue of informed consent. Children, handicapped individuals, and those who are temporarily incapacitated physically or mentally may be biometrically 'marked' without giving informed consent. As an example, the European Union (EU) announced plans, between 2004 and 2006, to fingerprint children for identification purposes from as young as six years of age, and compulsorily from 12 years, for various identification purposes (e.g., passports and visas). Young children are unable to judge whether they want to be biometrically identified and are not able for many years to understand or object to the system. By that time, their information may be in a multitude of databases, with no way to erase it.

Lies

Some organizations and individuals simply lie. In the United States alone, it is estimated that $40 billion a year is lost to telemarketing scams. In the course of being defrauded, people often give out personal information that can be sold to others or distributed through computer networks more readily than ever before.

Cost-Effective Deployment and Change of Use

Economic factors often drive administrative decisions that result in loss of privacy. Many organizations install biometric identification systems with the promise that the information will not be shared with other departments or organizations. However, by the time the systems are installed, the cost of individual stations using the technology is often found to be cost prohibitive, resulting in gradual expansion of the system to spread the cost out over more stations and thus reduce individual costs. With time, these systems become centralized and the initial promise of data security and compartmentalization is broken, without due process or public input.

As has been stated about other technologies described in this book, biometric technology is not inherently good or bad, but it is a powerful technology with a high potential for abuse when placed in the wrong hands, used for the wrong purpose, or redistributed without the owner's knowledge or consent. Using a face scan or electronic fingerprint system is not the same as presenting a driver's license photo. The driver's license data supposedly reside in only one database (although previous comments indicate this is not always the case), whereas a facial recognition system, if used by retailers or less accountable organizations, for example, resides in multiple databases that can be sold and swapped and accessed by employees at will. There is very little protection in this type of system.

Often data stored by small startup companies is acquired through mergers and acquisitions by larger companies. The larger companies do not always recognize or care about the safeguards promised by the company that originally gathered the data. As one example, a number of small software companies changed hands in the 1990s and, to the horror of their clients, the clients' private email communications showed up on the Internet and could be accessed by anyone through Google. Once information is 'in the wild' on the Net, you can't take it back, and the same could happen to sensitive biometric data.

Some privacy rights advocates are concerned that iris or retinal scans might be used to assess health information. An employer who uses iris or retinal scans (or both) might be tempted to analyze data for diabetes, glaucoma, cataracts or other signs of ill health when deciding on medical insurance benefits, raises, promotions, or who to lay off. There are even some fingerprint patterns that can indicate certain genetic characteristics or health conditions. In order to try to prevent abuse, some companies have developed systems that delete the original reference image after it has been coded into an electronic pattern, but there is no guarantee that all programmers and vendors will do this, especially if buyers express a willingness to pay for systems that do not add this security layer.

It was only a matter of time before people began lobbying for biometrics and RF/ID chips to be used to verify registered voters at voting booths.

It is hoped that all the issues regarding biometrics are discussed before systems are put in place and that the convenience of some of these systems doesn't overshadow good judgment in evaluating all the factors associated with their use.

10. Resources

Inclusion of the following companies neither constitutes nor implies an endorsement of their products and services and, conversely, does not imply their endorsement of the contents of this text.

10.a. Organizations

American Chemical Society (ACS) - Provides members with technical and educational information resources, professional development assistance, industry advocacy, awards, and insurance programs. Supports over 30 specialty divisions, including Analytical Chemistry (ANYL) and Chemical Toxicology (TOXI), which have their own publications in addition to the ACS publications. www.acs.org

American Society for Investigative Pathology (ASIP) - A society for biomedical scientists who investigate the mechanisms of disease. The discipline uses a variety of structural, functional, and genetic techniques, applying the research results to the diagnosis and treatment of disease. ASIP supports professional development and education of its members. asip.uthscsa.edu

Association for Biometrics (AfB) - A nonprofit organization providing services to government, academic, and industry members, while promoting technologies related to biometrics. www.afb.org.uk

Biometric Consortium - A U.S. government group of over 200 academic, private sector, and government members involved in research, developing, testing and evaluating biometric technologies. www.biometrics.org

Biometric Systems Lab - A biometrics research lab located at the University of Bologna, Cesena, Italy. www.csr.unibo.it/research/biolab/bio_home.html

Center for International Security and Cooperation (CISAC) - A multidisciplinary community within Stanford University's Institute for International Studies dedicated to research and training in issues of national security. www.stanford.edu/group/CISAC

Center for Security Systems - This is a research-development-applications center for creating technologies that aid in national security. It includes dedicated laboratories for sensors, image processing, alarms, communications, and biometrics. www.sandia.gov

Chemical Science and Technology Laboratory (CSTL) - Within the U.S. *Department of Commerce*, the CSTL is one of seven NIST measurement and standards laboratories. It works to promote U.S. economic growth by working with industry to develop and apply technology, measurements, and standards. It includes five divisions: analytical chemistry, biotechnology, physical and chemical properties, process measurements, and surface and microanalysis science. www.cstl.nist.gov

European Association for Biometrics (EAB) - This organization joined a consortium with the *National Computer Security Association* in 1997.

Federal Bureau of Investigation (FBI) - The FBI provides many print resources and services for law enforcement, along with a number of federal databases. The *National Crime Information Center* (NCIC) 2000 provides single fingerprint matching and mug shot data available to more than 80,000 criminal justice agencies. www.fbi.gov

International Association for Identification (IAI) - A nonprofit, professional organization for professionals engaged in forensic identification and scientific examination of physical evidence. The IAI provides a range of education and certification programs including latent fingerprint examination, crime scene certification, forensic artist, etc. Descended from the International Association for Criminal Identification, founded in 1915. www.theiai.org/

International Biometric Industry Association (IBIA) - This is a nonprofit trade organization established in 1998 to advance, advocate, defend, and support the international biometrics industry. IBIA also publishes newsletters and bulletins. www.ibia.org

International Biometric Society (IBS) - Founded in 1947, the IBS advances the subject-matter sciences related to biometrics. Members include biologists, statisticians, and others applying statistical techniques to research data. stat.tamu.edu/Biometrics

Lightning Powder Company, Inc. Supplies chemicals, including powders, to crime scene investigators and provides informational articles on their Web site on fingerprint technology. www.redwop.com

National Institute of Standards and Technology (NIST) - NIST is an agency of the U.S. Department of Commerce's Technology Administration, established in 1901 as the National Bureau of Standards and renamed in 1988. It aids industry in developing and applying technology, measurements, and standards through four major programs. The Chemical Science and Technology Laboratory is one of seven NIST measurement and standards laboratories. www.nist.gov

Southern California Association of Fingerprint Officers (SCAFO) - A nonprofit organization founded in 1837 to support professional identifiers. It now includes members in more than 50 law enforcement agencies. The Web site has an excellent fingerprint bibliography. www.scafo.org

SWGFAST - The Scientific Working Group on Friction Ridge Analysis, Study, and Technology was founded as the result of a 1995 FBI meeting of latent print examiners. SWGFAST (formerly TWGFAST)

provides guidelines, discussions of analysis methods and protocols, and support for the latent print professional community. Information is available through the FBI Laboratory. www.fbi.gov

U.S. Army Biometrics - This is within the Office of the Secretary of the Army Director of Information Systems C4. www.army.mil/biometrics

10.b Print

The author has endeavored to read and review as many mentioned resources as possible or to seek the recommendations of colleagues. In a few cases, it was necessary to rely on publishers' descriptions on books that were very recent, or difficult to acquire. It is hoped that the annotations will assist the reader in selecting additional reading.

These annotated listings may include both current and out-of-print books and journals. Those which are not currently in print are sometimes available in local libraries and second-hand book stores, or through interlibrary loan systems.

Bace, Rebecca Gurley, *Intrusion Detection*, Indianopolis: MacMillan Technical, 2000, 339 pages. History and developmental treatment of intrusion detection devices, legal issues, and other concepts. It is not a how-to book but rather an aid in making administrative decisions.

Bodziak, W., *Footwear Impression Evidence*, New York: Elsevier, 1990 and Boca Raton: CRC Press, 2000. The CRC Press edition includes added information on barefoot evidence and the O. J. Simpson trial.

Elashoff, Robert, *Perspectives in Biometrics*, New York: Academic Press, 1975.

Jain, Anil; Boole, Ruud; Pankanti, Sharath, Editors, Biometrics, Personal Identification in Networked Society: Personal Identification in Networked Society, from the Kluwer International Series in Engineering, Kluwer Academic Publishers, 1999.

Jain, L. C.; Halici, U.; Hayashi, I.; Lee, S.B., *Editors, Intelligent Biometric Techniques in Fingerprint and Face Recognition*, Series on Computational Intelligence, Boca Raton: CRC Press, 1999, 480 pages. This book has received mixed reviews.

Kovacich, Gerald L.; Boni, William C., *High Technology Crime Investigator's Handbook*, Butterworth-Heinemann, 1999, 350 pages. Not specifically on biometrics but a good general reference and practical guide to forensics using new technologies.

U.S. General Accounting Office, *Technology Assessment: Using Biometrics for Border Security*, GAO-03-174, Nov. 2002.

Articles

Adcock, J. M., The Development of Latent Fingerprints on Human Skin: The Iodine-Silver Plate Transfer Method, *Journal of Fingerprint Science*, 1977, V.22(3), pp. 599-604.

Almog, Joseph; Hirshfeld, Amiram; Klug, J. T., Reagents for the Chemical Development of Latent Fingerprints: Synthesis and Properties of Some Ninhydrin Analogues, *Journal of Fingerprint Science*, Oct. 1982, V.27(4), pp. 912-917. Almog has written a series of articles in JFS on this topic.

Ashbaugh, David R., Ridgeology - Modern Evaluative Friction Ridge Identification, *JFI*, 1991, V.41(1), pp. 16-64. Ashbaugh has written a series of articles in JFI on ridge patterns and identification.

Augibe, Frederick T.; Costello, James T., A New Method for Softening Mummified Fingers, *Journal of Fingerprint Science*, April 1985, V.31(2), pp. 726-731.

Baniuk, Krystyna, Determination of Age of Fingerprints, *Forensic Science International*, 1990, V.46, pp. 133-137.

Blank, Joseph P., The Fingerprint That Lied, *Reader's Digest,* Sept. 1975, pp. 81-85.

Brooks, Andrew J., Jr., Techniques for Finding Latent Prints, *Fingerprint and Identification Magazine,* Nov. 1972, pp. 3-11.

Burt, Jim A.; Menzel, E. Roland, Laser Detection of Latent Fingerprints: Difficult Surfaces, *Journal of Fingerprint Science,* April 1985, V.13(2), page 364-370.

Candela, Gerald T.; Grother, Patrick J.; Watson, Craig I.; Wilkinson, R. Allen; Wilson, Charles L., Public Domain PCASYS: PCASYS–A Pattern-level Classification Automation System for Fingerprints, *National Institute of Standards and Technology.*

Chemical Formulas and Processing Guide for Developing Latent Prints, *FBI Training Material,* 1994 (revised).

Chen, Hans H., Bio-Code Systems Promise New Age in Security: But Are They a Threat to Privacy, *APBnews.com,* 10 Oct. 1999. Details some of the uses and surprising abuses of biometrics and computer databases.

Chen, Hans H., Machines that Mesaure Your Body Parts, *ABPnews.com,* 10 Oct. 1999. Describes basic biometric devices with illustrations of some common systems.

Couto, Joe, Ontario farmer challenges government over photo ID, *Christian Week,* V.13(10). Describes Ontario's identification which includes the capability to transmit over networks and wireless communications systems.

Creer, Ken, Operational Experience in the Detection and Photography of Latent Fingerprints by Argon Ion Laser, *Forensic Sciences International,* 1983, V.23, pp. 149-160.

Duff, J. M.; Menzel, E. R., Laser-Assisted Thin-Layer Chromatography and Luminescence of Fingerprints: An Approach to Fingerprint Age Determination, *Journal of Forensic Science,* 1978, V.23(1), pp. 129-134.

Faulds, Henry, On the Skin-Furrows of the Hand, *Nature,* 28 Oct. 1880.

Feldman, M. A.; Meloan, C. E.; Lambert, J. L., A New Method for Recovering Latent Fingerprints from Skin, *Journal of Forensic Science,* Oct. 1982, V.27(4), pp. 806-811.

Fincher, Jack, Lifting 'latents' is now very much a high-tech matter, *Smithsonian,* Oct. 1989, pp. 201-218.

German, Edward R., The Admissibility of New Latent Print Detection Techniques in U.S. Courts, *Identification News,* Oct. 1986, pp. 12-13.

Grimoldi, Giuliana; Lennard, Christopher J.; Margot, Pierre A., 'Liquid Gloves' and Latent Fingerprint Detection, *JFI,* 1990, V.40(1), pp. 23-27.

Here's looking at you: One hospital's experience with biometrics, *HIPAA and Health Information Security.* Describes how biometric technology and access to electronic medical records may be integrated.

Lee, Henry C.; Gaensslen, R. E., Cyanoacrylate Fuming, *Identification News,* June 1984, pp. 8-14.

Menzel, E. Roland, Pretreatment of Latent Prints for Laser Development, *Forensic Science Review,* June 1989, V.1(1), pp. 44-66. Menzel has written several articles for a variety of publications including *Analytical Chemistry and Journal of Forensic Science.*

Nutt, Jim, Chemically Enhanced Bloody Fingerprints, *FBI Bulletin,* Feb. 1985, pp. 22-25.

Putting Your Finger on the Line: Biometric Identification Technology, *Framed,* Dec. 1997, Issue 34. Describes biometric technology and the controversy over the implementation of biometrics in the NSW prison system.

Scroggins, Steve, National ID card system threatens freedom, 18 July 1998. Describes the controversy surrounding a 1996 Department of Public Safety announcement to collect fingerprints for drivers' licenses and the implications of such practices.

Waver, David E., Photographic Enhancement of Latent Prints, *JFI*, 1988, V.38(5), pp. 189-196.

Wellborn, Stanley N., Foolproof ID: Opening Locks With Your Body, *U.S. News & World Report*, 17 Dec. 1984. Describes a variety of types of biometric systems and where and how they are used.

Woodward, John D., Biometric Scanning, Law & Policy: Identifying the Concerns–Drafting the Biometric Blueprint, *University of Pittsburgh Law Review, 1997*.

Journals

The Biometric Digest, a weekly email publication.

Biometric Technology Today, an Elsevier Advanced Technology publication.

Biometrics, a journal of the International Biometric Society. stat.tamu.edu/Biometrics/

Biometrics Bulletin, published by the American Statistical Association.

Biometrics in Human Services User Group Newsletter, provides biometric news, findings, and ideas.

Chemical Formulas and Processing Guide for Developing Latent Prints, FBI Training Material, 1994 (revised).

Framed, A quarterly magazine of justice action.

HIPAA and Health Information Security, a publication for health care professionals tasked with information security.

The Print, the professional journal of the Southern California Association of Fingerprint Officers (SCAFO). Six issues per year are available to members of SCAFO.

Privacy Digest, an online journal and news source listing international and local stories related to privacy invasion and erosion of freedom. It discusses technology, international agreements, and laws related to cybercrime and privacy. www.privacydigest.com

Voice ID Quarterly, information available through *jmarkowitz@pobox.com*.

10.c. Conferences and Workshops

Many of these conferences are annual events that are held at approximately the same time each year, so even if the conference listings are outdated, they can still help you determine the frequency and sometimes the time of year of upcoming events. It is very common for international conferences to be held in a different city each year, so contact the organizers for current locations.

Many of these organizations describe the upcoming conferences on the Web and may also archive conference proceedings for purchase or free download.

The following conferences are organized according to the calendar month in which they are usually held.

Biometrics Summit, Implementing Practical Applications in Biometrics. The *Global Biometrics Summit* is an international conference that is also available as a virtual conference on CD.

SmartCard, annual conferences since the late 1980s.

Experimental Biology, annual meeting of the American Society for Investigative Pathology (ASIP).

California State Division International Association for Identification, annual training seminar since ca 1916.

International Association for Identification, annual professional education conference on forensic identification and investigation.

International Biometric Conference, international conference held annually since ca 1983.

Biometric Consortium Conference, annual fall conference sponsored by the Biometric Consortium. www.biometrics.org/

Defending Cyberspace, CardTech SecurTech conference.

Carnahan Conference on Electronic Crime Countermeasures, covers a variety of topics including fingerprint technologies and publishes the proceedings.

10.d. Online Sites

The following are interesting Web sites relevant to this chapter. The author has tried to limit the listings to links that are stable and likely to remain so for a while. However, since Web sites do sometimes change, keywords in the descriptions below can help you relocate them with a search engine. Sites are moved more often than they are deleted.

Another suggestion, if the site has disappeared, is to go to the upper level of the domain name. Sometimes the site manager has simply changed the name of the file of interest. For example, if you cannot locate www.goodsite.com/science/uv.html *try going to* www.goodsite. com/science/ *or* www.goodsite.com/ *to see if there is a new link to the page. It could be that the filename* uv.html *was changed to* ultraviolet.html, *for example.*

Department of Social Services - The Connecticut Department of Social Services has information on the DSS Biometric Identification Project and many links to biometric publications. www.dss.state. ct.us/digital/dipubs2.htm

Fingerprinting Identification - This is an illustrated FBI educational site that introduces concepts and procedures related to fingerprinting technologies. There are also links to other pages, including polygraph testing. www.fbi.gov/kids/crimedet/finger/finger.htm

Southern California Association of Fingerprint Officers (SCAFO) - A nonprofit organization founded in 1837 to support professional identifiers. The Web site has an excellent, extensive fingerprint publications bibliography which, in some cases, includes links to the full text of articles. There is also general educational information about fingerprint science. www.scafo.org/

10.e. Media Resources

FBI Files, a weekly television series on the Discovery Channel that follows FBI investigations and describes various forensic techniques used in the solving of crimes.

Forensic Science, from the History Channel *Modern Marvels* series. This shows traces the history of forensic sciences using fingerprints, DNA profiles, fiber analysis and other technologies. VHS, 50 minutes. Cannot be shipped outside the U.S. and Canada.

Mission Impossible, a 1996 Paramount Pictures feature film starringa Rason: Tom Cruise as Ethan Hunt. A thriller that illustrates a few biometric technologies, including voice recognition and finger imaging.

Police Technology, from the History Channel *Modern Marvels* series. This show traces the history of the police department, from breakthroughs like fingerprint technology to modern forensic methods. VHS, 50 minutes. Cannot be shipped outside the U.S. and Canada.

Scene of the Crime, from the Arts & Entertainment *Scene of the Crime* series. This shows followscrime lab personnel as they gather evidence of a crime including fingerprints and other physical evidence. VHS, 50 minutes. Cannot be shipped outside the U.S. and Canada.

11. Glossary

Titles, product names, organizations, and specific military designations are capitalized; common generic and colloquial terms and phrases are not.

AFIS	Automated Fingerprint Identification System
assay	to analyze for one or more specific components
ATM	automatic teller/transaction machine
BAAPI	Biometric Authentication Application Programmers Interface, proprietary product of TrueTouch that is used by a number of third parties developing biometrics identification products
BIT	biometric identification technology
CCD	Consular Consolidated Database, a capacity of consular posts in foreign nations, as established by the Department of State, to capture electronic records of immigrant and nonimmigrant visas, including digitized photos and fingerprints (DoS)
CCH	computerized criminal history
DSV	dynamic signature verification
extraction	the assessment and processing of information or characteristics that are relevant to detection or identification
goats	individuals who, by lack of cooperation, eccentricity, or unremarkable personal characteristics tend not to be recognized by a detection or identification system
IAFIS	Integrated Automated Fingerprint Identification System, standardized FBI 10-finger print-matching system of rolled fingerprints. By the early 2000s, there were over 47 million sets of prints in the database that could be accessed from all 50 states.
IDENT	Automated Biometric Fingerprint Identification system piloted by the Immigration and Naturalization Service (INS) in 1995. As of the early 2000s, it included about 4.5 million prints of noncitizens.
identity discovery	assessing identification parameters and seeking to discover identity through investigation of the parameters and matching of the data and analysis with available information
identity verification	assessing identification parameters and confirming identity through matching with file information
IQS	image quality specifications (see IAFIS)
iris	the colored portion at the front of the eye surrounding the pupil but inside the whites of the eye which contains a unique pattern that is retained throughout life as long as there are no serious health problems or accidents
latent	hidden, not immediately or obviously discernible as faint stains or fingerprint oils
minutiae	the small details found in finger images which include ridges, patterns, valleys, and branching structures (bifurcations)
MIU	Mobile Imaging Unit, as used in squad cars for mobile identification and verification
SEERS	National Security Entry-Exit Registration System, a program in which certain foreign nationals who travel to the U.S. on nonimmigrant visas are required to register and deregister with the Department of Homeland Security upon arrival and departure, as well as being photographed and fingerprinted, with the information linked to the IDENT system.
PARIS	Pennsylvania Automated Recipient Identification System
PIN	personal identification number
platen	a surface on which a body part is placed in order to be imaged or otherwise scanned or detected
retina	the imaging portion of the back of the eye on which light is directed by the structures at the front of the eye, which contains a unique pattern of blood vessels

speech recognition the capability to recognize the content of a spoken communication (see voice recognition)

voice recognition the capability to recognize the presence of a voice or a voice that belongs to a particular individual, but not the content of a communication (see speech recognition)

US-VISIT the U.S. Visitor and Immigrant Status Indicator Technology administrated by Homeland Security

Biochemical Surveillance

14

Animals

1. Introduction

Animals have many senses humans don't have. Dolphins can communicate with sonar, dogs can smell and hear better than humans, birds can sense magnetic fields and see frequencies that we can't see, and moths and other insects can detect odors across remarkable distances.

Animals also have the ability to go places humans can't go. They can fly through forests, withstand cold temperatures without special equipment, and squeeze into tiny holes that would prevent human access.

A number of species are willing to aid humans with surveillance tasks—investigators have teamed up with mammals, birds, reptiles, and insects. Sea lions can perform a variety of tasks, from finding stowaways, criminals, and lost children to seeking out mines and shipwrecks. They've even been trained to attach deactivating devices to mines. The use of animals in surveillance and search and rescue operations can be highly effective.

Animals can be used as 'passive' sensing organisms as well. Their natural calls can be effective surveillance "alarms". As examples, dogs and birds will sound warning calls if they

The team of Federal Emergency Management Agency (FEMA) National Urban Search and Rescue (USAR) Response System dogs and their handlers. Sunny, Hawk, Louis, Ditto, Max, Duke, Miranda, and Guiness were deployed from four different states to Oklahoma City in May 1999 to aid in finding and aiding victims devastated by a tornado. [FEMA 1999 news photo, released.]

detect intruders. Similarly, natural animal behaviors, such as the scattering of rabbits or deer might indicate the presence of a large predator or an approaching storm or fire.

Animals as Detectors of Contamination

Animal and insect responses to stimuli or toxins can also be surveillance indicators. Animals, birds, and insects have been used in prospecting and industrial applications. From the earliest days of mining, canaries were used as 'chemical indicators' due to their sensitivity to vapors in mine shafts—if a canary was affected by gases in a mine shaft, the environment might be unsafe for humans, especially if there were prolonged exposure.

Some animal surveillance is of a less direct nature. Amphibians are sensitive to changes in the environment. Bear and eagle populations drop when salmon stocks are depleted. Fish disappear when streams and rivers are contaminated by industrial spills. Birds die when a new virus arrives in an area, as when the West Nile virus unexpectedly turned up in New York's Central Park region.

Insects can also be used to monitor environmental trends or changes. Spiders have been used to analyze the effects of drugs because drugs can selectively change the way they weave their webs. With further research, it may be possible to identify the quantity and toxicity of substances to which a spider has been exposed, which can provide information about the surrounding environment in the case of chemical spills or airborne pollutants.

Animal Protection and Conservation

Animals themselves need to be surveilled for their protection and ours. Poachers regularly kill endangered species and sell them for black market folk remedies or exotic foods. Gorillas are killed to turn their hands into ashtrays, game animals are taken out of season, elephants slaughtered for their tusks, endangered whales harpooned for meat or glands, rare cheetahs for their pelts. Using surveillance technologies to protect animals helps ensure their survival and ours, since we are part of a complex ecostructure that begins to die when essential parts of it disappear.

Bird flu and "mad cow" disease illustrate how important it is to surveil animal health. Many animals carry pathogens that can be transmitted to humans. HIV may have originated in monkeys. 'Mad cow disease' may be transmitted through the meat of cows, even if cooked. Paralytic shellfish poisoning can cause paralysis and may indicate water pollution. Salmonella may become endemic in chickens that are raised too closely together. Global surveillance and regular inspections are important health tools, and devices that can detect dangerous organisms are constantly under development. (See the Chemical Surveillance chapter for information on chemical-detection strategies.)

Animals that are exported to other countries must be inspected for diseases or injury but they also need to be checked as "vessels" for smuggling. Illegal drugs, gems, and other items are sometimes inserted into their digestive systems or other body cavities to hide them. Cocaine, heroin, and other drugs have been found in snakes, goats, dogs, snails, deer, and fish tanks. Thus, animal surveillance refers not only to the use of animals as detection systems, but also to the practice of inspecting animals and animal carcasses for the presence of controlled substances. This exploitation of animals is possible because surveillance of animals was generally restricted to quarantining them to detect diseases. This is slowly changing now that the abuse is becoming evident.

Animal Treatment as a Psychological Indicator

The way people treat animals has recently come under scrutiny. Psychologists have reported that those who abuse animals appear more likely to abuse people—information that may be relevant to a criminal investigation or may act as a warning sign to family members who are worried about the potential for violence among a relative and wish to seek psychiatric or legal assistance.

Animal surveillance is a broad topic and the smuggling and health aspects are touched on in other chapters, particularly in Chemical Surveillance. This chapter primarily discusses how animals (especially dogs) are used to aid humans in surveillance and, to some extent, covers animals as 'detection systems' based upon their reactions to chemical or physical stimuli. For reasons of space and scientific complexity, the less common aspects of animal surveillance are not covered in detail.

2. Kinds and Variations

There are four main aspects to animal surveillance:

animals as indicators - The behavior and vocalizations of animals, birds, and insects can warn of contamination, impending danger, or the presence of certain people or objects of interest. This knowledge is valuable when assessing contamination from industrial environments or accidents, or suspected chemical warfare. It can help when traversing unfamiliar terrain or when hunting for game. It can even alert a homeowner to the possibility of a prowler on the property.

animals as research subjects - Studying the flight patterns, chemical reactions, mating behaviors, hunting abilities, and sensory organs of animals teaches us about animals and about ourselves and leads to new sciences. Planes, helicopters, balloons, and many other technologies were suggested to our imaginations through our observations of animals, their physiology, and behaviors.

animal treatment as indicators - Observations of the ways in which certain people treat animals can sometimes aid in forensic 'profiling' in violent crimes investigations. A statistical correlation has been found between violent criminals behaviors, particularly in serial killers, and their treatment of animals. Cruelty to animals can't predict if a person is going to commit a violent crime, but violent criminals are often found to have been cruel to animals.

animal assistants - Animals are often enlisted to perform surveillance tasks. Millions of people have watchdogs to help surveil their property and homes. Dogs and dolphins have been used to find individuals or objects above ground and underwater, and cats can be trained to retrieve small objects in small spaces. K-9 police dog units have helped to find lost individuals or suspects who have fled the scene of a crime, including car thieves, bank robbers, and those who have committed assault.

The two most common animals specifically trained for animal surveillance are dogs and dolphins, so a large proportion of this chapter discusses them.

Canine Detectives

Dogs are the most favored surveillance assistants because they are eager for human companionship and often equally eager to help and to receive the praise and appreciation they get for a job well done. They are known for their extraordinarily good scent senses. Their sense of smell is thousands of times better than that of humans. The long muzzle of a dog provides an elongated nasal passage which, if unfolded, has a much greater surface area than that of a person. Dogs further have instinctive pack and watchdog abilities. Like birds, they will warn of intruders and sound a vocal alarm and, in some cases, will menace the intruder and prevent access to the property.

Dolphin Assistants

Dolphins, like dogs, are intelligent, social, and curious about people. Some of them even enjoy and seek out the company of humans. They have advanced natural sonar capabilities, large brains, and the ability to interpret two-dimensional images in a way that is superior to dogs and similar to humans. Dolphins can actually learn from viewing a videotape, an ability that has not been demonstrated by dogs and cats. They have been used in a variety of commercial and military applications for locating mines, wrecks, and vessels.

3. Context

Animal Warning Systems

Animals are used in situations where humans can't always be on the alert, where animals have better senses, or where humans can't go. Thus, animals are used to guard property when the owners are sleeping or away from the premises, to seek out lost, kidnapped, or murdered victims with their keen noses, or to attach neutralization charges to mines in areas that are hard to reach or hazardous for humans.

In the natural environment, animals are used indirectly for surveillance. We listen to their calls and monitor their reactions in order to sense danger ourselves. The frenetic behavior of bees, beetles, birds, and rodents can indicate a forest fire or impending storm. Elephants can hear and feel infrasonic vibrations and know when a large animal, vehicle, earthquake, or storm might be approaching.

Animal Use in Research and Testing

Animals are sometimes used in situations where there may be dangers to human lives and a tradeoff choice is made. In spite of continuing controversy, animals are used in scientific research and cosmetics testing, somewhat like the canary in the coal mine that may die if exposed to poisonous gases, but whose sacrifice, albeit involuntary, prevents the loss of human life. More and more, however, scientists are finding ways to synthetically duplicate animal sensory systems with detection technologies for use in situations where animals were once used. The use of animals for scientific research, especially in medicine, will probably never cease completely, however, as drug testing for new medicines is almost always tested on animals (typically mice and rats) before they are tested on humans.

Animals are sometimes used in testing situations where there may be dangers to humans or to the animals themselves. There is controversy over some aspects of animal testing, especially for cosmetics testing, but it is likely that testing will continue, in spite of objections, in medical research and pharmacy. Mice and rats are often used for medical experiments and birds are sometimes used to determine contamination, as they are sensitive to many substances. Many of the early concerns about the dangers of DDT pesticide were raised when observers noticed that birds were having fewer young and laying thin-shelled eggs because of direct exposure to the pesticide and indirect exposure through insects that were contaminated.

Space research has many rewards, but also some hazards. If life is discovered on other planets, even the most primitive viruses or bacteria could pose a danger to life on Earth if not handled carefully. Astronauts and their equipment must go through decontamination. Materials that are brought back from space, such as lunar rocks, are processed to determine if they might endanger any life on Earth. We don't want to risk exposure to pathogens that the organisms on our planet have never encountered and might be unable to counteract.

Materials brought back from the Apollo 12 space mission to the Moon were processed and inoculated into birds and rodents in order to ensure that they did not pose a danger to life on Earth. E. Landrum Young, of the Brown & Root Northrop facility, is shown here inoculating a bird in the quarantine cabinet (left) and examining inoculated mice in the Animal Laboratory, in 1969. [NASA/JSC December news photos, released.]

Research of Animal Physiology and Behavior

The study of animal physiology and behavior has led to many important scientific discoveries and has provided ideas and solutions for many different technologies. The study of dolphin sonar has taught us much about designing electronic sonar. The study of animals has even been suggested as a way to learn how humans might adapt to living in novel environments outside our planet. As an example, studies of the hibernation patterns of bears may yield information on how humans could better withstand the rigors of long space voyages without suffering from muscle wasting and other physical complications.

The observation of biological systems to gain ideas for designing new technological surveillance devices is not new. Studying wing designs and flight patterns of birds has provided a better understanding of aerodynamics and given rise to various models of aircraft.

The Indian River Lagoon, near the *Kennedy Space Center*, has a rich and diverse population of birds and also harbors dolphins, manatees, and many species of fish. Animals are frequently studied to learn more about locomotion, flight, and biological senses that are not found in humans and sometimes are directly enlisted to help, as in the case of pigeons, dolphins, sea lions, and dogs. [NASA/KSC news photos, released.]

Left: Scientist-Astronaut Owen K. Garriott is taking video footage of two spiders, Arabella and Anita, from the Skylab 3 space platform. In the weightlessness of space, the scientists are interested in observing how spiders cope with their new environment and build their webs, contributing to our knowledge of living and working in space. Right: A photo-reproduction of the color video transmission of Arabella, one of the two Skylab spiders. [NASA/JSC August 1973 news photo, released.]

The study of animal physiology and movement has led to some innovative design strategies for designing mobility systems for sensor-equipped robots that can navigate different terrains. Inventors have devised robots that bounce (so they can get up stairs), robots that move like snakes (so they can slither under things), and robots that move like insects or birds. The movement and gripping abilities of chameleons have been studied and applied to designs for robots that could walk along pipes and navigate through small spaces or hazardous environments [G. Immega, 1987].

Animal Assistants

Sometimes animals are used to surveil the environment on behalf of humans who have visual or auditory handicaps. Seeing eye dogs, hearing dogs, and dogs that can lift phones, open doors, and fetch objects, regularly assist in improving the quality of life for many people.

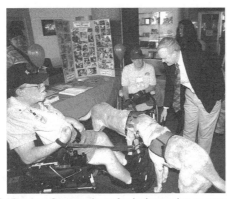

Left: Canine Companions for Independence was a vendor at the 1999 Disability Awareness and Action Working Group (DAAWG) Technology Fair that was held at the *Kennedy Space Center* (KSC). Vendors at the fair were highlighting technologies, both natural and man-made, that could assist in improving mobility, hearing, vision, and help people deal with 'silent' (less visually obvious) disabilities. These same technologies, which provide a way to enhance human senses, can also be used in surveillance for hazards and safety monitoring. The director of KSC, Roy Bridges, is shown petting a Labrador Retriever. Right: Nancy Shaw and her Golden Retriever 'dog in training' for the Canine Companions. [NASA/KSC news photo, NASA/GRC news photo by Tom Jares, released.]

Every human has a unique personality, and so does every animal. Dogs that are trained for one task are sometimes found to be unsuitable for the task, but do well at others due to individual differences in temperament and interests. Dogs that are trained as seeing or hearing dogs are sometimes found to be better at tracking or surveilling (and vice versa) and are sometimes reassigned to new handlers and trainers to continue their education in tasks that better match their abilities.

Dolphins and sea lions are intelligent, playful animals that have been used for many surveillance purposes, but dolphins are favored over sea lions for some types of tasks because sea lions have a tendency to suddenly leave—to play or chase fish.

4. Origins and Evolution

The importance of various kinds of animals as a source of food in early hunting and gathering cultures has been historically documented in cave paintings that are tens of thousands of years old, but some of the animals depicted in art also had mystical significance, as did cats in ancient Egypt. Large, black Siamese cats in Tibet are said to have acted as sentries, surveilling for intruders while guarding the crown jewels. Other animals served practical purposes, as well, such as dogs, horses, and monkeys. The partnership between humans and animals for companionship and protection has ancient roots. We'll never know exactly how long animals have been helping humans (and vice versa), but they may have been keeping amiable company for more than 100,000 years.

Dogs in Military Forces and Law Enforcement

Dogs of war existed thousands of years ago. Many surviving Roman artworks depict dogs and wolves. The Rottweiler breed is said to have descended from the dogs that accompanied the Romans on their invasions.

History is full of stories of animal helpers, sentries, and messengers. Horses, elephants, camels, and carrier pigeons have all aided in transporting not only individual travelers, but couriers, spies, and messengers, as well. Sometimes, in times of war, only animals and birds could get past the battle lines with a message.

Formal Animal Organizations

Historically, there are many stories of the strong bond between humans and animals, but there are also some unsettling stories of irresponsibility and cruelty. In the 1860s, the Society for the Prevention of Cruelty to Animals, one of the oldest animal rights organizations, began working in America to prevent animal abuses, to find homes for lost animals, and to provide education and medical care. It has since become a prominent and outspoken leader in the humane treatment of animals. Numerous other organizations now support this cause.

The formal use of dogs in modern police work originated in Nordic countries in the early 1900s, although hunting and tracking dogs had been used informally before that time for hundreds of years in many parts of the world. Dogs have long been used to find lost individuals or fleeing fugitives.

During World War I, thousands of dogs were used as sentries, scouts, pack-dogs, and as messengers throughout France and Germany, so it is not surprising that German and Belgian shepherds are still favored for many police and military tasks.

Seeing-Eye Dogs

The tradition of formal training schools for dogs was well-established in Europe by the

1920s, but was not common in America at the time. Once news circulated, however, training and breeding programs began to spread. In 1927, Dorothy H. Eustis wrote a column in the Saturday Evening Post describing how German shepherds were trained at a special school in Potsdam, Germany, and used to aid the blind. The Seeing Eye organization was established not long after, in 1929, to provide education and seeing-eye dogs to blind people. This organization is still active and other organizations for the blind and for training dogs were soon established soon.

Dogs were not the only animals that were of interest to scientists and animal handlers. The invention of the airplane had sparked a great deal of interest in human flight and a wide variety of successful and unsuccessful designs had been tried. Thus, the study of avian aerodynamics (the mechanics of bird flight) became more structured and objective in the early part of the century. Special equipment was developed in the 1930s to better study the mechanics and flying abilities of birds.

This historical photo from the NASA Langley Research Center shows an aerodynamics research system which is studying the flight characteristics of a seagull in motion on the carriage of the towing tank. [NASA/LRC 1932 news photo, released.]

Dogs in War

World War II was a time when new animal regiments of dogs and dolphins were established and new canine organizations formally founded.

During World War II, many thousands of dogs were used in the same way they were used in World War I, as sentries, messengers, and attack dogs. Sometimes they were also used in ways that were cruel— explosives were strapped to their backs and they were told to charge into the enemy lines, where the explosives would detonate.

While the British and American forces did not employ dogs to the same extent as the continental Europeans, they still recognized the value of dogs in military operations. Consequently, the British Royal Air Force *Guard Dog School* was established, in 1941.

The Royal Australian Air Force (RAAF) first introduced dogs into its security practices in 1943, using savage, untrained animals and, unfortunately, sometimes handling them callously.

Following the War, many dog units were disbanded, but a number of the military dog units were turned into training centers, for example, the British RAF *Guard Dog School* continued as the RAF *Police Dog Training Centre*.

In 1946, the *Guide Dog Foundation for the Blind* began providing guide dogs free of charge to blind people through its Guide Dog Foundation.

While the emphasis on dogs after the war wasn't as strong as it had been during the war, it was recognized that dogs could play a vital role in many peacetime activities.

In 1954, a RAAF *Dog Training Centre* was established, becoming the RAAF *Police Dog Training Centre* two years later. The purpose of the center was to breed and train dogs for security work, primarily using German shepherds, Doberman pinschers, and Labrador retrievers.

In 1958, the Australian Navy decided on the German shepherd as its sole breed for its training programs. The breed was selected for its protective coat, intelligence, and loyalty. Donated German shepherds of suitable temperament and health were accepted into the program.

Animals in Early Space Research

Since the beginning of space flight, animals have been involved in research into the special problems involved in working and living outside Earth's protective ozone/atmospheric envelope. Monkeys, apes, dogs, rodents, and spiders are examples of creatures that have been to space and back again.

A month after the launch of the unmanned space capsule Sputnik I, the Russian Federation sent up Sputnik II with a dog on board, to test the feasibility of sending people into space and the possible effects of radiation and other unanticipated hazards to health.

Stimulated by the launch of the Russian Sputniks, the U.S. soon embarked on its own space program.

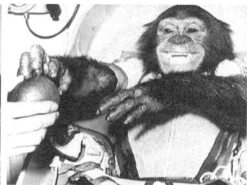

Chimpanzee Ham is recovered after a successful suborbital flight in a Mercury MR-2 space capsule. Ham, still strapped into his flight couch, which had been inserted in the space capsule, eagerly accepts an apple. [NASA/JSC January 1961 photos, released.]

Left and Middle Left: Enos with his handler before the flight of the MA-5, in his spacesuit and protective wrist tethers prior to the space flight. Middle Right: Enos holding hands with his handler as he is prepared in the flight couch that is designed to easily slip in and out of the space capsule. Right: Enos returns after orbiting the Earth twice and landing in the ocean waters south of Bermuda. [NASA/JSC news photos December 1961, released.]

In 1961, a chimpanzee named Ham was sent into a suborbital flight around Earth in a space capsule, to help us learn more about how animals and people react to the space environment and how we can develop technologies to protect their health. Later that same year, chimpanzee Enos successfully orbited the Earth twice in the Mercury-Atlas 5 space capsule. Chimpanzees were used partly for their intelligence and partly because they are almost genetically identical to humans. The physiological similarity between chimps and humans is much greater than most people realize and putting a chimpanzee in space was the closest thing researchers could do to putting a person in space. The return of the chimps indicated that human space flight was possible.

Dog platoons were added to U.S. military forces in the 1960s, including the 47th at Fort Benning and the 59th Infantry Platoon Scout Dog (IPSD), in 1968. The dogs were trained at the *Scout Dog Training School*. Some of the dogs were flown out by helicopter to serve forces in the Vietnam war. Dogs aided in patrols, the detection of snipers, and the detection of booby traps. Afterward, many were euthanized and buried in South Vietnam, in 1971.

The 1970s and 1980s - Increased Interest in Canine Programs

Many new canine programs were initiated or tested in the 1970s and 1980s. In addition to the already familiar seeing-eye dogs, larger numbers of hearing dogs, narcotics-sniffing dogs, and human-bonding dogs made the rounds of airports, borders, ships, and hospitals, and aided professionals in carrying out their duties more effectively.

In 1972, the designation of *Drug Detector Dog* and non-medical drug detection capabilities were added to the Royal Australian Air Force (RAAF) canine teams. The dogs were used not only for local policing, but for assisting in state and federal law enforcement.

During the 1970s when interest in canine programs increased, there was also interest in enlisting dolphins and sea lions, largely due to the research and educational information provided by Jacques Yves Cousteau and the Cousteau Society.

In 1982, a New York City bomb squad member joined a forensic chemist from the federal *Bureau of Alcohol, Tobacco, and Firearms* to demonstrate that a dog could be trained to seek out accelerant odor better than existing mechanical detectors. The ability to detect accelerant was useful not only in sniffing out bombs, but also in sniffing out chemicals at an arson site.

Space exploration and research continued in the 1980s with a series of Space Shuttle missions—animals were sometimes included in the scientific research payloads. In 1985, two squirrel monkeys and a number of rodents went along for the ride in the Shuttle Challenger Spacelab-3 mission.

The Mid-1980s - Health Care, Arson, and Transportation Security

Animals make good companions and many people respond positively to their presence. The U.S. Army initiated a human-animal bonding program in 1985, with two dogs who could soothe and entertain with a variety of tricks. Dogs and other animals have been found to have an especially therapeutic effect in hospitalized patients and tenants of geriatric/convalescent homes.

In 1985, the police role of the RAAF canine division was formally recognized and the RAAF Police and Security Service formed. The first female RAAF handler was graduated that year.

In 1986, the *Connecticut State Police Department* began a canine arson program with a black Labrador retriever that learned to detect more than a dozen different accelerant odors including solvents and lighter fluids. This established the viability of canine units for arson programs elsewhere.

In 1988, tragedy struck when Pan Am Flight 103 exploded over Lockerbie, Scotland. This horrific bombing resulted in tightened airport security, in many nations, and dogs were used in some places to search for explosives and other suspicious objects.

In 1988, the *Office of Fire Prevention and Control* for New York State began using dogs to help investigate fires. The office currently uses Labrador retrievers in its work.

We might think that the cruel use of dogs to deliver explosives ended with the end of World War II, but this hasn't been the case. In 1989, Israeli forces are said to have used Belgian shepherds to deliver remote-controlled explosives and gas into enemy bunkers in Lebanon. When the dogs reached the intended target, the explosives were remotely detonated.

The 1990s - Narcotics and Explosives Detection

In 1990, the County of Santa Barbara initiated their *Canine Drug Detection* program. The dogs are trained to detect a variety of types of illegal substances by searching buildings, vehicles, residences, and containers.

The *Australian Customs Service* had traditionally used dogs from animal shelters, but as the number of suitable dogs declined, due to responsible spaying and neutering programs, an internal breeding program was initiated. Labrador retrievers were chosen for the program to carry out a variety of enforcement-related tasks.

During the mid-1990s, there were some terrorist bombings of U.S. personnel and property, and dogs were used in search and rescue operations to help locate victims in the collapsed rubble. Dogs are usually trained either for the detection of live victims or the location of corpses. It was discovered that dogs, like humans, can become depressed and lacking in enthusiasm for their work if they come across too many dead victims. Handlers would sometimes allow them to find a live, planted 'victim' toward the end of a day in order to leave them with a positive experience and a happier frame of mind.

In 1996, President Clinton established a White House commission to study air travel security and also declared the investigation and prevention of church arsons to be a national priority. In June 1996, the *National Church Arson Task Force* (NCATF) was formed. In investigations spanning incidents occurring from January 1995 to September 1998, almost 700 cases of arson or bombings in houses of worship were investigated. A variety of law enforcement agencies cooperated in the subsequent efforts, which utilized a number of surveillance devices and techniques including the use of explosives- and accelerant-sniffing dogs. More than 300 arrests and more than 200 convictions resulted from these investigations.

In June 1998, the U.S. *Department of Defense* (DoD) participated in Working Dog trials and were awarded 12 out of 31 awards in the two-day competition. In September 1998, the *Office of the Assistant Secretary of Defense* announced that four German shepherd dogs had been added to the *Defense Protective Service* (DPS) which is essentially the police force to the U.S. Pentagon, which houses over 24,000 workers. The dogs would be used to help maintain security on a more regular basis at the DoD headquarters. Previously, dogs had been brought in, as needed, by military branches. The four dogs and their handlers were graduated from the Air Force police dog training school with a specialty in explosives detection.

Bee Studies

The DoD wasn't limiting its research on animal surveillance to dogs. Through a DARPA project, it began studying whether honeybees could be equipped with tiny radio frequency tags to enable them to be used to detect land mines as part of a larger effort called the *Controlled Biological and Biometric Systems* program, to determine some of the possible ways in which insects, reptiles, and crustaceans might be used in surveillance tasks. The *Pacific Northwest National Laboratory* and the *University of Montana* were assisting in fitting the bees with radio tags. The tags can help track the exit of a bee from the hive, its direction of travel, points where it may have landed, and the time that has elapsed. Chemical analysis of returning bees can lead to information about where they may have been and what they may have encountered, which might include industrial or warfare residues.

Bees have been used by other researchers in the past to collect information on pollutants or trace materials. Their bodies tend to absorb contaminants to the point where they are registered as valid data collection means with the *Environmental Protection Agency*. Some species of moths and wasps could perhaps also be used in similar experiments.

Sandia National Laboratory embarked on another ambitious project using bees. It is studying whether they can be trained to sniff out TNT and associate the smell with food. TNT residues sometimes seep from land mines and contaminant the surrounding area. The bees appear to learn readily, traveling up to 100 yards to the source after just a few hours of training.

The Late 1990s - Continued Security Concerns

In the late 1990s, the human population was reaching six billion, and never before had the Earth's populace been so mobile. With concerns about terrorism, violence, and smuggling on the increase, the interest in using dogs and other animals in ways that could improve security and detection continued.

In 1998, the CIA provided assistance to the Fairfax County and Vienna police departments in searching for explosives that might be present due to the high-profile Kasi trial. Dogs that were members of the K-9 Explosive Ordinance Disposal assisted with this security operation.

The U.S. Federal Aviation Advisory, in efforts to increase airport security and lower the threat of terrorism, recommended, among other things, that high-traffic airports have at least two explosives-sniffing dog teams, in conjunction with bomb-detecting equipment.

Electronic Sensors

New technologies are often developed from ideas that come from nature. Electronic 'sniffers' based on a dog's nose have been tried a few times in the past, but their use in surveillance is now considered a higher research priority because electronics has improved and made new types of devices not only possible, but more economical.

In the late 1990s, the American Forces press service announced that the *Department of Defense* was working on a three-year research project to attempt to electronically duplicate some of the sensing abilities of a dog's nose, which could aid in detecting explosives, such as those found in mines. It was hoped this would supplement and perhaps even overcome some of the limitations of other methods currently used to search for mines such as ground-penetrating radar, infrared sensors, and metal detectors. If the project were successful, it could signal the end of the use of dogs for the detection of land mines. It could also create technologies that could 'smell' a wide variety of substances for use in thousands of industrial and security applications. It remains to be seen how well these sniffers will work and how long it will take for the technology to become advanced enough to be practical.

In 2001, Electronic Sensor Technology, Inc., announced a fast, sensitive, and versatile electronic sniffing device called the zNose. It was initially promoted as a way to sense the environment, for scientists to study it, and to have potential applications for monitoring air quality in buildings, and sensing substances in the food and beverage and medical industries. The potential for detecting materials used in chemical warfare were also mentioned, but were not overly emphasized. By 2005, however, the zNose was widely promoted for use as a Homeland Security sensing technology, as a first responder system for detecting airborne pathogens, explosives, and chemical threats. (See the Chemical Surveillance chapter for more information on chemical sensing.)

5. Description and Functions

Dogs are the most common animal used in surveillance tasks, with tens of thousands of trained dogs enlisted worldwide.

Dogs as Surveillance Workers

Much of the surveillance work carried out by dogs is based upon their willingness to keep watch and their ability to smell.

The ability to pick up scents varies somewhat with the breed of dog. Bloodhounds are well known for their ability to scent people. The nose of a dog is long, with a much greater surface area than that of humans, and is divided into specialized chambers to distinguish and classify many hundreds of individual smells.

There are air-scenting dogs and near-scenting dogs, depending upon the breed and training of the dog:

air-scenting - These dogs can pick up a scent on the wind, usually of a lost or fugitive person who hasn't been gone for too long. Depending upon the wind and individual circumstances, a dog can scent a human up to about 550 ± 400 meters. Air-scenting is usually done on a leash unless the dog is specifically trained to seek, attack, and hold an intruder, in which case it may work off-leash.

near-scenting dogs - These dogs are trained to discriminate particular scents on items that are close to the dog's nose. Near-scenting dogs may be able to pick up a scent on the ground, but near-scenting is more commonly used in narcotics and explosives detection, in which the dog is taken close to the item being checked, which can include vehicles, dressers, school lockers, suitcases, clothing, luggage, or appliances. Most proximity-scenting is carried out on a leash, although some dogs are well-disciplined and well-trained and can be relied upon off-leash.

Selecting a Dog Breed

Breeders become very attached to their favorite breeds, but it should be remembered that almost any breed of dog can be trained for surveillance tasks if the animal is individually evaluated for an alert temperament, sufficient intelligence, and a willingness to work. Many professional dog trainers and official K-9 units get their dogs from local humane societies, which include both purebreed and mixed-breed animals. Mixed breeds, in some cases (there are no guarantees without X-rays and medical examinations), are less subject to genetic abnormalities such as hip dysplasia or deafness (as is found in some dalmations), but again, there are exceptions. A mixed breed of two genetically inbred breeds may be just as prone to genetic abnormalities as a purebreed. There is no hard and fast rule about whether a specific breed or a mixed breed is better for surveillance tasks in general, though some breeds can be more suitable for specific branches of surveillance.

Some breeds of dogs are favored for particular tasks because they have been bred for their willingness to work, their congenial dispositions, or for certain types of retrieval or scent surveillance. Some have also been bred for their willingness to work in or around water (not all dogs like water). For police work, the psychological advantage of using a dog that has dignity and a certain aloof air is also taken into consideration.

Genetics can strengthen particular traits in animals, but the way dogs are bred and judged for shows does not guarantee intelligence, unless the breeder and his or her dogs are regularly involved in and breeding for obedience trials. Not every individual animal has a good temperament even if the breed, in general, is docile or affectionate by nature. Even when a specific breed is chosen, each animal should be individually evaluated. A puppy with good potential will look up and try to understand when spoken to, rather than looking away or running off. High-strung dogs may be difficult to train and don't always calm down as they get older.

There aren't substantial differences between male and female dogs but there are some, including a greater tendency on the part of male dogs to wander or fight. These tendencies can be suppressed somewhat by neutering the dog at a young age. The dog trainer who trained the various dogs for the Lassie television series in the 1950s was once asked if there was a difference between the sexes. He replied that he felt the females were smarter, but the males were more willing to take risks, like crossing rough rivers.

Bloodhounds, golden retrievers, Labrador retrievers, Belgian malinois, German shepherds (Alsatians), and border collies are particularly favored for surveillance work. Beagles are used for some types of contamination detection. St. Bernards were traditionally bred for searching for victims of avalanches in the high mountains of the European Alps. Labs and Retrievers are particularly favored as *seeing-eye* dogs. Seeing-eye dogs are trained to surveil surroundings in order to protect and provide cues for their blind companions. The large Newfoundland dogs were bred to find and aid seafarers who had gone overboard from their boats.

Dog Training

The best time to start training a dog is when it is five weeks old. A dog that is trained young can be remarkably well-trained by the time it is four months old and often will have fewer 'bad habits' to try to undo later. However, for professional purposes this is often not practical or possible. Puppies want to chew, run, pee, and play, and most trainers in professional positions find it better for their purposes to train a dog that is between six and 18 months, when the dog has settled down and is more prone to listen. The ideal compromise is to board the dog with a good owner/handler while the dog is young, so it learns to listen and to respond to praise and instructions and then to give it professional training for a particular task when it has settled and matured.

Dogs are usually trained to indicate a 'find' by a behavior or a sound (or both). If a dog is used for surveillance, such as tracking down snipers or patroling a military zone, a dog may be taught to sit, to return, to grasp clothing, or to crouch, in order to indicate a 'hit'. For regular police work, search and rescue work, or forensic searches (e.g., for a body), the dog may be taught to sound an alarm by barking or whining. The breed may determine which type of task is best suited to the dog. Husky breeds tend to be very vocal, for example—they like to 'talk' and may not be the best breed for silent alarm tasks.

Dogs can be taught with treats as rewards for their work, but it is best to teach them with praise. Sometimes a dog that is taught with praise can be given a special treat for an especially good training session or action. Playing vigorously with the dog after a good training session is also a good reward.

Scenting dogs get tired, just as people get tired, when they are doing work that requires concentration and skill. It is best not to use a scenting dog for more than half an hour at a time and to give it breaks. Patrolling dogs have longer duty periods. Dogs are descended from wolves, which have good stamina. Most of the breeds used for surveillance work have good stamina and can run or stay on patrol for long periods of time. If a dog is fatigued and needs a break, a good handler will notice and respond appropriately.

Dolphins

Dolphins are air-breathing, highly intelligent mammals that live about 30 years, depending upon conditions, and whether they are wild or captive. They live in tight-knit groups and form strong social bonds. They only have about one offspring every two years, so they cannot quickly or easily replenish their numbers when their members succumb to injury or illness.

Dolphins are cetaceans, which also includes whales and porpoises. They are cooperative when they are 'in the mood' or if they know there will be a reward of food or something they desire. Otherwise, like human youngsters, they are usually more interested in socializing and playing with each other than in doing work.

Dolphins, like people, take a few years to mature, so dolphins are usually a few years old before people begin training them. It takes about three years to train a dolphin to be proficient at detection work.

Sea Lions

Sea lions may not be as intelligent as dolphins, but they are still bright animals, trainable, and have a number of good sensing abilities. They don't have the sophisticated sonar of dolphins, but they have excellent hearing and an amazing ability to navigate in murky water. This enables them to find objects under the water that make noise, like pingers that may be attached to wrecks or flight recording devices. Like dolphins, sea lions need to be motivated and rewarded for their work or they will take time out for unscheduled play.

Birds

Birds have been used as surveillance partners because they can fly, they are cooperative (birds often strongly bond to the people who raise them), they can live a long time. Some, like pigeons, have strong homing instincts. Carrier pigeons, once extremely numerous, were used for sending messages, but were ruthlessly hunted to extinction around 1914. Parrots and cockatoos are intelligent problem solvers that sometimes live longer than humans. Birds can be taught to pick out certain objects or colors and to place or retrieve various types of objects. They also make good 'watchdogs'. A cockatoo, for example, has a very strident alarm call that can be heard for some distance, that it uses to signal intruders (or people it doesn't like).

6. Applications

Search and Rescue

Dogs, with their keen sense of smell, are especially valuable in seeking out individuals who have run away, been lost, kidnapped, or become endangered by a hazardous situation or environment. Hazardous situations include collapsed buildings, train wrecks, chemical spills, avalanches, floods, and earthquakes.

Dogs can sometimes fit through holes that are too small for people, can swim in water that is too cold for people without wetsuits, and can smell individuals who may be unconscious or confused and unable to call out for help.

Two search dogs and their handlers en route to a Coast Guard Air Station after finding the body of Ned Rasmussen on Uganik Island. Rasmussen had been wearing camouflage clothing which made it difficult for searchers to locate the body. Mack, the German shepherd on the left, handled by Jennifer Brevik, was able to locate the body by smell. On the right is Jean Adams with her golden retriever Meeko. [U.S. Coast Guard 1999 news photo by Keith Alholm, released.]

Members of the Federal Emergency Management Agency National and Urban Search and Rescue (USAR) response systems, use dogs to aid in search and rescue operations. Left: A USAR worker with a trained golden retriever at an Oklahoma tornado site. Right: A USAR Task Force member and a retriever search through the debris of the Murrah Building after the 1995 Oklahoma City bombing that collapsed the building with many people inside. [Federal Emergency Management Agency 1999 and 1995 news photos, released.]

Accelerant, Explosives, and Firearms Sniffing

Dogs have been trained to sniff for explosive materials, which makes them able to locate bombs or land mines.

Since 1986, the U.S. Bureau of Alcohol, Tobacco, and Firearms (ATF) has been training dogs to detect accelerant, a telltale sign of arson. Since 1992, it has also had a canine team trained to detect explosives, explosives residues, and post-blast forensic evidence. The ATF canines can also detect firearms and ammunition that are buried or otherwise hidden. The ATF canine program is operated together with several organizational branches, including the National Response Team, the Explosives Technology Branch, ATF Labs, Certified Explosives

Specialists, and the ATF Firearms Branch and Tracing Center.

The ATF has also worked with the U.S. *Department of State* and the Connecticut State police to train and provide explosives-detecting dogs that can be sent overseas to aid foreign countries with programs to combat terrorism. The ATF also offers training to other federal and state agencies, and local law enforcement agencies.

U.S. ATF canines are trained to detect a variety of explosives scents, as well as hidden firearms and ammunition. The dogs are trained and certified for their jobs. [ATF Canine Operations Branch news photos, released.]

Terrorist threats at events that attract large numbers of people have also been a concern after bombings at sports events triggered calls for tighter security. New Orleans police canines were used to sniff under each seat and around public areas before the kickoff of the Super Bowl in the Louisiana Superdome, to seek explosives that might have been hidden in anticipation of the crowds. Dogs are also used to surveil other events in which large numbers of people congregate, such as religious meetings, pageants, and music concerts.

Left: A Massachusetts State Police collector's patch depicting a bomb detection dog, in this case, a shepherd. Right: A vehicle in Saudi Arabia is searched by Jacky and Snr. Airman Tammy Kirksey, who served as an Explosives Detector Dog Handler with the Security Police during Operation Southern Watch. [Classic Concepts photo, used with permission; U.S. DoD 1996 news photo by R. M. Heileman, released.]

Left: Jupiter, a bomb-detecting dog, with handler SSgt. Leach checks vehicles entering a compound at the Rafha Airport in northern Saudi Arabia. Right: Members of the Bureau of Alcohol, Tobacco, and Firearms (ATF) sift through the debris at the scene of a fire in Colorado, seeking evidence of arson, with the assistance of a dog trained to detect explosives and accelerant. [U.S. Army 1991 Airborne Corps archive photo by LaDona S. Kirkland; ATF news photo, released.]

Mica, a black Labrador retriever with the New York State Office of Fire Prevention and Control (OFPC) aids handler Randi Shadic in seeking out accelerants that may have been used to burn down this wood-frame building. [OFPC news photo, released.]

In February 2006, the British military opened a kennel in east Africa. as part of the International Mine Action Training Centre (IMATC), to offer landmine detection training using dogs to African forces involved in demining operations.

Narcotics, Pest, and Agricultural Sniffing

The chief responsibility of canines associated with most justice departments and customs organizations is sniffing for narcotics or explosives. However, there are also dogs trained for recovering victims or bodies from accident or disaster sites and dogs trained to identify contaminated foods or other substances that might pose a health risk.

Food Substance Sniffing

Dogs can be trained to search for a general category of substances or for specific substances. Since tainted food products or foreign species of plants can pose health risks if imported illegally, dogs are used to aid customs officials in inspecting luggage. The U.S. Beagle Brigade is specially trained to sniff luggage for the presence of meat products. In one year alone, almost 1/4 million pounds of illegally imported meat and related animal products are seized and confiscated in the U.S.

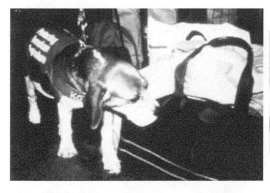

A member of the U.S. Department of Agriculture's 'Beagle Brigade', which aids the Animal and Plant Health Inspection Service detect illegal imports that might threaten the health of domestic agricultural products. The dogs are trained to detect fruits, meats, nuts, and other regulated plant materials. [U.S.D.A. news photos, released.]

Drug Sniffing

In 1992, The California Highway Patrol (CHP) acquired a number of drug-sniffing dogs and put 11 on the road in 1993 to participate in a national effort, sponsored by the U.S. Drug Enforcement Administration (DEA) to locate drugs and weapons carried on the nation's highways. By 1999, there were more than 40 dogs on the force.

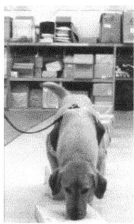

Left: The Australian Ministry of Justice canine section is used to detect illicit drugs in western Australian prisons. There are currently six dog teams that include dogs that are trained either as drug-detection dogs or as multipurpose dogs. Right: An Australian Customs Service Detector Dog Unit (DDU) aids in locating smuggled narcotics that may be hidden in luggage, cargo containers, vehicles, and other vessels, or carried on a person's body. The program began in 1969. [© 2000 Australian Ministry of Justice photo and 2000 Australian Custom Service photo, used as per copyright specifications.]

The customs agents of many countries use dogs to sniff for narcotics, smuggled items, and stowaway humans.

Top Left and middle: This 86-foot oil-supply vessel was found to have 86 kilos of cocaine hidden in the hull. Bottom Left and Right: Dogs sniff vessels to make a positive detection. In this case, a dog hit on a ship, so a hole was drilled into the hull. The dust that came out with the drill bit was put in a sensitive chemical identifier and the blue color indicated cocaine. The hull was then opened to reveal individually packed kilos of cocaine. Top Right: A portion of more than a ton of cocaine that was detected and seized aboard another marine vessel. [U.S. Customs news photos by James Tortelotte and top right photo by Todd Reeves, released.]

The detection of narcotics smuggling at various borders is an ongoing challenge for law enforcement agents. In half a year, in Arizona, more than 100,000 pounds of various narcotics were seized. Dogs are used to detect and locate narcotics hidden in vehicles and containers. Large amounts of undeclared currency are sometimes found with the narcotics.

Left: U.S. Customs sponsors a "Canine of the Month" section on their Website that features pictures and descriptions of the various dogs that aid enforcement officers in carrying out their duties. This golden retriever named "Zulu", was stationed in San Juan, Puerto Rico, in the late 1990s. "Zulu" had been working with the Customs service since 1990. Golden Retrievers are popular because of their mild, friendly dispositions, their desire to please, and their pride in their work and their handlers. Right: "Rufus" is another golden retriever who began working with Customs in 1990. Rufus was stationed in San Diego at the time this picture was composed. [U.S. Customs 2000 news photos, released.]

The following examples give a picture of how high technology and trained canine narcotics-sniffers are used to detect and locate hidden drug shipments:

- In February 2000, canine "Bo" sniffed a Dodge sedan driven by a man and woman from Mexico City. The dog signaled a positive hit in the vehicle floorboards. A drill was used to check through the floor and cocaine dust was positively identified with a chemical test. Opening the compartment revealed 91 pounds of cocaine.

- Another Dodge sedan with one male occupant was referred for a secondary inspection. The Customs Inspectors used a high-tech density-measuring device to detect irregularities in the vehicle's bumper area. Canine "Boyka" produced positive alerts for the bumper and dashboard. Inspection revealed 50 pounds of cocaine in the bumper, dashboard, and quarter panels.

- In San Diego, a pleasure boat was detained while a narcotics-detecting canine from the San Clemente Border Patrol station was summoned. The dog made a positive alert that resulted in the discovery of a hidden compartment retrofitted with an electronic latching device designed to make it look like a sealed compartment. More than 2,000 pounds of marijuana were found wrapped in cellophane bags in the compartment.

- When a Laredo truck driver came through the Colombia import lot hauling an empty trailer, a post-primary canine inspection by "Cowboy" resulted in a positive alert to the front wall of the trailer and the fifth wheel area. X-ray scanning then confirmed irregularities in the front wall. On inspection, almost 4,000 pounds of hidden marijuana were discovered. The seizure was a cooperative effort between Customs Special Agents and Laredo Police Narcotics Unit Officers.

- In March 2000, 13 canines and their handlers graduated from a two-week training session at the *Customs Canine Enforcement Training Center*. The training taught the dogs to detect MDMA "Ecstasy", which is being increasingly used by young people at 'raves' (parties). MDMA is primarily methamphetamine.

Left: Spc. Colleen Neubest and narcotics-sniffing Dutch shepherd "Fedor" disembarking from an aircraft in Bosnia, in August 1998. Right: Dr. Dunn of the *Armed Forces Institute of Pathology* veterinary department is shown here with a microscope. The IoP provides veterinary support for about 1,500 military working dogs in the U.S. *Department of Defense* plus others with the *Federal Aviation Administration*, *Customs and Border Patrol*, and the *Secret Service*. [Armed Forces Press Service news photo, ca 2000 by John Lasky; DoD news photo by Rudi Williams; U.S. DoD news photo, released.]

Pest Inspection

In the mid-1980s, a termite-sniffing dog named Sidney joined the Magic Exterminating Company in Flushing, Queens. Dogs have been trained to sniff walls for the presence of termites and other possible pests for an exterminating company in San Francisco, as well. Alabama-based *FSI Canine Academy* trains about three or four termite-sniffing dogs a year. After going through a rigorous training program, the bug-finding dogs have a 'hit' rate two or three times better than a human inspector. Human inspectors have traditionally depended upon visual cues for locating termites (although checking the density of materials with an electronic device is now also possible). Dogs are useful because they can fit into crawl spaces and behind small fixtures more easily than humans.

Pipeline Management and Hazardous Materials Detection

Dogs have been trained to detect Tekscent, a nontoxic chemical that is used as a detection aid in finding leaks in pipelines. Dogs are used by companies like Imperial Resources Limited in Calgary, Alberta to find leaky pipes buried as deep as 15 feet.

Dogs have also been used by the Swedish *Environmental Protection Agency* to sniff out shelves, instruments, pipes, sewage systems, and fissures for mercury contamination. Dogs can detect mercury in quantities as low as 1 milligram of mercury in normal temperatures. In 1998-1999, more than 1,300 kilograms of mercury were recovered from about 1,100 school sites. This is only a small proportion of the total amount of mercury, however. Between 1994 and 1999, the two dogs, Ville and Froy, sniffed out more than three tons of the highly toxic substance.

Wildlife Detection

Almost everyone is familiar with hunting dogs that have been trained to seek or pursue specific species of animals or birds, especially pheasant, duck, deer, wolves, or bears. Dogs can also sniff out endangered species or wildlife that are monitored for conservation purposes.

Dogs can also be trained to seek out individuals suspected of poaching, or animals that have been partially harvested and the remaining carcasses abandoned (gutpiles, hides, heads, etc.). They may also locate contraband carcasses in vehicles, campsites, or storage containers.

Dogs are able to detect the feces of target species so they can be studied and monitored and so sampling in the field can be done more selectively and efficiently. The dogs can find specific species so samples can be taken back to a lab for further analysis, including DNA profiling. With DNA profiling, the specific individual that left the feces can be identified for noninvasive tracking, without the use of radio collars or tags, or in instances where the radio collar has been lost. They can been trained to find ferrets, brown tree snakes, or the threatened indigo snake. The data from these studies are valuable in census-taking, wildlife management and conservation activities. Working Dogs for Conservation is one organization dedicated to the use of dogs for wildlife sensing.

Dogs have even been trained to search for undesirable plants, so they can be eradicated and could, theoretically be trained to seek out endangered plants, as well.

Underwater Surveillance

Both dolphins and sea lions are willing to surveil underwater environments for their trainers. Sea lions can hear well and navigate well in dark water and dolphins are extremely intelligent and have remarkable sonar systems that allow them to see objects inside other objects (described more fully in the Sonar Surveillance chapter). Dolphins can help locate underwater wrecks, divers in distress, minefields and individual mines, lost items, and more. Sometimes they can even place and retrieve specific items.

This dolphin is a member of the Mobil Unit 3 team out of San Diego, California, shown here on location in Lithuania with Staff Sgt. Fahs (left) and Col. Linn of the U.S. Air Force (right). The dolphins were deployed in Lithuania as part of Operation Baltic Challenge, a peace support exercise in the Baltics that included the U.S. and 11 European nations. [U.S. DoD 1998 news photo by Eduardo Guajardo, 55th Signal Company, released.]

In the U.S. Navy Mk 4 Mod 0 program, Pacific bottlenose dolphins have been trained to attach neutralization charges to the mooring cables of buoyant mines.

Health Sensing

It has been said that humans are 'walking bags of hormones', and this is largely true. Hormones circulate constantly, in different proportions, in our blood streams. A dog, with its sensitive nose, can detect subtle changes in a person's mood or health, based upon smells. This information may be useful in training dogs to detect liver and stomach problems and possibly even cancers.

Commercial Products

This is not a complete list of all vendors, nor does inclusion of a company imply endorsement of the quality of their products or services. The following is intended to provide examples only, to give an idea of the kinds of products that may be purchased on the market related to this topic.

Dog Training Scents

To train an animal to recognize and respond to particular scents, it helps to have selective examples of those scents for the animal to smell. There are commercial companies that create containers with specific kinds of odors that would be indicative of chemicals related to arson, smuggling, or explosives surveillance and package them for use for training dogs.

Sigma Chemical Company - Produces pseudoscents that simulate odors ranging from narcotics to a fresh corpse, or the smell of a person in distress.

Dog Training Equipment

Verschoorpak - A Dutch dog sport equipment company that sells bite-training cuffs and suits. It is Florida-based, with equipment produced in The Netherlands.

Dog Training Schools

Note, this is not a comprehensive list, but rather a few examples. Some of these training schools also supply dogs.

Command Dog College - Okotoks, Alberta. Labrador retrievers have learned to detect Tekscent.

The K9 Center - A two-acre dog training facility located in Richmond, B.C., Canada. It is managed by Eden & Ney Associates which provides information on police dogs and related law enforcement issues. The K9 Academy for Law Enforcement trains dog handlers.

Lynnwoods Kennels, Inc. - Located on 12 acres, this facility provides law enforcement canine handler courses including narcotics, explosives, accelerant, and cadaver detection along with patrol work handling. Located in Fremont, Ohio.

Southern Star Ranch K9 Training Center - A 100-acre Texas-based ranch that trains dogs to sniff out accelerants that might indicate arson, for law enforcement and insurance companies. Licenced by the U.S. Justice Department.

Ohio Township Training Center - A one-acre plot within a 52-acre park allocated as a training area for dogs to serve a number of police departments. The facility would provide training for narcotics and explosives detection and search and rescue activities.

In 1995, the U.S. Fish & Wildlife Service initiated the Search and Find canine program in collaboration with Special Agents and Inspectors in San Diego. Dogs trained for search and rescue tasks were taught to detect both live and dead smuggled wildlife destined for the San Diego and Los Angeles areas. [U.S. Fish & Wildlife Service news photos, released.]

Commercial Dog Sniffing Services

Commercial services require a license from the Drug Enforcement Administration.

Detector Dogs Against Drugs and Explosives (DDADE) - Founded in 1997 to provide narcotics and explosives sniffing services to businesses, schools, and private homes.

Located in Virginia with franchises in other states. The fee can range from $250 to $1,000 per search.

InterQuest Detection Canines - Established in the late 1970s, InterQuest conducts over 10,000 searches a year. About one in five searches leads to detection of drugs or explosive substances. The company also supplies dogs for law enforcement, school security, and commercial applications. InterQuest is a Houston-based multistate service.

Dog Suppliers

Many dogs that are trained for sniffing and search and rescue tasks are recruited from local humane societies and some are specifically bred for certain tasks. When China was interested in acquiring a number of drug-detection dogs, they chose dogs trained by the Australian Customs service. Twenty of the dogs were scheduled to be delivered to work in drug-detection tasks in southwest China startings in 2007.

Police Service Dogs International (PSDI) supplies patrol, tracking, detection, and multiple-purpose dogs of the German shepherd and Belgian malinois breeds. Canine Protection Services is a division of PSDI. Based in Massachusetts.

7. Problems and Limitations

Narcotics-Sniffing False Positives

False positives are somewhat common in drug sniffing operations. Dogs have excellent senses of smell and drug smoke and residues linger on clothing for some time, sometimes so strongly that human noses can pick up the scent. Thus, for every 100 people who might be identified by a dog as potentially carrying drugs, usually only about a third are found by subsequent searches by people to have drugs in their possession at the time of being 'sniffed'. This may be a mistake on the part of the dog—the person may have been playing with dogs or smell of something else that attracted the dog's attention. But there are also instances in which the person threw away the drugs or previously smoked or ingested drugs (or was in close proximity with others who had been using drugs), such that the smell lingers on the hands or clothing. In fact, people who do not smoke marijuana but who regularly socialize with those who do, sometimes have detectable amounts of marijuana in their bloodstream from second-hand smoke. While there hasn't been much research on this, dogs may be able to detect these trace blood levels if the person is perspiring or otherwise excreting body fluids.

More research is needed to determine the specific reasons for false positives. In the meantime, sniffing dogs will sometimes single out individuals based on their recent activities, rather than actual possession of substances at the time of the search.

Relocation of Animals

Animals are no different from people when it comes to susceptibility to infections and parasites, especially if they travel to foreign regions. If a dog or dolphin is transported, it may be vulnerable to tropical diseases, parasitic infections from worms in the water or food, or spirochetes transmitted by insects. Dolphins are specialized to survive in the ocean in which they were born—Atlantic dolphins can't survive in Pacific waters. The environments are different enough that it would take them generations, perhaps thousands of years, to adapt. They can't be moved to oceans anywhere in the world. Thus, adaptability and exposure to pathogens have to be considered when animals accompany humans to foreign regions.

8. Restrictions and Regulations

There are a number of professional and regulatory guidelines for the handling and care of animals, some of which specifically protect working animals.

Service Animal Protection

The regulations regarding the use of animals in law enforcement vary in wording from state to state but, in general, they safeguard animals from harm by levying fines and prison sentences for anyone who engages in malicious obstruction or harm to an animal acting in the line of duty. In some cases, persons are even barred from entering the area of control of a police service animal. Fines vary, but tend to be up to a few thousand dollars and prison terms range from approximately 6–18 months. Some statutes are written specifically for police service dogs and others are generic for police service animals. Since horses are used in many areas for police work, particularly in Canada, the generic wording is intended to cover different species of animals that might be brought into service in the future. This is not unreasonable, considering that many species of animals have been used in other parts of the world as messengers (e.g., carrier pigeons) or as patrol or 'watchdog' animals, including geese, ostriches, llamas, crows, cockatoos, and Siamese cats.

Legal Searches

Regulations regarding personal privacy and constitutional freedoms influence what an animal can or cannot search. Since dogs are often used to seek out controlled substances, explosives, and other contraband, the search of luggage, vehicles, possessions, school lockers, and closets is common. The search of people is also common, but there may be restrictions to using an animal to search a human, in addition to laws that apply to the search of a bag or cupboard.

Animal Handling and Protection

A number of humane societies that provide information on the proper care of animals and regulations governing the owning and use of animals have been listed in the resources section.

9. Implications of Use

The use of animals brings up some issues that are unique to this 'biological technology' that are different from using electronic or mechanical devices for surveillance.

Individual Temperament and Intelligence

Handling an animal is obviously different from handling an inanimate video camera, robot, or other surveillance 'device.' Unlike a piece of equipment, you can't stuff it in a briefcase when it isn't being used. It requires care and attention 24 hours a day. Unlike learning to use a tape recorder, a 15-minute training session isn't sufficient to make the trainee an expert in using the 'technology'. Trainers and handlers need years of experience. Even if the animal is a tiny firefly that is observed for its chemical reactions to stimuli, an animal requires humane, ethical treatment.

The more intelligent the animal, the more its personal attributes have to be considered, and the more experience it takes to work with the animal. Not every person has the temperament to be a good animal handler and not every animal has the temperament to carry out surveillance work. Some people believe animals shouldn't be used in surveillance work at all, with dogs a possible exception because they naturally enjoy sniffing out things, and protecting their owners and property.

Handling an animal is different from handling equipment. Animals have individual personalities, emotional needs, and intelligence. Dogs may not match humans in higher-level abstract or conceptual thinking skills, but they are extremely similar to us emotionally. While we can never confirm this with absolute certainly, their body language and reactions to situations suggest that they experience happiness, sadness, grief, anger, annoyance, and pretty much the range of emotions that we intuit a child feels and that we personally experience and acknowledge in adults.

Most canine handlers in professional positions are 'on duty' 24 hours a day, in that a dog requires continual care and desires companionship. There is a strong bond between the handler and the dog which is important to the partnership and also necessary for the emotional and physical well-being of the animal—dogs have highly social natures. Handling, training, and managing a dog includes home care, in many cases, above and beyond what the handler's non-canine-handling colleagues may be required to do in the execution of their duties. Home care includes feeding, grooming, exercise, the administration of medicines, when appropriate, and maintenance training. In some instances, this time is compensated in a variety of ways, depending on local and national labor laws and arrangements are made between the employer and the employee.

Dolphins are similar to dogs in that they are interested in humans and sometimes seek out their company, provided they are not threatened or unduly confined. The study and training of dolphins have resulted in many surprises. Dolphins have large and complex brains, sophisticated sonar and communication systems, and the ability to communicate to us what they are seeing inside various objects. They also appear to recognize each others as individuals and to have a sense of 'self'. We don't yet know how 'smart' they are relative to us, because they are different from humans in many essential ways, but preliminary studies indicate they may have a wealth of abstract thinking ability and mental talents as yet not fully understood.

Unfortunately, there are still many people who think that dolphins are fish. They're not aware that dolphins typically have only one offspring every two years or so, resulting in a dozen or so 'children' in a lifetime, as opposed to a fish that can lay hundreds or thousands of eggs every years. Given that fish stocks are declining, it can be seen why dolphins and whales, which have a harder time replenishing their numbers, are declining at alarming rates, in spite of protections, more humane nets, and export restrictions. The training and use of dolphins in surveillance activities have aided us in understanding more about the abilities and lifestyles of these remarkable mammals and will hopefully have a secondary benefit in educating the public in the need to observe and preserve the dolphins and their habitat.

Animal Rights

The use of animals in surveillance in the service of humans often provokes comment or criticism from animal rights advocates. Animal rights advocates are concerned for the safety, dignity, care, lifestyle, lifespan, and nonexploitive relationship of animals and people. Any attempts to confine animals, to conduct tests on them, to expose them to danger they may not understand, or to have them engage in activities that may reduce the quantity or quality of their lives will continue to be the subject of scrutiny and opposition. It is important for anyone training or handling animals to be aware of the concerns of animal rights groups, and the laws related to the keeping of animals, in order to work within our complex social structure to achieve ends that are of mutual benefit.

Rights activists are not opposed to every aspect of the use of animals in surveillance. Dogs clearly enjoy interacting with humans, engaging in field trips, and many of them love to work, requiring only appreciation and good care in return. They are indispensable partners in locating avalanche victims, missing children, people buried in earthquake rubble, and the bodies of those who have died under natural or suspicious circumstances. Given the enormous benefit we derive from their eager assistance, these kinds of activities raise few concerns as long as the animal receives good care.

10. Resources

Inclusion of the following companies neither constitutes nor implies endorsement of their products and services and, conversely, does not imply their endorsement of the contents of this text.

10.a. Organizations

American Rescue Dog Association (ARDA) - This is the oldest air-scenting search dog organization, founded in 1972. The organization has been involved in the development of standards and training for air-scenting in different conditions. ARDA developed the first National Search and Rescue Dog Directory for the National Association for Search and Rescue. www.ardainc.org/

American Society for the Prevention of Cruelty to Animals (ASPCA) - One of the most prominent animal rights and educational organizations. They sponsor adoption programs, spaying and neutering, hospitals, and investigations into cruelty to animals. www.aspca.org/

Animal and Plant Health Inspection Service (APHIS) - A department of the U.S. Department of Agriculture that is tasked with ensuring the health and care of animals and plants, agricultural productivity and competitiveness. APHIS uses dogs to aid in detecting illegal animal and plant matter. www.aphis.usda.gov

Dogs Against Drugs/Dogs Against Crime National Law Enforcement K9 Association - Established in 1989, this is a not-for-profit organization working with local and foreign law enforcement agencies to aid them in operations and, if necessary, funding for special purpose dogs and K9 equipment. www.dadac.com

Humane Society of the United States (HSUS) - Like the ASPCA, the HSUS has worked for many years toward educating the public about the humane treatment of animals. It has also disseminated research and other educational information, much of which is accessible through the Web site. www.hsus.org

National Association for Search and Rescue (NASR) - A nonprofit membership association for professionals involved in search and rescue, disaster assistance, and emergency medicine. The site includes a dog section which includes the National Search and Rescue Dog Directory. www.nasar.org

National Narcotic Detector Dogs Association (NNDDA) - Provides information, certification standards, events, etc. www.nndda.org

National Police Canine Association (NPCA) - Assists and promotes police service dogs and educational programs for the prevention and detection of crime. Further promotes task-related minimum standards and certification programs. www.npca.net

Nordic Police Dog Union (NPDU) - sponsors championships, publications, and activities related to the training and handling of police dogs. home3.inet.tele.dk/nphu/

North American Police Work Dog Association (NAPWDA) - Provides workshop information, training tips, and certification information. www.napwda.com

Northern Arizona Police Department (NAPD) - This site not only has illustrated information about police service dogs, but has a very large list of related links on the Web. www.nau.edu/~naupd/k9.htm

Official Site of the 47th IPSD "Paw Power" - This is a Scout Dog platoon founded in 1968, with photos, links, history, and communications of members of the platoon. www.geocities.com/Pentagon/4759

Royal Australian Air Force Police Dog Handlers Association - Established in 1996.

The Seeing Eye - Since 1929, this organization has promoted the training of dogs to aid the blind. Their Web site includes Dog Guide laws and has a number of articles and electronic books on working dogs that can be downloaded. www.seeingeye.org

U.S. Customs Service - This is the primary enforcement agency, within the Department of the Treasury, which protects national borders through extensive air, land, and marine interdiction and its own intelligence branch. Customs detects smuggling and enforces U.S. criminal and import/export laws. Canine sniffers can be used to aid in the detection of narcotics, explosives, and other illegal imports. www.customs.ustreas.gov

U.S. Fish and Wildlife Service (FWS) - The FWS is involved in conserving and monitoring American wildlife resources, issues hunting and fishing licenses, and provides educational publications and programs. Dogs are used to detect and deter smuggled wildlife. www.fws.gov

Working Dog Foundation - Incorporated as a nonprofit organization in 1995 to promote a positive public image of police working dogs, to operate the New Hampshire Police K-9 Academy, to promote minimum standards, and to aid organizations wishing to establish police dog units. www.workingdog.org

10.b. Print Resources

These annotated listings include both current and out-of-print books. Those which are not currently in print are sometimes available in local libraries and second-hand bookstores, or through interlibrary loan systems.

American Search and Rescue Dog Association, *Search and Rescue Dogs: Training Methods,* New York: Howell Book House, 1991, 208 pages.

Bulanda, Susan; Luther, Luana, Editor, *Ready! The Training of the Search and Rescue Dog,* Doral Publishing, 1995, 170 pages.

Burman, John C., D*og Tags of Courage: The Turmoil of War and the Rewards of Companionship,* MSG, 2000, 300 pages. Scout dogs stories from Vietnam vets.

Button, Lou, *Practical Scent Dog Training, Alpine Publications,* 1990. Techniques tested with Mountain Canine Corps search and rescue.

Duet, Karen Freeman; Duet, George, *The Business Security K-9: Selection and Training,* New York: MacMillan and Company, 1995.

Eden, Robert S., *Dog Training for Law Enforcement,* Calgary: Detselig Enterprises Ltd., 1985.

Gerritsen, Desi; Haak, Rudd, *K9 Search and Rescue,* Temeron, 185 pages.

Howorth, Peter C., *Whales and Dolphins Shorelines of America: The Story Behind the Scenery,* Las Vegas: KC Publications, 1995.

Johnson, Glen R., *Tracking Dog, 5th edition,* Barkleigh Productions, 2003, 214 pages. Reprint of classic with theory and methods.

Johnson, Glen R.; Patterson, Gary, *Training Police Dogs,* San Francisco: The Millbrook Press, 1989.

Kaldenbach, Jan, *K9 Scent Detection,* includes case files and dog requirements for The Netherlands.

Kearney, Jack, *Tracking: A Blueprint for Learning How,* revised edition, Pathways Press, 1999. Military and search and rescue reference for tracking people.

Koehler, William R., *The Koehler Method of Dog Training,* IDG Books Worldwide, 1996, 190 pages. Originally published in several editions by Howell Book House (New York).

Koehler, William R., *The Koehler Method of Guard Dog Training: An Effective and Authoritative Guide for Selecting, Training and Maintaining Dogs in Home Protection and Police Work,* IDG Books Worldwide, 1977. Matching the dog breed to the purpose, equipment, methods.

Koehler, William R., *The Koehler Method of Training Tracking Dogs,* Howell Books, 1984 (note that Glen Johnson's book on training tracking dogs has received stronger reviews).

Lemish, Michael G., *War Dogs: A History of Loyalty and Heroism,* Washington, D. C., Brassey's, Inc., 1999, 304 pages.

Mistafa, Ron, *K9 Explosive Detection,* Temeron Books, 1998, 189 pages. Training methods for different dogs, choosing a dog, general requirements, certification, training with explosives.

McKenzie, Stephen A., *Decoys and Aggression: A Police K9 Training Manual,* Temeron, 1996, 88 pages. Types of natural aggression, learning the body language, how to be a decoy for bite training.

Morgan, Paul B., *K-9 Soldiers: Vietnam and After,* Central Point, Or.: Hellgate Press, 1999, 220 pages.

Mullican, Jr., Herbert, Editor, *A Guide to the Technical Literature on Police and Military Working Dogs, Special Canine Services,* Frederick, Maryland. Bibliography that categorizes over 1,000 articles, publications, reports, and films.

Pearsall, Milo; Verbruggen, Hugo, *Scent: Training to Track,* Loveland, Co.: Alpine Publications, 1982. Scent science, tracking theory, and method.

Rapp, Jay, *How to Organize a K-9 Unit and Train Dogs for Police Work,* Denlinger's, 1979.

Reed, Don et al., *The Dolphins and Me,* Scholastic Book Services, Sierra Club, 1990.

Robicheaux, Jack; Jons, John A. R., *Basic Narcotic Detection Dog Training,* K9 Concepts, 1990. Selecting and training a dog.

Rosenthal, Richard, *K-9 Cops: Stories from America's K-9 Police Units,* Pocket Books, 1997, 360 pages.

Seguin, Marilyn; Caso, Adolph, Editor, *Dogs of War: And Stories of Other Beasts of Battle in the Civil War,* Branden Books, 1998. Juvenile reading level.

Simpson Smith, Elizabeth, *A Dolphin Goes to School: The Story of Squirt, a Trained Dolphin,* Morrow, 1986. Out of print.

Sjrotuck, William G., *Scent and the Scenting Dog,* Barkleigh Productions, Inc., 1972.

Sorg, Marcella H.; Rebmann, Andrew; David, Edward, *Cadaver Dog Handbook: A Forensic Guide for Training, Handling, and Searching,* CRC Press, 2000.

Thorpe, Samantha Glen; Pesaresi, Mary B., *Search and Rescue,* Fawcett Books, 1997, 197 pages. The story of German Shepherds with the Virginia Dept. of Emergency Services on rescue missions.

Tolhurst, Bill, *Police Textbook for Dog Handlers,* New York: Tolhurst, 1991, 85 pages. Trailing, cadaver recovery, accelerant training, crime scene canines. The author is a researcher and practitioner.

Tweedie, Jan, *On the Trail! A Practical Guide to the Working Bloodhound and Other Search and Rescue Dogs,* Alpine Publications, 1998. A practical professional guide to search and rescue record-keeping and tracking of humans.

White, Joseph J.; Luther Luana, Editor, *Ebony and White: A Story of the Canine Corps,* Wilsonville, Or.: Doral Publications, 1996, 168 pages. Canine corps from World War I and onward, including the Vietnamese war.

Articles

Albertson, Mike, Use of Military Working Dogs in Peace Support Operations, U.S. Army *Call*, 1998. This is a multipart article on many aspects of dog use in military operations.

Immega, Guy, Tension Actuator Load Suspension System, *United States Patent #4,826,206,* filed 9 Feb. 1987. Suspension and sensor systems applicable to robotics.

National Church Arson Task Force, Second Year Report for the President, Washington, D.C., October 1998.

Stone, Paul, Creatures Feature Possible Defense Applications, Washington, D.C., *American Forces Press Service.* An article about how creative thinking can come up with ways to utilize the unique characteristics and talents of a variety of critters to aid in surveillance tasks. It reports on the Controlled Biological and Biomimetic Systems project. Stone has also written a story on how bees might be used to detect mines.

Yarnall, Donn, My Dog Won't Out! 1999. Debate about training approaches, control, and aggressive instincts with the Belgian Malinois as the example.

Journals

American Academy on Veterinary Disaster Medicine Newsletter, journal of the AAVDM.

Journal of the National Association of Dog Obedience Instructors, Landing, NJ.

Nordic Police Dog Magazine, ships to about 61 countries.

The Canadian Search and Rescue Magazine, quarterly, free to the search and rescue community.

10.c. Conferences and Workshops

Many of these conferences are annual events that are held at approximately the same time each year, so even if the conference listings are outdated, they can still help you determine the frequency and sometimes the time of year of upcoming events. It is very common for international conferences to be held in a different city each year, so contact the organizers for current locations.

Many of these organizations describe the upcoming conferences on the Web and may also archive conference proceedings for purchase or free download.

The following conferences are organized according to the calendar month in which they are usually held.

International Police K9 Conference. Topics include beginning and advanced training techniques, tracking, tactics, problem-solving, muzzle work, deployment, and others.

Narcotic Detection Dog Seminar.

USPCA National Seminar, in conjuunction with the NPCA, the USPCA gives regular seminars and certification courses. www.uspcaregion25.org/training_seminars.htm

Police K9 Conference. Restricted to law enforcement personnel.

APDT Educational Conference, annual conference since the early 1990s.

DAD/DAC National Law Enforcement K9 Association offers regular training seminars and workday events through the U.S. www.daddac.com/

PLES.com provides regular seminars and certification programs for Drug K9 handlers and supervisors. www.ples.com/k9.htm

Southern Police Canine, Inc. conducts about one seminar per month ranging in length from two days to about eleven days, with various local police representatives or contact people from SPC. Some of these include certification. www.southernpolicecanine.com/seminar.htm

10.d. Online Sites

Albert Heim Foundation for Canine Research. This interesting site looks at the historical and physiological development of canines including archaeological skulls and other specimens and dog-related exhibits. The site is of particular interest to scientific researchers interested in the evolutionary history of canines. nmbe0.unibe.ch/abtwt/ahst.html

Tightlines. This site has a number of informative illustrated articles by writers/trainers on the subject of training and handling of tracking dogs. It also lists Tightlines seminars. The emphasis is on German Shepherds but the concepts apply to the training of many breeds. www.memlane.com/business/tightlines

Leerburg Dog Training. This company specializes in the breeding and training of service dogs and provides a series of training videos. Their illustrated Web site includes almost 200 free-access dog training articles on a wide variety of obedience, dog-care, and training topics. leerburg.com

K9 Case Law. Eden & Ney Associates Inc. includes several sections of interest on their Police Dog Web site, including training and standards information, links to police dog case law, and links to various canine agencies. www.policek9.com

10.e. Media Resources

Animals in the Service of the Military, an America's Defense Monitor series show that describes many types of animals, from elephants and horses to dogs and dolphins, that have served military forces over the centuries. It describes some of their abilities, but also some of their limitations and controversy over their use and handling, with dolphin programs as examples. Produced in 1991, available on video.

In the Wild: Dolphins with Robin Williams, an introduction and overview of dolphin habits, personalities, and abilities, NTSC format, color VHS.

K-9 Cop, a color VHS video in a Police Tactics series. How police dogs are prepared and drilled for attack and crime scene evidence hunting; 40 minutes, E&E Productions, Connecticut. The author has not viewed this video and cannot attest to its quality.

News Report on Dogs' Abilities to Detect Chemicals, part of CNN's Science and Technology Week programming which first aired in November 1996 and was subsequently made available to its global networks. It describes the research of an Auburn University professor into the use of dogs for detecting chemicals in explosives.

Search Dog Training: How to Get Started, Handlers, selecting a dog, basic training. By Stopper and Watts on VHS.

Tracking Fundamentals, video and book (Ganz and Boyd) that outline a training program, mapping, turn techniques, etc., 1992, VHS, 42 minutes.

Training Police Service Dogs, VHS video in Leerburg's extensive Dog Training series, available since 1979. This is an intermediate-level training video which covers equipment, handling, targeting of bites, and off-leash control; 2 hours, Leerburg Video Productions, Wisconsin. The author has not viewed this video and cannot attest to its quality, but notes that the producers specialize in the breeding of working dogs for police work, search and rescue, and obedience.

War Dogs: America's Forgotten Heroes, poignant video showing the canine heroes that aided soldiers in Vietnam. Donors to the Memorial Fund receive the video, VHS. This was shown on the Discovery Channel. www.war-dogs.com

11. Glossary

Titles, product names, organizations, and specific military designations are capitalized; common generic and colloquial terms and phrases are not.

AKC	American Kennel Club
CD	companion dog
CDX	companion dog excellent
CU	canine unit
EDC	explosives-detecting canine
HDC	human-detecting canine
K9	canine
mutt	mixed-breed dog
MWD	military working dog
NDC	narcotics-detecting canine
SAR	search and rescue
SPCA	Society for the Prevention of Cruelty to Animals
TD	tracking dog
TDX	tracking dog excellent
UDT	utility dog with tracking dog title
UDTX	utility dog title with tracking dog excellent title
UDX	utility dog excellent

Biochemical Surveillance

Genetics

1. Introduction

Few technologies are more controversial, more intimate, or more telling than DNA testing and analysis. DNA contains a type of genetic 'blueprint'* of our bodies describing our individual characteristics, our strengths and weaknesses, our health, our ancestral legacy, our race, and patterns that are partly mirrored in our living relatives. These essential attributes of our organic structure are coded into most of the cells in our bodies. Recently scientists have developed ways of unraveling this code and identifying or inferring important characteristics of an individual from tiny hair roots, teeth, or drops of saliva or blood.

DNA evidence is very powerful, sometimes damning, information, yet it is relatively easy to collect. You don't need advanced electronics skills and you don't have to set up complex equipment to gather saliva, or hairs from a brush or a comb—gloves, clean tweezers, and a sterile pouch are sometimes enough. The easy acquisition of DNA samples, compared to many other kinds of data, makes this technology a significant information-gathering tool.

---

*The genetic blueprint analogy is useful but not precise as it implies that the pieces of the genetic puzzle are stored in microscopic boxes that you can open and assemble like building blocks. The growth of an organism is a sophisticated chain of events, though, which is incompletely mapped out in the germ stage. The unraveling of the 'message' contained in DNA, and the process of cell reproduction is more like a bootstrapped, self-modifying computer program than an erector set, but for understanding DNA sampling and analysis from a lay point of view, the blueprint term provides a familiar starting point, and readers seeking a better understanding can consult references listed at the end of this chapter. The above photo illustrates gel electrophoresis being used to reveal gene sequences. [Courtesy of Pacific Northwest National Laboratory.]

DNA analysis is a recent science and its use is not yet strongly regulated outside of law enforcement applications. When new technologies arise, there is usually a 'window of opportunity' during which public access is open and unlicensed, and during which certification standards for laboratory procedures are lenient. When automobiles first appeared, there were few traffic laws, and newer vehicles like snowmobiles, are loosely regulated compared to cars. DNA testing and analysis are still in this 'honeymoon' phase. In contrast, there are many laws safeguarding individuals from having their phones tapped, and there are some laws controlling the use of DNA in law enforcement, but there are as yet few regulations protecting individuals from having their DNA sampled without their knowledge. Samples can easily be sent through the mail to a commercial lab for analysis, and the resulting profiles can be stored in a computer database indefinitely.

North Americans live in a society that values freedom—excessive regulation is considered to be contrary to this philosophy. In addition, lawmaking may lag, not because it significantly hinders individual freedoms, but because a technology is not understood by the ordinary individual and thus not immediately understood as being highly vulnerable to abuse. A DNA test can cost less than a night on the town and almost anyone can take a DNA sample and send it to a private company for analysis. Consequently, in our capitalist economy, DNA collection and analysis firms are now offering a wide range of services that may be used without the DNA donor's knowledge or permission. We can't anticipate all the problems that may result from this open system, but in terms of surveillance, there are currently novel opportunities for information gathering.

DNA technology makes it possible to identify plants and animals. With this information it is possible to track, monitor, and sometimes hold them accountable for their whereabouts and actions. DNA proponents and opponents are lining up on both sides of a privacy controversy that is likely to remain for a long time. Law enforcement officials have been lobbying for broader DNA collection powers, while private rights advocates have been lobbying for greater protection for individuals, whether or not they have been convicted of criminal activities.

DNA sampling and analysis are becoming easier and less inexpensive, yet there is an enormous amount of information that can be derived from a small amount of tissue. Consequently, a DNA profile is vulnerable to abuse on an unprecedented scale. Some futurists have proposed that this will lead toward a highly stratified society based on DNA characteristics, with genetically 'superior' humans procuring the best opportunities. Is this unlikely? Consider the fact that the Police Superintendents Association in England has called for the entire population to be DNA sampled. Consider also that current immigration regulations require that all applicants submit to medical tests, including HIV screening. Blood is routinely drawn for these tests. It may only be a matter of time before immigration officials suggest that we add DNA profiling to the blood tests already being conducted.

On the electronics side, administrators are promoting the development of computer network connections between government agencies. Many new centrally funded government databanks are being established. The growth of the Internet has motivated agencies to standardize their databases so there can be widespread sharing of data. We may be less than five years from a system in which immigration files, police files, and motor vehicle license files are fully cross-referenced, aiding law enforcement officials, but at the same time, blurring the line between private citizens and convicted felons. In a standardized database, they all look the same, and all data entry systems are vulnerable to file corruption, data entry error, spying, or sabotage by hackers or disgruntled employees. There are known instances where hackers have accessed sensitive data in banks and government agencies, and 'published' the information on the Net.

Once this type of information is uploaded, it is impossible to 'get it back' or quell its redistribution. It only takes minutes for digital data to be replicated and broadcast to millions of computers around the world.

A stratified society based on DNA characteristics has been chronicled in the movie *Gattaca*. Even though Gattaca depicts a futuristic society, much of the technology in the film is already available. The prediction that DNA will be used to control access to jobs, benefits, even mates, is not far-fetched. There are already more than 6,000 firms that report that they require DNA samples as a condition of employment, and insurance companies have already begun assessing the technology as a screening tool. In spite of the possible repercussions, DNA profiling is likely to be a growth industry, given that the motivations for obtaining individual DNA profiles are many: paternity confirmation, genetic testing of newborns for treatable and untreatable diseases, identification of remains, and public safety.

It is important for everyone seeking to understand surveillance technologies to grasp patterns of legislation related to DNA profiling and the basic concepts of DNA collection and analysis. DNA testing has been in general commercial use since the mid-1990s and is spreading rapidly. It is now regularly used in law enforcement and social service activities (and no doubt in espionage), in ways that may eventually impact every person on the planet.

1.a. What is DNA and How is it Used?

DNA is a microscopic, information-carrying structure that is contained in nearly all living cells. More than one type of cell structure contains DNA. Although it degrades over time, DNA persists for some time after the death of the organism and, depending upon where in the cell it is found, it can sometimes be recovered days or centuries later. When subjected to an exhaustive analysis, DNA can tell us a great deal about an individual's familial relations, physical characteristics, and health. The research in this area is recent and it is expected that our knowledge of DNA's relationship to biological evolution, growth, and health will increase dramatically in the coming decades.

The source of DNA

In sexual reproduction, a new individual develops from the coming together of genetic material from a male and a female. The *genes* in the DNA influence the form and characteristics of the new organism. As it develops, some genes are expressed and others are suppressed, and some interact in complex ways not yet fully understood.

Some of the better-known areas of the human genome, such as gender and certain physical characteristics, can be predicted with a high degree of probability from a small sampling of DNA. As such, DNA testing is quickly being adapted in investigative and surveillance applications as a means of determining information about a specific individual (human, animal, or plant) and where the person may have been. DNA samples taken in espionage operations or at crime sites are gaining acceptance in courts of law and scientific circles as compelling, if not definitive, evidence.

Using DNA

There is great concern about the ethics and legality of sampling genetic codes–and rightly so. A full understanding of DNA requires a scientific background. As our society becomes more complex, it becomes harder to convey the ramifications of a new technology in lay language. In legal debates about the admissibility, regulation, and constitutionality of DNA sampling, the DNA information is frequently compared to an inked fingerprint. This is a misunderstanding of the phrase *DNA fingerprinting* and an alarming comparison. Comparing a *fingerprint* to a *DNA print* or *profile* in terms of information content is like comparing a house number with the whole history of a house, its inhabitants, contents, structure, materials, and more.

In this volume, the phrase *DNA fingerprinting* is used to specifically mean matching a minimal type of DNA representation used for *identification only*. DNA fingerprinting refers to the use of what has sometimes been called *junk DNA**, which is a form of 'noncoding' DNA, to discern a pattern that is compared to a reference pattern. (DNA analysis prepared for paternity testing should be called a DNA *parental profile*, not a *DNA fingerprint*.)

1.b. Beyond Human DNA - Meta-Analysis

In the context of surveillance, when blood samples are drawn for DNA analysis, other medical information or organisms may sometimes be derived from the sample.

The information potential of DNA samples goes far beyond identification of an individual and his or her physical characteristics. In cryptologic surveillance one learns that much of the *intelligence* that can be derived from a coded message involves information outside the content of the message itself–information such as when the message was sent, how it was sent, who sent it, etc., all of which can be exceedingly important.

The same principle of macro-evaluation applies to intelligence derived from DNA testing. For example, one could observe the DNA of other organisms in a sample of a person's blood. Further information about the activities and of the donor and where he may have been can sometimes be deduced. For example, if particular strains of malaria or HIV, or other detectable pathogens, are found in the sample and can be analyzed and traced to a matching or related population in some part of the world, information about the individual's travels or activities may be revealed. In the surveillance industry, this secondary information may be of greater relevance than the particular genetic makeup of the individual sampled.

For example, someone smuggling narcotics out of South America might deny ever having been to that continent, but may carry a particular strain of organism found there, perhaps malaria or other types of infections. Conventional blood tests may reveal general information about the organism without a DNA test, but the specificity of the DNA used in conjunction with a reference database might someday pinpoint where the person contracted the organism with a precision not possible with current blood tests. This extrapolation of information from a sample may not interest a judge in a paternity case, but may greatly interest a government surveilling a suspected international spy or smuggler.

1.c. DNA Matching

DNA matching is the most basic level of DNA analysis, in which a DNA pattern is processed in a lab or with a mobile kit and compared to reference samples to see if any of the references correspond to the same pattern as the sample in question. In law enforcement, this reference-comparison process is already used for fingerprints and mug shots. In wildlife research, unique skin markings are commonly used to create photographic databases of whales and dolphins so they can be identified if they are seen later in other regions.

---

* While *DNA fingerprinting* in and of itself may not be considered invasive of a person's rights, *junk DNA* is not as innocent as the name may imply, and the sample from which it was derived may remain for an extended period in cold storage in some laboratory. There may also be an image or fuller analysis kept in a computer database. These related sources of information may be vulnerable to future use or abuse.

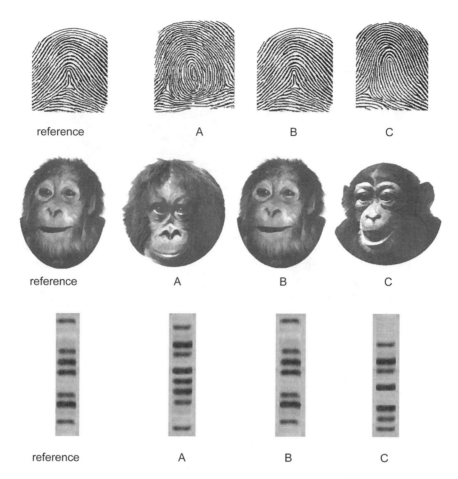

reference   A   B   C

reference   A   B   C

reference   A   B   C

The phrase DNA matching should be used instead of DNA fingerprinting to refer to the most basic identification procedures. In the top series of images, a traditional fingerprint is compared with three others to determine whether there is a match. In the second series, a 'mug shot' is matched by seeking out identifying facial features. In the third series, an X-ray 'photograph' of DNA fragments is compared to the other patterns. A full DNA analysis provides a broader range of information and should be termed profiling rather than fingerprinting. [Classic Concepts illustrations ©2000, used with permission.]

2. Kinds and Variations

2.a. Nuclear DNA

DNA is the abbreviation for *deoxyribonucleic acid,* a material that is found in living, reproducing organisms. DNA is in nucleated cells throughout the body, and the DNA pattern in a living person's saliva, for example, is the same as that in his or her hair roots or portions of the blood. There is no DNA in mature red blood cells, but it is present in white blood cells. Nuclear DNA is in chromosomes, which in turn are in the nucleus of a cell. Nuclear DNA comes from both the mother and the father, which is why it is preferred for many types of DNA analysis. As cells decay, the DNA contained in those cells decays as well. Nuclear DNA is particularly subject to decay and is not found in bodies that have been dead for some time unless they have been specially preserved.

Basic Cell Structure

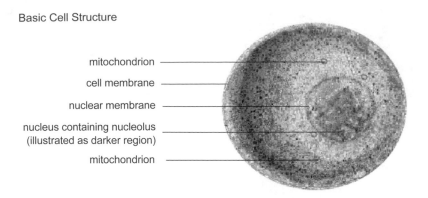

mitochondrion

cell membrane

nuclear membrane

nucleus containing nucleolus
(illustrated as darker region)

mitochondrion

2.b. Mitochondrial DNA

Mitochondrial DNA (mtDNA) is found in multiples in any given cell and, depending on conditions, may survive much longer than nuclear DNA. It can be recovered from structures that decay more slowly, like bones and teeth. Mitochondrial DNA comes almost entirely from the mother and does not provide information about the father.* Mitochondrial DNA does not contain as many bases (less than 17,000 in humans) as nuclear DNA, but mtDNA exists in more copies in the cell than nuclear DNA. The patterns derived from its analysis are not as broad or unique as those derived from nuclear DNA, but their specificity and usefulness can be improved with contextual information. Mitochondrial DNA's survivability makes it an important research tool for identification within a limited population, for analyzing ancient racial lineages, for studying historic population growth in humans, and for determining the maternal relationships in threatened species.

With the exception of identical twins, each person's DNA is unique and different, though patterns that link relatives with a high degree of probability can be detected. If two generations are available for testing, a great deal can be discovered about familial relationships. A genetic pattern, derived from analyzing a DNA sample, is called a *genotype* and can be symbolically represented in a number of ways. A diagram resembling a bar code, called an *autoradiograph*, is commonly used to represent the patterns in a DNA sample, especially for basic *DNA profiling*.

For general understanding, the DNA molecule is symbolically illustrated as a double helix with connecting strands resembling a spiral staircase. DNA contains a large number of *base pairs*, that is, a pair of bonded *nucleotides* on opposite strands of the DNA. The bases are commonly called A, T, G, and C. Base pairs can be C-G or A-T. There are many millions of these and unique sequences are found in individuals. It would be too time-consuming to examine all the base pairs in a sample, but enough is known about the patterns and repetitions in base pairs to select portions of the DNA that are known to vary from one person to the next and to concentrate on examining these.

Since these patterns vary less among family members than they do among unrelated individuals, they are useful for the identification or exclusion of familial kinships. The likelihood of two individuals being related is usually expressed as a percentage probability. In paternity tests, the offspring or father is generally said to be 100% excluded from being related or is related with a probability that usually ranges from about 99.6% t0 99.9%.

---

*There appears to be a cell-death process that occurs at conception that reduces the mtDNA contribution from the father to virtually nothing.

Basic Mitochondrian Structure

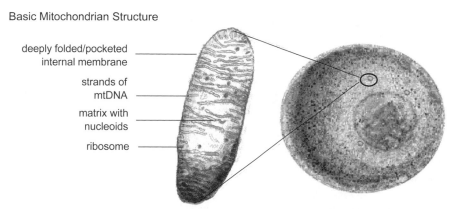

deeply folded/pocketed
internal membrane

strands of
mtDNA

matrix with
nucleoids

ribosome

The infolded structure on the left is a cutaway, simplified view of a mitochondrion showing its relative location in the cell on the right (multiple mitochondria exist outside the cell nucleus). Mitochondria are mobile, flexible cell organelles. In mammals, they are passed down through the maternal line (human sperm contains only about 100 mitochondria, compared to about 100,000 from the mother). They are usually elongated, though some may be round. The convoluted inner structure is a greatly infolded membrane. Associated with the membrane is a matrix containing nucleoids (free areas within the matrix), which in turn contain bodies of DNA. This mitochondrial DNA (mtDNA) is useful, as it may degrade more slowly than nuclear DNA. [Classic Concepts diagrams, pp. 15-6 and 15-7 ©1999, used with permission.]

Details of the structure of the molecule and the various components of DNA can be found in biological texts and forensic texts on DNA sampling and analysis and are not described in depth in this reference. Instead, this chapter focuses mainly on concepts and procedures essential to the basic understanding of *DNA identification* as it pertains to surveillance. Those involved in the surveillance industry typically turn the samples over to specialists (or a computer program) for analysis.

There are many ways to represent genetic coding. DNA sampling results are symbolically or photographically represented, usually in ladder-like bands that are created by processing the DNA fragments so that they are sorted according to size. The portions selected for basic identification of an individual are those that tend to vary widely from one person to another. This information is statistically and experimentally derived.

The human genome includes over three billion ($3x10^9$) base pairs. This is a number greater than most of us can conceptualize. Obviously not all of these are 'active' in the creation of an individual, but exist as potentialities. Of more than three billion base pairs in a human cell, approximately 15,000 genes are expressed at any one time from a typical tissue sample. We are still trying to unravel the complete genetic code and The Human Genome Project (HGP) was initiated as an international collaborative project specifically for this purpose.

The Human Genome Project has the goal of determining the complete human genetic sequence. While this information does not wholly predict or describe an individual's genetic makeup, it does provide significant information about general organization in humans, and particular patterns of organization in related humans or humans with similar characteristics or afflictions.

DNA patterns exist in all living, growing organisms. Thus, plants, fish, birds, dogs, horses, and people all have DNA characteristics distinct to their species. The DNA pattern in a controlled plant species confiscated in a smuggling operation can potentially be traced back to its source. Recovered children, who have been kidnapped at a very young age, can be reunited with their biological parents. DNA varies from individual to individual, but inherited characteristics can be mapped in family lines, providing information about sibling and parental relationships.

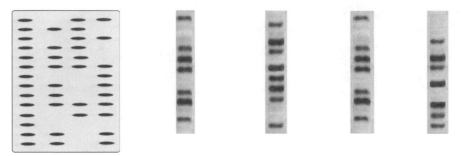

There are many ways to represent DNA patterns, but the most common is a ladder-like film of dark and light bands resembling a bar code, called an autoradiograph. In a common process called gel electrophoresis, DNA fragments are transferred from a gel surface to a membrane, subjected to radioactive probing, and recorded on an X-ray autoradiograph for viewing and storage. Common applications include DNA parental profiling for determining paternity or other relationships, or DNA matching, comparing several patterns to a specific reference pattern for a match. [Classic Concepts diagram ©1999, used with permission.]

3. Context

DNA testing is recent, and all the possible applications of the technology have not yet been developed. It is not difficult to procure samples for DNA testing; a cheek swab, drop of blood, glob of saliva or semen, or a handful of hair roots or feathers is often sufficient for basic testing. While contamination is possible, the use of sterile gloves and a sterile swab for each sample, and storage in a sterile envelope, vial, or other container, can substantially reduce the risk of contamination. Since collection procedures can be quickly learned, many people can be trained to obtain and handle noninvasive samples, particularly cheek swab samples.

Although DNA sampling can be learned by layworkers, invasive procedures, like drawing blood, should be carried out by trained professionals. DNA database entry is usually carried out by skilled or semi-skilled workers. DNA analysis is usually handled by forensic scientists with at least a Bachelor of Science degree and a background in Chemistry and Biology and extra training pertinent to DNA analysis. However, portable 'briefcase' systems sometimes provide preliminary computer analyses onsite, which can later be verified by fuller analyses by trained specialists in a lab.

The stability of typical DNA samples is relatively good. Uncomplicated storage procedures can preserve them for weeks or months. More sophisticated embalming or freezing techniques can preserve them for years and sometimes for centuries. Mitochondrial DNA survives longer than nuclear DNA, and while it only carries information from the maternal line, this information may be decades or centuries old, and can be very useful.

The following information on DNA testing is skewed toward commercially available or patented technologies, as these are the ones most likely to be accessible to surveillance professionals. Human DNA-testing is also emphasized and information on wildlife testing for law enforcement or conservation purposes is surveyed in less detail. For a more abstract theoretical background and pure research in this area, see the numerous resources listed at the end of the chapter.

Assessing Samples

Before looking at the information revealed by specific gene sequences, the species from which a sample has originated should be known or discovered. Murder suspects are sometimes apprehended on the basis of pet hairs found on their clothing. A crime scene may yield samples from spattered blood but, until it is analyzed, it is not usually known if the blood came from a victim, a witness, a criminal, or the family dog.

Research exists on the DNA patterns of many species — viral, botanical, avian, primate, etc. This knowledge, combined with various analytic techniques, including *protein electrophoresis, enzyme-linked immunosorbent assay,* or *radioimmunoassay* processes, allows one species to be distinguished from another. These techniques can be cumbersome, however, as they require that separate tests be performed on the suspected species to exclude it as a contributor. In 1995, Hershfield described a method of isolating nucleic acid from a biological sample and determining the interspersion pattern of repeats of a sequence in the isolated nucleic acid. This pattern could then be compared to known patterns of the microsatellite nucleotide sequence in selected organisms to determine the species.

Once the basic characteristics of a sample have been identified, and contaminants removed or neutralized, it can be further processed. There are two distinct methods of DNA testing and analysis in the field of forensics that have been widely adopted as they are more efficient than historic means of processing DNA:

- **polymerase chain reaction** (PCR) - This is currently the most commonly used group of methods for preparing a sample for analysis. PCR is less discriminatory than RFLP (described next) but it is a useful form of analysis that can be carried out with smaller samples that may not be in perfect condition, e.g., crime scene samples. The DNA molecule is extracted and replicated, thus providing copies of the original DNA in a process called *amplification* to assist in 'seeing' and evaluating the characteristics of the sample. Care is taken to try to amplify regions in the DNA that show greater variability among individuals, known as *polymorphic* regions. This is generally done in a test tube that is subjected to cycles of heat and cold. With PCR, results can be obtained fairly quickly. Larger, purer samples yield better results. Depending on the results, PCR may be followed up by RFLP. Amplification may be hindered by certain substances in the sample, such as hemoglobin in whole blood. Such substances need to be filtered or inactivated before replication can take place.

After initial processing, a sample can be *typed*. There are commercial kits and typing *strips* for examining genetic loci. *DQ alpha* is a common typing process.

- **restriction fragment length polymorphism** (RFLP) - This is a highly discriminatory analysis requiring a somewhat larger sample that is in good condition. The DNA molecule is extracted and cut or *restricted* at specific pre-determined sites by using an enzyme. The resulting fragments are separated by a process called gel electrophoresis in which they are sorted by applying an electric current to the gel medium through which the fragments are drawn. At this point a *blot* is usually made, and a *probe* applied to reveal the patterns in the DNA (blots and probes are described in more detail later). Traditionally, the sizes of the DNA fragments are determined manually, though automated procedures are now favored. The final pattern is represented on a membrane as a 'bar code', which is usually transferred to X-ray film but may also be digitally recorded and displayed on a monitor or high-resolution printouts. Analysis may take several weeks. RFLP may be used in situations where speed is not essential, or where a more accurate appraisal is important. It can also be used as a follow-up to the quicker PCR analysis, depending upon the circumstances.

Commercial improvements to basic DNA processes are always being sought, motivated in part by the huge backlog of unprocessed samples stored in law enforcement evidence archives. By the mid-1980s, researchers had developed a way to increase the concentration of a segment of target DNA without the need to purify or synthesize nucleic acid sequences unrelated to the desired sequence [Mullis et al., 1987]. A decade later, some of the problems of using replicated materials were overcome and specific target DNA could be detected without amplification through site-specific enzymatic cleaving [Dahlberg et al., 1992]. Many aspects of laboratory amplification and detection were improved at this time.

Surveillance occurs on a global scale, with samples often originating in foreign countries. The use of the PCR method can have secondary benefits in cases where export restrictions prevent transport of the original sample. In other words, one of the synthetic PCR-replicated samples may be sent instead. A thermal cycler is a portable device for replicating a DNA sample for international export without violating regulations.

DNA technology is still young, and these methods are some of the earliest best solutions to efficient processing of samples. However, they are not without their drawbacks.

In PCR, there is always the danger of amplifying non-target fragments, depending upon the quantity of the sample, the number of cycles of amplification performed, and environmental factors such as temperature [Erlich, 1989].

Specific nucleic acid sequences are usually detected by a process called a *hybridization reaction* or simply *hybridization*. While this process has become an important tool, it is not a perfect one. A mixture of DNA may only yield a low concentration of the sought-after target sequences; probe and target sequences are not necessarily perfect complements of one another and do not yield all possible probe-target complexes. Nevertheless, it remains a useful tool.

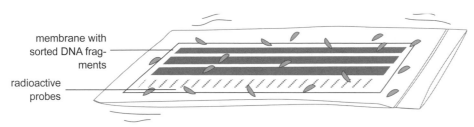

membrane with sorted DNA fragments

radioactive probes

Radioactive probing of DNA fragments can help locate target sequences. In many common 'blotting' procedures, the membrane containing the DNA fragment pattern can be reprobed with different markers. Radioactive probes are gradually being superseded by fluorescent probes. [Classic Concepts diagram ©1999, used with permission.]

There are many ways to process DNA samples. Not all will be discussed here, but examples of some common techniques are given to provide a basic understanding of what happens to a collected sample at the processing lab.

Southern Blot

The Southern blot is a well-known process of sample preparation, evolved from Edwin M. Southern's technique of identifying specific nucleic acid sequences. He developed these techniques in the mid-1970s. Blotting provides a stable replica of the distribution of the DNA fragments within a gel medium (after electrophoresis), and information on sequence organization.

A Southern blot involves separating out the DNA from other materials in the nucleus of a cell. This can be done with pressure, or with chemicals. The DNA is then chemically 'cut' into pieces of different sizes using enzymes with consistent effects on the DNA. A process of *gel electrophoresis* is then used to sort the DNA pieces by size. An electrical charge is applied to the mix to attract the DNA fragments. Smaller fragments travel more readily through the gel medium, leaving the longer fragments behind. The DNA is then denatured by heat or chemicals into single strands. The DNA strands are then mounted on a membrane in a process called 'blotting' for handling and analysis.

A radioactive DNA strand can be used as a 'probe' to aid in the analysis of the DNA sample. Subsequent exposure to X-ray film reveals areas in which the radioactive probe binds, indicating the occurrence and frequency of patterns within the sample. The fixation required for this process may take several hours to several days. While still a well-known technique, radioactive probes are giving way to other devices, such as fluorescent probes.

Short Tandem Repeat (STR) Analysis

Repeat markers such as *short tandem repeats* (STRs) were first described in the mid-1980s. STRs are end-to-end repeating blocks of DNA with less variability in the length of the fragments. STR is sometimes known as STMS (sequence-tagged microsatellite site). Use of these structures for nuclear DNA analysis is faster and less complicated than conducting a RFLP analysis.

In STR, particular areas of DNA selected for high population variability are examined for patterns and unique characteristics that distinguish one person from another, or which bear similarities to others in a given population. In STR, a small area in the DNA chain is targeted and amplified using the PCR reaction. In the mid-1990s, systems for analyzing more than one STR at the same time were developed. A couple of years later, second generation multiplex (SGM) was introduced, which utilizes six different areas of DNA. This provides a higher level of discrimination than previous systems when used in conjunction with database population profiles.

Mass spectrometry is one of the more recent technologies to utilize STR fragments. STR analysis is sometimes appropriate in situations where there is not enough sampled material in sufficiently good condition to perform a Southern blot.

Many companies are now using STR systems for paternity testing and crime suspect DNA-print matching.

4. Origins and Evolution

Our understanding and use of DNA technology are very recent. Humans walked the Earth for about a million years without understanding how our bodies expressed specific physical traits. Even when scientific strides were made during the Renaissance, social and religious prohibitions prevented the hands-on study of the human reproductive system (or any tampering with its functions), for many more decades. The systematic study of cell reproduction is less than 300 years old, and was at first grasped in only the most general way by leading philosophers and scientists.

General public awareness of the biological mechanics of sexual reproduction is less than 100 years old, and public awareness and commercial use of DNA did not become widespread until the mid-1990s.

Due to social prohibitions on human experimentation, early scientists turned to other forms of life to study cellular structure and reproduction. They may not have realized it at first, but plants, insects, and marine animals are excellent experimental subjects; their cells can be quite large and many engage in sexual reproduction. Scientists have unraveled many important genetic fundamentals by studying plants and animals. The vessel cradling the genetic code, the *nucleus*, is present in some form in all these living things.

Processing DNA

This is a much simplified illustration of a gel electrophoresis and blotting process. Lab techniques vary, and newer methods are beginning to supersede gel electrophoresis, but this has traditionally been a widespread means of processing DNA.

A gel layer is subjected to electrophoresis to cause the DNA fragments to separate (smaller fragments move more readily through the gel).

The fragments are 'blotted' onto a membrane. Good contact between the layers is important, as is the prevention of air bubbles.

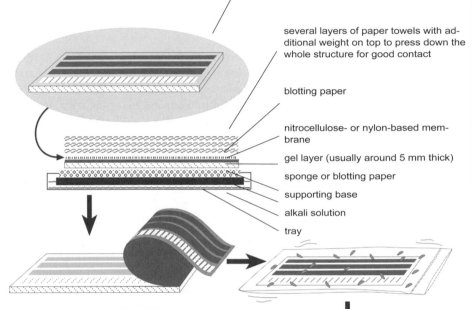

several layers of paper towels with additional weight on top to press down the whole structure for good contact

blotting paper

nitrocellulose- or nylon-based membrane

gel layer (usually around 5 mm thick)

sponge or blotting paper

supporting base

alkali solution

tray

The membrane bonded with the DNA is removed from the layer of gel and subjected to probing in a buffer medium to hybridize the DNA.

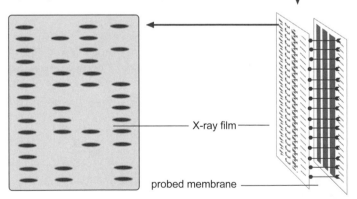

X-ray film

probed membrane

After probing, the membrane bonded with the DNA is placed adjacent to X-ray film, where the radioactive decay exposes the film to form a 'picture' of the placement of the fragments. The membrane can then be cleaned and reprobed and the imaging repeated several times.

DNA fragments are 'sorted' by gel electrophoresis and blotting and embedded in a medium so that the results can be probed and recorded on X-ray film for handling, analysis, and storage. [Classic Concepts diagram ©1999, used with permission.]

An overview of historical milestones in cell biology and genetics, and application of this information to DNA profiling, are provided next to give a backdrop to DNA sampling and the evolution of the technology. A summary chart of milestones is provided for quick reference. Following this is a practical section on collecting and handling DNA sample materials.

Understanding Cell Structure

Conducting DNA surveillance hinges on collecting useful cell samples. Thus, it is helpful to have a basic understanding of cell structure.

Most cell structures, particularly DNA, are too small to see with the unaided eye. Many of them are transparent, making them hard to recognize or distinguish one from another. Our detailed understanding of cell structure can be attributed to the invention of the microscope, in the 1600s, by Galileo (adapted from the telescope), Kepler, and Malpighi. Important strides in microscopy were made by Robert Hooke (1635-1703) and Antonie van Leeuwenhoek (1632-1723) in the late 1600s.

Van Leeuwenhoek greatly improved the single-lens microscope and used it to identify and study spermatozoa. The microscope is an indispensable laboratory tool which is routinely used for cell biology (cytology) and DNA sample assessment and preparation.

Antony van
Leeuwenhoek

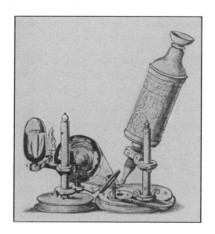 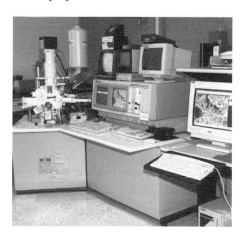

Left: A historic 1665 single-lens microscope, designed by Robert Hooke, was a significant invention that permitted scientists to discover microscopic cell structures and tiny living organisms. Right: Modern versions of the microscope are still indispensable laboratory tools, especially the electron scanning microscope. NASA uses microscopes extensively in space-related research, which includes searching for signs of life in samples from asteroids and other planets. [Carpenter, "The Microscope and its Revelations", copyright expired by date; NASA/LRL news photo, released.]

A few scientists stand out as significantly ahead of their time in studying reproductive processes. Regnier de Graaf (1641-1673) described human testicles toward the end of his short life. He also studied female reproductive structures and coined the term *ovary*. The Netherlands botanist Rudolph Jacob Camerarius (1665-1721) observed in 1694 in *"De sexu plantarum epistola"* that plants can reproduce sexually. Another forerunner, Lazzaro Spallanzani (1729-

1799), experimented with frog semen and discovered that filtering the liquid could prevent its ability to fertilize frogs' eggs, thus indicating that something contained in the liquid, not the liquid itself, must be responsible for fertilization. It took about 100 years for general scientific understanding to catch up with these discoveries. The evolution of DNA science from that time on follows a fairly orderly history from discovering basic cell structures and functions in the early 1800s, to developing theories to explain and predict heritability in living organisms in the mid-1800s, to developing commercial technologies in the 1980s. Trailing behind are ethical and social structures to understand, contain, and regulate the technology. These are ongoing developments.

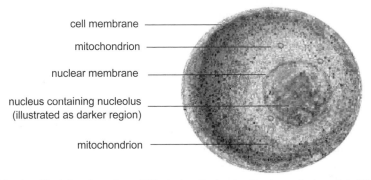

cell membrane
mitochondrion
nuclear membrane
nucleus containing nucleolus
(illustrated as darker region)
mitochondrion

This simplified drawing of a cell illustrates the basic structures that contain DNA. Historically, genetic structures could not be studied in detail until microscopes and staining techniques were developed to see the tiny, transparent contents of cells. [Classic Concepts diagram ©1999, used with permission.]

Once the microscope came into general use, great strides were made in cytology. In the early 1800s, Lorenz Oken (1779-1851) observed that organic beings originate from and consist of vesicles or cells. In 1831, Robert Brown (1773-1858) published *"On the organs and modes of fecundation in Orchideae and Asclerpiadae"* describing an important cell structure associated with fundamental building blocks. He observed an opaque area that he called an 'areola, or nucleus of the cell'. Matthias Jakob Schleiden (1804-1881) enlarged on the work of Brown and emphasized the importance of the nucleus as essential to the structure of the cell, calling it a *cytoblast* in his 1838 paper "Beitrage zur Phytogenesis". Within a year, a significant paper by Theodor Schwan, a professor at the University of Louvain, called attention to the importance of nuclei in animal tissues.

Most of these scientists didn't fully comprehend the role of the nucleus, thinking that perhaps new organisms budded from its surface, but the observations of Oken, Brown, Schleiden, and Schwan represent important pioneering steps in unraveling the mysteries of inherited traits within living organisms through the discovery of basic cell structure.

Charting Cell Structure and Exploring its Functions

Looking at further historical discoveries in detailed cell structure helps clarify the aspects of the cell related to DNA. This understanding of cells is useful in the DNA sampling process, as not all cells in the body contain nuclear DNA (as examples, hair shafts and mature red blood cells do not contain nuclear DNA).

Some of our understanding of cells comes from plant studies. In 1835, a German botanist, Hugo von Mohl (1805-1872), observed viscous fluid with a granular texture within cells that were sometimes in motion (motion in cells had been observed for about 60 years). He observed filamentous streams within the structures.

Around this time scientists noticed, with the help of microscopes, that the cell wall was not always present in small organisms, but that the same kinds of substances could be found within cell walls. They realized this was a clue, and subsequently directed their attention to these structures.

In 1848, Hofmeister observed the process of cell division in plants, and rod-like bodies within their cells.

By 1860, German physiologist Rudolph Virchow (1821-1902) had asserted a principle of the continuity of life by cellular division, roughly "All cells arise from cells", thus making an important contribution by establishing that the cell is a key reproductive unit, with the potential to generate a new organism.

New Theories and Procedures to Understand Heritability

Now that basic cell structure and reproduction were being unraveled, scientists developed theories for the hows, whens, and whys of cell reproduction, and how these might apply to our understanding of human origins. Many of these theories were met with disbelief or strong opposition, particularly in religious circles.

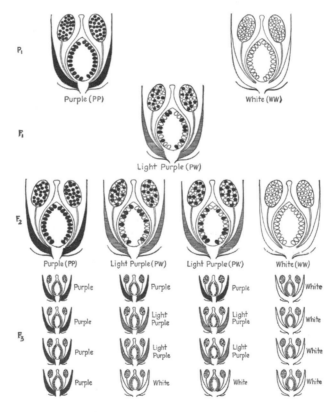

Gregory Mendel's work in plant inheritance had a strong influence on genetics research. This 1922 interpretation of Mendel's Law of Inheritance illustrates color traits in plants that are passed from the parents (P_1) onto their offspring through three successive generations (F_1, F_2, and F_3). [From *Botany Principles and Problems*, Sinnott, 1923, copyright expired by date.]

In 1859, British naturalist Charles Darwin (1809-1882) introduced a significant work, *The Origin of Species,* and sought to explain examples of natural selection and inheritance that he had observed in nature. Other scientists, such as Erasmus Darwin (1731–1802, Charles Darwin's grandfather), and one of Charles Darwin's contemporaries, Alfred Russel Wallace (1823–1913), developed evolutionary theories, but it was Charles Darwin's writings that managed to touch a sensitive chord, and he is the best remembered theorist of the time.

In 1866, Gregory Mendel (1822-1884), an Austrian cleric, published an important paper "Experiments in Plant Hybridization" in which he credits the earlier work of various researchers Gärtner, Pisum, et al., but points out the lack of a general theory, and the importance and difficulty of framing statements of general laws governing inheritance in organic forms. In his paper he describes his long experiments with plants and offers important observations regarding the stable patterns of heritability of traits among the offspring. Subsequent scientific research included studies of patterns of heritability in humans. Researchers now wanted to know where this information was stored and how it could be accessed and perhaps controlled. The works of Mendel came to the attention of the mainstream of science when they were independently rediscovered, at the turn of the century, by Hugo DeVries, Erich Von Tschermak, and Carl Correns.

Cell Division and Fertilization

The equal contributions of the mother and the father to the genetic makeup of human offspring is an important consideration in DNA analyses, particularly as they relate to parental profiles. Understanding basic cell division can help clarify this aspect of DNA testing.

By the late 1860s, there were many scientists who accepted the dual role of sperm and egg cells in reproduction, but most of them still assumed the sperm was performing some sort of chemical stimulation of the female egg, and did not yet credit the genetic material inserted by the sperm drilling into the egg. Microscopes were a big boon to cell research, but there were limits because they were unable to show transparent structures.

The brothers Oscar Hertwig (1849-1922) and Richard Hertwig (1850-1937) conducted genetic research together at a marine station in France. Around 1876/1877, Oscar Hertwig described the fertilization process as 'conjugation of two different sexual nuclei' from the male and the female, thus bringing the theories one step closer to our present understanding. In 1879, Hermann Fol, also working with marine animals, graphically illustrated the process of a sperm penetrating an egg, based upon his observations of starfish.

In spite of mounting scientific evidence to support heritability of traits from one generation to the next through eggs and sperm, there were still people who opposed the idea of nuclear continuity through descendants. Many still thought of family traits as being inherited in some mysterious manner through blood, a belief that resulted in terms like *blood relatives* and *bloodlines*. It was not until William Henry Perkin (1838-1907) made great strides in the synthetic creation of dyes, that microscopy provided the means for the next stage of discovery, starting around the 1870s.

Eduard Adolf Strasburger (1844-1912) made essential contributions to cell science when he used dyes to 'illuminate' various cell structures and plant cells undergoing mitosis. This allowed him to observe the *union of nuclei* during the process of fertilization.

In 1879, Walther Flemming (1843-1905) was studying dividing cells and observed the formation of what would later be termed *chromosomes*, though he didn't yet understand their function. Microscopic dyes were the next step in revealing previous unseen transparent structures. Dyes enabled Flemming to observe the progressive stages of cell division and, in 1882 he published his observations in *Cell Substance, Nucleus, and Cell Division*. This important work described cell mitosis and the role played by chromosomes in cell division. In 1888, the chromatic threads in the cell nucleus were termed *chromosomes* by Waldeyer.

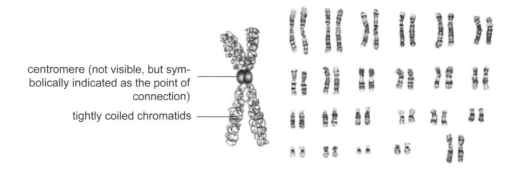

centromere (not visible, but symbolically indicated as the point of connection)

tightly coiled chromatids

These simplified drawings illustrate the general configuration and shape of human chromosome pairs. While roughly the same shape, they differ in size, and are typically sorted by size for identification purposes.

New scientific discoveries always seem to generate controversy. In the 1880s, August Friedrich Leopold Weismann (1834-1914) challenged the *blending and rejuvenation* theories of reproduction. In 1885, he introduced his theories in "Die Continuität des Keimplasmas als Grundlage einer Theorie der Vererbung" (The Continuity of Germ Plasm as the Foundation of a Theory of Transmission) and, in 1892, in *Das Keimplasma - eine Theorie der Vererbung,* describing the continuity of life through the reproducing nature of cells through chromosomes. He also described how genetic material in the egg and sperm are halved and then become a full germ plasm again when combined with one another in the fertilized egg.

Up to this time, many people wanted to believe that a mother's biology determined the sex of her offspring. Henry VIII had several wives killed because they did not bear him a male heir. Clarence E. McClung (1870-1946), an American biology professor, wrote a key paper that described how chromosomes were involved in sex determination in a species of grasshopper. He was assisted in his research work by Walter Stanborough Sutton.

The Rise of Modern Genetics

In the early 1900s, many of the key concepts and techniques in the emerging science of genetics were being developed and put into practical use.

Thomas Hunt Morgan (1866-1945) had enormous influence in early genetics research, not only through his own research, but also through the research of his students, who were inspired by his example. Morgan did some important work in the study of inheritance in fruit flies (Drosophila Melanogaster), confirming Hugo De Vrie's (1848-1935) observations from plants that mutations can occur, and that inheritance is not absolutely scientifically predictable. Morgan's experiments established the important role of chromosomes in heredity and motivated researchers to locate particular gene locations. In 1901, he authored "Regeneration" and, in 1926, "The Theory of the Gene". In 1933, he became the first biologist to win a Nobel Prize.

In 1902, Walter Sutton observed the process of cell division and published a paper on chromosome morphology. The following year, Theodor Boveri (1862-1915) and Sutton independently observed that each gamete receives only one chromosome from each original pair and Sutton proposed that the offspring get genetic material from each parent. Within the context of Mendel's studies, he suggested that heritable 'factors' were located on chromosomes and called them *genes*.

In 1902, William Bateson (1861-1926) authored "Mendel's Principles of Heredity: A Defence". In 1905, he made an important demonstration that some characteristics are not independently inherited, thus introducing the concept of 'gene linkage' which led to projects to map the genes and describe their order and relationships. Bateson coined the term *genetics*.

In 1905, Edmund Beecher Wilson (1856-1939) and Nellie Stevens proposed that separate X and Y chromosomes determine the sex of a human offspring. Thus, two X chromosomes resulted in a female and a Y chromosome from the father resulted in a male, an observation that was subsequently replicated by other scientists. Wilson had previously written "The Cell in Development and Inheritance", in 1896.

Archibald Garrod made some important connections between the work of Mendel and the biochemical pathways of reproduction and speculated on inborn metabolic causes for human diseases that swept away superstitions about 'bad air' and 'impure thoughts' as primary causes of illness and birth defects.

In the early 1900s, structures and nomenclature were specified in more detail, and processes explored more closely. In 1909, Danish botanist Wilhelm Johannsen (1857-1927) described a distinction between genotypes and phenotypes. In 1903, he first mentions the terms 'gene', 'genotyp', 'phenotyp' in the context of the breeding of beans.

In 1909, Archibald Garrod described the heritability of four metabolic diseases and interest in the speciality continues today, with the study of inheritance of serious diseases being a priority in many DNA research labs.

Genetics was introduced as a field of study at Columbia University in 1910.

Thomas Hunt Morgan demonstrated, in 1915, that genes are responsible for the transmission of traits. His research group also established the existence of 'sex-linked' traits, traits that are passed onto the offspring by one parent or the other or both, but which manifest in a particular gender (examples include hemophilia, pattern baldness, and color blindness). Morgan promoted a chromosome theory of heredity.

Genetics and Gene Mapping

Technology and discovery go hand in hand. The desire to know motivates us to create better tools for 'seeing', and better tools result in unanticipated discoveries. Just as the microscope played an essential role in the understanding of cell structure and function, X-ray technologies played an important role in the discovery of DNA.

Just as dyes enabled us to see previously invisible cellular components, X-ray crystallography extended our 'vision' by revealing structures through the diffraction of light. This technology was developed in large part by the father-son team of William Henry Bragg (1862-1942) and William Lawrence Bragg (1890-1971). Practical use of their invention came in the mid-1930s, and X-ray examinations are now used to reveal patterns in DNA. In 1915, the Braggs received a Nobel Prize for their work.

In 1913, Alfred Henry Sturtevant (1891-1970), a student of Thomas Morgan, constructed the first gene map by analyzing fruit fly matings and their results. This made it possible to follow and predict patterns of heritability.

With the introduction of genetics courses into universities, the pace of research accelerated, and the desire for sharing various discoveries increased. In 1916, the *Genetics* scientific journal was established by George Harrison Shull, a Princeton genetics professor, to meet this need.

In 1918, Herbert M. Evans declared that human cells contain 48 chromosomes (he was almost right).

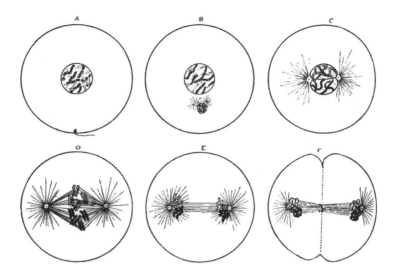

The process of fertilization and cell meiosis, as depicted in 1922. A) The sperm penetrates the egg. B) Having dropped its tail, the sperm moves toward the egg's nucleus. C) Interaction of the nuclei from sperm and egg. D) The splitting of chromosomes from sperm and egg. E) Separation of the chromosomes. F) Cell division. At the time this was drawn, genetics had recently been introduced to university curricula, and the nature of the gene was beginning to be explored and described. [Henry Holt and Co., copyright expired by date.]

In 1921, Hermann Joseph Muller (1890-1967), another of Thomas Morgan's students, described the nature of the gene in prescient theoretical paper. In the mid-1920s, he discovered that X-rays could increase mutation rates in fruit flies, indicating that the occurrence of mutations could be influenced in somewhat predictable ways. Muller's work was honored in 1946 with a Nobel Prize.

Genetics Development and Social Impact

In terms of its historical significance to surveillance, one of the most important landmarks in the application of the knowledge of genetics to human freedom and political control occurred in 1924, when the U.S. Immigration Act used genetics as a political lever to exclude poorly educated immigrants from southern and eastern Europe stating 'genetic inferiority' as a justification. This clear example of the abuse of political power is the reason why many people fear the potential misuse of DNA sampling and genetics databases.

In 1926, Thomas Hunt published *The Theory of the Gene* (Yale University Press) describing the physical basis for genetics—based upon breeding experiments and his observations with optical microscopes. Around this time, scientists increasingly peered into the intimate structure of DNA and its influence on physical traits.

In 1931, Barbara McClintock demonstrated that gene order in chromosomes can change by rearrangements, and that particular traits in maize are related to genetic distribution. In 1947 she reported on 'transposable elements', now known as 'jumping genes'.

The increased understanding of heritability of traits resulted in public attention being focused on people with undesirable traits, as interpreted by political powers of the time. As a result of this line of thinking, by 1931, compulsory sterilization laws had been implemented by the U.S. Government. Two years later, Germany instituted even more pervasive sterilization, labelling individuals as 'defective'. These historical precedents underline the importance of carefully evaluating the legal ramifications of DNA technology.

In 1934, Desmond Bernal demonstrated that very large molecules could be studied with X-ray crystallography.

Toward the end of World War II, Maclyn McCarty, Oswald Avery, and Colin MacLeod were conducting research on pneumonia bacteria. They published their findings in the *Journal of Experimental Medicine* on 1 February 1944. This paper discussed thread-like DNA fibers that carried hereditary information, an important precedent, promoting further research into the characteristics and components of these biological fibers.

Visualizing and Modeling DNA

With the evolution of radio electronics, and the introduction of transistors in the 1940s, many new instruments that could aid scientists in detecting the structure and function of cells were developed over the next several decades. One of the most significant inventions around this time was Vladimir Zworykin's (1889-1982) adaptation of the electron microscope, originally developed in Germany in the early 1930s. It quickly became a practical and exceedingly important tool for research in molecular biology and biochemistry related to genetics. During the 1950s, scientists began to delve deeply into the detailed components of cell structures.

In 1950, Erwin Chargaff (1905-) reported the proportional components of DNA, now known as Chargaff's Rules. He observed the relationship between A and T (adenine and thymine) and between G and C (guanine and cytosine). His research was important to subsequent models of the structure of DNA. Practical applications of inheritance in breeding were developed at about the same time. The principles of artificial insemination were known and had been described in print by the 1920s, but it was not until 1950 that artificial insemination became commercially routine in livestock breeding.

In 1952, Alfred Day Hershey and Martha Chase stated that genes consist of DNA, thus establishing a link between DNA and heritability. The electron microscope made it possible to study the anatomical structures of cells, including the *ribosomes*.

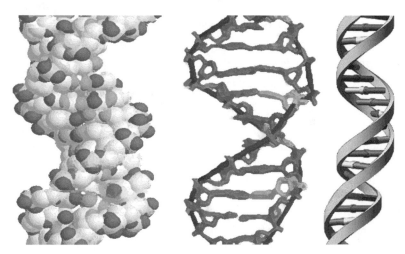

Three symbolic representations of the double helix structure of a DNA molecule. Forming a hypothetical model of DNA was an important step toward understanding the molecule, and synthesizing it in the laboratory. [Classic Concepts diagrams ©1998, used with permission.]

In the 1950s, Oliver Smithies developed a starch gel electrophoresis process for separating proteins, which eventually led to practical tools for processing DNA for analysis, such as the Southern blot, which is now commonly used.

In London, Maurice Hugh Frederick Wilkins (1916-) and Rosalind Elsie Franklin (1920–1958) conducted X-ray diffraction experiments that revealed the presence of two curving, related strands forming the basic structure of a DNA molecule. The use of X-ray diffraction was another important milestone in research and practical applications.

In the 1950s, James Dewey Watson (1928-), and Francis Harry Compton Crick (1916-) built on the work of Wilkins and Franklin and the chemical data of Chargaff, and published a more comprehensive descriptive model for DNA's double-helix structure. This important tool for visualizing DNA earned Watson, Crick, and Wilkins a 1962 Nobel Prize.

One way to confirm a biochemical model is to replicate it in the lab. Despite the complexity and microscopic scale of DNA, Arthur Kornberg (1918-) confirmed Watson and Crick's theoretical description of DNA and successfully synthesized it in 1956, earning a 1959 Nobel Prize.

In 1956, Tijo and Levan stated that humans have 46 chromosomes rather than 48 as reported by Evans in 1918.

Laboratory Tools and Manipulation of DNA

When Arthur Kornberg synthesized DNA in the lab, he opened up vast new roads in DNA research with practical applications in surveillance technology and general investigative sciences. The late 1960s to the late 1970s was a time during which laboratory synthesis and manipulation of DNA made it possible to develop general procedures for replicating and studying DNA, and set the stage for practical DNA profiling methods.

In 1958, Matthew S. Meselson and Franklin W. Stahl confirmed the 'splitting' of the double helix shape lengthwise to allow nucleotides to link with each half of the chain, forming two duplicates of the original structure. Important lab techniques now began to stem from such discoveries. Hybridization processes, which are now used for detection of specific nucleic acid sequences, were observed and reported in 1960 by Mannur and Lane and Doty et al. By 1970, it was possible to isolate discrete fragments of DNA.

Also in the late 1950s, Crick and his co-workers were exploring a 'messenger' molecule, ribonucleic acid (RNA), that is essential to transmitting DNA information.

In 1968, Victor A. McKusick described the repetitive nature of DNA and defined the first gene place. In 1970, Hamilton Smith and Daniel Nathans discovered a DNA-cutting enzyme. Enzymes which can 'cut' the DNA in certain locations and *ligasen* which can rejoin DNA, became important laboratory tools. Cutting DNA is now an essential procedure in laboratory processing of DNA.

In 1973, Herbert Boyer (1936-) and Stanley Cohen (1922-) created the first recombinant DNA molecule using the restriction endonuclease *EcoRI* and the plasmid *pSC101*, opening up a new level of genetic engineering and biotechnology.

1975 saw the practical application of immobilization techniques used in conjunction with restriction enzymes. Development of processing and analytic tools such as the Southern blot were underway.

Frederick Sanger (1918-) and Alan R. Coulson (1947-) presented a practical gene-sequencing technique that employed *dideoxynucleotides* and *gel electrophoresis*. In 1977, Walter Gilbert (1932-) and Allan M. Maxam presented a gene-sequencing technique that employs cloning and gel electrophoresis. Gel electrophoresis, a means of sorting DNA fragments in a gel medium, is still widely practiced. Modern DNA profiling was falling into place.

Applying Laboratory Methods to Unraveling the Genome

The most significant genetic research of the last twenty years has been the unraveling of the structure and function of individual human genes, and organizing this knowledge to better understand gene expression, interaction, and influence. The motivation to chart the human genome is in large part fed by our desire to eradicate or cure psychologically or physically painful conditions or life-threatening illnesses of a genetic origin. However, in the process of exploring the medical aspects of human genes, we are also uncovering many more complex social issues. Understanding the human genome is closely allied to DNA analysis/DNA surveillance, as the results of an analysis can be related to the growing body of work elucidating the genome, thus uncovering characteristics of a socially volatile nature.

Fred Sanger is credited as the first person to sequence an entire human gene, in 1978, and DNA evidence was first introduced into courtrooms at about the same time. In 1980 he was co-awarded a Nobel Prize with Paul Berg and Walter Gilbert.

Exploring the Human Genome

In 1980, David Botstein et al., suggested the use of DNA information as markers for exploring the human genome. Gene mapping became an important goal and tool of DNA researchers. At about the same time, automated sequencing was developed.

In 1983, Kary Banks Mullis (1944-) developed the polymerase chain reaction (PCR), an important tool for amplifying a DNA molecule which is now widely used both in laboratories and in portable DNA analysis kits.

In 1984, Alec J. Jeffreys of Leicester University developed *DNA 'fingerprinting.'* This was initially used in forensics in the U.K., in 1986, for investigating violent crimes. By 1987, DNA technology had made its mark on the legal system, a suspect had been convicted of a violent crime with the assistance of DNA evidence. The following year, in the United States, DNA was used to assist in convicting a criminal of rape. From this time on, there has been increasing acceptance of DNA evidence in court proceedings.

In the late 1980s, *fluorophore labels* began to supersede *radioisotope labels* for labeling DNA fragments for the purpose of DNA base sequencing. The fluorophore method permits analysis of DNA up to a length of about 500 base pairs (above this length, analysis is difficult due to reduced signal intensity). About the same time, the FBI began investigating the feasibility of using mitochondrial DNA for human-identity testing.

The Human Genome Project - HGP

In October 1990, the historically significant Human Genome Project (HGP) was officially launched, with James Watson as the first director. This project was initiated with the goal of mapping the entire human genome, with projections of completing the task to the 'working draft' stage sometime in 2002.

By 1993, French researchers had completed a physical map of human chromosomes. A year later, an international team published the first linkage map of the human genome.

In the early 1990s, following the work of Alec Jeffreys, practical methods for 'fingerprinting' a DNA sample, that is, determining whether it matched or didn't match a given example, were developed, with more precise and efficient quantification procedures for DNA profiling arising a few years later.

Ranajit Chakraborty, a professor of population genetics, designed DNA-sampling strategies that were used in analyzing hundreds of DNA-typing databases. He manages a repository of forensic DNA databases for more than six countries including India, Australia, Brazil, etc.

Some Key Developments in the Evolution of DNA Technology	
late 1400s	Leonardo da Vinci observes that skin color is influenced by both the mother and the father and Michelangelo writes letters to his brother advising him to find a wife with a good constitution and good traits that can be passed on to the children. These facts indicate that at least an intuitive understanding of heritability existed among the educated by Renaissance times.
mid-1600s	De Graaf studies reproductive anatomy and coins the term *ovary*.
late 1600s	Microscope technology is advanced by Hooke and Leeuwenhoek. Leeuwenhoek uses his microscope to study spermatozoa.
mid-1700s	Spallanzani studies frog semen and discovers that something contributing to fertilization can be filtered out of the semen.
1831	Robert Brown observes a cell nucleus.
1835	Hugo von Mohl observes a granular texture in cells.
1848	Hofmeister observes plant cell division and rod-like structures within the cells.
1859	Charles Darwin publishes "The Origin of Species".
1866	Gregory Mendel describes stable patterns of heritability in plants.
1879	Hermann Fol observes and graphically illustrates a sperm fertilizing an egg.
1902	William Bateson researches 'gene linkage.' Sutton proposes that genetic material is inherited from each parent.
early 1900s	X-ray crystallography is developed.
1916	The "Genetics" scientific journal is launched.
1920s	Thomas Morgan experiments with fruit flies and establishes the role of chromosomes in inheritance.
1940s	The emergence of modern microscopes, electronics (especially the transistor) and electron microscope technology.
1950s	Watson and Crick create a conceptual model of DNA.
1956	Kornberg confirms the Watson-Crick theoretical model and synthesizes DNA.
late 1950s	Technologies are developed for manipulating DNA.
1960	DNA profiling techniques are refined and hybridization is introduced.
late 1960s	Gene-cutting enzymes are applied to DNA sequencing techniques.
late 1970s	DNA evidence is introduced into court room trials, but is initially treated with some skepticism.
1980s	DNA profiling lab techniques are developed and refined. Acceptance in courtrooms is increasing, though not yet secure.
1990	The Human Genome Project is established.
mid-1990s	DNA testing becomes commercially available to public consumers. Privacy and law enforcement issues associated with DNA are debated.
1999	Human chromosome 22 is decoded by a joint international effort. Iceland turns over its genetic heritage to a private company.

By June 1996, after about four years of investigating protocol and validity, the FBI began conducting mitochondrial DNA forensic examinations.

By December 1999, an international team of researchers had sequenced an entire human chromosome (#22) containing 33 million base pairs as part of the Human Genome Project. The chromosome was found to have over 500 genes, with about 42% noncoding DNA. This significant achievement will be remembered as a milestone.

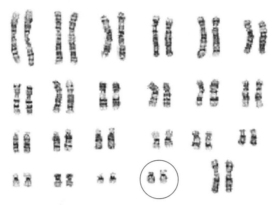

The first human chromosome to be fully decoded was chromosome #22, one of the smallest. This historic feat was accomplished by the end of 1999 in a collaborative international effort. [Classic Concepts diagram ©1999, used with permission.]

Consumer Access and Awareness

By the late 1980s, DNA profiling was a commercial commodity and an accepted law enforcement tool. By the mid-1990s, the proliferation of DNA testing, and the increased use of DNA information in the legal system, sparked public concern about privacy and the ethical uses of DNA.

As a result of public outcry from civil rights organizations, the *Genetics Confidentiality and Nondiscrimination Act* was introduced to the Senate in 1996. It was developed to safeguard the ownership of genetic information and the privacy of individuals. Similar bills were introduced in other areas. These legal restrictions have a direct influence on surveillance activities, limiting their scope and availability.

In the mid- to late-1990s, gel electrophoresis methods were improved by using capillaries rather than glass plates. Simultaneously, techniques using successive hybridization that did not require electrophoresis were developed. Commercially available, reasonably priced DNA testing was offered to consumers around this time.

Commercial services proliferated in the early 1990s, and the motivation to produce faster, cheaper, more portable DNA analysis tools increased in the mid- to late-1990s. Realtime analysis of DNA fragments in DNA sequencing processes became commercially available using a fluorescent process around 1998. Approximately 10,000 to 20,000 bases per day could be analyzed with newer systems. Acceptance of automated systems was quick, given that there was a significant backlog of samples in law enforcement storage facilities that had not yet been analyzed.

Consumer acceptance of DNA testing rose dramatically in the late 1990s, in large part due to paternity testing on daytime talk shows. As with many historical developments, television has had a substantial impact on public awareness (that is not to say that television has provided a well-rounded portrayal, but rather that it focuses public attention on specific aspects). In conjunction with this, the cost of a basic paternity test gradually dropped below $300 (for two samples—the alleged father and offspring). DNA research in the commercial sector is now heavily focused on automation and cost-saving measures. Law enforcement officials actively use the technology, and DNA databases are springing up everywhere. Home kits for personal DNA banking are in development.

A portable, battery-operated DNA analysis system has been developed at the Lawrence Livermore National Laboratory. Dean Hadley tests the suitcase-sized system for use in a variety of genetics profiling applications including forensics, paternity testing, remains identification, food and water analysis, and tests for pathogenic bacteria on the battlefield. [LLNL news photo ca 2000, by Joseph Martinez, released.]

By the late 1990s, realtime field analysis tools, such as those developed by the Lawrence Livermore Laboratory, made it possible to analyze a sample outside of the lab, a boon in situations where transport is difficult or impossible, or where confidentiality is desired.

Commercializing a Genetic Heritage

The genetic heritage of the people of Iceland has been of particular interest to those studying genetics and those seeking to promote its ethical use. More than a little controversy has been associated with the distribution of Icelandic DNA. Since the Vikings colonized Iceland in the 9th century, the settlement has been one of the more isolated and, hence, more genetically pure of the Scandinavian regions. In addition to this, the people of Iceland are organized and tidy and keenly committed to keeping detailed and careful records. A universal health-care system has been in place since the early part of the 20th century, and genealogical research is avidly pursued. A large percentage of the population, numbering a little over a quarter of a million, is able to trace its lineage back many generations and many Icelanders are able to read the old texts.

In 1998, Iceland's parliament passed the *Icelandic Health Sector Database Act* establishing a centralized database of the Icelandic peoples' family history and medical information. The government then decided to include genetic information, as well. Seeking to preserve and promote its genetic legacy, the Icelandic parliament then contracted to give access to medical and DNA records from the populace, to a biotechnology company named DeCODE Genetics, Inc., that was incorporated in the state of Delaware in 1996. DeCODE was given exclusive marketing rights for the information for a period of 12 years and permission to create the Icelandic Health Sector Database (IHD) by the Icelandic Ministry of Health.

In January 2000, deCODE announced that it was planning to publish a comprehensive reference called *The Book of Icelanders*. The primary focus of the company was developing drugs and DNA-based diagnostics based upon heritable common diseases. In March, there were public accusations of DeCODE having made contributions to a number of political parties with the implication that this was to influence approval of the project. DeCODE admitted having made contributions, but the parties denied having been influenced in any way and the publicity didn't change their plans to go ahead with a public offering of shares. In July 2000, DeCODE made public stock offerings on Nasdaq. It ran into problems, however, when a Class Action Complaint for federal securities law violations was filed in a New York district court, alleging that the "Defendants herein participated in a scheme to improperly enrich themselves through the manipulation of the aftermarket trading in deCode common stock following the IPO". The class action suit was amended in April 2002.

By the end of 2003, the company announced that it had identified 15 genes involved in 11 common diseases and located genes associated with 30 common diseases. It entered into an exclusive license with Bayer to develop and commercialize a compound that inhibits a protein to decrease production of leukotrienes as a means of remediating heart disease.

Concerns about privacy were raised right from the beginning of the project. In response, one of the 'informatics' tools developed by DeCODE was the Identity Protection System (IPS), marketed in alliance with IBM. Under the auspices of the Icelandic Data Protection Authority, the IPS is employed to automatically anonymize the clinical/genetic data.

In November 2003, the Icelandic Supreme Court made a landmark decision regarding the constitutionality of mandatory inclusion in the Iceland Health Sector Database. The plaintiff sought recognition of her right to prohibit the transfer of medical information on her deceased father.

> It was revealed in the course of proceedings that extensive information concerning people's health is entered into medical records, e.g. medical treatment, life-style and social conditions, employment and family circumstances, together with a detailed identification of the person that the information concerns. It was recognised as unequivocal that the provisions of Paragraph 1 of Article 71 of the Constitution applied to such information-protection of privacy in this respect.... In light of these circumstances, and taking into account the principles of Icelandic law concerning the confidentiality and protection of privacy, the Court concluded that the right of R in this matter must be recognised, and her court claims, therefore, upheld.

> [*Ragnhildur Guðmundsdóttir vs The State of Iceland* (No. 151/2003), 27 November 2003.]

In granting permission to create the IHD in 2000, the Icelandic government mandated a review of the IHD data encryption and protection protocols. DeCODE reported that as of March 2004, the review had not been completed and that they were having second thoughts about proceeding with this aspect of the project. In their words:

> When and if this review and issuance of related security certification is completed, we will evaluate whether and when, if at all, to proceed with the development of the IHD in light of our priorities and resources at that time. In light of our current business plans and priorities, we do not expect the IHD to be a material aspect of our business in the near future.

That is not to say the DeCODE data is wide open. The company described it as being "personally non-identifiable and held under encrypted identifiers generated by the Icelandic government's Data Protection Authority".

While genetics research may provide important information for the study of human origins, genetic patterns, and medical conditions, it also has significant ethical ramifications and raises concerns about what someone might do with the knowledge that a person has certain genetic weakness or latent illness or what happens to ethics when a public company becomes profit-oriented.

Similar efforts sprang up in other areas. In California, in 2000, DNA Sciences, Inc., launched a Website to recruit DNA donors to provide samples for a 'gene trust' for research into genetic illnesses, with James D. Watson as a director. DNA Sciences, Inc., was formerly Kiva Genetics, Inc., a provider of genetics information to academics and the pharmaceutical industry. Things did not go as planned. Hugh Rienhoff, CEO and chairman of DNA Sciences, Inc., left his positions with the company in 2001. In May 2003, Genaissance Pharmaceuticals, Inc., a public company, announced that a U.S. bankruptcy judge had approved their acquisition of "substantially all the assets of DNA Sciences, Inc.", including the facility, the patent estate, and the domain name (dna.com), which now redirects to Clinical Data, Inc.

At the same time that Watson was on the board of directors of DNA Sciences, he was president (and later chancellor) of the Cold Spring Harbor Laboratory, in New York.

All these mixed allegiances and staffing changes are typical of the corporate world and are, in themselves, important to consider with respect to private information. What happens to data in a commercial 'gene trust' if the staff moves on, is busy elsewhere, or if the company is acquired by another firm with different priorities and safeguards?

Subsequently, in a far more controversial move, Watson eschewed concerns about privacy invasion and encouraged Americans to offer their unique genetic patterns to a national database. News reports indicated that Watson had said this could help combat crime and terrorism and that the benefits would outweigh threats to privacy. The author has tried to editorial as little as possible in this book, allowing readers to read the facts and decide for themselves, but is of the opinion that there are not enough safeguards in place to protect data derived from emerging technologies, that stockpiling the DNA profiles of honest citizens is not a proven way to fight crime and, in fact, reduces the distinction between criminals and honest people, and that information like centralized digitized DNA profiles cannot be adequately safeguarded from hacking or covert use or resale, and may possibly be abused if they fall into the hands of future administrations. If you read the historical section of this chapter, you will see that past administrations have used genetic information to try to bar people from the country, to try to assert genetic superiority, and to discriminate in other ways—and this was before more detailed modern DNA profiles were available. To hand over sensitive genetic information on a national scale is highly unwise, because Constitutional mechanisms are too generalized to prevent specific forms of abuse by unknown people who may hold power in future administrations and there is no way to prevent this kind of information from getting onto the Net through human error or malicious distribution.

National Databases

Concerns about privacy don't seem to have prevented political administrations, health care firms, entrepreneurs, scientists, and law enforcement entities from amassing DNA information.

By February 2006, the U.K. Parliamentary Office of Science and Technology announced that the U.K.'s National DNA Database (NDNAD) included profiles of more than three million individuals. The database was first established in 1995 in England and Wales. Scotland and Northern Ireland maintain their own records and submit duplicates to the NDNAD. Legislative changes made it possible for the extensive expansion of the database by allowing the police to take DNA samples, without the person's consent, from "anyone arrested and detained in police custody with a recordable offence" (note that 'intimate' samples, like blood, still require consent). The database also includes samples from volunteer donors and from crime scenes. Note that individuals don't have to be convicted of a crime to be sampled, only to have been arrested for a crime (which they may not have committed). The samples themselves are not full profiles, but may provide limited information on medical propensities and ethnicity.

One of the concerns raised when the system changed from sampling those who were charged with crimes to those who had been arrested for a crime (but not necessarily charged), was that black and other ethnic minorities were found to be disproportionately represented. Under Scottish law, the sample is not retained if the person is acquitted. The U.K. Home Office, however, is in favor of keeping them. Those who voluntarily donate 'elimination samples' must provide written consent. Unfortunately, unlike Scottish law, in which volunteer consent can be withdrawn and the sample removed, the NDNAD consent is irrevocable—it cannot be withdrawn.

Public perception of the U.K. databases is often inaccurate. Many people assume that those represented in the database are all criminals, but some are voluntary donors and many are people who were never formally charged with a crime or who were acquitted. It has been argued that this contravenes articles in the *European Convention on Human Rights* (1998). When the case was heard and appealed, the House of Lords dismissed the appeals.

In 2003, the White House announced that juvenile offenders and adults who have been arrested (but not necessarily convicted) would be added to the FBI's national DNA database under a proposal by the G.W. Bush administration. Up until that time, only convicted adults could be added to the database.

As of January 2003, there were about 1.3 million DNA samples in the database, which primarily represented convicted adults. By 2006, the FBI had more than 3.3 million DNA profiles in its Combined DNA Index System (CODIS).

Thus, in the U.K. and U.S., it has been amply demonstrated that the original mandate almost never stays the same. As soon as a system is in place, those who find it useful seek to expand it. In both the U.K. and the U.S., concerns about the lack of public input into decisions about national databases have been raised. In the case of complex issues and sciences that are still evolving, like DNA science, some of the ethical considerations and potential dangers cannot be known in advance and the lack of democratic input into the expansion of surveillance tools should be of concern not only to private citizens but to law enforcement and other agencies seeking to use these tools in responsible ways.

5. Description and Functions

DNA samples will vary in their quality, purity, and quantity. What follows is a general overview of procedures for sampling and laboratory evaluation.

5.a. Permissions

Regulations for DNA use vary. Citizens and medical practitioners are not significantly restricted in their use of DNA. The DNA collection of some armed forces personnel is mandatory. DNA collection by law enforcement officials is regulated much the same way as other search procedures. Thus, depending on the situation and the person doing the collecting, the first step in getting a sample may be to inform the donor of his or her rights, and to get permission from the donor or donor's guardian for a sample, or to procure a court order or 'search warrant' for taking samples from an involuntary donor.

If a person being sampled is involved in a family or legal dispute, it is particularly important to verify the identity of the donor with *all* of the following: picture ID, a signature, and preliminary questioning. There are documented cases of imposters, some of them look-alikes, showing up to provide DNA samples for paternity disputes or on behalf of fugitive criminals. It will never be known how many of these imposters have gotten away with this fraud, and there are still clinics that need to improve their screening and verification procedures.

5.b. Collecting Samples

This section gives a general overall description of tissue collection, as there are variations in requirements from one lab to the next. For a practical understanding, it is recommended that a course of instruction be taken that includes hands-on demonstrations. Samples can be divided into two general categories: those that are given voluntarily or with permission of a guardian, and those that are collected without consent of the individual (individuals who are dead; who have committed a crime; who are being surveilled without their knowledge; or plants or animals which cannot give informed consent).

When tissue donations are given voluntarily, it is customary to use a blood draw or buccal (cheek) swab or, in the case of a rape victim, it might be a vaginal or anal swab. Blood collection or other invasive procedures should be done by trained professionals. For full analysis, it is valuable to take samples from more than one body site. It is often necessary to take the sample in the presence of a witness, with signatures, for it to qualify as admissible evidence. In the case of involuntary samples, DNA tissue is especially subject to contamination. It is recommended that sterile gloves be worn, and sterile glass vials, tissues, swabs, and envelopes be used whenever possible, with fresh gloves for each sample taken. Materials that can contaminate samples include chemical residues, dirt, grease, and some dyes (e.g., denim dyes).

It must always be remembered that human body fluids may contain dangerous viral or bacterial materials that can be contracted or spread by the person taking the sample (tuberculosis, HIV, streptococcus, etc.). Exercise good judgment in acquiring and handling body fluids. In the case of large fluid samples such as urine or mouthwashes, it may not be practical or necessary to transport and store the entire sample, a few ounces may be sufficient.

Changes occur in our cellular structure as we age or are affected by various illnesses. Genetic material is quite stable and resilient, but mutations can occur. *Germ line* mutations are likely to occur throughout an individual as this is a type of mutation that happens early in the reproductive process or growing stages of a new organism. *Somatic cell* mutations can happen during reproduction, or later in development, with the changes present only in cells developing after the mutation. Thus, a somatic cell mutation may not be present in all body cells, and a sample taken from a cell mutation site could differ from a sample from another part of the body. For practical purposes, these mutations are rare, and surveillance professionals gathering samples can concentrate their attention on getting as many good quality samples as possible, leaving micro-analysis of the tissues to a trained lab professional.

When sampling plant tissues, it is generally easier to obtain enough material to satisfy the needs of the lab. Collect as much as is practical and store it in a sterile container. Depending on the type of plant, samples can be dried, vacuum-sealed, or frozen, if they need to be transported. Animal tissues may be more difficult to collect in sufficient quantity, especially those that are associated with violent crimes such as murder, and rape, as they may be scarce or difficult to see. Blood traces can sometimes be located with chemical reagents, but the chemicals that reveal the stains may also alter them so that they can't be processed for DNA. In some cases, ultraviolet light can be used to reveal stains and other traces of body fluids. It is vital to collect as much good tissue as possible, as not all the samples may be from the same individual, and some may be of poor quality. Sometimes more than one test needs to be performed. Skin scrapings from a window sill or from under a victim's fingernails should be collected, but the chances of getting a good sample are slim. The tissue is usually too sparse, or too contaminated to be useful, but it may be the only thing available, and therefore should still be collected with care.

Scrapings can be put on clean dry paper which is folded before placing it in an envelope. Avoid putting the debris directly in the envelope, if possible.

Body Fluids

The saliva on an envelope, cigarette, or stamp may not be sufficient for a sample, but it should be collected nevertheless. If the saliva is associated with a large object, it is usually not necessary to send the whole object (unless it is to be used as evidence), but rather to cut out the relevant section with a margin of a few inches, and transport the portion containing the sample. Saliva samples kept in a wet condition tend to deteriorate rapidly; freezing may be appropriate; and air drying seems to be effective in prolonging the usefulness of the sample.

Blood and semen on a blanket or clothing can be treated in a similar fashion. As long as the entire object is not needed as evidence, the sample can be cut from the garment with a border of several inches and then packaged for transport. Blood is usually collected wet and transported wet or frozen. Body tissues are usually frozen. Certain types of preservatives are sometimes used with blood samples. Formaldehyde should not be used for tissues collected for DNA analysis.

Stains can be swabbed using a sterile swab lightly moistened with distilled water. The sample should be air dried with good ventilation before placing it in a container.

If a large object which has been spattered with fluid or tissue is being transported, and it is not practical to remove or cut out the sample, cover the sample area with clean dry paper securely taped around the edges. If there is a danger of exposure to moisture, put the whole object in a waterproof container or bag before transport. Nonabsorbent objects should generally be kept at room temperature.

In most cases, when sampling animal tissues, it is preferable to collect nuclear DNA, which has genetic material from both maternal and paternal lines, e.g., hair roots contain nuclear DNA, while hair shafts do not. Nuclear DNA degrades quickly, so blood drops and semen should be collected as soon as possible. For hair samples, it is desirable to get about 80 head hairs and about 50 pubic or other body hairs, but information can occasionally be derived from very small samples of five or more hairs. Contrary to popular belief, single hairs, or a few skin scrapings are rarely sufficient for a lab analysis, at least not at the present time. If good nuclear DNA samples are not available, sometimes mitochondrial DNA (mtDNA) can be analyzed. The remains of victims of transportation disasters, wars, fires, or individuals from past civilizations are analyzed using mtDNA, usually from teeth or bones.

Bones and Teeth

Bone samples of a few inches in length are best if taken from long bone and are usually transported at room temperature (unless fragments of tissue are attached to the bone, in which case it may be best to freeze the tissue). Teeth can be sent at room temperature, or the pulp removed and sent refrigerated.

Fetal cell samples for genetic testing must include a sample of the mother, to distinguish the child's DNA from that of the mother.

Do not expose samples to heat or bright sunlight, as this may degrade them, and air dry with sufficient ventilation (a low-power fan may be used) to avoid fungal or bacterial contamination.

Each sample should be carefully labelled with the probable source, the collection site, the date and time of collection, the method of collection, and the initials or name of the collector, taking care not to contaminate the sample with the labeling materials. If there are unusual conditions present, such as extreme temperatures or possible chemical contamination in an industrial setting, these should be noted as well.

5.c. Transporting Samples

Sterile blotters, swabs, envelopes and glass vials are generally provided for the temporary storage and transport of DNA samples to a lab or storage facility. Labs often provide these in kit form.

In the absence of professional storage containers, clean containers made of glass or stainless steel can be used, with paper or plastic bags as a second choice. If contaminating foodstuffs or other substances are present in a container, clean it thoroughly, add water and microwave for a few minutes and pour out as much of the water and debris as possible. Swab with alcohol and air dry. This isn't a perfect solution, but it can help reduce the level of unwanted materials prior to putting in the sample. Hair samples can be transported in a clean envelope with the flap folded or taped (not licked).

It is usually desirable to purify and dry samples before transport, if the technical expertise and equipment are available at the collection site.

Plastic is not recommended for storing samples, as chemicals in the plastic may leach into the sample, but acid-free solid plastic or vacuum-pack bags can be used for some types of samples if glass or stainless steel containers are not available.

Samples are usually transported dried or frozen. If they have been stored for a while before transport, the general rule of thumb is to transport them in the same state they have been stored in up to the time of transport. In other words, if they are frozen, keep them frozen (using dry ice if needed); if they have been kept liquid for a while after collection, then keep them liquid (rather than freezing them); and if they are air-dried, keep them air-dried. It is very important to prevent rehydration of dried samples, or thawing of frozen samples until they are ready to be used.

Samples are usually frozen at a temperature of about -20°C although temperatures of -7° to -15°C have been successful with some substances.

Unless a sample is being analyzed onsite, it must be transported to a testing facility. Portable testing kits for use at a scene, such as a crime scene, are now available and will increase in use as the technology improves and prices drop. However, portable kits may provide only a minimal, preliminary evaluation. If an onsite analysis is carried out, it is generally followed up with lab tests for confirmation or further analysis.

Transport of liquids or frozen materials may require specialized facilities, whereas the transport of hair samples through the regular postal system is practical and convenient in some circumstances. Potentially infectious or contagious materials that may be present in blood samples require special care and supervision to prevent them from contacting outside agents.

Blood can be dried onto filter paper, but without carefully controlled conditions, some breakdown of the DNA is likely to occur with this process. It is especially important to protect dried blood drops from rehydration.

Sometimes samples are purified before they are transported, but more commonly purification occurs after transport to a lab. Purification has been accomplished with a number of materials including silica gel or glass particles. Gordon, Stimpson, and Hsieh (1993) have developed a process to 'pull' the liquid through a filtering system while still retaining the desired cells, so that transport can be more efficiently carried out and on-site centrifuging avoided. Padhye, York, and Burkiewicz (1995) have suggested an improved method of using a *mixture* of silica gel and glass particles in combination with an aqueous solution of chaotropic salts to isolate DNA or RNA.

5.d. Assessing Samples

When received at a lab, the samples will be stored until they are ready for assessment and analysis. The following steps in the assessment and preparation of DNA fragments are a brief generic description only, as the processing of samples varies, depending on the type of tissue, the age of the tissue, its quantity, and the type of analysis being performed.

The technician will first read the notes associated with a sample, and give it a visual inspection. The notes can help indicate whether it is a mixed or relatively pure sample, and should describe the origin of the sample. The notes may also indicate the possible presence of contaminants or unusual environmental conditions at the point of origin. This may be followed by a microscopic examination to confirm the composition of the sample and whether there is sufficient good tissue to continue with processing. It is important to determine whether it is a pure or mixed sample and, if mixed, identify and isolate the individual components. This is followed by the extraction of a portion of the sample from its container or substrate.

Hair samples may be cleaned with detergent in an ultrasonic bath to remove contaminating residues. The tissues are then usually combined with an extraction solution and ground up to release the DNA from the surrounding cellular material.

Bones (including teeth) are cleaned and sometimes sanded. The inside of the tooth is generally used. A sample is extracted and finely ground. The resulting powder or extract is treated with a chemical solution to release the DNA.

Blood samples stored on blots can be cut away or punched from the blotting medium and the balance of the sample returned to storage for future use. Chemical sterilization and filtering can then be carried out.

Materials may be centrifuged to encourage separation of the various materials into layers. DNA remains soluble in the top layer and other cellular components are filtered out. The DNA sample is then subjected to a purification process ready for further processing, usually by the common PCR or RFLP methods.

If the results are successful, they are recorded or the process may be repeated until a good result occurs. Remaining pieces of tissue are stored for future reference.

5.e. Analyzing Samples

If you have not already done so, it is valuable to read the sections on PCR, RFLP, and Southern blotting (Section 3.a.) to understand the different means by which DNA fragments are separated, bonded, and recorded for visual analysis by a trained professional. Once records of the sample have been made, these records can be stored in files, or scanned and recorded in a computerized database, which may include a software analytical system.

When processing and sequencing DNA, the information is evaluated in the context of the situation. For example, if the samples were submitted for paternity testing, the DNA patterns of the alleged parent and child will be evaluated to see whether they 'match' (share key characteristics) with a high degree of probability.

Many sequencing methods, subprocedures, and apparatus have been developed, and there are likely to be more as our understanding of the science improves. They are too numerous to discuss in depth here, but some evolutionary examples that provide an overall understanding of basic methods and some of their advantages and disadvantages include

- *Sanger (dideoxy) sequencing method* - a well-known, frequently used method that uses enzymatic elongation procedures with chain-terminating nucleotides. The size of DNA fragments is determined by gel electrophoresis. Careful preparation of samples is required and the process is somewhat time-consuming [Sanger et al., 1977].

- *Maxam-Gilbert sequencing method* - Chemical cleavage to generate fragments that are randomly cleaved. Chemical reactions which exhibit specificity of reaction to generate nucleotide-specific cleavages. Careful preparation of samples is required [Maxam and Gilbert, 1977 and 1980].

These two important earlier methods generate fragments that are ordered by length. They have limitations related to errors and to the number of DNA segments that can be processed at one time. A number of alternative methods have been explored, including hybridization.

The creation of clones of the original DNA is an important aspect of sequencing. A larger sample provides more material for processing, and clones can sometimes be exported in circumstances where the original sample cannot. There are several ways to replicate or amplify and sequence DNA samples:

- *shotgun method* - a technique for digesting a DNA sample at random by ultrasonic vibration, preparing DNA fragments by subcloning, sequencing each fragment, and using overlaps to determine the full-length base sequence. The full-length base sequence is determined before the portion of the DNA that corresponds to the extracted DNA is known [Maniatis et al., 1989]. Analysis of a DNA length 10 to 20 times longer than the length of the DNA being sequenced must be obtained, a process requiring substantial time and care.

- *primer-walking method* - a huge intact DNA is used as a sample. The base sequence is first determined. For each sequencing, a primer is then synthesized to determine the DNA sequence of a contiguous portion. The process starts at one end of the sample and is sequential. While more efficient than the shotgun technique, it nevertheless requires very careful preparation of a primer for each sequencing process. The purification process is time-consuming [Matsunaga et al., 1997].

- *nested-deletion method* - fragments from a DNA sample are enzymatically digested, yielding different-sized fragments. The fragments are then sequenced according to longer length. A priming site is obtained in the process, and thus does not need to be repeated for each sequencing process. As in the primer walking method, the purification process is time-consuming [Maniatis et al., 1989].

Since the implementation of these three general methods in the late 1980s, other processes intended to streamline and simplify sequencing processes have been experimentally introduced, for example:

- *fragment-walking method* - direct sequencing of a DNA sample that is digested with a restriction enzyme. An oligomer with a known sequence is then ligated with the DNA fragment's terminus to recognize each DNA fragment in the mixture. A set of primers is used for a sequencing reaction for discriminating a complementary base sequence. After determining the base sequence of each DNA fragment, the base sequences of respective DNA fragments are reconstructed to obtain the overall base sequence. This has advantages over previous techniques in that it does not require cloning, but may yield digested fragments with long DNA sequences.

Sometimes techniques are combined in order to improve efficiency. For example, Matsunaga et al. (1997) have proposed the use of the fragment-walking method as a means to prepare DNA fragments for use with the nested-deletion method, followed by DNA sequencing. This hybrid approach streamlines the process by bypassing some of the operations for nested-deletion subcloning, and some of the difficulties in fragment-walking fragment connection.

By the mid-1990s there was widespread interest in commercializing and streamlining DNA testing techniques, opening up access outside of academic research to local law enforcement officials, federal agents, private detectives, and general consumers. Some other developments of commercial interest during the 1990s include

- Drmanac et al. - a non-gene-specific hybridization DNA sequencing method using probes (1993). Drmanac has more recently been involved with developing a system to read a base multiple times to improve the accuracy of known and novel DNA substitutions, deletions, and insertions.

- Rothberg, Deem, and Simpson with U.S. Government funding - developed a computerized apparatus for use in conjunction with a data library for identifying, quantifying, and classifying DNA sequences without sequencing. Distinctive signals (e.g., optical signals detected from fluorochrome labels) from short DNA sequences (or their absence), taken together, are used to identify a particular DNA sequence in conjunction with information from the data library. This is important for deriving information from mixed samples or those of unknown origin. Previous probe methods were somewhat effective for single samples, but cumbersome for mixed samples (1995).

- Jones - an iterative and regenerative method for DNA sequencing in discrete intervals using restriction enzyme and hybridization procedures. This is suitable for creating offset collections of DNA segments for providing continuous sequence information over long intervals and for sequencing large sets of segments (1996).

- Kambara and Okano - a gel electrophoresis method and apparatus for DNA and protein detection and DNA base sequencing using a photodetection system. The object of this method is to improve the efficiency of DNA sequencing by allowing two or more samples to be processed at once without sacrificing sensitivity. Improving efficiency over previous single-sample methods has important commercial and research implications (1996).

- Fodor, Solas, and Dower - de novo (new) sequencing of unknown polymer sequences for verification of known sequences and for mapping homologous segments within a sequence. Through automation, sample preparation is streamlined and speed and accuracy of results are enhanced over previous methods (1997).

One of the most common motivations for DNA analysis is paternity testing. Since fathering a child involves tremendous emotional, financial, and social implications and responsibilities, the confirmation or exclusion of paternity has become a substantial industry, and an important tool in legal proceedings. There are a number of ways to identify paternity. Love (1994) has described a means of detecting the presence or absence of multiple nucleic acid sequences. In this method, the multiple presence polymorphic (MPP) probes in a series of separate hybridization tests result in a pattern unique to an individual. This personal identification pattern (PIP) can be compared and contrasted with others within the test population to positively exclude or include within a high degree of probability, the likelihood of paternity.

While gel electrophoresis processing is still widely practiced, non-gel systems employing PCR samples and short tandem repeat (STR) sequences, micropellicular matrix separation of DNA fragments, and computerized displays are now available. These automated systems can create profiles in minutes, as compared to earlier technologies that took hours or days.

This portable, computerized DNA analyzer from Lawrence Livermore National Laboratories (LLNL) is contained in a briefcase, and can be used in the field to perform polymerase chain reaction (PCR) DNA analyses. The PCR chamber and analysis equipment are located on the compartment on the right side of the case. [LLNL news photo, released.]

In the late 1990s, the Lawrence Livermore National Laboratory (LLNL) created a 'suitcase' DNA lab that makes it possible to analyze a sample outside of the lab, a boon in situations where transport of the sample is difficult or impossible. This new level of miniaturization makes it possible to do testing and analysis in remote environmental sites, war zones, or crime scenes in politically unstable environments. Accurate measurements can be attained with flow cytometry, a general-purpose laser light-scattering diagnostic tool for assessing and categorizing biological cells and their components. This technology can be used for detecting biological warfare agents and cell contents such as DNA. The Department of Defense Armed Forces Institute of Pathology is using LLNL's Microtechnology Center's DNA analyzer for identifying human remains in the field, and testing for pathogenic bacteria and other contaminants.

5.f. Storing DNA Samples

Databanking of the information attained from a sample may have legal ramifications. It may be necessary to obtain permission to store the information or, as in the cases of samples taken from American service members, there may be policy stipulations as to how long the information may be retained, such as 50 years beyond the length of the term of service.

Destruction of DNA information after a specified period of time may be necessary to safeguard privacy. Currently, sample information provided to federal authorities usually reverts back to the state authorities who supplied it, with the onus on the state to destroy the original sampled tissues. As such, there is no centralized monitoring of the actual destruction of the materials, or the method of destruction (usually by incineration, or encapsulation and deposit in a landfill). Procedures for information and sample storage are still evolving with growing support for federal supervision.

In some situations it may be mandatory or ethical to convey the results of the analysis to the donor, whether or not the sampling was voluntary.

DNA samples are stored in clinical settings, forensic laboratories, research facilities, commercial 'DNA banks,' and homes. Samples may or may not be subjected to DNA extraction before storage. Specialized facilities may use special environments and preservation techniques such as transformed cell lines, or freezing, in order to maintain the samples over time.

The easiest method of storage, in the absence of storage equipment, is simple drying. Care should be taken not to contaminate the samples. Sterile containers and gloves are recommended, and in crime scenes in particular, it is recommended that a new pair of gloves be worn for each sample.

In cases of private sector DNA banking, the consumer is always faced with the problem of the firm going out of business, being taken over, or changing its product direction or corporate policies. In addition, private and public enterprises are subject to legislative inducements and restrictions that might require destruction of the samples after a specified date or length of time. There is also the lesser but provocative possibility of profit organizations using semi-skilled or unskilled labor to label and store samples, with the subsequent errors or abuse that may occur due to staff turnover, inexperience, or malicious mischief.

Solid storage and transport techniques are still being developed. One example is a patented process by L.A. Burgoyne, an Australian inventor. The use of a solid matrix medium with a protein denaturing agent and a free radical trap is used to protect against the degradation of DNA, and is suitable for blood samples. The free radical trap acts as a DNA-releasing agent to aid in releasing the DNA from the solid storage medium. Pathogenic organisms can be inactivated by the solid medium. The solid medium can be cellulose or synthetic in the form of a compressed pellet. If long-term storage is desired, it can be encased further in a protective material.

Uric acids or urate salts have found to be useful in long-term storage of purified DNA by providing buffering and free-radical protection. Once dried at room temperature, samples can be put into cold storage at about -15°C to -20°C.

Drying agents such as silica gels and dry sodium carbonate can be used in conjunction with cold storage to remove traces of acid vapors. Vacuum-packing can help reduce oxidation.

Equipment

Most of the equipment used for DNA processing and analysis is highly technical and is set up in a laboratory for use by trained professionals. A description of this equipment is outside the scope of this book and the reader is encouraged to contact labs directly for more detailed information.

In contrast, the equipment for collecting samples is relatively straightforward, and many labs will ship sampling kits that can be used by laypeople or professionals, depending on the type of sample. Buccal (cheek) swab kits can be used with a minimum of training. Sterile swabs, gloves, transport and storage containers should be used whenever possible.

FTA™ paper is a commercial dry reagent mixture commonly used for the transport and storage of blood samples. Small amounts of sample can be extracted by 'punching out' a portion from the medium. Sterile tissues, paper, and clean envelopes can be used for dry substances or those that have been collected wet and then air-dried. Dry ice can be used to keep samples cold if they have been previously frozen.

Sterile or well-cleaned tweezers, knives, or scrapers can be used to pick up dried samples to put them into a paper fold, though it is preferable to transport the sample with the object to which it is attached, when practical.

Syringes and pipettes can be used for wet samples. BD Vacutainers™ are commonly used to hold blood samples, and may come with substances that prevent coagulation or inhibit enzymes that could damage the sample. If a sample is scraped, the scraper must be thoroughly cleaned between each sample. If sterilization is not possible, cleaning with detergent and alcohol can help.

Clean scissors or a knife can be used to shorten hair root samples, if the shafts are long. It is usually not necessary to transport more than a few inches of the shaft.

6. Applications

DNA testing is a generic technology that can be applied in many aspects of life. It is used in law enforcement, poaching and wildlife conservation management, social services (especially child-support disputes and genetic counselling), crop evaluation, seed management, dog and livestock breeding, drug control, and many other areas in which information about the lineage, composition, or the whereabouts of an animal or plant is desired.

In this section, applications are described in general terms, followed by a selection of specific examples. It is not possible to list all the many applications of DNA analysis and new ones are constantly being developed, but there are enough listed here to convey the breadth of the field.

6.a. General Applications

Law Enforcement

Law enforcement officials have been quick to see the potential of DNA as evidence of both innocence and guilt. However, because of pre-existing laws concerning identification and search and seizure, the use of DNA technologies is more stringently regulated in law enforcement than anywhere else. Thus, there are limits for sampling, storage, and court admissibility of DNA evidence. Until recently, only crime scene samples, and samples from convicted criminals could be collected. However, in some jurisdictions, it is now lawful to take samples from suspects, in spite of challenges that this is an unconstitutional violation of a suspect's rights.

Not all suspects may be sampled. Currently, in most countries, suspects of violent crimes, primarily rape and murder, may be sampled. In some areas, burglary suspects may now also be sampled. DNA evidence can exonerate a falsely accused, innocent suspect so that limited resources are not expended in trying to apprehend and prosecute the wrong person. DNA can further be used to provide evidence or confirmation of more likely suspects.

A databank of original samples or a computer database of analyses may be available, in some cases, to determine if a suspect has previous offenses or if the DNA matches another sample from an unsolved crime. National databanks (e.g., CODIS) are being developed for reference. In the U.K., there have been instances where vast numbers of citizens have voluntarily provided DNA samples to exclude themselves as suspects and to aid in pinpointing a murder suspect. Since widespread testing expends large amounts of taxpayer's dollars, some citizens have supported the continued storage of these samples, rather than spending millions on retesting every time there is a violent crime. In other words, in law enforcement, DNA banking may increase as a result of economic pressures, in spite of concerns over privacy.

Certain law enforcement organizations have called for the DNA sampling of the entire populace, with the information to be kept in law enforcement files.

Attorney General Janet Reno has ordered a federal commission to study the legality of collecting DNA samples from millions of Americans who have never been convicted of any crime. This is currently an issue of substantial debate.

Breeding and Wildlife Conservation and Management

DNA allows animal lineages and pedigree documents to be verified and monitored to a degree of certainty never before possible.

Wildlife conservationists have found DNA sampling to be an invaluable tool in protecting the genetic diversity of threatened species. By breeding individuals that are as genetically different as possible, it is possible to strengthen a species, by improving resistance to disease and avoiding genetically related inherited abnormalities, especially those that are recessive and only express themselves when both parents have the traits. This can improve wild and captive breeding programs. Some rare species are privately owned, and access to these animals is sometimes difficult, but it is hoped that the owners will support DNA profiling for the sake of the continuation of the species.

DNA has been found valuable in determining gender in birds and animals. Some species are not easily sexed by visual inspection. Private and institutional bird breeders especially appreciate being able to know the sex of a bird before paying a high price for it or transporting it to a distant destination.

DNA sampling can be invaluable in providing evidence for the poaching of protected animals, whether wild, captive, or domesticated. A product on the market in another country can now potentially be traced to a specific herd or community of animals in the wild, or in a wildlife park or sanctuary. A decade ago, it was almost impossible to 'prove' this type o case, whereas it may now become routine.

With fish stocks declining dangerously, fishing boats may be subject to closer scrutiny. The sampling of fish debris found in the water of boats suspected of illegal activities can indicate whether certain species were present in the catches.

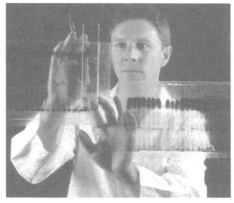

The U.S. Fish & Wildlife Service laboratory staff includes a genetics team which uses electrophoretic separation systems, protein analysis, and DNA analysis for a variety of scientific purposes, including species categorization, gender identification, and sample matching or exclusion. [U.S. Fish & Wildlife news photos, released.]

DNA sampling is used to chart the lineages of pedigreed animals, or animals bred for specific purposes like research. Litters with mixed parentage (i.e., more than one father) can be verified, as can fraudulently bred animals, or those which have been excessively line-bred. Fraud in the dog-breeding community has recently been revealed by DNA testing. 'Dogprints' designed to verify the parentage of an animal are becoming a routine part of the documentation that accompanies a purebred. Similar systems are being put in place for alpacas and cattle, and may soon be commonplace. In the past, brands and tattoos have been used to confirm the identity of an animal. Perhaps someday these will be replaced by tiny implantable microchips that can transmit stored DNA information about the animal's identity and characteristics.

In many areas, dog pedigree information has been confirmed or refuted with DNA evidence such as microsatellite (STR) DNA typing. Many cases of fabricated pedigrees (those in which one or both of the parents turn out to be unregistered dogs), or pedigrees listing the incorrect parent (usually the father) are now being investigated, and fraudulent breeders are being required to compensate purchasers of dogs from questionable litters.

Dog pedigrees can be used for other reasons not related to the breeding of dogs. It has been suggested that DNA could be used to locate dog owners who don't clean up after their dogs. Given the backlog of DNA testing needed for more important law enforcement activities, one would hope this suggestion of a 'poop patrol' remains a low priority compared to DNA testing for violent crimes. The idea, however, indicates that city database records for dog tags may someday be cross-referenced or supplemented with DNA information.

Archaeology and Anthropology

DNA analysis has been a tremendous boon to scientists studying our past history and evolution, and the evolution and characteristics of plants and extinct species. Since nuclear DNA degrades rapidly, mitochondrial DNA is usually used for these studies. Mitochondrial DNA (mtDNA) is passed through the maternal line and is more stable than nuclear DNA. Because skeletal remains last longer than soft tissue, and mtDNA can be found in bones, it is possible to sample mtDNA that is decades or hundreds of years old. This has produced some sensational and sometimes surprising discoveries in anthropology and archaeology.

DNA is also being proposed as a way to identify bodies in cemeteries whose origins are no longer known. By forming databases of family DNA patterns, a computer matching system may someday be able to identify an exhumed ancestor in minutes. Given the high degree of interest in charting family trees by amateur and professional genealogists all over the world, it is quite likely that DNA patterns will eventually be part of genealogical databases.

The exhumation of bodies is bound to yield some surprises. Throughout history it has been common for adoptions and out-of-wedlock pregnancies to be 'covered up' due to social stigmas against unmarried women and their children. As a result, many children have been raised by uncles and aunts as their own children. Often the offspring of a very young girl has been raised as a sibling by the young mother's parents, rather than as her child. Still other children have been parented by fathers that were not the husband of a married woman. Human curiosity about our past relations will result in at least a few unexpected revelations when people dig deeper into family histories.

The remains of victims of massacres or dire poverty may someday be returned to their families for services and reburial based on the results of DNA testing. There has been intense interest and research over the centuries in locating the bones and burial place of W. A. Mozart who was reportedly placed in a mass grave. Exhumation and DNA testing has already been used to verify the identity of a slaughtered Russian Czar (described in more detail later), and to study the descendants of Thomas Jefferson's family. Private detectives may someday find themselves investigating not just the extramarital affairs of living relatives, but those of dead relatives as well, especially where large inheritances are involved.

Tracking and Identification

DNA can be used as an identification tool for tracking wildlife, people, and plants. A sample taken in one part of the world can be matched with a reference specimen in another country in cases of poaching or smuggling.

In cases where the parentage of a child is unknown, DNA testing may be a way of alerting parents that they are related to kidnapped or adopted children. This type of identification can be carried out at a distance without a visual inspection. It can also be used when so much time has passed that it's difficult to recognize a child who was lost in infancy.

Identification of skeletal remains recovered from a crime scene, war zone, or search and rescue operation can be aided by DNA. Forensic anthropologists and dentists have long been involved in assisting with the identification of individuals who have been murdered or killed in a disaster. Now they have additional tools to carry out their work. A complete skeleton can reveal useful clues about sex, age, handedness, race, height, musculature, general health, and sometimes specific illnesses (e.g., rickets), and injuries. This provides information about characteristics which may be sufficient to identify an individual, but the skeleton may not be wholly recoverable, may be mixed with other skeletons, or may be incomplete. Even if a whole skeleton is found, identification may still be uncertain. If the teeth are present, dental work can help in a positive identification. But if the teeth are not present or have not been altered, DNA analysis and matching with an existing specimen, as in a service members database, can provide a positive ID. If an existing specimen is not available, matching with close family members can often provide a positive ID.

As DNA testing kits become more accurate and sensitive to smaller amounts of DNA, DNA may begin to be used to detect traces of unlawful entry by unauthorized persons in hazardous or politically sensitive buildings.

6.b. Specific Applications

There are thousands of ways in which DNA can be used to derive information about an individual or a situation, so not all can be described here. However, this section provides some specific examples, some of them high-profile, of how DNA has been used in practical applications.

Service Member Remains Identification

The U.S. Government has been collecting DNA samples from service members since 1992, with almost three million specimens being stored in Gaithersburg, Maryland. This information is gathered primarily for the identification of recovered remains. DNA collection from service members is now mandatory, even in cases where the member is not being sent to a war zone.

In 1995, service marines challenged this practice by disobeying an order to submit to DNA sampling under the threat of a court martial. In the lower courts, the action was not found to violate the plaintiffs' constitutional protections. A number of subsequent court cases that deal with very specific challenges to mandatory testing have occurred. In general, the plaintiffs appear to be losing, sometimes due to technical weaknesses in their preparations and arguments rather than the issues themselves, whereas in more widely ranging general constitutional studies, the higher courts, in some instances, are finding in favor of the plaintiffs. As of late 1999, compliance is being tracked over the Internet by U.S. Army leaders.

When the remains of soldiers are found in poor condition, sampling of the nuclear DNA is usually impossible, but mitochondrial DNA, which can be compared with the maternal line, may result in identification.

Left: The disinterred remains of the Tomb of the Unknown Soldier were carried to a waiting hearse at the Arlington National Cemetery en route to the Armed Forces Institute of Pathology. There, at the Walter Reed Army Medical Center, they underwent forensic testing, positively identifying them as the remains of U.S. Air Force 1st Lt. Michael Blassie. They had been interred for 14 years since recovery from Vietnam. Right: After being identified, the remains were buried in a service at the Jefferson Barracks National Cemetery. [U.S. DoD 1998 news photos by Helene C. Stikkel and Scott Seyer.]

The celebrated Unknown Soldier, interred since 1984, was identified in 1998 as Vietnam veteran 1st Lt. Blassie, an Air Force pilot, based on mitochondrial DNA evidence matching tissue donations from his female relatives. Since 1991 about 100 identifications have been made by the Armed Forces DNA Identification Laboratory (AFDIL), mostly of remains from Asian wars. A decision was made not to inter another service member in the tomb because officials are now confident that the identification of the remaining members can now be successfully undertaken.

Individuals seeking to identify service member remains can order a kit from AFDIL and go to a licensed health professional to take the sample, which can then be sent in for matching.

Royal Canadian Mounted Police (RCMP) forensic experts use wire trays to sift through the remains at a grave site in a village in Kosovo as U.S. Marines provide security. The remains were investigated to help determine war crimes, to establish evidence for breaches of international conventions. [U.S. DoD 1999 news photo by Craig J. Shell released.]

Identifying Political or Murder Victim Remains

DNA can be used to unearth information about historical incidents in which the facts have been lost or obscured.

One of the most publicized identification cases involves the use of DNA evidence to confirm that the bones buried in a mass grave after the murder of Czar Nicholas II of Russia in 1918, were indeed the remains of the Czar and some of the members of his family and staff, as reported in 1995. While tentative identification of bones and teeth had previously been undertaken, the DNA evidence produced in this case provided powerful corroborative evidence to existing information and oral reports. It has helped paint a clearer picture of political events of the time.

Top: A tent village is used as a base near the area of the alleged Kosovo massacres. Bottom: FBI investigators sift through the debris to recover remains from grave sites. In addition to human remains, the recovered evidence included shell casings and bullet fragments. The physical evidence was supplemented with photographic records and other scientific examinations. Autopsies of the deceased indicated that most had died from gunshot wounds. [FBI news photos, released.]

In spring 1999, at the request of the International Criminal Tribunal for the former Yugoslavia (ICTY), Federal Bureau of Investigation (FBI) agents carried out forensic examinations of burial sites identified by NATO to allegedly contain biological and chemical evidence of war crimes. The FBI investigators uncovered human remains of men, women, and children who had died from gunshot wounds to the head. A forensic anthropologist and four physicians from the Armed Forces Institute for Pathology (AFIP) conducted autopsies near the burial sites. Evidence to support allegations against Serbian forces was found.

This operation did more than establish evidence of war crimes. It also helped identify the remains of even badly decomposed bodies, so that after examination confirmation could be given to families who were unsure of the fate of their relatives. It also provided an opportunity for remains to be returned to the bereaved families.

Wrongfully Accused Exoneration

Since DNA evidence can be used to exclude a suspect or incarcerated individual, there is a movement to use this technology to free wrongfully convicted individuals, particularly those serving long jail sentences. Connors et al., have reported through the National Institute of Justice (1996) that a number of individuals, incarcerated mainly for cases of sexual assault and murder, have now been cleared by evidence showing their DNA did not match samples taken at the crime scene. Twenty-eight cases of released prisoners, having served an average of seven years in prison, were reviewed and analyzed for this report. The authors include commentary on the implications for public policy.

In "Law and Human Behavior", Wells et al. (1998) describe how DNA evidence exonerated individuals in 12 additional cases. The authors combined these with the cases cited in the Connors report, bringing the total cases under study to 40. Approximately half of the tissue samples examined came from semen, blood, or hair samples. The authors point out that about 90% of these cases included incorrect eyewitness identification testimony resulting in wrongful convictions. They make recommendations for improving eyewitness identification procedures in order to reduce the incidence of false selections.

The National Academy of Sciences has noted that "...the reliability of DNA evidence will permit it to exonerate some people who would have been wrongfully accused or convicted without it". Scheck and Neufeld report that in sexual assault cases referred to the FBI since 1989, DNA evidence excludes the primary suspect about 25% of the time.

Transportation Disasters

Families of individuals killed in disasters involving trains, planes, and other forms of transport have traditionally been identified from remains, documents, and dental records. However, there is often insufficient information to make a definite identification by traditional means. Katz et al. (1998) have reported the use of short tandem repeat (STR) and mitochondrial DNA (mtDNA) analysis to assist in identifying the remaining individuals from the 1997 Korean Airlines disaster.

In serious public transit accidents, including aircraft and train disasters, the remains may be seriously charred or crushed. Traditional visual inspection, fingerprinting, X-ray, and dental records-matching may be insufficient for identification. DNA can sometimes aid in these situations. The databanking of flight-crew DNA has been proposed and frequent flyers are being urged to voluntarily provide tissue samples.

Child Abductions

If at least one family member can be located, a child suspected of being abducted may be identified by means of DNA testing. A number of children abducted in El Salvador have been identified by a joint effort of the Physicians for Human Rights, and the Association in Search of Disappeared Children. In the case of civilian massacres, children of the victims are sometimes abducted and adopted out, or may be kept in servitude. Since many of these children were taken as infants, or as toddlers too young to remember their parents, DNA testing is the only way to verify their family relationships.

Young children have fewer distinguishing characteristics than adults and they grow and change quickly, so if a child is abducted and subsequently murdered, it can be very difficult to make a positive identification from the remains, especially if they are mutilated or badly decomposed. DNA can help provide definitive identification to a grieving family and allow them to come to terms with the loss of a child.

Pharmaceuticals

In the area of health care, there are many motivations for DNA testing. DNA can be used to predict some potentially devastating conditions and, eventually, drugs or pharmaceutical techniques may routinely correct some of these problems. Thus, those seeking help are likely to support pharmaceutical DNA-testing. This has an enormous potential for good, and also a frightening potential for abuse. Designer drugs may some day be used to target an individual's specific disease-causing agents, but may also be used for very specific and insidious types of chemical war agents. Individuals who test positive for serious illnesses may receive more effective health care, but they may also be denied access to jobs, insurance, and other important necessities for survival.

In war zones, DNA could be used to detect fevers, genetically altered micro-organisms, and other microscopic substances on face masks and other implements used in warfare. In turn, this information could be used to determine exposure risks.

It may not be long before DNA testing at birth becomes routine. There have even been suggestions that hospitals develop computerized databases of infants born with birth defects. The goal of health care is to study and pinpoint genes that cause these abnormalities, but how long will the information stay on file, and how much control will parents and the children tested have over the use of this DNA information?

6.c. Commercial Products

DNA technology is rapidly being commercialized. Many recent DNA patents are for faster, less expensive analysis methods that will speed up production and lower costs. The general public can purchase a variety of products and services, including sampling kits and, very recently, home DNA-banking kits. They can take noninvasive samples at home, or have blood sampled at a local clinic. Samples to be used as evidence require documentation as to the identity of the donor and the witness who oversaw the sample collection.

Genetic Counselling

DNA analysis is routinely used in genetics counselling, and research into organ growth and transplantation. The sample may be collected in a doctor's office or a hospital. DNA information is valuable in many situations related to childbirth. Couples planning to have a child can have their DNA screened for hereditary diseases and receive counselling based on the results. Once conception occurs, a fertilized egg may be examined when it reaches the blastula stage and selectively implanted in the mother if it shows no abnormalities or congenital diseases. If the pregnancy is underway, the fetus may be tested *in utero* to detect correctable abnormalities, since it is now possible to do some types of surgical correction during pregnancy. Even though genetics counselling is not a priority for surveillance professionals, there is still an indirect benefit from the commercialization of these services, as public demand for DNA analysis brings down the price of the technology in general, making it less expensive for other purposes.

Familial or Pedigree Testing

For routine paternity or animal pedigree tests, consumers can take samples to a walk-in laboratory or, in some cases, send them through the postal service. Outside of genetics counselling, the most common commercial services requested by private citizens are paternity tests, and pet and livestock pedigree documents. DNA records of children are also of interest to people whose offspring are at risk of developing a serious illness later in life (it may some day be possible to grow new organs from a person's own tissue). Parents may also want a DNA record of an adopted child for future reference. Parents of children at risk of being kidnapped may wish to store DNA records, especially those living in unstable political climates.

Costs for the more common DNA tests range from about $50 to $300. Costs for institutional storage of samples for future reference, medical procedures, or research vary greatly, depending on the purpose, the institution, and the length of time the DNA samples or records are stored.

Rosgen in the U.K., and PE AgGen in the U.S. are two prominent DNA service providers. PE AgGen, Inc. runs an animal DNA-profiling facility with locations in California and Europe. The company provides test kits, profiles, and storage for up to five years for a variety of pets and livestock breeds. The cost is about $50 for a dog profile. PE AgGen, originally developed as Zoogen at University of California, Davis, was purchased by PerkinElmer. The company is known for advocating a standardized marker system in the public domain so that registries can be free to use independent labs for profiles. www.pebio.com/ab/aggen/

Rosgen Ltd. is a major U.K. provider of animal DNA analysis services, primarily for dogs and livestock. Forensic casework DNA profiles are approximately $200 per sample, dogprints are around $50 each. Genotyping and sequence analysis are available on request. www.rosgen.co.uk/

Forensics

Law enforcement officials have access to a broader range of services than private citizens, including in-house labs and external specialized labs. In appropriate instances, they also have access to restricted state and national databanks, some of which are publicly funded (e.g., the Combined DNA Index System - CODIS). Lifecodes is a prominent accredited commercial laboratory providing services to law enforcement agencies. It manages one of the largest DNA databases in the world.

For a surveillance professional, access to services will depend on the person's status and affiliations with the law enforcement community. Association with law enforcement provides greater access to labs and databanks, but also involves greater restrictions on who can be sampled and how samples are gathered.

7. Problems and Limitations

Errors can happen in many aspects of DNA collection and profiling. The sample may come from the wrong donor, the processing may be compromised, files may be mislabeled or misfiled, and computer databases may be incorrectly recorded or accidentally or deliberately lost or corrupted. Safeguards against the misuse or abuse of DNA information should also to be taken into consideration. This section surveys some of the most common problems and limitations in the field of DNA analysis.

Sampling

The most common sampling errors involve contamination and a misunderstanding of how much tissue of a particular type must be collected. Fraud is another problem, as it's relatively easy to submit a fraudulent sample, or even a fraudulent donor, if there are no verification systems in place. Section 5 (Description) provides relevant information on gathering and transporting samples that should be understood in conjunction with this section.

DNA profiling is dependent on the quantity and quality of the samples. If the amount of tissue is too small, or has been contaminated by grease or dyes, it may be useless. Single hairs, hair shafts, or small skin scrapings (as might be found under fingernails) are not usually sufficient for nuclear DNA testing. A sufficient quantity of sample must be collected, and sterile or clean implements and storage containers must be used. It is important not to trample or overpopulate a crime scene, and to collect and store fluids and tissue as quickly as possible before they degrade. Important court cases, especially those involving murder, have been compromised by the shoddy collection of samples that might have led to a conviction.

One of the limitations of sampling is cost. In the U.K. there have been prominent cases in which a large part of the populace was sampled to apprehend a murderer, but such large-scale operations usually cost millions of dollars. There will always be a trade-off. Storing samples may compromise individual privacy and may prevent some people from voluntarily coming forward with samples. Not storing samples will likely result in repeatedly high costs, usually borne by taxpayers, for carrying out future investigations.

Deliberate substitution of a sample from another individual, or deliberate substitution of another individual for testing is not only possible, but has been documented in a number of paternity suits and even in a well-publicized murder case in which the murderer asked at least two people if they would donate a blood sample on his behalf. One of them agreed, thinking he was helping a friend, rather than considering that he might be harboring a criminal.

Processing

Many types of processing errors can occur in the lab. Samples can be mislabeled or misplaced, chemicals can be outdated or measured incorrectly. If the original sample is too small for repeat tests, an error or loss of a sample can be disastrous. Labs and law enforcement agencies are concerned about quality assurance in processing DNA samples, and certification programs for technicians and individual labs are being developed and implemented.

Gel electrophoresis and the creation of an autoradiograph on X-ray film are still common ways to process DNA fragments. The visual comparison of autoradiographs created on X-ray film can be subject to error if the intensity or exposure varies from one film to the next. Although time-consuming, creation of multiple films of each sample can help reveal processing errors, but does not overcome all the limitations of using X-ray film. More recent phosphor technologies are said to offer some improvements over older techniques.

Symbolic representations, or X-ray images of DNA patterns, are often filed, scanned, or entered into a database for long-term storage. Incorrect entry into databases can occur, with typographical or factual errors occurring as much as 30% of the time if good proofreading and quality assurance practices are not in place. In the private sector, the turnover in database entry jobs is moderately high, and it is repetitive and tedious work. A number of firms have hired prison workers in the past to save costs on various types of database entry. Also, insecure database entry systems provide opportunity for deliberate sabotage or tampering.

Storage

DNA degrades over time, particularly nuclear DNA. Mitochondrial DNA, as found in longer-lasting structures such as bones and teeth, may survive for centuries, but even mtDNA eventually degrades. Air-drying, vacuum-sealing, and freezing can help extend the life of a sample, but samples don't last indefinitely, and changes in humidity or temperature can affect storage conditions and lead to destruction or degradation.

It takes a financial investment to store samples in a long-term facility. Storage firms can have hardware failures (freezers breaking down, power outages, etc.) that compromise the samples. Private firms may be the target of business takeovers, sabotage, or bankruptcies. Home storage banks are seldom climate-controlled and frozen samples may thaw in a power outage without backup generators.

Libraries can provide hindsight clues as to what might happen to data in DNA banks as the demand for storage space increases. Historically, when shelf space limitations precluded the storage of large quantities of books, newspapers, and journals, libraries used microfilm as a means to keep the information on file. Anyone who's used a microfiche reader knows that the records of these documents are substandard compared to the originals. They are sometimes poorly exposed (too light or too dark), difficult to read, and usually awkward to handle. Now that microfiche readers are being sold at auctions due to better computer technology, but we no longer have the originals from which to rescan better images.

Librarians do the best they can with limited resources, and a microfiche is invaluable compared to nothing, but similar compromises could happen with DNA data, with more serious consequences. Just as it's difficult to store vast numbers of books, it's expensive and awkward to store large numbers of original DNA samples. There are already circumstances where DNA samples are being destroyed after they have been scanned or otherwise entered into a database. Thus, there is no way to reprocess the data or to check the computer record against a sample taken in a crime scene. This makes the data even more susceptible to tampering or corruption in the absence of a source of verification. The loss of an individual book is unfortunate, but does not usually have a serious impact. The loss, substitution, or corruption of DNA information, on the other hand, could have life or death consequences for a person involved in a medical or legal dispute, if too much faith is put into the computer data.

Population Demographics

Our understanding of DNA is still very incomplete. Practical use of DNA profiles is in part based on statistical assumptions. Since DNA analysis extends beyond the matching of one pattern to another, to analysis of statistical relationships with known or reference populations, we need databases of large numbers of profiles or profiles specific to the subgroup being investigated. Without them, an individual profile is often useless. DNA data acquisition requires cooperation, and attempts to encourage voluntary donations have sometimes met with strong resistance.

In 1999, in a controversial effort to get broader demographic information for African Americans, the Attorney General of Ohio requested that staff members from this ethnic group donate voluntarily. While the move may have been well-intended, it clearly was ill-conceived and was roundly criticized. Eventually the request was withdrawn. Thus, the development of useful population statistics that are needed for DNA pattern comparison and research is limited by concerns over privacy and discrimination. Most databases so far have been developed from mandatory testing of military service members and convicted or suspected felons.

Another limitation in developing reference databanks is the concern over commercial use of an individual's unique DNA pattern. One way to get people to agree to allow their DNA patterns to be entered into databanks is to keep the entries anonymous, but this means that if a particular pattern is found to have great commercial value, the person who owns or 'is' that pattern does not directly benefit, and may not even know his or her personal attributes are being exploited.

Discrimination

Discrimination in the business world with respect to race or health is rampant, in spite of equal opportunity laws. DNA tests could reveal medical conditions or mixed-race backgrounds that are not physically obvious, but which may result in a person being screened out of an existing or potential job. There are reports by the American Management Association that over 6,000 firms require genetic profiling from job candidates. While recent bills in the Senate are intended to protect private individuals from discrimination based on DNA patterns, it is unlikely that legislation against DNA discrimination will be enforced any more stringently than current legislation against other types of discrimination. The track record so far hasn't been good.

Presently, many health and life insurance companies require a potential insuree to have a medical examination and may try to refuse to insure someone with a debilitating or life-threatening condition. It is likely that insurance companies will want to require DNA sampling as part of that medical examination and may charge higher premiums on the basis of 'medical potentialities' for serious illnesses that may be indicated by their genetic makeup.

There are historical precedents for discrimination that provoke concerns over long-term storage spanning multiple political administrations. In the United States, by 1924, the Model Eugenical Sterilization Law, which discriminated against drug addicts, epileptics, criminals, and the insane, had been passed by 21 states. In 1927, the U.S. Supreme Court supported the widespread use of involuntary sterilization of 'genetically undesirable' individuals. If unanticipated political swings occur in the future, and massive DNA databanks are available, it is conceivable that DNA information might be expropriated for uses other than those for which it was originally collected.

8. Restrictions and Regulations

Before using a technology for surveillance purposes, the professional must be aware of the legal requirements and restrictions associated with the technology, and how its use could impinge on the rights, safety, or privacy of the organization or individuals being surveyed.

It is likely that laws related to genetic testing will never address all the issues in this complex science and will never fully protect everyone. Consider these three important social dynamics:

- the most significant scientific breakthroughs and their ultimate influence on society are rarely anticipated,

- opportunities for wealth often seduce entrepreneurs into ignoring the damaging consequences of their products, and

- citizens and lawmakers haven't sufficient time nor expertise to understand every aspect of genetic science.

Taken together, these factors have a substantial influence on lawmaking, with the result that genetics-related legislation will probably always be compensatory rather than proactive. This section gives a general overview of federal and state legislation related to DNA testing, issues which have only been under scrutiny for about 20 years. The ramifications of legislation are discussed further in Section 9 (Implications of Use).

Federal and state bills on health privacy have been debated since the early 1980s, with few resolutions and many delays. It is usually difficult to implement wholly new measures, so most of the DNA-related issues introduced to the Senate have taken the form of amendments and additions to existing legislation.

Many public officials have as yet only a novice understanding of DNA technology and its potential effect on society, which is not surprising, since the science is recent and ongoing. Many citizens have no knowledge of it at all. This is part of the reason why the various bills are moving slowly through legislative channels, with many as unresolved or shelved. As it stands, law enforcement officials are working with federal support to extend their jurisdictions and their databanks, employers and health insurance providers are on an honor system to not use DNA information to discriminate against employees and insurees, and free agents have few impediments to DNA collection and use.

National Legislative Concerns

Court challenges of DNA sampling have been relatively few, considering the significant number of people in our society who have already been sampled for one reason or another. In cases contesting the constitutionality of mandatory or nonvoluntary DNA sampling, the most commonly referenced amendments are the Fourth and the Fourteenth. The Fourth Amendment to the United States Constitution provides that

> "[t]he right of the people to be secure in their persons, houses, papers, and effects, against unreasonable searches and seizures, shall not be violated ...".

The Fourteenth Amendment to the Constitution provides for protection and due process of law. The First, Fifth, and Ninth Amendments also include explicit guarantees of privacy that are relevant to genetics testing.

In 1989, the National Institute of Justice initiated standards for DNA typing in agreement with the Office of Law Enforcement Standards (OLES) at the National Institute of Standards and Technology (NIST). Further standards were developed through the Technical Working Group on DNA Analysis Methods (TWGDAM).

The mid-1990s was a time when interest in DNA issues resulted in a number of new Acts and the establishment of DNA-related facilities and databanks.

In 1993, the *Privacy Protection Act* was introduced to the Senate to establish a Privacy Protection Commission. This sought to grant investigative powers to the Commission, but contained no similar grant of enforcement powers to the Commission itself. The Act charged the Commission with reporting violations of the Privacy Act, yet left private sector activities outside of its jurisdiction.

The *Violent Crime Control and Law Enforcement Act of 1994* included within it the *DNA Identification Act of 1994*. This act of Congress sought to protect genetic information derived from DNA samples held by law enforcement agencies for identification purposes. This law would not be affected by the Genetic Privacy Act. Related to this is the Violent Offender DNA Identification Act introduced five years later, in 1999 (described further below).

Also in 1994, the DNA Advisory Board (DAB) was established under the *DNA Identification Act of 1994* by the Director of the FBI. The DNA Identification Act was contained within the Omnibus Crime Control Act for quality assurance and proficiency testing standards and passed into law in 1995. It provided funding for forensic labs to improve the quality and availability of DNA analysis and for the FBI to establish a national DNA database called the *Combined DNA Index System* (CODIS). Funding was contingent on strict adherence to standards. Board members were appointed by the FBI Director from nominations proposed by the National

Academy of Sciences and professional societies of crime labs. Board recommendations resulted in the FBI's Quality Assurance Standards For Convicted Offender DNA Databasing laboratories which came into effect 1 April 1999. The DAB was slated to end in March 2000, but was extended and was still meeting in the summer of 2000. This effort included support from the Scientific Working Group DNA Analysis Methods (SWGDAM).

United Kingdom

In the U.K., the *Police and Criminal Evidence Act of 1984* (which included handling of fingerprint records) was further amended by the Criminal Justice and Public Order Act of 1994 to introduce compulsory sampling of bodily tissues and fluids, using reasonable force, if necessary, from anyone charged with 'a recordable offence'. Results of the samples, whether or not they are used in the case, would be used for a national DNA database. Later the mandate was changed to those who were arrested (and not necessarily charged or convicted) for a recordable offense and the result could be held indefinitely. Thus, powers to sample and keep samples were extended after the fact.

Canada

In June 1995, the Canadian Department of Justice released information on the amendment of the *Criminal Code and Young Offenders Act* which empowered a Peace Officer or other authorized warrant-holder to obtain bodily samples (hair, saliva, blood) for forensic DNA analysis from an individual reasonably believed to have committed a criminal offence, thereby broadening the powers of Canadian law enforcement officials to include suspects not proven guilty. It was stipulated that if subsequent investigation cleared the individual of wrongdoing, the results and the sample were to be destroyed. However, the judge would retain the power to veto this destruction.

United States

After Operation Desert Storm, in the early 1990s, the *Armed Forces Institute of Pathology* established a DNA registry and repository as a means to identify human remains, in addition to dental records (and fingerprints, when available). By 1998, more than a million records of armed service personnel and contract employees sent into hostile regions were on file, frozen in a warehouse in vacuum-sealed envelopes. The cards are normally retained for 50 years, the same length of time as military medical records [GIlbert, 1998].s

In 1994, the U.S. Congress passed the *Violent Crime Control and Law Enforcement Act*, which authorized the FBI to establish a national index (CODIS) of DNA samples from convicted offenders, crime scenes, crime victims, and unidentified human remains. DNA identification was gaining acceptance in various governmental departments.

More than a dozen bills were introduced to the U.S. Senate in 1996. These included the *Health Insurance Portability and Accountability Act of 1996*, designed to protect consumers from discrimination in receiving health insurance, includes provisions that would prevent companies from charging higher premiums for 'at risk' insurees.

In October 1998, the FBI announced the creation of the *National DNA Index System* (NDIS), a database through which DNA profiles could be compared in order to turn up investigative links.

Concerns over discrimination that might affect health care benefits or employment access were clearly growing in the mid-1990s, and a number of relevant bills were introduced in 1997, including the *Genetic Privacy and Nondiscrimination Act and the Genetic Confidentiality and Nondiscrimination Act* which would amend the *Public Health Service Act and the Employee Retirement Income Security Act of 1974*. These measures sought to prohibit health insurance providers from discrimination affecting eligibility, renewal, or benefits, and to prohibit employers from using genetic tests to discriminate against employee rights or benefits. The bills were similar in some ways, both encompassing broader issues than some of the other bills presented that same year, one of which stated:

> ...the circumstances under which DNA samples may be collected, stored, and analyzed, and genetic information may be collected, stored, analyzed, and disclosed, to define the rights of individuals and persons with respect to genetic information, to define the responsibilities of persons with respect to genetic information, to protect individuals and families from genetic discrimination, to establish uniform rules that protect individual genetic privacy, and to establish effective mechanisms to enforce the rights and responsibilities..

This bill stipulated specific safeguards, namely, that the person being sampled must not only give written consent, but must give fully informed written consent, putting an onus on the agency taking the sample to educate the individual about DNA before taking the sample. It was proposed in the bill that this Act take effect on 1 January 1999. In spirit, this would be a good safeguard. In practice, it is almost impossible to implement fully informed consent in large-scale screening activities. Also, 'informed consent' is usually carried out by handing the donor a complicated sheaf of small print with a place at the end for a signature that attests to the signer reading and understanding the contents of the document.

At about the same time that these bills were presented to the Senate, a number of states developed similar amendments to individual state privacy-related acts.

Employer Interests

One might think that a bill that protects the public against discrimination in the workplace or in the provision of health insurance would pass through Congress quickly and easily, but that hasn't been the case. A working citizen might consider it 'obvious' that employers should be barred from receiving confidential medical information from physicians or pharmaceutical companies, but employers have countered that as long as they are paying for employee medical benefits, they have a right to know about an individual's medical history, including the results of a DNA test. Given these conflicting interests, resolution to the benefit of all parties has not been achieved.

In 1997 there were amendments to the Criminal Evidence Act further extending law enforcement powers for obtaining non-intimate DNA samples, without consent, of violent offenders, sex offenders, mentally disordered offenders, and burglary offenders convicted prior to April 1995 and still serving their sentences.

In July 1999, the FBI announced the first 'cold hit' processed through the National DNA Index System (NDIS). A cold hit represents a DNA association between an offender or a crime without an investigative lead. Evidence linked six sexual assault cases occurring in the Washington, D.C. area through NDIS with three sexual assault cases occurring near Jacksonville, Florida. The Florida Department of Law Enforcement then notified the FBI Laboratory of the apparent association of the crimes in different areas. The FBI then turned up five more matches in the Washington, D.C. area and linked the crimes to a deceased individual, Leon Dundas. It appeared as if the system might be instrumental in helping solve violent serial crimes.

In 1998, the U.S. Bioethics Advisory Commission issued a report arguing for tighter controls on tissue and DNA sample banks.

In May 1999, an amendment was proposed to the DNA Detection of Sexual and Violent Offenders Act to add burglary to the list of crimes that require DNA sampling. Up to this time only felony sex offenses, murder, and indecent assault were eligible for mandatory testing.

The same year, the Violent Offender DNA Identification Act was proposed to eliminate local law enforcement agencies' backlogs of convicted offender samples or to have them reanalyzed with better technology, with help from the FBI, the results being recorded in the CODIS databank. This information is also intended to help form a statistical population database (with personal identifying information removed). The act further directed the expansion of the DNA Identification Index to include information on criminal offenses and acts of juvenile delinquency that would be considered violent crimes if committed by an adult.

International Organizations

By the late 1990s, international organizations were working out ethics and procedures for dealing with genetics research and technology on a global basis. Given the divergent interests and educational levels of the different countries, this is a major undertaking.

In 1997, the United Nations Educational, Scientific, and Cultural Organization (UNESCO), on behalf of almost 200 member states, adopted the *Universal Declaration on the Human Genome and Human Rights*, the first international statement of ethics in human genetics research. It is intended to safeguard the public in the area of scientific research, emphasizing consent and nondiscrimination within the context of fundamental human rights and freedoms.

State Legislative Concerns

At the state level, legislative concerns were similar to those at the federal level, with the addition of a number of challenges to the legality and constitutionality of mandatory or involuntary sampling of DNA. No clear direction is yet apparent as to whose rights take priority in courts of law. In some cases, law enforcement has successfully won broader powers, in some cases, citizens have won the right to refuse DNA testing. However, there does appear to be a general leaning in favor of empowering military and law enforcement agencies to acquire and store DNA.

In 1995, two marine service members under threat of court martial refused DNA testing. In court, they challenged the U.S. military's actions in requiring mandatory DNA samples of all service members. The military's principle motivation for testing was to aid in recovery of unidentifiable remains from armed conflicts, yet tests were being required of all service members, whether or not they were being sent to war zones. In initial court challenges, the military's actions in demanding the tests were upheld as lawful.

In 1996, Jason Aaron Boling challenged Roy Romer, Governor of the State of Colorado, on the constitutionality of Colorado Revised Statute 17-2-201(5)(g):

> ... which requires inmates convicted of an offense involving a sexual assault to provide the state with DNA samples before their release on parole, and the Department of Corrections' (DOC) policies implementing that statute...

The principal argument was that the statute violates the Fourth Amendment prohibition against unreasonable searches and seizures and plaintiff rights under the Fifth, Eighth, Ninth and Fourteenth Amendments. The plaintiff failed to support contentions that he was denied equal rights, and that the state might misuse the information, and his motions were denied.

The DNA Seizure and Dissemination Act enacted by the Commonwealth of Massachusetts has been challenged on constitutional and privacy grounds as well. The Act was put into place on 30 September 1997 to assist criminal justice and law enforcement agencies in identifying individuals and in deterring and discovering crimes. This would give wide-ranging powers affecting not only those convicted of a large number of statutory offenses, but also those convicted of an attempt or conspiracy to commit such acts. It would be incumbent not just on those incarcerated, but also on those on probation or parole. The results of all DNA analyses would become part of the state's DNA database. The Act further would permit officials to use 'reasonable force' in order to obtain samples from uncooperative donors, and the levying of fines and possible imprisonment for those refusing to provide samples.

The Act went into practical use in 1998. On 12 Aug. 1998, Massachusetts Superior Court Justice Isaac Borenstein challenged the DNA Seizure and Dissemination Act and ordered that "... defendants will not participate in the collection, analysis, and/or dissemination of DNA ..". and that "DNA samples already in the possession of the Director of the State Crime Laboratory shall remain sealed, impounded, segregated, and separately stored pending further notice by the Court". Judge Borenstein rejected a state bid for seizure of DNA from seven individuals, stating "... this Court would need to change the existing laws governing Fourth Amendment jurisprudence, and would not only exceed its authority, but adopt a rule that is new, extensive and unjustified under any analysis of constitutional doctrine".

In August 1999, the ACLU of Utah fought against policy that would take DNA samples from arrested individuals prior to conviction. They raised concerns about sampling individuals for minor offenses such as jaywalking, or civil disobedience acts of conscience that might occur at a political demonstration.

In October 1999, the New York State Senate and the Assembly passed a bill to revamp the New York State DNA database. This was significant in that it removed the distinction between criminals and those who have attempted a serious crime. It now covers violent felonies, serious nonviolent offenses (e.g., burglary), and some drug-related crimes. Offenders incarcerated prior to the establishment of the database in 1996 may now be sampled.

In some jurisdictions the onus is on an accused party to explicitly ask that DNA results and samples be destroyed after they have been acquitted or released. Many individuals are not aware of this fact, or that they may only have a few days in which to make the request.

As can be seen from this small sampling of examples, the DNA rights pendulum has not yet come to rest and there are many important issues still to be resolved.

In 2000, the *DNA Analysis Backlog Elimination Act of 2000* authorized the Attorney General to make grants to eligible states to carry out DNA sampling for inclusion in the FBIs Combined DNA Index/Identification System (CODIS). The act included authorization to use whatever means of restraint were reasonably necessary to obtain the sample from eligible felons. These acts and stipulations began to raise consitutional questions.

A 2002 case in a Ninth Circuit Court of Appeals, with an opinion by Judge Reinhardt and dissenting opinion by Judge O'Scannlain, challenged whether Fourth Amendment rights were violated by mandatory sampling of blood for DNA collection.

We hold that forced blood extractions pursuant to the DNA Analysis Backlog Elimination Act of 2000 violate the Fourth Amendment because they constitute suspicionless searches with the objective of furthering law enforcement purposes....

However intermingled with good intentions, DNA statutes, like the thermal imaging procedure struck down in Kyllo, 533 U.S. at 34-41, represent an alarming trend whereby the privacy and dignity of our citizens [are] being whittled away by [ ] imperceptible steps. Taken individually, each step may be of little consequence. But when viewed as a whole, there begins to emerge a society quite unlike any we have seen—a society in which government may intrude into the secret regions of man's life at will.

[Reinhart, United States of America vs Thomas Cameron Kincade, December 2002.]

In 2003, the G.W. Bush Administration announced a $1 billion initiative to improve the use of DNA in the criminal justice system. As part of this initiative, the *President's DNA Initiative* Website was established.

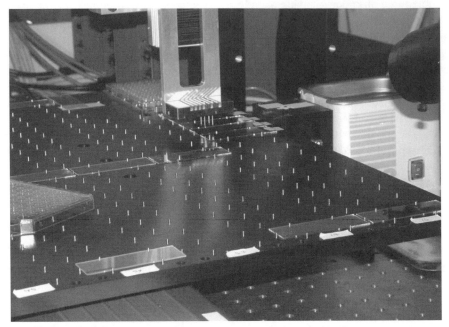

Automated Gene Preparation and Analysis. Sandia National Laboratories and the University of New Mexico cooperated in conducting genome microarray studies using a lab-assembled arrayer (shown here) and, later, a commercial machine. The instrument dips tiny metal pins into reservoirs of DNA (one gene per reservoir) and taps minute quantities onto up to 100 glass slides. Samples of RNA are then prepared and translated into DNA complements and tagged with a fluorescent marker and incubated above the spots of arrayed DNA. Matching cDNA and its corresponding gene are then competitively hybridized (bound) and the array is scanned with a fluorescent device that measures the concentration of the cDNA corresponding to each gene. The microarray is washed, dried, and read using red and green lasers to excite the fluorophores. The intensity of the emission peaks is interpreted into green and red using computer algorithms and the results analyzed in a number of ways, including through data mining. [Sandia National Labs news photo, 2002, released.]

9. Implications of Use

Accessibility

To a surveillance agent or law enforcement officer, the accessibility of DNA technology is based upon a number of factors, including ease of use, portability, cost, and the ease of surreptitious collection of samples.

It is relatively easy to collect voluntary DNA samples with cheek swabs, and only somewhat technical to collect blood samples. It's not quite as easy, but is reasonably routine, to collect involuntary or crime scene samples. Analyzing the samples requires technical expertise and equipment, but can be done by labs for a fee or, in some cases, with portable 'briefcase' systems costing several thousand dollars. DNA is moderately easy to collect surreptitiously from saliva (spit), semen (carelessly discarded condoms), and hair roots. The cost of a basic analysis on a couple of samples (as in a paternity test) is now about $300. Simple DNA profiling for one individual is less. A full analysis yielding more information, such as genetic defects, may cost several hundred to a couple of thousand dollars, but that is still reasonably low, and demand for genetic counselling will drive costs lower for similar, nongenetics services. Reduced costs of $200 for a complete genetic profile, $75 for a paternity test, and $25 for a livestock or endangered species profile were possible by 2003.

While sampling is relatively easy, access to DNA databases is somewhat restricted. Publicly funded national DNA databases are restricted mainly to government and law enforcement officials, but there are databases in research facilities that are open to scientists and instructors, and public databases containing generic information are available on the Internet. It is probable that individuals researching and publishing their genealogies will eventually make DNA patterns available to family members as part of their database histories. Many people's computer systems are wide open to anyone on the Internet when they are searching the Web, especially now that cable modems allow users to economically stay online all the time. Clearly, private individuals are not as security-conscious as corporate agencies (most of whom have firewalls, though even here there are exceptions). Thus, private databases may be vulnerable to unauthorized use.

Database entry is often carried out by lower-paid clerical workers who, in most circumstances, are only loosely supervised. While most of these people are hard-working and honest, they may still be the weakest security link in restricted databases. Low-paid employees are vulnerable to bribery, and young, low-paid employees and interns will sometimes hand over sensitive information because they are unaware of the possible consequences of their actions. Labs without careful verification measures may inadvertently sample a 'stand-in', that is, someone posing as the person who allegedly is the donor.

Given the ease and cost of sampling, in a couple of years, spying with DNA won't be substantially more difficult or more expensive than spying with video cameras, especially if the technical analysis is done with a briefcase lab, or is available for a reasonable fee through a private company that's not inclined to ask questions about the source of the sample or the purposes for which it might be used.

Ethics and Regulations

Ethics and regulations have been surveyed in earlier parts of this chapter, and a full discussion of ethics is too large a topic to be covered here since DNA is used in hundreds of fields of study. However, in terms of surveillance agents, regulations for the collection and use of DNA are fairly stringent in law enforcement, but not significantly restricted in the private sector.

DNA technology is not inherently good or bad. It is likely that a lot of good can come from pharmaceutical drugs designed to work cooperatively with the genetic makeup of individuals, giving them a chance at a longer life and better health. But the same technology can be applied to biological warfare that targets a specific family or ethnic group. Thus, DNA collection on behalf of hate- or protectionist-oriented groups may have some serious repercussions.

As described in earlier examples, most genetics-related bills are not comprehensive. They typically focus on protecting the confidentiality of information in the hands of institutions. Few seek to regulate individual or noninstitutional collection and use of DNA information. Some provisions protect physical access to the individual (e.g., the taking of blood), but few outside of law enforcement address the surreptitious gathering of spit, or hair from a brush, etc. Like any commodity, DNA may eventually be traded on the black market. If scientists find ways to inject DNA into developing embryos, people may buy particular attributes, or use illicitly collected DNA from famous personages, or unwitting parents. Just as there are professional killers willing to take a life for three- to five-thousand dollars, there will likely be professional

DNA collectors, willing to gather DNA for a fee. Taking a saliva sample or handful of hair roots from a person's home involves trespass and theft, but gathering it from a sidewalk, washroom, or other public place without the donor's knowledge is not that difficult. Even if it's illegal, it may be hard to prove when an involuntarily sampled donor doesn't know the crime has been committed.

Many legislative bills regarding national databanks are drafted on the assumption that anonymity is sufficient to safeguard the rights of individuals, but this reflects a simplistic understanding of DNA profiling. Anonymity may disappear when a sample shows a profile of unique interest to researchers or pharmaceutical companies. Who will safeguard the anonymous individual from exploitation of his or her unique genetic makeup for experimental or commercial purposes? What happens if an anonymous profile is later matched to one collected for another reason? What if the anonymous profile is evaluated in the context of close relatives, thus divulging the source of the DNA through deduction and inference? Currently there are no protections against these kinds of exploitation or intrusion.

Applications

The practical applications of DNA survcillance fit into roughly three categories: a) activities intended to benefit the 'greater good' (law enforcement, medicine, conservation, and defense); b) private information-gathering (by corporations, small detective agencies, and individuals); and c) sinister, underground, or black market surveillance (by unauthorized individuals or agencies).

Category a activities have been discussed throughout this chapter and are not repeated here, but *categories b and c* have not been covered in detail, and a few examples are cited to provide some perspective on less-sanctioned surveillance activities.

Private detectives and free agent spies are routinely hired to surveil corporations, allegedly cheating spouses, and individuals suspected of insurance fraud. This is currently done with hidden video cameras, radio transmitters, and sometimes with illegal phone taps. Given that phone call recording is rampant, in spite of tough legislation regulating its use, it's probable that surveillance agents will undertake clandestine collection of DNA in much the same way.

- Free agents collecting saliva, hair roots, or semen, and getting analyses from private labs, currently have few legislative hurdles to hinder them. They can conceivably collect and analyze DNA from motel sheets provided by the motel owners voluntarily or for a fee. They may sample a potential CEO or other employee, or a marriage hopeful who might want to know the genetic makeup of a potential spouse, perhaps a child that is suspected of being fathered through another woman, a deadbeat parent, or an adopted child.

- Supremacist groups might undertake 'racial screening' before admittance into their organizations.

- Enemies might sample a person for vulnerabilities, like a predisposition for alcoholism or heart disease, in order to discredit or harm a person.

- Underground research labs might design a 'smart virus' targeted specifically to be active in people with certain DNA patterns.

- Terrorists might target a specific species of plant, like a food crop, to be decimated by a designer gene that becomes active after the initial growing period.

When you are dealing with the building blocks of life, almost anything is possible and since many of these potential crimes are hard to trace, they will be equally hard to prosecute.

Strength of Evidence

Strength of evidence depends on its specificity, its accuracy, and admissibility in court. In the mid-1990s, there was a lot of debate about the strength and accuracy of DNA evidence. This is partly because other court-related tests are controversial, such as lie detector tests. From a scientific viewpoint, however, DNA is strong evidence, particularly in the areas of DNA *print matching* and DNA *family profiling,* which are relevant in many legal disputes. Courtrooms are gradually accepting DNA as 'good' evidence and supporting its credibility. Juries composed of lay people may not yet fully understand the technology, but gradually, as articles are written in the popular press, its applications and implications will be better understood. DNA evidence is now strong enough to exonerate falsely accused incarcerated individuals and is compelling enough to convict many murderers and rapists.

Freedom and Privacy

Freedom and privacy issues are probably the most important matters related to DNA surveillance and thus are covered in more detail. Some of the important implications of DNA sampling, storage, and use have already been discussed in Section 7 (Problems and Limitations), including discrimination and the logistics of sampling and storage, and the reader is encouraged to consult Section 7 in conjunction with this section.

The study and use of DNA technology is likely to grow in spite of concerns about privacy and commercial exploitation of an individual or family's unique attributes. There are so many stakeholders in the DNA goldrush: medical researchers, pharmaceutical companies, genetics counsellors, parents, potential parents, spies, and law enforcement officials, that the genie is already out of the bottle. With automation and commercialization, prices will continue to drop and DNA analysis will become readily available through portable suitcase labs. The cheaper it gets, the more people will use it—the more they use it, the cheaper it will get.

DNA analysis is a highly significant surveillance technology in terms of the amount of personal information that can be gathered and pinpointed to a particular location, or even to specific actions of an individual. Tissue samples can reveal not only characteristics of the individual and indirect information about his or her ancestors and siblings, but may also indicate medical conditions or pathogens that can betray a person's activities, general lifestyle choices, or specific whereabouts. This intimate information, taken together with data from other observations, can provide a composite intelligence picture that reveals much more than was ever before possible. Serial murders, for example, may be difficult to track when the killer is constantly on the move. Certain interstate murders might never before have been linked without careful studies of the modus operandus (MO) and sharing of notes with other jurisdictions. With DNA testing of crime scenes where blood or semen might be found, DNA analysis has the potential to immediately connect the murders, thus alerting law enforcement to cases in which cooperation between agencies would be fruitful.

DNA testing is favored over simple blood type testing in paternity suits, as it provides a far greater set of information, and requires less invasive sampling (suitable for children and animals). It can be performed on infants and in some cases, unborn children or their associated CVS and amniotic fluids. DNA sampling to detect genetic defects is already carried out in many hospitals and birth clinics. It's only a short step to sampling all children at birth, or even before the child is born. The information can be permanently stored, perhaps in a tiny implant on the underside of an arm or leg, one that might have an electronic lead or wireless transmitter for interfacing with a computer network.

The DNA sword is two-sided. Usually the easier it is to surveil, the easier it is to be surveilled. Clearly, law enforcement officials could do their jobs more efficiently if they had a key to every house in the community and, clearly, they could do better crime solving if they had a DNA file for every person in the nation, but is this a good idea or might it result in an insidious, irrevocable erosion of our free society? In the U.K., citizens have made voluntary donations of DNA samples to help fight crime, and police agencies have advocated sampling the entire population. There are tax incentives for permanent DNA banking of not just criminals, but ordinary people, as it's expensive to resample the populace each time a crime is committed. In some countries citizens won't even be given a choice. They'll be directed to submit a sample to the state, with strong penalties for noncompliance. Even in the U.S., when service marines refused to submit to mandatory DNA collection in the 1990s, the right of the U.S. military was given precedence over the concerns of the marines.

We should be both awed and concerned at the applications of this technology. Awed at the potential good from genetics counselling and the apprehension of violent criminals, and concerned about trading one set of freedoms for another without any guarantee of overall gain. If we look critically at computer technology, which developed in a flush of hope that automation would provide us with more free time and a better quality of life, we see that the trade-offs are sometimes greater than originally anticipated. Computers have spawned a global obsession with information-gathering, record-keeping, and publishing that now appears to be reducing free time, rather than increasing it. We are learning that computer use provides no guarantee of an improved quality of life except in some cases of medical applications, transportation, or communication with distant business associates or loved ones. In many other cases, quality of life has been lost.

Those particularly concerned about institutional and private abuse of DNA information have argued that DNA is a natural attribute of the person, subject to consent and privacy laws. Governments change and policies change but information databases tend to survive from one administration to the next. History has shown us that data collected for one purpose may be expropriated for other purposes by subsequent administrations. Those who handle the data, or turn it over to office colleagues or superiors, often don't know what it is or how it is used. Everyone involved with DNA technology should be aware that we may be handling 'data' that are more sensitive and more volatile than any technology ever devised.

10. Resources

Inclusion of the following companies does not constitute nor imply an endorsement of their products and services and, conversely, does not imply their endorsement of the contents of this text.

10.a. Organizations

American Academy of Forensic Sciences (AAFS) - The AAFS is a Colorado-based society with international activities dedicated to the application of science to law. Members include physicians, physical anthropologists, document examiners, criminologists, toxicologists, and others. www.aafs

American Association of Blood Banks - This organization has established standards for DNA use in paternity testing, and accreditation for laboratories.

American Board of Criminalistics (ABC) - Wisconsin-based organization that oversees a peer-developed, peer-reviewed certification program. www.criminalistics

American Civil Liberties Union (ACLU) - The ACLU is a prominent, nonpartisan individual rights advocate organization involved in legislation, litigation, and education on issues related to freedom in the United States. Established in 1920 by Roger Baldwin. www.aclu

American Society of Crime Laboratory Directors (ASCLD) - Founded in 1973 by the Director of the FBI Laboratory and the Director of the FBI, the ASCLD is a nonprofit professional organization concerned with the operations, standards, and ethics of crime laboratory operations. The ASCLD runs a voluntary Crime Laboratory Accreditation Program and publishes the ASCLD Newsletter. www.ascld

Armed Forces DNA Identification Laboratory & Repository - A subspecialty of the Center for Advanced Pathology (CAP) for consultation, education, and research. It processes identification of remains from Vietnam, Operation Desert Storm, Waco, etc. and collects and stores service member DNA samples, with more than 10,000 currently on file. www.cap

Armed Forces Institute of Pathology DNA Identification Laboratory (AFIP, AFDIL) - AFDIL is comprised of two lab. sections - mitochondrial DNA, used primarily for analyzing service member remains; and nuclear DNA, which supports the Office of the Armed Forces Medical Examiner in disaster investigation (explosions, air disasters, etc.). www.afip

A DNA Registry was established in 1992 to provide identification information for American service personnel and contractors sent to areas which may be dangerous to life. Blood specimens are collected (usually at induction centers), packaged, and frozen. Identifications are said to be based on the lengths of the repeated sequences, information sufficient to identify familial relationships, but not sufficient to provide in-depth medical information about the individual. However, with the original blood sample on file, it does not guarantee that different types of testing might not occur in the future. Cards are kept on file for the same length of time as other records, currently 50 years. Once service has been completed, an individual may request destruction of the DNA record.

Armed Forces Repository of Specimen Samples for the Identification of Remains (AFRSSIR) **Armed Forces Identification Review Board** (AFIRB) - A department involved in identification of the remains of service members, which now uses DNA technology to assist in this task. AFIRB was involved in identification of the memorialized Unknown Soldier.

Assistant Secretary of Defense for Health Affairs (ASDHA) - A federal agency responsible for establishment and enforcement of DNA sampling requirements. DNA samples superseded panographic samples as of 1 January 2000.

Association in Search of Disappeared Children - A humanitarian organization founded in 1994 to locate children who disappeared in the El Salvador civil war. It works in conjunction with Physicians for Human Rights in using DNA testing as an identification tool to help reunite families.

Bodymap - Web-accessible, grant-supported gene expression databases for humans and mice. It is based in Japan. bodymap.ims.u-tokyo.ac.jp/

Celera Genomics - One of the leading commercial companies in the process of unraveling the human genome and creating a base of genetics information and services for pharmaceutical, biotechnology, and academic research. www.celera

Combined DNA Index System (CODIS) - A national computer bank of DNA from all states including known and unidentified samples. Initiated October 1998, CODIS requires sampling from 13 gene loci that are specific to identification. (The FBI has had a typing lab in the Hoover Building in Washington, D.C. since 1989.) CODIS is a databank with profiles of evidence and convicted offenders. Access is restricted. Includes state (SDIS) and local (LDIS) indexes. Samples and records for convicted criminals are kept indefinitely.

Council for Responsible Genetics (CRG) - Founded in 1983, CRG is a nonprofit professional organization devoted to public awareness and monitoring of technological advances related to genetics. CRG advocates the responsible use of this technology, and publishes an information bulletin. www.gene-watch

Danish Centre for Human Genome Research - Web-accessible databases at the University of Aarhus for health and disease research. The protein files are extensively cross-referenced to other Internet databases (GenBank, UniGene, etc.). Includes still and video examples of procedures. biobase.dk/cgi-bin/celis/

DNA Advisory Board - A federal organization which sets standards for the interpretation of DNA evidence.

EarthTrust - An organization that supports the ongoing use of DNA testing to end whaling.

European DNA Profiling Group (EDNAP) - In 1991 EDNAP was accepted as a working group of the ISFH. Founded in October 1988 in London, it represents over a dozen European countries, providing law enforcement assistance based on cooperation and scientific data and decision-making. www.uni-mainz.de/FB/Medizin/Rechtsmedizin/ednap/ednap.htm

Federal Bureau of Investigation (FBI) - A U.S. federally funded investigative agency with a number of departments involved in DNA sampling, analysis, profiling and databasing. www.fbi.gov

Forensic Justice Project (FJP) - A national, Washington, D.C.-based effort to provide objective, expert evaluation and testimony, and to monitor and review cases handled by the FBI crime lab with the goal of protecting innocent people from the misuse or abuse of forensic sciences. www.forensicjustice.gov

The Forensic Science Service (FSS) - A U.K.-based, accredited group of laboratories employing over 1,200 scientists. Lab members examine samples, provide expert evidence in court, and engage in scientific research on advanced DNA techniques. www.forensic.gov.uk

Genetics Society of America - Established in 1931, this organization promotes academic research and recognition in the field of genetics, and publishes the monthly "Genetics" journal. www.faseb.org/genetics/gsa/gsamenu.htm

Genome MOT - A genome monitoring project which presents Web-accessible status reports for monitoring the progress of the many large genome sequencing projects, which includes the human genome. www.ebi.ac.uk/~sterk/genome-MOT/index.html

Genome Sequencing Center - Established in 1993 at the Washington University School of Medicine with a grant from the National Human Genome Research Institute of NIH. The teams each sequence about 7 million base pairs per year. The GSC works in conjunction with the Sanger Centre. genome.wustl.edu/gsc/index.shtml

Human Genome Diversity Project (HGDP) - A National Science Foundation-supported, international project proposed in the early 1990s to study global human variation, patterns of migration, and reproduction by collecting DNA samples from anonymous members of some 500 defined populations. Ethical and practical matters were being worked out before full-scale commencement of the project. THE HGDP is affiliated with HUGO.

Human Genome Organization (HUGO) - A nonprofit, non-governmental organization of scientists involved in the coordination of studies of human genetics, founded by Victor McCusick. HUGO is involved the Human Genome Diversity Project.

The Human Genome Project (HGP) - A highly significant international collaborative effort to 'map' the entire human genome. Established in 1991 by the Ministry of Education, Science, Sports and Culture (MESSC).

The Innocence Project - An effort which helps free people who have been jailed unjustly. DNA evidence is used to exonerate incarcerated individuals by excluding them as the perpetrators of a crime, usually involving rape or murder.

The Institute for Genomic Research (TIGR) - A not-for-profit research institute based in Rockville, Maryland. It conducts research on plant and animal cells and includes a large DNA sequencing lab. www.tigr

International Association of Forensic Science/Scientists (IAFS) - This organization conducts an annual international professional conference.

The International Association for Identification (IAI) - A large professional organization founded in 1915 for individuals around the world engaged in forensic identification, investigation, and scientific examination of physical evidence. The California-based IAI educates through conferences and workshops, encourages research, and promotes communication among members and affiliates. www.theiai

International Criminal Police Organization (INTERPOL) - An international cooperative organization of crime-fighting police authorities with members in almost 200 countries. Established in 1956 (descended from ICPC), INTERPOL is currently based in Lyon, France. www.kenpubs.co.uk/INTERPOL

INTERPOL European Working Party on DNA Profiling - A working group consisting of members from several countries, including Belgium, the Czech Republic, Germany, U.K., Norway, Spain et al. Its members discuss DNA profiling as an investigative tool and make recommendations concerning its use in criminal investigations.

Lawrence Livermore National Laboratory Forensic Science Center (LLNL FSC), **LLNL Human Genome Center** (HGC) - Research and commercialization of DNA sequencing techniques and equipment. The HGC has established a number of databanks and sequence tracking systems. LLNL has also been involved in the formation of a Joint Genome Institute for the U.S. Department of Energy with its three Genome Centers at Livermore, Berkeley, and Los Alamos. www-bio.llnl.gov/bbrp/genome/genome.html

Maxxam Analytics Inc. - Human DNA Department - recently inspected by the Standards Council of Canada in hopes of becoming accredited for paternity and forensic DNA analysis.

Medlantic Research Institute's Transplant and Immunogenetics Laboratories - Tissue typing for organ transplants, crossmatching cadavers and living recipients for compatibility, blood serology and viral screening. Designated as the central donor lab for the Washington Regional Transplant Consortium (WRTC) in 1990.

Mitochondrial DNA Concordance - Centered at the University of Cambridge, this database provides a cross-referenced list of single nucleotide substitutions in the two hypervariable segments of the mtDNA control region. The purpose is to provide information that is easier to analyze than that found in the GenBank, for example, which has been useful in the forensics field. shelob.bioanth.cam.ac.uk/mtDNA/

MITOMAP - A comprehensive human mitochondrial genome database that can be accessed through the Internet. The human mtDNA sequence was determined in the mid-1990s. infinity.gen.emory.edu/MITOMAP/

National Center for Biotechnology Information (NCBI) GenBank - A national, annotated, public databank of DNA sequences containing data submissions mainly from scientists. The GenBank is searchable over the Internet. GenBase is part of the International Nucleotide Sequence Database Collaboration project, and as such shares data with the DNA DataBank of Japan (DDBJ) and the European Molecular Biology Laboratory (EMBL). www.ncbi.nlm.nih.gov/Genbank/index.html

National Center for Genome Resources (NCGR) - A nonprofit center founded in 1994, arising from the Human Genome Project. NCGR supports the research, development, and application of knowledge systems in biology related to technologies that support humankind and the environment. The Center develops and applies computer-based systems for visualizing and analyzing data and disseminates them through collaborations with research institutions, both academic and commercial. Funding is from public and private sources. The center is one of two public repositories in the U.S. for human genomic data. Located in Santa Fe, New Mexico. www.ncgr

National DNA Advisory Board - Established in 1995, arising from the DNA Identification Act of 1994, to recommend quality assurance standards to the FBI.

National Human Genome Research Institute (NHGRI) - Originally established in 1989 as The National Center for Human Genome Research (NCHGR). NHGRI heads the Human Genome Project. It is one of two dozen centers comprising the National Institutes of Health (NIH). www.nhgri.nih

National Institute of Justice (NIJ) - This agency has a research branch awarding $5 million grants in 1999 to speed DNA chip technology.

National Institute of Justice Commission on DNA Evidence - Established as directed by the Attorney General to provide recommendations on the use of current and future DNA methods, applications, technologies, policies, and legal issues in the operation of the criminal justice system. Meetings are held about four times per year. www.ojp.usdoj.gov/nij/dna/welcome.html

National Center for Biotechnology Information (NCBI) - Established in 1988 as part of the National Library of Medicine as a recourse for molecular biology information. The NCBI conducts research in computational biology, creates public databases, develops software tools for analyzing genome data, and disseminates biomedical information. NCBI hosts the GenBank nucliotide sequence database. www.nih.gov   www.ncbi.nlm.nih.gov

National Research Council Committee from the National Academy of Sciences - Has been promoting the use of defined standards for proficiency testing, DNA laboratory work, and statistical calculations and issuing reports.

The Sanger Centre - Research in genomes, especially sequencing and analysis. The facility is jointly founded by the Wellcome Trust and the Medical Research Council. (See Wellcome Trust.) www.sanger.ac.uk/

Scientific Working Group DNA Analysis Methods (SWGDAM) - With support from the FBI, it includes members from more than 30 laboratories. Originally TWGDAM, it addresses issues related to statistical interpretation, DNA profiling, quality assurance and criteria. www.for-swg.org/swgdamin.htm

Selmar - DNA profiling services for law enforcement.

Stanford Human Genome Center (SHGC) - Research and education regarding the construction of high resolution radiation hybrid maps of the human genome, and sequencing of large, contiguous genomic regions. www-shgc.stanford.edu/

Technical Working Group on DNA Analysis Methods (TWGDAM) - Founded in 1988, TWGDAM is hosted and supported by the FBI Laboratory. It is now known as the Scientific Working Group on DNA Analysis Methods (SWGDAM).

UniGene - Unique Human Gene Sequence Collection. Records about 40,000 clusters of GenBank sequences representing the transcription products of distinct genes derived from an experimental system. Includes human, rat, and mouse data. www.ncbi.nlm.nih.gov/UniGene/index.html

United Nations Educational, Scientific and Cultural Organization (UNESCO) - Headquartered in France with about 60 field offices around the world, this multinational United Nations organization was founded in 1945 and currently represents almost 200 member nations. www.unesco.org

U.S. Army Central Identification Laboratory, Hawaii (CILHI) - Involved in the search, recovery, and identification of skeletal remains of American bodies lost in prior conflicts. Works through use of the Armed Forces Institute of Pathology (AFIP) labs.

U.S. Army Forces Command (FORSCOM) - Engaged in cataloguing all active soldiers by the end of 1999 and all reserve soldiers by the end of 2004, as per the Assistant Secretary of Defense for Health Affairs which is the agency responsible for the establishment and enforcement of DNA sampling requirements.

U.S. Fish & Wildlife Service - The Fish & Wildlife Service lab in Ashland, Oregon includes a Forensic Science Branch with Genetics and Criminalistics departments. These facilities aid the Service in prosecuting individuals 'beyond a reasonable doubt.' www.lab.fws

Wellcome Trust Genome Campus - Hinxton, near Cambridge, U.K. The campus houses the Sanger Centre, the European Bioinformatics Institute (EBI), a branch of the European Molecular Biology Laboratory (EMBL), and the Medical Research Council's Human Genome Mapping Project Resource Centre (HGMPRC).

10.b. Print Resources

The author has endeavored to read and review as many mentioned resources as possible or to seek the recommendations of colleagues. In a few cases, it was necessary to rely on publishers' descriptions on books that were very recent, or difficult to acquire. It is hoped that the annotations will assist the reader in selecting additional reading.

These annotated listings may include both current and out-of-print books and journals. Those which are not currently in print are sometimes available in local libraries and secondhand book stores, or through interlibrary loan systems.

Ballantyne, Jack; Sensabaugh, George; Witkowski, Jan, *DNA Technology and Forensic Science* (Banbury Report 32), New York: Cold Spring Harbor Lab Press, 1990, 368 pages. Discusses legal and policy issues, and the provision of DNA services.

Billings, Paul R., Editor, *DNA on Trial: Genetic Identification and Criminal Justice*, New York: Cold Spring Harbor Lab Press, 1992, 154 pages. Describes the history of DNA in trial courts, civil liberties, and public policy. Includes an analysis of decisions in various courts and whether juries should consider DNA evidence. Out of print.

Erlich, H. A., Editor, *PCR Technology: Principles and Applications for DNA Amplification*, New York: Stockton Press, 1989, 246 pages.

Evett, I.; Weir, B., *Interpreting DNA Evidence: Statistical Genetics for Forensic Scientists,* Sinauer, 1998, 278 pages.

Farley, Mark A.; Harrington, James J., Editors, *Forensic DNA Technology,* Chelsea, Mi.: Lewis Publishers, Inc., 1991. Includes contributions from some of the country's leading experts. Traces the underlying theory and historical development of DNA testing, legal admissibility requirements and courtroom utilization, as well as the history of DNA use in the clinical lab.

Fisher, Barry, A. J. *Techniques of Crime Scene Investigation, 6th edition,* Boca Raton, Fl.: CRC Press LLC, 1999. Describes field-tested techniques and procedures, and technical information concerning crime scene investigation. A comprehensive source used by police academies, community colleges, and universities. Recommended by three professional organizations, the International Association for Identification, the American Board of Criminalistics, and the Forensic Science Society, as a text to prepare for their certification examinations.

Frankel, Mark S.; Teich, Albert H., *Ethical and Legal Issues in Pedigree Research,* Washington, DC: AAAS, 1993, 216 pages. DNA and pedigree lineage issues. These are currently undergoing substantial change due to the availability of DNA information.

Frankel, Mark S.; Teich, Albert H., Editors, *The Genetic Frontier: Ethics, Law, and Policy,* Washington, DC: AAAS, 1994, 240 pages. Out of print.

Inman, Keith; Rudin, Norah, An Introduction to Forensic DNA Analysis, Boca Raton, Fl.: CRC Press LLC, 1997. Forensic DNA analysis theory and techniques explained so that professionals in other fields and laypeople can understand DNA analysis from sample collection to data interpretation. Describes the legal and scientific advantages and limitations of the various techniques.

Judson, Horace Freeland, The Eighth Day of Creation: Makers of the Revolution in Biology, New York: Cold Spring Harbor Lab Press, 1996. Historical information in molecular biology. Includes interviews with some of the scientists involved in the discovery of DNA.

Kitcher, Philip, The Lives to Come: The Genetic Revolution and Human Possibilities, New York: Simon & Schuster, 1996, 381 pages. Covers many aspects of DNA technology, including sequencing, fingerprinting, transcription, translation, restriction, cloning, etc. Includes a historical survey, an evaluation of its impact on our lives, and a good glossary.

Lagerkvist, Ulf, DNA Pioneers and Their Legacy, New Haven and London: Yale University Press, 1998, 156 pages.

National Institute of Justice, The National Institute of Justice and Advances in Forensic Science and Technology, a National Law Enforcement and Corrections Technology Center Bulletin, March 1998, 23 pages. Primarily important for providing background information on DNA procedures, legislation, quality assurance and application, but also includes information on latent fingerprinting and authorship of questioned documents. nlectc.org/pubs/

National Research Council, The Evaluation of Forensic DNA Evidence, Washington, D. C.: National Academy Press, 1996. Includes information on population genetics and statistics, case descriptions, and recommendations for handling samples.

Olby, Robert, The Path to the Double Helix, Seattle: University of Washington Press, 1974. A comprehensive historical presentation. Dover Publications has produced a 1994 reprint, 522 pages.

President's Commission for the Study of Ethical Problems in Medicine and Biomedical and Behavioral Research, Screening and Counseling for Genetic Conditions, Washington, D.C.: U.S. Government Printing Office, 1983.

Sayre, Anne, Rosalind Franklin and DNA, New York: W. W. Norton & Co., 1975, 221 pages.

Teich, Albert, Ethical and Legal Implications of Genetic Testing, Conference Proceedings: The Genome, Ethics, and the Law: Issues in Genetic Testing. Washington, D.C.: AAAS, 1992, 124 pages.

Watson, James D.; Stent, Gunther S., Editor, The Double Helix: A Personal Account of the Discovery of the Structure of DNA, New York: W. W. Norton & Co., 1980, 298 pages.

Zaborskey, Oskar R., An Evaluation of the Application of DNA Technology in Forensic Science, Washington, D. C.: National Research Council, National Academy Press, 1992, 185 pages.

Articles

Andrews, L. B.; Jaeger, A. S., Confidentiality of Genetic Information in the Workplace, American Journal of Law and Medicine, 1991, V.17, pp. 75-108.

Annas, George J., Privacy Rules for DNA Databanks: Protecting Coded 'Future Diaries', Journal of the American Medical Association, 1993, Nov. 17; 270(19), pp. 2346-2350. Discusses the wide-ranging information contained in DNA and how it relates to privacy, obligations, storage, and public policy and rule-making. Annas continues this discussion in Rules for gene banks: protecting privacy in the genetics age, a 1994 paper presented originally at a 1991 conference on Justice and the Human Genome.

Connors, Edward; Lundregan, Thomas; Miller, Neal; McEwen, Tom, Convicted by Juries, Exonerated by Science: Case studies in the use of DNA evidence to establish innocence after trial, NIJ Research Report, June 1996, 118 pages.

Croteau, Roger, Experts ID Dog DNA in Mauling, *San Antonio Express-News*, 14 May, 1998. Report in which saliva on a victim's clothes matched a dog accused of the mauling.

Dahlberg, James E.; Lyamichev, Victor I.; Brow, Mary A. D., Method of site specific nucleic acid cleavage, *U.S. Patent #5,422,253,* filed 7 Dec. 1992.

Drmanac, Radoje T.; Crkvenjakov, Radomir B., Method of sequencing genomes by hybridization of oligonucleotide probes, *U.S. Patent #6,018,041,* filed 29 July 1997 as a continuation of several other patent filings including *U.S. Patent #5,492,806,* filed 12 Apr. 1993.

Fodor, Stephen P. A.; Solas, Dennis W.; Dower, William J., Methods for nucleic acid analysis, *U.S. Patent #5,871,928,* filed 11 Jun. 1997.

Gill, P.; Jefferies, A. J.; Werrett, D. J. Forensic applications of DNA fingerprints, *Nature*, 1985, V.318, pp. 577-579.

Gill, P.; Ivanov, P. L.; Kimpton, C.; Piercy, R.; Benson, N.; Tully, G.; Evett, I.; Hagelberg, E.; Sullivan, K. Identification of the remains of the Romanov family by DNA analysis, *Nature Genetics*, 1994, V.6, pp. 130–135.

Ginther, C.; Issel-Tarver, L.; King, M. C., Identifying individuals by sequencing mitochondrial DNA from teeth, *Nature Genetics*, 1992, V.2, pp. 135–138.

Gordon, Julian; Stimpson, Donald; Hsieh, Wang-Ting, Method and apparatus for collecting a cell sample from a liquid specimen, *U.S. Patent #5,578,459,* filed 24 Nov. 1993.

Hershfield, Bennett, Method for identifying the species origin of a DNA sample, *U.S. Patent #5,674,687,* filed 29 Nov. 1995.

Isenberg, Alice R.; Moore, Jodi M., Mitochondrial DNA analysis at the FBI Laboratory, *Forensic Science Communications*, V.1(2), July 1999.

Jones, Douglas H., Iterative and regenerative DNA sequencing method, *U.S. Patent #5,858,671,* filed 1 Nov. 1996.

Kambara, Hideki; Okano, Kazunori, DNA sequencing method and DNA sample preparation method, *U.S. Patent #5,985,556,* filed 18 Sept. 1996.

Kass, N. E., The implications of genetic testing for health and life insurance, 1997, in Rothstein, M.A., Editor, *Genetic Secrets: Protecting Privacy and Confidentiality in the Genetic Era*, Yale University Press, pp. 299-316.

Katz, D.; Lee, D. et al, The Application of Mitochondrial DNA Sequence Analysis to the Identification of Remains from a Mass Disaster - KAL Flight 801, *Armed Forces DNA Identification Laboratory,* 1998.

Kusukawa, N.; Uemori, T.; Asada, K.; Kato, I., Rapid and reliable protocol for direct sequencing of material amplified by the polymerase chain reaction, *BioTechniques*, 1990, V.9, pp. 66-72.

Lewin, Roger Genes from a disappearing world, *New Scientist,* May 1993, V.138(1875), pp. 25-29. Discusses DNA samples and databanks, genome mapping, investigative methods, impact, risks and benefits, minority groups, and social discrimination. This article includes an inset Making the most of a human genetic atlas.

Love, Jack D., Identification and paternity determination by detecting presence or absence of multiple nucleic acid sequences, *U.S. Patent #5,645,990,* filed 22 Nov. 1994.

Maniatis, T. et al., Molecular Cloning, *Cold Spring Harbor Lab Report,* 1989, V.13, pp. 21-33.

Matsunaga, Hiroko; Okano, Kazunori; Kambara, Hideki, DNA sequencing method and reagents kit, *U.S. Patent #5,968,743,* filed 9 Oct. 1997.

Maxam, A.; Gilbert, W., A new method of sequencing DNA, *Proceedings of the National Academy of Sciences,* 1977, V.74, pp. 560-564.

Maxam, A. M.; Gilbert, W., Sequencing end-labeled DNA with base-specific chemical cleavages, *Methods in Enzymology,* 1980, V.65, pp. 499-560.

McEwen, Jean E.; Reilly, Philip R. A review of state legislation on DNA forensic data banking, *American Journal of Human Genetics*, 1994 June; V.54(6), pp. 941-958. Discusses confidentiality, criminal law, disclosure, DNA sampling, privacy issues, legislation, and advisory committees.

McEwen J.; McCarty K.; Reilly P., A survey of medical directors of life insurance companies concerning use of genetic information, *American Journal of Human Genetics*, 1993, V.53, p. 33.

Mirsky, Alfred E., The Discovery of DNA, *Scientific American,* June 1968, pp. 78-88. Provides an outline of major events and discoveries in the field.

Mullis, Kary B. et al., Process for amplifying, detecting, and/or cloning nucleic acid sequences, *U.S. Patent #4,683,195,* filed 28 July 1987.

Padhye, Vikas V.; York, Chuck; Burkiewicz, Adam, Nucleic acid purification on silica gel and glass mixtures, *U.S. Patent #5,658,548,* filed 7 Jun. 1995.

Rothberg, Jonathan Marc; Deem, Michael W.; Simpson, John W., Method and apparatus for identifying, classifying, or quantifying DNA sequences in a sample without sequencing, *U.S. Patent #5,871,697,* filed 24 Oct. 1995.

Sanger et al., DNA sequencing with chain terminating inhibitors, *Proceedings of the National Academy of Sciences*, USA, 1977, V.74(12), pp. 5463-5467.

Sensabaugh, George; Kay, D. H., Non-human DNA evidence, *Jurimetrics Journal,* 1998, V.38, pp. 1-16. Judging the soundness of new methods of DNA analysis, including samples from non-human sources, as evidence. The article is adapted from the 2nd edition of the U.S. Federal Judicial Reference Manual on Scientific Evidence.

Shapiro, E. D.; Weinberg; M.L., DNA Data Banking: The dangerous erosion of privacy, *Cleveland St. Law Review*, 1990, V.38, pp. 455, 477.

Tahara, T.; Kraus, J. P.; Rosenberg, L. E., Direct DNA sequencing of PCR amplified genomic DNA by the Maxam-Gilbert method, *BioTechniques,*1990, V.8, pp. 366-368.

Thompson, W. C., Evaluating the admissibility of new genetic identification tests: Lessons from the 'DNA War', *Journal of Criminal Law*, 1993, p. 22. The process and procedures related to DNA typing.

Wells, Gary; Small, Mark et al., Scientific Review of Lineups, White Paper, *Law and Human Behavior,* Dec. 1998.

Wilson, M. R.; Polanskey, D.; Butler, J.; DiZinno, J. A.; Repogle, J.; Budowle, B., Extraction, PCR amplification, and sequencing of mitochondrial DNA from human hair shafts, *BioTechniques*, 1995, V.18, pp.662–669.

Journals

Forensic Science Communications, a quarterly publication by FBI Laboratory personnel. It includes articles on document examination, imaging technologies, evidence and statistics, the recovery, handling and transport of evidence, and other topics related to forensic science (formerly Crime Laboratory Digest). www.fbi

Genewatch: The Bulletin of the Council for Responsible Genetics, a national newsletter which monitors the ethical, social, and biological impacts of biotechnology, six issues per year. www.gene-watch

Institute for Psychological Therapies Journal, official journal of the IPT, which deals with the private practice of clinical therapies for victims and alleged victims of violent crimes such as sexual abuse and murder. As such, the journal discusses issues related to DNA evidence and exoneration of wrongfully accused individuals. www.ipt-forensics

Journal of Forensic Sciences, the professional journal of the American Academy of Forensic Scientists (see listing under organizations). The JFS editorial base is at the University of Illinois at Chicago.

10.c. Conferences and Workshops

Many of these conferences are annual events that are held at approximately the same time each year, so even if the conference listings are outdated, they can still help you determine the frequency and sometimes the time of year of upcoming events. It is very common for international conferences to be held in a different city each year, so contact the organizers for current locations.

Many of these organizations describe the upcoming conferences on the Web and may also archive conference proceedings for purchase or free download.

The following conferences are organized according to the calendar month in which they are usually held.

The International Association of Forensic Scientists (IAFS) International Symposium on Forensic Sciences, sponsored by the Australian and New Zealand Forensic Science Society, annual since the mid-1980s,

Arizona State University SmithKline Beecham Symposium, Respecting Genetic Privacy, explored the relationship between laws, legal doctrines, and genetic privacy, information, and tissue samples.

Forensic e-Symposium—Human Identification, The Forensic Institute series. www.forensic.e-symposium.com

International Conference on Forensic Statistics, annual conference since 2000. /icfs.law.asu.edu/

International Symposium on Forensic Sciences, annual spring conference since ca 1990.

National Conference on the Future of DNA: Implications for the Criminal Justice System, annual conference since the mid-1990s.

Cambridge Healthtech Institute's DNA Forensics Conference, annual since the mid-1990s.

The Fourth Biennial Conference: International Perspectives on Crime, Justice and Public Order. This conference includes genetics topics.

The British Council International Seminar on Advancing the Scientific Investigation of Crime, annual since 1997.

International Congress on Forensic Genetics, annual since ca 1980.

International Conference on DNA Sampling and Human Genetic Research: Ethical, Legal, and Policy Aspects, international conference since 2000.

Forensic Bioinformatics Conference, annual national expert forum since ca 2001. www.bioforensics.com

TIGR Genome Sequencing and Analysis Conference, sponsored by the Institute for Genomic Research.

Conference on DNA Sampling - The Commercialization of Genetic Research: Ethical, Legal and Policy Issues. Re-presented at *The Canadian Bioethics Society Annual Conference: Reflections on a Decade of Bioethics,* annual conference.

10.d. Online Resources

DNA Tells All–Doesn't It? A site that specifically looks at crime investigation and other forensic aspects of DNA. This is part of a NISE project funded by the National Science Foundation. whyfiles.news.wisc.edu/014forensic/genetic_foren.html

The Gene School. An educational site with general information, applications, interactive tutorials, a glossary and other resources related to gene therapy, cloning, the Human Genome, agriculture, and medicine. library.advanced.org/tq-admin/day.cgi

National Institute of Justice Convicted Offender DNA Backlog Reduction Program. The site includes a program description and goals and information about legislation authorizing the program. www.ojp.usdoj.gov/nij/topics/forensics/dna/convicted/welcome.html

President's DNA Initiative - A site describing funding and plans to increase the use of DNA technology in the criminal justice system, established in 2003. www.dna.gov

10.e. Media Resources

Buried Secrets: Digging for DNA, from the History Channel *History's Mysteries* series. An examination of a number of important forensic techniques, including DNA matching, which determines identities of the famous dead, including the Unknown Soldier and descendants of Thomas Jefferson. VHS, 50 minutes. Cannot be shipped outside the U.S. and Canada.

The DNA Revolution, from the History Channel *20th Century with Mike Wallace* series. The show describes the history of our understanding of DNA and describes some of the current technologies and controversies concerning genetic engineering. VHS, 50 minutes. Cannot be shipped outside the U.S. and Canada.

Forensic Science: The Crime-Fighter's Weapon, from the History Channel *Modern Marvels* series. This traces the history and development of forensics and looks at various types of evidence (including DNA). VHS, 50 minutes. Cannot be shipped outside the U.S. and Canada.

Gattaca, feature film by Columbia Pictures with Jersey Film Production, an Andrew Niccol Film. A futuristic story that investigates the possible consequences of DNA testing to stratify and control movement within society. Worth watching as the implied consequences are not unprecedented, and much of the DNA technology illustrated in the film already exists.

The Wrong Man, from the History Channel *American Justice* series. This looks at the case of Rubin Hurricane Carter who was convicted of a crime he apparently did not commit. VHS, 50 minutes. Cannot be shipped outside the U.S. and Canada.

11. Glossary

Titles, product names, organizations, and specific military designations are capitalized; common generic and colloquial terms and phrases are not.

amplification	a process used to increase the number of copies of a specific fragment of DNA, usually to make it easy to test and analyze the sample
autoradiograph	a form of X-ray film image of DNA fragments that have been processed to yield certain patterns that can be studied and interpreted
Bp	base pair—pair of nitrogenous bases held together by weak bonds
chromosome	a replicating genetic structure in a cell which contains cellular DNA
cline	a continuous change in a trait, or the frequency of a trait, over space or time
clone	a cell or organism 'copied' from a specific ancestor such that it essentially retains the same genetic makeup as the ancestor
DNA	deoxyribonucleic acid, a molecule that contains encoded genetic information
DNA fingerprint	at its most basic, a DNA profile sufficient for basic identification
DNA profile	a DNA pattern that provides information about a person's genetic makeup that indicates gender, familial relationsp and may include information on racial characteristics and medical propensities
double helix	a geometric shape that is used to symbolize the physical relationship of bonded DNA strands
gene	a fundamental unit of heredity found in a chromosome
genome	the genetic material found in the chromosomes of a particular species of organism
marker	an identifiable location on a chromosome which can be monitored through subsequent generations
mutation	a heritable change in a DNA sequence that is not typical
PCR	polymerase chain reaction
RFLP	restriction fragment length polymorphism
RNA	ribonucleic acid, a chemical constituent of cells that has a structure similar to DNA

Surveillance
Technologies

Miscellaneous

Miscellaneous Surveillance

16

Magnetic

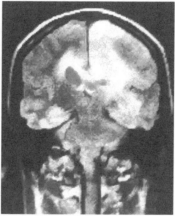

1. Introduction

Magnetic sensing is not as well known as other surveillance technologies (e.g., satellite surveillance or hidden microphones). It is nevertheless an important technology, used in law enforcement, geoscience, traffic analysis, search and salvage, hotel and conference management, and medical research, diagnostics, and treatment. It is also widely used for retail and premises security and to protect materials from theft in libraries.

Magnetic properties are inherent in many synthetic and natural substances. Some materials, like *lodestone,* can come out of the ground with good magnetic properties. Others can be influenced to acquire magnetic properties. When electricity is applied to substances with good conductivity, such as copper or iron, it is possible to turn them into *electromagnets.*

Biological organisms generate natural, low-level electrical currents and thus low-level magnetic activity is associated with living things, sufficient to be detected with sensitive devices. Magnetic surveillance is a versatile technology that can be used to detect both living and nonliving targets.

Very small magnetic sensors can be organized in arrays to sense over larger surface areas.

NASA space technology has contributed to the development of magnetic resonance imaging (MRI) systems which allow the internal structure of a body to be visualized as in this image of the human central nervous system. [NASA news photo, released.]

Another way to sense over a larger area is to make repeated readings at quick intervals and to combine the data. Large, sophisticated magnetic imaging machines are more costly than small discrete sensors and are used mainly in research and medical diagnostics.

Magnetic detection is often used in conjunction with other technologies to provide a 'bigger picture' of the phenomenon being investigated. By combining information from different kinds of sensing devices, a clearer picture of the sensed area may be obtained.

2. Kinds and Variations

The phrase *electromagnetic energy* should not be confused with the separate phenomena of 1) *electricity* and 2) *magnetism*. Electromagnetic energy is a collective term for radiant energies that include light, X-rays, gamma-rays, and radio waves. Electricity and magnetism are not part of the electromagnetic spectrum and, while they are closely related to one another, they are distinct phenomena—a fact that was not empirically established until the early 1800s.

electricity - a fundamental dynamic at the atomic level with attractive and repulsive properties. These attractive/repulsive properties/reactions are too small for us to see directly. However, they can be observed indirectly or mathematically inferred from physics experiments. At a macro level, electricity can be channeled in conductive materials to provide electrical power for activating and running surveillance devices. Electrical activity stimulates attractive properties that can be easily observed in some materials when they are rubbed. Electrical activity always has an associated magnetic field, which is why electricity and magnetism are almost always discussed together.

static electricity (sometimes called *stationary electricity*) - a phenomenon that can be observed in a glass rod, for example, if it is rubbed with silk so that it can briefly pick up small pieces of paper. While the glass can be made to exhibit a 'magnetic attraction' to the pieces of paper, it is not inherently magnetic in the same sense as 'lodestone', which is a natural magnet. In other words, lodestone can exhibit fairly strong magnetic properties without the application of electricity and without being rubbed, but the glass rod needs to be stimulated in order to exhibit attractive properties.

magnetism - a physical phenomenon consisting of an attractive or repulsive force inherent in an object or stimulated by the application of an electric current. Some substances have naturally strong magnetic characteristics and can retain these characteristics for a long time. These are called *permanent magnets*. Others can be stimulated to exhibit strong magnetic forces, for example, an iron nail wrapped with conductive wires can be stimulated by electricity to form an *electromagnet*. Some materials will retain their magnetism after electro-stimulation. Others lose the magnetism and only exhibit magnetic properties when the electricity is applied. Some natural structures, such as nerve impulses, exhibit very low-level electrical activity, and thus generate a low-level magnetic field associated with the electrical activity. Conversely, electrical activity can be stimulated in materials by moving a magnet over them.

All magnets have magnetic poles in pairs. By convention, they are designated as *north* and *south* poles. You cannot have one without the other. If you cut a magnet, each of the pieces will also have poles in pairs. The diagram on the following page illustrates magnetic *lines of force* emanating from the poles of a simple magnetic bar. Iron filings can be sprinkled around a magnet to make it easier to see the general direction of the magnetic lines of force associated with that magnet.

Lines of force will vary, depending on the shape, size, and strength of a magnet and neighboring magnetic forces. These lines are not fixed in one position. If you resprinkle the filings over the same magnet, they will line up in the same basic configuration but will be slightly different each time. The filings themselves have their own attractive, repulsive properties that cause them to orient themselves into lines. The actual force should be visualized as more of a continuous field than a series of lines, more like a light from a floodlight that radiates and extends out indefinitely but becomes gradually fainter as you move farther from the source of the light.

Whenever you have an electrical current running through an object such as an electric wire, there is a magnetic field associated with that current. The direction of that magnetic field, as the *axis of rotation,* is always the same in relation to the flow of the current. Since it's hard for many people to remember the correct direction, Fleming came up with a memory aid based on the right hand. By extending the thumb and fingers as shown in the diagram below, the thumb can be used to represent the current's direction of flow, with the curved fingers then indicating the rotational direction of the magnetic field associated with the current. Fleming's Rule is now commonly known as the *right-hand rule*, a memory aid that it is widely taught in introductory math and science classes.

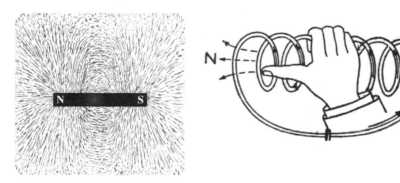

Left: An illustration of a theoretical bar magnet and its associated lines of force. Right: The 'right-hand rule' in which the thumb is used to indicate the direction of an electrical current. The directional rotation of the associated magnetic field is indicated by the direction in which the fingers are curving. [Illustrations ca. 1909 and ca. 1914, public domain by date.]

Polarity is the direction of the magnetic influence of a magnet, typically designated as *north* and *south* as shown in the above diagram. Since *fields of the same polarity repel* each other (try holding two bar magnets together in one direction and then in the opposite direction) and *fields of the opposite polarity attract* one another, polarity can be used to devise compasses, magnetic detection systems, and various types of technical instruments. There are lines of magnetic force emanating from the Earth in somewhat regular patterns in a north-south orientation. The north-seeking needle of a magnetic compass orients itself toward the Earth's magnetic north, near the North Pole, aiding in navigation. Polarity is not always fixed. Over very long periods of time, the Earth's polarity apparently can switch. On a smaller scale, the polarity of an electromagnet can be changed by altering the direction in which current is applied, a concept applied to the design of electric generators.

Magnetic sensing can be used as a *passive technology*, in which the sensor seeks out sources of magnetism without generating a magnetic field. It can also be used as an *active technology*, in which a source of magnetism is intentionally applied or inserted in materials in order to track their presence, movement, or other characteristics.

Various magnetic materials can be used, depending on the application. Metals, ceramics, and certain polymers have sufficient magnetic properties for a variety of magnetic sensing components.

Magnetic Access and Tracking Systems

Scientists have discovered that they can control vibrational frequencies by applying magnetic fields, thus, magnets can used in marking and tracking systems. This principle has been applied to the development of *tags* or *markers* that can be attached to articles and vehicles. A magnetized strip, patch, or tape can be combined in a small package with a resonator to give it a specific signal. The resonant frequency, and thus the signal, can be altered, or deactivated, to change it. This can be used in entry/exit systems. For example, materials can be 'checked out' of a store or library by deactivating them so they don't trip the exit sensor/alarm. The field can be reestablished when an article is returned.

Magnetic Security Systems

Counterfeiting with new high-resolution photocopy machines have become problems in recent years. Currency, stock certificates, sports and music concert tickets, coupons, and various other negotiable items are vulnerable to copy fraud. Magnetic security features have been developed for use in documents and other materials. Since magnetic particles can be quite small and magnetic sensors can be designed to be sensitive to trace amounts of these particles, it is practical to use magnetic properties in document surveillance systems. The magnetic or magnetizable particles can be applied as a coating after a document has been produced or they can be embedded in the fibers or inks during the fabrication of the materials themselves. The magnetic assessment of documents can be done in a number of ways, and may include

- detecting whether or not a magnetic force is present or absent,

- detecting a particular pattern of magnetism, like a magnetic watermark, in order to give the material a group magnetic 'signature', or

- detecting a random or ordered pattern of magnetism that has been mathematically analyzed and coded to produce a unique signature, somewhat like adding a checksum to a block of computer data.

3. Context

Magnetic Surveillance Strips

Magnets are easy to manufacture and control, so they work very well for tracking goods and maintaining inventory. Configurable magnetic strips and buttons are widely used to protect products in retail stores and materials in storehouses and libraries. Because magnets can be made small and inconspicuous, they have many surveillance applications.

Magnetic strips are also easy to attach to cards and keys and thus are used for many different kinds of access control through bank cards, credit cards, door control systems, computer access control, and hotel, trade show, and office key systems.

Computer floppy discs have magnetic surfaces in which the magnetic bits are arranged and rearranged to store data and instructions. The same process can be used to program and reprogram access cards, data cards, and electronic keys. This is an important aspect of surveillance that is discussed further in later sections because it means that an electronic data record of all aspects of the use of a data card or key can be tracked and stored in computer memory.

Access Devices

Magnetic strips provide a relatively easy way to design access switches for buildings and vehicles. Most of these switches have two separate components that slide or swing past one another. Depending on the system, when magnetic contact is broken, or if the state of the sensor changes, the device can trigger a light, an alarm, or an electrical impulse. These devices can be placed on doors, windows, safes, garages, and on appliances. Thus, if a door, window or lid is opened or an appliance is moved or lifted, the sensor reacts by sounding an alarm. Magnetic reed switches, which are commonly used for windows and doors, can often be wired in series. Some use radio frequencies to alert a central console if the magnetic sensor is triggered.

Magnetic Cards and Keys

Magnetic strips can be embedded in cards and keys to store data that can be read or written. When combined with 'smart card' technology, which may include a tiny programmable circuit board on the chip, the magnetic strip becomes an accessory to the programmable circuit.

Magnetic cards and keys are used to access accounts, vending machines, phones, doors, vehicles, and anything else that can be configured to accept card data. They can also be used to store data about individual employees, conference attendees, or clients at an amusement park or casino. They can serve as a reference for medical information or medications or as identification cards. They may also be combined with other technologies to track a person's movements.

Detection of Magnetic Materials

Many materials exhibit measurable natural magnetism, which makes it possible to locate them with magnetic sensors. A simple magnet can itself be a sensor. Many coin machines distinguish one coin from another based on attributes such as size, weight, and magnetic properties.

Magnetic sensors can be installed on bridges and causeways to monitor traffic patterns such as vehicle volume and speed.

Magnetic detectors are often used to locate hazardous, unexploded ordnances (military supplies) such as land mines and bombs. They may also help locate terrorist bombs in urban areas or on public transportation systems. Occasionally they are used to locate or identify explosive devices created by teenagers who have undertaken dangerous projects without realizing their full potential for harm. Thus, magnetic detectors are frequently used in law enforcement, public safety, and military contexts. They are also helpful in the geosciences, for studying terrain and various geophysical phenomena. Because they can help assess geophysical features, they are further used by forensic scientists and archaeologists to assess underground structures, terrain, and anomalies that may indicate bones, graves, artifacts, or other objects of interest.

Small-scale magnetic detectors have some limitations when used to sense weak magnetic differences. This is partly because portable systems are subject to interference from the Earth, which generates its own magnetic field. It is difficult to shield a portable detector from this interference. Thus, the data from small-scale magnetic detectors are rarely as distinct and clear as data from large-scale heavily shielded systems. Thus, small-scale magnetic detectors are often used in conjunction with other probes such as ground-penetrating radar and seismic detectors.

Data from multiple sensors are especially important in situations where incorrect readings could endanger lives. Thus, magnetic sensors may be used with acoustic and X-ray sensors in situations where multiple explosives are present. By cross-referencing the data from different systems it is sometimes easier to classify, confirm, and locate hazardous materials.

Magnetic sensors and electromagnetic sensors used together can be quite good at detecting certain classes of explosives, sometimes with a 90% detection rate. The problem is that they will also detect and flag many false alarms that can confuse the search. Thus, data are sometimes computer processed to try to reduce the number of false alarms.

Magnetic Resonance Imaging

Not all magnetic detectors are limited in their ability to detect objects of interest. With the right electronics, shielding, and equipment, extremely sensitive magnetic devices can be built.

Magnetic resonance imaging (MRI) technology is an advanced and expensive technology that is currently used mainly in the field of medical diagnostics and treatment monitoring, although it has also had some interesting archaeological applications, such as imaging the inside of a mummified corpse.

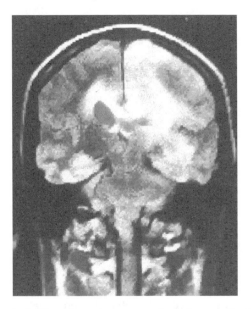 

Left: There have been a number of magnetic detection and satellite imaging technologies developed through various NASA space programs. Satellite imaging developed at the Stennis Space Center in the 1980s has contributed to Magnetic Resonance Imaging (MRI), which is now extensively used in medical diagnostics and patient monitoring. Right: Jerry Prince and Nael Osman, the engineers who invented the HARP-MRI system (shown in the background). [NASA 1994 news photo; Johns Hopkins University 1999 news photo by Keith Weller, released.]

MRI pictures can take some time to process, so engineers have been seeking ways to provide faster results. The images are especially suitable for assessing conditions such as heart attacks. The Harmonic Phase Magnetic Resonance Imaging (HARP-MRI) system is a step toward faster imaging, that can be more specifically used to measure cardiac function. The HARP-MRI system was developed at Johns Hopkins University and allows data to be processed in minutes rather than hours. It is the goal of the designers to continue to develop the technology until realtime images are possible, in which case a whole new dimension in imaging and diagnostics will open up.

Magnetoencephalography (MEG) is a subset of magnetic imaging that detects and images the magnetic activity of a brain. MEG differs from MRI in that it provides a near-realtime picture of the magnetic activity of a brain, whereas MRI provides an anatomical picture of the inner structure of the organ. The *Massachusetts Institute of Technology* (MIT) announced in 1997 that they had installed a MEG system built by the *Kanazawa Institute of Technology* (KIT).

In Section 2, the *right-hand rule* was introduced as a way to relate magnetism to the direction of the flow of electrical current in an object. Magnetic flow is also associated with biological nervous systems. The direction of current through neurons generates an associated magnetic field with certain predictable properties. A *biomagnetometer* is a sophisticated piece of equipment designed to detect and image this faint but distinctive magnetic field, allowing us to view a 'picture' of an organism's nervous system activity. The magnetic detection coils used in MEG machines are closely aligned and bathed in liquid helium to bring them down to extremely cold superconducting temperatures. Superconducting quantum interference devices (SQUIDs) make it possible to create the extremely sensitive magnetometers needed to monitor fine resolution activity as is found in the brain.

In operation, the coils are placed very close to the surface of the organism to record the electrical activity. The brain's electrical activity is detected by the coils by inducing a current which can then be converted into a printed or displayed image. The sensitivity of these systems is amazing. The coils can respond to magnetic energy as small as a single quantum. Because of this high sensitivity, the mechanism can only be used in an environment that is highly shielded—sources of interference such as the Earth's magnetic field, which is a billion times stronger than the signal from the brain that is being measured must be screened out.

MEG technology is used in several fields of study, including mapping the sensory and motor regions of the brain. By stimulating various parts of a body, the resulting magnetic activity in the corresponding part of the brain can be observed. At MIT, MEG technology is being used to study relationships between cognitive neuroscience and linguistics.

In 2004, Hitachi, Ltd., announced that they had used SQUID technology to develop a magneto-cardiograph system for screening for heart disease. The system can detect very weak magnetic fields "formed on the surface of the body as a results of cardiac activity". As the technology for magnetic resonance advances, less expensive machines that can be used for a wider variety of applications may be possible.

Magnetic Sensing and Common Sense

There are many useful ways to harness magnetic energy, but there are also tragic stories associated with magnetic detection of bombs and miness. Accidents occur even when the hazardous nature of an object is suspected. A moment of inattention seems to overcome people at times, with irrevocable results. One example is a police officer with bomb detection experience who was called to investigate a suspicious device found in a car. The object, suspected of being a bomb, was removed from the car and laid on the driveway. For some reason the officer picked up the box, which exploded, with tragic results. Another example is a soldier with land mine-detection experience working with a colleague and a mine-sniffing dog. The dog alerted to the presence of the mine and the service member, to the horror of his colleague, touched the mine, which exploded. In retrospect we can only assume these explosives professionals had a momentary lapse of reason. No matter which detector is used, whether magnetic, electrical, or biological, if the person handling the technology doesn't exercise a certain amount of care and attention at all times, the results can be tragic. Since magnetic technologies are often used to detect explosives, suspicious devices should never be handled until confirmatory equipment and procedures are used.

4. Origins and Evolution

Some of the origins of electricity and magnetism have already been covered in the Audio, Visual, and Radio Surveillance chapters and are not repeated here. You are encouraged to cross-reference the early discoveries described in other chapters, especially the Radio Surveillance chapter.

Thousands of years ago, the properties of highly magnetic materials were observed by the people of China and Greece. Magnetite, an oxide of iron, is a natural material that is known as *lodestone* when it comes out of the ground with natural magnetic properties. Magnetite can be readily magnetized. Magnetite was used in antiquity to make 'magic stones' and inspired the invention of compasses.

The attractive properties of amber and glass, when rubbed with silk or cotton, respectively, have been known for a long time and the phenomenon is colloquially known as *static electricity*. For hundreds of years, however, it was not known if magnetism and static electricity were the same or different phenomena.

In 1600, William Gilbert (1544-1603), an English physicist and physician, published "De magnete" based on his studies of magnetism and its characteristics. He made important distinctions between the attractive properties of amber and those of lodestone, a natural magnet. He further established the idea of the Earth having a large magnetic field.

Benjamin Franklin (1706-1790), American statesman and scientist, introduced the concept of positive and negative charges based upon his experiments with electricity. Franklin's experiments also suggested that *charge* is not something that is created at the point where materials are rubbed and 'electrified' but rather that a transfer of charge from one body to another takes place.

The 1700s and 1800s - Progress in Understanding Magnetism

In a relatively short time, beginning in the second half of the 18th century, significant progress in our theoretical and experimental understanding of magnetism was made by a number of gifted European physicists and mathematicians.

John Michell (1724-1793), an English geologist and cleric, invented a device called a *torsion balance* which he used to study gravitational attraction. In the mid-1700s, he demonstrated that magnetic poles exert attractive and repulsive forces on one another. Furthermore, he described a mathematical relationship between these forces and the distance between the poles. After his death, the apparatus came into the hands of Henry Cavendish, who continued experimenting with the device.

Joseph Priestley (1733-1804), a British chemist, provided experimental confirmation, around 1766, of the law that the force between electric charges varies inversely with the square of the distance between the charges.

Charles Augustin de Coulomb (1736-1806), a French physicist and engineer, independently confirmed various aspects of electricity and magnetism that had been studied by Michell and Priestley, by experimentally establishing the nature of the force between charges. He also invented a *torsion balance* apparatus, in 1777, consisting of a fine silver torsion wire suspended from the top of a tube with a horizontal carrier, with bodies of known electric charges at each end of the carrier suspended at the base of the wire. The whole thing was placed inside a protective housing to screen out outside influences, like wind. When the mechanism was displaced by attractive/repulsive forces, the amount of deflection could be observed through the glass housing. An arrow indicated the degree of torsion on a fixed scale. De Coulomb used the balance to accurately measure the force exerted by electrical charges and detailed his findings in a 1777 paper submitted to the Académie des Sciences and a series of memoirs submitted in the 1780s.

In 1819, Hans Christian Ørsted (1777-1851), a Danish physicist and educator, demonstrated to his students the effects of stimulating a magnetic needle with an electric current, an important milestone that helped us understand the relationship between electricity and magnetism. A unit for magnetic intensity was named the *oersted* in his honor.

In 1820, André-Marie Ampère (1775-1836) followed up on the research of Ørsted, trying to formulate a theory of electricity and magnetism that would bring together the two phenomena. He suggested that the electrical activities in atoms might be associated with magnetic fields. Ampère described electrodynamic forces in mathematical terms and his writings aided the subsequent development of magnet moving-coil instruments. In describing the relationship between the direction of magnetic rotation and current, he devised the 'left-hand rule' or Ampère's rule, which was originally based on the concept of a swimmer turning his head left while swimming in the direction of 'the current'. Early scientists weren't sure whether the phenomenon had an inherently left or right orientation. Later, Fleming contributed the 'right-hand rule' or Fleming's Rule, which is the one commonly used today.

While Ørsted and Ampère were adding to our knowledge of the relationship of electricity and magnetism, Michael Faraday (1791-1867), an English chemist and physicist, was conducting important experiments and creating the *Faraday magnet*, one of the earliest electromagnets. In 1831, he made a historic entry in his journal linking electricity and magnetism. Faraday put his newfound knowledge to practical use by creating a *dynamo*, which is essentially a pioneer version of the electrical generator that brought about the industrial revolution.

Now that some of the basic theory and mathematics related to magnetism had been established, inventors began applying the information to the development of new scientific instruments. Wilhelm Eduard Weber (1804-1891), a German physicist and associate of Karl Gauss, invented the electrodynamometer and developed sensitive *magnetometers* to detect weak magnetic fields.

Pioneer Work in Magnetic Resonance

There were many advances in detection, measurement, and imaging technologies in the first half of the 20th century. Many historic 'firsts' occurred at this time in theoretical and experimental physics, as well.

Not long after the concepts of superconductivity were developed, Walther Meissner (1882-1974) and Robert Ochsenfeld, German physicists, discovered the *Meissner effect,* also called *diamagnetism,* in 1933, whereby a superconductor expelled a magnetic field when cooling was applied at appropriate times. This opened up the possibility of magnetic 'levitation' which could be applied to technologies such as magnetic trains and magnetic bearings.

During the 1930s, display technologies began to improve as well, with cathode-ray tube (CRT) technologies leading to radar scopes, televisions and, eventually, computer monitors.

Scientists set the scene, at this time, for future magnetic resonance imaging by describing the phenomenon of nuclear magnetic resonance (NMR).

Isidor Isaac Rabi (1898-1988), from Rymanów, Poland (then Austro-Hungary), was brought to the United States the year after his birth by his family. He went on to study Chemistry at Cornel, later researching the magnetic properties of crystals for his PhD. He then studied in Europe for two years, coming into contact with some of the greatest pioneers in quantum physics, including Bohr, Schrödinger, Pauli, and Heisenberg. Upon his return to the U.S., he became a lecturer in physics at Columbia and carried out research on magnetic resonance based on molecular-ray models. By the late 1930s, Rabi had observed the absorption of radio frequencies by atomic nuclei and demonstrated the fundamentals of nuclear magnetic resonance (NMR). For his discoveries, he was awarded the Nobel Prize in physics in 1944.

In the mid-1940s, Felix Bloch (1905-1983) and Edward Mills Purcell (1912-1997) observed emission and absorption characteristics in nuclear magnetic phenomena and developed a nuclear induction method of measuring the magnetic fields of atomic nuclei. This was a remarkable achievement, considering the minute forces involved. For their work, they received a Nobel Prize in physics in 1952.

The Evolution of Electronics

The development of the computer was important to magnetic surveillance for two reasons. First, computing technology required memory to carry out more complex functions, leading to the evolution of magnetic memory, magnetic computer tapes, and magnetic discs. Second, the development of computer networks made it possible to process information stored on magnetic storage devices to develop sophisticated transaction, database, and surveillance systems.

Left: A magnetic core memory unit in a research laboratory in 1954 shows strides in the evolution of computers. Core memory provided important support to processing functions and later spawned other kinds of magnetic storage devices. Right: The inside of an IBM/UNIVAC computer, one of the first large-scale computer systems, at the *Lawrence Livermore National Lab* (LLNL). LLNL took delivery of the UNIVAC (Universal Automatic Computer) in 1952, a year after the Lab was founded. [NASA/GRC news photo by Walton; LLNL 1957 historical news photo, released.]

In 1947, the transistor was invented, which led to dramatic improvements in electronics, particularly in the areas of miniaturization and functional integration. However, it would take about a decade for transistor technology to become firmly established. In the meantime, vacuum-tube computers continued to evolve. The ENIAC and the UNIVAC weren't much more than advanced calculators, by current standards, but these advanced calculating capabilities were important, providing the basis for new discoveries in mathematics, physics, astronomy, engineering, and instrumentation. Within twenty years, full-scale software-programmable computing systems with magnetic memory and magnetic tape storage capabilities would become well established.

The development of *magnetic core memory* opened up a new world of computing capabilities. Core memory made it possible to develop reusable code, new programming languages, and compilers, innovations that were impractical to implement (or unreliable) on the earliest systems. Magnetic memory also contributed to the later evolution of miniature magnetic strip cards.

New miniaturization and memory storage technologies were well underway by the 1950s. They didn't have an immediate commercial impact, because of the huge financial investments that had been made in the first room-sized computers but, by the 1960s, the microelectronics industry was beginning to reach new markets and the limited-processing-capacity behemoth systems rapidly became obsolete.

As soon as magnetic storage systems were established in the computer industry, inventors sought other ways to use magnetic media to store information. Magnetic stripe cards began to emerge in the early 1960s and were almost immediately put to use in transit system checkpoints. Some cards were read-only—the same information was read from the card each time. Others were read-write systems, in which the data could be changed. This made it convenient to associate the data with a monetary value, such as is done with photocopy, phone, and public transit cards. The value on the card would be reduced each time it was used and could, in many cases, be 'recharged'.

It was not hardware alone that improved the state of the computer arts in the 1960s. Mathematics and software design evolved as well. Thus, both hardware and software contributed to steady improvements in the speed and efficiency of computer systems, and improved the process of creating magnetic resonance scan images. MRIs that took almost a day to scan on older systems could now be scanned in a few hours. Eventually, scanning speeds would improve to less than an hour and, by the year 2000, certain offshoots of magnetic imaging would be near-realtime.

Transaction Cards

Retailers have been extending credit to customers for hundreds of years. Trading goods stores and corner grocery stores commonly carried small accounts for good customers. By the 1920s, some companies were starting to create company charge cards that were engraved to protect them from forgery. These were the forerunners to the charge and credit cards that emerged in the 1950s. Public transit cards were in use by the 1960s, and automatic teller cards became prevalent in the 1970s and 1980s.

Standardization of magnetic stripe cards began to be established in the early 1970s. Up to that time, transit cards, for example, used proprietary data storage schemes. With the standardization of cards, broader use for other applications were possible and magnetic stripes became important for transactions associated with travel. By the 1980s, almost all transaction cards in western countries had a magnetic strip embedded into the surface.

By the mid-1980s, cards that could be used to access telephone service were developed and by the 1990s 'phone cards' with prepaid long-distance access available through many retail outlets.

Eventually the concept was extended to entry/exit cards and badges. By adding magnetic strips to employee cards, they could be designed to allow access to buildings or certain parts of buildings. Strips on conference badges could serve the same purpose, or could be coded to allow access to the specific seminars for which the conference attendee had paid.

Smart Cards

The real key to the evolution of 'smart cards' containing miniature circuits, was the invention of transistors in 1947 and the microcomputer revolution of the 1970s. These key technologies led to ultra-tiny computers capable of data processing that could store information on magnetic strips embedded in the cards. In 1974, the concept of the Smart Card was patented by Kunitaka Arimura and, in 1974, Roland Moreno also patented the idea. It was not long before companies saw their potential and began licensing the technology.

It was clear that widespread use of magnetic and computer cards was going to require some standardization, so that people could use their cards in a variety of locations when they were traveling. By the late 1980s, ISO standards for cards were being developed. By the early 2000s, faster and more sophisticated card reading and reprogramming machines were developed.

Space Science

Research and development in one industry often catalyzes inventions in other industries. This has been true of many aspects of the space program. Space is a difficult environment in which to live and work and large amounts of radiation that don't reach Earth's surface are prevalent in space. In order to put vehicles, sensing instruments, and people into space, new materials that can deflect heat and radiation, new protection gels, fine instrumentation, and a large variety of sensors were developed. The polymers and devices originally invented for space or extraplanetary exploration were later used to develop systems that have been useful in industrial and medical environments.

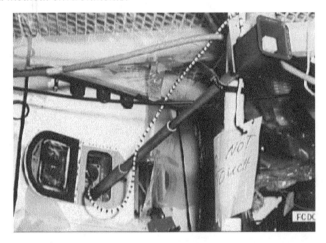

Shown here are the sensors and boom from the tri-axis magnetometer carried on board the Gemini 12 space flight in 1966. [NASA/JSC 1966 historic news photo, released.]

Left: This is a portable magnetometer designed to be used by the Apollo 14 crew to measure variations in the Moon's magnetic field at several different points. The device mounted on the tripod is a flux-gate magnetometer sensor head connected with a 50-foot flat ribbon cable that interfaces with an electronic data package. Right: A Lunar Portable Magnetometer (LPM) mounted on the Lunar Roving Vehicle on the Apollo 16 mission. [NASA/JSC 1970 and 1972 news photos, released.]

In 1966, the Gemini 12 space project included a *tri-axis magnetometer* (MSC-3) that was designed to monitor the direction and amplitude of the Earth's magnetic field relative to the spacecraft. Room on a space flight is very limited, so the instrument has to be designed to be light, portable, and to consume minimal power. It also has to be more sensitive than Earth-based magnetometers, since the magnetic forces on the Moon are only a fraction of those found on Earth.

Several Apollo space missions to the moon took measurements of the Moon's magnetic field using a Lunar Surface Magnetometer that was part of the ALSEP experimental package. The Moon's magnetic field is influenced not just by the Moon itself but also by external sources such as the Sun and the Earth. By taking a series of measurements over the time it takes the Moon to move through its orbits, it is possible to better distinguish the Moon's field from external forces.

In August 1999, work began on a magnetometer (MAG) calibration sequence that would be used on the Mars Global Surveyor spacecraft to help 1) characterize the Surveyor's magnetic signature and 2) to execute a series of solar array motions on the dark side (night side) of the orbit.

Magnetometers

Magnetometers are specialized instruments used for magnetic detection and imaging.

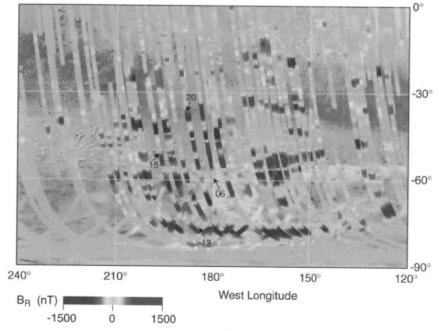

Magnetic mapping using data from a magnetometer. This is a magnetic map of a portion of Mars, near the Terra Cimmeria and Terra Sirenum regions. The pattern of the stripes is quite similar to patterns of magnetic mapping of the Earth's crust at the mid-oceanic ridges, indicating that the crust may have shifted in much the same way as the Earth's and that Mars may once have had a global magnetic pattern similar to what Earth has now. [NASA March 1999 news photo, released.]

Early magnetometers were bulky, which limited their use for field work or commercial surveillance. By the 1960s, components developed for the space program contributed to the science of portable, compact magnetometers. By the 1980s, miniaturization and improvements in electronics made it possible to develop more efficient, smaller systems, a trend that continued into the 1990s. By the mid-1990s, many magnetometers included data link capabilities so computers could be used to display results or to log activities. Thus, smaller, lower-power-consumption systems, such as demining magnetometers, magnetometric buoys, portable geoscience sensors, and commercial sensors, could now be constructed.

This magnetic suspension demonstration was part of the EAA AirVenture '99 exhibit in Wisconsin. If you look closely, you can see that the small object on the top of the cabinet on the right is suspended magnetically in the air without physical contact with the device underneath it. [NASA/Langley Research Center news photo by Donna Bushman, released.]

In 1996, UNICEF reported that there were about 110 million land mines distributed throughout more than sixty countries, many of them activated during conflicts that had long since been settled. These mines are often unintentionally triggered by vehicles, animals, and children's playtoys, blowing off limbs and sometimes causing death to innocent victims. Magnetic sensors are one of the means by which mine-detectors locate live mines.

Magnetic Tags

With increasing miniaturization, it became possible to create small-scale, cost-effective magnetic sensing instruments that could be used for inventory management, theft-deterrence, and wildlife-tracking. Libraries installed the systems to protect against book theft, with the practice becoming commonplace by the late 1980s. Retail stores began to adopt the systems for higher-priced items in the late 1980s and early 1990s and for smaller, easily shoplifted items like portable electronics and music CDs by the mid-1990s. Also, in the mid-1990s, magnetic systems for tracking wildlife were becoming practical. In conjunction with GPS technology and depth-recorders, tri-axial magnetometers could be used to track the movements and behaviors of marine mammals such as seals and whales.

Data Cards

By the 1990s, casinos, hotels, clubs, rental depots, and other businesses that catered to the public, began using magnetic cards and keys that would allow them to store and analyze the information provided when clients used the cards in various gambling and vending machines and in doors and vehicles. Soon universities and office complexes were installing magnetic locks and card readers as access control devices. Student cards could be designed so they not only opened doors, but could be used in vending machines, as well. By the end of the decade, the hardware to install these systems and the database software to track and log the activities of cardholders had become quite sophisticated.

Fundamental Research

Understanding magnetism at an atomic level enables us to influence the behavior of electrons and to use these atomic behaviors in practical applications. In 2005, Japanese and U.S. researchers described a polycrystalline substance with an interesting triangular arrangement of atoms that appears to prevent alignment of magnetic spins (the characteristic of electrons that produces magnetism), resulting in a 'liquid' magnetic state in which the spins fluctuate in a disorderly arrangement that does not produce an overall magnetic force. This may relate to the way electrons flow without resistance in superconducting materials [NIST Sept. 2005 News Release]. In 2006, scientists working at the National Institute of Standards and Technology (NIST) Center for Neutron Research, in collaboration with the University of Tennessee and Oak Ridge National Laboaratory, reported that magnetic fluctuations—in particular, peak magnetic excitations—were a key mechanism in pairing electrons and enabling resistance-free movement of electric current in high-temperature superconductors. Electrons are fundamental 'units' of energy and the ability to control them opens up many possibilities for new technologies that could work with conventional superconductors, with superconducting digital communications routers being one possible application.

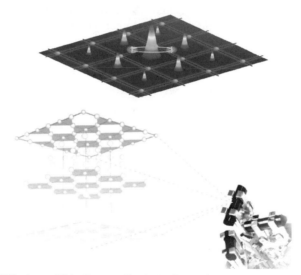

Magnetism and Electrons. This diagram illustrates how stimulation from a neutron probe can cause electrons in a superconducting medium to pair at the point of magnetic excitation. [NIST news photo, 2006, released.]

5. Description and Functions

Measures of Magnetic Field Strength

Some sensors measure the presence or absence of a magnetic field, some measure relative changes in a magnetic field over time, some measure the strength of a field in relation to other nearby objects, and some measure it on a standardized scale.

A *magnetometer* is a device that measures the strength of magnets and magnetic fields. It may also ascertain the direction and origin of the field in relation to the measuring device. A magnetometer is called a *Gaussmeter* in honor of Johann Karl Friedrich Gauss (1777-1855), a German mathematician and astronomer who investigated magnetism in conjunction with W.

Weber in 1831.

Several different unit systems are used to express the strength of a magnetic field and different aspects of magnetic flux density, so the conversions can be somewhat technical. This list is simplified to give an introductory idea of the kinds of units that are used:

gauss (G, Gs) - The Earth's magnetic field is about .25 to .5 gauss. Gauss tends to be used to measure magnetic flux density and magnetic induction (B). One gauss corresponds to 10^{-4} tesla.

gamma (λ) - One gamma corresponds to 10^{-9} tesla. This unit is useful in situations where very weak magnetic fields are being measured, as might be found on asteroids or moons. It is often used to express quantities in magnetic survey maps.

tesla (T) - The tesla, named after the physicist and inventor Nikola Tesla, tends to be used to measure magnetic flux density and magnetic induction. There are other units, including webers (W), ørsteds (Oe), and maxwells, that are applied to various aspects of magnetism. The *Resources* section at the end of this chapter lists online resources with tables of different magnetic measures and conversions.

A *geomagnetic field* is the overall magnetic field of a planet's or satellite's surface. It is comprised of the celestial body's internal field together with the magnetic fields within its atmosphere (if there is one). For reference, the Earth's magnetic field is around 300,000 gammas. In the Apollo space missions, the Moon's magnetic field was measured with sensitive lunar surface instruments and found to vary from about 6 gammas to about 313 gammas, depending upon where and when the measurements were taken.

Magnetic sensing may at times be combined with electrical assessments in a *resistivity survey*, which measures the relative differences in the electrical resistivity in materials such as soil. It may also be used in conjunction with ground-penetrating radar, a means of using radio waves to assess the relative densities of a stretch of ground or water, and seismic sensors, which measure pressure waves conducted through materials.

Magnetic Cards

Access and transaction cards are described here, while smart cards with processing capabilities are covered further in the Computer Surveillance chapter.

Access/transaction cards tend to come in three basic types:

proximity cards - These are cards that do not need to touch the sensor. Proximity cards are generally the simplest card and are used in situations where information is not needed. A proximity card is like a trip-switch. It can trigger a sensor so that a circuit can be turned off or a door or cash mechanism opened or closed. One of the advantages of proximity cards is that the contact surface does not readily get scratched or worn.

contact cards - These are cards that are held against a surface to be sensed or scanned. For some contact cards, the position of contact is standardized. They are usually used in access systems and typically do not convey much information, if any. They are similar to proximity cards.

swipe cards - These are cards that are either 'swiped' or pulled through a reader, or which are placed in a reader that has a built-in scanning 'swipe' mechanism to read and/or write data on the card. Swipe strips are commonly included on ATM cards, credit cards, and employee cards. There may be a significant amount of information stored on a swipe card, including account information, name and address, time, date, logging information, etc.

ID card printing machines may use a Magnetic Stripe Encoding Module (MSEM). Many of these card printing systems conform to ANSI/ISO standards. Magnetic encoders for standard cards fall into two general categories, so it is important to select the right kind of card to match the encoding scheme. The two common categories are

> *low coercitivity* - These data systems are common on transaction cards and are often brown in color. This is a less expensive method of encoding and is suitable for low to medium-high usage cards.

> *high coercitivity* - These data systems are more commonly used for higher-security applications and may be black in color. This kind of encoding may cost a little more, but it is generally more robust.

Most card readers are designed to read both types of cards.

The chief limitations of all magnetic cards are 1) the potential for loss or theft and 2) the potential for the magnetic data to be disturbed by proximity to other magnetic sources such as other magnetic cards (never carry them with the strips facing one another), magnets, computer monitors, or strong electrical sources.

EAS

Electronic Access Systems are those that use electronic or magnetic detection systems to sense or control in-and-out traffic to restricted areas. Many of them are more specifically *Electronic Exit Monitoring* (EEM) systems that are used widely in libraries and retail stores where managers are concerned about monitoring what people take out than what they bring in. Most people in the industry use EAS as an abbreviation for the more generic phrase *Electronic Article Surveillance* to describe the process of tagging individual articles for tracking or entry/exit monitoring.

For the purposes of this text, *Electronic Access Surveillance* is used to mean the use of electronic cards, keys, or door/window sensors to track or detect entry and/or exit. *Electronic Article Surveillance* is used to mean articles that have magnetic sensors attached or incorporated into their design that are subsequently detected or tracked through compatible sensors. There is sometimes overlap between these systems.

Libraries are big users of Electronic Article Surveillance (EAS). The theft of books by professors and students is apparently quite prevalent. Many libraries have rare or expensive books that are difficult or impossible to replace. Article surveillance helps cut down theft.

Retail stores now commonly have electronic and magnetic sensors in their doorways or in other areas where there are articles that are vulnerable to theft. These vary in design and in the technologies that are used, but many of them are like the electronic gates in libraries that sense magnetic disturbances and signal an alert or alarm when an active magnetic source is encountered.

There are two common tagging systems:

> *tags that can be attached/removed* - Removable tags are commonly used for more expensive items. Most of them remain activated all the time so that if an item is taken through a checkpoint, the alarm is sounded. If an article with the tag is purchased or legitimately removed, the tag itself is removed so that it doesn't trigger the checkpoint sensing device. These tags tend to be larger, palm-sized devices.

tags that can be activated/deactivated - Configurable tags are commonly embedded in an article or attached with a strip or sticker. These can be activated or deactivated as needed, usually by 'swiping' them through a source of magnetism/electricity. The checkpoint sensing device is usually set to sound an alarm if it senses an activated tag. These tags can be very small and may be hidden inside the item, or embedded in its design to prevent the patron from tampering with the security mechanism. Small security strips are sometimes called 'tattle tapes'.

Typically, products embedded with a magnetic device are 'swiped' to deactivate the sensor. Other 'high-ticket' items such as fur coats and diamond jewelry may have larger security tags that have to be removed before the customer can go through the exit without triggering an alarm. Radio frequency technologies are sometimes used together with magnetic fields to track removable-tag systems. The alarm is not necessarily heard by the customer or client. It may sound in a security area from which detectives or other personnel are dispatched to apprehend the person stealing or moving objects without proper authorization.

Types of EAS Systems

Not all access/article surveillance systems are based directly on magnetic devices, some use radio signals or acoustic signals, and many use a combination of these technologies. However, keep in mind that electricity and magnetism always occur together. Thus, access systems that rely on the generation of an electric current will always have an associated magnetic field, and many security systems rely indirectly on magnetic forces even if physical magnets are not present in the system.

The type of sensor and the sensitivity of the sensor will dictate how far apart they can be spaced in an entrance or exit checkpoint system. Low-sensitivity sensors may require more than one 'gate' as are sometimes seen at wide entrances in retail stores in shopping malls.

Most EAS systems work on the same general principles. One of the gates is a transmitter, one is a receiver, and the tag is sensitive to the emissions specific to the transmitter. Gates for wide entrances are sometimes paired or combined, but the principles are the same. The receiver expects to receive a certain signal to indicate a deactivated or activated tag, depending upon the system. Three common EAS schemes used in retail and library systems are

Acousto-magnetic systems - three-part active systems that consist of a transmitter, a receiver, and security tags. They work by transmitting a radio frequency pulse. When someone walks through the checkpoint, this pulse is received by a security tag, which responds with a single-frequency pulse as the transmitting pulse ends (so the signals don't clobber one another), which is then detected by the receiver. If the received pulse matches the frequency and timing characteristics expected (or doesn't match it, depending on the system), an alert is sounded.

Electromagnetic systems - three-part active systems that consist of a transmitter, a receiver, and security tags. The transmitter transmits a continuously varying field between the transmitter and receiver. Since there are always magnetic fields associated with electrical fields, a magnetic field is created between the transmitter and receiver with a shifting polarity. The strength of the field also varies. The magnetic field influences the field of a tag, as a person walks through the checkpoint, generating a signal that can be checked against the transmitter signal. If the received signal from the tag matches the harmonics and strength expected by the receiver from an active tag, an alert is sounded.

Microwave systems are three-part active systems that consist of a transmitter, a detector, and transceiving security tags. The transmitter generates modulated signals, usually over two frequency ranges. The signal interacts with the tag, which is a small palm-sized microwave transceiver, and the tag processes the signal, emitting its own characteristic signal when someone walks through the checkpoint. If the signal received by the detector is a specific frequency, indicating an non-deactivated tag, an alert is sounded.

It is estimated that as of 2000, there are close to one million EAS systems installed around the world.

Magnetic Surveying

A magnetic survey is a technique for taking reference measurements of the Earth's magnetic field so that magnetic structures can be distinguished from the background magnetism. A *proton magnetometer* may be used for this task.

Many human-made objects are fairly highly magnetic in relation to soil, water, plants, and other materials that might be present at a site that is being surveilled. Appliances, coffin hardware, ships, vehicles, jewelry, kiln-heated bricks and pottery, all have measurable magnetic properties that can be detected with appropriate tools. This makes magnetic surveys valuable for forensics, mine detection, and archaeological exploration, as well as earth sciences research.

6. Applications

Access Monitoring and Data Tracking

The use of cards with magnetic strips to access automatic teller/transaction machines (ATMs) is probably one of the most familiar uses of magnetic detection and data management. The same principles are used in credit cards, phone cards, and cards designed for vending machines. Similar systems can also be used for doors and restricted access areas in labs, industrial facilities, and offices. When linked to a computer database, more sophisticated data can be stored and processed, including frequency of use, times used, locations, etc.

Left: Point-of-purchase machines now commonly accept debit and credit cards with magnetic 'stripes' for transactions at numerous locations including banks, gas stations, supermarkets, and department stores. Middle: Cards generally fall into two categories: low coercitivity (top) and high coercitivity (middle and bottom), depending on the level of security desired. These are sometimes distinguished by brown or black magnetic surfaces. Right: Magnetic cards and keys are also used to open doors to buildings, offices, hotel rooms, and other restricted access areas. [Classic Concepts ©2000 photos, used with permission.]

Archaeology

In archaeology, magnetic detection is used along with resistivity surveys, and pulse radar surveys, to assess research locations and to explore data from surveys. *Archaeomagnetic dating* is a technique based on assessing the variation of the Earth's magnetic field as it changes through time. The Earth, as it spins, has a rotational wobble that repeats over a long period of time. Knowledge of this variation can help assess many types of artifacts such as historic kilns and other metallic and metal-smelting structures.

Detection of Land Mines and Bomblets

By the mid-1990s, it was estimated that more than 100 million active land mines were spread around the world. In addition to this, the U.S. forces dropped millions of 'bomblets' over the Laotian region during the conflicts in Vietnam, not all of which exploded on impact.

The detection and neutralization of land mines and bomblets is a continuing problem. Since many technologies are used to detect mines, it has also been mentioned in other chapters. Mine-detection is a process that involves determining a suspect region, finding individual mines, and dealing with the mines once they have been detected.

Demining can be especially challenging if it has to be done in forests or swampy areas, where there are many obstacles, objects, or conductive surfaces to interfere with demining equipment or detection devices. Magnetic detection is one of the technologies that has aided in this process. When added to the arsenal of probing sticks, bulldozers, ground-penetrating radars, sniffing dogs, and electronic sniffers, it provides one more tool to protect innocent civilians and armed service personnel from disfigurement or death.

One demining device of interest is the Meandering Winding Magnetometer (MWM) that was devised at MIT by James R. Melcher and his colleagues. The device makes it possible to sense the approximate size, shape, depth, and sometimes even the composition, of a buried metal object. MIT currently sponsors the Humanitarian Demining Project which has received funding by the U.S. *Department of Defense.*

Detection of Submarines

Ships and submarines alter the magnetic fields of the water in which they are located. Submarine-sensing magnetometers can be designed to indicate the presence of surface or submerged vessels. Using a magnetic submarine sensor from a ship or another sub would be difficult, since the interference from the deploying vessel would obscure the readings from other vessels. However, it is practical to consider the use of submarine sensors from helicopters and autonomous or remotely operated aerial vehicles. A magnetometer can be attached to a 'boom' in much the same way as a sonar towfish is used to troll waters looking for submerged vessels. It is not uncommon for sonar and a marine magnetometer to be used together. It can also be attached to a buoy and might be further equipped with radio transmitting capabilities to send readings to a satellite relay or nearby vessel.

Electronic Article Surveillance

Electronic security devices to track vehicles, books, and retail goods are now common.

Libraries and retail stores are the biggest users of electronic article security systems. By attaching a magnetic strip, 'tattle tape', button, film or other object that can be programmed or activated/deactivated, it is possible to detect the presence of an object and, in some cases, even its location and velocity. Most retail outlets use a combination of acoustic, magnetic, and electromagnetic (radio frequency) emissions to reduce the incidence of theft. 'Gates' in entrances typically incorporate transmitters and receivers that sense security tags that are attached to items or embedded in the items at the time of manufacture.

These photos show four slightly different electronic article surveillance systems that are used in the retail industry to prevent theft of clothing, jewelry, music CDs, and computer games. As the articles are paid for at the cash registers, the attached sensors are removed or deactivated so they don't trigger the exit systems when the customer leaves with paid items. Most systems consist of a transmitting 'gate' and a receiving 'gate'. Each retail outlet shown here also has visual surveillance camera systems installed inside the store. [Classic Concepts ©2000 photos, used with permission.]

Geophysical Sciences and Site Surveys

Magnetometers are used in hundreds of kinds of research and sensing projects, but are especially valuable for studying the Earth and its various geophysical structures. The Earth itself has an overall magnetic field that we can detect with a compass and some regions have mineral deposits with sufficient magnetism to interfere with the normal use of a compass. *Magnetic susceptibility* instruments aid in studying rocks and sediments. They have also been used to assess the magnetic characteristics of other bodies in our solar system, including the Moon, Mars, and Io, a moon of the planet Jupiter.

NASA and other organizations have developed quite a number of magnetometer devices, and there have been spinoffs of this science in other areas, including magnetic resonance imaging (MRI) which is used extensively in the medical field. Johns Hopkins has been active in the development of space-related magnetometers, including a high-sensitivity, wide-dynamic-range sensor called a *xylophone bar magnetometer*.

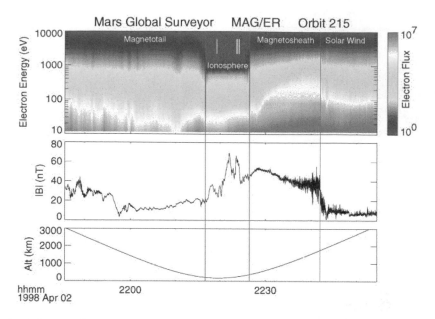

A chart of magnetic measurements taken as the Mars Global Surveyor spacecraft passed from the solar wind regime through the magnetic regimes of the planet Mars. [NASA June 1999 news diagram, released.]

Fluxgate magnetometers can be used to survey a site prior to installing sensitive magnetic equipment as might be used in a research lab or medical facility. They can also be used to provide reference data for Earth mapping projects and observatories.

Home, Business, and Vehicle Security

Magnetic strips, tapes, buttons, keys, and reed switches are all used to provide various degrees of security to homes, offices, retail products, library books, and vehicles. Most of these are two-piece systems that consist of a strip or card that generates a magnetic field or which holds magnetic data and a sensing system that detects (or fails to detect) the magnetic field or reads the magnetic data.

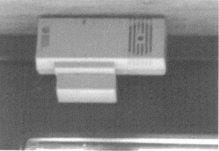

Two-piece sensors that are activated by the motion of a magnet breaking the contact or moving past the contact are common in home and office security systems. Since most of them are small and wireless (battery-operated), they can easily be placed on doors, windows, closets, and lids. Many of them use radio frequencies to send the alert to a central console that can be configured to sound an alarm. These Black & Decker magnetic components are common in homes and offices. [Class Concepts ©1999, used with permission.]

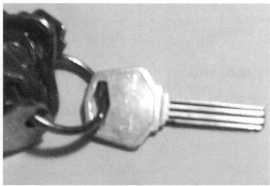

This looks like a conventional lock and key, but the key is magnetic (note the three dark indented strips), and the lock has an electronic storage capability for logging the times and dates when the lock is used, along with who has been accessing the door. To read the data, a portable computer, equipped with a cable that attaches to a data-reading key, is hooked into the locking mechanism. The data are then downloaded to the computer for analysis and storage. If a key is used inappropriately or there is some question about the user, the lock can be reprogrammed to reject access the next time the key is used. In more sophisticated systems, an alert can be sounded, or a video camera activated if someone questionable tries to gain unauthorized access. [Class Concepts ©1999, used with permission.]

When magnetic systems are used for exit monitoring, they are sometimes combined with other technologies, including radio-frequency signals or acoustic signals.

Shock Wave Detection and Research

The detection of shock waves is important for an understanding of geosciences, but it can also be a life-saving measure. Shock waves from tidal waves, nuclear or chemical explosions, volcanic eruptions, land slides, avalanches, or other significant events can often signal trouble before the trouble arrives. Shock wave laboratories use a variety of technologies to study and classify shock waves. Light gas guns are one of the devices used to launch projectiles at high speeds in order to generate and study impact craters and shock waves. These guns can use magnetic detection to measure the velocity of the projectile so the data can be cross-referenced with results from other tests.

Traffic Assessment

Magnetic sensors have been used in many types of traffic assessment. Depending upon their type and placement, they can be used to assess the presence, volume, and speed of cars and trains, and have been established in a fairly extensive network in the San Francisco Bay area, for example.

Intelligent Vehicle Systems, those that incorporate vehicles that can essentially drive themselves have been designed in a number of ways, but some use magnetic sensors to assess the location and proximity of vehicles to certain locations and to each other in order to automatically manage the traffic. Potentially, the information on a specific vehicle and its occupants would be put into a database to assess route preferences, speed preferences, schedules, priorities, etc. While this might be convenient and of interest to some, others might be concerned about the privacy issues inherent in this kind of system.

Underwater Surveillance

Many of the structures and objects that are sought underwater have strong magnetic fields that lend themselves to detection with various magnetic devices. Towed magnetometers and diving magnetometers are available for salvage and search and rescue operations. These systems are often interfaced to computer systems for analysis and display. Deep-tow and shallow-tow systems are usually sold as separate items as the deep-tow devices (e.g., 2,000 meters or more) require special shielding to protect them from the high pressures that exist at great depths. Some marine magnetometers can also be used for certain land-based operations.

7. Problems and Limitations

The most significant problems with magnetic sensing devices are background interference from other magnetic sources such as the Earth, false alarms triggered by magnetic items other than those being monitored, and the loss or theft of items containing information that is stored magnetically. There are also limitations associated with the amount of time it takes to read and interpret data in highly sensitive magnetic arrays with many elements. Finally, there are always problems with improper maintenance of machines or use by insufficiently trained personnel.

Magnetic Interference

Background interference can be overcome to some extent by carrying out a 'survey' of an area, that is, a series of readings, sometimes over an extended period of time, that provide references from which to interpret the data of interest. Shielding is another way in which background interference can be reduced. Highly sensitive machines such as magnetic resonance imaging (MRI) machines are heavily shielded to screen out the Earth's magnetic interference, as well as interference from other building structures and appliances.

Databases of the magnetic properties of various materials can aid in reducing false alarms, and in identifying specific kinds of objects, as in archaeological and forensic surveys.

One interesting source of cosmic magnetic interference is a newly discovered celestial body called a *magnetar.* A magnetar, once considered only theoretical, is a rapidly spinning core that remains after a supernova. The magnetar generates an enormous magnetic field, discharging huge amounts of charged particles that can affect other celestial bodies for quite a distance. Just as Sun spots can disrupt radio communications on our planet, discharges from magnetars can interfere with sensitive orbiting communications satellites.

False Alarms

False alarms are common in magnetic sensing systems. Many objects other than special tags, or the object being sought, can trigger a magnetic sensor.

False alarms are usually handled either by seeking out the items that are triggering the alarm (as in airport, library, or retail security systems) or by using several different kinds of sensors and making a determination on the basis of multiple inputs.

Theft and Tampering

There are two aspects of using magnetic tags and tapes for electronic article surveillance that are somewhat difficult to overcome:

- Removal of the magnetic strip tag prior to shoplifting an item, enabling a thief to pass through a sensor checkpoint undetected. Manufacturers have tried to minimize this problem by embedding the tag in the product or hiding it underneath the visible layers of a product in a process called *source tagging.* Source tagging also saves the retailer time, as it is not necessary to individually tag items before placing them on the shelves.

- Deactivation of the tag could allow it to be shoplifted or taken out of a library without being properly checked out.

Fortunately, theft and tampering take extra effort and equipment, and hence is less frequent. As magnetic films, inks, and papers are devised, it becomes more difficult for people to defeat the security systems.

Detection Speed and Processing

Speed (or the lack of it) is a limitation and an important aspect of magnetic detection in a number of technologies.

- Towed magnetometers, as are used from aircraft or boats, need to be able to resolve and store or transmit the data at whatever speed the host vessel is traveling.

- Intelligent vehicle transportation systems need to be able to respond to inputs at the speed of traffic and may need to be able to recognize individual events or vehicles.

- Magnetic resonance imaging systems need to be able to scan as quickly as possible without loss of data for the sake of the comfort of the person being scanned and for economic reasons related to the cost of operating the equipment. Improvements in both software algorithms and hardware have greatly decreased the time it takes to do a scan from many hours to several hours to less than an hour, so the technology has greatly improved over the last two decades. By 2000, adaptations of MRI technology were operating in near-realtime.

8. Restrictions and Regulations

Transaction Cards

Cards with magnetic stripes for local use can be custom-designed with proprietary data formats, but if the cards are to be used in standardized card writers and readers or on Internet-based computer systems, they must conform to certain standards in order to ensure compatibility. Various organizations have established data standards for transactions cards with ANSI/ISO standards being prevalent.

The use of cards for electronic transactions and fees associated with their use are regulated by such acts as the *Electronic Fund Transfer Act* and the *ATM Fee Reform Act of 1999*. These kinds of legislation not only regulate the amount and frequency of fee assignments, but also contain stipulations about notifying the customer in various ways, such as on an ATM display screen or on the account printout associated with a transaction.

Electromagnetic and Magnetic Interference

Some electronic article surveillance systems are hybrid systems that use more than one kind of sensing technology. Radio-frequency sensors and magnetic sensors are often combined, thus creating a potential for both magnetic and radio emissions that might influence the surrounding environment.

Because magnetic and electromagnetic fields can interfere with the operation of nearby devices, there are some regulations for electronic article systems (EASs) that limit the magnetic field strength and detection sensitivity of the tags, the sensors, and transmitters. A clothes rack full of articles that have magnetic tags could potentially affect customers with small electronic devices and there have been reports that certain EAS checkpoint sensors may interfere with medical pacemakers and other bionic medical devices.

Some EAS systems are passive detectors that light up or sound a local alarm. These are not likely to cause problems with pacemakers. However, active sensing systems and those which generate radio frequencies may cause problems, particularly those in the microwave wavelengths. Some stores have signs warning pacemaker users that the security system may have an affect on their pacemakers. In some areas, these signs are required.

In the United States, the *Federal Communications Commission* (FCC) is the primary agency regulating emissions standards. Those with an interest in manufacturing or selling EAS systems have petitioned the FCC to increase maximum allowable standards, but these requests have usually been denied.

See also the Restrictions and Regulations section in the Computer Surveillance chapter, since some of these apply to computerized 'smart card' surveillance technologies.

9. Implications of Use

Most aspects of magnetic sensing are subject to a minimum of controversy. Magnetometers that are used in geosciences, search and salvage, and medical applications generally contribute to our knowledge and quality of life. Some can even save lives. Magnetic strips that are used to protect inventory in stores and libraries are not often challenged or questioned, since the owners have a right to protect their investment and magnetic monitoring is far less obtrusive than many other methods.

However, there are a few aspects of magnetic sensor use that are likely to become controversial, mainly because they can link with computer processing techniques to create extensive and targeted databases.

Magnetic Cards - Use and Abuse

The most common problem with the use of magnetically coded cards is that they can be lost or stolen. It is inconvenient and problematic to replace ID and credit cards, and fees associated with cards go up when the incidence of theft and unauthorized use rises.

Since most card swipe machines are now standardized, it is not difficult for thieves to acquire the machines (or access to the machines in their places of work) and to write their own cards or to read information off the cards that might help them locate your house, your job, your bank account, or your magnetic-access home safe.

Some vendors have proposed using biometric identification systems along with the cards in order to reduce the incidence of theft and unauthorized use.

A growing problem with card abuse is that they hold increasing amounts of information and the systems that swipe the cards are not necessarily secure. In 2005, MasterCard International announced that information on more than 40 million credit cards might have been stolen in a May 2005 security breach through the third-party processing company, CardSystems Solutions. The FCC took the matter under consideration and reprimanded CardSystems for inadequate security measures and ordered that they improve security and submit to regular audits.

Electronic Article Surveillance

Magnetic strips in clothing, library books, and other retail or loan properties are now common (as are radio-frequency id chips). The surveillance checkpoints at the entrances of retail stores are usually based upon magnetic or radio-frequency surveillance technologies. Magnetic gates in most stores work with *dumb tags*, that is, magnetizable tags that are either activated or deactivated. However, a number of inventors have been working on *smart tags* that can be coded or assigned with group signatures or unique signatures (or both). This has obvious benefits for retailers, as it can potentially be used to aid in the management and automation of inventory. But it may spell trouble for consumer privacy. Consider this potential scenario:

A retailer attaches smart tags to all the items in a retail store in a mall. The store is equipped with magnetic exit monitors that can read the smart tags. Since most systems are standardized so they can be sold to thousands of retailers, other stores in the mall have installed compatible smart tag systems. The smart tag has two kinds of information, the on/off system that tells whether the consumer has paid for the item and an inventory system. The store also has a video surveillance camera and a computer database. A customer enters the store, purchases a number of products, has them deactivated by the cashier and pays for the purchase. The inventory information on the card is not deactivated and is entered in the computer database. The customer now heads for the exit and, as she or he passes through the checkpoint, the video image is triggered to snap a still shot of the person's face and add it to the database, along with a record of the items carried out by the consumer, which include not only the purchases just made (which can be distinguished by the computer), but the purchases made in all the other stores. The whole thing is automatically processed and the retailer can now read data on the computer screen that shows pictures, the patrons' names (since most people pay with checks or credit cards), the date and time of past and present purchases, and items purchased in other establishments, without the customer knowing all this information has been gathered and automatically combined. Over time, the vendor builds a 'data picture' of what the person buys, how often he or she shops, and the total value of the goods purchased.

The technology to carry out this kind of surveillance *profiling* already exists. The only real hindrance at the moment is designing checkpoints that can read the data fast enough, as the patron passes through the gates. It may be possible to produce this kind of system within two or three years.

There has already been a court case in which a store tried to use the information in a shopping database to discredit a shopper. The customer had slipped on a spill in a supermarket and decided to sue the establishment for his injuries. The supermarket looked up his shopping record in their database and argued that it showed that he frequently bought liquor, implying that he may have slipped because he was intoxicated. In this case, the court didn't admit the shopping record evidence, but it shows how quickly the retailer attempted to use personal data to protect its business interests, regardless of whatever rights of privacy the consumer may have had.

There are currently no safeguards to protect consumers from being monitored and manipulated based on their shopping profiles. Advertisers are use supermarket 'member' cards to print out ads on cash register tapes that are targeted toward the buying habits of individual shoppers. The same concept can be used by a retailer to flash a targeted commercial at a consumer as he walks out the door.

Consumer protection organizations in the 1970s were concerned about the potential for abuse from 'subliminal advertising'. But now we have the potential for far greater abuse from overt advertising targeted to specific people, especially young or technologically naive shoppers who aren't aware that they are being manipulated.

Personal Information and Magnetic Profiling

The increasing practice of issuing readable and/or writable magnetic cards and keys that are registered to a particular person or organization and that can be tracked through a databank or computer network extends to other businesses besides retail stores. Some organizations are building substantial data profiles on individual customers.

Casinos, hotels, and some trade shows now issue magnetic gambling cards, keys, or access cards to their clients. They typically request the person's name, address, and even personal interests before issuing the card. Sometimes occupation and gender are also requested. Each time the card is used, the information about the transaction is transmitted through a data network and entered in the firm's central database.

It doesn't take long for a gaming establishment, for example, to develop detailed economic and psychological profiles of individuals. Instead of using coins, clients purchase 'game time' on magnetic rewritable cards. It's convenient. They don't have to carry around heavy coins or constantly plug the machines. The gaming machines are networked to a central computer database. The data are processed to reveal how often people gamble, how much they gamble, which machines they prefer, their gender, names, addresses, and sometimes more. Since software is easy to enhance and modify, there's no reason why the information couldn't be cross-referenced to data on relatives or friends. The gaming establishment, trade show, or hotel knows where clients come from, what vehicles they drive, and what facilities they use within the premises.

There are honest businesses and there are dishonest businesses. A small percentage of dishonest employers and business owners now have the technological tools to determine where a patron lives and whether he or she is poor or wealthy, young or old, and married or single. Any employee with access to the database also knows that the patron is not at home while using the local services.

Honest hotel owners and casinos will assert that they have no intention of using the information for anything other than the comfort of the patron while gambling. In some cases this will be true. However, there are no social or legal guarantees, at the present time, to prevent them from providing the information to closely allied business contractors, marketers, mail-order sellers, and others. Just as catalog companies sell their mailing lists to generate additional revenue, retailers and casino owners may sell the mailing lists and profiles to other firms to generate additional revenue.

The Monitoring of Employees

A further cause for concern is the fact that even an honest business does not have absolute control over the actions of all its employees. When data obtained from magnetic access cards are fed into computers alongside images from video cameras, an employee who might be a potential thief, killer, or stalker, unknown to the employer, has a great deal of sensitive information that could enable him or her to find and harm an innocent victim. A now-famous case from 1989 involved two Swiss banking employees who were offered a large sum for aiding foreign tax authorities in decoding magnetic tape contained in the bank customers' data. Thus, it has already been demonstrated that people we generally trust, such as government officials and employees of banking institutions, have been known to abuse sensitive information for profit or political gain. There are also numerous documented cases of programmers leaving an employer and taking all the computer data with them to start new companies, without the employer being aware that data have been stolen.

In order to build up the confidence of their patrons, casinos and hotel owners using client databases will have to set up stringent safeguards to protect privacy and safety. Law enforcement agencies and privacy rights groups will have to recognize that this is an area that is particularly vulnerable to abuse and lobby for safeguards to protect individual rights. Unfortunately, the users of magnetic cards can't protect themselves, because they have little technical understanding of the technology and consequently don't even know how vulnerable they are and how easily the information can be stored and analyzed without their knowledge.

10. Resources

10.a. Organizations

These organizations are related to the industry and have information of relevance to this chapter. No endorsement of these companies is intended nor implied and, conversely, their inclusion does not imply their endorsement of the contents of this document.

Association for Payment Clearing Services (APACS) - The U.K. national standards body for transaction cards, including those with magnetic stripes, located in London.

Bartington Instruments - A commercial vendor of a variety of surveillance equipment including a line of magnetometers and gradiometers that are especially applicable to the earth sciences and medical and geophysical site surveys. Vehicle-detector surveillance systems are also available. Based in Oxford, England.

Billingsley Magnetics - A laboratory in a magnetically 'quiet' region which is equipped to assemble and characterize magnetic sensors. The president has a background with NASA, NOAA, and the private sector. Products include ultraminiature and high temperature magnetometers, medical gradiometers, and custom applications. Based in Brookeville, Maryland.

Bioelectromagnetics Society (BEMS) - An independent, nonprofit organization of biological scientists, engineers, and medical practitioners established in 1978 to study the interactions of non-ionizing radiation with biological systems. bioelectromagnetics.org

Electronic Funds Transfer Association (EFTA) - An inter-industry trade association advocating the use and advancement of electronic payment systems, located in Virginia. www.efta

Francis Bitter National Magnet Laboratory (FBML) - Established in 1961 at MIT to research state-of-the-art magnetic technologies. The Center for Magnetic Resonance was further established within the FBML in the early 1970s. web.mit.edu/fbml/cmr/

Geomagnetism/Ørsted-Satellite Group - A geomagnetic research group at the *Niels Bohr Institute* Department of Geophysics, in Copenhagen, Denmark. www.ggfy.ku.dk

IEEE Magnetics Society - A society with numerous chapters around the world. It supports and sponsors engineering research and applications in magnetism. yara.ecn.purdue.edu/~smag/

International Card Manufacturers Association (ICMA) - A worldwide nonprofit association of manufacturers that serves the dynamic plastic card industry and related industries, located in New Jersey. www.icma

International Organization for Standardization (ISO) - A significant international standards organization, located in Geneva, Switzerland. www.iso.ch

Magnetic Materials Program - A National Institute of Standards and Technology (NIST) program to obtain scientific measurements of key magnetic properties and fundamental research of magnetic characteristics, particularly for new materials. Thus, NIST seeks to accelerate the use of advanced magnetic materials by the industrial sector. www.msel.nist.gov/magnetic.html

Marine Magnetics Corporation - A commercial vendor and renter of magnetic and gradiometer marine and land exploration devices and equipment. Based in Ontario, Canada.

National High Field Magnetic Field Laboratory (NHMFL) - Funded by the National Science Foundation, the U.S. Dept. of Energy, and the State of Florida, the lab conducts and supports research in high magnetic fields and instrumentation. There are three labs, including the Pulsed Field Facility at Los Alamos National Laboratory. www.lanl.gov/orgs/mst/nhmfl/welcome.html

National Institute of Standards and Technology (NIST) - An agency of the U.S. Department of Commerce Technology Administration, established in 1901 as the National Bureau of Standards, located in Maryland. www.nist.gov/

Topical Group on Magnetism and its Applications (GMAG) - A special interest group of the American Physical Society. There is discussion of the science of magnetism and also in magnetic recording technologies that are used in the computer industry. www.aps.org/units/gmag/index.html

10.b. Print

Arnold, J. Barto, *Marine Magnetometer Survey of Archaeological Materials Found Near Galveston, Texas,* Austin: Texas, 1987, 53 pages.

Asimov, Isaac, *Understanding Physics: Light, Magnetism, and Electricity,* New York: New American Library, 1969, 249 pages.

Bond, Clell L., *Palo Alto Battlefield: A Magnetometer and Metal Detector Survey,* Texas A&M University, Cultural Resources laboratory, Sept. 1979, 63 pages.

Chikazumi, S., *Physics of Magnetism,* New York: Wiley, 1964, 554 pages.

Davy, Humphry, *Further Researches on the Magnetic Phænomena Produced by Electricity; With Some New Experiments on the Properties of Electrified Bodies in Their Relations to Conducting Powers and Temperature,* London, 1821.

Davy, Humphry, *On a New Phenomenon of Electro-Magnetism,* London, 1823.

Jianming, Jin, *Electromagnetic Analysis and Design in Magnetic Resonance Imaging,* Boca Raton, Fl.: CRC Press, 1998, 304 pages. An introduction to MRI with an analysis and survey of the components of a system, the magnet and coils. Includes analytical and numerical methods for analyzing electromagnetic fields in biological objects.

Maxwell, J. C. (Clerk-Maxwell, James), *A Treatise on Electricity and Magnetism,* Oxford: Clarendon Press, 1873.

Morrish, A. H., *The Physical Principles of Magnetism,* New York: Wiley, 1965, 680 pages.

Smart, J. S., *Effective Field Theories of Magnetism,* Philadelphia: W. B. Saunders Co., 1966, 188 pages.

Articles

Coulomb, C., First and Second Memoirs on Electricity and Magnetism, *Memoires de l'Academie des Sciences for 1785, 1788,* Institut de France.

Rezai, Ali R.; Mogilner, Alon, Introduction to Magnetoencephalography, NYU Medical Center Department of Neurosurgery.

Sieber, Ulrich, Computer Crime and Criminal Information Law: New Trends in the International Risk and Information Society, prepared for the European Commission, January 1998. This describes many kinds of risks associated with computer data, including data found on magnetic tapes and on smart cards. It can be downloaded at europa.eu.int/ISPO/legal/en/comcrime/sieber.doc .

Journals

BioElectroMagnetics Journal, published by Wiley-Liss, Inc. for the Bioelectromagnetics Society and the European Bioelectromagnetics Association.

Bulletin of Magnetic Resonance, a journal of the International Society of Magnetic Resonance.

Geophysical Journal International, published for the Deutsche Geophysikalische Gesellschaft, the European Geophysical Society, and the Royal Astronomical Society. It is a leading solid earth geophysics journal covering theoretical, computational, and observational geophysics.

IEEE Transactions on Magnetics, sponsored by the IEEE Magnetics Society, with articles on magnetic materials, magnetism, numerical methods, recording media, magnetic devices.

Magnetic Resonance Imaging, an Elsevier Science publication. International multidisciplinary journal dedicated to research and applications. www.elsevier.nl/

Magnetic Resonance Quarterly, a publication of Raven Press.

Magnetic Resonance Review, by Gordon Breach Publishers.

Solid State Nuclear Magnetic Resonance, Elsevier Science publication.

10.c. Conferences and Workshops

Many of these conferences are annual events that are held at approximately the same time each year, so even if the conference listings are outdated, they can still help you determine the frequency and sometimes the time of year of upcoming events. It is very common for international conferences to be held in a different city each year, so contact the organizers for current locations.

Many of these organizations describe the upcoming conferences on the Web and may also archive conference proceedings for purchase or free download.

The following conferences are organized according to the calendar month in which they are usually held.

Joint MMM-Intermag Conference, annual conference since the mid-1990s.

Card-Tech Secure-Tech. This conference took place some time ago (1995), but the Conference Proceedings printed after the event are of interest.

Symposium on Magnetic Materials for Magnetoelectronic Devices, an annual international symposium since the mid-1990s. research.ihost.com/symposiummmpd/

Physics of Magnetism, international conference.

ICM, international conference on magnetism.

Biomag, international conference on biomagnetism, since ca 1988.

Hermann von Helmholtz Symposium: New Frontiers and Opportunities in Biomagnetism, the von Helmholtz Symposia are held in Europe and cover a range of technological topics.

Applied Superconductivity Conference.

International Symposium on Magnetic Materials, Processes, and Devices, annual symposium of the Electrochemical Society, Inc., since the mid-1990s.

Asia Pacific Magnetic Recording Conference, an annucal international conference. apmrc2006.dsi.a-star.edu.sg/

10.d. Online Sites

The following are interesting Web sites relevant to this chapter. The author has tried to limit the listings to links that are stable and likely to remain so for a while. However, since Web sites do sometimes change, keywords in the descriptions below can help you relocate them with a search engine. Sites are moved more often than they are deleted.

Another suggestion, if the site has disappeared, is to go to the upper level of the domain name. Sometimes the site manager has simply changed the name of the file of interest. For example, if you cannot locate www.goodsite.com/science/uv.html *try going to* www.goodsite.com/science/ *or* www.goodsite.com/ *to see if there is a new link to the page. It could be that the filename* mgntc.htm *was changed to* magnetic.html, *for example.*

AIM, Inc. This international trade association for manufacturers and providers of automatic identification products has a Web site with a high proportion of educational content, including information on card systems, electronic article surveillance, standards organizations with addresses, glossaries for each subject area, and conversion charts and illustrations. It's a pleasure to come across a site like this. Recommended. www.aimglobal

The Basics of Electronic Commuications. This is aimed at children and youth, but can be enjoyed by all. It includes colorful, well-illustrated introductory information on sound, light, electricity, magnetism, and other technology-related phenomena. The characteristics of magnets are discussed as is the relationship between electricity and magnetism. park.org/Japan/NTT/DM/html_st/ST_menu_4_e.html

Electricity and Magnetism. IPPEX Interactive has a Quicktime interactive introduction to electricity and magnetism. IPPEX provides online pages on matter, electricity, magnetism, energy, and fusion. There is also an opportunity to ask a scientist a question through email. ippex.pppl.gov/ippex/module_4/intro.html

Magnet Facts. A short list of some of the things we know and some of the things we don't know about magnets. There are also links to information on types of magnets and magnetic coils. www.technicoil.com/magnetism.html

Magnetic Units and Symbols Conversion Charts. A useful set of tables for converting the units used for expressing various aspects of magnetism and associated symbols. Sponsored by Miller at Iowa State University. www.public.iastate.edu/~miller/tables/convert2.htm

TravInfo® System. A traveler information system for the Bay Area that detects and reports traffic flow and speed, especially over well-traveled bridge routes. The system is based in part on the data from magnetic sensors and data about and from individual sensors can be downloaded from the site. The system allows the user to check travel information for a specific route. There is also information on the scope of the project when it is fully implemented. The program is based on an Intelligent Transportation System (ITS) Field Operational Test (FOT) approved by the Federal Highway Administration. www.travinfo.org/ www.erg.sri.com/travinfo/

10.e. Media Resources

Many science and technology museums have exhibits relating to the history and science of magnetism, too many to list, so here are just a few examples.

Museum of Science and Industry. Located in Chicago, Illinois, this extensive exhibition space includes an "Idea Factory" a learn-through-play section that allows youngsters to observe and test the basic concepts of mechanics, light, color, and magnetism.

Science Center of the Americas. The Miami Museum of Science and the Smithsonian Institution are developing America's first international science center. The Hands-on Hall of Science will feature a number of science and technology exhibits, including biomedicine, telecommunications, and others.

Whipple Museum of the History of Science. Located at the University of Cambridge, in the U.K., the exhibits include magnetic materials, lodestones, bar magnets, and a very rare amplitude compass.

11. Glossary

Titles, product names, organizations, and specific military designations are capitalized; common generic and colloquial terms and phrases are not.

ABA track	ANSI/ISO standardized data track #2 encoded in BCD format
AIDC	automatic identification and data capture, e.g., as is accomplished with cards with magnetic strips
air gap	a nonactive section or break in a magnetic surface circuit
ATB	automatic ticketing and boarding, a magnetically coded ticket system used by airlines and other forms of public transportation to expedite passenger boarding
ATM	automatic teller/transaction machine
bulk eraser	a strong magnetic 'scrambling' unit used to 'remove' the ordered magnetic patterns (but not the magnetic character) of a magnetic data medium such as a floppy disc or coded card
CENELEC	Comité Européean de Normalisation Electrotechnique, a European telecommunications regulating authority
degaussing	the process of demagnetizing a substance or system. Some systems build up a magnetic charge over time (e.g., computer monitors) that can eventually interfere with the functioning of the system and must be periodically degaussed to remove the source of the interference.
doping	the process of embedding tiny amounts of magnetic or other materials to increase the conductivity or magnetizability of a substance
EAS	electronic article surveillance, electronic access surveillance, electronic access system
EM	electronically magnetized/magnetizable
f	a symbol for magnetic flux
FCC	Federal Communications Commission, the primary U.S. body for radio frequency transmissions and emissions regulation
Gilbert	a centimeter-gram-second (CGS) unit of magnetomotive force
Henry	a unit and associated symbol (H) for magnetic field strength or magnetic inductance
MCCL	magnetic-code computer lock
NMR	nuclear magnetic resonance
Oersted	or Ørsted, a unit and associated symbol (Oe) for magnetic intensity
ordnance	(not to be confused with ordinance, which is a decree) military supplies and equipment such as weapons, land mines, vehicles, etc.
remote sensing	sensing at a distance, which is usually, though not always, non-destructive (remote-sensing of biological specimens with X-ray technology may have destructive effects)
sampling, probabilistic	a mathematical technique used in forensics and archaeology to interrelate small samples to larger populations or amounts, in other words, mathematically extrapolating information from what is at hand from information that is already statistically known or calculated
tesla	a standard international (SI) unit and associated symbol (T) for magnetic flux density
UOD	unexploded ordnance detection, the detection of undetonated explosives such as land mines or bombs
UXO	unexploded ordnance, see UOD
Weber	a standard international (SI) unit and associated symbol (Wb) for magnetic flux

Miscellaneous Surveillance

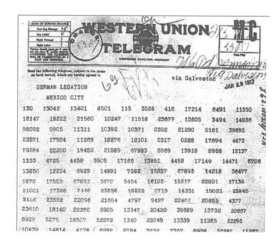

Cryptologic

1. Introduction

Cryptology is the study of the discovery and analysis of coded and hidden information. *Cryptologic surveillance* involves seeking, analyzing, and decoding coded messages. The decoding process is more specifically known as the field of *cryptanalysis*. Cryptologic surveillance is a subset of cryptology, the part of the process that occurs after the devising of codes and the encoding of messages.

Some cryptologic surveillance overlaps with chemical surveillance. For example, a message written in lemon juice will become less visible as it dries and can only be seen if heated or chemically treated.

This chapter introduces the discovery of messages that have been hidden to obscure the *existence* of the message and messages that have been encoded with a cipher or other means of obscuring the *content* of a message. Sometimes visual, chemical, and cryptologic surveillance methods are used together. For example, a message that has been encoded to defy comprehension may also have been hidden, making it necessary to engage in a two-step process to 1) discover and 2) decode the message.

The broader term *cryptology* is used here instead of *cryptography* because not all coded messages are graphically transcribed in words or pictures. Some may be transmitted in radio or television broadcasts, may be hidden in music, encoded into the words of a public speech, or may engage senses other than those of sight or hearing. Prearranged signals are a form of coding that can be used to signal danger in adverse situations or to signal approval in private social situations. For example, a cook fixing dinner in an occupied war zone might cook onions to signal others in the vicinity that an intruder is hiding nearby. When others near the kitchen, the smell of onions will tip them off. Or imagine a female spy signaling a yes/no, safe/not-safe situation to another agent at a dinner party. If she's wearing cologne, the answer is 'yes' or 'safe,' if she's not wearing cologne (or a flower, or a dress of a certain color) the answer is 'no' or 'not-safe'. These types of one-time, simple-answer codes are very difficult to detect and interpret. Cryptologic technologies tend to be used to decode longer communications, such as written messages and radio broadcasts and, now that personal computers are linked via the Internet, email and other digital communications.

Cryptology is a highly specialized technical field and those who are hired by businesses or governments to detect and decode messages tend to be problem-solvers with a knack for crosswords, logic puzzles, and games like Chess and Go. They are often generalists with multidisciplinary interests and the ability to 'think outside the box', a talent that aids in proposing novel solutions.

Cryptography and hidden writing frequently involve the use of chemical, mechanical, and computer technologies. A portion of chemical surveillance deals with ways to detect and reveal hidden writing, and ultraviolet light can sometimes make it easier to see writing (or ancient drawings) on rocks, paper, or other materials.

Cryptologic surveillance is an important aspect of surveillance technology, but it is a very specialized topic that is only broadly introduced in this chapter. There are thousands of books on cryptographic technique, machines, and history, and this chapter does not seek to repeat the information but rather to provide an overview within the general context of technological surveillance with an emphasis on cryptanalysis.

If you are interested in learning more about cryptographic surveillance or learning some ciphering or deciphering techniques, you are encouraged to consult the resources described at the end of this chapter.

2. Kinds and Variations

Not all codes are secret; some are created for convenience, to save space, or for use with electronic devices (e.g., bar codes). In the context of this text, emphasis is on the detection and discovery of codes that are intended to be private or secret.

There are two basic aspects of cryptologic surveillance that involve detecting and deciphering messages. These are hidden messages and coded messages:

hidden messages - messages obscured from view by size, shape, color, visibility, context, or other attributes. A message may have been shrunk to microscopic size and hidden in a punctuation mark on the page of a telephone book or integrated with the pixels in an image. It might be transcribed with a material that is visible only under ultraviolet or infrared light or which is revealed only by a change in temperature or the application of chemical reagents. It may be split across a series of billboards flanking a long stretch of highway. It may be seen through a crack in the rocks only at a particular time of day. Hidden messages may or may not be coded.

coded messages - messages that are obscured by transforming the information content into another system or medium. Letters may be substituted for other letters, numbers may be substituted for letters, colors may be substituted for letters, or tonal values may be substituted for directions. One language may be substituted for another, math formulas may be substituted for musical notes, bird calls may be used to communicate outdoors. There are more sophisticated encryption systems that depend on more complicated encoding procedures than just substituting one form of information or one unit of information for another, such as key codes, and these are generally favored for sensitive documents or computer communications. Coded messages may or may not be hidden.

There are many different kinds of codes and ways to obscure messages, and not all are listed here, but a few examples include semaphore, map codes, rock codes (rocks piled to indicate direction or a message), alphabetic codes (e.g., Morse), tape codes (e.g., colored tapes on trees), machine-readable codes, and binary codes.

Invisible Ink

Invisible ink is relatively easy to make and has been used for centuries. Lemon juice is readily available for writing 'hidden' messages. During the American Revolution, writers would mix ferrous sulfate and water and write 'between the lines' of otherwise innocent-looking correspondence, to avoid interception.

There are three common ways to make invisible ink visible:

- heating, with a flame or light bulb,

- chemical manipulation with a chemical reagent, such as sodium carbonate, or

- illumination, often with a specific kind of light (e.g., ultraviolet).

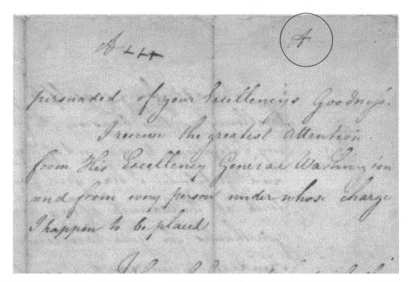

In this letter from John André to Henry Clinton, dated September 1780, the letter "A" has been inscribed at the top of the last page, indicating that acid rather than heat should be used to reveal the message. [Clements Library Clinton Collection, copyright expired by date.]

Depending upon how a message was created, using the wrong method to reveal the message may damage the chemistry and make it impossible to ascertain the contents. During the

Revolution, John André, head of intelligence for the British commander-in-chief, Henry Clinton, used an initial on his letters to indicate how they should be read. The letter 'F' stood for 'fire' and the letter 'A' for 'acid.'

Substitution Codes

The most common type of alphabetic cipher, and the easiest to generate, is a *substitution code* in which one unit or block of information is substituted for another unit of information. Substitution codes need not be one-to-one. A simple alphabetic substitution code in which the letters of the alphabet are substituted with another set of letters from the same alphabet is extremely easy to 'break' or decode. When spaces between words are eliminated, and groups of letters such as 'ing' or 'sch' are replaced with a single letter or group of letters with a different number of individual units, it becomes more difficult to decipher (though not much more difficult if you are an experienced decoder). By writing the whole message backward and adding nonsense words, it is even harder to break the code, but the general concept is still fairly simple. In spite of the ease with which substitution codes can be broken, they are still widely used in personal and business correspondence. School children often code messages to prevent classmates (and teachers) from discovering their contents.

A simple alphabetic 'substitution code' is one of the easiest and most popular means to hide a message. Spy and code stories from World War I resulted in many popular fiction and nonfiction books and commercial products that described how to create and read codes. This Orphan Annie™ ring and accompanying book were released to consumers in the mid-1930s to make it easy for them to write a message with a substitution code. The center ring rotates relative to the outer ring to set the 'key' letter relationship so that the rest of the code can be read easily off the outer dial. [Classic Concepts photos copyright 1999, used with permission.]

Key Lookup Codes

Some codes require the possession of a 'key' with which to decrypt and subsequently interpret a coded message. The coded message is said to be *encrypted*. A key code can be somewhat difficult to break if the format or content of the key is not known. Some of the better computer encryption schemes are based on key systems and they are now widely used to protect electronic data, including voice and computer communications.

Decoding Methods

There are thousands of decoding methods, just as there are thousands of codes, and these are discussed in a little more detail in Section 5 (Description and Functions). However, four basic foundation methods are mentioned here, as they apply to many aspects of cryptanalysis.

databases, tables, and precedence - a data archive of information related to a problem can sometimes be used to help solve it. For example, if a building collapses, a look at the data on similar buildings might reveal that there is a particular structural flaw common to those buildings. Another example is opening moves in chess. Over the decades, it has been noticed that certain opening moves lead to a higher probability of a good outcome and that Grandmaster chess players tend to favor them and have them memorized. Every once in a while someone comes up with a novel opening, but at the highly competitive international level, the openings have standardized somewhat for the first few moves to the point where they have been given names. Knowing them can give a recreational player an advantage over someone of otherwise similar playing ability. Thus, a database of common scenarios is combined with heuristic problem-solving strategies. Similarly, secret service agents have vast storehouses of information about the cultures, habits, and priorities of other nations to make it easier to interpret their actions and motives. Without this information, a move on the part of an apparently hostile nation that was not intended to be aggressive might be misinterpreted by a different culture and lead to war. For example, in many western nations, the approach to perceived hostilities is to "shoot first and ask questions later". In contrast, there are some eastern cultures in which one must apologize before the other side is willing to enter into negotiations to sort out a problem. Obviously, if the opposition doesn't know this, it could lead to serious misunderstandings. Thus, a knowledge base is an important aspect of decision-making and of carrying out cryptanalysis on foreign communications.

brute force - the process of finding a solution by trying every apparent possibility. For example, one way to find out if 17 is a prime number (a number that can only be divided by one and itself) is to multiply together all the combinations of integers between 2 and 16 and see if the result of any of the calculations is 17. In other words, it's a long process that eventually yields results, but isn't necessarily the most efficient way, or may only be an efficient way up to a certain point. When you want to find out if 9,342,154,673 is a prime number, brute force is going to take a long, long time, even with optimization methods. Thus, strategy and elimination are sometimes used to solve certain problems.

selected attack - problem-solving developed to deal with a specific problem. For example, in computer decryption algorithms, sometimes the nature and even the method used to encrypt a message are known. In this case, the most effective way to break the code is with mathematical techniques known to work for the specific decryption system. Often, when cryptanalysts discover a formula or technique that works for a specific code, they will publish it to make the community aware of the limitations of the code. In some cases, if the body using the encryption technique finds a way to break it, or if a secret service agency doesn't want it known that they have broken it, they will keep the solution classified. Certain key encryption algorithms can be broken if the keys are short. In some cases, the same encryption system might be much stronger and more difficult to break, not by changing the nature of the encryption, but by making the key longer.

heuristics - a problem-solving process in which logic, elimination, and exploration are used to follow 'fruitful' paths of inquiry—those which are more likely to result in a solution. Artificial intelligence programs strive to incorporate effective heuristics. Many of the historic computer chess programs used brute force and were slow and fairly easy to beat. Gradually, however, programmers added problem-solving heuristics (along with a database of strong opening moves) which, instead of calculating every possible move, would calculate moves based upon knowledge and experience of what would more likely be strong moves. These might include strategies such as advancing pawns, crowding a Queen, using the bishop and knight in tandem, etc. Heuristics often are based on an intuitive human element and sometimes are based on algorithmic/mathematical discoveries of fruitful paths of exploration. To solve present-day computer encryption systems, a knowledge of probability, statistics, and geometry helps hone heuristic intuition.

The above methods are almost always used together to different degrees, depending on the nature of the problem to be solved.

Fields of Study

The study of coding and its related fields has been divided up in a number of ways. Some of the more common fields of study within cryptology are

- coding theory, codes, and applied coding,
- general information theory and mathematics related to cryptology, and
- decoding, cryptanalysis, cryptologic surveillance.

3. Context

Cryptology is applied to many disciplines, from computer compression schemes to secure military communications, converting plain data to other less easily recognized forms has many advantages, not all of which are related to privacy. However, cryptologic surveillance is almost always used in the context of secrecy. Whether it's a child sending a secret message to a classmate two desks away, or a war department sending a coded message to troops on an aircraft carrier, the basic premise is the same, the sender wants only the intended recipient to be able to receive and decipher the message. Thus, the most common contexts for cryptologic surveillance involve the deciphering or discovery of a hidden or coded message intended for someone else.

Cryptologic surveillance was once limited to secret service and military agents, for apprehending messages from hostile individuals, forces, or nations. But, with the advent of encoded communications devices, cryptology is now practiced by a larger segment of the population wanting to access the communications of others. On the Internet, and now on cellular phones and other electronic products, encryption is a common security measure and surveillance to make sure the security measures work is important. There are also a number of people who like to decode these communications for the challenge of breaking the code, and some who do it for material gain, or for reasons of curiosity or vengeance.

4. Origins and Evolution

The use of secret messages, and the discovery and deciphering of these messages, have been pivotal in decisions made during global conflicts, including World Wars I and II. But their origins stretch back much farther than that, probably farther than we can know. Coded and hidden messages have allowed friends and lovers to share secrets for centuries. They have been an integral part of political intrigue; they provided a means for Mary Queen of Scots to escape her imprisonment and for Queen Elizabeth to prevent her later escape when Mary was again imprisoned. They now form an essential aspect of computer communications and have created more consternation in government policy discussions than almost any other single aspect of computer security.

Secret writing includes many different kinds of hidden and open messages which may or may not be ciphered. Surveillance techniques to detect and decipher secret messages have evolved and become more sophisticated as the secret writings themselves have evolved. Surprisingly, it is only in the last few decades that reasonably secure encryption techniques have been developed. In prior ages, the systems were simple and often effective, but they were also rather easy to decrypt by someone clever enough and determined enough to do so.

Writing itself used to be 'secret' for the simple reason that very few people were taught to read or write. For many centuries, reading was a skill known only by clergy, rulers, their elder statesmen, and a tiny portion of the population who learned to read spontaneously without formal instruction (estimated today at about two percent, but in those days people had less exposure to written works and less opportunity to study them, and so the percentage was probably much less).

When messages were drawn in sand or carved in large blocks of stone, there was no easy way to transport them, but pigments for body adornment have been used for thousands of years and early messengers may have carried symbolic messages on their backs, heads, or buttocks or other less-visible parts of their anatomy if they wanted to escape detection.

We can guess that some secret writings were probably modifications of regular writing, letters or words that were written backward or upside down or in a slightly different style or color to indicate that certain letters were more significant than others. There are examples of Egyptian hieroglyphics that differ somewhat from the norm. This may have been a way to imbue them with special meaning or perhaps it was a way to obscure or encode a message or call attention to particular aspects of a message.

Historic clay tablets weren't very portable or flexible, but once papyrus came into use, new systems were developed, such as rolling or folding the papyrus so a message could only be read if the paper was reassembled (if torn apart) or refolded in its original configuration (there are biblical references that indicate rolling may have been used).

Hidden writing has probably been in use longer than ciphered writing. Certain aspects of practical chemistry were well understood in antiquity, given the sophisticated embalming techniques of the Egyptians. Since writing was already in use and pigments for body adornment were common at the time, there is a possibility that someone noticed that marks made with lemon juice disappeared as they dried. It takes only a little heat to make them reappear, so heating may have been one of the first examples of a surveillance technique to reveal secret writing. When hidden writing and cryptography were first used is not known, but they may have roots in ancient history. Those clever enough to invent a system of hidden writing probably sought ways to detect and decipher the hidden writing of others.

The Roman Emperor Julius Caesar was apparently supportive of the use of codes for communications. The "Caesar Cipher" is a substitution cipher Caesar is said to have used to write to Cicero. These basic Caesarian ideas continued right through the Middle Ages and were not substantially challenged with more sophisticated systems until Renaissance times.

'Seals' to protect the confidentiality of correspondence have been in use for many centuries by rulers and others of higher social stature. The royal seal, stamped onto a document with heated wax, was intended to safeguard the contents both by the implication that snooping would have dire consequences and also by the physical seal that would make it hard to pry it apart to look inside. The seal was one of the first 'tamper' devices. It may seem like a rather loose security system, since one could pry it open, rewrite the message (changing it, if desired), and reheat the seal on the backside to attach it to the new document, but both opportunity and forgery skills were necessary and resources for the average person were in short supply in the Middle Ages.

The historic use of tunnels, secret drawers, and hidden compartments has been documented in many texts. The people who built and used secret implements may also have used secret writings and hidden some of those writings in secret places.

Journals have been written in code for many years and the privacy of journals designed for teenagers is still loosely guarded with locking flaps and keys.

The use of a *nom de plume*, a pen name, or the anonymous authoring of a message or publication, are two of the simplest ways to obscure identity, and were common in the Middle Ages when death or excommunication could result from expressing 'heretical' thoughts. Anonymous publications sometimes had the identity of the author encoded into the text, chapter headings, or the binding of the book and there may be many of these that have not yet been detected. Men sometimes used pen names to protect themselves from political censure and women commonly used male pen names right up until the 1970s, because of social censure against women writers. There have been many periods of history where censure against certain political, scientific, or religious ideas was so strong that entire volumes were written in code.

Hidden messages within a piece of writing have a long history. The first (or second, etc.) letter of each paragraph, or of each sentence, or before or after each punctuation mark, for example, could be designed to be assembled into a encrypted message, which then might be deciphered forward or backward or using a key. Sir John Trevanion is said to have escaped confinement in a castle by receiving a tip from a jailer in this way.

During the Renaissance, when art and culture blossomed in western Europe, a German monk named Johannes Trithemius (1462-1526) penned a document called "Steganographia", based upon a system of hiding a message and providing a 'clue' or key to aid in locating and deciphering the message. To this day, the term steganography is used to indicate messages hidden within other messages, often without any indication that they are present.

Many persons have stated that Leonardo da Vinci wrote backward, sometimes called 'mirror writing', to hide secrets of his inventions and tools of warfare, but the author has doubts about this explanation. Mirrors existed in the Renaissance that easily allowed backward-writing to be read, and many artists and virtually all traditional typesetters can read backwards fairly easily, so it's not an uncommon skill and thus not especially secret. Given that da Vinci was left-handed and writing was done in those days not with ballpoint pens but with slow-drying quill-pen ink that smeared easily, it is more likely that the resourceful da Vinci simply found it cleaner and more comfortable to write in the natural direction for the left hand. Further support for this theory is that Leonardo's images, which accompanied his text, were often more explanatory than the text itself and he made no attempt to obscure those. Da Vinci was a politically astute genius who earned his reputation and his living by promoting his ideas and designs to rulers and dignitaries of the time, so it doesn't seem likely that he would have taken extensive steps to hide his efforts.

Henry VIII, on the other hand, was always trying to get away with something that might provoke the ire of his detractors or the populace. He was a highly manipulative statesman, who managed to take several wives despite religious prohibitions against doing so. Thus, it's not surprising that he apparently made significant use of cryptographic services during his reign.

The Elizabethan Age and Beyond

There were a number of scholarly writings in the 1500s and 1600s about cryptology and its application to affairs of state.

Queen Elizabeth I, in the 1500s, was a wily ruler and determined to hold onto her crown during a time when rulers were often deposed or executed within weeks or a few short years of the start of their reigns. The Queen's security system, established by Sir Francis, Earl of Walsingham (ca. 1531-1590), was probably a significant factor in her long reign and set a precedence for secret service departments for generations to come. Walsingham got his training and experience in foreign intelligence under William Cecil, Baron Burghley, and put the knowledge to use as Elizabeth's Joint Secretary of State.

Queen Elizabeth I retained Walsingham in this position for many years. He was a meticulous and dedicated man who made ready use of spies and various means to cipher and decipher messages that might affect the monarchy, including those of the imprisoned Mary Queen of Scots. Secret messages had enabled Mary to escape imprisonment on a prior occasion years before, but Walsingham was not so easily fooled as her previous jailers. His administrative and surveillance resources were excellent. He not only gathered evidence against her from her correspondence, but claims to have retrieved dozens of different cipher systems from her premises. This eventually led to Mary's execution and the rounding up of a number of her followers, based on the written evidence provided by his surveillance efforts.

Walsingham is also known for having uncovered news of an impending attack by the Spanish Armada upon England. His secret service included dozens of agents. Thomas Phelippes was one of the most talented, able to solve codes in multiple languages.

Cryptography was a familiar tool at least by Renaissance times. In 1565, Italian Giovanni Baptista della Porta published a simple but effective cryptographic table consisting of thirteen 'key' letters, with an alphabet in which the lowest line in the table moved one position to the right for each pair of capitals.

A number of texts, systems, and counter arguments were published at this time.

- Blaise de Vigenère (1523-1596) wrote about ciphers and secret writing and created a system that remained unbroken for centuries.

- In 1588, Timothy Bright wrote a book called *The Arte of Shorte, Swifte and Secret Writing,* which detailed a system of shorthand. While only loosely a cipher system, shorthand could provide a certain measure of light security, if the system were not understood by prying eyes. While shorthand systems are easy to decipher, they still provide a deterrent to the lazy or slow.

- Francis Bacon (1561-1626) had a particular interest in cryptology and referred to it in his writings.

- John Wilkins (1641-1666) published *Mercury the Secret and Swift Messenger,* in 1641, cautioning against weak ciphers. He wrote of a number of systems and developed at least one of his own.

Armand Jean du Plessis, Cardinal and Duke of Richelieu (1585-1642) was the First Minister of France for almost twenty years and is said to have engaged in secret correspondence to bring about his political ends.

Technology and Cryptology

Fine craftsmanship was prevalent during and after the Renaissance. There were many ingenious mechanical devices developed in the decades that followed. Music boxes, coded looms, and concepts for calculating machines were being invented and it was only a matter of time before someone applied the idea of clockworks or cylinders to the creation of ciphering machines.

In the 1700s, the fathers of democracy, Thomas Jefferson, Benjamin Franklin, and George Washington, all were interested in hidden writing and cryptography. During the American

Revolution, written communications were routinely surveilled by both the British and Americans. Runners and riders were intercepted and the contents of their mailbags searched, often turning up war-related letters. Many of the British documents captured by George Washington are now archived in the U.S. Library of Congress.

To reduce the possibility of messages being intercepted, secret communications were used by George Washington's networks of spies when he commanded the Continental Army. Troop movements and other news were forwarded to General Washington through a system of secret correspondence. Invisible ink made it possible for hidden messages to be written between the lines of other innocent-looking correspondence. His secretary, Alexander Hamilton, used secret inks, codes, and symbols to conceal information from British eyes. James Lovell, a Boston school teacher who is considered the 'father of American cryptanalysis' provided George Washington with news about British troop activities.

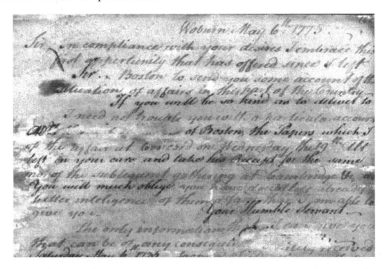

This detail of a three-page letter, written by Benjamin Thompson in 1775, actually carries two messages, a longer visible message and a short invisible message. The visible letter begins "Sir, In compliance with your desires ...". However, interspersed within it is a message written in invisible ink (now made visible) that states "Sir | If you will be so kind as to deliver to | Mr. of Boston, the Papers which I | left in your care, and take his Receipt for the same, | You will much oblige | Your Humble Servant | [erased] | Saturday May 6th 1775". [Clements Library Gold Star Collection historical document, copyright expired by date.]

Navajo band members were also reported to have been used as messengers. Since the Navajos had no written language, and their spoken language was known only to a few (and didn't resemble English), there was pretty effective 'encryption' inherent in transmitting a message through spoken Navajo.

American and British spies in the Revolutionary War used both simple and elaborate schemes for hiding messages and the paper on which they were written. They were often rolled or folded and hidden inside bullet casings, small silver balls (which could be swallowed), quill pens, and buttons. The methods of detecting these messages ranged from simple searches, to the use of emetics to cause the suspected messenger to vomit up anything that might have been swallowed. It may be that some of these messages were never delivered or discovered and are still hidden in garments of the era (or buried with their bearers). With new X-ray and other scanning technologies, it may now be possible to examine preserved relics in museums without taking them apart or otherwise damaging them.

Some of the methods of conveying secret messages during the Revolution are revealed by John André in a four-page letter to Joseph Stansbury, in May 1779, in which he writes about codes, hidden writing, key words in conversation, and keeping the messengers ignorant of the contents of their messages:

> You will leave me a long book similar to yours. Three Numbers make a Word the 1st is the Page and the 2d the Line the third the Word a comma is placed between each Word when only the first letter of the line is wanted in order to compose a Word not in the book, the number representing the Word will be + Unit with a stroke across. In writings to be discover'd by a process F is fire and A acid.
>
> In general information, as to the Complesion of Affairs an Old Woman's health may be the Subject. The Lady might write to me at the Same time with one of her intimates She will grasp who I mean, the latter remaining ignorant of interlining & sending the letter. I will write myself to the friend to give occasion for a reply. This will come by ~~any~~ a flag of truce, exchang'd Officer & @ every messenger remaining ignorant of what they are charg'd with, the letters may talk of the Meschianza & other nonsense.
>
> You will take your mysterious notes from this letter and burn it or rather leave it Sealed for me with -------
>
> [Excerpt transcribed from the original John André letter in the Clements Library Clinton Collection.]

The Journals of the Continental Congress notes from Friday, 7 April 1786, mention the use of cipher for correspondence in a letter of December 1785 from John Adams to John Jay that was referred back to the Secretary for Foreign Affairs to report. The report was regarding the attitude of Great Britain toward the United States, recorded as being transcribed in cipher No. 84, VI, folio 43, with a translation in folio 51.

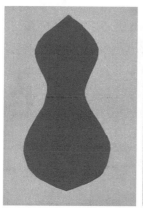  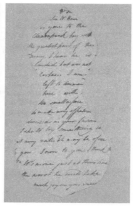

Sometimes a mask was used to create a seemingly innocent message with a hidden message incorporated into the main part of the text. Henry Clinton wrote a letter to John Burgoyne dated 10 August 1777 that used this technique. The mask is shown on the left, the innocent-seeming message in the middle, and the intended message through the mask on the right. Careful composition of the words in the letter had to be considered when writing the text. If it was awkward, rather than natural, it might create suspicion. [Clements Library Gold Star Collection, copyright expired by date.]

The Jefferson cylinder was developed in the 1790s, consisting of three dozen discs, each with a random alphabet, which could be organized in different ways to change the code. This cylinder somewhat resembled the metal cylinders that were coded with pegs or bars to make music boxes.

In 1830, a passionate French teenager named Evariste Galois (1811-1832) wrote a paper in mathematics that was published posthumously after he died in a duel. This remarkable paper led to further research in finite fields called Galois fields, which were to become important years later in disciplines such as cryptology, coding theory, and specialized areas of geometry.

In 1833, Kerckhoff wrote *The Handbook of Applied Cryptography,* which discussed some generally preferred characteristics of an effective cryptosystem. Kerckhoff shed light on a way to decipher an 'unbreakable' system devised by de Vigenère in the 1500s. Kerckhoff principles are still mentioned today.

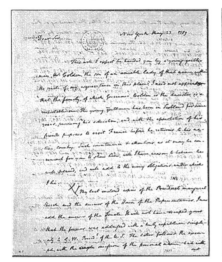

A letter from James Madison to Thomas Jefferson in 1789 clearly shows a portion written in cipher. Note the cipher digits on the second page, with the translation penned above the numbers. The bottom image shows the lookup table used to determine which words correspond with the numbers in a letter. A closer look indicates that it is a very orderly cipher, with numbers assigned to the words alphabetically, which actually weakens the security of the cipher. If some of the numbers are decoded and the orderly pattern of the numbers is discovered, it is much easier to decode the rest of the message than if the numbers had been assigned to words on a random basis. Jefferson made use of a number of different ciphers. [Library of Congress James Madison and Thomas Jefferson Papers, public domain by date.]

In the 1840s, Edgar Allen Poe, poet and writer of dark and mysterious literature, showed himself to be an aficionado of cryptology. In his story, "The Gold Bug", he developed a mystery around a secret message that may have spurred interest in this fascinating subject on the part of the public.

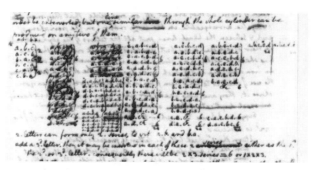

A portion of Thomas Jefferson's correspondence describing a cipher wheel system. [Library of Congress Thomas Jefferson Papers, public domain by date.]

When Allan Pinkerton, the famous detective, and Samuel Morse Felton, apparently uncovered a secessionist plot to assassinate President Lincoln, Pinkerton sent Felton a coded message "Plums delivered nuts safely", to indicate that he had arranged the successful transport of President Lincoln through Baltimore. Pinkerton and his colleagues were actively engaged in intercepting and interpreting enemy communications during the Civil War.

World War I

Sometimes a single encrypted message can be a pivotal document during a War.

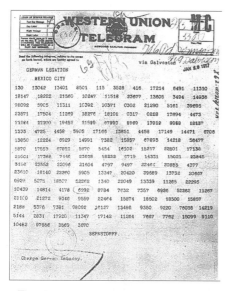
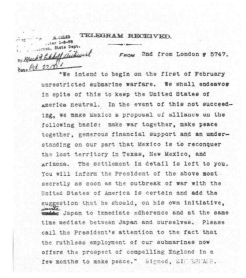

The Zimmermann telegram described the German intent to break the Sussex Pledge and engage in unrestricted submarine warfare. It further entreated Mexico to join the Germans against the U.S. in exchange for U.S. territory. The breaking of this code by British agents was a pivotal piece of intelligence drawing the then-neutral U.S. into World War I. The coded telegram on the left is signed Bernstorff, the German ambassador in Washington. [Clements Library Gold Star Collection, copyright expired by date.]

The European nations were engaged in a significant armed conflict between 1914 and 1917. Under Woodrow Wilson, the United States remained neutral until early 1917. When Germany broke a pledge (the Sussex Pledge) to limit submarine warfare, the United States severed its diplomatic ties with Germany. The conflict came to be known as the Great War and, after World War II broke out, was retroactively called World War I.

In January 1917, British cryptographers deciphered a telegram from Arthur Zimmermann (1864-1940), the German Foreign Minister, to von Eckhardt, the German Minister to Mexico. The content of the message entreated Mexico to join in the German cause, in exchange for U.S. territory. In February, the British revealed the telegram to President Woodrow Wilson and it was published in the press in March. On 6 April 1917, the U.S. formally declared war on Germany and its allies.

Following the War, a German engineer named Arthur Scherbius (1878-1929) developed a cryptographic machine and tried to interest the German Navy in the invention. The Navy declined, so Scherbius marketed it instead to businesses such as banks that wanted to secure their communications. It was named "Enigma". The German forces evidently changed their minds about the Enigma machine, adapting modified versions for the Navy, Army, and Air Force during the 1920s and 1930s.

The cracking of the German code in World War I inspired a wave of enthusiasm for spies and ciphers in the post-War years. Three examples of simple alphabetic substitution coders/decoders sold or given away to consumers are shown here. They consist of rings that could be turned in relation to each other to create different codes based on the same principle (they are just packaged in different ways). Left: A simple cardboard dial. Right: A collectible metal "Captain Midnight" decoder. Bottom: A "Spy King" decoder box with the code ring built in on the left side. [Classic Concepts photos copyright 1999, used with permission.]

Thus, the Enigma was known before the outbreak of World War II, and Polish and French cryptanalysts had already broken a portion of the Enigma code by the late 1920s. This would later aid British cryptologists, although it would still be a challenge, with the Germans making changes to the system during the War.

America After World War I

Herbert O. Yardley (1889-1958) was a telegrapher for the U.S. State Department during World War I. After the war, Yardley established a cryptanalytic bureau whose job it became to decipher codes in foreign diplomatic correspondence. This organization came to be known as the "Black Chamber" or MI-8 and made history by decrypting enciphered Japanese correspondence. In 1929, Yardley was directed to shut down operations and turn the unit resources over to William F. Friedman (1891-1969) and the Signal Intelligence Service. Friedman, a Russian immigrant who had set up a cryptology school with his wife, became director in 1930. In response, the disenchanted Yardley wrote "The American Black Chamber" describing the detailed activities of his former unit. The second edition was suppressed by the U.S. government, but not before the first edition reached other nations, who promptly changed their codes.

Friedman had his work cut out for him. Not only had the foreign codes been changed, but foreign nations were now especially wary. Nevertheless, Friedman's unit succeeded, a few years later, in decoding the new Japanese code, called the Purple Code. This allowed the U.S. to decipher many of the Japanese communications during the war.

The *National Security Agency* sponsors the National Cryptologic Museum, which includes, among other exhibits, historical information and artifacts related to Yardley and the Black Chamber (left) and a variety of cipher machines, including some from other countries. Shown here is a Jade machine, in the same family of encryption machines as the famous Japanese Purple machine cracked by U.S. Cryptologists. A portion of a Purple machine is also on exhibit. [National Security Agency Cryptologic Museum news photos, released.]

In August 1940, U.S. Army Intelligence engaged in a decryption effort named MAGIC to break the Japanese diplomatic code, making it possible to access the content of a radio message sent on 7 Dec. 1941, from Tokyo to the Japanese Embassy in Washington, D.C. Thus, eight hours before the Japanese bombed Pearl Harbor, news of the break in diplomatic relations between Japan and the U.S. was intercepted, but the political policies necessary to quickly process and respond to this information and avert the bombing were not in place or, apparently, were ineffective.

"Prior to the attack, the Americans had broken several Japanese naval and diplomatic codes and ciphers and also had intelligence indicating that an action was imminent. Lacking was the accurate and timely processing and evaluation of the available information, which in a well-organized and smoothly running organization would have been a matter of routine. Pearl Harbor stimulated the development of an efficient secret service and led to the establishment of the CIA in its present form". [Dr. Georg Walter, "Secret Intelligence Services", August 1964.]

The U.S. Navy *Radio Intelligence Section* was responsible for providing communications intelligence on the activities of the Japanese Navy. It had to intercept the radio communication, decipher them, translate the text, and forward the results to the appropriate authorities. The various intercept stations, dotted around the Pacific, exchanged information, data, codes, and keys. The Navy had two categories, Traffic Intelligence (TI), more general information about the communication, and Decryption Intelligence (DI), more specific information derived from the content of the message itself. In spite of the gradual losses of some of its intelligence stations, the Navy learned about Japanese plans to attack Port Moresby and Tulagi, prior to the Battle of the Coral Sea. While U.S. losses were high in the Battle of the Coral Sea, it was a turning point and a setback for the Japanese and weakened their position in the Battle of Midway that followed. After the Battle, even those skeptical about the value of cryptologic data had to accede that it had provided essential information strongly affecting the outcome.

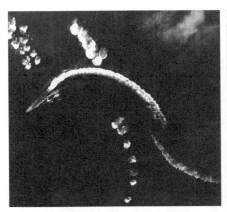 

Left: An aerial surveillance photo shows the Japanese aircraft carrier Hiryu maneuvering to avoid bombs dropped by the U.S. Air Force B-17 bombers. Naval intelligence cryptanalysts had advance notice of the 'surprise' attack and were able to rally forces to counter the Japanese offensive. Right: Burning oil tanks are seen in the background after the Japanese air attack on Sand Island, Midway, in June 1942. [U.S. Navy historical photos, released.]

The War on the British Front

Alan Turing (1912-1954), the shy and eccentric father of much of modern computing became part of the elite group of code-breakers working on the German Enigma code, and its successors such as the Fish codes, at Bletchley Park. The cryptanalysts at Bletchley Park were there under the umbrella of the British Foreign Office's Government Code and Cypher School and their activities were kept secret as much as possible.

During the War, Turing and Johann (John) von Neumann (1903-1957), the brilliant founder of game theory, were involved in developing concepts of programmable computers which were used in efforts to crack the German codes. The highly classified Colossus, an advanced Boolean logic calculator that preceded the ENIAC, was based in part on their ideas.

The British had a code machine of their own, called the British Typex, which was similar to the Enigma, but apparently more complex and presumably more secure. The Typex was apparently never fully put into service.

Cryptology had a significant impact on the outcome of World War II. Allied knowledge of the content of intercepted communications about German movements and plans gave them a decisive advantage.

An Unfortunate Enigma

On a side note, on 1 April 2000, a rare Abwehr Enigma G312 four-rotor cipher machine, disappeared from the mansion at Bletchley Park, where scientists and codebreakers broke the German codes used in World War II. Because the machine disappeared on April 1st, it was a puzzle as to whether it was a temporary prank or theft. The mystery was cleared up, however, when a number of ransom demands were made to the Bletchley Park Trust.

In October 2000, the rare antique was sent by the thief to a BBC newscaster, minus three of the four vital rotor wheels. Another ransom note for the missing parts soon followed. Fortunately, the thief was apprehended after contacting the Sunday Times about the missing parts, and later pleaded guilty. Wikipedia has an extensive entry on the Enigma machine that includes circuit diagrams and a mathematical description of the machine's encrypting/decrypting schemes.

Postwar Developments

The defeat at Pearl Harbor and the failure of U.S. officials to process the information obtained by cracking the Japanese code and intercepting radio communications illustrated the need for better lines of communication with the cryptologic unit. In 1945, the *U.S. State Department*, Army and Navy forces, acknowledged that a better system was needed. As a result, the Coordinator for Joint Operation (CJO) was established in 1946 to coordinate cryptologic services. The *Joint Intelligence Bureau*, which replaced the Joint Intelligence Committee in 1946, became a primary agency for Central Intelligence Analysis.

In 1947, General Hoyt Vandenberg of the U.S. Air Force endorsed a separate Air Force cryptologic intelligence organization to provide independent support to the evolving Strategic Air Command. A 1948 agreement between the Air Force and the Army Security Agency set out a mobile and tactical role for the U.S. AFSS. In June 1948, the *Air Force Security Group* was established to oversee the transfer of resources to the as-yet unestablished Air Force cryptologic agency. In spring 1949, Secretary of Defense Louis Johnson announced the *Armed Forces Security Agency* (AFSA) to oversee the separate cryptologic efforts.

One of the groups that is currently most active in gathering and interpreting historical information on intelligence which includes cryptologic information is the International Intelligence History Study Group. Members of IIHSG are engaged in compiling the manuals and cover sheets related to the ciphering methods used by the Wehrmacht (the German armed forces), as a complete set had not been reconstructed as of 1996. Gilbert Bloch, who suggested the project, has also obtained a monthly list of Enigma settings, "Sonder-Maschinenschlüssel BGT". These efforts are gradually building a clearer picture of one of the most historically significant uses of cryptology in warfare.

Computers, Electronics, and Cryptology

Gigantic, expensive, vacuum-tube computer systems with limited calculating abilities began to emerge in the 1940s and 1950s. For the first twenty years, they were used almost exclusively in government applications, scientific research, and big business settings. The ENIAC computer, used by the U.S. Army to calculate ballistics tables and trajectories, beginning in 1946, weighed 30 tons.

Computers changed cryptology and cryptanalysis in some fundamental ways. Not only could more complex or extensive encryption and decryption techniques be applied, but codes of an entirely different character could be developed, compared to those devised by hand or with mechanical tools.

Another consideration in early computing was the cost. Since only a privileged few had access to computing resources, it put encryption and decryption in the hands of those with power or money. It was not until the transistor was invented in 1947, and electronics gradually came into the hands of the general public, that discussions over who could encrypt their messages and who couldn't became controversial.

The development of computers and computer networks also provided a way for private citizens to more closely follow the actions and decisions of government policy makers and secret service agencies. Computer networks greatly facilitated global communications, but individuals bent on crime could now access information and collaborate with others in distant states in unprecedented ways. Dealers in illegal shipments, and money launderers, could carry on their business on a grand scale, using encrypted telecommunications technologies to avoid detection.

Law enforcement agents were concerned about keeping up with criminal communication venues, and lobbied to update wiretapping authorizations to include electronic communications. At about the same time, computers began to infiltrate mainstream entertainment media. Movies like MGM's *WarGames* (1983) increased concerns about 'hacking' (and made it a household word). Five years later, President Reagan signed a bill intended to safeguard declassified information, such as census and tax records, and the *National Bureau of Standards* was tasked with developing software to protect computers from illegal entry. The *National Security Agency* sought empowerment to decrypt foreign telecommunications. Private citizens started voicing concerns about how general policies might affect basic freedoms. As computing moved into the late 1980s and early 1990s, controversy followed every new evolution in encryption technology and every step on the part of the government to gain access to private communications.

The 1990s - Secure Communications and Controversy

By the 1990s, businesses were computerized, business software was sold with various security features, and the electronic encryption of radio and data communications was commonplace. National security experts were alarmed at the potential for criminals to carry out clandestine planning without intervention by law enforcement agencies and the general public was alarmed because the NSA and the FBI wanted broader access and open-ended wiretapping authorization in order to be able to intercept secure electronic communications. This important struggle is discussed at some length in the historical sections of Chapter 1 (Introduction and Scope) and Chapter 2 (Audio Surveillance) and is still ongoing.

By 1990, the U.S. *National Security Agency* (NSA) had established a lab for developing special-needs computer resources, such as electronic chips and circuit boards. This served two purposes. It allowed the NSA to design and construct circuits that weren't obtainable through commercial channels and it ensured the secrecy of the specialized circuits and the state-of-the-art of NSA resources. The NSA was developing one of the most extensive computer facilities in the country, which was used, in part, for encryption and decryption development.

In 1993, the Clinton Administration announced the *Escrowed Encryption Initiative* (EEI), commonly known as the *Clipper Initiative,* after it was developed as a hardware device called the *Clipper Chip.* The Clipper Chip was a hotly contested system incorporating a classified, secret-key encryption algorithm called SKIPJACK that was implemented in an Escrowed Encryption Standard (EES). The controversy over the EEI arose partly because of a Law Enforcement Access Field (LEAF) incorporated into the system. LEAF was intended to enable access by law enforcement agents to otherwise secure communications. The keys were intended to be deposited with the National Institute of Standards and Technology and the U.S. Treasury's Automated Systems Division. Thus, a court-authorized 'wiretapping' could be deciphered through obtaining the key information. The computing population didn't like the system and debated it over the Internet and through elected representatives.

Transfers of Regulatory Authority

Two key events occurred in 1996 with regard to encryption. The Government relented to public pressure and presented a modified Clipper Chip plan. It also reclassified encryption technology. Up to this point, encryption was treated pretty much the same as a physical weapon. The export of the technology was regulated under Arms Export acts and regulations. However, regulatory control for non-military encryption technology was transferred to the U.S. Department of Commerce *and new regulations were issued at the close of the year, to be administered by the* Bureau of Export Administration. *Categorical distinctions for different types of encryption were now being formally recognized.*

Thus, the government came back with a modified plan (Clipper II) with a change in the escrow agencies with input from users. The debate still didn't die down. A government draft in 1996 (called Clipper III by some) proposed a key management infrastructure based on a voluntary key escrow plan. At this point, users would be permitted to choose their encryption algorithms. Public-key certificates would be issued by a Certification Authority. In some cases, if certain requirements were met, self-escrow might be an acceptable option. These were significant adjustments to the original plan, but there was still opposition. A Technical Advisory Committee was formed, but after many meetings, a comfortable agreement between government and business was not reached.

In 1997, the *Department of Defense* made an announcement that the *National Security Agency* (NSA) would not be implementing EES in favor of a *key recovery* system. It was published as the *Electronic Data Security Act of 1997*. The act was amended only two months later, as a response to further criticisms. A number of bills and options were presented over the next two years with no general consensus. Monetary incentives in the form of tax credits for developing and producing recoverable cryptography were even proposed to try to come to a resolution or compromise.

By summer 1998, the U.S. *Department of Commerce* restrictions on software exports were challenged at several court levels, with courts initially ruling that the government export regulations on encryption were constitutional.

By this time, programmers and hardware designers had developed ways to integrate the Internet with public telephone systems and started promoting Internet phone systems. The traditional phone carriers were alarmed and opposed the systems, which essentially threatened to make long distance calls over the Internet free of charge. Internet phone systems also blurred the line between regular phone regulations and regulations for electronic communications. Law enforcement agents were more concerned than ever about their ability to intercept communications and sought to adjust the scope of 'wiretapping' to encompass electronic technologies while still facing continued opposition.

Encryption's Soft Underbelly

At the 1998 EPIC Cryptography Conference, Robert Litt spoke on behalf of the Attorney General's office regarding the strength of DES and reiterated the need for a law enforcement 'back door' to access encrypted communications. Shortly thereafter, government assertions regarding security faced an even stronger challenge from cryptanalysts.

In July 1998, the Electronic Frontier Foundation announced that its EFF Data Encryption Standard Cracker (ESC) machine had cracked the Digital Encryption Standard (DES) in less than three days, substantially faster than a network of computers that had previously established a record of 39 days.

Undersecretary William Reinsch responded to this announcement a month later on behalf of the *Department of Commerce* stating law enforcement's point of view, which included the following statement regarding the practicality of the code-breaking success:

> "... Spending 56 hours breaking a single message in a situation where those making the attempt knew where the message was and, presumably, knew it was in English, is not analogous to the real-time problems facing law enforcement. At the same time, this is a fast-moving sector, and recent developments in it, including EFF's own efforts, demonstrate a need to continually review our policy in light of such changes. In fact, such a review is underway right now, but like previous reviews, it will continue to be based on the same fundamental principles that underlie our current policy..."

> [William A. Reinsch, United States Department of Commerce letter to Barry Steinhardt, President, the Electronic Frontier Foundation, 26 August 1998.]

Half a year later, the Encryption Standard Cracker, designed by John Gilmore at a cost of under $250,000, had cracked a message encoded by RSA Data Security in less than a day. This strengthened the skepticism surrounding the security of DES and U.S. Government advocacy of its use. The ESC machine had cracked the code in conjunction with the nonprofit Distributed. net for the DES Challenge, which was an annual event hosted by RSA Data Security, offering cash prizes for successful attacks and larger prizes for fast attacks.

Data Security and International Software Export Regulations

The Technical Advisory process on electronic encryption continued and the group released a report. Before the deadline for comments, in September 1999, a proposed draft of the *Cyberspace Electronic Security Act of 1999* was released. Once again, it contained provisions for access to encryption keys by law enforcement agents. The FBI would be funded to aid law enforcement agencies to deal with increasing use of encryption for criminal purposes.

While the public and the government were trying to balance information access with privacy, U.S. computer vendors were trying to maintain leadership in the global marketplace. For several years there had been pressure from the U.S. software-marketing community for looser restrictions on encrypted products. Vendors had persistently argued that U.S. export restrictions had a negative effect on their international competitive edge. On the other hand, law enforcement officials feared that covert communications of an illegal nature would be facilitated by stronger encryption and argued against the change. The law enforcement community was further concerned that the detection and collection of evidence to convict in cases of criminal activities, both domestic and foreign, would be greatly impeded by increased encryption.

U.S. businesses wanted to export stronger encryption schemes. The U.S. public didn't want the government to have special privileged access to any kinds of messages, and the various security agencies felt their power to carry out their responsibilities were being eroded. Trade-offs were proposed. Certain government parties suggested that encryption export laws could be relaxed if law enforcement had encryption keys or other methods of accessing encrypted data. A bargaining tone had been set.

Meanwhile, the European Union was continuing its task of uniting Europe into a common market that would exceed the population size of the U.S. once it was fully integrated. At the same time, India and several European nations began to export stronger encryption technologies than were provided by American vendors. The competitive pressure was cited in further U.S. encryption debates.

In July 1999, a House Armed Services Committee hearing listened to concerns by the Attorney General and the Director of the FBI regarding the proposed relaxation of encryption restrictions on electronic communications-related products. Law enforcement officials warned that the ability to carry out investigations of illegal activities would be severely hampered if the U.S. sanctioned high-level encryption on a variety of consumer products, ranging from cell phones to computer software. Officials were particularly concerned that their ability to detect and investigate criminal activities such as international terrorism, economic fraud, narcotics and weapons trafficking, and related violent crimes, would be greatly curtailed.

Export Regulations Adjustments

Widespread criticism had broken out in Europe in the mid-1990s when it was discovered the U.S. software for export was only weakly protected, with most of the encryption key information provided to the *National Security Agency* (NSA) so that they could easily 'crack' the codes, if needed. Several European nations discovered that their documents were far less secure than they had assumed at the time of purchase.

Then, in 1998, Signitron, a Calcutta-based company, announced that it had developed a 448-bit encryption package suitable for desktop computers. The market was globalizing. Signitron's package was fast and reasonably priced, offering excellent privacy—unencrypted data was never stored on the working hard drive. Commercial encryption in the U.S. for personal computers was typically 32- or 64-bit at the time. Stronger encryption had been developed, but the government imposed export restrictions on anything 128 bits or stronger. Signitron estimated it would take "at least 2 trillion years to crack the[ir encryption] code". This put pressure on U.S. developers to export more efficient, more secure encryption schemes for desktop systems.

In September 1999, President Clinton responded to corporate concerns and relaxed restrictions on the export of privacy-protecting software, a move to 'level the playing field' in international data security.

The FBI and others in the intelligence community opposed the move. How could law enforcement keep up with criminal activity if they couldn't intercept or examine criminal evidence?

The White House countered the FBI and others by promising encryption keys would be stored with third parties and law enforcement agents could obtain them through a court order similar to requesting wiretap. The situation could still be complicated however, by two-key systems if the person holding the second key refused to divulge information. It may also partly have been a face-saving gesture in response to criticism from the European Union.

Whatever the motivations for the Clinton announcement, encryption was now officially sanctioned by the White House, pending submission of encrypted products to the U.S. *Department of Commerce* for one-time review prior to export. Thus, regulations for products with up to 64-bit key lengths were relaxed and those with over 64 bits could obtain a license exemption, provided they were not shipped to a short-list of countries suspected of terrorist leanings.

These regulatory changes were intended to address the needs of vendors shipping products on physical media. The guidelines didn't fully answer questions about multinational collaborative projects or software distribution through the Web. Presumably vendors couldn't permit software downloads by foreign buyers without a *Department of Commerce* review—a bureaucratic hurdle that could hurt small software development companies and startup businesses, and a logistical problem, since foreigners could use proxy servers to simulate domestic downloads.

The world had changed. In the past, if people wanted to send encrypted messages, they would have to labor over codes or use machines like the Enigma to slowly construct and deconstruct messages. Now it was possible to buy inexpensive, off-the-shelf packages with strong encryption to automate the process—encryption and de-cryption was no longer a time-intensive endeavor, but a basic convenience.

China also has an interest in developing encryption technologies and was one of the nations submitting an encryption scheme to the ISO group setting global encryption standards. In 2006, the American 802.11i standard was selected over the Chinese scheme, known as WAPI. China expressed concern that the selection was politically motivated, while others claimed that China had released too little information about the scheme for the international community to be comfortable with establishing it as a standard. Despite the setback to setting a world standard, the Chinese announced the creation of a group of companies to promote WAPI, which is said to be more secure than 802.11i.

Spying on Consumers

In the 2000s, new problems associated with encryption began to arise. As the capability to build encryption into consumer telecommunications devices and digital data formats became easier, they were incorporated into products in a number of ways that weren't obvious to consumers and which caused concern among privacy advocates.

Retailers have always been eager to gather information on customers and their buying habits so they can target their marketing dollars as effectively as possible. In the 1990s, more sophisticated tracking systems and less expensive storage devices made it possible to monitor the buying patterns of individual consumers on an unprecedented scale. By the early 2000s, serial numbers or other identifying data was encrypted into commonly purchased products such as cell phones, printers, and digital cameras—information that could lead directly to the registered owner. Many consumers took for granted that their cell phone calls were secure and private. When service providers were subpoenaed for information related to specific customers' calls, they discovered, after-the-fact, that their assumptions were wrong. In some cases, messages had been logged and stored. Similarly, many consumers weren't aware that some digital cameras and desktop printers were secretly encoded with identifying information. Patterns were routinely inserted into digital images and physical printouts that could identify the specific unit that created the image or hardcopy, in such a way that they could not be easily detected or removed.

From a law enforcement perspective, it's easier to track counterfeiters and those who create ransom notes on their printers, if the documents are encoded. From a consumer point of view, having one's printouts secretly coded raises issues of privacy.

Effectiveness of Encryption

How secure are encryption schemes? Encryption based upon 448-bit keys is very strong indeed. But recent history has taught us that every time a newer, stronger form of encryption is developed, someone finds a way to crack it. This is partly due to individual enterprise, and partly because computer processing speeds and storage capacities continue to increase, providing better tools for decryption than were available when the original encryption schemes were developed. Even if a new system is strong enough to withstand decryption attempts, that still doesn't mean it's secure. Many individual devices or components can be reproduced with the right equipment---which makes the encryption aspect irrelevant. In other words, if a credit card contains encrypted data that can't be decrypted, a thief may try other strategies, like getting access to the machinery that makes the card. Encryption protects only the data. It doesn't protect the mechanism that creates the data or take into account the personal habits (or integrity) of the person carrying the medium upon which the data is stored. It also doesn't protect the object from theft or change the fact that employees may have access to protected information and equipment and that people can be bribed.

5. Description and Functions

Some people write coded messages to protect the privacy of their correspondence, or simply because it is personally convenient. Beatrix Potter, the author of the Peter Rabbit books, used a sort of 'coded' writing in her journal, but it may have been a way that was comfortable for her to write, rather than a deliberate attempt to obscure. As mentioned earlier, Leonardo da Vinci may have written backward for reasons of convenience rather than secrecy.

Businesses encrypt their messages and electronic financial transactions to ensure privacy and protection against theft and fraud. This is especially true for Internet transactions and other computer network electronic transfers, particularly those used in the banking business.

Governments encrypt their communications to prevent spying, aggression, economic disadvantage, and terrorist acts. They also rely heavily on encryption for creating, sending, and decrypting general communications. Communications encryption systems are commonly based upon 'key' encryption, which is difficult to break if the keys are sufficiently long.

Since this chapter focuses on cryptologic surveillance, the discovery and discernment of coded messages, it doesn't delve into the coding of the messages themselves. The reader is encouraged to consult specialized references for information on coding and encryption.

Discovering Messages

Hidden messages may be discovered deliberately or accidentally. Checking the pockets of a slain or captured messenger might reveal a hidden message. Rifling the drawers of someone suspected of stealing might unearth a coded diary.

The fact that a message is obscured is not necessarily a secret. We all know that sensitive government communications are classified, obscured, secured, and often encrypted. Users who buy spread-spectrum wireless phones usually know they are getting a certain amount of security built into their private or business communications.

Discerning Message Contents

Sometimes people decipher messages because they are nosy and curious about personal relationships. Some do it to get a 'news scoop' or to steal sensitive business information (e.g., insider stock information). Some people are temperamentally drawn to solving puzzles and 'mysteries'. Others are hired specifically to discover hidden messages or break codes.

The motive for deciphering a message often determines how hard a person will try to decipher it. In the case of personal or business-related information, people may give up fairly easily. It usually isn't worth the weeks or months it takes to unravel a message that may yield nothing of interest, and not everyone has the skill and patience to succeed. Those who like to solve mysteries usually devote only as much time as they feel they can spare. Those who have a passion for code-breaking or do it in the course of their work may spend months or years unraveling more sophisticated problems.

There is no one method for breaking a code, because there are tens of thousands of ways to make a code. It takes persistence, patience, logic, insight, intuition, and sometimes a bit of luck, to find and decipher hidden and encrypted messages. Deciphering the Enigma codes in World War II would have been far more difficult if the French and Polish hadn't known something about the machines and if Enigma machines had not been found on a German U-boat and a German trawler.

Code-breaking takes talent and dedication, but it also requires a good knowledge of the various schemes that have been used over the years. These days, with sophisticated computer programs and advanced techniques, it is almost essential to have good mathematical skills to deal with the kind of programs that now handle a majority of encrypted communications. Codes are becoming harder to break, but 'unbreakable' codes are often shown to be fallible by brilliant sleuths who put their minds and computer resources into cracking them.

Some codes can be broken in more than one way. Obviously, in breaking a simple substitution code, the easiest way is to find the original coding ring or table. When this isn't possible, analyzing the patterns of letters until familiar landmarks appear can eventually lead to reconstruction of the code—it just takes longer. Finding a written translation by the recipient of the coded letter is a third option and, while some people might laugh at this solution, it is often the best, as the human element is usually the weakest link in any code. In other words, looking on a person's computer monitor for a sticky note with a password written on it is often easier and faster than trying to 'break in' to a secure account by trying thousands of passwords. When Richard Feynman started working at the Los Alamos laboratory, he took a certain delight in breaking into 'secure' file areas, but he discounted the human element, at first, until he was convinced that it was, indeed, an important aspect of security. A knowledge of human nature is important in all aspects of cryptology—birthdays, names, and information about loved ones are frequently used as passwords or 'keys'. A knowledge of sociology and psychology makes it easier to to see the weak links in schedules, personal traits, and administrative policies, in order to gain unauthorized access to accounts. In fact, this may be the only way to gain access considering that strong computer encryption makes many communications virtually inaccessible by any means.

Cryptanalysis Strategy

One of the key aspects to breaking a code is finding a weak element. The human element was mentioned in the previous paragraph as it is often the weakest link, but sometimes the human element is not available (or not cooperative), or the code is intercepted far from its origin or destination. In these cases, other weak elements can be sought. Here are just a few examples:

- The weak element might be the language in which it is written. For example, the English language has many consistent patterns. Sentences are constructed in certain ways, each letter of the alphabet has a particular frequency of use in normal correspondence, and certain words like 'the' or 'a' or 'I' are common giveaways. Sometimes parts of an intercepted message aren't even encoded. If a time or date or important name is spelled out, it can provide a clue to the content of the entire message. Thomas Jefferson and his correspondents would sometimes encode only part of a message, betraying information about sentence style, possible content, etc., through the unencoded portions of the text.

- The weak element might be tidiness. Lookup tables that are neatly numbered or alphabetized or organized according to categories are security weaknesses. If small parts of a message with random lookups are decoded, the decoded parts don't give away much about the rest of the message. If ordered lookups are used, however, decoding small parts can provide significant clues as to the structure of the rest of the message.

- Sometimes the weak element is sloppiness or laziness. Many codes derive their security from the fact that the key or code changes every time it is used. Thus, decoding one message may not provide any information about subsequent messages. However, if the coder (or the computer algorithm) fails to change the code each time, the next message may give away the structure of the previous message.

- The overt nature of the message itself might be the weak element. Even before a message is decoded, its situation or basic characteristics will often give away the general nature of the contents. For example, if a teacher sees a piece of paper being passed between two giggling teenagers of the opposite sex a few days before the school dance, she can make a pretty good guess as to what it might say. Similarly, if the Germans in World War II sent a short coded telegraphic message to an Axis country after just sinking an Allied submarine, cryptanalysts could make a pretty good guess as to what its contents might be before it is even decoded. This, in turn, can provide 'Rosetta Stone' clues to future correspondence. Further, the length of the message, the material on which it is transcribed, whether it is typed or written, may all give away clues as to its origin or intent. The archetypal ransom note is often depicted with newspaper letters pasted on a sheet of paper, intended to prevent the kidnapper's handwriting from being analyzed. However, the source of the letters, the type of paper, the skill of the paste job, and even the brand of glue and who sells it, can betray the kidnapper's location and personal characteristics.

- One of the weaknesses in computer encryption systems is that many rely on 'random' number generators. Pseudorandom numbers are the rule on computers and thus represent a security weakness that can sometimes be replicated by using a similar system. Many computer encryption technologies use pseudorandom numbers at some point in their processing.

Data Encryption Approaches

It's impossible in this limited space to include all the encryption techniques now used in electronic communications, but it is worthwhile to mention a few. Various methods of breaking these systems have been attempted and have sometimes succeeded. Both differential and linear cryptanalysis have been used, and other attacks such as interpolation. Two general categories include *stream encryption* (small substitutions that change as they go) and *block encryption* (larger blocks of substitutions that include some of the key encryption techniques).

Key encryption is accomplished with a string of data that determines the mapping of the unencrypted data to the encrypted data to make it possible to encrypt or decrypt that data. Some schemes have two keys, often a *public key* and a *private key* known only to the user. Keys have different lengths which are usually related to the 'strength' of the encryption. Key encryption is one of the most prevalent means of encrypting electronic data for security purposes. (Compression algorithms also encrypt data, but their primary purpose is not usually to obscure the meaning of the data but rather to create a means to store or transmit the data more efficiently.)

In general, stronger encryption means longer encryption and/or decryption times, though this is not true in every instance, as more efficient algorithms are sometimes devised. The choice of encryption techniques depends very much on the need for convenience and the required level of security. Personal letters are less sensitive than classified government documents, for example, and function well with lower levels of encryption and single keys (or no keys).

Encryption techniques can be symmetric, in which the encryption and decryption use the same key or in which the encryption and decryption processes require about the same amount of time. They can also be asymmetric, with different keys, or substantially different encryption/decryption times. For purposes of security, a method that is quickly encrypted and slowly decrypted is favored for some purposes. For general correspondence, however, slow decryption is an inconvenience, as are multiple keys.

Encryption systems can be *deterministic* or *nondeterministic*. One that generates the same result each time, given the same key, is deterministic and is generally not as strong as one that generates a different result, given the same key. However, true randomness in computer operations is not usually the rule and many systems are deterministic.

Encryption algorithms can be reversible or irreversible. A reversible scheme is one in which data can be recovered back to its unencrypted state. Irreversible schemes cannot be recovered, but since they tend to be used as authentication or tamper mechanisms rather than as message recovery mechanisms, they are valuable for certain tasks.

Some specific encryption schemes of interest include

AES - IA-8314: Advanced Encryption Standard. A new, stronger encryption algorithm intended by the *National Institute of Standards and Technology* (NIST) Computer Security Division to replace DES. The project was initiated in January 1997 as a standard laboratory network protocol. Da Vinci is an example of a shareware encryption system based upon AES. CipherMax is an example of a commercial product that provides 256-bit AES encription.

APKC (Absolute Public Key Cryptography) - This system was patented in August 2006. It is a public key cryptographic system and method that offers two-way communication security even when a private key is revealed. APKC can support mobile devices with low processing power and short keys. The keys have two or more components, a random number is bound to each of the components of the public key, the message is encrypted into the same number of cipher versions as components, and the ciphers are delivered to the destination in source routing or hop-by-hop routing with a small time gap. All ciphered version are mathematically manipulated to reconstruct the original message. This method prevents an attack at an intermediary IP router because all cipher versions must be available and the original message cannot be obtained even if the attacker knows the private key. [U.S. Patent # 7,088,821]

Blowfish - Designed by Bruce Schneier and first presented in 1994. Widespread. Standard in OpenBSD. A symmetric block cipher that accepts a variable-length key from 32 bits up to a maximum of 448 bits. It can be used as a replacement for DES or IDEA. Small block size, good speed, uncomplicated interface. Introduced in *Dr. Dobb's Journal* in Apr. 1994, with source published in *Dr. Dobbs Journal*, Sept. 1995. Variations on Blowfish have been developed, including Blowfish Updated Re-entrant Project (BURP) by Geodyssey Limited. Blowfish is unpatented and license-free.

Discussions on weaknesses and attempted breaks of Blowfish include papers by Serge Vaudenay and Vincent Rijmen.

DEAL - Data Encryption Algorithm with Larger blocks. A 128-bit block cipher based on DES (DEA), but which is intended to overcome some of its weaknesses. DEAL allows for key sizes of 128, 192, and 256 bits. It is intended to deter 'matching ciphertext attacks' that can be used against DES.

DES (DEA) - IA-8307: Standard Unclassified Data Encryption Protocol, approved for use in 1977. A 64-bit block cipher that accepts a 64-bit key, of which 56 bits are active. Established as a standard in Dec. 1993. With current computing systems, it is possible to break 64-bit systems and thus DES is no longer considered sufficient for high-security applications. To overcome this weakness, DES may be encrypted multiple times with multiple independent keys to improve security somewhat (e.g., Triple-DES). It is expected to be superseded by AES.

932

Kerberos - A client-server sign-on model in which cryptographic keys are exchanged. Similar to SESAME, but focused on UNIX systems. See SESAME.

MEAS (Manansala Encryption and Authentication System) - A means to tighten Internet security and authentication for applications such as online credit card purchases such that private knowledge from the user is used to encrypt the data upon receipt, by the user, of an authentication code consisting of fa long random number. The receiver (e.g., a credit card verification agency) generates the private knowledge ky to decipher the message and can thus verify the identity of the person using the card for the online purchase without the credit card number itself being transmitted. This overcomes some of the weaknesses in digital public key systems and the lack of individual identification of the persons establishing a Website based on digital certificates (or impersonating digital certificates).

PGP - Pretty Good Privacy. Developed by Philip Zimmermann, based upon Blowfish technology, PGP provides a measure of privacy and authentication for data communications. It is widely used for email communications and freely distributable for noncommercial use. Acquired by Network Associates in 1998.

RSA - Designed by RSA Data Security, Inc. A well-recognized, patented asymmetric public-key/private-key cryptography standard named for its developers, Rivest, Shamir, and Adelman.

SESAME - Secure European System for Applications in a Multi-vendor Environment. A scalable, client-server, sign-on, distributed access control system using digitally signed Privilege Attribute Certificates supporting cryptographic protection of remote applications and communications between users. Uses authentication tokens. It is an Open Systems protocol that provides components for computer product developers. SESAME is partly funded by the European Commission. It originated in the late 1980s as part of the Open Systems Standards work of the European Computer Manufacturers Association and was beta released in 1994. It is similar to Kerberos.

Solitaire - Designed by Bruce Schneier. An output-feedback mode stream cipher (key generator). A physical system, though it can work on computers. Not fast, but intended to be highly secure when used for small messages, even against those who understand the algorithm. Security is dependent on keeping the key secret. To maintain security, the same key should never be used for more than one message and the deck must be reshuffled after encryption. The system requires diligence, for if an error is made in the process of encryption, everything beyond that point will be incorrect.

Twofish - Designed by B. Schneier, J. Kelsey, D. Whiting, D. Wagner, C. Hall, and N. Ferguson at Counterpane Labs. A 128-bit block cipher that accepts a variable-length key up to a maximum of 256 bits. It can be implemented in hardware in 14000 gates. Counterpane has extensively tried to break Twofish, reporting that the best attack "breaks 5 rounds with $2^{22.5}$ chosen plaintexts and 2^{51} effort". Selected as a finalist for the Advanced Encryption Standard (AES). The source code is license-free, uncopyrighted, and free.

Yarrow - Designed by Bruce Schneier and John Kelsey. A pseudorandom number generator intended to be secure compared to the native 'random' number generators incorporated into compilers or computer hardware. The system is unpatented and license-free.

There are many others, including FEAL, variations on LOKI, MacGuffin, MAGENTA, SAFER, SSL, TEA, library encryption (based upon a five-rotor Enigma), etc. Information on these schemes and attacks against them can be found on the Net and in Print and Online Resources listed at the end of the chapter.

6. Applications

Cryptanalysis is intensely interesting, challenging, and rewarding to those who enjoy it and inscrutable and mystical to those who don't have the mindset for this particular kind of problem-solving. For this reason, code-breaking, beyond the basic alphabetic substitution code, is only practiced by a very small portion of the population. However, that small portion can have enormous power and influence over the affairs of state if they happen to find weaknesses in classified domestic or foreign codes or if they are unscrupulous enough to surveil private business or financial dealings.

The application of cryptanalysis is much like the playing of chess. It takes time, energy, talent, and practice. It is also very specific, very technical, and thus beyond the scope of this book in terms of specifics.

Code-breaking was a pivotal surveillance technology affecting the outcome of World War II and it may be a pivotal technology in the competitive economic race that is playing itself out on the global stage.

While encryption is broadly used in many consumer products, code-breaking is mainly used in the government secret services and, to some extent, by the computing and mathematical communities in the pursuit of knowledge and better encryption/decryption algorithms and systems.

7. Problems and Limitations

Inconvenience

One of the biggest problems with encryption systems is that they add time to any task. Some aspects of encryption can be automated. Email can be automatically encrypted when sent and decrypted when received, which certainly adds a good measure of security to messages en route, but the computer on which the email is stored may still be wide open. If the computer is password-protected, it's safer, but then you have to remember the password and a hundred other passwords for all the different services and features offered by banks, retailers, and various Web sites. Anyone who uses the Net regularly has at least a dozen passwords and some have hundreds. Putting them in a database isn't secure. Writing them down isn't secure. Memorizing them is beyond the capabilities of most people. The only reasonable solution is an algorithmic password and even that can have its disadvantages, depending on how it is designed, because many sites impose odd password requirements that might conflict with the algorithm.

The Human Element

While advanced mathematics and more powerful computers have made it possible to develop encryption systems that are difficult and, in some cases, virtually impossible to break, it will never be 100% possible to ensure the security of information before it gets encoded or after it has been decoded. Encryption is not a panacea. At best it is like locking a door—it will deter opportunists or those who are lazy—it won't deter the persistent and determined.

No encryption system can be considered to be completely secure because people make mistakes or have personal agendas—they lose keys, they forget passwords, they insert 'back doors', they get vengeful, they get sloppy, they may have to tell the truth in court, or they may accept bribes. An increase in the level of encryption in communications systems will probably result in spies shifting their focus from the technology back to the people who handle the messages rather than trying to decrypt them.

Arrogance

One of the problems with 'secure' systems is that the designers of new encryption systems are almost always certain they are unbreakable until someone breaks them. The events of the past ten years have revealed some remarkable feats of cryptanalysis that many encryption experts didn't anticipate. This has certainly advanced the science, but it has also taught us the lesson that no matter how sure you are, you can't be absolutely sure there isn't some way to solve what at first appears to be an intractable problem.

Politics

One of the problems with encryption systems is well-known to law enforcement and secret service agents. They feel that they are cut out of the communications loop if they do not have access to encryption keys or other means to access seized or surveilled information. Unfortunately, normal law-abiding citizens rarely are comfortable with giving access to these agents, feeling that past abuses with wiretapping and other surveillance activities on the part of government officials suggest that current administrators have the potential to engage in corrupt or unethical activities. These issues are discussed to some extent in Chapters 1 and 2 and they are as yet unresolved.

Another aspect of politics is balancing the needs of national security with the global competitive needs of businesses. U.S. security professionals are concerned about the U.S. supplying other nations with strongly encrypted business and communications tools. Other countries, on the other hand, will supply them if the U.S. does not, thus leading to a loss of momentum and business on the part of American software vendors. This issue has been hotly debated, with the result that export requirements were relaxed at the end of the 1990s to allow U.S. companies to retain a competitive advantage. Some agencies strongly opposed this move. At one point, the *National Security Agency* was given access to a portion of encryption keys in order to reduce the amount of time it might take to decode a communication. This issue, too, met with resistance and criticism from foreign nations and American privacy advocates.

8. Restrictions and Regulations

There are important export laws restricting the sale of encrypted products and encryption schemes, especially those intended to safeguard national security. This is a very difficult situation to legislate. It is relatively easy to say you can't ship an encrypted physical product overseas, but what if users download the product off the Internet? What if you have a note on the site saying foreigners can't download it but U.S. residents can? What if you post the source code to the encryption system for anyone to download, or perhaps you post just the theory? Is that considered an export of the technology? What about emailing the same code? If you collaborate with a mathematician in a foreign country on developing encryption technology, what export restrictions apply?

If you have questions about the legality of encryption matters, it is recommended that you consult sites specifically devoted to cryptology legal issues.

Some Acts of particular interest, in order of date, include

Computer Fraud and Abuse Act of 1986. Public Law 99-474. USC Title 18, Part 1 - Crimes, Chapter 47 - Fraud and False Statements. Defines fraudulent computer-related activities such as access without authorization or in excess of authorization and describes penalties. Cites the Atomic Energy Act of 1954.

Electronic Communications Privacy Act of 1986 (ECPA). USC Title 18, Part I - Crimes, Chapter 121 - Stored Wire and Electronic Communications and Transactional Records Access. Defines unlawful access to stored communications and associated penalties.

Computer Security Act of 1987. Intended to improve the security and privacy of sensitive information in Federal computer systems by establishing minimum acceptable security practices. Amends the Act of 3 March 1901 to assign standards and guidelines development to the *National Bureau of Standards.* Stipulates mandatory periodic training requirements for persons managing or using federal systems containing sensitive information.

Encrypted Communications Privacy Act (ECPA). Introduced in March 1996 to provide a higher legal level of security for computer network transmissions and protection of civil liberties by preventing unlimited access to computer records by the government. Affirms Americans' rights to use any encryption strength in order to maintain a competitive advantage in the world marketplace. Proposed barring of government-mandated key recovery or key escrow encryption allowing user to choose their method of encryption and protect the privacy of online communications and data files. Affects export restrictions on the export of strongly encrypted products to allow U.S. firms to compete in the global marketplace, meeting the demands of customers.

The lawmaking process is often slow, requiring three or more years for definitions and priorities to be sorted out before they pass through all the bureaucratic hurdles. In addition to the congressional acts and restrictions mentioned above, there have been many proposed and amended acts since 1997, and some which have passed into law. They are not discussed in detail here, but here are some for study and review. They are in various stages of completion, review, or rejection, which illustrates the complexity of the process of integrating technological change and law.

In October, 1998, the *Digital Millennium Copyright Act* was passed into law. The act was implemented, in part, to comply with treaties signed two years earlier at the World Intellectual Property Organization conference. In summary, it makes it illegal to circumvent anti-piracy protections and to sell or otherwise distribute code-cracking devices, except for testing and research purposes. Because of opposition from libraries and academics, there are provisions in the act to exempt nonprofit libraries and educational institutions in some circumstances. Service providers are also exempted from liability for normal transmission of data over networks by their clients/users. However, service providers are still expected to police their users' Websites and to remove materials that violate copyrights. While copyrights and encryption are two distinct issues, they are interrelated in that encryption is often used to protect copyright materials and cryptanalysis may be used to intercept protected data or, sometimes, to determine if copyrights are being violated.

Other acts of relevance to cryptography include:

Cyberspace Electronic Security Act of 1999. September proposal. Intended to protect the growing use of encryption for the protection of privacy and confidentiality by businesses and individuals, while still providing a means for law enforcement to obtain evidence to investigate and prosecute criminals using encryption to hide criminal activity. This would further provide funding to the FBI's *Technical Support Center* to establish a new cryptologic unit. The bill does not seek to regulate domestic use or sale of encryption.

Secure Public Networks Act (SPNA). Introduced to extend existing regulations.

Security and Freedom Through Encryption (SAFE) Act. Introduced to facilitate the removal of computer export restrictions, it was almost immediately amended to make substantial changes.

Internet Integrity and Critical Infrastructure Protection Act of 2000 (IICIPA). This bill was introduced to "enhance the protections of the Internet and the critical infrastructure of the United States, and for other purposes". This significant bill deals very specifically with hacking, unsolicited email, protection for privacy and confidentiality of various digital communications venues, online fraud, and international computer crime enforcement.

Cyber Security Enhancement Act (CSEA). Passed by the House of Representatives in July 2002 by an overwhelming majority, this act reduced the privacy of stored communications on commercial networks, allowing a service provider to disclose information if the provider had reasonable suspicion that the communication concerned a serious crime. It also set penalties for those involved in computer crimes, especially if the crimes resulted in death. It further increased penalties for those surreptitiously listening to unencrypted cellular phone communications, thus imposing new restrictions on amateur radio enthusiasts who had previously enjoyed 'open season' in terms of listening in on the airwaves. While the CSEA does not focus on encryption, but rather on matters of privacy, it is relevant in terms of general trends in balancing privacy concerns with meted responsibility.

9. Implications of Use

The struggle for balance between the need for public freedom and privacy and the need for access to evidence of criminal activity in order to safeguard public safety is one of the more potent and intractable debates facing the U.S. government and its citizens. In another age and time, the government might have implemented encryption restrictions with impunity, but times have changed and personal input into government decisions is at an all-time high, due to accessibility on the Internet, as is the rate of technological change. It has never been easier to use secure communications and it has never been harder to gain outside access to those communications. Whether there is a solution within the current social structure is debatable. Perhaps new technology that has nothing to do with electronic communications needs to be developed to combat crime. Or perhaps the solution is closer to home. Perhaps another approach is to devote more resources to family support and efforts to remove some of the motivations for crime in the first place, as we will reach a point, perhaps soon, when the difficulty of breaking strong encryption and returning the genie to the bottle may exceed the difficulty of implementing a whole different class of preventive measures throughout the world.

Technological Enhancements

One of the more interesting implications of cryptanalysis is that a system designed to be more secure than previous systems sometimes opens the door to security weaknesses in previous systems. For example, it has been proposed by scientists exploring quantum computing that quantum computers may be capable of better security than previous systems but that the quantum computers might make it much easier to break existing codes. In other words, in advancing the technology, it may obsolete all existing systems and all the documents created on those older schemes.

Thus, there may come a time when nations who have stockpiled formerly undecipherable documents might gain access to a new computing scheme that makes the old encryption methods weaker and the once-secure documents vulnerable to access by unauthorized users. This is not unlike what has happened with DNA technology. Rapists who went free ten years ago for lack of evidence can now be prosecuted, on the basis of sperm samples stored in evidence lockers, using new DNA techniques. Emailers who sent messages in the 1990s and early 2000s without fear of capture or reprisal might find themselves in a situation of being unable to deny their guilt some time in the future.

10. Resources

Inclusion of the following companies neither constitutes nor implies endorsement of their products and services and, conversely, does not imply their endorsement of the contents of this text.

10.a. Organizations

American Cryptogram Association (ACA) - Established in 1929 to further the professional and recreational strategy/gaming aspects of cryptography for the enjoyment and elucidation of its members. The ACA publishes The Cryptogram, a bimonthly journal. www.und.nodak.edu/org/crypto/

Americans for Computer Privacy (ACP) - A coalition of businesses and association in a wide variety of computer-related industries. The group provides information on legislations, facts relating to encryption, FAQs, and terms. www.computerprivacy.org

Consolidated Cryptologic Program (CCP) - Provides personnel to staff the Community Open Source Program within the CIA. Personnel are also drawn from the General Defense Intelligence Program, the Central Intelligence Agency Program, and others.

International Association for Cryptologic Research (IACR) - A nonprofit scientific organization to promote research in cryptology and related fields. www.iacr.org

International Financial Cryptography Association (IFCA) - An international organization dedicated to advancing the theory and practice of financial cryptography and related fields. www.ifca.ai

Security Technology Research Group (STRG) - STRG promotes interdisciplinary, interdepartmental research on all aspects of cryptology theory and applications. Sponsored by the Department of Computer Science and Electrical Engineering at the University of Maryland. STRG sponsors regular seminars. www.cs.umbc.edu/www/crypto/

UCL Crypto Group - A department of the *Microelectronics Laboratory* that brings together researchers with backgrounds in microelectronics, telecommunications, mathematics, and computer science to study and report on security technology. www.dice.ucl.ac.be/cryptos

10.b. Print

Babson, Walt, *All Kinds of Codes,* New York: Four Winds Press, 1976.

Bauer, Friedrich, *Decrypted Secrets: Methods and Maxims of Cryptology,* Berlin: Springer-Verlag, 1997, 448 pages. A technical reference on mathematical aspects of cryptology.

Biham, Eli; Shamir, Adi, *Differential Cryptanalysis of the Data Encryption Standard,* Springer-Verlag, 1993. Out of print.

Brazier, John R. T., *Possible NSA Decryption Capabilities,* discussion draft, U.K., June 1999. This study estimates the costs and capabilities that could make it possible for the NSA to break certain difficult encryption codes, based partly on work on a DES cracker by the Electronic Frontier Foundation. It describes the hardware, controllers, and other aspects of building a machine to accomplish this task.

Electronic Frontier Foundation, *Cracking DES: Secrets of Encryption Research, Wiretap Politics & Chip Design,* O'Reilly & Associates, 1998, 272 pages. How, after 20 years, the DES was attacked, with implications for political encryption efforts.

Flaconer, J., *Rules for Explaining and Deciphering All Manner of Secret Writing, Plain and Demonstrative,* London, 1692.

Foster, Caxton, *Cryptanalysis for Microcomputers,* Hayden, 1982.

Friedman, William F., *Elements of Cryptanalysis: Military Cryptanalysis,* Laguna Hills, Ca.: Aegean Park Press, 1984. This has been published in a series of at least four parts, each on different types of systems.

Gaines, Helen Fouche, *Cryptanalysis: A Study of Ciphers and their Solutions,* New York: Dover Publications, 1939 and 1956. This book can still be found, despite its original publication date. Descriptions of ciphers and methods of attack.

Hinsley, Francis; Knight, R. C.; Thomas, Edward E., *British Intelligence in the Second World War,* multiple volumes, London: Cambridge University Press, 1982.

Hinsley, F. H.; Stripp, Alan, Editors, *Codebreakers: The Inside Story of Bletchley Park,* Oxford: Oxford University Press, 1994, 321 pages.

Junken, Jeremiah S.; Kline, Gary, Editor; Simons, Peter, Editor, *PGP: A Nutshell Overview,* 1994.

Kahn, David, *The Codebreakers: The Comprehensive History of Secret Communication from Ancient Times to the Internet,* New York: Scribner, 1996, revised edition, 1181 pages. The author is a member of the U.S. Intelligence Community Editorial Board. It is a history of cryptography which doesn't require a technical background to enjoy.

Kerckhoff, *The Handbook of Applied Cryptography,* 1883.

Kullback, Solomon, *Statistical Methods in Cryptanalysis,* Laguna Hills, Ca.: Aegean Park Press, 1976, a reprint from a 1938 publication.

Marks, Leo, *Between Silk and Cyanide: A Codemaker's War, 1941-1945,* New York: Free Press, 1999. A personal account from the Bletchley Park days of a young man who became part of the Special Operations Executive (SOE) as head of communications. It includes new details reflecting Marks' contribution and those of others, with a history of the SOE woven through the narrative.

Menezes, A. J.; van Oorschot, P. C.; Vanstone, S. A., *Handbook of Applied Cryptography,* Boca Raton, Fl.: CRC Press, 1997. A good technical reference to the subject.

Meyer, Carl; Matyas, S. M., *Cryptography: A New Dimension in Computer Data Security,* New York: Wiley, 1982.

Nelson, N., *Codes,* New York: Thomson Learning, 1993.

Sarnoff, J.; Ruffins, R., *The Code and Cipher Book,* New York: Charles Scribner's Sons, 1975.

Schneier, Bruce, *Applied Cryptography,* John Wiley & Sons, 1996. The author is a practitioner and author of well-known encryption schemes.

Schomburg, Bernd prepared a study on behalf of the Bundesamt on the achievements of German cryptology up to 1945. Unfortunately, most of these documents have been lost.

Simmons, Gustavus, *Contemporary Cryptology,* IEEE Press, 1992, 640 pages. A survey of cryptology and its mathematical relationships.

Sinkow, Abraham, *Elementary Cryptanalysis: A Mathematical Approach,* The Mathematical Association of America, 1980.

Smedley, William T., *The Mystery of Francis Bacon,* London, Robert Banks, 1912, 196 pages.

Smith, Michael; *Station X: Decoding Nazi Secrets,* London: TV Books Inc., 2000, 240 pages. Mostly anecdotes which have appeared in other texts. The breezy style does not include a strong analytical history.

Stallings, William, *Cryptography and Network Security: Principles and Practice,* Prentice Hall, 1998. Introductory text and implementation reference for public-key cryptography, ciphers, and others.

Stinson, D. R., *Cryptography: Theory and Practice,* Boca Raton, Fl.: CRC Press, 1995.

Tuchman, Barbara, *The Zimmermann Telegram,* New York: MacMillan, 1978, 244 pages. A historical account of British intelligence efforts to decipher German codes during World War I that is suitable for secondary educational programs.

Wayner, Peter, *Disappearing Cryptography,* Boston: Academic Press Professional, 1996, 295 pages.

Yardley, Herbert O., *The American Black Chamber,* Indianapolis: Bobbs-Merrill, 1931, 375 pages.

Zimmermann Telegram, the coded and decoded Zimmermann telegrams from World War II are archived in the General Records of the U.S. Department of State, Record Group 59 as Decimal File 862.2021/82A and 862.20212/69.

Articles

Den Boer, B, Cryptanalysis of F.E.A.L., *Advances in Cryptology - EUROCRYPT '88 Proceedings,* Springer-Verlag, 1988, pp. 275-280. Lists some means of attacking the FEAL algorithm.

Diffie, W.; Hellman, M., Exhaustive Cryptanalysis of the Nbs Data Encryption Standard, *Computer,* pp. 74–84, 1977.

Diffie, W.; Hellman, M. E., Privacy and Authentication: An Introduction to Cryptography, *Proceedings of IEEE,* March 1979, V.67(3), pp. 397-427.

Kelsey, J.; Schneier, B.; Wagner, D., Key-Schedule Cryptanalysis of IDEA, G-DES, GOST, SAFER, and Triple-DES, *Advances in Cryptology: CRYPTO'96, LNCS 1109,* Springer-Verlag, 1996, pp. 237–251.

Krebs, Gerhard, Radio decoding in the Pacific during World War II, IIHSG 1996 annual conference presentation. (See next listing.)

Leiberich, Otto, provided a personal description of the development of information technology security in the Federal Republic of Germany at the IIHSG 1996 annual conference presentation. This included information on the development of cryptography.

Pearson, P. Cryptanalysis of the Ciarcia Circuit Cellar Encryptor, *Cryptologia*, 1988, V.12(1), pp. 1-9. For those who remember *Byte Magazine* when it was promoted as the "Small Systems Journal", Steve Ciarcia wrote the popular Circuit Cellar electronics column.

Retter, C., A key search attack on MacLaren-Marsaglia systems, *Cryptologia*, 1985, V.9, pp. 114–130.

Schneier, Bruce, A self-study course in block-cipher cryptanalysis, Counterpane Internet Security, Inc. A paper that presents the literature related to block-cipher cryptanalysis as an introduction to breaking new algorithms. For those interested in the practice and technical aspects of cryptanalysis, it lists many references for the cryptanalysis of specific encryption schemes.

Siegenthaler, T., Decrypting a class of stream ciphers using ciphertext only, *IEEE Transactions on Computers*, 1985, C–34, pp. 81–85.

Ulbricht, Heinz, The Enigma coding machine - Enigma 95, IIHSG 1996 annual conference presentation. The science and math of radio decoding in World War II. Enigma is used as a backdrop for how coding can be enhanced with personal computers.

van der Meulen, Michael, Werftschlüssel: A German Navy Hand Cipher System, a multipart article in *Cryptologia* beginning in 1995, V.XIX(4), pp. 349–364.

Journals

ACM Transactions on Information and System Security, published by the Association for Computing Machinery. www.acm.org/pubs/tissec/

Advances in Cryptology, regularly published proceedings of the annual international Cryptology Conference (e.g., Crypto 2000).

Antenna, Newsletter of the Mercurians, in Society for the History of Technology. Includes a range of articles about systems for encryption.

Cipher, an electronic newsletter of the Technical Committee on Security & Privacy of the IEEE. Back issue archive is available online. www.issl.org/cipher.html

Crypto-gram Newsletter, an email newsletter from Bruce Schneier, developer of Blowfish and Twofish encryption schemes.

The Cryptogram, a bimonthly journal of the American Cryptogram Association, which takes a strategic/gaming approach to cryptography, published for 70 years.

Cryptolog, published by Naval Cryptologic Veterans Association (NCVA). www.usncva.org/clog/

Cryptologia, a quarterly scholarly journal established in 1977 and published at but not by the U.S. Government.

Designs, Codes and Cryptography, An international journal available electronically through Kluwer Online.

IACR Newsletter, published electronically by the International Association for Cryptologic Research. www.iacr.org/

Journal of Cryptology, published by Springer-Verlag. The official journal of the International Association for Cryptologic Research.

10.c. Conferences and Workshops

Many of these conferences are annual events that are held at approximately the same time each year, so even if the conference listings are outdated, they can still help you determine the frequency and sometimes the time of year of upcoming events. It is very common for international conferences to be held in a different city each year, so contact the organizers for current locations.

Many of these organizations describe the upcoming conferences on the Web and may also archive conference proceedings for purchase or free download.

The following conferences are organized according to the calendar month in which they are usually held.

Financial Cryptography Conference, an annual conference since the mid-1990s, sponsored by the International Financial Cryptography Association.

International Workshop on Coding and Cryptography, organized by INRIA and sponsored by the Écoles de Coëtquidan.

Network and Distributed System Security Symposium, annual symposium since the early 1990s.

Theory of Cryptography Conference, annual conference since 2003. research.ihost.com/tcc06/

RSA Conference, A large crypto and data security conference sponsored by RSA Security, Inc., which developed the RSA system.

Information Hiding Workshop, annual international conference since the mid-1990s.

Eurocrypt. Proceedings from previous years are available online.

History of Cryptography Symposium, sponsored in 1998 by The British Society for the History of Mathematics. www.dcs.warwick.ac.uk/bshm/

Crypto, annual conference (since 1980) organized by the IACR and IEEE and Computer Science at the University of California, Santa Barbara.

AFITC, Air Force Information Technology Conference. ossg.gunter.af.mil/

Asiacrypt, annual international conference on the theory and application of cryptology and information security. www.lois.cn/Asiacrypt2006/index.htm

10.d. Online Sites

The following are interesting Web sites relevant to this chapter. The author has tried to limit the listings to links that are stable and likely to remain so for a while. However, since Web sites do sometimes change, keywords in the descriptions below can help you relocate them with a search engine. Sites are moved more often than they are deleted.

Another suggestion, if the site has disappeared, is to go to the upper level of the domain name. Sometimes the site manager has simply changed the name of the file of interest. For example, if you cannot locate www.goodsite.com/science/uv.html *try going to* www.goodsite. com/science/ *or* www.goodsite.com/ *to see if there is a new link to the page. It could be that the filename* uv.html *was changed to* ultraviolet.html, *for example.*

Crypto Law Survey. This site, compiled by Bert-Jaap Koops, provides an extensive set of links to cryptography that are not used for digital signatures (see next listing). It includes import/export and domestic control links and links to laws of a large list of countries. The U.K. and U.S. links include further details on bills and regulations and bills presented before the U.S. Congress. cwis.kub.nl/~frw/people/koops/lawsurvy.htm

Cryptography Conferences. Mihir Bellare maintains an online list, with links, of about 100 cryptography conferences, dating back to 2000. www-cse.ucsd.edu/~mihir/confs.html

Cryptography.org Archive. This site has both ftp and http access to many directories of source code and notes for major encryption technologies in PC, Amiga, and other formats. Includes Blowfish, BeOS crypto support, DES, Enigma, HASH, IDEA, etc. You must be eligible in terms of location and residence to download files. www.cryptography.org/

Digital Signature Law Survey. This site is like the one above, except that it covers cryptography related to digital signatures, providing information and links to legislation in a list of countries. There are also U.N. information, descriptions of research projects and policy statements. The site is compiled by Simone van der Hof. cwis.kub.nl/~frw/people/hof/DS-lawsu.htm

Glossary of Cryptographic Terms. A good list of cryptographic concepts and terms including some specifically related to cryptanalysis. It also includes some links of interest to other cryptology sites. www.identification.de/crypto/cryterms.html

Introduction to Cryptography. An basic, illustrated introduction to the main concepts and historical events associated with cryptography.
www.cs.adfa.oz.au/teaching/studinfo/csc/lectures/classical.html

Spy Letters of the American Revolution. This is a wonderful educational exhibit from the Clements Library at the University of Michigan. It contains scanned letters from the Revolution that illustrate not only military strategies and secrets of the time, but also the methods that were used to convey them, including codes and hidden writing. The images have been scanned at high resolution so that the text of the messages can be read in conjunction with the transcriptions supplied on the site. The methods to hide the messages and sometimes the paper on which they were written are explained. A site worth visiting. www.si.umich.edu/spies/

Standard Specification for Public-Key Cryptography. This describes the IEEE P1363 project for issuing standards for public-key cryptography, the method commonly used for electronic data communications. It includes links to completed documents and those in progress.
grouper.ieee.org/groups/1363/index.html

Steganography. An introduction to the practice of hiding data or information inside another message, communication, or entity in order to obscure its presence. Includes links and information on the Steganography Mailing List. www.iks-jena.de/mitarb/lutz/security/stegano.html

UCL Crypto Group. A department of the Microelectronics Laboratory that brings together researchers with backgrounds in microelectronics, telecommunications, mathematics, and computer science to study and report on security technology. www.dice.ucl.ac.be/cryptos

Wiretapped.net. An Australian Web site devoted to archiving downloadable information and source code on cryptographic development resources, computer security, privacy, and network-related operations. The archive categories are divided into security-related files, cryptography-related files, and audio files. www.wiretapped.net/

UCL Crypto Group Call for Papers. The UCL Crypto Group sponsors a Web page that lists upcoming conferences and symposiums devoted to cryptographic topics. Several years are listed at any given time. www.dice.ucl.ac.be/crypto/callforpapers

10.e. Media Resources

The National Cryptologic Museum has been open to the public since December 1993 and is sponsored by the National Security Agency. Exhibits dedicated to people and devices that played important roles in the history of cryptology. It includes a rare book collection, cipher machines and wheels, and other information of interest. www.nsa.gov/museum/

Das Boot (The Boat) is a 1981 film by Wolfgang Petersen set inside a German U-boat (submarine). There are a number of scenes featuring a four-rotor Kriegsmarine Enigma. The popular American film *U-571* shows an enigma machine. The filmmakers took some pains to be authentic with some aspects of the production (e.g., many features of the submarine itself), but the Americanization of the Enigma capture is Hollywood fantasy. Decryption of the Enigma was accomplished by researchers in Bletchley Park, England, in 1940, before America entered the war, after U-33 was attacked off the coast of Scotland and, later, after apprehension of a German trawler carrying two Enigma machines.

War Games (1983) is a light, fun MGM film featuring Matthew Broderick as a young computer hacker who breaks into a secured system and ends up hunted as a result. While it takes some liberties for the sake of entertainment, the movie was a good reflection of increasing fears about data security at the time, and accurately portrayed that fact that many hackers are curious teenagers with time on their hands who break into secured systems without a full understanding of the possible consequences of their actions.

11. Glossary

Titles, product names, organizations, and specific military designations are capitalized; common generic and colloquial terms and phrases are not.

ADARS	Airborne Digital Audio Recording System. An RC-135V/W recording system used by cryptologic intelligence crew.
AFCO	Air Force Cryptologic Office
AFELTP	Air Force Exportable Language Training Program. Training program for cryptologic linguists.
ANNULET	a cryptologic maintenance system
ASTW	312TRS exportable computer-based training cryptologic program
attack	an attempt to discern, break, or penetrate a communication or encrypted data by an entity (usually human or machine) not intended or authorized to have access to the information
CCP	Consolidated Cryptologic Program
CHAINWORK	A cryptologic maintenance course
cryptanalysis	the principles and practice of decoding an encrypted message without knowledge of the encryption technique or key
cryptography	the analysis of encoded written communications
cryptology	the study and analysis of encoded information
CTAC	Cryptologic Training Advisory Committee
CTAP	Cryptologic Training Appraisal Program
CTC	Cryptologic Training Council
CTEP	Cryptolinguistic Training and Evaluation Program
CTS	Cryptologic Training System
DES	Data Encryption Standard, an encryption standard developed by the U.S. government in the 1970s to establish an official method for government use
ECAC	Encyphered Communications Analysis Course

ECB	electronic code book
encryption	the process of coding a communication in order to obscure its contents or to facilitate its storage or administration
key	a data packet, number, word, or other reasonably compact information entity which is intended to provide a quick means to encrypt and/or decrypt a communication by mapping to data in a unique way. It is often used to protect the security of electronic communications, just as a physical key will allow a locked door to be quickly locked or unlocked.
key escrow	a key 'bank' or data storage vehicle, that is, a databank to hold keys so that authorized personnel can access the information for communications or security requirements
MECCAP	Middle Enlisted Cryptologic Career Advancement Program
NCS	National Cryptologic School
PGP	Pretty Good Privacy, an encryption and authentication system developed by Philip Zimmermann in the early 1990s, based on Blowfish technology
plaintext	an uncoded text message
RSA	an encryption algorithm published by Ron Rivest, Adi Shamir, and Len Adleman
S-box	substitution box or table
SCE	Service Cryptologic Element
Stealth	a stenographic software program that strips out RSA headers and other identifying crypto marks so that PGP-encrypted communications may masquerade as or be imbedded in other types of files (e.g., graphics)
steganography	the practice of hiding data inside other data in order to obscure its existence
symmetric key	an encryption system in which the encryption key and the decryption key are the same
TRANSEC	transmission security (see steganography)
USCS	U.S. Cryptologic Service (formerly USSS)

Miscellaneous Surveillance

Computers

1. Introduction

Computers are changing the surveillance industry in dramatic ways. There are three aspects of computers that make them particularly suitable for surveillance tasks: 1) microelectronic components can be used to design, test, and automate surveillance devices, 2) they can quickly process enormous quantities of data, and 3) they can be interconnected to share information. *Computer network surveillance* in the sense of administering and monitoring a network is not covered in this text because it is a broad and technical field—specialized sources should be consulted. *Computerized surveillance*, using computers to control electronic components and to process data is introduced in this chapter because it is extremely prevalent and is carried out by people with skills ranging from novice to expert.

Computers have caused a dramatic shift in information access. Up until the early 1980s, it was difficult for private citizens to look up the name, address and phone number of someone hundreds of miles away. Even if they called Directory Assistance, only the phone number, not the street address, was given out by the phone company. Reverse directories that provided a name for a phone number, were hard to find. Now there are dozens of reverse directories on the Internet, and sites that provide phone numbers, ages, occupations, the names of neighbors

Programmers rewiring the ENIAC computer in 1946. Programs were executed with physical wires to connect the circuits. Changes to the program necessitated rewiring. [U.S. Army historical photo, public domain.]

and, sometimes, even social security numbers—most of it for free, and the rest for reasonable fees. By searching, one can also find census, obituary, and other private information. Thus, computers allow home users to access information that was once available only to law enforcement personnel and private detectives.

These dynamics are changing the way people think about privacy and, at the same time, creating windows of opportunity for organizations and individuals to spy. Surveillance professionals are having to reinvent their roles, shifting from 'insiders' and 'information hunters' to consultants and information organizers and brokers. This chapter introduces some of the more common ways in which computers are used to conduct surveillance and to process data from a variety of surveillance devices.

The Power of Networking

There is so much information available on the Internet that it is possible to profile the lives, preferences, and personal habits of millions of people who don't even have computer access and who may never have logged onto the Internet.

There two primary sources of network computer intelligence are 1) the *Internet* and 2) *local area networks* (mainly in the workplace and government institutions). With tens of millions of people regularly connected to each other on the Internet, the amount of information that is shared and can be gleaned from the Net is orders of magnitude greater than what can be derived from local networks (though these are still important sources of corporate intelligence). The impact of this information exchange will probably be greater than anyone has predicted and we will see its effects unfolding for many years to come.

Surveillance of and with computer networks, whether global or local, is prevalent and, in many cases, does not require sophisticated tools. Computer surveillance is used for many legitimate purposes, for many marginal or gray area purposes, and for many illegal or unethical purposes. Network surveillance is regularly used to log user activities, to profile people, to find lost relatives, old high-school friends, deadbeat parents, criminals, potential customers, and payment skippers. Internet search engines, and the links to which they point, used as intelligence-gathering tools by a wide variety of people, including private detectives, corporate and government spies, law enforcement agents, social services workers, charity organizers, stalkers, smugglers, terrorists, thieves, and illicit dealers.

Information Sources

This chapter describes some of the more common strategies and techniques used to locate information using computer networks. It emphasizes Internet surveillance and workplace surveillance, the two most common sources of surveillance information.

It further focuses on those aspects of networking that use commonly available applications. This chapter assumes that you have mastered the basic skills of logging onto a computer network, running a browser, and doing a basic search on one of the major search engines. If you do not already have these skills, there are hundreds of books that cover the basics.

In the computer age, trying to conduct a surveillance-related business without computer skills is like trying to publish a book with a pen instead of a word processor. The old ways of conducting surveillance are not obsolete, but by themselves are complete.

2. Kinds and Variations

Basic Terms and Concepts

Here are some of the most basic terms and concepts related to networking that will aid in understanding the contents of this chapter (a few more can also be found in the Glossary at the end of the chapter). The more technical terms (and network topologies) have been de-emphasized, to keep this chapter as straightforward and user-oriented as possible. If you have a technical background, you should seek out specialist sources. If you are somewhat familiar with computers, this chapter will provide an introductory overview of computer surveillance.

Here are some of the most common terms associated with computer surveillance. They are organized conceptually, rather than alphabetically:

network - two or more computers with a communications link between them

intranet - an internal network as would be found in an office or a school

internet (with a small 'i') - networks joined externally, such as those in separate buildings

Internet - a worldwide network of millions of computers all linked together through wired and wireless connections, routers, and servers.

World Wide Web, the Web, WWW - a somewhat self-contained client-server software interface/environment that runs on networks. The Web supports graphics, files, text, and other structures that provide document display and simple user interactions mainly through a markup language called *HTML*.

Web browser - a software application that interacts with a Web server to enable users to access Web-compliant files on the World Wide Web. To most people using browsers like FireFox, Safari, Opera, and Internet Explorer, the Web appears as a graphical window within their computer desktop environment.

Web browser plugin - a third party software program that works together with a Web browser to provide added capabilities, usually graphics, sound, and animation programs. Web browsers (and servers) are evolving to support technologies that are conceptually similar to interactive television. Plugins serve to extend the technology while allowing browsers to remain essentially the same, resulting in a modular approach to providing content on the Web.

cookie - a software marker or identifying piece of information that may potentially be accessed by other computer programs. Browser cookies are sometimes used by Web vendors to streamline customer service and purchasing functions. In most cases, a cookie can be disabled in order to safeguard privacy, though there is nothing right now that stops a large (or small) vendor from hiding cookies in their commercial applications programs and not telling their customers about their presence. When this happens, the computing community usually hears about it from watchdog agencies before the vendor admits to the fact.

back door - a means of accessing a software program that is intentionally (and usually clandestinely) left by the programmer. It will never be known just how many back doors there are in computer software, as they can be difficult to detect. Some are there for legitimate reasons, for allowing the programmer or network manager to quickly access a program for maintenance or troubleshooting—others are there so that un-ethical or disgruntled employees or ex-employees can access the program or system without the knowledge of others.

search engine - a software program that allows you to enter keywords, titles, or concepts in order to search and retrieve information from one or more archives on a computer or network. There are about a dozen prominent search engines on the World Wide Web.

firewall - a form of data security system designed to keep certain users or data out (or in) of a specific computer, group of computers, or wide area network. For example, a classroom or corporate network that is connected to the Internet may have a firewall to prevent outside Internet users from accessing the school computers. A secure firewall may even prevent outside users from being able to easily detect the presence of the school system on the Internet. Each individual computer may also have a firewall.

router (or *switcher*) - an electronic peripheral device that controls or aids in controlling the flow of data traffic in a network. Switchers will often have four, eight, sixteen, or more ports into which individual computers or peripherals can be connected with a network cable, usually an Ethernet cable. Switchers are becoming so sophisticated that there is now overlap between routers and switchers and the distinctions between them are disappearing. Routers range from simple consumer routers for managing traffic between a few machines to *routing distribution networks*, city-block-sized buildings with tens of thousands of connections.

media space - an environment wired with media connections, usually video and sound, to support communication or collaboration between individuals in different locations. Media spaces are used to facilitate remote communication between workplace departments, different company branches, collaborating businesses, and teleworkers. Videoconferencing is commonly used in media spaces.

3. Context

The Internet versus The Web

One of the most important concepts to grasp in computer network surveillance is that the Internet and the World Wide Web (WWW) are not the same thing.

A *Web browser* is a software tool that displays information dispensed by *Web servers*. Imagine going to a food complex with a dozen restaurants offering food ranging from fast food to fine dinners. The various kitchens are full of ingredients, cooking equipment, and chefs who are capable of providing a vast number of meals, but there are no signs or menus. Experienced patrons know the best chefs and can walk into the back rooms and let them know which meals to prepare. However, this is logistically difficult for first-time visitors or people without a culinary background. The Internet is similar to a food complex with no signs or menus. There are a lot of people behind the scenes, many resources, and there could be many more meals cooked from the basic ingredients than people actually order. Experts could come into the food court and pick a chef and a group of ingredients and ask for custom meals. However, most visitors to the food court would be too inexperienced and too shy to ask for custom meals, and many would be just too busy to wait.

The Internet is somewhat like the unadorned food complex without waiters or menus. It has a wealth of resources that experts can readily use, but is confusing for people without computer expertise. There are strange codes, unfamiliar procedures, and hundreds of commands. Along comes a menu system. By organizing some basic dishes and putting up colorful billboards, the restaurants in the food complex can immediately advertise their style of food and the various meals that are available to the general public. The Web provides a similar means of presenting goodies on the Internet. The World Wide Web enables less technical users to display and dispense information about what is available on their sites in a readily accessible form so the person without a technical background can access the resources. Just as a restaurant can hire waiters to present menus and information on daily specials, a *Web server* can dispense information on the various resources that are on the Internet in Web format. Just as a restaurant patron uses his eyes and ears to read the menu and listen to the waiter describe the specials, the *Web browser* provides an interface that allows a computer to display the various sights and sounds on the Web that are 'served up' to the browser. This is called a *client-server* system.

Thus, the Web is only a subset of the Internet. In fact, in the mid-1990s, it was a very tiny subset of the Internet, but over time, it has grown. The Web doesn't let you see the whole Internet, it doesn't let you talk directly with the people behind the scenes, and it is a somewhat prepackaged environment compared to the Internet as a whole, but it is colorful, sometimes relevant, dynamic, and much easier to use for people who don't have a background in computers. Most of the Web is point-and-click. In other words, if you can use a pointing device like a mouse, you can use the Web.

As far as surveillance is concerned, there are thousands of data repositories that are not Web-accessible, or which are indirectly Web-accessible, that are important information sources. There are also thousands of newsgroup feeds and dozens of realtime chat and email discussion lists that are not necessarily listed or directly accessible via the Web. This is gradually changing, as the Web evolves and programmers get better at integrating Sun Java™ routines into Internet browsers, but at the present time, it is important to note that you need a variety of techniques to access all the different sources of public information.

Since this is an introductory text, it concentrates on readily accessible Web tools. For delving deeper into the Internet, it is recommended that you learn some *command line* skills and *UNIX* skills from some of the many good books that are available. Then you can go in the back room of the Internet, 'behind' the Web, and order your own custom meals.

Search in the Broader Context

The single most important skill that a surveillant using the Web needs to acquire is the ability to use search engines effectively. A search engine is a tool for finding specific types of Web pages. There are billion of 'pages' on the Web, and there is no easy way to find the appropriate ones without using a search engine. If you are looking for sites that list names and addresses, a Web search engine can pick out the relevant sites and list them for you with a short description. With practice, it's possible to use search engines to narrow your search to a few hundred or a few dozen of the best sources.

First you need to understand Uniform Resource Locators (URLs). In simple terms, these are Web addresses. Just as you have a house number and a phone number to identify your physical location, a site on the Web needs a URL to identify the location at which the information is stored. URLs follow standard formats. The one most relevant to this topic is the location of a basic Web page which will look something like this:

```
http://www.taylorandfrancis.com
```

Typing this URL into a Web browser takes you to the publisher's site. From there you can click on links to the resources listed at the back of each chapter. It will also take you to some of the major search engines on the Web.

There are hundreds of Web search engines but there are a number that are particularly useful and popular sites including:

Name	Web Address (URL)	Notes
AlltheWeb	www.alltheweb.com	Fast and uncluttered search engine.
AltaVista	www.altavista.com	Good advanced search capabilities, large list of sites, language translation.
Ask Jeeves	www.askjeeves.com	Will accept phrase and sentence queries and narrow down hits to the best choices from each source.
DogPile	www.dogpile.com	A research that aggregates search engines.
Excite	www.excite.com	General search, weather, stocks.
Google	www.google.com	Good search engine with over a billion indexed pages.
Google Groups	groups.google.com	A good resource for word-of-mouth information. (It used to be a superb source of newsgroup dialog but is now somewhat diminished by flamers, graphics, and ads.)
InfoSeek (Go.com)	www.go.com	Web pages, newsgroups, individuals.
Lycos	www.lycos.com	General search, maps, news, names.
Search Thingy	www.searchthingy.com/	Lets you select from search sites.
Sleuth	www.sleuth.com	Access to Internet databases in general categories.
Starting Point	www.stpt.com	Web and Internet, news, includes advanced search.
Web Crawler	www.webcrawler.com	Quick, simple.
Yahoo	search.yahoo.com	Organized by categories.

Note that search engines will log your searches and that this information could conceivably be surrendered to authorities if the search engine provider is served with a subpoena. In August 2006, AOL freely distributed three months-worth of people's searches on a Website intended for researchers. The site was closed soon after the information was made available for download due to serious concerns about privacy posted on blogs by people who had downloaded and examined the materials.

Sources of information on the Web are diverse and plentiful, so it is usually best to tackle information-gathering with a specific focus. Everyone is trying to hold your attention on the Web, so it is very easy to get lost or distracted while searching links.

The kinds of surveillance-related information that can readily be found on the Web include

- the names, addresses, and phone numbers of people and businesses, particularly those who have listed numbers in North American or Western European paper directories.

Age and occupation are sometimes also available, as are census and marriage records and genealogical histories. Much more information is available for a fee.

- personal Web pp. and profiles of people who are frequent Web users
- detailed maps of roads, businesses, recreational services, and transportation systems from most communities in developed nations, particularly North America
- commercial and government satellite photographs
- environmental and geological statistics and images
- lists of library holdings and educational publication lists
- lists of business products and services
- announcements of new technologies and their inventors
- patents and trademarks
- lists of people who have made significant achievements in sports, science, and business
- newspaper and magazine articles about specific people or technologies
- discussion groups of hobbyists, professionals, and advocates for specific causes
- government and private statistical compilations and population demographics
- proceedings and positions of state and federal government political bodies
- unclassified activities of the armed services and White House personnel

Private detectives make regular use of the Web to locate 'skips' (people who are avoiding payment of bills, child support, or who are seeking to avoid the law), adoptees and their families, and to track the activities of individuals and corporations.

Marketing agents use the Web to plot demographics, consumer buying trends, and to assess the competition. They also amass postal and electronic mailing lists.

Entrepreneurs use the Web to assess existing patents and inventions and to determine whether other companies are already making or distributing products similar to those they are considering developing. They also use Web information to develop business plans and communicate with potential investors, employees, and business associates.

Stock investors, brokers, and venture capitalists extensively use the Web to assess market trends, business activities, new ventures, and competitive activities.

Intelligence agents use the Web to track foreign communications, to uncover crimes, to detect and apprehend stalkers, poachers, smugglers, and terrorists, and to assemble information of a general nature that might be relevant in later contexts.

Using Good Tools

An enormous amount of information can be gathered with a simple browser and a dialup Internet account, but if searching the Net is part of a professional service or task, then an up-to-date browser and a fast-access account (T-1, DSL, ISDN, or cable modem) can save an agency, and a client, a lot of time and money. Many Web sites are now enhanced by graphics (or hopelessly cluttered by them). Many now have animated ads that greatly slow down the speed at which a page will load. When downloading files or images, the difference between a fast access account and a regular modem for downloading a 10-MByte file can range from a couple of minutes to over an hour.

Privacy, Security, and Anonymity

It is not easy to remain anonymous on the Web. There are many extremely intelligent computer techies who love to solve puzzles. One of those puzzles is finding ways to detect the activity of a browser and the person using it. They have also worked out ways to get your email address. Don't believe it if someone says you don't give out personal information when using a browser if your identity is taken out and your 'cookies' disabled. If you use an email program that is built into the browser and someone sends you email with a built-in URL that you can just click instead of typing, that automated click can provide information automatically. As browsers become more sophisticated, it becomes easier to build 'smart features' into the browser that can be controlled offsite. Anonymous proxy servers can mask Web surfing to some extent, but there are often other ways to determin a user's identity.

Many users of IBM-licensed/Microsoft-based personal computers don't realize that when they get their DSL or cable modem linkup they are visible to other people on the Internet unless they set up their system with certain security settings or a firewall. This information can make it possible for hackers to access your files.

Browsers leave trails. If you wish to search or email anonymously (or avoid the inevitable backwash of junk email that results from leaving messages while searching the Web), you need to disable cookies, remove the 'Identity' information from your browser, use particular servers, and often set up other security. This is beyond the technical expertise of most casual computer users.

A *cookie* is a token or ID that can be passed from computer to computer or from one software process to another on one or more systems to keep track of a user. It's a way to keep records of processes or activities. If you regularly log onto a site such as an online book store or forum, you may have noticed that they will ask if you want to be 'remembered' for subsequent transactions (so you don't have to type in your name and other information every time). This information is derived from a cookie and is stored in the vendor's database. When you log back onto the system at a later time, the software checks your browser cookies and calls up the previously stored information, sometimes greeting you by name. Cookies are used in many Web 'shopping cart' systems. Some sites are even set up to track a browser as it leaves a site. In some cases, the remote site will deposit a *reverse cookie* on your site, which creates a security gap through which hackers could potentially compromise your data. If cookies are turned on, then it's best to accept cookies for the serving site only.

It is difficult to completely cover your tracks and protect your privacy online. Some individuals choose to disable the majority of identity tags in their browsers as a sort of minimal protection and leave it at that. For high security, aliases and anonymous servers are necessary to preserve privacy. An anonymous server is a system set up to forward messages without identity tags to protect the identity of the sender. For example, a frightened witness to a violent crime might send a tip to law enforcement officials through an anonymous server. A refugee from a foreign country might use an anonymous server to communicate with relatives in the home country in order not to reveal his whereabouts to a repressive government. A news journalist in a war zone might use an anonymous server to send information to his or her editor in another region. Anonymous server issues are described in more detail in specialized books on preserving privacy on the Internet.

Information-Gathering Strategy

As has been mentioned, the Internet is a big place. If you were an archaeologist who just located an ancient 15-acre underground ruin of passages in an area that was going to be bull-dozed in three weeks, you would understand the importance of priorities and good searching strategies. Time and resources are never unlimited, and it is best to approach information-gathering on the Net in the same way. A typical search for a topic on a Web search engine can yield anywhere from 8,000 to 8,000,000 'hits', that is, sites with information pertaining to your query. It is recommended that you learn to use the Power Search features provided with every major search engine, as this will help narrow the searches down to several hundred hits without sacrificing quality.

Decide what you need to know and then doggedly and patiently seek it rather than trying to see everything there is to see. There are times when a wider viewpoint is valuable, but much of the time it is best to focus clearly on a specific goal. Avoid the temptation to explore side-routes. When scanning the 'hits' that are provided by search engines, read the descriptions and select the ones that look most relevant, don't just go down through the list clicking each one unless the number of hits is small and you are having trouble finding the information you need.

The Context of the Internet

The Internet is a powerful source of information as long as the information-seeker under-stands the particular character of the Internet. At the present time, users of the Internet do not represent a microcosm of human society. If the world is a cake, then the Internet is not a slice of cake, but rather a spoonful of whipped cream icing.

There are many sources of information on the general population that can be found on the Internet, but not all surveillance intelligence is based on looking up names and addresses. Some of it is based on studying Web pp. and articles written by the people who comprise the Net community. This is not a typical cross-section of society.

There are many research companies that have studied the demographics of the people using the Internet. They have discovered that, statistically, Internet users as a group are predominantly affluent, male, white, well-educated, young to middle-aged, and generally Liberal and Libertarian in their political leanings. They have also discovered that, as huge as the Internet has become, many users are only passive listeners (called *lurkers*), browsing the Web once in a while, or are private emailers or private messagers who interact exclusively with small social groups. These people leave indirect footprints on the Web. Much can still be discovered about them, but it must be done through clandestine technical surveillance or through secondary sources (which are, in fact, numerous).

Software and content providers on the Net are themselves a specialized society. Those who are the most visible and active on the Net usually have a strong technical background or a strong political agenda. Those who produce the vast majority of information and programs that define the character of the Web only represent a tiny fraction of the human population, about 0.1%. This produces a bias in the medium itself. Marshall McLuhan made a historic, often-quoted statement regarding television in the 1960s when he stated that "The medium *is* the message". The concept holds as true today for the Net as it does for television.

Given this select group of people creating and relating on the Net, it can be seen that the opinions of people on the Internet and their behavior and demographics are not likely to represent the opinions of people as a whole. Thus, detectives and other professional surveillants who are gathering statistics and profiles on the Internet, need to be aware of the skewed distribution in the following areas:

Economics - Until recently, computers were not cheap. Even now, many laptops and workstations cost a couple of thousand dollars, over two month's pay for someone earning minimum wage. While prices have dropped substantially, many people still can't afford computers or Internet access fees. Those with higher incomes are more apt to purchase and use computers and be connected to the Net.

Program Providers - The majority of programmers are male (around 85% as of 2000) and there are indications that they write programs to reflect their personal interests and economic goals (e.g., video games). These personal and commercial leanings are apparent on the Web as well, though diversity is increasing.

Traditional Stereotypes - Computer network technology grew out of electronics developments from the 1940s to the mid-1970s, a time during which women were actively prevented from entering electronics-related professions. Women and minorities still have difficulty getting upper level management positions in the technology industries (less than 3% as of 2000). The age curve is also a factor in Internet use–many retired people don't use computers and never will, but there is also a trend for more retired people to use the Net when they are offered reading, gaming, and other recreational opportunities online. Minority populations often have less Net access due to job discrimination that results in lower incomes. As computers become less expensive and Internet connections are more broadly installed in public schools and libraries, these stereotypes should gradually diminish.

Recreation Time - In families with two working parents, men spend more time on computers while at home. This may be related to differences in personal interests generally and statistics indicating that women still handle the majority of child care, shopping, cooking, and other household chores that might take time away from computer activities in the home.

One other factor of relevance to surveillance is the tradition of women taking their husband's names when they marry and of adoptees being given the names of their adopted parents. Surveillance professionals using the Net to find felons, skips, lost family members, or adoption families usually find it easier to locate primary source data on young to middle-aged white males than on other demographic groups. Information from city governments and community groups about property assessments, marriages, and legal proceedings are gradually becoming accessible on the Web and may make it easier to locate information on female subjects in the future.

Language Considerations

There are other aspects of the Internet which directly affect information-seekers. One of the most significant of these is that most computer applications are designed in English.

The majority of commercial computer programming languages evolved in America. This results in a certain amount of cultural-centricity in the technology. Just as most opera terms are in Italian and most ballet and cooking terms are in French, most computer terms are in English, where a large portion of the trade originates.

That is not to say computer technology is essentially American. A great deal of computer development has occurred in Europe and Asia. There is a strong interest in computer technology and mathematical algorithms in India and business owners and governments in developing nations would gladly use computers if they could afford them.

Consequently, in spite of a strong interest in computers throughout the world, Europeans generally program in English. Because Asian written languages are difficult to interpret into keyboards and computer syntaxes, many Asians also program in English. Many search engines have language translators that will translate a German, French, Italian, or Spanish Web pp. into English text (some also translate other languages into English).

What this means for a surveillance professional is that if English is your primary language and you live in the United States, Canada, or the United Kingdom, you have an advantage in using the Internet.

Surveillance in the Context of the Workplace

Up to now, most of the discussion has been about Internet surveillance. However, local computer networks and computerized badges and access systems are also prevalent in workplaces that may or may not have Internet access.

Computer surveillance in the workplace takes many forms, from *access surveillance* for monitoring entries and exits, to *keystroke logging* to monitor activities on individual computers, to *tracking* an employee throughout a complex, including trips to the washroom. With the help of computers, a complete record of every person's movements is now theoretically possible.

It is easy to see how the computerization of tracking and monitoring technologies is of interest to employers. They are concerned about preventing employee malingering, game-playing, personal emailing, fraud, and theft, all of which contribute to a loss of productivity and profits. Honest employees, on the other hand, are concerned that they are being discriminated against due to the activities of a small minority and further concerned about being surveilled and recorded without there being policies in place to allow an employee to review his or her file or to provide feedback on the manager, to correct misconceptions or errors, or to have transcripts destroyed after a reasonable amount of time.

The current workplace climate is one of increased surveillance through a variety of technologies, including entry access, electronic badge tracking, video surveillance, phone and email accounting systems, keystroke recorders, and phone call recorders. And 'workplace' surveillance doesn't stop at the office. With many more people working from home, workplace surveillance now often includes a video camera installed in the home that is networked to a main office. Employees have also tried to implement tracking systems that follow an individual outside of work. Privacy advocates have strongly criticized this practice.

Surveys indicate more than half of employers currently use some form of electronic monitoring, with phone logging, storage and retrieval of email, and video monitoring listed as examples [AMAI 1999]. Telemarketing firms and computer software suppliers are particularly known for monitoring sales and technical support calls.

Electronic Performance Monitoring (a form of workplace surveillance) involves the use of computer technology to evaluate the speed, efficiency, and effectiveness characteristics of an employee. Since EPM occurs most frequently in jobs that lend themselves to monitoring, predominantly phone work and clerical work performed on computers, EPM is indirectly biased to monitoring women, who fill the majority of receptionist, word processing, and data entry jobs. Since managers rarely perform keyboard-intensive activities, they are less subject to this kind of monitoring. As software programs become more sophisticated and able to monitor loans and brokerage transactions, which are statistically handled more often by men, the gender bias may decrease, but management biases may remain.

A number of concerned agencies have been studying workplace surveillance and have made some initial recommendations regarding legislation and corporate policies. Their recommendations include, but are not limited to, the following:

> *disclosure* - Full disclosure must be given by the employer to the employees of surveillance device locations and operations, including the purpose of the surveillance, when it is active, how records are stored, who is authorized to see them or distribute them, how long they are kept, in what manner they are distributed outside the company, and how they are destroyed.

> *consent* - An employee must give informed written consent for collection of or access to the employee's personal data or for covert surveillance of that specific individual, except under very exceptional circumstances related to 'grievous criminal activity'.

review - Mechanisms for employee review of his or her records must be put in place, with a process for the employee to correct incorrect information.

bias removal - In systems where human monitoring through video cameras exists, particularly if those cameras can be remotely aimed, strict guidelines as to objective and nonvoyeuristic surveillance must be established and upheld. This is to prevent reported abuses in the areas of minority prejudice, voyeuristic preference for videotaping young women, and prejudice against monitoring men and youth in high crime areas.

balance - Two-way surveillance between workers and management should be instituted. In other words, if employees are to be held accountable for their actions with respect to the employer/managers, a balancing mechanism needs to be in place to hold employers accountable for their actions with respect to employees. Many upper level managers resist the concept, thinking they should be 'above scrutiny' or somehow exempt, but given the powerful nature of surveillance data and the need for trust and collaboration between employers and employees for maximum workplace efficiency, these concerns need to be addressed and balanced for everyone's mutual benefit and for the long-term health of the company.

change of use prevention - Data collected for one purpose should not later be used for another purpose without the explicit written consent of those surveilled. In other words, information from employees collected for company demographics cannot later be used for marketing without employee consent. Similarly, cameras aimed at cafeterias or parking lots for 'safety of the employees' cannot later be used for employee evaluation, litigation, or other purposes, unless employees were informed in advance and agreed that they might be used for these purposes.

free zones - It is important to establish surveillance-free zones where worker privacy is guaranteed in order to ensure and promote trust and the emotional, mental, and physical well-being of employees.

Many workplace-related surveillance systems (phone-logging devices, video cameras, magnetic access devices, etc.) are described in other chapters.

4. Origins and Evolution

When a new generic technology is invented, it may take a while before its surveillance potential is recognized and used. In the case of computers, their early development and their use for surveillance are so closely linked as to be almost inseparable. Almost as soon as computers were invented they were being used to create and break ciphers, to store information on people and governments, and to help solve problems related to economics, warfare, and information brokerage.

Using Machines to Gather Data

Mechanical calculating machines have existed at least since the 1600s, when inventors like Wilhelm Schickard (1592-1635) and Blaise Pascal (1623-1662) created some of the first devices that could facilitate financial transactions and tax assessments.

The use of mechanical calculators for the large-scale collection of information on people began with Herman Hollerith's (1860-1929) invention of a tabulating machine for counting national census data. In a sense, this was the beginning of the computer database, foreshadowing one of the uses for which computers would be adapted 100 years later. The Hollerith machine could read punched cards using an electrical sensor. Following publicity about his invention,

Hollerith established the Tabulating Machine Company, which evolved into the International Business Machines Corporation, now known as IBM.

Computers have only been around since the middle of the 20th century, and personal computers didn't become consumer items until the mid-1970s, but the mechanisms for keystroke monitoring, and other aspects of worker surveillance have been in existence at least since the early part of the century.

Cyclometers were invented for counting keystrokes on typewriters, which provided a means to monitor the efficiency of clerical workers. Similarly, the work of telephone operators was monitored for decades. Operators were required to keep track of call statistics on cards they organized on their desks. Thus, the concept of monitoring both workers and client activities was already prevalent before computers were invented. So perhaps it shouldn't surprise us that computers were quickly adapted for surveillance purposes each time they evolved new capabilities.

Development of Electronics

Electronics were at the heart of the development of computers and a great deal of radio technology contributed to the evolution of early computers. Vacuum tubes and printed circuit boards were two important technologies that contributed to the development of computers and other electronic devices (the early history of vacuum tubes started with the *Fleming valve* and the de Forest *Audion* are described in the Radio Surveillance chapter). Circuit board design emerged gradually through the 1920s and onward.

A great deal of hardware design related to computers occurred within the U.S. military forces, especially the Army, where it was needed for ballistics targeting tables and complex calculations related to navigation and the aiming and firing of advanced weapons. Since many electronics inventions arose out of these activities, the development of printed circuit boards has generally been assumed to have been developed in the 1940s. However, an investigation of radio fabrication shows that at least a few radios from the late 1920s already had printed circuit board fabrications, indicating that the practice was invented earlier, even if it did not become prevalent until the 1950s.

Left: An example of a 1928 circuit board that was discovered in a commercial cabinet radio by Jonathan Winter, the proprietor of the American Antique Radio Museum. The copper traces have been blasted onto the underside of the circuit, with connections to the upper side through basic nuts and bolts (the wire jumpers were later additions). During the 1940s, the U.S. Army developed photolithographic processes for creating circuit boards for use in electronics and later in computers. Right: A modern printed circuit board can support tiny traces and computer chips. [Classic Concepts photos ©1998, used with permission.]

The Early History of Practical Computers

In Germany, in the 1930s, a brilliant young engineer named Konrad Zuse (1910-1995) developed a general purpose calculating and computing machine in his parents' apartment. The *Zuse Z1* had a mechanical memory storage and was programmed by punching instructions onto film. Later, with help from Helmut Schreye, electronic relays and vacuum tubes were added. By 1942, Zuse was developing the Z4, which was demonstrated in April 1945.

In Britain, the Colussus computer, involved in decyphering the German Enigma code, was being developed under complete secrecy during 1943. It became operational in 1944, a fact that wasn't publicly revealed until almost three decades later.

Meanwhile, in America, John Vincent Atanasoff (1903-1995) and Clifford E. Berry were collaborating on the development of the *Atanasoff-Berry computer* (ABC). It was prototyped in 1939 after two years of design development. It used rotating drum capacitors to refresh memory so that data wouldn't be lost. It is significant for the fact that the data and memory were implemented as separate functions. Data were entered into the computer with punch cards, a system that was still in use on many computers in the early 1980s.

Another computer developed in the late 1930s and early 1940s was the *Harvard Mark I*, an automatic relay computer constructed by Howard Aiken with support from IBM engineers. It could run long calculations from punched paper tape.

The *Electrical Numerical Integrator and Calculator,* better known as the ENIAC, was developed by John W. Mauchly and J. Presper Eckert at the University of Pennsylvania under the guidance of John Brainerd. The ENIAC was dedicated in 1943 and unveiled in 1946. When complete it weighed over 30 tons and included nearly 20,000 individual vacuum tubes.

Thus, the first gigantic, expensive, vacuum-tube computer systems were first introduced in the 1940s. They were little more than advanced calculating machines by current standards, costing millions of dollars, yet their capabilities far exceeded previous technologies. For the first two decades, because of their cost, they were used almost exclusively for large-scale government, science, and business uses. The average homeowner was barely aware of their existence until television shows began to feature computers and robots, in the late 1950s and 1960s.

Left: The ENIAC computer, constructed in 1945 and rolled out in February 1946, was essentially a giant advanced calculator that was used for a variety of U.S. Army projects including ballistics calculations, scientific computing, weather prediction, and thermal ignition studies. Right: Programmers are shown changing the coding by physically rearranging the wires. [U.S. Army historical photos, public domain.]

Grace Murray Hopper (1906-1992) was a mathematician, physicist, and pioneer computer scientist who developed the COBOL programming language in the days when computers were programmed by rearranging wires. More important, however, was the fact that she realized that code could be reused, that routines could be self-referencing, even though many detractors scoffed at the idea at the time. Nevertheless, Hopper championed the basic concept, even though the technology was not yet in place, and her ideas established a foundation for other programming languages, which would come into their own when circuit boards and binary electronics superseded the old physically wired systems.

The early computers were limited in capabilities, difficult to maintain and program and yet, in spite of their limitations, contributed fundamental calculations and tables that built a foundation for scientific advancement and future computing. Computers bootstrapped their own evolution by contributing to our understanding of physics and mathematics and to the development of electronics and more powerful computers.

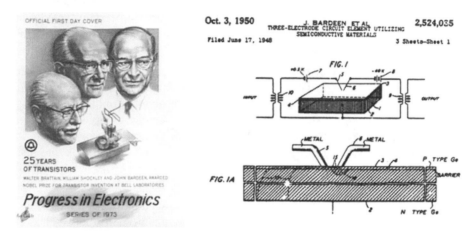

Left: The inventors of the transistor were commemorated in this U.S. Postal Service first day cover. Right: A portion of the original patent for the transistor, which was registered by and subsequently granted to the inventors from Bell Laboratories. [Cover from the collection of the author, used with permission; U.S. Patent and Trademark Office, public domain.]

The invention of the transistor at the Bell Laboratories in 1947, following the end of the World War II, was one of the most significant developments in all of electronics history and had a big impact on the design of electronics from that point on. Not only could computers be made smaller and less expensive, but they could be made to perform functions that were not practical on the giant, tube-based systems. Compilers, assembly languages, and other aspects of programming continued to evolve, which, when combined with new systems, would herald a new age of data processing and information exchange.

Innovative Ideas

When microcomputer electronics became a reality, many visionary thinkers were excited by the technology. Even though microcomputers didn't become established until the mid-1970s, a number of inventors had already envisioned a world of small computers by the 1960s. In 1968, Jürgen Dethloff proposed the idea of using a plastic card as the medium to support microelectronics, essentially describing a 'smart card' a decade before they became technologically practical. A few years later, Roland Moreno's idea of associating a personal identification number (PIN) with digital information led to the development of *magnetic stripe cards* that are now familiar transaction tools.

The Birth of Computer Networks

Computer networks, particularly the Internet, have had a profound effect on computer surveillance. There are several factors that brought about the birth of computer networks:

decreased size and price - As room-sized computers became smaller and less expensive, it became possible for organizations to own more than one. This facilitated experimentation and the evolution of networking hardware and software.

advanced calculating concepts - The idea of hooking computers together was seen as a way to collectively increase computing power over what a single machine or several individual machines could accomplish. If a computing task could be 'farmed out' to several computers and the data brought together through automation after the individual tasks were 'solved', this could greatly speed up certain types of tasks. This idea of achieving more efficient *distributed computing* provided motivation for interconnecting computers.

time-sharing - As word about computers spread, more organizations became interested in accessing computer services and were willing to pay for it. Time-sharing capabilities were developed so that individuals equipped with 'dumb terminals' could pay for computer time on large-scale systems. The idea of creating 'smart terminals' was not far behind. Network protocols were developed to facilitate dumb and smart terminal connectivity, eventually making it possible for multiple users to hook into a single system.

human social factors - The pervasive desire for people to intercommunicate was also a factor in the evolution of networking. The desire to send messages through computers motivated the invention of specific network communications tools such as email and newsgroups.

By the early 1970s, the ARPANET, the forerunner to the Internet, was being developed by the U.S. *Defense Advanced Research Projects Agency* (DARPA). Modems to enable computers to communicate through phone lines were developed and improved at about the same time.

Microcomputer Electronics

When microcomputers as we know them were invented in the early 1970s, no one paid much attention. The Kenbak-1 rose and sank in 1971 with barely a whisper. It wasn't until the *Altair* was developed in 1974 and advertised and featured as a hobbyist kit in the January 1975 issue of *Popular Electronics* that the world suddenly took notice. Within two years, CRT screens and keyboards were standard peripherals and the Apple and TRS-80 lines of computers were launched. Microcomputers soon became a tool of daily life.

After microcomputers caught on, hardware developers began selling modem cards that could be inserted in a computer to allow it to interface with a phone line to share data with systems. These early modems were slow, only 300 bits per second, but that didn't deter hobbyists from using them. *Bulletin board systems* (BBSs), in which individuals could dial up a computer that was being used as a *server* for games and message boards became highly popular by the early 1980s and modem speeds increased to 1200 bits per second.

During the 1970s, Gary Kildall (1942-1994) developed CP/M, the first significant widely distributed microcomputer operating system. Other operating systems joined CP/M, including AppleDOS, TRS-DOS, and L-DOS, giving consumers a variety of options on their personal computers. The public eagerly began buying the new computers, paying almost $5,000 for a full system with some software, a modem, and a printer.* By the late 1970s, there was a fierce competitive battle between Apple, Radio Shack, and IBM to capture the microcomputer market.

A historic milestone in computing history occurred when Microsoft bought an operating system from a programmer named Tim Paterson, who had created a version of CP/M adapted from Kildall's CP/M manual. Microsoft sold this to IBM and it came to be known as PC-DOS. PC-DOS was then adapted by Microsoft as MS-DOS and sold in competition to IBM's operating system (much to IBM's surprise as they assumed they had purchased exclusive rights to the software). Kildall continued to develop CP/M, which evolved into Digital Research's DOS (DR DOS). Thus, the competition for a major computer and a major operating system for home and business markets heated up and the microcomputer revolution was underway.

By the early 1980s, microcomputer sales were booming and computer networks were getting established. By the mid-1980s computers had found their way into many homes and schools. Since schools were particularly interested in networking connectivity, so that teachers could monitor student activities and students could work on group projects, Apple Computer built networking capabilities into their Macintosh computers right from the start. At the same time, Kildall's company was developing some of the first effective connectivity technologies for IBM-licensed technologies. On a bigger scale, DARPA was developing large-scale networks for military and scientific work. For home computers, dialup BBS systems were thriving, with thousands of users now sharing information and home-brew software through telephone networks. The speed of modems had now increased to 2400 bits per second.

By the mid-1980s FidoNet emerged as a way to automatically transfer files from one computer to another through phone lines. Thus, by 1984, networks were evolving rapidly on many fronts and eventually the technologies would converge into one big distributed network that came to be known as the *Internet*.

Tiny Computers

As computer electronics devices dropped in size and price, developers continued to capitalize on the new technologies and seek ways to incorporate computing capabilities into new media. By the late 1980s, pen computers, schedulers, advanced programmable calculators, and smart cards began to emerge. In terms of surveillance, tiny computers provided a way to unobtrusively access data in the field and, when small radio modems emerged, to send that data to a central facility or individual colleague.

Magnetic stripe cards provided a way for the issuer of the card to track the movements and habits of the person using the card. Not all smart card issuers used them for this purpose, but the technology was in place and the possibility now presented itself. By the mid-1990s, retailers recognized the surveillance potential of 'member cards' and began offering discounts and other enticements to get customers to carry membership cards. It was now possible to track a person's shopping habits, including what kinds of products they typically purchased, how often they shopped, and how much they spent.

The 1990s - Increasing Integration and Processing Capabilities

The most significant advancements in computer surveillance in the 1990s, were profiling systems and computer-processed video systems.

---

*In 1980, that was nearly the cost of a new car, a hefty price considering that computers operated only at 4 MHz and had only about 8 kilobytes (not Megabytes) of RAM. They stored information on tapes, not hard drives, and the monitors could only display 320 *x* 200 pixels.

Identity profiling

Profiling databases are systems that combine computer networks and magnetic stripe or other access devices to develop profiles on the attributes and actions of the people who use the technologies. These systems made it possible for supermarkets to profile shopping preferences and to target mailings and cash receipt advertising to the individual consumer. They allow financial institutions to profile client transactions and marketing professionals to use Web sites to profile potential customers. In many cases the customers have no knowledge or control over what is being done with the personal information that was collected. Profiling is one of the areas over which privacy advocates have expressed special concern.

> .... There exists a massive wealth of information in today's world, which is increasingly stored electronically. In fact, experts estimate that the average American is "profiled" in up to 150 commercial electronic databases. That means that there is a great deal of data–in some cases, very detailed and personal–out there and easily accessible courtesy of the Internet revolution. With the click of a button it is possible to examine all sorts of personal information, be it an address, a criminal record, a credit history, a shopping performance, or even a medical file...

> [Mr. Kohl, speaking on a bill to establish the Privacy Protection Study Commission, in 1999.]

Audio/Video Integration with Computer Processing

Computer-processed video systems are video camera feeds in which the image data is analyzed by computer algorithms. These systems can identify faces, individuals, actions, movement, and a host of other triggers. By 2000, the systems were so sophisticated, they could even recognize people who were wearing hats, glasses, or beards. Some were integrated with audio sensors that could detect the sound of a traffic accident, aim a camera at the source of the sound, and record video images related to the accident.

In 1997, the American Management Association released survey statistics on midsize and large member companies, reporting that almost two-thirds were using some form of monitoring or surveillance. Worker productivity and accountability were cited as motivations for monitoring. However, contradictory views were expressed by the U.S. Office of Technology Assessment, which reported that no reliable evidence yet supports the contention that monitoring increases production, but that stress does appear to increase when employees are monitored.

In 1999, the American Management Association reported that 27 percent of large U.S. corporations check employee email on a routine basis.

Surveillance was increasing, becoming prevalent in the workplace, and the kinds of devices used for surveillance were also increasing in variety and sophistication. In light of these changes, there were many legislative changes proposed to regulate use of new technologies.

5. Description and Functions

The three main areas of computer surveillance that have been introduced here have been surveillance of information on the Internet, computer profiling, and workplace surveillance. The technical aspects of computer surveillance including hacking, firewalls, proxies, and packet sniffing are outside the scope of this volume, but there are many excellent references on computer security that deal with these issues. Computer processing is also described here in a little more detail.

5.a. Internet Surveillance

A large proportion of information that is surveilled on the Internet is *open source* information, which is information that is freely available. Web pages, chat rooms, search engines, and USENET newsgroups are all examples of open sources. Additional useful sources of information include

discussion lists - unmoderated, moderated, or by-invitation discussion groups in which members participate in sharing information, data, and opinions. Discussion lists cover every conceivable topic and number in the tens of thousands. The best known public discussion lists are on USENET and are known as newsgroups. These are established on a vote process. Unfortunately, due to junk and advertising messages (which are not permitted), many newsgroups have gone to moderated status and some have simply died because the noise-to-signal ratio from irresponsible posters became too high. As a result a number of Web and email discussion lists have emerged to take the place of some of the better USENET groups of the past (there are still good computer newsgroups, but the general ones suffer from problems). Other sources of discussion lists include

www.liszt.com

www.onelist.com which is now www.egroups.com

yellow and white pp. lists - free or commercial sources of 'phone book' style information on individuals and companies. Some also include email addresses. Some of the best known sites include

www.411.com

www.411.ca

www.infospace.com

www.freeyellow.com

www.whitepages.com

www.yellowpages.com

maps and travel locators - free or commercial sources of terrain and street maps (satellite images are discussed in the Aerial Surveillance chapter). Some of the popular sources include

www.earth.google.com   3D rendering of the Earth's surfaces, including selectable geographical markers for landmarks, transportation corridors, and more

www.mapquest.com   city and street maps

www.maps.com   world atlas, topographic maps

www.usgs.gov   terrain and satellite maps

5.b. Computerized Profiling

Computerized *profiling* is the process of collecting data on a person's individual characteristics, personal information, or activities by monitoring their actions through voluntary reporting, access devices, or other surveillance technologies that can be interfaced with a database. It also applies to records kept on individuals convicted of criminal offenses, which include personal information, violations, fingerprints, and other law enforcement records.

The most common devices for profiling include magnetic stripe cards (identity cards, ATM cards, supermarket or sports club membership cards, etc.) and video cameras. Other systems, such as biometric fingerprints and iris scans are now also combined with computer databases.

Law enforcement, national security, and customs agencies use a wide variety of databases to profile visitors to the country, foreign agents, criminal suspects, convicted offenders, and prisoners. These databases are increasingly being linked together through the Internet to allow agencies to cooperate in cases that involve more than one state.

In the past, law enforcement agents working in patrol cars had to radio in driver's license information or license plates to ask the dispatcher if there were any outstanding warrants or other problems with a particular individual. New systems are now put in place in which patrol officers can put the data directly into a mobile unit and call up any relevant data from a central database.

Joint automated booking stations (JABS) are multimedia information systems based on a DEA-Rome Laboratory pilot project to enable Federal law enforcement agencies to share information more effectively.

5.c. Work Monitoring Systems

The most prevalent forms of workplace monitoring using computer electronics include:

- the use of access cards or keys that log an individual's entry and exit patterns and sometimes also their movements around a complex,

- the monitoring of keyboard input, use of computer applications, Internet activities, and email, and

- telephone activities, especially telemarketing calls, sales follow-up calls, and technical support calls.

"The range of occupations susceptible to electronic monitoring is surprisingly wide. Some of the positions most likely to be monitored include: word processors, data-entry clerks, telephone operators, customer service representatives, telemarketers, insurance claims clerks, mail clerks, supermarket cashiers, and bank proof clerks [OTA, 1987; ILO, 1993]. Professional and technical workers may believe that their work is too complex to be monitored successfully. However, sophisticated groupware applications and work-flow tracking systems provide an abundance of information on the status of an electronic document as it is "passed" from one professional to another. As well, electronic mail and scheduling applications provide additional potential for surveillance of employee activities and communications [Clement, 1988; ILO, 1993; Piller, 1993; Allen, 1994]".

[Susan Bryant, Electronic Surveillance in the Workplace, *Canadian Journal of Communication*, V.20(4).]

5.d. Computer Processing

A technology that has been mentioned in other chapters is the capability of computers to manually or automatically analyze or clean up information from other surveillance devices. This is usually an expensive 'last resort' process, but in cases of murders or kidnappings, the technology can mean the difference between life or death for vulnerable victims if the perpetrator is convicted. Two of the more common computer processing technologies include:

data processing - the process of enhancing an image or other data (e.g., sound) to clarify the source of the data or the identity of an object or person. For example, an image may be sharpened, or the colors or tonal values adjusted to make details more clear. Image processing can be used to clarify faint writing on a wall or a photo of a faint footprint in the mud. It can also be used to change the apparent age of a person in a photo, such as child lost for several years (age progression), or to remove or add a beard or glasses. It may further be used to clean up a poor audio signal taped from a phone call or to raise or lower the tone of a voice of someone who was using a voice changing device to hide his or her identity (e.g., a kidnapper).

data fusion - the process of combining data from more than one source. For example, video footage of a license plate may not show the numbers, but fusion of several frames of video, choosing the best parts of each frame and combining them, can sometimes yield a recognizable number.

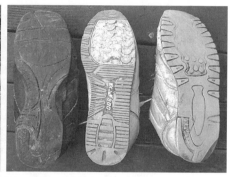

The photo of the footprint on the left was difficult to see and no plaster cast was made of the print. In the photo in the middle, the image was processed with a common image processing software program to make the outline and ridges easier to identify and thus easier to match with its mate on the right. [Classic Concepts ©1999, used with permission.]

Computer processing is not limited to data we can see or hear. Patterns that are invisible to humans can sometimes be readily interpreted by special devices and computer processing. For example, infrared images called *thermograms* can be used to identify individuals. When various types of data are combined and analyzed by the computer, it becomes almost impossible to hide your identity through disguises or changes in haircuts and facial hair.

Computer processing is not used only for examining evidence or identifying criminal suspects or unauthorized intruders, it may also be used to manage access systems for employees or those authorized to enter restricted areas.

6. Applications

This section does not constitute a complete list of all vendors, nor does it imply an endorsement of the quality of their products or services. The following are intended as educational examples only, to provide an introduction to the types of products that may be purchased on the market related to this topic.

Computer Spy Systems

Computer spy systems are products designed to monitor the various activities of a computer user, including keystrokes, password entries, applications, specific Web pages, Internet resources, and other files or data that are accessed or executed. Some of these systems are intended to protect the security of a computer from unauthorized use, others are used to monitor workplace productivity or to collect evidence of unauthorized, incorrect or illegal use. Some are used to monitor the activities of children for parental guidance or educational purposes. Occasionally computer 'spyware' programs will steal or 'grab' passwords to allow someone to later access to a system without authorization. Examples with a variety of purposes and capabilities include

Data-Trak, Inc. *WinSpy* - A Windows-based network surveillance software tool that allows a computer to monitor several others on a network. The software on the monitored systems will take screen captures of the desktop environment and send them in encrypted format to the WinSpy console which then displays the JPEG-format screen shots of the selected computers being monitored. The monitoring program is password-protected.

Codex Data Systems *DIRT* - Data Interception by Remote Transmission. A powerful surveillance tool sold only to authorized law enforcement and government agencies, this product is intended to help users read encrypted messages. The system sends email to a target machine with embedded security software that will open up the system for covert access and monitor keystrokes to capture information and encryption keys that are subsequently transmitted to the sender when the target computer is online.

Omniquad *Detective* - A tool to retroactively construct the historical actions on the target computer system, including content downloaded from the Internet. System images can be viewed in a slideshow format. The software has advanced search capabilities.

MicroSpy (U.K.) *MicroSpy* - This consists of a microchip concealed in a short length of cable which is attached to the computer being investigated. It is designed to monitor keyboard characters and store them as they are typed, up to 1,000 keystrokes. The device can be retrieved and be later connected to the computer running the software and it will automatically upload the capture information (address, phone numbers, passwords). The device can then be reset and used again. Used ito investigate embezzlement, tax evasion, computer fraud, and unauthorized access. U.K. patent. IBM-licensed compatibles.

Olivetti *Pandora* - A system for viewing remotely through video cameras associated with each Pandora station. The software includes Peek (employee approval) and Spy (no approval) modes.

Intrusion detection systems are based on the premise that people may try to access an unauthorized computer, one that has been left unattended, or programs or data, without authorization. They detect an intrusion or anomalous use of a system, optionally log the activities on the computer, and optionally email or otherwise alert the owner or authorized user of the computer. They may also optionally shut down the system or prompt the user for authorization if anomalous activities are detected. They may even be programmed to turn on other detection devices such as video cameras or audio listening devices.

Ghost Keylogger - A computer security surveillance tool that monitors whether someone is accessing a secured computer. It can be used to monitor an off-limits computer or to monitor the computer activities of children. It records every keystroke to an optionally encrypted file that can be saved or sent to a specified email address. It further logs the title of the active application and the time of its use.

Tracking Systems

Tracking systems frequently combine radio or infrared transmitters with computer databases. There are now realtime programs that plot the location of a person within the vicinity of a monitored premises and display the information on a picture of the premises on the computer monitor. Depending on the system, the computer can show current location and recent path. These systems are useful for trucking and taxi services for safety, efficiency, and dispatch purposes. They can also track workers in hazardous areas or visitors to a secured area. Similar systems are used in some wildlife tracking programs.

> *Active Badge* - A worn-badge system that contains an infrared transmitter that transmits a unique 48-bit word every fifteen seconds so that the location of the badge (and the person wearing it) can be tracked. Badge information in a database can include security clearance and personal information. Radio transmitters can be used in much the same way and don't require line-of-site connections.

Video Peripherals

Computers are now commonly used to control, manage, and poll inputs from video surveillance or biometric identity systems.

> AITech *WaveWatcher-TV* - A hardware peripheral card for Windows-based IBM-licensed compatibles that allows realtime 21-bit full-motion video to be displayed in a window within the Windows interface environment. It is positionable and sizable from full-screen to icon-size. It takes inputs from three independent NTSC or PAL video sources and three independent audio sources. It can be used with VCRs and closed-circuit television feeds for video conferencing and video surveillance. Individual frames can be captured, stored, and edited such that a VCR or camcorder can be used as a scanner.

Computer Database Systems

There are hundreds of database systems related to surveillance activities, including systems that hold mug shots, fingerprints, data on employees and contractors, and lists of suspected foreign agents or terrorists. This is just a tiny selection of the many systems available:

> *Digital Justice Solution*™ - by Printrak International Inc. This system combines realtime automated fingerprint identification, computerized criminal history, a mug shot, and document storage and retrieval capabilities. As of 1999, over 80 law enforcement agencies were linked to the system, with an average time of identification of about 15 minutes.

> *MADRID-LE*™ - by Electronic Warfare Associates, Inc. This is a relational database system designed for law enforcement applications to support information analysis, storage, and retrieval of large volumes of investigative data from numerous databases.

> *TrueID*™ - by Image Data, LLC. This is a means to enable a staff member to check the picture of a customer on file when he or she is making a transaction. First, picture ID is presented to the vendor, who inserts the photo into a scanner where it is transmitted and mapped into a customer database. Later, when a transaction takes place, the identity device can be queried on the customer and the image that is stored in the database is securely transmitted to the device and displayed on a small screen to the cashier or teller. The teller then verifies the image with the person seeking to make the transaction.

Other resources include the *National Crime Information Center* and the National Law Enforcement Telecommunications System. Many states have criminal history and sex offender databases, in addition to which there is a National Sex Offender database.

Government Alliances

In 1995, the *National Security Agency* (NSA), the *Defense Information Systems Agency* (DISA), and the *Advanced Research Projects Agency* (ARPA) signed a memorandum of agreement to cooperate in computer system security research and development. The *Information Systems Security Research–Joint Technology Office* (ISSR–JTO) was created to support these efforts and to aid in safeguarding data in *Department of Defense* (DoD) information systems.

7. Problems and Limitations

It is difficult to discuss problems associated with computer surveillance when the field is so diversified and still rapidly evolving. In most cases, the problems are social rather than technological. Technologists have made remarkable progress in solving individual problems, including increased storage capacity, speed, and connectivity. The rate of change still appears to be increasing and the capabilities of computers are far beyond what most people envisioned ten years ago.

The most prevalent problems appear to be associated with finding exactly what you want and managing the veritable flood of information that is currently available. Improved search algorithms, intelligent agents to located and filter information on the surveillant's behalf, and good prioritizing to separate the good information from the rest are probably the most important advancements that could be made to improve the efficiency of computer surveillance.

8. Restrictions and Regulations

Several regulations related to privacy in general have already been listed in the Chapter 1 and 2 and can be cross-referenced. Some that are more specific to computers in order of date include

Electronic Communications Privacy Act of 1986 (ECPA) - Updates the *Crime Control Act of 1968* to protect digital communications from interception and disclosure. It requires a court order for Federal agents to conduct a 'wiretap' on electronic communications, including data, video, and audio, from unauthorized interception.

Computer Security Act of 1987 - Public Law 100-235. Sets standards for security and mandates for sensitive systems. The act created the Computer System Security and Privacy Advisory Board (CSSPAB) as a public advisory committee. The CSSPAB was to identify security issues and provide advisement.

Computer Matching and Privacy Protection Act of 1988 - An amendment to the *Privacy Act of 1974* that expressly regulates matching of data from different databases (federal, state, local). It requires notification of matches and provides an opportunity for the findings to be challenged. This has been further amended.

Telecommunications Infrastructure Act of 1993 - Prohibits telecommunications carriers from disclosing subscriber information of a personal nature.

Wiretapping

There are strict laws against the interception of electronic communications, as included in the *U.S. Wiretap Act*, and amended by the *Electronic Communications Privacy Act of 1986*.

These are covered in some detail in the Audio Surveillance chapter. There are some exceptions for employers who, as providers of email systems, may retrieve stored messages, but only in the normal course of employment.

Workplace Monitoring

With the increase of computer monitoring and logging of employee activities, sometimes down to individual keystrokes, there has been concern by labor organizations about protection of employee rights within an increasingly surveillant work world. One example of this is the consideration and publication of a report on *Telecommunications and Privacy in Labour Relationships* by the *European Union Data Protection Commissioners* released in 1996. The paper discusses various data-collection methods used in the workplace and their potential to generate data on employee activities. The second part includes a number of recommendations for the respect of privacy in the workplace. The third part includes specific applications of the report recommendations to information technologies and telecommunications. At about the same time, the *International Labour Organisation* was also discussing a draft Code of Practice with regard to privacy. These developments indicate the trend toward assessing workplace policies in the light of new technologies.

9. Implications of Use

Workplace Surveillance with Computers

There are a number of problems associated with computer surveillance and assessment of workplace performance. Computers tend to be used to surveil activities that are easy to quantify. Since management activities are less quantifiable than production line work or data entry activities, there is a stronger focus on monitoring the activities of workers with less seniority or authority, establishing a double-standard in the workplace beyond what already exists. All people make mistakes. If an employee who is not surveilled makes a mistake, there is often no record of the event and there are opportunities for the individual to remedy the mistake. In computer surveillance systems, every mistake, every hesitation, every learning step can be recorded with chilling accuracy, making it possible to create 'justifications' for layoffs, firings, or blocked promotions or raises that affect only certain workers and which may be unfair in the broader context of the workplace.

> Today 40 million American workers are under surveillance at the office. Women make up 85 percent of that number, as they tend to occupy customer-service and data-entry positions, which are more commonly scrutinized.
>
> [Brad Marlowe, "You Are Being Watched", *ZDNet*, Dec. 1999.]

While Marlowe is correct in pointing out the preponderance of surveillance in traditional office jobs staffed by women, men are not exempt from workplace surveillance. Truck drivers, cab drivers, travelling sales representatives, and many production line workers are male and are also likely to come under heavier surveillance as the technologies become easier to install and monitor.

Computer surveillance has the potential to increasingly stratify society, separating workers from executives more than in the past through the use of one-way monitoring that provides data to the executive that are not available to the worker. This dichotomy of documentation would become especially apparent in legal suits. Courtroom judgments are based largely on concrete evidence. If there is a dispute between an employer and an employee, the employer has the opportunity to gather reams of 'substantive data' on an employee through computer surveillance, in the form of computer logs and video tapes, while the employee may have little or no proof to prosecute or defend against actions of an employer.

In some cases, employers are justifiably concerned about employee abuse of computing facilities. If employees use workplace computing systems to make purchases of illegal goods or are downloading illegal images and the actions are traced to the employer's premises, there may be a problem defending the actions and determining who, in fact, was responsible for the illegal actions. This, in part, has motivated employers to put workplace surveillance in place. In some states, employers must inform employees before using surveillance tools, but this is not a universally requirement.

Balancing Costs and Losses

Computerized surveillance is changing the workplace. Rather than being structured on trust and communication among employees and managers, offices are increasingly being equipped with video cameras, electronic access and tracking technologies, and computer activity logs. This 'automation' does not come free. Not only can it be expensive to install and maintain all the equipment, but it can be unprofitable to constantly monitor, archive, and search and retrieve the enormous volumes of data that can be recorded. Employers have to weigh the losses traceable to a small percentage of dishonest employees against the cost of buying, maintaining, and monitoring workplace surveillance equipment. They must further assess the impact on productivity that may occur if employees feel they are being controlled by intimidation rather than being motivated by trust.

Data Mining and Personal Privacy

The use of computer networks for data mining creates a problem for individuals who are unaware of the extensive amount of information about them that can be collected and disseminated without their knowledge. It is difficult or impossible to have your name removed from many of the lists on the Internet. Some of the more reputable directory sites allow you to log on and 'unlist' your number, but as soon as they update their data from newly released phone directories, about once a year, your name is usually back and you have to do it again. Less reputable directory sites sell your information and may not provide a way to have your name removed. Even if you remove it, it may already have been replicated by a hundred or a million computers around the world and then the information is 'out there' or 'in the wild' and impossible to recall. If you don't have a computer and network account to spot the exploitation in the first place, you're out of luck.

Powerful Programs in the Wrong Hands

There are a number of software programs that have been developed for law enforcement officials and government agents to capture information, files, and keystrokes on target machines. Many of the companies selling these products are careful to qualify the buyers so the software doesn't get into the wrong hands. However, not all companies are so discriminating in distributing software and piracy of software is rampant on the Internet.

If software algorithms designed for covert law enforcement agents were to get into the hands of criminals, it would give them the tools to examine the contents of millions of computer systems. How do users know that the technology is not used by the suppliers themselves to make covert examinations of networked computers? Law enforcement agents have to operate under certain legal restraints related to wiretapping and often require authorizations to use this type of tool. However, unauthorized users of the software accept no such similar constraints. There are currently few laws to protect the public from this kind of spying, because they rarely know their systems have been violated.

Computers Integrated with Video Systems

Many computer cams have been integrated with the Web so that Internet users anywhere can log onto a Web site and remotely control the swivel-and-zoom cameras to watch any aspect of the scene that they choose. This is wonderful for schools, zoos, and museums that want to give shut-ins, children, and other Internet patrons a chance to see the animals, educational materials, sports events, cultural events, and exhibits, but what about the cams that are mounted on campuses and downtown buildings? Could a stalker use the cameras to chart the movements of a coed walking home from classes at the same time every day? Could a corporate spy use them to monitor the customers walking into or out of a store each day?

Public Sales of 'Private' Information

As was mentioned in earlier sections, a number of Public Safety Departments have sold or attempted to sell driver's license photos to outside private agencies without public input or approval. In some cases, the public found out and had the deals stopped but in one case, at least, the judge ruled that driver's license photos are not protected by privacy laws (e.g., South Carolina). Since there is a precedence now for government agencies to exploit public information, does this open the door for these agencies to sell other kinds of information kept in databases? (Note that the *Driver's Privacy Protection Act of 1997* regulates how records may be released and how the recipients of records may subsequently distribute the information.)

10. Resources

10.a. Organizations

Endorsement of these companies is neither intended nor implied and, conversely, their inclusion does not imply their endorsement of the contents of this document.

American Management Association International (AMAI) - This development and training organization considers matters of employee testing and monitoring and periodically issues the results of surveys, including surveys on Electronic Monitoring and Surveillance. www.amanet.org/

Computer Emergency Response Team (CERT) - Established in 1988 by the Defense Advanced Research Projects Agency (DARPA) to respond to security problems in networked computers and computer networks, particularly the Internet. CERT supplies technical assistance to aid in protecting the digital infrastructure. www.cert.org/

Computer Operations, Audit, and Security Technology (COAST) - This is now part of CERIAS. It is a cooperative project, multiple-lab computer security research established at the Computer Sciences Department at Purdue University. www.cerias.purdue.edu/coast/coast.html

Computer Security Institute (CSI) - Since 1974, CSI has been dedicated to serving and training computer security professionals. www.gocsi.com

High Tech Crime Investigation Association (HTCIA) - An international association which promotes education, research, and discussion about the investigation of high technology crime. www.htcia.org/

National Computer Security Association (NCSA) - Security assurance services for Internet-connected companies. ICSA publishes Information Security Magazine. www.icsa.net

National Computer Security Center (NCSC) - Originally established by the Department of Defense (DoD) for certifying various computer systems for security.

10.b. Print

Bentham, Jeremy, *The Works of Jeremy Bentham,* New York: Russell & Russell, 1962.

Casey, Eoghan, *Digital Evidence and Computer Crime,* Academic Press, 2000, 279 pages. Presents technical and legal concepts discussing the application of computer forensics.

Chantico Publications, *Combating Computer Crime: Prevention, Detection, Investigation,* New York: McGraw-Hill, 1992.

Conly, Catherine H., *Organizing for Computer Crime Investigation and Prosecution,* Washington, D. C.: U.S. Dept. of Justice, 1989, 124 pages.

Icove, David, *Computer Crime: A Crimefighter's Handbook,* O'Reilly & Associates, 1995.

Ilgun, Koral, USTAT - A *Real-time Intrusion Detection System for UNIX,* master's thesis, November 1922, U. C. Santa Barbara.

Judson, Karen, *Computer Crime: Phreaks, Spies, and Salami Slicers,* Enslow, 1994.

Power, Richard, *Current and Future Danger: A CSI Primer on Computer Crime & Information Warfare,* San Francisco: Computer Security Institute, 1995.

Rule, James B., *Private Lives and Public Surveillance: Social Control in the Computer Age,* New York: Schocken Books, 1974, 382 pages.

Stephenson, Peter, *Investigating Computer-Related Crime,* Boca Raton, Fl.: CRC Press, 2000.

Zuboff, Shoshana, *In the Age of the Smart Machine: The Future of Work and Power,* New York: Basic Books, 1988.

Articles

Aeillo, John R., Computer-Based Work Monitoring: Electronic Surveillance and its Effects, *Journal of Applied Social Psychology,* V.23, pp. 499-507.

American Management Association International (AMAI), More U.S. Firms Checking E-Mail, Computers Files, and Phone Calls, 14 April 1999. A report on an annual survey of 1,054 member organizations conducted from Jan. to Mar. 1999. Reports on the prevalent and increasing trend of electronic workplace monitoring.

Attewell, Paul, Big Brother and the Sweatshop: Computer Surveillance in the Automated Office, *Sociological Theory,* V.5, pp. 87-99.

Balitis, John J., Jr.; Silvyn, Jeffrey S., Big Brother at Work: Supervision: Employers' Electronic Monitoring of Employees in the Workplace Raises Federal and State Liability Issues, *Daily Journal,* April 1998.

Boehmer, Robert G., Artificial Monitoring and Surveillance of Employees: The Fine Line Dividing the Prudently Managed Enterprise from the Modern Sweatshop, *DePaul Law Review,* V.41, pp. 739-819.

Bryant, Susan, Electronic Surveillance in the Workplace, *Canadian Journal of Communication,* Papers, V.20(4). Describes increasing prevalence of monitoring and of power of the corporation over the individual worker along with legislative and public policy concerns.

Burgess, John, 'Active Badges' Play Follow the Worker: Computerized Trackers Spark Worries about 'Big Brother,' *The Washington Post,* 8 October 1992.

Bylinsky, Gene, How Companies Spy on Employees, *Fortune,* 1991, V.124, pp. 131-140.

Denning, Dorothy E., An Intrusion Detection Model, *IEEE Transactions on Software Engineering,* Feb. 1987, Number 2, p. 222.

DeTienne, Dristen Bell; Nelson, T. Abbot, Developing an Employee-Centered Electronic Monitoring System, *Journal of Systems Management,* 1993, V.44, pp. 12-16.

The End of Privacy: The surveillance society, Editorial, *The Economist,* 5 Jan. 1999.

Gandy, Oscar H., Jr., The Surveillance Society: Information Technology and Bureaucratic Social Control, *Journal of Communications,* 1989 V.39, pp. 61-76.

Griffith, Terri L., Teaching Big Brother to be a Team Player: Computer Monitoring and Quality, *Academy of Management Executive,* 1993, V.7, pp. 73-80.

Iadipaolo, Donna Marie, Monster or Monitor? Have Tracking Systems Gone Mad? *Insurance & Technology,* 1992, V.17, pp. 47-54.

Jewett, Christina, Your Life: Private as a Postcard, *Indiana Daily,* 13 Oct. 1999.

Levy, Michael, Electronic Monitoring in the Workplace: Power Through the Panopticon, Impact of New Technologies Web Server, *Library and Information Studies,* UC Berkeley, 1993. The author looks at the impact of workplace monitoring with quotes from a number of significant writings on the subject at the time.

Lunt, Teresa F., A survey of intrusion detection techniques, *Computers and Security,* 1993, V.12, pp. 405-416.

Marlowe, Brad, You Are Being Watched, *ZDNet,* Dec. 1999. Workplace monitoring.

Marx, Gary T., Let's Eavesdrop on Managers, *Computerworld,* 20 April 1992, p. 29.

National Security Agency, Information Security and Privacy in Network Environments, *Office of Technology Assessment OTA-TCT-606,* U.S. Government Printing Office, Sept. 1994.

Nitzberg, Sam, Emerging security issues involving the presence of microphones and video cameras in the computing environment, *ACM SIGSAC Security Audit & Control Review,* 1996, V.14 (3), pp. 13-16.

Nussbaum, Karen, Workers Under Surveillance, *Computerworld,* 6 January 1992, p. 21.

Piller, Charles, Bosses with X-Ray Eyes, *MacWorld,* 10 July 1993, pp. 118-123.

Rome Laboratory Law Enforcement Technology Team, Transferring Defense Technology to Law Enforcement, *The New Horizon,* April 1996.

Schwartau, Winn, DIRT Bugs Strike, *Network World,* July 1998. Describes the Data Interception by Remote Transmission system which is sold only to authorized government and law enforcement personnel.

Sundaram, Aurobindo, An Introduction to Intrusion Detection, *Crossroads.*

Vitone, Philip, Reflections on Surveillance, *Canadian Journal of Communication,* V.19(1).

Journals

Note, the U.S. Navy has a very good list of computer communications-related journals at chacs.nrl.navy.mil/ieee/cipher/readers-guide/journals.html

Canadian Journal of Communication, includes articles on privacy and computer surveillance. www.cjc-online.ca/

Cipher, a newsletter of the IEEE Computer Society's TC on Security and Privacy.

Computer Underground Digest, a weekly electronic journal available without a subscription fee. It can be found on the Web. comp.society.cu-digest (Google Groups)

Crossroads, an electronic publication of the ACM that discusses various computer-related topics. www.acm.org/crossroads/

Information Security Magazine, published by ICSA for Internet-connected security assurance services professionals.

Journal of Computer Security, quarterly journal of research and developments.

Law Enforcement & Corrections Technology News Summary, a publication of the National Law Enforcement and Corrections Technology Center of the National Institute of Justice.

The Risks Digest, Forum on Risks to the Public in Computers and Related Systems, sponsored by the ACM Committee on Computers and Public Policy.

10.c. Conferences and Workshops

Many of these conferences are annual events that are held at approximately the same time each year, so even if the conference listings are outdated, they can still help you determine the frequency and sometimes the time of year of upcoming events. It is very common for international conferences to be held in a different city each year, so contact the organizers for current locations.

Many of these organizations describe the upcoming conferences on the Web and may also archive conference proceedings for purchase or free download.

The following conferences are organized according to the calendar month in which they are usually held.

NDSS, The Internet Society network and distributed system security symposium.

NetSec, network security technical conference.

Safecomp, international conference on computer safety, security, and reliability. www.safecomp.org/

ICICS, international conference on information and communications security, since the late 1990s.

International Conference of Data Protection Commissioners, longstanding international conference. www.privacyconference2005.org/

National Computer Security Conference, jointly organized by the National Institute of Standards and Technology (NIST) and the National Security Agency (NSA) for attendees from private industry and the government.

10.d. Online Sites

The following are interesting Web sites relevant to this chapter. The author has tried to limit the listings to links that are stable and likely to remain so for a while. However, since Web sites do sometimes change, keywords in the descriptions below can help you relocate them with a search engine. Sites are moved more often than they are deleted.

Another suggestion, if the site has disappeared, is to go to the upper level of the domain name. Sometimes the site manager has simply changed the name of the file of interest. For example, if you cannot locate www.goodsite.com/science/uv.html *try going to* www.goodsite.com/science/ *or* www.goodsite.com *to see if there is a new link to the page. It could be that the filename* uv.html *was changed to* ultraviolet.html, *for example.*

Electronic Monitoring & Surveillance. A survey published online by the American Management Association International. This site includes charts, tables, discussion, and a summary of key findings. When viewing this site keep in mind that there was a large upswing in the installation of video cameras in 1999 that may change the demographics somewhat from those published in this 1997 survey. www.amanet.org/survey/elec97.htm

10.e. Media Resources

Criminals in Cyberspace, an Arts & Entertainment program from the 20th Century with Mike Wallace series. It provides a look into cybercrime and the types of terrorist activities that can occur on computer systems. VHS, 50 minutes. May not be shipped outside the U.S. and Canada.

Cybersex Cop, an Arts & Entertainment program from the Investigative Reports series which discusses pornography and the impact of pornography on the Internet, including the arrest of a child pornographer. VHS, 50 minutes. May not be shipped outside the U.S. and Canada.

11. Glossary

Titles, product names, organizations, and specific military designations are capitalized; common generic and colloquial terms and phrases are not.

CCH	computerized criminal history
EPM	electronic performance monitoring
firewall	a computer security configuration intended to selectively or completely limit access to a system or process
gateway	a transmission connection between dissimilar networks which may or may not have security features incorporated into the system
IDS	intrusion detection system
packet sniffer	a technical tool for analyzing and monitoring digital data packets that are transmitted over a computer network. A packet sniffer can be used to covertly moonitor communications.
proxy	a system or software agent intended to act on behalf of clients and which can act as a server or client for processes associated with security

Surveillance Technologies

Index

Symbols